General relativity

An Einstein centenary survey

General relativity

An Einstein centenary survey
edited by
S. W. Hawking
Professor of Gravitational Physics
University of Cambridge

W. Israel
Professor of Physics
University of Alberta

Cambridge University Press
Cambridge
London   New York   Melbourne

Published by the Syndics of the Cambridge University Press
The Pitt Building, Trumpington Street, Cambridge CB2 1RP
Bentley House, 200 Euston Road, London NW1 2DB
32 East 57th Street, New York, NY 10022, USA
296 Beaconsfield Parade, Middle Park, Melbourne 3206, Australia

© Cambridge University Press 1979

First published 1979

Printed in Great Britain by
J. W. Arrowsmith Ltd, Bristol BS3 2NT

Library of Congress Cataloguing in Publication Data
Main entry under title:
General relativity
Includes index
1. General relativity (Physics) 2. Black holes (Astronomy)
I. Hawking, S. W. II. Israel, W.
QC173.G46 530.1'1 78-62112
ISBN 0 521 22285 0

Contents

	page
List of contributors	xi
Preface	xv

1. An introductory survey   1
   S. W. Hawking and W. Israel

1.1	Historical background	1
1.2	The field equations	9
1.3	Cosmology	11
1.4	Gravitational collapse	15
1.5	Quantum gravity	21
1.6	Future prospects	23

2. The confrontation between gravitation theory and experiment   24
   C. M. Will

2.1	Introduction	24
2.2	Principles of equivalence and the foundations of gravitation theory	26
2.3	Post-Newtonian gravity in the solar system	40
2.4	Gravitational radiation as a tool for testing gravitation theory	62
2.5	Stellar-system tests: the binary pulsar	70
2.6	Gravitation in the universe: the influence of global structure on local physics	84
2.7	Summary	88

3. Gravitational-radiation experiments   90
   D. H. Douglass and V. B. Braginsky

3.1	Introduction	90
3.2	Characteristics of gravitational radiation	94

Contents

		page
3.3	Sources of gravitational radiation	97
3.4	Detection of gravitational radiation	119
3.5	Prospects for the future	135

4. The initial value problem and the dynamical formulation of general relativity   138
A. E. Fischer and J. E. Marsden

4.1	Canonical formalism	140
4.2	The constraint manifold	158
4.3	The abstract Cauchy problem and hyperbolic equations	168
4.4	The Cauchy problem for relativity	183
4.5	Linearization stability of the vacuum Einstein equations	194
4.6	The space of gravitational degrees of freedom	202

5. Global structure of spacetimes   212
R. Geroch and G. T. Horowitz

5.1	Introduction	212
5.2	What is the topology of our universe?	217
5.3	Is our universe singular?	255
5.4	How noticeably singular is our universe?	269
5.5	Conclusion	288
5.6	Appendix	289

6. The general theory of the mechanical, electromagnetic and thermodynamic properties of black holes   294
B. Carter

6.1	Introduction	294
6.2	The evolution of the horizon	302
6.3	Local properties of a stationary horizon	314
6.4	Energy and angular momentum transport in a black hole background	328
6.5	Electromagnetic effects in a black hole background space	336
6.6	The total mass and angular momentum	352
6.7	Uniqueness and no-hair theorems	359

		page
7.	**An introduction to the theory of the Kerr metric and its perturbations** S. Chandrasekhar	370
7.1	The tetrad formalism	371
7.2	The Newman–Penrose formalism	375
7.3	Tetrad transformations and related matters	383
7.4	The Kerr metric and the perturbation problem	391
7.5	The solution of Maxwell's equations	404
7.6	Gravitational perturbations	411
7.7	The solution of Dirac's equation	425
7.8	The potential barriers round the Kerr black hole and the problem of reflection and transmission	429
8.	**Black hole astrophysics** R. D. Blandford and K. S. Thorne	454
8.1	Introduction	454
8.2	On the character of research in black hole astrophysics	457
8.3	Isolated holes produced by collapse of normal stars	461
8.4	Black holes in binary systems	469
8.5	Black holes in globular clusters	481
8.6	Black holes in quasars and galactic nuclei	485
8.7	Primordial black holes	494
8.8	Concluding remarks	502
9.	**The big bang cosmology – enigmas and nostrums** R. H. Dicke and P. J. E. Peebles	504
9.1	Introduction	504
9.2	Enigmas	504
9.3	Nostrums and elixirs	510
10.	**Cosmology and the early universe** Ya B. Zel'dovich	518
10.1	Introduction	518
10.2	The average matter density in the universe	520
10.3	The lepton era	522
10.4	The hadron era	523
10.5	The quantum era and its effect	526

Contents

		page
11.	**Anisotropic and inhomogeneous relativistic cosmologies** *M. A. H. MacCallum*	533
11.1	Introduction	533
11.2	Spacetime symmetries	536
11.3	Spatially-homogeneous anisotropic metrics	542
11.4	Inhomogeneous metrics	563
11.5	Physics of the models	570
11.6	Constraints and inferences	576
12.	**Singularities and time-asymmetry** *R. Penrose*	581
12.1	Introduction	581
12.2	Statement of the problem	582
12.3	Singularities: the key?	611
12.4	Asymmetric physics?	635
13.	**Quantum field theory in curved spacetime** *G. W. Gibbons*	639
13.1	Introduction	639
13.2	Basic notions	640
13.3	Applications	663
13.4	Conclusion	679
14.	**Quantum gravity: the new synthesis** *B. S. DeWitt*	680
14.1	Introduction	680
14.2	The quantum ether	683
14.3	The back reaction	698
14.4	The one-loop approximation	702
14.5	The full quantum theory	720
14.6	Conclusion	743
15.	**The path-integral approach to quantum gravity** *S. W. Hawking*	746
15.1	Introduction	746
15.2	The action	749

		page
15.3	Complex spacetime	752
15.4	The indefiniteness of the gravitational action	757
15.5	The stationary-phase approximation	762
15.6	Zeta function regularization	766
15.7	The background fields	771
15.8	Gravitational thermodynamics	778
15.9	Beyond one loop	782
15.10	Spacetime foam	785

16. Ultraviolet divergences in quantum theories of gravitation 790
S. Weinberg

16.1	Introduction	790
16.2	Renormalizable theories of gravitation	792
16.3	Asymptotic safety	798
16.4	Physics at ordinary energies	809
16.5	Dimensional continuation	814
16.6	Gravity in $2+\varepsilon$ dimensions	822
16.7	Appendix. Calculation of b	828

References 833

Index 903

Contributors

R. D. Blandford
Department of Theoretical Astrophysics, California Institute of Technology, Pasadena, California 91125, USA

V. B. Braginsky
Department of Physics, Moscow State University, Moscow 117234, USSR

B. Carter
Groupe d'Astrophysique Relativiste, Observatoire de Paris, 92 Meudon, Paris, France

S. Chandrasekhar
Laboratory for Astrophysics and Space Research, The University of Chicago, The Enrico Fermi Institute, 933 East 56th Street, Chicago, Illinois 60637, USA

B. S. DeWitt
Department of Physics, The University of Texas at Austin, Austin, Texas 78712, USA

R. H. Dicke
Department of Physics, Joseph Henry Laboratories, Jadwin Hall, PO Box 708, Princeton University, Princeton, New Jersey 08540, USA

D. H. Douglass
Department of Physics and Astronomy University of Rochester, River Campus Station, Rochester, New York 14627, USA

A. E. Fischer
Division of Natural Sciences, University of California, Santa Cruz, California 95064, USA

Contributors

R. P. Geroch
Laboratory for Astrophysics and Space Research, The University of Chicago, The Enrico Fermi Institute, 933 East 56th Street, Chicago, Illinois 60637, USA

G. W. Gibbons
Department of Applied Mathematics and Theoretical Physics, Silver Street, Cambridge CB3 9EW, UK

S. W. Hawking
Department of Applied Mathematics and Theoretical Physics, Silver Street, Cambridge CB3 9EW, UK

G. T. Horowitz
Laboratory for Astrophysics and Space Research, The University of Chicago, The Enrico Fermi Institute, 933 East 56th Street, Chicago, Illinois 60637, USA

W. Israel
Theoretical Physics Institute, Department of Physics, University of Alberta, Edmonton T6G 2J1, Canada

M. A. H. MacCallum
Department of Applied Mathematics, Queen Mary College, University of London, Mile End Road, London E1 4NS, UK

J. E. Marsden
Division of Natural Sciences II, University of California, Santa Cruz, California 95064, USA

P. J. E. Peebles
Department of Physics, Joseph Henry Laboratories, Jadwin Hall, PO Box 708, Princeton University, Princeton, New Jersey 08540, USA

R. Penrose
University of Oxford, Mathematical Institute, 24–29 St. Giles, Oxford OX1 3LB, UK

K. S. Thorne
Department of Theoretical Astrophysics, California Institute of Technology, Pasadena, California 91125, USA

Contributors

S. Weinberg
Department of Physics, Harvard University, Cambridge, Massachusetts 02138, USA

C. M. Will
Department of Physics, Stanford University, Stanford, California 94305, USA

Ya B. Zel'dovich
Institute for Cosmic Research, Space Research Institute, USSR Academy of Science, Profsoyuznaya 88, Moscow 117485, USSR

Preface

To a physicist of the time, the year 1879 seemed uneventful. He may have read with some scepticism and not much excitement an article by Josef Stefan in the *Sitzungsberichte der Akademie der Wissenschaften in Wien*. Stefan suggested, on rather flimsy empirical grounds, that the energy radiated by a perfectly absorbing body was proportional to the fourth power of its temperature. A new fundamental constant had quietly slipped into physics. But it took another twenty years and the insight of Planck to recognize this.

Probably our physicist would have greeted with more interest the appearance that year of volume VIII of the ninth edition of the *Encyclopaedia Britannica*, containing the article 'Ether' by James Clerk Maxwell. But he could not have known of a letter from Maxwell to D. P. Todd Esq., Director of the Nautical Almanac Office in Washington, asking whether anomalies had been observed in the eclipse times of Jupiter's satellites. For, he wrote, 'they afford the *only* method, as far as I know, of getting any estimate of the direction and magnitude of the velocity of the sun with respect to the luminiferous medium ... In the terrestrial methods of determining the velocity of light, the light comes back along the same path again, so that the velocity of the earth with respect to the ether would alter the time of the double passage by a quantity depending on the square of the earth's velocity to that of light, and this is quite too small to be observed'. Maxwell's letter found its way into *Nature* shortly after his death in November 1879, and caught the eye of a young instructor at the US Naval Academy, Albert A. Michelson. The letter is headed Cavendish Laboratory, Cambridge, 19 March 1879 and is reproduced at the end of the preface. Five days earlier, on 14 March, Albert Einstein was born in Ulm. Thus the twin streams that were to grow into quantum theory and relativity came into being at the same moment as the man who, for a critical decade, was to be their surest oarsman.

From its inception in 1916, general relativity was considered an extraordinarily difficult theory. Its development in the last twenty years

Preface

has been so phenomenal that a book like this cannot avoid many pages of technicalities. Yet the main impression one always carries away is a sense of wonder at the magical, strangely rational universe which has opened to us through the genius of one man. No-one can recall without a thrill his first encounter with this Carollian world where space is curved, time a fourth dimension and honest witnesses blithely disagree on the most elementary questions of what happened when and where. Perhaps this was the most precious gift that Albert Einstein bequeathed to our century: that he restored to a world of billiard balls its aura of depth and mystery.

Even apart from the fact that it is now a hundred years since Einstein's birth, this is a very appropriate time to review his greatest achievement, the General Theory of Relativity. In the last twenty years there has been tremendous progress, both in theoretical understanding and in experimental verification, which has transformed general relativity from an intellectual curiosity into an everyday tool for astrophysicists, something whose effects have to be taken into account in such practical matters as the design of satellite navigational systems. We now have a fairly complete picture of the theory, at least up to the point at which it would have to be modified to take into account quantum gravitational effects. Most of the articles in this survey describe different aspects of this classical, i.e. non-quantum, theory. However the classical theory also predicts that there will be spacetime singularities, regions where the curvature of spacetime becomes unboundedly large. In such situations one would expect quantum gravitational effects to play a dominant role. At present, this is an area of intense research activity, some of which is described in the last few articles in this volume. The ultimate aim is to realise Einstein's dream of a complete and consistent theory that would unify all the laws of physics.

The aim in preparing this volume was to provide a survey of the current state of research in general relativity, which would be accessible to the non-specialist. Inevitably, some topics, such as the initial value problem or perturbations of the Kerr metric, are complicated and involve a lot of formalism and equations. However the non-expert reader should have little difficulty with the introductory survey, with the articles on experimental tests and the detection of gravitational radiation, the article on the astrophysics of black holes and the articles by Geroch and Horowitz, and Penrose.

We, the editors, are very grateful to the authors who wrote these articles at short notice, to our secretaries, Judy Fella and Mary Yiu, to Ian

Bailey and Richard Mably for help in the preparation of the index, and to the sub-editors at the Cambridge University Press who carried out the mammoth task of preparing the manuscript for the printers.

S. W. HAWKING
W. ISRAEL

August 1978

"On a Possible Mode of Detecting a Motion of the Solar System through the Luminiferous Ether." By the late Prof. J. Clerk Maxwell. In a letter to Mr. D. P. Todd, Director of the *Nautical Almanac* Office, Washington, U.S. Communicated by Prof. Stokes, Sec. R.S.

Mr. Todd has been so good as to communicate to me a copy of the subjoined letter, and has kindly permitted me to make any use of it.

As the notice referred to by Maxwell in the *Encyclopædia Britannica* is very brief, being confined to a single sentence, and as the subject is one of great interest, I have thought it best to communicate the letter to the Royal Society.

From the researches of Mr. Huggins on the radial component of the relative velocity of our sun and certain stars, the coefficient of the inequality which we might expect as not unlikely, would be only something comparable with half a second of time. This, no doubt, would be a very delicate matter to determine. Still, for anything we know *à priori* to the contrary, the motion might be very much greater than what would correspond to this; and the idea has a value of its own, irrespective of the possibility of actually making the determination.

In his letter to me Mr. Todd remarks, "I regard the communication as one of extraordinary importance, although (as you will notice if you have access to the reply which I made) it is likely to be a long time before we shall have tables of the satellites of Jupiter sufficiently accurate to put the matter to a practical test."

I have not thought it expedient to delay the publication of the letter on the chance that something bearing on the subject might be found among Maxwell's papers.

(Copy.)

Cavendish Laboratory,
Cambridge,
19th March, 1879

Sir,

I have received with much pleasure the tables of the satellites of Jupiter which you have been so kind as to send me, and I am encouraged by your interest in the Jovial system to ask you if you have made any special study of the apparent retardation of the eclipses as affected by the geocentric position of Jupiter.

I am told that observations of this kind have been somewhat put out of fashion by other methods of determining quantities related to the velocity of light, but they afford the *only* method, so far as I know, of getting any estimate of the direction and magnitude of the velocity of the sun with respect to the luminiferous medium. Even if we were sure of the theory of aberration, we can only get differences of position of stars, and in the terrestrial methods of determining the velocity of light, the light comes back along the same path again, so that the velocity of the earth with respect to the ether would alter the time of the double passage by a quantity depending on the square of the ratio of the earth's velocity to that of light, and this is quite too small to be observed.

But if J E is the distance of Jupiter from the earth, and l the geocentric longitude, and if l' is the longitude and λ the latitude of the direction in which the sun is moving through ether with velocity v, and if V is the velocity of light and t the time of transit from J to E,

$$J E = [V - v \cos \lambda \cos (l - l')] t.$$

By a comparison of the values of t when Jupiter is in different signs of the zodiac, it would be possible to determine l' and $v \cos \lambda$.

I do not see how to determine λ, unless we had a planet with an orbit very much inclined to the ecliptic. It may be noticed that whereas the determination of V, the velocity of light, by this method depends on the differences of J E, that is, on the diameter of the earth's orbit, the determination of $v \cos \lambda$ depends on J E itself, a much larger quantity.

But no method can be made available without good tables of the motion of the satellites, and as I am not an astronomer, I do not know whether, in comparing the observations with the tables of Damoiseau, any attempt has been made to consider the term in $v \cos \lambda$.

I have, therefore, taken the liberty of writing to you, as the matter is beyond the reach of any one who has not made a special study of the satellites.

In the article E [ether] in the ninth edition of the " Encyclopædia Britannica," I have collected all the facts I know about the relative motion of the ether and the bodies which move in it, and have shown that nothing can be inferred about this relative motion from any phenomena hitherto observed, except the eclipses, &c., of the satellites of a planet, the more distant the better.

If you know of any work done in this direction, either by yourself or others, I should esteem it a favour to be told of it.

     Believe me,
      Yours faithfully,
   (Signed)   J. CLERK MAXWELL
D. P. Todd, Esq.

1. Introductory survey

S. W. HAWKING AND W. ISRAEL

1.1 Historical background

To the ancient Greeks it appeared self-evident that there was a preferred velocity or standard of rest which all particles would take up if not propelled by some external force. This preferred velocity led to the concept of Absolute Space or, more strictly, the ability to establish that the events at different times occurred at the same point in space. Such was the regard for ancient learning, and the low status of observational evidence, that this view persisted right up to the end of the sixteenth century, when Galileo showed it to be mistaken. Galileo's ideas and observations were codified by Newton in his laws of mechanics, the first of which states that a body continues to move uniformly in a straight line unless acted on by external forces. This meant one had to abandon the concept of Absolute Space, as there was no way in which one could measure it. However, Galileo and Newton retained the concept of Absolute Time, or the ability to establish the coincidence in time of events at different spatial positions. Absolute Time was defined to be the time appearing in Newton's laws of motion. It could be measured by well-regulated clocks, all of which agreed within the limits of the technology of that period. An absolute value of time could be given measured from the creation of the world, which was estimated by Bishop Usher to have occurred in 4004 B.C.

The Newtonian concept of the structure of space and time remained unchallenged until the development of electromagnetic theory in the nineteenth century, principally by Faraday and Maxwell. Maxwell showed that electromagnetic waves ought to propagate at a speed of 186 000 miles per second, the velocity of light. But what did this mean, if time but not space was absolute? What was the velocity relative to? In an attempt to overcome this paradox the concept of a luminiferous aether was introduced as the medium in which electromagnetic waves propagate. This was, in fact, a return to absolute space. However, a famous series of experiments by Michelson and Morley failed to detect

Chapter 1. Introductory survey

any motion of the earth relative to the frame defined by the aether. It is not clear whether Einstein knew of the Michelson–Morley experiment when in 1905 he proposed the special theory of relativity as an answer to the paradox (see Holton, 1972). In this theory neither space nor time was absolute but they were just regarded as coordinates or labels on a four-dimensional spacetime continuum or manifold. The structure of spacetime was represented by the Minkowski metric, which determined the proper distance or proper time interval between two events.

In special relativity the spacetime metric is a fixed quantity. It was Einstein's greatest stroke of genius to realize that it could be given a dynamical role. After a number of unsuccessful attempts he arrived at the general theory of relativity in 1915. In this theory the metric of spacetime was no longer flat as in special relativity, but was curved or distorted by the matter and energy contained in spacetime. Four years later, in 1919, an eclipse expedition confirmed that, as predicted by Einstein, light passing near the sun is deflected through a small angle by the gravitational field. The world woke up to the fact that it lived in a curved universe.

Although general relativity was recognized as a major conceptual revolution, it was generally felt to be of little practical significance because it was thought that gravitational fields could never be so strong that there would be much difference between its predictions and those of the much simpler Newtonian theory. The first evidence that this view was mistaken came in the early 1920s with the discovery, by Slipher and Hubble, that the light from distant galaxies was systematically shifted towards the red, i.e. to longer wavelengths. The farther a galaxy was from us, the greater the redshift. This was interpreted as indicating that the galaxies were receding from us with velocities proportional to their distances. About this time, Friedmann discovered quite independently that the field equations of general relativity admitted solutions representing a universe which is homogeneous and isotropic in spatial directions but which is expanding with time. Since then, further observations and theoretical work have confirmed that a Friedmann solution is a very good approximation to the large-scale structure of the universe, at least at the present epoch.

A feature of the Friedmann solutions that was not taken too seriously at the time was that they had a singular epoch in the past at which all matter of the universe was concentrated into a single point. It was felt that conditions in the real universe would never get so extreme. A related discovery whose significance was also not properly appreciated at the time was made by Chandrasekhar and Landau in the early 1930s; they showed that, according to Newtonian gravitational theory, there ought to be a maximum mass of the order of one solar mass for a cold star. On the one hand this work pointed to the possibility of having very condensed objects with strong gravitational fields and, on the other hand, it was fairly easy to show that the nonlinearity of the field equations of general relativity merely made matters worse and reduced the maximum mass. The question then arose: what happens to a star of more than the maximum mass when it has exhausted its nuclear fuel and can no longer maintain its temperature? The answer was provided by Oppenheimer in 1939: the star would collapse down to a single point under its own gravity. Most people at the time, including Einstein and Eddington, were unwilling to accept this. Einstein in fact wrote a paper in 1939 claiming that collapsed objects could not exist (Einstein, 1939).

Very little work was done on general relativity in the 1940s and 50s. This was partly because most physicists were primarily occupied with atomic and nuclear physics and partly because it was felt that general relativity was not of much physical importance except in cosmology, and cosmology was regarded as an area in which wild theoretical speculation was unfettered by any possible observations. This attitude began to change in the late 1950s as new technology greatly increased the range of observations. A significant landmark was the first Texas symposium on relativistic astrophysics in 1963. It took place shortly after the discovery of quasars, very distant and compact objects which emitted very large fluxes of light and radio waves. From the first it was clear that if they were really at the distances they appeared to be (and this was a subject of some controversy) they must certainly contain very large concentrations of mass and strong gravitational fields, and they might even be similar to the gravitational collapsed objects that Oppenheimer had described. Interest in such objects was now thoroughly aroused.

In 1965 a new and surprising discovery was made. Penzias and Wilson, working on the development of a satellite communication system, found that there was a noise in their microwave receiver which could not be removed or accounted for by any terrestrial, solar system or galactic source. In the end they realized that it must be cosmological in origin. Subsequent measurements by other people showed that it had a spectrum of thermal radiation at a temperature of about 3 K. It was interpreted as a relic of the dense phase of the universe and had in fact been predicted by Gamow and his coworkers back in 1948, though not

Chapter 1. Introductory survey

many people took their model seriously at the time. Gamow's calculations and subsequent refinements thereof, which were based on the Friedmann model of the universe, predicted that, about one hundred seconds after the initial singularity, about one quarter of the original protons and neutrons in the universe should have changed into helium and a small amount of deuterium. These abundances seem to agree with observation and it is very difficult to account for so much helium in any other way.

Thus it appears that the Friedmann models are a good approximation, at least back to the first one hundred seconds. What about earlier times? Did the universe really have a singular origin at which the density was infinite? In the Friedmann models the assumption of spatial isotropy meant that the relative motion of a pair of particles had to be purely radial and the assumption of spatial homogeneity excluded the possibility of pressure gradients. It was therefore not surprising that the world-lines of all particles should intersect at a single point in the past. However, in a realistic model one would expect that there would be some transverse velocities and inhomogeneities. As one went back in time, these might grow large and cause the particles to miss each other, giving a nonsingular 'bounce' instead. This was the reason that the initial singularity of the Friedmann models had not been taken very seriously. However, between 1965 and 1970, Penrose and Hawking proved a number of theorems which showed fairly conclusively that there must have been a singularity if general relativity was correct. This conclusion was also reached by Lifshitz, Khalatnikov and Belinsky in 1969.

The theorems of Penrose and Hawking also showed that a singularity should occur in the collapse of a burnt-out star even if it was not exactly spherical (as Oppenheimer had assumed). However, in this case, there seemed to be a phenomenon which was called 'cosmic censorship': the singularity occurred in a region of spacetime from which neither light nor anything else could escape to the rest of the universe. Such regions were christened 'black holes' by Wheeler, who also inspired much of the work on them. Stimulated by a theorem of Israel in 1967, and earlier results of Doroschkevich, Zel'dovich and Novikov, he conjectured that 'a black hole has no hair'. By this he meant that a black hole would rapidly settle down to a stationary state which was determined by only three parameters – mass, angular momentum and electric charge – and was otherwise independent of the details of the body that collapsed to form the black hole. These stationary states would be the Kerr–Newman family of solutions. This conjecture was later proved by the combined work of Israel, Carter, Hawking and Robinson. It was of considerable practical importance because it meant that one could predict the structure of the gravitational field of any black hole and so one could construct models of astrophysical objects thought to contain black holes and compare their properties with observation.

The search for black holes was greatly encouraged by the discovery in 1968 of rapidly pulsing radio sources or 'pulsars', as they became known. These were interpreted as rotating neutron stars, bodies of about the mass of the sun but a radius of only 10 km, composed mainly of neutrons packed together as tightly as in an atomic nucleus. They had first been proposed by Zwicky in 1934 and their theory had been extensively investigated by Wheeler and his school. In neutron stars the gravitational potential would be a significant fraction of unity, so it would not be a great extrapolation to consider the existence of objects in which the potential attained unity, i.e. black holes. Of course, by their very definition, black holes could not emit any light (apart from an effect which will be mentioned below), so they could not be observed directly. However, they could be detected by their gravitational influence on nearby matter.

In 1972, a rapidly fluctuating X-ray source called Cygnus X-1 was identified with a binary system consisting of a normal giant star of about fifteen solar masses and an unseen companion whose mass was estimated to be at least four solar masses, more likely six. This fitted very nicely with a model proposed by Shakura and Sunyaev and by Pringle and Rees, in which matter was dragged or blown off a normal star and fell towards the companion. As it approached, it would develop spiral motion and get very hot, emitting X-rays. For this model to work, the companion had to be a compact object such as a white dwarf, a neutron star or a black hole. The lower limit of four solar masses seemed to rule out the first two possibilities, so one was led to the conclusion that Cygnus X-1 probably contains a black hole. There are alternative models not involving a black hole, but they all seem rather far-fetched.

During the same period that qualitatively new phenomena such as black holes were coming within the range of observation, there was a steady progress in the quantitative verification of general relativity within the solar system. Einstein's theory came through these tests with flying colours, but many rival theories of gravity were eliminated and the few that remain seem rather contrived.

Another area in which experimental interest was aroused was gravitational radiation. Einstein had shown in 1918 that the linear

Chapter 1. Introductory survey

approximation to the field equations admitted wavelike solutions propagating at the speed of light. For a long time there was disagreement about whether gravitational waves really had any physical existence and whether they carried energy away from the system that emitted them. This was finally resolved in 1962 by Bondi, Sachs, Newman and Penrose, and their collaborators, who showed that the mass of a system (measured at infinity) would decrease if the system emitted gravitational waves. However, the extreme weakness of the gravitational interaction made these waves very difficult to detect. Nonetheless, the observational search for them was inaugurated by Weber, using antennae consisting of massive aluminium bars suspended in vacuum. Although the first hopes of detecting waves now appear too optimistic, there is some confidence that improvements in technology will make this possible within the next twenty years.

General relativity is a purely classical theory. Yet, in the 1920s and 30s, it was found that all other fields seemed to be governed by quantum laws. Paradoxically, Einstein, who had played an important role in the development of quantum theory, could never really accept its degree of uncertainty or chance. His feelings were expressed in his statement, 'God does not play dice'. However, this did not stop other people from trying to quantize general relativity. At the time this seemed a rather academic exercise, because it was felt that even classical general relativistic effects would be very small, so quantum ones should be quite unmeasurable. The situation changed, however, when it was realized that the universe probably had a singular origin. The strong gravitational fields near the singularity could have created a significant number of particles. Indeed, in 1969 Zel'dovich suggested that particle creation in an initially anisotropic and inhomogeneous universe could have damped out the anisotropies and inhomogeneities to give the very highly isotropic universe we observe today.

Another situation in which strong gravitational fields would be expected is in the gravitational collapse of an object to form a black hole. In 1969 Penrose proposed a classical gedanken experiment which showed that energy could be extracted from a rotating black hole. The wave analogue of this process was called 'super-radiance': incident waves in certain modes would be amplified (rather than absorbed) by a rotating black hole, and would carry away some of the black hole's rotational energy. In 1971, Zel'dovich pointed out that, on a quantum level, one could regard this amplification as stimulated emission, and therefore there ought to be a corresponding rate of spontaneous emission in these modes, produced by the vacuum fluctuations of the field under consideration. This was confirmed by a detailed calculation by Unruh in 1973. About the same time, Hawking discovered that there should be an additional spontaneous emission in all modes, even if the black hole was not rotating. This emission would be thermal and would be just as if the black hole were an ordinary body with a temperature inversely proportional to its mass. This fitted remarkably well with some analogues between black holes and thermodynamics which had been discovered earlier by Bardeen, Carter, Hawking and Bekenstein. In particular, it was now clear that black holes should have an intrinsic entropy which would be proportional to the area of the event horizon, the boundary of the black hole. This entropy could be interpreted as corresponding to the internal states of the black hole, which would be unobservable from the outside of the black hole because of the 'no-hair' theorem. The radiation emitted could be regarded as having tunnelled out of an unobservable region of spacetime inside the black hole. Its random thermal character would then correspond to a lack of knowledge of the internal behaviour of the black hole.

The work on particle creation in cosmology and black holes described above was performed in the semi-classical approximation, in which the gravitational field is represented by a given, unquantized, curved metric and the matter fields are treated quantum mechanically in this metric. However, the success with black holes reawakened interest in efforts to quantize the gravitational field itself. The difficulty is that the perturbation methods which had been successful in quantum electrodynamics and Yang–Mills theories do not seem to work for gravity. One possibility is to adopt some non-perturbative approach which reflects the fact that the gravitational field can have many different global structures. Another is to introduce additional fields and invariances which might cancel the infinities appearing in perturbation theory. A promising development along these lines is supergravity, which was proposed in 1976 by Ferrara, van Nieuwenhuizen and Freedman, and independently by Deser and Zumino. However, the situation on quantum gravity is still very unclear, though the large amount of effort that is currently being applied encourages the hope that significant progress will be made.

In this brief historical sketch, we have not had space to mention the work of many people who have made important contributions. In the remaining sections of this article we shall attempt to survey some of the main areas of work in general relativity in the last sixty years or so. This will serve as an introduction to the following articles.

Chapter 1. Introductory survey

1.2. The field equations

In general relativity, spacetime is represented by a four-dimensional manifold M on which there is a metric g_{ab}. In the neighbourhood of a point $p \in M$ one can choose local coordinates x^a ('local inertial coordinates') such that, at p, the components g_{ab} take on their Minkowski-space values $g_{ab} = \text{diag}(1, 1, 1, -1)$ and $\partial g_{ab}/\partial x^c = 0$. This means that locally spacetime has the same structure as in special relativity, and all the experimentally observed special relativistic effects, such as time-dilation and increase of mass with velocity, are predicted. In particular, the time registered by a clock is determined by the metric interval along its world-line, and light moves along geodesics of zero length. However, on a larger scale, significant departures from the Minkowski metric may occur, and these are interpreted as representing the gravitational influence of the matter and energy in the spacetime. At this stage one has what is called a 'metric' theory of gravity (see chapter 2 by Will). To complete the theory one has to provide field equations which relate the metric to the matter distribution. One would expect that these equations would involve the matter only through its energy–momentum tensor, because experiments first carried out by Galileo and subsequently refined by Eötvös, Dicke, Braginsky and Kreuzer indicate that gravity couples only to the energy content of a body. If one introduces no additional geometrical structure into spacetime apart from the metric itself, and if one requires that the field equations should not involve derivatives higher than the second, one is led uniquely to the Einstein equations (Einstein, 1915; Hilbert, 1915),

$$R_{ab} - \tfrac{1}{2} g_{ab} R + \Lambda g_{ab} = 8 \pi G T_{ab}.$$

Here, R_{ab} is the Ricci tensor of the metric g_{ab}, R is the Ricci scalar, T_{ab} is the energy–momentum tensor and G is Newton's constant of gravitation. The 'cosmological term' Λg_{ab} (Λ a constant) will be discussed in the next section. Its effects can be ignored in most situations.

It is not immediately obvious that these very geometrical field equations should have the properties one would require:

(i) They should reproduce the results of Newtonian gravitational theory in the limit of low masses, large separations and low velocities.

(ii) They should determine the metric up to the freedom to make coordinate transformations, given initial data on some spacelike surface.

Einstein essentially demonstrated property (i) in his original 1915 paper by examining the linear approximation to the field equations.

From this and the next approximation (the post-Newtonian), he obtained his three famous tests of general relativity: the gravitational redshift, the bending of light by the sun, and the advance of the perihelion. He also showed that the linearized approximation was consistent with property (ii), though a full derivation did not come till later. At present we can show that suitably differentiable data determine a unique solution for a finite time interval from the initial surface (see chapter 4 by Fischer and Marsden). The differentiability requirements, however, are still higher than seems physically reasonable, and there is no proof of the existence of a solution for all time. This latter point is closely connected with the occurrence of singularities, discussed below.

In his original papers, Einstein assumed as an additional postulate that small, low-mass, isolated, uncharged bodies (test bodies) move along geodesics of the metric g_{ab}. It was later shown, however, that this postulate was a consequence of the local conservation of the energy–momentum tensor, which in turn was a consequence of the field equations. To analyse the motion of the bodies whose gravitational mass could not be neglected, approximation schemes are necessary. There are two main methods, the slow-motion, and the fast-motion approximations. The first was introduced by Einstein, Infeld, and Hoffmann in 1938 to describe the motion of a system of supposedly point masses under their mutual gravitational influence. In it, the perturbations are expanded in a power series in the velocities which are taken to be of the same order as the square-root of the gravitational binding energy. Subsequent work on gravitational collapse showed that the picture of point masses moving along timelike world-lines is too idealized, but the scheme was extended by Chandrasekhar to deal with continuous distributions of matter, and by D'Eath to deal with black holes, which are the nearest approximation to point masses. In the fast-motion scheme one expands in powers of the gravitational binding energy, but this is no longer assumed to be related to the velocities, which may be large. This makes it much harder to keep track of the different orders of quantities in the perturbation expansion, and the accuracy of the results obtained is often unclear.

In general relativity the energy–momentum tensor obeys the equation

$$T^{ab}{}_{;b} = 0$$

as a consequence of the field equations. These express the local conservation of energy and momentum in local Minkowskian coordinates. However, the curvature of the metric in general prevents them from

Chapter 1. Introductory survey

being integrated to give global conservation laws. The physical reason is that T^{ab} represents the energy and momentum of the matter field only, and does not include a contribution from the gravitational field. In fact, there is no locally defined quantity which measures the energy and momentum of the gravitational field, because the field can always be transformed to zero locally by choosing local inertial coordinates. This non-localizability of gravitational energy caused much confusion in the early days of general relativity. One method of dealing with it, proposed by Einstein, was to introduce for each given set of coordinates x^a an object called the pseudo-tensor $t_a{}^b$, supposed to represent the gravitational energy–momentum. This satisfied the equations

$$\frac{\partial}{\partial x^b}(t_a{}^b + T_a{}^b) = 0.$$

These could be integrated to give global conservation laws. By a coordinate transformation the quantity $t_a{}^b$ could be made zero at any given point, but Einstein showed that the integrated energy and momentum would be unchanged, provided the metric approached the Minkowski metric in a suitable way at spatial infinity. Later work on such asymptotically flat spacetimes showed that this total energy and momentum, as measured at infinity, could be defined simply from the asymptotic form of the metric on an asymptotically flat, spacelike, initial hypersurface. One would expect that the matter fields would give a positive contribution to the mass as measured at infinity. But because gravity is attractive one would expect the gravitational potential energy to give a negative contribution. It is therefore not obvious that the total mass should be positive, though one of the major unproved conjectures of general relativity is that it is always positive for non-singular initial data. What seems to happen is that, if one considers a sequence of initial-data sets in which the matter is concentrated into smaller and smaller regions, an event horizon forms and the matter collapses into a black hole with a positive mass before the negative gravitational potential energy exceeds the positive matter contribution.

The work on the initial-value problem shows that the Einstein equations form a quasi-linear hyperbolic system. This means that the gravitational effects of a change in the matter distribution propagate along bicharacteristics of the system, which are in fact the null geodesics and coincide with the paths of light rays. In other words, gravitational disturbances propagate at the velocity of light. For a long time, however, there was disagreement whether those 'gravitational waves' carried energy away from a bounded system in an asymptotically flat space. It was difficult to answer this using the definition of mass in terms of spacelike surfaces, because later spacelike surfaces would intersect the gravitational waves and therefore the mass defined on them would also include the energy of the waves. This problem was overcome by Bondi, Metzner, Van den Burg, and Sachs in 1962 by introducing the concept of mass defined on an outgoing null surface. They were able to show that the mass associated with successive null surfaces decreased by a positive amount that was a measure of the gravitational radiation that escaped to infinity between the null surfaces. It is, however, still an unsolved problem to relate this mass to the one defined on spacelike surfaces. One would expect that the spacelike mass would be the limit of the null mass when the null surface is taken in the infinite past. Another problem that has not been completely cleared up is to relate the decrease of mass at infinity to the motion of the bodies producing the gravitational radiation. One would expect some sort of radiation reaction as in electrodynamics, but attempts to derive this from the approximation schemes mentioned earlier have produced conflicting results.

1.3 Cosmology

Newton's theory of gravity was formulated to deal with a finite collection of bodies, the solar system. However, problems arose when one tried to apply it to the whole universe: if the material content of the universe were finite, why did it not all fall together to make one big body? At the time, and indeed until the 1920s, no-one considered the possibility that the bodies in the universe could be moving systematically towards or away from each other. Newton therefore suggested that the universe contained an infinite number of bodies, uniformly distributed and at rest relative to each other, and that they would not fall towards each other because there would be no preferred point for condensation. This concept of an infinite, uniform and static universe raised a number of problems. First of all, the gravitational potential would be infinite, and the gravitational force on a body would not be well defined. In an attempt to overcome this difficulty it was suggested that the Newtonian law of attraction should be weakened at large distances. This could be done by adding a 'cosmological term' to Poisson's equation

$$\nabla^2 \Phi = 4\pi G \rho + \Lambda \Phi$$

where Φ is the gravitational potential, Λ is the 'cosmological constant'

Chapter 1. Introductory survey

and ρ the material density. Another problem, first pointed out by Halley, later Olbers, and nowadays referred to as 'Olbers' paradox', was that if the universe had existed in its present state for an infinite time, almost every line of sight would end on the surface of a star, and therefore the sky ought to be as bright as the surface of the sun. Because of the finite velocity of light, this difficulty could be overcome by postulating that the stars had been in their fixed positions for ever, but that they had begun to shine only at some finite time in the past. However, it was not clear why they should suddenly have started to shine after all that time.

Newton's theory was criticized by Bishop Berkeley because it endowed space with a structure of its own. Although there was no absolute standard of rest since all uniform linear velocities were equivalent, there was an absolute zero of acceleration (that in which no external forces acted) and an absolute zero of rotational velocity (that in which there were no centrifugal forces). This structure of space would exist even if the universe contained no matter. Berkeley claimed that such absolute properties of space were metaphysical and that one should deal only with the motion of matter relative to the other matter in the universe. Berkeley's ideas were taken up by Mach, and are now generally known as 'Mach's Principle'. Einstein was strongly influenced by Mach and believed that general relativity incorporates his Principle. However, this is not the case, because the Einstein equations admit many solutions, including the flat Minkowski metric, which contain no matter at all. In fact, by making the metric a dynamical field, Einstein removed the distinction between matter and the metric, which embodies the structure of spacetime. On a modern view, one would say that a graviton, an excitation of the metric, is just as much an elementary particle as a 'matter quantum' like a photon or an electron.

Einstein's first attempt to construct a general relativistic cosmological model was made in 1917, when the universe was still regarded as being static. In order to balance the gravitational attraction of the matter in the universe he introduced the 'cosmological term' Λg_{ab} into the field equations. This was similar to the Newtonian cosmological term. However, whereas the Newtonian model had been infinite in both space and time, Einstein's model was infinite in time, but its spatial cross-sections were finite but unbounded, and were in fact three-dimensional spheres. Einstein believed that this was the only cosmological model with positive Λ which was non-singular. However, in 1919 de Sitter obtained another solution in which the density of matter was zero and in which test particles receded from each other with ever-increasing velocities. About this time evidence began to accumulate which indicated that nearby galaxies were receding from us. De Sitter's solution was not a satisfactory cosmological model because it contained no matter, but in 1922 Friedmann showed that the Einstein equations admit spatially isotropic, homogeneous solutions representing a uniform distribution of expanding matter. This class of solutions fitted the observations of the time, and indeed all subsequent measurements of the large-scale structure of the universe. The Friedmann models did not suffer from Olbers' paradox, because the expansion of the universe redshifted the light from distant galaxies and thereby diminished its intensity. They also had a beginning at a singularity a finite time in the past. They did not require a cosmological term to balance the gravitational attraction of the matter, because the matter was not stationary. The motivation for the cosmological term therefore disappeared, and it is now generally ignored. Measurements of distant galaxies place an upper limit of 10^{-66} cm^{-2} on $|\Lambda|$.

Although the spatially homogeneous and isotropic Friedmann models were very successful in describing the large-scale structure of the universe, they obviously do not reflect the local irregularities, such as stars and galaxies. To describe these, one wants to study deviations from the Friedmann models. This was first done by Lifshitz in 1946 using a linearized approximation. He found that small density enhancements would grow, but only rather slowly. It is therefore difficult to understand why the universe should be so homogeneous and isotropic on a large scale, and yet contain initial perturbations which were sufficiently large to have grown into galaxies by the present epoch. We refer the reader to chapter 9 by Dicke and Peebles. A class of deviations from the Friedmann models which can be analysed beyond the linear approximation are those which are anisotropic but spatially homogeneous. They were studied first by Taub and later, extensively, by Misner. The observed isotropy of the cosmic microwave background radiation places extremely low upper limits on any large-scale anisotropy at the present epoch. Possible reasons that have been suggested to explain why the universe should be so isotropic now, even if it began in a chaotic state, include particle creation in the very early universe and neutrino viscosity at a slightly later epoch. We refer the reader to chapters 10 and 11 by Zel'dovich and MacCallum.

All the Friedmann models without a cosmological term had a singular origin at a finite time in the past. If the cosmological constant was

Chapter 1. Introductory survey

positive and sufficiently large, it was possible to obtain a model without a singularity in which a contracting phase was followed by a 'bounce' and then an expansion. However, it was fairly easy to show this model was inconsistent with observations. There was considerable unwillingness to accept the idea that the universe and time itself had a beginning. For this reason Bondi, Gold and Hoyle suggested a steady state model in 1948. This was a Friedmann model in which the separation between neighbouring particles increased exponentially with time so that the rate of expansion was the same at all times. As the particles moved apart, new matter was created at a steady rate to maintain the density at a constant value. This 'continual creation' required either a violation of the local conservation of energy or the introduction of a field with negative energy density whose creation balanced the creation of matter with positive density. Both these alternatives seem rather undesirable because they might allow the rate of creation to run away. Nevertheless the model had the apparently attractive feature of not having a beginning while at the same time avoiding Olbers' paradox by expanding and redshifting the light from distant galaxies. However a series of counts of radio-sources in the late 1950s and early 1960s showed it to be incompatible with observations: there were more radio-sources in the past than there are now.

The conclusion that the universe is evolving was reinforced by the discovery in 1965 of microwave background radiation with a thermal spectrum at 3 K. It is very difficult to account for this radiation except as a remnant from an earlier hot, dense phase of the universe. Such a phase would occur if the Friedmann models remained a good approximation back to about one hundred seconds after the initial singularity. But was there really a singularity which was the beginning of the universe and of time itself? This question was studied by Lifshitz and Khalatnikov (1963). They constructed a power series approximation to what they thought was the most general solution of the Einstein equations with a singularity. They found it contained one fewer arbitrary function than a fully general solution. From this they concluded that singularities would not occur in a general solution but only in those with some special symmetry like the Friedmann models. However they later realized that their solution was not the most general and that one could indeed obtain singular solutions with the full number of arbitrary functions (Lifshitz and Khalatnikov, 1970). This suggests that singularities are possible in general solutions but it does not answer the question of whether they are inevitable under certain conditions.

Moreover the validity of their power series expansion is not completely clear.

A very different approach to the occurrence of singularities was initiated by Penrose (1965). He used global geometrical constructions to show that, under certain assumptions, a singularity was inevitable in the gravitational collapse of a star. These global methods were extended and applied to cosmology by Hawking and Geroch. The basic idea was to obtain a contradiction to the postulate that the universe is singularity-free by showing that if it were, on the one hand there should be a timelike geodesic which maximized the length of timelike curves between a certain pair of points, and on the other hand the timelike geodesic could be perturbed to give a slightly longer curve. The culmination of this work was a general-purpose singularity theorem of Hawking and Penrose (1970). It, and the earlier theorems, did not depend on the detailed nature of the material content of the universe but only on certain reasonable inequalities on the energy–momentum tensor which expressed the property that gravity should always be attractive. The observation of the microwave background radiation indicates that the universe contains enough matter to satisfy the main condition of the Hawking–Penrose theorem. Thus, according to classical general relativity at least, the universe should contain a singularity which would be a beginning of time for at least some timelike or null geodesic. The generality of the theorem does not allow one to conclude that all the timelike or null geodesics have a beginning a finite distance in the past, but it seems likely that this is the case in generic solutions. The theorem also does not prove that the incompleteness of the geodesics is caused by the curvature becoming unboundedly large. Indeed there are some examples in which this does not happen. However it seems that these are special cases and Clarke (1975) has shown that, at least in certain cases, a general solution will have a curvature singularity. For more detail on global methods and singularities we refer the reader to chapters 5 and 12).

1.4 Gravitational collapse

The theorems of Penrose and Hawking also imply that a singularity is inevitable in the gravitational collapse of a star once it has passed a certain critical stage. This 'point of no return' occurs when the gravitational field becomes so strong that it drags back any further light emitted by the star. It can be characterized by the appearance of a 'closed

Chapter 1. Introductory survey

trapped surface', a spacelike 2-surface such that both the outgoing and the ingoing families of future-directed null geodesics orthogonal to it are converging. This means that if one imagined that the 2-surface emitted an instantaneous flash of light, both the outgoing and ingoing wavefronts would be decreasing in area. The effect of any matter or curvature of spacetime that the wavefronts encountered would be to make them decrease in area even more rapidly. Thus the outgoing wavefront would shrink to zero in a finite time (more precisely, a finite affine distance) trapping within it all the material of the star. This cannot get out because it cannot travel outwards faster than light. Obviously something goes wrong and what happens is that a singularity occurs.

A cosmological singularity in the past is bad enough because it implies that there was a time before which nothing was defined and no questions could be asked, but a singularity in the future, such as is predicted to occur in gravitational collapse, is even more disturbing because it means that time itself comes to an end as would any helpless observer who was foolish enough to follow the collapse of the star. Near the predicted singularity the gravitational field would presumably get very large and as yet unknown quantum gravitational effects would be expected. Thus the singularity theorems show that classical general relativity predicts its own breakdown. However it seems that we may be shielded from the full effects of a breakdown because it may always occur in a region of spacetime, a black hole, which is not visible to an external observer, at least classically. This conjecture was given the name 'the cosmic censorship hypothesis' by Penrose. Without it we could be in for a surprise every time a star in the galaxy collapsed.

Despite considerable effort, the cosmic censorship hypothesis remains undecided within the framework of classical general relativity. The problem is complicated by the fact that one does not wish to consider inessential singularities caused by bad choices of coordinates or matter singularities, such as the intersection of the flow lines of a fluid, that are nothing to do with gravity but merely reflect the failure of a phenomenological description of the material content. However, the hypothesis is supported by perturbation and computer calculations and by the failure of a number of attempts to obtain contradictions from it. It is therefore generally believed and forms the basis of all work on black holes.

From this hypothesis and the assumption that the energy density of matter is non-negative (which should hold in the classical regime), it can be shown that the surface area of the event horizon, the boundary of a

Gravitational collapse

black hole, can never decrease with time (Hawking, 1971; an earlier version of this result was obtained by Christodoulou, 1970). Moreover when two black holes collide and merge together to form a single black hole, the area of the event horizon around the final black hole is greater than the sum of the areas of the horizons of the initial black holes. One would therefore expect that a system undergoing gravitational collapse or containing colliding black holes would eventually settle down to a stationary black hole state with the emission of a certain amount of gravitational radiation. By the 'no-hair theorem' referred to earlier, these stationary states are described by the Kerr–Newman family of solutions which depend only on three parameters; the mass, the angular momentum, and the electric charge of the black hole. (See chapter 6.) A real black hole interacting with the rest of the universe will not be in an exactly stationary state but in many cases it can be treated as a small perturbation from a Kerr solution. It is a remarkable property of these solutions that the perturbation equations in them can be solved by separation of variables (see chapter 7). This allows detailed calculations of many processes to be made. Situations in which a small perturbation treatment is not valid are the collapse of an asymmetric star or the collision of two black holes. One would like to know how much gravitational radiation would be given off in such cases because they seem to be the most likely sources to be detectable (see chapter 3). Computer calculations of axisymmetric collisions indicate that the amount of gravitational radiation emitted is rather small (Smarr, 1977) but there are suggestions that the amount may be much larger in the more realistic non-axisymmetric case.

The observational search for black holes really got off the ground with the launch of the X-ray satellite Uhuru in 1971. At present the best candidate is Cygnus X-1 which was described earlier, but there are a number of similar systems for which the observational data are less complete. Uhuru also discovered X-ray sources which seem to be at the centres of globular clusters of stars. It is thought that these might be black holes of about one thousand solar masses which are accreting gas lost from the stars in the cluster. On a large scale, it seems likely that black holes of a million solar masses or more may be present in galactic nuclei and in quasars, but there is not yet much direct observational evidence for this. We refer the reader to chapter 8.

As mentioned above, the area of the event horizon of a black hole has the property that it can only increase and not decrease with time. This led Bekenstein in 1972 to suggest that it might be connected with the

Chapter 1. Introductory survey

thermodynamic quantity, entropy, which measures the degree of disorder of a system or one's lack of knowledge of it. He pointed out that the 'no-hair theorem' implied that a very large amount of information about a star was irretrievably lost when it collapsed to form a black hole and he claimed that the area of the event horizon was a measure of this unobservable information which could be regarded as the entropy of the black hole. Other analogies between classical black holes and thermodynamics were found by Bardeen, Carter and Hawking (1973) but there was an apparently insurmountable obstacle to attributing a finite entropy to a black hole because that would imply that it should have a finite temperature and should be able to remain in equilibrium with thermal radiation at the same temperature. However this seemed impossible because the black hole would absorb some of the radiation but, by its very definition, it would not be able to emit anything in return.

The paradox remained until Hawking (1974) discovered that applying quantum mechanics to matter fields in the background geometry of a black hole metric led to a steady rate of particle creation and emission to infinity. The emitted particles would have a thermal spectrum with a temperature proportional to the surface gravity of the black hole, which is a measure of the strength of the gravitational field at the event horizon and which is inversely proportional to the mass. This emission would enable the black hole to remain in equilibrium with thermal radiation at the same temperature.

There are a number of different ways of explaining the thermal radiation from black holes but the following is probably the easiest to understand. The Uncertainty Principle implies that 'empty' space is filled with pairs of 'virtual' particles and antiparticles which appear to gather at some point of spacetime, move apart and then come together again and annihilate each other. They are called virtual because unlike 'real' particles they cannot be observed directly with a particle detector but their indirect effects have been measured in a number of experiments such as the Lamb shift and the Casimir effect. If a black hole is present, one member of a pair may fall into the hole leaving the other without a partner with whom to annihilate. The forsaken particle or antiparticle may follow its mate into the black hole but it may also escape to infinity where it will appear to be a particle or antiparticle emitted by the black hole. Equivalently, one can think of the member of the pair that fell into the black hole (say, the particle) as an antiparticle travelling backwards in time and coming out of the hole. When it reached the point at which the particle–antiparticle pair first appeared, it would be scattered by the gravitational field into an antiparticle travelling forwards in time and going out to infinity. Thus one can think of the emitted radiation as having come from inside the black hole and having quantum mechanically tunnelled through the potential barrier around the hole created by the gravitational field, a barrier that could not be surmounted classically. By the 'no-hair theorem' an external observer's knowledge of the region from which the radiation came is limited to its total mass, angular momentum, and electric charge. For this reason the black hole emits with equal probability every configuration of particles with the same total energy, angular momentum and charge (though not all these configurations escape to infinity with equal probability because the potential barrier around the black hole depends on the energies and angular momenta of individual particles and may reflect them back into the hole). Thus it is possible for the black hole to emit a television set or Charles Darwin, but the number of configurations corresponding to such exotic possibilities is very small. The overwhelming probability is for the emitted particles to have an almost thermal spectrum. The fact that the radiation is thermal means that it has an extra degree of randomness or unpredictability over and above that normally associated with quantum mechanics. In classical mechanics one can make definite predictions of both the position and velocity of a particle. In ordinary quantum mechanics one can definitely predict *either* the position *or* the velocity but not both. Alternatively one can predict one combination of position and velocity. Thus, roughly speaking, one's ability to make definite predictions is cut by half. However in the case of particles emitted by a black hole, one can definitely predict neither the position nor the velocity. All that one can predict is the *probabilities* that the particles will be emitted in certain modes. This loss of predictability seems to be associated with the breakdown that one would expect at a spacetime singularity. One could interpret it by saying that new random information is entering the region of the universe that we observe from the interior of the black hole.

For solar-mass black holes the predicted temperature is only about 10^{-7} K so the radiation would be completely negligible and would be swamped by the 3 K microwave background. However there might also be a population of very much smaller 'primordial black holes' which might have been formed by the collapse of inhomogeneities in the hot early stages of the universe. A black hole of 10^{15} g will have a radius of about 10^{-13} cm and would be emitting energy at a rate of about 6000

Chapter 1. Introductory survey

megawatts, mainly in gamma rays, neutrinos and electron–positron pairs. The emitted energy would be balanced by a flux of negative energy through the event horizon into the black hole. The event horizon would therefore decrease in area, thus violating the classical law that it could only increase. However one would still have a generalized second law of thermodynamics which states that the entropy outside black holes plus a certain multiple of the area of black hole horizons never decreases.

The emission from black holes has been derived from the semi-classical approximation in which the metric of spacetime is treated classically and the matter fields are treated quantum mechanically on the given spacetime background (see chapter 13). This should be a good approximation for black holes which are large compared to the Planck mass of about 10^{-5} g but it will fail when the black hole mass gets down to this value. So far there is no good theoretical framework with which to treat the final stages of a black hole but there seem to be three possibilities:

(i) The black hole might disappear completely, leaving just the thermal radiation that it emitted during its evaporation.

(ii) It might leave behind a non-radiating black hole of about the Planck mass.

(iii) The emission of energy might continue indefinitely creating a negative-mass naked singularity.

The third possibility seems implausible because it would lead to runaway solutions. The second possibility also seems rather improbable and it would imply that there might be very large numbers of dead Planck-mass black holes which might have a mass density higher than the limits set by cosmological observations. The first possibility, that black holes disappear completely, seems the most natural. If it holds, one must have non-conservation of baryon number from the following argument. Suppose a star consisting mainly of baryons collapses to form a black hole. The temperature of the black hole will be very low initially, so it will radiate most of its energy in zero-rest-mass particles like neutrinos, gravitons and photons. Only when the mass of the black hole gets down to about 10^{15} g will the temperature become high enough to emit significant numbers of baryons and antibaryons. Even if there were some mechanism for favouring the emission of baryons over antibaryons at this stage (and none is known) it would be too late because there would be insufficient energy left to emit the original number of baryons. Thus, if the black hole disappears completely, there will be a violation of the usual law of conservation of baryon number.

Quantum gravity

The only hope of observing thermal radiation from black holes would seem to be if there were significant numbers of primordial black holes with masses less than about 10^{17} g. Whether this is the case or not depends on conditions in the early universe about which little is known. Measurements of the cosmic gamma-ray background around 100 MeV place an upper limit of about 200 per cubic light year for the average density of black holes in the mass range around 10^{15} g. To do better than this one would have to try to detect individual black holes by the burst of very rapid emission that they would produce at the end of their life. This might require very large gamma-ray detectors in space.

1.5 Quantum gravity

So far the work that has been described in this chapter on classical general relativity, and even on quantum theory on a fixed spacetime background is fairly well understood and complete. We, the editors, are reasonably confident that the articles on these subjects in this volume will remain reasonable reviews for some time to come and that what further developments there are will merely add to and not replace what we already have. The same cannot be said for the three articles on quantum gravity by DeWitt, Weinberg and Hawking. At the moment it is not clear what form a consistent quantum theory of gravity would take or whether one even exists at all. However, there is a great deal of work going on in this area and some progress is being made.

There are a number of different approaches to the quantization of gravity, some of which are described in the three articles in this volume. Following the successful use of Feynman diagrams in quantum electrodynamics, it was natural to try similar techniques in gravity. However, the naive application of Feynman rules gave results which violated unitarity whenever there was a closed graviton loop. Feynman, DeWitt, Fadeev and Popov showed how this lack of unitarity arose from the gauge freedom in general relativity which corresponds to the fact that the same geometry can be represented by many different metrics in different coordinate systems. One can restore unitarity by introducing a correction term which can be represented by additional Feynman graphs containing closed loops of 'fictitious' or 'ghost' particles.

Even with the ghost particles, the closed loops cause difficulties because, as usual in field theory, they diverge. There are three types of divergence which go quartically, quadratically and logarithmically in a cut-off parameter. The first two do not cause any problem because they

Chapter 1. Introductory survey

can be discarded without introducing any ambiguity. However this is not true of the logarithmic term: one is left with an undetermined multiple of some function of the external parameters of the Feynman graph. In other field theories like quantum electrodynamics or Yang–Mills theory one can absorb this undetermined quantity into a redefinition or renormalization of the coupling constant which then becomes a function of the energy at which it is measured. However this will not work in general relativity because there is no dimensionless coupling constant. There are currently several different opinions as to what to do about this awkward problem. One school sticks to the traditional Feynman rules but seeks to introduce additional matter fields which will cancel out the embarrassing logarithmic divergences. A particularly promising development along these lines is supergravity in which the graviton is related to all the other physical fields, bosons and fermions, by certain supersymmetric transformations. These theories have been shown to have no logarithmic divergences at one loop and probably also at two loops in perturbation theory around flat space (the same is true of pure gravity but not gravity coupled to ordinary, non-supersymmetric matter). However it seems that there will probably be divergences at three loops and there also are divergences at one loop for perturbations around metrics which are not topologically trivial.

A second school of thought also uses perturbation theory around flat space. It does not seek to cancel all the logarithmic divergences but rather to find a criterion for determining the ambiguous quantities to which they give rise. An approach on these lines is described in the article by Weinberg. He considers (in a formal sense) general relativity in $2+\varepsilon$ dimensions. He shows that if the undetermined quantities in this theory have certain definite values, the theory takes a particularly simple form at high energies. He then argues that this requirement of 'asymptotic safety' should also determine the ambiguous quantities even when ε is 2, i.e. in four dimensions.

A third school maintains that perturbation theory around flat space is bound to break down and that one cannot hope to obtain a consistent theory of quantum gravity without taking account of the many different topologies that the gravitational field can have. Some success along these lines has already been obtained by deriving the entropy associated with black holes from a path-integral formulation in which the gravitational field itself is quantized. This agrees strikingly with the results of the semi-classical approximation. However a lot more remains to be done to show that this approach leads to a successful theory. More details are contained in chapter 15.

1.6 Future prospects

The main qualitative features of the classical theory of general relativity now seem to be well understood. While this chapter was being prepared a result was reported which effectively resolves one of the major outstanding questions, the positive energy conjecture. The main global problem now remaining is the cosmic censorship hypothesis which is the basis for all work on black holes. On the other hand a lot remains to be done on the quantitative side. We do not yet have reliable estimates of how much gravitational radiation would be emitted in a gravitational collapse or how radiation reaction forces modify the motion of bodies in strong gravitational fields. It is likely that in the future classical general relativity will become mainly a 'service subject' for astrophysicists, like electrodynamics. The main areas of interest will probably centre around black hole models of quasars and processes in the early universe.

There is currently a great deal of interest in the problem of combining gravity and quantum mechanics in a consistent theory. The eventual hope is to find a Grand Unified Theory which will incorporate all particles and all interactions. This would realize the cherished dream of Einstein's later years, but in a form rather different from that he envisaged.

2. The confrontation between gravitation theory and experiment†

CLIFFORD M. WILL‡

2.1 Introduction

For over half a century, the general theory of relativity has stood as a monument to the genius of Albert Einstein. It has altered forever our view of the nature of space and time, and has forced us to come to grips with the issues of the birth and ultimate fate of the universe. Yet despite its great influence on scientific thought, general relativity was supported initially by very meager empirical evidence. For nearly forty-five years, the validity of the theory was confirmed by data that, by today's standards, would be considered qualitative at best. Three 'classical tests' formed the backbone of general relativity: the excess perihelion shift of Mercury, in agreement with the theory's prediction, but attributable at least in part to other possible causes; the deflection of light by the Sun, measured to be anywhere between one half and twice its predicted value; and the gravitational redshift, observed in spectral lines of white dwarfs again to be anywhere between one half and twice its predicted value, and moreover suspected to be not a true test of general relativity anyway.

But the astronomical and technological revolution of the 1960s and 70s has brought about a confrontation between general relativity and experiment at unprecedented levels of accuracy. Technological advances in atomic clocks, radar and laser ranging to planets and artificial satellites, long-baseline radio interferometry, low-noise motion sensors, to name only a few, have made the systematic high-precision testing of gravitational theory almost routine. Measurements to accuracies of fractions of a per cent and better of the miniscule effects predicted by general relativity for the solar system are not unheard of.

The desire to perform such high-precision tests of gravitation theories, moreover, was intensified during the past two decades by

---
† Supported in part by the National Aeronautics and Space Administration (NSG 7204 S1) and the National Science Foundation (PHY 76-21454).
‡ Alfred P. Sloan Foundation Research Fellow.

Introduction

discoveries of exotic astronomical objects in which relativistic gravitational effects appear to play a crucial role, discoveries such as quasars, pulsars, compact X-ray sources, and the cosmic microwave background, that heightened the theoretical astrophysicist's need for confidence in one theory of gravitation to be used in model-building. That need was further heightened by the introduction of several competing theories of gravitation, principally the Brans–Dicke theory, that made forceful claims to be viable alternatives to general relativity. During the 1960s and 70s the amount of activity and degree of sophistication that went into the construction of alternative theories of gravity made general relativity at times appear to be just one among many equal competitors. So much so that a 'theory of gravitation theories' was developed during the late 1960s and early 70s to study and classify all theories of gravitation in as unbiased a manner as possible. Pioneered by Robert H. Dicke and Kenneth Nordtvedt, Jr, this 'theory of theories' could also be put to powerful use analyzing the new high-precision experiments, and suggesting future experiments made possible by further technological advances.

To date, general relativity has passed every experimental test it has been put to, and many of its competitors have fallen by the wayside, but the confrontation of gravitation theory with experiment is far from over. Solar-system tests will continue to improve in precision, with major advances in sight during the 1980s. But tests of gravitation theory need not be confined to the solar system. New arenas for the confrontation may soon be available, arenas where experimenters can probe aspects of gravitation theory unobservable in the standard solar-system tests. These include gravitational radiation tests, stellar-system tests (the binary pulsar), cosmological tests, and laboratory tests. One can only guess whether general relativity will continue to survive each new confrontation, but whether it does or not, the depth of understanding of gravitation and its effects that we have gained because of general relativity will in itself remain as a lasting tribute to the insight of Albert Einstein.

We begin this review (section 2.2) by analyzing the foundations of gravitation theory – principles of equivalence, the fundamental criteria for the viability of a gravitational theory, and the experiments that support those criteria. One of the central conclusions of section 2.2 is that the correct, viable theory of gravity must in all probability be a 'metric' theory. In section 2.3 we focus attention on solar-system tests, using a 'theory of theories' known as the parametrized post-Newtonian

Chapter 2. Confrontation between gravitation theory and experiment

(PPN) formalism that encompasses most metric theories of gravity and that is ideally suited to the solar-system arena. In section 2.4 we discuss gravitational radiation as a possible tool for testing gravitational theory. The binary pulsar, a new, 'stellar-system' testing ground is studied in section 2.5 In section 2.6 we describe tests of gravitation theory in a cosmic arena. A summary is given in section 2.7.

Throughout this review we adopt the units and conventions of Misner, Thorne and Wheeler (1973, hereafter referred to as MTW). Portions of this paper are based on the author's 1972 Varenna lectures 'The theoretical tools of experimental gravitation' (Will, 1974a, hereafter referred to as TTEG). For detailed discussion of the mathematical methods and calculational tools that are only touched on here, the reader is referred to TTEG, and to MTW chapters 38–40. Other useful reviews of this subject include Richard (1975), Brill (1973), Nordtvedt (1972), Will (1972, 1974b), and Dicke (1964a, b). For a thorough review of the experimental situation up to 1961, see Bertotti *et al.* (1962).

2.2 Principles of equivalence and the foundations of gravitation theory

The Principle of Equivalence has played an important role in the development of gravitation theory. Isaac Newton regarded this principle as such a cornerstone of mechanics that he devoted the first paragraphs of the *Principia* to a detailed discussion of it. He also made mention there of the pendulum experiments he had performed to verify it. Einstein used the principle as a basic element of general relativity. Yet it is only relatively recently that we have gained a deeper understanding of the significance of the Principle of Equivalence for gravitation theory and for experiment. Largely through the work of Robert H. Dicke, we have come to view the Principle – or better, Principles – of Equivalence, and experiments such as the Eötvös experiment, the gravitational redshift experiment, and so on, as probes more of the foundations of gravitation theory than of general relativity itself. This viewpoint is part of what has come to be known as the Dicke framework (section 2.2.1); it allows one to discuss at a very fundamental level the nature of spacetime and gravity. Within it one asks questions such as: Do all bodies respond to gravity with the same acceleration? Is space locally isotropic in its intrinsic properties? What types of fields, if any, are associated with gravity – scalar fields, vector fields, tensor fields...? Out of this framework has come a set of fundamental criteria that any potentially viable gravitation theory should satisfy (section 2.2.2), and a way of viewing the key experiments that form the empirical basis for these criteria (section 2.2.3). These criteria have important consequences for gravitation theory according to a conjecture based on the work of the late Leonard Schiff. Schiff's conjecture (section 2.2.4) states that any theory of gravity that satisfies the fundamental criteria for viability must necessarily be a metric theory. Schiff's conjecture and the Dicke framework have spawned a number of concrete theoretical formalisms (such as the $TH\epsilon\mu$ framework, section 2.2.5) for comparing and contrasting metric and non-metric theories of gravity, for analyzing the experiments that test the criteria for viability, and for proving Schiff's conjecture (section 2.2.5).

2.2.1 The Dicke framework

The Dicke framework for analyzing experimental tests of gravitation was expounded in Appendix 4 of Dicke's 'Les Houches lectures' (Dicke, 1964a). It makes two main assumptions about the nature of gravity:

(i) Spacetime is a four-dimensional manifold, with each point in the manifold corresponding to a physical event. The manifold need not *a priori* have either a metric or an affine connection.

(ii) The equations of gravity and the mathematical entities in them are to be expressed in a form that is independent of the particular coordinates used, i.e. in covariant form.

The Dicke framework is particularly useful for designing and interpreting experiments that ask what types of fields are associated with gravity (Dicke, 1964a; Thorne and Will, 1971; TTEG § 2; MTW § 38.7). For example, there is strong evidence from elementary-particle physics for at least one symmetric, second-rank tensor field that reduces to the Minkowski metric η when gravitational effects can be ignored. The Hughes–Drever experiment rules out the existence of more than one second-rank tensor field, both coupling directly to matter, and various ether-drift experiments rule out a vector field coupling directly to matter. (These experiments do not rule out, however, vector and tensor fields which couple only to gravity or to matter's self-gravitational energy (Will and Nordtvedt, 1972).) No experiment has been able to rule out or reveal the existence of a scalar field, although several experiments have set limits on specific scalar–tensor theories (see section 2.3).

Chapter 2. Confrontation between gravitation theory and experiment

However, this is not the only powerful use of the Dicke framework.

2.2.2 Criteria for the viability of gravitation theory

The general, unbiased viewpoint embodied in the Dicke framework has permitted gravitation theorists to formulate a set of fundamental criteria that any gravitation theory should satisfy if it is to be viable. Two of these criteria are purely theoretical, while four are based on experimental evidence.

(i) It must be complete: that is, it must be capable of analyzing from 'first principles' the outcome of every experiment of interest. It is not enough for the theory to *postulate* that bodies made of different material fall with the same acceleration. The theory must incorporate and mesh with a complete set of electromagnetic and quantum mechanical laws, which can be used to calculate the detailed behavior of bodies in gravitational fields. This demand of completeness does not extend, of course, to areas such as quantum gravity (see chapters 14, 15 and 16), strong and weak interaction theory, singularities (see chapters 5 and 12), and so on, where the theory even in special or general relativity cannot be regarded as fully developed or complete.

(ii) It must be self-consistent. A gravitation theory is consistent if its prediction for the outcome of every experiment is unique, i.e. if when one calculates the prediction by two different methods, one always gets the same results, for example the bending of light computed either in the geometrical optics limit of Maxwell's equations or in the zero-rest-mass limit of the motion of test particles.

(iii) It must be relativistic: that is, in the limit as gravity is 'turned off' compared to other physical interactions, the nongravitational laws of physics must reduce to the laws of special relativity.

(iv) It must have the correct Newtonian limit: that is, in the limit of weak gravitational fields and slow motions, it must reproduce Newton's laws (to an appropriate level of accuracy) as observed for example in Cavendish experiments, planetary motion, and stellar structure.

TTEG § 3.2 and Thorne *et al.* (1971) give partial lists of theories that violate one or more of these four criteria.

(v) It must embody the Weak Equivalence Principle (WEP): which states that *if an uncharged test body is placed at an initial event in spacetime and is given an initial velocity there, then its subsequent worldline will be independent of its internal structure and composition* (see

Thorne *et al.*, 1973, for detailed discussion and definitions). This principle has sometimes been called the Universality of Free Fall (UFF) (MTW, § 38.3).

(vi) It must embody the Universality of Gravitational Redshift: which states that *the gravitational redshift between a pair of identical ideal clocks at two events in spacetime is independent of their structure and composition* (Will, 1974c).

Before addressing the theoretical implications of these criteria, we turn to their experimental justification.

2.2.3 Experiments that probe gravity's roots

A variety of experiments, both in the laboratory and in the solar system provide the empirical backbone of criteria (iii) through (vi).

(a) Tests of special relativity

A host of experiments in the high-energy laboratory have verified and reverified the validity of special relativity in the limit when gravitational effects can be ignored. These experiments range from direct tests of time-dilation to tests of esoteric predictions of Lorentz-invariant quantum field theories (see for example, MTW § 38.4; Thorne and Will, 1971; Mansouri and Sexl, 1977a, b).

The Hughes–Drever, or 'Isotropy of Inertial Mass' experiment (Hughes, *et al.*, 1960; Drever, 1961) can be regarded as a test of special relativity (as well as a test for a second cosmic tensor field (Peebles and Dicke, 1962; Peebles, 1962)). That experiment placed a limit of 2×10^{-16} eV on the relative shift between the $m = \frac{3}{2}$ and $m = \frac{1}{2}$ levels (split in an external magnetic field) of the $1p_{3/2}$ nuclear energy level of ^{7}Li. The total binding energy in those levels is about 1 MeV so the experimental limit represented a few parts in 10^{22}. If there were a violation of Lorentz invariance in electrodynamics for instance, then because of the Earth's motion through the preferred universal rest-frame singled out by the broken Lorentz invariance, just such a splitting would occur. If δ_L is a dimensionless parameter that symbolizes the 'deviation' from Lorentz invariance (for a concrete form of δ_L see section 2.2.5), then the limit placed on δ_L by the Hughes–Drever experiment is given by

$$|\delta_L| < 10^{-13} (w/30 \text{ km s}^{-1})^{-2}, \qquad (2.1)$$

(Nordtvedt and Haugan, 1978) where w is the velocity of the Earth relative to the universal frame.

Chapter 2. Confrontation between gravitation theory and experiment

(b) The Newtonian limit

Massive amounts of empirical data support the validity of Newtonian gravitation theory (NGT), at least as an approximation to the 'true' relativistic theory of gravity. Observations of the motions of planets and spacecraft agree with NGT down to the level (parts in 10^8) at which post-Newtonian effects can be observed. Laboratory Cavendish experiments provide weaker support for NGT for small separations between gravitating bodies. However NGT has recently come under fire. It has been suggested that a massive short-range component to the gravitational interaction could be present, leading to modifications to NGT that can be described most conveniently by a gravitational 'constant' $G(r)$ that depends on the separation between gravitating bodies (Wagoner, 1970; Fujii, 1971, 1972; O'Hanlon, 1972). Long (1976) has recently reported experimental results that indicate a variation in $G(r)$ over distances between 5 and 30 cm of size

$$\varepsilon \equiv G^{-1} r \, dG/dr \approx (20 \pm 5) \times 10^{-4}, \quad (1\sigma)$$

where the error is one standard deviation (1σ). Attempts to perform similar experiments are in progress at other laboratories (Newman *et al.*, 1977; Paik *et al.*, 1977), while Mikkelson and Newman (1977) have carried out a detailed study of the limits placed on ε for distances larger than 100 cm, by a variety of astronomical observations. Further experiments and observations will be needed to strengthen our faith in the Newtonian limit.

(c) Tests of the Weak Equivalence Principle

A direct test of WEP is the comparison of the acceleration of laboratory-size objects of different composition. Many high-precision experiments to test WEP have been performed, from the pendulum experiments of Newton, Bessel and Potter, to the classic torsion-balance measurements of Eötvös, Dicke, Braginsky and their collaborators (see Dicke, 1964*a* for a detailed discussion of the experimental problems). Experiments to test WEP for individual atoms and elementary particles have been inconclusive or inaccurate with the exception of neutrons (Koester, 1976) (for the case of electrons, see Fairbank *et al.*, 1974). Table 2.1 discusses various experiments and quotes the limits that they set on the difference in acceleration *a* between objects of different

Table 2.1. *Tests of the Weak Equivalence Principle*

Experiment	Reference	Method	Substances tested	Limit on $|\eta|$
Newton	Newton (1686)	Pendula	Various	10^{-3}
Bessel	Bessel (1832)	Pendula	Various	2×10^{-5}
Eötvös	Eötvös, Pekár and Fekete (1922)	Torsion balance	Various	5×10^{-9}
Potter	Potter (1923)	Pendula	Various	2×10^{-5}
Renner	Renner (1935)	Torsion balance	Various	2×10^{-9}
Princeton	Roll, Krotkov and Dicke (1964)	Torsion balance	Aluminum and gold	10^{-11}
Moscow	Braginsky and Panov (1972)	Torsion balance	Aluminum and platinum	10^{-12}
Koester	Koester (1976)	Free fall	Neutrons	3×10^{-4}
Stanford	Worden (1976)	Magnetic suspension	Niobium, Earth	2×10^{-5}
Orbital[a]	Worden and Everitt (1974)	Free fall in orbit	Various	10^{-15} to 10^{-18}

[a] Experiments yet to be performed.

Chapter 2. Confrontation between gravitation theory and experiment

composition (A and B), given by

$$\eta \equiv (a_A - a_B)/\tfrac{1}{2}(a_A + a_B).$$

Future improved tests of WEP must reduce noise due to thermal and seismic effects, and so may have to be performed in space using cryogenic techniques. Hoped-for limits on η in such experiments range between 10^{-15} and 10^{-18} (Worden, 1976; Worden and Everitt, 1974).

(d) *Tests of UGR*

Gravitational redshift measurements are tests of UGR. A typical experiment measures the frequency or wavelength shift $z = \Delta\nu/\nu = -\Delta\lambda/\lambda$ between identical atomic frequency standards at different locations in a gravitational potential as a function of the potential difference ΔU. It is convenient for most purposes to view these experiments as measuring a parameter α given by

$$z = (1+\alpha)\Delta U. \qquad (2.2)$$

If UGR is satisfied, α is independent of the nature of the clocks being studied; moreover, theoretical arguments (Schiff's conjecture, see sections 2.2.4, 2.2.5) suggests that if UGR is satisfied, α must be zero. Table 2.2 summarizes the important redshift experiments that have been performed since 1960 (see Bertotti *et al.*, 1962, for a review of pre-1960 experiments) and notes a few that may be performed in the coming years. The first and most famous high-precision test was the Pound–Rebka–Snider experiment that measured the frequency shift of γ-ray photons from ^{57}Fe as they ascended or descended the Harvard Tower. The high accuracy was obtained by making use of the Mössbauer effect to achieve a narrow resonance line. Other experiments measured the shift of solar spectral lines in the Sun's gravitational field. Several tests have involved atomic clocks transported aloft on aircraft, rockets and satellites.

Recently however, a new era in redshift experiments has been ushered in with the development of frequency standards of ultrahigh stability – parts in 10^{15} to 10^{16} over averaging times of 10 to 100 seconds and longer. Examples are hydrogen-maser clocks (Vessot, 1974), superconducting – cavity stabilized oscillator (SCSO) clocks (Stein, 1974; Stein and Turneaure, 1975), and cryogenically cooled monocrystals of dielectric materials such as silicon and sapphire (McGuigan and

Table 2.2. *Tests of the Universality of Gravitational Redshift*

Experiment	Reference	Method	Limit on $|\alpha|$
Pound–Rebka–Snider	Pound and Rebka (1960) Pound and Snider (1965)	Fall of photons from Mössbauer emitters	10^{-2}
Brault	Brault (1962)	Solar spectral lines	5×10^{-2}
Jenkins	Jenkins (1969)	Crystal oscillator clocks on GEOS-1 satellite	9×10^{-2}
Snider	Snider (1972)	Solar spectral lines	6×10^{-2}
Jet-Lagged Clocks (*A*)	Hafele and Keating (1972*a*, *b*)	Cesium beam clocks on jet aircraft	10^{-1}
Jet-Lagged Clocks (*B*)	Alley *et al.* (1977)	Rubidium clocks on jet aircraft	2×10^{-2}
NASA-SAO Rocket Redshift Experiment	Vessot and Levine (1976)	Hydrogen maser on rocket	2×10^{-4}
Null Redshift[a] Experiment	Turneaure and Will (1975), Will (1977*a*)	Hydrogen maser versus SCSO	10^{-2}
Close Solar Probe[a]	Nordtvedt (1977)	Hydrogen maser or SCSO on satellite	10^{-6}

[a] Experiments yet to be performed.

Chapter 2. Confrontation between gravitation theory and experiment

Douglass, 1977). The first such experiment was the National Aeronautics and Space Administration/Smithsonian Astrophysical Observatory (NASA–SAO) Rocket Redshift Experiment that took place in June 1976. A hydrogen-maser clock was flown on a rocket to an altitude of about 10 000 km and its frequency compared to a similar clock on the ground. The experiment took advantage of the high-frequency stability of hydrogen-maser clocks (parts in 10^{15} over 100-second averaging times) by monitoring the frequency shift as a function of altitude. A sophisticated data acquisition scheme (Vessot, 1974) accurately eliminated all effects of the first-order Doppler shift due to the rocket's motion, while tracking data were used to determine the payload's location and velocity (to evaluate the potential difference ΔU, and the second-order Doppler shift). A crude, preliminary analysis of the data has yielded a limit of 200 parts per million on α (Vessot and Levine, 1976). (By strange coincidence, the Scout rocket that carried the maser aloft stood 22.6 meters in its gantry, almost exactly the height of the Harvard Tower.) In an interplanetary version of this experiment, a stable clock (H-maser or SCSO clock) would be flown on a spacecraft in a very eccentric solar orbit (closest approach ~4 solar radii); such an experiment could test α to a part in 10^6 (Nordtvedt, 1977) and could conceivably look for 'second-order' redshift effects (Jaffe and Vessot, 1976).

Advances in stable clocks have also made possible a new type of redshift experiment that is a direct test of UGR: a 'null' gravitational redshift experiment that compares two different types of clocks, side by side in the same laboratory. One such experiment (Turneaure and Will, 1975; Will, 1977a) would compare a hydrogen-maser and a SCSO clock and search for a daily variation in their relative frequencies as the Earth's rotation carries the laboratory in and out of the Sun's gravitational potential. Such a variation would have the form

$$\nu_H/\nu_{SCSO} = A[1 - 3 \times 10^{-13}(\alpha_H - \alpha_{SCSO})\cos(2\pi t/1 \text{ solar day})],$$

where A is the constant average ratio between the two clock frequencies (Will, 1977a). A limit of a part in 100 on the composition dependent coefficient $\alpha_H - \alpha_{SCSO}$ may be possible.

2.2.4 Principles of equivalence and Schiff's conjecture

In section 2.2.2 we described two principles of equivalence, WEP and UGR, that must be embodied in any potentially viable theory of gravity.

Principles of equivalence

A third principle, the Einstein Equivalence Principle (EEP), also plays an important role in understanding the foundations of gravitation theory. It states that (i) *WEP and UGR are valid*, and (ii) *the outcome of any non-gravitational experiment performed in a local freely falling frame is independent of where and when in the universe it is performed, and independent of the velocity of the frame* (see Thorne et al., 1973, for discussion). It is a straightforward leap from EEP to the postulates of a *metric theory of gravity*: (i) *spacetime is endowed with a metric $g_{\mu\nu}$*; (ii) *the world-lines of test bodies are geodesics of that metric*; (iii) *EEP is satisfied, with the non-gravitational laws in any freely falling frame reducing to those of special relativity*. Thus EEP makes a fundamental distinction between metric theories of gravity, and non-metric theories, those that fail to satisfy one or more of the metric postulates.

Schiff's conjecture states that *any theory of gravity that obeys the fundamental criteria for viability* (section 2.2.2) *necessarily obeys EEP*. This form of Schiff's conjecture is an embellished classical version of his original 1960 quantum mechanical conjecture (Schiff, 1960a). His interest in the conjecture was rekindled in November 1970 by a vigorous argument with Kip S. Thorne at a conference on experimental gravity held at the California Institute of Technology. Unfortunately, his untimely death in January 1971 cut short his renewed effort.

If Schiff's conjecture is correct, then Eötvös experiments and gravitational redshift experiments may be seen as the direct empirical foundation for EEP, hence for the interpretation of gravity as a geometrical, curved-spacetime phenomenon. For this reason, much effort has recently gone into 'proving' Schiff's conjecture within specific mathematical frameworks that encompass both metric and non-metric theories of gravity, yet that are simple enough to permit concrete computations.

2.2.5 Metric and non-metric theories of gravity: the $TH\epsilon\mu$ framework

The first successful attempt to prove Schiff's conjecture was made by Lightman and Lee (1973a). They developed a framework called the $TH\epsilon\mu$ formalism that encompasses all metric theories of gravity and many non-metric theories (table 2.3). It restricts attention to the behavior of charged particles (electromagnetic interactions only) in an external static spherically symmetric (SSS) gravitational field, described by a potential U. It characterizes the motion of the charged particles in the external potential by two arbitrary functions $T(U)$ and $H(U)$, and

Chapter 2. Confrontation between gravitation theory and experiment

Table 2.3. *The THεμ framework (Lightman and Lee, 1973a)*

A. *Lagrangian for particles with mass m_{0a}, charge e_a*

$$L = \sum_a \int [-m_{0a}(T - Hv_a^2)^{1/2} + e_a A_\mu(x_a^\nu) v_a^\mu] \, dt$$
$$+ (8\pi)^{-1} \int [\varepsilon \mathbf{E}^2 + \mu^{-1} \mathbf{B}^2] \, d^3x \, dt.$$

B. *Gravitational field*: $U(x) = M/r$ where M is a constant and r is Cartesian coordinate distance from source.

C. *Arbitrary functions*: $T(U)$, $H(U)$, $\varepsilon(U)$, $\mu(U)$; EEP is satisfied if and only if $\varepsilon = \mu = (H/T)^{1/2}$ for all U; far from source of field, functions have values T_∞, H_∞, ε_∞, μ_∞ determined by present cosmological boundary conditions; can rescale coordinates and charges so that (today), $T_\infty = H_\infty = \varepsilon_\infty = 1$. Hughes–Drever experiment constrains μ_∞ by $|\varepsilon_\infty \mu_\infty T_\infty / H_\infty - 1| < 10^{-13}$.

D. *Weak-field expansion and non-metric parameters*: (assume $T_\infty = H_\infty = \varepsilon_\infty = \mu_\infty = 1$):

$T = 1 - 2U + 2\beta U^2 + \ldots$,

$H = 1 + 2\gamma U + \frac{3}{2}\delta U^2 \ldots$,

$\varepsilon = 1 + (1 + \gamma - \Gamma_0)U + \frac{1}{2}(3 + 2\gamma - 2\beta - \gamma^2 + \frac{3}{2}\delta - \Gamma_1)U^2 + \ldots$,

$\mu = 1 + (1 + \gamma - \Lambda_0)U + \frac{1}{2}(3 + 2\gamma - 2\beta - \gamma^2 + \frac{3}{2}\delta - \Lambda_1)U^2 + \ldots$

Parameters: γ, β, δ = analogues of PPN parameters (section 2.3.2); (Γ_0, Λ_0), (Γ_1, Λ_1), ..., measure deviations from EEP at each order in U.

---

characterizes the response of electromagnetic fields to the external potential (gravitationally modified Maxwell equations) by two functions $\varepsilon(U)$ and $\mu(U)$. The forms of T, H, ε and μ vary from theory to theory, but every metric theory satisfies

$$\varepsilon = \mu = (H/T)^{1/2} \tag{2.3}$$

for all U. Conversely, every theory within this class that satisfies (2.3) can have its electrodynamic equations cast into 'metric' form. Lightman and Lee (1973a) then calculated explicitly the rate of fall of a 'test' body made up of interacting charged particles, and found that the rate was independent of the structure of the body (WEP) if and only if (2.3) was satisfied. In other words WEP⇒EEP and Schiff's conjecture was verified, at least within the restrictions built into the formalism (for the converse EEP⇒WEP, see Nordtvedt, 1970c). Will (1974c) derived a gravitationally modified Dirac equation from the $TH\varepsilon\mu$ formalism, determined the gravitational redshift experienced by a variety of atomic clocks, and found that the redshift was independent of the nature of the clock (UGR) if and only if (2.3) was satisfied; in other words UGR $\Rightarrow$ EEP, and another aspect of Schiff's conjecture was verified.

The heart of the $TH\varepsilon\mu$ formalism is the 'metric meshing law', (2.3), so denoted because it imposes a constraint on the relative manner in which the electromagnetic laws and the equations of motion of particles 'mesh' with gravitational fields if those laws are to be 'metric'. Experimental tests of the equivalence principle can then be regarded as tests of the metric meshing law, at least insofar as electromagnetic effects are concerned. Because such experiments are performed in the solar system where gravitation is weak ($U < 10^{-5}$), the weak-field version of the meshing law is of more direct interest. The functions T, H, ε, and μ may be expanded in powers of U using arbitrary parameters (table 2.3). Pairs of parameters Γ_0 and Λ_0, Γ_1 and Λ_1, and so on, measure the deviation from the metric meshing law (2.3) at each order in U. Every metric theory of gravity has $\Gamma_i = \Lambda_i = 0$, while every non-metric theory has at least one Γ_i or Λ_i non-zero.

In this weak-field limit, the rate of fall of a composite spherical test body has the form (Lightman and Lee, 1973a; Haugan and Will, 1977)

$$a = (m_P/m)\nabla U,$$
$$m_P/m = 1 - \Gamma_0(E_e/m) - \Lambda_0(E_m/m) + [\text{terms involving } \Gamma_1, \Lambda_1] + \ldots, \quad (2.4)$$

where m_P and m are the passive gravitational and inertial masses of the test body, and E_e and E_m are the electrostatic and magnetostatic self-energies of the body (see Haugan and Will, 1977, for details). Eötvös experiments place limits on the WEP-violating terms in (2.4), and ultimately place limits on the non-metric parameters Γ_0, Λ_0, Γ_1, Λ_1, etc., given by

$$|\Gamma_0| < 4 \times 10^{-10}, \quad |\Lambda_0| < 6 \times 10^{-6},$$
$$|\Gamma_1| < 4 \times 10^{-2}, \quad |\Lambda_1| < 600.$$

These limits are sufficiently tight to rule out a number of non-metric theories of gravity thought previously to be viable, including theories due to Belinfante and Swihart, Naida and Capella (Lightman and Lee, 1973a).

Chapter 2. Confrontation between gravitation theory and experiment

The gravitational redshifts experienced by photons emitted by various kinds of clocks have the form of (2.2) (Will, 1974c, 1977a) with

$$\alpha = \begin{cases} -3\Gamma_0 + \Lambda_0 & \text{[hydrogen hyperfine transition, H-maser clock],} \\ -\tfrac{1}{2}(3\Gamma_0 + \Lambda_0) & \text{[electromagnetic mode in cavity, SCSO clock],} \\ -2\Gamma_0 & \text{[phonon mode in solid, high } Q \text{ dielectric crystal;} \\ & \text{principal transition in hydrogen].} \end{cases}$$

Thus the NASA–SAO Rocket Redshift Experiment sets a limit on the parameter combination $|3\Gamma_0 - \Lambda_0| < 2 \times 10^{-4}$; a null-redshift experiment comparing hydrogen-maser and SCSO clocks (section 2.2.3d) would set a limit on $|\alpha_H - \alpha_{SCSO}| = |\tfrac{3}{2}(\Gamma_0 - \Lambda_0)|$.

The Hughes–Drever experiment (section 2.2.3a) can also be analyzed within the $TH\varepsilon\mu$ framework. The $TH\varepsilon\mu$ Lagrangian shown in table 2.3 is locally Lorentz invariant as long as $\varepsilon\mu T/H = 1$; a violation of this constraint leads to an inertial-mass anisotropy of a kind ruled out by the experiment. A detailed calculation (Nordtvedt and Haugan, 1978) leads to the constraint (see 2.1)

$$\delta_L = |\varepsilon\mu T/H - 1| < 10^{-13},$$

where we have used $w = 30 \text{ km s}^{-1}$.

The $TH\varepsilon\mu$ framework has many built-in restrictions – to SSS fields, to electromagnetic interactions, to a specific Lagrangian form. Several attempts have been made to relax some of these restrictions. Ni (1977) abandoned the restriction to SSS fields, but continued to work with a specific electromagnetic-particle Lagrangian, though of a more general form than the $TH\varepsilon\mu$ Lagrangian. He verified that Schiff's conjecture is valid as long as there is no pseudo-scalar gravitational field that couples to the Maxwell field. In the presence of a non-metric pseudo-scalar field, test bodies still obey WEP, but experience anomalous torques that couple to moments of their internal electromagnetic energy distribution. The $TH\varepsilon\mu$ formalism is a special case of Ni's equations.

Nordtvedt (1975a) has employed a gedanken experiment that assumes only conservation of energy and local Lorentz invariance to illustrate the close connection between WEP and UGR. In this experiment a system in a quantum state A decays to state B, emitting a quantum of frequency ν. The quantum falls a height H in an external gravitational field and is shifted to frequency ν', while the state B falls with acceleration g_B. At the bottom, state A is rebuilt out of state B, the quantum of frequency ν' and the kinetic energy $m_B g_B H$ that state B has gained during its fall. The energy left over must be exactly enough, $m_A g_A H$, to raise state A to its original location. (The assumption of local Lorentz invariance permits the inertial masses m_A and m_B to be identified with the total energies of the bodies.) If g_A and g_B depend on that portion of the internal energy of the states that was involved in the quantum transition from A to B according to

$$g_A = g(1 + \alpha E_A/m_A), \quad g_B = g(1 + \alpha E_B/m_B), \quad E_A - E_B = h\nu,$$

(violation of WEP), then by conservation of energy, there must be a corresponding violation of UGR of the form (to lowest order in $h\nu/m$)

$$z = (\nu' - \nu)/\nu = (1 + \alpha)gH.$$

In the $TH\varepsilon\mu$ framework, this relationship has been verified by Haugan (1978) for coulomb electrostatic internal energies and for fine-structure and hyperfine internal energies, provided Lorentz invariance holds, that is, provided $\varepsilon\mu T/H = 1$. When WEP is imposed ($\alpha = 0$), one recovers the classic 'derivation' of the standard gravitational redshift employed by Dicke (1964a) and others.

Attempts to relax the $TH\varepsilon\mu$ framework's restriction to electromagnetic fields are hampered by the uncertainties and the technical complexities of the theories of weak and strong interactions. However, Haugan and Will (1976) have shown that if weak interactions cause violations of WEP of the form

$$m_P/m = 1 + \Gamma_w(E_w/m),$$

where E_w is the contribution of weak interactions to the energy of the nucleus, and Γ_w is a parameter measuring the violation of WEP, then the Moscow version of the Eötvös experiment places the limit $|\Gamma_w| < 5 \times 10^{-3}$; in other words, weak interactions obey WEP to better than a part in 100. This refuted earlier claims that those experiments did not test weak-interaction effects (Worden and Everitt, 1974; Worden, 1976; Chapman and Hanson, 1971; Chiu and Hoffman, 1964; Dicke, 1964a).

One subject that is important for the validity of EEP but that lies outside the restrictions of the present form of the $TH\varepsilon\mu$ framework is the constancy of the fundamental non-gravitational constants over cosmological timescales (we delay discussion of the gravitational 'constant' until section 2.6.1). Such constancy is a direct test of that portion of EEP (section 2.2.4) which states that 'the outcome of any

Chapter 2. Confrontation between gravitation theory and experiment

non-gravitational experiment...is independent of *when* it is performed'. We shall not review the various theories and proposals originating with Dirac that permitted variable fundamental constants (for a detailed review and references, see Dyson, 1972), rather we shall cite the most recent observational evidence (table 2.4). The observations range from comparisons of spectral lines in distant galaxies and quasars (Wolfe, Brown and Roberts, 1976), to measurements of isotopic abundances of elements in the solar system (Dyson, 1972), to laboratory comparisons of atomic clocks (Stein, 1974). Recently, Shylakhter (1976) has made significant improvements in the limits on variations in the electromagnetic, weak and strong coupling constants by studying fission yields in the 'Oklo Natural Reactor', a sustained fission reactor that evidently occurred in Gabon, Africa nearly 2 billion years ago. By comparing the nuclear cross-sections inferred from the abundances of the yield isotopes with currently measured cross-sections he obtained the limits shown in the last column of table 2.4.

2.2.6 Summary

The strong experimental support for the fundamental criteria for viability of gravitation theories, coupled with theoretical arguments such as Schiff's conjecture, serve to bolster our faith that, whatever theory of gravity is correct, it must be a metric theory. For the remainder of this review we shall make this assumption. It must be kept in mind, however, that this assumption rests on potentially unsteady ground, and continued experimental and theoretical effort is crucial to maintaining a solid foundation for gravitation theory.

2.3 Post-Newtonian gravity in the solar system

When the postulates of metric theories of gravity (section 2.2.4) are examined, one notices a crucial feature: no matter how complex the theory, no matter what additional gravitational or cosmological fields it deals with, matter and non-gravitational fields respond only to the metric $g_{\mu\nu}$. The role of the other fields which a given theory may contain can only be that of helping to generate the spacetime curvature associated with the metric. Matter may create these fields, and the fields plus the matter may generate the metric, but they cannot act back directly on the matter. The matter responds only to the metric. From this point of view, the metric $g_{\mu\nu}$ becomes the primary theoretical entity, and all that

Table 2.4. Limits on cosmological variation of non-gravitational constants

Constant k	Limit on $\dot{k}/k$ per Hubble time 2×10^{10} yr ($H_0 = 55$ km s^{-1} Mpc^{-1})	Method	Reference	Limit from Oklo reactor (Shlyakhter, 1976)
Dimensional constants:				
$\hbar c$	2×10^{-2}	Energy versus wavelength for old and young photons	Solheim, Barnes and Smith (1976); Baum and Florentin-Nielsen (1976)	
Dimensionless constants:				
Fine-structure constant:				
$\alpha = e^2/\hbar c$	4×10^{-4}	^{187}Re β-decay rate over geological time	Dyson (1972)	10^{-7}
	8×10^{-2}	MgII fine-structure and 21 cm line in radio-source at $z = 0.5$	Wolfe, Brown and Roberts (1976)	
	8×10^{-2}	SCSO clock versus cesium beam clock	Turneaure and Stein (1976)	
Weak interaction constant:				
$\beta = g m^2 c/\hbar^3$	2	^{187}Re, ^{40}K decay rates	Dyson (1972)	2×10^{-4}
Electron–proton mass ratio:				
m_e/m_p	1	Mass shift in quasar spectral lines ($z \sim 2$)	Pagel (1977)	
Proton gyromagnetic factor:				
$g_p m_e/m_p$	10^{-1}	MgII, 21 cm line	Wolfe, Brown and Roberts (1976)	
Strong interactions:				
g_s^2	8×10^{-2}	Nuclear stability	Davies (1972)	8×10^{-9}

distinguishes one metric theory from another is the particular way in which matter generates the metric.

The comparison of metric theories of gravity with each other and with experiment becomes particularly simple when one takes the weak-field, slow-motion, post-Newtonian limit. In this limit, one learns that theories differ from one another only in the values they predict for a set of ten dimensionless parameters – so-called PPN parameters (section 2.3.1). Using the resulting parametrized post-Newtonian (PPN) formalism, one can study and classify competing metric theories of gravity (section 2.3.2). Because gravity in the solar system is weak, the PPN formalism is ideally suited to analyzing solar-system experiments that can detect post-Newtonian effects. Each experiment may then be regarded as a measurement of the appropriate PPN parameter or combination of parameters, and thereby as a test of the predictions of competing theories. The important experiments include light-deflection and time-delay experiments (section 2.3.3), lunar-laser-ranging tests of the Nordtvedt effect (section 2.3.4), perihelion-shift measurements (section 2.3.5), geophysical and planetary tests of preferred-frame and preferred-location effects (section 2.3.6), gyroscope and other precession experiments (section 2.3.7) and laboratory experiments (section 2.3.8).

2.3.1 The parametrized post-Newtonian (PPN) formalism

In the weak-field, slow-motion limit, the spacetime metric $g_{\mu\nu}$ predicted by nearly every metric theory of gravity has the same structure. It can be written as an expansion about the Minkowski metric ($\eta_{\mu\nu}$ = diag $(-1, 1, 1, 1)$) in terms of dimensionless gravitational potentials of varying degrees of smallness. These potentials are constructed from the matter variables (table 2.5) and from the Newtonian gravitational potential

$$U(\mathbf{x}, t) = \int \rho(\mathbf{x}', t)|\mathbf{x} - \mathbf{x}'|^{-1} \, \mathrm{d}^3 x'.$$

The 'order of smallness' is determined according to the rules $U \sim v^2 \sim \Pi \sim p/\rho \sim 0(2)$, $v^i \sim |\mathrm{d}/\mathrm{d}t|/|\mathrm{d}/\mathrm{d}x| \sim 0(1)$, and so on. A consistent post-Newtonian limit requires determination of g_{00} correct through $0(4)$, g_{0i} through $0(3)$ and g_{ij} through $0(2)$ (for details see TTEG § 4.2). The only way that one metric theory differs from another is in the numerical values of the coefficients that appear in front of the metric potentials. The PPN formalism inserts parameters in place of these coefficients,

Table 2.5. *The parametrized post-Newtonian formalism*

A. *PPN parameters*:
    $\gamma, \beta, \xi, \alpha_1, \alpha_2, \alpha_3, \zeta_1, \zeta_2, \zeta_3, \zeta_4$.

B. *Metric*:
$$g_{00} = -1 + 2U - 2\beta U^2 - 2\xi\Phi_w - (\zeta_1 - 2\xi)\mathcal{A} + (2\gamma + 2 + \alpha_3 + \zeta_1 - 2\xi)\Phi_1$$
$$+ 2(3\gamma - 2\beta + 1 + \zeta_2 + \xi)\Phi_2 + 2(1 + \zeta_3)\Phi_3 + 2(3\gamma + 3\zeta_4 - 2\xi)\Phi_4$$
$$- (\alpha_1 - \alpha_2 - \alpha_3)w^2 U - \alpha_2 w^i w^j U_{ij} + (2\alpha_3 - \alpha_1)w^i V_i,$$
$$g_{0i} = -\tfrac{1}{2}(4\gamma + 3 + \alpha_1 - \alpha_2 + \zeta_1 - 2\xi)V_i - \tfrac{1}{2}(1 + \alpha_2 - \zeta_1 + 2\xi)W_i$$
$$- \tfrac{1}{2}(\alpha_1 - 2\alpha_2)w^i U - \alpha_2 w^j U_{ij},$$
$$g_{ij} = (1 + 2\gamma U)\delta_{ij}.$$

C. *Metric potentials*:
    For detailed definitions see TTEG § 4 and Will (1973).

D. *Differences between this version and the TTEG version*
    1. Adoption of MTW signature $(-1, 1, 1, 1)$ and index convention (Greek indices run 0, 1, 2, 3; Latin run 1, 2, 3).
    2. New symbol for Whitehead parameter: ξ instead of ζ_w as in Will (1973).
    3. Modified conservation-law parameters incorporating effects of Whitehead term (see Lee *et al.*, 1974)
    $$(\zeta_1)_{\text{NEW}} = (\zeta_1)_{\text{OLD}} + 2\xi, \quad (\zeta_2)_{\text{NEW}} = (\zeta_2)_{\text{OLD}} - \xi,$$
    $$(\zeta_4)_{\text{NEW}} = (\zeta_4)_{\text{OLD}} + \tfrac{2}{3}\xi.$$

---

parameters whose values depend on the theory under study. In the current version of the PPN formalism, ten parameters are used, chosen in such a manner that they measure or indicate general properties of metric theories of gravity (table 2.6). The parameters γ and β are the usual Eddington (1922)–Robertson (1962)–Schiff (1967) parameters used to describe the 'classical' tests of general relativity; ξ is nonzero in any theory of gravity that predicts preferred-location effects such as a galaxy-induced anisotropy in the local gravitational constant G_L (also called 'Whitehead' effects (Will, 1971*d*, 1973); ξ corresponds to ζ_w of Will, 1973); $\alpha_1, \alpha_2, \alpha_3$ measure whether or not the theory predicts post-Newtonian preferred-frame effects (Will and Nordtvedt, 1972; Nordtvedt and Will, 1972); $\alpha_3, \zeta_1, \zeta_2, \zeta_3, \zeta_4$ measure whether or not the theory predicts violations of global conservation laws for total momentum (Will, 1971*c*). In table 2.6, we show the values these parameters

Chapter 2. Confrontation between gravitation theory and experiment

Table 2.6 *The PPN parameters and their significance*

Parameter (see table 2.5)	What it measures, relative to general relativity[a]	Value in general relativity	Value in semi-conservative theories	Value in fully conservative theories
γ	How much space-curvature is produced by unit rest mass?	1	γ	γ
β	How much 'non-linearity' is there in the superposition law for gravity?	1	β	β
ξ	Are there preferred-location effects?	0	ξ	ξ
α_1		0	α_1	0
α_2	Are there preferred-frame effects?	0	α_2	0
α_3		0	0	0
ζ_1	Is there violation	0	0	0
ζ_2	of conservation	0	0	0
ζ_3	of total	0	0	0
ζ_4	momentum?	0	0	0

[a] These descriptions are valid only in the standard PPN gauge, and should not be construed as covariant statements. For examples of the misunderstandings that can arise if this caution is not heeded, especially in the case of β, see Deser and Laurent (1973), and Duff (1974).

take: (i) in general relativity; (ii) in any theory of gravity that possesses conservation laws for total momentum, called 'semi-conservative' (any theory that is based on an invariant action principle is semi-conservative (Lee *et al.*, 1974)); and (iii) in any theory that in addition possesses six global conservation laws for angular momentum, called 'fully conservative' (such theories automatically predict no post-Newtonian preferred-frame effects). Semi-conservative theories have five free PPN parameters ($\gamma, \beta, \xi, \alpha_1, \alpha_2$) while fully-conservative theories have three (γ, β, ξ). Table 2.6 summarizes the basic ingredients and formulae of the PPN formalism; for further details see MTW § 39 and TTEG § 4.

A primitive version of the PPN formalism was devised and studied by Eddington (1922), Robertson (1962) and Schiff (1967). It treated the solar-system metric as that of a spherical non-rotating Sun and idealized the planets as test bodies moving on geodesics of the metric. Schiff (1960*b*) generalized the metric to incorporate rotation (Lense–Thirring effect, section 2.3.7). But the pioneering development of the full PPN formalism was performed by Kenneth Nordtvedt, Jr (1968b), who studied the post-Newtonian metric of a system of gravitating point masses. Will (1971a) generalized the formalism to incorporate matter described by a perfect fluid (see also Baierlein, 1967). A unified version of the PPN formalism was then presented by Will and Nordtvedt (1972), and the preferred-location or Whitehead parameter was added by Will (1973). The differences between the version of the formalism used here and the Will–Nordtvedt version (as summarized in TTEG § 4) are shown in row D of table 2.5. Other versions of the PPN formalism have been developed to deal with point masses with charge (Will, 1976a) and fluid with anisotropic stresses (MTW § 39).

2.3.2 A zoo of competing theories of gravity

One of the important applications of the PPN formalism is the comparison and classification of alternative metric theories of gravity. The population of viable theories has fluctuated in recent years as new effects and tests have been discovered, largely through the use of the PPN framework. Examples are the Nordtvedt effect (Nordtvedt, 1968c), preferred-frame tests (Will, 1971d, Nordtvedt and Will, 1972), and preferred-location tests (Will, 1971d, 1973). These new tests eliminated many theories thought previously to be viable. The theory population has also fluctuated as new, viable theories have been invented.

The present population of theories may be divided into two broad theoretical classes. The first class consists of theories of gravity that contain only dynamical gravitational fields, while the second class consists of theories that possess prior geometry, i.e. that contain absolute or non-dynamical fields, such as flat background metrics or cosmic time fields. These concepts are discussed in detail by Thorne et al. (1973) and briefly in MTW (§ 17.6) (see also Anderson, 1967). It must be emphasized that theories with prior geometry can still be generally covariant, in keeping with the second of the basic assumptions of the Dicke framework (section 2.2.1). For instance, a flat background metric can be completely specified by the covariant statement that its Riemann tensor vanishes everywhere, and a cosmic time coordinate t may be specified by the covariant statement that it is a scalar field whose gradient is timelike and covariantly constant.

We now give a very brief description of the currently viable metric theories of gravity and summarize their PPN parameter values (table

Table 2.7. *Metric theories of gravity and their PPN parameter values*

Theory	Arbitrary functions or constants	Parameters fixed by present cosmological conditions	PPN parameters						Can Parameters be made equal to GRT?
			γ	β	ξ	α_1	α_2	(α_3, ζ_i)	
(a) *Purely dynamical theories*									
(i) General relativity $(\mathbf{g})$:	None	None	1	1	0	0	0	0	
(ii) Scalar–tensor $(\mathbf{g}, \phi)$:									
BWN	$\omega(\phi)$	ϕ_0	$\dfrac{1+\omega}{2+\omega}$	$1+\Lambda$	0	0	0	0	No[b]
Bekenstein	$\omega(\phi), r, q$	ϕ_0	$\dfrac{1+\omega}{2+\omega}$	$1+\Lambda$	0	0	0	0	No[b]
Brans–Dicke	ω	ϕ_0	$\dfrac{1+\omega}{2+\omega}$	1	0	0	0	0	No[b]
(iii) Vector–tensor $(\mathbf{g}, \vec{K})$:									
Will–Nordvedt	None	K_0	1	1	0	0	$K_0^2(1+\tfrac{1}{2}K_0^2)^{-1}$	0	Yes, for special cosmology $K_0 = 0$

(b) *Theories with prior geometry*

(iv) Bimetric theories:

Theory	Variables								Agrees with cosmology?
Rosen (g, η)	None	c_0, c_1	1	1	0	0	$(c_0/c_1)-1$	0	Yes, for special cosmology ($c_0 = c_1$)
Lightman–Lee (g, h, η)	a, f, k	ω_0, ω_1	γ'^c	β'	0	α'_1	α'_2	0	Yes, for $(a, f, k, \omega_0, \omega_1) = (\tfrac{1}{4}, -\tfrac{5}{64}, \tfrac{1}{16}, 0, 0)$
Rastall ($g, \eta, \phi, \bar{K}$)	None		1[a]	1	0	0	0	0	Yes

(v) Stratified theories:

Theory	Variables								Agrees with cosmology?
Ni ($g, \eta, t, \phi, \bar{K}$)	$e, f_1(\phi), f_2(\phi),$ $f_3(\phi), a, b, d$		a[a]	b	0	$-4(a+1+\tfrac{1}{2}e)$	$-(d+1)$	0	Yes, for $(a, b, d, e) = (1, 1, -1, -4)$

---

[a] A complete match between the post-Newtonian limit and appropriate cosmological boundary conditions has not been carried out. The values of the PPN parameters shown assume that **g** is asymptotically equal to **η**.

[b] See discussion in section 2.3.2.

[c] In the Lightman–Lee theory, γ', β', α'_1 and α'_2 are complicated functions of a, f, k, ω_0 and ω_1.

Chapter 2. Confrontation between gravitation theory and experiment

2.7). For further details of the theories and of the methods of computing their post-Newtonian limits see Ni (1972) and TTEG § 5 and references therein; for the more recent theories we provide primary references.

(a) Purely dynamical theories

(i) General relativity: the metric **g** is the sole dynamical field and the theory contains no arbitrary functions or parameters (we ignore the cosmological constant λ).

(ii) Scalar–tensor theories: contain the metric **g**, a scalar field ϕ, and an arbitrary coupling function $\omega(\phi)$ that determines the relative 'strength' of the scalar field. The most general version is the Bergmann (1968)–Wagoner (1970)–Nordtvedt (1970b) – (BWN) – theory, with $\omega(\phi)$ arbitrary. Special cases are Bekenstein's (1977) theory ($\omega(\phi)$ a particular function of ϕ with arbitrary coefficients r and q) and Brans–Dicke (1961) theory ($\omega(\phi)$ a constant). The parameters that enter the post-Newtonian limit are

$$\omega = \omega(\phi_0), \quad \Lambda = [(d\omega/d\phi)(3+2\omega)^{-2}(4+2\omega)^{-1}]|_{\phi_0},$$

where ϕ_0 is the value of ϕ today far from the solar system, as determined by appropriate cosmological boundary conditions. In Brans–Dicke theory, the larger the value of ω, the smaller the effects of the scalar field, and in the limit $\omega \to \infty$, the theory becomes indistinguishable from general relativity in all its predictions. In the BWN–Bekenstein formulations, the function $\omega(\phi)$ could be sufficiently pathological that for the present value of the scalar field ϕ_0, ω is very large, and Λ is very small (theory's predictions almost identical to general relativity today), but that for past or future values of ϕ, ω and Λ could take on values that would lead to significant differences in cosmological models, for instance (Bekenstein, 1977).

(iii) Vector–tensor theories: contain the metric **g** and a vector field $\vec{K}$. In the Will–Nordtvedt (1972) version the parameter in the post-Newtonian limit is the present value of K_0 far from the solar system.

(b) Theories with prior geometry

(iv) Bimetric theories: contain dynamical scalar, vector or tensor fields, and a non-dynamical background metric $\boldsymbol{\eta}$. In Rosen's bimetric theory (Rosen, 1973, 1974, 1977a; Rosen and Rosen, 1975) the physical metric **g** is the dynamical field, the metric $\boldsymbol{\eta}$ is Riemann-flat (recently, Rosen (1978) has generalized η to spacetimes of constant curvature). The theory has no arbitary parameters, however the post-Newtonian limit must be matched asymptotically to a cosmological model in order to determine the relative values of the asymptotic components of $g_{\mu\nu}$ and $\eta_{\mu\nu}$. For a homogeneous isotropic cosmology, this matching yields two parameters c_0 and c_1 given by (Lee et al., 1976)

$$c_0 + 3c_1 = \eta^{\mu\nu} g_{\mu\nu}, \quad c_0^{-1} + 3c_1^{-1} = \eta_{\mu\nu} g^{\mu\nu}, \tag{2.5}$$

where the expressions on the right are to be evaluated far from the solar system. This meshing between local gravitation and global structure can have important consequences for the viability of theories of this type. For instance the PPN parameter values are identical to those of general relativity, except for α_2 which depends on c_0 and c_1. There exist cosmological models in Rosen's theory for which $\alpha_2 \approx 0$ (Rosen, 1978), but those models may be in violation of cosmological observations. We discuss this point further in section 2.6.2.

In the Lightman–Lee (1973b) bimetric theory, the dynamical field is a tensor $\boldsymbol{h}$; the physical metric is then obtained algebraically from $\boldsymbol{h}$ and a flat background metric $\boldsymbol{\eta}$. The theory contains three arbitrary constants a, f, k, and two cosmological matching parameters ω_0 and ω_1. For an appropriate choice of these values $(a, f, k, \omega_0, \omega_1) = (\frac{1}{4}, -\frac{5}{64}, \frac{1}{16}, 0, 0)$ the PPN parameters are identical to those of general relativity.

In Rastall's (1976, 1977a, b) bimetric theory, the dynamical fields are a scalar ϕ and a vector $\vec{K}$, the physical metric is obtained algebraically from ϕ, $\vec{K}$ and a flat background metric $\boldsymbol{\eta}$. The current version of the theory has no arbitrary constants (Rastall, 1977b), however the problem of matching to a realistic cosmological model has not been solved. Rastall (1977b) has found that the PPN parameters are identical to those of general relativity if the physical metric is assumed to be asymptotically equal to $\boldsymbol{\eta}$.

(v) Stratified theories: contain a mechanism for singling out preferred space-slices ('strata') in the universe. The most common mechanism is a non-dynamical cosmic time function t whose tangent four-vector is timelike and covariantly constant with respect to some metric. In Ni's (1973) theory, t and a flat background metric $\boldsymbol{\eta}$ are non-dynamical fields, a scalar field ϕ and a vector field $\vec{K}$ are dynamical. The physical metric $\boldsymbol{g}$ is constructed algebraically from these fields. The theory has three arbitrary functions of ϕ and one arbitrary parameter; in the post-Newtonian limit with physical metric asymptotically flat (a complete match to a cosmological model has not been carried out) there

Chapter 2. Confrontation between gravitation theory and experiment

are four arbitrary parameters a, b, d and e. They can be chosen so that the PPN parameters are identical to those of general relativity.

These theories and their PPN parameter values are summarized in table 2.7. All the theories shown are based on invariant action principles (Lagrangian-based), hence all are semi-conservative, i.e. have zero values for the parameters α_3, ζ_1, ζ_2, ζ_3 and ζ_4 (denoted α_3, ζ_i in table 2.7).

We have not discussed at all a variety of theories that have been shown to be unviable. These include a class of stratified theories that predict $\alpha_1 \approx -8$ in violent disagreement with geophysical measurements (section 2.3.6), a class of 'quasi-linear' theories including Whitehead's (1922) theory that predict galaxy-induced anisotropies in the local gravitational constant in violent disagreement with Earth-tide measurements (section 2.3.6), and conformally flat theories that predict $\gamma = -1$ in violent disagreement with light-deflection and time-delay measurements (section 2.3.3). Nor have we discussed theories yet to be invented.

2.3.3 Light-deflection and time-delay experiments: probing curvature in the solar system

The bending of light rays by the Sun, and the added delay in the round-trip travel time of a radar signal which passes the Sun both measure the PPN parameter γ. A light ray (or photon) which passes the Sun at a distance d is deflected by an angle

$$\delta\theta = \tfrac{1}{2}(1+\gamma)(4m_\odot/d)[(1+\cos\theta)/2] \qquad (2.6)$$

(Shapiro, 1967; Ward, 1970; TTEG § 7.2), where $m_\odot$ is the mass of the Sun and θ is the angle between the Earth–Sun line and the incoming direction of the photon. For a grazing ray, $d = R_\odot$, $\theta \approx 0$, and

$$\delta\theta \approx \tfrac{1}{2}(1+\gamma)1.75'',$$

independent of the frequency of light.

A radar signal sent across the solar system beyond the Sun to a planet or satellite and returned to the Earth suffers an additional non-Newtonian delay in its round-trip travel time; for a ray which passes close to the Sun it is given by (Shapiro, 1964; TTEG § 7.2)

$$\delta t \approx \tfrac{1}{2}(1+\gamma)[250 - 20\ln(d^2/r)]\,\mu\text{s},$$

where d is the distance of closest approach of the ray in solar radii, and r is the distance of the planet or satellite from the Sun, in astronomical units.

Measurements of these two effects have given us our most precise direct measurements of the parameter γ to date.

The prediction of the bending of light by the Sun was one of the great successes of Einstein's general relativity. Eddington's confirmation of the bending of optical starlight observed during a solar eclipse in the first days following World War I helped make Einstein famous. However, the experiments of Eddington and his coworkers had only 30 per cent accuracy, and succeeding experiments weren't much better: the results were scattered between one half and twice the Einstein value, and the accuracies were low (for reviews, see Richard, 1975; Merat et al., 1974; Bertotti et al., 1962). The most recent optical measurement, during the solar eclipse of 30 June 1973 illustrates the difficulty of these experiments. It yielded a value

$$\tfrac{1}{2}(1+\gamma) \approx 0.95 \pm 0.11$$

(Texas Mauritanian Eclipse Team, 1976; Jones, 1976). The accuracy was limited by poor seeing (caused by a dust storm just prior to the eclipse, and by clouds and rain during the follow-up expedition in November 1973) that drastically reduced the number of measurable star images. There were also variable scale changes between eclipse- and comparison-field exposures. Recent advances in photoelectric and astrometric techniques may make possible optical deflection measurements without the need for solar eclipses (Hill, 1971).

The development of long-baseline radio-interferometry has altered this situation. Long-baseline and very-long-baseline (VLBI) interferometric techniques have the capability in principle of measuring angular separations and changes in angles as small as 3×10^{-4} seconds of arc. Coupled with this technological advance is a series of heavenly coincidences: each year, groups of strong quasi-stellar radio-sources pass very close to the Sun (as seen from the Earth), including the group 3C273, 3C279, and 3C48, and the group 0111+02, 0119+11 and 0116+08. By measuring the relative bending of the signals from the quasars within each group, radio-astronomers over the past decade have been able to measure the coefficient $\tfrac{1}{2}(1+\gamma)$ in (2.6), which has the value unity in general relativity. Their results are shown in figure 2.1.

One of the major sources of error in these experiments is the solar corona, which bends radio-waves much more strongly than it bent the

Chapter 2. Confrontation between gravitation theory and experiment

visible light rays which Eddington observed. Improvements in dual frequency techniques have improved accuracies by allowing the coronal bending, which depends on the frequency of the wave, to be measured separately from the gravitational bending, which does not. Fomalont and Sramek (1977) provide a thorough review of these experiments, and discuss the prospects for improvement.

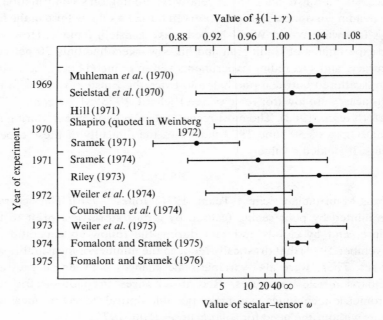

Figure 2.1. Results of radio-wave deflection experiments 1969–75.

The 'time-delay' effect was not predicted by Einstein; it was 1964 when this effect was discovered by Shapiro (1964) as a theoretical consequence of general relativity and of other theories of gravity (see also Muhleman and Reichley, 1964). In the following decade, attempts have been made to measure this effect using radar ranging to targets passing through 'superior conjunction' (target on the far side of the Sun; radar signals passing close to the Sun). Three types of targets have been employed: planets such as Mercury and Venus, used as passive reflectors of the radar signals; spacecraft such as Mariner 6 and 7 used as active retransmitters of the radar signals, and combinations of planet and spacecraft ('anchored' spacecraft), such as the Mariner 9 Mars orbiter, and the Viking landers and orbiters. Detailed analyses of the measured

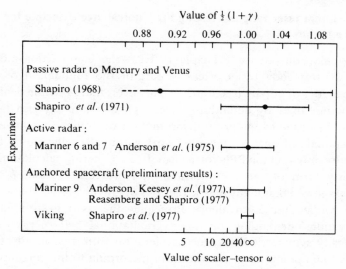

Figure 2.2. Results of time-delay experiments 1968–77.

round-trip travel times have yielded results for the coefficient $\frac{1}{2}(1+\gamma)$ shown in figure 2.2.

Here, as in the radio-wave deflection measurements, the solar corona causes uncertainties because of its slowing down of the radar signal; again dual frequency ranging helps reduce these errors. Other major sources of systematic errors are (i) the perturbing effects of non-gravitational forces such as solar wind, radiation pressure and unbalanced attitude-control forces on the trajectory of spacecraft, and (ii) the degrading effect of planetary topography on the ability to locate a planet within 0.5 km by means of passive radar. The use of 'anchored' spacecraft such as the Mariner 9 orbiter and the Viking orbiters and landers is designed to reduce both these error sources. In fact, preliminary analysis of data from Viking has yielded approximately 0.5 per cent accuracy for $\frac{1}{2}(1+\gamma)$, and may ultimately reach 0.1 per cent (Shapiro et al., 1977). For detailed reviews of the experimental problems and prospects see Anderson (1974) and Shapiro (1972).

From these results, we can conclude that the coefficient $\frac{1}{2}(1+\gamma)$ must be within, at most, one per cent of unity. Most of the theories shown in table 2.7 can select their adjustable parameters or cosmological boundary conditions with sufficient freedom to meet this constraint. Scalar-tensor theories must have $\omega > 48$ to be within one per cent or $\omega > 23$ to be within 2 per cent of unity.

Chapter 2. Confrontation between gravitation theory and experiment

2.3.4 Lunar laser ranging: testing the Equivalence Principle for massive bodies

The Weak Equivalence Principle (WEP) – the composition-independence of test-body trajectories – is part of the foundation of every metric theory of gravity (section 2.2). However, that principle deals only with bodies that, by definition, have negligible internal gravitational energy. In order to understand the motion of bodies that have *finite* gravitational self-energy, one must work within a specific metric theory of gravity or within the PPN formalism. In a pioneering calculation using his early form of the PPN formalism, Nordtvedt (1968a, b, 1969, 1971a) showed that many metric theories of gravity predict that massive bodies violate the Equivalence Principle – that is, fall with different accelerations depending on their gravitational self-energy. For a spherically symmetric body, the acceleration from rest in an external gravitational potential U has the form (according to the version of the PPN formalism shown in table 2.5) (cf. equation (2.4))

$$a = (m_P/m)\nabla U,$$
$$m_P/m = 1 - \eta(E_g/m), \qquad (2.7)$$
$$\eta = 4\beta - \gamma - 3 - \tfrac{10}{3}\xi - \alpha_1 + \tfrac{2}{3}\alpha_2 - \tfrac{2}{3}\zeta_1 - \tfrac{1}{3}\zeta_2,$$

where E_g is the gravitational self-energy of the body. This violation of the Massive-Body Equivalence Principle is known as the 'Nordtvedt effect' (the possibility of such an effect was first noticed by Dicke (1964b); see also Dicke (1969), Will (1971a)). The effect is absent in general relativity ($\eta = 0$) but present in Brans–Dicke theory [$\eta = 1/(2+\omega)$]. The existence of the Nordtvedt effect does not violate the results of laboratory Eötvös experiments, since for laboratory-sized objects, $E_g/m \sim 10^{-27}$, far below the sensitivity of current experiments. However, for astronomical bodies, E_g/m may be significant (10^{-5} for the Sun, 10^{-8} for Jupiter, 4.6×10^{-10} for the Earth), and in fact it is an 'Eötvös experiment' using solar-system bodies that has yielded a significant test of the Nordtvedt effect. If the Nordtvedt effect is present ($\eta \neq 0$) then the Earth should fall toward the Sun with a slightly different acceleration than the Moon ($E_g/m \sim 0.2 \times 10^{-10}$). This perturbation in the Earth–Moon orbit leads to a polarization of the orbit that is directed toward the Sun as it moves around the Earth–Moon system, as seen from Earth. This polarization represents a perturbation in the Earth–

Moon distance of the form

$$\delta r \approx 9.2\eta \cos(\omega_0 - \omega_s)t \text{ m}, \tag{2.8}$$

where ω_0 and ω_s are the angular frequencies of the orbits of the Moon and Sun around the Earth (see Nordtvedt, 1968c, 1973; Will 1971b; TTEG § 7.7, for detailed derivations).

Since August 1969, when the first successful acquisition was made of a laser signal reflected from the Apollo 11 retroreflector on the Moon, the Lunar Laser Ranging Program has made regular measurements of the round-trip travel times of laser pulses between McDonald Observatory in Texas and the lunar retroreflectors, with accuracies of 1 ns (30 cm) (see Bender *et al.*, 1973, for a review). Two independent analyses have been made of data taken between 1969 and 1975; after accurately taking into account a large Newtonian perturbation of the form $\delta r \sim 110 \cos(\omega_0 - \omega_s)t$ km, upon which the Nordtvedt effect (2.8) would be superimposed, both analyses found no evidence for the Nordtvedt effect. Their results for η were

$$\eta \approx \begin{cases} 0.00 \pm 0.03 & [\text{Williams } et\, al., (1976)], \\ 0.001 \pm 0.015 & [\text{Shapiro } et\, al., (1976)]. \end{cases}$$

This represents a limit on a possible violation of WEP for massive bodies of 7 parts in 10^{12} (cf. table 2.1). For Brans–Dicke theory these results force a lower limit on the coupling constant ω of 29 (2σ, Shapiro result).

Improvements in the measurement accuracy and in the theoretical analysis of the lunar motion may improve this limit by an order of magnitude (Williams *et al.*, 1976), while a comparable test of the Nordtvedt effect may be possible using the Mars–Sun–Jupiter system (Shapiro *et al.*, 1976). Nordtvedt (1968a, 1970a, 1971a, b) has discussed other potentially observable effects of such equivalence principle violations.

2.3.5 Perihelion shifts

The explanation of the 'anomalous' 43 seconds per century advance of the perihelion of Mercury was another of the early successes of general relativity, yet today the interpretation of this effect is shrouded in uncertainty and disagreement. The *measured* perihelion shift is accurately known: after the perturbing effects of the other planets have been accounted for, the remaining perihelion shift is known (*a*) to a

Chapter 2. Confrontation between gravitation theory and experiment

precision of about one per cent from optical observations of Mercury during the past three centuries (Morrison and Ward, 1975), and (*b*) to about 0.5 per cent from radar observations during the past decade (Shapiro *et al.*, 1976).

But the predicted effect has two major contributions, one from relativistic gravity and one from a possible oblateness of the Sun that leads to a modified solar gravitational potential. In terms of the PPN formalism the predicted advance per orbit $\Delta\omega$ for an orbit with semi-major axis *a*, and eccentricity *e* is

$$\Delta\omega = \frac{6\pi m_\odot}{a(1-e^2)}\left\{\tfrac{1}{3}(2+2\gamma-\beta)+J_2\left[\frac{R_\odot^2}{2m_\odot a(1-e^2)}\right]\right\}, \qquad (2.9)$$

where $R_\odot$ is the mean solar radius, and J_2 is a dimensionless measure of the solar quadrupole moment given by $J_2 = (C-A)/m_\odot R_\odot^2$ where C and A are the moments of inertia about the Sun's rotation axis and equatorial axis, respectively. For Mercury, the predicted shift, in seconds of arc per century, is

$$\dot\omega_{\mathrm{\breve{Q}}} = 42.95\lambda_\mathrm{p},$$

$$\lambda_\mathrm{p} = [\tfrac{1}{3}(2+2\gamma-\beta)+0.0003\,(J_2/1\times 10^{-7})],$$

where J_2 is shown normalized by the approximate value it would have if the Sun rotated uniformly with its observed surface angular velocity (Anderson, Colombo *et al.*, 1977). Measurements of the orbit of Mercury alone are incapable at present of separating the effects of relativistic gravity and of solar quadrupole moment in the determination of λ_p. Thus, in two recent analyses of radar distance measurements to Mercury, J_2 was *assumed* to have a value corresponding to uniform rotation (effect on λ_p negligible), and the PPN parameter combination was estimated. The results were

$$\tfrac{1}{3}(2+2\gamma-\beta) \approx \begin{cases} 1.005 \pm 0.020 & \text{[1966--71 data: Shapiro } et\,al.\text{ (1972)]}, \\ 1.003 \pm 0.005 & \text{[1966--76 data: Shapiro } et\,al.\text{ (1976)]}. \end{cases}$$

The origin of the uncertainty that clouds the interpretation of perihelion-shift measurements is a series of experiments performed in 1966 by Dicke and Goldenberg (see Dicke and Goldenberg, 1974, for a detailed review). Those experiments measured the visual oblateness or flattening of the Sun's disk and found a difference in the apparent polar and equatorial angular radii of $\Delta R = 43.3'' \pm 3.3'' \times 10^{-3}$. By taking into account the oblateness of the surface layers of the Sun caused by centrifugal flattening, this oblateness signal can be related to J_2 by (see Dicke, 1974)

$$J_2 \approx \tfrac{2}{3}(\Delta R/R_\odot) - 5.3 \times 10^{-6},$$

which gives ($R_\odot = 959''$)

$$J_2 \approx 2.47 \pm 0.23 \times 10^{-5} \quad \text{[Dicke and Goldenberg, 1974]}.$$

On the other hand, measurements made in 1973 by Hill and collaborators yielded $\Delta R = 9.2'' \pm 6.2'' \times 10^{-3}$ or

$$J_2 \approx 0.10 \pm 0.43 \times 10^{-5} \quad \text{[Hill et al., 1974]},$$

(see also Hill and Stebbins, 1975). The disagreement between these two results remains unresolved.

One of the major difficulties in relating visual solar oblateness results to J_2 is that a considerable amount of complex solar physics theory must be employed. There is, however, a way of determining J_2 unambiguously, namely by probing the solar gravity field at different distances from the Sun, thereby separating the effects of J_2 from those of relativistic gravitation through their different radial dependences. One method would compare the perhelion shifts of different planets. But the perihelion shifts of Venus, Earth and Mars are not known to sufficient accuracy, although Shapiro *et al.*, (1972) have pointed out that several more years of radar observations of the inner planets may permit such a comparison. Another method would take advantage of Mercury's orbital eccentricity ($e \sim 0.2$) and search for the different *periodic* orbital perturbations induced by J_2 and by relativistic gravity. The accuracy required for such measurements would necessitate tracking of a spacecraft in orbit around Mercury, but preliminary studies have shown that J_2 could be determined to within a few parts in 10^7 (Anderson, Colombo *et al.*, 1977; Wahr and Bender, 1976). Finally, a spacecraft in a high-eccentricity solar orbit with perihelion distance of four solar radii could yield a measurement of J_2 with a precision of a part in 10^8 (Nordtvedt, 1977; Anderson, Colombo *et al.*, 1977). Such missions would also lead to improved determinations of γ and β.

2.3.6 Tests of preferred-frame and preferred-location effects

Some theories of gravity predict that the outcomes of local gravitational experiments may depend on the velocity of the laboratory relative to the mean rest-frame of the universe (preferred-frame effects) or on the

Chapter 2. Confrontation between gravitation theory and experiment

location of the laboratory relative to a nearby gravitating body (preferred-location effects). Preferred-frame effects are governed by the values of the PPN parameters α_1, α_2 and α_3, and preferred-location effects are governed by ξ (see table 2.6). The most important such effects are (*a*) variations and anisotropies in the locally measured value of the gravitational constant G_L, and (*b*) anomalous perihelion shifts of the planets.

(a) *Variations and anisotropies in* G_L

The local value of the gravitational constant may be determined by measuring the force of attraction between two gravitating bodies (Cavendish experiment). However, the precision of laboratory Cavendish experiments is not high enough to detect post-Newtonian effects (see Rose *et al.*, 1969). On the other hand, a high-precision gravimeter on the surface of the Earth is a form of Cavendish experiment, one that is particularly sensitive to *variations* in G_L, and one that, because of advances in gravimeter technology, can search for post-Newtonian effects. According to the PPN formalism, the predicted form of G_L is (Will, 1971*d*, 1973; Nordtvedt and Will, 1972; TTEG § 6.5)

$$G_L = 1 - \left[4\beta - \gamma - 3 - \zeta_2 - \xi\left(3 + \frac{I}{mR^2}\right)\right] U_{\text{ext}}$$
$$+ \xi\left(1 - \frac{3I}{mR^2}\right) U_{\text{ext}}(e_e \cdot e_r)^2 + \frac{1}{2}\left[\alpha_3 - \alpha_1 + \alpha_2\left(1 - \frac{I}{mR^2}\right)\right] w_\oplus^2$$
$$- \frac{1}{2}\alpha_2\left(1 - \frac{3I}{mR^2}\right)(w_\oplus \cdot e_r)^2,$$

where U_{ext} is the Newtonian gravitational potential of an external body such as the Sun or the galaxy and e_e a unit vector directed from the Earth toward that body, $w_\oplus$ is the Earth's velocity relative to the mean rest-frame of the universe, e_r is a unit vector joining the gravimeter to the center of the Earth, and I, m and R are the Earth's spherical moment of inertia, mass and radius, respectively.

Because of the Earth's eccentric orbital motion, the external potential U_{ext} produced by the Sun and galaxy varies on a yearly basis by only a part in 10^{10}, too small to be detected with confidence by Earth-bound gravimeters; on the other hand, the anisotropic term varies by $U_{\text{ext}} \sim U_{\text{galaxy}} \sim 5 \times 10^{-7}$ because of the variation in $e_e \cdot e_r$ as the Earth rotates relative to e_e. In the preferred-frame terms, the Earth's velocity $w_\oplus$ is made up of a uniform velocity w (~ 300 km s^{-1}) of the solar system relative to the preferred frame, and the Earth's orbital velocity v (~ 30 km s^{-1}) around the Sun. Thus the dominant variations in G_L are given by

$$\Delta G_L / G_L \approx (\tfrac{1}{2}\alpha_2 + \alpha_3 - \alpha_1) w \cdot v - \tfrac{1}{2}\xi U_{\text{ext}}(e_e \cdot e_r)^2$$
$$+ \tfrac{1}{4}\alpha_2 [(w \cdot e_r)^2 + 2(w \cdot e_r)(v \cdot e_r)], \tag{2.10}$$

where we have used the fact that for the Earth $I \approx \tfrac{1}{2} m R^2$.

The $w \cdot v$ and $(w \cdot e_r)(v \cdot e_r)$ terms in (2.10) lead to a yearly variation in the strength of G_L that causes a change in the Earth's moment of inertia, which in turn leads to variations in the Earth's sidereal rotation rate Ω. For a velocity w of 300 km s^{-1} in the direction of the Sun's motion through the cosmic microwave radiation: $\alpha = 165° \pm 9°$, $\delta = 6° \pm 10°$ (Smoot et al., 1977), the value adopted throughout this discussion, the variations have amplitude

$$(\Delta\Omega/\Omega)_{\text{PPN}} \approx 9(\tfrac{2}{3}\alpha_2 + \alpha_3 - \alpha_1) \times 10^{-9}.$$

Nordtvedt and Will (1972) and Rochester and Smylie (1974) have shown that observations of variations in the length of the day coupled with estimates of the atmospheric wind contribution to those variations set a limit on the predicted effect given by

$$|\tfrac{2}{3}\alpha_2 + \alpha_3 - \alpha_1| < 0.02.$$

The anisotropic terms in (2.10) lead to variations in gravimeter measurements and deformations of the elastic Earth that have the same characteristics as solid-Earth tides. The dominant predicted tides occur at angular frequencies Ω (diurnal sidereal), 2Ω (semi-diurnal sidereal) and $2\Omega - \omega$ (semi-diurnal) where ω is the angular frequency of the Earth's orbit. The amplitudes of the tides are roughly parts in 10^8 for the preferred-frame tides and parts in 10^7 for the preferred-location tide. Warburton and Goodkind (1976) have analyzed an 18-month record of gravimeter data obtained with a superconducting gravimeter (accuracy of measurements: $\sigma(\Delta G/G) \approx 10^{-11}$) at Piñon Flat California, and found that for the above choice of w, they could set upper limits on both α_2 and ξ given by

$$|\alpha_2| < 10^{-3}, \quad |\xi| < 10^{-3},$$

an order of magnitude improvement over the crude limits set by Will (1971d). Other effects of these variations in G_L are discussed by Nordtvedt and Will (1972) and TTEG § 7.4.

Chapter 2. Confrontation between gravitation theory and experiment

(b) **Anomalous perihelion shifts**

Preferred-frame and preferred-location effects can produce anomalous perihelion shifts for the planets given in seconds of arc per century by (Nordtvedt and Will, 1972; Will, 1973; and TTEG § 7.3)

$$\dot{\omega}_{\xi} \approx -104\alpha_1 + 70\alpha_2 + 12 \times 10^4 \alpha_3 + 63\xi,$$
$$\dot{\omega}_{\oplus} \approx -165\alpha_1 + 9\alpha_2 + 21 \times 10^5 \alpha_3 + 14\xi. \quad (2.11)$$

By comparing these anomalous shifts with the measured shifts for Mercury and the Earth, and taking into account the limits on the parameters set by geophysical measurements, one obtains from (2.11) a limit on α_3 and thereby a limit on α_1 given by

$$|\alpha_3| < 2 \times 10^{-6}, \quad |\alpha_1| < 2 \times 10^{-2}.$$

Lunar-laser-ranging data can also be used to limit a large number of preferred-frame perturbations in the lunar orbit (Nordtvedt and Will, 1972; Nordtvedt 1973).

2.3.7 The gyroscope experiment and tests of the dragging of inertial frames

Since 1960, much effort has been directed toward testing theories of gravitation using an orbiting superconducting gyroscope (Schiff, 1960b, c; Everitt, 1974; Lipa et al., 1974). The object of the experiment is to measure the precession of the gyroscope's spin axis relative to the distant stars as the gyroscope orbits the Earth. Two relativistic effects are potentially observable. One of these is the 'geodetic' precession, produced in part by the space curvature around the Earth. For a polar orbit with the gyroscope axis in the plane of the orbit, the geodetic precession produces a secular rotation of the spin axis of amplitude

$$\text{geodetic precession} \sim \tfrac{1}{3}(1 + 2\gamma) \times (7'' \, \text{yr}^{-1}).$$

The other is the Lense–Thirring precession or 'Dragging of inertial frames' produced by the Earth's rotation (see MTW §§ 19.2, 33.4, for detailed discussion). For a polar orbit with the spin axis normal to the orbital plane, the dragging of inertial frames produces a secular rotation of the spin axis with amplitude

$$\text{Lense–Thirring precession} \sim \tfrac{1}{8}(4\gamma + 4 + \alpha_1) \times (0.05'' \, \text{yr}^{-1}). \quad (2.12)$$

The experimental goal is 10^{-3} seconds of arc per year accuracy. Detailed derivations of formulae for the gyroscope experiment are given by Wilkins, (1970); MTW, §40.7; TTEG, § 6.4.

A variant of the gyroscope experiment has recently been proposed by Van Patten and Everitt (1976), in which the 'gyroscope' is itself a satellite in Earth orbit. The dragging of inertial frames causes the plane of the orbit to rotate about an axis parallel to the Earth's rotation axis. For a polar-orbiting satellite 800 km above the Earth's surface the rotation rate is $\sim \frac{1}{8}(4\gamma + 4 + \alpha_1) \times (0.18'' \, \text{yr}^{-1})$. In order to eliminate the effects of other sources of rotation (such as the quadrupole moment of the Earth) two satellites counter-rotating in nearly identical orbits are necessary. With the use of drag-free satellites, Van Patten and Everitt estimate that a one per cent experiment may be possible using two and a half years of orbit data.

2.3.8 Laboratory experiments to test post-Newtonian gravity

Because the gravitational force is so weak, most tests of post-Newtonian effects in the solar system have required the use of the Sun and planets as sources of gravitation. One disadvantage of such experiments is that the experimenter has no control over the sources, and so is unable to manipulate the experimental configuration to test or improve the sensitivity of the apparatus or at the very least, to repeat the experiment. Despite this disadvantage of solar-system-sized experiments, the weakness of post-Newtonian gravity has prohibited decent laboratory experiments, with one exception.

That exception is the Kreuzer experiment (Kreuzer, 1968) that compared the active and passive gravitational masses of fluorine and bromine. Kreuzer's experiment used a Cavendish balance to compare the Newtonian gravitational force generated by a cylinder of Teflon (76 per cent fluorine by weight) with the force generated by that amount of a liquid mixture of trichloroethylene and dibromomethane (74 per cent bromine by weight) that had the same passive gravitational mass as the cylinder (namely, the amount of liquid displaced by the cylinder at neutral buoyancy). Kreuzer's central conclusion was that the forces were the same, and hence that the ratio of active mass (m_A) to passive mass (m_P) for fluorine and bromine were the same to 5 parts in 10^5 (Kreuzer, 1968), that is,

$$\left| \frac{(m_A/m_P)_{\text{fluorine}} - (m_A/m_P)_{\text{bromine}}}{(m_A/m_P)_{\text{bromine}}} \right| < 5 \times 10^{-5}. \qquad (2.13)$$

(For further discussion of Kreuzer's experiment see Gilvarry and Muller, 1972, and Morrison and Hill, 1973.)

Will (1976a) has analyzed this experiment using a version of the PPN formalism that deals with charged point masses and has shown that the electrostatic binding energy E_e of a body makes a contribution to m_A/m_P of the form

$$m_A/m_P = 1 + \tfrac{1}{2}\zeta_3(E_e/m_P). \tag{2.14}$$

Because the nuclei of fluorine and bromine contain different fractions of electrostatic binding energy, (2.13) and (2.14) yield a limit on ζ_3 given by

$$|\zeta_3| < 6 \times 10^{-2}.$$

This generalizes and corrects a previous result of Thorne *et al.* (1971).

Recently, Braginsky *et al.* (1977) have suggested that advancing technology may make several laboratory post-Newtonian experiments possible in the coming decade. The progress that makes such experiments feasible is the development of sensing systems with very low levels of dissipation, such as torque-balance systems made from fused quartz or sapphire fibers at temperatures $\leqslant 0.1$ K, massive dielectric monocrystals cooled to millidegree temperatures, and microwave resonators with superconducting walls. Among the experimental possibilities they cite are a measurement of the spin–spin coupling of two rotating bodies, a search for time variations of the gravitational constant (section 2.6), a search for preferred-frame and preferred-location effects, and a measurement of the dragging of inertial frames by a rotating body.

2.3.9 Summary

General relativity has passed every solar-system test of post-Newtonian gravity with flying colors. Yet so have several alternative theories, so it is clear that the solar system cannot be the sole arena for the confrontation between gravitational theory and experiment. Thus, to refine this confrontation, one must look to predictions of theory outside the post-Newtonian regime. It is to these new regimes – gravitational radiation, stellar-system tests and cosmological tests – that we now turn.

2.4 Gravitational radiation as a tool for testing gravitation theory

Solar-system tests of post-Newtonian gravity have played an important role in delineating which metric theories of gravity are unviable, and

Gravitational radiation as a tool for testing gravitation theory

which theories may be viable. Yet such experiments are not the entire story, because they probe only a limited portion, the post-Newtonian limit, of the whole space of predictions of gravitational theories. This is underscored by the fact that there are several metric theories of gravity that are completely different in their formulations yet that can be made to agree completely with general relativity in the post-Newtonian limit, and thereby with current solar-system tests (see table 2.7). The problem of testing these theories has thus forced interest away from the post-Newtonian approximation, toward new areas of 'prediction space', new possible testing grounds.

One new testing ground is gravitational radiation. Recent work has shown that metric theories of gravitation may differ from each other and from general relativity in their gravitational-radiation predictions in at least three important ways: (i) they may predict different polarization states for generic gravitational waves (section 2.4.1); (ii) they may predict that the speed of weak gravitational radiation may differ from that of light (section 2.4.2); and (iii) they may predict different multipolarities, i.e. monople, dipole, quadrupole, etc., of gravitational radiation emitted by given sources (section 2.4.3). The use of polarization and speed as tests requires the regular detection of gravitational radiation, a prospect that may be far off (see chapter 3). On the other hand, the multipolarity of gravitational waves can be studied by analyzing the back-influence of the emission of radiation on the source (radiation reaction) for different multipoles. One example is the change in the period of a two-body orbit caused by the change in energy of the system as a result of the emission of gravitational radiation. Such a test may already be possible in the binary pulsar PSR 1913+16 (see sections 2.5.2a and 2.5.2b).

2.4.1 Polarization of gravitational waves

General relativity predicts that weak gravitational radiation has two independent states of polarization, the '+' and '×' modes, to use the language of MTW (§ 35.6), or the +2 and −2 helicity states, to use the language of quantum field theory. However, general relativity is probably unique in that prediction; every other known, viable metric theory of gravity predicts more than two polarizations for the generic gravitational wave. In fact, the most general weak gravitational wave that a theory may predict is composed of *six* modes of polarization, expressible in terms of the six 'electric' components of the Riemann tensor

Chapter 2. Confrontation between gravitation theory and experiment

R_{0i0j} that govern the driving forces in a detector (Eardley *et al.*, 1973; Eardley, Lee and Lightman, 1973).

For a weak, plane gravitational wave propagating in the z direction, the amplitudes of the six polarization modes may be written as two real functions $\Psi_2(u)$, $\Phi_{22}(u)$ and two complex functions $\Psi_3(u)$, $\Psi_4(u)$ of retarded time $u = t - z$ (the notation is that of the Newman–Penrose formalism). These functions are given by

$$\Psi_2 = -\tfrac{1}{6}R_{z0z0}, \quad \Psi_3 = \tfrac{1}{2}(-R_{x0z0} + iR_{y0z0})$$
$$\Psi_4 = R_{y0y0} - R_{x0x0} + 2iR_{x0y0}, \quad \Phi_{22} = -(R_{x0x0} + R_{y0y0}).$$

Figure 2.3 shows the action of each mode on a sphere of test bodies; Ψ_4 and Φ_{22} are purely transverse, Ψ_2 is purely longitudinal and Ψ_3 is mixed. General relativity permits only the two Ψ_4 modes, while scalar–tensor theories permit the Ψ_4 and the Φ_{22} modes. Theories can be classified according to the number and type of modes they permit for the generic wave. For instance, the most general class, II_6 has $\Psi_2 \neq 0$, and all six modes may be present; class III_5 has $\Psi_2 \equiv 0$, $\Psi_3 \neq 0$ and five modes may be present. For these two classes the amplitudes of the remaining modes are observer-dependent, i.e. they are not invariant under Lorentz transformations that leave the wavevector unchanged ('little group' E(2)). Class N_3 has $\Psi_2 \equiv \Psi_3 \equiv 0$, three modes ($\Psi_4$, Φ_{22}) present; N_2 has $\Psi_2 \equiv \Psi_3 \equiv \Phi_{22} \equiv 0$, two modes ($\Psi_4$) present; 0_1 has $\Psi_2 \equiv \Psi_3 \equiv \Psi_4 \equiv 0$, one mode ($\Phi_{22}$) present. For these three classes the amplitudes that are present are 'little group' invariant; only for the classes (N_2, N_3, 0_1) is it possible to describe the waves by means of states of definite helicity (± 2 for Ψ_4, 0 for Φ_{22}). Table 2.8 shows the classes of the currently viable theories of gravity discussed in section 2.3.2.

Because the six electric components of the Riemann tensor determine the most general wave, then an appropriately designed gravity-wave detector can uniquely determine the six amplitudes (see figure 2.3), and thereby the class of the wave, *as long as the direction to the source is known*. The direction must be determined by means of gravitational-wave interferometry, or by association of the gravitational wave event with an optical display, as for example, from a supernova, or by some other method. If the direction is unknown, the class of the wave may be limited, but not determined uniquely (observables: 6 driving forces; unknowns: 6 amplitudes, 2 direction cosines). A detailed strategy for determining or limiting the class of the waves is spelled out by Eardley, Lee, and Lightman (1973). If the class of the waves is assumed *a priori* to

Gravitational radiation as a tool for testing gravitation theory

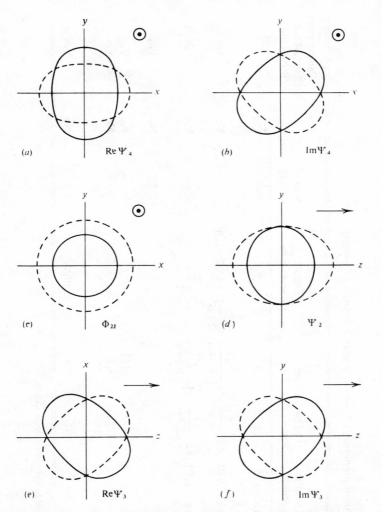

Figure 2.3. The six polarization modes of a weak, plane, null-gravitational wave permitted in any metric theory of gravity. Shown is the displacement that each mode induces on a sphere of test particles. The wave is propagating in the $+z$ direction and has time dependence $\cos \omega t$. The solid line is a snapshot at $\omega t = 0$, the broken line one at $\omega t = \pi$. There is no displacement perpendicular to the plane of the figure.

be N_2, N_3 or 0_1 (waves of definite helicity), then both the class and the direction may be determined simultaneously by an appropriate single detector (Paik, 1977; Wagoner and Paik, 1977). If the observed class of a wave is more general (more polarizations present) than that predicted by a given theory, then that theory is unviable; however, if the observed class is less general than the predicted class, no conclusion can be reached

Table 2.8. *Gravitational-radiation properties of currently viable theories of gravity*

Theory	E(2) class	Definite helicity?	PM parameters κ_1	κ_2	Dipole parameter κ_D	Sign of energy	$c_g = c_{em}$?	Can be equal to GRT in PPN limit?
General relativity	N_2	Yes (± 2)	12	11	0	+	Yes	
Scalar–tensor:								
BWN–Bekenstein	N_3	Yes ($\pm 2, 0$)	$12 - \dfrac{5}{+\omega}$	$11 - \dfrac{45}{8+4\omega}$	$\dfrac{2}{2+\omega}[1+4\Lambda(2+\omega)]^2$	+	Yes	No
Brans–Dicke	N_3	Yes ($\pm 2, 0$)	$12 - \dfrac{5}{2+\omega}$	$11 - \dfrac{45}{8+4\omega}$	$\dfrac{2}{2+\omega}$	+	Yes	No
Vector–tensor:								
Will–Nordtvedt	III_5	No	a	a	a	a	No	Yes
Bimetric:								
Rosen	II_6	No	$-21/2$	$-23/2$	$-20/3$	$+/-$	No	Yes
Lightman–Lee	II_6	No	$-21/2$	$73/8$	$-125/3$	$+/-$	No	Yes
Rastall	II_6	No	a	a	a	a	No	Yes
Stratified:								
Ni	II_6	No	-18	-19	$-400/3$	$+/-$	No	Yes

[a] Calculations have not been performed to determine these values.

because the nature of a particular astrophysical source (a symmetry, for instance) may cause a polarization state to be absent that would have been present in a generic situation.

2.4.2 Speed of gravitational waves

Some theories of gravity predict the same propagation speed for gravitational waves c_g as for light c_{em} (see table 2.8) but others predict a difference that in weak gravitational fields is typically

$$(c_g - c_{em})/c_{em} \sim U/c_{em}^2,$$

where U is the local Newtonian gravitational potential. For our region of the galaxy, or in the field of the Virgo cluster $U/c_{em}^2 \sim 10^{-7}$, so by comparing the arrival times for light and for gravitational waves from a discrete event such as a supernova in the Virgo cluster a limit could be set on the speed difference (see Eardley et al., 1973) of

$$|c_g - c_{em}|/c_{em} < 10^{-9} \times (\text{time-lag precision})/(1 \text{ week}).$$

2.4.3 Multipolarity of gravitational waves: radiation from binary systems

It is common knowledge that general relativity predicts that the lowest multipole emitted in gravitational radiation is quadrupole, in the sense that, if a multipole analysis of the gravitational field in the radiation zone is performed in terms of tensor spherical harmonics, then only the transverse-traceless spin 2 harmonics $\vec{T}^{E2}{}_{lm}$ and $\vec{T}^{B2}{}_{lm}$ with $l \geq 2$ are present (see Thorne, 1977 for discussion and notation). For material sources, this statement can be reworded in terms of appropriately defined multipole moments of the matter and gravitational field distribution within the near-zone surrounding the source (Thorne, 1977): the lowest source multipole that generates gravitational radiation is quadrupole. For slow-motion, weak-field sources, such as binary star systems, quadrupole radiation is in fact the dominant multipole. For a binary system with total mass $m = m_1 + m_2$, reduced mass $\mu = m_1 m_2 / m$, orbital separation r and velocity v, quadrupole radiation leads to a loss of orbital energy at a rate (Peters and Mathews, 1963)

$$\frac{dE}{dt} = -\left\langle \frac{\mu^2 m^2}{r^4} \frac{8}{15} (12v^2 - 11\dot{r}^2) \right\rangle, \tag{2.15}$$

Chapter 2. Confrontation between gravitation theory and experiment

where $\dot{r} = dr/dt$, and where angular brackets denote an average over an orbit. This loss of energy results in a decrease in the orbital period P given by Kepler's third law,

$$\dot{P}/P = -\tfrac{3}{2}E^{-1}\,dE/dt$$

(see however Ehlers *et al.*, 1976, for an opposing viewpoint on this conclusion). Quadrupole radiation also leads to a decrease in the angular momentum of the system, and to a corresponding decrease in the eccentricity of the orbit (see Wagoner, 1975, for references and a summary of the formulae). Faulkner (1971) has pointed out that these effects of quadrupole gravitational radiation may play an important role in the evolution of ulstrashort-period binary systems. But probably the most promising test of the existence of quadrupole radiation will be provided by observations of period changes $\dot{P}$ in the binary pulsar (section 2.5.2*a*).

Unlike general relativity, however, nearly every alternative theory of gravity predicts the presence in gravitational radiation of all multipoles – monopole and dipole, as well as quadrupole and higher multipoles (Eardley, 1975; Will and Eardley, 1977; Will, 1977*b*). For binary star systems, these additional multipole contributions have two effects on the energy-loss-rate formula (2.15): (*a*) modify the numerical coefficients in (2.15); and (*b*) generate an additional term (produced by *dipole* moments) that depends on the self-gravitational binding energy of the stars. The resulting formula for dE/dt may be written in a form that contains dimensionless parameters whose values depend on the theory under study. Two parameters κ_1 and κ_2 are denoted PM parameters because they refer to that part of dE/dt that corresponds to the Peters–Mathews (1963) result for general relativity. A parameter κ_D refers to the dipole self-gravitation contribution. The result is

$$\frac{dE}{dt} = -\left\langle \frac{\mu^2 m^2}{r^4}[\tfrac{8}{15}(\kappa_1 v^2 - \kappa_2 \dot{r}^2) + \tfrac{1}{3}\kappa_D \mathfrak{S}^2]\right\rangle, \qquad (2.16)$$

(Will, 1977*b*) where $\mathfrak{S}$ is the difference in the self-gravitational binding energy per unit mass between the two objects in the binary system (see Eardley, 1975; or Will and Eardley, 1977, for a more precise definition of $\mathfrak{S}$).

Table 2.8 shows the predicted values of κ_1, κ_2 and κ_D for those theories whose values have been computed. We note the surprising result that for all the theories except scalar–tensor theories, the dipole radiation carries negative energy, i.e. increases the energy of the system

($\kappa_D < 0$), and that the PM radiation may carry either positive or negative energy, depending on the theory and on the nature of the orbit. It could be argued (and presumably will be argued by some) that this prediction alone should be sufficient grounds to judge each such theory unviable. However, this is a theorist's constraint that has little experimental foundation in the case of gravitation, and so we will restrict attention to observational evidence for or against such an effect.

The only theory shown in table 2.8 that automatically predicts no dipole gravitational radiation is general relativity. Scalar–tensor theories of the BWN–Bekenstein type can also avoid dipole radiation if for example the function $\omega(\phi)$ is given by $\omega(\phi) = \frac{1}{2}(4-3\phi)/(\phi-1)$, then $1+4\Lambda(2+\omega) = 0 = \kappa_D$. In fact, in this case, the locally measured gravitational constant G_L is truly constant, and the theory predicts no Nordtvedt effect ($4\beta - \gamma - 3 = 0$) (Barker, 1978). For the other theories in table 2.8, there may be a close relationship between the presence or absence of dipole radiation, the sign of the energy carried by the waves, and the E(2) class of the theory. General relativity predicts waves of the least general class (N_2) (except for the uninteresting class 0_1) and of definite helicity (± 2). This results from the fact that the full Einstein field equations may be written in the form (MTW § 20; Epstein and Wagoner, 1975)

$$\Box \theta^{\mu\nu} = -16\pi\tau^{\mu\nu}, \quad \tau^{\mu\nu}{}_{,\nu} = 0,$$

where $\theta^{\mu\nu}$ is related to the spacetime metric, and $\tau^{\mu\nu}$ is a stress–energy pseudo-tensor for matter and gravitational fields, and where the coordinates have been specialized (gauge choice) to satisfy $\theta^{\mu\nu}{}_{,\nu} = 0$. But there is an additional gauge freedom permitted by the conservation of $\tau^{\mu\nu}$ ($\tau^{\mu\nu}{}_{,\nu} = 0$) that eliminates all but two helicity states for $\theta^{\mu\nu}$. In field-theoretic language, the waves have definite helicity because they couple to a conserved current. But it is precisely the conservation law $\tau^{\mu\nu}{}_{,\nu} = 0$ that also eliminates contributions to $\theta^{\mu\nu}$ from monopole and dipole moments of sources, however relativistic. Thus it is perhaps not surprising that theories of gravity whose waves are more general than N_2 predict dipole radiation. In purely dynamical theories, at least one of the additional dynamical fields does not couple to a conserved current (otherwise definite helicity states would result), and so dipole radiation may be permitted. The exceptions are scalar–tensor theories (N_3), where the scalar field automatically has definite helicity zero, despite the fact that it may couple to a non-conserved source (Will, 1977b). In prior-geometric theories, the presence of non-dynamical fields destroys

Chapter 2. Confrontation between gravitation theory and experiment

the gauge invariance of the dynamical fields that contribute to the physical metric, and so permits monopole and dipole moments of sources to contribute. It is also perhaps not surprising that such theories predict negative or indefinite sign for the emitted energy, as, again in quantum field theoretic terms, definite helicity, quantizability, and positive-definiteness of energy go hand-in-hand.

For systems containing compact objects such as neutron stars ($\mathfrak{S} \sim E_g/m \sim 0.5$) the dipole contribution in (2.16) may dominate over the PM contribution ($v^2 \sim \dot{r}^2 \sim 10^{-6}$, for $v \sim 200$ km s^{-1}) if $\kappa_D \neq 0$. The most promising testing ground for the existence of dipole gravitational radiation may be the binary pulsar (section 2.5.2b). Such a test may provide crucial evidence for or against a wide class of gravitation theories that are indistinguishable from general relativity in solar-system tests.

2.5 Stellar-system tests: the binary pulsar

It was the summer of 1974. Russell Hulse and Joseph Taylor were carrying out a systematic survey for new pulsars at the Arecibo Observatory in Puerto Rico. During that survey, they detected 50 pulsars, of which 40 were not previously known, and made a variety of observations, including measurements of their pulse periods to an accuracy of one microsecond. But one of these pulsars, denoted PSR 1913+16, was peculiar: besides having a pulsation period of 59 ms – shorter than that of any known pulsar except the one in the Crab Nebula – it also defied any attempts to measure its period to ±1 μs, by making 'apparent period changes of up to 80 μs from day to day, and sometimes by as much as 8 μs over 5 minutes' (Hulse and Taylor, 1975). Such behavior is uncharacteristic of pulsars, and Hulse and Taylor rapidly concluded that the observed period changes were the result of Doppler shifts due to orbital motion of the pulsar about a companion. By the end of September 1974, Hulse and Taylor had obtained an accurate velocity curve of this 'single line spectroscopic binary'. By a detailed fit of this curve to a Keplerian two-body orbit, they obtained the following elements of the orbit of the system: K_1, the semi-amplitude of radial velocity variation of the pulsar relative to the center of mass of the system; P_b, the period of the binary orbit, corrected for the motion of the observatory; e, the eccentricity of the orbit, ω the longitude of periastron at a chosen epoch (September 1974); $a_1 \sin i$, the projected semi-major axis of the pulsar orbit where i is the inclination of the orbit relative to the plane of the sky; and $f_1 = (m_2 \sin i)^3/(m_1 + m_2)^2$, the mass function, where m_1 and m_2 are the mass of the pulsar and companion. In addition, they obtained the 'rest' period P_p of the pulsar, corrected for orbital Doppler shifts at a chosen epoch. The results are shown in the middle column of table 2.9 (Hulse and Taylor, 1975).

Table 2.9. *Parameters of PSR 1913+16*

Parameter	Value from period data (summer 1974)	Value from arrival-time data (9/74–10/75)
K_1 (km s^{-1})	199 ± 5	
P_b (s)	$27\,908 \pm 7$	$27\,906.980 \pm 0.002$
e	0.615 ± 0.010	$0.617\,17 \pm 0.000\,05$
ω (degrees)	179 ± 1	178.861 ± 0.007
$a_1 \sin i$ (cm)	$(6.96 \pm 0.13) \times 10^{10}$	$(7.0043 \pm 0.0004) \times 10^{10}$
$f_1 (m_\odot)$	0.13 ± 0.01	$0.131\,26 \pm 0.000\,02$
P_p (s)	$0.059\,030 \pm 1$	$0.059\,029\,995\,272 \pm 5$
Reference	Hulse and Taylor (1975)	Taylor *et al.* (1976)

However, at the end of September 1974, the observers switched to an observation technique that yielded vastly improved accuracy. That technique measures the arrival times of individual pulses (as opposed to the period or the difference between adjacent pulses) and compares those times to arrival times predicted using the best available pulsar and orbit parameters (Blandford and Teukolsky, 1976). The parameters are then improved by means of a least-squares analysis of the arrival-time residuals. The results of this analysis using data up to October 1975 are shown in the right-hand column of table 2.9 (Taylor *et al.*, 1976).

The discovery of PSR 1913+16 caused considerable excitement in the relativity community (to say nothing of the editorial offices of the *Astrophysical Journal Letters*), because it was realized that the system could provide a new laboratory for studying relativistic gravity. Post-Newtonian orbital effects would have magnitudes of order $v^2 \sim K_1^2 \sim 5 \times 10^{-7}$, $m/r \sim f_1/a_1 \sin i \sim 3 \times 10^{-7}$, a factor of ten larger than the corresponding quantities for Mercury, and the shortness of the orbital period ($\sim$8 hours) would amplify any secular effect such as the periastron shift. This expectation was confirmed by the announcement in December 1974 (Taylor, 1975) that the periastron shift had been measured to be 4.0 ± 1.5 degrees per year (compare with Mercury!). Moreover, the system appeared to be a 'clean' laboratory, unaffected by

Chapter 2. Confrontation between gravitation theory and experiment

complex astrophysical processes such as mass transfer. The pulsar radio signal was never eclipsed by the companion, placing limits on the geometrical size of the companion, and the dispersion of the pulsed radio signal showed little change over an orbit, indicating an absence of dense plasma in the system, as would occur if there was mass-transfer from the companion onto the pulsar. These data effectively ruled out a main-sequence star as a companion: although such a star could conceivably fit the geometrical constraints placed by the eclipse and dispersion measurements, it would produce an enormous periastron shift ($>5000°$ yr^{-1}) generated by tidal deformations due to the pulsar's gravitational field (Masters and Roberts, 1975; Webbink, 1975). Another suggested companion was a helium main-sequence star, which could accommodate the geometrical and periaston-shift constraints. However, estimates of the distance to the pulsar (5 kpc: Hulse and Taylor, 1975) and extinction along the line of sight (~ 3.3 mag: Davidsen et al., 1975) indicated that such a helium star would be brighter by at least two magnitudes than limits ($m_v \geqslant 23$, $m_R \geqslant 21$) placed on any optical object associated with the pulsar (Kristian et al., 1976; Roberts et al., 1976). Other possible companions are the condensed stellar objects: white dwarf, neutron star or black hole.

Attempts to further delineate the nature of the companion involved constructing scenarios for the formation and evolution of the system (see table 2.10). The favored scenario appears to be evolution from an X-ray binary phase whose end product is two neutron stars (Flannery and van den Heuvel, 1975; Webbink, 1975; Smarr and Blandford, 1976). However alternative scenarios have been constructed that lead to white dwarf companions (Van Horn et al., 1975, Smarr and Blandford, 1976), black hole companions (Webbink, 1975; Bisnovatyi-Kogan and Komberg, 1976; Smarr and Blandford, 1976) and helium star companions (Webbink, 1975; Smarr and Blandford, 1976). The nature of the companion is crucial for discussion of various relativistic and astrophysical effects in the system, to which we now turn.

The binary pulsar generated excitement among relativists because it can play two rather different roles: (i) it is a 'clean' laboratory with potentially large relativistic effects in which tests of gravitation theory may be possible (section 2.5.2), and (ii) *it is the first known system in which relativistic gravity can be used as a practical tool for the determination of astrophysical parameters* (section 2.5.1). Because the latter is a novel and quite unexpected role for relativistic gravitation we address it first.

Table 2.10. *A bibliography on the binary pulsar*

A. *Observations and astrophysics*

Pulsar observations
  Hulse and Taylor (1975)
  Taylor (1975)
  Taylor et al. (1976)

Evolutionary scenarios
  Flannery and van den Heuvel (1975)
  Webbink (1975)
  Van Horn et al. (1975)
  Van den Bergh (1975)
  Wheeler (1976)
  Smarr and Blandford (1976)
  Bisnovatyi-Kogan and Komberg (1976)

Astrophysical effects
  Ozernoi and Shishov (1975)
  Balbus and Brecher (1976)
  Shapiro and Terzian (1976)

Other observations
  Hjellming and Gibson (1975)
  Chanan et al. (1975)
  Davidsen et al. (1975)
  Bernacca et al. (1975)
  Kristian et al. (1976)
  Nather et al. (1976)
  Van Citters and Rybski (1977)

Data analysis methods
  Wheeler (1975)
  Blandford and Teukolsky (1975, 1976)
  Epstein (1977)

Nature of the companion
  Masters and Roberts (1975)
  Ozernoi and Reinhardt (1975)
  Roberts et al. (1976)

B. *Relativistic gravity*

General
  Damour and Ruffini (1974)
  Brecher (1975)
  Brumberg et al. (1975)
  Esposito and Harrison (1975)
  Demianski and Shakura (1976)

Redshift-mass of pulsar
  Wheeler (1975)
  Blandford and Teukolsky (1975, 1976)

Tests of gravitation theory
  Eardley (1975)
  Nordtvedt (1975b)
  Will (1976b, 1977b)
  Will and Eardley (1977)

Periastron shifts
  Will (1975)
  Barker and O'Connell (1975)
  Nordtvedt (1975b)

Precession of pulsar spin
  Barker and O'Connell (1975)
  Zel'dovich and Shakura (1975)
  Hari Dass and Radhakrishnan (1975)
  Rudolf (1977)

Gravitational radiation
  Eardley (1975)
  Wagoner (1975, 1976)
  Will and Eardley (1977)
  Will (1977b)

Chapter 2. Confrontation between gravitation theory and experiment

2.5.1 Relativistic gravity as a tool for measuring astrophysical parameters

(a) The periastron shift

According to the PPN formalism, the rate of periastron precession of two bodies of mass m_1 and m_2 is given by (cf. (2.9))

$$\frac{d\omega}{dt} = \frac{6\pi m}{P_b a(1-e^2)}\left[\tfrac{1}{3}(2+2\gamma-\beta)+\tfrac{1}{6}(2\alpha_1-\alpha_2+\alpha_3+2\zeta_2)\frac{m_1 m_2}{m^2}\right], \tag{2.17}$$

where $m = m_1 + m_2$ is the total mass, and P_b is the orbital period. The second term in (2.17) may be significant when neither of the bodies has negligible mass compared to the other (for Mercury and the Sun, $m_1 m_2/m^2 \sim 2\times 10^{-7}$), however it is absent in any case for any fully-conservative theory of gravity ($\alpha_1 = \alpha_2 = \alpha_3 = \zeta_2 = 0$). For a nearly Keplerian orbit described by the relation $P_b/2\pi = a^{3/2} m^{-1/2}$, (2.17) may be written

$$\frac{d\omega}{dt} = \frac{3 m^{2/3}}{(P_b/2\pi)^{5/3}(1-e^2)}\left[\tfrac{1}{3}(2+2\gamma-\beta)+\tfrac{1}{6}(2\alpha_1-\alpha_2+\alpha_3+2\zeta_2)\frac{X}{(1+X)^2}\right],$$

where $X = m_1/m_2$. For the binary pulsar, P_b and e are known (table 2.9) hence, in degrees per year,

$$\frac{d\omega}{dt} = 2.11°\left(\frac{m}{m_\odot}\right)^{2/3}\left[\tfrac{1}{3}(2+2\gamma-\beta)+\tfrac{1}{6}(2\alpha_1-\alpha_2+\alpha_3+2\zeta_2)\frac{X}{(1+X)^2}\right].$$

Because the periastron shift depends on the unknown total mass m and mass ratio X, it is impossible to use this effect as a test of gravitation theory. Instead, it has proved more profitable to turn the tables: to assume a particular theory of gravity, and to use the measured periastron shift to determine the total mass of the system. For example, in general relativity the predicted shift is ($\gamma = \beta = 1$, $\alpha_1 = \alpha_2 = \alpha_3 = \zeta_2 = 0$)

$$(d\omega/dt)_{\text{GRT}} = 2.11°(m/m_\odot)^{2/3}\text{ yr}^{-1}.$$

Taylor *et al.*, (1976) have measured the periastron shift to be

$$(d\omega/dt)_{\text{Measured}} \approx 4.22 \pm 0.04° \text{ yr}^{-1},$$

so the total mass of the system is $m \approx 2.83 \pm 0.04 m_\odot$. To our knowledge, this is the first instance in which general relativity has been used as a

Stellar-system tests: the binary pulsar

practical tool to make a high-precision ($\sim$ one per cent) astronomical measurement.

However, two notes of caution must be sounded before the periastron-measured mass can be regarded as final.

(i) There may be non-relativistic sources of periastron shift in the system, principally from possible quadrupole deformations of the companion (analogous to the effect in (2.9) of a solar quadrupole moment). These deformations are of two types: tidal deformations induced by the pulsar's gravitational field – significant only if the companion is a helium star – and rotational deformations that can produce either advance or regression of the periastron depending on the orientation of the companion's spin axis – significant if the companion is a helium star or a rapidly rotating white dwarf. If the companion is a black hole, neutron star on non-rotating white dwarf, then only the relativistic periastron precession is present (for detailed discussion see Smarr and Blandford, 1976: see also Will, 1975; Roberts et al., 1976).

(ii) In alternative theories of gravitation such as those shown in table 2.7, the PPN parameters may agree with those of general relativity, yet the predicted shift may differ markedly from (2.17). Although the orbital motion may be described as post-Newtonian ($m/r \sim v^2 \ll 1$), the structure of the pulsar ($m/R \sim 0.5$) generally must be described by the fully relativistic gravitational equations of the theory (for examples of neutron star models in alternative theories, see Eardley, 1975; Rosen and Rosen, 1975; Caporaso and Brecher, 1977; Mikkelson, 1977; Will and Eardley, 1977; Rastall, 1977c). The result may be large relativistic modifications of the masses – active, passive, inertial, etc. – that appear in the Newtonian and post-Newtonian equations of motion. In general relativity, no such modifications occur, no matter now relativistic the source. For example, in Rosen's bimetric theory, Kepler's third law and the periastron prediction become (Will and Eardley, 1977)

$$(P_b/2\pi) = \mathfrak{G}^{-1/2} a^{3/2} m^{-1/2},$$

$$d\omega/dt = 6\pi \mathfrak{P} \mathfrak{G}^{-1} m / P_b a (1-e^2),$$

where $\mathfrak{G} = 1 - 4s_1 s_2/3$, $\mathfrak{P} = (1 - \frac{8}{9}(s_1 + s_2) + 28 s_1 s_2/27) \mathfrak{G} + 0(s^3)$, and where s_1 and s_2 are related to the self-gravitational binding energies per unit mass of the two bodies ($s_1 - s_2 = \mathfrak{S}$ in (2.16)). For neutron stars in Rosen's theory, s can be as large as 0.6. The resulting periastron-measured mass is then given by

$$m \approx (2.83 \pm 0.04) m_\odot \mathfrak{G}^2 \mathfrak{P}^{-3/2}.$$

Chapter 2. Confrontation between gravitation theory and experiment

For instance, if the companion is a neutron star of mass close to that of the pulsar, then $m \approx 6.96 m_\odot$ (Will and Eardley, 1977).

Keeping in mind these cautionary notes, we shall proceed under the assumption that the total mass of the system is around 2.83 solar masses.

(b) Gravitational redshift and second-order Doppler shift: a way to weigh the pulsar

Relativistic gravitational effects may also be tools that help perform the first precision measurement of the mass of a neutron star. The observed period of the pulsar is affected not only by its orbital velocity projected along the line of sight (first-order Doppler shift), but also by the magnitude of its velocity (second-order Doppler shift) and by the gravitational potential of the companion (gravitational redshift). These effects have the form

$$(P_p)_{obs}/(P_p)_{em} = 1 + \boldsymbol{n} \cdot \boldsymbol{v}_1 + \tfrac{1}{2} v_1^2 + m_2/r,$$

where $\boldsymbol{n}$ is a unit vector along the line of sight, $\boldsymbol{v}_1$ is the pulsar's velocity, and r is the distance between the pulsar and the companion. In terms of Keplerian orbital elements, this can be written

$$(P_p)_{obs}/(P_p)_{em} = 1 + K_1[\cos(\omega + \phi) + e \cos \omega] + B \cos \phi + C, \qquad (2.18)$$

where ϕ is the angle of the orbit measured from periastron, and C is an unmeasurable constant. The coefficients K_1 (see table 2.9) and B are given by

$$K_1 = \frac{2\pi a_1 \sin i}{P_b (1-e^2)^{1/2}}, \quad B = \frac{m_2^2 (m_1 + 2m_2) e}{m^2 a_1 (1-e^2)},$$

(note $B/K_1 \sim [m/a_1]^{1/2} \sim v$). A measurement of B could be combined with the measured values of the mass function f_1, the projected semi-major axis $a_1 \sin i$ and the periastron-measured total mass m to give unique values for m_1, m_2, a_1 and $\sin i$. For instance, if $m = 2.83 m_\odot$, then the mass of the pulsar would be given by

$$m_1 = 1.42 \, [3 - (1 + B/5 \times 10^{-7})^{1/2}] m_\odot.$$

But a determination of B would be totally unfeasible were it not for the presence of the relativistic periastron shift. It is the periastron shift ($\omega = \omega_0 + \dot\omega t$) that allows the term with amplitude K_1 in (2.18) to be separated from the B-term; without it, the two terms in (2.18) would be completely degenerate. This is because the first-order Doppler shift depends on the orientation of the orbit relative to the observer's line of sight, which changes as the periastron precesses, while the redshift second-order Doppler effects depend only on internal orbital variables which do not change. Since the periastron rotates at 4.22 degrees per year, a complete separation of these effects would be possible in about 23 years. However, Blandford and Teukolsky (1975, 1976) have shown that in practice, measurements of the arrival times of pulses to 1 ms accuracy made systematically over a period of about 5 years would yield a 10 per cent measurement of B (a one per cent measurement would require about 10 years). Such a measurement of the mass of a neutron star is an important piece of data in testing equations of state of matter at high densities.

(c) Precession of the pulsar's spin axis

If the pulsar is a rapidly rotating neutron star, it should experience the same precession effects on its spin axis as a gyroscope in orbit around the Earth (section 2.3.7). The dominant effect is the geodetic precession, given by

$$\langle d\mathbf{n}/dt \rangle = \mathbf{\Omega} \times \mathbf{n},$$

$$\mathbf{\Omega} = (3\pi/P_b)[m_2^2/ma(1-e^2)][\tfrac{1}{3}(2\gamma+1) + \tfrac{2}{3}(\gamma+1+\tfrac{1}{4}\alpha_1)(m_1/m_2)]\mathbf{n}_0,$$

(Barker and O'Connell, 1975; Hari Dass and Radhakrishnan, 1975; Rudolph, 1977) where γ, α_1 are PPN parameters, $\mathbf{n}$ is a unit vector parallel to the pulsar's spin axis, and $\mathbf{n}_0$ is a unit vector normal to the plane of the orbit. The magnitude of $\mathbf{\Omega}$ is about one degree per year for the binary pulsar (compare with an Earth-orbiting gyroscope, section 2.3.7); note however, that no precession occurs if the pulsar's spin axis is normal to the plane of the orbit.

If precession does occur, it could be viewed as a means to test the PPN parameters; however it may be more fruitful to use the relativistic precession as a means to probe the nature of the pulsar's emission mechanism. As the pulsar precesses, the observer's line of sight intersects the surface of the neutron star at different latitudes, thus it may be possible to obtain two-dimensional information on the shape of the emitted beam, as well as to study the variation of spectrum and polarization with latitude (Smarr and Blandford, 1976). Unfortunately, in most pulsar models, the radio pulses are emitted in a pencil beam, so the pulsar might disappear altogether.

Chapter 2. Confrontation between gravitation theory and experiment

2.5.2 The binary pulsar as a testing ground for relativistic gravity

(a) *Test for the existence of gravitational radiation*

General relativity predicts that the emission of quadrupole gravitational radiation causes the energy of a binary system to decrease according to (2.15). For a Keplerian orbit, (2.15) takes the form

$$\frac{dE}{dt} = -\frac{32}{5}\left(\frac{\mu}{m}\right)^2\left(\frac{m}{a}\right)^5\left(1+\frac{73}{24}e^2+\frac{37}{96}e^4\right)(1-e^2)^{-7/2}.$$

For the measured parameters of the binary pulsar, the ensuing change in the orbital period of the system is (Wagoner, 1975, 1976)

$$\frac{\dot{P}_b}{P_b} = -3.8\times 10^{-9}\frac{(m/2.83m_\odot)^{4/3}}{\sin i}\left[1-0.36\frac{(m/2.83m_\odot)^{-1/3}}{\sin i}\right]\text{yr}^{-1}.$$

The predicted period change is relatively insensitive to inclination angle except between 20 and 30 degrees ($0 < m_1 < 0.8 m_\odot$). Blandford and Teukolsky (1976) have shown that the error in a determination of $\dot{P}_b/P_b$ carried out over T years behaves roughly as $10^{-7}T^{-5/2}$ if the pulse arrival times can be measured with an accuracy of 10^{-3} s, and if phase coherence with the pulses can be maintained over that period. Thus a 10 per cent measurement of $\dot{P}_b/P_b$ may be possible in 10 years, by which time the individual masses and the inclination angle will presumably have been determined, thus making the prediction more definite.

A confirmation of this effect goes beyond testing merely the existence of gravitational radiation. It also tests whether gravitational radiation carries energy away from the source. Rosen (1977*b*) has recently suggested that, in view of the time-symmetry of the field equations for gravitational radiation in all theories of gravity, the solutions to those equations should also be symmetric, that is half-retarded plus half-advanced, as opposed to the conventional retarded, outgoing-wave solutions. Such symmetric solutions still represent gravitational radiation (R_{0i0j}) as measured by a laboratory detector, but they do not carry energy. Detection of $\dot{P}_b/P_b$ would rule out this possibility.

However, before this can be viewed as a reliable test for the existence of gravitational radiation, other possible sources of period changes must be accounted for. Since all the tests of gravitational theory discussed in this chapter involve detecting changes in both the orbital and the pulsar period, we review here sources of both (see table 2.11).

Stellar-system tests: the binary pulsar

Table 2.11. *Comparison of secular effects on periods in the binary pulsar*

Effect	$[\dot{P}/P]_{\text{Pulsar}}$ yr^{-1}	$|\dot{P}/P|_{\text{Orbit}}$ yr^{-1}
Tests of gravitation theory		
Gravitational radiation		
Quadrupole–general relativity	Negligible	2×10^{-9}
Dipole	Negligible	3×10^{-7}
Conservation-law breakdown	4×10^{-7}	4×10^{-7}
Competing effects		
Tidal dissipation		
Neutron star, black hole companion	Negligible	Negligible
White dwarf, helium star companion	Negligible	Uncertain[a]
Mass loss	Uncertain	$< 10^{-14}$
Galactic acceleration	2×10^{-13}	2×10^{-13}
Observed values (2σ)	$(4.6 \pm 0.2) \times 10^{-9}$ (Taylor *et al.*, 1976)	$(3.6 \pm 0.7) \times 10^{-9}$ (Taylor, 'Texas' symposium, Munich, 1978)

[a] See following discussion.

Tidal dissipation. Tides raised on the companion by the gravitational field of the pulsar will change orbital and rotational energies via viscous heating (the corresponding tides raised on the pulsar are negligible because of its small size). The rate of change of orbital period caused by this dissipation is given by (Smarr and Blandford, 1976; Will, 1976*b*)

$$\dot{P}_b/P_b \approx -3 \times 10^{-13} X(1+X)^2$$
$$\times \left(1 - \frac{P_b}{\tau}\right)\left(\frac{R}{10^5 \text{ km}}\right)^9 \left(\frac{m}{2.83 m_\odot}\right)^{-11/3} \langle\mu\rangle_{13} \text{ yr}^{-1},$$

where $\langle\mu\rangle_{13}$ is the mean coefficient of viscosity of the companion in 10^{13} g cm^{-1} s^{-1}, R and τ are the radius and rotational period of the companion. The absence of eclipses in the observations of the pulsar signal indicates that $R < 10^5$ km (Hulse and Taylor, 1975). If the source of viscosity is molecular, $\langle\mu\rangle_{13} < 10^{-10}$, and tidal dissipation yields utterley negligible changes in the orbital period. However, if the companion to the pulsar is a white dwarf or helium star, tidally driven shear (Press, Wiita and Smarr, 1975) or turbulence in the atmosphere (Balbus and Brecher, 1976) could lead to $\langle\mu\rangle_{13} \geqslant 1$. Magnetic viscosity in a white dwarf could make $\langle\mu\rangle_{13} \sim 10^5$ (Smarr and Blandford, 1976).

Chapter 2. Confrontation between gravitation theory and experiment

Galactic rotation. Because the galaxy does not rotate uniformly, the relative velocity along the line of sight between the solar system and the binary system changes as a function of time. The resulting changing Doppler shift produces changes in both the pulsar and orbital periods; however estimates of the location and distance of the pulsar combined with the conventional galactic rotation law yield

$$|\dot{P}_b/P_b| = |\dot{P}_p/P_p| \sim 2 \times 10^{-13} \text{ yr}^{-1}.$$

(Will, 1976b; Shapiro and Terzian, 1976).

Energy loss. The emission of energy of various forms (particles, electromagnetic radiation, etc.) from the pulsar results in a decrease in the pulsar's rotational kinetic energy, and thus to an increase in its period, given by

$$dE/dt = -I(2\pi/P_p)^2 \dot{P}_p/P_p,$$

where dE/dt is the rate of energy loss, and I is the moment of inertia of the pulsar. The orbital period also changes at a rate given by

$$\dot{P}_b/P_b = -\tfrac{1}{2}\dot{m}_1/m,$$

where m_1 is the mass of the pulsar. However, because $dE/dt \lesssim dm_1/dt$, we may write

$$\dot{P}_b/P_b \lesssim 1 \times 10^{-6} \, (m/2.83 m_\odot)^{-1} I_{45} (\dot{P}_p/P_p)$$

where $I_{45} = I/10^{45}$ g cm^2. Since the observed value for $\dot{P}_p/P_p$ is less than 10^{-8} yr^{-1} (table 2.11), then $\dot{P}_b/P_b$ due to energy loss must be less than 10^{-14} yr^{-1}.

(b) Test for the existence of dipole gravitational radiation

As we pointed out in section 2.4.3, most alternative theories of gravitation to general relativity predict the existence of dipole gravitational radiation. Since the magnitude of the effect in binary systems depends on the self-gravitational binding energies of the two bodies, the binary pulsar provides an ideal testing ground. In general relativity, neutron star binding energies can be as large as half their rest masses, and in other theories even larger, so the dipole effect, if present, would produce more rapid period changes than the general relativistic quadrupole effect.

Stellar-system tests: the binary pulsar

The predicted period change averaged over an orbit (cf. (2.16)) is given by (Eardley, 1975; Will, 1977b)

$$\dot{P}_b/P_b = -\kappa_D \mu \mathfrak{S}^2 (2\pi/P_b)^2 (1+\tfrac{1}{2}e^2)(1-e^2)^{-5/2}.$$

For the binary pulsar, this yields

$$\dot{P}_b/P_b = -(3.1 \times 10^{-7})\kappa_D (\mathfrak{S}/0.1)^2 (\mu/m_\odot)\,\text{yr}^{-1}.$$

The observed upper limit on $\dot{P}_b/P_b$ is $10^{-7}\,\text{yr}^{-1}$ (table 2.11). This datum can be used to set a limit on the parameter κ_D and thereby test the alternative theories listed in table 2.8, but because of the uncertainty in the masses of the components and the identity of the companion to the pulsar, a decisive test is not feasible at present. However, figure 2.4 illustrates the rough limits that would be set on κ_D for the various pulsar masses assuming the companion is (i) a neutron star or (ii) a white dwarf. Values of $\mathfrak{S}$ were computing using general relativistic neutron star models (Will, 1977b) The calculations that led to figure 2.4 were carried out through post-Newtonian order (Will, 1977b) and so could not take into account the extreme relativistic structure of the pulsar, hence the limits on κ_D shown should be regarded as merely approximate. A calculation that does take the fully relativistic structure of the neutron star into account was performed within the context of Rosen's bimetric theory by Will and Eardley (1977). The important differences in the result are (i) possibly larger periastron measured masses ($6.96 m_\odot$ for two equal neutron stars) and (ii) larger $\mathfrak{S}$-values for neutron star models in Rosen's theory.

Further observations of the system that tighten the limit on $\dot{P}_b/P_b$ and establish the identity and mass of the companion will provide crucial tests of what otherwise could be viable alternatives to general relativity.

(c) Test of post-Newtonian conservation laws

Some metric theories of gravitation predict that the center of mass of a binary system may 'self-accelerate' in the direction of the system's periastron, with an acceleration given by (Will, 1976b)

$$\boldsymbol{a}_\text{CM} = \frac{\pi m_1 m_2 (m_1 - m_2) e}{P m^{3/2} a^{3/2} (1-e^2)^{3/2}} (\alpha_3 + \zeta_2)\boldsymbol{n},$$

where $\boldsymbol{n}$ is a unit vector directed from the center of mass of the system to the point of periastron of m_1. Any semi-conservative theory of gravity has $\alpha_3 = \zeta_2 = 0$ and predicts no such effect, so the effect can occur only in

Chapter 2. Confrontation between gravitation theory and experiment

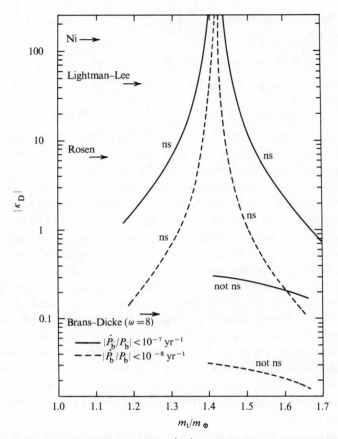

Figure 2.4. Limits on the dipole parameter $|\kappa_D|$ placed by observed 2σ upper limits on $|\dot{P}_b/P_b|$ of 10^{-7} yr^{-1} (present value) and 10^{-8} yr^{-1} (possible future value), as a function of pulsar mass. Two possible companions to the pulsar are considered: a neutron star (curve 'ns') and not a neutron star (curve 'not ns'). It is assumed that the companion makes negligible contributions to the periastron shift ($m = 2.83 m_\odot$) and to the orbital decay. Predicted values of $|\kappa_D|$ are indicated by arrows.

theories of gravity that lack global post-Newtonian momentum conservation laws. However, solar-system experiments to date (section 2.3) place limits only on the parameter α_3, namely $|\alpha_3| < 2 \times 10^{-5}$. No decent direct limit has been placed on ζ_2, although Shapiro and Teukolsky (1976) have used gravitational redshift data for white dwarfs to obtain the weak limit $|\zeta_2| < 100$. Other experiments such as lunar laser-ranging (section 2.3.4) test combinations of PPN parameters that contain ζ_2.

There are few currently known viable theories of gravitation that have $\zeta_2 \neq 0$. Any theory based on an invariant action principle is automatically semi-conservative (Lee et al., 1974); all the theories listed in table 2.7 fall into this class. Yet a nonzero value of ζ_2 may have important physical consequences; in addition to being a 'conservation law' parameter, ζ_2 is a rough measure of 'how much gravity is generated by gravitational energy'. Wagoner and Malone (1974) have shown that the contribution of self-gravitational energy to the active gravitational mass of a neutron star is sensitive to the value of ζ_2, and have recommended caution in attempts to discern the nature of a compact object (neutron star versus black hole) based solely on its Kepler-measured active gravitational mass. Thus an observed upper limit on an anomalous center-of-mass acceleration would test gravitational conservation laws directly, and provide a limit on the PPN parameter ζ_2.

In the binary pulsar, the observable effect of such an anomalous acceleration is a secular change in both the observed pulsar period and the observed obital period (changing overall Doppler shift), given by

$$\frac{\dot{P}_p}{P_p} = \frac{\dot{P}_b}{P_b} \approx -4 \times 10^{-7} \frac{X(1-X)}{(1+X)^2} \left(\frac{m}{2.83 m_\odot}\right)^{2/3}$$

$$\times (\alpha_3 + \zeta_2)\left(\frac{t}{1 \text{ yr}}\right) \text{yr}^{-1}, \quad t \leq 10 \text{ yr}$$

where t is the time since September 1974, where we have taken into account the effect of the periastron advance on the vector $\mathbf{n}$.

Observations by Taylor et al. (1976) up to September 1975 indicate that, for the pulsar itself

$$\dot{P}_p/P_p \approx (4.6 \pm 0.2) \times 10^{-9} \text{ yr}^{-1}.$$

The limit thereby placed on ζ_2 is

$$|\zeta_2| < 0.1[(1+X)^2/8X(1-X)](2.83 m_\odot/m)^{2/3}$$

Because the predicted effect increases linearly with time, repeated measurements of $\dot{P}_p/P_p$ will set tighter limits on ζ_2 (assuming the pulsar mass has been measured, thereby fixing the prediction). The most important competing astrophysical effect is energy loss from the pulsar, which presumably is the source of the finite 4.6×10^{-9} yr^{-1} value for $\dot{P}_p/P_p$.

In principle, the two effects could be separated because of their different time dependences – if the rate of energy loss is roughly constant, then $\dot{P}_p/P_p$ will be constant, whereas if there is a secular acceleration of the center of mass, $\dot{P}_p/P_p$ will change roughly linearly with time at the present epoch.

Chapter 2. Confrontation between gravitation theory and experiment

The problem of the secular acceleration of the center of mass of binary systems has a checkered history. Levi-Civita (1937) first pointed out that general relativity *predicted* a secular acceleration in the direction of the periastron of the orbit, and found a binary system candidate in which he felt the effect might one day be observable. Eddington and Clark (1938) repeated the calculation using de Sitter's (1916) n-body equations of motion. After first finding a secular acceleration of opposite sign to that of Levi-Civita, they then discovered an error in de Sitter's equations, and concluded finally that the secular acceleration was zero. Robertson (1938) independently reached the same conclusion using the Einstein–Infeld–Hoffman equations of motion, and Levi-Civita (1964) later verified that result. Now, we realize that only non-conservative theories of gravitation may predict a secular center-of-mass acceleration, and that measurements of the secular changes in period of the pulsar PSR 1913 + 16 may test such theories.

2.6 Gravitation in the universe: the influence of global structure on local physics

Since gravitation appears to be a long-range, purely attractive force, one might expect the global matter distribution in the universe to affect local gravitational physics. In fact from this 'Machian' point of view, it is a mystery that any metric theory of gravity can avoid such influence. But general relativity obviously avoids it, and as a consequence is said to satisfy the Strong Equivalence Principle (SEP) which states that (i) *WEP and UGR are valid*, and (ii) *the outcome of any gravitational or non-gravitational experiment performed in a local freely falling frame is independent of where and when in the universe it is performed, and independent of the velocity of the frame.* (Compare with EEP (section 2.2.4) which restricts attention to non-gravitational experiments.)

In order to understand why general relativity embodies SEP and why most alternative theories do not, one must consider the manner in which a local gravitational problem, such as the spacetime metric of the solar system, is solved when the local system is embedded in the real universe, as opposed to an asymptotically flat spacetime. The computation of the metric is separated into two parts: a cosmological solution, and a 'local' solution. From this viewpoint, the universe affects the local gravitational physics of the system by establishing the 'matching' or boundary conditions (at a boundary 'far' from the matter) for the various fields generated by the local system. Several conclusions can be reached:

(i) *A theory that contains solely a metric field yields local gravitational physics which is identical in all asymptotic Lorentz frames, and which is unaffected by the structure or evolution of the universe, i.e., embodies SEP.* All this follows from the invariance properties of the Minkowski metric η (the asymptotic form of g), the only field coupling the local system asymptotically to the universe, and from general covariance, which allows us to always find a coordinate system in which the metric takes this Minkowski form at the boundary between the universe and the local system. It has also been argued that any theory that embodies SEP predicts no Nordtvedt effect (MTW §§ 20.6, 40.9), no variations in the locally measured gravitational constant due to nearby matter (Will and Nordtvedt, 1972) and no dipole gravitational radiation (Eardley, 1975; Will and Eardley, 1977). General relativity falls into this class of theories.

(ii) *A theory that contains a metric field and a scalar field ϕ yields physics which is identical in all asymptotic Lorentz frames, but which may depend on universal structure and evolution.* These conclusions follow from the invariance of both η and ϕ under Lorentz transformations, but now ϕ may depend on universal structure and evolution. Scalar–tensor theories are of this type, and hence may violate SEP.

(iii) *A theory that contains, in addition to the metric, vector fields $\vec{K}$, or tensor fields C, or non-dynamical fields, t, η, or combinations of these, yields local gravitational physics which may depend on motion relative to a preferred universe rest-frame, and which may depend on the structure and evolution of the universe.* This follows because the asymptotic values of the additional fields are not necessarily invariant under Lorentz transformations, and take values determined by the cosmological structure. This is true even for a flat background metric η, as in bimetric theories (section 2.3.2b; Lee et al., 1976; Lightman and Lee, 1973b).

It thus appears that any alternative to general relativity that introduces additional fields, whether dynamical or not, can predict violations of SEP. Hence the influence of global structure on local gravitational physics can lead to observable consequences, including secular variations in the Newtonian gravitational constant (section 2.6.1), and cosmologically determined PPN parameter values (section 2.6.2). In the latter case, the limits placed on PPN parameters by solar-system tests can be used to constrain cosmological models in theories of this type. Cosmological models that are consistent with local gravitational physics may then violate global observational constraints such as the existence of the microwave background radiation.

Chapter 2. Confrontation between gravitation theory and experiment

2.6.1 Constancy of the gravitational constant

Most theories of gravity that violate SEP predict that the locally measured Newtonian gravitational constant may vary with time as the universe evolves. For the theories listed in table 2.7, the predictions for $\dot{G}/G$ can be written in terms of time derivatives of the asymptotic dynamical fields or of the asymptotic matching parameters. Other, more heuristic proposals for a changing gravitational constant, such as that due to Dirac cannot be written this way. Dyson (1972) gives a detailed discussion of these proposals. Where G does change with cosmic evolution, its rate of variation should be of the order of the expansion rate of the universe, i.e.

$$\dot{G}/G = \sigma H_0,$$

where H_0 is the Hubble expansion parameter whose value is $H_0 \approx 55 \text{ km s}^{-1} \text{ Mpc}^{-1} \approx (2 \times 10^{10} \text{ yr})^{-1}$, and σ is a dimensionless parameter whose value depends on the theory of gravity under study and on the detailed cosmological model. For very few theories has a systematic study of values of σ been carried out (for Brans–Dicke theory, see review and references in section 16.4 of Weinberg, 1972). However, several observational constraints can be placed on $\dot{G}/G$, using methods that include studies of the evolution of clusters of galaxies and of the Sun, observations of lunar occultations, planetary radar-ranging measurements, and yet-to-be performed laboratory experiments. The present status of these observations is summarized in table 2.12. With the exception of Van Flandern's (1976) lunar occultation measurements, all results are consistent with zero cosmic variation in G. Reasenberg and Shapiro (1976) have pointed out that, because the errors in the radar observations of $\dot{G}/G$ decrease as $T^{-5/2}$ where T is the time span of the observations, one can expect from this method an accuracy of $\Delta|\dot{G}/G| < 10^{-11} \text{ yr}^{-1}$ by 1985. Anderson, Keesey et al. (1977) and Wahr and Bender (1976) have shown that radar observations of Viking or of a Mercury orbiter over two-year missions could yield $\Delta|\dot{G}/G| \sim 10^{-12} \text{ yr}^{-1}$.

2.6.2 Cosmological tests of gravitational theories

Because general relativity satisfies SEP, observational cosmology involves measuring purely cosmic parameters such as the Hubble expansion parameter H_0, density parameter $\Omega_0 = 8\pi G\rho_0/3H_0^2$, where ρ_0 is the present mean density of the universe, deceleration parameter

Table 2.12. *Observations of the constancy of the gravitational constant*

$\sigma = \dot{G}/G \times (2 \times 10^{10} \text{ yr})$	Method	Reference
$|\sigma| < 8 \times 10^{-1}$	Evolution of galactic clusters	Dearborn and Schramm (1974) (see however Marchant and Mansfield, 1977)
$|\sigma| < 2$	Solar evolution	Chin and Stothers (1976)
$|\sigma| < 8 \times 10^{-1}$	Lunar occultations	Morrison (1973)
$\sigma = -(7.2 \pm 3.6) \times 10^{-1}$	Lunar occultations	Van Flandern (1975a, b, 1976)
$|\sigma| < 8$	Planetary radar	Shapiro *et al.* (1971)
$|\sigma| < 2$	Planetary radar	Reasenberg and Shapiro (1976)
[a]	Laboratory experiments	Braginsky and Ginzberg (1974), Ritter *et al.* (1976), Braginsky *et al.* (1977)

[a] Experiments yet to be performed.

q_0, and so on (see MTW § 29 for detailed definitions). For homogeneous isotropic cosmological models with vanishing cosmological constant, two of the above parameters, say H_0 and Ω_0, suffice to uniquely determine the model. (One must in all cases specify the equation of state of matter, or equivalently the density of radiation, in order to determine the past behavior of the model.) However in theories that violate SEP the influence of the global structure on local gravitational physics produces additional parameters that must be specified, for instance the value of the measured gravitational constant G relative to the 'bare' coupling constant G_* of the theory, the rate of change of G, or the values of some or all of the PPN parameters. For example, a unique cosmological model in scalar–tensor theories requires knowledge of H_0, Ω_0, ϕ_0, $\dot{\phi}_0$, and of the function $\omega(\phi)$, while in Rosen's bimetric theory (with flat background metric), a unique model is determined by H_0, Ω_0, $(c_0)_0$ and $(c_1)_0$ (section 2.3.2*b*) or alternatively by H_0, Ω_0, $(\dot{G}/G)_0$, and the PPN parameter $(\alpha_2)_0$ where the subscript (0) refers to present values (Caves, 1977). When the observed values for these parameters are used as boundary conditions, one may then ask whether cosmological models in these theories agree with observations such as the existence and isotropy of the microwave background radiation, the abundances of

Chapter 2. Confrontation between gravitation theory and experiment

helium and deuterium, observed values or limits on q_0, and so on (see chapter 9 for discussion). In Brans–Dicke theory for example, models can be constructed that agree with the observations, although the larger the value of ω (as required by solar-system experiments, section 2.3), the closer these models approximate general relativistic models. However, in Rosen's bimetric theory and in Ni's stratified theory, cosmological models that agree with present values of cosmological and local parameters predict in some cases a 'bounce' in the past, as opposed to a big bang. Caves (1977a) has shown that such a bounce typically occurs at densities and temperatures less than those required to ionize hydrogen. Hence, such models are hard pressed to produce the observed microwave background radiation, and cannot produce the observed helium abundance by means of primeval nucleosynthesis.

If the universe is anisotropic, theories that violate SEP may predict anisotropic PPN parameters (Nordtvedt, 1976). Nordtvedt (1975b, 1976) has used Earth-tide measurements and observations of the periastron shift in the binary pulsar to set limits on some of the anisotropies between parts in 10^4 and parts in 10^9. We note that recent observations of the microwave background indicate that the universe is isotropic to at least three parts in 10^4 (Smoot et al., 1977).

The universe itself may be the next exciting arena for the high-precision confrontation between gravitation theory and experiment.

2.7 Summary

Our discussion of the confrontation between gravitation theory and experiment began close to home, in the laboratory, where many of the foundations of gravitation theory are tested. But our viewpoint quickly widened from the laboratory to the solar system, to gravitational radiation from space, to the binary pulsar, to the cosmos. This widening viewpoint in experimental gravitation parallels the ever-widening horizons of astronomy: as technology advances, the measuring tools of the astronomer are able to probe farther and deeper, and with more precision. That probing reveals again and again the fundamental role that gravity plays in our universe, and makes the need for a correct theory of gravitation more acute. Yet the same technological advances that widen the astronomical horizons also provide the means to confront gravitation theory with experiment in new arenas and at ever more rigorous levels of precision. As the articles in this volume attest, Einstein's general relativity is favored by most gravitation theorists and

Summary

astrophysicists. But this is not enough. General relativity must be confronted with each new experimental fact and must stand or fall only on that basis. So far it has withstood every confrontation, but new confrontations, in new arenas, are on the horizon. Whether general relativity survives is a matter of speculation for some, pious hope for some, and supreme confidence for others. Regardless of one's theoretical prejudices, it can certainly be agreed that gravitation, the oldest known, and in many ways most fundamental interaction, deserves an empirical foundation second to none.

It is a pleasure to acknowledge helpful discussions with Bob Wagoner, Mark Haugan, John Anderson, Carl Caves and Joe Taylor. Thanks also go to Victoria LaBrie for her masterful typing of the manuscript.

3. Gravitational-radiation experiments

D. H. DOUGLASS AND V. B. BRAGINSKY†

3.1 Introduction

Einstein (1916) showed in the very early days of the general theory of relativity that there were weak-field solutions of the field equations that obeyed the wave equation. His interpretation of these solutions as gravitational radiation did not go unchallenged. Eddington (1922), for example, suggested that these solutions were coordinate changes which he characterized as 'propagating . . . with the speed of thought'. Among present-day theoretical physicists, however, there is a very strong consensus that gravitational radiation does exist and that the real questions concern only the sources and the strengths of the radiation. Nevertheless, the question of the existence of gravitational radiation must remain open until gravitational radiation is discovered experimentally and its properties (such as the velocity of propagation and its states of polarization) are determined.

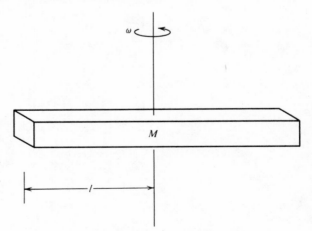

Figure 3.1. Rotating rod of length $2l$ and mass M.

† Research supported in part by the National Science Foundation, USA, and by the Ministry of Higher Education, USSR.

Introduction

How can gravitational waves be generated? Again it was Einstein (1918) who made the first calculations using the general theory. He showed that for a system of accelerating masses the analogue of electromagnetic dipole radiation does not exist. This is a direct consequence of the fact that in his theory the masses all have the same sign, or more precisely, that the ratio of the gravitational mass and inertial mass is a constant for all bodies, which is experimentally verified to a very high level of accuracy. Thus one expects the efficiency for the generation of gravitational radiation to be small. The analogue of electromagnetic quadrupole radiation does exist, however. A rod (mass M, length $2l$) rotating with angular velocity ω about an axis perpendicular to the rod axis has a time-varying quadrupole moment and will generate gravitational radiation (see figure 3.1). It is an easy calculation to show that the gravitational luminosity L is

$$L = \frac{128 G M^2 l^4 \omega^6}{45 c^5} = \frac{128 G}{45 c^5} \left(\frac{M}{l}\right)^2 v^6, \tag{3.1}$$

where G is the gravitational constant and c is the velocity of light. As one might expect, more radiation comes from a large massive rod rotating rapidly. Equation (3.1) is also expressed in terms of the velocity v ($= l\omega$) of the ends of the rod.

Let us estimate how much radiation one could generate with a manmade device of this kind. The first consideration is that rods tend to break when the velocity approaches some fraction of the velocity of sound v_s in the material. Thus we should choose a material such as beryllium, which has a large velocity of sound ($v_s \approx 1.7 \times 10^6$ cm s^{-1}). The quantity M/l in (3.1) is the linear density of the rod, which could perhaps be as large as 3×10^7 g cm^{-1}. (For $\omega \sim 10^3$, this results in $l \sim 1700$ cm and $M \sim 6 \times 10^{10}$ g.) For these values of parameters, which are admittedly optimistic, one estimates

$$L_{\text{man}} \sim 10^{-7} \text{ erg s}^{-1}.$$

This can be considered as some sort of upper limit for the gravitational radiation from a man-made generator based upon conventional materials and technology. When converted to a flux at distances in the wavezone of the generator, this is a discouragingly small number. This means that there is little prospect at the present time to perform a Hertz-type experiment whereby one generates gravitational radiation and detects that same radiation.

Chapter 3. Gravitational-radiation experiments

Therefore, if one wishes to observe gravitational radiation, some other generator, such as an astrophysical source, must be sought and appropriate detectors constructed. How strong can astrophysical sources be? We estimate this value by rewriting (3.1) in the form

$$L = \frac{32}{45} \frac{c^5}{G} \left(\frac{r_g}{l}\right)^2 \left(\frac{v}{c}\right)^6, \qquad (3.2)$$

where the gravitational length $r_g = 2GM/c^2$ has been introduced. For astrophysical sources one can envision objects of sizes comparable with their gravitational lengths $l \sim r_g$ and of velocities approaching that of light, $v \sim c$. Thus an upper limit on the luminosity of an astronomical source can be estimated by putting these values into (3.2):

$$L_{ast} < c^5/G \approx 4 \times 10^{59} \text{ erg s}^{-1}.$$

We see that astrophysical sources could be many tens of orders of magnitude more luminous than any man-made generators. Possible sources include rotating binary stars, vibrating stars, rotating stars, supernova explosions, black hole events in galactic nuclei, and various others that are enumerated and described later in this chapter. The variety and number of possible sources is so rich and the possible information about these sources from gravitational waves is so important that we certainly agree with the remark of Hawking (1972) – 'the construction of gravitational-radiation detectors may open up a whole new field of "gravitational astronomy" which could be as fruitful as radio-astronomy has been in the last two decades'.

The detection of gravitational radiation is just as inefficient as its generation – again because of the equivalence of gravitational and inertial masses. Since all particles undergo the same acceleration in a uniform gravitational field, one has to measure the gradient of the gravitational field in order to detect gravitational radiation. This leads to the consideration of mass quadrupole detectors, the simplest of which consists of two test masses separated by a distance l (see figure 3.2(a)). A gravitational wave interacting with these two test masses will change their separation by a small amount. Even for strong sources the displacement is not large. For example, a supernova explosion at the center of our galaxy would be expected to produce a fractional change in l of about 10^{-17}. For a separation of 100 cm this requires measuring a displacement of 10^{-15} cm. This is less than the diameter of a nucleus, so perhaps some idea is conveyed of the experimental difficulty of con-

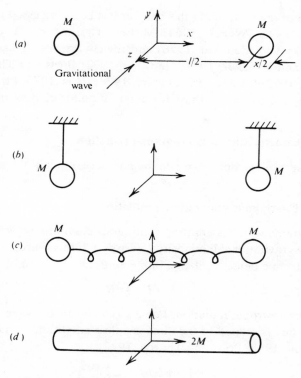

Figure 3.2. Diagram of gravitational-wave detectors: (*a*) two free masses, (*b*) 'almost'-free-mass detector, (*c*) resonant-mass–spring detector, (*d*) resonant-cylindrical-bar detector.

structing a gravitational-wave antenna to detect even the sources which might be considered strong. The first experimentalist to seriously consider the possibility of constructing a gravitational antenna was Weber (1961); he, in fact, proceeded with his students to construct several. These efforts and those of others will be discussed in later sections of this chapter, which consider and discuss the following.

The nature of gravitational radiation is discussed in section 3.2. This is followed in section 3.3 by a rather complete enumeration and classification of the possible astrophysical sources of gravitational radiation, along with estimates of the magnitudes. Extensive discussion of the possible detectors that have been considered and the intrinsic limitations are given in section 3.4; past and present efforts are described. The prognosis for the eventual discovery of gravitational radiation is given in section 3.5.

Chapter 3. Gravitational-radiation experiments

Much of the material in this chapter has been discussed previously in various books (Weber, 1964; Misner, Thorne and Wheeler, 1973; Braginsky and Manukin, 1976) and review articles (Braginsky, 1966; Ruffini and Wheeler, 1971; Hawking, 1972; Press and Thorne, 1972; Sciama, 1972; Misner, 1974; Papini, 1974; Rees, 1974; Pizzella, 1975). One is urged to consult these for different points of view and emphasis.

3.2 Characteristics of gravitational radiation

In this section we derive and define a variety of quantities.

3.2.1 Continuous gravitational radiation

For a source which is emitting continuous gravitational waves, one can define a gravitational luminosity L (erg s^{-1}). If this source is at a distance R(cm) from a detector, then a flux F (erg cm^{-2} s^{-1}) can be defined,

$$F = L/(4\pi R^2), \tag{3.3}$$

assuming isotropic radiation. For a monochromatic wave of frequency ω_g, the flux is related to the dimensionless amplitude h via the relation (Isaacson, 1968; Press and Thorne, 1972)

$$h \sim [16\pi GF/(c^3 \omega_g^2)]^{1/2}, \tag{3.4}$$
$$\sim 5.6 \times 10^{-20}(F/f_g^2)^{1/2},$$

where $f_g = \omega_g/(2\pi)$. This can be expressed in terms of L and R using (3.3):

$$h \sim \left(\frac{G}{\pi^2 c^3}\right)^{1/2} \frac{L^{1/2}}{f_g R}$$
$$\sim 5 \times 10^{-20} \left(\frac{10^{-3} \text{ Hz}}{f_g}\right)\left(\frac{100 \text{ pc}}{R}\right)\left(\frac{L}{10^{36} \text{ erg s}^{-1}}\right)^{1/2}. \tag{3.5}$$

3.2.2 Bursts of gravitational radiation

The above relations are correct for monochromatic radiation. For bursts of gravitational radiation it is convenient to define additional quantities. For an event involving an object of mass M, the energy given off in the form of gravitational radiation can be expressed as

$$E_g = \varepsilon M c^2, \tag{3.6}$$

where ε is the efficiency factor. Values of ε as high as 0.50 are possible (Hawking, 1971); however, various estimates for processes which are thought to represent astrophysical events are much less than this value. The burst is characterized by a time $\hat{\tau}$, from which one can define a characteristic frequency f_g,

$$f_g = 1/(2\pi\hat{\tau}), \qquad (3.7)$$

which characterizes the peak in the power spectrum. The width of the spectrum Δf would be of the order of f_g itself. For these events the gravitational energy E_g and the luminosity during the event are related by

$$L \sim E_g/\hat{\tau}. \qquad (3.8)$$

The energy flux I (erg cm^{-2}) is defined and used, and is

$$I \sim F\hat{\tau}.$$

The energy flux is frequently decomposed into its Fourier components

$$I = \int F_f(f)\,df$$

where $F_f(f)$ (erg cm^{-2} Hz^{-1}) is the intensity of the flux at f, which we will call the burst intensity. For the bursts we may assume

$$I \sim F_f(f)\,\Delta f \sim f F_f(f).$$

The burst intensity is defined because this is the fundamental quantity that a resonant detector measures – all other quantities would be derived or inferred from this.

Bursts from gravitational collapse

For a burst event involving gravitational collapse, there is a characteristic time $\hat{\tau}$ which is the time for the wave to propagate across the region of strong gravitation, d_s. (Although this characteristic time can be different from that of the burst duration, we will assume that they are the same.)

$$\hat{\tau} \sim d_s/c$$

We will assume the distance d_s to be approximately twice the gravitational length, $2GM/c^2$, as is frequently done, so that

$$\hat{\tau} \sim 4GM/c^3.$$

Chapter 3. Gravitational-radiation experiments

From this, one estimates the frequency f_g of the maximum in the spectrum to be

$$f_g \sim 1/(2\pi \hat{\tau}) \sim c^3/(8\pi GM) \qquad (3.9)$$

$$\sim (4 \times 10^3 \text{ Hz})(2M_\odot/M), \qquad (3.10)$$

where $M_\odot$ is the mass of the Sun. The dimensionless amplitude h can be expressed in terms of the characteristic frequency by using (3.5), (3.6), (3.7) and (3.8) and is

$$h \sim \frac{c}{\omega_g R}\varepsilon^{1/2}$$

$$\sim 2 \times 10^{-14}\left(\frac{1 \text{ Hz}}{f_g}\right)\left(\frac{10^4 \text{pc}}{R}\right)\left(\frac{\varepsilon}{0.01}\right)^{1/2}. \qquad (3.11)$$

where $\omega_g = 2\pi f_g$ has been used to obtain the last expression. One sees the somewhat remarkable result that this equation is independent of mass and the gravitational constant. If, in addition, one has an antenna at a fixed frequency, which is often the case, then the dimensionless amplitude that one would attempt to measure depends only on the distance. For example, if $f_g \sim 3 \times 10^3$ Hz, then

$$h \sim 7 \times 10^{-18}\left(\frac{10^4 \text{pc}}{R}\right)\left(\frac{\varepsilon}{0.01}\right)^{1/2}.$$

Alternatively, one can use (3.9) to express h in terms of the mass:

$$h \sim \frac{4GM}{Rc^2}\varepsilon^{1/2}$$

$$\sim 5 \times 10^{-18}\left(\frac{M}{2M_\odot}\right)\left(\frac{10^4 \text{pc}}{R}\right)\left(\frac{\varepsilon}{0.01}\right)^{1/2}.$$

The above expressions describing bursts from gravitational collapse, in the absence of a specific model or calculation, are only qualitative and could be in error by orders of magnitude. One can make better estimates (see Thorne, 1976). These relations, however, will serve well enough to estimate the magnitude of the strength of various events. The various definitions introduced in this section are listed in table 3.1. A classification (after Press and Thorne, 1972) of gravitational radiation according to characteristic frequency is given in table 3.2.

Sources of gravitational radiation

Table 3.1. *Symbols and definitions*

Monochromatic waves	
L	gravitational luminosity (erg s^{-1})
R	distance to detector (cm)
F	flux (erg cm^{-2} s^{-1}); $F = L/(4\pi R^2)$
ω_g	angular frequency of gravitational wave (rad s^{-1})
h	dimensionless amplitude; $h \sim [16\pi GF/(c^3 \omega_g^2)]^{1/2}$
Bursts	
E_g	gravitational energy in a burst (erg)
ε	efficiency factor (fraction of rest-mass energy converted to gravitational waves)
$\hat{\tau}$	characteristic time of burst (s)
f_g	characteristic frequency of burst (Hz); $f_g = 1/(2\pi\hat{\tau})$
L	luminosity during burst (erg s^{-1}); $L \sim E_g/\hat{\tau}$
I	energy flux of burst (erg cm^{-2}); $I \sim F\hat{\tau}$
F_f	burst intensity (erg cm^{-2} Hz^{-1}); $F_f \sim I/f_g$
Bursts from gravitationally collapsing objects of mass M	
$\hat{\tau}$	characteristic time (s); $\tau \sim 4GM/c^3$
f_g	characteristic frequency (Hz); $f_g \sim c^3/(8\pi GM)$
h	dimensionless amplitude; $h \sim c/(\omega_g R)\varepsilon^{1/2}$

3.3 Sources of gravitational radiation

In this section we shall consider many possible sources of gravitational radiation starting from the certain sources, the binary stars, and finishing with sources about whose existence one may speculate, such as pregalactic black hole events. For purposes of presentation we will group these sources into those producing continuous radiation, following a review by Douglass (1978), and those producing bursts, following a review by Thorne (1978).

3.3.1 Sources of continuous gravitational radiation

There are many potential sources of continuous gravitational radiation. We consider in this subsection radiation from rotating binary stars, vibrating stars, and rotating stars. We also consider the case of incoherent sources such as the W Uma stars.

Rotation of binary star systems

Consider a binary star system with components of mass M_1 and M_2 rotating at orbital frequency ω. This system will radiate gravitational

Table 3.2. *Gravitational-wave frequency bands*

Designation	Frequency (Hz)	Period (s)	Wavelength (cm)	Sources
ELF (Extremely low frequency)	10^{-7} to 10^{-4}	10^7 to 10^4	3×10^{17} to 3×10^{14} (0.1 pc to 20 AU)	Slow binary stars, black hole events $M > 10^8 M_\odot$
VLF (Very low frequency)	10^{-4} to 10^{-1}	10^4 to 10^1	3×10^{14} to 3×10^{11} (20 AU to 300 Earth diameters)	Fast binary stars, black hole events $M \sim 10^5$ to $10^8 M_\odot$, white dwarf vibrations
LF (Low frequency)	10^{-1} to 10^2	10^1 to 10^{-2}	3×10^{11} to 3×10^8 (300 Earth diameters to 0.3 Earth diameters)	Doubly compact binaries, black hole events $M \sim 10^2$ to $10^5 M_\odot$
MF (Medium frequency)	10^2 to 10^5	10^{-2} to 10^{-5}	3×10^8 to 3×10^5	Supernova explosions, neutron starquakes, CW (continuous wave) radiation from pulsars
HF (High frequency)	10^5 to 10^8	10^{-5} to 10^{-8}	3×10^5 to 3×10^2	Man-made?
VHF (Very high frequency)	10^8 to 10^{11}	10^{-8} to 10^{-11}	3×10^2 to 3×10^{-1}	Black-body cosmological?

Sources of gravitational radiation

radiation at frequencies $\omega_{g,n}$ which are harmonics of ω:

$$\omega_{g,n} = n\omega.$$

The total gravitational luminosity L_n contained in each harmonic is given by Peters and Mathews (1963):

$$L_n = \frac{32 G^{7/3}}{5c^5} \frac{(M_1 M_2)^2}{(M_1+M_2)^{2/3}} \omega^{10/3} g_n(e). \qquad (3.12)$$

The function $g_n(e)$ depends on the eccentricity e. If $e = 0$, $g_n = \delta_{n2}$ (δ_{n2} is the Kronecker delta function) and all the power is radiated at twice the rotational frequency. When $e \neq 0$, there is always some power in the higher harmonics. For large e the maximum power may be in a rather high harmonic (if $e = 0.7$, the maximum power is in the tenth harmonic, for example). Equation (3.12) can be expressed in the convenient form

$$L_n = (2.2 \times 10^{45} \text{ erg s}^{-1}) \left(\frac{m_1^2 m_2^2}{(m_1+m_2)^{2/3}}\right) \left(\frac{f}{1 \text{ Hz}}\right)^{10/3} g_n(e), \qquad (3.13)$$

where $m_1 = M_1/M_\odot$ and $m_2 = M_2/M_\odot$. The average radiated flux F and the dimensionless amplitude h at the Earth can be estimated from (3.3) and (3.4). Putting (3.13) into (3.3), and that into (3.4), one obtains

$$h = 2.4 \times 10^{-20} \left(\frac{10^4 \text{ pc}}{R}\right) \frac{m_1 m_2}{(m_1+m_2)^{1/3}} \left(\frac{f}{1 \text{ Hz}}\right)^{2/3}. \qquad (3.14)$$

The radiated flux and dimensionless amplitude from a rotating binary source depends, of course, on the angular orientation and polarization, so that for any particular direction those quantities will probably differ from the average value estimated from (3.3) and (3.4); the differences, however, are at most a factor of 4.

If the binary evolves as a consequence of radiation of gravitational radiation, then the separation of the two members will decrease and the frequency will increase with time, and so also will the gravitational luminosity. A measure of the lifetime of the binary due to gravitational radiation is the time τ for the star to reach zero separation (Peters, 1964),

$$\tau = \frac{5c^5 a^4}{256 G^3 M_1 M_2 (M_1+M_2)}, \qquad (3.15)$$

where a is the separation at time τ. We will find it convenient to express this lifetime in terms of the binary frequency. Using Kepler's third law,

$$a^3 = G(M_1+M_2)/\omega^2, \qquad (3.16)$$

Chapter 3. Gravitational-radiation experiments

(3.14) can be expressed as

$$\tau = \frac{5c^5(M_1+M_2)^{1/3}}{256 G^{5/3} M_1 M_2 \omega^{8/3}}$$

$$= (1.0 \times 10^5 \text{ s}) \frac{(m_1+m_2)^{1/3}}{m_1 m_2} \left(\frac{1 \text{ Hz}}{f}\right)^{8/3}. \quad (3.17)$$

Thus with the knowledge of a binary orbital frequency and masses one can estimate the lifetime due to gravitational radiation. For example, if $m_1 = 1.0$, $m_2 = 0.5$, $f = 10^{-4}$ Hz (values close to those of the binary ι Boo), then $\tau \sim 1 \times 10^{16}$ s (3×10^8 yr). The evolution of these binaries via gravitational radiation (i.e. decreasing separation, and increasing frequency and gravitational power) is only one possible mechanism. However, this process can proceed until one member tidally disrupts the other, leading to a 'catastrophic' ending to this process and the beginning of new processes. A reasonable estimate of the separation distance at which this occurs is four times the radius of the largest component. This condition determines the maximum frequency for which equation (3.17) applies. Table 3.3 shows the maximum frequency for various binary star systems obtained by putting 'typical' diameters of various types of stars into (3.16), assuming $m_1 \sim m_2 \sim 1$. Thus from the table one infers that observed binaries with orbital frequencies much greater than 2×10^{-5} Hz (period ~ 0.6 d) would almost certainly have a white dwarf, a neutron star or a black hole as one member. In addition, no binary, except a black hole–black hole, could have an orbital frequency much greater than 10^3 Hz.

Table 3.3. *Maximum binary rotational frequency*

Largest member of binary	Radius of largest member (cm)	Maximum orbital frequency (Hz)
'Ordinary' star	10^{11}	2×10^{-5}
White dwarf	10^9	2×10^{-2}
Neutron star	10^6	7×10^2

Since 50 per cent of the stars in our galaxy are thought to be binary stars, there are potentially a very large number of sources. Approximately 3000 eclipsing binaries are known; several hundreds of these are known well enough to determine the major orbital elements.

'Classical' binaries

Table 3.4 contains a short list of 'classical' binary star systems obtained from a list of eclipsing binaries (Gaposchkin, 1958). The masses and distances that were used to obtain estimates of the gravitational luminosity and the gravitational flux, and the dimensionless amplitude of the wave, arriving at the Earth, were compiled from the indicated references. In all cases it was assumed that the radiation occurs in the $n=2$ harmonic. One sees that ι Boo, which appears frequently in reviews on gravitational radiation, produces the largest gravitational flux, with a value of 10^{-10} erg cm^{-2} s^{-1} at a frequency of 10^{-4} Hz. However, it is to be noted that μ Sco has the largest h in the list, with a value $\sim 2 \times 10^{-20}$, which is four times that of ι Boo. It should also be

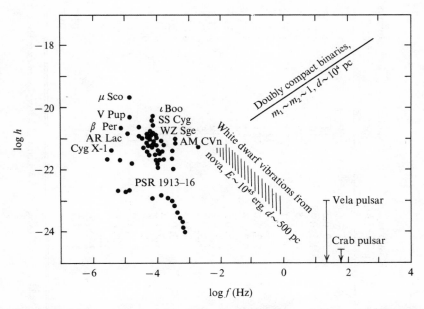

Figure 3.3. Dimensionless amplitude h of gravitational wave versus frequency f. The solid dots are the amplitudes of the sources (~40) listed in tables 3.4, 3.5 and 3.6. The shaded area in the middle is the amplitude estimated for vibrations of white dwarfs. The line of slope 2/3 is for doubly compact binaries. The vertical lines in the lower right are for the Vela and Crab neutron stars.

noted that the number of binaries above a certain amplitude or flux level increases rapidly as one decreases the amplitude or flux level. The dimensionless amplitudes of all these binaries are plotted in figure 3.3.

Table 3.4. 'Classical' binaries

Name	f_g (10^{-6} Hz)	Masses m_1, m_2[a] (units of $M_\odot$)	Distance[b] (pc)	GW Luminosity (erg s^{-1})	GW Flux at Earth (10^{-12} erg cm^{-2} s^{-1})	Dimensionless amplitude h (10^{-22})
RW Com	96	0.7, 0.6	200	1.2	0.3	3
ι Boo	86	1.0, 0.5	11.7	1.1	68.0	51
VW Cep	83	1.1, 0.4	16.7	0.7	23.0	30
YY Eri	72	0.8, 0.5	29	0.4	4.4	15
SW Lac	72	0.9, 1.1	73	1.9	3.2	13
W UMa	70	0.8, 0.6	110	0.5	0.4	5
AB And	70	1.7, 1.0	121	4.2	2.5	12
RZ Com	68	0.7, 0.6	64	2.3	5.0	17
U Peg	62	0.7, 0.6	170	0.3	0.1	3
TZ Cne	61	1.5, 0.8	140	1.5	0.7	7
AH Vir	56	1.3, 0.6	65	0.5	1.1	10
ER Ori	55	0.5, 0.3	62	0.03	0.1	3
V502 Oph	51	1.3, 0.5	51	0.29	1.0	10
UV Leo	38	1.3, 1.2	75	0.51	0.8	13
RT And	37	1.5, 1.0	54	0.4	1.2	16
GO Cyg	32	3.2, 1.7	230	2.1	0.4	10
YY Gem	28	0.6, 0.6	11.7	0.02	1.1	22
μ Sco	16	12, 12	109	51	38.0	210
V Pup	16	16.5, 9.7	520	59	1.9	46
AR Lac	12	1.3, 1.3	31	0.01	0.1	13
β Per	8.1	4.7, 0.9	30	0.01	0.1	21

[a] Masses were taken from Gaposchkin (1958), except for SW Lac, which were taken from Hack (1963).
[b] Distances from Gaposchkin (1958).

Sources of gravitational radiation

Binary cataclysmic variables

The cataclysmic variables are interesting systems which include the classical novae (N), the recurrent novae (RN), the dwarf novae (DN), and the nova-like (NL) variables. Many of these systems are binaries with orbital periods ranging from tens of minutes to several days. Table 3.5 contains a short list of these binary stars along with estimates of the gravitational luminosity, and the flux and dimensionless amplitude at the Earth due to the orbital motion. It is noted that AM CVn is the most luminous source in this list and has a flux of 2×10^{-10} erg cm^{-1} at a frequency of 2×10^{-3} Hz. However, when one looks at the dimensionless amplitudes, one sees that SS Cyg is largest with $h \sim 3 \times 10^{-21}$, a value comparable with ι Boo in table 3.4. The amplitudes are plotted in figure 3.3. As we will indicate below, some of these same binaries would be expected to emit gravitational radiation at higher frequencies due to non-radial vibrations of the white dwarf component. Since these binary stars are also rich in the electromagnetic signals that they emit, one could expect a combined analysis with any gravitational waves received to be quite rewarding.

Binary X-ray sources

Many of the X-ray sources have been discovered to be binary star systems. Again one would expect that if gravitational radiation could be detected from these sources, that when analyzed along with the X-ray data (and optical and radio-data, if available), significant progress could be made in the understanding of their nature. Table 3.6 contains a list of the known X-ray binaries. Compared with the binaries listed in tables 3.4 and 3.5, the fluxes and amplitudes here are smaller. However, technical innovations may improve to the point where these systems might be detectable. The binary pulsar PSR 1913−16 is particularly listed because of the fact that it has a large eccentricity ($e = 0.61$). This has the consequence that most of the gravitational radiation will be emitted in the harmonics of the orbital frequency. The harmonic with the most power is $n = 8$, but harmonic $n = 4$ has the largest dimensionless amplitude. Table 3.6 lists the fluxes and amplitudes expected at the various harmonics (only even harmonics are listed). The signal from this source, when optimally detected, is the equivalent of approximately 10 coherent sources of the same frequency. The amplitudes of these binaries are also plotted in figure 3.2.

Table 3.5. *Binary cataclysmic variables*

Name	Type[c]	GW frequency[d] (10^{-6} Hz)	Masses m_1, m_2 (units of $M_\odot$)	Distances[e] (pc)	GW luminosity (10^{30} erg s^{-1})	GW flux at Earth (10^{-12} erg cm^{-2} s^{-1})	Dimensionless amplitude h (10^{-22})
Am CVn[a]	NL or DN	1900	1.0, 0.041	100	300	240	5
WZ Sge[b]	RN	410	1.5, 0.12	75	24	37	8
EX Hya	DN	337	1.07, 0.16	170	14	4	6
VV Pup	NL	333	1.0, 0.16	1000	10	0.1	1
Z Cha	DN	312	1.0, 0.17	300	10	1	2
VW Hyi	DN	312	1.0, 0.17	150	10	4	4
TT Ari	NL	174	0.79, 0.38	170	5	2	6
V603 Agl	N	169	0.87, 0.40	290	5	0.7	2
RR Pic	N	160	1.0, 0.44	220	7	1	4
WW Cet	DN	145	1.0, 0.50	170	8	3	8
U Gem	DN	133	0.35, 0.53	250	1	0.1	2
SS Aur	DN	130	0.89, 0.57	280	3	0.4	2
DQ Her	N	120	0.83, 0.60	280	3	0.4	4
UX UMa	NL	118	1.0, 0.61	360	5	0.3	2
RX And	DN	109	1.02, 0.65	110	4	3	10
RW Tri	NL	100	1.0, 0.71	480	3	0.1	2
SS Cyg	DN	84	0.97, 0.83	30	2	20	30
Z Cam	DN	80	1.17, 0.86	170	3	0.9	8
EM Cyg	DN	79	0.67, 0.86	120	1	0.6	6
RU Peg	DN	63	0.97, 1.12	200	1	0.3	6
AE Aqr	NL	56	1.25, 1.18	250	2	0.2	6

[a] For AM CVn, masses from Faulkner, Flannery and Warner (1972). The distance was taken as 100 pc, which is based upon the parallax measurement of 0.010 seconds of arc of Vasilevskis *et al.* (1975).
[b] Masses from Warner (1976); distance estimate from Kraft (1962).
[c] Abbreviations: N, nova; NL, Nova-like; DN, dwarf nova; RN, recurrent nova.
[d] Masses and frequencies from Warner (1976). Exceptions: for RW Tri, UX UMa, WW Cet, RR Pic, VW Hyi, Z Cha and VV Pup m_1 was set equal to 1.0.
[e] Distances d were estimated according to the relation $M = V + 5 - 5\log_{10} d$, where M is the absolute magnitude and V is the visual magnitude. Both M and V were taken from Warner (1976). The exception is SS Cyg, where the distance of 30 pc corresponds to an estimate of the parallax of 0.032 seconds of arc reported by Strand (1948).

Table 3.6. *X-ray binaries*[a]

Name	Orbital period (d)	GW frequency (10^{-6} Hz)	GW luminosity (10^{30} erg s^{-1})	GW flux (10^{-15} erg cm^{-2} s^{-1})	Dimensionless amplitude h (10^{-22})
Cyg X-3	0.20	120	9	1	0.2
Sco X-1	0.79	29	0.1	10	2
Her X-1	1.7	14	0.04	0.01	0.2
Cen X-3	2.1	11	0.3	0.01	0.2
3U 1700-37	3.4	6.8	0.2	1	2.0
SMC X-1	3.9	5.9	0.5	0.001	0.2
Cyg X-1	5.6	4.1	1.0	1	4
3U 0900-40	9.0	2.6	0.01	0.1	2
PSR 1913+16[b]	0.323	Harmonic, n			
		2 70	0.6	0.2	0.12
		4 140	2.9	1.1	0.14
		6 220	5.8	2.1	0.12
		8 280	5.8	2.1	0.10
		10 360	4.8	1.6	0.07
		12 430	2.9	1.1	0.04
		14 500	2.0	0.7	0.03
		16 570	1.6	0.5	0.02
		18 640	1.0	0.35	0.02
		20 720	0.6	0.2	0.01

[a] Masses, orbital periods and distances were taken from Gursky and Schreier (1975), van den Heuvel (1976), and Stockman, *et al.* (1973).
[b] This is the binary pulsar with $e = 0.61$. Gravitational radiation will occur at high harmonics of the orbital frequency, with $n \sim 6$ to 8 containing the most.

Chapter 3. Gravitational-radiation experiments

Doubly compact binary stars

In this subsection we consider explicitly the case where both members of the binary are compact (neutron star or black hole), which should not imply that none exist in the lists above (the binary pulsar PSR 1913−16 could be one). Our interest, of course, is because the binary frequency can be high and hence the luminosity and dimensionless amplitude also. Lattimer and Schram (1976) estimate a lower limit on the birth rate of progenitors of doubly compact binaries in the Galactic plane as $2 \times 10^{-12} \mathrm{pc}^{-2} \mathrm{yr}^{-1}$. Clark and Eardley (1977), using an efficiency of 1 per cent, estimate the rate of birth of doubly compact binaries in our galaxy at $6 \times 10^{-6} \mathrm{yr}^{-1}$ ($1.9 \times 10^{-13} \mathrm{s}^{-1}$), and the rate within a sphere of radius 15 Mpc (to the Virgo cluster) of $1/80 \mathrm{yr}^{-1}$ ($4 \times 10^{-10} \mathrm{s}^{-1}$). Revised estimates by Clark, van den Heuvel and Sutantyo (1978) give $2.9 \times 10^{-4} \mathrm{yr}^{-1}$ ($9.2 \times 10^{-12} \mathrm{s}^{-1}$) in the Galaxy and $5.5 \times 10^{-3} \mathrm{yr}$ ($1.8 \times 10^{-10} \mathrm{s}^{-1}$) out to the Virgo cluster.

To consider the general problem, let Ω_R be the rate for a given radius R. If the death rate equals the birth rate, and is a constant, then we estimate the number N of deaths in the next interval of time t as

$$N \sim \Omega_R t.$$

If these systems evolve because of gravitational radiation, then t is the lifetime and (3.17) can be expressed in terms of the frequency of the binary:

$$N(f) \sim 1 \times 10^5 \frac{(m_1+m_2)^{1/3}}{m_1 m_2} \left(\frac{1 \text{ Hz}}{f}\right)^{8/3} \left(\frac{\Omega_R}{\mathrm{s}^{-1}}\right). \quad (3.18)$$

If $m_1 \sim m_2 \sim 1$, this becomes

$$N(f) \sim 1.3 \times 10^5 \left(\frac{1 \text{ Hz}}{f}\right)^{8/3} \left(\frac{\Omega_R}{\mathrm{s}^{-1}}\right).$$

This relation can be used to estimate the number of such binaries that one could expect to find with orbital frequency above a certain value. If $m_1 \sim m_2 \sim 1$, then (3.18) can be expressed as $f \sim 80 (\Omega_R/N)^{3/8}$. Using the Clark, van den Heuvel and Sutantyo estimates for Ω_R, one finds for the example of the Galaxy that one would expect to find one compact binary with a frequency greater than 5.8×10^{-3} Hz. Using (3.13) one predicts a gravitational luminosity greater than $5 \times 10^{37} \mathrm{erg\,s}^{-1}$, and placed at the Galactic center (10 kpc) this would produce a flux greater than $4 \times 10^{-9} \mathrm{erg\,cm}^{-2}\mathrm{s}^{-1}$ and an amplitude 6×10^{-22}. The amplitude–frequency relation, (3.14), for this case is plotted in figure 3.3. The second example

Sources of gravitational radiation

of the 15 Mpc sphere (to the Virgo cluster) leads one to predict one compact binary with $f \sim 2 \times 10^{-2}$ Hz, with gravitational luminosity greater than 6×10^{39} erg s^{-1}, with flux at the Earth greater than 2×10^{-13} erg cm^{-2} s^{-1}, and with an amplitude greater than 10^{-24}. The binary pulsar PSR 1913+16 is a candidate for a doubly compact object. The flux and dimensionless-amplitude estimates in these two examples are sufficiently high that one should be conscious of the possibility of radiation from doubly compact binaries in any search for gravitational radiation. The doubly compact binaries will evolve to the point where one will tidally disrupt the other and will emit a burst of radiation (Clark and Eardley, 1977). This burst will be considered a little later.

W UMa stars

The total amount of (incoherent) gravitational radiation from rotation of all of the binary stars and the (incoherent) flux reaching the Earth has been considered by Mironovskii (1966). Since the gravitational luminosity is proportional to $f^{10/3}$, where f is the rotational frequency, one expects the close binaries to dominate the spectrum. The most populous class of close binaries is the W Uma type of star system. Mironovskii estimates that there are 10^8 such binary stars in our galaxy. Of the entries in table 3.4, 12 (i.e. more than half) are, in fact, W UMa stars (ι Boo, VW Cep, YY Eri, SW Lac, W UMa AB And, RZ Com, U Peg, TZ Cne, AH Vir, ER Ori, and V502 Oph). The flux from the direction of the Galactic center is estimated to be

$$F \sim 10^{-7} \text{ erg cm}^{-2} \text{ s}^{-1}.$$

This is peaked at a frequency $f_g \sim 8 \times 10^{-5}$ Hz with a width of $\sim 0.5 \times 10^{-4}$ Hz, and corresponds to a spectral density of F at f_g of

$$S_F \sim 10^{-10} \text{ (erg cm}^{-2} \text{ s}^{-1})^2 \text{ Hz}^{-1}.$$

Using (3.4) for the dimensionless amplitude, the flux from the direction of the Galactic center produces an amplitude of

$$h \sim 10^{-19},$$

which corresponds to maximum spectral density of h of

$$S_h \sim 2 \times 10^{-34} \text{ Hz}^{-1}$$

peaked at $f \sim 8 \times 10^{-5}$ Hz in a band $\Delta f \sim 50 \times 10^{-6}$ Hz. The spectral-density curve for these stars is plotted in figure 3.4. The value of h from

Chapter 3. Gravitational-radiation experiments

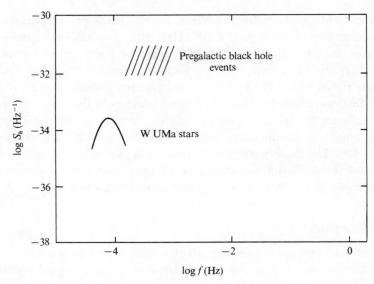

Figure 3.4. Spectral density of dimensionless amplitude S_h versus frequency f for incoherent sources. The gravitational radiation from the W UMa stars is discussed in section 3.3.1 and that from pregalactic black hole events in section 3.3.2.

this incoherent source in a narrow bandwidth Δf would then be given by $h \sim (S_h \Delta f)^{1/2}$. The estimate of the total power from all the W UMa stars in the Galaxy is 10^{35} erg s^{-1}. It is clear that this (incoherent) source is stronger than any single source thus considered, which could facilitate its detection. Problems could arise in detecting a single source whose frequency was within this band and whose direction was close to the Galactic center.

Vibrations of stars

Stars vibrating in non-radial modes can relax by radiating gravitational radiation. The lowest mode that can generate gravitational waves is the quadrupole ($l = 2$) mode. The general problem of vibrational modes of neutron stars has been considered by Thorne and Campolattaro (1967), Price and Thorne (1969), and Thorne (1969a, b); and that of white dwarfs by Van Horn and Savedoff (1976), Hansen, Cox and Van Horn (1977), Osaki and Hansen (1973), Brickhill (1975), as well as others. Among the various types of quadrupole modes the so-called f- and p-modes are the strong radiators of gravitational waves, with the f-mode being strongest. If these modes are excited by some mechanism,

Sources of gravitational radiation

then gravitational radiation will be emitted. If the star has a close companion or an accretion disk, then mechanisms for exciting these modes either periodically or continuously can be envisioned. Self-excitation via a Cepheid-variable mechanism is also possible.

White dwarfs

Fourier analysis of the light curves of some variable white dwarfs has yielded discrete frequencies in the range 10^{-1} to 10^{-3} Hz (Warner, 1976; Van Horn and Savedoff, 1977; Robinson, 1977). These are commonly interpreted as non-radial modes of vibration of g-type, which are not strong emitters of gravitational radiation when compared with the f- and p-modes. This is of course what one would expect: the f- and p-modes relax faster by radiating gravitational radiation and not by thermal emission. Also, there is no reason to assume that the 'gravitational-wave' modes are not excited. The frequencies of the f- and p-modes have been estimated by Osaki and Hansen (1973) for various white dwarf models. They occur at values larger than those for the g-modes (2×10^{-2} Hz to 1 Hz for the models considered). In addition, the frequencies of the f-, p- and g-modes approach each other as the luminosity of the star increases.

For novae, recurrent novae, and dwarf novae, an eruptive nova event occurs whereby the star increases its luminosity by many orders of magnitude to a value L_{max} corresponding to a kinetic energy E_k. Osaki and Hansen (1973) have considered the question of how much gravitational radiation a vibrating white dwarf will emit using equation of state and stellar models of Savedoff, Van Horn and Vila (1969). The amount of gravitational radiation for the f-modes, the efficient gravitational radiators, can be expressed in the following form for the models considered:

$$L_{GW} \sim (2 \times 10^{35} \text{ erg s}^{-1}) k(m) \left(\frac{\beta}{0.1}\right)\left(\frac{E_k}{10^{45} \text{ erg}}\right), \quad (3.19)$$

where $k(m)$ is a dimensionless constant which depends on the mass of the dwarf and the type of mode, and β is the fraction of the kinetic energy in the mode. For the strongest gravitational-wave mode (f-mode), they found $k(1.0) \sim 1$, and $k(0.4) \sim 2 \times 10^{-3}$.

One would expect the duration of time over which the gravitational radiation would remain strong to be the order of the time for the luminosity to drop by a factor of e (months to years for novae, ~ 10 d for

Chapter 3. Gravitational-radiation experiments

dwarf novae). This is not necessarily the same as the relaxation (or coherence) time τ^* of the oscillation, which would be given by

$$\tau^* = Q/(\pi f_g), \qquad (3.20)$$

where $f_g \sim 0.01$ to 0.1 for a white dwarf. If one assumes that the Q of the gravitational-wave mode is comparable to that of the observed g-modes, then one would estimate $Q \sim 300$. Thus one estimates a correlation time $\sim 10^3$ to 10^4 s for white dwarfs.

Novae. Nova events are eruptions of stars in which the absolute luminosity of the star reaches a value $\sim 10^5 L_\odot$ and the total energy approaches 10^{45} erg. The pre- and post-nova star is a white dwarf, which we will consider to be vibrating as well (and emitting gravitational waves as a consequence of this vibration). Although supernova events are more energetic, novae occur much more frequently. The estimates for our galaxy are 50 to 100 yr^{-1}. Approximately 200 have been observed optically with the average rate being ~ 2 yr^{-1}. In a list of 80 observed novae for which distances R have been deduced, 12 are found with $R < 1000$ pc and five for which $R < 500$ pc (Payne-Gaposchkin, 1957). Let us estimate the gravitational-wave properties of a white dwarf star of mass $M_\odot$ left vibrating by a nova explosion with an energy of 10^{45} erg. Assuming $\beta = 0.1$, we obtain $L_{GW} \sim 2 \times 10^{35}$ erg s^{-1} from (3.19). Assuming $R \sim 500$ pc, the estimated flux is $F \sim 7 \times 10^{-9}$ erg cm^{-2} s^{-1}. For an assumed frequency $\sim 10^{-2}$ Hz, an estimate of the dimensionless amplitude $h \sim 4 \times 10^{-22}$ is obtained. The duration of this signal could be months or years, and the coherence time would be estimated from (3.20) at 10^4 s. These estimates (which are conceivably optimistic) of flux and dimensionless amplitude obtained, are seen to be comparable with other continuous sources.

All novae are also thought to be members of close binaries. There are at least 13 nova stars which are confirmed members of binaries (T CrB, GK Per, AE Aqr, RW Tri, T Aur, UX UMa, DQ Her, RR Pic, VA Scl, TT Ari, VV Pup, AM CVn, and WZ Sge), and there are at least seven novae (VY Aqr, T CrB, RS Oph, T Pyx, WZ Sge, V1017 Sge, μ Sco) which recur after an interval of time that ranges from 10 to 100 yr. From these lists one sees that there are two (T CrB and WZ Sge) which are simultaneously binary novae and recurrent novae.

Dwarf novae. Dwarf novae are novae for which the recurrence time and the change in amplitude during eruption are much smaller than those of ordinary novae. There are approximately 120 known members of this class. Typical luminosities at maximum vary from 10 to $100L_\odot$, which recurrence times ~ 10 to 100 d. Even though the energy at maximum is less than that of a nova, dwarf novae may, in fact, be observable if they are closer than ordinary novae. One of the most studied is SS Cyg. It has a luminosity at maximum $\sim 16L_\odot$ (Warner, 1976) and a recurrence time ~ 50 d, from which we estimate an energy of the outburst of 10^{41} erg. Assuming $m \sim 1$ and using (3.19), one estimates $L_{GW} \sim 2 \times 10^{31}$ erg s^{-1}. The parallax has been estimated at 0.032 seconds of arc (Strand, 1948), from which one estimates a distance of 30 pc. Assuming this distance and $f_g \sim 10^{-2}$ Hz, one estimates a flux $F \sim 2 \times 10^{-10}$ erg cm^{-2} s^{-1} and a dimensionless amplitude $h \sim 10^{-22}$. These are relatively interesting values. There is a subclass of the dwarf novae, called Z Cam, for which the system sometimes stays in the 'on state' (i.e. high-luminosity state) for a period of time longer than the mean recurrence time. The duration of this 'on time' is not predictable and appears to be random. While in the 'on state', the Fourier analysis of the light curves reveals continuously excited frequencies which are identified with the non-radial modes of vibration. Presumably some mechanism is continuously supplying energy to these modes of vibration. And one could further presume that the modes that radiate gravitational waves are also supplied with energy. However, to attempt to estimate the amount of gravitational radiation would require going beyond the 'hand waving' given above to mere speculation. This section is concluded by noting that many of the dwarf novae are also members of binary systems (e.g. BV Cen, RU Peg, EM Cyg, Z Cam, SS Cyg, RX And, SS Aur, U Gem, WW Cet, VW Hyi, Z Cha, EX Hya, and V436 Cen).

Neutron stars

The vibration of neutron stars has been considered by Thorne and Compolattaro (1967), Thorne (1969a, b), and Price and Thorne (1969). Neutron stars have higher densities and smaller dimensions than white dwarfs. The vibrational frequencies occur at kilohertz values and would also be relatively sharp. The coherence time, however, would be expected to be less than 1 s even for modestly high Qs (100, say). These modes could be excited, for example, during the formation of the neutron star. Because of the narrowness of the resonance, one would need a broad-band detector, since the frequencies would be unknown. It

Chapter 3. Gravitational-radiation experiments

is conceivable that a narrow-band resonant detector (discussed below) could accidently have its frequency equal to the vibrational frequency of an excited neutron star. This, however, is highly improbable. It is interesting to note that Boriakoff (1976) has suggested that vibrations of a neutron star may be the explanation of a 1100 Hz periodicity found in the pulsar system AP 2016+28.

Rotation of stars

If a rotating star of frequency f has a 'transverse' quadrupole moment, gravitational radiation will be emitted. The gravitational luminosity is

$$L_{GW} = \frac{32G}{5c^5} \varepsilon^2 J^2 (2\pi f)^6$$

where the ellipticity ε is the difference between the equatorial and polar radii divided by the mean radius, J is the moment of inertia and f is the rotational frequency. Putting this equation into (3.5) one obtains

$$h = 4\pi^2 \left(\frac{32}{5}\right)^{1/2} \frac{G}{c^4} \frac{J\varepsilon f^2}{R}$$

$$= 8.1 \times 10^{-28} \left(\frac{J}{3 \times 10^{44} \text{ g cm}^2}\right)\left(\frac{\varepsilon}{10^{-6}}\right)\left(\frac{100 \text{ pc}}{R}\right)\left(\frac{f}{10 \text{ Hz}}\right)^2.$$

(3.21)

The difficulty in estimating h comes almost entirely from estimating ε. The latest estimates for the Crab and Vela pulsar are by Zimmerman (1977). He estimates $h \sim 10^{-29}$ to 10^{-26} for the Crab pulsar and 10^{-27} to 10^{-23} for the Vela pulsar. These are listed in table 3.7. It should be noted that although the amplitude increases as f^2, it also increases as R^{-1}, so that a close pulsar at a low frequency could have an amplitude comparable with a pulsar further away with a higher frequency. This suggests a study of the quantity f^2/R for all the pulsars. This was done for the 37 pulsars listed in Prentice and ter Haar (1969) and two were found with f^2/R values comparable with those of the Crab and Vela pulsars. Parameters for these two plus the Crab and Vela pulsars are given in table 3.7. If it is worth considering gravitational radiation from the Crab and Vela pulsars, then it is also worth considering radiation from CP 9050 and PSR 1451−68.

The estimates of dimensionless amplitude for the Crab and Vela pulsars are plotted in figure 3.3.

Table 3.7. *Gravitational radiation from some pulsars*

Pulsar	Period (s)	Distance[a] (pc)	f_g (Hz)	$\left(\dfrac{1700}{d}\right)\left(\dfrac{f}{60.6}\right)^2$	Dimensionless[b] amplitude h
NP 0532 (Crab)	0.033	1700	60.6	1.0	10^{-29} to 10^{-26}
NP 0835 (Vela)	0.089	400	22.5	0.58	10^{-27} to 10^{-23}
CP 0950	0.253	60	7.9	0.48	?
PSR 1451–68	0.227	150	8.8	0.24	?

[a] Distances from Prentice and ter Haar (1969).
[b] Zimmerman (1977) has estimated h for the Crab and Vela pulsars. There are no estimates for CP 0950 and PSR 1451–68.

Chapter 3. Gravitational-radiation experiments

3.3.2 Sources of bursts of gravitational waves

Gravitational collapse: order-of-magnitude estimates

In section 3.2 we gave an elementary discussion of a system undergoing gravitational collapse. In that discussion it was found that a characteristic frequency could be defined in terms of the mass. Furthermore, if one knew the efficiency for conversion of mass into energy, and the distance from the source, then the dimensionless amplitude h could be estimated. The dimensionless amplitude estimated in this way, (3.20), is plotted in figure 3.5 versus characteristic frequency f_g for various distances, assuming an efficiency of 0.01. Since the mass is a unique function of f_g, a mass scale is plotted also across the top of the figure. The various considerations that have been neglected (such as angular momentum, mass loss to infinity, neutrino production, etc.) seem to affect the efficiency factor the most. More exact calculations of various

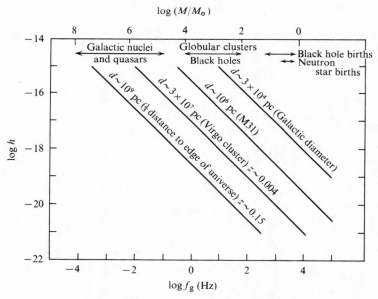

Figure 3.5. Dimensionless amplitude h versus characteristic frequency f_g for gravitational-collapse events for an assumed efficiency $\varepsilon \sim 0.01$ (see (3.11)). The various curves are for different assumed distances to the sources. Because of the unique relationship between mass and frequency, (3.10), a mass scale can also be used as an abscissa and is indicated along the top. Thus it is seen that if one knows the mass, the distance, and the efficiency, estimates of the frequency and the dimensionless amplitude can be made. The mass ranges of events involving neutron stars, black holes, globular clusters, galactic nuclei and quasars are also indicated.

Sources of gravitational radiation

models include these effects and will be discussed later. We now proceed to estimate the orders of magnitude of h of some possible sources ignoring these other effects.

Events with mass 1 to $10M_\odot$

This includes most of those events involving collapse of stars, including binary collapse. From (3.10) we see that the corresponding characteristic frequency will be in the medium-frequency range (10^2 to 10^5 Hz). This is the band that includes all of the Earth-bound resonant detectors that are being built. These events which may be occuring in our galaxy would have amplitudes in the range $h \sim 10^{-18}$ to 10^{-17}, again assuming an efficiency of 0.01. Such events include black hole formation, neutron star formation, and doubly compact binary collapses. The more accurate estimates of gravitational-radiation bursts from black holes and neutron stars (Thorne, 1976) and doubly compact binary collapse (Clark and Eardley, 1977) are found to be within these ranges.

One would expect, however, that the efficiency of production of gravitational waves during neutron star formation would be less than that for the formation of black holes, because of the larger size. In addition to the burst of waves during the formation, the neutron star may be left vibrating in its quadrupole mode after the event, which could relax by emitting gravitational radiation. Misner (1974) estimates that the efficiency could be as high as 0.05. The frequency of the radiation would be in the range 1000 to 3000 Hz and the width would be ~ 100 Hz or less.

In regard to the binary neutron star events, Clark and Eardley calculate a maximum h that is comparable with the estimate from (3.11) for $m_1 \sim m_2 \sim 1$. The estimated event rate is 10^{-2} yr^{-1} out to a distance of 15 Mpc. Also, they estimate that the maximum frequency of the gravitational radiation is $\sim 10^3$ Hz, slightly below the values of most of the resonant-bar detectors being constructed. The lower limit on the rate of these 1 to $10M_\odot$ events is given by the observed supernova rate, since these are thought to involve either a black hole or a neutron star. The supernova rate has been estimated at 0.03 yr^{-1} for our galaxy (Tamman, 1974). On the other hand, the birth rate of pulsars, which are assumed to be neutron stars, is estimated by Taylor and Manchester (1977) to be as high as ~ 0.16 yr^{-1}, five times the supernova rate.

An upper limit on these events can perhaps be inferred from the birth rate of stars of mass greater than $1.5M_\odot$. Zel'dovich and Novikov

Chapter 3. Gravitational-radiation experiments

(1971) estimate this birth rate at $7 \, \text{yr}^{-1}$. An experimentalist would be satisfied with an event rate of $7 \, \text{yr}^{-1}$, but would probably be discouraged if the rate were as low as $0.03 \, \text{yr}^{-1}$, and would be forced to increase the range of his detector to reach beyond the Galaxy. For example, at a distance of 10 Mpc (to M 101), Talbot (1976) estimates the supernova event rate at $1 \, \text{yr}^{-1}$ (and perhaps by inference a pulsar birth rate of $5 \, \text{yr}^{-1}$). The corresponding dimensionless amplitudes would be $h \sim 3 \times 10^{-21}$ to 3×10^{-20}, requiring a 300-fold increase in sensitivity.

Black hole events in globular clusters

Thorne (1976) has considered the possibility of black hole events occurring in globular clusters involving masses in the range 10^2 to $10^4 M_\odot$. From (3.10), or from figure 3.4, one estimates the frequency of the radiation to be in the range 10^0 to 10^2 Hz, and for clusters in the Galaxy $h \sim 10^{-16}$ to 10^{-14}, assuming $\varepsilon \sim 0.01$. These are close to the estimates of Thorne. He estimates the rate at 1 month^{-1} at a distance of 10^9 pc. At this distance one estimates a dimensionless amplitude of 3×10^{-21} to 3×10^{-19}.

Black hole events in galacti nuclei and quasars

Supermassive black holes (of mass $M \sim 10^6$ to $10^{10} M_\odot$) at the centers of galaxies have been postulated as a possible explanation for the quasars and other active galaxies. Thorne and Braginsky (1976) estimate the rate of events associated with these holes to be between $50 \, \text{yr}^{-1}$ and $1/300 \, \text{y}^{-1}$ out to distances within the observable universe. Choosing this distance 3×10^9 pc one would estimate $h \sim 10^{-16}$ at $f_g \sim 10^{-2}$ Hz. The more exact calculation by Thorne, which includes the correction for the redshift ($z \sim 2.6$), yields values close to these.

Pregalactic black hole events

Some have suggested that objects of mass 10^5 to $10^6 M_\odot$ may have condensed out of the primordial plasma prior to galaxy formation at a time when the redshift was $z \sim 30$ to 100 (Doroshkevich, Zel'dovich and Novikov, 1967; Peebles and Dicke, 1968). If these objects collapsed to form black holes, then a gravitational burst would result. Thorne (1978) considers objects of $M \sim 3 \times 10^5 M_\odot$, $z \sim 60$ and obtains $h \sim 10^{-18}$, $f_g \sim 5 \times 10^{-4}$ Hz. One obtains nearly the same estimate for h by choosing

$R \sim 3 \times 10^9$ pc in (3.11), and for f_g from (3.9), by allowing for a red-shifted frequency.

Thorne estimates that of the order of 100 such bursts may be passing the Earth at any one time. These waves would add incoherently producing an amplitude $h \sim (100)^{1/2} \times 10^{-18} \sim 10^{-17}$. This source of gravitational waves would be better described by a spectral density for h: $S_h \sim 2 \times 10^{-31}\,\text{Hz}^{-1}$ at $f_g \sim 5 \times 10^{-4}\,\text{Hz}$ over a frequency band of $5 \times 10^{-4}\,\text{Hz}$. The spectral density of this source is plotted in figure 3.4.

Gravitational collapse: the question of efficiency

The question of efficiency of conversion of matter into gravitational energy is of utmost importance. How closely one may approach the Hawking (1971) limit of $\varepsilon = \frac{1}{2}$ can only be determined for a particular process by a calculation. A number of calculations have been made for various systems which we now consider.

Collapse of single objects

Thuan and Ostriker (1974) considered the gravitational radiation emitted from a $1.4 M_\odot$ collapsing spheroid which was slowly rotating uniformly with no internal pressure. They found a maximum efficiency between 10^{-3} and 10^{-2}. Novikov (1975) has considered semi-quantitatively the same problem with internal pressure. He argues that most of the radiation occurs during the abrupt deceleration at the moment of maximum flattening, and that the total can be much more than Thuan and Ostriker estimate. Shapiro (1977), and Saeny and Shapiro (1977), have examined quantitatively the combined effects of internal pressure and rotation on spheroid collapse. They find that, for a wide range of initial configurations, the gravitational-radiation efficiency during the first full collapse and rebound never exceeds 10^{-2}. Whether these models are realistic approximations to real processes, or whether they are correct for larger masses and other changes in parameter, is a difficult question. Thorne, for example, believes that most estimates of ε are highly uncertain, but that for the strongest sources ε could be as high as 0.1 (Thorne, 1978).

Collapse of binary objects

The case of a small mass m directly falling into a Schwarzschild black hole of mass M has been calculated by Davis *et al.* (1971). They find

Chapter 3. Gravitational-radiation experiments

that the total gravitational radiation emitted is

$$E_g \approx 0.01(m/M)mc^2. \tag{3.22}$$

Detweiler (1977), on the other hand, has considered the corresponding case of a (rotating) Kerr black hole. He finds

$$E_g \sim 1.5(m/M)mc^2.$$

Thus an enhancement by a factor of 150 is possible when one considers angular momentum, which would almost certainly be present in any real problem.

The case where both masses are of the same order is much more difficult. The very ambitious and important calculation of two black holes is being carried out by Smarr (1976). He is considering the fully relativistic case of two colliding Schwarzschild holes. His calculations seem to be yielding efficiencies less than 0.001, and they could be much less. However, one might expect a considerable Detweiler enhancement factor when angular momentum is put into the problem. For the case of the collapsing binary (rotating) neutron star mentioned above, Clark and Eardley (1977) find $\varepsilon \sim 0.02$.

Neutron starquakes

Neutron stars have been observed to suddenly change their rotational frequency. A common interpretation is that a 'quake' has occurred and that a large amount of strain energy has been released (Pines, Shaham and Ruderman, 1974). Thorne (1978) estimates that the energy release could be 10^{45} erg. In a scenario similar to that discussed above for white dwarfs, one might expect some of this mechanical energy to be deposited into the quadrupole modes, which then relax by radiating gravitational radiation. The frequencies of vibration tend to lie in the range 10^3 to 10^4 Hz (Thorne, 1969a, b; Detweiler, 1975a, b). Furthermore, the Qs of these modes are high (10^2 to 10^3), implying relaxation times $\tau^* \sim 1$ s ($\tau^* \sim 2Q/\omega$); thus the radiation would be confined to a narrow frequency band. If this frequency is not known, then the probability of a resonant detector being at the correct frequency will be very small indeed. Thorne (1976) estimates the dimensionless amplitude as

$$h \sim 10^{-23}\left(\frac{\Delta E}{10^{45} \text{ erg}}\right)\left(\frac{3000 \text{ Hz}}{f_g}\right)\left(\frac{1 \text{ s}}{\tau^*}\right)^{1/2}\left(\frac{3 \text{ kpc}}{R}\right).$$

Detection of gravitational radiation

White dwarf quakes

The gravitational radiation emitted by a white dwarf after a quake (nova explosion) can be considered a burst also. However, since the wavetrain could last for days or weeks, this radiation was classified (somewhat arbitrarily) as continuous, and was discussed in section 3.3.1.

Gravitational scattering of two bodies

Zel'dovich and Polnarev (1974) have considered the possibility of clusters of 10^9 superdense stars (neutron stars or black holes) at the Galactic center. They estimated the frequency spectrum of gravitational radiation due to close flybys (i.e. scattering without capture). Under favorable conditions they found that for the case of black holes, pulses with dimensionless amplitude of the order of 10^{-20} could occur perhaps once a year. The peak in the frequency spectrum of the radiation would be in the kilohertz band or below. Compared to other possible sources of gravitational radiation, this one seems considerably weaker.

3.4 Detection of gravitational radiation

3.4.1 Amplitude of the signal

Free-mass- or almost-free-mass antenna

Consider two masses separated by a distance l interacting with a gravitational wave of amplitude $h(t)$. Assuming that the wave is optimally oriented and polarized (see figure 3.2), the relative displacement x for a gravitational wave $h = h_0 \, e^{i\omega_g t}$ is

$$x \approx h_0 \frac{\lambda_g}{\pi} \left(\sin \frac{\pi l}{\lambda_g} \right) e^{i\omega_g t}, \qquad (3.23)$$

where λ_g is the wavelength of the gravitational wave. This expression has the familiar $x \sim h_0 l$ limit when $l \ll \lambda_g/2$. One notes that the maximum amplitude occurs when the separation l has the value $\lambda_g/2$.

The almost-free-mass antenna consists of a pair of masses which can move 'freely' with respect to each other at frequencies greater than some relatively low frequency. Two pendulum masses interacting with a gravitational wave whose direction is normal to the plane containing the pendula, and whose frequency is greater than the pendulum frequency,

Chapter 3. Gravitational-radiation experiments

is such a case (see figure 3.2(*b*)). The free-mass- and the almost-free-mass antennae have the advantage that they can detect gravitational waves over a broad frequency band.

Resonant antenna

The thermal noise associated with the free-mass- or almost-free-mass antenna may be such a problem that one is forced to consider a resonant antenna. The conceptually simplest resonant antenna is one whereby two masses separated by a distance l are connected by a spring (figure 3.2(*c*)); thus one has a mechanical oscillator of resonant frequency ω_m, also characterized by an amplitude-relaxation time τ^*, which is related to the quality factor Q by

$$\tau^* = 2Q/\omega_m. \tag{3.24}$$

The cylindrical-bar detector (figure 3.2(*d*)), which most experimental groups are using, closely approximates this antenna type. The response of the detector to a gravitational wave $h_0 \, e^{i\omega_g t}$, where ω_g is equal to ω_m, is given approximately by

$$x \sim n h_0 \, e^{i\omega_g t},$$

where

$$n \sim \begin{cases} f_g t & t \ll \tau^* \\ Q & t \gg \tau^* \end{cases} \tag{3.25}$$

and where it is assumed that $l < \lambda_g/2$. The quantity n is the number of cycles of the wave interacting with the detector. If t is longer than τ^*, n has the value Q. For broad-band bursts of gravitational radiation one expects that $n \sim 1$, thus x for the resonant bar in this case is of the same order of magnitude as the expression for the displacement that one had for the free-mass- and almost-free-mass detectors. It will be shown below, however, that the noise competing with the signal can be reduced more easily for the resonant detectors.

3.4.2 Detection schemes under study and development

Mechanically resonant detectors

Almost all of the initial effort to detect gravitational radiation went into the construction of detectors of this type. The inventor and pioneer was

Detection of gravitational radiation

Weber (1961), who along with his students constructed the first gravitational-wave detector from a large cylinder of aluminum ($l \sim 150$ cm, $f_m \sim 1660$ Hz; see figure 3.6). His announcement of coincident detection of pulses (Weber, 1970a, b, c, 1971, 1972; Weber et al., 1973) on two of these antennae stimulated five to six other groups to attempt to confirm his observations by construction of similar resonant-cylinder antennae. Most of the antennae that have been constructed have involved cylinders of aluminum ($f \sim 1$-to-2×10^3 Hz, $M \sim 10^6$ g) similar to that of Weber. The sensitivity of these later detectors reached a level which is stated by those investigators to be a considerable improvement over that of the early Weber detectors. None of these other groups claimed to have discovered gravitational radiation (Braginsky et al., 1972; Drever et al., 1973; Levine and Garwin, 1973; Tyson, 1973; Billing et al., 1975; Douglass et al., 1975; Allen and Christodoulides, 1975). The most sensitive of these detectors reached a noise level corresponding to $x/l \sim 3 \times 10^{-17}$.

Various schemes for detecting the small displacement x, either directly or indirectly, have been used. One such scheme is represented

Figure 3.6. Photograph of Joseph Weber and an early aluminum cylindrical gravitational-wave detector.

Chapter 3. Gravitational-radiation experiments

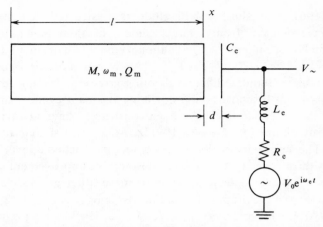

Figure 3.7. Diagram of the capacitance pick-off system. The antenna under the action of a gravitational wave undergoes a displacement x. This displacement changes the value of the capacitor C, which is detected as a change in some parameter in the *RLC* electrical circuit.

diagramatically in figure 3.7; the motion of the antenna moves one face of a capacitor which is part of an electromagnetic resonator. The motion of this face modulates some parameter of the circuit (frequency, voltage, phase) which one can measure to infer the displacement. It is noted for later reference that this sensor has a source of noise (i.e. the electrical noise associated with R_e). Assuming that one does have a method of detecting such small displacements, the limitation on the sensitivity is due to the noise from all the various sources.

Noise

The condition for detectability is that the displacement x_g due to the signal is larger than the noise x_n:

$$x_g > x_n.$$

It is sometimes convenient to express this condition in terms of an energy

$$E_g \sim M\omega_m^2 x_g^2 > E_n,$$

where E_n is the energy associated with the noise. (It should be always kept in mind, however, that one actually measures displacement and that one can also observe negative changes in displacement amplitude which correspond to negative energy changes.) The noise can be considered in two parts.

Detection of gravitational radiation

The Brownian-motion noise. The Brownian-motion noise of the antenna takes a simple form when expressed in terms of the energy:

$$E_{BM} \sim kT(2\tau_{meas}/\tau^*),$$

where k is Boltzmann's constant, τ_{meas} is the measurement time, and τ^* is the amplitude-relaxation time defined in (3.24). This equation is for the case where $\tau_{meas} \ll \tau^*$; it can also be expressed as

$$E_{BM} \sim kT\omega_m \tau_{meas}/Q. \tag{3.26}$$

In terms of the displacement, (3.26) becomes

$$x_{BM} \sim \left(\frac{kT\tau_{meas}}{M\omega_m Q}\right)^{1/2},$$

or, if one wishes, in terms of the strain,

$$\left(\frac{x}{l}\right)_{BM} \sim \left(\frac{kT\tau_{meas}}{M\omega_m l^2 Q}\right)^{1/2}. \tag{3.27}$$

For the cylindrical-bar detector there is a relation between ω_m and l (assuming the first longitudinal mode of vibration):

$$\omega_m l \approx \pi v_s,$$

where v_s is the velocity of sound of the material of the antenna. Thus (3.27) becomes

$$\left(\frac{x}{l}\right)_{BM} \sim \left(\frac{kT\omega_m \tau_{meas}}{\pi^2 M v_s^2 Q}\right)^{1/2}$$

$$\sim 3 \times 10^{-21} \left[\left(\frac{f_m}{10^3 \text{ Hz}}\right)\left(\frac{10^{17} \text{ erg}}{Mv_s^2}\right)\left(\frac{10^9}{Q}\right)\left(\frac{T}{1 \text{ K}}\right)\left(\frac{\tau_{meas}}{10^{-2} \text{ s}}\right)\right]^{1/2}. \tag{3.28}$$

Among the parameters available to the experimentalist, one sees that the Brownian-motion noise is made smaller by small T, and large Mv_s^2 and Q. (Although τ_{meas} can be made as small as the interaction time $\hat{\tau}$ without diminishing the signal, the experimenter cannot arbitrarily reduce τ_{meas} because the noise from the sensor also depends on τ_{meas}.) To estimate the magnitude of the Brownian-motion noise we consider the example of a single crystal of silicon (Douglass, 1976; McGuigan et al., 1978) or sapphire (Braginsky, 1977b) for which Qs of the order of 10^9 have been achieved. For both silicon and sapphire, $v_s \sim 10^6$ cm s^{-1},

Chapter 3. Gravitational-radiation experiments

and assuming $M \sim 3 \times 10^4$ g and $T \sim 1$ K, one estimates

$$\left(\frac{x}{l}\right)_{\text{BM}} \sim 5 \times 10^{-21} \left[\left(\frac{T}{1\,\text{K}}\right)\left(\frac{\tau_{\text{meas}}}{10^{-2}\,\text{s}}\right)\right]^{1/2}.$$

For an aluminum cylinder with $M \sim 5 \times 10^6$ g, $Q \sim 10^7$, $v_s = 5 \times 10^5$ cm s^{-1} one obtains

$$\left(\frac{x}{l}\right)_{\text{BM}} \sim 10^{-20} \left[\left(\frac{T}{1\,\text{K}}\right)\left(\frac{\tau_{\text{meas}}}{10^{-2}\,\text{s}}\right)\right]^{1/2},$$

a value comparable with the silicon or sapphire example. It is clear that it is technically feasible to reduce the Brownian-motion noise to the levels $x/l \sim 10^{-21}$ (which is necessary to be able to measure bursts from the Virgo cluster) by increasing M and Q, and decreasing T. It may be necessary, however, to cool below 1 K, which is an important technical consideration.

Sensor noise. The sensor introduces noise into the experiment in two ways. The first is directly from its own noise (for example, from the resistor R_e in figure 3.7) entering the measuring device following the sensor. This we call the 'direct noise'. Secondly, the sensor may act back on the antenna, which we will call the 'back reaction'. This analysis is also conveniently done in terms of an energy description.

The direct noise in the sensor is usually broad-band so that the noise is proportional to the frequency bandwidth Δf. In addition, it can be made smaller by increasing the energy P in the sensor, so that we can express the noise E_D relative to the signal as

$$E_D \propto \Delta f / P.$$

Since $\Delta f \sim \tau_{\text{meas}}^{-1}$,

$$E_D \propto 1/(P\tau_{\text{meas}}).$$

On the other hand, this same sensor noise produces a broad-band random force acting back on the antenna. The action of this force is mathematically the same as that of the Brownian-motion force. Also, increasing the power in the sensor makes the force larger, so as a result we have

$$E_{\text{BR}} \propto P\tau_{\text{meas}},$$

which is seen to be opposite to the action of the direct noise. Thus there

Detection of gravitational radiation

is some optimum set of sensor parameters that minimizes the sum of the direct and back reactions.

Let us assume that these two effects are independent (this is not proven and is not necessarily true):

$$E_{\text{sens}} \sim E_D + E_{BR}. \tag{3.29}$$

It is found that (3.29) can be expressed in the form

$$E_{\text{sens}} \sim E_s(\tau_{\text{op}}/\tau_{\text{meas}} + \tau_{\text{meas}}/\tau_{\text{op}}), \tag{3.30}$$

where τ_{op} and E_s are a characteristic time and a characteristic energy of the particular sensor.

The optimum strategy

The optimum strategy will be to reduce both the Brownian-motion noise and the sensor noise to levels below that from the expected signal. The exact criteria depend upon whether one is attempting to detect bursts or continuous radiation.

Bursts of gravitational radiation. The first criterion is to reduce the Brownian noise (see (3.28)) to acceptable levels. Since ω_m is fixed and τ_{meas} cannot be reduced below $\hat{\tau}$, one must increase Mv_s^2 and decrease T/Q. It appears that this can be done with present materials (high Q) and present technology (low T). The next criteria concern the sensor. The characteristic time τ_{op} appearing in (3.30) is determined by the parameters of the sensor. Although choosing $\tau_{\text{meas}} \sim \tau_{\text{op}}$ will minimize the sensor noise, if it is much larger than $\hat{\tau}$, the interaction time of the wave, the Brownian-motion noise will increase and one may need to decrease T or increase Q to compensate. Thus, there is the additional somewhat weaker requirement

$$\tau_{\text{op}} \sim \tau_g \sim 1/\omega_m. \tag{3.31}$$

Choosing $\tau_{\text{meas}} \sim \tau_{\text{op}}$, the minimum noise energy due to the sensor is

$$E_{\text{sens,min}} \sim 2E_s.$$

Expressed in terms of the strain this becomes

$$\frac{x}{l} \sim \left(\frac{2E_s}{\pi^2 M v_s^2}\right)^{1/2}. \tag{3.32}$$

The value of E_s and τ_{op} depend on the particular detection scheme. For

Chapter 3. Gravitational-radiation experiments

the capacitance pick-off scheme in figure 3.7, the result is (Braginsky, 1977a; Braginsky and Manukin, 1977)

$$E_D \sim \frac{3}{4} M \omega_m^2 \left(\frac{d}{V_\sim}\right)^2 \frac{2kT_e}{Q_e \omega_e C_e} \frac{1}{\tau_{\text{meas}}}$$

$$E_{BR} \sim \frac{2kT_e Q_e C_e}{3 \omega_e M} \left(\frac{V_\sim}{d}\right)^2 \tau_{\text{meas}},$$

where T_e, ω_e, Q_e, C_e are respectively the temperature, frequency, quality factor, and capacitance of the sensor. The spacing in the capacitor is d, and the voltage across the capacitor is $V_\sim$. It has been assumed in the above analysis that $2Q_e/\omega_e < \tau_{\text{meas}}$. The opposite case has been considered by Braginsky (1977a). The values of E_s and τ_{op} are

$$E_s \approx kT_e(\omega_m/\omega_e)$$

$$\tau_{\text{op}} = \frac{3M}{2Q_e} \frac{\omega_m}{\omega_e} \left(\frac{d}{V_\sim}\right)^2. \tag{3.33}$$

Let us estimate if condition (3.31) can be satisfied assuming a scheme based upon a superconducting cavity. Choose the parameters $M \sim 10^6$ g, $\omega_m \sim 10^4$ rad s^{-1}, $(d/V_\sim) \sim 10^{-4}$ cm V^{-1}, $c_e \sim 10^{-10}$ F, $Q_e \sim 10^8$; then $\tau_{\text{op}} \sim 10^{-3}$, which is of the order of τ_g for kilocycle bursts. Thus this scheme will work, in principle. Implied in the scheme is a very stable electromagnetic oscillator which, although technically feasible, has not yet been built (Braginsky, 1977b). However, the prospects that superconducting oscillators of sufficient stability can be built are quite good (Stein and Turneaure, 1973; Stein, 1975).

Now let us consider the energy E_s of (3.33). It would appear that one could make E_s as small as one wishes by lowering T_e or making ω_e larger. However, rewriting (3.33),

$$E_s \approx (kT_e/\hbar\omega_e)\hbar\omega_m,$$

(where $\hbar$ is Planck's constant divided by 2π) one sees that the factor $kT_e/(\hbar\omega_e)$ is the 'number' of thermal photons in the electromagnetic resonator. When this factor approaches 1, the electromagnetic resonator must be considered as a quantum system (and one notes also that the energy changes of the antenna are comparable to the quantum-mechanical spacings). Furthermore, one cannot use an ordinary instrument (such as a linear amplifier) to determine the quantum state of the electromagnetic resonator. Thus the minimum energy sensitivity achievable in the classical regime is

$$E_{s,\text{min}} \sim \hbar\omega_m. \tag{3.34}$$

Putting (3.34) into (3.32) one obtains a lower limit for this case on the measurable strain

$$\left(\frac{x}{l}\right)_s \sim \left(\frac{2\hbar\omega_m}{\pi^2 M v_s^2}\right)^{1/2}$$

$$\sim 1.1 \times 10^{-21} \left(\frac{f}{10^3 \text{ Hz}}\right)^{1/2} \left(\frac{10^6 \text{ g}}{M}\right)^{1/2} \left(\frac{10^6 \text{ cm s}^{-1}}{v_s}\right).$$

We conclude that one cannot measure a dimensionless amplitude h much less than 10^{-21} in the kilohertz band using ordinary instrumentation.

The above results obtained for the capacitance pick-off scheme are in fact much more general. Transducer schemes based upon the SQUID (superconducting quantum interference device) have been analyzed by Gusev and Rudenko (1977) and Hough et al., (1977). Their calculations lead to an energy-sensitivity limit which is very close to (3.34). Giffard (1976) has recently considered the case of an arbitrary linear transducer and derives equations close to (3.33) and (3.34). In his scheme the transducer is a parametric amplifier, where the factor (ω_m/ω_e) in (3.33) is identified as the inverse of the power gain, and T_e is the noise temperature of the following amplifier. Using the best possible linear amplifier (a maser, perhaps) the value of T_e can be no lower than $\hbar\omega_e/k$, which when put into (3.33) yields (3.34). Thus, to do better than this limit, one must consider the sensor and the antenna as quantum systems.

Braginsky (1977b) has considered the problem of determining the quantum state of an electromagnetic oscillator without disturbing that state (quantum non-demolition). Although this is a very difficult problem, there is no fundamental difficulty with a scheme based upon quantum non-demolition. (For example, in the flux quantization experiments, one can determine the integer describing the quantum state exactly without demolishing that state.) Of course, in order not to violate the uncertainty principle, one has to give up all information concerning the phase of the wavefunction. Unruh (1977a) has shown that quantum non-demolition schemes based upon a quadratic perturbation can work, but that schemes based upon linear perturbations will not. Quadratic schemes have been presented and discussed by Braginsky (1977c) and Unruh (1977b).

Continuous radiation. One may wish to construct an antenna to detect the continuous radiation expected from a known source, the Vela pulsar

Chapter 3. Gravitational-radiation experiments

for example. From (3.25) one sees that the signal from the detector increases with integration time so that one could expect to measure smaller dimensionless amplitudes h_0 from a continuous source than from a burst source. This is, in fact, true. However, increasing the measuring time also increases the Brownian-motion noise, as an inspection of (3.27) reveals. Thus we arrive at a necessary condition for the detection of continuous gravitational radiation:

$$h_0 > \left(\frac{kT}{M}\right)^{1/2} \frac{1}{l\omega_g Q} \tag{3.35}$$

$$\sim 4 \times 10^{-24} \left(\frac{T}{1\,\mathrm{K}}\right)^{1/2} \left(\frac{10^5\,\mathrm{g}}{M}\right)^{1/2} \left(\frac{100\,\mathrm{cm}}{l}\right) \left(\frac{100\,\mathrm{rad\,s^{-1}}}{\omega_g}\right) \left(\frac{10^9}{Q}\right),$$

where n was set equal to its maximum value of Q. Thus, using optimistic parameters for a resonator, one can barely get into the range of values for h_0 that Zimmerman (1977) assigns to this source ($h_0 \sim 10^{-27}$ to 10^{-23}). In addition to the back reaction, which will not be discussed, there are many other problems, such as tuning the frequency of the detector to take out the Doppler effect. Another problem is that resonant detectors of the cylindrical-bar type would be prohibitively large at these frequencies, so one is forced to consider a detector of the second class (Douglass, 1970) whose size can be much less than a wavelength of sound in the material (such as a tuning fork). The resonant frequency of such a detector can be made much lower than that of a cylinder of the same approximate size and mass by proper choice of dimensions; however, Qs approaching 10^{10}, which have been found for the first longitudinal modes of cylinders, have not been demonstrated for this type of resonator.

In spite of these problems, Hirakawa, Tsubono and Fujimoto (1977) have constructed a mechanical resonant detector at the frequency (60.2 Hz) expected for radiation from the Crab pulsar. They report an upper limit on the flux of 1.4×10^4 erg cm^{-2} s^{-1} ($h_0 < 1 \times 10^{-19}$). Even though this limit is many orders of magnitude above the limit set by the observed deceleration ($h_0 < 5 \times 10^{-25}$), one has to admire the conception and elegance of this experiment. In addition, the techniques developed will certainly be useful in constructing more sensitive detectors in this frequency band.

If one contemplates constructing resonant antennae to detect continuous gravitational radiation at still lower frequencies, then condition (3.35) becomes more difficult to satisfy. In addition, one encounters

Detection of gravitational radiation

the increasingly severe problem of low-frequency noise on the surface of the Earth. One could avoid the Earth noise by putting the detector in space. However, if one is forced to consider a space experiment, then there are other more attractive schemes, which we consider below.

Free-mass- and almost-free-mass detectors

The free-mass- and almost-free-mass detectors are particularly attractive because they are broad-band and can detect gravitational radiation at various frequencies. Furthermore, we see from (3.23) that larger signals can be obtained by increasing the separation l (provided $l < \lambda_g/2$). We now discuss two versions of this type of detector under active study and development.

Laser ranging

In this scheme the change in the separation x of the masses is detected by means of a laser. Most of the proposals thus far involve three masses, as indicated in figure 3.8, which is recognized as an optical interferometer. For an optimally oriented detector and for a polarized wave, a displacement x of mirror M_1 and $-x$ of mirror M_2 would occur. The

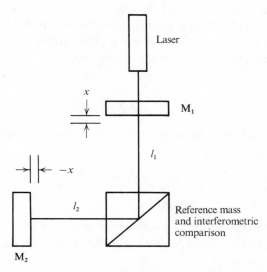

Figure 3.8. Diagram of 'almost'-free-mass antenna using optical detection. The gravitational wave propagates in or out of the plane, and is optimally polarized. The interaction causes a displacement x in arm 1 and a displacement $-x$ in arm 2.

Chapter 3. Gravitational-radiation experiments

first group to build such a system was Forward, Moss and Miller (1971), who achieved a displacement sensitivity of 1.3×10^{-12} cm in a 1 Hz bandwidth. The strain sensitivity x/l, assuming a length of 10^3 cm, would be 10^{-15}, which is within two orders of magnitude of that achieved by the resonant detectors. Examination of (3.23) shows that the signal x due to the gravitational wave can be made larger by making the arms of the interferometer larger, subject to the condition

$$l \leq \lambda_g/2. \tag{3.36}$$

For bursts, (3.36) is equivalent to the criterion that the light travel time must be less than a period of the wave. For continuous waves, if $l > \lambda_g/2$, one sees from (3.23) that the amplitude just oscillates with l. Relatively large values of l can be obtained with laboratory-sized apparatus (conceivably as large as 10^5 cm) by reflecting the light beam many times back and forth through the interferometer; with high-quality dielectric coatings one could perhaps expect 300 reflections. This would result in an effective separation of 3×10^7 cm, which is about five orders of magnitude larger than the resonant detectors. There are at least four groups presently studying similar schemes (at the University of Glasgow, MIT, the Max Planck Institute at Munich, and the Hughes Research Laboratories at Malibu, California).

We now consider the fundamental limitations of a laser-ranging antenna system. The most fundamental limitation is the uncertainty in displacement x_{ph} due to fluctuations in the number of photons N_{ph} during the time interval τ_{meas} of the measurement:

$$x_{ph} \sim \frac{\lambda_{ph}}{2\pi N_{ph}^{1/2}}, \tag{3.37}$$

where λ_{ph} is the photon wavelength. Since the number of photons detected during τ_{meas} is

$$N_{ph} \sim \frac{\varepsilon P_{ph} \tau_{meas}}{h \nu_{ph}},$$

where P_{ph} is the power of the photon generator, ε is an efficiency factor for detection of the photons, and ν_{ph} is the frequency, equation (3.37) can be expressed as

$$x_{ph} \sim \left(\frac{\hbar c \lambda_{ph}}{2\pi \varepsilon P_{ph} \tau_{meas}} \right)^{1/2} \tag{3.38}$$

Detection of gravitational radiation

$$\sim \left(\frac{\hbar c \lambda_{ph} \Delta f}{\pi \varepsilon P_{ph}}\right)^{1/2}, \tag{3.39}$$

where in (3.39) the bandwidth $\Delta f \sim 1/(2\tau_{meas})$ has been introduced.

Bursts of gravitational radiation. If we are interested in detecting gravitational-wave bursts, which from the discussion in section (3.3.2) are expected to be the strongest sources, then

$$\Delta f_{bur} \sim f_g,$$

and (3.39) becomes

$$x_{ph} \sim \left(\frac{\hbar c \lambda_{ph} f_g}{\pi \varepsilon P_{ph}}\right)^{1/2}.$$

Expressed in terms of a strain this becomes

$$\frac{x_{ph}}{l} \sim \left(\frac{\hbar c \lambda_{ph} f_g}{\pi l^2 \varepsilon P_{ph}}\right)^{1/2}.$$

The minimum possible strain sensitivity as a consequence of this noise source is given by choosing the maximum l allowed by (3.36):

$$\left(\frac{x_{ph}}{l}\right)_{min} \sim \left(\frac{4\hbar \lambda_{ph} f_g^3}{\pi c \varepsilon P_{ph}}\right)^{1/2}$$

$$\sim 4.5 \times 10^{-20} \left(\frac{\lambda_{ph}}{5 \times 10^{-5} \, cm}\right)^{1/2} \left(\frac{0.3}{\varepsilon}\right)^{1/2} \tag{3.40}$$

$$\times \left(\frac{10 \, W}{P_{ph}}\right)^{1/2} \left(\frac{f_g}{3 \times 10^3 \, Hz}\right)^{3/2}.$$

Although the limiting strain sensitivity in the kilohertz band is three orders of magnitude below the present sensitivity of resonant antennae, it would appear that in order to detect supernova explosions in the Virgo cluster ($h \sim 10^{-21}$) one would need a laser with a power $\sim 10^4$ W. Perhaps a more promising application of this scheme would be to attempt to detect gravitational bursts of lower characteristic frequency. The fundamental strain-sensitivity limit for bursts given above, (3.40), becomes less as one looks for lower-frequency radiation, 100 Hz for example. If it were possible to eliminate all other sources of noise at 100 Hz and to also satisfy (3.36), then the fundamental limit at this

Chapter 3. Gravitational-radiation experiments

frequency would be $x/l \sim 10^{-22}$ for a power of 100 W. However, this would require that $l \sim 1.5 \times 10^8$ cm, which is 10 times the distance estimated above as the practical limit for an Earth-based system. However, with a sensitivity of $h \sim 10^{-20}$ one could easily see black hole events in globular clusters in the Virgo cluster (see figure 3.4). At frequencies below 100 Hz it would be increasingly difficult to increase l to the limit allowed by (3.36); also, at frequencies below 100 Hz, one has for an Earth-based experiment the very difficult problem of ground vibrations and other environmental effects. In a space experiment these problems would not be present. It would also be relatively easy to choose any l that was desired.

Continuous radiation. The photon-noise limit for detection of continuous waves is given by (3.38). Expressing this in terms of a strain, and choosing the limit when $l \sim \lambda_g/2$ to obtain the minimum sensitivity, we find

$$\left(\frac{x_{\text{ph}}}{l}\right)_{\text{min}} \sim \left(\frac{2\hbar\lambda_{\text{ph}}f_g^2}{\pi c \varepsilon P_{\text{ph}}\tau_{\text{meas}}}\right)^{1/2}$$

$$\sim 2.0 \times 10^{-26}\left[\left(\frac{\lambda_{\text{ph}}}{5 \times 10^{-5}\,\text{cm}}\right)\right. \tag{3.41}$$

$$\left.\times \left(\frac{0.3}{\varepsilon}\right)\left(\frac{10\,\text{W}}{P_{\text{ph}}}\right)\left(\frac{10^6\,\text{s}}{\tau_{\text{meas}}}\right)\left(\frac{f_g}{100\,\text{Hz}}\right)^2\right]^{1/2}.$$

The strain sensitivity indicated by (3.41) suggests that the gravitational radiation expected from the Crab and Vela pulsar could not be detected by this scheme unless one had kilowatts of power available and integrated for years. If, on the other hand, one wished, for example, to detect the continuous radiation from the binary star system WZ Sge (see table 3.5), then putting in the frequency of the expected radiation ($f_g \sim 4 \times 10^{-4}$ Hz) one easily obtains a limit lower than the expected amplitude. This experiment would have to be done in space for two reasons. The separations required to achieve this sensitivity are of the order of 1 AU and the Earth's surface is extremely noisy at these frequencies. Although the problems of a space experiment are formidable, there appear to be no fundamental reasons why such a system would not work. However, there are myriad practical problems associated with *any* laser scheme, but these will not be discussed here. For further details of the 'practical' difficulties see Weiss (1972), Winkler (1976) and Drever (1977a).

Doppler ranging

In the Doppler-ranging scheme there are two free masses, e.g., two spacecraft, or the Earth and a spacecraft. One of the free masses has on it a master oscillator of frequency ν_0. The second mass has the capability of transmitting electromagnetic signals at a frequency ν which is close to ν_0. This signal can be generated from an oscillator on the second mass, or by transponding back a signal received from the oscillator on the first mass. One then has the observable

$$y = (\nu - \nu_0)/\nu_0.$$

If a gravitational wave interacts with the two masses and changes their separation, y will change because of the (first-order) Doppler effect.

Estabrook and Wahlquist (1975) and Anderson (1971, 1977) have shown that the magnitude of the Doppler signal y_g due to the gravitational wave of dimensionless amplitude h is given by the rather simple expression

$$y_g \sim h \sin(2\pi l/\lambda_g),$$

where λ_g is the wavelength of the gravitational wave (i.e. a signal of this magnitude modulates the frequency of the electromagnetic wave at the gravitational-wave frequency). For $l < \lambda_g/4$, the signal magnitude increases with separation of the spacecraft as expected. The magnitude is largest at a separation of $l = \lambda_g/4$, again as expected, and oscillates sinusoidally with separation for large l.

The relation $l \sim \lambda_g/4$ for the maximum signal defines for a particular frequency a minimum separation that must be achieved in order to obtain that largest signal. For example, the separation corresponding to 10 Hz is approximately 1 Earth diameter, and for 10^{-4} Hz the separation is about 10 AU. Thus the Doppler-ranging-of-spacecraft detection scheme will be of interest primarily for frequencies within this range.

There are many sources of noise that one has to consider and contend with. One unavoidable source is frequency noise in the master oscillator. Another is dispersion of the interstellar medium. The condition for detectability is that the signal y_g from the gravitational wave must be larger than the noise y_N from all other sources,

$$y_g > y_N. \tag{3.42}$$

We shall consider in our discussion primarily the noise from the master oscillator. This noise is easily measured, and can be described in terms of a spectral density of fractional frequency deviations S_y (Barnes *et al.*,

Chapter 3. Gravitational-radiation experiments

1971). The best oscillators have values of spectral density of $\sim 10^{-28} \text{ Hz}^{-1}$. However, it would not be unreasonable to expect that $S_y \sim 10^{-30} \text{ Hz}^{-1}$ could be achieved using hydrogen masers (Vessot, 1976) or superconducting-cavity-stabilized oscillators (Stein and Turneaure, 1973) in the frequency range 10^{-4} to 1 Hz. Considering only the oscillator noise, condition (3.42) becomes

$$y_g > (S_y \, \Delta f)^{1/2}, \tag{3.43}$$

where Δf is the bandwidth associated with the measurements.

Bursts of gravitational radiation. The most likely candidates for detection by this scheme are bursts from the nuclei from distant galaxies and quasars (Thorne and Braginsky, 1976; Thorne, 1978) discussed in section 3.3.2. We consider first the condition on the oscillator noise. To detect the burst, the bandwidth Δf in (3.43) must be of the order of, or larger than, the characteristic frequency f_g of the gravitational wave. Thus (3.43) becomes

$$y_g > f_g^{1/2} S_y^{1/2}.$$

For black hole events with mass between 10^6 and $10^7 \, M_\odot$, values of y_g would be between 10^{-17} and 10^{-16} for a burst with characteristic frequency of 10^{-3} Hz. The noise from present oscillators, $y_N \sim (10^{-3} \times 10^{-28})^{1/2} \sim 3 \times 10^{-16}$, is seen to be only one or two orders of magnitude larger than the expected signal. As mentioned above, one could expect that the stability of the oscillators could be improved considerably (i.e. smaller S_y). If this is done, one would also have to improve the instrument that does the comparison of the frequencies, the Doppler extractor (Wahlquist *et al.*, 1976). This is technically an easy thing to do. Wahlquist *et al.* also indicate that the noise from the interplanetary plasma can be reduced by going to a dual-frequency Doppler scheme (s-band and x-band, say). Estabrook and Wahlquist (1975) point out that the gravitational waveform is repeated three times in the Doppler signal. This feature may allow one to extract the gravitational wave in the presence of larger noise. Vessot, Levine and Nordtvedt (1977) have proposed a four-link Doppler system which would allow for a more complete 'modeling-way' of plasma noise.

Continuous radiation. There are many sources of continuous radiation in this frequency band that one might wish to attempt to detect by the Doppler-tracking technique. The amplitude of the strongest sources at frequencies of 10^{-4} to 10^{-3} Hz approaches $h \sim 10^{-20}$. Since many of these sources are continuous and coherent, one can reduce the noise by integrating for a long time. Thus the bandwidth of the measurement is $\sim 1/(2\tau_{\text{meas}})$, and equation (3.43) becomes

$$y_g > 2 \times 10^{-18} \left(\frac{S_y}{10^{-28} \text{ Hz}^{-1}} \right)^{1/2} \left(\frac{10^7 \text{ s}}{\tau_{\text{meas}}} \right)^{1/2}.$$

Again we see that present technology is within one or two orders of magnitude of having the required sensitivity to detect gravitational-wave sources. These sources, unlike burst sources, are known to exist. Furthermore, the positions and frequencies are known precisely, and the estimates of the dimensionless amplitudes are not likely to be in error by more than a factor of 10. Improvements in the oscillators and the Doppler extractor would be necessary for this case also. At some point, however, one will encounter the noise of the interplanetary medium (Woo, 1975). Since this noise decreases with increasing electromagnetic frequency, one can, in principle, reduce it by going to a higher master-oscillator frequency. Another possibility is to make Doppler-ranging measurements at two or more frequencies and to 'model' it out.

3.5 Prospects for the future

In this chapter we have shown that there is a large variety of different sources of gravitational radiation. There is considerable excitement about the possibility of detection of such radiation. Many groups are now attempting to build antennae sufficiently sensitive to detect these sources.

The room-temperature resonant antennae are at a sensitivity of $x/l \sim 10^{-17}$, which is close to that required to observe the gravitational-wave bursts expected from supernova explosions in our galaxy. Since the rate of these events is low, one would like to have about three more orders of magnitude improvement in displacement sensitivity in order to observe similar events in the Virgo cluster. The vigorous and ongoing world-wide efforts to develop more sensitive detectors almost guarantee considerable improvement. There are at least nine groups attempting to improve resonant antennae by combinations of high Q, large mass and low temperature (at the University of Maryland, the University of Rochester, the Moscow State University, the Louisiana State University, Rome University, Stanford University, Regina University, the University of

Chapter 3. Gravitational-radiation experiments

Figure 3.9. Photograph of the cryogenic gravitational-wave experiment at Louisiana State University.

Tokyo, and the University of Western Australia). Figure 3.9 shows a photograph of a 4800 kg aluminum cylinder at Louisiana State University being prepared for cooling to temperatures of 1 K or below. A photograph of a high-Q single crystal of silicon is shown in figure 3.10 ($M \sim 20$ kg, $l \sim 150$ cm, $f \sim 3000$ Hz). There seems to be little doubt that these efforts to reduce the Brownian-motion noise below the level of expected signals will be successful. However, one is less optimistic with regard to the sensor. As discussed in section 3.4.2, there are severe problems in constructing an optimum sensor; the solution of these problems represents a considerable challenge to the experimentalist and will take many years to overcome. Also, to reach the Virgo cluster one may have to build 'quantum non-demolition' detectors. This, without a doubt, would require more than a few years.

The development of free-mass- or almost-free-mass detectors is being carried out by at least five groups. Initial efforts with prototype devices on the surface of the Earth are being carried on by groups at MIT, the University of Glasgow, the Max Planck Institute at Munich, and the

Prospects for the future

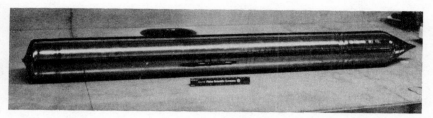

Figure 3.10. Photograph of a high-Q single crystal of silicon ($M \sim 20$ kg, $l \sim 150$ cm, $f \sim 3000$ Hz).

Hughes Research Laboratories at Malibu. These detectors presumably could be put into space later so that the large separations which are necessary in order to reach the ultimate limit of this scheme could be achieved. The proposed free-mass Doppler-tracking experiments of Estabrook and Wahlquist at the Jet Propulsion Laboratory at Pasadena also look promising, especially if one or two orders of magnitude of improvement in sensitivity can be obtained.

It appears to these authors that the prospects for the unambiguous discovery of gravitational radiation in the near future is good. This will probably occur as a result of coincidence experiments with two or more groups at a few locations around the Earth. Very important experiments could then be done. The direction of the sources could be inferred from phase differences or arrival times; and correlation of pulses with optical observations could be used to determine the speed of propagation. Measurements of the polarization of the waves could be used to test various theories of gravitation and to infer something about the nature of the sources. Perhaps soon we will be able to probe phenomena associated with gravitational collapse, the interior of supernovae, black holes, and countless other astronomical events. The information that could be obtained about the universe in which we live might be enormous. We may be on the threshold of opening a new scientific field – gravitational-wave astronomy.

4. The initial value problem and the dynamical formulation of general relativity

ARTHUR E. FISCHER AND JERROLD E. MARSDEN[†]

In this chapter we discuss some of the inter-relationships between the initial value problem, the canonical formalism, linearization stability and the space of gravitational degrees of freedom. In the last decade, these topics have experienced a resurgence of interest as more advanced mathematical methods and viewpoints have begun to show the intimate relationships among these topics. At present, the literature regarding these areas of general relativity is a rapidly expanding body of knowledge.

Our purpose here is to present the current state of affairs from our own point of view. We shall use geometric methods developed by the authors to establish various connections between the above-mentioned topics. The main tools we shall use to develop this material are nonlinear functional analysis, an adjoint formalism for Hamiltonian field theories, and infinite-dimensional symplectic geometry. As we shall see, these tools and the topics we shall consider are naturally related.

For a more complete picture of the current state of affairs, the reader is urged to consult Choquet-Bruhat (1962), Arnowitt, Deser and Misner (1962), Hawking and Ellis (1973), Misner, Thorne and Wheeler (1973), Hanson, Regge and Teitelboim (1976), Kuchař (1976, 1977), Müller zum Hagen and Seifert (1976) and Choquet-Bruhat and York (1979).

Section 4.1 develops the Hamiltonian formalism for the dynamics of general relativity, usually called the ADM (Arnowitt, Deser and Misner) formalism. This is done using invariant concepts and the adjoint formalism developed by the authors. We show how to write the Einstein dynamical system explicitly in the compact form

Evolution equations $\begin{pmatrix} \frac{\partial g}{\partial \lambda} \\ \frac{\partial \pi}{\partial \lambda} \end{pmatrix} = J \circ [D\Phi(g, \pi)]^* \cdot \begin{pmatrix} N \\ X \end{pmatrix},$

Constraint equations $\Phi(g, \pi) = 0.$

[†] Research for this paper was partially supported by the National Science Foundation, USA, and the Science Research Council, UK.

This form of the equations is useful in understanding linearization stability and the space of gravitational degrees of freedom. We sketch how this adjoint formalism can be extended to all field theories which are minimally coupled to gravity.

The adjoint formalism leads naturally to the study of the constraint manifold in section 4.2; the main result in this section tells which points are manifold (regular) points and which are bifurcation (singular) points. We also show (using the adjoint formalism) that the constraint submanifold is in involution under the dynamical equations. The equations used to establish this result are equivalent to the Dirac canonical commutation relations.

With the dynamical formalism in hand, we discuss existence, uniqueness, and stability for the Cauchy problem in sections 4.3 and 4.4. In section 4.3 we summarize the general theory of hyperbolic initial value problems that we shall need for relativity. We give an abstract approach which gives as special cases existence and uniqueness results for first-order symmetric hyperbolic systems, second-order hyperbolic systems, or combinations of these systems. The theorems we present yield the sharpest known results with regard to differentiability. When applied in section 4.5, these theorems give the sharpest results regarding the existence and uniqueness theorems for the Cauchy problem of the empty space field equations (theorems 4.23 and 4.27). Some remarks are made to show how the abstract theory is applied to fields coupled to gravity.

Although considerable progress in the initial value problem has been made, the basic open problem of relating dynamical singularities (non-existence of 'all-time' solutions to the evolution equations) to singularities in the Hawking–Penrose sense remains unsolved.

Section 4.5 combines sections 4.2 and 4.4 to give conditions under which first-order perturbation theory is and is not valid and shows that perturbation series must be readjusted to be made consistent whenever a Killing vector is present. Necessary second-order conditions are given for a perturbation to be integrable. These results are due to the joint work of V. Moncrief and the authors.

Finally, section 4.6 discusses the elimination of gauges by a general reduction procedure for Hamiltonian systems. An application of this general procedure is then used to show that the space of gravitational degrees of freedom is generically an infinite-dimensional symplectic manifold. Thus, generically, the set of empty space geometries is an infinite-dimensional gravitational phase space without singularities. Our

Chapter 4. The initial value problem

general formalism is also applicable to fields minimally coupled to gravity, and with little extra effort it can also be shown that the space of degrees of freedom for fields and gravity is also generically a symplectic manifold.

For more information regarding the topics presented here, the reader may consult Arms (1977a, b), Arms, Fischer and Marsden (1975), Choquet-Bruhat, Fischer and Marsden (1978), Fischer and Marsden (1978a, b, c), Fischer, Marsden and Moncrief (1978), Hughes, Kato and Marsden (1976), Kato (1977), Marsden and Weinstein (1974), Moncrief (1975a, b, c, 1976, 1977) and Weinstein (1977).

The authors thank J. Arms, Y. Choquet-Bruhat, K. Kuchař, V. Moncrief, R. Palais, R. Sachs, and A. Taub for their helpful advice and S. Hawking and W. Israel for their kind invitation to contribute this chapter.

4.1 Canonical formalism

We begin by recalling the four-dimensional Lagrangian formalism in classical field theory for fields coupled to gravity. Following this we treat the '3+1' or dynamical approach.

The notation is as follows: V_4 is a smooth 4-manifold with connected, oriented, paracompact and Hausdorff included in the term 'manifold'. TV_4 denotes its tangent bundle. We also let

$L(V_4)$ = all smooth Lorentz metrics with signature $(-+++)$;

$S_2(V_4)$ = all smooth 2-covariant symmetric tensor fields on V_4.

Now let E be a vector bundle over V_4 with projection $\pi\colon E \to V_4$ and let its C^∞ sections be denoted $C^\infty(E)$. Often we take $E = T^r_s(V_4)$, the bundle of tensors with r contravariant indices and s covariant indices. However, it is important not to restrict one's attention merely to tensor field theories or else important field theories, such as Yang–Mills fields, will be excluded; see for example Hermann (1975), Arms (1977b), and references therein. Strictly speaking, Yang–Mills fields require the use of an affine bundle (bundle of connections on a principal bundle over V_4), but this does not affect the formalism in any significant way.

In coordinates, we write the components of $\varphi \in C^\infty(E)$ as φ_A where A denotes a set of multi-indices.

Let $\mathscr{D}(V_4)$ denote the (orientation-preserving) diffeomorphisms of V_4. For 'natural bundles' any $F \in \mathscr{D}(V_4)$ extends functorially to a bundle diffeomorphism $F_E: E \to E$ covering F; i.e., the diagram

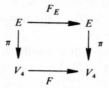

commutes and $(F \circ G)_E = F_E \circ G_E$. For $E = T^r_s(V_4)$, F_E is the usual transformation of tensors under F. Then *pull back* by F is defined on sections of E, and is given by

$$F^*: C^\infty(E) \to C^\infty(E); \quad \varphi \mapsto F_E^{-1} \circ \varphi \circ F = F^*\varphi$$

and its inverse, *push forward*, is defined by

$$F_*\varphi = F_E \circ \varphi \circ F^{-1}.$$

For the bundles associated with Yang–Mills fields, one also has transformation by an infinite-dimensional gauge group in addition to the notion of pull back and push forward by $\mathcal{D}(V_4)$.

Note that E may be a Whitney sum $E_1 \oplus E_2 \oplus \cdots E_k$ for k types of fields, so that our formalism is appropriate for interacting fields.

Let Ω denote the bundle of densities over V_4 (i.e., 4-forms) and write E^* for the *dual bundle* over V_4 whose fiber at x is

$$E^*_x \otimes \Omega_x$$

where E^*_x represents the vector space dual to E_x. Thus E^* is a bundle of vector densities over V_4. For example, if $E = T^r_s(V_4)$, $E^* = T^s_r(V_4) \otimes \Omega$ is the bundle of tensor densities of type $\binom{s}{r}$.

We have a natural L_2-pairing between $C^\infty(E)$ and $C^\infty(E^*)$ given by

$$(\varphi, \psi \otimes d\mu)_{L_2} = \int_{V_4} \psi(\varphi) \, d\mu,$$

where $\psi(\varphi) \, d\mu = \psi(\varphi) \otimes d\mu$, and we are assuming $\psi(\varphi)$ is $d\mu$-integrable. We speak of $C^\infty(E^*)$ as the *natural L_2 dual of $C^\infty(E)$*.

Let E and F be two bundles over V_4 and $A: C^\infty(E) \to C^\infty(F)$ be a linear operator. The *natural adjoint* of A is defined by

$$A^*: C^\infty(F^*) \to C^\infty(E^*), \quad (A^*(\tilde{\psi}), \varphi)_{L_2} = (\tilde{\psi}, A\varphi)_{L_2}.$$

for $\tilde{\psi} \in C^\infty(F^*)$, $\varphi \in C^\infty(E)$. We tacitly assume A^* exists.

Chapter 4. The initial value problem

If A is a differential operator, A^* will be computed in the usual way by integration by parts to give the adjoint differential operator. In general, A^* can be interpreted in the sense of unbounded operators (Kato, 1966).

For a bundle E over V_4 with dual bundle E^*, we take E to be the dual bundle of E^* so that $(E^*)^* = E$. Thus, in the example of $E = T_s^r(V_4)$, $E^* = T_r^s(V_4) \otimes \Omega$, and $(E^*)^* = T_s^r(V_4)$. Thus, with this convention, if

$$A: C^\infty(E) \to C^\infty(F^*),$$

then

$$A^*: C^\infty(F) \to C^\infty(E^*).$$

Consider now a Lagrangian density for a field theory coupled to gravity:

$$\mathscr{L}: L(V_4) \times C^\infty(E) \to C_d^\infty(V_4),$$

where $C_d^\infty(V_4) = \Omega$ is the bundle of scalar densities over V_4. Write $\mathscr{L}(g, \varphi) = \mathscr{L}_{\text{grav}}(g) + \mathscr{L}_{\text{fields}}(g, \varphi)$ where $\mathscr{L}_{\text{grav}}(g) = (1/16\pi)R(g)\,d\mu(g)$ and where $R(g)$ is the scalar curvature of g and $d\mu(g) = (-\det g_{\alpha\beta})^{1/2} dx^0 \wedge dx^1 \wedge dx^2 \wedge dx^3$ is the volume element associated with $g \in L(V_4)$; later we shall designate such a g as $^{(4)}g$.

If we demand that the action integral

$$S(g, \varphi) = \int_{\mathscr{D}} [\mathscr{L}_{\text{grav}}(g) + \mathscr{L}_{\text{fields}}(g, \varphi)] \, d\mu(g)$$

be stationary for any bounded open region $\mathscr{D} \subset V_4$ with smooth boundary and for any variation h of g and variation ψ of φ vanishing on the boundary, we get

$$0 = \int_{\mathscr{D}} [D\mathscr{L}_{\text{grav}}(g) \cdot h + D_g \mathscr{L}_{\text{fields}}(g, \varphi) \cdot h + D_\varphi \mathscr{L}_{\text{fields}}(g, \varphi) \cdot \psi] \, d\mu(g)$$

for all h, ψ vanishing on $\partial\mathscr{D}$, where D, D_g, D_φ denote the (Fréchet) derivatives, and partial derivatives with respect to g and φ, respectively. Note that the variation h is in $S_2(V_4)$, the space of symmetric 2-covariant tensor fields on V_4, and ψ is in $C^\infty(E)$.

In terms of natural adjoints, this condition becomes the Euler–Lagrange equations:

$$[D\mathscr{L}_{\text{grav}}(g)]^* \cdot 1 + [D_g \mathscr{L}_{\text{fields}}(g, \varphi)]^* \cdot 1 = 0$$

and

$$[D_\varphi \mathscr{L}_{\text{fields}}(g, \varphi)]^* \cdot 1 = 0$$

where 1 is the constant function 1 in the space of real-valued functions, the dual space to the densities $C_d^\infty(V_4)$. These equations are equivalent to the usual way of writing the Euler–Lagrange equations (if $\mathscr{L}_\text{fields}$ are assumed to depend on the k-jet of g, φ):

$$\frac{\delta \mathscr{L}_\text{grav}}{\delta g} + \frac{\delta \mathscr{L}_\text{fields}}{\delta g} = 0$$

$$\frac{\delta \mathscr{L}_\text{fields}}{\delta \varphi} = 0.$$

Now, as in Lichnerowicz (1961), we have

$$DR(g) \cdot h = \Delta \operatorname{tr} h + \delta\delta h - h \cdot \operatorname{Ric}(g)$$

where

$\Delta = $ Laplace–Beltrami operator on scalars; $\Delta f = -f_{;\alpha}{}^{;\alpha}$

tr = trace; tr $h = h^\alpha{}_\alpha$

$\delta h = -\operatorname{div} h = -h_\alpha{}^\beta{}_{;\beta}$

$\delta\delta h = $ double divergence $= h^{\alpha\beta}{}_{;\alpha;\beta}$

$\operatorname{Ric}(g) = $ Ricci tensor of $g = R_{\alpha\beta}$

and

$$D[d\mu(g)] \cdot h = \tfrac{1}{2}(\operatorname{tr} h)\, d\mu(g).$$

Thus,

$$D\mathscr{L}_\text{grav}(g) \cdot h = \frac{1}{16\pi}[\Delta \operatorname{tr} h + \delta\delta h - \operatorname{Ein}(g) \cdot h]\, d\mu(g),$$

where

$$\operatorname{Ein}(g) = \operatorname{Ric}(g) - \tfrac{1}{2} g R(g)$$

is the Einstein tensor of g (i.e., $G_{\alpha\beta} = R_{\alpha\beta} - \tfrac{1}{2} g_{\alpha\beta} R$). Since the integral of $\Delta \operatorname{tr} h + \delta\delta h$ vanishes for variations h that vanish on $\partial \mathscr{D}$, it follows that

$$[D\mathscr{L}_\text{grav}(g)]^* \cdot 1 = -\frac{1}{16\pi}[\operatorname{Ein}(g)]^\# d\mu(g)$$

where $^\#$ means indices raised by g.

We let (see Hawking and Ellis, 1973, section 3.3)

$$\mathscr{T}(g, \varphi) = 2 \frac{\delta \mathscr{L}_\text{fields}}{\delta g} = 2[D_g \mathscr{L}_\text{fields}(g, \varphi)]^* \cdot 1 \in S_d^2(V_4),$$

Chapter 4. The initial value problem

where $S_d^2(V_4) = S^2(V_4) \otimes \Omega$ denotes the space of 2-contravariant symmetric tensor densities on V_4.

Let $T(g, \varphi) = \mathcal{T}(g, \varphi)^*$, the dual tensor in $S_2(V_4)$ induced by the metric. Thus, $\mathcal{T} = T^\# d\mu(g)$ and T is the usual symmetric energy-momentum tensor associated with $\mathscr{L}_{\text{fields}}(g, \varphi)$.

The field equations now read:

$$\text{Ein}(g) = 8\pi T(g, \varphi) \quad (\text{i.e., } G_{\mu\nu} = 8\pi T_{\mu\nu})$$

and

$$\frac{\delta \mathscr{L}_{\text{fields}}}{\delta \varphi} = 0.$$

If one wishes to obtain field theories which are well posed, there are severe restrictions on the possible choices of $\mathscr{L}_{\text{fields}}$. For example, if we have a tensor theory and $\mathscr{L}_{\text{fields}}$ depends on derivatives of g, e.g. on the covariant derivative $\nabla \varphi$ of φ, then T in general will depend on second derivatives of g and on second derivatives of the fields φ. Likewise the equations for the fields will depend on second derivatives of the metric as well as second derivatives of the fields. In such circumstances, one may not have a well-defined system of hyperbolic equations (see Kuchař, 1976). Thus, one usually requires *minimal coupling*, i.e., $\mathscr{L}_{\text{fields}}$ depends only on point values of g. For the scalar, electrodynamic, and Yang–Mills fields, where $\mathscr{L}_{\text{fields}} = (1/16\pi) F \cdot F \, d\mu(g)$ and $F = dA + [A, A] = $ {curvature of a connection field A}, this presents no difficulties, as these systems are minimally coupled. For minimally coupled tensor field theories, one is able to classify the *natural* differential operators that may occur; see Palais (1959), Nijenhuis (1951) and Terng (1976).

Now we turn our attention to the Dirac-ADM dynamical formulation. We develop this material using modern symplectic geometry and an intrinsic version of the Dirac theory of constraints; see Abraham and Marsden (1978). Since gravitation plays a distinguished role, we shall discuss it first. Then we shall make some comments on the case of fields coupled to gravity.

As above, let V_4 be a four-dimensional manifold with Lorentzian metric $^{(4)}g$ which is oriented and time-oriented. We write $^{(4)}g$ to avoid confusion with Riemannian metrics g to be introduced later. Let M be a *compact* oriented three-dimensional manifold,† and let $i: M \to V_4$ be an

---

† The Hamiltonian formalism for the non-compact case is rather different. See Regge and Teitelboim (1974) and Choquet-Bruhat, Fischer and Marsden (1978). The existence and uniqueness theory discussed in sections 4.3 and 4.4 is valid in either case.

embedding of M such that the embedded manifold $i(M) = \Sigma$ is spacelike; i.e., the pull back $i^*(^{(4)}g) = g$ is a Riemannian metric on M. Let $C^\infty_{\text{space}}(M; V, ^{(4)}g)$ denote the set of all such spacelike embeddings. As in Ebin and Marsden (1970), this is a smooth manifold. Let k denote the second fundamental form of the embedding, defined at $m \in M$, for X, $Y \in T_m M$, by the usual formula

$$k_m(X, Y) = -^{(4)}g \circ i(m) \cdot ((T_m i \cdot Y), {}^{(4)}\nabla_{(T_m i \cdot X)} {}^{(4)}Z_\Sigma \circ i(m))$$

where $^{(4)}Z_\Sigma \circ i(m)$ is the forward-pointing unit timelike normal to Σ at $i(m)$. Thus $k_{ij} = -Z_{i;j}$ (where ';' denotes covariant differentiation using $^{(4)}g$; covariant differentiation using g is denoted with a vertical bar).

Let $\pi = \pi' \otimes d\mu(g)$ be a 2-contravariant tensor density, whose tensor part π' is defined by $\pi' = [(\operatorname{tr} k)g - k]^\sharp$, where $^\sharp$ indicates the contravariant form of a covariant tensor with indices raised by g; similarly, $^\flat$ denotes the covariant form of a contravariant tensor. In the Hamiltonian formulation of Arnowitt, Deser and Misner, k plays the role of a velocity variable and π is its canonical momentum. Note that $\pi^{\text{ours}} = \pi^{\text{ADM}} d^3 x$. When we discuss the space of gravitational degrees of freedom in section 4.6, it is useful to know that if $(V_4, {}^{(4)}g)$ is globally hyperbolic with a Cauchy surface diffeomorphic to M, then any spacelike embedding of M in V_4 is also a Cauchy surface (see Hawking and Ellis, 1973; and Budic, Isenberg, Lindblom and Yasskin, 1978).

Now suppose we have a curve in $C^\infty_{\text{space}}(M; V_4, {}^{(4)}g)$; i.e., a curve i of spacelike embeddings of M into $(V_4, {}^{(4)}g)$. The λ-derivative of this curve defines a one-parameter family of vector fields $^{(4)}X_{\Sigma_\lambda}$ on the embedded hypersurfaces by the equation

$$\frac{di_\lambda}{d\lambda} = {}^{(4)}X_{i_\lambda} = {}^{(4)}X_{\Sigma_\lambda} \circ i_\lambda : M \to TV_4$$

(see figure 4.1). The normal and tangential projections of $^{(4)}X_{\Sigma_\lambda}$ define a curve of functions $N_\lambda = {}^{(4)}X_\perp : M \to \mathbb{R}$ and vector fields $^{(4)}X_\| = X_\lambda : M \to TM$ on M by the equation

$$^{(4)}X_{\Sigma_\lambda} \circ i_\lambda(m) = {}^{(4)}X_\perp(\lambda, m){}^{(4)}Z_{\Sigma_\lambda} \circ i_\lambda(m) + T_m i_\lambda \cdot {}^{(4)}X_\|(\lambda, m)$$

where $^{(4)}Z_{\Sigma_\lambda}$ is the forward-pointing unit timelike normal to Σ_λ. If $N_\lambda > 0$, then the map

$$F: I \times M \to V_4; \quad (\lambda, m) \mapsto i_\lambda(m)$$

is a diffeomorphism of $I \times M$ onto a tubular neighborhood of $i_0(M) = \Sigma_0$, if the interval $I = (-\beta, \beta)$ is chosen small enough. In this case, we call

Chapter 4. The initial value problem

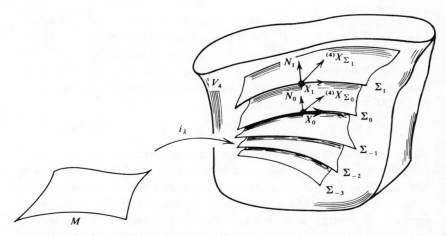

Figure 4.1. Spacelike embeddings of M into V_4 with the normal and tangential decomposition of the generator $^{(4)}X_{\Sigma_\lambda}$.

either the curve i_λ or the embedded spacelike hypersurfaces $\Sigma_\lambda = i_\lambda(M)$ a *slicing* of V_4.

The functions N_λ and the vector fields X_λ are the *lapse functions* and *shift vector fields* of Arnowitt, Deser and Misner (1962) and Wheeler (1964).

Using $F: I \times M \to V_4$ as a coordinate system for a tubular neighborhood of Σ_0 in V_4, coordinates (x^i), $i = 1, 2, 3$ on M, and $(x^\alpha) = (\lambda, x^i)$, $\alpha = 0, 1, 2, 3$ as coordinates on $I \times M$, the pulled back metric $F^{*(4)}g$ is

$$(F^{*(4)}g)_{\alpha\beta}\, dx^\alpha\, dx^\beta = -(N^2 - X_i X^i)\, d\lambda^2 + 2X_i\, dx^i\, d\lambda + g_{ij}\, dx^i\, dx^j$$

where $g_{ij} = (g_\lambda)_{ij}$, and $g_\lambda = i_\lambda^{*\ (4)}g$.

Let k_λ be the curve of second fundamental forms for the embedded hypersurfaces $\Sigma_\lambda = i_\lambda(M)$, and let π_λ be their associated canonical momenta.

The following theorem contains the basic geometrodynamical equations due to Lichnerowicz (1944), Choquet-Bruhat (1952), Dirac (1959, 1964), and Arnowitt, Deser and Misner (1962).

Theorem 4.1

Let the vacuum Einstein field equations $\mathrm{Ein}\,(^{(4)}g) = 0$ *hold on* V_4. *Then for each one-parameter family of spacelike embeddings* $\{i_\lambda\}$ *of* V_4, *the induced metrics* g_λ *and momentum* π_λ *on* Σ_λ *satisfy the following equations*:

Evolution equations
$$\begin{cases} \dfrac{\partial g}{\partial \lambda} = 2N[(\pi') - \tfrac{1}{2}g(\operatorname{tr}\pi')] + L_X g, \\ \dfrac{\partial \pi}{\partial \lambda} = -2N[\pi' \times \pi' - \tfrac{1}{2}(\operatorname{tr}\pi')\pi']\,d\mu(g) \\ \qquad + \tfrac{1}{2}Ng^{\#}[\pi' \cdot \pi' - \tfrac{1}{2}(\operatorname{tr}\pi')^2]\,d\mu(g) \\ \qquad - N[\operatorname{Ric}(g) - \tfrac{1}{2}R(g)g]^{\#}\,d\mu(g) \\ \qquad + (\operatorname{Hess} N + g\Delta N)^{\#}\,d\mu(g) + L_X\pi, \end{cases}$$

Constraint equations
$$\begin{cases} \mathcal{H}(g,\pi) = [\pi' \cdot \pi' - \tfrac{1}{2}(\operatorname{tr}\pi')^2 - R(g)]\,d\mu(g) = 0, \\ \mathcal{J}(g,\pi) = 2(\delta_g \pi) = -2\pi_i{}^j{}_{|j} = 0. \end{cases}$$

Conversely, if i_λ is a slicing of $(V_4, {}^{(4)}g)$ such that the above evolution and constraint equations hold, then ${}^{(4)}g$ satisfies the (empty space) field equations.

Our notation in the theorem is as follows: $(\pi' \times \pi')^{ij} = (\pi')^{ik}(\pi')_k{}^j$; $\pi' \cdot \pi' = (\pi')^{ij}(\pi')_{ij}$; $\operatorname{Hess} N = N_{|i|j}$; $\Delta N = -g^{ij} N_{|i|j}$; and $L_X \pi = (L_X \pi')\,d\mu(g) + \pi'(\operatorname{div} X)\,d\mu(g)$ is the Lie derivative of the tensor density $\pi = \pi'\,d\mu(g)$; note, $L_X\,d\mu(g) = (\operatorname{div} X)\,d\mu(g)$. The Ricci tensor $R_{\mu\nu}$ of ${}^{(4)}g$ is denoted $\operatorname{Ric}({}^{(4)}g)$ and that of g by $\operatorname{Ric}(g)$; $R(g)$ is the scalar curvature of g. We write $\operatorname{Ein}(g) = \operatorname{Ric}(g) - \tfrac{1}{2}R(g)g$, the Einstein tensor of g.

A sketch of the proof of theorem 4.1 is given after theorem 4.3.

The twelve first-order evolution equations for (g, π) correspond to the six second-order equations ${}^{(4)}G_{ij} = 0$, while the other four Einstein equations ${}^{(4)}G^{00} = 0$ and ${}^{(4)}G^0{}_i = 0$ appear as the constraint equations. More explicitly, in coordinates determined by a slicing i_λ, ${}^{(4)}Z_\Sigma$ has components ${}^{(4)}Z_\alpha = (-N, 0)$. If we define the 'perpendicular–perpendicular' and 'perpendicular–parallel' projections of the Einstein tensor by

$$ {}^{(4)}G_{\perp\perp} = Z_\alpha Z_\beta {}^{(4)}G^{\alpha\beta} = N^2\,{}^{(4)}G^{00} $$

and

$$ {}^{(4)}G^{\perp}{}_i = -Z_\alpha {}^{(4)}G^\alpha{}_i = N\,{}^{(4)}G^0{}_i, $$

then

$$ \mathcal{H}(g, \pi) = -2\,{}^{(4)}G_{\perp\perp}\,d\mu(g) $$

Chapter 4. The initial value problem

and

$$\mathcal{J}(g, \pi)_i = 2\,{}^{(4)}G^\perp{}_i\,d\mu(g).$$

The evolution equations of this theorem are well posed, as is shown in section 4.4.

In the formulation of theorem 4.1, the lapse and shift are regarded as freely specifiable. In the 'thin sandwich' formulation, one regards g and $\dot{g}$ as Cauchy data, expresses π as a function of $(\dot{g}, N, X)$ and solves for N and X from the constraint equations

$$\mathcal{H}(g, \pi(\dot{g}, N, X)) = 0$$

$$\mathcal{J}(g, \pi(\dot{g}, N, X)) = 0;$$

see Misner, Thorne and Wheeler (1974). Upon linearizing, it is easy to see that this is not an elliptic system, so even if it is solvable, there will be some technical problems; in particular, regularity must fail. Thus, the thin sandwich formulation is rejected by most workers. For other difficulties with the thin sandwich formulation, see Christodoulou and Francaviglia (1978).

It is important to recognize various combinations of terms in the ADM evolution equations as Lie derivatives, and we have done so in the way theorem 4.1 is written. It is also useful to write the quadratic algebraic part of $\partial\pi/\partial\lambda$ as

$$S_g(\pi, \pi) = -2\{\pi' \times \pi' - \tfrac{1}{2}(\mathrm{tr}\,\pi')\pi'\}\,d\mu(g) + \tfrac{1}{2}g^{\#}\{\pi' \cdot \pi' - \tfrac{1}{2}(\mathrm{tr}\,\pi')^2\}\,d\mu(g).$$

This is the spray of the DeWitt metric, i.e., the terms in the evolution equation quadratic in π' (see below, and Fischer and Marsden, 1972a). Thus the terms in the evolution equation for π may be interpreted as follows:

$$\frac{\partial \pi}{\partial \lambda} = NS_g(\pi, \pi) - N\,\mathrm{Ein}\,(g)^{\#}\,d\mu(g) + (\mathrm{Hess}\,N + g\,\Delta N)^{\#}\,d\mu(g) + L_X\pi.$$

geodesic spray of the DeWitt metric	forcing term of the scalar curvature potential	'tilt' term due to non-constant N	'shift' term due to nonzero shift

See DeWitt (1967), Fischer and Marsden (1972a), and Kuchař (1976) for more information regarding the geometric interpretation of this equation.

In order to understand these equations in terms of a symplectic structure on a cotangent bundle, we must introduce the following spaces. Let $\mathcal{M}$ denote the space of C^∞ Riemannian metrics on M, and $\mathcal{D} = \mathcal{D}(M)$ the diffeomorphism group of M. We let $\mathcal{M}^{s,p}$ with $s > n/p$ denote the Riemannian metrics of Sobolev class $W^{s,p}$; the diffeomorphisms and other maps and tensors of class $W^{s,p}$ being denoted similarly. For ease of notation, however, we shall restrict to the C^∞ case in this section.

Let $T\mathcal{M} \approx \mathcal{M} \times S_2$ denote the tangent bundle of $\mathcal{M}$, where, as above, S_2 is the space of C^∞ 2-covariant symmetric tensor fields on M, and S_d^2 is the space of C^∞ 2-contravariant symmetric tensor densities on M. Define $T^*\mathcal{M} \approx \mathcal{M} \times S_d^2 = \{(g, \pi) | g \in \mathcal{M}, \pi \in S_d^2\}$. We shall refer to $T^*\mathcal{M}$ as the 'L_2-cotangent bundle to $\mathcal{M}$'. For $k \in T_g\mathcal{M} \approx S_2$, $\pi \in T_g^*\mathcal{M} \approx S_d^2$, there is a natural L_2-pairing

$$(\pi, k)_{L_2} = \int_M \pi \cdot k,$$

as explained above. Thus $T^*\mathcal{M}$ as defined is a subbundle of the 'true' contangent bundle. Since $T^*\mathcal{M}$ is open in $S_2 \times S_d^2$, the tangent space of $T^*\mathcal{M}$ at $(g, \pi) \in T^*\mathcal{M}$ is $T_{(g,\pi)}(T^*\mathcal{M}) \approx S_2 \times S_d^2$.

We now show that $T^*\mathcal{M}$ carries a natural symplectic structure in which the evolution equations of the theorem are Hamiltonian. In order to include the lapse function and shift vector field into this scheme, it is necessary to develop the notion of a generalized Hamiltonian system.

On $T^*\mathcal{M}$ we define the globally constant symplectic structure

$$\Omega = \Omega_{(g,\pi)}: T_{(g,\pi)}(T^*\mathcal{M}) \times T_{(g,\pi)}(T^*\mathcal{M}) \to \mathbb{R}$$

as follows: for $(h_1, \omega_1), (h_2, \omega_2) \in T_{(g,\pi)}(T^*\mathcal{M}) = S_2 \times S_d^2$,

$$\Omega_{(g,\pi)}((h_1, \omega_1), (h_2, \omega_2)) = \int_M \omega_2 \cdot h_1 - \omega_1 \cdot h_2.$$

Let

$$J = \begin{pmatrix} 0 & I \\ -I & 0 \end{pmatrix}: S_d^2 \times S_2 \to S_2 \times S_d^2$$

be defined by

$$(\omega, h) \mapsto J\begin{pmatrix} \omega \\ h \end{pmatrix} = \begin{pmatrix} h \\ -\omega \end{pmatrix},$$

so that

$$J^{-1} = \begin{pmatrix} 0 & -I \\ I & 0 \end{pmatrix}: S_2 \times S_d^2 \to S_d^2 \times S_2, (h, \omega) \mapsto (-\omega, h).$$

Chapter 4. The initial value problem

Then

$$\Omega((h_1, \omega_1), (h_2, \omega_2)) = \int_M \left\langle J^{-1}\begin{pmatrix} h_1 \\ \omega_1 \end{pmatrix}, (h_2, \omega_2) \right\rangle.$$

We shall return to J shortly.

Let $C^\infty = C^\infty(M; \mathbb{R})$ denote the smooth real-valued functions on M;

C_d^∞ = smooth scalar densities on M;

$\mathscr{X}$ = smooth vector fields on M;

and

Λ_d^1 = smooth one-form densities on M.

Consider the functions

$$\mathscr{H}: T^*\mathcal{M} \to C_d^\infty; \quad (g, \pi) \mapsto \mathscr{H}(g, \pi) = [\pi' \cdot \pi' - \tfrac{1}{2}(\operatorname{tr} \pi')^2 - R(g)] \, d\mu(g);$$

$$\mathscr{J} = 2\delta: T^*\mathcal{M} \to \Lambda_d^1; \quad (g, \pi) \mapsto 2(\delta_g \pi) = -2\pi_i^{\ j}{}_{|j};$$

and

$$\Phi = (\mathscr{H}, \mathscr{J}): T^*\mathcal{M} \to C_d^\infty \times \Lambda_d^1; \quad (g, \pi) \mapsto (\mathscr{H}(g, \pi), \mathscr{J}(g, \pi)).$$

At this point it is necessary to compute the derivatives of $\mathscr{H}$, $\mathscr{J}$, and Φ and their natural adjoints. The results are collected in the following.

Proposition 4.2

Letting $(g, \pi) \in T^*\mathcal{M}$, $(h, \omega) \in T_{(g,\pi)}(T^*\mathcal{M}) = S_2 \times S_d^2$ *and* $(N, X) \in C^\infty \times \mathscr{X}$, *the derivatives of* $\mathscr{H}, \mathscr{J}, \Phi$

$$D\mathscr{H}(g, \pi): S_2 \times S_d^2 \to C_d^\infty,$$

$$D\mathscr{J}(g, \pi): S_2 \times S_d^2 \to \Lambda_d^1,$$

$$D\Phi(g, \pi): S_2 \times S_d^2 \to C_d \times \Lambda_d^1$$

and their natural adjoints

$$[D\mathscr{H}(g, \pi)]^*: C^\infty \to S_d^2 \times S_2,$$

$$[D\mathscr{J}(g, \pi)]^*: \mathscr{X} \to S_d^2 \times S_2,$$

$$[D\Phi(g, \pi)]^*: C^\infty \times \mathscr{X} \to S_d^2 \times S_2$$

are given as follows:

$$D\mathcal{H}(g, \pi) \cdot (h, \omega) = -S_g(\pi, \pi) \cdot h + [\text{Ein}(g) \cdot h - (\delta\delta h + \Delta \text{ tr } h)] \, d\mu(g)$$
$$+ 2[(\pi') - \tfrac{1}{2}(\text{tr } \pi')g] \cdot \omega;$$

$$[D\mathcal{H}(g, \pi)]^* \cdot N = \{[-NS_g(\pi, \pi) + (N \text{ Ein}(g) - (\text{Hess } N + g \, \Delta N))^{\#}]$$
$$\otimes d\mu(g), 2N[(\pi')^{\flat} - \tfrac{1}{2}(\text{tr } \pi')g]\};$$

$$D\mathcal{J}(g, \pi) \cdot (h, \omega) = -2[\omega_i{}^j{}_{|j} + h_{ik}\pi^{kj}{}_{|j} + \pi^{jl}(h_{ij|l} - \tfrac{1}{2}h_{jl|i})];$$

$$[D\mathcal{J}(g, \pi)]^* \cdot X = (-L_X\pi, L_Xg);$$

$$D\Phi(g, \pi) \cdot (h, \omega) = (D\mathcal{H}(g, \pi) \cdot (h, \omega), D\mathcal{J}(g, \pi) \cdot (h, \omega));$$

and

$$[D\Phi(g, \pi)]^* \cdot (N, X) = [D\mathcal{H}(g, \pi)]^* \cdot N + [D\mathcal{J}(g, \pi)]^* \cdot X$$
$$= \{[-NS_g(\pi, \pi) + (N \text{ Ein}(g) - (\text{Hess } N + g \, \Delta N))^{\#}]$$
$$\otimes d\mu(g) - L_X\pi, 2N[(\pi')^{\flat} - \tfrac{1}{2}(\text{tr } \pi')g + L_Xg]\}.$$

The proof is a slightly long, but straightforward, computation.

As is shown in Arnowitt, Deser and Misner (1962), the evolution equations of theorem 4.1 are Hamilton's equations with Hamiltonian $N\mathcal{H} + X \cdot \mathcal{J}$, i.e.,

$$\frac{\partial g}{\partial \lambda} = \frac{\delta}{\delta \pi}(N\mathcal{H} + X \cdot \mathcal{J}),$$

$$\frac{\partial \pi}{\partial \lambda} = -\frac{\delta}{\delta g}(N\mathcal{H} + X \cdot \mathcal{J}).$$

Using the symplectic structure on $T^*\mathcal{M}$ defined by

$$J = \begin{pmatrix} 0 & I \\ -I & 0 \end{pmatrix} : S_d^2 \times S_2 \to S_2 \times S_d^2, (\omega, h) \mapsto J\begin{pmatrix} \omega \\ h \end{pmatrix} = \begin{pmatrix} h \\ -\omega \end{pmatrix},$$

and the correspondence

$$\left(\frac{\delta}{\delta g}(N\mathcal{H} + X \cdot \mathcal{J}), \frac{\delta}{\delta \pi}(N\mathcal{H} + X \cdot \mathcal{J})\right) = [D\Phi(g, \pi)]^* \cdot \begin{pmatrix} N \\ X \end{pmatrix},$$

the Hamiltonian equations in theorem 4.1 can be written in a very compact way.

Chapter 4. The initial value problem

Theorem 4.3

The Einstein system, defined by the evolution equations and constraint equations of theorem 4.1 can be written as

Evolution equations $\dfrac{\partial}{\partial \lambda}\begin{pmatrix} g \\ \pi \end{pmatrix} = J \circ [D\Phi(g, \pi)]^* \cdot \begin{pmatrix} N \\ X \end{pmatrix},$

Constraint equations $\Phi(g, \pi) = (\mathcal{H}(g, \pi), \mathcal{J}(g, \pi)) = 0,$

where (N, X) are the lapse function and shift vector field associated with the slicing, and where $[D\Phi(g, \pi)]^* \cdot \begin{pmatrix} N \\ X \end{pmatrix}$ is given by proposition 4.2.

Sketch of proof of theorems 4.1 and 4.3. The Lagrangian density which generates the empty space Einstein equations is

$$\mathcal{L}_{\text{grav}}(^{(4)}g) = \frac{1}{16\pi} R(^{(4)}g) \, d\mu(^{(4)}g)$$

where $d\mu(^{(4)}g) = (-\det {}^{(4)}g)^{1/2} \, d^4x = N(\det g)^{1/2} \, d^3x \, d\lambda = N \, d\lambda \, d\mu(g)$. A computational part of the proof, which we shall not do, is to show that $\mathcal{L}_{\text{grav}}$ can be written in the following $(3+1)$-dimensional form (see equation 7-3.13 in Arnowitt, Deser and Misner, 1962, and equations 21–90 in Misner, Thorne and Wheeler, 1974)

$$16\pi \mathcal{L}_{\text{grav}}(^{(4)}g) = NR(^{(4)}g) \, d\mu(g) \, d\lambda$$

$$= \left[\pi^{ij} \frac{\partial g_{ij}}{\partial \lambda} - N\mathcal{H}(g, \pi) - X \cdot \mathcal{J}(g, \pi) \right] d\lambda$$

$$- 2[\pi^i{}_j X^j - \tfrac{1}{2} X^i \operatorname{tr} \pi + (\operatorname{grad} N)^i \, d\mu(g)]_{,i} \, d\lambda$$

$$- \left(\frac{\partial}{\partial \lambda} \operatorname{tr} \pi \right) d\lambda.$$

Here i_λ is a slicing of V_4 so that V_4 can be identified with $I \times M$. Note that our $\pi = \pi' \, d\mu(g) = \pi'(\det g)^{1/2} \, d^3x = \pi^{\text{ADM}} \, d^3x$ contains the d^3x term to complete $(\det g)^{1/2}$ to a volume element on M. Similarly, the volume element $d\mu(^{(4)}g)$ conains $d^4x = d^3x \, d\lambda$, explaining the overall multiplicative factor $d\lambda$.

Set $\beta = \beta^i = -2[\pi^i{}_j X^j - \tfrac{1}{2} X^i \operatorname{tr} \pi + (\operatorname{grad} N)^i \, d\mu(g)]$, a vector density on M; note that $\beta^i{}_{,i} = \beta^i{}_{|i} = \operatorname{div} \beta$. The action for gravity can be written as

Canonical formalism

$$16\pi S_{\text{grav}}(^{(4)}g) = 16\pi \int_{V_4} \mathscr{L}_{\text{grav}}(^{(4)}g)$$

$$= 16\pi \int_I d\lambda \int_M \left[\pi \cdot \frac{\partial g}{\partial \lambda} - N\mathscr{H}(g,\pi) - X \cdot \mathscr{J}(g,\pi) \right]$$

$$+ 16\pi \int_I d\lambda \int_M \left(\text{div } \beta - \frac{\partial}{\partial \lambda} \text{tr } \pi \right).$$

Integrating the div β term to zero on M, and dropping the total time derivative term

$$\int_{I=[a,b]} d\lambda \int_M \frac{\partial}{\partial \lambda} \text{tr } \pi = \int_M (\text{tr } \pi)_{\lambda=b} - \int_M (\text{tr } \pi)_{\lambda=a}$$

as a constant that will not enter into the variation of S_{grav}, we have

$$16\pi S_{\text{grav}}(^{(4)}g) = 16\pi \int_I d\lambda \int_M \left(\pi \cdot \frac{\partial g}{\partial \lambda} - N\mathscr{H} - X \cdot \mathscr{J} \right)$$

$$= 16\pi \int_I d\lambda \int_M \left[\pi \cdot \frac{\partial g}{\partial \lambda} - \Phi(g,\pi) \cdot \binom{N}{X} \right].$$

Varying the action with respect to $^{(4)}g$ in the direction $^{(4)}h$ which vanishes on $\{a\} \times M$ and $\{b\} \times M$ induces a variation (h, ω) of (g, π) which also vanishes on each end manifold $\{a\} \times M$ and $\{b\} \times M$. Thus, taking the extremum of the action for an arbitrary variation (h, ω) vanishing on the end manifolds $\{a\} \times M$ and $\{b\} \times M$ gives

$$0 = 16\pi \, dS_{\text{grav}}(^{(4)}g) \cdot {}^{(4)}h = 16\pi \int_I d\lambda \int_M \left(\omega \cdot \frac{\partial g}{\partial \lambda} + \pi \cdot \frac{\partial h}{\partial \lambda} \right)$$

$$- 16\pi \int_I d\lambda \int_M \left\langle D\Phi(g,\pi) \cdot (h,\omega), \binom{N}{X} \right\rangle$$

$$= 16\pi \int_I d\lambda \int_M \left(\omega \cdot \frac{\partial g}{\partial \lambda} - \frac{\partial \pi}{\partial \lambda} \cdot h \right) + 16\pi \left[\int_M (\pi \cdot h)_{\lambda=b} - \int_M (\pi \cdot h)_{\lambda=a} \right]$$

$$- 16\pi \int_I d\lambda \int_M \left\langle (h,\omega), [D\Phi(g,\pi)]^* \cdot \binom{N}{X} \right\rangle$$

$$= 16\pi \int_I d\lambda \int_M \left\langle (h,\omega), \left[\left(-\frac{\partial \pi}{\partial \lambda}, \frac{\partial g}{\partial \lambda} \right) - [D\Phi(g,\pi)]^* \cdot \binom{N}{X} \right] \right\rangle$$

where the term involving the total time derivative $\int_I d\lambda \int_M (\partial/\partial \lambda)(\pi \cdot h)$ integrates to zero in the λ-variable by virtue of the vanishing of h on the

Chapter 4. The initial value problem

end manifolds. Since the variation (h, ω) was arbitrary, we conclude that

$$\left(-\frac{\partial \pi}{\partial \lambda}, \frac{\partial g}{\partial \lambda}\right) = [D\Phi(g, \pi)]^* \cdot \binom{N}{X}$$

so that

$$J\begin{pmatrix} -\frac{\partial \pi}{\partial \lambda} \\ \frac{\partial g}{\partial \lambda} \end{pmatrix} = \begin{pmatrix} \frac{\partial g}{\partial \lambda} \\ \frac{\partial \pi}{\partial \lambda} \end{pmatrix} = J \circ [D\Phi(g, \pi)]^* \cdot \binom{N}{X}. \blacksquare$$

We now give a few additional details on the Hamiltonian structure of the adjoint equation in theorem 4.3.

Let $F: T^*\mathcal{M} \to R$ be a real-valued function on $T^*\mathcal{M}$ that comes from a density $\mathcal{F}: T^*\mathcal{M} \to C_d^\infty$; i.e.,

$$F(g, \pi) = \int_M \mathcal{F}(g, \pi).$$

Then the *Hamiltonian vector field* of F,

$$X_F: T^*\mathcal{M} \to T(T^*\mathcal{M})$$

is defined by

$$dF(g, \pi) \cdot (h, \omega) = \Omega(X_F(g, \pi), (h, \omega))$$

where Ω is the symplectic structure on $T^*\mathcal{M}$.

Proposition 4.4

The Hamiltonian vector field X_F is given by

$$X_F(g, \pi) = J \circ [D\mathcal{F}(g, \pi)]^* \cdot 1.$$

Proof. $\Omega(X_F(g, \pi), (h, \omega)) = -\int \langle X_F(g, \pi), J^{-1}(h, \omega) \rangle,$

and so

$$dF(g, \pi) \cdot (h, \omega) = \int D\mathcal{F}(g, \pi) \cdot (h, \omega)$$

$$= \int \langle [D\mathcal{F}(g, \pi)]^* \cdot 1, (h, \omega) \rangle$$

$$= -\int \langle J \circ [D\mathcal{F}(g, \pi)]^* \cdot 1, J^{-1}(h, \omega) \rangle \quad (J^* = -J)$$

$$= \Omega\{J \circ [D\mathcal{F}(g, \pi)]^* \cdot 1, (h, \omega)\}. \blacksquare$$

Canonical formalism

In particular, if $F = \int N\mathcal{H} + X \cdot \mathcal{J} = \int \langle (N, X), \Phi \rangle$, then

$$X_F(g, \pi) = J \circ [D(N\mathcal{H} + X \cdot \mathcal{J})]^* \cdot 1$$

$$= J \circ [D\Phi(g, \pi)]^* \cdot \binom{N}{X}$$

showing that the Einstein evolution equations are Hamilton's equations on the symplectic manifold $T^*\mathcal{M}$ with Hamiltonian density $N\mathcal{H} + X \cdot \mathcal{J}$.

Now suppose $F_1, F_2: T^*\mathcal{M} \to \mathbb{R}$ are real-valued functions on $T^*\mathcal{M}$ that arise from densities $\mathcal{F}_1$ and $\mathcal{F}_2$, respectively. Then their Poisson bracket,

$$\{F_1, F_2\}: T^*\mathcal{M} \to \mathbb{R},$$

is defined by

$$\{F_1, F_2\}(g, \pi) = \Omega(X_{F_1}(g, \pi), X_{F_2}(g, \pi)),$$

where X_F is the Hamiltonian vector field for F.

Proposition 4.5

The Poisson bracket $\{F_1, F_2\}$ defined above is given by

$$\{F_1, F_2\}(g, \pi) = \int \langle [D_g\mathcal{F}_1(g, \pi)]^* \cdot 1, [D_\pi\mathcal{F}_2(g, \pi)]^* \cdot 1 \rangle$$

$$- \int \langle [D_\pi\mathcal{F}_1(g, \pi)]^* \cdot 1, D_g\mathcal{F}_2(g, \pi) \cdot 1 \rangle.$$

Proof.

$$\{F_1, F_2\}(g, \pi) = \Omega(X_{F_1}(g, \pi), X_{F_2}(g, \pi))$$

$$= -\int \langle X_{F_1}(g, \pi), J^{-1} \circ X_{F_2}(g, \pi) \rangle$$

$$= -\int \langle J \circ [D\mathcal{F}_1(g, \pi)]^* \cdot 1, J^{-1} \circ J \circ [D\mathcal{F}_2(g, \pi)]^* \cdot 1 \rangle$$

$$= -\int \langle [D\mathcal{F}_1(g, \pi)]^* \cdot 1, J^* \circ [D\mathcal{F}_2(g, \pi)]^* \cdot 1 \rangle$$

$$= \int \langle [D\mathcal{F}_1(g, \pi)]^* \cdot 1, J \circ [D\mathcal{F}_2(g, \pi)]^* \cdot 1 \rangle$$

Chapter 4. The initial value problem

$$= \int \langle ([D_g \mathcal{F}_1(g, \pi)]^* \cdot 1, [D_\pi \mathcal{F}_1(g, \pi)]^* \cdot 1)$$
$$\times \begin{pmatrix} [D_\pi \mathcal{F}_2(g, \pi)]^* \cdot 1 \\ -[D_g \mathcal{F}_2(g, \pi)]^* \cdot 1 \end{pmatrix} \rangle$$
$$= \int \langle [D_g \mathcal{F}_1(g, \pi)]^* \cdot 1, [D_\pi \mathcal{F}_2(g, \pi)]^* \cdot 1 \rangle$$
$$- \int \langle [D_\pi \mathcal{F}_1(g, \pi)]^* \cdot 1, [D_g \mathcal{F}_2(g, \pi)]^* \cdot 1 \rangle.$$

This result may be written in 'physics notation' as

$$\{F_1, F_2\} = \int \left(\frac{\delta \mathcal{F}_1}{\delta g} \frac{\delta \mathcal{F}_2}{\delta \pi} - \frac{\delta \mathcal{F}_1}{\delta \pi} \frac{\delta \mathcal{F}_2}{\delta g} \right).$$

Now consider the case when $F_2 = \int N \mathcal{H} + X \cdot \mathcal{J}$. Then, from the proof above,

$$\{F, N\mathcal{H} + X \cdot \mathcal{J}\}(g, \pi) = \int \langle [D\mathcal{F}(g, \pi)]^* \cdot 1, J \circ [D(N\mathcal{H} + X \cdot \mathcal{J})]^* \cdot 1 \rangle$$
$$= \int \left\langle [D\mathcal{F}(g, \pi)]^* \cdot 1, J \circ [D\Phi(g, \pi)]^* \cdot \binom{N}{X} \right\rangle$$
$$= \int D\mathcal{F}(g, \pi) \cdot \begin{pmatrix} \frac{\partial g}{\partial \lambda} \\ \frac{\partial \pi}{\partial \lambda} \end{pmatrix}$$
$$= \int \frac{\mathrm{d}}{\mathrm{d}\lambda} \mathcal{F}(g, \pi)$$
$$= \frac{\mathrm{d}}{\mathrm{d}\lambda} F(g, \pi).$$

What this means is the following. Let $(g(\lambda), \pi(\lambda))$ be a solution of the Einstein evolution equations with lapse and shift $N(\lambda)$, $X(\lambda)$. Let $F(\lambda) = F(g(\lambda), \pi(\lambda))$. Then

$$\frac{\mathrm{d}F}{\mathrm{d}\lambda} = \{F, N\mathcal{H} + X \cdot \mathcal{J}\}.$$

Thus, as expected, a Poisson bracket with the Hamiltonian $\int N\mathcal{H} + X \cdot \mathcal{J}$ generates λ-derivatives of $F(g(\lambda), \pi(\lambda))$ where $(g(\lambda), \pi(\lambda))$ is the flow with initial data $(g(0), \pi(0))$ and lapse and shift $(N(\lambda), X(\lambda))$.

Actually, the form of the Einstein equations as they appear in theorem 4.3 can be extended to include field theories coupled to gravity.

This extended form is at the basis of a covariant formulation of Hamiltonian systems (Kuchař, 1976a, b, c; Fischer and Marsden, 1976, 1978a, b). For example, the canonical formulation of the covariant scalar wave equation $\Box \phi = m^2\phi + F'(\phi)$ on a spacetime $(V_4 = I \times M, {}^{(4)}g)$ in terms of a general lapse and shift is as follows.

Consider the Hamiltonian

$$\mathcal{H}(g;\phi,\pi) = \{\tfrac{1}{2}[(\pi'_\phi)^2 + |\nabla\phi|^2 + m^2\phi^2] + F(\phi)\}\,d\mu(g)$$

for the scalar field (the background metric is considered as implicitly given for this example). We can construct a 2-contravariant symmetric tensor density $\mathcal{T}$ obtained by varying $\mathcal{H}(g;\phi,\pi_\phi)$ with respect to g:

$$\mathcal{T} = -2[D_g\mathcal{H}(g;\phi,\pi_\phi)]^* \cdot 1,$$

and a one-form density $\mathcal{J}(\phi,\pi_\phi)$ from the relationship

$$\int \langle X, \mathcal{J}(\phi,\pi_\phi)\rangle = \int \langle \pi_\phi, L_X\phi\rangle,$$

so that $\mathcal{J}(\phi,\pi_\phi) = \pi_\phi \cdot d\phi$. This condition expresses $\mathcal{J}$ as the conserved quantity for the coordinate invariance group on M (Fischer and Marsden, 1972). If we set $\Phi = (\mathcal{H}, \mathcal{J})$, then the Hamiltonian equations of motion for ϕ in a general slicing of the spacetime with lapse N and shift X are

$$\frac{\partial}{\partial\lambda}\begin{pmatrix}\phi\\\pi_\phi\end{pmatrix} = J \circ [D\Phi(g;\phi,\pi_\phi)]^* \cdot \begin{pmatrix}N\\X\end{pmatrix},$$

exactly as for general relativity. A computation shows that this system is equivalent to the covariant scalar wave equation given above. Here $D\Phi(g;\phi,\pi_\phi)$ is the derivative of Φ with respect to the scalar field and its canonical momentum π_ϕ.

If we couple the scalar field with gravity by regarding the scalar field as a source, the equation for the gravitational momentum $\partial\pi/\partial\lambda$ in theorems 4.1 and 4.3 is altered by the addition of the term $\tfrac{1}{2}N\mathcal{T}$, and the equation for $\partial g/\partial\lambda$ is unchanged. The constraint equations become

$$\mathcal{H}_{\text{grav}}(g,\pi) + \mathcal{H}_{\text{scalar}}(g;\phi,\pi_\phi) = 0 \quad \text{and} \quad \mathcal{J}_{\text{grav}}(g,\pi) + \mathcal{J}_{\text{scalar}}(\phi,\pi_\phi) = 0.$$

More generally, if one considers the total Hamiltonian $\mathcal{H}_T = \mathcal{H}_{\text{grav}} + \mathcal{H}_{\text{fields}}$ and a total universal flux tensor $\mathcal{J}_T = \mathcal{J}_{\text{grav}} + \mathcal{J}_{\text{fields}}$, and $\Phi_T = (\mathcal{H}_T, \mathcal{J}_T)$, and if the non-gravitational fields are non-derivatively

Chapter 4. The initial value problem

(minimally) coupled to the gravitational fields, the general form of the coupled equations is

$$\frac{\partial}{\partial \lambda}\begin{pmatrix} g \\ \pi \\ \phi_A \\ \pi^A \end{pmatrix} = J \circ [D\Phi_T(g, \pi; \phi_A, \pi^A)]^* \cdot \begin{pmatrix} N \\ X \\ \psi \end{pmatrix},$$

$$\Phi(g, \pi; \phi_A, \pi^A) = (\Phi_T(g, \pi; \phi_A, \pi^A), \Phi_{\text{deg}}(g, \pi; \phi_A, \pi^A)) = 0.$$

Here, ϕ_A represents all non-gravitational dynamical fields, π^A their conjugate momenta, and $\Phi_{\text{deg}} = 0$ represents additional constraints due to degeneracies in $^{(4)}\mathscr{L}$, and ψ are the corresponding non-dynamical (degenerate) fields. These results provide a unified covariant Hamiltonian formulation of general relativity coupled to other Lagrangian field theories and in fact allow the empty space case to be extended formally to the non-derivative coupling case. The proof that the description of fields coupled to gravity can be given in the 'L_2-adjoint formulation' as above is based on the work of Kuchař (1976a, b, c; 1977), who, in his milestone series of papers, gives in detail the canonical formulation for covariant field theories, a study initiated by Dirac (see Dirac, 1964, and the references therein). We refer to Arms (1977a, b) for the realization of this formulation for Yang–Mills fields.

The formalism of this section can be extended to the case where M is non-compact. This case has many technical problems, but there is one basic difference: the fall-off rate for asymptotically flat metrics is not fast enough to allow integration by parts. This has led Regge and Teitelboim (1974) to conclude that the proper Hamiltonian actually generating the evolution equations contains an additional surface integral term corresponding to the mass. Thus, in the asymptotically flat case, the mass can be interpreted as the 'true' generator of the evolution equation after the constraints $\Phi = 0$ are imposed. These ideas are discussed in Choquet-Bruhat, Fischer and Marsden (1978).

4.2 The constraint manifold

Let $C_{\mathscr{H}} = \{(g, \pi) \in T^*\mathscr{M} | \mathscr{H}(g, \pi) = 0\}$ denote the set of solutions of the Hamiltonian constraint and let $C_\delta = \{(g, \pi) \in T^*\mathscr{M} | \mathscr{J}(g, \pi) = -2\pi_i{}^j{}_{|j} = 0\}$ denote the set of solutions of the divergence constraint. Thus $\mathscr{C} = \mathscr{C}_{\mathscr{H}} \cap \mathscr{C}_\delta \subset T^*\mathscr{M}$ is the constraint set for the vacuum Einstein system.

Two important facts about $\mathcal{C}_\mathcal{H} \cap \mathcal{C}_\delta$ are that the constraints are maintained by the evolution equations for any choice of lapse function and shift vector field, and that generically, $\mathcal{C}_\mathcal{H} \cap \mathcal{C}_\delta$ is a smooth submanifold of $T^*\mathcal{M}$.

From the spacetime point of view, maintenance of the constraints is equivalent to the contracted Bianchi identities, differential identities generated by the covariance of the four-dimensional field equations. This maintenance in time of the constraints is necessary for the consistency of the evolution and constraint equations.

The manifold nature of $\mathcal{C}_\mathcal{H} \cap \mathcal{C}_\delta$, while of intrinsic interest, is the key to understanding the linearization stability of the field equations, as we shall see.

We begin by noting that the Hamiltonian and momentum functions are covariant with respect to the infinite-dimensional gauge group $\mathcal{D}(M)$ of diffeomorphisms of M. That is, for any $\eta \in \mathcal{D}(M)$ and $(g, \pi) \in T^*\mathcal{M}$,

$$\mathcal{H}(\eta^*g, \eta^*\pi) = \eta^*\mathcal{H}(g, \pi),$$

and

$$\mathcal{J}(\eta^*g, \eta^*\pi) = \eta^*\mathcal{J}(g, \pi),$$

and hence

$$\Phi(\eta^*g, \eta^*\pi) = \eta^*\Phi(g, \pi).$$

Here η^* denotes the usual pull back of tensors.

If η_λ is a curve in $\mathcal{D}(M)$ with η_0 = identity, and we define the vector field X by

$$X = \left(\frac{d}{d\lambda}\eta_\lambda\right)_{\lambda=0},$$

then differentiating the relations above in λ and evaluating at $\lambda = 0$ gives the infinitesimal version of covariance:

$$D\mathcal{H}(g, \pi) \cdot (L_Xg, L_X\pi) = L_X[\mathcal{H}(g, \pi)],$$

and

$$D\mathcal{J}(g, \pi) \cdot (L_Xg, L_X\pi) = L_X[\mathcal{J}(g, \pi)],$$

and hence

$$D\Phi(g, \pi) \cdot (L_Xg, L_X\pi) = L_X[\Phi(g, \pi)].$$

Similar identities are generated by the gauge invariance of Yang–Mills fields.

Chapter 4. The initial value problem

The next theorem computes the rate of change of $\mathcal{H}$ and $\mathcal{J}$ along a solution of the evolution equations for a general lapse and shift. The infinitesimal covariance accounts for the Lie derivatives in the resulting formulae.

Theorem 4.6

For an arbitrary lapse $N(\lambda)$ and shift $X(\lambda)$, let $(g(\lambda), \pi(\lambda))$ be a solution of the Einstein evolution equations

$$\frac{\partial}{\partial \lambda}\begin{pmatrix} g \\ \pi \end{pmatrix} = J \circ [D\Phi(g, \pi)]^* \cdot \begin{pmatrix} N \\ X \end{pmatrix}.$$

Then $(\mathcal{H}(\lambda), \mathcal{J}(\lambda)) = \{\mathcal{H}(g(\lambda), \pi(\lambda)), \mathcal{J}(g(\lambda), \pi(\lambda))\}$ satisfies the following system of equations:

$$\frac{d\mathcal{H}}{d\lambda} = \frac{1}{N} \operatorname{div}(N^2 \mathcal{J}) + L_X \mathcal{H}$$

and

$$\frac{d\mathcal{J}}{d\lambda} = (dN)\mathcal{H} + L_X \mathcal{J}.$$

If, for some λ_0 in the domain of existence of the solution, $(g(\lambda_0), \pi(\lambda_0)) = (g_0, \pi_0) \in \mathcal{C}_{\mathcal{H}} \cap \mathcal{C}_\delta$ (that is, $\Phi(g_0, \pi_0) = 0$), then $(g(\lambda), \pi(\lambda)) \in \mathcal{C}_{\mathcal{H}} \cap \mathcal{C}_\delta$ for all λ for which the solution exists.

Remark. It follows (from uniqueness theorems in the next section) that if a solution of the evolution equations intersects $\mathcal{C}_{\mathcal{H}} \cap \mathcal{C}_\delta$, it must lie wholly within $\mathcal{C}_{\mathcal{H}} \cap \mathcal{C}_\delta$.

Proof. The infinitesimal covariance of $\mathcal{H}$ is used as follows:

$$\frac{d\mathcal{H}(g, \pi)}{d\lambda} = D\mathcal{H}(g, \pi) \cdot \left(\frac{\partial g}{\partial \lambda}, \frac{\partial \pi}{\partial \lambda}\right)$$

$$= D\mathcal{H}(g, \pi) \cdot \left\{ J \circ [D\Phi(g, \pi)]^* \cdot \begin{pmatrix} N \\ X \end{pmatrix} \right\}$$

$$= D\mathcal{H}(g, \pi) \cdot \{J \circ [(D\mathcal{H}(g, \pi))^* \cdot N + (D\mathcal{J}(g, \pi))^* \cdot X]\}$$

$$= D\mathcal{H}(g, \pi) \cdot \{J \circ [(D\mathcal{H}(g, \pi))^* \cdot N + (-L_X \pi, L_X g)]\}$$

$$= D_g \mathcal{H}(g, \pi) \cdot \{[D_\pi \mathcal{H}(g, \pi)]^* \cdot N\}$$

$$\quad - D_\pi \mathcal{H}(g, \pi) \cdot \{[D_g \mathcal{H}(g, \pi)]^* \cdot N\} + L_X \mathcal{H}(g, \pi).$$

The constraint manifold

The first two terms in the expression for $\partial \mathcal{H}/\partial \lambda$ involve a rather tedious computation. The results are

$$D_g\mathcal{H} \cdot [(D_\pi\mathcal{H})^* \cdot N] - D_\pi\mathcal{H} \cdot [(D_g\mathcal{H})^* \cdot N] = -\frac{1}{N}\delta[N^2\delta(2\pi)].$$

Thus we arrive at

$$\frac{d\mathcal{H}}{d\lambda} = -\frac{1}{N}\delta[N^2\delta(2\pi)] + L_X\mathcal{H}$$

$$= \frac{1}{N}\operatorname{div}(N^2\mathcal{J}) + L_X\mathcal{H}.$$

The evolution equation for $\mathcal{J}(g, \pi)$ follows from infinitesimal covariance of $\Phi(g, \pi)$ as follows:

Let $Y \in \mathcal{X}$ be any vector field on M (independent of λ). Then

$$\frac{d}{d\lambda}\int \langle Y, \mathcal{J}(g, \pi)\rangle = \int \left\langle Y, \frac{d\mathcal{J}(g, \pi)}{d\lambda}\right\rangle$$

$$= \int \left\langle Y, D\mathcal{J}(g, \pi) \cdot \left(\frac{\partial g}{\partial \lambda}, \frac{\partial \pi}{\partial \lambda}\right)\right\rangle$$

$$= \int \left\langle Y, D\mathcal{J}(g, \pi) \cdot \left\{J \circ [D\Phi(g, \pi)]^* \cdot \binom{N}{X}\right\}\right\rangle$$

$$= -\int \left\langle D\Phi(g, \pi) \cdot \{J \circ [D\mathcal{J}(g, \pi)]^* \cdot Y\}, \binom{N}{X}\right\rangle \quad (J^* = -J)$$

$$= -\int \langle D\Phi(g, \pi) \cdot (L_Y g, L_Y \pi), (N, X)\rangle$$

$$= -\int \langle L_Y\Phi(g, \pi), (N, X)\rangle \quad \text{(infinitesimal covariance of } \Phi\text{)}$$

$$= -\int NL_Y\mathcal{H}(g, \pi) + \langle X, L_Y\mathcal{J}(g, \pi)\rangle$$

$$= \int (L_Y N)\mathcal{H} + \int \langle L_Y X, \mathcal{J}\rangle \quad \text{(integration by parts)}$$

$$= \int Y(dN)\mathcal{H} - \int \langle L_X Y, \mathcal{J}\rangle$$

$$= \int Y(dN)\mathcal{H} + \int \langle Y, L_X\mathcal{J}\rangle.$$

Since Y is arbitrary,

$$\frac{d\mathcal{J}}{d\lambda} = (dN)\mathcal{H} + L_X\mathcal{J}. \qquad \blacksquare$$

In terms of the Poisson brackets introduced in the previous section, we can rewrite theorem 4.6 as follows.

Chapter 4. The initial value problem

Theorem 4.7

Given $N_1, N_2: M \to \mathbb{R}$, $X_1, X_2: M \to TM$, and

$$F_1 = \int (N_1 \mathcal{J} + X_1 \cdot \mathcal{J}): T^*\mathcal{M} \to \mathbb{R}$$
$$F_2 = \int (N_2 \mathcal{H} + X_2 \cdot \mathcal{J}): T^*\mathcal{M} \to \mathbb{R},$$

then

$$\{F_1, F_2\} = \int (L_{X_1} N_2 - L_{X_2} N_1) \mathcal{H}$$
$$+ \int \langle (N_1 \operatorname{grad} N_2 - N_2 \operatorname{grad} N_1), \mathcal{J} \rangle + \langle L_{X_1} X_2, \mathcal{J} \rangle,$$

and, in particular,

$$\left\{ \int N_1 \mathcal{H}, \int N_2 \mathcal{H} \right\} = \int \langle N_1 \operatorname{grad} N_2 - N_2 \operatorname{grad} N_1, \mathcal{J} \rangle$$
$$\left\{ \int N \mathcal{H}, \int X \cdot \mathcal{J} \right\} = -\int (L_X N) \mathcal{H}$$
$$\left\{ \int X_1 \cdot \mathcal{J}, \int X_2 \cdot \mathcal{J} \right\} = \int \langle L_{X_1} X_2, \mathcal{J} \rangle.$$

The verification that these relations are equivalent to theorem 4.6 is straightforward. We refer to these relationships as the *Dirac canonical commutation relations*.

The following infinitesimal version of theorem 4.1 will be important in understanding and interpreting a splitting due to Moncrief (1976), and in understanding the construction leading to the space of gravitational degrees of freedom (section 4.6).

Proposition 4.8

Let $(g, \pi) \in \mathscr{C}_\mathscr{H} \cap \mathscr{C}_\delta$. Then

$$\operatorname{range} \{ J \circ [D\Phi(g, \pi)]^* \} \subset \ker D\Phi(g, \pi).$$

Proof. Let $(h, \omega) \in \operatorname{range} \{ J \circ [D\Phi(g, \pi)]^* \}$, and $(N, X) \in C^\infty \times \mathscr{X}$ be such that $(h, \omega) = J \circ [D\Phi(g, \pi)]^* \cdot (N, X)$. Let $(N(\lambda), X(\lambda))$ be an arbitrary lapse and shift such that $(N(0), X(0)) = (N, X)$. Let $(g(\lambda), \pi(\lambda))$ be the solution to the evolution equations with lapse and shift $(N(\lambda), X(\lambda))$ and with initial data $(g, \pi) \in \mathscr{C}_\mathscr{H} \cap \mathscr{C}_\delta$. Since $\Phi(g, \pi) = 0$, by theorem 4.6,

The constraint manifold

$\Phi(g(\lambda), \pi(\lambda)) = 0$ for all λ for which the solution exists. Hence,

$$0 = \frac{d}{d\lambda}\Phi(g(\lambda), \pi(\lambda))\Big|_{\lambda=0} = D\Phi(g(\lambda), \pi(\lambda)) \cdot \left(\frac{\partial g(\lambda)}{\partial \lambda}, \frac{\partial \pi(\lambda)}{\partial \lambda}\right)\Big|_{\lambda=0}$$

$$= D\Phi(g(\lambda), \pi(\lambda))$$

$$\cdot \left\{J \circ [D\Phi(g(\lambda), \pi(\lambda))]^* \cdot \binom{N(\lambda)}{X(\lambda)}\right\}\Big|_{\lambda=0}$$

$$= D\Phi(g, \pi) \cdot \left\{J \circ [D\Phi(g, \pi)]^* \cdot \binom{N}{X}\right\}$$

$$= D\Phi(g, \pi) \cdot (h, \omega).$$

Hence, $(h, \omega) \in \ker D\Phi(g, \pi)$. ∎

We now examine the manifold structure of the constraint set $\mathscr{C}_{\mathscr{H}} \cap \mathscr{C}_{\delta}$. We introduce the following conditions on $(g, \pi) \in T^*\mathscr{M}$:

$C_{\mathscr{H}}$: If $\pi = 0$, then g is not flat;

C_{δ}: If for $X \in \mathscr{X}(M)$, $L_X g = 0$ and $L_X \pi = 0$, then $X = 0$;

C_{tr}: tr π' is a constant on M.

We consider the constraints one at a time; first, the Hamiltonian constraint.

Proposition 4.9

Let $(g, \pi) \in \mathscr{C}_{\mathscr{H}}$ satisfy condition $C_{\mathscr{H}}$. Then $\mathscr{C}_{\mathscr{H}}$ is a C^∞ submanifold of $T^*\mathscr{M}$ in a neighborhood of (g, π) with tangent space

$$T_{(g,\pi)}\mathscr{C}_{\mathscr{H}} = \ker D\mathscr{H}(g, \pi).$$

The proof relies on some facts about elliptic operators and Sobolev spaces. We briefly recall the relevant facts (see Palais, 1965; Berger and Ebin, 1969, for proofs).

Let Ω be an open bounded region of $\mathbb{R}^n$ with smooth boundary. For any C^∞ function f from $\mathbb{R}^n$ to $\mathbb{R}^m$, we define the $W^{s,p}(\Omega, \mathbb{R}^m)$ norm of f to be

$$\|f\|_{W^{s,p}} = \sum_{0 \leq \alpha \leq s} \|D^\alpha f\|_{L_p(\Omega)}$$

where D^α is the total derivative of f of order α and $\|\ \|_{L_p(\Omega)}$ denotes the

Chapter 4. The initial value problem

usual L_p norm on Ω: $\|g\|_{L_p(\Omega)} = (\int_\Omega |g(x)|^p \, dx)^{1/p}$. By definition, $W^{s,p}(\Omega, \mathbb{R}^m)$ is the completion of $C^\infty(\Omega, \mathbb{R}^m) = \{$restrictions of C^∞ functions on $\mathbb{R}^n$ to $\Omega\}$ with respect to this norm.

We shall shorten $W^{s,p}(\Omega, \mathbb{R}^m)$ and similar expressions to $W^{s,p}$ when there is little chance of confusion.

For a compact manifold M with no boundary and a vector bundle E over M, $W^{s,p}(E)$ shall denote the space of all sections of E that are of class $W^{s,p}$ in some (and hence every) covering of M by charts. For real-valued functions we shall just write $W^{s,p}$, but for other tensor bundles we shall make up special notations for $W^{s,p}(E)$, such as $\mathcal{M}^{s,p}$ for the $W^{s,p}$ space of Riemannian metrics.

In case $p = 2$ the spaces $W^{s,p}$ are denoted H^s. In this case, and only in this case, do we get Hilbert spaces.

Now suppose we have two vector bundles E and F, over the same manifold M, and a linear differential operator D of order k,

$$\mathrm{D}: C^\infty(E) \to C^\infty(F).$$

A linear differential operator of order k is a map such that for given charts on E and F (and hence for all charts), the operator takes the form $\mathrm{D} = \sum_{|\alpha| < k} a_\alpha(x) \mathrm{D}^\alpha$, where $\mathrm{D}^\alpha = \partial^{|\alpha|} / \partial x_1^{\alpha_1}, \ldots, \partial x_n^{\alpha_n}$ is a partial derivative in a chart U for M, $\alpha = (\alpha_1, \ldots, \alpha_n)$, $|\alpha| = \sum_{i=1}^n \alpha_i$, and $a_\alpha(x)$ is a linear function from the model space for the fiber E_x to the model space for the fiber F_x over $x \in U$. We can regard D as a map between Sobolev spaces:

$$\mathrm{D}: W^{s+k,p} \to W^{s,p}.$$

D has an L_2-adjoint D* defined as usual by the equation

$$(\mathrm{D}f, g)_{L_2} = (f, \mathrm{D}^* g)_{L_2}; \text{ that is, } \int_M \langle \mathrm{D}f, g \rangle \, d\mu = \int_M \langle f, \mathrm{D}^* g \rangle \, d\mu,$$

where $d\mu$ is some preferred volume element such as that associated with a metric: $d\mu(g) = [\det(g_{ij})]^{1/2} \, dx^1 \wedge \cdots \wedge dx^n$, and $\langle , \rangle$ is an inner product on the fibers. This structure is not needed if one uses natural adjoints.

A differential operator D is *elliptic* if it has injective (principal) symbol. For each x in M and for each $\xi \in T_x^* M =$ the fiber of the cotangent bundle, the *symbol* $\sigma_\xi(\mathrm{D})$ is a linear map from the fiber E_x to the fiber F_x. In the expression of D in charts, $\sigma_\xi(\mathrm{D})$ is obtained by substituting the components of $\xi \in T_x^* M$ for the corresponding partial derivatives in the terms involving the highest-order derivatives. Thus, for each coordinate on F_x, $\sigma_\xi(x)$ is a homogeneous kth degree polynomial in the components of ξ. For example, the symbol of the ordinary Laplacian $\nabla^2 = \sum_{i=1}^{2} \partial^2/\partial x_i^2$ is $\sigma_\xi(\nabla^2) = \|\xi\|^2$.

For elliptic operators we have the following basic splitting theorem.

Fredholm alternative: Theorem 4.10

If either D *or* D* *is elliptic, then* $W^{s,p}(F) =$ range D$\oplus$ker D*, *where the sum is an* L_2 *orthogonal direct sum.*

Proof of proposition 4.9. Consider the map $\mathcal{H}: T^*\mathcal{M} \to C_d^\infty$; $(g, \pi) \mapsto \mathcal{H}(g, \pi)$. We shall show that under condition $C_\mathcal{H}$,

$$D\mathcal{H}(g, \pi): T_{(g,\pi)}(T^*\mathcal{M}) = S_2 \times S_d^2 \to T_{\mathcal{H}(g,\pi)}C_d^\infty = C_d^\infty$$

is surjective with splitting kernel so that $\mathcal{H}$ is a submersion at (g, π). Using Sobolev spaces and the implicit function theorem, and then passing to the C^∞ case via a regularity argument, it follows that $\mathscr{C}_\mathcal{H} = \mathcal{H}^{-1}(0)$ is a smooth submanifold in a neighborhood of (g, π).

From theorem 4.10 it follows that $D\mathcal{H}(g, \pi)$ is surjective provided that its L_2-adjoint

$$[D\mathcal{H}(g, \pi)]^*: C^\infty \to S_d^2 \times S_2$$

$$[DH(g, \pi)]^* \cdot N = \{-NS_g(\pi, \pi) + [N \text{ Ein } (g) - \text{Hess } N - g \, \Delta N]^\# \, d\mu(g),$$
$$2N[(\pi')^\flat - \tfrac{1}{2}(\text{tr } \pi')g]\}$$

is injective and has injective symbol.

The symbol of $[D\mathcal{H}(g, \pi)]^*$ is

$$\sigma_\xi[D\mathcal{H}(g, \pi)]^* = [(-\xi \otimes \xi + g\|\xi\|^2)^\# \, d\mu(g), 0]:$$
$$\mathbb{R} \to ((T_x^*M \otimes T_x^*M)_{\text{sym}} \, d\mu(g), (T_xM \otimes T_xM)_{\text{sym}})$$

for $\xi \in T_x^*M$. For $s \in \mathbb{R}$, $\xi \neq 0$, $(-\xi \otimes \xi + g\|\xi\|^2)s = 0$ implies, by taking the trace, $2\|\xi\|^2 s = 0$ so $s = 0$, so that the symbol is injective.

Any $N \in \ker [D\mathcal{H}(g, \pi)]^*$ satisfies
 (i) $-NS_g(\pi, \pi) + [N \text{ Ein } (g) - \text{Hess } N - g \, \Delta N]^\# \, d\mu(g) = 0$
 (ii) $2N[(\pi') - \tfrac{1}{2}(\text{tr } \pi')g] = 0$.

Taking the trace of (ii) gives $N(\text{tr } \pi') = 0$ and so from (ii) again $N\pi = 0$. Thus, from (i),
 (iii) $N \text{ Ein } (g) - \text{Hess } N - g \, \Delta N = 0$.

Chapter 4. The initial value problem

From the trace of (iii)
$$2 \Delta N + \tfrac{1}{2} R(g) N = 0.$$
However, from $\mathcal{H}(g, \pi) = 0$ and $N\pi = 0$, it follows that $NR(g) = 0$. Hence
$$\Delta N = 0$$
and so $N =$ constant.

If $\pi \neq 0$, then $N\pi = 0$ implies $N = 0$, since N is constant. Thus $[D\mathcal{H}(g, \pi)]^*$ is injective and hence $D\mathcal{H}(g, \pi)$ is surjective.

If $\pi = 0$, then from (iii), $N \text{Ein}(g) = 0$ since N is constant and so $N \text{Ric}(g) = 0$. Thus, if $N \neq 0$, then $\text{Ric}(g) = 0$ and hence g is flat, since $\dim M = 3$. But a flat g and $\pi = 0$ is ruled out by condition $C_\mathcal{H}$. Hence $N = 0$, and again $D\mathcal{H}(g, \pi)$ is surjective. ∎

Proposition 4.11

If $(g, \pi) \in \mathcal{C}_\delta = \{(g, \pi) | \mathcal{J}(g, \pi) = 0\} \subset T^*\mathcal{M}$ satisfies condition C_δ, then $\mathcal{C}_\delta$ is a smooth submanifold of $T^*\mathcal{M}$ in a neighborhood of (g, π) with tangent space
$$T_{(g,\pi)} \mathcal{C}_\delta = \ker [D\mathcal{J}(g, \pi)].$$

Proof. The adjoint of the derivative of $\mathcal{J}(g, \pi)$ is given by
$$[D\mathcal{J}(g, \pi)]^* \cdot X = (-L_X \pi, L_X g).$$
The symbol is injective (from its injectivity in the second component alone). The kernel of $[D\mathcal{J}(g, \pi)]^*$ is $\{X | L_X \pi = 0,\ L_X g = 0\}$ so that injectivity of $[D\mathcal{J}(g, \pi)]^*$ is exactly condition C_δ. The result then follows by the implicit function theorem as in proposition 4.9. ∎

To show that the intersection $\mathcal{C} = \mathcal{C}_\mathcal{H} \cap \mathcal{C}_\delta$ is a submanifold of $T^*\mathcal{M}$, we need additional restrictions because there may be points at which the intersection is not transversal. At this point it is necessary to assume that (g, π) satisfies the condition $\text{tr } \pi' =$ constant.

Theorem 4.12

Let $(g, \pi) \in \mathcal{C}_\mathcal{H} \cap \mathcal{C}_\delta$ satisfy the conditions $C_\mathcal{H}$, C_δ, and C_{tr}. Then the constraint set $\mathcal{C} = \mathcal{C}_\mathcal{H} \cap \mathcal{C}_\delta$ is a C^∞ submanifold of $T^*\mathcal{M}$ in a neighborhood of (g, π) with tangent space
$$T_{(g,\pi)}\mathcal{C} = \ker D\Phi(g, \pi)$$
where $\Phi = (\mathcal{H}, \mathcal{J})$.

The constraint manifold

Proof. We want to show $D\Phi(g, \pi) = (D\mathcal{H}(g, \pi), D\mathcal{J}(g, \pi))$ is surjective for $(g, \pi) \in \mathcal{C}$ and satisfying the given conditions. The adjoint is

$$[D\Phi(g, \pi)]^*: C^\infty \times \mathcal{X} \to S_d^2 \times S_2,$$

$$(N, X) \mapsto [D\Phi(g, \pi)]^* \cdot (N, X) = [D\mathcal{H}(g, \pi)]^* \cdot N + [D\mathcal{J}(g, \pi)]^* \cdot X.$$

For $\xi \in T_x^* M$, $\xi \neq 0$, the symbol of this map, $\sigma_\xi [D\Phi(g, \pi)]^*$, $\xi \in T_x^* M$, may be shown to be injective, as above (see, however, remarks on various types of ellipticity in Fischer and Marsden, 1975b). Thus it remains to show that $[D\Phi(g, \pi)]^*$ is injective. Let $(N, X) \in$ ker $[D\Phi(g, \pi)]^*$. Then, from the formula for $[D\Phi(g, \pi)]^*$, we have

(i) $-NS_g(\pi, \pi) + [N \text{ Ein } (g) - (\text{Hess } N + g \Delta N)]^\# d\mu(g) - L_X \pi = 0$

and

(ii) $2N[(\pi') - \frac{1}{2}(\text{tr } \pi')g] + L_X g = 0.$

Taking the trace of (i) and (ii), we get:

(iii) $-\frac{N}{2}\mathcal{H}(g, \pi) + 2(\Delta N) d\mu(g) + \text{tr } L_X \pi = 0$

and

(iv) $-N \text{ tr } \pi' + 2 \text{ div } X = 0.$

Now $\text{tr } L_X \pi = X \cdot d \text{ tr } \pi - \pi \cdot L_X g + (\text{div } X)(\text{tr } \pi)$, since $L_X \pi = (L_X \pi') \otimes d\mu(g) + \pi' \otimes (\text{div } X) d\mu(g)$ (in coordinates, $(L_X \pi)^{ij} = X^k \pi^{ij}{}_{|k} - \pi^{ik} X^j{}_{|k} - \pi^{jk} X^i{}_{|k} + X^k{}_{|k} \pi^{ij}$).

Since $\mathcal{H}(g, \pi) = 0$, (iii) reduces to

(v) $2(\Delta N) d\mu(g) + X \cdot d \text{ tr } \pi - \pi \cdot L_X g + (\text{div } X)(\text{tr } \pi) = 0.$

Using (ii) and (iv) to eliminate $L_X g$ and div X, respectively, in (v) gives

(vi) $2(\Delta N) + X \cdot d \text{ tr } \pi' - \pi' \cdot L_X g + (\text{div } X)(\text{tr } \pi')$

$= 2 \Delta N + 2N\pi' \cdot [(\pi')^\flat - \frac{1}{2}(\text{tr } \pi')g] + \frac{N}{2}(\text{tr } \pi')(\text{tr } \pi') + X \cdot d \text{ tr } \pi'$

$= 2 \Delta N + 2N\pi' \cdot \pi' + \frac{N}{2}(\text{tr } \pi')^2 + X \cdot d \text{ tr } \pi'$

$= 2 \Delta N + 2N[\pi' \cdot \pi' - \frac{1}{4}(\text{tr } \pi')^2] + X \cdot d \text{ tr } \pi' = 0.$

Now note that the coefficient of N, namely $P(\pi', \pi') = \pi' \cdot \pi' - \frac{1}{4}(\text{tr } \pi')^2 = [\pi' - \frac{1}{2}(\text{tr } \pi')g] \cdot [\pi' - \frac{1}{2}(\text{tr } \pi')g]$, is positive-definite.

Chapter 4. The initial value problem

Thus, if tr π' is a constant, (vi) becomes

$$2 \Delta N + 2P(\pi', \pi')N = 0$$

which implies $N = 0$ unless $\pi' = 0$, in which case $N = $ constant. In this case, from (i), Ein $(g) = 0$ and so Ric $(g) = 0$, i.e., g is flat since dim $M = 3$. However, the case $(g_F, 0)$, where g_F is flat, is excluded by condition $C_{\mathcal{H}}$. Thus, $\pi' \neq 0$ and $N = 0$. Then, by (i) and (ii), $L_X g = 0$ and $L_X \pi = 0$, which, by condition $C\delta$, implies $X = 0$. Thus $(N, X) = (0, 0)$ and so $[D\Phi(g, \pi)]^*$ is injective, under conditions $C_{\mathcal{H}}$, C_δ, and C_{tr}. The result of the theorem then follows by the implicit function theorem. ∎

Remark. That one must impose the condition tr π' is a constant to show that the intersection $\mathscr{C}_{\mathcal{H}} \cap \mathscr{C}_\delta$ is a manifold is an annoying feature of the analysis. One might suspect that under conditions $C_{\mathcal{H}}$ and C_δ alone, the system (i) and (ii) is injective. The difficulty is that in the system, say (vi) and (ii) for (N, X), the $X \cdot d$ tr π' coupling term seems to be sufficient to prevent one from showing uniqueness for this system. The results of Moncrief, discussed in section 4.5, will shed light on this point.

In Choquet-Bruhat, Fischer and Marsden (1978), and Marsden and Tipler (1979) the existence of hypersurfaces with tr π' equal to a constant is discussed. Thus these preferred hypersurfaces will be the place to check conditions $C_{\mathcal{H}}$ and C_δ.

4.3 The abstract Cauchy problem and hyperbolic equations

This section summarizes the general theory for hyperbolic initial-value problems that we shall need for relativity. The complete proofs are technical and lengthy, so only the ideas will be given. The papers of Kato (1975a, b, 1977) and Hughes, Kato and Marsden (1977) can be consulted for details. The present abstract approach is preferred since it gives as special cases both first-order symmetric and second-order hyperbolic, or combinations of these systems. Moreover, it yields the sharpest known results with regard to differentiability.

We shall begin with the linear case, then treat the nonlinear. We give a result on differentiability of the time t map for later use and then explain how the results apply to hyperbolic systems.

It is necessary to assume the reader is familiar with linear semigroup theory; see, for example, Hille and Phillips (1967), Yosida (1974), or Marsden and Hughes (1978).

The abstract Cauchy problem and hyperbolic equations

If X is a Banach space, $G(X, M, \beta)$ denotes the set of generators A of C_0 semigroups $e^{tA} = U(t)$ on X satisfying

$$\|U(t)\| \leq M e^{t\beta}, \quad t \geq 0;$$

i.e., by the Hille–Yosida theorem, $\lambda - A$ is one-to-one and onto X and

$$\|(\lambda - A)^{-n}\| \leq \frac{M}{(\lambda - \beta)^n}, \quad n = 1, 2, \ldots, \lambda > \beta.$$

If $M = 1$, we say A is *quasi-accretive* or $U(t)$ is *quasi-contractive*. This is the class of linear semigroups of importance to us. We recall that for $\varphi \in D(A)$, the domain of A, $U(t)\varphi = \varphi(t)$ lies in $D(A)$ as well and satisfies the evolution equation

$$\frac{d}{dt}\varphi(t) = A\varphi(t), \tag{4.1}$$

where $\varphi(\cdot)$ is regarded as a map of $[0, \infty)$ to X for purposes of computing the time derivative.

Let X, Y be Banach spaces, $Y \subset X$ with the inclusion continuous and dense. Let $U(t, s)$ be a family of bounded operators on X defined for $0 \leq s \leq t \leq T$; here $[0, T]$ is a conveniently chosen time interval; T could be arbitrarily large. Let $A(t)$ be a family of linear generators on X, $Y \subset D(A(t))$, $0 \leq t \leq T$. We call $U(t, s)$ a family of (strong) *evolution operators* for A if

(i) $U(s, s) = 1$ and $(t, s) \mapsto U(t, s)$ is strongly continuous in X;
(ii) $U(t, s)U(s, r) = U(t, r)$, $0 \leq r \leq s \leq t \leq T$;
(iii) $U(t, s)$ is a bounded operator of Y to Y and is strongly continuous in (t, s);
(iv) $(\partial/\partial t)U(t, s)\varphi = A(t)U(t, s)\varphi$, $\varphi \in Y$ (forward differential equation) and each side is strongly continuous in (t, s) with values in $B(Y, X)$ (the bounded operators from Y to X) and $\partial/\partial t$ is taken in the X-norm.

If we differentiate (ii) with respect to s at $s = r$, and use (iv), we formally get the backwards differential equation:

$$\frac{\partial}{\partial s} U(t, s)\varphi = -U(t, s)A(s)\varphi$$

for $\varphi \in Y$. If we write (iv) as an integral equation in time, this is easy to prove; write

$$U(t, s)\varphi = \varphi + \int_s^t A(\tau) U(\tau, s)\varphi \, d\tau$$

Chapter 4. The initial value problem

and use the identity

$$\frac{1}{h}[U(t, s+h)\varphi - U(t, s)\varphi]$$

$$= U(t, s+h)\left[\frac{\varphi - U(s+h)\varphi}{h}\right] + \frac{1}{h}[U(s+h, s)\varphi - \varphi]$$

$$- \frac{1}{h}\int_s^{s+h} A(\tau)U(\tau, s+h)\varphi \, d\tau.$$

A family $A(t) \in G(X, M, \beta)$ (for M, β fixed) is called *stable* if for any $s_j \geq 0$ and $0 \leq t_1 \leq \cdots \leq t_k \leq T$,

$$\exp[s_k A(t_k)] \exp[s_{k-1} A(t_{k-1})] \ldots \exp[s_1 A(t_1)]$$
$$\leq M \exp[\beta(s_1 + \ldots + s_k)]$$

or, equivalently,

$$\|(\lambda - A(t_k))^{-1} \cdots (\lambda - A(t_1))^{-1}\| \leq \frac{M}{(\lambda - \beta)^k}, \quad \lambda > \beta.$$

If $A(t) \in G(X, 1, \beta)$, then $A(t)$ is clearly stable. If we let X_t denote X with a new norm $\| \|_t$ depending on t in an exponential fashion:

$$\|\varphi\|_t \leq \|\varphi\|_s \, e^{c|t-s|}, \quad s, t \in [0, T]$$

and if $A(t) \in G(X_t, 1, \beta)$, then $A(t)$ is stable in X_{t_0} with $M = e^{2cT\beta}$; see Kato (1970, Proposition 3.4). The same reference, Proposition 3.5, shows that a bounded perturbation of a stable family is stable.

In the following theorem, $L_*^\infty([0, T]; B(Y, X))$ denotes the (equivalence class of) *strongly* measurable essentially bounded functions from $[0, T]$ to $B(Y, X)$ and $\text{Lip}_*([0, T]; B(Y, X))$ denotes the strong indefinite integrals of functions in $L_*^\infty([0, T]; B(Y, X))$.

Theorem 4.13 (Kato, 1973)

Assume
  (i) *$A(t)$ is a stable family of generators in X, $0 \leq t \leq T$;*
  (ii) *$Y \subset X$, with continuous dense inclusion and $D(A) \supset Y$;*
  (iii) *there is a family $S(t): Y \to X$ of isomorphisms (onto) such that*

$$S(t)A(t)S(t)^{-1} = A(t) + B(t),$$

where $B(t) \in B(X)$, a bounded operator on X, and
 (a) $t \mapsto S(t)$ lies in $\text{Lip}_*([0, T]; B(Y, X))$
 (b) $t \mapsto B(t)$ lies in $L^\infty_*([0, T]; B(X))$;
(iv) $t \mapsto A(t) \in B(Y, X)$ is norm continuous.
Then there is a unique family of strong evolution operators for A.

See Kato (1973) for the proof.

The case where the domain of $A(t)$ is constant in time is much simpler. Here we assume $D(A(t)) = Y$ and that $A \in \text{Lip}_*([0, T]; B(Y, X))$. Then (iii) will hold with $B = 0$ and

$$S(t) = \lambda - A(t), \quad \lambda > \beta.$$

However, for the hyperbolic problems we wish to consider, the domains need not be constant. The constant domain case was the subject of the original work of Kato (1966); see also Yosida (1974).

The inhomogeneous problem

$$\frac{\partial u}{\partial t} = A(t)u + f(t), \quad u(0) = \varphi$$

can be treated by a clever trick of Kato (1977). Namely, we suspend the equation on $X \times \mathbb{R}$ and consider the equivalent homogeneous problem

$$\frac{\partial}{\partial t}\begin{pmatrix} u \\ k \end{pmatrix} = \hat{A}(t)\begin{pmatrix} u \\ k \end{pmatrix}, \quad u(0) = \varphi, k(0) = 1$$

where

$$\hat{A}(t) = \begin{pmatrix} A(t) & f(t) \\ 0 & 0 \end{pmatrix}.$$

Then the theorem above may be applied to $\hat{A}$.

In many nonlinear problems it is often convenient to consider associated (time-dependent) linear Cauchy problems and for these, the theorem above is applicable.

To illustrate how the theorem applies we consider the two cases that mainly concern us, namely first-order symmetric hyperbolic and second-order hyperbolic systems. We shall treat these on $\mathbb{R}^m$, but due to the hyperbolicity of the equations, the results can be localized and therefore applied to compact manifolds as well; see Hawking and Ellis (1973).

We first consider first-order symmetric hyperbolic systems of Friedrichs (1954); see also Fischer and Marsden (1972b) and Kato

Chapter 4. The initial value problem

(1975*a*, *b*, 1977). The form is

$$a_0(t,x)\frac{\partial u}{\partial t} = \sum_{j=1}^m a_j(t,x)\frac{\partial u}{\partial x^j} + a(t,x)u, \qquad (4.2)$$

where $u(t,x) \in \mathbb{R}^N$, a_j, a are real. We assume
  (i) there are constant matrices a_j^∞, a^∞ such that

$$a_j - a_j^\infty, a - a^\infty \in C([0,T], H^0(\mathbb{R}^m)) \cap L^\infty([0,T], H^s(\mathbb{R}^m)),$$
$$j = 0, 1, \ldots, m$$

$$a_0 - a_0^\infty \in \mathrm{Lip}\,([0,T], H^{s-1}(\mathbb{R}^m)).$$

Here $H^s(\mathbb{R}^m)$ is the usual Sobolev space on $\mathbb{R}^m$ (with range unspecified) and $s > (m/2)+1$.
  (ii) a_j are symmetric matrices;
  (iii) $a_0(t,x) \geq cI$ for some $c > 0$.

Theorem 4.14

Under these conditions, the hypotheses of theorem 4.13 *are satisfied with*

$$X = L^2(\mathbb{R}^m) = H^0(\mathbb{R}^m)$$
$$Y = H^{s'}(\mathbb{R}^m), \quad 1 \leq s' \leq s$$
$$S(t) = (1-\Delta)^{s'/2}$$
$$A(t) = a_0(t,\cdot)^{-1}\left[\sum_{j=1}^m a_j(t,\cdot)\frac{\partial}{\partial x^j} + a(t,\cdot)\right]$$

(the closure of this operator on C_0^∞), i.e., (4.2) generates a strong evolution system in L^2 which maps $H^{s'}$ to $H^{s'}$ (regularity).

Warning. The domain of $A(t)$ need not be $H^1(\mathbb{R}^m)$; e.g., the a_j may vanish.

The idea of the proof is as follows. If we put on X the energy norm

$$\|\varphi\|_t^2 = \int_{\mathbb{R}^m} [a_0(t,x)\varphi(x)] \cdot \varphi(x)\,dx,$$

we find that

$$A(t) \in G(X_t, 1, \beta)$$

with

$$\beta = \sup_{t,x} \left| a(t,x) - \frac{1}{2} \sum_{j=1}^{m} \frac{\partial a_j}{\partial x^j}(t,x) \right|,$$

which is finite by the Sobolev inequalities. The key idea here is the estimate

$$\|[\lambda - A(t)]^{-1}\|_t \geq \frac{1}{(\lambda - \beta)},$$

which, by the Schwarz inequality, is implied by

$$\langle [\lambda - A(t)]\varphi, \varphi \rangle_t \geq (\lambda - \beta)\|\varphi\|_t^2.$$

The latter is readily proved using integration by parts and the symmetry of a_j. The stability of $A(t)$ results from the fact that the norms $\|\;\|_t$ depend exponentially on t. The hardest part is to prove that

$$B(t) = [S, A(t)]S^{-1}$$

is a bounded operator in X, where $[,]$ is the commutator. One writes the commutator out explicitly; the key estimate boils down to an estimate on the commutator

$$\left[S, \frac{\partial a_j}{\partial x^i} \right].$$

The required estimates on this commutator use a lengthy but relatively straightforward series of Sobolev-type estimates. Details may be found in Kato (1975b) for $s' = 1$, the general case being similar.

Remark. Results of this type for (4.2) already appear in early work of Friedrichs (1954) and Courant and Hilbert (1962). However, sharp differentiability hypotheses, which are crucial for nonlinear problems, were never clearly spelled out. An intermediate attempt was given in Fischer and Marsden (1972b) and the formulation was then sharpened and clarified by Kato (1975a). The present unified scheme, suggested by Hughes, Kato, and Marsden (1977) is due to Kato.

Next, we consider second-order hyperbolic systems. The form is

$$a_{00}(t,x)\frac{\partial^2 u}{\partial t^2} = \sum_{i,j=1}^{m} a_{ij}(t,x)\frac{\partial^2 u}{\partial x^i \partial x^j} + 2\sum_{i=1}^{m} a_{0i}(t,x)\frac{\partial^2 u}{\partial t \partial x^i} + a_0(t,x)\frac{\partial u}{\partial t}$$

$$+ \sum_{i=1}^{m} a_i(t,x)\frac{\partial u}{\partial x^i} + a(t,x)u \qquad (4.3)$$

Chapter 4. The initial value problem

where, again, $u(t, x) \in \mathbb{R}^N$, $a_{\alpha\beta}$, a_α, a are $N \times N$ matrix functions and we assume: $s > \frac{1}{2}m + 1$ and

(i) there are constant matrices $a_{\alpha\beta}^\infty$, a_α^∞, a^∞ such that
$$a_{\alpha\beta} - a_{\alpha\beta}^\infty, a - a^\infty \in \text{Lip}\,([0, T]; H^{s-1}(\mathbb{R}^m)) \subset L^\infty([0, T]; H^s(\mathbb{R}^m));$$

(ii) $a_{\alpha\beta}$ is symmetric;

(iii) $a_{00}(t, x) \geq cI$ for some $c > 0$;

(iv) strong ellipticity; there is an $\varepsilon > 0$ such that
$$\sum_{i,j=1}^M a_{ij}(t, x)\xi_i\xi_j \geq \varepsilon \left(\sum_{j=1}^m \xi_j^2 \right)$$

(a matrix inequality) for all $\xi = (\xi_1, \ldots, \xi_m) \in \mathbb{R}^m$.

Theorem 4.15

Under these conditions, the hypotheses of theorem 4.13 *are satisfied with*
$$X = H^1(\mathbb{R}^m) \times H^0(\mathbb{R}^m),$$
$$Y = H^{s'+1}(\mathbb{R}^m) \times H^{s'}(\mathbb{R}^m), \quad 1 \leq s' \leq s,$$
$$S = (1-\Delta)^{s'/2} \times (1-\Delta)^{s'/2},$$
$$A(t) = \begin{pmatrix} 0 & I \\ a_{00}^{-1}\left[\sum a_{ij}\dfrac{\partial^2}{\partial x^i \partial x^j} + \sum a_j \dfrac{\partial}{\partial x^j} + a \right] & a_{00}^{-1}\left[2\sum a_{0j}\dfrac{\partial}{\partial x^j} + a_0 \right] \end{pmatrix}$$

(*the closure of this operator on* C_0^∞), *i.e.*, (4.3) *generates a strong evolution system in* X *which maps* Y *to* Y.

Here we have written (4.3) in the usual way as a system in $(u, \dot{u})$, first order in time.

One uses the norm
$$\|(\varphi, \dot\varphi)\|_t^2 = \int_{\mathbb{R}^m} \left[\sum_{i,j=1}^m a_{ij}(t, x)\frac{\partial \varphi}{\partial x^i} \cdot \frac{\partial \varphi}{\partial x^j} + c\varphi \cdot \varphi + a_{00}(t, x)\dot\varphi \cdot \dot\varphi \right] dx$$

where the constant c is chosen sufficiently large. By Gårdings inequality, this gives an equivalent norm on X (this uses strong ellipticity). It is then straightforward to get the estimate
$$\|[\lambda - A(tS)]^{-1}\|_t \geq \frac{1}{(\lambda - \beta)}$$

by showing, as before, that

$$\langle [\lambda - A(t)]u, u \rangle_t \geq (\lambda - \beta)\|u\|_t^2.$$

One can also show, as in Yosida (1974) that $\lambda - A(t)$ is one-to-one and onto X, so $A(t) \in G(X_t, 1, \beta)$, and, as above, $\|\cdot\|_t$ varies exponentially with t, so $A(t)$ is a stable family.

Again, the proof of boundedness of $B(t)$ requires estimates on commutators; for details see Hughes, Kato and Marsden (1977).

For later use in the nonlinear problem (and in lemma 4.22), it is crucial to have sharp differentiability assumptions on the coefficients as stated here.

Remark. Since the abstract theorem includes both (4.2) and (4.3) as special cases, it is clear that coupled systems of such equations can be handled in a similar way. This is important for certain types of matter fields coupled to the gravitational field.

Now we turn to the nonlinear problem. As above, let X and Y be Banach spaces, with Y densely and continuously included in X. Let $W \subset Y$ be open, let $T > 0$ and let $G: [0, T] \times W \to X$ be a given mapping. A nonlinear evolution equation has the form

$$\dot{u}(t) = G(t, u(t)), \quad \text{where } \dot{u} = \frac{du}{dt}. \tag{4.4}$$

If $s \in [0, T]$ and $\Phi \in W$ are given, a solution curve (or integral curve) of G with value ϕ at s is a map $u(\cdot) \in C^0([s, T], W) \cap ([s, T], X)$ such that (4.4) hold on $[s, T]$ and $u(s) = \phi$.

If these solution curves exist and are unique for ϕ in an open set $U \subset W$, we can define evolution operators $F_{t,s}: U \to W$ that map $u(s) = \phi$ to $u(t)$. We say (4.4) is *well posed* (or is *Cauchy stable*) if $F_{t,s}$ is continuous (in the Y-topology on U and W) for each t, s satisfying $0 \leq s \leq t \leq T$. We remark that joint continuity of $F_{t,s}(\phi)$ in (t, s, ϕ) follows under general hypotheses (Chernoff and Marsden, 1974). Furthermore, if one has well-posedness for short time intervals, it is easy to obtain it for the maximally extended flow.

Well-posedness can be difficult to establish in specific examples, especially for 'hyperbolic' ones. The continuity of $F_{t,s}$ from Y to Y cannot in general be replaced by stronger smoothness conditions such as Lipschitz or even Hölder continuity; a simple example showing this, namely $\dot{u} + uu_x = 0$ in $Y = H^{s+1}$, $X = H^s$ on $\mathbb{R}$, is given in Kato (1975a). A discussion of these smoothness questions is given below.

Chapter 4. The initial value problem

The most thoroughly studied nonlinear evolution equations are those giving rise to nonlinear contraction semigroups generated by monotone operators (Brezis, 1973). These sometimes have evolution operators defined on all of X. This is not typical of hyperbolic problems, where $F_{t,s}$ may be defined only in Y, may be continuous from Y to Y, be differentiable from Y to X, and be Y-locally Lipschitz from X to X, without being X-locally Lipschitz from X to X or Y-locally Lipschitz from Y to Y, as is shown by the example above.

Specializing (4.4), we shall consider the quasi-linear abstract Cauchy problem

$$\dot{u} = A(t, u)u + f(t, u), \quad 0 \leq t \leq T, \quad u(0) = \varphi \qquad (4.5)$$

where u takes values in X and $A(t, u)$ is an (unbounded) linear operator depending on the unknown u in a nonlinear fashion. We include f for completeness, although it can be omitted by using Kato's suspension trick mentioned above.

Here are our assumptions.

We start from four (real) Banach spaces

$$Y \subset X \subset Z' \subset Z,$$

with all the spaces reflexive and separable and the inclusions continuous and dense. We assume that

(Z') Z' is an interpolation space between Y and Z; thus if $U \in B(Y) \cap B(Z)$, then $U \in B(Z')$ with $\|U\|_{Z'} \leq c \max\{\|U\|_Y, \|U\|_Z\}$; $B(Y)$ denotes bounded operators on Y.

Let $N(Z)$ be the set of all norms in Z equivalent to the given one $\|\ \|_Z$. Then $N(Z)$ is a metric space with the distance function

$$d(\|\ \|_\mu, \|\ \|_\nu) = \log \max\{\sup_{0 \neq z \in Z} \|z\|_\mu/\|z\|_\nu, \sup_{0 \neq z \in Z} \|z\|_\nu/\|z\|_\mu\}.$$

We now introduce four functions, A, N, S, and f on $[0, T] \times W$, where $T > 0$ and W is an open set in Y, with the following properties:

For all $t, t', \ldots, \in [0, T]$ for all $w, w', \ldots \in W$, there is a real number β and there are positive numbers $\lambda_N, \mu_N, \ldots$ such that the following conditions hold.

(N) $N(t, w) \in N(Z)$, with
$d(N(t, w), \|\ \|_Z) \leq \lambda_N$,
$d(N(t', w'), N(t, w)) \leq \mu_N(|t' - t| + \|w' - w\|_X)$.

(S) $S(t, w)$ is an isomorphism of Y onto Z, with
$\|S(t, w)\|_{Y,Z} \leq \lambda_S$, $\|S(t, w)^{-1}\|_{Z,Y} \leq \lambda'_S$,
$\|S(t', w') - S(t, w)\|_{Y,Z} \leq \mu_S(|t'-t| + \|w'-w\|_X)$.

(A1) $A(t, w) \in G(Z_{N(t,w)}, 1, \beta)$, where $Z_{N(t,w)}$ denotes the Banach space Z with norm $N(t, w)$. This means that $A(t, w)$ is a C_0-generator in Z such that $\|e^{\tau A(t,w)} z\| \leq e^{\beta \tau} \|z\|$ for all $\tau \geq 0$ and $z \in Z$.

(A2) $S(t, w) A(t, w) S(t, w)^{-1} = A(t, w) + B(t, w)$, where $B(t, w) \in B(Z)$, $\|B(t, w)\|_Z \leq \lambda_B$.

(A3) $A(t, w) \in B(Y, X)$, with $\|A(t, w)\|_{Y,X} \leq \lambda_A$ and
$\|A(t, w') - A(t, w)\|_{Y,Z'} \leq \mu_A \|w' - w\|_{Z'}$
and with $t \mapsto A(t, w) \in B(Y, Z)$ continuous in norm.

(f1) $f(t, w) \in Y$, $\|f(t, w)\|_Y \leq \lambda_f$, $\|f(t, w') - f(t, w)\|_{Z'} \leq \mu_f \|w' - w\|_Z$,
and $t \mapsto f(t, w) \in Z$ is continuous.

Remarks. (i) If $N(t, w) = \text{const} = \|\ \|_Z$, condition (N) is redundant. If $S(t, w) = \text{const} = S$, condition (S) is trivial. If both are assumed, and $X = Z' = Z$, we have the case of Kato (1975b).

(ii) In most applications we can choose $Z' = Z$ and/or $Z' = X$.

(iii) The paper of Hughes, Kato and Marsden (1977) had an additional condition (A4) which was then shown to be redundant in Kato (1977).

Theorem 4.16

Let (Z'), (N), (S), (A1) to (A3), and (f1) be satisfied. Then there are positive constants ρ' and $T' \leq T$ such that if $\phi \in Y$ with $\|\phi - y_0\|_Y \leq \rho'$, then (4.5) has a unique solution u on $[0, T']$ with

$$u \in C^0([0, T']; W) \cap C^1([0, T']; X).$$

Here ρ' depends only on λ_N, λ_S, λ'_S, and $R = \text{dist}(y_0, Y \setminus W)$, while T' may depend on all the constants β, λ_N, μ_N, ... and R. When ϕ varies in Y subject to $\|\phi - y_0\|_Y \leq \rho'$, the map $\phi \mapsto u(t)$ is Lipschitz continuous in the Z'-norm, uniformly in $t \in [0, T']$.

To establish well-posedness, we have to strengthen some of the assumptions. We assume the following conditions:

(B) $\|B(t, w') - B(t, w)\|_Z \leq \mu_B \|w' - w\|_Y$.
(f2) $\|f(t, w') - f(t, w)\|_Y \leq \mu'_f \|w' - w\|_Y$.

Chapter 4. The initial value problem

Theorem 4.17

Let (Z'), (N), (S), (A1) *to* (A3), (B), (f1), *and* (f2) *be satisfied, where* $S(t, w)$ *is assumed to be independent of w. Then there is a positive constant* $T'' \leq T'$ *such that when* ϕ *varies in Y subject to* $\|\phi - y_0\|_Y \leq \rho'$, *the map* $\phi \mapsto u(t)$ *given by theorem 4.16 is continuous in the Y-norm, uniformly in* $t \in [0, T'']$.

Remark. As in Kato (1975b), one can prove a similar continuity theorem when not only the initial value ϕ but also the functions N, A, and f are varied, i.e., the solution is 'stable' when the equations themselves are varied. It appears, on the other hand, that the variation of S is rather difficult to handle.

The theorem thus guarantees the existence of (locally defined) maps

$$F_{t,s}: Y \to Y$$

which are continuous in all variables. We have

$$F_{s,s} = \text{Id}$$
$$F_{t,s} \circ F_{s,r} = F_{t,r}$$

as in the linear case. We speak of $F_{t,s}$ as the *evolution operators generated by the equation* (4.5). The general notion of evolution operators for (4.4) is defined in an analogous manner.

The idea behind the proof of theorem 4.17 is to fix a curve $v(t)$, $v(0) = \phi$ in Y and to let $u(t)$ be the solution of the 'frozen coefficient problem'

$$\dot{u} = A(t, v)u + f(t, v), \quad u(0) = \phi$$

which is guaranteed by theorem 4.13. This defines a map $\Phi: v \mapsto u$ and we look for a fixed point of Φ. In a suitable function space and for T' sufficiently small, Φ is in fact a contraction, so has a unique fixed point.

However, it is not so simple to prove that u depends continuously on ϕ and detailed estimates from the linear theory are needed. The proof more or less has to be delicate since the dependence on ϕ is not locally Lipschitz in general. For details of these proofs, we refer to Kato (1975b, 1977) and Hughes, Kato and Marsden (1977).

The continuous dependence of the solution on ϕ leads us naturally to investigate if it is smooth in any sense. This is important for studying the relationship between nonlinear theories and their linearization. The following results are taken from some unpublished notes of Dorroh and Marsden.

First, we give the notion of differentiability appropriate for the generator G of (4.4). Let X and Y be Banach spaces with $Y \subset X$ continuously and densely included. Let $U \subset Y$ be open and $f: U \to X$ be a given mapping. We say f is α-*differentiable* if for each $x \in U$ there is a bounded linear operator $Df(x): Y \to X$ such that

$$\frac{\|f(x+h)-f(x)-Df(x)\cdot h\|_X}{\|h\|_X} \to 0$$

as $\|h\|_Y \to 0$. If f is α-differentiable and $x \mapsto Df(x) \in B(Y, X)$ is norm continuous, we call f C^1 α-*differentiable*. Notice that this is *stronger* than C^1 in the Fréchet sense. If f is α-differentiable and

$$\|f(x+h)-f(x)-Df(x)\cdot h\|_X/\|h\|_X$$

is uniformly bounded for x and $x+h$ in some T neighborhood of each point, we say that f is *locally uniformly α-differentiable*.

Most concrete examples can be checked using the following proposition.

Proposition 4.18

Suppose $f: U \subset Y \to X$ is of class C^2, and locally in the Y topology

$$x \mapsto \frac{\|D^2 f(x)(h, h)\|_X}{\|h\|_Y \|h\|_X}$$

is bounded. Then f is locally uniformly C^1 α-differentiable.

This follows easily from the identity

$$f(x+h)-f(x)-Df(x)\cdot h = \int_0^1 \int_0^1 D^2 f(x+sth)(h, h)\, ds\, dt.$$

Next, we turn to the appropriate notion for the evolution operators.

A map $g: U \subset Y \to X$ is called β-*differentiable* if it is α-differentiable and $Dg(x)$, for each $x \in U$, extends to a bounded operator X to X.

Chapter 4. The initial value problem

β-differentiable maps obey a chain rule. For example, if $g_1: Y \to Y$, $g_2: Y \to Y$ and each is β-differentiable (as maps of Y to X) and are continuous from Y to Y, then $g_2 \circ g_1$ is β-differentiable with, of course,

$$D(g_2 \circ g_1)(x) = Dg_2[g_1(x)] \circ Dg_1(x).$$

The proof of this fact is routine. In particular, one can apply the chain rule to $F_{t,s} \circ F_{s,r} = F_{t,r}$ if each $F_{t,s}$ is β-differentiable. Differentiating this in s at $s = r$ gives the backwards equation for $x \in Y$:

$$\frac{\partial}{\partial s} F_{t,s}(x) = -DF_{t,s}(x) \cdot G(x).$$

Then differentiation in r at $r = s$ gives

$$DF_{t,s}(x) \cdot G(x) = G[F_{t,s}(x)],$$

the flow invariance of the generator.

We leave it to the reader to supply rigorous proofs of these claims following the hint from the linear case.

For the following theorem we assume these hypotheses: $Y \subset X$ is continuously and densely included and $F_{t,s}$ is a continuous evolution system on an open subset $D \subset Y$ and the X-infinitesimal generator $G(t)$ of $F_{t,s}$ has domain† D. Also, we assume:

(H_1) $G(t): D \subset Y \to X$ is locally uniformly C^1 α-differentiable. Its derivative is denoted $D_x G(t, x)$ and is assumed strongly continuous in t.

(H_2) For $x \in D$, $s \ge 0$, let $T_{x,s}$ be the lifetime of x beyond s, i.e., $\sup\{t \ge s | F_{t,s}(x) \text{ is defined}\}$. Assume there is a strongly continuous linear evolution system $\{U^{x,s}(\tau, \sigma): 0 \le \sigma \le \tau \le T_{x,s}\}$ in X whose X-infinitesimal generator is an extension of $\{D_x G(t, F_{t,s}x) \in B(Y, X); 0 \le t \le T_{x,s}\}$; i.e., if $y \in Y$,

$$\left. \frac{\partial}{\partial \tau} U^{x,s}(\tau, \sigma) \cdot y \right|_{\tau=\sigma} = D_x G(\tau, F_{\tau,s}(x)) \cdot y.$$

Theorem 4.19 (*J. R. Dorroh*)

Under the hypotheses above, $F_{t,s}$ is β-differentiable at x and in fact,

$$DF_{t,s}(x) = U_{x,s}(t, s).$$

---

† As in the linear case, $G(t)$ may have an extension to a larger domain, but we are only interested in $G(t)$ on D here.

Proof. Define $\varphi_t(x, y) = \varphi(t, x, y)$ by

$$G(t, x) - G(t, y) = D_x G(t, y) \cdot (x - y) + \|x - y\|_X \varphi_t(x, y)$$

(or zero if $x = y$) and notice that by local uniformity, $\|\varphi(t, x, y)\|_X$ is uniformly bounded if x and y are Y-close. By joint continuity of $F_{t,s}(x)$, for $0 < t < T_{x,s}$, $\|\varphi(t, F_{t,s}y, F_{t,s}x)\|_X$ is bounded for $0 \leq s \leq T$ if $\|x - y\|_Y$ is sufficiently small.

By construction, we have the equation

$$\frac{d}{dt} F_{t,s}(x) = G[F_{t,s}(x)], \quad 0 \leq s \leq t \leq T_{x,s}, x \in D.$$

Let

$$w(t, s) = F_{t,s}(y) - F_{t,s}(x)$$

so that

$$\frac{\partial w(t, s)}{\partial t} = G(t, F_{t,s}(y)) - G(t, F_{t,s}(x))$$

$$= D_x G(t, F_{t,s}(x)) w(t, s) + \|w(t, s)\|_X \varphi(t, F_{t,s}y, F_{t,s}x).$$

Since $D_x G(t, F_{t,s}x) \cdot w(t, s)$ is continuous in t, s with values in X, and writing $U = U_{x,x}$ we have the backwards differential equation:

$$\frac{\partial}{\partial \sigma} U(t, \sigma) w(\sigma, s) = U(t, \sigma) \frac{\partial w(\sigma, s)}{\partial \sigma} - U(t, \sigma) D_x G(\sigma, F_{\sigma,s}(x)) \cdot w(\sigma, s)$$

$$= U(t, \sigma) \cdot \|w(\sigma, s)\|_X \varphi(\sigma, F_{\sigma,s}(y), F_{\sigma,s}(x)).$$

Hence, integrating from $\sigma = s$ to $\sigma = t$,

$$w(t, s) = U(t, s)(y - x) + \int_s^t U(t, \sigma) \|w(\sigma, s)\|_X \varphi(\sigma, F_{\sigma,s}(y), F_{\sigma,s}(x)) \, d\sigma.$$

Let $\|U(\tau, \sigma)\|_{X,X} \leq M$, and $\|\varphi[\sigma, F_{\sigma,s}(y), F_{\sigma,s}(x)]\|_X \leq M_2, 0 \leq s \leq \sigma \leq \tau \leq T$. Thus, by Gronwall's inequality,

$$\|w(t, s)\|_X \leq M_1 e^{M_1 M_2 T} \|y - x\|_X = M_3 \|y - x\|_X.$$

In other words,

$$\frac{\|F_{t,s}(y) - F_{t,s}(x) - U(t, s)(y - x)\|_X}{\|y - x\|_X} \leq M_1 M_3 \int_s^t \|\varphi(\sigma, F_{\sigma,s}(y), F_{\sigma,s}(x))\|_X \, d\sigma.$$

From the bounded convergence theorem, we conclude that $F_{t,s}$ is β-differentiable at x and $DF_{t,s}(x) = U(t, s)$. ($\varphi(t, F_{t,s}(y), F_{t,s}(x))$ is strongly measurable in s since $\varphi(x, y)$ is continuous for $x \neq y$.) ∎

Chapter 4. The initial value problem

This completes our description of the abstract nonlinear theory. Next, we state how the nonlinear existence and uniqueness theorem applies to quasi-linear equations of types (4.2) and (4.3).

First, we consider the first-order case:

$$a_0(t, x, u)\frac{\partial u}{\partial t} = \sum_{j=1}^{m} a_j(t, x, u)\frac{\partial u}{\partial x^j} + a(t, x, u). \qquad (4.6)$$

We assume

(i) $s > \frac{1}{2}m + 1$ and a_α, a are of class C^{s+1} in the variables t, x, u (possibly locally defined in u);
(ii) the linear conditions (i), (ii), (iii) of theorem 4.14 hold locally uniformly in u.

Theorem 4.20

Under these conditions, theorems 4.16, 4.17 and 4.19 hold for (4.6), i.e., (4.6) generates a unique local evolution system $F_{t,s}$ in $X = H^{s-1}(\mathbb{R}^m)$ with $Y = H^s(\mathbb{R}^m)$ and $Z = Z' = L^2(\mathbb{R}^m)$; $F_{t,s}$ maps Y to Y continuously and, for t, s fixed, is β-differentiable as a map of Y to X.

The full details of the proof require a lengthy discussion of Sobolev space estimates to verify the hypotheses, but it is relatively straightforward. See Kato (1975a, b) for details. We note that one may also choose $X = Z = Z' = L^2(\mathbb{R}^m)$, $Y = H^s(\mathbb{R}^m)$, but the choices in theorem 4.20 are appropriate for theorem 4.19. Again, length and their technical nature preclude giving details of how theorem 4.19 applies to (4.6). It is again a semi-routine Sobolev space exercise.

For the second-order case, we proceed as follows. Consider

$$a_{00}(t, s, u, \nabla u)\frac{\partial^2 u}{\partial t^2} = \sum_{i,j=1}^{m} a_{ij}(t, x, u, \nabla u)\frac{\partial^2 u}{\partial x^i \partial x^j}$$
$$+ 2\sum_{j=1}^{m} a_{0i}(t, x, u, \nabla u)\frac{\partial^2 u}{\partial t \partial x^i} + a(t, x, u, \nabla u). \qquad (4.7)$$

Here

$$\nabla u = \left(\frac{\partial u}{\partial x^1}, \ldots, \frac{\partial u}{\partial x^m}, \frac{\partial u}{\partial t}\right).$$

We assume

(i) $a_{\alpha\beta}$, a are of class C^{s+1} in all variables (possibly locally defined in u);

(ii) the linear conditions (i), (ii), (iii), (iv) of theorem 4.15 hold locally uniformly in u.

Theorem 4.21

(i) *If* $s > \frac{1}{2}m + 1$, *theorems* 4.16, 4.17 *and* 4.19 *hold for* (4.7) *with*

$$X = H^s(\mathbb{R}^m) \times H^{s-1}(\mathbb{R}^m),$$
$$Z = Z' = H^1(\mathbb{R}^m) \times H^0(\mathbb{R}^m),$$
$$Y = H^{s+1}(\mathbb{R}^m) \times H^s(\mathbb{R}^m),$$

i.e., (4.7) *generates a unique local evolution system* $F_{t,s}: Y \to Y$ *which is continuous and for fixed* t, s *is* β-*differentiable from* Y *to* X.

(ii) *If* $a_{\alpha\beta}$ *do not depend on* ∇u, *then the same conclusions hold with* $s > \frac{1}{2}m$.

For details of the proof, see Hughes, Kato and Marsden (1977).

As we shall see in the next section, case (ii) is the case relevant for general relativity. Note that if $m = 3$, solutions $(u, \dot{u})$ will lie in $Y = H^r \times H^{r-1}$ where $r > 2.5$. For example, in this case, (4.7) gives a well-posed problem for u in H^3. (Notice that u is only C^1 in this case and need not be C^2.) For hyperbolic systems, theorems 4.20 and 4.21 are the sharpest known results, although these problems have been considered by a large number of authors,† such as Choquet-Bruhat (1952, 1962), Courant-Hilbert (1962), Dionne (1962), Frankl (1937), Krzyzanski and Schauder (1934), Leray (1953), Lichnerowicz (1967), Lions (1969), Petrovskii (1937), Schauder (1935), and Sobolev (1939).

4.4 The Cauchy problem for relativity

We shall begin with the vacuum problem and then go on to consider gravity coupled to other fields. We begin by reviewing the classic work of Lichnerowicz (1944) and Choquet-Bruhat (1952) and the introduction of harmonic coordinates. We shall be brief since this is described in Choquet-Bruhat (1962) and in Hawking and Ellis (1973). Our main result is that for H^s spacetimes with $s > 2.5$, there is a satisfactory existence theorem 4.23 and uniqueness theorem 4.27 for

---

† For relativity, some partial results in H^3 were indicated by Hawking and Ellis (1973, p. 251).

Chapter 4. The initial value problem

the Cauchy problem. These results are the sharpest presently known for the Cauchy problem.

We work on $\mathbb{R}^4$ for simplicity and because of the hyperbolicity, with no essential loss of generality. For empty space relativity, one searches for a Lorentz metric $g_{\mu\nu}(t, x^i)$ whose Ricci curvature $R_{\mu\nu}$ is zero; i.e., $g_{\mu\nu}(t, x^i)$ must satisfy the system

$$R_{\mu\nu}\left(t, x^i, g_{\mu\nu}, \frac{\partial g_{\mu\nu}}{\partial x^\alpha}, \frac{\partial^2 g_{\mu\nu}}{\partial x^\alpha \partial x^\beta}\right) = -\frac{1}{2} g^{\alpha\beta} \frac{\partial^2 g_{\mu\nu}}{\partial x^\alpha \partial x^\beta} - \frac{1}{2} g^{\alpha\beta} \frac{\partial^2 g_{\alpha\beta}}{\partial x^\mu \partial x^\nu}$$

$$+ \frac{1}{2} g^{\alpha\beta} \frac{\partial^2 g_{\alpha\nu}}{\partial x^\beta \partial x^\mu} + \frac{1}{2} g^{\alpha\beta} \frac{\partial^2 g_{\alpha\mu}}{\partial x^\beta \partial x^\nu}$$

$$+ H_{\mu\nu}\left(g_{\mu\nu}, \frac{\partial g_{\mu\nu}}{\partial x^\alpha}\right)$$

$$= 0,$$

where $H_{\mu\nu}(g_{\mu\nu}, \partial g_{\mu\nu}/\partial x^\alpha)$ is a rational combination of $g_{\mu\nu}$ and $\partial g_{\mu\nu}/\partial x^\alpha$ with denominator $\det g_{\mu\nu} \neq 0$. Note that the contravariant tensor $g^{\mu\nu}$ is a rational combination of the $g_{\mu\nu}$ with denominator $\det g_{\mu\nu} \neq 0$.

Let $G_{\mu\nu} = R_{\mu\nu} - \frac{1}{2} g_{\mu\nu} R$ be the Einstein tensor, where $R = g^{\alpha\beta} R_{\alpha\beta}$ is the scalar curvature. Then, as is well known, $G^0{}_\mu$ contains only first-order time derivatives of $g_{\mu\nu}$. Thus $G^0{}_\mu(0, x^i)$ can be computed from the Cauchy data $g_{\mu\nu}(0, x^i)$ and $\partial g_{\mu\nu}(0, x^i)/\partial t$ alone, and therefore $G^0{}_\mu(0, x^i) = 0$ is a necessary condition on the Cauchy data in order that a spacetime $g_{\mu\nu}(t, x^i)$ have the given Cauchy data and satisfy $G_{\mu\nu} = 0$, which is equivalent to $R_{\mu\nu} = 0$.

The existence part of the Cauchy problem for the system $R_{\mu\nu} = 0$ is as follows.

Let $(\mathring{g}_{\mu\nu}(x^i), \mathring{k}_{\mu\nu}(x^i))$ be Cauchy data of class $(H^s(\Omega), H^{s-1}(\Omega))$, $s \geq 3$, such that $\mathring{G}^0{}_\mu(x^i) = 0$. Let Ω_0 be a proper subdomain, $\bar{\Omega}_0 \subset \Omega$. Find an $\varepsilon > 0$ and a spacetime $g_{\mu\nu}(t, x^i)$, $|t| < \varepsilon$, $(x^i) \in \Omega_0 \subset \Omega$ such that
  (i) $g_{\mu\nu}(t, x^i)$ *is H^s jointly in* $(t, x^i) \in (-\varepsilon, \varepsilon) \times \Omega_0$;
  (ii) $(g_{\mu\nu}(0, x^i), \partial g_{\mu\nu}(0, x^i)/\partial t) = (\mathring{g}_{\mu\nu}(x^i), \mathring{k}_{\mu\nu}(x^i))$;
  (iii) $g_{\mu\nu}(t, x^i)$ *has zero Ricci curvature.*

The system $R_{\mu\nu} = 0$ is a quasi-linear system of ten second-order partial differential equations for which the highest-order terms involve mixing of the components of the system. As it stands, there are no known theorems about partial differential equations which can be applied to resolve the Cauchy problem. However, as was first noted by

Lanczos (1922) (and in fact by Einstein himself (1916b) for the linearized equations) the Ricci tensor simplifies considerably in harmonic coordinates, i.e., in a coordinate system for which the contracted Cristoffel symbols vanish, $\Gamma^\mu = g^{\alpha\beta}\Gamma^\mu_{\alpha\beta} = 0$. In fact, an algebraic computation shows that

$$R_{\mu\nu} = -\frac{1}{2}g^{\alpha\beta}\frac{\partial^2 g_{\mu\nu}}{\partial x^\alpha \partial x^\beta} + \frac{1}{2}g_{\mu\alpha}\frac{\partial \Gamma^\alpha}{\partial x^\nu} + \frac{1}{2}g_{\nu\alpha}\frac{\partial \Gamma^\alpha}{\partial x^\mu} + H_{\mu\nu}$$

so that in a coordinate system for which $\Gamma^\mu = 0$,

$$R_{\mu\nu} = R^{(h)}_{\mu\nu} = -\frac{1}{2}g^{\alpha\beta}\frac{\partial^2 g_{\mu\nu}}{\partial x^\alpha \partial x^\beta} + H_{\mu\nu}.$$

The operator $-\frac{1}{2}g^{\alpha\beta}(\partial^2/\partial x^\alpha \partial x^\beta)$ operates the same way on each component of the system $g_{\mu\nu}$ so that there is no mixing in the highest-order derivatives. Thus the normalized system $R^{(h)}_{\mu\nu} = 0$ is considerably simpler than the full system. In fact, the system $R^{(h)}_{\mu\nu} = 0$ has only simple characteristics so that $R^{(h)}_{\mu\nu} = 0$ is a strictly hyperbolic system.

The importance of the use of harmonic coordinates and of the system $R^{(h)}_{\mu\nu} = 0$ is based on the fact that it is sufficient to solve the Cauchy problem for $R^{(h)}_{\mu\nu} = 0$; this remarkable fact, discovered by Choquet-Bruhat (1952), is based on the observation that the condition $\mathring{\Gamma}^\mu(x^i) \equiv \mathring{g}^{\alpha\beta}(x^i)\mathring{\Gamma}^\mu_{\alpha\beta}(x^i) = 0$ is propagated off the hypersurface $t = 0$ for solutions $g_{\mu\nu}$ of $R^{(h)}_{\mu\nu} = 0$. This is established in the next lemma.

Lemma 4.22

Let $(\mathring{g}_{\mu\nu}(x^i), \mathring{k}_{\mu\nu}(x^i))$ be of Sobolev class (H^s, H^{s-1}) on Ω, $s > \frac{1}{2}n + 1$, $n = 3$, and suppose that $(\mathring{g}_{\mu\nu}(x^i), \mathring{k}_{\mu\nu}(x^i))$ satisfies
  (i) $\mathring{\Gamma}^\mu(x^i) = 0$,
  (ii) $\mathring{G}^0_\mu(x^i) = 0$.
If $g_{\mu\nu}(t, x)$, $|t| < \varepsilon$, $x \in \Omega_0$, Ω_0 a proper subdomain, $\bar{\Omega}_0 \subset \Omega$, is an H^s-solution of

$$R^{(h)}_{\mu\nu} = -\tfrac{1}{2}g^{\alpha\beta}(\partial^2 g_{\mu\nu}/\partial x^\alpha \partial x^\beta) + H_{\mu\nu} = 0,$$

$$(g_{\mu\nu}(0, x), \partial g_{\mu\nu}(0, x)/\partial t) = (\mathring{g}_{\mu\nu}(x^i), \mathring{k}_{\mu\nu}(x^i)),$$

then $\Gamma^\mu(t, x^i) = 0$ for $|t| < \varepsilon$, $x \in \Omega_0$.

Proof. Let $g_{\mu\nu}(t, x^i)$ satisfy (i), (ii), and $R^{(h)}_{\mu\nu} = 0$. Then a straightforward computation shows that $\Gamma^\mu(t, x^i) = g^{\alpha\beta}(t, x^i)\Gamma^\mu_{\alpha\beta}(t, x^i)$ satisfies

Chapter 4. The initial value problem

$\partial \Gamma^\mu(0, x^i)/\partial t = 0$. From $G^{\mu\nu}{}_{;\nu} = 0$ and $R^{(h)}_{\mu\nu} = 0$, Γ^μ is shown to satisfy the system of linear equations

$$g^{\alpha\beta} \frac{\partial^2 \Gamma^\mu}{\partial x^\alpha \partial x^\beta} + A^{\beta\mu}_\alpha\left(g_{\mu\nu}, g^{\mu\nu}, \frac{\partial g_{\mu\nu}}{\partial x^\lambda}\right) \frac{\partial \Gamma^\alpha}{\partial x^\beta} = 0.$$

This linear system is of the form (4.3) for which a uniqueness and existence theorem holds. Thus, by theorem 4.15, $\Gamma^\mu(0, x^i) = 0$ and $\partial \Gamma^\mu(0, x^i)/\partial t = 0$ imply $\Gamma^\mu(t, x^i) = 0$. ∎

According to the lemma, an H^s-solution of $R^{(h)}_{\mu\nu} = 0$ with prescribed Cauchy data is also a solution of. $R_{\mu\nu} = 0$ (since $\Gamma^\mu(t, x) = 0 \Rightarrow R^{(h)}_{\mu\nu} = R_{\mu\nu}$), provided that the Cauchy data satisfies (i) $\mathring{\Gamma}^\mu = 0$ and (ii) $\mathring{G}^0{}_\mu = 0$. As mentioned above, (ii) is a necessary condition on the Cauchy data for a solution $g_{\mu\nu}(t, x)$ to satisfy $R_{\mu\nu} = 0$. If (i) is not satisfied, then a set of Cauchy data can be found whose evolution under $R^{(h)}_{\mu\nu} = 0$ leads to an H^s-spacetime which, by an H^{s+1}-coordinate transformation, gives rise to a spacetime with the original Cauchy data (see theorem 4.26 below and Fischer and Marsden, 1972b).

From theorem 4.21 we conclude that Cauchy data of class (H^s, H^{s-1}) has an H^s-time evolution for $s > 2.5$ and Cauchy stability holds.

We can also prove this result by reducing the strictly hyperbolic system $R^{(h)}_{\mu\nu} = 0$ to a quasi-linear symmetric hyperbolic first-order system. This will be outlined below.

Theorem 4.23

Let Ω be an open bounded domain in $\mathbb{R}^3$ with Ω_0 a proper subdomain, $\bar{\Omega}_0 \subset \Omega$, and let $(\mathring{g}_{\mu\nu}(x), \mathring{k}_{\mu\nu}(x))$, $(x^i) \in \Omega$, $0 \leq \mu$, $\nu \leq 3$, $1 \leq i \leq 3$, be of Sobolev class (H^s, H^{s-1}), $s > 2.5$. Suppose that $\mathring{\Gamma}^\mu(x^i) = 0$ and $\mathring{G}^0{}_\mu(x) = 0$. Then there exists an $\varepsilon > 0$ and a unique Lorentz metric $g_{\mu\nu}(t, x)$, $|t| < \varepsilon$, $(x^i) \in \Omega_0$ such that

(i) *$g_{\mu\nu}(t, x^i)$ is jointly of class H^s;*
(ii) *$R^{(h)}_{\mu\nu}(t, x^i) = 0$;*
(iii) *$(g_{\mu\nu}(0, x^i), \partial g_{\mu\nu}(0, x^i)/\partial t) = (\mathring{g}_{\mu\nu}(x^i), \mathring{k}_{\mu\nu}(x^i))$.*

From lemma 4.22, this $g_{\mu\nu}(t, x^i)$ also satisfies $R_{\mu\nu}(t, x^i) = 0$. Moreover, $g_{\mu\nu}(t, x^i)$ depends continuously on $(\mathring{g}_{\mu\nu}(x^i), \mathring{k}_{\mu\nu}(x^i))$ in the (H^s, H^{s-1}) topology. If $(\mathring{g}_{\mu\nu}(x^i), \mathring{k}_{\mu\nu}(x^i))$ is of class (C^∞, C^∞) on Ω, then $g_{\mu\nu}(t, x^i)$ is C^∞ for all t for which the solution exists.

See below for a discussion of solutions on all of $\mathbb{R}^3$ with spatial asymptotic conditions.

We have already indicated how this follows directly from theorem 4.21 and lemma 4.22. To give another proof using theorem 4.20, we reduce the system $R^{(h)}_{\mu\nu} = 0$ to a first-order system by introducing the ten new unknowns $k_{\mu\nu} = \partial g_{\mu\nu}/\partial t$ and the thirty new unknowns $g_{\mu\nu,i} = \partial g_{\mu\nu}/\partial x^i$ and considering the quasi-linear first-order system of fifty equations:

$$\partial g_{\mu\nu}/\partial t = k_{\mu\nu},$$

$$g^{ij}\left(\frac{\partial g_{\mu\nu,i}}{\partial t}\right) = g^{ij}\frac{\partial k_{\mu\nu}}{\partial x^i}, \qquad (4.8)$$

$$-g^{00}\frac{\partial k_{\mu\nu}}{\partial t} = 2g^{0j}\frac{\partial k_{\mu\nu}}{\partial x^j} + g^{ij}\frac{\partial g_{\mu\nu,i}}{\partial x^j} - 2H_{\mu\nu}(g_{\mu\nu}, g_{\mu\nu,i}, k_{\mu\nu}).$$

We are considering $H_{\mu\nu}$ as a polynomial in $g_{\mu\nu,i}$ and $k_{\mu\nu}$, and rational in $g_{\mu\nu}$ with denominator $\det g_{\mu\nu} \neq 0$. At first, we extend our initial data to all of $\mathbb{R}^3$, say, to equal the Minkowski metric outside a compact set, and consider the system (4.8) on $\mathbb{R}^3$. Note that the Cauchy data need not satisfy the constraints $G^0{}_\mu = 0$ during the transition.

The matrix g^{ij} has inverse $g_{jk} - (g_{j0}g_{k0}/g_{00})$, i.e., $g^{ij}[g_{jk} - (g_{j0}g_{k0}/g_{00})] = \delta^i_k$, so that the second set of thirty equations can be inverted to give

$$\partial g_{\mu\nu,i}/\partial t = \partial k_{\mu\nu}/\partial x^i. \qquad (4.9)$$

For $g_{\mu\nu}$ of class C^2, (4.9) implies

$$g_{\mu\nu,i} = \partial g_{\mu\nu}/\partial x^i,$$

so that the system (4.8) is equivalent to $R^{(h)}_{\mu\nu} = 0$.

Let

$$u = \begin{pmatrix} g_{\mu\nu} \\ g_{\mu\nu,i} \\ k_{\mu\nu} \end{pmatrix}$$

be a fifty-component column vector, where $g_{\mu\nu,i}$ is listed as

$$\begin{pmatrix} g_{00,1} \\ \vdots \\ g_{33,1} \\ \vdots \\ g_{00,3} \\ \vdots \\ g_{33,3} \end{pmatrix}$$

Chapter 4. The initial value problem

Let $0^{10} = 10 \times 10$ zero matrix, $I^{10} = 10 \times 10$ identity matrix, and let $A^0(u) = A^0(g_{\mu\nu}, g_{\mu\nu,i}, k_{\mu\nu})$ and $A^j(g_{\mu\nu}, g_{\mu\nu,i}, k_{\mu\nu})$ be the 50×50 matrices given by

$$A^0(g_{\mu\nu}, g_{\mu\nu,i}, k_{\mu\nu}) = \begin{pmatrix} I^{10} & 0^{10} & 0^{10} & 0^{10} & 0^{10} \\ 0^{10} & g^{11}I^{10} & g^{12}I^{10} & g^{13}I^{10} & 0^{10} \\ 0^{10} & g^{12}I^{10} & g^{22}I^{10} & g^{23}I^{10} & 0^{10} \\ 0^{10} & g^{13}I^{10} & g^{23}I^{10} & 0^{10} & 0^{10} \\ 0^{10} & 0^{10} & 0^{10} & 0^{10} & -g^{00}I^{10} \end{pmatrix}$$

$$A^j(g_{\mu\nu}, g_{\mu\nu,i}, k_{\mu\nu}) = \begin{pmatrix} 0^{10} & 0^{10} & 0^{10} & 0^{10} & 0^{10} \\ 0^{10} & 0^{10} & 0^{10} & 0^{10} & g^{j1}I^{10} \\ 0^{10} & 0^{10} & 0^{10} & 0^{10} & g^{j2}I^{10} \\ 0^{10} & 0^{10} & 0^{10} & 0^{10} & g^{j3}I^{10} \\ 0^{10} & g^{1j}I^{10} & g^{2j}I^{10} & g^{3j}I^{10} & 2g^{j0}I^{10} \end{pmatrix}$$

and let $B(g_{\mu\nu}, g_{\mu\nu,i}, k_{\mu\nu})$ be the fifty-component column vector given by

$$B(g_{\mu\nu}, g_{\mu\nu,i}, k_{\mu\nu}) = \begin{pmatrix} k_{\mu\nu} \\ 0^{30} \\ -2H_{\mu\nu}(g_{\mu\nu}, g_{\mu\nu,i}, k_{\mu\nu}) \end{pmatrix}$$

where 0^{30} is the thirty-component zero column vector.

Note that $A^0(u)$ and $A^i(u)$ are symmetric, and that $A^0(u)$ is positive-definite if $g_{\mu\nu}$ has Lorentz signature. A direct verification shows that the first-order quasi-linear symmetric hyperbolic system

$$A^0(u)(\partial u/\partial t) = A^j(u)(\partial u/\partial x^j) + B(u)$$

is just the system (4.8). From theorem 4.20 we conclude that for Cauchy data

$$\mathring{u}(x^i) = \begin{pmatrix} \mathring{g}_{\mu\nu}(x^i) \\ \mathring{g}_{\mu\nu,i}(x^i) \\ \mathring{k}_{\mu\nu}(x^i) \end{pmatrix}$$

of Sobolev class H^{s-1}, $s - 1 > \frac{1}{2}n + 1$, there exists an $\varepsilon > 0$ and a solution

$$u(t, x^i) = \begin{pmatrix} g_{\mu\nu}(t, x^i) \\ g_{\mu\nu,i}(t, x^i) \\ k_{\mu\nu}(t, x^i) \end{pmatrix}$$

of class H^{s-1}. By Sobolev's lemma, $u(t, x^i)$ is also of class C^2, and so, by the second set of equations of (4.5), $g_{\mu\nu,i} = \partial g_{\mu\nu}/\partial x^i$. Since $(g_{\mu\nu,i}, k_{\mu\nu}) = (\partial g_{\mu\nu}/\partial x^i, \partial g_{\mu\nu}/\partial t)$ is of class H^{s-1}, $g_{\mu\nu}(t, x^i)$ is in fact of class H^s. The continuous dependence of the solutions on the initial data follows from the general theory.

To recover the result for the domain Ω from the result for $\mathbb{R}^n$, we can use the standard domain of dependence arguments; see Courant and Hilbert (1962).

Since Ω is bounded, $(\overset{\circ}{g}_{\mu\nu}, \overset{\circ}{k}_{\mu\nu})$ of class C^∞ implies that the solution is in the intersection of all the Sobolev spaces and hence is C^∞; again, we are using the general regularity result about symmetric hyperbolic systems.

From lemma 4.22, the $g_{\mu\nu}(t, x^i)$ so found satisfy the field equations $R_{\mu\nu} = 0$.

While the second-order approach gives $s > 2.5$, e.g., $s = 3$ (see theorem 4.21(ii)) the first-order approach as it stands only gives $s > 3.5$, e.g., $s = 4$. It can be refined, but it requires a knowledge of the special structure of the equations and ellipticity. For these reasons, the second-order methods seem more attractive.

For the case of asymptotic conditions, some care must be exercised. Spacetimes which are spatially like $1/r$ will not be of class H^s. Fix a background spacetime $g^b_{\alpha\beta}$ with prescribed fall-off to the Minkowski metric at ∞. For example, a specified mass will determine the coefficient of $1/r$; $g^b_{\alpha\beta}$ could be a Schwarzschild-type solution with the singularity at $r = 0$ smoothed out.

We let our variables be $u_{\alpha\beta} = g_{\alpha\beta} - g^b_{\alpha\beta}$ and solve for $u_{\alpha\beta}$. Although $g_{\alpha\beta}$ will not be in H^s itself, $u_{\alpha\beta}$ will be.

Assume the following conditions on $g^b_{\alpha\beta}$:

$$g^b_{\alpha\beta} \in C^{s+1}_b(\mathbb{R}^3, \mathbb{R}), \quad \dot{g}^b_{\alpha\beta} \in H^s(\mathbb{R}^3, \mathbb{R})$$

and (4.10)

$$\frac{\partial g^b_{\alpha\beta}}{\partial x^i} \in H^s(\mathbb{R}^3, \mathbb{R}), \quad 0 \leq \alpha, \beta \leq 3, \quad 1 \leq i \leq 3.$$

In the variables $u_{\alpha\beta}$, the equations (4.8) are of the form (4.7).

The coefficients of the second-order terms do not involve derivatives of u, so only $s > \frac{1}{2}n$ is required.

Let us write $H^s_{g^b_{\alpha\beta}}$ for the space of $g_{\alpha\beta}$ such that $g_{\alpha\beta} - g^b_{\alpha\beta} \in H^s$, topologized accordingly. Then theorem 4.21 yields:

Chapter 4. The initial value problem

Theorem 4.24

Let (4.10) hold. Then, for $s > 1.5$ and initial data in a ball about $(g^b_{\alpha\beta}, \dot{g}^b_{\alpha\beta})$ in $H^{s+1}_{g^b_{\alpha\beta}} \times H^s_{g^b_{\alpha\beta}}$, (4.8) have a unique solution in the same space for a time interval $[0, T']$, $T' > 0$. The solution depends continuously on the initial data in this space (i.e., it is well posed or 'Cauchy stable') and smoothly in the sense of theorem 4.19.

Thus, with the asymptotic conditions subtracted off, $H^3 \times H^2$ initial data generates a piece of H^3 spacetime in a way which depends continuously on the initial data. If T' is allowed to be large, the Lorentz character of $g_{\alpha\beta}$ could be lost or a singularity could develop.

A by-product of the proof is regularity; i.e., if existence holds in $H^{s+1} \times H^s$ on $[0, T']$ and the initial data is smoother, then so is the solution on the *same* interval $[0, T']$. Thus C^∞ initial data gives C^∞ solutions.

An interesting problem is to determine whether or not the spacetime generated by initial data satisfying (4.10) is large enough to include asymptotic boosts. An examination of the proofs shows that the time of existence increases at least logarithmically at spatial infinity, so the proofs as they stand do not seem to give an affirmative answer.

We now show that any two H^s-spacetimes, $s > 2.5$, which are Ricci flat and which have the same Cauchy data are related by an H^{s+1}-coordinate transformation. The key idea is to show that any H^s-spacetime when expressed in harmonic coordinates is also of class H^s. This in turn is based on an old result of Sobolev (1963); namely, that solutions to the wave equation with (H^s, H^{s-1}) coefficients preserve (H^{s+1}, H^s) Cauchy data, a result implied by theorem 4.15. We can give an alternative proof of this result using the well-known result that any single second-order hyperbolic equation can be reduced to a system of symmetric hyperbolic equations (see Fischer and Marsden, 1972*b*). The result follows:

Lemma 4.25

Let $s > 2.5$ and $(\psi_0(x), \dot{\psi}_0(x))$ be of Sobolev class (H^{s+1}, H^s) on $\mathbb{R}^3$. Then there exists a unique $\psi(t, x)$ of class H^{s+1} that satisfies

$$g^{\mu\nu}(t, x)\left(\frac{\partial^2 \psi}{\partial x^\mu \partial x^\nu}\right) + b^\mu(t, x)\left(\frac{\partial \psi}{\partial x^\mu}\right) + c(t, x)\psi = 0$$

$$\left(\psi(0, x), \frac{\partial \psi(0, x)}{\partial t}\right) = (\psi_0(x), \dot{\psi}_0(x)),$$

The Cauchy problem for relativity

where $g^{\mu\nu}(t, x)$ is a Lorentz metric of class H^s, $b^\mu(t, x)$ is a vector field of class H^{s-1}, and $c(t, x)$ is of class H^{s-1}.

We can now prove that when one transforms an H^s-spacetime to harmonic coordinates, it stays H^s.

Theorem 4.26

Let $g_{\mu\nu}(t, x)$ be an H^s-spacetime, $s > 2.5$. Then there exists an H^{s+1}-coordinate transformation $\bar{x}^\lambda(x^\mu)$ such that

$$\bar{g}_{\mu\nu}(\bar{x}^\lambda) = \frac{\partial x^\alpha}{\partial \bar{x}^\mu}(\bar{x}^\lambda)\frac{\partial x^\beta}{\partial \bar{x}^\nu}(\bar{x}^\lambda)g_{\alpha\beta}[x^\mu(\bar{x}^\lambda)]$$

is an H^s-spacetime with $\bar{\Gamma}^\mu(\bar{t}, \bar{x}) = \bar{g}^{\alpha\beta}\bar{\Gamma}^\mu_{\alpha\beta}(\bar{t}, \bar{x}) = 0$.

Proof. To find $\bar{x}^\lambda(x^\mu)$ consider the wave equation

$$\Box \psi = -g^{\alpha\beta}\left(\frac{\partial^2 \psi}{\partial x^\alpha \partial x^\beta}\right) + g^{\alpha\beta}\Gamma^\mu_{\alpha\beta}\left(\frac{\partial \psi}{\partial x^\mu}\right) = 0,$$

and let $\bar{t}(t, x)$ be the unique solution of the wave equation with Cauchy data $t(0, x) = 0$, $\partial \bar{t}(0, x)/\partial t = 1$, and let $\bar{x}^i(t, x)$ be the unique solution of the wave equation with Cauchy data

$$\bar{x}^i(0, x) = x^i, \quad \frac{\partial \bar{x}^i}{\partial t}(0, x) = 0.$$

For $g_{\mu\nu}$ of class H^s, Γ^μ is of class H^{s-1}, so $\bar{t}(t, x)$ and $\bar{x}(t, x)$ are H^{s+1}-functions and in fact by the inverse function theorem for H^s-functions (Ebin, 1970), $(\bar{t}(t, x), \bar{x}(t, x))$ is an H^{s+1} diffeomorphism in a neighborhood of $t = 0$.

Since $\Box \bar{x}^\mu(t, x) = 0$ is an invariant equation,

$$\Box \bar{x}^\mu = -\bar{g}^{\alpha\beta}\frac{\partial^2 \bar{x}^\mu}{\partial \bar{x}^\alpha \partial \bar{x}^\beta} + \bar{g}^{\alpha\beta}\bar{\Gamma}^\nu_{\alpha\beta}\frac{\partial \bar{x}^\mu}{\partial \bar{x}^\nu} = \bar{g}^{\alpha\beta}\bar{\Gamma}^\mu_{\alpha\beta} = 0$$

in the barred coordinate system, so $\bar{x}^\mu$ is a system of harmonic coordinates. ∎

Remark. This theorem may be regarded as a special case of the general theory of harmonic maps (Eells and Sampson, 1964).

Chapter 4. The initial value problem

As a simple consequence of lemma 4.25 we have the following uniqueness result for the Einstein equations:

Theorem 4.27

Let $g_{\mu\nu}(t, x)$ and $\bar{g}_{\mu\nu}(t, x)$ be two Einstein flat H^s-spacetimes with $s > 2.5$ and such that $(g_{\mu\nu}(0, x), \partial g_{\mu\nu}(0, x)/\partial t) = (\bar{g}_{\mu\nu}(0, x), \partial \bar{g}_{\mu\nu}(0, x)/\partial t)$. Then $g_{\mu\nu}(t, x)$ and $\bar{g}_{\mu\nu}(t, x)$ are related by an H^{s+1}-coordinate change in a neighborhood of $t = 0$.

Proof. From lemma 4.25 there exist H^{s+1}-coordinate transformations $y^\mu(x^\alpha)$ and $\bar{y}^\mu(x^\alpha)$ such that the transformed metrics

$$(\partial x^\alpha/\partial y^\mu)(\partial x^\beta/\partial y^\nu)g_{\alpha\beta} \text{ and } (\partial x^\alpha/\partial \bar{y}^\mu)(\partial x^\beta/\partial \bar{y}^\nu)\bar{g}_{\alpha\beta}$$

satisfy $R^{(h)}_{\mu\nu} = 0$. Since the Cauchy data for $g_{\mu\nu}$ and $\bar{g}_{\mu\nu}$ are equal, the transformed metrics also have the same Cauchy data. By uniqueness,

$$(\partial x^\alpha/\partial y^\mu)(\partial x^\beta/\partial y^\nu)g_{\alpha\beta} = (\partial x^\alpha/\partial \bar{y}^\mu)(\partial x^\beta/\partial \bar{y}^\nu)\bar{g}_{\alpha\beta}.$$

Since the composition of H^{s+1}-coordinate changes is also H^{s+1}, $\bar{g}_{\alpha\beta}$ is related to $g_{\alpha\beta}$ by an H^{s+1}-coordinate change in a neighbourhood of $t = 0$. ∎

The local existence and uniqueness theorems 4.23 and 4.27 can be globalized in the same spirit that one studies maximal integral curves for systems of ordinary differential equations. This leads to the following theorem of Choquet-Bruhat and Geroch (1969).

Theorem 4.28

Fix a compact manifold M and let $(g_0, \pi_0) \in \mathcal{C}_\mathcal{H} \cap \mathcal{C}_\delta = \mathcal{C}$ (the solutions of the constraint equations). Then there is a spacetime $(V_4, {}^{(4)}g_0)$ and a spacelike embedding $i_0: M \to V_4$ such that:
  (i) *Ein $({}^{(4)}g_0) = 0$;*
  (ii) *the metric and conjugate momentum induced on $\Sigma_0 = i_0(M)$ is (g_0, π_0);*
  (iii) *Σ_0 is a Cauchy surface;†*
  (iv) *$(V_4, {}^{(4)}g_0)$ is maximal (i.e., cannot be properly and isometrically embedded in another spacetime with properties (i), (ii), and (iii)).*

---

† So that $(V_4, {}^{(4)}g)$ is globally hyperbolic (Hawking and Ellis, 1973, Proposition 6.6.3), and hence any compact spacelike hypersurface is Cauchy (Budic, Isenberg, Lindblom and Yasskin, 1977).

This spacetime $(V_4, {}^{(4)}g_0)$ *is unique in the sense that if we have another* $(V'_4, {}^{(4)}g'_0)$ *with* (i)–(iv) *holding, there is a unique diffeomorphism* $F: V_4 \to V'_4$ *such that*
   (i) $F^* \, {}^{(4)}g'_0 = {}^{(4)}g_0$ (F *is isometric*) *and*
   (ii) $F \circ i_0 = i'_0$.

The proof is conveniently available in Hawking and Ellis (1973). The uniqueness of F uses the fact that an isometry is determined by its action on a frame at a point. The linearized version of this result is needed in the next section (see Fischer and Marsden, 1978a, for details).

Theorem 4.29

Let $(V_4, {}^{(4)}g_0)$ *be a vacuum spacetime, i.e.,* $\mathrm{Ein}({}^{(4)}g_0) = 0$ *with a compact Cauchy surface* $\Sigma_0 = i_0(M)$ *and with induced metric and canonical momentum* $(g_0, \pi_0) \in \mathscr{C}_\mathscr{H} \cap \mathscr{C}_\mathscr{S}$. *Let* $(h_0, \omega_0) \in S_2 \times S_d^2$ *satisfy the linearized constraint equations, i.e.,*

$$D\Phi(g_0, \pi_0) \cdot (h_0, \omega_0) = 0.$$

Then there exists an ${}^{(4)}h_0 \in S_2(V_4)$ *such that*

$$D \, \mathrm{Ein}({}^{(4)}g_0) \cdot {}^{(4)}h_0 = 0$$

and such that the linearized Cauchy data induced by ${}^{(4)}h_0$ *on* Σ_0 *is* (h_0, ω_0).

If ${}^{(4)}h'_0$ *is another such solution, there is a unique vector field* ${}^{(4)}X$ *on* V_4 *such that*

$${}^{(4)}h'_0 = {}^{(4)}h_0 + L_{{}^{(4)}X} \, {}^{(4)}g_0$$

and ${}^{(4)}X$ *and its derivative vanish on* Σ_0.

Remarks (*a*). The linearized Cauchy data is defined in the same manner as the (g, π) are defined. In fact, if ${}^{(4)}g(\rho)$ is a curve of Lorentz metrics tangent to ${}^{(4)}h$ at ${}^{(4)}g_0$, then

$$(h_0, \omega_0) = \left(\left. \frac{\partial g(\rho)}{\partial \rho} \right|_{\rho=0}, \left. \frac{\partial \pi(\rho)}{\partial \rho} \right|_{\rho=0} \right)$$

where $(g(\rho), \pi(\rho))$ are the Cauchy data induced on Σ_0 from ${}^{(4)}g(\rho)$.

(*b*). One can view harmonic coordinates as a technical tool in which to verify the abstract theory in section 4.3. However, once this is done, well-posedness follows in any gauge. For example, one can give a

Chapter 4. The initial value problem

coordinate-free treatment of hyperbolic systems (see Marsden, Ebin and Fischer, 1972 p. 247). Furthermore, for numerical calculations, work of Smarr and others indicates that maximal slices or slices of constant mean curvature may be more useful than harmonic coordinates.

The abstract theory given in section 4.3 (see theorem 4.16) applies to fields coupled to gravity as well as to pure gravity. There are several points to be noted however (cf. Hawking and Ellis, 1963, Section 7.7).

(i) The fields should be minimally coupled to gravity so the hyperbolic character of the equations for the gravitational field is not destroyed.

(ii) The energy–momentum tensor must be a smooth function (not necessarily polynomial) of $^{(4)}g$, $^{(4)}\varphi$.

(iii) For fixed $^{(4)}g$, the (linearized) matter equations should be well posed. This is needed so that hypothesis (A1) of theorem 4.16 can be verified.†

The other conditions of theorem 4.16 are of a technical nature, but cannot be ignored (they sharpen and are needed to verify condition (b), p. 254, of Hawking and Ellis, 1973). For examples of coupled systems and existence theory done by direct methods, see Choquet-Bruhat (1962).

For systems coupled to gravity, the results on uniqueness and global Cauchy developments given above for the vacuum equations carry over in routine fashion.

4.5 Linearization stability of the vacuum Einstein equations

Linearization stability concerns the validity of first-order perturbation theory. The idea is the following. Suppose we have a differentiable function F and points x_0 and y_0 such that $F(x_0) = y_0$. A standard procedure for finding other solutions to the equation $F(x) = y_0$ near x_0 is to solve the linearized equation $DF(x_0) \cdot h = 0$ and assert that $x = x_0 + \rho h$ is, for small ρ, an approximate solution to $F(x) = y_0$. Technically, this assertion may be stated as follows: there exists a curve of exact solutions $x(\rho)$ for small ρ such that $F(x(\rho)) = y_0$, $x(0) = x_0$, and $x'(0) = h$. If this assertion is valid, we say F is *linearization stable* at x_0. It is easy to give examples where the assertion is false. For instance, in two dimen-

---

† As noted by Hawking and Ellis (1973), this can be roughly described by saying that 'the null cones of the matter equations coincide with (as in the Einstein–Maxwell system) or lie within the null cone of the spacetime metric'.

sions $F(x_1, x_2) = x_1^2 + x_2^2 = 0$ has no solutions other than $(0, 0)$, although the linearized equation $DF(0, 0) \cdot (h, k) = 0$ has many solutions. Thus it is a non-vacuous question whether or not an equation is linearization stable at some given solution. Intuitively, linearization stability means that first-order perturbation theory is valid near x_0 and there are no spurious directions of perturbation.

The question of linearization stability is important for relativity. In the literature it was often assumed that solutions to the linearized equations do in fact approximate solutions to the exact equations. However, Brill and Deser (1973) indicated that for the flat three-torus, with zero extrinsic curvature, there are solutions to the linearized constraint equations which are not approximated by a curve of exact solutions. They gave a second-order perturbation argument to show that subject to the condition tr $\pi = 0$, there are no other nearby solutions to the constraint equations, except essentially trivial modifications, even though there are many non-trivial solutions to the linearized equations (see Fischer and Marsden, 1975a, for a complete proof). It is analogous to and is proved by techniques similar to the following *Isolation Theorem* in geometry (Fischer and Marsden, 1975b).†

Theorem 4.30

If M is compact and g_F is a flat metric on M, then there is a neighborhood U_{g_F} of g_F in the space of metrics $\mathcal{M}$ such that any metric g in the neighborhood U_{g_F} with $R(g) \geq 0$ is flat.

The proof amounts to a version of the Morse lemma adapted to infinite-dimensional spaces with special attention needed because of the coordinate invariance of the scalar curvature map.

The results on linearization stability are due, independently, to Choquet-Bruhat and Deser (1972) for flat space, and Fischer and Marsden (1973a, 1974, 1975a) for the general case of empty spacetimes with a compact hypersurface. The methods used are rather different. O'Murchadha and York (1974a) generalized the Choquet-Bruhat and Deser method to the case of spacetimes with a compact hypersurface; see Choquet-Bruhat and York (1979). Results for Robertson–Walker spacetimes were proved by D'Eath (1975) and results

---

† This result has recently been globalized by Schoen and Yau as a special case of their solution to the mass problem in relativity. For example, they prove that on the three-torus, any metric with $R(g) \geq 0$ is flat.

Chapter 4. The initial value problem

for gauge theories coupled to gravity have been obtained by Arms (1977). The flat space result is:

Theorem 4.31

Near Minkowski space, the Einstein empty space equations $\text{Ein}\,(^{(4)}g) = 0$ *are linearization stable.*

In this theorem, one must use suitable function spaces with asymptotic conditions and asymptotically flat spacetimes. We will only consider the compact case in this article; see Choquet-Bruhat, Fischer and Marsden (1978) for the non-compact case.

We begin by defining linearization stability for the empty space Einstein equations.

Let $\text{Ein}\,(^{(4)}g_0) = 0$. An *infinitesimal deformation* of $^{(4)}g_0$ is a solution $^{(4)}h \in S_2(V_4)$ of the linearized equations

$$D\,\text{Ein}\,(^{(4)}g_0) \cdot {}^{(4)}h = 0.$$

The Einstein equations are *linearization stable at* $^{(4)}g_0$ (or $^{(4)}g_0$ is *linearization stable*) if for every infinitesimal deformation $^{(4)}h$ of $^{(4)}g_0$, there exists a C^1 curve $^{(4)}g(\rho)$ of exact solutions to the empty space field equations (on the same V_4),

$$\text{Ein}\,[^{(4)}g(\rho)] = 0,$$

such that $^{(4)}g(0) = {}^{(4)}g_0$ and $\partial^{(4)}g(0)/\partial\rho = {}^{(4)}h_0$.

This definition has to be qualified slightly to be strictly accurate. Namely, for any compact set $D \subset V_4$, we only require $^{(4)}g(\rho)$ to be defined for $|\rho| < \varepsilon$ where ε may depend on D. The reason for this is that $^{(4)}g(\rho)$ will be developed from a curve of Cauchy data $(g(\rho), \pi(\rho))$ and so $^{(4)}g(\rho)$ will be uniformly close to $^{(4)}g_0$ on compact sets for $|\rho| < \varepsilon$, but not on all of V_4 in general.

Since we are fixing the hypersurface topology M here, all Cauchy developments lead topologically to the same spacetime $V_4 \approx \mathbb{R} \times M$, so fixing V_4 is not a serious restriction. Topological perturbations are, of course, another story.

Using the linearized dynamical Einstein system, linearization stability of the Einstein equations is equivalent to linearization stability of the constraint equations, as we shall see below. In fact, linearization stability of a well-posed hyperbolic system of partial differential equations is equivalent to linearization stability of any nonlinear constraints present.

Linearization stability of the vacuum Einstein equations

In terms of the linearized map $D\Phi(g, \pi)$, we can give necessary and sufficient conditions for the constraint equations

$$\Phi(g, \pi) = 0$$

to be linearization stable at (g_0, π_0); that is, if $(h, \omega) \in S_2 \times S_d^2$ satisfies the linearized equations

$$D\Phi(g_0, \pi_0) \cdot (h, \omega) = 0,$$

then there exists a differentiable curve $(g(\rho), \pi(\rho)) \in T^*\mathcal{M}$ of exact solutions to the constraint equations

$$\Phi(g(\rho), \pi(\rho)) = 0$$

such that $(g(0), \pi(0)) = (g_0, \pi_0)$ and

$$\left(\frac{\partial g(0)}{\partial \rho}, \frac{\partial \pi(0)}{\partial \rho} \right) = (h, \omega).$$

The main result follows:

Theorem 4.32

Let $\Phi = (\mathcal{H}, \mathcal{J}): T^*\mathcal{M} \to C_d^\infty \times \Lambda_d^1$ *be defined as in section 4.2 so* $\mathcal{C}_\mathcal{H} \cap \mathcal{C}_\delta = \Phi^{-1}(0)$. *Let* $(g_0, \pi_0) \in \mathcal{C}_\mathcal{H} \cap \mathcal{C}_\delta$. *The following conditions are equivalent*:

(i) *the constraint equations*

$$\Phi(g, \pi) = 0$$

are linearization stable at (g_0, π_0);

(ii) $D\Phi(g_0, \pi_0): S_2 \times S_d^2 \to C_d^\infty \times \Lambda_d^1$ *is surjective*;

(iii) $[D\Phi(g_0, \pi_0)]^*: C^\infty \times \mathcal{X} \to S_d^2 \times S_2$ *is injective*.

Remark. In section 4.2 we listed some sufficient conditions in order for (ii) to be valid, namely the conditions $C_\mathcal{H}$, C_δ, and C_{tr}.

Proof of theorem 4.32. In section 4.2 we showed that $[D\Phi(g_0, \pi_0)]^*$ is elliptic. Thus, the equivalence of (ii) and (iii) is an immediate consequence of the Fredholm alternative.

(ii) implies (i). The kernel of $D\Phi(g_0, \pi_0)$ splits by the Fredholm alternative. Thus the implicit function theorem implies that near (g_0, π_0), $\Phi^{-1}(0)$ is a smooth manifold. Here one must use the Sobolev spaces and

Chapter 4. The initial value problem

pass to C^∞ by a regularity argument, as in Fischer and Marsden (1975*b*). Since any tangent vector to a smooth manifold is tangent to a curve in the manifold, (i) results.

(i) implies (iii). This is less elementary and will just be sketched. Assume (i) and that $[D\Phi(g_0, \pi_0)]^* \cdot (N, X) = 0$, but $(N, X) \neq 0$. We will derive a contradiction by showing that there is a necessary second-order condition on first-order deformations (h, ω) that must be satisfied in order for the deformation to be tangent to a curve of exact solutions to the constraints. Thus, let (h, ω) be a solution to the linearized equations, and let $(g(\rho), \pi(\rho))$ be a curve of exact solutions of

$$\Phi(g(\rho), \pi(\rho)) = 0 \tag{4.11}$$

through (g_0, π_0) and tangent to (h, ω). Differentiating (4.11) twice and evaluating at $\rho = 0$ gives

$$D\Phi(g_0, \pi_0) \cdot (g''(0), \pi''(0)) + D^2\Phi(g_0, \pi_0) \cdot ((h, \omega), h, \omega)) = 0 \tag{4.12}$$

where

$$g''(0) = \frac{\partial^2 g(0)}{\partial \rho^2} \quad \text{and} \quad \pi''(0) = \frac{\partial^2 \pi(0)}{\partial \rho^2}.$$

Contracting (4.12) with (N, X) and integrating over M, the first term of (4.12) gives

$$\int \langle (N, X), D\Phi(g_0, \pi_0) \cdot (g''(0), \pi''(0)) \rangle$$
$$= \int \langle [D\Phi(g_0, \pi_0)]^* \cdot (N, X), (g''(0), \pi''(0)) \rangle = 0,$$

since $(N, X) \in \ker [D(g_0, \pi_0)]^*$.

Thus the first term of (4.12) drops out, leaving the necessary condition

$$\int \langle (N, X), D^2\Phi(g_0, \pi_0) \cdot ((h, \omega), (h, \omega)) \rangle = 0, \tag{4.13}$$

which must hold for all $(h, \omega) \in \ker D\Phi(g_0, \pi_0)$. An argument like that in Bourguignon, Ebin and Marsden (1975) can be used to show that (4.13) is a non-trivial condition (see Arms and Marsden, 1979, and Fischer, Marsden and Moncrief, 1978). ∎

The procedure for finding a second-order condition when linearization stability fails is quite general. See Fischer and Marsden (1975*a, b*) for other applications.

From the linearization stability of the constraint equations, we can deduce linearization stability of the spacetime, and vice versa, as follows.

Theorem 4.33

Let $(V_4, {}^{(4)}g_0)$ be a vacuum spacetime which is the maximal development of Cauchy data (g_0, π_0) on a compact hypersurface $\Sigma_0 = i_0(M)$.
Then the Einstein equations on V_4,

$$\text{Ein}\,({}^{(4)}g) = 0,$$

are linearization stable at ${}^{(4)}g_0$ if and only if the constraint equations

$$\Phi(g, \pi) = 0$$

are linearization stable at (g_0, π_0).

In particular, if conditions $C_{\mathscr{H}}$, C_δ, and C_{tr} hold for (g_0, π_0), then the Einstein equations are linearization stable.

Proof. Assume first that the constraint equations are linearization stable. Let ${}^{(4)}h_0$ be a solution to the linearized equations at ${}^{(4)}g_0$ and let (h_0, ω_0) be the induced deformation of (g, π) on Σ_0. Now (h_0, ω_0) satisfies the linearized constraint equations. By assumption, there is a curve $(g(\rho), \pi(\rho)) \in \mathscr{C}_{\mathscr{H}} \cap \mathscr{C}_\delta$ tangent to (h_0, ω_0) at (g_0, π_0).

By the existence theory for the Cauchy problem, there is a curve ${}^{(4)}g(\rho)$ of maximal solutions on $V_4 \approx \mathbb{R} \times M$ of $\text{Ein}\,({}^{(4)}g(\rho)) = 0$ and with Cauchy data $(g(\rho), \pi(\rho))$. By theorems 4.19 and 4.24 ${}^{(4)}g(\rho)$ will be, for a given choice of lapse and shift, a smooth function of ρ in the sense of theorem 4.19 or in the usual C^∞ sense. As earlier, for any compact set $D \subset V_4$ and $\varepsilon > 0$, there is a $\delta > 0$ such that ${}^{(4)}g(\rho)$ is within ε of ${}^{(4)}g_0$ (using any standard topology) on D.

Using the uniqueness results for the linearized and full Einstein system, one can transform the curve ${}^{(4)}g(\rho)$ by diffeomorphisms so that ${}^{(4)}h_0$ is its tangent at $\rho = 0$. See Fischer and Marsden (1978a) for details. ∎

Moncrief (1975a) has proved that for $(g, \pi) \in \mathscr{C}_{\mathscr{H}} \cap \mathscr{C}_\delta$, the map $[D\Phi(g, \pi)]^*$ is injective if and only if a spacetime ${}^{(4)}g$ generated by (g, π) has no (non-trivial) Killing vector fields ${}^{(4)}Y$ (i.e., $L_{{}^{(4)}Y}\,{}^{(4)}g = 0$ implies ${}^{(4)}Y = 0$); together with theorems 4.32 and 4.33, Moncrief's result then gives necessary and sufficient conditions for a spacetime with compact Cauchy spacelike hypersurfaces to be linearization stable.

Chapter 4. The initial value problem

Moncrief's result still does not give necessary and sufficient conditions for $[D\Phi(g, \pi)]^*$ to be injective in terms of the (g, π) (the conditions $C_{\mathcal{H}}$, C_δ, and $\mathfrak{C}_{tr}$ are sufficient but not necessary), but bypasses the condition tr π' = constant, apparently rendering it much less important.

Theorem 4.34 (Moncrief, 1975a)

Let $^{(4)}g$ be a solution to the empty space field equations $\mathrm{Ein}\,(^{(4)}g) = 0$. *Let $\Sigma_0 = i_0(M)$ be a compact Cauchy hypersurface with induced metric g_0 and canonical momentum π_0. Then* $\ker [D\Phi(g_0, \pi_0)]^*$ *(a finite-dimensional vector space) is isomorphic to the space of Killing vector fields of $^{(4)}g$. In fact,*

$$(Y_\perp, Y_\parallel) \in \ker [D\Phi(g_0, \pi_0)]^*$$

if and only if there exists a Killing vector field $^{(4)}Y$ of $^{(4)}g$ whose normal and tangential components to Σ_0 are $Y_\perp$ and $Y_\parallel$.

See Coll (1977) and Fischer and Marsden (1978a) for alternative proofs to the one given by Moncrief.

As an important corollary of this result, we observe that *the condition* $\ker [D\Phi(g_0, \pi_0)]^* = \{0\}$ *is hypersurface independent* (since it is equivalent to the absence of Killing vector fields, which is hypersurface independent). The condition is also obviously unchanged if we pass to an isometric spacetime.

Putting all this together yields the main linearization stability theorem.

Theorem 4.35

Let $^{(4)}g_0$ be a solution of the vacuum field equations $\mathrm{Ein}\,(^{(4)}g_0) = 0$. *Assume that the spacetime $(V_4, {}^{(4)}g_0)$ has a compact Cauchy surface Σ_0. Then the Einstein equations on V_4*

$$\mathrm{Ein}\,(^{(4)}g) = 0$$

are linearization stable at $^{(4)}g_0$ if and only if $^{(4)}g_0$ has no Killing vector fields.

We conclude this section by briefly examining the case in which $^{(4)}g_0$ is not linearization stable. The goal is to find necessary and sufficient conditions on a solution $^{(4)}h$ of the linearized equations so that $^{(4)}h$ is

Linearization stability of the vacuum Einstein equations

tangent to a curve of exact solutions through $^{(4)}g_0$. The necessary conditions will be derived; for sufficiency, see Fischer, Marsden and Moncrief (1978).

In theorem 4.32 we showed that if $^{(4)}h$ is tangent to a curve of exact solutions and $(N, X) \in \ker [D\Phi(g_0, \pi_0)]^*$, then

$$\int_{\Sigma_0} \langle (N, X), D^2\Phi(g_0, \pi_0) \cdot ((h, \omega), (h, \omega)) \rangle = 0.$$

Following Moncrief (1976), we can re-express this second-order condition in terms of the spacetime, just as the condition $\ker D\Phi(g_0, \pi_0) = \{0\}$ was so re-expressed. See Fischer and Marsden (1978a) and Fischer, Marsden and Moncrief (1978) for alternative proofs.

Theorem 4.36 (Moncrief, 1976).

Let Ein $(^{(4)}g_0) = 0$, and let $^{(4)}h \in S_2(V_4)$ satisfy the linearized equations

$$D \text{ Ein } (^{(4)}g_0) \cdot {^{(4)}h} = 0.$$

Let $^{(4)}Y$ be a Killing vector field of $^{(4)}g_0$ (so that $^{(4)}g_0$ is linearization unstable). Let Σ_0 be a compact Cauchy hypersurface and let $(Y_\perp, Y_\parallel)$ be the normal and tangential components of $^{(4)}Y$ on Σ_0. Then a necessary second-order condition for $^{(4)}h$ to be tangent to a curve of exact solutions is

$$\int_{\Sigma_0} \langle D^2 \text{ Ein } (^{(4)}g_0) \cdot (^{(4)}h, {^{(4)}h}), \;(^{(4)}Y_{\Sigma_0}, {^{(4)}Z_{\Sigma_0}}) \rangle \, d\mu(g_0)$$

$$= \int_{\Sigma_0} \langle (Y_\perp, Y_\parallel), D^2\Phi(g_0, \pi_0) \cdot ((h, \omega), (h, \omega)) \rangle = 0. \tag{4.14}$$

If Ein $(^{(4)}g_0) = 0 = D \text{ Ein } (^{(4)}g_0) \cdot {^{(4)}h_0}$, then $D^2 \text{ Ein } (^{(4)}g_0) \cdot (^{(4)}h, {^{(4)}h})$ has zero divergence (Taub, 1970). Thus, if $^{(4)}Y$ is a Killing vector field, then the vector field

$$^{(4)}W = {^{(4)}Y} \cdot [D^2 \text{ Ein } (^{(4)}g_0) \cdot (^{(4)}h, {^{(4)}h})]$$

also has zero divergence. Thus the necessary second-order condition

$$\int_{\Sigma_0} \langle {^{(4)}W}, {^{(4)}Z_{\Sigma_0}} \rangle \, d\mu(g_0) = 0$$

on first-order deformations is independent of the Cauchy hyper-surface on which it is evaluated. The integral of $^{(4)}W$ over a Cauchy hypersurface then represents a conserved quantity for the gravitational field,

Chapter 4. The initial value problem

constructed from a solution $^{(4)}h$ of the linearized equations and from a Killing vector field $^{(4)}Y$. The interesting and important feature of this conserved quantity of Taub, as shown by theorem 4.36, is that unless it is zero, the first-order solution $^{(4)}h$ from which $^{(4)}W$ was constructed is not tangent to any curve of exact solutions. Thus, for spacetimes which are not linearization stable, Taub's conserved quantity plays the central role in testing whether or not perturbations $^{(4)}h$ are spurious (i.e., are not tangent to any curve of exact solutions).

4.6 The space of gravitational degrees of freedom

We now review some results of symplectic geometry that provide a basis for a unified description of the various splittings that occur in general relatively (Arms, Fischer and Marsden, 1975). These results are based on a general reduction of phase spaces for which there is an invariant Hamiltonian system under some group action (Marsden and Weinstein, 1974). A further application of these results leads to the construction of the symplectic space of gravitational degrees of freedom (Fischer and Marsden, 1978b).†

Background references for the material in this section are Abraham and Marsden (1978), Chernoff and Marsden (1974), and Marsden (1974).

Let P be a manifold and Ω a symplectic form on P; that is, Ω is a closed (weakly) non-degenerate two-form. For relativity, P will be $T^*\mathcal{M}$ and Ω will be the canonical symplectic form J^{-1}, as described in section 4.1.

Let G be a topological group which acts canonically on P; that is, for each $g \in G$, the action of g on P, $\Phi_g : p \mapsto g \cdot p$, preserves Ω. Assume there is a *moment* Ψ for the action. This means the following: Ψ is a map from P to $\mathfrak{g}^*$, the dual to the Lie algebra $\mathfrak{g} = T_e G$ of G, such that

$$\Omega(\xi_P(p), v_p) = \langle d\Psi(p) \cdot v_p, \xi \rangle$$

for all $\xi \in \mathfrak{g}$, where ξ_P is the corresponding infinitesimal generator (Killing form) on P, and $v_p \in T_p P$. Another way to define Ψ is to require that for each ξ, the map $p \mapsto \langle \Psi(p), \xi \rangle$ be an energy function for the Hamiltonian vector field ξ_P. This concept of a moment is an important

---

† It should be noted that in the case of compact Cauchy surfaces, the space of gravitational degrees of freedom has had all of the dynamical degrees of freedom factored out. For some purposes this may be undesirable and a less severe identification may be wanted. (See York, 1972, and Fischer and Marsden, 1977.)

The space of gravitational degrees of freedom

geometrization of the various conservation theorems of classical mechanics and field theory, including Noether's theorem.

It is easy to prove that if H is a Hamiltonian function on P with corresponding Hamiltonian vector field X_H, i.e., $dH(p) \cdot v = \Omega_p(X_H(p), v)$, or equivalently, $i_{X_H}\Omega = dH$, and if H is invariant under G, then Ψ is a constant of the motion for X_H; i.e., if F_t is the flow of X_H, then $\Psi \circ F_t = \Psi$.

As an example, consider a group G acting on a configuration space Q. This action lifts to a canonical action on the phase space T^*Q. The moment in this case is given by

$$\langle \Psi(\alpha_q), \xi \rangle = \langle \xi_Q(q), \alpha_q \rangle,$$

where α_q belongs to T^*Q. If G is the set of translations or rotations, Ψ is linear or angular momentum, respectively. As expected, Ψ is a vector, and the transformation property required of this vector is equivariance of the moment under the co-adjoint action of G on $\mathfrak{g}$; that is, the diagram

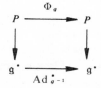

must commute. We shall consider only equivariant moments.

There are several classical theorems concerning reduction of phase spaces. In celestial mechanics, there is Jacobi's elimination of the node, which states that in a rotationally invariant system, we can eliminate four of the variables and still have a Hamiltonian system in the new variables. Another classical theorem of Hamiltonian mechanics states that the existence of k first integrals in involution allows a reduction of $2k$ variables in the phase space. Both of these theorems follow from a theorem of Marsden and Weinstein (1974) on the reduction of phase space.

To construct this reduced space, let $\mu \in \mathfrak{g}^*$ and set

$$G_\mu = \{g \in G | \mathrm{Ad}^*_{g^{-1}} \mu = \mu\}.$$

Consider $\Psi^{-1}(\mu) = \{p | \Psi(p) = \mu\}$. The equivariance condition implies that G preserves $\Psi^{-1}(\mu)$, so we can consider $P_\mu = \Psi^{-1}(\mu)/G_\mu$. In the case that $\Psi^{-1}(\mu)$ is a manifold (e.g., μ is a regular value) and G acts freely and properly on this manifold, we have:

Chapter 4. The initial value problem

Theorem 4.37

P_μ inherits a natural symplectic structure from P, and a Hamiltonian system on P which is invariant under the canonical action of G projects naturally to a Hamiltonian system on P_μ.

In Jacobi's elimination of the node, G is SO(3), so $\mathfrak{g}$ is $\mathbb{R}^3$ and the co-adjoint action is the usual one. Thus the isotropy subgroup G_μ of a point μ in $\mathbb{R}^3$ is S^1. If n is the dimension of P, then $\Psi^{-1}(\mu)$ is the solution set for three equations so the dimension of $\Psi^{-1}(\mu)/G_\mu$ is $n - 3 - 1 = n - 4$. For k first integrals in involution, G is a k-dimensional abelian group, so the co-adjoint action is trivial and $G_\mu = G$. Thus the dimension of $\Psi^{-1}(\mu)/G$ is $n - 2k$. Another known theorem that follows from theorem 4.37 is the Kostant–Kirillov theorem which states that the orbit of a point μ of $\mathfrak{g}^*$ under the adjoint action is a symplectic manifold.

Now we shall show how to obtain a general splitting theorem for symplectic manifolds, one piece of which is tangent to the reduced space P_μ (Arms, Fischer and Marsden, 1975). This includes the splitting theorems for symmetric tensors as a special case.

A splitting theorem for a symplectic manifold P requires a positive-definitive but possibly only weakly non-degenerate metric, or other such structure to give a dualization. This is so that orthogonal complements may be defined. Suppose we know, say from the Fredholm theorem, that

$$T_p P = \text{range } (T_p \Psi)^* \oplus \ker T_p \Psi$$

(here $(T_p \Psi)^*$ is the usual L_2-adjoint). Of course, in finite dimensions this is automatic. Define

$$\alpha_p : \mathfrak{g}_\mu \to T_p P;\ \xi \mapsto \xi_P(p)$$

where $\mathfrak{g}_\mu$ is the Lie algebra of G_μ. Suppose we also have the splitting

$$T_p P = \text{range } \alpha_P \oplus \ker \alpha_P^*.$$

There is a general compatibility condition between these two splittings, namely range $\alpha_p \subset \ker T_p \Psi$, which follows readily from equivariance. In fact,

$$\text{range } \alpha_p = T_p(G \cdot p) \cap \ker T_p \Psi.$$

This compatibility condition implies the finer splitting:

$$T_p P = \text{range } (T_p \Psi)^* \oplus \text{range } \alpha_p \oplus (\ker T_p \Psi \cap \ker \alpha_p^*), \quad (4.15)$$

The space of gravitational degrees of freedom

i.e.,

$$T_pP = \text{range } (T_p\Psi)^* \oplus T_p(G_\mu \cdot p) \oplus \ker T_p\Psi/[Tp(G_\mu \cdot p)].$$

Note that the third summand is the tangent space to P_μ. The geometric picture is given in figure 4.2. For the purposes of this figure we number the

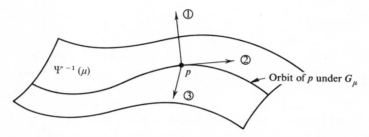

Figure 4.2. The geometry of a general symplectic decomposition.

summands in the previous decomposition as

$$T_pP = ① \oplus ② \oplus ③,$$

where

① belongs to range $(T_p\Psi)^*$, the orthogonal complement of the tangent space to the level $\Psi^{-1}(\mu)$;

② belongs to range α_p, the tangent space to the orbit of p under G_μ;

③ is in $(\ker T_p\Psi \cap \ker \alpha_P^*)$, and is the part of the decomposition which is tangent to the reduced symplectic manifold.

② and ③ together are $\ker T_p\Psi$, the tangent space to $\Psi^{-1}(\mu)$.

A basic splitting of Moncrief (1975b) can be viewed as a special case of this result. We choose $P = T^*\mathcal{M}$ and the 'group' is $G = C^\infty_{\text{space}}(M; V_4, {}^{(4)}g)$, the spacelike embeddings of M to Cauchy hypersurfaces in $(V_4, {}^{(4)}g)$, an Einstein flat spacetime which is the maximal development with respect to some Cauchy hypersurface $\Sigma \subset V_4$.

Although G is not a group, it is enough like a group for the analysis to work.† G 'acts' on (g, π) as follows (see figure 4.3). Let $(V_4, {}^{(4)}g, i_0)$, Ein $({}^{(4)}g) = 0$, be a maximal development which has (g_0, π_0) as Cauchy data on an embedded Cauchy hypersurface $\Sigma_0 = i_0(M)$, $i_0: M \to V_4$ (i_0 is like an origin for C^∞_{space}). Then $i \in C^\infty_{\text{space}}(M; V_4, {}^{(4)}g)$ maps (g_0, π_0) to

† One uses the more general reduction procedure described in Weinstein (1977) and Abraham and Marsden (1978).

Chapter 4. The initial value problem

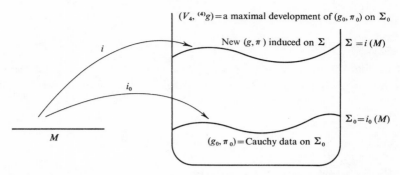

Figure 4.3. Representation of the 'action' of the space of embeddings on the space of Cauchy data.

the (g, π) induced on the hypersurface $\Sigma = i(M)$. The set of all such (g, π) define the orbit of (g_0, π_0) in $\mathscr{C}_{\mathscr{H}} \cap \mathscr{C}_{\mathscr{S}}$. These orbits are disjoint, and so define an equivalence relation, $\sim$, in $\mathscr{C}_{\mathscr{H}} \cap \mathscr{C}_{\mathscr{S}}$.

Although this is not an action (since C^∞_{space} is not a group), it has well-defined orbits and the symplectic analysis above applies (Fischer and Marsden, 1978b). Using the adjoint form of the Einstein evolution system, the moment of 'this action' on a tangent vector $^{(4)}X_\Sigma \in T_{i_0} C^\infty_{\text{space}}(M; V_4, {}^{(4)}g)$ with lapse N and shift X is computed to be

$$\Psi_{(g,\pi)}({}^{(4)}X) = \int N \mathscr{H}(g, \pi) + X \cdot \mathscr{J}(g, \pi).$$

Here the $^{(4)}X_\Sigma$ or the (N, X) can be thought of as belonging to the 'Lie algebra' of C^∞_{space}.

Since $\Psi^{-1}(0)$ is precisely the constraint set $\mathscr{C}_{\mathscr{H}} \cap \mathscr{C}_{\mathscr{S}}$, we choose $\mu = 0$, so $G_\mu = G$. From the equations of motion, we find that

$$\alpha_{(g,\pi)}: \mathfrak{g} \to T_{(g,\pi)}(T^*\mathcal{M})$$

is given by

$$(N, X) \mapsto J \circ [D\Phi(g, \pi)]^* \cdot \binom{N}{X},$$

so the symplectic decomposition (4.15) becomes

$$T_{(g,\pi)} T^* \mathcal{M} = \{\text{range } [D\Phi(g, \pi)]^*\}^* \oplus \text{range } \{J \circ [D(g, \pi)]^*\}$$
$$\oplus \ker D\Phi(g, \pi) \cap [\ker D\Phi(g, \pi) \circ J]^*$$

which is Moncrief's splitting. Elements of the first summand infinitesimally deform (g, π) to Cauchy data which do not satisfy the constraint equations. Elements of the second summand infinitesimally deform

The space of gravitational degrees of freedom

(g, π) to Cauchy data that generate an isometric spacetime, and elements of the third summand infinitesimally deform (g, π) in the direction of new Cauchy data that generate a non-isometric solution to the empty space field equation; see figure 4.4 and compare with figure 4.2.

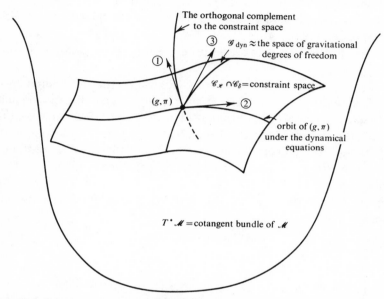

Figure 4.4. Symplectic decomposition applied to the Einstein equations to construct the space of gravitational degrees of freedom.

This third summand represents the tangent space to the reduced space $P_\mu \approx \mathscr{C}_{\mathscr{H}} \cap \mathscr{C}_\delta / \sim$. This quotient by the equivalence relation described above is naturally isomorphic to the space of gravitational degrees of freedom,

$$\mathscr{G}(V_4) = \mathscr{E}(V_4)/\mathscr{D}(V_4),$$

namely the set of maximal solutions to the vacuum Einstein equations

$\mathscr{E}(V_4) = \{^{(4)}g \mid \text{Ein}\,(^{(4)}g) = 0$, and such that $(V_4, {}^{(4)}g)$ is the maximal development of the Cauchy data on some Cauchy hypersurface$\}$

modulo the spacetime diffeomorphism group $\mathscr{D}(V_4)$. This is the space of isometry classes of empty space solutions of the Einstein equations, or the space of gravitational degrees of freedom since the coordinate gauge group has been factored out.

We call the representation of $\mathscr{G}(V_4)$ described here the *dynamical representation* since one uses the canonical formulation to define $P_\mu \approx$

Chapter 4. The initial value problem

$\mathscr{C}_{\mathscr{H}} \cap \mathscr{C}_{\delta}/\sim$. See York (1971), Choquet-Bruhat and York (1979), and Fischer and Marsden (1977) for a conformal representation of $\mathscr{G}(V_4)$.

As we have emphasized, in the case of compact hypersurfaces, one identifies all the (g, π) which occur on slicings in an Einstein flat maximal spacetime. In the non-compact case one does *not* do this, as is explained in Regge and Teitelboim (1974) and Choquet-Bruhat, Fischer and Marsden (1978).

A few further remarks are in order regarding the decomposition of $T_{(g,\pi)}(T^*\mathcal{M})$.

Set $\mathfrak{g}_{(g,\pi)} = \ker D\Phi(g, \pi) \cap [\ker D\Phi(g, \pi) \circ J]^*$, the third summand in the decomposition above. The summand $\mathfrak{g}_{(g,\pi)}$ generalizes the classical transverse traceless (TT) decomposition of Deser (1965) and Brill and Deser (1968). Indeed for $\pi = 0$ and $R(g) = 0$, Moncrief's decomposition reduces to two copies of the Berger and Ebin (1972) splitting. If moreover, Ric $(g) = 0$ (so that g is flat), we regain the original Brill–Deser splitting.

Now suppose $(h, \omega) \in \mathscr{G}_{(g,\pi)}$. Then (h, ω) satisfies the following equations:

$$D\Phi(g, \pi) \cdot (h, \omega) = 0, \qquad (4.16)$$

and

$$[D\Phi(g, \pi) \circ J] \cdot (h, \omega)^* = D\Phi(g, \pi) \cdot ((\omega')^\flat, -h^\sharp \, d\mu(g)) = 0. \qquad (4.17)$$

Written out in terms of the constraint functions $\mathscr{H}$ and $\mathscr{J}$, these equations are

$$D\mathscr{H}(g, \pi) \cdot (h, \omega) = 0,$$
$$D\mathscr{H}(g, \pi) \cdot ((\omega')^\flat, -h^\sharp \, d\mu(g)) = 0,$$
$$D\mathscr{J}(g, \pi) \cdot (h, \omega) = 0,$$
$$D\mathscr{J}(g, \pi) \cdot ((\omega')^\flat, -h^\sharp \, d\mu(g)) = 0.$$

These equations, eight conditions on twelve functions of three variables, formally leave four functions of three variables as parameters of the space $\mathscr{G}_{(g,\pi)}$. Formally, $\mathscr{G}_{(g,\pi)}$ is the tangent space to the space of gravitational degrees of freedom, which is parametrized by four functions of three variables.

Moreover, there is a certain 'symplectic symmetry' in the summand $\mathscr{G}_{(g,\pi)}$, reflected in (4.16) and (4.17) above: if $(h, \omega) \in \mathscr{G}_{(g,\pi)}$, then $J \circ (h, \omega)^*$ is also in $\mathscr{G}_{(g,\pi)}$. We shall refer to this symmetry as J-invariance of $\mathscr{G}_{(g,\pi)}$.

Proposition 4.38

The (weak) symplectic form Ω on $S_2 \times S_d^2$ naturally induces a weak symplectic form Ω' on any J-invariant subspace of $S_2 \times S_d^2$. In particular, $\mathcal{G}_{(g,\pi)}$ is a (weak) symplectic linear space.

Proof. The symplectic form Ω on $S_2 \times S_d^2$ defined by

$$\Omega((h_1, \omega_1), (h_2, \omega_2)) = \int_M \langle J^{-1}(h_1, \omega_1), (h_2, \omega_2) \rangle$$

defines by the same formula an antisymmetric bilinear form Ω' on $\mathcal{G}_{(g,\pi)}$ (or any other J-invariant subspace of $S_2 \times S_d^2$). One has to show Ω' is non-degenerate. Thus suppose for $(h_1, \omega_1) \in \mathcal{G}_{(g,\pi)}$,

$$\int_M \langle J^{-1}(h_1, \omega_1), (h_2, \omega_2) \rangle = 0$$

for all $(h_2, \omega_2) \in \mathcal{G}_{(g,\pi)}$. Since $\mathcal{G}_{(g,\pi)}$ is J-invariant,

$$J \circ \begin{pmatrix} h_1 \\ \omega_1 \end{pmatrix}^* \in \mathcal{G}_{(g,\pi)}.$$

Thus letting

$$\begin{pmatrix} h_2 \\ \omega_2 \end{pmatrix} = -J \circ \begin{pmatrix} h_1 \\ \omega_1 \end{pmatrix}^*,$$

and since $J^* = -J$, we have

$$0 = -\int \langle J^{-1}(h_1, \omega_1), J \circ (h_1, \omega_1)^* \rangle = \int \langle (h_1, \omega_1), (h_1, \omega_1)^* \rangle$$

$$= \int_M (h_1 \cdot h_1 + \omega_1' \cdot \omega_1') \, d\mu(g).$$

Thus $(h_1, \omega_1) = 0$ so that Ω' is non-degenerate. ∎

Proposition 4.38 is a special case of the following general result of symplectic geometry (see Weinstein, 1977).

Theorem 4.39

Let (V, Ω) be a (weak) symplectic vector space and $W \subset V$ a subspace. Let $W_\Omega^\perp = \{v \in V | \Omega(v, \omega) = 0 \text{ for all } \omega \in W\}$ and assume W is co-isotropic, i.e., $W_\Omega^\perp \subset W$. Then $W/W_\Omega^\perp$ is, in a natural way, a (weak) symplectic vector space.

Chapter 4. The initial value problem

Proof. Denote an element of $W/W_\Omega^\perp$ by $w + W_\Omega^\perp$. Define Ω on the quotient by $\Omega(w_1 + W_\Omega^\perp, w_2 + W_\Omega^\perp) = \Omega(w_1, w_2)$. Since $\Omega(w_i, W_\Omega^\perp) = 0$, $i = 1, 2$ this is well defined. On the other hand, if $\Omega(w_1 + W_\Omega^\perp, w_2 + W_\Omega^\perp) = 0$ for all $w_2, w_2 \in W_\Omega^\perp$, then $w_2 + W_\Omega^\perp$ is the zero element of the quotient. ∎

Proposition 4.38 follows as a corollary by letting $V = S_2 \times S_d^2 = T_{(g,\pi)}T^*\mathcal{M}$, Ω as given, and $W = \ker D\Phi(g, \pi)$. Then,

$$\begin{aligned}W_\Omega^\perp &= \{(h, \omega) | \Omega((h, \omega), W) = 0\} \\ &= \{(h, \omega) | J^{-1}(h, \omega) \text{ is orthogonal to } \ker D\Phi(g, \pi)\} \\ &= \{(h, \omega) | J^{-1}(h, \omega) \in \text{range } [D\Phi(g, \pi)]^*\} \\ &= \text{range } \{J \circ [D\Phi(g, \pi)]^*\} \subset W.\end{aligned}$$ ∎

The symplectic structure on $\mathcal{G}$ described above may be important for the problem of quantizing gravity. The symplectic structure presented here is probably implicit in the work of Bergmann (1958), Dirac (1959), and DeWitt (1967). The present formulation, however, allows one to be rather precise and geometric. First of all, it may allow one to use the Segal or Kostant–Souriau quantization formalism to carry out a full quantization or a semi-classical quantization. Secondly, the approach presented here enables one to show that near metrics $^{(4)}g$ in $\mathcal{E}(V_4)$ with no isometries (and hence no spacetime Killing vector fields), $\mathcal{G} = \mathcal{E}(V_4)/\mathcal{D}(V_4)$ is a smooth manifold and is locally isomorphic, in a natural way, to $\mathcal{C}_\mathcal{H} \cap \mathcal{C}_\delta/\sim$, and thus carries a canonical symplectic structure.† Thus, in the neighborhood of Einstein flat spacetimes without Killing vector fields, the space $\mathcal{G} = \mathcal{E}(V_4)/\mathcal{D}(V_4)$ of gravitational degrees of freedom is itself a symplectic manifold, or if you prefer, a gravitational *phase space* without singularities, each element of which represents an empty space geometry. Note that $\mathcal{G}$ is (generically) a symplectic manifold even though it is not a cotangent bundle. We conjecture that $\mathcal{G}$ can actually be stratified into symplectic manifolds, similar to the stratification of superspace; see Fischer (1970) and Bourguignon (1975). The singularities in $\mathcal{G}$ occur near spacetimes with symmetries, and these are of a conical nature (Fischer, Marsden and Moncrief, 1978). Moncrief has emphasized in his 1978 Gravity Research

---

† An interesting point here is that $\mathcal{C}_\mathcal{H} \cap \mathcal{C}_\delta$, although (generically) a submanifold of $T^*\mathcal{M}$, does not have a natural symplectic structure induced from $T^*\mathcal{M}$, since the tangent space of $\mathcal{C}_\mathcal{H} \cap \mathcal{C}_\delta$ is not J-invariant. One must pass to the quotient manifold $\mathcal{C}_\mathcal{H} \cap \mathcal{C}_\delta/\sim$ in order to get a symplectic structure induced by the symplectic structure of $T^*\mathcal{M}$.

Foundation essay that these singularities have an important effect on quantization procedures (for instance, for de Sitter spacetimes).

The methods we have employed to analyze gravity, being based on the L_2-adjoint formalism, carry over directly to fields minimally coupled to gravity, and in particular to Yang–Mills fields. In the latter case, one divides out not by $\mathscr{D}(V_4)$, but by the larger group of equivariant bundle diffeomorphisms (i.e. gauge transformations covering diffeomorphisms of spacetime). Using the methods presented here, we can show that the space of degrees of freedom for fields and gravity, for fields minimally coupled to gravity, is, generally, a symplectic manifold; see Fischer and Marsden (1978*b*, *c*).

Finally we remark that we hope that the geometric methods presented here help to unfold some of the inter-relationships that exist between general relativity, differential geometry, functional analysis, nonlinear partial differential equations, infinite-dimensional dynamical systems, symplectic geometry, and the theory of singularities. Certainly, all of these areas of mathematics (and others) will have to make their contribution to the study of gravitational theory before the final analysis is in.

5. Global structure of spacetimes†

ROBERT GEROCH AND GARY T. HOROWITZ‡

5.1. Introduction

A model for the physical world in classical general relativity begins, of course, with a spacetime – a four-dimensional manifold (representing the physical events) endowed with a metric of Lorentz signature (representing the results of measurements of spatial distances and elapsed times). Further, one has certain additional fields on this manifold (representing other physical phenomena) satisfying the appropriate differential equations. Finally, these fields and the metric are tied together by Einstein's equation. The 'global structure' refers to certain features of such models, including, for example, the issue of what is the underlying manifold, the qualitative behavior of the light-cones, the possibilities for orientations, causal structure and possible causality violations, the existence and properties of spacelike surfaces, the meaning and implications of determinism, etc. As this list suggests, the features of interest refer primarily to just the underlying spacetime. 'Global structure' may not be an entirely apt description: perhaps 'qualitative structure', 'kinematical structure', or 'structure prior to, for example, the curvature and detailed local behavior of the differential equations' would be better.

Global structure has become popular largely in the last decade or so, although its mathematics, as well as most of the physical ideas on which it is based, was around long before. It is a rather small, and by now a relatively fixed, body of ideas. In particular, although its techniques continue to find significant application to important physical problems in

---

† Supported in part by the National Science Foundation, under contract number PHY-76-81102.
‡ National Science Foundation predoctoral fellow.

Introduction

relativity, rather few substantial additions to the body of the subject itself have occurred in recent years.

The original impetus for the study of global structure was the early singularity theorems. These results have the feature that, in order to say things which are true, one has to have some degree of control over the large-scale structure of one's spacetime. One must, among other things, rule out certain, rather pathological, possibilities. This 'ruling out' took the form of, first, isolating careful statements of the unwanted behavior; second, understanding the physics of these statements; and finally, figuring out what would rule out what. The subject retains some of this flavor today. It has very much the character of a 'service subject', i.e. it consists largely of techniques designed to be useful in physical problems in various areas of relativity. In particular, there are very few results which one would regard as exclusively within the domain of global structure, and yet which could stand alone on their intrinsic physical interest. Various problems in relativity require these techniques to a greater or lesser extent. Their influence is very strong, for example, in the study of singularities or black holes, somewhat more moderate in the analysis of exact solutions or asymptotic structure, and comparatively weak in the post-Newtonian approximations or the analysis of the interaction of gravitational and acoustic waves.

One could argue that there is a second reason – besides the applicability of its techniques to specific physical problems – for interest in global structure. There is a sense that one is isolating features which are, not only more qualitative, but perhaps also more 'fundamental'. Imagine compiling a list of the structural ingredients of general relativity, with an estimate for each of a probability that it will be regarded as an essential part of our understanding of space, time, and gravitation fifty years from now. Perhaps 'Einstein's equation' would be assigned a rather small probability compared with, say, 'manifold and metric'. And yet one could easily imagine that both of these could have disappeared, while some remnant of 'causal structure' remains. In short, one is studying space, time, and gravitation with perhaps a bit more distance from the details of general relativity.

The elements of global structure can be divided into three classes: the definitions; the theorems, which show that various relationships hold between these definitions; and the examples, which show that various relationships do not hold.

The definitions are certainly important, but fortunately they are relatively easy, and surprisingly few in number. Indeed, a repertoire of ten

Chapter 5. Global structure of spacetimes

standard definitions will suffice for almost anything one wishes to do with global structure: space- and time-orientable, past and future, stable causality, slice, domain of dependence, Cauchy surface, Cauchy horizon, singular, asymptotically flat, and event horizon.

The theorems, as it turns out, possess a somewhat anomalous status within the subject. It is normally the case, at least in mathematics, that one is able to isolate a relatively small number of key theorems from which the network of relationships within the body of the subject follows more or less directly. The pattern of relationships within global structure, by contrast, seems to be somewhat more diffuse. One by and large lacks 'key theorems', and instead has a large number of true statements, all of about equal utility and such that they do not all follow easily from any yet isolated small number. This structure of the pattern of theorems in turn affects the structure of the proofs. Most proofs start from scratch, and thus tend to be rather long and to have a somewhat technical character. Furthermore, what is called a proof is often a reduction of the statement of the theorem to be proved to various 'intuitively obvious', usually local, assertions about curves, surfaces, metrics, etc. Again, one has been unable to compile a short but sufficiently inclusive list of such local assertions, so that they could then be conveniently subject to full proof. The role of key theorems has been assumed instead by 'methods of proof', for indeed there are a limited number of these, which do occur over and over. In fact, the following eight methods of proof will suffice for most results in this subject: 'Introduce a timelike vector field (and, usually, then consider its integral curves).' 'Carry information about closed curves and compare with what one had at the beginning.' 'Connect timelike or null curves together, and smooth the corners, to obtain new curves.' 'Find a sequence of points with some property and ask for an accumulation point.' 'Choose a sequence of points which approach the point in question (usually from a timelike direction) and try to carry some property to the limit.' 'Ask for the timelike curve of maximal length between two points, or between a point and a surface.' 'Take the domain of dependence, and study the Cauchy horizon.' and 'Find a sequence of timelike curves with some property, and ask for an accumulation curve.' An attempt to reorganize the theorems of this subject along more conventional lines (say, with just one key theorem to isolate all that can be done with each method of proof) might be an interesting project.

We come, finally, to the third class of elements of global structure, the examples. These, too, have some curious features. First, their role could hardly be over-emphasized. There is a vast number of plausible-sounding statements within global structure which are in fact false. Probably, we are still somewhat over-conditioned to Minkowski spacetime. Since it is easy to waste most of one's time trying to prove such statements, the examples can serve as time-savers. The examples, although they are particularly numerous in this subject, are in practice rather easy to obtain. Indeed, the following seven techniques (or, more often, combinations of two or more) will normally produce the desired example: 'Check the known solutions.' 'Tip the light-cones in some way.' 'Take a covering space or product.' 'Isolate what makes an example fail in a local region, and push that region off to infinity.' 'Introduce a conformal factor.' 'Patch spacetimes together across boundary surfaces' and 'Cut holes of various types in given spacetimes.' In fact, the construction of examples seems to be so much easier than retaining or constructing theorems that one's normal initial thought on being confronted with a statement about global structure is not 'Is this the theorem proved by...?', or 'How would I go about making a proof?', but rather 'Can I find a counterexample?'. If one cannot, then a good first guess is that the statement is true. A further feature common to these examples is that they are almost invariably spacetimes with little or no direct physical interest. In particular, Einstein's equation, with some reasonable matter source, is often neither mentioned nor satisfied. This, however, in no way diminishes their significance, but rather merely reflects an aspect of the purpose of the examples: to gain an understanding of what will work and what will not, of what needs to be and can be ruled out – and not to provide physically realistic cosmological models. One's lack of concern with Einstein's equation in these examples is a reflection of the experience that things which can happen in the absence of this equation can usually also happen in its presence.

This paper is a brief survey of global structure. At least three circumstances intervene to make such a project difficult: the fact that the subject is divided into numerous small, and somewhat independent, topics; the fact that its contact with physics is often through its application to specific external problems; and the existence of a number of excellent surveys of this subject, with which we could hardly hope to compete directly. Our response to these circumstances is to emphasize certain aspects of the subject at the expense of others. The emphasis here will be on the flavor and the physics of global structure. That is we give the main definitions and what they are intended to capture, the methods of proof and what they will prove, and the techniques for

Chapter 5. Global structure of spacetimes

examples and where they will work. These items are discussed, in so far as possible, from the standpoint of the physical phenomena one is trying to capture, eliminate, or relate to other phenomena. We have also tried to include a fair amount of solid information as to what is true, what is false, and what is open. The price one pays is the almost complete absence of the more technical aspects of the subject. This material is certainly important, but is easily accessible elsewhere.[1] We avoid even the mention of more refined versions of various definitions, concentrating instead on what seems to be the most useful version. We do not attempt to formulate the strongest statement of a theorem or example when there is available a simpler statement with most of the ideas and utility. We give no full proofs anywhere, but rather 'sketches' in a few sentences to illustrate the structure of the argument.

The discussion is divided into three parts, with each organized around the application of global structure to a broadly defined physical problem. The definitions, techniques, examples, and so on are developed as needed. In section 5.2, 'What is the topology of our universe?', we consider such topics as the underlying manifold, the qualitative behavior of the light-cones, causal structure, and determinism. It is here that most of the basic ideas of the subject are developed. Although all of these topics do make some contact with the issue of the topology of the universe in its broadest sense, there is necessarily more emphasis on mathematics here than in the other two sections. In section 5.3, 'Is our universe singular?', we discuss the famous singularity theorems. These results are certainly the most striking examples of the application of global techniques. There are a number of such theorems on the market, all with a common thread: a certain local calculation involving the convergence of a geodesic congruence, and suitable controls over the global structure of one's spacetime. Finally, in section 5.4, 'How noticeably singular is our universe?', we discuss the issue of cosmic censorship, i.e. that of whether or not one expects under certain circumstances that surviving observers will be able to detect singular behavior in spacetime. The problem here is to formulate a suitable conjecture (presumably, to the effect that under certain circumstances singular behavior will not be observable), and to prove it. It seems certain that global structure will play an important role in this, currently rather active, question. It is particularly useful, in order to make effective use of global structure in particular contexts, to be able to decide quickly and efficiently what is true and what is false. The Appendix contains a list of statements for which this decision is to be made.

5.2 What is the topology of our universe?

The manifold

General relativity requires, among other things, an underlying four-dimensional manifold for spacetime. What is the appropriate manifold for our universe? In this section, we shall discuss what information is available in answer to this question, as well as certain related broad, qualitative aspects of spacetimes.

The most direct mathematical attack on our question might begin by asking for a classification of all 4-manifolds. Such a classification would at least provide one with an overview of the possibilities with which one must contend. The plan would then be to use such a classification as a framework, with the intention of then attempting to eliminate various of the classes by other considerations. Unfortunately, a program along these lines does not appear at present to be very promising. On the one hand, it is easy to construct an enormous variety of possible 4-manifolds; on the other, we are far from having anything like a 'classification' of them all.

At the other extreme is what might be called the experimental approach. One begins with the observation that the question 'What is the underlying manifold of our universe?' is genuinely an experimental one in that, for example, if one had access to all regions of our spacetime then one could determine the answer. While we of course do not have such access, one might hope that, since at least some observations have been made on our universe, some information might be obtained about its underlying manifold. This program, unfortunately, does not appear to be very promising either, for the observational effects due to the underlying manifold seem to be completely dominated by the observational effects due to the fields on that manifold. For example the solutions representing collapsing spherical dust clouds, the Gödel solution, the static fluid ball solutions, the Kasner solutions, the plane-wave solutions, certain Weyl solutions representing two masses held apart by wires stretching to infinity, the open Friedman solutions, and Minkowski spacetime all have the same underlying manifold, $\mathbb{R}^4$. The maximally extended positive-mass Schwarzschild solution ('a wormhole connecting two asymptotically flat regions'), the negative-mass Schwarzschild and certain Weyl solutions ('a singularity as a source for an asymptotically flat spacetime'), and the Reissner–Nordström solution all have the same underlying manifold, $S^2 \times \mathbb{R}^2$. (Of course, the fact that the underlying manifolds of so many familiar solutions are products of spheres and real

Chapter 5. Global structure of spacetimes

lines could hardly be taken as evidence that our universe has so simple a topology!) In these examples, then, spacetimes which obviously have quite different physical characteristics none the less have the same topology. Even such gross features as singular behavior, or the presence of one or many asymptotically flat regions, does not seem to be recorded in the topology. In short, at least as far as physical effects are concerned, the geometry dominates the topology.

There are, none the less, physical arguments for the elimination of a few classes of manifolds – although these consist essentially of a few pathologies which are normally permitted mathematically under the term 'manifold'.[2] Thus one would perhaps not wish to admit a manifold with boundary as a candidate for the underlying manifold of our universe (for the boundary would represent physically an 'edge' to spacetime, while such edges have never been observed); or a non-Hausdorff manifold (i.e. a manifold in which there are two points which cannot be separated by disjoint neighborhoods, such behavior would perhaps violate what we mean physically by 'distinct events'); or a non-connected manifold (for by no stretch of the imagination could communication ever be carried out between the separate connected components, so physically 'our universe' *is* connected); or a non-paracompact manifold (i.e. some connected component cannot be covered by a countable collection of coordinate patches; we shall see the physical reason for this exclusion shortly). In light of these remarks, we shall assume hereafter that, unless otherwise stated, 'manifold' means without boundary, Hausdorff, connected, and paracompact.

The conclusion from all this, then, is that these considerations yield very little information indeed about the underlying manifold of our universe. But general relativity requires of course more than merely a manifold: there must also be a metric of Lorentz signature. One might therefore impose this additional condition – the existence of such a metric – in the hope that it will significantly reduce the number of manifolds which must be dealt with. It turns out, unfortunately, that very little additional information is obtained even by this criterion, for 'practically every' four-dimensional manifold admits *some* metric of Lorentz signature.

In order to discuss this issue in more detail, we shall first need a simple and important tool. We begin by noting that every manifold† M admits[3]

† If a manifold admits a Lorentz metric, it is paracompact. A manifold admits a positive-definite metric if and only if it is paracompact. These facts are the reason we earlier assumed paracompactness.

What is the topology of our universe?

some positive-definite metric g^+_{ab}. The proof consists of first choosing, for each of a countable collection of coordinate patches which cover M, a metric positive-definite in the interior of the patch and zero outside, and then taking the sum of these metrics, possibly after rescaling so that the sum will converge to some metric on M. This same argument will not work, of course, in the Lorentz case, for the sum of two metrics of Lorentz signature need not have that signature. Indeed, the essence of the Lorentz case is this non-additivity – the fact that a Lorentz metric has a 'directional character', which can become intertwined with the topology of the underlying manifold. This directional character is given expression as follows.[4,5] Let manifold M admit Lorentz metric g_{ab}, and fix a positive-definite metric g^+_{ab} on M. Fix a point p of M, and consider the function, on the set of all nonzero tangent vectors at p, whose value at ξ^a is $(g_{ab}\xi^a\xi^b)/(g^+_{mn}\xi^m\xi^n)$. This function must have a minimum, say at $\bar{\xi}^a$. (We are essentially simultaneously diagonalizing the matrices representing the components of our two metrics.) This $\bar{\xi}^a$ will of course be timelike, and unique up to a nonzero factor. Since p is arbitrary, we conclude that a Lorentz metric on M gives rise to a nowhere-vanishing, timelike vector field, ξ^a, unique up to a factor at each point of M – what is called a *direction field* on M. Now suppose, conversely, that M is endowed with a direction field ξ^a. Choose any positive-definite g^+_{ab} on M, and set $g_{ab} = g^+_{ab} - 2g^+_{am}\xi^m g^+_{bn}\xi^n(g^+_{pq}\xi^p\xi^q)^{-1}$, noting that this is a Lorentz metric and that it is independent of the scaling of ξ^a. In this way a direction field on M leads to a Lorentz metric on M. We conclude, then, that manifold M admits a Lorentz metric if and only if it admits a direction field. The advantage of direction fields over Lorentz metrics is that the former are usually easier to work with geometrically.

We now return to the question of which manifolds admit Lorentz metrics. A manifold is said to be *compact* if every sequence of its points has an accumulation point. For example, S^2 and $S^1 \times S^1$, the sphere and torus, are compact, while $\mathbb{R}^2$ and $S^1 \times \mathbb{R}$, the plane and cylinder, are not. We claim, first, that every non-compact manifold M admits a direction field. To see this, first choose a direction field defined at all points of M except possibly for some isolated points at which it cannot be defined, e.g. because of 'dipole-like' behavior.[6] By non-compactness one can find sequences of points on M having no accumulation points. Now introduce a sequence of diffeomorphisms on M which 'shift' these isolated singular points to successive points along such a sequence. Taking the limit, one obtains, since the sequence has no accumulation point, a direction field defined everywhere on M. We conclude, then,

Chapter 5. Global structure of spacetimes

that every non-compact manifold M admits some metric of Lorentz signature. In the compact case, the situation is more complicated: some manifolds, such as the 4-sphere S^4 and $S^2 \times S^2$, do not admit any Lorentz metric; while others, such as $S^1 \times S^3$ and $S^1 \times S^1 \times S^2$, do. (The following simple result is sometimes helpful in deciding the issue for compact manifolds: a product of two compact manifolds admits a direction field if and only if at least one of the factors does.) However, as we shall see shortly, there are physical reasons why compact manifolds are of rather little interest in relativity.

We conclude, then, that even the criterion of the existence of a Lorentz metric sheds little light on the issue of what is the underlying manifold of our universe. But of course one is interested physically, not in just *any* Lorentz metric, but rather one which seems to have some of the qualitative physical features of our own universe. Thus, one might hope to obtain additional, stronger, criteria on the underlying manifold. We give an example of such an argument. In everyday units, the 'average curvature of the universe' is, at least in our own local region, rather small. But within the class of all four-dimensional manifolds, there are many which seem intuitively to be 'highly contorted'. The idea, then, would be to formulate some kind of argument by which certain manifolds are eliminated from consideration on the grounds that every Lorentz metric on such a manifold has greater 'average curvature' than we see in our universe. It is clear that a great deal of technical work would have to be done to formulate a careful argument along these lines, but we can at least consider this issue in a rough way by asking the following, much simpler, question: 'Which 4-manifolds admit a flat metric?' Those manifolds which do not would at least be candidates for elimination by a more refined argument. But it turns out that a surprising variety of 4-manifolds survive even this criterion. A large class of examples can be obtained by what is the most useful and important construction in this subject.

Let M, g_{ab} be a spacetime, and let C be any closed (i.e. such that any limit point, if it exists, of a sequence of points of C is itself in C) subset of M. Then $M - C$ (the manifold which results by removing from M the points of C) with the metric induced by g_{ab}, is, if connected, itself a spacetime. (That C be closed was necessary to ensure that $M - C$ even be a manifold.) This simple fact is the basis for an enormous number of examples: one removes from a given spacetime (often Minkowski) some points or other regions in order to obtain a spacetime having certain properties. There is also a closely related construction in which, instead of 'cutting holes', one 'patches together' different regions of spacetimes. Let M, g_{ab} be a spacetime with boundary consisting of two disjoint three-dimensional manifolds, B_1 and B_2. Then one obtains a new four-dimensional manifold by identifying B_1 and B_2 under some diffeomorphism between these manifolds. The original metric will define a metric on this new manifold, provided the identification is such that the metric joins smoothly. (Of course, one can in this way join two distinct spacetimes to obtain one.) Most often, these boundaries are created by removing some region which is not closed from a given spacetime. The spacetimes which result from these constructions are, in almost every case, physically unrealistic for various reasons. The point of the construction, however, is not normally to construct physically realistic cosmological models, but rather to demonstrate by means of *some* example that a certain assertion is false, or that a certain line of argument cannot work.

We return now to the question of which manifolds admit flat metrics. One might think, for example, that $S^3 \times R$ – the underlying manifold of the closed Friedmann models – would be a good initial candidate. In this spacetime, after all, 'curvature seems to be necessary to permit the spatial sections to be bent around to form three-spheres'. So, one might expect that $S^3 \times R$ admits no flat metric. But it does: removing a point from Minkowski spacetime, we obtain the manifold $S^3 \times R$, with a flat metric. Similarly, the underlying manifold of the maximally extended Schwarzschild spacetime is $S^2 \times R^2$: 'curvature seems necessary in order to have a worm-hole connecting two universes.' But $S^2 \times R^2$ also admits a flat metric: remove a straight line from Minkowski spacetime. Finally, $S^1 \times S^1 \times S^1 \times R$ admits a flat metric. Remove from Minkowski spacetime the region given, in the usual coordinates, by $|x| > 1$ or $|y| > 1$ or $|z| > 1$ (figure 5.1). Now identify pairs of boundary points of the form $(t, 1, y, z)$ and $(t, -1, y, z)$, and similarly for y and z. That is to say, we identify the spatial sections to obtain 3-toruses. These are of course very simple manifolds. A few examples are known of non-compact, four-dimensional manifolds which admit no flat metric.† For other manifolds the question is, as far as we are aware, largely open. We conclude, then, that even the demand that the underlying manifold of our universe admit some flat metric seems to give little information as to what that manifold is.

---

† These turn out to be certain manifolds which do not admit any spinor structure. See reference 7.

Chapter 5. Global structure of spacetimes

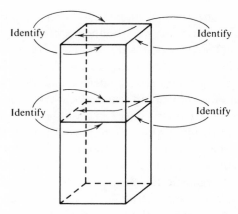

Figure 5.1. A Flat metric exists on $S^1 \times S^1 \times S^1 \times \mathbb{R}$. The figure shows a subset of Minkowski spacetime (one spatial dimension suppressed), with opposite vertical planes identified.

We consider one final physical criterion on the underlying manifold. We see in our universe electrons and neutrinos. But such fermions are described in spacetime by spinor fields, while in certain spacetimes spinor fields cannot even be defined as global objects. However, this criterion – that the manifold is such that spinor fields do make sense – also turns out to be too weak to be of much use. Indeed, it is not even easy to give an example of a manifold in which spinor fields cannot be defined globally.

We conclude this section with one example of a precise meaning for 'broad, qualitative features of a Lorentz metric'. Fix a four-dimensional manifold, M. We say that Lorentz metric g_{ab} on M can be *continuously deformed* (through Lorentz metrics) to Lorentz metric g'_{ab} on M if there exists a 1-parameter family, $g_{ab}(\lambda)$, with λ in the closed interval $[0, 1]$, of Lorentz metrics on M, jointly continuous in its dependence on the parameter and on the point of M, such that $g_{ab}(0) = g_{ab}$, and $g_{ab}(1) = g'_{ab}$. (Note that we demand that the intermediate fields also be Lorentz metrics: otherwise, any two Lorentz metrics could be continuously deformed to each other. This is just one of several possible, closely related, notions.) Clearly, 'has the same qualitative features as' could be taken to mean 'can be continuously deformed to'. Given this notion, one can ask, among other things, which Lorentz metrics on a given manifold can be continuously deformed to which others. It is convenient, as with so many issues of this type, to replace Lorentz metrics by direction fields. Define 'can be continuously deformed

What is the topology of our universe?

to' similarly for positive-definite metrics and for direction fields. We first note that any two positive-definite metrics can be continuously deformed to each other by additivity. (Use the family $\lambda g_{ab}^+ + (1-\lambda) g_{ab}'^+$ of intermediate metrics, noting that these are indeed positive-definite.) That is, there are no 'topological entanglements' with positive-definite metrics. But it now follows that one Lorentz metric can be continuously deformed to another if and only if their associated direction fields can be so deformed, for one can characterize the Lorentz metric in terms of a positive-definite metric and a direction field. But one can often determine by inspection, at least in simple examples, whether or not one direction field can be continuously deformed to another. Consider, for example, the manifold $\mathbb{R}^4$. It seems reasonable intuitively – and is in fact true – that any direction field on this manifold can be continuously deformed to one 'pointing along the t-direction'; and so in particular any two direction fields can be deformed to each other. It follows, then, that any two Lorentz metrics on $\mathbb{R}^4$ – e.g. that of Minkowski spacetime and the plane waves – can be continuously deformed to each other. By contrast consider (now passing to two dimensions, to simplify the example) the two-dimensional annulus: the open disk in the plane with a hole in the center removed (figure 5.2). A direction field on this manifold can be described by the number of times, and the direction in which, the field rotates through angle π (relative, say, to the vertical in the figure) during one clockwise trip about the annulus. In this way, a direction field defines a (positive, negative, or zero) integer n. One

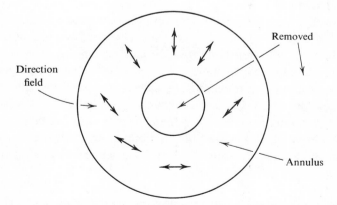

Figure 5.2. Two-dimensional annulus with a direction field. Both the region interior to the inner circle and that exterior to the outer circle (including the circles themselves) are removed from the plane. A direction field with $n = 1$ is shown.

Chapter 5. Global structure of spacetimes

convinces oneself that one direction field can be continuously deformed to another if and only if they have the same integer. Thus, in this example, the classes of metrics up to deformations are characterized by the integers. Similarly, for example, there is but one† deformation class on $S^2 \times \mathbb{R}^2$ (so, in this sense, the maximally extended positive-mass Schwarzschild solution, although a wormhole, is 'qualitatively the same' as the negative-mass solution), two on $S^1 \times \mathbb{R}^3$, four on $S^1 \times S^1 \times \mathbb{R}^2$, and one for each integer on $S^3 \times \mathbb{R}$. More detailed results regarding the possibilities for continuous deformations are available in the presence of suitable boundary conditions.

The conclusion to be drawn is that, at this general, qualitative level, it seems to be difficult to obtain any solid information about the underlying manifold of the universe. We shall shortly be more specific, and will be able to say more.

Orientations

Let M, g_{ab} be a spacetime. Fix a point p of M. Then the set of all timelike vectors at p can be divided into two classes, one of which could (for the moment arbitrarily) be designated the set of future-directed timelike vectors, the other the set of past-directed. (More precisely, call two timelike vectors at p equivalent if their inner product is negative. Then there are precisely two equivalence classes.) Similarly, the set of all triads of unit, mutually orthogonal, spacelike vectors at p can be divided into two classes, which could be designated the left-handed and right-handed triads.

Physically, the designation of future- and past-directed timelike vectors corresponds to a choice of a direction for the arrow of time; the designation of left- and right-handed triads to a choice of spatial parity. These designations can of course be made at any point of any spacetime, and, by continuity, throughout any sufficiently small neighborhood of any point. Questions of orientability arise when one asks whether or not such designations can be made globally over the entire spacetime manifold. We briefly describe what the possibilities are, how they are characterized, and the types of physical inferences which can be drawn.

---

† The number of deformation classes is computed as follows. For a manifold admitting four everywhere-independent vector fields, a property possessed by all of these manifolds, Lorentz metrics up to continuous deformations are classified by homotopy classes of continuous mappings from the manifold to projective 3-space.

What is the topology of our universe?

A spacetime is said to be *time-orientable*[4] if a designation of which timelike vectors are to be future-directed and which past-directed can be made at each of its points, where this designation is continuous from point to point over the entire manifold. Clearly, if a spacetime is time-orientable, then there are exactly two such designations, the original, and that obtained by reversing everywhere the roles of past and future. Each of these two designations is called a *time-orientation*. Similarly for *space-orientable* and *space-orientation*. A few facts follow immediately from the definitions. For example, given two Lorentz metrics on a manifold such that one can be continuously deformed to the other, then the first spacetime is time-orientable (respectively, space-orientable) if and only if the second is. One easily proves this by choosing, for example, a time-orientation on the first spacetime, and carrying this designation continuously through the continuous deformation. For the second property, fix a spacetime and a corresponding timelike direction field. Then the spacetime is time-orientable if and only if the timelike direction field can be replaced by a timelike vector field (i.e. one can choose at each point one of the two oppositely directed unit timelike vectors along the given 'direction', such that this choice is continuous from point to point).

The above definitions of orientable are, as it turns out, not in general the most useful. In practice, one is often faced with such questions as whether or not a given spacetime is time- or space-orientable, how to construct various counterexamples, etc. It is normally inconvenient to spell out in detail which timelike vectors are to be future-directed, which triads are to be left-handed, etc. Fortunately there is available a much simpler characterization. It rests essentially on the observation that what happens in the entire spacetime depends only on what happens on passage about certain closed curves. Fix a spacetime M, g_{ab}, and fix a point p. Consider any (not necessarily timelike) closed curve, γ, in M, beginning and ending at p. Fix a time-orientation at p, and carry this choice continuously about γ from point to point, until one returns to p. (Since γ need not be timelike, 'carry' is in a more mathematical than physical sense!) One thus obtains, on the return to p, a final time-orientation at p, which will be either the same or the opposite as that with which one began. One describes the situation by saying that the curve γ is, respectively, time-preserving or time-reversing. Similarly for space-preserving and space-reversing. The whole point of introducing these closed curves is the following fact: a spacetime is time-orientable (respectively, space-orientable) if and only if every closed curve through

Chapter 5. Global structure of spacetimes

p is time-preserving (respectively, space-preserving). The 'only if' part is immediate because, for example, a time-orientable spacetime must certainly have every closed curve time-preserving. The proof of the converse runs as follows. Choose a time-orientation at the base point p, and then define a time-orientation at any other point q by carrying the choice at p continuously along some curve joining p and q. The resulting time-orientation at q will be unambiguous, for, given another curve joining p and q, it can be combined with the original curve to obtain a closed curve through p – while, by hypothesis, every closed curve is time-preserving. Thus, one obtains a time-orientation at each point q, and so on the entire spacetime.

Thus in order to decide whether or not a spacetime is time-orientable, one has only to 'test' time-orientation about all the closed curves through a given point p. But this, it might be thought, is not much of an advantage, for there will of course be a vast number of curves to test. It turns out, fortunately, that not even that many curves need even be tested, as we now show. Call two closed curves, γ and γ', through p *homotopic*[8] if γ can be continuously deformed to γ', i.e., if there exists a 1-parameter family, γ_λ with λ in $[0, 1]$, of closed curves through p, jointly continuous in λ and the curve-parameter, such that $\gamma_0 = \gamma$ and $\gamma_1 = \gamma'$. Any two closed curves through the origin in the plane, for example, are homotopic, while a closed curve in the annulus which 'goes once around the hole' is not homotopic to one which does not. We now claim: for γ and γ' homotopic, either they are both time-reversing or both time-preserving. Indeed, this is immediate, since a continuous deformation cannot result in the discontinuous change from time-preserving to time-reversing. Thus, in order to decide whether or not a spacetime is time-orientable, one has only to test enough closed curves so that every such curve is homotopic to one of the tested curves. Again, similarly for space-orientable.

A few examples will illustrate this method. Again we pass to two dimensions to facilitate drawing pictures. Consider the manifold R^2, the plane, and let the point p be the origin. On this manifold any two closed curves through p are homotopic, or, equivalently, any closed curve is homotopic to a closed curve lying in an arbitrarily small neighborhood of p. (A suitable continuous deformation is that which shrinks the curve radially toward the origin.) But clearly, given a sufficiently small neighborhood of p, any closed curve lying entirely in this neighborhood must be time- and space-preserving. We conclude, then, that every space-time based on R^2 (and similarly for R^4) must be time- and space-

What is the topology of our universe?

orientable. As a second example, consider the two-dimensional annulus. Characterize any closed curve on this manifold by the integer n giving the net number of times the curve goes clockwise around the 'hole'. Then two curves are homotopic if and only if they have the same value of n. Thus, in order to decide whether or not a given spacetime based on this manifold is time-orientable, one has only to test one closed curve for each value of n. Thus, the first example of figure 5.3 is time-orientable; the second is not, for the closed curve with $n = 1$ in that example is time-reversing. Taking the product of the examples of figure 5.3 with a spacelike, R^2, to obtain four-dimensional examples, the first is space-orientable while the second is not.

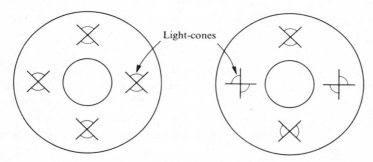

Figure 5.3. Two spacetimes based on the two-dimensional annulus. The first is time-orientable; the second is not.

As these examples illustrate, whether or not a spacetime is time- or space-orientable depends not just on the underlying manifold but also on its particular Lorentz metric. One can ask, none the less, whether there are *any* connections between the underlying manifold and these orientations. In what sense, in short, does the manifold topology interact with the 'light-cone topology', as expressed through orientability? It turns out that there are two such senses. The first is an easy generalization of the example above of the plane. A manifold M will be said to be *simply connected* if any two closed curves through p are homotopic (noting that this property is independent of the choice of point p). Simple connectivity is of course a property only of the underlying manifold. As examples, R^1, R^2, ... and S^2, S^3, ... are all simply connected, while S^1 and the annulus are not. The product of two manifolds is simply connected if and only if each factor is. Repeating the argument we used for the plane, we see that every spacetime based on a simply connected manifold is both time- and space-orientable. Thus, for

Chapter 5. Global structure of spacetimes

example, not only are the Schwarzschild and closed Friedmann spacetimes time- and space-orientable but also *any* spacetimes based on their underlying manifolds ($S^2 \times R^2$ and $S^3 \times R$, respectively). The second relation between the orientations and the topology of the manifold involves a combination of time- and space-orientations. We say that closed curve γ through p is *total-orientation-preserving* if either it is both time- and space-preserving, or neither. It then turns out that, although of course individually 'time-preserving' and 'space-preserving' refer to the particular Lorentz metric, the combination 'total-orientation-preserving' is independent of that metric. Indeed, one could have defined 'total-orientable' initially and directly, in analogy with time and space-orientable, but without reference to the metric. (The set of all tetrads of independent tangent vectors at p can be divided into two classes ... .) The previous notions for time- and space-orientable go through also with total-orientable. The R^ns and S^ns are total-orientable, and a product of manifolds is total-orientable if and only if each factor is. The standard example of a non-orientable manifold is of course the Möbius strip. We thus conclude, for example, that any spacetime based on the Möbius strip crossed with R^2 (to make the manifold four-dimensional) either fails to be time-orientable or fails to be space-orientable.

There are essentially no other relations between the three types of orientations than those described above. It is immediate from the definitions, for instance, that no spacetime can be orientable in two of these three senses but not in the third. There are examples to show that any other combination is indeed possible. By an argument similar to that used previously in connection with the existence of Lorentz metrics, one can also show the following. Any non-compact manifold can be the underlying manifold for a time-orientable spacetime, and also for a space-orientable spacetime. In addition, if M is not simply connected, then there exists a spacetime based on M which is not time-orientable, and one which is not space-orientable. In effect, 'anything not obviously forbidden can occur'.

We next turn to the physical aspects of the issue of orientations. One asks in particular whether there are any physical arguments which might suggest that realistic models of our own universe must be time-orientable, or must be space-orientable. A typical argument might run as follows. We observers, in our own local region of spacetime, perceive a preferred future time-direction. Furthermore, there is agreement between different observers as to which time-direction this is. Suppose,

What is the topology of our universe?

then, that one universalizes these local experiences – i.e. one imagines that there could be local observers in all regions of the universe, that each observer would perceive a preferred future time-direction, and that there would be agreement among these observers. One would then conclude that a physically realistic model of our universe must be time-orientable. The analogous argument for space-orientable seems to be somewhat less convincing, for, although one can certainly distinguish locally two space-orientations, neither seems to be preferred physically to quite the extent that 'future' is distinguished over 'past'. There is in fact available what is essentially a more quantitative version of these same arguments. Instead of 'perceived preferred time-direction', one uses the observed, local, symmetry violations of elementary-particle physics.[9] The violations of these symmetries permit one, by local experiments, to prefer certain simultaneous choices of future time-direction, right-handed spatial parity, and particles as opposed to anti-particles, over various other simultaneous choices. These locally determined preferences, if extrapolated globally, permit one to impose restrictions on the orientation-structure of one's spacetime. Suppose, for example, that parity-reversal invariance was broken in local experiments. It would *not* follow that one could, by local experiments, single out preferred 'right-handed' frames, but rather only that such frames could be singled out given a preferred 'future' time-direction and a choice as to which are the particles and which anti-particles. These arguments of course require the introduction of an additional, non-geometrical, 'orientation', which essentially asks whether particles and anti-particles are interchanged on passage about a closed curve. It turns out[10] that the strongest conclusion to be drawn from such arguments, using the presently observed symmetry violations in elementary-particle physics, is that our spacetime must be total-orientable. One cannot conclude from this, for example, that our spacetime must be separately time- and space-orientable.

There is a second line of argument which touches on this question of what are the appropriate orientation properties of a physically realistic model of our universe. The conclusion of this argument is that no physical possibilities would be lost by demanding that one's spacetime be space- and time-orientable. It runs as follows. Let M, g_{ab} be a spacetime, and fix a point p of M. We construct a new spacetime, $\tilde{M}$, $\tilde{g}_{ab}$. Each point of M will define several (possibly one, possibly a finite number, possibly an infinite number) of points of $\tilde{M}$, as follows. Consider pairs (q, γ), where q is a point of M, and γ is a curve (not

Chapter 5. Global structure of spacetimes

necessarily timelike) in M from p to q. Call two such pairs, (q, γ) and (q', γ'), equivalent if $q = q'$ and there is a continuous deformation (keeping the endpoints fixed) of γ to γ'. (This construction, of course, is a mild generalization of the notion of homotopic closed curves.) The set of equivalence classes we denote by $\tilde{M}$. Thus, a point of $\tilde{M}$ is essentially 'a point of M and a curve from p to that point, up to continuous deformations of the curve'. The metric $\tilde{g}_{ab}$ on $\tilde{M}$ is obtained as follows. Consider a point $\tilde{q}$ of $\tilde{M}$ (represented by point q of M and curve γ in M from p to q), and a nearby point $\tilde{q}'$ of $\tilde{M}$ (represented by nearby point q' of M, and nearby curve γ'). We define the interval from $\tilde{q}$ to $\tilde{q}'$ in $\tilde{M}$ to be just the interval from q to q' in M. The resulting spacetime $\tilde{M}, \tilde{g}_{ab}$ is called the *universal covering space (-time)* of M, g_{ab}. By construction, $\tilde{M}$, $\tilde{g}_{ab}$ is always locally indistinguishable from M, g_{ab}, but globally it can be quite different. Suppose, for example, that M, g_{ab} is simply connected. Then any two curves in M with the same endpoints can be continuously deformed to each other, and so we shall have (q, γ) equivalent to (q', γ') provided only that $q = q'$. Thus, in this case, the universal covering spacetime will always be identical to the original spacetime. Let, as a second example, M, g_{ab} be the flat, two-dimensional annulus (figure 5.4). For the point q of the figure, curve γ can be continuously deformed

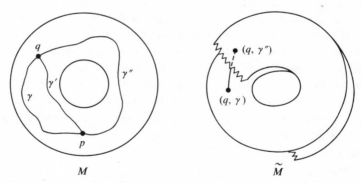

Figure 5.4. The universal covering space $\tilde{M}$ of the two-dimensional annulus. Each point of M defines an infinite number of points of $\tilde{M}$, for the spiral continues indefinitely. The metric on each manifold is flat.

to γ', and so (q, γ) and (q, γ') are equivalent, i.e. they define the same point of $\tilde{M}$. On the other hand, (q, γ) and (q, γ'') are not equivalent, i.e. they define different points of $\tilde{M}$. More generally, a curve which reaches q from p after winding around the 'hole' n times will be deformable to a curve which winds around the hole n' times if and only if $n = n'$. Each point q of M, therefore, will give rise to an infinite number (one for each integer n) of points of $\tilde{M}$. The structure of $\tilde{M}$, $\tilde{g}_{ab}$, then, will be as shown in the figure. Essentially, we have just unwrapped the annulus.

The relevance of the universal covering space to the issue of orientability stems from the following fact: the universal covering spacetime of any spacetime is simply connected. (Indeed, the universal covering spacetime was constructed by just 'unwrapping' the original spacetime about every closed curve not continuously deformable to a point.) Therefore, the universal covering spacetime is always time-orientable and space-orientable. But there is a sense in which the universal covering spacetime is physically indistinguishable from the original spacetime. Since the only effect of taking the universal covering spacetime is to possibly produce several copies of each local region in the original spacetime, the 'local physics' remains the same. Thus, so the argument would go, one could never discover by local experiments (which, after all, are the only ones we can perform) whether or not our spacetime is simply connected, and so nothing is lost by always assuming simple connectivity and hence time- and space-orientability.

It is also possible to introduce coverings which are intermediate between the original spacetime and its universal covering spacetime. Fix a point p of M, g_{ab}, and also a collection of closed curves through p which satisfy the following conditions: (i) two homotopic curves are either both in or both out of the collection; (ii) a closed curve in a sufficiently small neighborhood of p is in the collection; (iii) for τ and τ' in the collection, the new closed curve obtained by first going around τ and then τ' is in the collection; (iv) for τ in the collection, so is the curve obtained by going around in the opposite direction. Now, for (q, γ) and (q', γ') as above, regard these as equivalent if the closed curve through p obtained by going from p to q along γ and from q back to p along γ' is in this collection. Thus, if one's collection of curves is all closed curves through p, one obtains the original spacetime; if it is all closed curves homotopic to a curve in a small neighborhood of p, one obtains the universal covering spacetime. An intermediate possibility would be to let the collection consist of all time-preserving closed curves. The resulting covering spacetime would then be time-orientable, but not necessarily space-orientable.

Covering spacetimes find occasional use, both in proofs and in counterexamples. As an example of the former, one frequently needs time-orientability in an argument. One passes to a covering spacetime, completes one's argument there, and then carries the result back to the

Chapter 5. Global structure of spacetimes

original spacetime. An example of the latter is the following. It is easy to find flat metrics on the manifold R^4 which are different from the Minkowski metric, e.g. by removing that portion of Minkowski spacetime outside of an open ball. These spacetimes, however, are *extendible*, i.e. they can be embedded as a proper open subset of a larger spacetime. Are all flat spacetimes based on R^4 either Minkowski spacetime or extendible? The answer is no. Remove from Minkowski spacetime a 2-plane (so the resulting manifold has topology $S^1 \times R^3$), and take the universal covering spacetime.

Causal structure. I – Domain of influence[†]

Let M, g_{ab} be a spacetime with fixed time-orientation. For p and q any two points of M, we say that p *precedes*[11] q (chronologically) if there exists a future-directed timelike curve which begins at p and ends at q. In Minkowski spacetime, for example, with p and q labeled in the usual coordinates by (t, x, y, z) and (t', x', y', z'), respectively, this relation means $(t - t') < -[(x - x')^2 + (y - y')^2 + (z - z')^2]^{1/2}$. Physically, the points p and q represent events, while the relation 'p precedes q' means that some signal could be sent from p to be received later at event q. (One might think it more appropriate, since signals can also travel along null curves in relativity, to also allow such curves in the definition of 'precede'. It turns out, however, that to do so adds nothing substantial to the physical content, while somewhat complicating the mathematics.) The relation above is the central one of what is called the causal structure of spacetimes. One is dealing with an aspect of the metric – signal propagation – which on the one hand has an obvious and direct physical significance, and on the other seems to capture an aspect of spacetimes more fundamental than, say, the curvature of the metric. The importance of causal structure to the study of the global properties of spacetimes could hardly be over-emphasized. A surprising fraction of all global arguments in general relativity are most naturally formulated within this framework.

There are two basic properties of the relation 'precedes'. The first is: if p precedes q and q precedes r, then p precedes r. The proof consists of drawing the future-directed timelike curves from p to q and from q to r, joining them at q, and there rounding of the corner (if any) of the joined curve, to obtain a smooth future-directed timelike curve from p to r.

[†] We assume throughout this section that all spacetimes are time-oriented.

What is the topology of our universe?

The second property requires a definition. For p any point of M, the *past* (or past domain of influence) of p, denoted $I^-(p)$, is the set of all points which precede p, and similarly the *future* of p, $I^+(p)$, is the set of all points which p precedes. In Minkowski spacetime, for example, these are just the interiors of the past and future light-cones of p. The second property of the relation 'precedes' is that $I^-(p)$ and $I^+(p)$ are always open sets. (That is, any point of $I^-(p)$ or $I^+(p)$ has a neighborhood entirely within that set.) The proof is again an elementary property of timelike curves. Let q be in $I^+(p)$, fix a future-directed timelike curve from p to q, and fix a point q' just prior to q along this curve (figure 5.5). Consider now the points q'' which can be reached from p by a future-directed timelike curve which begins at p, coincides with our original curve until q', and there branches off to reach q''. The points q'' so obtained all lie within $I^+(p)$, and include a neighborhood of q.

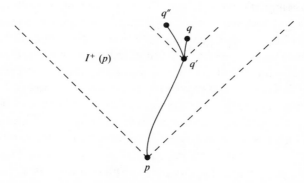

Figure 5.5. $I^+(p)$ is an open set. The set of all points q'' reached by future-directed timelike curves from q' form a neighborhood of q which lies inside $I^+(p)$.

Various other facts about the relation 'precedes' are direct consequences of these, or of the definitions. For example: (i) point p is in $I^-(q)$ if and only if q is in $I^+(p)$; (ii) for p in $I^-(q)$, $I^-(p)$ is a subset of $I^-(q)$; (iii) if a sequence $p_1, p_2, \ldots$ of points approaches p and p is in $I^-(q)$, then q is in $I^+(p_i)$ for some i; (iv) if some future-directed timelike or null curve joins p to q, then $I^+(q)$ is a subset of $I^+(p)$ (for r in $I^+(q)$, join the future-directed timelike or null curve from p to q with the future-directed timelike curve from q to r, round off the corner at q, and possibly vary the curve slightly to obtain a timelike curve from p to r (figure 5.6)); (v) $I^-(p)$ is precisely the union of all sets of the form $I^-(q)$ with q in $I^-(p)$.

Chapter 5. Global structure of spacetimes

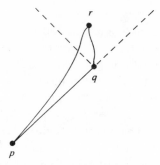

Figure 5.6. If a future-directed timelike or null curve joins p to q, then $I^+(q)$ is a subset of $I^+(p)$.

The pasts and futures have two further properties, which turn out to be quite useful in many causal arguments. The first (say, for the future) is the following: if γ is a future-directed timelike or null curve from point q, and γ does not enter $I^+(q)$, then γ is a null geodesic. (This is clearly true, for example, in Minkowski spacetime.) The proof in general begins by noting that, just as for Minkowski spacetime, the curve must certainly be a null geodesic near q. Suppose, then, that there were a point p on this curve beyond which it ceases to be a null geodesic. Then the curve would have to enter $I^+(p)$, i.e. there would have to be a point r on the curve γ and in $I^+(p)$. But now, by property (iv) above, this r would also have to be in $I^+(q)$, which contradicts γ not entering $I^+(q)$. The second property is essentially a corollary of the first. This property (say, for the past) is that for any point p the boundary of $I^-(p)$ must be null, in the following sense: through any point $q \neq p$ of the boundary of $I^-(p)$, there passes a future-directed null geodesic which remains entirely within this boundary; and furthermore this geodesic either cannot be assigned any future endpoint, or else can be assigned future endpoint p. To prove this, let q be as above, and choose a sequence of points $q_1, q_2, \ldots$ in $I^-(p)$ which approach q (figure 5.7). From each q_i, draw a future-directed timelike curve to p. Let γ be an accumulation curve of this sequence, so γ is everywhere in either $I^-(p)$ or its boundary, and is everywhere either timelike or null. Now γ cannot enter $I^+(q)$, for this would imply that q is in $I^-(p)$, and hence would violate q being on the boundary of $I^-(p)$. Hence, γ must be a null geodesic. But γ cannot enter $I^-(p)$ either, for this would again imply that q is in $I^-(p)$. Hence, γ must be on the boundary of $I^-(p)$. Finally, if this null geodesic γ could be assigned any future endpoint other than p itself, one could

What is the topology of our universe?

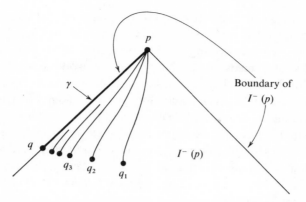

Figure 5.7. Construction of a null geodesic on the boundary of $I^-(p)$. The sequence of points $q_1, q_2, \ldots$ approach q, and γ is the accumulation curve of timelike curves from q_i to p.

repeat the argument starting from that endpoint, and thus obtain an extension of that geodesic within the boundary of $I^-(p)$.

We have given these arguments in some detail because they represent examples of some important methods of proof in causal structure. Consider the second. We must, first, take advantage of the hypothesis that q is on the boundary of $I^-(p)$. This is done by choosing the sequence $q_1, q_2, \ldots$ in $I^-(p)$ and approaching q. We must next take advantage of the fact that the q_i are in $I^-(p)$, which is done by considering a sequence of timelike curves from the q_i to p. We now have a point q which is 'almost' in $I^-(p)$ (namely, on its boundary), and a sequence of points q_i which 'approximate' q, and which are actually in $I^-(p)$, a feature which is made explicit by the timelike curves from the q_i to p. The next step is to carry this feature of the approximations, the q_i, to the point they approximate, q, i.e. to take the accumulation curve. We thus obtain our candidate for the null geodesic. The rest is brute force: showing that this candidate has no choice but to be the desired null geodesic. Such limiting arguments tend to work when one has or seeks some point or curve which 'almost has' some desirable property (e.g., in this case, q is almost in $I^-(p)$, and γ is to be almost a timelike curve to p).

The subject of causal structure has the curious feature that there is a seemingly endless stream of plausible-sounding statements within its framework which are actually false. Frequently, these are exactly the statements one would like to have true in order to prove something interesting. The decision as to which statements are false is normally

Chapter 5. Global structure of spacetimes

made by means of counterexamples. Furthermore, the ease with which these examples can be obtained endows them with various other uses. For instance if one were carrying out the argument of the previous paragraph for the first time one would probably stop, having obtained the candidate curve γ for the desired null geodesic, to check by means of a quick search for counterexamples that it is actually a good candidate. Having found no counterexamples, one would go ahead and try to complete the proof by showing that γ indeed is the desired null geodesic. In any case, the ability to decide quickly and efficiently what is true and what is false is a most useful skill in making effective use of causal ideas in physical problems. We here give just a few examples of false statements. There are many others in the Appendix.

(1) If q is on the boundary of $I^-(p)$, then the future-directed null geodesic from q in the boundary of $I^-(p)$ actually reaches p. Remove the origin from Minkowski spacetime, and let p and q lie on opposite sides of that removed point along a null geodesic which in the original Minkowski spacetime passed through the origin.

(2) If $I^-(p)$ contains $I^-(q)$, then $I^+(q)$ contains $I^+(p)$. Let the spacetime be Minkowski spacetime with the half-plane $t = 0$, $x \geq 0$ removed. Let the coordinates of p be $(1, -1, 0, 0)$, and q $(-1, 1, 0, 0)$ (figure 5.8).

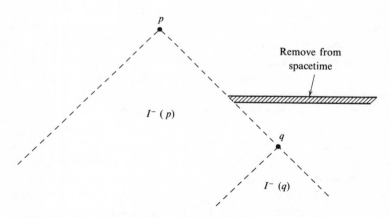

Figure 5.8. Two-dimensional Minkowski spacetime with the positive x-axis removed. $I^-(p)$ contains $I^-(q)$, but $I^+(q)$ does not even intersect $I^+(p)$.

(3) If $I^-(p) = I^-(q)$, then $p = q$. Let the spacetime be that portion of Minkowski spacetime between $t = -1$ and $t = +1$, and identify each boundary point of the form $(-1, x, y, z)$ with $(+1, x, y, z)$. Then the past

What is the topology of our universe?

of every point in the resulting spacetime is the entire spacetime (figure 5.9).

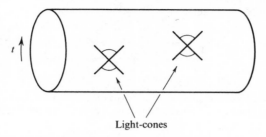

Figure 5.9. Strip of two-dimensional Minkowski spacetime with $t=1$ identified with $t=-1$. The past of every point is the entire spacetime.

(4) If q is not in $I^-(p)$, then there exists a maximally extended past-directed timelike or null curve from q which never enters $I^-(p)$. Remove a hole from Minkowski spacetime as illustrated in figure 5.10 and let points p and q be as shown.

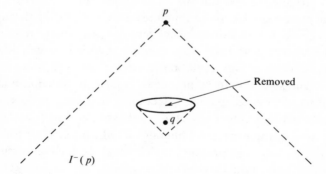

Figure 5.10. Two-dimensional Minkowski spacetime with a hole removed. Point q is not in $I^-(p)$, yet every maximally extended past-directed timelike or null curve from q enters $I^-(p)$.

(5) If q is simultaneously on the boundary of both $I^+(p)$ and $I^-(p)$, then $q=p$. From Minkowski spacetime, remove two identical horizontal slits, as shown in figure 5.11. Identify the bottom edge of the slit marked A with the top edge of B. The point p has past and future as indicated in the figure. Choose q as shown.

Chapter 5. Global structure of spacetimes

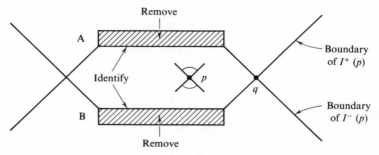

Figure 5.11. Two-dimensional Minkowski spacetime with two slits removed and their edges identified. Point $q \neq p$ is on the boundary of both $I^+(p)$ and $I^-(p)$.

We turn next to the possibilities involving causality violations in spacetimes. The most obvious manifestation of such a violation would be the existence of a closed timelike curve through, say, event p, i.e. in terms of our causal relation, the existence of a point p which precedes itself. Example 3 above, as well as such familiar spacetimes as the Gödel, Taub–NUT, and extended Kerr, make it clear that general relativity, as it is normally formulated, does not prohibit closed timelike curves. A very different question is whether or not one wishes, on physical grounds, to rule out such anomalous causal behavior.[12] Physically, the existence of a closed timelike curve, say passing through point p, would mean that an individual at event p would be able in principle to influence his own past. One might argue that endowing individuals with this ability violates our most basic conceptions of how the world operates, and so it is entirely proper to impose, as an additional condition for physically acceptable spacetimes, that they possess no such causality violations. On the other hand, physical theories frequently suggest new and unexpected phenomena, which are later found to be realized physically. The present may very well be such an occasion, and so perhaps one should take the possibility of causality violations seriously.

It might be of interest to see if one could make the issue of whether or not it is physically appropriate to rule out causality violations more quantitative. One approach along these lines might be the following. Consider, as an example, Maxwell's equations with zero sources. We, in our own local region of the universe, observe that these equations are the only constraints on electric and magnetic fields, for this feature is perhaps the key one which made possible the original discovery of these equations. Now make universal this observation, i.e. demand that, given

What is the topology of our universe?

any point p of spacetime and a solution of Maxwell's equations in a neighborhood of p, there exists some Maxwell solution defined over the entire spacetime which reduces in some neighborhood to the given local solution. This, then, is a property which may or may not be satisfied in a given spacetime, and which one might demand on physical grounds. This property is satisfied, for example, in Minkowski spacetime. There are, however, examples of spacetimes in which it is not satisfied, and typically these are precisely the spacetimes with causality violations. Taub–NUT spacetime is a good example. What happens here is that, given a Maxwell field in a small neighborhood in the Taub-part of the spacetime, the resulting radiation fields would normally 'collect' near the Misner boundary, resulting in a Maxwell field which is badly behaved at that boundary. On the other hand, there are also spacetimes which have causality violations, but in which none the less local Maxwell fields can always be made global. Consider, for example, the spacetime which results from Minkowski spacetime by excising the regions with $|x| \geq 1$, $|y| \geq 1$, or $|z| \geq 1$, and identifying $t = 0$ and $t = 10$ (figure 5.12).

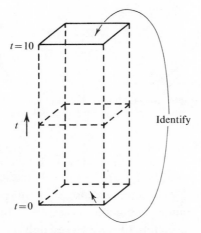

Figure 5.12. Tube of Minkowski spacetime (one spatial dimension suppressed) with top and bottom identified. This spacetime has closed timelike curves, yet every local Maxwell field can be made global.

That this spacetime imposes no restrictions on local Maxwell fields is an immediate consequence of Huygen's principle, for the radiation from a local Maxwell field will 'go off the edge of the spacetime' before it is able to traverse a time-cycle. The idea would be to see if one could show that, in some sense, it is only in very special cases that spacetimes with

Chapter 5. Global structure of spacetimes

causality violations permit every local Maxwell field to be made global. One would thus argue that in the 'generic' case with causality violations this property fails, and therefore causality violation is physically unacceptable.

One might imagine that the issue of whether or not a given spacetime has causality violations would be clear-cut: either there are closed timelike curves or there are not. However, matters are not so simple, as the following example will show. Take the region of Minkowski spacetime between $t = 0$ and $t = 1$, and identify points represented by $(0, x, y, z)$ with $(1, x + 1, y, z)$ (figure 5.13). While this spacetime has no

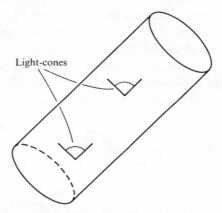

Figure 5.13. Strip of two-dimensional Minkowski spacetime between $t = 0$ and $t = 1$, with the point $(0, x)$ identified with $(1, x + 1)$. This spacetime has no closed timelike curves, but there are closed null curves.

closed timelike curves, it does have closed null curves, and so, by any reasonable standard, would have to be regarded as causally anomalous. Furthermore, even if one demands also that there be no closed null curves, there still remain ambiguous examples. There are spacetimes which have neither closed timelike nor closed null curves, but which none the less violate what is called the strong causality condition: there is a point p of the spacetime such that through every sufficiently small neighborhood of p some timelike curve passes more than once. (In other words, there are 'almost closed' timelike curves, which begin very close to p and also end very close to p.) Again, it would seem inappropriate to regard such spacetimes as having satisfactory causal behavior. In fact, there is at least a countably infinite collection of different meanings of 'causally badly behaved'.

What is the topology of our universe?

This unsatisfactory state of affairs was resolved by the introduction of the 'right' causality condition. A spacetime M, g_{ab} is said to be *stably causal*[13] if there exists a (nowhere-vanishing) timelike vector field t_a such that the Lorentz metric on M given by $g_{ab} - t_a t_b$ admits no closed timelike curves.† To see what this means, let us consider the geometry at a single point p. The null cone of the Lorentz metric $g_{ab} - t_a t_b$ is more 'opened out' (isotropically about the vector t^a) than is the null cone of g_{ab} (figure 5.14). The amount of opening out depends on 'how large' t_a

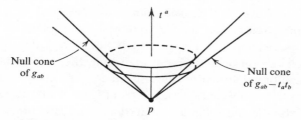

Figure 5.14. The null cones of the metric $g_{ab} - t_a t_b$ are more 'opened out' than those of the metric g_{ab}.

is. The effect of opening out the null cones in this way is, of course, that more curves are taken as timelike, and in particular the likelihood of finding a closed timelike curve is increased. The definition requires then that, in order to be stably causal, a spacetime must not only admit no closed timelike curves, but it must even have the property that one can open out the null cones by a finite amount at each point, where this 'amount' varies smoothly but otherwise arbitrarily over the manifold, all without creating any closed timelike curves. In particular, a spacetime which has closed timelike curves, or closed null curves, or which violates strong causality – or, indeed, which violates any of the causality conditions – will necessarily violate stable causality. Minkowski spacetime, by contrast, is stably causal: choose for t_a any unit, constant, timelike vector field. Similarly, the Schwarzschild, Friedmann, Weyl, Reissner–Nordström, and plane-wave solutions all satisfy stable causality. Stable causality seems to capture completely one's intuitive idea of having no causal anomalies.

There are a number of useful consequences of stable causality. We give three examples.

Let M, g_{ab} be a stably causal spacetime. Then the topology of M follows already from the causal structure, in the following sense.[11] We

† Our signature is $(-, +, +, +)$.

Chapter 5. Global structure of spacetimes

have seen that, even without any causality conditions, sets of the form $I^+(p)$ and $I^-(p)$ are open. By the properties of open sets, it follows that arbitrary unions of finite intersections of sets of this form are also open. We now try to reverse this process. We *define* a collection of subsets of M, consisting of all unions of finite intersections of sets of the form $I^+(p)$ or $I^-(p)$. It turns out that, in the stably causal case, the sets so defined are precisely the open sets of the underlying manifold M. (To prove this, it suffices to show that very small open sets can be obtained by such unions of intersections. But sets of the form $I^+(p) \cap I^-(q)$, for p near q and preceding q, will do if the spacetime is stably causal. Without stable causality, these sets can be 'too big' to reproduce the manifold topology, e.g. as in the example of figure 5.9.) It turns out[14] that, by a more complicated version of this type of argument, the differentiable structure and conformal structure can also be determined from the causal structure in the stably causal case. Here, at least, is one sense in which causal structure is 'more fundamental' than other structures. A second consequence of stable causality is this: every stably causal spacetime is non-compact.[11] Indeed, suppose that M were compact. Fix a timelike vector field on M, fix a point p, and draw the maximally extended integral curve of this field, beginning at p. Choose a sequence of points along this curve, and let, by compactness, q be an accumulation point. Then through every sufficiently small neighborhood of q there would pass a timelike curve (namely our integral curve) more than once, violating stable causality. Thus, compact spacetimes are always subject to causality violations – and it is for this reason that such spacetimes are rarely considered in relativity. The third consequence of stable causality is the most surprising. By a *time-function* on spacetime M, g_{ab}, we mean a smooth scalar field t whose gradient is strictly timelike. For example, the Minkowski coordinate 't' is a time-function. In the general case, a time-function represents essentially the assignment of a 'time' (the value of t) to every point of spacetime, in such a way that this time is strictly increasing along every timelike curve (i.e. just the property one would expect of a global time). It is immediate that, if a spacetime admits a time-function, then it has no closed timelike curves, for t is strictly increasing along timelike curves. It is almost as easy to see that the existence of a time-function implies stable causality: set $t_a = \nabla_a t$ and then note that the gradient of t is still timelike with respect to the new metric, $g_{ab} - t_a t_b$, and so this new metric could hardly admit closed timelike curves. The third result is that the converse of this is also true: a spacetime admits a time-function if and only if it is stably causal.[13] Thus,

What is the topology of our universe?

in these 'causally well-behaved' spacetimes, one can always introduce a time-function. Although this result does perhaps give one confidence that stable causality is the 'right' condition, it is normally the case in practice that the time-function has little direct physical significance. In particular, the time-function is never even close to being unique, and the spacelike surfaces of constant t need not be connected (and, in many examples, cannot be chosen to be connected).

Causal structure. II – Slices†

Our discussion of causal structure so far has involved primarily timelike curves, i.e., physically, 'all time at one point of space'. The dual notion is 'all space at one instant of time', i.e. spacelike three-dimensional submanifolds.

We begin with the observation that 'spacelike, three-dimensional submanifold' does not fully capture the physical notion we are after. Consider, for example, the disk in Minkowski spacetime given by $t = 0$, $x^2 + y^2 + z^2 < 1$. This is certainly a spacelike, three-dimensional submanifold, but it is 'small and local', hardly representing 'all space at one instant of time'. We are interested in objects more global. The appropriate and most useful notion is that of a *slice*: a spacelike, three-dimensional submanifold of spacetime which in addition is closed as a subset of the manifold M. The disk above, for example, is not a slice, because, for example, the point represented by $(0, 1, 0, 0)$ is a limit point, but is not itself in the submanifold. (Just including the boundary will not help. The region given by $t = 0$, $x^2 + y^2 + z^2 \leq 1$ is also not a slice, for it is not even a submanifold.) Examples of slices in Minkowski spacetime include the entire plane $t = 0$ and the hyperboloid given by $t = (x^2 + y^2 + z^2 + 1)^{1/2}$.

Which spacetimes admit slices? Certainly not all: the Gödel spacetime is a well-known example. However, the Gödel spacetime also violates stable causality. This example rather suggests that the presence or absence of slices has something to do with whether or not stable causality is satisfied. There is one result along these lines: every stably causal spacetime admits a slice and, indeed, for each point p a slice passing through p. (To prove this, introduce a time-function and consider the spacelike, three-dimensional submanifolds given by $t = c$, a constant. But these submanifolds are also closed, and hence slices, for, by

---

† We assume throughout this section that all spacetimes are time-oriented.

Chapter 5. Global structure of spacetimes

continuity of t, any limit-point of a sequence of points at which $t = c$ also has $t = c$.) The converse of this result is false. The example of figure 5.9 admits slices, and in fact a slice through every point, although this spacetime is certainly not stably causal. There is apparently not known any simple condition for the existence of a slice in a spacetime which violates stable causality.

What topologies are possible for slices in a spacetime? In the absence of further conditions, there is no restriction: any three-dimensional manifold can be the underlying manifold of a slice in some spacetime. (Given three-dimensional S, let $M = S \times \mathbb{R}$, and impose on this manifold a product metric, with the Ss spacelike and the $\mathbb{R}$s timelike.) In certain examples (e.g., Minkowski spacetime) all connected slices have the same topology. But this property does not hold in general. Consider, for example, Minkowski spacetime with I^+ (origin), as well as its boundary, removed (figure 5.15). In this spacetime, there are some slices which are

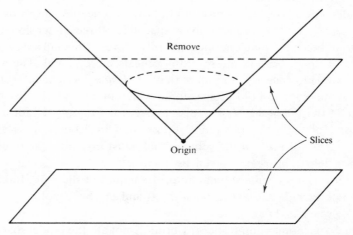

Figure 5.15. Minkowski spacetime (one spatial dimension suppressed) with the future of the origin, as well as its boundary, removed. Two slices are shown; one has topology $\mathbb{R}^2$, the other topology $S^1 \times \mathbb{R}$.

$\mathbb{R}^3$, and others which are $S^2 \times \mathbb{R}$. Similarly the closed Friedmann model admits slices which are S^3, $\mathbb{R}^3$, $S^2 \times \mathbb{R}$, $S^1 \times \mathbb{R}^2$ as well as many other topologies. In fact, there even exist spacetimes which can be covered by a family of slices all of which are topologically identical, and also by a second family all topologically identical – but such that the slices in the two families are of different topologies! An example (in two dimensions, to simplify the figure) is a 'short' cylinder with light-cones vertical and

What is the topology of our universe?

slices as shown in figure 5.16. One might have hoped that the topology of the slices would provide a physically useful characterization of spacetimes – e.g. that one could define a 'closed universe' to be a spacetime having compact slices. However, the examples above make it clear that, because of the enormous variety of slices normally possible even in very tame spacetimes, such a characterization is unlikely to be useful in practice.

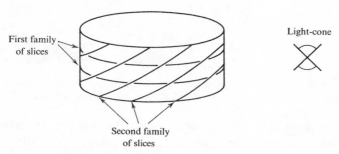

Figure 5.16. A two-dimensional spacetime which can be covered by two different families of slices, one of topology S^1 and the other $\mathbb{R}$.

Are there any results which assert that, under certain circumstances, two slices *must* have the same topology? An easy result to the contrary is that, given any two three-dimensional manifolds, a spacetime can be found in which there are two slices having the respective topologies of these manifolds. In short, with no additional conditions, anything is possible. One example of a positive result along these lines is the following.[15] Let a spacetime have two compact slices, S and S', and let there be a compact subset of the manifold whose boundary consists of S and S' (figure 5.17). Let, furthermore, stable causality hold. Then S and S' are diffeomorphic. For the proof, one first introduces a timelike vector field. Start at any point of S, and follow the corresponding integral curve of this field. If this curve remained always in the compact region between S and S' then it would have to eventually come near itself, by compactness, which would violate stable causality. Hence this curve must terminate on S'. So there results a mapping from S to S' which, one shows, is the required diffeomorphism. Physically, the statement is that in a closed universe (in a rather strong sense!) with no causality violations, the 'topology of space' cannot change.

In addition to these topological features of slices, various causal features can also be distinguished. Perhaps the most important is the following: a slice is said to be *achronal* if no point of that slice precedes

Chapter 5. Global structure of spacetimes

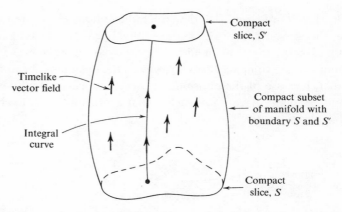

Figure 5.17. Two compact slices S and S' must be diffeomorphic if there exists a compact subset of the manifold with boundary S and S', and stable causality holds.

any other point of the slice. For example, the plane or hyperboloid in Minkowski spacetime are achronal, while none of the slices of the spacetime of figure 5.9 are achronal. As these examples suggest, failure of achronality is often associated with failure of stable causality. Indeed, every stably causal spacetime admits achronal slices, for, since the time-function is increasing along timelike curves, any slice of constant time-function must be achronal. However, certain stably causal spacetimes can also admit slices which are not achronal. Consider, for example, the 'spacelike ribbon' in Minkowski spacetime shown in figure 5.18. This is a spacelike submanifold but, since it is not closed, not a slice. Now, however, we remove the points of its edge from the spacetime itself. We thus obtain a new spacetime in which S, now closed, is indeed a slice, and which retains from the original Minkowski spacetime stable causality. This slice is of course not achronal. There also exist spacetimes which are not stably causal, but still admit achronal slices.

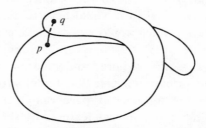

Figure 5.18. Spacelike submanifold of Minkowski spacetime (one spatial dimension suppressed) which is not achronal, e.g. because p precedes q.

What is the topology of our universe?

The conclusion, then, is that the interaction between the topology and causal structure of the underlying spacetime on the one hand, and of the slices on the other, is rather weak.

Causal structure. III – Domain of dependence†

The basic notion of the domain of influence, 'precede', reflects the physical question of whether or not event p can influence event q by means of a signal. A closely related question asks whether information given in some region of spacetime will determine the physical situation in some other region. It is immediately clear that the mathematics appropriate to this question will not be, as with the domain of influence, a relation between single points of spacetime. One knows, e.g. for plane electromagnetic waves in Minkowski spacetime, that no amount of physical information at a single point will suffice to determine the physical situation at some other point. Indeed, in order to 'determine' what will happen at point p one needs to 'register all signals which could influence the physics at p'. The appropriate ledger on which to register this information, as it turns out, is a slice.

The definition which captures these physical ideas is the following. Let M, g_{ab} be a spacetime with fixed time-orientation, and let S be an achronal slice. By the *future domain of dependence* [16,17] of S, $D^+(S)$, we mean the set of all points p such that every past-directed timelike curve from p which cannot be assigned a past endpoint meets S. The idea of the definition is this. Regard signals in relativity as traveling along timelike curves. By demanding that all timelike curves reaching p also meet S, one is ensuring that all signals which could influence the physics at p are registered on S. Hence, the physical situation in $D^+(S)$ will be completely determined by information on S. (Again, including null curves with timelike curves, while perhaps somewhat more appropriate physically, adds little to the physical content and complicates the mathematics.)

We give some examples. First consider Minkowski spacetime. If S is given, in the usual coordinates, by $t=0$, then clearly $D^+(S)$ is just the set of all points with $t \geq 0$. If S is the hyperboloid $t = (r^2+1)^{1/2}$ (where $r = (x^2+y^2+z^2)^{1/2}$) then $D^+(S)$ is given by $t \geq (r^2+1)^{1/2}$. Now let S be the 'other' hyperboloid, $t = -(r^2+1)^{1/2}$ (figure 5.19). Then $D^+(S)$ is given by $-r \geq t \geq -(r^2+1)^{1/2}$, as illustrated in the figure. The point p, for

---

† We assume throughout this section that all spacetimes are time-oriented.

Chapter 5. Global structure of spacetimes

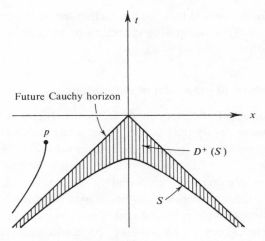

Figure 5.19. Future domain of dependence of the past hyperboloid in Minkowski spacetime (two spatial dimensions suppressed). Point p, for example, is not in $D^+(S)$ because of the past-directed timelike curve shown.

example, is not in $D^+(S)$, because of the past-directed timelike curve shown. Suppose next that the spacetime is Minkowski spacetime with the origin removed, and S is given by $t=0$ (respectively, $t=-1$). Then $D^+(S)$ is given, respectively, by $0 \leq t \leq r$ or $-1 \leq t \leq r$ (figure 5.20). The point p, for example, is in neither domain of dependence, because of the timelike curve from p which 'enters the hole at the origin', and thus does not meet S. Finally, the future domain of dependence of an achronal slice in Taub–NUT spacetime, as shown in figure 5.21, is in the 'Taub-part' of the space-time. Here p fails to be in $D^+(S)$ because the closed timelike curve through p, remaining in the 'NUT-part', fails to meet S.

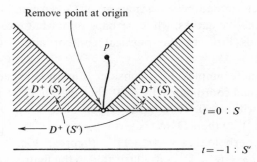

Figure 5.20. Future domain of dependence of two slices in Minkowski spacetime with the point at the origin removed (two spatial dimensions suppressed). Point p, for example, is in neither domain of dependence because of the past-directed timelike curve shown.

What is the topology of our universe?

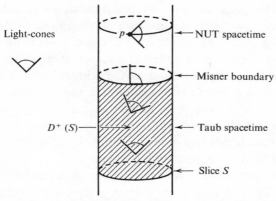

Figure 5.21. Future domain of dependence of a slice in Taub–NUT spacetime (two spatial dimensions suppressed). Point p is not in $D^+(S)$ because of the closed timelike curve passing through p.

One sees in these examples how the domain of dependence works. 'Holes in the spacetime' (e.g., in more physical examples, singular regions) leave a 'shadow', removed from the domain of dependence, which corresponds physically to the fact that information could, at least in principle, be emitted from the hole, influencing the physics in the 'shadow region' without that information having been registered on S. In the last example one sees that regions with causality violations must also normally be deleted from the domain of dependence, for closed timelike curves allow signals to propagate without being registered on S.

Interchanging the roles of 'past' and 'future' in the above, we obtain the *past domain of dependence*, $D^-(S)$. The *domain of dependence* of S is the union of $D^+(S)$ and $D^-(S)$.

There are available a number of results, involving global aspects of initial-value formulations in general relativity, which make more precise the sense in which physical data on an achronal slice determine the physical situation at points of its domain of dependence. Consider, as an example, Maxwell fields with zero source. The initial data, to be specified on S, are the values of the electric and magnetic fields, each with vanishing divergence. The result in this case is that such data on S determine one and only one smooth Maxwell field in the domain of dependence of S. The analogous result for Einstein's equation (say, with vanishing stress-energy) is somewhat more difficult to state, since of course in this case the 'solution' of the equation is just the metric, while that metric is part of the spacetime. One introduces the notion of an initial-data set: a three-dimensional manifold S with a positive-definite

Chapter 5. Global structure of spacetimes

metric q_{ab} (ultimately, the induced spatial metric on S) and a symmetric p_{ab} (ultimately, the extrinsic curvature of S), subject to the usual constraint equations of general relativity. The statement in this case is then the following:[18] given such an initial-data set, there exists a spacetime satisfying Einstein's equation and having achronal slice S, such that the data induced on S from the spacetime are the originally given initial data and such that every point of that spacetime is in the domain of dependence of S. Furthermore, one can choose that spacetime to be maximal (i.e. such that any other spacetime satisfying these conditions can be embedded in it) and then it is unique.

The elementary properties of the domain of dependence include the following.[17] The past and future domain of dependence of any achronal slice S are both closed, and, of course, each includes S itself. Furthermore, for p in $D^+(S)$, the 'region between p and S', i.e. all qs which precede p and are to the future of S, is in $D^+(S)$. (Were such a q not in $D^+(S)$, then some past-directed timelike curve from q would fail to meet S. Join this curve with a past-directed timelike curve from p to q and round off the corner at q to obtain a past-directed timelike curve from p which fails to meet S – contradicting p in $D^+(S)$ (figure 5.22).)

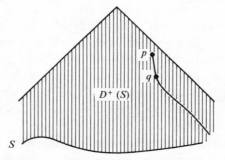

Figure 5.22. If p is in $D^+(S)$, then all points q to the future of S which precede p are also in $D^+(S)$.

There is a second class of particularly useful properties of the domain of dependence. The main one is that the domain of dependence is 'internally causally compact', in the following sense: for p and q in the interior of the domain of dependence of achronal slice S, the closure of $I^-(p) \cap I^+(q)$ is compact (figure 5.23). (Consider, for example, the case with p and q in $D^+(S)$. Were the set above not compact, one could find a sequence of points in $I^-(p) \cap I^+(q)$ without accumulation point. Draw, for each of these points, a past-directed timelike curve from p to that point, and take the accumulation curve of this sequence. This past-

What is the topology of our universe?

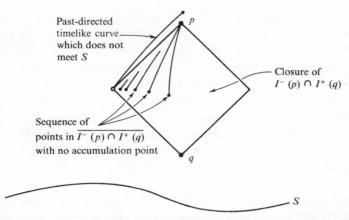

Figure 5.23. If p, q are in the interior of $D^+(S)$, then the closure of $I^-(p) \cap I^+(q)$ must be compact.

directed timelike or null curve from p cannot meet S. Now move p slightly into the future, and 'tilt' this curve, to obtain a past-directed timelike curve from this point which does not meet S – a contradiction.) This property reflects our earlier observation that the domain of dependence excludes 'holes'. A second property is essentially a corollary of the first. For p and q, with p preceding q, in the interior of the domain of dependence of S, there exists a timelike geodesic of maximal length from p to q. We first remark that it is false in general that whenever p precedes q there is a timelike geodesic from p to q. For example, let the spacetime be Minkowski spacetime with the origin removed, and let p and q be given by $(-1, 0, 0, 0)$ and $(+1, 0, 0, 0)$, respectively. To show that such a maximal geodesic does exist when p and q are in the domain of dependence of an achronal slice, one proceeds as follows. Consider the set of all timelike or null curves joining p and q, and consider the function 'length of the curve' on this set. One shows by compactness, that this function must achieve its maximum, which is the desired geodesic. Similarly, one shows that, for p in the interior of the domain of dependence of S, there is a timelike geodesic of maximal length from p to S. As we shall see in the following section, these properties play an important role in the singularity theorems.

Just as the notion of 'precedes' leads to the formulation of an additional condition which may be imposed on spacetimes – lack of causality violations – so the domain of dependence leads to a condition – roughly speaking, lack of 'determinism violation'. Let M, g_{ab} be a spacetime, and

Chapter 5. Global structure of spacetimes

S an achronal slice. This S is said to be a *Cauchy surface*[17] for the spacetime if every point of M is in the domain of dependence of S. For example, the 3-plane $t = 0$ in Minkowski spacetime is a Cauchy surface, while the hyperboloid is not. The positive-mass Schwarzschild and Friedmann spacetimes possess Cauchy surfaces, while Minkowski spacetime with the origin removed, the negative-mass Schwarzschild, the maximally extended Reissner–Nordström, most Weyl, the Kerr, the Taub–NUT, and the plane-wave spacetimes do not. Physically, the existence of a Cauchy surface means that there is a single slice such that information on that slice suffices to determine the physics throughout the entire spacetime.

The demand of the existence of a Cauchy surface, as it turns out, is an extremely strong condition on spacetimes – in particular much stronger than the various conditions we have discussed previously. A few examples of the consequences of this condition will illustrate this point. If a spacetime has a Cauchy surface, then it must be stably causal. Furthermore, one can even choose the time-function such that each of the achronal slices of constant t is itself a Cauchy surface of the spacetime, so in particular there must pass a Cauchy surface through every point of the spacetime. In addition, all of these slices must be topologically identical, i.e. diffeomorphic. As an example of the arguments, we prove the last property. Fix a timelike vector field on the spacetime and consider two of our slices, S and S', each a Cauchy surface. We now define a mapping from S to S' which sends each point p of S to that point of S' reached by the integral curve of our vector field passing through p. (There must exist such a point, since S' is a Cauchy surface; and it must be unique, by achronality of S'.) This mapping is smooth; its inverse exists (reversing the roles of S and S'), and so we have produced our diffeomorphism from S to S'. What all these properties say, then, is that a spacetime with a Cauchy surface has a very 'tame' global structure: in particular, there can be no causal anomalies. An interesting consequence of the last property above is that the underlying manifold of a spacetime with Cauchy surface must be of the form $S \times \mathbb{R}$, where S is some three-dimensional manifold. But this is a condition only on the underlying manifold. Thus, there are numerous 4-manifolds (e.g. $\mathbb{R}^4$ with two points removed) which are known not to admit any Lorentz metric such that the resulting spacetime possesses a Cauchy surface!

One thinks of a spacetime with a Cauchy surface as being 'predictive', in that all the physics in such a spacetime can be determined from suitable initial conditions. On the other hand, there are of course many spacetimes without Cauchy surfaces, and one thinks of these as subject to uncontrollable influences entering the spacetime, e.g. from singular regions. The fact that the second possibility is available in general relativity complicates the treatment within the theory of a number of physical questions, for implicit in many of these questions is the idea of prediction. Consider, for example, the collapse of a star of a given mass and angular momentum. What fraction of the initial mass is emitted as gravitational radiation? Since this is essentially a problem of predicting the future from initial conditions, it is possible that the answer from general relativity would be, not the fraction, but rather the assertion that there are no spacetimes representing this phenomenon which have Cauchy surfaces.

In what sense is the existence of a Cauchy surface a physically reasonable requirement? The situation is very much like that regarding stable causality: one lacks a firm argument one way or the other, and so is reduced to the judgement as to whether or not non-predictive spacetimes should be ruled out. However, there does seem to exist at least the potential for more solid evidence. One might try to show that spacetimes which do not admit Cauchy surfaces are in some sense 'unstable', and hence, presumably, of little physical interest. Consider a spacetime, and an achronal slice S which is not a Cauchy surface. For simplicity, consider test electromagnetic fields. Given initial data for Maxwell's equations on S, we of course obtain a unique smooth Maxwell field in the domain of dependence of S. (So, in particular, if S were a Cauchy surface we would have a smooth Maxwell field in the entire spacetime.) Returning to the general case, one can now ask whether the field so obtained is everywhere well-behaved, or whether this field becomes increasingly large as one approaches the boundary of the domain of dependence. The latter alternative would be a good indication that, when this situation is treated within the full context of general relativity, the resulting spacetime would become singular along this boundary, 'cutting off' the region of the spacetime outside of the domain of dependence. There are examples which indicate that this second alternative might indeed occur, e.g. the maximally extended Reissner–Nordström spacetime. If one could show that it does occur in the 'generic' spacetime then one would have an argument that the existence of a Cauchy surface is perhaps a physically reasonable requirement. The argument itself would have to be formulated with some care, for there are certainly spacetimes, e.g. Minkowski spacetime, which admit Cauchy surfaces and also achronal slices which are not Cauchy surfaces.

Chapter 5. Global structure of spacetimes

Given an achronal slice in a spacetime, one usually does not know ahead of time whether that slice is a Cauchy surface. One normally wishes to prove that it is, in order to make available all the strong consequences of the existence of a Cauchy surface. It is useful for such arguments to have available some tool which will pinpoint a slice's failure to be a Cauchy surface. An appropriate tool is the notion of a Cauchy horizon. Let S be an achronal slice in spacetime M, g_{ab}. The *future Cauchy horizon*[17] of S is the collection of all points p which are in $D^+(S)$, but which precede no points of $D^+(S)$. The future Cauchy horizon, in other words, is the 'future boundary' of $D^+(S)$. Similarly for the *past Cauchy horizon*. For example for S, the hyperboloid in Minkowski spacetime as shown in figure 5.19, the future Cauchy horizon is the surface shown. It is immediate from the definition that the future Cauchy horizon is always an achronal set. It is also always closed, a consequence of the fact that $D^+(S)$ is closed while the horizon includes no point preceding a point of $D^+(S)$. The sense in which the Cauchy horizon of S 'pinpoints the failure of S to be a Cauchy surface' is the following: an achronal slice S is a Cauchy surface if and only if its future and past Cauchy horizons are both empty, which is immediate from the fact that the Cauchy horizon is the boundary of the domain of dependence.

Thus the mere existence of a non-empty Cauchy horizon means that one's original slice is not a Cauchy surface. If one wishes to prove that a given slice is a Cauchy surface, one supposes that there is a non-empty Cauchy horizon, and tries to use various properties of this horizon to obtain a contradiction. The utility of this horizon, then, depends essentially on its having accessible properties. By far the most important such property is that the Cauchy horizon is null, in the following sense. Let p be a point of the future Cauchy horizon of S. Then there exists a maximally extended past-directed null geodesic from p which always remains on that future Cauchy horizon. To prove this choose a sequence, $p_1, p_2, \ldots$, of points which approach p from the future. Since p is on the horizon, none of these can be in $D^+(S)$, and so there exists a sequence of maximally extended, past-directed timelike curves, $\gamma_1, \gamma_2, \ldots$, from these respective points, all of which fail to meet S. Let γ be an accumulation curve, a past-directed timelike or null curve from p (figure 5.24). Then this γ cannot enter $I^-(p)$, for otherwise one of the γ_i would, whence this γ_i would have to meet S. So, γ is a null geodesic. This γ must be in $D^+(S)$, since p is. Furthermore, it must remain on the Cauchy horizon, for if a point to the future of a point of γ were in

What is the topology of our universe?

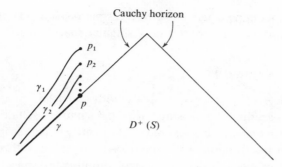

Figure 5.24. Construction of a null geodesic on the Cauchy horizon. The points p_i approach p, and γ is an accumulation curve of the maximally extended past-directed timelike curves γ_i from p_i which never enter $D^+(S)$.

$D^+(S)$, then this point would also be to the future of a point of one of the γ_i, whence that γ_i would have to meet S. This γ is the desired null geodesic.

5.3 Is our universe singular?

Numerous exact solutions of Einstein's equation, including, for example, the Schwarzschild, Kerr, Weyl, Friedmann, and Reissner–Nordström spacetimes, are known to be in some sense singular. It was long felt (or at least suspected) that the singular character of these exact solutions was merely a consequence of their special symmetries. Indeed, already in Newtonian gravitation certain highly symmetric solutions (e.g. those for a collapsing spherical dust cloud) are singular. It turns out, however, that the status of the singular solutions in general relativity is quite different: 'generic' classes of spacetimes in general relativity are singular; this singular character seems to be an essential feature of the theory. The theorems to this effect are remarkable: their applicability is rather general (because, for example, they use Einstein's equation only in a very limited way), and their proofs are quite simple. These singularity theorems require certain limitations on the global structure of the spacetime under consideration. In fact, it was perhaps largely because of these theorems that global techniques have seen the development they have. The theorems thus provide a good example of the applications of the various techniques of the previous section.

We here discuss these results. This section is divided into three parts. In the first we consider the definition of a singular spacetime; in the

Chapter 5. Global structure of spacetimes

second, the theorems which show that certain spacetimes are singular; and in the third, various properties of singular spacetimes.

Singular spacetimes

In most areas of physics one deals with particles, fields, etc. on Minkowski spacetime. There 'singularity' has a natural definition: a point of the spacetime at which some physical quantity becomes infinite (e.g. the origin for the Coulomb solution in electrodynamics). But in general relativity the primary field under consideration is the spacetime metric itself. How is one to characterize singular behavior of this field?

Inspection of the various exact solutions might suggest some definition along the following lines: a singularity is a point at which the components of the metric are badly behaved (e.g. become infinite, oscillate indefinitely, etc.). Such attempts, unfortunately, suffer from at least two difficulties. First, it seems to be extremely difficult, when dealing with metric components, to disentangle genuine geometrical information about the spacetime from 'gauge' information contained in the chart-choice. For example, the metric $-t^{-2} dt^2 + dx^2 + dy^2 + dz^2$ might appear to be 'singular at $t = 0$', while this is of course just Minkowski spacetime; the Schwarzschild spacetime, in the usual coordinates, might seem to be singular at $r = 2m$, while this is just a consequence of an inappropriate choice of coordinates. Although it is certainly clear that the behavior of metric components has *something* to do with singular behavior, it seems difficult to put one's finger on exactly what. A second difficulty with such definitions, closely related to the first, is the issue of what is the 'point' they refer to. One may have in mind its characterization by certain limiting values of the coordinates of some chart, but this again seems to depend in an essential way on the choice of chart. In fact, the isolation of what are to be called the 'singular points' of a spacetime is apparently a difficult technical problem, to which we shall return shortly. Although one could imagine a start toward a resolution of the first difficulty, e.g. by replacing components by scalar invariants, this would hardly help for the second.

The key idea of what is now widely accepted as the most fruitful definition of a singular spacetime is the following.[19] General relativity, as it is usually formulated, requires a manifold with a smooth Lorentz metric. This formulation leaves no room for points of the manifold at which the metric is singular. Indeed, it is even hard to see how one could modify the theory to admit such 'singular points', for it is only through

Is our universe singular?

the metric that one acquires the ability to identify the individual points of the manifold as events. One cannot isolate, as additional physical events, points at which the metric itself is badly behaved. In short, it seems to be a necessary part of general relativity that all 'singular points' have been excised from the spacetime manifold. How can one tell, from the spacetime itself, that 'a singularity has been excised'? Consider, as an example, the Schwarzschild spacetime. Investigation of the topology will never allow one to decide, for the topology in this example is $S^2 \times \mathbb{R}^2$, and there are many Lorentz metrics on this manifold which by no stretch of the imagination would be called singular. Neither will investigation of the causal structure, etc. A feature which does distinguish this spacetime from similar ones which one would regard as non-singular is the following. In the Schwarzschild spacetime, there are geodesics ('running into the singularity') which are maximally extended and which none the less have the property that their affine parameters do not extend over the entire range from $-\infty$ to $+\infty$.

These remarks suggest the following definition. A spacetime is said to be *timelike* (respectively *null*) *geodesically incomplete* if it contains a maximally extended timelike (respectively, null) geodesic whose affine parameter does not assume values in the full range from $-\infty$ to $+\infty$. The proposal is to make 'singular' precise as geodesic incompleteness. That this is physically reasonable is suggested not only by the discussion above, but also by the fact that timelike geodesics represent the paths of freely falling observers, while their affine parameters are local time as recorded by these observers. Thus, the existence of an incomplete timelike geodesic means that there could be a freely falling observer who experiences only a finite time – certainly a rather 'singular' occurrence.

There is, however, a potential problem with this definition. Although it is certainly true that 'excised singularities' result in geodesic incompleteness, one could also excise well-behaved regions from spacetimes (e.g. the origin from Minkowski spacetime) and again produce geodesic incompleteness. Are we to call such spacetimes singular too? Physically, one would prefer not to, since in particular there appears to be no physical mechanism that would result in such a region of spacetime being removed. In some sense, one should never have allowed the freedom to excise regions of spacetime in the first place. However, it seems to be difficult to formulate a criterion which distinguishes between geodesic incompleteness which results from excision of a 'real singularity' as opposed to a 'regular region'. For example, inextendibility alone will

Chapter 5. Global structure of spacetimes

not serve as such a criterion. The universal covering space of Minkowski spacetime with a 2-plane removed is geodesically incomplete, inextendible, and yet would not seem physically to be very singular. Fortunately, this problem does not interfere with the singularity theorems, for the following reason. The form of a typical theorem is: 'If a spacetime satisfies ..., then it is geodesically incomplete'. Now, if one wishes to maintain that the spacetime in question is geodesically incomplete simply because of the removal of some well-behaved regions from the spacetime, then one is invited to restore those regions. But this new spacetime will normally again satisfy the conditions of the theorem, and so must still be geodesically incomplete. In short, it is the form of the theorems themselves which renders viable what is clearly too inclusive a definition.

Neither timelike nor null incompleteness implies the other. In fact, there are even several other inequivalent notions of incompleteness, of which we mention two as examples. A spacetime is said to be bounded-acceleration incomplete if it possesses a maximally extended timelike curve (say, future-directed from an initial point) with bounded acceleration and finite total length. A spacetime is said to be bundle-incomplete[20] if it admits a timelike curve of finite length, where now this length is defined by introducing a parallel-transported orthonormal frame along the curve and taking, for the squared-norm of the tangent vector of the curve, the sum of the squares of the inner products of this tangent vector and the vectors in the tetrad. These definitions of course have the same idea as that of geodesic incompleteness, namely to identify certain curves which, while maximally extended, are not 'as long as similar curves in Minkowski spacetime'. While the obvious implications – geodesically incomplete implies bounded-acceleration incomplete implies bundle-incomplete – are true, there are examples which show that their converses are all false. These other possibilities serve to illustrate that 'has an excised singularity' can have many facets in relativity. None, with the exception of geodesic incompleteness, seems to have found any significant applications.

Singularity theorems

Each of the proofs which show that certain spacetimes must be singular consists of two parts: a local geometrical part, whose substance is that there is an irreversible tendency for geodesics to converge toward each other; and a global part, in which this 'tendency to converge' leads to geodesic incompleteness. We consider these two parts in turn.

Let M, g_{ab} be a time-oriented spacetime, and let S be a slice. We introduce a function t defined near S by the conditions that $t = 0$ on S and that $\nabla_a t$ be unit, future-directed, and timelike. Setting $t^a = \nabla^a t$, we have $t^m \nabla_m t^a = t^m \nabla^a t_m = 0$, where the first equality uses that t_a is a gradient, the second that t_a is unit. Thus, t^a is a geodesic vector field, namely the field of tangents to the timelike geodesics emanating normally from S. Set $c = -\nabla_m t^m$, the convergence of our field. We now ask for the rate of change of this convergence along the geodesics:†

$$t^a \nabla_a c = -t^a \nabla_a \nabla_b t^b = R_{ab} t^a t^b - t^a \nabla_b \nabla_a t^b$$
$$= R_{ab} t^a t^b + (\nabla_b t_a)(\nabla^a t^b).$$

The two terms in this last expression are now dealt with as follows. The first is rewritten, using Einstein's equation, as $(T_{ab} - \tfrac{1}{2} T g_{ab}) t^a t^b$, where $T = T^m{}_m$. We now demand that this expression be non-negative (for every timelike t^a), a condition on the stress-energy of the matter in the spacetime called the *energy condition*. More on this shortly. For the second term, we first note that $\nabla_a t_b$ is symmetric, and has vanishing contraction with t^a. But our second term is the trace of the square of $\nabla_a t_b$, which (since $\nabla_a t_b$ is essentially a symmetric three-dimensional tensor in the space orthogonal to t^a, and since the metric in that space is positive-definite) cannot be less than one-third the square of its trace. (Proof: the square of its trace-free part must be non-negative. Expand.) This second term, then, must be at least $\tfrac{1}{3} c^2$. Thus, our equation becomes

$$t^a \nabla_a c \geq \tfrac{1}{3} c^2. \tag{5.1}$$

This equation states that the convergence has an irreversible tendency to increase along the geodesics, and further that the rate of increase must itself become larger as c increases. There results a 'runaway effect' which normally results in c becoming infinite. It is immediate from (5.1) in particular that, if $c = c_0 > 0$ at some initial point of a geodesic, then c must become infinite at least by a further distance $3/c_0$ along the geodesic. That c becomes infinite is the signal that nearby geodesics have begun to cross. Equation (5.1) can of course be obtained directly from the equation for geodesic deviation. There are analogous results –

† The Riemann tensor is defined by $\nabla_{[a} \nabla_{b]} k_c = \tfrac{1}{2} R_{abc}{}^d k_d$, and the Ricci tensor by $R_{ab} = R_{amb}{}^m$.

Chapter 5. Global structure of spacetimes

to the effect that nearby geodesics must cross – for null geodesics, and also for timelike geodesics under somewhat different conditions. We shall state other results as needed.

The physical meaning of this little calculation is the following. The number $T_{ab}t^a t^b$ is of course the local energy density of matter, as seen by an observer with 4-velocity t^a. The energy condition, however, refers to the combination $(T_{ab} - \frac{1}{2}Tg_{ab})t^a t^b$, which is essentially this local energy density, but with an additional term involving the stresses. All matter normally considered in relativity turns out to satisfy this energy condition, i.e. turns out to have $(T_{ab} - \frac{1}{2}Tg_{ab})t^a t^b$ non-negative for all timelike t^a. For the Maxwell field, for example, this combination is the sum of the squares of the electric and magnetic fields as determined by the observer with 4-velocity t^a. For a perfect fluid, the energy condition requires that both $\rho + p$ and $\rho + 3p$, where ρ is the density and p the pressure, be non-negative. Thus, for $\rho > 0$, this condition would be violated only for a large *negative* pressure. The energy condition requires, in short, that a certain 'effective local energy density of matter' be non-negative. What (5.1) means, in effect, is that gravitation is then attractive. Gravitational effects, in the presence of the energy condition, serve to draw the geodesics (free particles) toward each other, causing the convergence c to increase.

One would of course have to impose *some* restriction on the stress-energy of matter in order to obtain any singularity theorems, for with no restrictions Einstein's equation has no content. One might have thought, however, that only a detailed specification of the stress-energy at each point would suffice, e.g. that one might have to prove a separate theorem for each combination of the innumerable substances which could be introduced into spacetime. It is the energy condition which intervenes to make this subject simple. On the one hand it seems to be a physically reasonable condition on all types of classical matter, while on the other it is precisely the condition on the matter one needs for the singularity theorems.

All the singularity theorems, then, start from the energy condition and its tendency to force geodesics to cross, and end with geodesic incompleteness. One gets from one to the other by exercising suitable control over the global structure of the spacetime, i.e. additional conditions which, together with the crossings of geodesics, will imply incompleteness. Of course, one hopes also that these additional global conditions can be justified by the physical situation under consideration. That such conditions are indeed needed can be seen from the following

Is our universe singular?

example. Consider the future-directed timelike geodesics emerging normally from the hyperboloid in Minkowski spacetime given by $t = -(r^2+1)^{1/2}$. These all meet at the origin, although of course there is no singular behavior of the spacetime there.

We now consider a few examples of singularity theorems.[21]

Let M, g_{ab} be a spacetime, and let S be a Cauchy surface. Let it be the case that the convergence, at S, of the timelike geodesics emanating normally from S is bounded below by some positive constant c_0. Physically, S represents 'the entire universe at one instant of time', and the condition on the convergence means that the universe is, at 'time S', contracting at a rate bounded below. (Alternatively, one could time-reverse, and deal with an expanding universe.) We claim that these properties together with the energy condition imply that the spacetime must be timelike geodesically incomplete. Suppose, for a contradiction, that this spacetime were geodesically complete. Draw a future-directed timelike geodesic from S, go along it a distance greater than $3/c_0$ from S, and there select point p. Since p is in the future domain of dependence of S, there will exist a timelike curve from p to S of maximal length; by construction, this length will exceed $3/c_0$. This curve will have to be a geodesic (for otherwise it could be lengthened by straightening it out), and will have to meet S normally (for otherwise it could be lengthened by moving slightly its point of intersection with S). But, since the convergence on S exceeds c_0, the geodesics emerging normally from S will begin to cross by at least distance $3/c_0$, and in particular our geodesic γ will be so crossed before it reaches p. This crossing of γ by another geodesic means that we can further lengthen γ, by rounding off the corner at the crossing point as shown in figure 5.25. We thus

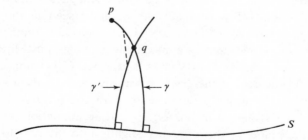

Figure 5.25. If a geodesic is crossed, then it cannot be a maximal geodesic. The distance from S to p along γ is essentially the distance from S to q along γ' plus the distance from q to p along γ. But this latter curve is a broken geodesic, which can be lengthened by rounding off the corner as shown. Hence, γ could not have been the maximal curve from S to p.

Chapter 5. Global structure of spacetimes

contradict γ being the longest timelike curve from p to S. We conclude, therefore, that this spacetime – satisfying the energy condition and having a Cauchy surface the convergence of whose normals is bounded below – must be timelike geodesically incomplete.

While the above is perhaps the simplest singularity theorem, it is also, at least from a physical viewpoint, one of the weakest. Its weakness stems from two of its hypotheses. First, we demand that S be a Cauchy surface for the spacetime. As discussed in the previous section, we do not have strong evidence that the existence of a Cauchy surface is a physically reasonable restriction on spacetimes. Second, we must demand that the convergence of the normals from the slice S be bounded away from zero. While this is certainly one way of saying 'contracting universe', it is in fact a rather strong condition, for it must be possible to choose S so that this 'contraction' is not less than some c_0 at every point of S. It is, for example, far from clear that our own universe possesses the time-reverse of such an S. The next two theorems show how these respective two conditions can be weakened.

Consider first the condition that S be a Cauchy surface. It will not do, first, to simply drop this condition. Consider, for example, Minkowski spacetime, and let the slice S be the usual hyperboloid. In this case the convergence of the normals of S is bounded away from zero, and the energy condition is satisfied, and yet this spacetime is of course geodesically complete. Thus, we must replace the condition that S be a Cauchy surface by other conditions, and it turns out that two are needed: that S be compact (i.e. one has a 'closed universe'); and that the spacetime be stably causal. The proof now proceeds as follows. Consider the future domain of dependence of S, $D^+(S)$, and then its future Cauchy horizon, H. This H must be either compact or non-compact. Suppose first that H is compact. As we have seen in the previous section, it follows from the fact that H is a Cauchy horizon that through every point of H there passes a maximally extended past-directed null geodesic which remains always in H. But, since H is compact, such a curve would have to come back arbitrarily near to itself – clearly a violation of stable causality. Suppose, then, that H is non-compact. Let $p_1, p_2, \ldots$ be a sequence of points, without accumulation point, in H. Displace the points of this sequence slightly into the past to obtain a sequence, $q_1, q_2, \ldots$, of points in the interior of $D^+(S)$, also without accumulation point. Since the qs are in the interior of $D^+(S)$, we can apply our earlier argument to obtain, for each q_i, a maximal timelike curve, γ_i, to S. Each γ_i will be a timelike geodesic, which meets S

Is our universe singular?

normally, say at point s_i (figure 5.26). Now we use compactness of S: the s_i on S must have an accumulation point, say s. We now conclude that the timelike geodesic from s normal to S cannot have length greater than $3/c_0$, for if it did a point on this geodesic would provide an accumulation point for the q_i. We conclude, then, that this spacetime must be timelike geodesically incomplete.† What we have done, then, is retained the energy condition, the slice S, and the demand that the convergence of its normals be bounded below, while replacing the condition that S be a Cauchy surface by the conditions that S be compact and that the spacetime satisfy stable causality.

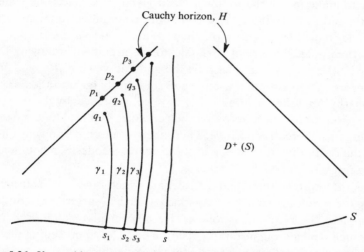

Figure 5.26. If a stably causal spacetime satisfies the energy condition and admits a compact slice whose normals are everywhere converging, then it must be timelike geodesically incomplete.

We consider next the relaxation of the condition that the convergence of the normals of S be bounded below. Again, we cannot simply drop this condition – i.e. claim that any spacetime with a Cauchy surface, which satisfies the energy condition, is geodesically incomplete – for Minkowski spacetime is a counterexample. Again, we must replace the condition on the convergence by other conditions, and again it turns out

---

† In fact the assumption of stable causality is not necessary. Were the Cauchy horizon compact, then the function which assigns to each point of that horizon the length of the maximal geodesic to S would have to assume a minimum. But this would contradict the fact that this function must be strictly decreasing along past-directed null geodesics in the horizon.

Chapter 5. Global structure of spacetimes

that two are needed: that S be compact, and that the spacetime be generic. A spacetime is said to be *generic* if every timelike or null geodesic possesses a point at which $\xi_{[a}R_{b]mn[c}\xi_{d]}\xi^m\xi^n$ is nonzero, where ξ^m denotes the tangent vector to that geodesic. This condition demands, roughly speaking, that every such geodesic encounter some 'effective curvature' somewhere. It is regarded as physically reasonable because one would expect that, even if a given spacetime is not generic, a small perturbation in the metric would result in a spacetime which is. The presence of the 3 K blackbody radiation in our own Universe rather suggests that it is generic. We shall need the property of being generic in order to make use of the following fact: given, in a spacetime which is generic and satisfies the energy condition, a complete timelike or null geodesic, then some nearby timelike or null geodesic meets that one more than once. The proof is similar to the argument following (5.1). Let γ be the geodesic, and let p be the point at which $\xi_{[a}R_{b]mn[c}\xi_{d]}\xi^m\xi^n$ is nonzero. Choose a point q on γ and sufficiently far into the past that the nearby geodesics meeting γ at q have convergence, by the time they reach p, sufficiently near zero that the curvature at p is able to make the convergence thereafter positive. There then ensues the 'runaway effect', with the result that at least one of these nearby geodesics meets γ again in the future.

We are to prove, then, that a spacetime which is generic, satisfies the energy condition, and has compact Cauchy surface S, must be incomplete. This is done as follows. Fix any timelike geodesic τ in the spacetime, and a sequence of points $a_1, a_2, \ldots$, each preceding the next and extending indefinitely into the future along this geodesic, and also a sequence, $b_1, b_2, \ldots$, into the past. Since we have Cauchy surface S, there is for each i a timelike geodesic, γ_i, from b_i to a_i of maximal length (figure 5.27). Characterize each geodesic by the point at which it meets S, and its tangent direction there. Since S is compact, we can find an accumulation point and tangent direction, where the direction is either timelike or null. Let γ be the geodesic which meets S at this point with this direction. We suppose, for contradiction, that this γ were complete. But now, since our spacetime is generic and satisfies the energy condition, some nearby geodesic must meet γ more than once. But this γ is essentially a limit curve of the γ_i, and so some nearby geodesic must meet one of the γ_i more than once. Finally, 'meeting more than once' would mean that we could lengthen that γ_i by rounding off corners, which contradicts our construction of γ_i as the curve of maximal length from b_i to a_i.

Is our universe singular?

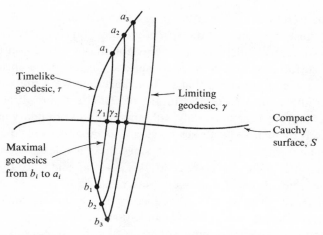

Figure 5.27. A spacetime which is generic, satisfies the energy condition, and admits a compact Cauchy surface must be geodesically incomplete.

It turns out, in fact, that one can combine these two arguments to eliminate simultaneously the conditions that the spacetime admit a Cauchy surface and that the convergence of the normals be bounded away from zero. That is to say: a spacetime which is stably causal, which is generic, which satisfies the energy condition, and which has compact slice S, must be incomplete. The proof consists essentially of using the argument of figure 5.26 to choose appropriate sequences within the domain of dependence of S, and then the argument of figure 5.27, within that domain of dependence, to obtain incompleteness.

The thrust of these results is that a closed universe, i.e. a spacetime having a compact slice, satisfying certain additional conditions one might regard as physically reasonable for our universe (i.e. is stably causal, generic, and satisfies the energy condition) must be singular. These results all deal with singular cosmological models. There is also a class of results which refer to the gravitational collapse of smaller objects, e.g. stars, within the universe. A *trapped surface* in a spacetime is a two-dimensional, compact, spacelike surface T such that both classes of future-directed null geodesics emerging from S orthogonally have positive convergence at S. For the 2-spheres of spherical symmetry in the Schwarzschild spacetime, for example, those with $r < 2m$ are trapped surfaces. The existence of a trapped surface is taken to mean that gravitational collapse has proceeded beyond a certain point. The next result indicates that under certain circumstances this point is in fact

Chapter 5. Global structure of spacetimes

a point of no return, i.e. the creation of a trapped surface means that the collapse must continue and the spacetime must be singular. More precisely, a spacetime satisfying the energy condition, with a non-compact Cauchy surface, S, and a trapped surface, T, must be incomplete. Suppose, for contradiction, that such a spacetime were null complete. Consider the boundary of the future of T. The future-directed null geodesics leaving T orthogonally will remain in this boundary until they are crossed by nearby geodesics leaving T orthogonally, and after this they must enter the future of T. But, since these geodesics are assumed complete, since their convergence at T is positive, and since the energy condition holds, each of these geodesics must be so crossed in a finite affine distance. (This is a consequence of the equation analogous to (5.1), but for the null case.) We conclude that the boundary of the future of T, since it is made up of finite segments of null geodesics, is compact. Now introduce a timelike vector field, and consider the mapping from this boundary back to the Cauchy surface S obtained by projecting along its integral curves. This mapping will have to be one-to-one (for no point of this boundary can precede itself) from this compact boundary to the non-compact S, which is a contradiction. We conclude, therefore, that our assumed null completeness must fail. As before, the existence of a Cauchy surface can be replaced by stable causality and the generic property.

The theorems discussed above are indicative of the types of results which are available to the effect that broad classes of spacetimes must be singular. The situation, in general terms, is as follows. It appears that closed universes (i.e. those possessing a compact achronal slice) are normally singular. For open universes, by contrast, there seem to be 'generic' spacetimes which are singular, and also generic ones which are not. Generally speaking, an open universe needs a catalyst, e.g. a trapped surface or a slice with convergence bounded away from zero, before one can conclude that it must be singular.

Properties of singular spacetimes

The conclusion from all this, then, is that large classes of solutions of Einstein's equation are singular. This state of affairs naturally gives rise to a variety of possible viewpoints. On the one hand, one might regard the frequency of singular spacetimes as a reminder that general relativity has only a finite range of applicability. The experimental basis for the detailed theory is not at all strong, and in particular we have virtually

Is our universe singular?

no evidence that it is applicable in strong gravitational fields. One might, for example, look for a more refined theory, in which the problem of singular behavior disappears. Unfortunately, the fact that the theorems use Einstein's equation only in a very weak sense – essentially, only to conclude that gravitation is attractive – rather suggests that the construction of a new theory, without singular behavior, may be difficult. We are apparently very far from either a reasonable alternative to general relativity which avoids singular behavior or some result to the effect that no such theory exists. Alternatively, one might try to argue that these results will turn out not to be applicable to our own universe, e.g. because there is available some matter which violates the energy condition and which becomes important precisely in regions of high density. Indeed, there are indications that the inclusion of quantum effects will result in a contribution to the stress-energy of exactly this type. Finally, one might take the position that the theory is correct, and that our universe really is singular. Our task is not to try to eliminate this behavior somehow but rather to adapt our physical ideas to it.

It would presumably be helpful, under any of these viewpoints, to obtain some more detailed information as to the structure of singular spacetimes. One program along these lines would be to look for some construction which, given a singular spacetime, yields 'singular points', presumably attached to the spacetime manifold as some sort of boundary. That is to say, one would like to be able to use the word 'singularity'. Then the physics of the spacetime near these singular points could be studied locally. An example will illustrate why such a construction must be formulated with some care. Consider, for example, a Weyl solution with its 'line singularity'. Suppose that there has been attached to this spacetime in some direct way, e.g. by examination in a chart, such singular points. Consider now an example of some physical statement one may wish to make, for example, that curves γ_1 and γ_2 in figure 5.28 reach the same singular point. We show qualitatively that, at least in this example, there are alternative choices of the singular points such that these curves reach different singular points. We first rewrite this spacetime, e.g. by replacing the coordinate ρ, representing distance from the axis, by $\rho + 1$, so that its singularity appears to be a cylinder, as shown. Each of our original singular points now defines a circular cross-section of this cylinder. We next choose a different set of circular cross-sections, and finally recollapse the cylinder to a line, but now with each of our new cross-sections collapsed to a point. Each of these operations was a smooth transformation on our original spacetime, and

Chapter 5. Global structure of spacetimes

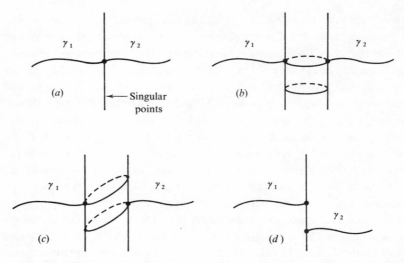

Figure 5.28. A naive specification of a singular boundary can be ambiguous. (*a*) Two curves which appear to approach the same singular point. (*b*) Same spacetime rewritten so that the singular region becomes a cylinder. (*c*) New cross-sections chosen for the cylinder. (*d*) Cylinder recollapsed to a line.

so the final spacetime represents precisely the same physical situation as the original. However, the final spacetime is so presented that one would not regard the curves γ_1 and γ_2 as approaching the same singular point. In short, 'two curves in the spacetime approach the same singular point' makes no sense without a detailed specification of how those singular points are to be constructed. Similar problems are involved, for example, in the discussion of the behavior of scalar invariants as one approaches singular points from various 'directions'. For example, certain Weyl solutions have the property that the scalar invariant $R_{abcd}R^{abcd}$ approaches a finite limit as one approaches the 'singular point' along the axis, but grows without bound as one does so within the equatorial plane (i.e., in the usual Weyl coordinates, $z = 0$). But, for example, one can introduce a new system of coordinates such that, on approaching the 'singular point' along any radial direction with respect to these new coordinates, this scalar invariant remains finite (figure 5.29).

Although various attempts have been made to provide a construction for singular points, none so far has been particularly successful. One involves the use of equivalence classes of incomplete geodesics,[22] but this seems to be both complicated and somewhat contrived; another involves the introduction of a positive-definite metric on the bundle of

Is our universe singular?

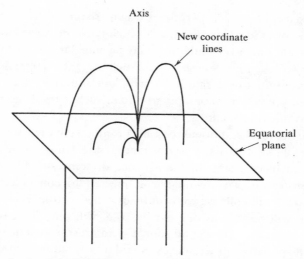

Figure 5.29. A new choice of radial coordinate for a Weyl solution in which the scalar invariant $R_{abcd} R^{abcd}$ remains finite as one approaches the origin along any radial direction.

frames of the spacetime,[20] but this turns out to yield in certain examples non-Hausdorff spaces,[23] i.e. it gives unphysical answers in such examples. The failure of these attempts at least leaves the suggestion that we may simply be asking the wrong question. Perhaps, for example, singular behavior is to be understood, not by means of local singular points, but rather as a more global property of the entire spacetime.

Certain aspects of singular spacetimes can be discussed even in the absence of any construction of singular points. One such aspect – perhaps the most important – is the subject of the next section.

5.4 How noticeably singular is our universe?

In the previous section we found that a large class of physically reasonable spacetimes must be singular, i.e. geodesically incomplete. These results pose a potential problem for general relativity. On the one hand, geodesic incompleteness is often associated with an inability to predict the future from the present, for uncontrollable influences are thereby able to enter the spacetime affecting the physical situation; on the other, the existence of some determinate relationship between present and future could be viewed as a characteristic feature of physical theories. In this section, we shall discuss a body of ideas for how one might circumvent this problem. The idea is to obtain some result to the effect that the

Chapter 5. Global structure of spacetimes

'generic' spacetime, or all spacetimes which are physically reasonable in some sense, or at least all spacetimes which could be regarded as viable models for our universe, while they may be singular can be so only in a rather benign sense. This sense is to be such that their singular character will not interfere too dramatically with our intuitive ideas of determinism and prediction. Attempts to obtain such a result go under the name of cosmic censorship.[24]

There are several additional reasons for interest in the issue of cosmic censorship. First, this issue is relevant to the study of gravitational collapse and black holes. In the absence of some suitable version of cosmic censorship, we are unable at present to conclude that nonspherical stars will collapse to form black holes, or that once formed a black hole will behave as one expects (i.e. will only increase in size, never bifurcate, etc.). A second source of interest in cosmic censorship involves the possible consequences should it turn out that no such result is available for our universe. There might then arise, for example, the possibility of observing directly 'nearby' regions of spacetime with large curvature and the effects which would arise within such regions. One might thus be able to very much expand the range over which general relativity could be tested experimentally, or observe directly the effects of quantum gravity associated with high curvature.

Is some version of cosmic censorship true in general relativity? This question is open and currently under active investigation. The problem is to either find some precise statement of cosmic censorship and prove it, or else show, presumably by means of a counterexample, that there is no true statement which seems to capture cosmic censorship.

Spacetimes which, in some sense to be made precise, have the property that certain observers can detect that their spacetime is singular (i.e. can perceive directly the singular character) are said to be nakedly singular. Nakedly singular spacetimes already have a somewhat nonpredictive character. A very strong result of cosmic censorship would thus rule out, on some physical grounds, all nakedly singular spacetimes. But such a result, as we shall see shortly, appears to be hopeless. Some nakedly singular spacetimes are worse than others, and a more realistic goal would be to try to isolate and rule out only the more unpleasant ones. We begin by considering three families of examples. These will both clarify the meaning to be attached to 'nakedly singular' and illustrate the kinds of nakedly singular spacetimes one does and does not wish to avoid.

How noticeably singular is our universe?

Consider first the closed Friedmann solutions. These spacetimes are certainly singular, and in fact they are both future geodesically incomplete (because of the final collapse at late epochs) and past geodesically incomplete (because of the initial big bang). The future incompleteness could not of course be directly perceived by any surviving observer, while the past incompleteness could be detected by all observers. It is only because of the latter that these spacetimes would be regarded as nakedly singular. For similar reasons, one would regard the maximally extended, positive-mass Schwarzschild solution as nakedly singular. The singular character of these examples, however, is not the sort of thing one could reasonably rule out under cosmic censorship. These spacetimes were already singular at the beginning, i.e. they have 'singular initial conditions'. While observers can certainly detect the singular character, they do not have direct experimental access to such regions. Some closely related examples include the Schwarzschild solution with $m<0$ and the Kerr solution with $a^2>m^2$. These spacetimes, too, would have to be regarded as nakedly singular by any reasonable definition. Now, however, the spacetimes are 'singular at all times', so in particular observers could experience directly the regions of high curvature. This behavior, while clearly unpleasant from the point of view of prediction, is also not what one wishes to rule out. Since our universe presumably arose from a big bang, one could hardly hope to exclude spacetimes which are initially singular, or those for which the singular character persists from the early epoch to the present.

The sort of behavior one *would* wish to rule out by cosmic censorship is more the following. Consider a spacetime which admits a non-singular slice. ('Non-singular' in some sense to be made precise; e.g. for a spacetime asymptotically flat at spatial infinity[19] one might demand that the slice be topologically R^3 and that it be asymptotically a 3-plane.) Physically, one thinks of this slice as representing a 'time at which the spacetime is non-singular'. Suppose furthermore that the future evolution of the spacetime, e.g. because of gravitational collapse, is such that the spacetime is nakedly singular to the future of that slice. Now, one could not attribute the singular character of the spacetime to its being either 'initially singular' or 'persistently singular from an early epoch'. Such a spacetime, subject to certain other conditions, might be taken as a counterexample to cosmic censorship. We can now see already why one might believe that some version of cosmic censorship might be true. Suppose that one has a spacetime up to a non-singular slice, and that one attempts to 'cause' that spacetime to be nakedly singular to the

Chapter 5. Global structure of spacetimes

future of that slice, e.g. by permitting a spherically symmetric cloud of negative-mass dust† to collapse to form, at late times, the negative-mass Schwarzschild solution; or by permitting a cloud of positive-mass dust with large angular momentum relative to its mass to collapse to form, at late times, a Kerr solution with $a^2 > m^2$. But, in either of these cases, there is a tendency for the collapse not to occur at all: in the first, because the gravitational effects of the negative mass are repulsive; in the second, because the effective centrifugal effects of the large angular momentum are repulsive. Thus, although Schwarzschild with $m < 0$ and Kerr with $a^2 > m^2$ are nakedly singular spacetimes, one does not seem to have a viable physical mechanism by which one could 'create' the objects represented by these spacetimes within an otherwise non-singular spacetime.

The second family of examples will all admit a non-singular slice and will be singular to the future of this slice, but it will not be clear whether one should call the future *nakedly* singular in a strong sense. Consider first the maximally extended Kerr spacetime with $a^2 < m^2$. While this spacetime is singular, only a certain class of observers to the future of a non-singular slice are able to detect this fact. Indeed, the only such observers are those who fall into the black hole; observers who remain in the asymptotic region do not have observational access to the singular character. As a second example, consider the Taub–NUT spacetime. This spacetime again admits a non-singular slice (in the Taub part) whose future is singular. Now, however, all observers who are able to detect the singular character are directly involved with it. It is not clear that one would wish even to call this spacetime nakedly singular.

The third family of examples will all admit a non-singular slice, and will be clearly nakedly singular to the future of this slice. The issue now will be whether or not these spacetimes are to be regarded as 'physically reasonable'. Consider first a ball of spherically symmetric dust in Newtonian theory. Arrange the initial conditions so that the outer shells of dust begin collapsing inward and the inner shells begin expanding outward, so these shells cross each other at a certain radius and later time. When the shells meet, the density becomes infinite. There are, as it turns out, analogous solutions in general relativity, where 'infinite density' results in a singular spacetime, and in fact one nakedly

---

† Of course, this use of negative-mass dust is already somewhat unphysical. Indeed, we shall see later that one can easily construct counterexamples to cosmic censorship using such matter.

singular.[25] One would like to argue that such examples do not represent a violation of cosmic censorship because they are 'unphysical', i.e. that 'normal matter' will not execute this shell-crossing behavior. But it turns out, unfortunately, to be difficult to formulate such an argument precisely. In particular, the mere demand that there be a nonzero pressure (say, as a function of density) will not do, for it is known that shell-crossing can occur with a large variety of equations of state. Similarly, an attempt to replace the energy condition† (that $(T_{ab} - \frac{1}{2}Tg_{ab})t^a t^b \geq 0$ for all timelike t^a) by some stronger condition does not look too promising, for it seems very unlikely that there is any other condition which involves only the stress-energy, which is satisfied by what one would regard as 'normal matter', and which would prevent shell-crossing. Thus, while one has the feeling that there is *something* unphysical about shell-crossing, it appears to be difficult to put one's finger on exactly what. The situation is further complicated by examples which seem to be halfway between shell-crossing and collapse. Consider, for example, an infinite dust-filled cylinder in general relativity, with the dust initially at rest. On evolution, the dust will collapse toward the axis, and the resulting space-time will be nakedly singular.[26] This example has some aspects of the shell-crossing examples above (a Newtonian effect), and some of the collapse of a spherical dust cloud to a black hole (a general relativistic effect). A similar example is the following. Consider a collapsing spherical dust cloud together with outgoing radiation, which we shall take, to simplify the discussion, to be a null fluid. This radiation carries away mass from the object as it collapses. One can now carefully adjust the rate at which mass is being carried away so that the radius of the star, even as it collapses, remains always greater than its Schwarzschild radius.[27] But the density can still become infinite, and so one obtains a nakedly singular spacetime.

The significance of these examples – particularly the last two – is not yet clear. It could be that they are just the beginning of a series of examples which ultimately show that there is in fact no result in general relativity having connotations of cosmic censorship. Alternatively, they may somehow carry the key to obtaining the 'correct' statement of cosmic censorship.

---

† It is not difficult to construct counterexamples to any version of cosmic censorship if no conditions are placed on the stress-energy of matter. For example, remove the origin from Minkowski spacetime and introduce a conformal factor which is one outside a neighborhood of the origin and badly behaved near the origin. Set $T_{ab} = G_{ab}$.

Chapter 5. Global structure of spacetimes

We turn now to a discussion of four possible approaches to cosmic censorship. Undoubtedly, there are many others. All four – a causal approach, an asymptotic approach, a stability approach, and an evolutionary approach – will make frequent use of the examples above.

Causal approach

Cosmic censorship essentially refers to the question of which observers can detect, by means of a signal, that their spacetime is singular, i.e. to the causal structure of singular spacetimes. It is natural, therefore, to try to formulate cosmic censorship within this framework.

We first consider a possibility for making precise the notion of 'detect'. There is of course a natural meaning for 'q can detect p' when p and q are points of a spacetime, namely there exist a future-directed timelike curve from p to q, i.e. that p precede q. How is one to characterize a point p at which an observer could detect that his spacetime is singular? One possibility would be to ask whether there exists a past-incomplete timelike or null geodesic with future endpoint p, for the geodesic could be thought of as the world line of a 'signal from the singularity' to p. Unfortunately, such a characterization is subject to two closely related defects. First, the detection is essentially of the effects of 'initially singular' spacetimes (e.g. for the closed Friedmann models every point p would satisfy this characterization) while this is not the type of phenomenon of interest for cosmic censorship. Second, the characterization refers only to past incompleteness, while of course it is the future-incomplete spacetimes which are singular in the more physical sense. A more appropriate characterization is the following. Imagine an observer traveling along a future-incomplete timelike geodesic, continuously emitting a signal. A second observer would be able to detect that his spacetime is future incomplete if he receives the signal emitted by the first observer for a while but then notices that it ends abruptly. More precisely, we define N to be the set of all points p in the spacetime such that there exists a point q which precedes p and whose past contains a future-incomplete timelike or null geodesic. Thus N represents the region of spacetime in which observers could detect, in the sense above, that their universe is singular. Other versions are possible, e.g. with geodesic incompleteness replaced by bundle, or some other, incompleteness. A slightly different notion results if one replaces 'incomplete timelike or null geodesic' by 'maximally extended timelike or null curves which cannot be assigned future endpoints'. The result is

How noticeably singular is our universe?

to enlarge N to include points at which observers could detect 'naked infinities', such as occur, for example, in the extended Kerr spacetime with $a^2 < m^2$ and anti-de Sitter spacetime.

We give some examples of this set N. Clearly, N is empty for any non-singular spacetime. Further, N is also empty for the positive-mass Schwarzschild spacetime. For the negative-mass Schwarzschild spacetime, however, N is certainly not empty, and in fact consists of the entire spacetime. Finally, consider Minkowski spacetime with the point at the origin removed. For this spacetime, N consists of the future of the origin (figure 5.30). This last example illustrates an important issue

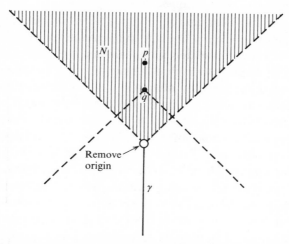

Figure 5.30. The set N for Minkowski spacetime with the point at the origin removed (two spatial dimensions suppressed). The point p, for example, is in N since q precedes p and contains in its past a future-incomplete timelike geodesic.

concerning the formulation of any statement of cosmic censorship. Since we have taken 'singular' to mean geodesically incomplete, nakedly singular spacetimes are easily created by simply removing points from otherwise well-behaved spacetimes. One would not, of course, wish to regard such spacetimes as counterexamples to cosmic censorship. Unfortunately, although one perhaps has a good intuitive idea of what it is that one wants to avoid, it seems to be difficult to formulate a precise condition to rule out such examples. Demanding that one's spacetime be maximally extended, for example, will not do, for there are maximally extended, nakedly singular spacetimes which are everywhere flat, e.g. the universal covering space of Minkowski spacetime with a 2-plane excised.

Chapter 5. Global structure of spacetimes

The basic property of this set N is that it is a future set, i.e. $N = I^+(N)$ (= the union of the futures of the points of N). Indeed, p is in N if and only if there is a point q which precedes p such that $I^-(q)$ contains a future-incomplete timelike or null geodesic. But this holds if and only if there is a point r which precedes p and is in N. In particular, since $I^+(N)$ is always an open set, it follows that the set N is always open. A second consequence is that through any point of the boundary of N there passes a past-directed null geodesic without past endpoint which remains entirely within this boundary. The proof is essentially the same as that of the fact that the boundary of the past of a point is null (i.e. take an accumulation curve of a sequence of timelike curves).

We now return to cosmic censorship. The idea would be that, since this N isolates the nakedly singular character of the spacetime, cosmic censorship is to be formulated as the statement that N cannot be 'too large'. The problem with such a formulation, of course, is that it is difficult to spell out precisely what 'too large' is to mean. Suppose, for example, that one tried to obtain a result to the effect that, for all physically reasonable spacetimes, N must be empty. Then one's criterion for physical reasonableness must exclude, for example, spacetimes with shell-crossing and Taub–NUT spacetime – for in these N is certainly not empty. There is, as it turns out, a criterion which does exclude these examples: in any spacetime with a Cauchy surface, N is empty. (Let S be the Cauchy surface. Since every future-incomplete timelike or null geodesic must meet S, N must be strictly to the future of S. But now, were N not empty, a past-directed null geodesic in the boundary of N could not meet S, which would contradict S being a Cauchy surface.) Unfortunately, the existence of a Cauchy surface is a rather strong condition, and in particular it would be difficult to argue that this condition should be satisfied for all 'physically reasonable' spacetimes (see discussion in 'Causal Structure III'). In any case, we are able to conclude that no spacetime with a Cauchy surface should be regarded as nakedly singular, and in particular that no such spacetime could be taken as a counterexample to cosmic censorship.† Two observations are of interest in connection with this conclusion. The first is that certain of the singularity theorems have as a hypothesis the existence of a Cauchy surface. Thus these theorems, which show that one's spacetime must be singular, are only applicable in the case when the spacetime is not

---

† Of course there do exist spacetimes in which N is empty yet there is no Cauchy surface, e.g. anti-de Sitter spacetime.

nakedly singular. The second is that there is an additional result – essentially a weakened converse of the above – connecting the structure of N with the existence of a Cauchy surface: if a spacetime is such that not only N, but even the expanded N (with future-incomplete timelike or null geodesics replaced by all maximally extended future-directed timelike or null curves) is empty, then that spacetime admits a Cauchy surface. That is to say a spacetime which is neither nakedly singular nor possesses 'naked infinities' must have a Cauchy surface, and of course conversely.

This causal approach, then, takes one only part of the way to a formulation of cosmic censorship. One is able to obtain a set, N, which effectively isolates the nakedly singular character of one's spacetime. This N furthermore has certain desirable properties which may be useful in proofs. What one is not able to do in this approach is isolate a suitable sense of how small one may demand that N will be.

Asymptotic approach

Whereas in the causal approach above one was concerned with the detection by any observer that his spacetime is singular, in the asymptotic approach one makes distinctions between observers. Consider, as an example, the Kerr spacetime with $a^2 < m^2$. This spacetime is nakedly singular, and in particular N is not empty. One would not, however, wish to regard this spacetime as a counterexample to cosmic censorship, for it is only the 'nearby' observers who can detect the singular behavior, not the 'distant' observers. The idea, then, is to sharpen this distinction between nearby and distant.

There is in fact available already a tool in general relativity – that of asymptotic structure – which can be used to obtain a suitable distinction between distant and nearby observers. We briefly describe this tool.

A spacetime M, g_{ab} is said to be *asymptotically flat*[28] (at null infinity) if there exists a manifold with boundary, $\tilde{M} = M \cup I$, consisting of M with boundary I attached, together with a smooth Lorentz metric $\tilde{g}_{ab}$ and a smooth scalar field Ω on $\tilde{M}$, such that:

(1) On M, $\tilde{g}_{ab} = \Omega^2 g_{ab}$.
(2) On I, Ω vanishes, its gradient is nonzero and null, and its second derivative vanishes.
(3) The boundary I consists of a past and future part, I^- and I^+ respectively, each $S^2 \times \mathbb{R}$ with the $\mathbb{R}$s the null generators and each complete.

Chapter 5. Global structure of spacetimes

The motivation and meaning of this definition is as follows. We wish to capture the idea that the spacetime represents an 'isolated system', far from which the spacetime is nearly Minkowskian. This definition does so in a way which avoids potentially awkward limits 'as $r \to \infty$'. This is done, first, by introducing the conformal factor Ω, and, by the first condition, the conformally scaled metric $\tilde{g}_{ab}$. By means of this scaling one can 'bring null infinity into a finite region', and attach 'additional ideal points at null infinity'. These points at infinity are the points of I. Thus $\tilde{M}$ represents the spacetime with its points at infinity attached. That Ω vanish on I, in the second condition, guarantees that the necessary rescaling ultimately be by an infinite amount, i.e. it states that 'infinity is far away'. The rest of the second condition gives more detail as to the required asymptotic behavior of Ω. It essentially requires that Ω fall to zero asymptotically 'as $1/r$, where r is a typical radial function'. Note also that this second condition demands that, on I, the gradient of Ω be normal to I; hence, that I be a null boundary. Finally, the third condition requires that I have 'the correct size and shape'. Since I is null, each point of I is either reached by future-directed timelike curves in the spacetime, or by past-directed. The subsets of I so distinguished are denoted, respectively, I^+ and I^-. (Physically, the distinction is between going to infinity in future or past null directions, respectively.) That each of these parts of I be topologically $S^2 \times \mathbb{R}$ requires that they have the 'shape' of null infinity for, for example, Minkowski spacetime. Finally, completeness refers to the vector field $\tilde{\nabla}^a \Omega$ on I. It is a further consequence of condition (2) that completeness or incompleteness of this vector field is independent of the specific choice of Ω. The requirement that $\tilde{\nabla}^a \Omega$ be complete effectively ensures that I be 'as large' as the boundary at infinity for Minkowski spacetime.

As an example we claim that the Schwarzschild spacetime (and, as a special case, Minkowski spacetime) is asymptotically flat. In the usual coordinates, the metric takes the form

$$ds^2 = -(1-2m/r)\,dt^2 + (1-2m/r)^{-1}\,dr^2 + r^2(d\theta^2 + \sin^2\theta\,d\varphi^2).$$

Replacing r and t by new coordinates given by $x = r^{-1}$ and $u = t - r - 2m \ln(r - 2m)$, this becomes

$$ds^2 = x^{-2}\{2\,du\,dx - x^2(1-2mx)\,du^2 + (d\theta^2 + \sin^2\theta\,d\varphi^2)\}.$$

Now let $\tilde{M}$ be the manifold with boundary consisting of the points of M together with additional points (of I^+) labeled by $x = 0$ in this chart and,

278

reversing the sign of t, attach I^- similarly. Set $\Omega = x$, so that the metric $d\hat{s}^2$ is given by the curly brackets above. The verification of the conditions for asymptotic flatness is straightforward (figure 5.31).

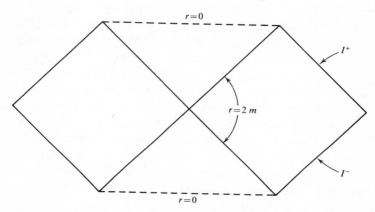

Figure 5.31. The $r-t$ plane of the Schwarzschild solution with null infinity shown.

The important feature of this definition of asymptotic flatness is that it yields a boundary surface I of points 'at null infinity'. Thus various asymptotic questions, such as that of the character of emitted radiation, reduce to local questions about I. Further, as we shall see in a moment, the surface I itself is useful for certain global arguments. It should be remarked, however, that it seems rather unlikely that a spacetime representing our own universe should be asymptotically flat. In particular, the presence of an initial big bang would make unlikely the existence of past null infinity I^-. None the less, the notion of asymptotic flatness remains a most useful approximation for 'isolated systems', i.e. subsystems within the universe on which the effects of the rest of the universe are taken as negligible.

We now return to the issue of cosmic censorship. Recall that in the causal approach we introduced, for any spacetime, the set N consisting of all points at which observers could detect that their spacetime is future incomplete. The plan was to formulate cosmic censorship as the statement that N cannot be 'too large' in a suitable sense. But it turned out to be difficult to specify what too large was to mean, and in particular the demand that N be empty seemed too strong. We are now, however, in a position to formulate a weaker requirement. Let us

Chapter 5. Global structure of spacetimes

restrict consideration to spacetimes which are asymptotically flat. We wish to permit N to be not empty, but demand that it contain no points 'in the asymptotic region'. This is made precise as follows: demand that the closure of N in $\tilde{M}$ not intersect I^+. (A point in this intersection would mean that certain 'asymptotic observers' could detect that their spacetime is singular.) The resulting version of cosmic censorship – that in any asymptotically flat spacetime the closure of N has empty intersection with I^+ – does eliminate some of our earlier examples. It is satisfied, for example, for the Kerr spacetime with $a^2 < m^2$; and it excludes such examples as Taub–NUT spacetime, which are not even asymptotically flat. Unfortunately, a number of our examples violate even this version, e.g. the negative-mass Schwarzschild spacetime, the shell-crossing models, and Minkowski spacetime with the point at the origin removed. (All of these are asymptotically flat, and in all the closure of N has non-empty intersection with I^+.) These examples are of course also unphysical in some sense, but apparently this sense is not captured by the asymptotic structure.

We remarked earlier that cosmic censorship has a bearing on certain issues involving the structure of black holes. Indeed, this relationship to black holes was the original impetus for the study of cosmic censorship. Since it turns out that what is needed is essentially an asymptotic version of cosmic censorship we discuss this relationship here.

Let M, g_{ab} be an asymptotically flat spacetime. Then the past of future null infinity I^+ represents physically the set of all events from which an observer could escape to the asymptotic region. The boundary of this past is called the *event horizon*[29] of the spacetime. A spacetime which has an event horizon (i.e. for which the past of future null infinity is not the entire spacetime, i.e. in which certain observers can never escape to the asymptotic region) is said to possess a black hole. The study of spacetimes which possess a black hole reduces essentially to the study of event horizons. These horizons turn out to have a number of properties which make them useful for various arguments. One such property, an immediate consequence of its definition as the boundary of a past, is that through each point of the horizon there passes a maximally extended future-directed null geodesic which remains always in the horizon and never reaches I^+. These null geodesics are called the generators of the horizon. A second property, called the area theorem, asserts essentially that the convergence of these generators can never be positive. Consider first the case in which each of these null geodesic generators is future-complete. Then the result is immediate for, were the convergence any-

How noticeably singular is our universe?

where positive, it would (by the energy condition, completeness, and the equation analogous to (5.1) for the null case) eventually have to become infinite along that geodesic. But this means that our geodesic is crossed by nearby geodesics on the horizon and hence that this geodesic does not remain on the horizon: a contradiction. This argument of course does not work in the absence of completeness, for only with this condition does (5.1) imply infinite convergence. Unfortunately, there is no good physical reason for believing that the null geodesic generators of the horizon should be future complete. It turns out, however, that one can still obtain the area theorem, even without the assumption of completeness, provided that one invokes a suitable version of cosmic censorship. Let the convergence of the generators be positive at some point p of the horizon. Draw a small spacelike 2-surface A in a neighborhood of p such that its boundary is on the horizon, it is otherwise in the past of I^+, and such that the convergence of its outgoing null normals is positive (figure 5.32). Now consider the boundary of the

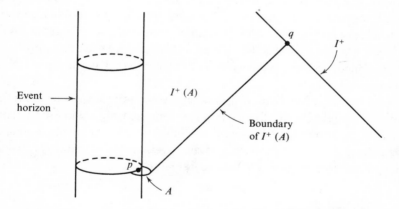

Figure 5.32. The event horizon in an asymptotically flat spacetime. If the convergence of the null generators ever became positive, say at p, one could attach a spacelike 2-surface A such that the convergence of the outgoing null normals from A is everywhere positive, and then obtain a contradiction.

future of this A. By construction, through every point of this boundary, there is a maximally extended past-directed null geodesic which either remains always in this boundary, or else remains in this boundary until it meets A. We now require a certain version of cosmic censorship. What is needed is that all of these null geodesics do in fact meet A, i.e. that

Chapter 5. Global structure of spacetimes

these null geodesics, which certainly lie in the past of I^+ and are thus certainly 'visible' to external observers, do not 'terminate in singular behavior'. But now, under this additional hypothesis, we can obtain a contradiction. Fix a point q where the boundary of the future of A meets I^+, and consider the past-directed null geodesic γ from q which remains in the boundary of the future of A. This geodesic is certainly future complete, while by our cosmic censorship assumption it meets A. But this is a contradiction since the null geodesics from A begin with positive convergence, whence by (5.1), the energy condition, and completeness they must cross, whence our γ could hardly remain on the boundary of the future of A. We conclude, then, that in a spacetime which is asymptotically flat, which satisfies the energy condition and which satisfies cosmic censorship in the sense above, the convergence of the generators of the horizon must be everywhere non-positive. There is also a similar result to the effect that, under the same assumption of cosmic censorship, no asymptotic observers can detect the trapped surfaces, i.e. all trapped surfaces must be inside the horizon.

Several attempts have been made to find counterexamples to cosmic censorship using these results.[30] One tries, for example, to find a generic spacetime with trapped surfaces and no event horizon, or with an event horizon which does not satisfy the area theorem. All such attempts have so far failed. Indeed, there is some evidence[31] that there are no such counterexamples.

To summarize, the causal approach gives one access to a notion of nakedly singular, while the asymptotic approach represents the beginning of a distinction between various types of nakedly singular spacetimes. The two approaches taken together do not, however, result in a precise formulation of cosmic censorship.

Stability approach

In the above two approaches to cosmic censorship, we have been dealing essentially with individual spacetimes. That is, we have tried to formulate a statement to the effect that any spacetime having certain properties cannot be nakedly singular in too severe a sense. However, it may turn out that certain specific spacetimes, such as possibly Taub–NUT, will survive all such criteria. The present approach is based on an attempt to formulate cosmic censorship, not as a statement about individual spacetimes, but rather as a statement about certain classes. Specifically, one seeks a result to the effect that counterexamples are

How noticeably singular is our universe?

rather 'sparse' among all spacetimes – that the 'generic' spacetime must satisfy some version of cosmic censorship.

Some examples will illustrate the general idea. Consider first a spherically symmetric fluid which collapses to form a black hole, Schwarzschild outside. The resulting spacetime would not be regarded as violating cosmic censorship since the presence of the horizon prevents asymptotic observers from detecting the singular character of their spacetime. Now suppose that this spacetime is perturbed, e.g. as follows. Consider a slice in this spacetime, and change its initial data slightly. It is known[32] that the perturbed spacetime will also have a horizon, i.e. that it will also satisfy cosmic censorship. Consider next, by contrast, the Taub–NUT spacetime. This spacetime could perhaps be regarded as nakedly singular, for there are incomplete timelike geodesics in the Taub part, while these geodesics could certainly be seen by observers in the NUT part. Now fix a slice in the Taub part of this spacetime, and again introduce a perturbation, say the introduction of some dust, on this slice. One finds[33] that the dust accumulates near the Misner boundary, causing this null surface to become singular. Thus, the nakedly singular character of this spacetime does not seem to be stable under small perturbations. A closely related example is the Reissner–Nordström spacetime. This spacetime is certainly nakedly singular, for observers who enter the wormhole to re-emerge in the 'next Universe' can detect the singular behavior. Again, however, it is known[34] that certain perturbations in the spacetime will become singular on the inner horizon, i.e. will prevent these observers from detecting the singular behavior.

What happens in these last two examples is that the perturbation field tends to 'build up' at the Cauchy horizon of one's slice, causing that horizon to disappear into singular behavior in the perturbed spacetime. The idea, then, would be to try to show that what happens in these examples must happen more generally – that the introduction of perturbations in any spacetime which would be regarded as violating cosmic censorship will essentially change the character of that spacetime. Indeed, as discussed earlier, it does not appear to be completely out of the question to obtain the much stronger result that failure to have a Cauchy surface is unstable in this sense.

One is here dealing with a sense of 'stability' somewhat different from, for example, that involved in stable causality. Whereas in the latter one is concerned essentially with 'kinematical stability' (i.e. what happens when one opens up the light-cones a bit), in the former the

Chapter 5. Global structure of spacetimes

issue is 'dynamical stability', what happens under evolution in time. All this suggests that evolution is relevant to cosmic censorship, an idea that will be developed in the next, and final, approach.

Evolutionary approach

While each of the above approaches to cosmic censorship provides some insight into the type of statement one is looking for and the types of examples one must be wary of, all of these approaches flounder on a common class of examples, namely certain spacetimes obtained by excising regions from otherwise well-behaved spacetimes. The problem, of course, is that while the excision of certain regions is permitted by general relativity, the result is often a nakedly singular spacetime. In short, general relativity permits the capricious introduction of singular behavior. The present approach is designed, among other things, to avoid this class of counterexamples. The idea is to use the initial-value formulation in general relativity to detect the excision of certain regions. The result is that we shall here be able to isolate at least some precise statements which are not clearly false and which do have connotations of cosmic censorship. Unfortunately, these statements appear to be very difficult to prove.

We first recall the initial-value formulation of general relativity. Let M, g_{ab} be a spacetime, let there be given on M the various fields representing the matter, subject to the appropriate equations, and let Einstein's equation be satisfied. Let S be a slice in this spacetime. Then there is induced on S, from the metric g_{ab}, a positive-definite induced metric q_{ab} and a symmetric extrinsic curvature p_{ab}, and, from the fields representing the matter, various additional fields. These fields on S must satisfy the appropriate constraint equations. (For example for the matter field, the Maxwell field, the corresponding fields on S will be the electric and magnetic fields, and the constraints will include the conditions that the electric and magnetic fields be divergence-free.) These remarks suggest the following definition.[18] An initial-data set is a three-dimensional manifold S with a positive-definite metric q_{ab}, a symmetric tensor field p_{ab}, and possibly other tensor fields representing the initial data for the matter, subject to the appropriate constraint equations. The main theorem of the initial-value formulation is this. Given any initial-data set, there exists a spacetime with matter fields corresponding to the initial data on S, such that Einstein's equation and the equations for the matter fields are satisfied, together with a Cauchy surface S in this spacetime such that the data induced on S from the spacetime agree with the original initial data on S. Furthermore, one can find such a spacetime which is maximal with respect to its properties (i.e. which cannot be embedded in any larger spacetime with the same properties) and then it is unique. What this means is that one can uniquely and maximally evolve the initial data on S to obtain a spacetime in which S must (for uniqueness) be a Cauchy surface.

As an example, let the initial-data set be that which arises from the 3-plane $t=0$ in Minkowski spacetime. Then the maximal evolution (whose existence is guaranteed above) would be the whole of Minkowski spacetime. Note that one loses the option of cutting holes in this maximal evolution: it is fixed already by the initial-data set. If, on the other hand, one begins with an initial-data set which is destined to produce a singular spacetime, then the maximal evolution will be 'cut off' in response to the singular behavior, for only in this way can S be a Cauchy surface in the final spacetime. In short, these maximal evolutions distinguish in a sense between spacetimes which are 'genuinely singular' and those which are singular merely because some region has been excised. The idea is to exploit these features to obtain a statement of cosmic censorship. We consider two examples. In each, the final statement will assert that under certain circumstances the maximal evolution is 'sufficiently large'.

The first draws on some ideas from the asymptotic approach. Define an initial-data set to be asymptotically flat if its maximal evolution is a spacetime which satisfies all the conditions for asymptotic flatness except possibly for the completeness condition. That is, we require that it be possible to attach to the spacetime an 'initial piece' of null infinity, but not necessarily an entire null infinity. Consider now the following statement: for any asymptotically flat initial-data set, topologically $\mathbb{R}^3$, its maximal evolution is an asymptotically flat spacetime (i.e. it even satisfies completeness of null infinity). This statement, we claim, captures a sense of cosmic censorship. That the initial-data set be topologically $\mathbb{R}^3$ ensures that the evolution is not singular already on S; that S be asymptotically flat ensures that one deals with an isolated system, and in particular that any singular behavior of the evolution must be due to the system itself and not external influences. Suppose, then, that this maximal evolution were singular, say to the future of S. It certainly cannot be nakedly singular to the future of S, for S must be a Cauchy surface for its evolution. Furthermore, the statement asserts that this evolution must be sufficiently large that it includes the entire

Chapter 5. Global structure of spacetimes

asymptotic regime, so in particular asymptotic observers can live out their entire lives within this maximal evolution. What this statement means, then, is that asymptotic observers will forever be unaffected by any singular behavior of the spacetime. But this is a version of cosmic censorship. We have essentially just re-expressed our earlier statement (that the closure of N have empty intersection with I^+), but now in a way which avoids the counterexample obtained by removing a point from Minkowski spacetime.

The second statement of cosmic censorship also refers to the size of the maximal evolution, without using asymptotic structure. Consider, for motivation, an initial-data set whose maximal evolution is extendible to the future of S, as shown in figure 5.33. This extended spacetime

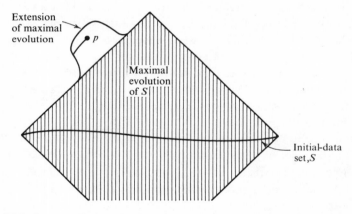

Figure 5.33. An extendible maximal evolution. From any point p in an extension, there exists a maximally extended past-directed curve which does not meet S.

cannot, by definition of the maximal evolution, have S as a Cauchy surface. That is, from a point such as p in the extension there must exist a maximally extended past-directed timelike curve which cannot be assigned a past endpoint, and which fails to meet S. In this rather mild sense the extended spacetime must be nakedly singular. One might therefore imagine formulating cosmic censorship as the assertion that every maximal evolution is inextendible, i.e. that, once the maximal evolution is completed, it is not possible to add any 'extra regions' as vantage points from which observers could detect that their spacetime is singular to the future of S. But this assertion, unfortunately, is false. Let S be the initial-data set induced on a small, spacelike disk in Minkowski spacetime. Then its maximal evolution is a double cone. This evolution is of course extendible, e.g. to Minkowski spacetime. What is necessary, clearly, is to distinguish between extensions which arise because S was chosen 'too small', and those which arise otherwise. One formulation of this distinction is the following. Suppose that, for p in an extension of the maximal evolution of an initial-data set, as in figure 5.33, the intersection of the past of p with S is compact. Under these circumstances, the existence of the extension could certainly not be attributed to S being too small, for the data on a compact region of S already suffices to ensure that such an extension will be available. That is, the evolution must have been stopped because of singular behavior which would be detectable by observers traveling into the extension. These remarks suggest the following statement of cosmic censorship: given any point p in any extension of the maximal evolution of an initial-data set, the closure of the intersection of $I^-(p)$ with S must be non-compact. Unfortunately, even this statement is false. Let S be a slice in Taub–NUT spacetime, as shown in figure 5.21, so the maximal evolution is Taub space. Let the extension be the entire Taub–NUT spacetime, and let q be a point on the Misner boundary. Then the intersection of $I^-(q)$ with S is all of S, which of course is compact. We may at least avoid this example by demanding that S itself be non-compact.

We have so far ignored the issue of what matter sources are to be permitted. Since in particular we have not ruled out shell-crossing, all of the statements of cosmic censorship in this section are, as presently formulated, false. A temporary expedient would be to rule out such examples by restricting consideration to the case of pure gravitation, i.e. with the stress-energy zero. Although this additional condition is obviously too strong physically, its imposition at least allows one to deal with the more global aspects of cosmic censorship at a non-trivial level while avoiding the more local issues of the detailed conditions to be placed on the matter. This special case might also be regarded as reasonable because one expects that, at the late stages of gravitational collapse, the gravitational field will dominate the matter.

We thus have:

Conjecture 5.1. The maximal evolution of every asymptotically flat, vacuum initial-data set, topologically $\mathbb{R}^3$, is an asymptotically flat spacetime.

Chapter 5. Global structure of spacetimes

Conjecture 5.2. For p any point in an extension of the maximal evolution of any non-compact initial-data set S, $I^-(p) \cap S$ has non-compact closure.

These conjectures are, as far as we are aware, open. Although each captures a sense of cosmic censorship, the senses seem to be somewhat different. For instance, if one could find an asymptotically flat initial-data set S whose maximal evolution is singular to the future of S, but such that the large curvature creates gravitational radiation which goes out to infinity and whose intensity becomes infinite, then one would have a counterexample to the first conjecture but not the second. (The intense radiation would cut off the evolution, and prevent it from being asymptotically flat, while it may also prohibit any extension of that evolution (figure 5.34).) Conversely, an asymptotically flat initial-data

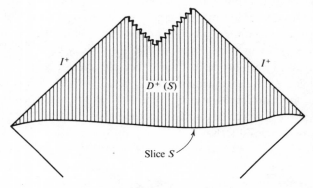

Figure 5.34. A spacetime which is extendible (so Conjecture 5.2 is satisfied), but in which $D^+(S)$ is not asymptotically flat (so Conjecture 5.1 fails).

set S whose maximal evolution is also asymptotically flat, but such that the maximal evolution could be suitably extended 'inside the black hole', in a region unable to communicate with future null infinity, might be a counterexample to the second conjecture, but not the first (figure 5.35).

5.5 Conclusion

The ingredients of global structure have by now become rather stable; its ideas have been incorporated into the arsenal of the working relativist. The focus is no longer on the internal mechanism of the subject

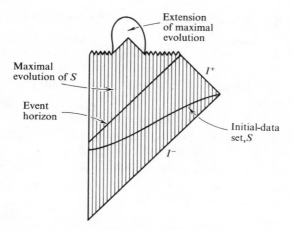

Figure 5.35. An extendible maximal evolution which satisfies Conjecture 5.1 but violates Conjecture 5.2.

itself but rather on its applications to external physical questions. Is there, for example, a good physical reason for demanding stable causality or the existence of a Cauchy surface? Is there some theorem expressing cosmic censorship? How, more generally, are singular spacetimes to be dealt with? In short, global structure has now become a rather mature subject.

Appendix

Prove or find a counterexample to the following assertions. Take all spacetimes to be four-dimensional, and, for assertions involving causal structure, time-oriented. All manifolds are to be Hausdorff, paracompact, connected, and without boundary.

Manifolds, metrics, orientation

(1) A manifold admits a Lorentz metric if and only if its universal covering manifold does.
(2) Every flat spacetime admits a nonzero Killing field.
(3) Call two Lorentz metrics on M equivalent if there is a diffeomorphism from M to M which sends one to the other. Then each of two equivalent metrics can be continuously deformed to the other.
(4) For M, a manifold, and p, a point of M, M with p removed is not diffeomorphic with M.

Chapter 5. Global structure of spacetimes

(5) If spacetime M, g_{ab} has vanishing Weyl tensor, then $\Omega^2 g_{ab}$ is flat for some Ω.
(6) If, for C a closed subset of M, $M-C$ admits a Lorentz metric, then so does M.
(7) For S, a two-dimensional manifold, and p, a point of S, $S-p$ is not simply connected.
(8) The product of two simply connected manifolds is simply connected.
(9) If, for S and S', three-dimensional manifolds, $S \times \mathbb{R}$ and $S' \times \mathbb{R}$ are diffeomorphic then S and S' are diffeomorphic.
(10) No compact spacetime is extendible.
(11) Every simply connected 4-manifold which is a product admits a Lorentz metric, except $S^2 \times S^2$.
(12) A spacetime is extendible if and only if its universal covering spacetime is.
(13) A non-simply connected manifold which admits a Lorentz metric admits one which is neither time-orientable nor space-orientable.
(14) A flat spacetime is time-orientable.
(15) If each of two spacetimes can be embedded isometrically as an open subset of the other, then the spacetimes are isometric.
(16) The sum of two complete vector fields is complete.

Domain of influence

(17) If the futures of two points do not intersect, then neither do their pasts.
(18) A non-empty subset of a spacetime equal to its own future and its own past is the entire spacetime.
(19) A non-compact manifold admits a stably causal Lorentz metric.
(20) A stably causal spacetime admits a time-function t with all submanifolds t = constant connected.
(21) A manifold which admits a Lorentz metric admits a stably causal Lorentz metric.
(22) Every compact spacetime admits a closed timelike curve through every point.
(23) A spacetime admitting a closed timelike curve admits a closed null curve.
(24) For p a point of a spacetime, the past of the future of the past ... of p is the entire spacetime.
(25) Fix a point p of a spacetime. We say that p has index n if the past of the future of the past ... (n times) of p is the entire spacetime, while

this fails for $(n-1)$. Every spacetime has finite index. For every positive integer, there exists a spacetime with that integer as index. Every compact spacetime has finite index.

(26) No spacetime has exactly one closed timelike curve; exactly one closed null curve.

(27) Every local Maxwell field in the example of figure 5.9 can be extended to a global field.

(28) The set of points of a spacetime through which there pass closed timelike curves is open.

(29) If there passes a closed timelike curve through p, and q precedes p, then there passes a closed timelike curve through q.

(30) There exists a spacetime with points p and q such that these can be joined simultaneously by timelike, null, and spacelike geodesics.

(31) If the past of p intersects the future of q, then the past of p contains q.

(32) Any two points of a spacetime can be joined by a spacelike curve.

(33) For any timelike vector field t_a in Minkowski spacetime, $g_{ab} - t_a t_b$ admits no closed timelike curves.

(34) In a stably causal spacetime, two points with the same future are the same.

(35) If the future of p is the entire spacetime, then through every point of that spacetime there passes a closed timelike curve.

(36) If a spacetime violates stable causality, then some open subset of that spacetime with compact closure violates stable causality.

(37) If M, g_{ab} is stably causal then, for some timelike t_a, $M, g_{ab} - t_a t_b$ is stably causal.

(38) If the past of p contains the future of q, then there is a closed timelike curve through p.

(39) If a stably causal spacetime has an extension, then it has a stably causal one.

Slices, domain of dependence

(40) Every compact slice in a spacetime with Cauchy surface is itself a Cauchy surface.[35]

(41) No compact spacetime admits an achronal slice.

(42) Let S be a three-dimensional manifold, and set $M = S \times R$. Then there exists a Lorentz metric on M which admits a Cauchy surface.

(43) The future Cauchy horizon is achronal.

Chapter 5. Global structure of spacetimes

(44) Through a point of the future Cauchy horizon there passes a *unique* past-directed null geodesic remaining in that horizon.

(45) All Cauchy surfaces for a spacetime are diffeomorphic.

(46) No spacetime with non-compact Cauchy surface admits a compact slice.

(47) Through a point of the future Cauchy horizon there passes a future-directed null geodesic remaining, at least for a while, in that horizon.

(48) If all connected slices in a spacetime are achronal, then it admits a Cauchy surface.

(49) In a spacetime with Cauchy surface, every connected slice is achronal.

(50) Let S be an achronal slice not a Cauchy surface. Then the past of the future Cauchy horizon of S includes S.

(51) Let S be an achronal slice. Then the future domain of dependence of any slice in $D^+(S)$ is itself in $D^+(S)$.

(52) Every maximally extended, past-directed null geodesic from a point of $D^+(S)$ meets S.

(53) For p and q in $D^+(S)$, with p preceding q, there is a timelike geodesic joining p and q.

(54) If the non-empty future Cauchy horizon of S is compact, so is S.

(55) Two slices having the same domain of dependence are the same.

(56) If two points of a stably causal spacetime can be joined by neither a timelike nor null curve, then some connected achronal slice contains them both.

(57) If p precedes q in a spacetime with Cauchy surface, then there is a Cauchy surface having p in its past and q in its future.

(58) If S is a Cauchy surface, then for no p is S in $I^-(p)$.

(59) If through every point of a spacetime there passes an achronal slice, then that spacetime is stably causal.

(60) A simply connected spacetime which has a closed timelike curve through every point admits no slices.

(61) Every maximally extended null geodesic meets a Cauchy surface.

(62) No null curve can meet an achronal slice more than once.

(63) For S a Cauchy surface, every point of the spacetime is on S, in its future, or in its past.

(64) If the closure of $I^-(p)$ meets achronal slice S compactly, then p is in the future domain of dependence of S.

(65) A spacetime with orientable Cauchy surface is space-orientable.

Appendix

Singular spacetimes

(66) Every spacetime is conformally related to one which is timelike and null geodesically complete.

(67) Two points, one of which precedes the other, in a geodesically complete spacetime can be joined by a timelike geodesic.

(68) If a spacetime has a compact Cauchy surface with converging normals, and satisfies the energy condition, then *every* timelike geodesic is incomplete.

(69) Every non-compact manifold admits a geodesically complete Lorentz metric.

(70) Every compact spacetime is geodesically complete.

(71) In a geodesically complete spacetime, every maximally extended timelike curve has infinite length.

(1) false, (2) false, (3) false, (4) false, (5) false, (6) false, (7) false, (8) true, (9) false, (10) true, (11) true, (12) false, (13) false, (14) false, (15) false, (16) false, (17) false, (18) true, (19) true, (20) false, (21) false, (22) false, (23) true, (24) true, (25) false, true, true, (26) true, false, (27) true, (28) true, (29) false, (30) true, (31) true, (32) true, (33) false, (34) true, (35) false, (36) false, (37) true, (38) true, (39) true, (40) true, (41) false, (42) true, (43) true, (44) false, (45) true, (46) true, (47) false, (48) false, (49) false, (50) false, (51) true, (52) false, (53) false, (54) true, (55) false, (56) false, (57) true, (58) true, (59) false, (60) true, (61) true, (62) true, (63) true, (64) false, (65) true, (66) false, (67) false, (68) true, (69) true, (70) false, (71) false.

6. The general theory of the mechanical, electromagnetic and thermodynamic properties of black holes

BRANDON CARTER

6.1 Introduction

6.1.1 Preface

It has long been well known to astronomers, and has (more recently) come to be widely accepted (for reasons to be summarized in section 6.6) even by specialists in general relativity, that, according to this theory, an equilibrium state of a pure vacuum black hole must be described by one of the Kerr (1963) solutions. In view of this state of affairs it is not surprising that a large proportion of the work that has been carried out on the theory of black holes has been devoted to the study of the special properties of the Kerr solutions, taking advantage of their exceptionally convenient properties (which are related to the type-D character of the Weyl tensor) of separability of variables, not only in the Hamilton–Jacobi and scalar wave equations (Carter, 1968a, b), but also in higher-spin wave equations (Teukolsky, 1972, 1973; Chandrasekhar, 1976; Page, 1976) as well as the existence of explicit analytic forms (Linet, 1977) for the corresponding Green functions. (See chapter 7 by Chandrasekhar.)

This chapter is intended to provide a reasonably comprehensive account of the comparatively small body of mathematically well-established results in the theory of *general* (i.e. not necessarily Kerr) black hole states. There are several reasons for being interested in this more general theory. To start with, many results that were first derived laboriously, using heavy algebra, from the explicit form of the Kerr solutions can be obtained by simpler and more elegant means in a more general context. A more fundamental reason is the obvious fact that the use of a more general approach is inescapable if one wishes to consider *dynamic* processes of gravitational collapse and the formation of black holes. Moreover, even after the achievement of stationary or quasi-stationary *equilibrium* (which is believed to be attainable on astronomically short timescales), there may well remain *external rings* of (slowly accreting) matter, which can easily be conceived (at least in the context of galactic nuclei and quasar theory, though perhaps not for ordinary, stellar-mass black holes) to be sufficiently massive to give rise to *significant deformations* from the standard Kerr geometry. Finally, even if one is concerned only with *unperturbed pure vacuum equilibrium states*, it is still logically necessary to *start* with a general formulation if one is interested in seeing how it may be *proved* that these states must belong to the Kerr family. (A general approach is necessary *a fortiori* if one is concerned to close the minor subsisting loopholes that prevent one being able to claim, even today, that the proof of uniqueness is absolutely watertight.)

Although this chapter is intended to be reasonably self-contained, it is nevertheless assumed that the reader will have at least some slight previous familiarity with the subject, such as might be obtained by reading the chapter by Hartle (1973*b*) in the Banff Summer School proceedings, or (as a minimum) the relevant sections of Zel'dovich and Novikov (1971) and of Misner, Thorne and Wheeler (1973). However it is *not* required that the reader should be familiar with more specialized accounts such as are given in chapter 9 of Hawking and Ellis (1973), or in the relevant chapters in the proceedings of the 1972 Les Houches Summer School (Bardeen, 1973; Carter, 1973*b*; Hawking, 1973*a*). Indeed the present survey could be read as both an introduction and an epilogue to the more specialized works that have just been cited since it provides both an overview of their content, as well as an account of the (relatively limited) progress that has been made in this field since 1973.

This chapter aims not just at describing the results that have been achieved, but also offers reasonably rigorous (by theoretical physical, if not pure mathematical, standards) and complete proofs wherever it is possible to do so without recourse either to calculations and arguments that would be disproportionately long and unwieldy (in relation to the importance or unexpectedness of the results) or to technically specialized methods (of global spacetime topology, Newman–Penrose formalism, etc.) that might be unfamiliar to a reader with a general background in theoretical physics. It is fortunate that results lacking a reasonably simple and elegant proof, and for whose demonstration scrupulous readers will be referred elsewhere, usually turn out to have a high degree of intrinsic plausibility, so that such results should not in general be too difficult to accept on trust. (A classic example in a quite different field is the four-colour theorem, described by Appel and Wolfgang,

Chapter 6. Mechanical, electromagnetic and thermodynamic properties

1977. This state of affairs arises quite naturally: results that are *neither* plausible in advance *nor* easy to prove tend to remain undiscovered.)

The plan of this review is as follows. We shall complete this introductory section by giving a brief account of the basic mathematical concept of a black hole in a general dynamical context. Section 6.2 contains a more detailed examination of the properties of the horizon with particular reference to situations in which the black hole is allowed to tend asymptotically towards a stationary final equilibrium state. The following section provides a more specialized description of the properties of the horizon in an *exactly* stationary state. Sections 6.4 and 6.5 are effectively concerned with *quasi*-stationary states in which weak – i.e. gravitationally passive – external (in particular electromagnetic) fields are considered to act in a stationary background space, giving rise to phenomena such as super-radiance. Section 6.6 discusses the mass of a black hole and the law governing its variation in a transition between nearby *exactly* stationary states. Finally, section 6.7 summarizes the results that, taken together, provide the justification for the belief that a stationary, asymptotically flat black hole state is fully determined by its mass and angular momentum when external matter and (non-gravitational) fields are absent, or by its mass, angular momentum, and electric charge if electromagnetic fields are allowed for. (The mathematical but unphysical possibility that the hole might possess a net magnetic monopole moment is ignored throughout.)

6.1.2 Global characterization of a black hole and the second law

Throughout this work we shall be dealing exclusively with four-dimensional time-oriented spacetime manifolds that are *asymptotically flat* (meaning that they are weakly asymptotically simple in the sense of Penrose, 1967).

We shall be primarily concerned with the *domain of outer communications* $\langle\!\langle \mathcal{J} \rangle\!\rangle$ of the manifold (as defined by Carter, 1971b), and with the horizon $\mathcal{H}^+$ forming its future boundary. The domain $\langle\!\langle \mathcal{J} \rangle\!\rangle$ is the *intersection* of the regions – denoted respectively by $I^-(\mathcal{J}^+)$ and $I^+(\mathcal{J}^-)$ in the notation of Penrose (1965) – from which it is possible to construct a respectively future- or past-oriented timelike line (and hence respectively to send or receive a signal) to or from arbitrarily large asymptotic distances (i.e. to $\mathcal{J}^+$ or from $\mathcal{J}^-$ in the Penrose's conformally extended manifold). The boundary of the domain of outer communications, $\mathcal{H} = \partial \langle\!\langle \mathcal{J} \rangle\!\rangle$, can obviously be decomposed as a union of two parts, $\mathcal{H}^+ \subset \partial I^-(\mathcal{J}^+)$ and $\mathcal{H}^- \subset \partial I^+(\mathcal{J}^-)$. In a time-symmetric model (such as the Schwarzschild and Kerr solutions) both $\mathcal{H}^+$ and $\mathcal{H}^-$ would be present, but in a physically realistic gravitational collapse situation, starting from well-behaved initial conditions, the past horizon $\mathcal{H}^-$ would be an empty subset.

Throughout the work to be described it is, of course, to be taken for granted that the part of spacetime under consideration, including at least the domain $\langle\!\langle \mathcal{J} \rangle\!\rangle$, is *causally well behaved* in the sense that it contains no closed timelike or null lines. (It is hardly necessary to justify such a condition physically; the question of whether it entails any restriction mathematically has recently been examined by Tipler (1976).) (The topological conditions imposed in section 6.6 imply that $\langle\!\langle \mathcal{J} \rangle\!\rangle$ is also future-complete in the sense that no future-directed timelike or null geodesic without future endpoint in $\langle\!\langle \mathcal{J} \rangle\!\rangle$ can have a finite affine length, unless it crosses the boundary $\mathcal{H}^+$.) For some purposes, notably the theorems of Hawking that are enuciated in this section, it is necessary to make the stronger assumption that the domain of communications is *future-predicable*, meaning that there is a well-behaved spacelike hypersurface without edges (or at least without edges in $\langle\!\langle \mathcal{J} \rangle\!\rangle$) that is a Cauchy surface for the part of $\langle\!\langle \mathcal{J} \rangle\!\rangle$ lying to its future, in the sense that any inextensible past-directed timelike line from any point in $\langle\!\langle \mathcal{J} \rangle\!\rangle$ to the future of the hypersurface must intersect it once and once only. This is the condition that Hawking (1972) refers to as *asymptotic predictability*; it can be interpreted as a form of the cosmic censorship condition to the effect that there are no naked singularities, i.e. singularities visible at large distance that are formed subsequent to the initial time defined by the spacelike hypersurface. It was conjectured by Penrose (1969) that such naked singularities could not be formed in a 'physically reasonable' model, but despite the efforts of Gibbons (1972) and Penrose (1974) there has been no real progress towards either confirming or invalidating this cosmic censorship hypothesis, except for the work of Yodzis, Seifert and Müller zum Hagen (1973), who have shown that effectively naked singularities may form in matter obeying an extremely soft equation of state of the kind that might be obeyed by very high-density matter in the (unlikely) event that a Hagedorn-type model of hadronic behaviour is valid. It is to be remarked that the presence of such 'pressure-failure' singularities would in any case invalidate the letter, but not the spirit, of Penrose's hypothesis, since the formation of this kind of singularity owes nothing to gravity but would arise in essentially the same way even in a Minkowskian background

Chapter 6. Mechanical, electromagnetic and thermodynamic properties

space (quite unlike the more serious kind of spacetime singularities whose formation subsequent to gravitational collapse is predicted by the famous theorem of Penrose (1965) and by the subsequent work of Hawking (see Hawking and Ellis, 1973)). One would therefore hope to be able to interpret the pressure-failure singularities as merely representing a technical difficulty within the framework of a classical spacetime treatment, rather than as representing a fundamental breakdown of the latter. (Before leaving this question, it is relevant to mention that Hawking's (1975) particle-creation mechanism implies the possiblity of a fundamental breakdown of classical spacetime in the form of a quantum singularity that would be visible to exterior observers *independently* of whether or not Penrose's *classical* cosmic censorship hypothesis is valid.)

Whether or not the cosmic censorship condition is obligatory, we are always at liberty to impose it as a mathematical restriction, and when it holds we shall have what may be described as a well-behaved black hole situation. In these circumstances the whole of the region exterior to $\langle\!\langle \mathcal{J} \rangle\!\rangle$ may be described as the *hole*, the term *black* hole being reserved to describe that part of the hole lying on or to the future of $\mathcal{H}^+$. In the standard collapse situation, for which the past horizon $\mathcal{H}^-$ is absent, the distinction does not arise: the black hole covers the entire hole region. (In the case of a 'primordial' hole for which $\mathcal{H}^-$ might be non-empty the term 'white hole' has been used for regions outside $\langle\!\langle \mathcal{J} \rangle\!\rangle$ lying on or to the past of $\mathcal{H}^-$. The region outside $\langle\!\langle \mathcal{J} \rangle\!\rangle$ which is neither in the future of $\mathcal{H}^+$ nor in the past of $\mathcal{H}^-$ has no particular name and is of little practical interest being neither accessible nor visible from outside.)

In much – indeed most – of what follows we shall be concerned with the special properties of the *horizon*, $\mathcal{H}^+$, which in the standard collapse situation forms the entire boundary of the hole. Such a black hole boundary horizon is a special case of the class of achronal boundaries, the study of which was pioneered by Penrose (1965) and widely applied by Hawking in the context first of singularity theorems and later of black hole theorems (see chapter 6 of Hawking and Ellis (1973) for a systematic introduction). The horizon of a black hole is not just an *achronal* boundary, as characterized by the property that no two distinct points on it can be joined by a timelike line, but has the more special property that (with respect to $I^+(\mathcal{J}^-)$ or with respect to the entire spacetime in the standard collapse situation wherein $\mathcal{H}^-$ is absent) it forms a *global past horizon* in the sense (Carter, 1971a) that *from each point of $\mathcal{H}^+$ there is a future-directed null geodesic without future endpoint lying* *entirely within* $\mathcal{H}^+$. This can easily be seen, using lemmas given originally by Penrose (1965) and Hawking (1967) (see Hawking and Penrose, 1970), from the fact that from any point just outside $\mathcal{H}^+$ (despite its achronality property) there is an inextensible future-directed timelike line without future endpoint that remains always outside (in fact extending out into the asymptotically flat region).

The property that has just been described is shared, more generally, with the *boundary of the past* of an arbitrary spacetime subset, *except* in so far as points lying on the boundary of the original subset itself are concerned, and *except also* for the fact that in the more general case the future-directed null geodesic generators of the boundary of the past may have future endpoints where they reach the boundary of the original subset itself (though not elsewhere). Moreover, if two of the null geodesic generators intersect – or have a point of convergence – at some point on the boundary, their further extensions will leave the boundary forever. (Extensions to the past of such a point will immediately enter the interior of the past of the original subset. A future intersection or convergence point may only occur on the boundary of the original subset itself, and extension beyond would lead into its future or at least onto the boundary of its future.) It is of course evident that analogous properties hold with future and past interchanged. These properties were originally utilized by Penrose (1965) in the particular case of the boundary of the future of a *closed trapped surface*, i.e. a topologically spherical, spacelike 2-surface with the property that the divergence, $\theta^{(0)}$, of the family of outgoing null geodesics starting from it is *everywhere negative* on it, i.e. precisely the opposite of what would be the case for an ordinary 2-sphere in flat spacetime. Penrose observed that the fundamental evolution equation (whose derivation and form will be described below) for $\theta^{(0)}$ as a function of an *affine* parameter τ implied

$$\frac{d\theta^{(0)}}{d\tau} \leq -\tfrac{1}{2}(\theta^{(0)})^2 \qquad (6.1)$$

for any dynamical field equations that are 'sufficiently reasonable' for the condition

$$R_{ab}l^a l^b \geq 0 \qquad (6.2)$$

to be satisfied, where R_{ab} are the Ricci components and l^a are the components of the null-tangent vector. Since (6.1) clearly implies that if $\theta^{(0)}$ once becomes negative it will necessarily diverge to $-\infty$ within a finite affine distance, Penrose could argue that the null geodesic generators of

Chapter 6. Mechanical, electromagnetic and thermodynamic properties

the boundary of the future of a closed trapped surface could only have finite affine length, and hence that the connected part of the boundary of the future must be compact, which is quite unimaginable (and also as he and Hawking (1970) subsequently showed, mathematically impossible under very general conditions) unless spacetime is torn off short by some kind of singularity (whose actual nature is not yet clearly established).

Hawking (1972) has pointed out (cf. Hawking and Ellis, 1973) that a corollary to this line of reasoning is the conclusion that, if the asymptotic predictability condition referred to above is satisfied, *any closed trapped surface* to the future of the Cauchy hypersurface *must lie entirely outside the domain of outer communications*, $\langle\langle \mathscr{J} \rangle\rangle$, since otherwise part of the future of the closed trapped surface, and hence also of the boundary of the future, would extend out to $\mathscr{J}^+$, so that contrary to the above conclusion there would exist boundary generators of infinite length, necessarily with past endpoints (since otherwise they would intersect the Cauchy hypersurface) which could only occur on the closed trapped surface. Thus the presence of a closed trapped surface is a very useful criterion that provides a sufficient – though not a necessary – indication that a region of spacetime lies within a (black) hole. Such a local criterion (local at least in terms of *time* if not of space) is valuable in view of the awkward teleological nature of the black hole horizon, i.e. the fact that its precise locality at a given time (i.e. on a given spacelike hypersurface) cannot be determined without complete knowledge of the entire future evolution of spacetime. It is for this reason that Hawking refers to the outer boundary of the region containing closed trapped surfaces on a given spacelike hypersurface as the *apparent horizon* (relative to that hypersurface): it may not coincide with the intersection of the true horizon with the hypersurface, but it is at least guaranteed *not to lie outside it*.

The same kind of argument that applies to the boundary of the future of a closed trapped surface can *a fortiori* be applied – in the time-reversed sense – to $\mathscr{H}^+$, the horizon itself. Although the null geodesic generators of $\mathscr{H}^+$ may have past endpoints, we have seen that they can never have future endpoints. Hence although their divergence $\theta^{(0)}$ may blow up towards $+\infty$ as one goes back in a past sense, it may never go to $-\infty$ in the future sense, which (by (6.1)) means that it can never even become negative at all, i.e. we must always have

$$\theta^{(0)} \geq 0. \qquad (6.3)$$

Now $\theta^{(0)}$ governs the rate at which the area $d\mathscr{S}$ of a 2-surface element d$\vec{\mathscr{S}}$ with components d$\mathscr{S}^{ab}$ in the horizon varies as the element undergoes transport towards the future along the generators according to the equation

$$\frac{d}{dt}(d\mathscr{S}) = \theta \, d\mathscr{S} \tag{6.4}$$

where θ is the divergence with respect to an *arbitrary* (not necessarily affine) transport parameter t, which will be related to the *affine* divergence $\theta^{(0)}$ by

$$\theta = \dot{\tau}\theta^{(0)} \tag{6.5}$$

with

$$\dot{\tau} = \frac{d\tau}{dt}. \tag{6.6}$$

Since new generators may come into existence, but existing ones can never terminate towards the future, the area

$$\mathscr{A} = \oint d\mathscr{S} \tag{6.7}$$

of any spacelike 2-surface (e.g. formed by the intersection of a spacelike hypersurface with $\mathscr{H}^+$) will satisfy the purely kinematic condition

$$\frac{d\mathscr{A}}{dt} \geq \oint \theta \, d\mathscr{S} \tag{6.8}$$

when the 2-surface is transported towards the future. (This would be a strict equality if the number of generators were conserved.) Hence provided the dynamic inequality (6.2) does indeed hold, so that (6.1) and (6.3) are satisfied, one can immediately deduce that we shall have

$$\delta\mathscr{A} \geq 0 \tag{6.9}$$

for any transport *towards the future* (so that $\dot{\tau}$ is everywhere positive) of a closed spacelike 2-surface $\mathscr{S}$ within the horizon. This result (derived originally by Hawking, 1972) has been known since its discovery as the 'second' law of black hole mechanics in view of its obvious analogy with the second law of thermodynamics (to the effect that the entropy S of an isolated system must satisfy $\delta S \geq 0$) even though it was not at first realized just how deep the analogy actually is.

Chapter 6. Mechanical, electromagnetic and thermodynamic properties

6.2 The evolution of the horizon

6.2.1 Notation conventions and the kinematics of (timelike and) null congruences

In preparation for a closer examination of the situation (which will include an explicit derivation of the inequality (6.1)), I shall recapitulate some of the basic kinematic properties of congruences of timelike and (most specially) null curves and of the corresponding fields of tangent vectors $\vec{l}$, as defined in terms of a parametrization with respect to an arbitrary time coordinate, t, by

$$l^a = \frac{dx^a}{dt} \tag{6.10}$$

where x^a ($a = 0, 1, 2, 3$) are local spacetime coordinates. We start with a demonstration of the well-known fact that the null generators of any hypersurface that is itself null must necessarily be geodesic. One may see this using the identity

$$\vec{l} \cdot (\boldsymbol{\pi} \wedge \boldsymbol{\Omega}) = (\vec{l} \cdot \boldsymbol{\pi})\boldsymbol{\Omega} - \boldsymbol{\pi} \wedge (\vec{l} \cdot \boldsymbol{\Omega}) \tag{6.11}$$

(which holds for the *contraction* of any vector $\vec{l}$ with the exterior product of any covector $\boldsymbol{\pi}$ with any differential form $\boldsymbol{\Omega}$ of arbitrary order) in conjunction with the *irrotationality condition*

$$l \wedge \partial l = 0 \tag{6.12}$$

which is the Frobenius orthogonality requirement that is necessary and sufficient for the local existence of a hypersurface tangent to a 1-form l, where we use the symbol ∂ to denote Cartan's operation of exterior differentiation. This operation may be defined in terms of the *covariant differentiation* operator, ∇, by

$$\partial \equiv \nabla \wedge \tag{6.13}$$

but its effect on a *differential form* (i.e. a covariant tensor field that is antisymmetric under all possible interchanges of pairs of indices) may be specified *independently* of any affine connection in terms of the operator $\partial_a \equiv \partial/\partial x^a$ of *partial differentiation* (with respect to the local coordinates x^a) by

$$(\partial \boldsymbol{\Omega})_{abcd\ldots} = \partial_a \Omega_{bcd\ldots} - (\partial_b \Omega_{acd\ldots} + \partial_c \Omega_{bad\ldots} + \partial_d \Omega_{bca\ldots} + \ldots). \tag{6.14}$$

So as to obtain dimension- and order-independent formulae such as (6.11) we are employing the standard Cartan convention as described, for example, in Misner *et al.* (1973) for the normalization of the exterior product operation. Some authors, e.g. Hawking and Ellis (1973), use a different convention that would require the insertion of algebraic correction factors dependent on the order of Ω in formulae such as (6.11) as well as in the important Lie differentiation formula (6.20) quoted below. The dot appearing in (6.11) is just the standard physicist's notation for contraction of adjacent indices: i.e. $\vec{v}\cdot$ represents what pure mathematicians commonly denote by $i(\vec{v})$ which, apart from being more cumbersome, has the disadvantage of being liable to misinterpretation in situations where $i((-1)^{1/2})$ is involved. Similar possibilities of misinterpretation arise from the regrettably widespread use of d instead of the boundary/coboundary symbol ∂ for exterior differentiation: by using the latter symbol for this purpose we remain free to use the symbol d in the traditional manner to denote surface elements, as in equations (6.4) and (6.7), or to denote infinitesimal changes and displacements as in the expression

$$ds^2 = d\vec{x} \cdot \mathbf{g} \cdot d\vec{x} = g_{ab}\, dx^a\, dx^b \qquad (6.15)$$

defining the infinitesimal length ds associated with an infinitesimal displacement d$\vec{x}$ with components dx^a in terms of a covariant metric tensor $\mathbf{g}$. We shall consistently use the same alphabetic symbol with a distinguishing overhead arrow or printed boldface to denote corresponding contravariant and covariant tensors related by index lowering and raising operations as performed by contractions with $\mathbf{g}$ or with its inverse $\vec{g} = (\mathbf{g})^{-1}$. Thus in the case in question (before we embarked on this digression on notation) the covector l (representing the normal, i.e. the tangent *form*, to the null hypersurface under consideration) is to be interpreted as being related to the *null-tangent vector* (as specified by (6.10)) by

$$l = \mathbf{g} \cdot \vec{l} \qquad (6.16)$$

(i.e. in terms of components, $l_a = g_{ab}l^b$), so that the nullity property will be expressed simply by the condition that the squared norm of $\vec{l}$ vanishes, i.e.

$$|\vec{l}|^2 \equiv \vec{l} \cdot l = 0. \qquad (6.17)$$

If now we substitute l in place of π, and ∂l in place of Ω in (6.11), we see by (6.12) that we shall have

$$l \wedge (\vec{l} \cdot \partial l) = 0. \qquad (6.18)$$

Chapter 6. Mechanical, electromagnetic and thermodynamic properties

Now the fact that the squared magnitude $|l|^2$ satisfies (6.17) on the hypersurface implies that its gradient form is, if not zero, at least orthogonal to the hypersurface (i.e. parallel to l). Hence, as an immediate consequence, we obtain the condition

$$l \wedge (\vec{l} \pounds\, l) = 0, \qquad (6.19)$$

as may be seen by substituting l in place of Ω in the general *Cartan formula*

$$\vec{l} \pounds\, \Omega = \vec{l} \cdot \partial \Omega + \partial(\vec{l} \cdot \Omega) \qquad (6.20)$$

which holds for the *Lie derivative* with respect to a vector field $\vec{l}$ of a *differential form* Ω of *arbitrary order* (see, for example, Hicks (1971); Choquet-Bruhat (1968)). (This is another example of a formula which would require awkward dimension- and order-dependent correction coefficients if one were to use a non-standard normalization covention for exterior multiplication and differentiation.) Condition (6.19) merely expresses the obvious fact the null cotangent form must remain parallel to itself when dragged along the congruence. However if we now take a suitable linear combination of (6.18) and (6.19) we obtain as a less trivial consequence the *geodesic equation* in the form

$$l \wedge \dot{l} = 0, \qquad (6.21)$$

expressing the condition that $\dot{l}$ is proportional to l, where we use a superior dot to denote covariant differentiation with respect to the parameter t, which in this example gives

$$\dot{l} \equiv (\vec{l} \cdot \nabla) l. \qquad (6.22)$$

Although trivial as it stands, the property (6.19) has significant implications, since it is a necessary and sufficient condition for it to be possible to give an objective meaning to the concepts of the *expansion* and *shear* of a null congruence, concepts whose description and application to the context of black hole horizons will take up the remainder of this section. The concepts of expansion and shear – as measurements of rates of changes of infinitesimal orthogonal distances to neighbouring curves – present no ambiguities when measured along a *timelike* curve. However, infinitesimal distances from a *null* curve can be given a well-defined meaning only for those infinitesimally neighbouring curves that lie in the tangent null hyperplane through the curve, and hence *rates of change* of these distances can be defined in an invariant manner only if such curves *remain* in the tangent null hyperplane as they propagate, i.e. only if the tangent null hyperplane is transported parallel to itself, which is precisely the condition expressed by (6.19). Under these circumstances – but not otherwise – one has a solid foundation in terms of which it is meaningful to discuss expansion and shear of a null congruence. The basic requirement for this (i.e. (6.19)) will be satisfied automatically for geodesics generating any individual null hypersurface (such as a black hole horizon) and, since

$$\vec{l} \pounds \, l \equiv \vec{l} \cdot (\vec{l} \pounds \, g), \qquad (6.23)$$

the requirement (6.19) will also be satisfied on a locus of nullity of a congruence of *Killing vectors*, as characterized by

$$\vec{l} \pounds \, g_{ab} \equiv \nabla_a l_b + \nabla_b l_a = 0, \qquad (6.24)$$

(which takes care of cases such as 'ergosurfaces' or 'speed of light cylinders', where respectively an asymptotically stationary or a corotating Killing vector becomes null) although in such a case the expansion tensor will trivially be zero. However for a congruence that is *everywhere* null so that $|\vec{l}|^2$ is not only zero on the surface but *also* has vanishing gradient, i.e.

$$\partial |l|^2 = 0, \qquad (6.25)$$

which by (6.20) implies

$$\dot{l} = \vec{l} \pounds \, l, \qquad (6.26)$$

then the condition (6.19) will *not* hold automatically but must be imposed as a restriction, which is equivalent to *requiring* that the congruence be geodesic.

The properties of null-geodesic congruences have been explored by numerous workers, the properties most relevant to the present discussion having been most notably developed by Sachs (1962) and by Newman and Penrose (1962) (see the review of Pirani, 1965). We shall start by describing *general null-geodesic* congruences, before restricting our attention to the *irrotational* case, as characterized by (6.2), which is relevant to black hole horizons. A slightly different analysis, to be described in the next section on *stationary* black hole theory, is relevant when the generators of the horizon are considered as belonging to a (non-geodesic) *Killing vector* congruence, and it is in order to be able to relate these two types of analysis that we work throughout in terms of a *general* parametrization instead of simplifying the formulae by the use of an *affine* parametrization as is usually done in other contexts.

Chapter 6. Mechanical, electromagnetic and thermodynamic properties

In order to bring out the analogy with the more familiar timelike case, we shall proceed as far as possible with a formalism that is applicable both to a *general* (not necessarily geodesic) *timelike* congruence as well as to the limiting case of a *null-geodesic* congruence.

To this end we start by introducing a vector $\vec{\beta}$ with components β^a and an orthogonal projection tensor γ with components $\gamma^a{}_b$ characterized by

$$\vec{l} \cdot \vec{\beta} = -1 \tag{6.27}$$

$$\gamma \cdot \vec{\beta} = \gamma \cdot \vec{l} = \vec{l} \cdot \gamma = 0 \tag{6.28}$$

$$\gamma \cdot \gamma = \gamma. \tag{6.29}$$

In the case of a *timelike* vector $\vec{l}$ these properties together with the supposition that γ is of maximum rank, i.e. three, are sufficient to specify γ and $\vec{\beta}$ uniquely: they will be given explicitly by

$$\beta^a = -(\vec{l} \cdot \vec{l})^{-1} l^a \tag{6.30}$$

$$\gamma^a{}_b = g^a{}_b + \beta^a l_b. \tag{6.31}$$

In the *null case*, however, such a solution obviously fails, which implies that γ can be at most of rank *two* so that there will be considerable ambiguity of choice, which can be partially restricted by the imposition of the additional condition that $\vec{\beta}$ be null, i.e.

$$|\vec{\beta}|^2 = 0. \tag{6.32}$$

For *any* $\vec{\beta}$ satisfying (6.27) and (6.32) we shall have a well-defined solution for γ given by

$$\gamma^a{}_b = g^a{}_b + \beta^a l_b + l^a \beta_b. \tag{6.33}$$

Conversely, we may start by fixing γ, by choosing a particular 2-plane in the null tangent hyperplane into which it projects, and there will then be a well-defined solution for β satisfying (6.27) and (6.32).

Although the fundamental parameter t with which we shall work will not be required to be affine, the significance of the results may be made clearer by introducing an alternative parameter, τ say, that is required to be affine, with a corresponding *affine tangent vector*, $\vec{u}$ say, given by

$$u^a = \frac{dx^a}{d\tau} \tag{6.34}$$

in terms of which our original tangent vector, $\vec{l}$, will be given by

$$\vec{l} = \dot{\tau} \vec{u} \tag{6.35}$$

where a dot denotes (covariant) differentiation with respect to t. (In the timelike case we could take τ to measure proper time so that $\vec{u}$ becomes a unit vector.) We now introduce the *acceleration* $\dot{u}$ of the congruence, as defined with respect to the *coordinate time*, t (as opposed to the affine or proper time, τ) which is given by

$$\dot{u} = \vec{l} \cdot \nabla u. \qquad (6.36)$$

This quantity will of course be *zero* in the null geodesic case, and it will in general satisfy

$$\vec{l} \cdot \dot{u} = 0. \qquad (6.37)$$

The *absolute* (physical) acceleration that would be felt by a particle moving on one of the *timelike* lines is given by a covector a which is expressible by

$$a_a = \beta^d(\nabla_d l_a + \beta_a l^c \nabla_d l_c). \qquad (6.38)$$

If τ measures proper time so that $\vec{u}$ is a unit vector and $l \cdot l = -\dot{\tau}^2$ then we shall have $\dot{u} = \dot{\tau} a$. In the null geodesic case, the quantity a defined by (6.38) has a quite different interpretation that has nothing to do with the acceleration, but it will in any case satisfy

$$\vec{l} \cdot a = 0. \qquad (6.39)$$

Whatever case we are concerned with we shall have

$$\dot{\tau}\dot{u} = -(\vec{l} \cdot l)a \qquad (6.40)$$

and hence we can make the decomposition

$$\dot{l} = \kappa l - (\vec{l} \cdot l)a \qquad (6.41)$$

where the coefficient κ, which will play a central role in the following discussion, is effectively a measure of the extent to which $\vec{l}$ is *non-affine*. It is given explicitly by

$$\kappa = (\ln \dot{\tau})^{\cdot} = \ddot{\tau}/\dot{\tau} = -\vec{\beta} \cdot \dot{l}. \qquad (6.42)$$

We can now go on to decompose the covariant derivative of l in the form

$$\nabla_a l_b = v_{ab} - l_b(\kappa \beta_a + \gamma_a{}^d \beta^c \nabla_d l_c) - l_a a_b \qquad (6.43)$$

which is valid for a general timelike congruence, as well as for a null-geodesic congruence, with

$$v_{ab} = \gamma_a{}^c \gamma_b{}^d \nabla_c l_d = \omega_{ab} + \theta_{ab} \qquad (6.44)$$

Chapter 6. Mechanical, electromagnetic and thermodynamic properties

where
$$\omega_{ab} = v_{[ab]}, \qquad \theta_{ab} = v_{(ab)} \qquad (6.45)$$

using square and round brackets to denote symmetrization and antisymmetrization respectively. It is evident from the form of (6.43) that the vanishing of the *rotation tensor*, ω, is a necessary and sufficient condition for the Frobenius hypersurface orthogonality condition (6.12) to be satisfied in both the null-geodesic and timelike cases. It can equally be seen in both cases that the tensor θ_{ab} introduced in this way is interpretable as an expansion rate tensor since, for infinitesimally neighbouring members of the congruence separated by a relative displacement vector $d\vec{x}$, the rate of change of their separation will be given by

$$\tfrac{1}{2}(ds^2)^{\cdot} = d\vec{x} \cdot \boldsymbol{\theta} \cdot d\vec{x} = \theta_{ab}\, dx^a\, dx^b. \qquad (6.46)$$

Having established this, it follows that it is appropriate to describe its *trace*,

$$\theta = \theta_a{}^a = \gamma^{ab}\nabla_a l_b, \qquad (6.47)$$

as representing the *divergence of the congruence*, since it evidently gives the fractional *volume*, or in the null case the *area*, expansion rate. (This provides the justification for (6.4).) It is to be noted however that the *divergence of the vector* $\vec{l}$ contains an additional contribution allowing for possible variations in its length, being given (as can be seen by taking the trace of (6.43)) by

$$\nabla \cdot \vec{l} = \theta + \kappa. \qquad (6.48)$$

A similar term involving κ also appears when (6.43) is contracted with itself, giving

$$(\nabla^a l^b)(\nabla_b l_a) = \kappa^2 + \theta_{ab}\theta^{ab} - 2\omega^2 \qquad (6.49)$$

where ω is the *angular velocity* scalar defined by

$$2\omega^2 = \omega_{ab}\omega^{ab} \qquad (6.50)$$

(whose vanishing is necessary and sufficient for the Frobenius orthogonality condition (6.12) to be satisfied).

Up to this point all our numbered formulae from (6.34) onward (as well as (6.27) to (6.29)) have been equally valid for both timelike and null-geodesic congruences. We are now ready to introduce the important concept of the shear rate tensor $\boldsymbol{\sigma}$, which is defined as the trace-free part of the expansion rate tensor, and hence given by the decomposition

$$\theta_{ab} = (\gamma^c{}_c)^{-1}\theta\gamma_{ab} + \sigma_{ab} \tag{6.51}$$

where the projection tensor trace $\gamma^c{}_c$ is equal to its *rank*, which is three in the timelike case, but only two in the null-geodesic case. The shear scalar (normalized so as to agree with traditional usage in Newtonian fluid mechanics) is defined by

$$\tfrac{1}{2}\sigma^2 = \sigma_{ab}\sigma^{ab}. \tag{6.52}$$

In terms of these quantities, and using (6.43), one can now evaluate the terms in the identity

$$l^b\nabla_b(\nabla_a l^a) \equiv \nabla_a(l^b\nabla_b l^a) - (\nabla^a l^b)(\nabla_b l_a) - R_{ab}l^a l^b \tag{6.53}$$

that is derivable for a completely arbitrary vector field from the defining relations

$$(\nabla_a\nabla_b - \nabla_b\nabla_a)l^c = R^c{}_{dab}l^d \tag{6.54}$$

and

$$R_{ab} = R^c{}_{acb} \tag{6.55}$$

for the Riemann and Ricci tensors. One thus obtains the generalized Raychaudhuri equation

$$\dot{\theta} - \kappa\theta = -(\gamma^c{}_c)^{-1}\theta^2 - \tfrac{1}{2}\sigma^2 + 2\omega^2 + \nabla_a(\vec{l} \cdot la^a) - R_{ab}l^a l^b \tag{6.56}$$

in a form valid for arbitrarily parametrized *null-geodesic* or *timelike* congruences.

For the irrotational *null*-geodesic congruences with which we are primarily concerned this reduces to

$$\dot{\theta} + \kappa\theta = -8\pi\mathcal{D} \tag{6.57}$$

where we introduce a quantity $\mathcal{D}$ (which will be shown to be interpretable, in the case of a black hole horizon, as a *dissipation rate* per unit surface area) defined by

$$8\pi\mathcal{D} = \tfrac{1}{2}(\theta^2 + \sigma^2) + R_{ab}l^a l^b. \tag{6.58}$$

Obviously the terms in $\mathcal{D}$ will all be non-negative provided the inequality (6.2) is satisfied. This provides the justification for the inequality (6.1) on which the second (area increase) law (6.9) was based (as may be seen either by choosing t to be equal to τ, or else by noting that one has

Chapter 6. Mechanical, electromagnetic and thermodynamic properties

quite generally $\dot\theta - \kappa\theta \equiv \dot\tau \theta^{(0)}$ (where $\theta^{(0)}$ is the expansion rate as measured in terms of τ instead of t)).

6.2.2 Approach to equilibrium: the Hartle–Hawking formula

It is plausible to suppose that if the amount of energy available for accretion into a black hole is limited, then after a certain time the oscillations of the external field will be damped out by gravitational radiation, and the hole will settle down asymptotically towards a final *stationary equilibrium* state (whose properties will be discussed in the subsequent sections). If these conditions are satisfied, the values of the expansion rate θ must obviously tend to *zero* in the final state, and hence the differential equation (6.57) may be replaced by a corresponding integral formula

$$\theta = \int_t^\infty e^{\kappa t - \kappa' t'}(8\pi\mathcal{D}' + \kappa'\theta't')\,dt' \qquad (6.59)$$

where quantities valued at the intermediate time t' are distinguished by a prime.

We are now ready to examine in more detail the way in which the area $d\mathcal{S}$ of a surface element of the black hole increases with time. By (6.4) the change in $d\mathcal{S}$ between two times, t_0 and t_1 say, will be given by

$$\ln\left(\frac{d\mathcal{S}_1}{d\mathcal{S}_0}\right) = \int_{t_0}^{t_1} \theta\,dt \qquad (6.60)$$

where we introduce the use of suffices 0 and 1 to denote quantities evaluated at the times t_0 and t_1 respectively. If now we substitute (6.59) and interchange the order of the integration, in the manner first employed by Hartle and Hawking (1972b) in a slightly more restricted analysis (in which deviations from stationarity were assumed to be at all times small), we obtain

$$\ln\left(\frac{d\mathcal{S}_1}{d\mathcal{S}_0}\right) = \left\{\int_{t_0}^{t_1} dt' \int_{t_0}^{t'} dt + \int_{t_1}^{\infty} dt' \int_{t_0}^{t_1} dt\right\} e^{\kappa t - \kappa' t'}(8\pi\mathcal{D}' + \kappa'\theta't'). \qquad (6.61)$$

This formula could be made to appear simpler if we chose to set κ and $\dot\kappa$ to zero by using an *affine* parametrization. However, in any stationary or temporarily stationary situation it is preferable for purposes of physical interpretation to work with the natural parametrization determined by the stationary symmetry group. In terms of such a symmetry group parametrization we shall have a value of κ which (though obviously constant along the generators by the group invariance property) will *not* in general be zero. One of the principal results of the following section will in fact be to show that for any stationary black hole state the appropriate value of κ is not only constant along the individual generators but actually has the *same positive value* on all of them. If now we wish to study the change in κ as the hole evolves between two distinct states of approximate stationary equilibrium, then it is obviously desirable to work with the parametrizations and values of κ appropriate to such states. If we arrange the parametrization in such a way as to ensure that κ is always positive, with a *constant* value, κ_1 say, during the time interval from t_0 to t_1, and if we choose t_0 as the time origin (i.e. set $t_0 = 0$), the formula (6.61) simplifies to give

$$\ln\left(\frac{\mathrm{d}\mathcal{S}_1}{\mathrm{d}\mathcal{S}_0}\right) = \frac{8\pi}{\kappa_1}\int_0^{t_1}\mathrm{d}t(1-\mathrm{e}^{-\kappa_1 t})\mathcal{D} + \frac{(\mathrm{e}^{\kappa_1 t}-1)}{\kappa_1}\int_{t_1}^{\infty}\mathrm{d}t\,\mathrm{e}^{-\kappa t}(8\pi\mathcal{D}+\dot{\kappa}\theta t). \quad (6.62)$$

The second term on the right-hand side of this formula has a teleological (as opposed to causal) character: it shows that the behaviour of the horizon during the time interval from t_0 to t_1 depends on what will happen subsequent to the time t_1. This rather bizarre feature is a consequence of the teleological way in which a black hole is defined as a region from which light cannot escape to infinity: it results from the fact that local information can never guarantee the possibility of escape – there is always the possibility of being engulfed by a future infall of matter. However, the teleological term in (6.62) can be eliminated, and the first term simplified, if we take the limit (in which the source rate term, $\mathcal{D}$, is supposed to be switched on) for a limited period which is both preceded and followed by periods that are sufficiently long (compared with the characteristic time $\tau \sim 1/\kappa_1$) during which $\mathcal{D}$ (and $\dot{\kappa}$) is held equal to zero. More explicitly, if the intervals during which $\mathcal{D}$ is held equal to zero are made to follow immediately after the initial time $t_0 = 0$ and immediately after the final time t_1 at which the overall change is to be measured, then we obtain the limit

$$\ln\left(\frac{\mathrm{d}\mathcal{S}_1}{\mathrm{d}\mathcal{S}_0}\right) = \frac{8\pi}{\kappa_1}\int_0^{t_1}\left(\frac{1}{16\pi}\theta^2 + \frac{1}{16\pi}\sigma^2 + \frac{1}{8\pi}R_{ab}l^a l^b\right)\mathrm{d}t \quad (6.63)$$

writing in the value of $\mathcal{D}$ explicitly in the bracket on the right. In these circumstances it follows from (6.7) that the value of the *total* area at the

Chapter 6. Mechanical, electromagnetic and thermodynamic properties

later time t will be given by

$$\mathcal{A}_1 = \oint d\mathcal{S}_1 \geq \oint d\mathcal{S}_0 \exp \frac{8\pi}{\kappa_1} \int_{t_0}^{t_1} dt\, \mathcal{D} \qquad (6.64)$$

with *equality* provided that there exists no caustic from which new generators enter the horizon during the time interval under consideration. The formulae (6.63) and (6.64) are generalizations to *finite* area changes of the formulae originally derived by Hawking and Hartle in the approximately stationary limit for which the source rate, $\mathcal{D}$, is supposed to be infinitesimal, so that the $d\mathcal{S}_1$ is only infinitesimally larger than $d\mathcal{S}_2$, in which case (6.63) may be replaced by

$$\delta(d\mathcal{S}) = \frac{8\pi}{\kappa} d\mathcal{S} \int_{t_0}^{t_1} \left(\frac{1}{16\pi} \sigma^2 + \frac{1}{8\pi} R_{ab} l^a l^b \right) dt, \qquad (6.65)$$

where we have dropped the suffix (1) on κ in view of the fact that, since we are now dealing with an approximately stationary situation, no ambiguity in the value of κ is likely to arise, at least to first order. Since the expansion rate, θ, itself will necessarily be of the same order as the change in $d\mathcal{S}$, the term in θ^2 has been dropped as relatively negligible in these circumstances. If now we transform from a kinematic to a dynamic presentation by introducing the Einstein equations

$$R_{ab} - \tfrac{1}{2} R g_{ab} = 8\pi T_{ab} \qquad (6.66)$$

then we obtain the final integral form of the Hartle–Hawking formula as

$$\delta\mathcal{A} = \oint \frac{8\pi}{\kappa} d\mathcal{S} \int_{t_0}^{t_1} \left(\frac{1}{16\pi} \sigma^2 + T^a{}_b l^b l_a \right) dt. \qquad (6.67)$$

Now we shall see in the next section that in a stationary situation, with the natural stationary-group parametrization, the vector $\vec{\Pi} \equiv T \cdot \vec{l}$ with components $T^a{}_b l^b$ may be interpreted as a flux of conserved material energy as defined with respect to a certain natural rest-frame defined by the black hole, and hence, since we have

$$T^a{}_b l^b l_a\, dt\, d\mathcal{S} = \Pi^a\, d\Sigma_a, \qquad (6.68)$$

where $d\Sigma_a$ are components of the normal hypersurface element corresponding to $dt\, d\mathcal{S}$, it follows that the second term on the right-hand side of (6.67) can be interpreted as being equal to the flux of this material energy across the horizon, with a proportionality factor $8\pi/\kappa$. Taking account of this, one sees that the expression (6.67) has a highly suggestive form. If the area is thought of as analogous to entropy, as is suggested by the 'second law' (6.9), then the quantity $8\pi/\kappa$ plays the role corresponding to *temperature*. The analogy thus brought to light was first taken seriously by Bekenstein (1973) who suggested that a black hole should really be thought of as having an entropy equal to its area in terms of some suitable chosen units with order of unity values for the Dirac–Planck constant, $\hbar$, and the Boltzmann constant, k.

The analogy with entropy generation in an ordinary body can be extended further by remarking that the energy flux term in (6.67) is augmented by an effective dissipation term proportional to the square of the shear rate tensor, just as would be the case in an ordinary viscous fluid. As was observed by Hartle and Hawking, this effect may be accounted for quantitatively by attributing a surface *shear* viscosity η_v of magnitude $\eta_v = (16\pi)^{-1}$. Furthermore, one can see from the more general formula (6.63) that the horizon should similarly be thought of as having a surface *dilatation* viscosity ζ_v (the analogue of a fluid *bulk* viscosity) of the *same* magnitude, i.e. $\zeta_v = (16\pi)^{-1}$.

The analogy between the black hole horizon and an ordinary material medium can be pursued even further in electromagnetic contexts. Let us decompose the energy momentum tensor into a material part, indicated by a suffix M, and an electromagnetic part indicated by a suffix F, i.e.

$$T^a{}_b = T_M{}^a{}_b + T_F{}^a{}_b, \qquad (6.69)$$

where the electromagnetic part is given in terms of the electromagnetic field 2-form $\boldsymbol{F}$ by the standard Maxwellian expression

$$T_F{}^{ab} = (1/4\pi)(F^a{}_c F^{bc} - \tfrac{1}{4} F_{cd} F^{cd} g^{ab}). \qquad (6.70)$$

Then in consequence of the nullity of $\vec{l}$ on the horizon we have

$$T^{ab}{}_b l^b l_a = T_M{}^a{}_b l^b l_a + (1/4\pi) E^{Ha} E^H{}_b \qquad (6.71)$$

where $\vec{E}^H$, the effective *electric field* vector on the horizon, is defined by

$$\vec{E}^H = \vec{F} \cdot \boldsymbol{l} \qquad (6.72)$$

and thus evidently satisfies the necessary and sufficient condition

$$\vec{E}^H \cdot \boldsymbol{l} = 0 \qquad (6.73)$$

to be tangent to the horizon. It can now be seen that the purely electromagnetic contribution to the energy flux, i.e. the second term in (6.71), is exactly the same as the dissipation term that would arise from an electric field of the form (6.72) if the black hole horizon were thought

Chapter 6. Mechanical, electromagnetic and thermodynamic properties

of as having a finite resistivity, corresponding quantitatively to a *surface resistivity* coefficient ρ_R, given numerically by $\rho_R = 4\pi$.

Recent work by Znajek (1977) and Damour (1978) has revealed that in a *stationary* situation the analogy with a finite conductor (of which our analysis so far provides only a hint) can in fact be taken very literally indeed, as will be shown in section 6.4.

6.3 Local properties of a stationary horizon

6.3.1 The energy condition

Despite its brevity, the preceding section has covered a substantial part of the little that is known about *dynamic* black hole theory. From this point onward we shall restrict our attention to the more thoroughly explored terrain of *stationary* black hole theory. Apart from the fact that it is more amenable to mathematical analysis, this more specialized domain derives its *physical* – and more particularly its astrophysical – interest from the conjecture that whenever a black hole is formed in circumstances where there is only a limited amount of matter available for accretion then the vibrations in its external field will ultimately be damped out by gravitational radiation. So (provided it is not subject to any external stimulation mechanism) the black hole will settle down asymptotically towards an equilibrium state that is *stationary*. It is stationary in the sense of being invariant under the action of a group whose generator, $\vec{k}$, which must of course satisfy the Killing equations

$$\nabla_{(a} k_{b)} = 0, \tag{6.74}$$

is such as (*a*) to be *timelike* at sufficiently large asymptotic distances and (*b*) such that on the horizon, $\mathcal{H}^+$, (which, being invariant, must contain $\vec{k}$ as a – necessarily *spacelike or null* – tangent vector) it transports any given spacelike 2-section $\mathcal{S}(0)$ towards the future over a family of congruent 2-surfaces, $\mathcal{S}(t)$ say, where t is a *group parameter*, which can be used as an ignorable time coordinate, such that

$$k^a \frac{\partial}{\partial x^a} = \frac{\partial}{\partial t}. \tag{6.75}$$

Hawking has shown (see Hawking and Ellis (1973) proposition 9.3.3) that subject to asymptotic predictability the second condition (*b*) (including the global existence on $\langle\!\langle \mathcal{J} \rangle\!\rangle$ of the well-behaved ignorable time coordinate, t) follows from the first postulate (*a*); but despite having a high degree of intuitive plausibility, the postulate that there is an asymptotically approached final state satisfying condition (*a*) remains conjectural, with a rather similar status to Penrose's cosmic censorship conjecture in so much as it not only lacks any semblance of a mathematical proof, but also has been worded in a deliberately vague manner so as to maximize the chances that it may one day be justified. It can be maintained nevertheless that the conjecture of *ultimate asymptotic stationarity* is on a considerably stronger footing than the cosmic censorship hypothesis, and merits correspondingly much greater confidence since it is supported on the one hand by the perturbation analyses of vacuum (Schwarzschild and Kerr) black hole states (using the specialized methods described in chapter 7 of this book by Chandrasekhar) which have been carried out by Visveshwara (1970), Press and Teukolsky (1973), Wilkins and Hartle (1974), Stewart (1975) and others; while on the other hand analogous behaviour is well known and understood, experimentally and theoretically, in many other branches of physics (in contrast with cosmic censorship which is a novel and as yet unobserved phenomenon).

Since the surface area, $\mathcal{A}$, of a spacelike 2-section of the horizon in the postulated *stationary final state* is *unchanged* as $\mathcal{S}$ is transported toward the future, the non-negativity (on which the second law (6.9) was based) of the terms in (6.58) allows us to draw the strong conclusion that they will all have to be strictly *zero*, i.e. we must have

$$\theta = 0, \qquad \sigma = 0 \qquad (6.76)$$

and

$$R_{ab}l^a l^b = 0 \qquad (6.77)$$

on an equilibrium-state horizon.

In view of the vital role it plays in the reasoning of our entire analysis, it is appropriate at this stage to comment on the *non-negativity* of the last-mentioned term, which is based on a *physical* (and therefore experimentally verifiable or contradictable) assumption, unlike that of the preceding terms (whose non-negativity holds as a purely kinematic identity).

For the purely *electromagnetic* contribution (6.70) to the classical energy–momentum tensor it is a standard algebraic exercise to demonstrate that for *any* timelike or null vector, $\vec{l}$, the associated energy–momentum flux, $\vec{\Pi}$, defined by

$$\Pi_F{}^a = -T_F{}^{ab} l_b, \qquad (6.78)$$

Chapter 6. Mechanical, electromagnetic and thermodynamic properties

must satisfy the inequalities

$$\Pi_F{}^a\Pi_{Fa} \leq 0, \quad \Pi_F{}^a l_a \leq 0. \tag{6.79}$$

Furthermore (6.79) will hold as *strict* inequalities, (with $<$ in place of $\leq$) except in special circumstances occuring only when either the vector $\vec{l}$ or the Maxwell field $\boldsymbol{F}$ is *null*. Explicitly, the only conditions under which they can hold as strict equalities are as follows:

$$\Pi_F{}^a\Pi_{Fa} = 0 \Leftrightarrow \begin{cases} \text{either } F^{ab}F_{ab} = 0 = (*F^{ab})F_{ab} \\ \text{or } l^a l_a = 0 \end{cases} \tag{6.80}$$

$$\Pi_F{}^a l_a = 0 \Leftrightarrow l \wedge (\boldsymbol{F} \cdot \vec{l}) = 0 \Leftrightarrow \begin{cases} \vec{\Pi}_F \wedge \vec{l} = 0 \\ \text{and } l^a l_a = 0. \end{cases} \tag{6.81}$$

The latter condition (6.81) is an expression of the requirement that $\vec{l}$ be a *principal null vector* of $\boldsymbol{F}$ (i.e. one of its – necessarily null – nonzero eigenvalue characteristic vectors) which can lie in one of two possible directions, except in the case where $\boldsymbol{F}$ is itself *null* (in the sense that it satisfies the first alternative condition of (6.81)), in which case there is only one possible direction for a vector, $\vec{l}$, satisfying (6.81).

Now although one cannot give any analogous mathematical proof without restricting oneself to some very explicitly specified kind of matter, nevertheless it is probably safe to assume that the *non-electromagnetic* energy–momentum contribution, $T_M{}^{ab}$, obeys even stricter inequalities. As far as we know today, the electromagnetic field is the *only* long-range classical field that is *fundamental* (in the sense that it arises simply by taking the classical limit of a zero rest-mass bosonic quantum field) except for *gravitation* itself, which latter is treated as giving no direct contribution to the energy–momentum tensor (at least in the standard formulation of Einstein's theory). Thus whenever a classical continuous field treatment is appropriate (i.e. in any circumstances to which the conventional Einstein field equations might be applicable), the non-electromagnetic contribution to the energy–momentum tensor, $T_M{}^{ab}$, must represent the effect of a *compound* medium (which might represent an ordinary fluid or solid, or perhaps a plasma, or some kind of distribution of very short wavelength gravitational or electromagnetic radiation) in which the continuum description represents a suitable averaging over a possibly very complicated underlying microscopic structure. All our physical experience suggests the law that for any such *compound* medium, the corresponding *energy–momentum vector* as defined with respect to an *arbitrary* rest frame will be *strictly timelike and future-directed*, i.e. the quantity

$$\Pi_M{}^a = -T_M{}^{ab} l_b \tag{6.82}$$

will satisfy

$$\Pi_M{}^a \Pi_{Ma} < 0, \quad \Pi_M{}^a l_a < 0 \tag{6.83}$$

for any *timelike* vector, $\vec{l}$, unless of course $T_M{}^{ab}$ is zero. Moreover the attainment of limiting special cases is rendered unlikely by the averaging over microscopic substructure, and hence it would seem fairly safe to assume that (6.83) will still hold as strict inequalities even when $\vec{l}$ is null. We shall therefore take it as a fundamental postulate, which we shall refer to as the *energy axiom*, that (6.83) hold for arbitrary timelike *or* null vectors, $\vec{l}$. (Hawking and Ellis (1973) have listed a wide range of more or less restrictive energy inequalities that may be involved for various uses, but the condition that has just been described will suffice for our present purposes.)

By combining (6.83) with (6.79) one can deduce that the corresponding *total* energy–momentum vector, as given by

$$\Pi^a = -T^{ab} l_b \tag{6.84}$$

will satisfy

$$\Pi^a \Pi_a \leq 0, \quad \Pi^a l_a \leq 0 \tag{6.85}$$

(which is what Hawking and Ellis (1973) call the *dominant* energy condition) for an *arbitrary* timelike or null vector, $\vec{l}$, and that neither of these relations can hold as an equality *unless* $T_M{}^{ab}$ vanishes and one or other of the conditions (6.80), (6.81) is satisfied.

If we now restrict ourselves to the case for which the vector $\vec{l}$ is *null*, as will be the horizon generators with which we are actually concerned, then substitution of the Einstein equations in the last inequality of (6.85) leads immediately to the positivity condition (6.2) on which so much of our reasoning has been based. Conversely substitution of the Einstein equations in our conclusion (6.77) leads to the equality

$$T^{ab} l_a l_b = 0 \tag{6.86}$$

which must be satisfied by the null generators of a horizon in the *stationary case*. By the axiom (6.83) we can see that this strict inequality can hold only if we have

$$T_M{}^{ab} = 0 \tag{6.87}$$

Chapter 6. Mechanical, electromagnetic and thermodynamic properties

on the horizon, (i.e. there can be no material accretion into the black hole in a *strictly* stationary state) and the theorem (6.81) then implies that the electric field $\vec{E}^H$ defined on the horizon by (6.72) *must be parallel to* $\vec{l}$, i.e.

$$\vec{E}^H \wedge \vec{l} = 0, \tag{6.88}$$

which tells us that $\vec{l}$ must in fact be a principal null vector of $\boldsymbol{F}$ on the horizon. (Later on, in section 6.4, we shall consider *quasi*-stationary accretion processes in which the Einstein equations are required to be satisfied only *approximately*, with accreting matter and electromagnetic fields treated as non-selfgravitating perturbations, so that the conditions (6.87) and (6.88) may be somewhat relaxed.)

Combining (6.81) with (6.87) we deduce also that *the total momentum vector*, $\vec{\Pi}$, *must be parallel to the tangent vector*, $\vec{l}$, or equivalently that the Ricci tensor must satisfy

$$R_{ab}l^b = R^{(0)}_{(0)}l_a \tag{6.89}$$

for some scalar $R^{(0)}_{(0)}$ on the horizon. The previous condition (6.77) from which (6.89) was originally derived can be recovered again from (6.89) as a special case.

6.3.2 The angular velocity and electromagnetic potential of a stationary black hole

It can be seen from the discussion of section 6.2.1 that the conclusions (6.76) that θ and σ must be zero on a stationary horizon imply that the flow defined by the null-tangent vector, $\vec{l}$, on the horizon is effectively *rigid*, so that $\vec{l}$ generates an *isometry* of the *intrinsic* geometry of $\mathcal{H}^+$. This conclusion has been extended by Hawking (1972) (see Hawking and Ellis, 1973) to give what I shall refer to as the *strong rigidity theorem* to the effect that, when suitably normalized, $\vec{l}$ must actually coincide with a *Killing vector* generator of an *isometry of the full four-dimensional geometry*, not just the intrinsic geometry of the horizon. This result – whose proof is much too elaborate to be recapitulated here – is a generalization of an earlier and technically simpler theorem (Carter, 1969, 1971*b*) to the same effect in the more restrictive context of a spacetime that is assumed in advance to be axisymmetric with purely circular flow of any external matter that may be present. (A complete derivation of this more readable restricted theorem is given by Carter, 1973*b*; a more mathematically abstract version relevant to global contexts other than those of an ordinary asymptotically flat black hole has also been given by Carter, 1978a.)

In the axisymmetric case with which the present analysis is primarily concerned, there is a second isometry generated by a vector $\vec{m}$, satisfying the Killing equations

$$\nabla^{(a}m^{b)} = 0 \tag{6.90}$$

with *closed* (topologically circular) and hence (in order to maintain causality) *necessarily spacelike* trajectories. Its normalization is customarily fixed by the requirement that the corresponding group parameter, φ say, which can be introduced as an ignorable coordinate such that

$$m^a \frac{\partial}{\partial x^a} = \frac{\partial}{\partial \varphi}, \tag{6.91}$$

has the *standard period* 2π. In an *asymptotically flat* spacetime such an axisymmetric Killing vector, $\vec{m}$, must necessarily commute with the stationary Killing vector, $\vec{k}$ (as introduced in the previous section), whenever both are defined; i.e. we must have

$$\vec{m} \,\pounds\, \vec{k} \equiv (\vec{m} \cdot \nabla)\vec{k} - (\vec{k} \cdot \nabla)\vec{m} = 0. \tag{6.92}$$

(See Carter (1970) but note that there is an omission, in the statement of proposition 5 and the proof of proposition 6 of this reference, of the qualification that the vector fields under consideration should be complete.) This commutation property makes it possible to use coordinates in which the stationary group parameter, t, and the axisymmetry group parameter, φ, are introduced as *simultaneously ignorable* coordinates.

Assuming that there is not a third independent Killing vector field (which could in fact be present only in the very special case of spherical symmetry), the Killing vector field predicted by the strong rigidity theorem to coincide with $\vec{l}$ on the horizon must simply be a linear combination of $\vec{k}$ and $\vec{m}$, i.e., subject to suitable normalization of $\vec{l}$ it must have the form

$$\vec{l} = \vec{k} + \Omega^H \vec{m} \tag{6.93}$$

where, in order for the combination itself to satisfy the corresponding Killing equations

$$\nabla^{(a}l^{b)} = 0, \tag{6.94}$$

Chapter 6. Mechanical, electromagnetic and thermodynamic properties

the quantity Ω^H so defined must be constant throughout space. Since the horizon must itself be invariant under the symmetry group action, and hence will contain $\vec{k}$ and $\vec{m}$ as tangent vectors so that they obey

$$\vec{k} \cdot \mathbf{l} = \vec{m} \cdot \mathbf{l} = 0, \tag{6.95}$$

it follows that we can eliminate the scalar Ω^H and the vector $\vec{l}$ from (6.93) and thus obtain, as a corollary to the strong rigidity theorem, the conclusion that the quantity

$$\rho^2 \equiv (\vec{k} \cdot \mathbf{m})^2 - |\vec{k}|^2 |\vec{m}|^2 \tag{6.96}$$

must satisfy

$$\rho^2 = 0 \tag{6.97}$$

on the horizon. The more primitive and restricted approach described by Carter (1973b) proceeds in the reverse direction: the condition (6.94) is derived *first* – in fact it is shown that the quantity ρ^2 must be non-negative throughout the closure of the domain of outer communication $\langle\!\langle \mathcal{J} \rangle\!\rangle$, being allowed to be zero only on the symmetry axis where the vector $\vec{m}$ itself vanishes, and, in accordance with (6.97), on the horizon. It is then shown as a consequence – indeed it is virtually obvious – that the null generator must be a linear combination of the form (6.93), the coefficient Ω^H being obviously interpretable as representing the local *angular velocity* of the horizon ($\nu^H = \Omega^H/2\pi$ being the number of times per unit of the ignorable coordinate time, t, that a null generator makes a complete rotation around the horizon). Finally it is proved as a not quite trivial corollary (using an argument given by Carter, 1969, 1972) that the rotation is *rigid* in the sense that Ω^H must be uniform over the horizon, i.e.

$$\vec{\xi} \cdot \nabla \Omega^H = 0 \tag{6.98}$$

for any vector $\vec{\xi}$ satisfying the tangency condition

$$\vec{\xi} \cdot \mathbf{l} = 0 \tag{6.99}$$

which brings us back to what, in the Hawking approach, was the initial conclusion, namely the result that $\vec{l}$ coincides with a Killing vector field.

In the more general situation where the spacetime is not assumed in advance to be axisymmetric the more powerful Hawking approach still goes through, but if the second Killing vector, $\vec{m}$, does not exist, the strong rigidity theorem leaves us no choice but to conclude that $\vec{l}$

coincides with $\vec{k}$ the stationary Killing vector itself, i.e. we obtain

$$\vec{k} = \vec{l} \qquad (6.100)$$

in which case the black hole is described as *non-rotating*, since this is the same situation that obtains in the axisymmetric case in the limit when the angular velocity vanishes, i.e.

$$\Omega^H = 0. \qquad (6.101)$$

This – at first surprisingly strong – conclusion can be interpreted as implying that whenever a black hole is perturbed by a non-axisymmetric external perturbing system, viscous drag effects and gravitational radiation will come into play in such a way as to reduce the departures from axisymmetry and/or the rate of rotation of the black hole until either an axisymmetric or a non-rotating stationary state is attained. Such effects have been studied in detail by Press (1972) in the case of an external perturbation produced by a hypothetical scalar field, by Pollock (1976); Pollock and Brinkmann (1977); King and Lasota (1977); Znajek (1977) and Damour (1978) in the more realistic case of electromagnetic perturbations (of the kind to be discussed in section 6.5) and finally by Hartle (1973*a*) and Prior (1977) in the more delicate case of gravitational perturbations.

Having defined the angular velocity of the black hole it may be mentioned that one can also define a corresponding (non-uniform) surface 3-*velocity* vector

$$\vec{v}_H = \Omega^H \vec{m} \qquad (6.102)$$

whose magnitude, $|\vec{v}_H|$, represents the *proper* distance (relative to the stationary frame defined by $\vec{k}$) through which the hole rotates per unit of the *improper* ignorable time, t (i.e. the time defined with respect to an observer at infinity). Although $|\vec{v}_H|$ is *not* an ordinary 3-velocity (which would measure a rate of proper displacement with respect to *proper* time in some well-defined rest-frame), it has been pointed out by Damour (1978) that it has properties analogous to those one would expect for the local 3-velocity of rotation of an ordinary material body in at least the particular case of the *Kerr* solutions, where $|\vec{v}_H|$ is found to increase monotonically from zero on the poles to a *maximum on the equator* of the horizon, this maximum being *always less than one* except in the physically unattainable limit of the *extreme* $(a = m)$ Kerr solutions in which $|\vec{v}_H|$ attains the value of unity (representing in some sense 'rotation at the speed of light') on the equator, the unattainability of this limit

Chapter 6. Mechanical, electromagnetic and thermodynamic properties

resulting from the application of the 'zeroth law' of black hole mechanics as described in section 6.3.3. It is an unsolved problem (to which the inequality of Hajicek (1975) may be relevant) whether $|v_H|$ would behave in the same way for externally perturbed black holes.

By comparison with the effort required to establish that a stationary black hole has a uniform angular velocity, Ω^H, it is very easy to show that it has uniform effective electric potential, Φ^H, specified as the limit on the horizon of a globally defined, corotating electrostatic potential $\Phi_{(0)}$ where the latter is defined modulo an additive constant as the solution of

$$\partial \Phi_{(0)} = E_{(0)}. \quad (6.103)$$

The corotating electric field, $E_{(0)}$, appearing here is defined in terms of the corotating Killing vector, $\vec{l}$ (given by (6.93)) as

$$E_{(0)} = F \cdot \vec{l} \quad (6.104)$$

the integrability of (6.103) being obtained using the condition that, being invariant under the actions generated by $\vec{k}$ and (in the axisymmetric case) also by $\vec{m}$, the system as a whole must also be invariant under the action generated by the constant coefficient linear combination, $\vec{l}$, which in the particular case of the electromagnetic field means

$$\vec{l} \, \pounds \, F = 0. \quad (6.105)$$

Working this out using the Cartan formula (6.20) we see from the first of the Maxwell equations

$$\partial F = 0 \quad (6.106)$$

$$\nabla \cdot \vec{F} = 4\pi \vec{j} \quad (6.107)$$

($\vec{j}$ being the electric current vector) that we shall have

$$\partial E_{(0)} = 0 \quad (6.108)$$

which is a necessary and sufficient condition for at least local integrability of (6.103). Using the notation Φ^H and E^H for the values on the horizon of the fields $\Phi_{(0)}$ and $E_{(0)}$, we see that the gradient of Φ^H in the horizon will satisfy

$$\vec{\xi} \cdot \nabla \Phi^H = \vec{\xi} \cdot E^H \quad (6.109)$$

for any vector $\vec{\xi}$ satisfying the condition of tangency to the horizon. Since we have already derived the boundary condition (6.88) to the effect that E^H is parallel to l and therefore orthogonal to $\vec{\xi}$, we immediately obtain the required result

$$\vec{\xi} \cdot \nabla \Phi^H = 0. \tag{6.110}$$

This result also shows immediately that the field $\Phi_{(0)}$ can be *globally* defined, at least in the neighbourhood of the horizon. If we are willing to postulate that the domain of outer communications $\langle\!\langle \mathcal{J} \rangle\!\rangle$ is *simply connected* there can in any case be no doubt that $\Phi_{(0)}$ will be globally well defined everywhere on it, and the constant Φ^H can then be uniquely defined by fixing the constant of integration in the standard way by the requirement that $\Phi_{(0)}$ should *tend to zero* at sufficiently large asymptotic distances.

6.3.3 The decay constant: the third and zeroth laws

In order for the stationary equilibrium states to be of practical interest it is not sufficient to know merely that they are approached as asymptotic final states: it is also necessary that the timescales characterizing the convergence to the asymptotic final state should be reasonably short by the relevant practical standards. However the available evidence strongly suggests that this will in fact be the case in most practical situations. The analysis of the previous section provides a clear hint that in terms of a coordinate time, t, chosen so as to be consistent with the stationary symmetry group parametrization in the asymptotic final state, the convergence will be characterized by a timescale, τ, given by the inverse of the corresponding value of κ, i.e. by the formula $\tau \sim \kappa^{-1}$. The idea that the appropriate value of κ characterizes the rate of exponential decay of deviations from the final equilibrium state is confirmed by the results of the perturbation analyses mentioned in section 6.3.1.

It follows from these considerations that stationary black hole states with zero κ (which in the Kerr solutions corresponds to the limit $J^2 = M^4$) can be regarded as representing a *physically inaccessible* limit.

It is already evident from the Hartle–Hawking formula (6.67) that in terms of a thermodynamic analogy in which the black hole is thought of as having an entropy S_H proportional to its surface area, i.e. such that

$$\mathcal{A} = \alpha S_H, \tag{6.111}$$

then it must be thought of as having a temperature, Θ^H, proportional to the decay constant, κ, with

$$\Theta^H = (\alpha/8\pi)\kappa \tag{6.112}$$

Chapter 6. Mechanical, electromagnetic and thermodynamic properties

where α is a proportionality constant that cannot be calculated within the framework of the purely classical theory to which the present analysis is restricted. (The quantum particle creation theory of Hawking (1975) has shown that the relations (6.111) and (6.112) represent more than a mere analogy, but should be taken quite literally, with the proportionality factor given – in terms of the Planck and Boltzmann constants – by $\alpha = 4\hbar c/k$ in accordance with an earlier order of magnitude argument of Bekenstein (1973b). A thorough discussion is given in chapter 13 by Gibbons.)

The possibility of interpreting the decay constant in this way makes it natural to describe the fact that the characteristic timescale (given by $\tau \sim \kappa^{-1}$) becomes infinitely large in the limiting case of a black hole for which κ is infinitesimally small, as the *third law* of black hole mechanics (cf. Bardeen, Carter and Hawking, 1973). (One can obtain an analogue of $\tau \sim \Theta^{-1}$ in ordinary thermodynamics by applying the uncertainty principle to energy and time.)

Having chosen to normalize $\vec{l}$ so as to satisfy (6.93) (which corresponds to using the stationary group parameter as the variable t in the definition (6.10)), we see that the corresponding value of the decay constant κ will be given by

$$(\vec{l} \cdot \nabla)l = \kappa l. \qquad (6.113)$$

Using the antisymmetry of $\nabla_a l_b$, which follows from Killing's equation (6.94), one can see that κ can be specified equivalently by the alternative expression

$$\nabla |\vec{l}|^2 = 2\kappa l \qquad (6.114)$$

and that (by the orthogonality condition (6.12)) one also has the more explicit expression

$$\kappa^2 = \tfrac{1}{2}(\nabla_a l_b)(\nabla^b l^a). \qquad (6.115)$$

Now any Killing vector field, $\vec{l}$, always satisfies

$$|\vec{l}|^2 (\nabla_a l_b)(\nabla^a l^b) + \tfrac{1}{2} |\nabla |\vec{l}|^2|^2 = 2|\vec{\omega}^\dagger|^2, \qquad (6.116)$$

where $\vec{\omega}^\dagger$ is the vector defined in terms of the alternating tensor $\vec{\varepsilon}$ by

$$\vec{\omega}^\dagger = *(l \wedge \partial l). \qquad (6.117)$$

Since the vanishing of this vector, $\vec{\omega}^\dagger$, is equivalent to the Frobenius hypersurface orthogonality condition, which is necessarily satisfied on the horizon (cf. (6.12)), it follows from (6.115) that κ can be interpreted

Local properties of a stationary horizon

as being given by

$$\frac{(\nabla_a |\vec{l}|^2)(\nabla^a |\vec{l}|^2)}{|\vec{l}|^2} \to 4\kappa \qquad (6.118)$$

in the limit on the horizon.

This last formula in its turn suggests yet another interpretation of the quantity κ. If we use the somewhat vague and conjectural 'third law' as a justification for leaving out of account the special case for which κ is zero, we can see from (6.114) that the square of the Killing vector, $|\vec{l}|^2$, changes sign across the horizon. Moreover, by the remarks in the previous section, one knows – at least in the axisymmetric (rotating) case – that the quantity $|\vec{l}|^2$ is necessarily positive just within the horizon (where ρ^2 is negative) which implies that the vector $\vec{l}$ must be *timelike* in a finite region extending from *just outside* the horizon out to a timelike hypersurface – the 'speed of light cylinder' – beyond which the velocity of corotation would be faster than light. (Whether $\vec{l}$ would necessarily be timelike outside the horizon in the *non-rotating* case is an important, but not yet completely resolved question which will be discussed further in section 6.6.) Within the region where $\vec{l}$ is timelike we can construct a timelike unit vector field, $\vec{u}$, parallel to it, obeying

$$\vec{l} = \dot{\tau}\vec{u} \qquad (6.119)$$

where $\dot{\tau}$ is the derivative of *proper* time, τ, with respect to the stationary symmetry group parameter t along the Killing vector trajectories. The time dilatation (or 'mean redshift') factor will be given by

$$\dot{\tau} = (-|\vec{l}|^2)^{1/2} \qquad (6.120)$$

so that one has $|\vec{u}|^2 = -1$.

The acceleration, $\dot{u}$, of this unit vector with respect to the group parameter (i.e. the ignorable time coordinate) t will be given in terms of the *proper* acceleration a by

$$\dot{u} = (\vec{l} \cdot \nabla)u = \dot{\tau}a. \qquad (6.121)$$

It can now be seen from (6.81) that κ represents the magnitude of this vector $\dot{u}$ in the limit on the horizon, where one obtains

$$\dot{u}_a \dot{u}^a \to \kappa^2. \qquad (6.122)$$

(It is evident that the magnitude of the *proper* acceleration diverges to infinity on the horizon. One sees that κ can be interpreted as the finite but improper acceleration corresponding to Damour's finite but

Chapter 6. Mechanical, electromagnetic and thermodynamic properties

improper 3-velocity 4-vector (6.102) on the horizon.) It is for this reason that the decay constant of a stationary black hole, κ, is frequently referred to as the surface gravity of the horizon.

The thermodynamic analogy that we have been developing would lead one to guess immediately that κ (like an ordinary temperature) should be *uniform* over the horizon in any state of equilibrium. (In a non-equilibrium state there is in any case no unambiguous way of fixing the normalization of κ, just as in an ordinary thermodynamic situation there is no unique way of defining the temperature except in the thermal equilibrium limit.) The fact that the uniformity of κ is not just a special property of the Kerr solutions came to light first as a step in the proof of the no-hair theorem (Carter, 1971b, 1973b) to be discussed in section 6.6, in the restricted context of *axisymmetric* black holes. It turned up again in a more general context as a step in the proof of the strong rigidity theorem (Hawking, 1972) whose content has already been described in section 6.3.2. In view of its importance for the thermodynamic analogy I shall present the general proof (in the form given by Bardeen *et al.*, 1973).

We start from the condition that the rotation vector as defined by (6.117) must vanish on the horizon, which is equivalent to the requirement that the curl of the normal form l should satisfy

$$\partial l = 2q \wedge l \qquad (6.123)$$

for some 1-form, q, on the horizon; q can be fixed uniquely by requiring that it should also satisfy $\vec{\beta} \cdot q = 0$, where $\vec{\beta}$ is the null vector transverse to the horizon that was introduced by (6.27) and (6.28). It follows that for any vector fields $\vec{\xi}, \vec{\eta}$ lying in the horizon, i.e. such that $\vec{\xi} \cdot l = \vec{\eta} \cdot l = 0$, we can derive (using the Killing equation (6.94)) the condition

$$\xi^a \eta^b \nabla_a l_b = \xi^a l^b (q_a l_b - q_b l_a) = 0. \qquad (6.124)$$

This expresses the fact that the stationary horizon is *extrinsically flat* (and thus *geodesically generated*) since by differentiating (6.123) we can see that (6.124) is equivalent to

$$\vec{l} \cdot \{(\vec{\xi} \cdot \nabla)\vec{\eta}\} = 0 \qquad (6.125)$$

which shows that the vector field $(\vec{\xi} \cdot \nabla)\vec{\eta}$ must lie within the horizon. Furthermore by differentiating (6.124) and using (6.123) again twice in succession we also find that

$$\xi^a \eta^b \nabla_c \nabla_b l_a = 0 \qquad (6.126)$$

on the horizon. Now it follows directly from the fact that l satisfies the Killing equation (6.94) that its second derivatives can be expressed in terms of the Riemann tensor by

$$\nabla_c \nabla_b l_a = R_{abc}{}^d l_d. \tag{6.127}$$

This implies that (6.126) is equivalent to the condition that

$$R_{abcd} l^b \xi^c \eta^d = 0 \tag{6.128}$$

for any vectors $\vec{\xi}$, $\vec{\eta}$ lying in the horizon. This allows us to derive a corresponding condition on the Ricci tensor, which we evaluate by contracting the Riemann tensor with the metric in the decomposed form

$$g^{ab} = \gamma^{ab} - \beta^a l^b - l^a \beta^b \tag{6.129}$$

thereby obtaining

$$R_{ab} l^b = R_{abcd}(l^b \beta^c - \gamma^{bc}) l^d \tag{6.130}$$

and hence, by (6.128),

$$R_{ab} l^b \xi^b = R_{abcd} \xi^a l^b \beta^c l^d \tag{6.131}$$

for any vector $\vec{\xi}$ lying in the horizon.

To demonstrate the uniformity of κ we start from the expression

$$\kappa = -\beta^a l^b \nabla_b l_a \tag{6.132}$$

(which is derived immediately from (6.113) using the normalization condition (6.27) for the transverse null vector, $\vec{\beta}$). Using the decomposition (6.123) with (6.132) and (6.131) we can obtain the relation

$$\xi^c \nabla_c \kappa = -R_{ab} l^b \xi^a \tag{6.133}$$

for any vector $\vec{\xi}$ lying in the horizon.

So far our reasoning in this section has been purely kinematic. However we now use the dynamic boundary condition (6.89) (obtained from the Einstein equations and the energy condition) to obtain finally the required conclusion

$$\xi^c \nabla_c \kappa = 0. \tag{6.134}$$

We shall terminate our analysis of intrinsic properties of the horizon at this point. For further interesting details the reader is referred to Hawking and Ellis (1973) and Hajicek (1975, 1976).

Chapter 6. Mechanical, electromagnetic and thermodynamic properties

6.4 Energy and angular momentum transport in a black hole background

6.4.1 The canonical and metric energy–momentum flux vectors in a free-field theory

We have already remarked that, apart from gravitation itself, the electromagnetic field is the only known long-range classical field that is fundamental (in the sense that it is directly derivable from a zero rest-mass bosonic quantum field) and it is for this reason that it is given special attention here. Nevertheless we shall start off by drawing some general conclusions that would be valid for any field whose equations of motion are derivable from a generally covariant second-order Lagrangian scalar function, Λ_F say, that, in addition to the metric components g_{ab}, depends also on a set of field components A_a and their derivatives $\partial_a A_r$, where r is not necessarily a covariant vectorial index (although this is what we shall in fact suppose when we specialize to the particular electromagnetic case later on) but may label a quite general set of scalar or tensorial components. From any such scalar function one may derive a set of Noetherian identities in the manner described by Trautman (1965) starting from the obvious fact that the Lie derivative of Λ_F with respect to an *arbitrary* vector field $\vec{\xi}$ may be expanded in the form

$$\vec{\xi} \, \pounds \, \Lambda_F = \frac{\partial \Lambda_F}{\partial g_{ab}} \vec{\xi} \, \pounds \, g_{ab} + \frac{\partial \Lambda_F}{\partial A_r} \vec{\xi} \, \pounds \, A_r + D_F{}^{ra} \vec{\xi} \, \pounds \, \partial_a A_r \qquad (6.135)$$

using the abbreviation

$$D_F{}^{ra} \equiv \frac{\partial \Lambda_F}{\partial(\partial_a A_r)} \qquad (6.136)$$

for what will be seen to be a generalization of the electromagnetic displacement tensor. Introducing the further abbreviation

$$j_F{}^r \equiv |g|^{-1/2} \partial_a(|g|^{1/2} D^{ra}) - \frac{\partial \Lambda_F}{\partial A_r} \qquad (6.137)$$

(for what is effectively the field source current), together with

$$T_F{}^{ab} \equiv 2 \frac{\partial \Lambda_F}{\partial g_{ab}} + \Lambda_F g^{ab} \qquad (6.138)$$

(for what is evidently the *metric* energy–momentum tensor), and

$$\mathcal{P}_F{}^a(\xi) \equiv D_F{}^{ra} \vec{\xi} \, \pounds \, A_r - \Lambda_F \xi^a \qquad (6.139)$$

(for the canonical energy–momentum flux as defined with respect to $\vec{\xi}$), we may rewrite the identity (6.135) in the more interesting form

$$\tfrac{1}{2}T_F^{ab}\vec{\xi}\,\pounds\,g_{ab} - j_F^r\vec{\xi}\,\pounds\,A_r + \nabla_a \mathscr{P}^a(\vec{\xi}) \equiv 0. \tag{6.140}$$

Now for any tensorial quantities A_r we can always express the Lie derivative in the form

$$\vec{\xi}\,\pounds\,A_r \equiv \xi^a \nabla_a A_r + A_{r\ a}^{\ b}\nabla_b \xi^a \tag{6.141}$$

where the coefficients $A_{r\ a}^{\ b}$ are appropriate tensorial components with *homogeneous linear algebraic* dependence on the A_r. It follows that the terms in the identity (6.140) all have homogeneous linear dependence on $\vec{\xi}$ and its first and second derivatives. Moreover, since $\vec{\xi}$ is arbitrary, it follows that the quantities ξ^a, $\nabla_b \xi^a$, $\nabla_{(c}\nabla_{b)}\xi^a$ can be chosen *independently* at any given point, which implies that the coefficients of each of these quantities must vanish separately. The last of these conditions gives the identity

$$S_F^{(cb)}{}_a \equiv 0 \tag{6.142}$$

where we make the abbreviation

$$S_F^{cb}{}_a \equiv D_F^{rc} A_{r\ a}^{\ b} \tag{6.143}$$

for what may be described as the *spin density* tensor. Setting the coefficient of $\nabla_a \xi^b$ equal to zero gives

$$T_F^{\ b}{}_a \equiv \mathscr{T}_F^{\ b}{}_a - \nabla_c S_F^{cb}{}_a + j_F^r A_{r\ a}^{\ b} \tag{6.144}$$

where the *canonical* energy–momentum tensor, $\mathscr{T}_F$, is defined by

$$\mathscr{T}_F^{\ b}{}_a \equiv -D_F^{rb}\nabla_a A_r + \Lambda_F g^b{}_a. \tag{6.145}$$

Finally, setting the coefficient of ξ^a equal to zero gives

$$\nabla_b \mathscr{T}_F^{\ b}{}_a \equiv -j_F^r \nabla_a A_r + \tfrac{1}{2}R^d{}_{abc}S^{bc}{}_d. \tag{6.146}$$

If we use (6.144) to convert this into terms of the metric energy–momentum tensor, the Riemann tensor term goes out and we are left with the divergence law

$$\nabla_b T_F^{\ b}{}_a \equiv \nabla_b(j_F^r A_{r\ a}^{\ b}) - j_F^r \nabla_a A_r. \tag{6.147}$$

The canonical energy–momentum *flux* vector is clearly related to the canonical energy–momentum *tensor* by

$$\mathscr{P}_F^a(\xi) \equiv -\mathscr{T}_F^{\ a}{}_b \xi^b + S_F^{ab}{}_c \nabla_b \xi^c \tag{6.148}$$

Chapter 6. Mechanical, electromagnetic and thermodynamic properties

and it can therefore be expressed in terms of the *metric* energy–momentum *flux* vector,

$$\Pi_F{}^a(\xi) \equiv -T_F{}^a{}_b \xi^b \qquad (6.149)$$

by

$$\mathcal{P}_F{}^a(\xi) = \Pi_F{}^a(\xi) + \nabla_b(S^{ab}{}_c \xi^c) + j_F{}^r A_r{}^a{}_b \xi^b. \qquad (6.150)$$

By the antisymmetry condition (6.142) it follows that their divergences are related simply by

$$\nabla_a \mathcal{P}_F{}^a(\xi) \equiv \nabla_a \Pi_F{}^a(\xi) + \nabla_a(j_F{}^r A_r{}^a{}_b \xi^b). \qquad (6.151)$$

Let us now impose the condition that the fields are *source free* in the sense that they obey the field equations obtained by treating the function Λ_F as a Lagrangian scalar, namely

$$j_F{}^r = 0. \qquad (6.152)$$

Then we see from (6.147) that the metric (but *not* in general the canonical) energy–momentum tensor will satisfy the *covariant conservation law*

$$\nabla_b T_F{}^{ob} = 0 \qquad (6.153)$$

and hence that the geometric energy–momentum flux will satisfy

$$\nabla_b \Pi_F{}^b(\xi) = -T^{ab} \nabla_a \xi_b \qquad (6.154)$$

which shows that the metric energy–momentum flux will be rigorously conserved whenever $\vec{\xi}$ is a Killing vector. It follows that the *same* applies to the canonical energy–momentum flux, since the last term in (6.150) vanishes when the field equations are satisfied. These results may be expressed briefly as the following theorem:

$$\left. \begin{array}{c} \vec{j}_F = 0 \\ \text{and} \\ \vec{\xi} \,\pounds\, \mathbf{g} = 0 \end{array} \right\} \Rightarrow \left\{ \begin{array}{c} \nabla \cdot \vec{\Pi}_F(\xi) = 0 \\ \text{and} \\ \nabla \cdot \vec{\mathcal{P}}_F(\xi) = 0. \end{array} \right. \qquad (6.155)$$

Although the canonical energy–momentum tensor is in general less useful than its metric analogue, the canonical energy–momentum *flux* is for many purposes more convenient than the metric one, particularly in situations where $\vec{\xi}$ generates a symmetry not just of the geometry but also of the fields, since in these conditions it is parallel to $\vec{\xi}$ and conserved *independently* of whether the field equations (6.152) are satisfied,

Energy and angular momentum transport

since by (6.139) we have

$$\vec{\xi} \pounds A_r = 0 \Rightarrow \vec{\mathscr{P}}_F(\xi) = -\Lambda_F \vec{\xi} \qquad (6.156)$$

while evidently by (6.140)

$$\left.\begin{array}{c} \vec{\xi} \pounds A_r = 0 \\ \text{and} \\ \vec{j} \pounds \mathbf{g} = 0 \end{array}\right\} \Rightarrow \nabla \cdot \vec{\mathscr{P}}_F(\xi) = 0. \qquad (6.157)$$

6.4.2 The angular momentum: energy transport ratio

Now let us consider the application of these results to the case where there is a stationary Killing vector, $\vec{k}$, and (or) an axisymmetry Killing vector $\vec{m}$, in terms of which we may define a canonical energy and (or) angular momentum flux vectors

$$\vec{\mathscr{E}} = \mathscr{P}_F(\vec{k}), \qquad \vec{\mathscr{J}} = -\mathscr{P}_F(\vec{m}). \qquad (6.158)$$

In view of the fact that it is the metric energy–momentum tensor that is fundamental – in the sense of being directly coupled to gravitation – it is appropriate to define the *total* mass–energy flux, ΔM, and (or) the total angular momentum flux, ΔJ, across a hypersurface Σ by

$$\Delta M = -\int_\Sigma \Pi^a(k) \, d\Sigma_a, \qquad \Delta J = \int_\Sigma \Pi^a(m) \, d\Sigma_a \qquad (6.159)$$

where any set of infinitesimal displacement vectors $d_{(1)}\vec{x}, d_{(2)}\vec{x}, d_{(3)}\vec{x}$ are considered as generating a normal surface element $d\Sigma$ defined by

$$d\Sigma = *(d_{(1)}\vec{x} \wedge d_{(2)}\vec{x} \wedge d_{(3)}\vec{x}). \qquad (6.160)$$

It follows from (6.150) by the generalized Green's theorem that when the *source-free* field equations (6.152) are satisfied we shall have contributions

$$\Delta M_F = -\int_\Sigma \mathscr{E}_F{}^a \, d\Sigma_a + \tfrac{1}{2} \oint_{\mathscr{S}=\partial\Sigma} S_F{}^{ab}{}_c k^c \, d\mathscr{S}_{ab} \qquad (6.161)$$

and

$$\Delta J_F = -\int_\Sigma \mathscr{J}_F{}^a \, d\Sigma_a + \tfrac{1}{2} \oint_{\mathscr{S}=\partial\Sigma} S_F{}^{ab}{}_c m^c \, d\mathscr{S}_{ab} \qquad (6.162)$$

where $\mathscr{S}$ is the boundary of Σ and the surface element $d\mathscr{S}$ is defined in

Chapter 6. Mechanical, electromagnetic and thermodynamic properties

terms of corresponding infinitesimal displacements $d_{(2)}\vec{x}$, $d_{(3)}\vec{x}$ by

$$d\mathscr{S} = *(d_{(2)}\vec{x} \wedge d_{(3)}\vec{x}). \qquad (6.163)$$

It is evident that the 2-surface integral terms will drop out if we are considering only the effect of a finite wave packet and Σ is taken sufficiently large for its boundary to lie everywhere outside the packet. Another interesting situation is the one where we consider a wave that is periodic with respect to the ignorable coordinate time, t, as characterized by the condition

$$\vec{k} \cdot \nabla t = 1 \qquad (6.164)$$

and where we consider a 3-surface Σ generated by $\vec{k}$ with *compact* space sections bounded by a pair of *congruent* 2-surfaces $\mathscr{S}_1$ and $\mathscr{S}_2$ on which t takes two different constant values, t_1 and t_2 say. Then, provided the difference $t_1 - t_2$ is an *integral multiple*, n, say, *of the period* $2\pi/\omega$, the contributions of $\mathscr{S}_1$ and $\mathscr{S}_2$ to the integral over the entire boundary $\mathscr{S} = \mathscr{S}_1 \cup \mathscr{S}_2$ will *cancel*, so that we obtain

$$\Delta M_\mathrm{F} = -\int_\Sigma \mathscr{S}_\mathrm{F}{}^a \, d\Sigma_a = -\int_{t_1}^{t_1+(2n\pi/\omega)} dt \int_{\mathscr{S}(t)} \mathscr{E}_\mathrm{F}{}^a k^b \, d\mathscr{S}_{ab} \qquad (6.165)$$

and analogously for ΔJ_F, where $\mathscr{S}(t)$ is a time-dependent, compact, two-dimensional section and where we have used the substitution $d_{(1)}\vec{x} = \vec{k} \, dt$ in (6.160) to obtain

$$d\Sigma_a = (k^b \, dt) \, d\mathscr{S}_{ab}. \qquad (6.166)$$

Hence we deduce that the *coordinate time-averaged* energy and angular momentum *flux rates* will be given by

$$\langle \dot{M}_\mathrm{F} \rangle = -\left\langle \int_\mathscr{S} \mathscr{E}_\mathrm{F}{}^a k^b \, d\mathscr{S}_{ab} \right\rangle \qquad (6.167)$$

$$\langle \dot{J}_\mathrm{F} \rangle = -\left\langle \int_\mathscr{S} \mathscr{J}_\mathrm{F}{}^a k^b \, d\mathscr{S}_{ab} \right\rangle. \qquad (6.168)$$

A general field will of course depend on both the ignorable time coordinate, t, and the ignorable periodic angle coordinate, φ, as characterized by

$$\vec{m} \cdot \nabla \varphi = 1 \qquad (6.169)$$

which can be made compatible with t (in the sense that both are simultaneously ignorable) provided $\vec{k}$ and $\vec{m}$ satisfy the commutation condition (6.92) which as we have remarked will necessarily be the case under conditions of asymptotic flatness. However an interesting special case is that of a field which is independent of some linear combination, $\varphi - \Omega^F t$ say, of the geometrically ignorable coordinates, that is to say, in more invariant terminology, a field for which there is some Killing vector combination

$$\vec{l}^F = \vec{k} + \Omega^F \vec{m} \qquad (6.170)$$

such that

$$\vec{l}^F \pounds A_r = 0. \qquad (6.171)$$

Under these circumstances – in which it is obviously appropriate to describe the field as *rigidly rotating* with *constant angular velocity* Ω^F – the theorem (6.157) clearly implies that we shall have

$$\vec{\mathscr{E}}_F - \Omega^F \vec{\mathscr{J}}_F = -\Lambda_F \vec{l}^F \qquad (6.172)$$

and hence that

$$\int_{\mathscr{S}} \mathscr{E}_F{}^a l^{Fb} \, d\mathscr{S}_{ab} = \Omega^F \int_{\mathscr{S}} \mathscr{J}_F{}^a l^{Fb} \, d\mathscr{S}_{ab} \qquad (6.173)$$

for any 2-surfaces $\mathscr{S}$. If now we restrict our attention to 2-surfaces that have *circular* sections in the sense that they are invariant under the axisymmetry action generated by $\vec{m}$, i.e. 2-surfaces to which $\vec{m}$ is a tangent vector so that $m^a \, d\mathscr{S}_{ab}$ will vanish, then the right-hand sides of (6.167) and (6.168) will be unaffected by replacing $\vec{k}$ by $\vec{l}^F$ so that (6.171), will evidently imply

$$\langle \dot{M}_F \rangle = \Omega^F \langle \dot{J}_F \rangle. \qquad (6.174)$$

Since the field is necessarily periodic in φ, the fact that it is also assumed to be periodic in t with fundamental angular frequency ω implies that it can be expressed in the form

$$A_r = \sum_{n,m} \tilde{A}_{rnm} \, e^{im\varphi - in\omega t} \qquad (6.175)$$

for some finite or infinite set of integers m and n, where $\tilde{A}_{rnm}$ depend only on the non-ignorable coordinates (r and θ say), i.e. in more invariant language they are characterized by

$$\vec{k} \pounds \tilde{A}_{rnm} = \vec{m} \pounds \tilde{A}_{rnm} = 0 \qquad (6.176)$$

so that we have

$$i\vec{k} \pounds A_r = \omega \sum_{n,m} n\tilde{A}_{rnm} \, e^{im\varphi - in\omega t}$$

$$i\vec{m} \pounds A_r = -\sum_{n,m} m\tilde{A}_{rnm} \, e^{im\varphi - im\omega t} \, . \qquad (6.177)$$

Chapter 6. Mechanical, electromagnetic and thermodynamic properties

Evidently the invariance condition (6.170) will be satisfied if and only if there is some *fixed* integer m such that

$$\omega = \Omega^F m \qquad (6.178)$$

and such that $A_{mm'}$ is zero except when $m' = nm$, so that (6.175) can be reduced to the form

$$A_r = \sum_n A_{rn} \, e^{in(m\varphi - \omega t)}. \qquad (6.179)$$

In particular it is evident that there will *always* be an appropriate angular velocity Ω^F satisfying (6.178) for any *pure mode*, i.e. one that is simply harmonic in both t and φ with corresponding quantum numbers ω and m. The foregoing work thus provides a rigorous basis for the oft-quoted but seldom rigorously justified assertion that the ratio of energy to angular momentum transport is ω/m, since for any wave of the form (6.179) the equations (6.174) and (6.178) imply that the *average* flux rates of energy and angular momentum will satisfy

$$m \langle \dot{M}_F \rangle = \omega \langle \dot{J}_F \rangle \qquad (6.180)$$

whenever the *source-free* field equations are satisfied. In particular the formula (6.180) may be applied to a 2-sphere at asymptotically large distance – in which case $\dot{M}$ and $\dot{J}$ would measure rates of influx 'from infinity'. They can also be applied on the horizon of an axisymmetric black hole – in which case they would measure rates of flux into the hole. Now it follows from (6.161) and (6.162) that the total flux rates (with respect to the stationary group parameter, t) on the horizon will be given *exactly* by

$$\dot{M} = \int_{\mathscr{S}} T^a{}_c k^c l^b \, d\mathscr{S}_{ab} \qquad (6.181)$$

$$\dot{J} = -\int_{\mathscr{S}} T^a{}_c m^c l^b \, d\mathscr{S}_{ab} \qquad (6.182)$$

where $\vec{l}$ is the null Killing vector on the horizon that was introduced in the previous section. (Actually since $\vec{m}$ is known to be tangent to the horizon we could replace $\vec{l}$ indifferently by $\vec{l}^F$ – when this latter is defined – or for that matter by $\vec{k}$ without affecting the validity of (6.182).) Since we may write

$$d\mathscr{S}_{ab} = 2\beta_{[a} l_{b]} \, d\mathscr{S} \qquad (6.183)$$

for any 2-surface element on the horizon (with β as introduced by

Energy and angular momentum transport

(6.27)) these formulae may also be expressed as

$$\dot{M} = \int_{\mathscr{S}} T^a{}_b k^b l_a \, \mathrm{d}\mathscr{S} \qquad (6.184)$$

$$\dot{J} = -\int_{\mathscr{S}} T^a{}_b m^b l_a \, \mathrm{d}\mathscr{S} \qquad (6.185)$$

which implies that the contribution of the field to the 'dissipation rate' on the horizon – as evaluated in terms of the analogy revealed by the Hartle–Hawking equation (6.67) – will be given by

$$(\kappa/8\pi)\dot{\mathscr{A}} = \int_{\mathscr{S}} T_{ab} l^a l^b \, \mathrm{d}\mathscr{S} = \dot{M} - \Omega^H \dot{J}. \qquad (6.186)$$

Hence for a pure vacuum wave of the form (6.179) the time-averaged rate of contribution to the dissipation will be given by the equivalent expressions

$$(\kappa/8\pi)\langle\dot{\mathscr{A}}_F\rangle = \langle\dot{M}_F\rangle - \Omega^H \langle\dot{J}_F\rangle$$
$$= (\Omega^F - \Omega^H)\langle\dot{J}_F\rangle$$
$$= (1 - (m/\omega)\Omega^H)\langle\dot{M}_F\rangle. \qquad (6.187)$$

Now for any fields obeying the energy axiom described in the previous section, the left-hand side is non-negative. It follows that in the non-rotating case, i.e. when Ω^H is zero, $\langle \dot{M}_F \rangle$ must also be non-negative, i.e. the hole can only accrete energy. However if the hole is *rotating*, let us say in a sense such that Ω^H is positive, then there will be a subset of modes characterized by

$$0 < (\omega/m) = \Omega^F < \Omega^H \qquad (6.188)$$

for which $\langle \dot{M}_F \rangle$ will be negative, so that there will be a net *outflux* of energy from the hole, i.e. incoming radiation from infinity is *amplified* as it is reflected by the black hole. This is the phenomenon that is now well known by the name of *super-radiance* to which attention was first drawn by Misner (1972), Press and Teukolsky (1972), and Zel'dovich (1972). Its possibility provided the first suggestion that there could be spontaneous creation of particles by the corresponding quantum fields, but the subsequent discovery of the Hawking effect has rather diminished its significance in this context because it is now predicted that there should be spontaneous creation in all modes, not just the super-radiant ones.

The relation (6.187) can also be used to draw other conclusions. By the same argument that was used to derive (6.181) (or by direct reading

Chapter 6. Mechanical, electromagnetic and thermodynamic properties

off from (6.178)) we can see that we have

$$\left.\begin{array}{l}\vec{k} \pounds A_r = 0 \\ \text{i.e. } \omega = 0\end{array}\right\} \Rightarrow \left\{\begin{array}{l}\langle \dot{M} \rangle = 0 \\ \text{and } (\kappa/8\pi)\langle \dot{\mathscr{A}} \rangle = -\Omega^H \langle \dot{J} \rangle\end{array}\right. \quad (6.189)$$

and similarly

$$\left.\begin{array}{l}\vec{m} \pounds A_r = 0 \\ \text{i.e. } m = 0\end{array}\right\} \Rightarrow \left\{\begin{array}{l}\langle \dot{J} \rangle = 0 \\ \text{and } (\kappa/8\pi)\langle \dot{\mathscr{A}} \rangle = \langle \dot{M} \rangle\end{array}\right. \quad (6.190)$$

when the *source-free* field equation (6.152) holds. Since as, we have already remarked, the energy axiom implies that the integrand of (6.186) is non-negative, we can deduce that it will in fact be zero, i.e. we must satisfy the boundary conditions

$$T_F{}^{ab} l_a l_b = 0 \quad (6.191)$$

on the horizon whenever the A_r are themselves *both stationary and axisymmetric* (in the sense that the conditions (6.189) and (6.190) are satisfied) and *provided* that the source-free field equations are satisfied. (We shall consider what happens for a stationary axisymmetric field when source currents are present later on in this section.) We can also deduce that (6.191) will hold for a *non-rotating* black hole whenever the A_r are *static* in the weak sense of satisfying just the condition (6.189) whether or not the axisymmetry condition (6.190) is satisfied (and whether or not the background space is static in the strict sense referred to in section 6.7.1). It is to be noticed that the derivation of (6.191) obtained in this way is quite different from the derivation of (6.86) in the previous section, in so much as it does *not* depend on the gravitational field equations (whether Einstein's or those of any alternative theory) that may or may not be satisfied by the background space. Nevertheless (6.189) provides some physical intuition into the reasons underlying Hawking's strong rigidity theorem (i.e. the deduction that a stationary black hole state must be either non-rotating or axisymmetric).

6.5 Electromagnetic effects in a black hole background space

6.5.1 Fluxes and boundary conditions

The results described in the previous section are valid for quite general sets of scalar or tensorial fields obeying field equations derived from a covariant Lagrangian variation principle. The electromagnetic case with which we are primarily concerned is obtained simply by treating r as an

Electromagnetic effects

ordinary covariant coordinate index, so that A_r are interpreted as coordinate components of a 1-form $\boldsymbol{A}$ in terms of which the electromagnetic 2-form is constructed in the standard manner by taking the exterior derivative, i.e.

$$\boldsymbol{F} = \partial \boldsymbol{A}. \tag{6.192}$$

The standard Lagrangian scalar for the electromagnetic field is given by

$$\Lambda_F = (1/16\pi) F_{ab} F^{ba} \tag{6.193}$$

and hence by (6.136) the corresponding displacement tensor is obtained as

$$D_F{}^{ab} = (1/4\pi) F^{ab} \tag{6.194}$$

so that by (6.137) we see that the corresponding field equations (with allowance for the possible presence of a source current) are given by

$$\nabla_b D_F{}^{ab} = j_F{}^a. \tag{6.195}$$

By comparing (6.141) with the standard expression for the Lie derivative of a covector we see that in the electromagnetic case we shall have explicitly

$$A_c{}^b{}_a = A_a g^b{}_c \tag{6.196}$$

and hence that the spin-density tensor will be given by

$$S_F{}^{cb}{}_a = (1/4\pi) A_a F^{bc}. \tag{6.197}$$

The geometric energy–momentum tensor will of course be given by the Maxwellian expression (6.70) that we have already utilized and the corresponding conservation identity (6.147) can be reduced to

$$\nabla_b T_F{}^b{}_a = j_F{}^b F_{ba} \tag{6.198}$$

using the source-conservation law

$$\nabla_b j_F{}^b = 0 \tag{6.199}$$

obtainable from (6.195).

The geometric energy flux vector $\vec{\Pi}_F(k)$ is related to the *conserved* canonical energy flux vector of the field, $\vec{\mathscr{E}}_F$, by

$$\Pi_F{}^a(k) = \mathscr{E}_F{}^a + \nabla_b(F^{ab} A_c k^c) - j_F{}^a A_c k^c. \tag{6.200}$$

It can be seen from the explicit forms of these vectors (using the antisymmetry of $\boldsymbol{F}$) that they satisfy the useful orthogonality conditions

$$(\vec{k} \cdot \boldsymbol{F}) \cdot \vec{\Pi}_F(k) = 0 = (\vec{m} \cdot \boldsymbol{F}) \cdot \vec{\Pi}_F(m). \tag{6.201}$$

Chapter 6. Mechanical, electromagnetic and thermodynamic properties

However, when source currents are present the gauge-invariant geometric energy and angular momentum fluxes (unlike their gauge-dependent canonical counterparts) *need not* be separately conserved but satisfy

$$\nabla \cdot \vec{\Pi}_F(k) = -j_F \cdot \nabla A_t$$
$$\nabla \cdot \vec{\Pi}_F(m) = -j_F \cdot \nabla A_\varphi \qquad (6.202)$$

where we have introduced the abbreviations

$$A_t = \vec{k} \cdot \boldsymbol{A}, \qquad A_\varphi = \vec{m} \cdot \boldsymbol{A} \qquad (6.203)$$

for the pair of scalars that are in fact the same as the ignorable coordinate components of the vector potential in a system characterized by the conditions (6.164) and (6.169).

It can now be seen that if the field itself satisfies the stationarity condition (6.189), which is equivalent to

$$\vec{k} \pounds \boldsymbol{A} = \boldsymbol{F} \cdot \vec{k} - \nabla A_t = 0, \qquad (6.204)$$

then we shall have

$$\vec{\Pi}(k) \cdot \nabla A_t = 0 \qquad (6.205)$$

and similarly with $\vec{k}$ replaced by $\vec{m}$. This shows that (as noticed by Damour, 1975) the geometric energy or angular momentum flux vectors lie on the hypersurfaces on which respectively A_t or A_φ is constant in any respectively stationary or axisymmetric situation.

Let us now look more closely at the behaviour of the electromagnetic field on the horizon. We have already had occasion to refer to the natural electric field vector $\vec{E}^H$ which is expressible on the horizon by

$$\boldsymbol{E}_{(0)} = \boldsymbol{F} \cdot \vec{l} \to \boldsymbol{E}^H \qquad (6.206)$$

where the (improper) corotating electric field, $\boldsymbol{E}_{(0)}$ satisfies (6.103) with the potential given by

$$\Phi_{(0)} = \boldsymbol{A} \cdot \vec{l} = A_t + \Omega^H A_\varphi. \qquad (6.207)$$

(The *proper* corotating electric field, as defined with respect to the rest-frame determined by $\vec{l}$, would be obtained by substituting the unit vector $\vec{u}$ given by (6.119) in place of $\vec{l}$ in (6.206), and hence it would differ from the improper electric field by the mean redshift factor $\dot{\tau}^{-1}$ which diverges on the horizon.) The analogous *magnetic* field $\vec{B}^H$ on the horizon as given by

$$\vec{B}_{(0)} = (*\boldsymbol{F}) \cdot \vec{l} \to \boldsymbol{B}^H \qquad (6.208)$$

Electromagnetic effects

will automatically satisfy the relations

$$|\vec{B}^H|^2 = |\vec{E}^H|^2 \tag{6.209}$$

$$\vec{B}^H \cdot \vec{E}^H = 0 \tag{6.210}$$

as a direct algebraic consequence of the fact that l is null on the horizon, and it is obvious that both $\vec{E}^H$ and $\vec{B}_H$ must lie in the horizon, since by the antisymmetry of F and $*F$ they evidently satisfy

$$\vec{E}^H \cdot l = 0 = \vec{B}^H \cdot l. \tag{6.211}$$

To see how the less obvious properties (6.209), (6.210) come about we introduce an orthogonal basis of unit vectors $\vec{e}_{(i)}$ on the horizon ($(i) = 2, 3$) such that the orthogonal projection tensor introduced in equation (6.33) takes the form

$$\gamma^a{}_b = e_{(2)}{}^a e_{(2)b} + e_{(3)}{}^a e_{(3)b} \tag{6.212}$$

which clearly requires

$$\vec{e}_{(i)} \cdot e_{(j)} = \delta_{ij}, \qquad \vec{e}^{(i)} \cdot l = 0 = \vec{e}_{(i)} \cdot \boldsymbol{\beta}. \tag{6.213}$$

The fact that $\vec{l}$ is null allows us (subject to an appropriate choice of orientation) to express its adjoint in the form

$$*\vec{l} = l \wedge e_{(2)} \wedge e_{(3)} \tag{6.214}$$

and hence to derive the expression

$$B^H{}_a = 3 l_{[a} e_{(2)b} e_{(3)c]} F^{bc} \tag{6.215}$$

for the magnetic field components. Using the obvious notation

$$B^H{}_{(i)} = \vec{e}_{(i)} \cdot \boldsymbol{B}^H, \qquad E^H{}_{(i)} = \vec{e}_{(i)} \cdot \boldsymbol{E}^H \tag{6.216}$$

we can express the result of contracting (6.215) with the basis vectors by the relations

$$B^H{}_{(3)} = E^H{}_{(2)}, \qquad B^H{}_{(2)} = -E^H{}_{(3)}. \tag{6.217}$$

If we bear in mind (6.211), the relations (6.209), (6.210) will now be obvious, and we shall also have

$$|\vec{E}^H|^2 = (E^H{}_{(1)})^2 + (E^H{}_{(2)})^2 \tag{6.218}$$

and similarly for $\vec{B}^H$.

6.5.2 The surface current in the horizon

It has recently been pointed out by Damour (1978) that in addition to the electric and magnetic field vectors that have just been described, a

Chapter 6. Mechanical, electromagnetic and thermodynamic properties

black hole can also be thought of as having a certain naturally defined *effective surface current 4-vector*, $\vec{\sigma}_H$, to which there corresponds a naturally defined (but fictitious) 4-current density distribution,

$$\vec{j}_H = \vec{\sigma}_H \delta(\lambda), \tag{6.219}$$

where δ denotes a Dirac distribution and λ is any function that is positive everywhere within but nowhere outside the domain of exterior communications, and zero on the horizon $\mathcal{H}^+$, with its gradient adjusted in magnitude so as to satisfy the requirement

$$\vec{\beta} \cdot \partial \lambda = -1. \tag{6.220}$$

Damour's fictitious current distribution is characterized first and foremost by the requirement that it should 'complete the circuit' formed by the real currents outside the black hole, so that if the currents within the black hole are imagined to be suppressed and replaced by $\vec{j}_H$ the total field will still be conserved, i.e. we shall have

$$\nabla \cdot (\vec{j}_H + \vec{j}_F H(\lambda)) = 0, \tag{6.221}$$

where H is the Heaviside function, whose derivative is the Dirac distribution, which implies that it satisfies

$$\partial H(\lambda) = \delta(\lambda) \partial \lambda. \tag{6.222}$$

The standard solution to this problem – which is a familiar one in other electrodynamic contexts – is to take

$$j_H{}^a = D_F{}^{ab} \delta(\lambda) \partial_b \lambda \tag{6.223}$$

(where the vacuum displacement tensor is given by (6.194)). This solution is not unique (obviously (6.221) would still be satisfied if the distribution (6.223) were augmented by an arbitrary divergence-free vector field) but it is distinguished from other non-standard solutions by the property that it correctly gives the apparent surface charge density that is derived from the application of *Gauss's law*, according to which the *total charge* Q^H enclosed within a boundary 2-surface, $\mathcal{S}$, is given (using the normalization conventions of (6.163) and (6.183)) by

$$Q^H = -\tfrac{1}{2} \oint_{\mathcal{S}} D_F{}^{ab} \, \mathrm{d}\mathcal{S}_{ab}. \tag{6.224}$$

This can be seen (with the aid of (6.183)) to be expressible in terms of an effective surface charge density $\sigma_F^{(0)}$ in the form

$$Q^H = \int \sigma_F^{(0)} \mathrm{d}\mathcal{S} \tag{6.225}$$

Electromagnetic effects

where (for any pair of null normals satisfying (6.27)) $\sigma_F^{(0)}$ is given by

$$\sigma_F^{(0)} = -\beta_a D_F{}^{ab} l_b \tag{6.226}$$

which is *precisely the same* as the effective surface charge density that would be derived by considering the flux of the current (6.233) across an element of hypersurface Σ cutting across the discontinuity hypersurface $\lambda = 0$. This can be demonstrated conveniently by arranging for Σ to contain $\vec{\beta}$ as a tangent vector where it meets the discontinuity surface, so that the tangent element (6.217) may be taken to have the particular form

$$-d\boldsymbol{\Sigma} = *(\vec{e}_{(2)} \wedge \vec{e}_{(3)} \wedge \vec{\beta}) \, d\mathcal{S} \, d\lambda \tag{6.227}$$

where $\vec{e}_{(2)}$ and $\vec{e}_{(3)}$ are mutually orthogonal unit vectors within the common intersection 2-surface $\mathcal{S}$. Using the fact that $\vec{\beta}$ is null as well as orthogonal to $\vec{e}_{(2)}$ and $\vec{e}_{(3)}$, we obtain

$$*(\vec{e}_{(2)} \wedge \vec{e}_{(3)} \wedge \vec{\beta}) = -\boldsymbol{\beta} \tag{6.228}$$

and hence

$$-\int_\Sigma j_H{}^a \, d\Sigma_a = -\iint_\Sigma j_H{}^a \beta_a \, d\mathcal{S} \, d\lambda = \int_\mathcal{S} \sigma_F^{(0)} d\mathcal{S} \tag{6.229}$$

where

$$\sigma_F^{(0)} = -\beta_a D_F{}^{ab} \partial_b \lambda. \tag{6.230}$$

The agreement between the two expressions (6.226) and (6.230) for $\sigma_F^{(0)}$ follows from the fact that since l and $\partial\lambda$ are both orthogonal to the 2-surface $\mathcal{S}$ (on which Σ intersects the surface, $\lambda = 0$) and since they both satisfy (6.220) their difference – if any – must be parallel to the null form $\boldsymbol{\beta}$.

The procedure that we have just described is completely standard in the case where the surface $\lambda = 0$ is a *timelike one*, which might represent the boundary of an ordinary conducting medium. However, in a case such as that of the black hole horizon with which we are concerned here, for which the hypersurface $\lambda = 0$ is *null*, which by (6.220) implies that we actually have an equality,

$$\partial\lambda = l, \tag{6.231}$$

an ambiguity arises when we wish to go on from the definition (6.223) of the *distribution* $\vec{j}_H$ to the definition of a corresponding *surface* current

Chapter 6. Mechanical, electromagnetic and thermodynamic properties

density $\vec{\sigma}_F$, since a rescaling of the form $\lambda \to \alpha \lambda$ would have no effect on the definition (6.223) but *would* affect the value of σ_F that would be read out from (6.219). In the timelike case the relation between the current distribution and the corresponding current density is customarily fixed by normalizing λ so that its gradient has unit length, which makes it possible to interpret the 4-vector $\vec{\sigma}_F$ read out from (6.219) as representing the rate of transport of charge per unit proper time. In the null case there is no unambiguous way of fixing the normalization by purely *local* considerations – which would allow the null vector l appearing in (6.231) to be rescaled at will – so that although the distribution $\vec{j}_H$ would be perfectly well defined even for an *expanding* black hole of the kind described in section 6.2, the surface density current $\vec{\sigma}_F$ would not. In the *stationary* case we have seen however that $\vec{l}$ is canonically determined as a Killing vector with normalization fixed by asymptotic boundary conditions, and in this case the prescription (6.220), which is equivalent by (6.231) to

$$\sigma_F{}^a = D_F{}^{ab} l_b, \qquad (6.232)$$

does in fact give a uniquely determined result which evidently obeys an Ohm-type law of the form

$$\vec{E}^H = 4\pi \vec{\sigma}_F \qquad (6.233)$$

where the surface resistivity of 4π agrees with the conductivity formula that was deduced in section 6.2.2. This surface-current vector can be interpreted as representing the rate of transport of charge per unit *not* of proper time (which does not exist in a null hypersurface) but of the canonical *group parameter*, t. To see this, consider the flux across a hypersurface, Σ, which intersects the horizon, $\mathcal{H}^+$, in a 2-surface formed by the null generators passing through a curve, c. Then if we choose the basis vector, $\vec{e}_{(3)}$, to be tangent to c we may evaluate the expression (6.217) for a surface element where Σ intersects the horizon by setting

$$d\vec{x}_{(1)} = \vec{\beta} \, d\lambda, \qquad d\vec{x}_{(2)} = \vec{l} \, dt, \qquad d\vec{x}_{(3)} = e_{(3)} \, ds \qquad (6.234)$$

where ds is the metric length of the spacelike element $d\vec{x}_{(3)}$. Since we have

$$*(\vec{\beta} \wedge \vec{l} \wedge \vec{e}_{(3)}) = e_{(2)} \qquad (6.235)$$

we obtain

$$\int j_H{}^a \, d\Sigma_a = \iiint \vec{\sigma}_F \cdot e_{(2)} \, \delta(\lambda) \, d\lambda \, ds \, dt$$
$$= \int dt \int_c ds \, \sigma_F{}^{(2)} \qquad (6.236)$$

using the notation

$$\sigma_F^{(i)} = \vec{\sigma}_F \cdot \vec{e}_{(i)} \quad (6.237)$$

so that we have

$$\vec{\sigma}_F = \sigma_F^{(0)}\vec{l} + \sigma_F^{(2)}\vec{e}_{(2)} + \sigma_F^{(3)}\vec{e}_{(3)}. \quad (6.238)$$

The expression (6.236) shows that $\sigma^{(2)}$ represents the rate with respect to τ of transport across unit length in the direction of $\vec{e}_{(3)}$ within the horizon relative to the frame determined by the null generators (and similarly $\sigma^{(3)}$ gives the rate of transport across a unit length in the direction of $\vec{e}_{(2)}$). We see that the electromagnetic contribution to the 'dissipation rate' over the horizon will be given by

$$(\kappa/8\pi)\dot{\mathcal{A}} = \int T_F{}^{ab} l_a l_b \, \mathrm{d}\mathcal{S} = \int \vec{\sigma}_F \cdot \boldsymbol{E}^H \, \mathrm{d}\mathcal{S}. \quad (6.239)$$

Let us now consider the case when the black hole is not only stationary but also axisymmetric, and let us take $\vec{e}_{(3)}$ parallel to the axial Killing vector, so that we have

$$\vec{m} = |\vec{m}|\vec{e}_{(3)} \quad (6.240)$$

where the magnitude $|\vec{m}|$ represents the circumferential radius of the black hole at the latitude under consideration, which is obtainable in terms of metric components in any system characterized by (6.114), from the formula

$$|\vec{m}|^2 = g_{\varphi\varphi}. \quad (6.241)$$

When the basis vector $\vec{e}_{(3)}$ is fixed by (6.240), the component $\sigma^{(2)}$ represents the *azimuthal* current density in the horizon, which can be interpreted as the rate with respect to t of transport of charge across unit length of a line of latitude in the horizon with respect either to the corotating frame determined by $\vec{l}$ and *also* with respect to the fixed (non-rotating) frame determined by $\vec{k}$ (since with $\vec{e}_{(3)}$ given by (6.240) replacement of $\vec{l}$ by $\vec{k}$ has no effect on (6.235)). In this sense the *azimuthal* component, $\sigma^{(2)}$, has a more absolute significance than the axial surface current component, $\sigma^{(3)}$. The concept of an effect azimuthal surface current on the horizon plays a key role in the more specialized analysis given by Znajek (1978) for the case when the electromagnetic field, F, is itself stationary and axisymmetric. Throughout the remainder of this section we shall restrict attention to the case when the electromagnetic field stationarity and axisymmetry conditions ((6.189) and (6.190)) are in fact satisfied, in which case it follows immediately

Chapter 6. Mechanical, electromagnetic and thermodynamic properties

(Carter, 1969) from the commutation condition (6.92) that the axial component

$$E_{(0)\varphi} \equiv \vec{m} \cdot \boldsymbol{F} \cdot \vec{l} = \vec{m} \cdot \boldsymbol{F} \cdot \vec{k} \tag{6.242}$$

of the electric field $\boldsymbol{E}_{(0)}$ defined with respect to the corotating Killing vector $\vec{l}$ by (6.206) will satisfy

$$E_{(0)\varphi} = 0. \tag{6.243}$$

This may be shown easily by repeated use of the Cartan formula (6.20) which gives

$$\partial(\vec{m} \cdot \boldsymbol{F} \cdot \vec{k}) \equiv (\vec{m} \pounds \boldsymbol{F}) \cdot \vec{k} + \boldsymbol{F} \cdot (\vec{h} \pounds \vec{k}) + \vec{m} \cdot (\vec{k} \pounds \boldsymbol{F} - \vec{k} \cdot \partial \boldsymbol{F}) = 0 \tag{6.244}$$

since all the Lie derivatives are zero by the assumed symmetry conditions and $\partial \boldsymbol{F}$ is zero by Maxwell's equation. Thus one sees that $\vec{m} \cdot \boldsymbol{F} \cdot \vec{k}$ must be uniform through spacetime, and hence – since it evidently vanishes on the axis of symmetry where the axisymmetry Killing vector $\vec{m}$ itself is zero – that it is zero everywhere in accordance with (6.243). Since $\boldsymbol{E}_{(0)}$ tends to $\boldsymbol{E}^{\mathrm{H}}$ on the horizon, it can be seen by (6.216), (6.217) that we must therefore have

$$E^{\mathrm{H}}_{(3)} = 0 = B^{\mathrm{H}}_{(2)} \tag{6.245}$$

there in the stationary axisymmetric case.

Let us now see what can be deduced by applying an exactly analogous line of reasoning to the adjoint form $*\vec{F}$ with components given by

$$*F_{ab} = \tfrac{1}{2}\varepsilon_{abcd}F^{cd}. \tag{6.246}$$

Since the second Maxwell equation (6.107) can be written in terms of purely exterior derivatives as

$$\partial(*\vec{F}) = 4\pi * \vec{j}_{\mathrm{F}}, \tag{6.247}$$

where $*\vec{j}_{\mathrm{F}}$ is the 3-form with components

$$(*\vec{j}_{\mathrm{F}})_{abc} = \varepsilon_{abcd} j_{\mathrm{F}}^{\ d}, \tag{6.248}$$

one sees that in the *source-free* case one can again draw the conclusion (Carter, 1969) that the axial component

$$B_{(0)\varphi} \equiv \vec{m} \cdot (*\vec{F}) \cdot \vec{l} = \vec{m} \cdot (*\vec{F}) \cdot \vec{k} \tag{6.249}$$

of the corresponding field. $\boldsymbol{B}_{(0)}$ as given by (6.208) will have to be zero everywhere. However, when source currents are present $B_{(0)\varphi}$ need not vanish, since substitution of $*\vec{F}$ in place of $\boldsymbol{F}$ in (6.244) leads only to the weaker restriction that

$$\partial B_{(0)\varphi} = 4\pi \vec{k} \cdot (*\vec{j}_F) \cdot \vec{m}$$
$$= 4\pi *(\vec{m} \wedge \vec{k} \wedge \vec{j}_F). \qquad (6.250)$$

This relation allows us (following Blandford and Znajek, 1977) to make the deduction that the scalar $B_{(0)\varphi}$ is proportional to the total *current*, I, passing through any axisymmetric 2-surface $\mathscr{S}$ (i.e. any 2-surface generated by $\vec{m}$) such that its boundary $\partial\mathscr{S}$ consists of a circle passing through the point at which $B_{(0)\varphi}$ is evaluated, the current being defined as the *total charge* flux per unit of the globally defined *ignorable time* coordinate, t, across the invariant hypersurface (i.e. the stationary axisymmetric hypersurface generated by both $\vec{k}$ and $\vec{m}$) passing through $\mathscr{S}$. To show this we start by remarking that (with the aid of (6.166)) this current I can be expressed as

$$I = -\int_{\mathscr{S}} j^a k^b \, d\mathscr{S}_{ab}$$
$$= \int_{\mathscr{S}} *(d_{(2)}x \wedge d_{(3)}\vec{x} \wedge \vec{k} \wedge \vec{j}_F) \qquad (6.251)$$

where the surface element $d\mathscr{S}$ is given by (6.163). Since $\mathscr{S}$ is generated by the axisymmetry Killing vector, $\vec{m}$, it is possible to substitute

$$d_{(3)}\vec{x} = \vec{m} \, d\varphi, \qquad (6.252)$$

where φ is the axisymmetry group parameter, and hence to obtain

$$I = \int_{\mathscr{S}} d\varphi *(d_{(2)}\vec{x} \wedge \vec{m} \wedge \vec{k} \wedge \vec{j}_F)$$
$$= 2\pi \int_c *(d\vec{x} \wedge \vec{m} \wedge \vec{k} \wedge \vec{j}_F) \qquad (6.253)$$

where the last integral is taken over any curve c in $\mathscr{S}$ starting on the axis of symmetry and ending on a point of the circular boundary $\partial\mathscr{S}$, in such a way that the 2-surface $\mathscr{S}$ may be thought of as generated by sweeping the curve c round its endpoint on the axis under the action of the axisymmetry group. Now the formula (6.250) shows that $dB_{(0)\varphi}$, the infinitesimal change of $B^{(0)}{}_\varphi$ associated with a tangent vector $d\vec{x}$ along c, is expressible by

$$dB_{(0)\varphi} = d\vec{x} \cdot \partial B_{(0)\varphi}$$
$$= 4\pi *(d\vec{x} \wedge \vec{m} \wedge \vec{k} \wedge \vec{j}_F) \qquad (6.254)$$

Chapter 6. Mechanical, electromagnetic and thermodynamic properties

so that (6.253) reduces to

$$I = \tfrac{1}{2} \int_c \mathrm{d}B_{(0)\varphi}. \tag{6.255}$$

Hence, since $B_{(0)\varphi}$ must be zero at the initial point of c (on the axis where $\tilde{m}$ is zero) we are able finally to conclude that

$$B_{(0)\varphi} = 2I. \tag{6.256}$$

This result – a general relativistic version of the historic Ampère law – is valid everywhere, and in particular can be applied on the *horizon* $\mathcal{H}^+$ on which $\boldsymbol{B}_{(0)}$ tends to the limit $\boldsymbol{B}_\mathrm{H}$, so that by (6.263) we shall have

$$B_{(0)\varphi} = |\tilde{m}| B_{(3)}{}^\mathrm{H} \tag{6.257}$$

where it is to be recalled that, by (6.241), $|\tilde{m}|$ represents the circumferential cylindrical radius of the latitude under consideration. It can thus be seen from (6.233) and (6.217) that, as one would expect, Znajek's surface current, I, is related to $\sigma_\mathrm{F}^{(2)}$, azimuthal component of Damour's surface current density $\vec{\sigma}_\mathrm{F}$, by

$$I = 2\pi |\tilde{m}| \sigma_\mathrm{F}^{(2)} \tag{6.258}$$

(where $2\pi|\tilde{m}|$ is the circumference at the latitude in question). It is to be remarked that the fact that I vanishes on the south as well as (by definition) on the north pole of the horizon reflects the fact that the total charge influx

$$\Delta Q = -\int_\Sigma j_\mathrm{F}{}^a \, \mathrm{d}\Sigma_a = -\int_{t_1}^{t_2} \mathrm{d}t \int j^a k^b \, \mathrm{d}\mathcal{S}_{ab} = \int_{t_1}^{t_2} I \, \mathrm{d}t \tag{6.259}$$

across the whole of the horizon in any time interval $t_1 \leq t \leq t_2$ must be *zero*, since any variation of the total charge, Q, would obviously be incompatible with stationarity.

We can now see from (6.186) that the electromagnetic contribution to the dissipation rate on the horizon will be given, in the case when the field is stationary and axisymmetric, by

$$(\kappa/8\pi)\mathcal{A}_\mathrm{F} = \int_\mathcal{S} T_\mathrm{F}{}^{ab} l_a l_b \, \mathrm{d}\mathcal{S}$$

$$= \int_c I E^\mathrm{H}{}_{(2)} \, \mathrm{d}s \tag{6.260}$$

$$= \tfrac{1}{2} \int_c B_{(0)\varphi} B^\mathrm{H}{}_{(3)} \, \mathrm{d}s$$

where ds denotes the *proper length* differential along a curve c extending from the north to the south pole in the horizon. Since we clearly must have

$$T_F{}^{ab} m_a l_b = (1/4\pi) E^H{}_a (F^{ab} m_b) \tag{6.261}$$

on the horizon, and since substitution of (6.212) and use of (6.245) and (6.203) gives

$$\begin{aligned} E^H{}_a (F^{ab} m_b) &= E^H{}_a (\beta^a l_c + \gamma^a{}_c) F^{cb} m_b \\ &= E^H{}_{(2)} \vec{e}_{(2)} \cdot \nabla A_\varphi, \end{aligned} \tag{6.262}$$

we find that the corresponding rate of transfer of angular momentum can also be expressed in terms of the magnetic field by

$$\begin{aligned} \dot{J}_F &= -\oint_{\mathscr{S}} T_F{}^{ab} m_a l_b \, d\mathscr{S} \\ &= -\tfrac{1}{2} \int_c B_{(0)\varphi} \vec{e}_{(2)} \cdot \nabla A_\varphi \, ds \\ &= -\tfrac{1}{2} \int_c B_{(0)\varphi} \, dA_\varphi = \tfrac{1}{2} \int_c A_\varphi \, dB_{(0)\varphi}. \end{aligned} \tag{6.263}$$

Finally a similar argument gives

$$\begin{aligned} \dot{M}_F &= (\kappa/8\pi) \dot{\mathscr{A}}_F + \Omega^H \dot{J}_F \\ &= \oint_{\mathscr{S}} T_F{}^{ab} k_a l_b \, d\mathscr{S} \\ &= \tfrac{1}{2} \int_c B_{(0)\varphi} \, dA_t = -\tfrac{1}{2} \int_c A_t \, dB_{(0)\varphi}. \end{aligned} \tag{6.264}$$

6.5.3 Degenerate electromagnetic fields

For a given differential form (i.e. an antisymmetric covariant tensor field) F of any order p, a congruence of curves can appropriately be described as *flux* (conserving) *lines* if the integral of F across a p-surface segment depends only on the subset of curves of the congruence passing through the segment, the necessary and sufficient condition for this being that the Lie derivative of F with respect to *any* field $\vec{\xi}$ of tangent vectors to the congruence should be zero. It can be seen fairly easily from the Cartan formula (6.20) that a congruence generated by some given tangent vector field, $\vec{\xi}$, will behave as a set of flux lines in this sense

Chapter 6. Mechanical, electromagnetic and thermodynamic properties

if and only if it satisfies

$$\vec{\xi} \cdot \boldsymbol{F} = 0 \qquad (6.265)$$

$$\vec{\xi} \cdot \partial \boldsymbol{F} = 0. \qquad (6.266)$$

(It is obvious from the Cartan formula (6.20) that these conditions are sufficient, and it can easily be checked that they are also necessary.) In general there need not be any nonzero solutions to the system of equations (6.25) and (6.266) but in favourable conditions there may be a family of solutions, $\vec{\xi}$, forming a multidimensional subspace at each point, and in such a case there is a non-trivial but not too difficult standard theorem (see, for example, Choquet-Bruhat, 1968) to the effect that the subspaces are necessarily *integrable* to form a locally well-behaved family of *flux* (conserving) *surfaces* with the property that the integral of $\boldsymbol{F}$ over a p-dimensional segment depends only on the particular subset of flux surfaces that the segment intersects. Now in the case of a *Maxwell* field, the condition (6.266) holds automatically for arbitrary $\vec{\xi}$ and hence (6.265) alone is a necessary condition for $\vec{\xi}$ to be the generator of a flux line. Since any (nonzero) solution to (6.265) is a zero eigenvalue characteristic vector, one sees that (nonzero) solutions will exist if and only if the component matrix F_{ab} is *degenerate* in the sense that its determinant is zero, which is equivalent to the requirement

$$F_{ab}(*F^{ab}) = 0. \qquad (6.267)$$

Since (by its antisymmetry) the component matrix F_{ab} necessarily has *even* rank, it follows that if (6.267) is satisfied the solutions to (6.265) will necessarily span a *two*-dimensional tangent subspace, and hence (by the theorem referred to above) that they will generate a two-dimensional family of surfaces which can appropriately be described as *magnetic* flux 2-surfaces, provided they are timelike. (In the spacelike case they would be interpretable as surfaces of constant electric potential. The null case arises only if the other scalar invariant, $F_{ab}F^{ab}$, is also zero, which occurs for plane waves.) For a general electromagnetic field, for which (6.267) does not hold, the concept of magnetic flux can be given no *invariant* meaning and 'magnetic flux lines' cannot be defined without reference to some particular frame. It is, therefore, fortunate that the condition (6.267) can be justified on physical grounds as at least a reasonably good approximation in a fairly wide range of circumstances, of which the best known is the case of a *perfect magnetohydrodynamic medium* in which the electric field $\boldsymbol{E}$, defined with respect to the unit flow vector of the medium by

$$E = F \cdot \vec{u}, \qquad |u|^2 = -1, \qquad (6.268)$$

is *zero*, so that $\vec{u}$ itself is a solution of (6.265). Under these conditions the magnetic field vector defined by

$$\vec{B} = (*F) \cdot u \qquad (6.269)$$

will also be a solution of (6.265) and therefore it, together with $\vec{u}$, will generate the magnetic flux surfaces which will obviously be 'frozen in' in the sense of being dragged along by the flow (because they have $\vec{u}$ itself as a tangent vector). A quite different situation in which (6.267) may be a good approximation occurs in the case of *extremely strong* magnetic fields (which might be built up by accretion of coherently magnetized material into a black hole) for which electric breakdown of the vacuum by diverse mechanisms can be expected to reduce any electric field *parallel* to the magnetic field (as defined with respect to an arbitrary base, $\vec{u}$) to a comparatively very small (though not strictly zero) value, so that the invariant

$$\vec{B} \cdot \vec{E} \equiv \tfrac{1}{2}(*F_{ab})F^{ab} \qquad (6.270)$$

becomes, if not strictly zero, at least *negligibly small* by comparison with the other invariant

$$|\vec{B}|^2 - |\vec{E}|^2 \equiv \tfrac{1}{2}F_{ab}F^{ab} \qquad (6.271)$$

(so that one obtains $|(*F_{ab})F^{ab}| \ll F_{ab}F^{ab}$). In such a situation it will often be possible (although it is not essential for the application that follows) to assume that the field is *force-free*, in the sense that the total electromagnetic current $\vec{j}_F$ is *itself* one of the null eigenvalue characteristic vectors of F, i.e.

$$F \cdot \vec{j}_F = 0. \qquad (6.272)$$

(This will be particularly likely to hold in low-density situations, where the individual charged particles involved can spiral many times around the magnetic field direction before being scattered, so that the contributions of transverse components of the current will cancel out in any macroscopic averaging.)

Now it can be seen from (6.204) that *any* vector $\vec{\xi}$ satisfying the requirement (6.265) for tangency to the magnetic 2-surfaces will also satisfy

$$\vec{\xi} \cdot \nabla A_t = \vec{\xi} \cdot \nabla A_\varphi = 0, \qquad (6.273)$$

Chapter 6. Mechanical, electromagnetic and thermodynamic properties

which shows that A_t and A_φ are each constant on the magnetic flux 2-surfaces. Since they are each constant also on the invariant 2-surfaces generated by $\vec{k}$ and $\vec{m}$ we can go on to deduce that the *loci of constant A_t and of constant A_φ* must *coincide* with each other, i.e.

$$(\nabla A_t) \wedge (\nabla A_\varphi) = 0$$
$$\nabla A_t = 0 \Leftrightarrow \nabla A_\varphi = 0 \quad (6.274)$$

and these loci must *contain the magnetic flux 2-surfaces as* subspaces. Thus (except in the special case when A_t and A_φ are both constant, so that the magnetic flux lines actually coincide with the invariant 2-surfaces) the common loci on which A_t and A_φ are constant will form a congruence of 3-surfaces containing *both* the magnetic flux 2-surfaces *and* the invariant 2-surfaces. This implies that there must be a well-defined family of (one-dimensional) curves that are common to *both* the magnetic flux of 2-surfaces *and* the invariant 2-surfaces, and which are therefore generated by a field of vectors, $\vec{\xi}$, satisfying (6.265), and having the form

$$\vec{\xi} = \vec{k} + \Omega^F \vec{m}, \quad (6.275)$$

where the scalar Ω^F represents the *angular velocity* of rotation of the magnetic field. (In the special case when A_t and A_φ are both constant – so that the magnetic field vector is purely axial – there will still exist vectors of the form (6.275) satisfying (6.265), but the value of Ω^F will no longer be unique.) Since substitution of (6.275) in (6.265) and use of the stationarity and axisymmetry conditions (in the form of (6.204)) immediately gives

$$\nabla A_t = -\Omega^F \nabla A_\varphi \quad (6.276)$$

we can deduce – again provided A_t and A_φ are not both uniform throughout – that Ω^F is itself constant over the common hypersurfaces of constant A_t and A_φ and hence more particularly that it is *constant over each magnetic flux 2-surface*. (This is the relativistic generalization of Ferraro's law of isorotation – cf. Carter (1978*b*) for a discussion in a more general dynamical context.)

Blandford and Znajek (1977) have used the angular velocity Ω^F so defined to express the formula for the dissipation rate and the rate of absorption of energy and angular momentum in forms analogous to those we obtained for *non*-stationary/axisymmetric fields in our earlier discussion of the super-radiance phenomenon. Thus (6.276) allows us

Electromagnetic effects

immediately to replace (6.263) by

$$\dot{M}_F = -\tfrac{1}{2}\int_c \Omega^F B_{(0)\varphi}\, dA_\varphi \qquad (6.277)$$

so that we obtain for the dissipation rate

$$(\kappa/8\pi)\dot{\mathscr{A}}_F = \tfrac{1}{2}\int_c (\Omega^H - \Omega^F) B_{(0)\varphi}\, dA_\varphi. \qquad (6.278)$$

Znajek pointed out that not only the combination (6.218) but also the separate energy and angular momentum fluxes (6.277) and (6.263) will tend to zero in the corotating limit. Since we can use (6.93) to write

$$\vec{l} = \vec{v} + (\Omega^H - \Omega^F)\vec{m} \qquad (6.279)$$

so that (6.265) implies that we have

$$\boldsymbol{E}_{(0)} = (\Omega^H - \Omega^F)\nabla A_\varphi, \qquad (6.280)$$

and hence by (6.217) that on the horizon

$$B_{(0)\varphi} = |\vec{m}|(\Omega^H - \Omega^F)\vec{e}_{(2)} \cdot \nabla A_\varphi, \qquad (6.281)$$

using the fact that contraction of (6.215) with the transverse null vector, $\vec{\beta}$, gives

$$\vec{e}_{(2)} \cdot \nabla A_\varphi = -\vec{\beta} \cdot \boldsymbol{B}^H \qquad (6.282)$$

we may finally write

$$\dot{J}_F = \tfrac{1}{2}\int_c (\Omega^F - \Omega^H)(\vec{\beta} \cdot \boldsymbol{B}^H)^2 |\vec{m}|\, ds \qquad (6.283)$$

$$\dot{M}_F = \tfrac{1}{2}\int_c \Omega^F(\Omega^F - \Omega^H)(\vec{\beta} \cdot \boldsymbol{B}^H)^2 |\vec{m}|\, ds \qquad (6.284)$$

$$\dot{\mathscr{A}}_F = \tfrac{1}{2}\int_c (\Omega^F - \Omega^H)^2 (\vec{\beta} \cdot \boldsymbol{B}^H)^2 |\vec{m}|\, ds \qquad (6.285)$$

for any stationary, axisymmetric, degenerate field. As Blandford and Znajek have pointed out, this implies that there will be a super-radiance-type phenomenon, that is to say *negative* $\dot{M}_F$ (meaning extraction of electromagnetic energy from the rotational energy of the hole) if and only if Ω^F has the *same* sign as, but *smaller* magnitude than, Ω^H, in accordance with the inequalities (6.188). It is the opinion (not just) of the present author that this is by far the most promising kind of

Chapter 6. Mechanical, electromagnetic and thermodynamic properties

mechanism for extraction of energy from black holes in models of astrophysical phenomena such as quasars.

6.6 The total mass and angular momentum

6.6.1 Mass and angular momentum decomposition formulae for stationary axisymmetric systems

The standard way to define the total mass, M, and angular momentum, J, in a stationary axisymmetric asymptotically flat system proceeds via the requirement that there should exist suitable asymptotically Minkowskian coordinates x^1, x^2, x^3, x^4, such that the stationary Killing vector has the form $k^a = \delta^a_4$, in which the metric components are given by

$$g_{44} = -1 + (2M/r) + O(1/r^2)$$
$$g_{4\nu} = (2J/r^3)m_\nu + O(1/r^3) \qquad (6.286)$$
$$g_{\mu\nu} = [1 + (2M/r)]\delta_{\mu\nu} + O(1/r^2),$$

with μ, ν running over the values 1, 2, 3 and with $r^2 = \delta_{\mu\nu}x^\mu x^\nu$, where $\delta_{\mu\nu}$ is the Kronecker delta. Following Komar (1959) one may express the quantities M and J so defined in a coordinate-independent manner by the formulae

$$\lim_{\infty} \oint \nabla^a k^a \, d\mathscr{S}_{ab} = 8\pi M \qquad (6.287)$$

$$\lim_{\infty} \oint \nabla^a m^a \, d\mathscr{S}_{ab} = -16\pi J \qquad (6.288)$$

where the integral is taken over a spacelike 2-sphere $\mathscr{S}$ (with $d\mathscr{S} = (e_{(3)} \wedge e_{(4)}) \, d\mathscr{S}$ for the same mutually orthogonal spacelike and timelike normals $e_{(3)}$ and $e_{(4)}$) and where the limit is taken as the 2-sphere taken out to arbitrarily large asymptotic distances. One can then use the identities

$$\Box k^a = -R^a{}_b k^b, \qquad \Box m^a = -R^a{}_b m^b \qquad (6.289)$$

(where $\Box \equiv \nabla_c \nabla^c$) which hold in consequence of the Killing equations (6.74), (6.90) to convert these expressions to the form

$$M = M_{\text{GS}} + M_{\text{H}} \qquad (6.290)$$

$$J = J_{\text{GS}} + J_{\text{H}} \qquad (6.291)$$

The total mass and angular momentum

using Stokes' theorem, where the external *gravitational source* contributions are given by

$$M_{GS} = (1/4\pi) \int_\Sigma R^a{}_b k^b \, d\Sigma_a \qquad (6.292)$$

$$J_{GS} = -(1/8\pi) \int_\Sigma R^a{}_b m^b \, d\Sigma_a \qquad (6.293)$$

(Σ is any spacelike hypersurface extending from spacelike infinity into some bounding 2-surface $\mathscr{S}$ on the horizon). In the absence of a black hole (the situation originally envisaged by Komar) the inner boundary would not occur and the M_H and J_H would be unnecessary, but in the black hole case the horizon gives contributions

$$M_H = (1/8\pi) \oint_\mathscr{S} \nabla^a k^b \, d\mathscr{S}_{ab} \qquad (6.294)$$

$$J_H = -(1/16\pi) \oint_\mathscr{S} \nabla^a m^b \, d\mathscr{S}_{ab} \qquad (6.295)$$

with $d\mathscr{S}$ given on the horizon by the expression (6.183). (Note that in Bardeen, Carter and Hawking (1973), Carter (1973a, b) and Hawking and Ellis (1973) a convention differing by a factor of 2 was used for the normalization of $d\mathscr{S}$. The present convention is in better accord with the standard Cartan scheme as used by Misner, Thorne and Wheeler and as advocated in section 6.2.1.)

With the natural definitions (6.294) and (6.295) it is evident from (16.93) that (as originally pointed out by Bardeen, Carter and Hawking, 1973) one will have

$$M_H - 2\Omega^H J_H = (1/8\pi) \oint_\mathscr{S} \nabla^a l^b \, d\mathscr{S}_{ab}, \qquad (6.296)$$

where $\vec{l}$ is the standard corotating Killing vector that coincides with the null generator on the horizon. Combining this result with the expression (6.183) for $d\mathscr{S}$ and with the expression (6.132) for κ we immediately obtain the generalized *Smarr Formula*

$$M_H = 2\Omega^H J_H + (\kappa/8\pi)\mathscr{A} \qquad (6.297)$$

so called because it was first noticed by Smarr (1973) in the particular case of the *pure vacuum* Kerr solutions where the contributions M_{GS} and J_{GS} are absent, so that M_H and J_H coincide with the total values M and J.

Chapter 6. Mechanical, electromagnetic and thermodynamic properties

A further refinement (cf. Carter, 1973b) can be made when the external source contributions are partly attributable to an electromagnetic field. Using the Einstein equations

$$R^a{}_b = 8\pi(T^a{}_b - \tfrac{1}{2}T^c{}_c g^a{}_b) \tag{6.298}$$

and making the decomposition

$$T^{ab} = T_F{}^{ab} + T_M{}^{ab}, \tag{6.299}$$

where $T_F{}^{ab}$ is the Maxwellian contribution defined by (6.70) and $T_M{}^{ab}$ is the resultant of all other material sources, we see that it is appropriate to decompose the external source contributions in the form

$$M_{GS} = M_F + M_M \tag{6.300}$$

$$J_{GS} = J_F + J_M \tag{6.301}$$

with

$$M_M = \int_\Sigma (2\, T_M{}^a{}_b k^b - T_M{}^c{}_c k^a)\, d\Sigma_a \tag{6.302}$$

$$J_M = -\int_\Sigma T^a{}_b m^b\, d\Sigma_a \tag{6.303}$$

(using the assumption that Σ is invariant under the axisymmetry action, so that $m^a\, d\Sigma_a$ vanishes, to simplify the latter expression) and with

$$M_F = \int_\Sigma 2\Pi_F{}^a(\vec{k})\, d\Sigma_a \tag{6.304}$$

$$J_F = -\int_\Sigma \Pi_F{}^a(\vec{m})\, d\Sigma_a \tag{6.305}$$

(using the tracelessness of $T_F{}^a{}_b$).

Since we are now assuming that the system as a whole, including the electromagnetic field in particular, is invariant under the actions generated by the Killing vectors, we can obtain the formula

$$-\vec{\Pi}_F(\vec{m}) = (\vec{m}\cdot \mathbf{A})\vec{j} + \Lambda_F \vec{m} + \nabla\cdot\{(\vec{m}\cdot \mathbf{A})\vec{D}_F\} \tag{6.306}$$

directly from (6.139) and (6.150), using the condition that $\vec{m}\, \pounds\, \mathbf{A}$ is zero. Hence (again using the assumption that Σ is invariant) we are led by Green's theorem to make the decomposition

$$J_F = J_C + J_{FH} \tag{6.307}$$

where the *electromagnetic source* contribution arising directly from the external *charge current distribution* is given by

$$J_C = -\int_\Sigma (m^c A_c) j^a \, d\Sigma_a \qquad (6.308)$$

and where the horizon contribution to the electromagnetic angular momentum is given by

$$\begin{aligned} J_{\text{FH}} &= \tfrac{1}{2} \oint_{\mathscr{S}} (m^c A_c) D_F{}^{ba} \, d\mathscr{S}_{ab} \\ &= \oint_{\mathscr{S}} (m^c A_c) \sigma_F^{(0)} \, d\mathscr{S} \end{aligned} \qquad (6.309)$$

where $\sigma_F^{(0)}$ is the surface charge density as given by (6.226). Due to the fact that the hypersurface Σ will certainly not be invariant under the action generated by $\vec{k}$, the treatment of the mass contribution is more complicated because the term $\Lambda_F \vec{k}$ in the formula obtained by substituting $\vec{k}$ for $\vec{m}$ in (6.306) does not simply drop out, but must itself be decomposed into the sum of a source contribution and a divergence. Using the explicit form (6.193) of Λ_F we finally obtain the suitable decomposed expression

$$-\vec{\Pi}_F(\vec{k}) = (\vec{k} \cdot \mathbf{A}) \vec{j} - (\vec{j} \cdot \mathbf{A}) \vec{k} + \nabla \cdot \{(\vec{k} \cdot \mathbf{A}) \vec{D}_F + \tfrac{1}{2} \vec{k} \wedge (\vec{D}_F \cdot \mathbf{A})\} \qquad (6.310)$$

for the electromagnetic energy flux vector. This leads to the corresponding decomposition

$$M_F = M_C + M_{\text{FH}}, \qquad (6.311)$$

where the electromagnetic source contribution from the external charge current distribution is given by

$$M_C = \int \{2(k^b A_b) j^a - (j^b A_b) k^a\} \, d\Sigma_a \qquad (6.312)$$

(which can be seen to bear a close analogy to (6.302)) and where the horizon contribution to the electromagnetic mass is given by

$$\begin{aligned} M_{\text{FH}} &= \oint_{\mathscr{S}} \{k^c A_c D_F{}^{ab} + k^a D_F{}^{bc} A_c\} \, d\mathscr{A}_{ab} \\ &= -\oint_{\mathscr{S}} (l^c A_c - 2\Omega^H m^c A_c) \sigma_F^{(0)} \, d\mathscr{S}. \end{aligned} \qquad (6.313)$$

The derivation of the last step of (6.313) depends on the use of the strict boundary condition

$$\vec{D}_F \cdot \mathbf{l} = \sigma_F^{(0)} \vec{l} \qquad (6.314)$$

Chapter 6. Mechanical, electromagnetic and thermodynamic properties

(obtained from (6.88) and (6.238)) which is necessary for exact stationarity, since (in contrast with the less restrictive test field situation described in section 6.5) we are now treating the electromagnetic field as a source for gravity via the Einstein equations, so that the considerations of section 6.3 will apply. Using the fact that the considerations of section 6.3 also imply that the electromagnetic potential,

$$\Phi^H = -\vec{l} \cdot \mathbf{A}, \tag{6.315}$$

must be uniform over the horizon, and that Ω^H is uniform in any case, we can evaluate the final integral in (6.313) so as to obtain the result

$$M_{FH} = 2\Omega^H J_{FH} + \Phi^H Q_H \tag{6.316}$$

which is the electromagnetic analogue of the Smarr formula (6.297). Let us now regroup the terms in our decomposition in the form

$$M = M_{BH} + M_{MC} \tag{6.317}$$

$$J = J_{BH} + J_{MC} \tag{6.318}$$

where the *black hole* contributions, as defined in terms of surface integrals on the horizon, are grouped together to form

$$M_{BH} = M_H + M_{FH} \tag{6.319}$$

$$J_{BH} = J_H + J_{FH} \tag{6.320}$$

and where the terms expressible in terms of the external *material* and *charge* current distribution are grouped together to form

$$M_{MC} = M_M + M_C \tag{6.321}$$

$$J_{MC} = J_M + J_C. \tag{6.322}$$

It can be seen from (6.297) and (6.316) that the contributions of the first group will satisfy the formula

$$M_{BH} = 2\Omega^H J_{BH} + \Phi^H Q_H + (\kappa/4\pi)\mathcal{A} \tag{6.323}$$

whose validity was also noticed first by Smarr (1973) in the particular case of the Kerr–Newman *source-free* Einstein–Maxwell solutions, in which the external contributions M_{MC} and J_{MC} vanish so that M_{BH} and J_{BH} coincide with the *total* values M and J.

6.6.2 Mass variation: the first law

We shall now consider an infinitesimal variation – denoted by δ – between two neighbouring states of an asymptotically flat black hole, introducing a tensor with covariant components h_{ab} to represent the variation of the metric itself, so that

$$\delta g_{ab} = h_{ab}, \qquad \delta g^{ab} = -h^{ab}. \tag{6.324}$$

It can be seen that in an asymptotically Cartesian system of the form (6.286) we shall have

$$h_{ab} = (2M/r)\delta_{ab} + O(1/r) \tag{6.325}$$

and hence that the variation of the asymptotic mass, which we intend to evaluate, will be expressible by

$$4\pi\delta M = \lim_{\infty} \oint k^a \nabla^{[c} h^{b]}{}_c \, d\mathscr{S}_{ab}. \tag{6.326}$$

Whenever one wishes to consider the variation of locally defined quantities – such as tensor fields – on a manifold, ambiguities may arise from the freedom to adjust the mapping between the new and original manifolds. We shall use this gauge freedom to adjust the mapping so as to preserve the stationary and axisymmetry actions, so that the Killing vectors, $\vec{k}$ and $\vec{m}$, remain unchanged, i.e.

$$\delta k^a = 0, \qquad \delta m^a = 0 \tag{6.327}$$

and we shall also take the horizon to lie in the same position in the new and perturbed states (which is consistent with (6.327) because the horizon is itself invariant). However the angular velocity of the horizon may change, and therefore the Killing vector, $\vec{l}$, specified by (6.93) may also change, its variation being given by

$$\delta l^a = m^a \delta \Omega^{\mathrm{H}}. \tag{6.328}$$

It follows that the covariant tangent form, l, on the horizon will have a variation given by

$$\delta l_a = h_{ab} l^b + m_a \, \delta \Omega^{\mathrm{H}} \tag{6.329}$$

but since l remains normal to the horizon its direction will be unaffected, so that we shall have

$$l \wedge \delta l = 0. \tag{6.330}$$

Combining this property with the group invariance property

$$\vec{l} \, \pounds \, \delta l = 0 \tag{6.331}$$

one can deduce that the variation of the decay constant, κ, will be given

Chapter 6. Mechanical, electromagnetic and thermodynamic properties

by

$$\delta\kappa = \tfrac{1}{2}l^a\nabla^b h_{ab} - l_a\beta_b(\nabla^a m^b)\,\delta\Omega^{\mathrm{H}} \qquad (6.332)$$

whence

$$\mathcal{A}\,\delta\kappa = \oint (\delta\kappa)\,\mathrm{d}\mathscr{S}$$
$$= \oint l^b \nabla^{[c} h^{a]}{}_c\,\mathrm{d}\mathscr{S}_{ab} - 8\pi J_{\mathrm{H}}\delta\Omega^{\mathrm{H}}. \qquad (6.333)$$

Now by choosing $\mathscr{S}$ to be invariant under the axisymmetry action, which implies

$$m^a\,\mathrm{d}\mathscr{S}_{ab} = 0 \qquad (6.334)$$

we see that we can obtain

$$\oint l^b \nabla^{[c} h^{a]}{}_c\,\mathrm{d}\mathscr{S}_{ab} = \oint k^b \nabla^{[c} h^{a]}{}_c\,\mathrm{d}\mathscr{S}_{ab}$$
$$= \int_\Sigma k^a \nabla_b \nabla^{[b} h^{c]}{}_c\,\mathrm{d}\Sigma_a - 4\pi\delta M \qquad (6.335)$$

using Green's theorem and (6.326). Thus we obtain

$$4\pi\delta M + 8\pi J_{\mathrm{H}}\,\delta\Omega^{\mathrm{H}} + \mathcal{A}\,\delta\kappa = \int_\Sigma k^a \nabla_b \nabla^{[b} h^{c]}{}_c\,\mathrm{d}\Sigma_a. \qquad (6.336)$$

Now the integrand on the right-hand side is familiar from the variation principle derivation of Einstein's equations which uses the relation

$$\nabla_b \nabla^{[b} h^{c]}{}_c + \tfrac{1}{2}G^{cd}h_{cd} + \tfrac{1}{2}\|g\|^{-1/2}\,\delta(R\|g\|^{1/2}) = 0 \qquad (6.337)$$

where

$$G^{ab} = R^{ab} - \tfrac{1}{2}Rg^{ab} \qquad (6.338)$$

is the Einstein tensor, so that, using

$$\delta(\mathrm{d}\Sigma_a) = \|g\|^{-1/2}(\delta\|g\|^{1/2})\,\mathrm{d}\Sigma_a = \tfrac{1}{2}h^c{}_c\,\mathrm{d}\Sigma_a \qquad (6.339)$$

we obtain

$$4\pi\,\delta M + 8\pi J_{\mathrm{H}}\,\delta\Omega^{\mathrm{H}} + \mathcal{A}\,\delta\kappa = -\tfrac{1}{2}\int G^{cd}h_{cd}k^a\,\mathrm{d}\Sigma_a - \tfrac{1}{2}\delta\int Rk^a\,\mathrm{d}\Sigma_a. \qquad (6.340)$$

Since it follows from the formulae (6.292) and (6.297) that the *total* mass will satisfy

$$M - 2\Omega^{\mathrm{H}}J_{\mathrm{H}} - (k/4\pi)\mathcal{A} = (1/4\pi)\int R^a{}_b k^b\,\mathrm{d}\Sigma_a \qquad (6.341)$$

we obtain for the *total mass variation* the corresponding formula

$$\delta M - \Omega^H \delta J_H - (\kappa/8\pi)\delta\mathscr{A} = (1/8\pi)\delta\int G^a{}_b k^b \, d\Sigma_a \\ - (1/16\pi)\int G^{cd} h_{cd} k^a \, d\Sigma_a. \quad (6.342)$$

In the case of a variation between pure vacuum states the right-hand side of (6.340) will vanish, and we shall be left with the formula that was originally derived by Bekenstein (1973a) for the particular case of the pure vacuum Kerr solutions by simply differentiating the known formula for their mass M as a function of J_H and $\mathscr{A}$. Simple expressions for the matter contributions on the right-hand side of (6.342) have been given for the case of a perfect fluid by Bardeen (1973) and with allowance for an electromagnetic field by Carter (1973b). (See also Carter 1973a, 1975, 1978b.) The extension of the vacuum mass variation law to a variation principle has been described by Hawking (1973).

6.7 Uniqueness and no-hair theorems

6.7.1 The non-rotating case

The purpose of this final section is to provide a brief account of the progress that has been achieved towards establishing the truth of the widely accepted dogma that in the absence of external material and electric charge sources an asymptotically flat black hole equilibrium state is completely characterized, at least as far as continuous variations are concerned, by just three parameters which might, for example, be taken to be the intensive (locally defined but necessarily uniform) quantities Ω^H, Φ^H, κ as described in section 6.3. Alternatively one might prefer the three corresponding *extensive* quantities J_{BH}, Q_H, $\mathscr{A}$ to which they were shown in section 6.6 to be conjugate, or again one might prefer to work with the three quantities M, J, Q that are measurable as surface integrals at infinity. Demonstrations that at least in suitable restricted (e.g. static or axisymmetric) circumstances these parameters are indeed sufficient to characterize any continuous variations have come to be widely known as 'no-hair' theorems, although in other mathematical contexts analogous results would usually be described in less colourfully metaphoric language as no-bifurcation theorems. Even stronger results, describable as *uniqueness theorems*, have been conjectured, and at least partially proved, to the effect that an appropriate

Chapter 6. Mechanical, electromagnetic and thermodynamic properties

combination of not more than three parameters is sufficient to characterize the states *completely*, not just in so far as continuous variations are concerned. I shall not attempt to describe the full technical details of the work that has been done, but merely to present statements (without proof) of the key steps as a series of lemmas, using just sufficient rigour to make it clear where the main logical gaps remain.

The starting point (logically, though not chronologically) is the strong rigidity theorem already referred to in section 6.3.2 whose content, as relevant to the present context, may be described as follows:

Lemma 6.1 (Hawking (1972); Hawking and Ellis (1973))

The domain of outer communications $《g》$ of an asymptotically flat asymptotically predictable black hole equilibrium state will be either *non-rotating* or *axisymmetric* provided that the vacuum Einstein or source-free Einstein–Maxwell equations are satisfied.

(Although it is not relevant for the subsequent reasoning in the present section, it is to be mentioned that the conclusion should remain true in the presence of external sources obeying almost any reasonably well-behaved field equations. The Hawking argument is based on the assumption of *analyticity* of the stationary field solutions, as justified by the work of Müller zum Hagen (1970), but I see no reason to expect that the conclusion should necessarily be invalidated by such comparatively trivial analyticity failures as would arise from ordinary thermodynamic phase discontinuities in external matter sources.)

The next requirement for the subsequent lines of reasoning, which however cannot yet be presented as a lemma, is the following.

Condition 6.2

Subject to the conditions of lemma 6.1 and to the exclusion of the physically unattainable zero κ limit (and perhaps as a consequence also of other, as yet unformulated, restrictions) the domain of outer communications has the topology of the product of a 2-sphere (which might naturally be given coordinates θ and φ) and a 2-plane (which might naturally be given coordinates r and t) while the future horizon, $\mathcal{H}^+$, has the topology of the product of a 2-sphere with a line.

Hawking has made some progress towards establishing this condition by showing (Hawking (1972); Hawking and Ellis (1973)) that the topology of each connected component of the horizon must be the product of a 2-sphere with the line, thereby ruling out the most obvious way in which condition 2 might have been conceived as failing – namely due to the hole having a toroidal or otherwise multiply connected horizon. However it has not yet been possible *rigorously* to exclude the possibility of such physically implausible configurations as that of a pair of coaxial disjoint corotating black holes, despite the evidence that the mutual (Lense–Thirring effect) repulsion generated by the rotation should be insufficient to counterbalance the basic (monopole) attractive force between them, except perhaps in the extreme zero limit. (The coaxial configuration has been excluded in the non-rotating *static* limit by Gibbons (1974), while at the other extreme two- or many-hole solutions have been shown by Hartle and Hawking (1972a) actually to be *possible* in the zero κ limit with *electrostatic* repulsion balancing gravitational attraction.)

In order to make further progress we must follow up two separate lines of argument originating in the two alternative predictions made by lemma 6.1. We first consider the *non-rotating* case, for which one would like to be able to show that the solution is *static* in the sense that the stationary Killing vector, $\vec{k}$ (which in this case coincides with $\vec{l}$) is *hypersurface orthogonal* (which can be the case only if the Frobenius condition (6.12) is satisfied *everywhere*) and that the *magnetic* field $\vec{B}_{(0)}$ as defined by (6.198) is everywhere zero.

If staticity could be established somehow then it would *follow* from the global extension (Carter 1973b, 1977) of a local result originally due to Vishveshwara (1968) that we should also have:

Condition 6.3

In the *non-rotating* case predicted by lemma 6.1, and subject to the condition 6.2, the scalar $|\vec{k}|^2$ is *negative* everywhere within the domain of outer communications (and of course zero on $\mathcal{H}^+$). It was shown by Hawking (1972) in the vacuum case, and by Carter (1973b) in the electromagnetic case that *if* condition 6.3 can be established by *other means* then one can *establish staticity* by extending a theorem given originally by Lichnerowicz, 1955. Formally one has:

Lemma 6.4 (Hawking and Ellis (1973); Carter (1973b))

In the *non-rotating* case predicted by lemma 6.1, and subject to condition 6.3, the domain of outer communications will be *static*.

Chapter 6. Mechanical, electromagnetic and thermodynamic properties

In view of this valuable result it would obviously be desirable to establish condition 6.3 directly. A certain amount of progress towards this objective has been achieved by the work of Hajicek (1973, 1975). The next step, which is also in many ways (particularly in the electromagnetic case) the most difficult, was actually carried out earlier (in the historical sense) than its logical predecessors, by Israel, subject to a comparatively minor (though by no means trivial) technical reservation later removed by Müller zum Hagen, Robinson and Seifert (1973, 1974) and recently in a more concise and elegant manner by Robinson (1977). We have:

Lemma 6.5 (Israel (1967, 1968); Müller zum Hagen et al. (1973, 1974); Robinson (1977))

Subject to the conditions (and of course the conclusions) of lemma 6.4, the domain of outer communications must necessarily be spherically symmetric.

Lemma 6.6 (Birkhoff (1923); Hoffmann (1933))

Subject to the conditions (and of course the conclusions) of lemma 6.5, the domain of outer communications is given by the *Reissner–Nordström* or (in the absence of an electromagnetic field) by the *Schwarzschild* solution, and is therefore determined *uniquely* by the parameters M and (in the presence of an electromagnetic field) Q.

6.7.2 The axisymmetric case

Let us now go back and consider the second alternative possibility allowed by lemma 6.1, namely the *axisymmetric case*. At first things go more smoothly than in the static case, because the analogues of condition 6.3 and lemma 6.4 can *both* be proved, albeit in the inverse order. We have first the analogue of lemma 6.4 to the effect that the solution must have *circular symmetry* (in the sense discussed in detail by Carter, 1973b) meaning that the 2-surfaces generated by $\vec{k}$ and $\vec{m}$ (which would be characterized as surfaces of constant r and θ in a coordinate system of the standard kind with ignorable φ and t) are *orthogonal* to another family of 2-surfaces (which it is obviously convenient to take as surfaces of constant φ and t). Formally we have:

Lemma 6.7 (Papapetrou (1966); Carter (1969))

In the *axisymmetric* case predicted by lemma 6.1 (and independently of condition 6.2) the domain of outer communications will have *circular symmetry*.

We can then obtain the analogue of conditions 6.3 as a *conclusion*, obtained as a global generalization (cf. Carter, 1978a) of a local result given originally by Carter, 1969. We have:

Lemma 6.8 (Carter (1971b, 1973b))

Subject to condition 6.2 and the conditions (and of course the conclusion) of lemma 6.3, the quantity ρ^2 (as defined by (6.119)) is positive throughout the domain of outer communications except on the axis of symmetry where it is zero. It is also zero on the horizon.

The next step uses the Morse theory of *harmonic* functions (Morse and Heinz, 1945) as applied to the quantity ρ to establish:

Lemma 6.9 (Carter (1971b, 1973b))

Subject to the conditions (and conclusion) of lemma 6.8, the domain of outer communications can be covered globally by a manifestly stationary–axisymmetric *ellipsoidal* system of coordinates λ, μ, φ, t (the two last being *ignorable*) such that the metric takes the form

$$ds^2 = \Xi\left\{\frac{d\lambda^2}{\lambda^2-c^2}+\frac{d\mu^2}{1-\mu^2}\right\}+X\,d\varphi^2+2W\,d\varphi\,dt-V\,dt^2 \quad (6.343)$$

where

$$\rho^2 = (\lambda^2-c^2)(1-\mu^2) \quad (6.344)$$

with

$$c = M - 2\Omega^H J - \Phi^H Q \quad (6.345)$$

the north and south poles of the symmetry axis being given by $\mu = \pm 1$, while the horizon, $\mathcal{H}$, is given by the limit $\lambda \to c$.

The following step relies heavily on the work of Ernst (1968a, b) who showed that for a metric of the form (6.343) the source-free Einstein–Maxwell equations can be expressed as a partial differential system of

Chapter 6. Mechanical, electromagnetic and thermodynamic properties

equations for just four unknowns from which all the other field variables can be obtained by explicit integrations. One of these unknowns is usually taken to be the coefficient V, but for the present purpose it turns out to be more convenient to exploit the algebraic (though not geometric) symmetry between V and X, and to use the latter instead, since X has the advantage of being known (by the causality axiom that is implicit in the asymptotic predictability condition) to be *positive* except on the symmetry axis, whereas V is known to change sign on an 'ergosurface' surrounding the horizon. The other Ernst variables Y, B, E say (using the notation of Carter, 1973b and Robinson, 1974, 1975), are such that in the non-electromagnetic case only X and Y remain, E and B being zero, while in the electrostatic (non-rotating, non-magnetic) case only X and E remain, Y and B being zero.

Not only do these Ernst variables obey simple field equations that turn out to be obtainable from a *simple, non-negative, Lagrangian density*, L, but they also turn out to admit an extremely simple and convenient formulation of the *global boundary conditions* for the black hole problem. One obtains:

Lemma 6.10 (Carter (1970, 1973b))

Under the conditions of lemma 6.9 the solution of the full set of field equations and boundary conditions is uniquely determined by a solution in the two-dimensional domain

$$-1 < \mu < 1, \qquad c < \lambda < \infty \tag{6.346}$$

with metric

$$ds^2 = \frac{d\lambda^2}{\lambda^2 - c^2} + \frac{d\mu^2}{1 - \mu^2} \tag{6.347}$$

of the reduced system specified by the variational integral $\int L \, d\lambda \, d\mu$ with

$$L = \frac{|\nabla X|^2 + |\nabla Y + 2(E\nabla B - B\nabla E)|^2}{2X^2} + 2\frac{|\nabla E|^2 + |\nabla B|^2}{X} \tag{6.348}$$

subject to the boundary conditions that the variables X, Y, E, B and their derivatives be *bounded*, that $\partial E/\partial \lambda$, $\partial B/\partial \lambda$, $\partial Y/\partial \lambda$, $\partial Y/\partial \mu + 2(E \, \partial B/\partial \mu - B \, \partial E/\partial \mu)$ and X (but not $\partial X/\partial \mu$) must tend to *zero* on the axes $\mu \to \pm 1$, and that as $\lambda \to \infty$

$$E = -Q\mu + O(1/\lambda)$$
$$B = O(1/\lambda)$$
$$Y = 2J\mu(3-\mu^2) + O(1/\lambda) \quad (6.349)$$
$$\lambda^{-2}X = (1-\mu^2)(1+O(1/\lambda)).$$

(It is to be remarked that the chosen variables have the convenient property that no special conditions other than boundedness are necessary on the horizon where $\lambda \to c$.)

Up to this stage the analysis has proceeded more smoothly in the axisymmetric than in the non-rotating case. However the ultimate step (the analogue of lemmas 6.5 and 6.6) to a simple *uniqueness* theorem is lacking except in the *non-electromagnetic* case (for which $E = B = 0$) and even this has been achieved only very recently by Robinson (1975) who has shown:

Lemma 6.11 (Robinson, 1975)

Under the conditions of lemma 6.10 and subject to the restriction that E and B are zero, there is *at most one* possible solution of the reduced system (and hence a correspondingly unique geometry for the domain of outer communications) for each pair of values of the parameters c and J (which are in fact the only ones remaining in the formulation of the problem).

Subject to the restriction

$$c^4 > 4J^2 \quad (6.350)$$

solutions do exist, which are of course the black hole metrics of Kerr (1963), which are usually thought of as being specified by M and J rather than by c and J, the relation between the two formulations being expressible by

$$M = \tfrac{1}{2}\{(c^2 + 2J)^{1/2} + (c^2 - 2J)^{1/2}\}. \quad (6.351)$$

Prior to 1975 the best result, for the non-electromagnetic case, was a *no-hair theorem* (Carter 1971b, 1973b) to the effect that *continuous variations* of the solutions are completely determined by the corresponding continuous variations of c and J, (so that the solutions could

Chapter 6. Mechanical, electromagnetic and thermodynamic properties

at least be classified in discrete 2-parameter families, the question remaining whether the Kerr solutions were the only such family) and even today the best result for the full electromagnetic case is a corresponding extended no-hair theorem provided by:

Lemma 6.12 (Robinson (1974))

Under the conditions of lemma 6.10, continuous variations of solutions of the reduced system are fully determined by continuous variations of the three parameters c, J, Q.

This lemma supersedes the more specialized results, based on perturbations of the Kerr and Kerr–Newman solutions by Ipser (1971), Wald (1972) and Bose and Wang (1974)).

This implies that the solutions of the full system can be classified as belonging to *discrete families* each of which is completely characterized by (at most) the *three parameters c, J, Q*. At the date of writing however there still remains a shadow of doubt as to whether the well-known solutions discovered by Newman *et al.* (1965) are the only such family.

Having mentioned that some doubt still exists, I hasten to emphasize that the no-hair theorems available since 1971 in the non-electromagnetic case and since 1974 in the general case are quite sufficient to justify – with at least the degree of rigour usually considered acceptable in physics – the assumption by any practically minded astrophysical theorist that any (external source free) black hole equilibrium-state solution with which he may be concerned belongs to the Kerr or Kerr–Newman families. (The important *physical* doubts that should be entertained are whether the Einstein or Einstein–Maxwell equations themselves really are valid in a highly nonlinear black hole situation.) The argument that discrete non-Kerr–Newman families (if they exist) will almost certainly be irrelevant to practical physical situations is based on the fact that – *unlike the well-behaved Kerr–Newman family* – they are known to have the *pathological* property that they *cannot be varied continuously* (by quasi-stationary angular momentum loss mechanisms of the kind considered in section 6.4) to the *non-rotating spherical limit* nor (by analogous electric discharge processes) to the *electrically neutral* limit. The impossibility of completely carrying out ordinary spin-down and discharge processes for non-Kerr–Newman families (if they exist) follows immediately from the lemmas 6.5 and 6.11 respectively, which

Uniqueness and no-hair theorems

show that the *only* (electromagnetic or neutral) *static solutions*, and the *only* (static or rotating) *neutral* solutions, belong to the well-behaved 3-parameter family of Kerr–Newman. To avoid a physical paradox one is therefore obliged to conclude that the non-Kerr–Newman families – if they exist – must certainly be *unstable* at least for sufficiently low values of their allowed angular momentum and charge parameters.

6.7.3 The divergence identities

It would take far too much space to describe the proofs of all the earlier lemmas quoted in this section, but the proof of lemma 6.10 is so short (consisting essentially of just one – albeit rather long – line) and so elegant that I shall present it in full.

Let us first define the restricted Lagrangian

$$L_G = \frac{|\nabla X|^2 + |\nabla Y|^2}{2X^2} \qquad (6.352)$$

to which (6.348) reduces when the electromagnetic variables E, B are set equal to zero, so that the corresponding pure vacuum field equations take the form

$$G_X \equiv \frac{\delta L(X, Y)}{\delta X} = 0, \qquad G_Y \equiv \frac{\delta L(X, Y)}{\delta Y} = 0. \qquad (6.353)$$

The proof of lemma 6.10 can now be obtained directly by inspection from the identity

$$\rho \left| \bar{X}\left(\frac{\nabla Y}{X}\right)^{\cdot} - \dot{Y}\overline{\left(\frac{\nabla X}{X}\right)} \right|^2 + \rho \left| \dot{X}\overline{\left(\frac{\nabla Y}{X}\right)} - \dot{Y}\overline{\left(\frac{\nabla X}{X}\right)} \right|^2 + \rho \left| \dot{Y}\frac{\nabla Y}{X} + X\overline{\left(\frac{\nabla X}{X}\right)^{\cdot}} \right|^2$$

$$= \langle X \rangle^2 \nabla \left\{ \rho \nabla \left(\frac{\dot{X}^2 + \dot{Y}^2}{\langle X \rangle^2} \right) \right\} - \dot{Y}(X^2 G_Y)^{\cdot} + \dot{Y}^2 \overline{(XG_X)}$$

$$- \bar{X}\dot{X}(XG_X)^{\cdot} \qquad (6.354)$$

where for any function or functional $f(X, Y)$ and any two pairs of fields (X_1, Y_1), (X_2, Y_2) we use the abbreviated notation

$$\dot{f} \equiv f(X_1, Y_1) - f(X_2, Y_2) \qquad (6.355)$$

$$\bar{f} \equiv f(X_1, Y_1) + f(X_2, Y_2) \qquad (6.356)$$

$$\langle f \rangle^2 \equiv f(X_1, Y_1) f(X_2, Y_2). \qquad (6.357)$$

It can be seen that whenever we substitute two pairs of fields (X_1, Y_1), (X_2, Y_2) that are both *solutions* of the system, the right-hand side of

Chapter 6. Mechanical, electromagnetic and thermodynamic properties

(6.354) will reduce to the first term which is proportional to a *divergence*. Moreover when integrated over the whole domain the divergence gives just a surface integral which conveniently turns out to *vanish* by the boundary conditions. Since the factor multiplying the divergence is strictly positive, and since the terms on the left-hand side are manifestly *non-negative*, it follows that each of these terms must *vanish separately*. This leads (almost) immediately to the required conclusion that the two solutions (X_1, Y_1) and (X_2, Y_2) must *coincide*.

The identity (6.354) on which the Robinson lemma 6.11 is based is a generalization of a simpler identity (Carter, 1971b) to which it reduces when the two pairs of functions are supposed to be at most *infinitesimally* distinct so that the dot operation (6.355) is interpreted as no longer as a finite difference but as a *differential* – or as a derivative with respect to a continuous variation parameter. This original identity – which can be read off directly from (6.354) simply by ignoring the presence of the bars and angle brackets – was sufficient to establish the non-electromagnetic *no-hair* theorem, by excluding the existence of nonzero solutions of the linearized system of equations for perturbations leaving c and J fixed. (It is a general principle of bifurcation theory of non-linear systems that the ruling out of the existence of linearized perturbations is sufficient to exclude the possibility of a corresponding bifurcating continuous sequence of exact solutions. The converse does *not* hold: it is quite possible that a linearized solution may exist *without* implying the existence of a corresponding continuous sequence of exact solutions.) Robinson's lemma 6.12 is based on a generalization of the linearized version of (6.354) (as read off by ignoring the bars and angle brackets) to an analogous identity that applies to the full Lagrangian (6.348) including the variables E and B as well as X and Y. This electromagnetic identity is too complicated to be quoted without considerable risk of a copying error (the interested reader is referred to the original paper of Robinson, 1974) and its discovery required a veritable algebraic *tour de force* (I found it hard enough to find even the comparatively simple linearized version of (6.354)), and the fact that such convenient identities exist at all remains mysterious. No one has yet succeeded in generalizing the linearized electromagnetic identity analogous to (and including as a special case) the nonlinear (but non-electromagnetic) identity (6.354). If such a generalization exists it will certainly be extremely complicated, since even the linearized identity of Robinson (1974) equates (when the field equations are satisfied) a divergence to the sum of no less than *nine distinct* non-negative terms. I

Uniqueness and no-hair theorems

feel strongly that there must be a deep but essentially simple mathematical reason why the identities found so far should exist, and I would conjecture that the generalization required to tie up the problem completely will not be constructed until *after* the discovery of such an underlying explanation, which would presumably show one how to construct the required identities *directly* (without recourse to the algebraic trial and error method to which Robinson and I were obliged to resort).

I shall complete this discussion by drawing attention to a secondary application of Robinson's nonlinear identity (that does not seem to have been noticed before) as a means of providing a *short-cut proof of the Israel theorem* subject to the *restriction* (which is not needed in the more complicated full-scale Israel theorem) that *axisymmetry* is postulated in advance. This application is based simply on noticing that in the static case, for which Y and B are zero, the full electromagnetic Lagrangian (6.348) reduces to the form

$$L_S = \frac{|\nabla X|^2}{2X^2} + 2\frac{|\nabla E|^2}{X} \qquad (6.358)$$

and that the subsequent substitution $X = Z^2$ then leads to the form

$$\tfrac{1}{4}L_S = \frac{|\nabla K|^2 + |\nabla E|^2}{2K^2} \qquad (6.359)$$

in which the right-hand side has exactly the same form as in (6.352) with K and E respectively in place of X and Y. (This symmetry between the gravitational and electrostatic subcases of the Ernst equations has been previously exploited for quite different purposes by Geroch (1974) and by Diaz Alonso and Catenacci (1976).) It follows that simply by substituting K, E in place X, Y in (6.354) we obtain an identity which can be used to establish the conclusions of lemmas 6.5 and 6.6 in the axisymmetric case.

7. An introduction to the theory of the Kerr metric and its perturbations†

S. CHANDRASEKHAR

Introduction

The discovery by Kerr (1963) of the 2-parameter family of metrics associated with his name is one of the principal landmarks in the development of the general theory of relativity. It ranks alongside Schwarzschild's discovery (1916) of the solution associated with his name, within a month of the founding of the theory by Einstein. Schwarzschild's solution is in fact included in the Kerr family as a limiting member.

The special significance of Kerr's solution for astronomy derives from the theorems of Carter (1971) and Robinson (1975) which establish its uniqueness for an exact description of black holes that occur in nature.‡ But Kerr's solution has also surpassing theoretical interest: it has many properties which have the aura of the miraculous about them. These properties are revealed when one considers the problem of the reflection and transmission of waves of different sorts (electromagnetic, gravitational, neutrino, and electron waves) by the Kerr black hole. It is with this basic problem, in the theory of the perturbations of the Kerr metric, that we shall principally be concerned in this account.

In large measure, the disclosure of the many striking features of Kerr geometry followed from the fortunate circumstance that one had, ready at hand, the formalism of Newman and Penrose (1962) which is peculiarly adapted to its study. We shall, accordingly, begin with an elementary account of the Newman–Penrose formalism and then proceed to the study of the basic problems in the theory of the perturbations of the Kerr metric. (Indeed, the entire account will be elementary and may not appeal to a sophisticated reader.)

---

† This essay is a selection and an expansion of some of the material covered in a set of four Hermann Weyl Lectures I gave at the Institute for Advanced Study, School of Mathematics (Princeton, New Jersey, USA) in October 1975.
‡ These aspects of Kerr's solution are considered by Carter in chapter 6.

7.1 The tetrad formalism

7.1.1 The tetrad representation

In the standard tetrad formalism, one sets up, at each point of spacetime, a basis of four contravariant vectors

$$e_{(a)}{}^i \quad (a = 1, 2, 3, 4), \tag{7.1}$$

where enclosure in parenthesis distinguishes the tetrad indices from the tensor indices. Associated with the contravariant vectors (7.1), we have the covariant vectors

$$e_{(a)i} = g_{ik} e_{(a)}{}^k, \tag{7.2}$$

where g_{ik} denotes the metric tensor. In addition, we also define the inverse, $e^{(b)}{}_i$ of the matrix $e_{(a)}{}^i$ (with the tetrad index labelling the rows and the tensor index labelling the columns) so that

$$e_{(a)}{}^i e^{(b)}{}_i = \delta_{(a)}{}^{(b)} \quad \text{and} \quad e_{(a)}{}^i e^{(a)}{}_j = \delta^i{}_j, \tag{7.3}$$

where the summation convention with respect to the indices of the two sorts, independently, is assumed (here and elsewhere). Further, as a part of the definition, we shall also assume that

$$e_{(a)}{}^i e_{(b)i} = \eta_{(a)(b)} = \text{constant} \tag{7.4}$$

where $\eta_{(a)(b)}$ is a constant symmetric matrix with signature $(+, -, -, -)$. One most often supposes that $\eta_{(a)(b)}$ is diagonal and Minkowskian so that the vectors $e_{(a)}$ form an orthonormal basis; but we shall not suppose this. And finally, let $\eta^{(a)(b)}$ be the inverse of the matrix $\eta_{(a)(b)}$ so that

$$\eta^{(a)(b)} \eta_{(b)(c)} = \delta^{(a)}{}_{(c)}. \tag{7.5}$$

As a consequence of the foregoing definitions, we have the identities,

$$\eta_{(a)(b)} e^{(a)}{}_i = e_{(b)i}, \quad \eta^{(a)(b)} e_{(a)i} = e^{(b)}{}_i; \tag{7.6}$$

and most importantly

$$e_{(a)i} e^{(a)}{}_j = g_{ij}. \tag{7.7}$$

Given any tensor, we can project it onto the tetrad frame to obtain its tetrad components. Thus,

$$\left. \begin{array}{l} A_{(a)} = e_{(a)i} A^i = e_{(a)}{}^j A_j, \\ A^{(a)} = e^{(a)}{}_j A^j = e^{(a)j} A_j; \\ A^i = e_{(a)}{}^i A^{(a)} = e^{(a)i} A_{(a)}. \end{array} \right\} \tag{7.8}$$

and

Chapter 7. The Kerr metric and its perturbations

More generally, we have

$$T_{(a)(b)} = e_{(a)}{}^i e_{(b)}{}^j T_{ij} = e_{(a)}{}^i T_{i(b)},$$

and (7.9)

$$T_{ij} = e^{(a)}{}_i T_{(a)j} = e^{(a)}{}_i e^{(b)}{}_j T_{(a)(b)}.$$

It is clear from the foregoing formulae that we can pass freely from the tensor to the tetrad components, and vice versa; and also that we can raise and lower the tetrad indices with $\eta^{(a)(b)}$ and $\eta_{(a)(b)}$ even as we can raise and lower the tensor indices with the metric tensor.

7.1.2 Directional derivatives; Ricci rotation-coefficients

The contravariant vectors, $e_{(a)}$, considered as tangent vectors, define the directional derivatives

$$e_{(a)} = e_{(a)}{}^i \frac{\partial}{\partial x^i}; \tag{7.10}$$

and we shall write

$$\phi_{,a} = e_{(a)}{}^i \frac{\partial \phi}{\partial x^i} = e_{(a)}{}^i \phi_{,i}, \tag{7.11}$$

where ϕ is any scalar function. More generally, we define

$$A_{(a),(b)} = e_{(b)}{}^i \frac{\partial}{\partial x^i} A_{(a)} = e_{(b)}{}^i \frac{\partial}{\partial x^i} e_{(a)}{}^j A_j$$

$$= e_{(b)}{}^i [e_{(a)}{}^j A_{j,i} + A_k e_{(a)}{}^k{}_{,i}]. \tag{7.12}$$

In the second line of the foregoing equation, we can replace the ordinary ('comma') derivatives by covariant ('semi-colon') derivatives (for symmetric connections, which we shall always assume). We, thus, obtain

$$A_{(a),(b)} = A_{j;i} e_{(a)}{}^j e_{(b)}{}^i + e_{(a)}{}^k{}_{;i} A_k e_{(b)}{}^i$$

$$= A_{j;i} e_{(a)}{}^j e_{(b)}{}^i + e_{(a)k;i} e_{(b)}{}^i e_{(c)}{}^k A^{(c)}. \tag{7.13}$$

With the definition,

$$\gamma_{(c)(a)(b)} = e_{(c)}{}^k e_{(a)k;i} e_{(b)}{}^i, \tag{7.14}$$

we can write

$$A_{(a),(b)} = e_{(a)}{}^i A_{i;j} e_{(b)}{}^j + \gamma_{(c)(a)(b)} A^{(c)}. \tag{7.15}$$

The tetrad formalism

The quantities $\gamma_{(c)(a)(b)}$, which we have defined in (7.14), are the so-called *Ricci rotation-coefficients*. These coefficients are antisymmetric in the first pair of indices:

$$\gamma_{(c)(a)(b)} = -\gamma_{(a)(c)(b)} \qquad (7.16)$$

– a fact which may be verified by expanding the identity

$$\eta_{(a)(b);k} = [e_{(a)i}e_{(b)}{}^i]_{;k} = 0. \qquad (7.17)$$

In view of this antisymmetry, it is clear that there are 24 independent rotation coefficients.

We can write equation (7.15) alternatively in the form

$$e_{(a)}{}^i A_{i;j} e_{(b)}{}^j = A_{(a),(b)} - \eta^{(n)(m)} \gamma_{(n)(a)(b)} A_{(m)}. \qquad (7.18)$$

The quantity on the right-hand side of equation (7.18) is called the *intrinsic derivative* of $A_{(a)}$ in the direction (b) and written $A_{(a)|(b)}$; thus,

$$A_{(a)|(b)} = A_{i;j} e_{(a)}{}^i e_{(b)}{}^j. \qquad (7.19)$$

This notion of intrinsic derivatives is readily extended. Thus, if R_{ijkl} is some fourth-order tensor, we write

$$R_{(a)(b)(c)(d)|(f)} = R_{ijkl;m} e_{(a)}{}^i e_{(b)}{}^j e_{(c)}{}^k e_{(d)}{}^l e_{(f)}{}^m. \qquad (7.20)$$

Now expanding

$$R_{(a)(b)(c)(d),(f)} = [R_{ijkl} e_{(a)}{}^i e_{(b)}{}^j e_{(c)}{}^k e_{(d)}{}^l]_{;m} e_{(f)}{}^m \qquad (7.21)$$

and replacing the covariant derivatives of the different basis vectors by the respective rotation-coefficients, we find (analogous to (7.18))

$$R_{(a)(b)(c)(d)|(f)} = R_{(a)(b)(c)(d),(f)} - \eta^{(n)(m)} [\gamma_{(n)(a)(f)} R_{(m)(b)(c)(d)}$$
$$+ \gamma_{(n)(b)(f)} R_{(a)(m)(c)(d)} \qquad (7.22)$$
$$+ \gamma_{(n)(c)(f)} R_{(a)(b)(m)(d)} + \gamma_{(n)(d)(f)} R_{(a)(b)(c)(m)}].$$

Finally, it is important to observe that the evaluation of the rotation coefficients does not require the evaluation of covariant derivatives (and, therefore, of the Christoffel symbols). For, defining

$$\lambda_{(a)(b)(c)} = e_{(b)i,j} [e_{(a)}{}^i e_{(c)}{}^j - e_{(a)}{}^j e_{(c)}{}^i], \qquad (7.23)$$

and rewriting it in the form

$$\lambda_{(a)(b)(c)} = [e_{(b)i,j} - e_{(b)j,i}] e_{(a)}{}^i e_{(c)}{}^j, \qquad (7.24)$$

we observe that we can replace the ordinary derivatives of $e_{(b)}$ by the

Chapter 7. The Kerr metric and its perturbations

covariant derivatives and write

$$\lambda_{(a)(b)(c)} = [e_{(b)i;j} - e_{(b)j;i}]e_{(a)}{}^i e_{(c)}{}^j \qquad (7.25)$$
$$= \gamma_{(a)(b)(c)} - \gamma_{(c)(b)(a)}.$$

By virtue of this last relation, we have

$$\gamma_{(a)(b)(c)} = \tfrac{1}{2}[\lambda_{(a)(b)(c)} + \lambda_{(c)(a)(b)} - \lambda_{(b)(c)(a)}]; \qquad (7.26)$$

and as is manifest from equation (7.23) the evaluation of $\lambda_{(a)(b)(c)}$ requires only the evaluation of ordinary derivatives.

7.1.3 The commutation relations and the structure constants

The commutator,

$$[e_{(a)}, e_{(b)}], \qquad (7.27)$$

plays an important role in the subsequent developments. The commutator, being a tangent vector itself, can be expanded in terms of the same basis, $e_{(a)}$. We can, accordingly, write

$$[e_{(a)}, e_{(b)}] = C^{(c)}{}_{(a)(b)} e_{(c)}. \qquad (7.28)$$

The coefficients, $C^{(c)}{}_{(a)(b)}$, in this expansion, are the *structure constants*.

The structure constants can be expressed in terms of the rotation coefficients as follows. Consider the effect of the commutator on a scalar function f. We have

$$\left.\begin{aligned}
[e_{(a)}, e_{(b)}]f &= e_{(a)}{}^i [e_{(b)}{}^j f_{,j}]_{,i} - e_{(b)}{}^i [e_{(a)}{}^j f_{,j}]_{,i} \\
&= [e_{(a)}{}^i e_{(b)}{}^j{}_{;i} - e_{(b)}{}^i e_{(a)}{}^j{}_{;i}] f_{,j} \\
&= [-\gamma_{(b)}{}^j{}_{(a)} + \gamma_{(a)}{}^j{}_{(b)}] f_{,j} \\
&= [-\gamma_{(b)}{}^{(c)}{}_{(a)} + \gamma_{(a)}{}^{(c)}{}_{(b)}] e_{(c)}{}^j f_{,j}.
\end{aligned}\right\} \qquad (7.29)$$

Comparison with equation (7.28) shows that

$$C^{(c)}{}_{(a)(b)} = \gamma^{(c)}{}_{(b)(a)} - \gamma^{(c)}{}_{(a)(b)}. \qquad (7.30)$$

7.1.4 The Ricci and the Bianchi identities

Projecting the Ricci identity

$$e_{(a)i;k;l} - e_{(a)i;l;k} = R_{mikl} e_{(a)}{}^m, \qquad (7.31)$$

The Newman–Penrose formalism

where R_{mikl} is the Riemann tensor, onto the tetrad frame, we have

$$R_{(a)(b)(c)(d)} = R_{mikl}e_{(a)}{}^m e_{(b)}{}^i e_{(c)}{}^k e_{(d)}{}^l$$
$$= \{-[\gamma_{(a)(f)(g)}e^{(f)}{}_{;i}e^{(g)}{}_k]_{;l} + [\gamma_{(a)(f)(g)}e^{(f)}{}_{;i}e^{(g)}{}_l]_{;k}\}e_{(b)}{}^i e_{(c)}{}^k e_{(d)}{}^l,$$
(7.32)

and expanding the quantities in the square brackets, on the right-hand side, and replacing, once again, the covariant derivatives of the basis vectors by the respective rotation-coefficients, we obtain

$$R_{(a)(b)(c)(d)} = -\gamma_{(a)(b)(c),(d)} + \gamma_{(a)(b)(d),(c)} + \gamma_{(a)(b)(f)}[\gamma^{(f)}{}_{(c)(d)} - \gamma^{(f)}{}_{(d)(c)}]$$
$$+ \gamma_{(a)(f)(c)}\gamma^{(f)}{}_{(b)(d)} - \gamma_{(a)(f)(d)}\gamma^{(f)}{}_{(b)(c)}.$$
(7.33)

Because of the antisymmetry of the rotation coefficients in the first pair of their indices and the manifest antisymmetry of the operation by which the tetrad components of the Riemann tensor are constructed from them, it is evident that we can write 36 equations of the kind (7.33).

If we identify R_{ijkl} in equation (7.20) with the Riemann tensor, the Bianchi identity,

$$R_{ij[kl;m]} = 0,$$
(7.34)

expressed in terms of intrinsic derivatives takes the form (cf. equation (7.22))

$$0 = R_{(a)(b)[(c)(d)|(f)]}$$
$$= \sum_{[(c)(d)(f)]} \{R_{(a)(b)(c)(d),(f)}$$
$$- \eta^{(n)(m)}[\gamma_{(n)(a)(f)}R_{(m)(b)(c)(d)} + \gamma_{(n)(b)(f)}R_{(a)(m)(c)(d)}$$
$$+ \gamma_{(n)(c)(f)}R_{(a)(b)(m)(d)} + \gamma_{(n)(d)(f)}R_{(a)(b)(c)(m)}]\}.$$
(7.35)

7.2 The Newman–Penrose formalism

7.2.1 The null basis and the spin coefficients

The Newman–Penrose formalism differs from the tetrad formalism described in section 7.1 only in the manner of choice of the basis vectors: instead of an orthonormal basis, the choice is made of a complex null-basis $(l, n, m, \bar{m})$ where l and n are two real null-vectors

Chapter 7. The Kerr metric and its perturbations

and m and $\bar{m}$ are a pair of complex conjugate null-vectors; and the null vectors are, further, assumed to satisfy the orthogonality relations

$$l \cdot n = 1, \quad m \cdot \bar{m} = -1, \quad l \cdot m = l \cdot \bar{m} = n \cdot m = n \cdot \bar{m} = 0. \tag{7.36}$$

There are several advantages to the Newman–Penrose choice of the basis vectors. One of them is the fact that, since the basis is complex, it will suffice to write down only half the number of equations. However, its special advantage for the study of the Kerr metric (and, more generally, algebraically special fields) derives from the Goldberg–Sachs theorem which we prove in section 7.3.4.

The null basis, which underlies the Newman–Penrose formalism, can be derived from an orthonormal basis, (e_t, e_x, e_y, e_z), where e_t is timelike and e_x, e_y, and e_z are spacelike, in the manner

$$\begin{aligned} e_1 = l = (e_t + e_z)/\sqrt{2}, \quad & e_2 = n = (e_t - e_z)/\sqrt{2}, \\ e_3 = m = (e_x + ie_y)/\sqrt{2}, \quad & e_4 = \bar{m} = (e_x - ie_y)/\sqrt{2}. \end{aligned} \tag{7.37}$$

For the basis so chosen,

$$\eta_{ab} = \eta^{ab} = \begin{vmatrix} 0 & 1 & 0 & 0 \\ 1 & 0 & 0 & 0 \\ 0 & 0 & 0 & -1 \\ 0 & 0 & -1 & 0 \end{vmatrix}; \tag{7.38}†$$

and the covariant basis is given by

$$e^1 = e_2 = n, \quad e^2 = e_1 = l, \quad e^3 = -e_4 = -\bar{m}, \quad \text{and} \quad e^4 = -e_3 = -m. \tag{7.39}$$

The basis vectors, considered as directional derivatives, are designated by special symbols:

$$e_1 = e^2 = D, \quad e_2 = e^1 = \Delta, \quad e_3 = -e^4 = \delta, \quad \text{and} \quad e_4 = -e^3 = \delta^*. \tag{7.40}$$

And the various Ricci rotation-coefficients, now called *spin coefficients*, are also designated by special symbols:

$$\left.\begin{aligned} \gamma_{311} &= \kappa; & \gamma_{314} &= \tilde{\rho}; & \tfrac{1}{2}(\gamma_{211} + \gamma_{341}) &= \varepsilon; \\ \gamma_{313} &= \sigma; & \gamma_{243} &= \mu; & \tfrac{1}{2}(\gamma_{212} + \gamma_{342}) &= \gamma; \\ \gamma_{244} &= \lambda; & \gamma_{312} &= \tau; & \tfrac{1}{2}(\gamma_{214} + \gamma_{344}) &= \alpha; \\ \gamma_{242} &= \nu; & \gamma_{241} &= \pi; & \tfrac{1}{2}(\gamma_{213} + \gamma_{343}) &= \beta. \end{aligned}\right\} \tag{7.41}$$

---

† Here and in the sequel we shall suppress the distinguishing parenthesis for the tetrad indices so long as there is no ambiguity as to what is intended.

Note that by interchanging the indices 3 and 4 we obtain the complex conjugates of the respective quantities.

7.2.2 The representation of the Weyl and the Ricci tensors

The Weyl tensor is the trace-free part of the Riemann tensor; and its tetrad components are given by

$$C_{abcd} = R_{abcd} + \tfrac{1}{2}(\eta_{ac}R_{bd} - \eta_{bc}R_{ad} - \eta_{ad}R_{bc} + \eta_{bd}R_{ac}) \\ - \tfrac{1}{6}(\eta_{ac}\eta_{bd} - \eta_{ad}\eta_{bc})R, \tag{7.42}$$

where R_{bc} denotes the tetrad components of the Ricci tensor and R is the curvature scalar:

$$R_{bc} = \eta^{ad}R_{abcd} \quad \text{and} \quad R = \eta^{ab}R_{ab}. \tag{7.43}$$

The fact that C_{abcd} is trace-free requires

$$\eta^{ad}C_{abcd} = C_{1bc2} + C_{2bc1} - C_{3bc4} - C_{4bc3} = 0; \tag{7.44}$$

and we must also require

$$C_{1234} + C_{1423} + C_{1342} = 0. \tag{7.45}$$

These requirements, when written out explicitly, yield the following relations:

$$C_{3114} = C_{4113} = C_{3224} = C_{4223} = C_{1332} = C_{1442} = 0, \tag{7.46}$$

and

$$\left. \begin{array}{l} C_{1212} = C_{3434}, \quad C_{1231} = C_{1334}, \quad C_{1241} = C_{1443}, \\ C_{1232} = C_{2343}, \quad C_{1242} = C_{2434}, \\ C_{1342} = \tfrac{1}{2}(C_{1212} - C_{1234}) = \tfrac{1}{2}(C_{3434} - C_{1234}). \end{array} \right\} \tag{7.47}$$

In the Newman–Penrose formalism, the ten independent components of the Weyl tensor are represented by the five complex scalars,

$$\left. \begin{array}{l} \Psi_0 = -C_{1313} = -C_{pqrs}l^p m^q l^r m^s, \\ \Psi_1 = -C_{1213} = -C_{pqrs}l^p n^q l^r m^s, \\ \Psi_2 = -C_{1342} = -C_{pqrs}l^p m^q \bar{m}^r n^s, \\ \Psi_3 = -C_{1242} = -C_{pqrs}l^p n^q \bar{m}^r n^s, \\ \Psi_4 = -C_{2424} = -C_{pqrs}n^p \bar{m}^q n^r \bar{m}^s. \end{array} \right\} \tag{7.48}$$

Besides these scalars, representing the Weyl tensor, we also define the

Chapter 7. The Kerr metric and its perturbations

following scalars representing the Ricci tensor:

$$\left.\begin{array}{ll} \Phi_{00} = -\tfrac{1}{2}R_{11}; & \Phi_{01} = \Phi_{10}^{*} = -\tfrac{1}{2}R_{13} = -\tfrac{1}{2}R_{14}^{*}, \\ \Phi_{11} = -\tfrac{1}{4}(R_{12}+R_{34}); & \Phi_{12} = \Phi_{21}^{*} = -\tfrac{1}{2}R_{23} = -\tfrac{1}{2}R_{24}^{*}, \\ \Phi_{22} = -\tfrac{1}{2}R_{22}; & \Phi_{02} = \Phi_{20}^{*} = -\tfrac{1}{2}R_{33} = -\tfrac{1}{2}R_{44}^{*}, \end{array}\right\} \quad (7.49)$$

and

$$\Lambda = R/24 = (R_{12} - R_{34})/12.$$

7.2.3 The commutation relations and the structure constants

We shall now proceed to write down the explicit forms of the various equations of the theory.

We consider first the commutation relation (equation (7.30))

$$[e_{(a)}, e_{(b)}] = (\gamma_{cba} - \gamma_{cab})e^{c} = C^{c}{}_{ab}e_{c}. \quad (7.50)$$

As an example consider

$$\begin{aligned}[][\Delta, D] = [n, l] = [e_2, e_1] &= (\gamma_{c12} - \gamma_{c21})e^{c} \\ &= -\gamma_{121}e^{1} + \gamma_{212}e^{2} + (\gamma_{312} - \gamma_{321})e^{3} + (\gamma_{412} - \gamma_{421})e^{4} \quad (7.51) \\ &= -\gamma_{121}\Delta + \gamma_{212}D - (\gamma_{312} - \gamma_{321})\delta^{*} - (\gamma_{412} - \gamma_{421})\delta, \end{aligned}$$

or, on substituting for the spin coefficients their named symbols given in equations (7.41), we obtain

$$\Delta D - D\Delta = (\gamma + \gamma^{*})D + (\varepsilon + \varepsilon^{*})\Delta - (\tau^{*} + \pi)\delta - (\tau + \pi^{*})\delta^{*}. \quad (7.52)$$

In similar fashion, we find

$$\delta D - D\delta = (\alpha^{*} + \beta - \pi^{*})D + \kappa\Delta - (\tilde{\rho}^{*} + \varepsilon - \varepsilon^{*})\delta - \sigma\delta^{*}, \quad (7.53)$$

$$\delta\Delta - \Delta\delta = -\nu^{*}D + (\tau - \alpha^{*} - \beta)\Delta + (\mu - \gamma + \gamma^{*})\delta + \lambda^{*}\delta^{*}, \quad (7.54)$$

$$\delta^{*}\delta - \delta\delta^{*} = (\mu^{*} - \mu)D + (\tilde{\rho}^{*} - \tilde{\rho})\Delta + (\alpha - \beta^{*})\delta + (\beta - \alpha^{*})\delta^{*}. \quad (7.55)$$

By expressing the foregoing relations in the manner of (7.50), we find that the structure constants are related to the spin coefficients as in the accompanying tabulation:

$$\left.\begin{array}{lll} C_{21}^{1} = (\gamma + \gamma^{*}); & C_{31}^{3} = -(\tilde{\rho}^{*} + \varepsilon - \varepsilon^{*}); & C_{32}^{4} = \lambda^{*}; \\ C_{21}^{2} = (\varepsilon + \varepsilon^{*}); & C_{31}^{4} = -\sigma; & C_{43}^{1} = \mu^{*} - \mu, \\ C_{21}^{3} = -(\tau^{*} + \pi); & C_{32}^{1} = -\nu^{*}; & C_{43}^{2} = \tilde{\rho}^{*} - \tilde{\rho}, \\ C_{21}^{4} = -(\tau + \pi^{*}); & C_{32}^{2} = \tau - \alpha^{*} - \beta; & C_{43}^{3} = \alpha - \beta^{*}, \\ C_{31}^{1} = \alpha^{*} + \beta - \pi^{*}; & C_{32}^{3} = \mu - \gamma + \gamma^{*}, & C_{43}^{4} = \beta - \alpha^{*}, \\ C_{31}^{2} = \kappa. \end{array}\right\} \quad (7.56)$$

7.2.4 The Ricci identities

The explicit form which the Ricci identities (7.33) take in the Newman–Penrose formalism can now be written down. Thus considering the (1313)-component of (7.33), we have (cf. (7.48)),

$$-\Psi_0 = C_{1313} = R_{1313} = \gamma_{133,1} - \gamma_{131,3}$$
$$+ \gamma_{133}(\gamma_{121} + \gamma_{431} - \gamma_{413} + \gamma_{431} + \gamma_{134})$$
$$- \gamma_{131}(\gamma_{433} + \gamma_{123} - \gamma_{213} + \gamma_{231} + \gamma_{132});$$
(7.57)

or, substituting for the directional derivatives and the spin coefficients their named symbols, we obtain

$$D\sigma - \delta\kappa = \sigma(3\varepsilon - \varepsilon^* + \tilde{\rho} + \tilde{\rho}^*) + \kappa(\pi^* - \tau - 3\beta - \alpha^*) + \Psi_0.$$
(7.58)

As we have noted in the context of the standard tetrad formalism (section 7.1.4) we can write down a total of 36 equations by considering the various different cases of equation (7.33). But in the context of the Newman–Penrose formalism, it will suffice to write down only half the number of equations (by omitting to write down the complex conjugate of an equation). We list below a set of 18 equations (equations 7.59) as originally derived by Newman and Penrose (1962); and we have indicated in each case the component of the Riemann tensor which gives rise to the equation.

Equations (7.59)

$$D\tilde{\rho} - \delta^*\kappa = (\tilde{\rho}^2 + \sigma\sigma^*) + \tilde{\rho}(\varepsilon + \varepsilon^*) \qquad [R_{1314}] \qquad (a)$$
$$- \kappa^*\tau - \kappa(3\alpha + \beta^* - \pi) + \Phi_{00},$$

$$D\sigma - \delta\kappa = \sigma(\tilde{\rho} + \tilde{\rho}^* + 3\varepsilon - \varepsilon^*) \qquad [R_{1313}] \qquad (b)$$
$$- \kappa(\tau - \pi^* + \alpha^* + 3\beta) + \Psi_0,$$

$$D\tau - \Delta\kappa = \tilde{\rho}(\tau^* + \pi) + \sigma(\tau^* + \pi) + \tau(\varepsilon - \varepsilon^*) \qquad [R_{1312}] \qquad (c)$$
$$- \kappa(3\gamma + \gamma^*) + \Psi_1 + \Phi_{01},$$

$$D\alpha - \delta^*\varepsilon = \alpha(\tilde{\rho} + \varepsilon^* - 2\varepsilon) + \beta\sigma^* - \beta^*\varepsilon - \kappa\lambda \qquad [\tfrac{1}{2}(R_{3414} - R_{1214})] \quad (d)$$
$$- \kappa^*\gamma + \pi(\varepsilon + \tilde{\rho}) + \Phi_{10},$$

$$D\beta - \delta\varepsilon = \sigma(\alpha + \pi) + \beta(\tilde{\rho}^* - \varepsilon^*) - \kappa(\mu + \gamma) \qquad [\tfrac{1}{2}(R_{1213} - R_{3413})] \quad (e)$$
$$- \varepsilon(\alpha^* - \pi^*) + \Psi_1,$$

Chapter 7. The Kerr metric and its perturbations

$$D\gamma - \Delta\varepsilon = \alpha(\tau + \pi^*) + \beta(\tau^* + \pi) - \gamma(\varepsilon + \varepsilon^*) \qquad [\tfrac{1}{2}(R_{1212} - R_{3412})] \quad (f)$$
$$\qquad - \varepsilon(\gamma + \gamma^*) + \tau\pi - \nu\kappa + \Psi_2 + \Phi_{11} - \Lambda,$$

$$D\lambda - \delta^*\pi = (\tilde{\rho}\lambda + \sigma^*\mu) + \pi(\pi + \alpha - \beta^*) - \nu\kappa^* \qquad [R_{2441}] \qquad (g)$$
$$\qquad - \lambda(3\varepsilon - \varepsilon^*) + \Phi_{20},$$

$$D\mu - \delta\pi = (\tilde{\rho}^*\mu + \sigma\lambda) + \pi(\pi^* - \alpha^* + \beta) \qquad [R_{2431}] \qquad (h)$$
$$\qquad - \mu(\varepsilon + \varepsilon^*) - \nu\kappa + \Psi_2 + 2\Lambda,$$

$$D\nu - \Delta\pi = \mu(\pi + \tau^*) + \lambda(\pi^* + \tau) + \pi(\gamma - \gamma^*) \qquad [R_{2421}] \qquad (i)$$
$$\qquad - \nu(3\varepsilon + \varepsilon^*) + \Psi_3 + \Phi_{21},$$

$$\Delta\lambda - \delta^*\nu = -\lambda(\mu + \mu^* + 3\gamma - \gamma^*) \qquad [R_{2442}] \qquad (j)$$
$$\qquad + \nu(3\alpha + \beta^* + \pi - \tau^*) - \Psi_4,$$

$$\delta\tilde{\rho} - \delta^*\sigma = \tilde{\rho}(\alpha^* + \beta) - \sigma(3\alpha - \beta^*) + \tau(\tilde{\rho} - \tilde{\rho}^*) \qquad [R_{3143}] \qquad (k)$$
$$\qquad + \kappa(\mu - \mu^*) - \Psi_1 + \Phi_{01},$$

$$\delta\alpha - \delta^*\beta = (\mu\tilde{\rho} - \lambda\sigma) + \alpha\alpha^* + \beta\beta^* - 2\alpha\beta \qquad [\tfrac{1}{2}(R_{1234} - R_{3434})] \quad (l)$$
$$\qquad + \gamma(\tilde{\rho} - \tilde{\rho}^*) + \varepsilon(\mu - \mu^*) - \Psi_2 + \Phi_{11} + \Lambda,$$

$$\delta\lambda - \delta^*\mu = \nu(\tilde{\rho} - \tilde{\rho}^*) + \pi(\mu - \mu^*) + \mu(\alpha + \beta^*) \qquad [R_{2443}] \qquad (m)$$
$$\qquad + \lambda(\alpha^* - 3\beta) - \Psi_3 + \Phi_{21},$$

$$\delta\nu - \Delta\mu = (\mu^2 + \lambda\lambda^*) + \mu(\gamma + \gamma^*) \qquad [R_{2423}] \qquad (n)$$
$$\qquad - \nu^*\pi + \nu(\tau - 3\beta - \alpha^*) + \Phi_{22},$$

$$\delta\gamma - \Delta\beta = \gamma(\tau - \alpha^* - \beta) + \mu\tau - \sigma\nu - \varepsilon\nu^* \qquad [\tfrac{1}{2}(R_{1232} - R_{3432})] \quad (o)$$
$$\qquad - \beta(\gamma - \gamma^* - \mu) + \alpha\lambda^* + \Phi_{12},$$

$$\delta\tau - \Delta\sigma = (\mu\sigma + \lambda^*\tilde{\rho}) + \tau(\tau + \beta - \alpha^*) \qquad [R_{1332}] \qquad (p)$$
$$\qquad - \sigma(3\gamma - \gamma^*) - \kappa\nu^* + \Phi_{02},$$

$$\Delta\tilde{\rho} - \delta^*\tau = -(\tilde{\rho}\mu^* + \sigma\lambda) + \tau(\beta^* - \alpha - \tau^*) \qquad [R_{1324}] \qquad (q)$$
$$\qquad + \tilde{\rho}(\gamma + \gamma^*) + \nu\kappa - \Psi_2 - 2\Lambda,$$

$$\Delta\alpha - \delta^*\gamma = \nu(\tilde{\rho} + \varepsilon) - \lambda(\tau + \beta) + \alpha(\gamma^* - \mu^*) \qquad [\tfrac{1}{2}(R_{1242} - R_{3442})] \quad (r)$$
$$\qquad + \gamma(\beta^* - \tau^*) - \Psi_3.$$

7.2.5 The Bianchi identities

In writing down the Bianchi identities (7.35) we shall restrict ourselves to the case of empty space so that we do not need to distinguish between the Weyl and the Riemann tensor.

As an example, consider the (13)-component of (7.35). We have

$$R_{1313|4} + R_{1334|1} + R_{1341|3} = 0. \tag{7.60}$$

We readily verify that the terms derived from R_{1341} vanish (cf. (7.46)). The two remaining terms give

$$\begin{aligned}R_{1313|4} &= R_{1313,4} - \eta^{nm}(\gamma_{n14}R_{m313} + \gamma_{n34}R_{1m13} + \gamma_{n14}R_{13m3} + \gamma_{n34}R_{131m}) \\ &= R_{1313,4} - 2(\gamma_{214} + \gamma_{344})R_{1313} + 2\gamma_{314}(R_{1213} + R_{4313}) \quad (7.61)\\ &= -\delta^*\Psi_0 + 4\alpha\Psi_0 - 4\tilde{\rho}\Psi_1,\end{aligned}$$

and (similarly),

$$R_{1334|1} = D\Psi_1 + 2\varepsilon\Psi_1 - 3\kappa\Psi_2 + \pi\Psi_0. \tag{7.62}$$

We thus obtain

$$\delta^*\Psi_0 - D\Psi_1 = (4\alpha - \pi)\Psi_0 - 2(2\tilde{\rho} + \varepsilon)\Psi_1 + 3\kappa\Psi_2. \tag{7.63}$$

The remaining Bianchi identities can be evaluated in similar fashion. We list below a complete set.

Equations (7.64)

$$D\Psi_1 - \delta^*\Psi_0 = -3\kappa\Psi_2 + 2(\varepsilon + 2\tilde{\rho})\Psi_1 + (\pi - 4\alpha)\Psi_0, \quad (a)$$

$$D\Psi_2 - \delta^*\Psi_1 = -2\kappa\Psi_3 + 3\tilde{\rho}\Psi_2 + 2(\pi - \alpha)\Psi_1 - \lambda\Psi_0, \quad (b)$$

$$D\Psi_3 - \delta^*\Psi_2 = -\kappa\Psi_4 - 2(\varepsilon - \tilde{\rho})\Psi_3 + 3\pi\Psi_2 - 2\lambda\Psi_1, \quad (c)$$

$$D\Psi_4 - \delta^*\Psi_3 = -(4\varepsilon - \tilde{\rho})\Psi_4 + 2(2\pi + \alpha)\Psi_3 - 3\lambda\Psi_2, \quad (d)$$

$$\Delta\Psi_0 - \delta\Psi_1 = (4\gamma - \mu)\Psi_0 - 2(2\tau + \beta)\Psi_1 + 3\sigma\Psi_2, \quad (e)$$

$$\Delta\Psi_1 - \delta\Psi_2 = \nu\Psi_0 + 2(\gamma - \mu)\Psi_1 - 3\tau\Psi_2 + 2\sigma\Psi_3, \quad (f)$$

$$\Delta\Psi_2 - \delta\Psi_3 = 2\nu\Psi_1 - 3\mu\Psi_2 - 2(\tau - \beta)\Psi_3 + \sigma\Psi_4, \quad (g)$$

$$\Delta\Psi_3 - \delta\Psi_4 = 3\nu\Psi_2 - 2(\gamma + 2\mu)\Psi_3 - (\tau - 4\beta)\Psi_4. \quad (h)$$

Chapter 7. The Kerr metric and its perturbations

7.2.6 Maxwell's equations

In the Newman–Penrose formalism, the Maxwell tensor F_{ij} is replaced by the three complex scalars

$$\begin{aligned}\phi_0 &= F_{13} = F_{ij}l^i m^j, \\ \phi_1 &= \tfrac{1}{2}(F_{12}+F_{43}) = \tfrac{1}{2}F_{ij}(l^i n^j + \bar{m}^i m^j), \\ \text{and} \qquad \phi_2 &= F_{42} = F_{ij}\bar{m}^i n^j; \end{aligned} \qquad (7.65)$$

and Maxwell's equations (expressed in terms of intrinsic derivatives), namely,

$$F_{[ab|c]} = 0 \quad \text{and} \quad \eta^{nm} F_{an|m} = 0 \qquad (7.66)$$

are replaced by the equations,

$$\begin{aligned}\phi_{1|1} - \phi_{0|4} &= 0, \quad \phi_{2|1} - \phi_{1|4} = 0, \\ \phi_{1|3} - \phi_{0|2} &= 0, \quad \phi_{2|3} - \phi_{1|2} = 0.\end{aligned} \qquad (7.67)$$

The explicit forms of these equations are readily found. Thus,

$$\begin{aligned}\phi_{1|1} &= \tfrac{1}{2}[F_{12,1} - \eta^{nm}(\gamma_{n11}F_{m2} + \gamma_{n21}F_{1m}) \\ &\quad + F_{43,1} - \eta^{nm}(\gamma_{n41}F_{m3} + \gamma_{n31}F_{4m})] \\ &= \phi_{1,1} - (\gamma_{131}F_{42} + \gamma_{241}F_{13}) \\ &= D\phi_1 + \kappa\phi_2 - \pi\phi_0;\end{aligned} \qquad (7.68)$$

and, similarly,

$$\phi_{0|4} = \delta^*\phi_0 - 2\alpha\phi_0 + 2\tilde{\rho}\phi_1. \qquad (7.69)$$

The first of Maxwell's equations (7.67), therefore, takes the form

$$D\phi_1 - \delta^*\phi_0 = (\pi - 2\alpha)\phi_0 + 2\tilde{\rho}\phi_1 - \kappa\phi_2. \qquad (7.70)$$

And the remaining equations are

$$D\phi_2 - \delta^*\phi_1 = -\lambda\phi_0 + 2\pi\phi_1 + (\tilde{\rho} - 2\varepsilon)\phi_2, \qquad (7.71)$$

$$\delta\phi_1 - \Delta\phi_0 = (\mu - 2\gamma)\phi_0 + 2\tau\phi_1 - \sigma\phi_2, \qquad (7.72)$$

$$\delta\phi_2 - \Delta\phi_1 = -\nu\phi_0 + 2\mu\phi_1 + (\tau - 2\beta)\phi_2. \qquad (7.73)$$

7.3 Tetrad transformations and related matters

7.3.1 Tetrad transformations

Having chosen a tetrad frame – either an orthonormal frame as in the conventional tetrad formalism, or a complex null-frame as in the Newman–Penrose formalism – at each point in spacetime, we can subject the frame to Lorentz transformations (in some continuous manner through spacetime). Corresponding to the six parameters of the group of Lorentz transformations, we have six degrees of freedom to rotate a chosen tetrad frame to another. In considering the effect of such Lorentz transformations on the various Newman–Penrose quantities, we shall find it convenient to regard a general Lorentz transformation of the basis $(l, n, m, \bar{m})$ as made up of the following three classes of rotations.

(a) Rotations of class I which leave the vector l unchanged;
(b) rotations of class II which leave the vector n unchanged;
(c) rotations of class III which leave the directions of l and n unchanged; but rotate m (and $\bar{m}$) by an angle θ in the $(m, \bar{m})$-plane.

Associated with the foregoing three classes of rotation, we have the following explicit transformations (which, as may be readily verified, preserve the underlying orthogonality relations):

$$\text{I.} \quad l \to l, \; m \to m + al, \; \bar{m} \to \bar{m} + a^*l,$$
$$n \to n + a^*m + a\bar{m} + aa^*l, \tag{7.74}$$

$$\text{II.} \quad n \to n, \; m \to m + bn, \; \bar{m} \to \bar{m} + b^*n,$$
$$l \to l + b^*m + b\bar{m} + bb^*n, \tag{7.75}$$

and

$$\text{III.} \quad l \to A^{-1}l, \; n \to An,$$
$$m \to e^{i\theta}m, \; \bar{m} \to e^{-i\theta}\bar{m}, \tag{7.76}$$

where a and b are two complex functions and A and θ are two real functions.

The effect of a rotation of class I on the various Newman–Penrose quantities is readily found. Thus, considering the Weyl scalars Ψ_0 and Ψ_1, we have

$$-\Psi_0 = C_{1313} \to C_{pqrs}l^p(m^q + al^q)l^r(m^s + al^s)$$
$$= C_{pqrs}l^p m^q l^r m^s = -\Psi_0; \tag{7.77}$$

Chapter 7. The Kerr metric and its perturbations

and

$$-\Psi_1 = C_{1213} \rightarrow C_{pqrs}l^p(n^q + a^*m^q + a\bar{m}^q + aa^*l^q)l^r(m^s + al^s)$$
$$= C_{pqrs}l^p(n^q + a^*m^q)l^r m^s \qquad (7.78)$$
$$= C_{1213} + a^*C_{1313} = -(\Psi_1 + a^*\Psi_0),$$

where in the reductions for Ψ_1, apart from the obvious symmetries of the Weyl tensor, the fact that $C_{1413} = 0$ (cf. (7.47)) has been used. The effects on the remaining Weyl scalars are similarly found. We find

$$\left.\begin{array}{l}\Psi_0 \rightarrow \Psi_0, \quad \Psi_1 \rightarrow \Psi_1 + a^*\Psi_0, \quad \Psi_2 \rightarrow \Psi_2 + 2a^*\Psi_1 + (a^*)^2\Psi_0 \\ \Psi_3 \rightarrow \Psi_3 + 3a^*\Psi_2 + 3(a^*)^2\Psi_1 + (a^*)^3\Psi_0, \\ \Psi_4 \rightarrow \Psi_4 + 4a^*\Psi_3 + 6(a^*)^2\Psi_2 + 4(a^*)^3\Psi_1 + (a^*)^4\Psi_0.\end{array}\right\} \quad (7.79)$$

In a similar fashion, we find that the spin-coefficients are transformed as follows:

$$\text{Transformations (7.80)}$$

$$\kappa \rightarrow \kappa; \quad \sigma \rightarrow \sigma + a\kappa; \quad \tilde{\rho} \rightarrow \tilde{\rho} + a^*\kappa; \quad \varepsilon \rightarrow \varepsilon + a^*\kappa;$$
$$\tau \rightarrow \tau + a\tilde{\rho} + a^*\sigma + aa^*\kappa; \quad \pi \rightarrow \pi + 2a^*\varepsilon + (a^*)^2\kappa + \mathrm{D}a^*;$$
$$\alpha \rightarrow \alpha + a^*(\tilde{\rho} + \varepsilon) + (a^*)^2\kappa; \quad \beta \rightarrow \beta + a\varepsilon + a^*\sigma + aa^*\kappa;$$
$$\gamma \rightarrow \gamma + a\alpha + a^*(\beta + \tau) + aa^*(\tilde{\rho} + \varepsilon) + (a^*)^2\sigma + a(a^*)^2\kappa;$$
$$\lambda \rightarrow \lambda + a^*(2\alpha + \pi) + (a^*)^2(\tilde{\rho} + 2\varepsilon) + (a^*)^3\kappa + \delta^*a^* + a^*\mathrm{D}a^*;$$
$$\mu \rightarrow \mu + a\pi + 2a^*\beta + 2aa^*\varepsilon + (a^*)^2\sigma + a(a^*)^2\kappa + \delta a^* + a\mathrm{D}a^*,$$
$$\nu \rightarrow \nu + a\lambda + a^*(\mu + 2\gamma) + (a^*)^2(\tau + 2\beta) + (a^*)^3\sigma + aa^*(\pi + 2\alpha)$$
$$\quad + a(a^*)^2(\tilde{\rho} + 2\varepsilon) + a(a^*)^3\kappa + (\Delta + a^*\delta + a\delta^* + aa^*\mathrm{D})a^*.$$

And the scalars representing the Maxwell field become:

$$\phi_0 \rightarrow \phi_0, \quad \phi_1 \rightarrow \phi_1 + a^*\phi_0, \quad \phi_2 \rightarrow \phi_2 + 2a^*\phi_1 + (a^*)^2\phi_0. \quad (7.81)$$

The corresponding effects of a rotation of class II on the various Newman–Penrose quantities can be readily written down from the foregoing formulae, since the effect of interchanging l and n results in the transformation

$$\Psi_0 \rightleftarrows \Psi_4^*, \quad \Psi_1 \rightleftarrows \Psi_3^*, \quad \Psi_2 \rightleftarrows \Psi_2^*, \quad \phi_0 \rightleftarrows -\phi_2^*, \quad \phi_1 \rightleftarrows -\phi_1^*,$$
$$\kappa \rightleftarrows -\nu^*, \quad \tilde{\rho} \rightleftarrows -\mu^*, \quad \sigma \rightleftarrows -\lambda^*, \quad \alpha \rightleftarrows -\beta^*, \quad \varepsilon \rightleftarrows -\gamma^*, \quad (7.82)$$
$$\text{and} \quad \pi \rightleftarrows -\tau^*.$$

Tetrad transformations and related matters

In particular, the effect of a rotation of class II on the Weyl scalars is

$$\left.\begin{aligned}
\Psi_0 &\to \Psi_0 + 4b\Psi_1 + 6b^2\Psi_2 + 4b^3\Psi_3 + b^4\Psi_4, \\
\Psi_1 &\to \Psi_1 + 3b\Psi_2 + 3b^2\Psi_3 + b^3\Psi_4, \\
\Psi_2 &\to \Psi_2 + 2b\Psi_3 + b^2\Psi_4, \quad \Psi_3 \to \Psi_3 + b\Psi_4, \quad \Psi_4 \to \Psi_4.
\end{aligned}\right\} \quad (7.83)$$

And finally, we write down the effects of a rotation of class III on the Newman–Penrose quantities

Equations (7.84)

$$\Psi_0 \to A^{-2} e^{2i\theta} \Psi_0; \quad \Psi_1 \to A^{-1} e^{i\theta}\Psi_1, \quad \Psi_2 \to \Psi_2$$

$$\Psi_3 \to A\, e^{-i\theta}\Psi_3, \quad \text{and} \quad \Psi_4 \to A^2\, e^{-2i\theta}\Psi_4;$$

$$\phi_0 \to A^{-1} e^{i\theta}\phi_0, \quad \phi_1 \to \phi_1, \quad \phi_2 \to A\, e^{-i\theta}\phi_2;$$

$$\kappa \to A^{-2} e^{i\theta}\kappa, \quad \sigma \to A^{-1} e^{2i\theta}\sigma, \quad \tilde{\rho} \to A^{-1}\tilde{\rho}, \quad \tau \to e^{i\theta}\tau,$$

$$\pi \to e^{-i\theta}\pi, \quad \lambda \to A\, e^{-2i\theta}\lambda, \quad \mu \to A\mu, \quad \nu \to A^2 e^{-i\theta}\nu$$

$$\gamma \to A\gamma - \tfrac{1}{2}\Delta A + \tfrac{1}{2}iA\Delta\theta,$$

$$\varepsilon \to A^{-1}\varepsilon - \tfrac{1}{2}A^{-2}DA + \tfrac{1}{2}iA^{-1}D\theta,$$

$$\alpha \to e^{-i\theta}\alpha + \tfrac{1}{2}i\, e^{-i\theta}\delta^*\theta - \tfrac{1}{2}A^{-1} e^{-i\theta}\delta^*A,$$

$$\beta \to e^{i\theta}\beta + \tfrac{1}{2}i\, e^{i\theta}\delta\theta - \tfrac{1}{2}A^{-1} e^{i\theta}\delta A.$$

7.3.2 Geometric considerations

Consider the congruence of the null vectors l. By expanding

$$l_{i;j} = e^{(a)}{}_i \gamma_{(a)(1)(b)} e^{(b)}{}_j, \quad (7.85)$$

and giving the spin coefficients their named symbols, we find

$$\begin{aligned}
l_{i;j} = &(\varepsilon + \varepsilon^*)l_i n_j + (\gamma + \gamma^*)l_i l_j - (\alpha^* + \beta)l_i \bar{m}_j - (\alpha + \beta^*)l_i m_j \\
&- \kappa \bar{m}_i n_j - \kappa^* m_i n_j + \sigma \bar{m}_i \bar{m}_j + \sigma^* m_i m_j \\
&- \tau \bar{m}_i l_j - \tau^* m_i l_j + \tilde{\rho}\bar{m}_i m_j + \tilde{\rho}^* m_i \bar{m}_j.
\end{aligned} \quad (7.86)$$

From this equation it follows that

$$l_{i;j} l^j = (\varepsilon + \varepsilon^*)l_i - \kappa \bar{m}_i - \kappa^* m_i. \quad (7.87)$$

Accordingly, if $\kappa = 0$, *the l-vectors will form a congruence of null geodesics; and the geodesics will, in addition, be affinely parametrized if, by a rotation of class* III *(which will not affect the directions of l or an*

Chapter 7. The Kerr metric and its perturbations

initially vanishing κ), we make $\varepsilon = 0$. And when $\kappa = \varepsilon = 0$, (7.86) becomes

$$l_{i;j} = (\gamma + \gamma^*)l_i l_j - (\alpha^* + \beta)l_i \bar{m}_j - (\alpha + \beta^*)l_i m_j - \tau \bar{m}_i l_j \\ + \sigma \bar{m}_i \bar{m}_j + \sigma^* m_i m_j + \tilde{\rho} \bar{m}_i m_j + \tilde{\rho}^* m_i \bar{m}_j - \tau^* m_i l_j. \quad (7.88)$$

From this last equation we find

$$l_{[i;j]} = -(\alpha^* + \beta - \tau)l_{[i}\bar{m}_{j]} - (\alpha + \beta^* - \tau^*)l_{[i}m_{j]} + (\tilde{\rho} - \tilde{\rho}^*)\bar{m}_{[i}m_{j]},$$

and

$$l_{[i;j}l_{k]} = (\tilde{\rho} - \tilde{\rho}^*)\bar{m}_{[i}m_j l_{k]}. \quad (7.89)$$

*The congruence of the **l**s will therefore be hypersurface orthogonal* (i.e. proportional to a gradient) *if $\tilde{\rho}$ is real and will be equal to a gradient field if, in addition, $\alpha^* + \beta = \tau$.*

Further consequences of (7.88) are:

$$\left.\begin{aligned} \tfrac{1}{2}l^i{}_{;i} &= -\tfrac{1}{2}(\tilde{\rho} + \tilde{\rho}^*) = \theta^2 \quad \text{(say)}, \\ l_{[i;j]}l^{i;j} &= -\tfrac{1}{2}(\tilde{\rho} - \tilde{\rho}^*)^2 = \omega^2 \quad \text{(say)}, \\ l_{(i;j)}l^{i;j} &= 2(\theta^2 + |\sigma|^2). \end{aligned}\right\} \quad (7.90)$$

and

The quantities θ, ω, and σ we have defined are the so-called *optical scalars*; and they describe the expansion, the twist, and the shear, respectively, of the congruence.

A consideration, related to the foregoing, is the change which the tetrad experiences when it is subjected to an infinitesimal displacement ξ. The change is given by

$$\delta e_{(a)i} = e_{(a)i;j}\xi^j = e^{(b)}{}_i \gamma_{(b)(a)(c)} e^{(c)}{}_j \xi^j \\ = \gamma_{(b)(a)(c)} e^{(b)}{}_i \xi^c = \Omega_{(a)(b)} e^{(b)}{}_i \quad \text{(say)} \quad (7.91)$$

where

$$\Omega_{(a)(b)} = -\gamma_{(a)(b)(c)}\xi^{(c)}. \quad (7.92)$$

The tensor components of this *rotation matrix*, Ω, are given by

$$\Omega_{ij} = e^{(a)}{}_i e^{(b)}{}_j \gamma_{(b)(a)(c)}\xi^{(c)}. \quad (7.93)$$

Accordingly, for a displacement ξ in the direction of (c),

$$\tfrac{1}{2}\Omega_{ij(c)} = e^{(a)}{}_{[i} e^{(b)}{}_{j]} \gamma_{(b)(a)(c)}. \quad (7.94)$$

From this equation, we find that for a displacement in the direction of l

$$\tfrac{1}{2}\Omega_{ij(1)} = n_{[i}l_{j]}\gamma_{211} + \bar{m}_{[i}m_{j]}\gamma_{431} - n_{[i}m_{j]}\gamma_{411} \\ - l_{[i}m_{j]}\gamma_{421} - l_{[i}\bar{m}_{j]}\gamma_{321} - n_{[i}\bar{m}_{j]}\gamma_{311},$$
(7.95)

or giving the spin coefficients their named symbols, we have

$$\tfrac{1}{2}\Omega_{ij(1)} = (\varepsilon + \varepsilon^*)n_{[i}l_{j]} - \kappa n_{[i}\bar{m}_{j]} + \pi l_{[i}m_{j]} \\ - (\varepsilon - \varepsilon^*)\bar{m}_{[i}m_{j]} - \kappa^* n_{[i}m_{j]} + \pi^* l_{[i}\bar{m}_{j]}.$$
(7.96)

Consequently, if the l-congruence is one of geodesics affinely parametrized, (so that $\kappa = \varepsilon = 0$), and if, further, we make $\pi = 0$ by a rotation of class I (which will not affect l, nor κ or ε if they are initially zero), then $\Omega_{ij(1)} \equiv 0$; and the remaining basis vectors of the tetrad, n, m, and $\bar{m}$, will be parallelly propagated.

Again, considering a displacement in the direction of m, we have

$$\tfrac{1}{2}\Omega_{ij(3)} = (\alpha^* + \beta)n_{[i}l_{j]} + (\alpha^* - \beta)\bar{m}_{[i}m_{j]} + \lambda^* l_{[i}m_{j]} \\ - \sigma n_{[i}m_{j]} - \tilde{\rho}^* n_{[i}m_{j]} + \mu l_{[i}\bar{m}_{j]}.$$
(7.97)

From this last equation, it follows that the change in the direction of l (by a displacement in the direction of m) is determined solely by $\tilde{\rho}$ and σ: the Re $\tilde{\rho}$ measures the convergence of the congruence, while the Im $\tilde{\rho}$, the Re σ, and the Im σ measure, respectively, the deviation of l in the directions of $-im$, $\bar{m}$, and $i\bar{m}$; in particular, the Im $\tilde{\rho}$ measures the rotation (or twist) of the congruence while the Re σ and the Im σ measure the two independent shearing motions.

7.3.3 The Petrov classification and algebraically special fields

Let $\Psi_4 \neq 0$. (If it should be zero in the chosen frame, we can make it non-vanishing by a rotation of class I, so long as spacetime is not flat.) Consider a rotation of class II with parameter b. The Weyl scalars Ψ_0, Ψ_1, etc., become (cf. (7.83))

$$\Psi_0 \to \Psi_0 + 4b\Psi_1 + 6b^2\Psi_2 + 4b^3\Psi_3 + b^4\Psi_4, \\ \Psi_1 \to \Psi_1 + 3b\Psi_2 + 3b^2\Psi_3 + b^3\Psi_4, \quad \text{etc.}$$
(7.98)

It is clear that by a rotation of this class, we can make $\Psi_0 = 0$, if b is a root of the equation,

$$\Psi_0 + 4b\Psi_1 + 6b^2\Psi_2 + 4b^3\Psi_3 + b^4\Psi_4 = 0.$$
(7.99)

Chapter 7. The Kerr metric and its perturbations

This equation has always four roots and the corresponding new directions of l, namely, $l + b^*m + b\bar{m} + bb^*n$, are called the *principal null-directions* of the field. If some of the roots coincide, the field is said to be *algebraically special*. And the various ways in which the roots can coincide lead to the Petrov classification.

(a) *Petrov type I*. In this case, all four roots of (7.99) are distinct. Let them be b_1, b_2, b_3, and b_4. Then by a rotation of class II, with parameter $b = b_1$ (say), we can make Ψ_0 vanish. Next, by a rotation of class I, we can make $\Psi_4 = 0$ (without affecting the vanishing of Ψ_0). With Ψ_0 and Ψ_4 made to vanish in this manner, Ψ_1, Ψ_2, and Ψ_3 will be left non-vanishing; and of these three scalars, $\Psi_1 \cdot \Psi_3$, and Ψ_2 will be invariant to the remaining rotations of class III.

(b) *Petrov type II*. Let (7.99) allow two coincident roots: $b_1 = b_2 (\neq b_3 \neq b_4)$. In this case, besides (7.99), its derivative, namely,

$$\Psi_1 + 3b\Psi_2 + 3b^2\Psi_3 + b^3\Psi_4 = 0 \tag{7.100}$$

will also be satisfied when $b = b_1(=b_2)$. Therefore, by a rotation of class II, with parameter $b = b_1(=b_2)$, Ψ_0 and Ψ_1 will vanish simultaneously. Then, by a rotation of class I, we can make Ψ_4 vanish. Only Ψ_2 and Ψ_3 will be left non-vanishing; and of them only Ψ_2 will be invariant to rotations of class III.

(c) *Petrov type D*. If two pairs of roots coincide and $b_1 = b_2$ and $b_3 = b_4 (\neq b_1$ or $b_2)$, it is clear that by a rotation of class II, with parameter $b = b_1(=b_2)$ we can make Ψ_0 and Ψ_1 vanish simultaneously, while by a rotation of class I, we can also make Ψ_4 and Ψ_3 vanish. And Ψ_2 will be the only remaining non-vanishing scalar, invariant to rotations of class III.

(d) *Petrov type III*. If three of the roots coincide and $b_1 = b_2 = b_3 \neq b_4$, then by a rotation of class II, with parameter $b = b_1 (=b_2 = b_3)$ we can make Ψ_0, Ψ_1, and Ψ_2 vanish simultaneously; and by a rotation of class I, we can make Ψ_4 vanish. And Ψ_3 will be the only non-vanishing scalar.

(e) *Petrov type N*. If all four roots coincide and there is only one distinct root, b (say), then by a rotation of class II, with parameter b, we can make Ψ_0, Ψ_1, Ψ_2, and Ψ_3, all, vanish simultaneously; and Ψ_4 will be the only non-vanishing scalar.

Tetrad transformations and related matters

We shall now derive some simple conditions for l being a principal null-vector. Consider

$$Q_{ij} = C_{ipqj}l^p l^q = C_{i(1)(1)j}. \qquad (7.101)$$

We readily verify that Q_{ij} has the expansion

$$Q_{ij} = q_{11}l_i l_j - \Psi_1(l_i \bar{m}_j + l_j \bar{m}_i) - \Psi_1^*(l_i m_j + l_j m_i) + \Psi_0 \bar{m}_i \bar{m}_j + \Psi_0^* m_i m_j, \qquad (7.102)$$

where besides the definitions of Ψ_0 and Ψ_1, the fact that $C_{4113} = C_{3114} = 0$ (see (7.46)) has been used. From (7.102), we obtain the relations

$$Q_{i[j}l_{k]} = (\Psi_0^* m_i - \Psi_1^* l_i) m_{[j} l_{k]} + (\Psi_0 \bar{m}_i - \Psi_1 l_i) \bar{m}_{[j} l_{k]} \qquad (7.103)$$

and

$$l_{[h} Q_{i][j} l_{k]} = \Psi_0^* m_{[i} l_{h]} m_{[j} l_{k]} + \Psi_0 \bar{m}_{[i} l_{h]} \bar{m}_{[j} l_{k]}. \qquad (7.104)$$

Consequently, the conditions that l be a principal null-vector are

$$l_{[h} C_{i]pq[j} l_{k]} l^p l^q = 0 \quad \text{for } \Psi_0 = 0 \qquad (7.105)$$

and

$$C_{ipq[j} l_{k]} l^p l^q = 0 \quad \text{for } \Psi_1 = \Psi_0 = 0. \qquad (7.106)$$

7.3.4 The Goldberg–Sachs theorem

The theorem states that *in empty space, if the Riemann tensor is algebraically special and $\Psi_0 = \Psi_1 = 0$, then $\kappa = \sigma = 0$; and conversely.* In other words, the theorem relates algebraic speciality with the geodesic and shear-free character of the congruence formed by the principal null direction. The proof which follows is due to Newman and Penrose (1962).

To prove that, when $\Psi_0 = \Psi_1 = 0$, the congruence formed by the vector l is geodesic ($\kappa = 0$) and shear-free ($\sigma = 0$) is straightforward. Thus, when $\Psi_0 = \Psi_1 = 0$, the Bianchi identities (7.64a–h) give

Equations (7.107)

$$3\kappa \Psi_2 = 0, \qquad (a)$$

$$D\Psi_2 = -2\kappa \Psi_3 + 3\tilde{\rho} \Psi_2, \qquad (b)$$

$$D\Psi_3 - \delta^* \Psi_2 = -\kappa \Psi_4 - 2(\varepsilon - \tilde{\rho})\Psi_3 + 3\pi \Psi_2, \qquad (c)$$

$$3\sigma \Psi_2 = 0, \qquad (d)$$

Chapter 7. The Kerr metric and its perturbations

$$-\delta\Psi_2 = -3\tau\Psi_2 + 2\sigma\Psi_3, \qquad (e)$$

$$\Delta\Psi_2 - \delta\Psi_3 = -3\mu\Psi_2 - 2(\tau-\beta)\Psi_3 + \sigma\Psi_4. \qquad (f)$$

If $\kappa \neq 0$ and $\sigma \neq 0$, then it follows from (7.107a) and (7.107d) that $\Psi_2 = 0$. Then (7.107b) and (7.107e) show $\Psi_3 = 0$; and, finally, (7.107c) and (7.107f) show that $\Psi_4 = 0$. Thus, if $\kappa \neq 0$ and $\sigma \neq 0$, $\Psi_0 = \Psi_1 = 0$ implies that the remaining Weyl scalars must also vanish; and since we are dealing with empty spacetime, it follows that it is flat. Therefore, if spacetime is not flat, $\Psi_0 = \Psi_1 = 0$ implies that $\kappa = \sigma = 0$, i.e. the congruence of the principal null directions is geodesic and shear-free.

To prove the converse, namely that if $\kappa = \sigma = 0$, then $\Psi_0 = \Psi_1 = 0$, is somewhat less direct. First, we shall suppose that by a rotation of class III (which will not affect the vanishing of κ and σ), ε has been made to vanish. Then, the Ricci identities, (a), (b), (c), (e), and (k) of (7.59), when $\kappa = \sigma = 0$, become

(i) $D\tilde{\rho} = \tilde{\rho}^2$; (ii) $\Psi_0 = 0$; (iii) $D\tau = (\tau + \tilde{\pi}^*)\tilde{\rho} + \Psi_1$; (7.108)

(iv) $D\beta = \tilde{\rho}^*\beta + \Psi_1$; and (v) $\delta\tilde{\rho} = (\alpha^* + \beta)\tilde{\rho} + (\tilde{\rho} - \tilde{\rho}^*)\tau - \Psi_1$;

(7.109)

and the Bianchi identities (7.64) when $\kappa = \sigma = 0$ and $\Psi_0 = 0$ (by equation (ii) above) give

(vi) $\delta\Psi_1 = 2(2\tau + \beta)\Psi_1$; and (vii) $D\Psi_1 = 4\tilde{\rho}\Psi_1$. (7.110)

And, finally, the commutation relation (7.53) gives

$$\delta D - D\delta = (\pi^* - \alpha^* - \beta)D + \tilde{\rho}^*\delta \qquad (\kappa = \sigma = \varepsilon = 0). \quad (7.111)$$

From (7.80), it is clear that by a rotation of class I (which will not affect Ψ_0 nor the vanishing of κ, σ, or ε), we can arrange that $\tau = 0$, *provided $\tilde{\rho} \neq 0$.* (If, on the other hand, $\tilde{\rho}$ should vanish, it would follow from equation (v) that $\Psi_1 = 0$ which is the result that we wish to prove.) Assuming, then, that τ has been made to vanish, we have (cf. equations (vi) and (vii))

$$\delta \log \Psi_1 = 2\beta \quad \text{and} \quad D \log \Psi_1 = 4\tilde{\rho}; \qquad (7.112)$$

and, accordingly,

$$(D\delta - \delta D) \log \Psi_1 = 2D\beta - 4\delta\tilde{\rho}; \qquad (7.113)$$

or, on using the relations (iv) and (v), we have

$$(D\delta - \delta D) \log \Psi_1 = 2\beta\tilde{\rho}^* - 4(\alpha^* + \beta)\tilde{\rho} + 6\Psi_1. \qquad (7.114)$$

The Kerr metric and the perturbation problem

On the other hand, by applying the commutation relation (7.111) to $\log \Psi_1$, and making use of equations (vi) and (vii) (in (7.110)), we obtain

$$(D\delta - \delta D) \log \Psi_1 = (\pi^* - \alpha^* - \beta) D \log \Psi_1 + \tilde{\rho}^* \delta \log \Psi_1$$
$$= 4(\pi^* - \alpha^* - \beta)\tilde{\rho} + 2\beta\tilde{\rho}^*. \tag{7.115}$$

Equating now the right-hand sides of (7.114) and (7.115), we find

$$\Psi_1 = \tfrac{2}{3}\pi^* \tilde{\rho}; \tag{7.116}$$

whereas equation (iii) (in (7.108)) with $\tau = 0$, requires

$$\Psi_1 = -\pi^* \tilde{\rho}; \tag{7.117}$$

and this is clearly incompatible with (7.116) since we have assumed that $\tilde{\rho} \neq 0$. Hence $\Psi_1 = 0$; and the theorem stated follows.

A corollary to the Goldberg–Sachs theorem is that if the field is algebraically special and of Petrov type D, then the congruences formed by the two principal null-directions, l and n, must both be geodesic and shear-free, i.e. $\kappa = \sigma = \nu = \lambda = 0$ when $\Psi_0 = \Psi_1 = \Psi_3 = \Psi_4 = 0$; and conversely. The particular significance of these remarks for the study of the Kerr metric will become apparent in section 7.4.3.

7.4 The Kerr metric and the perturbation problem

7.4.1 The Kerr metric: an outline of its derivation

It is customary in accounts of the Kerr metric to posit it as a manna from Kerr and proceed from there without any ado. And if on occasion one feels that some explanation for this brusque procedure is called for, then one says: 'There is no constructive analytic derivation of the metric that is adequate in its physical ideas, and even a check of this solution of Einstein's equation involves cumbersome calculations' (Landau and Lifshitz, 1975). Actually, it is not difficult to physically motivate the particular form which the Kerr metric has in the most commonly used coordinates of Boyer and Lindquist (1967) and isolate the single place where the verification is 'cumbersome' (but, even then, not unduly so).

Now, it is known that a general stationary axisymmetric metric can be written in the form (in units in which $c = 1$ and $G = 1$)

$$ds^2 = e^{2\nu}(dt)^2 - e^{2\psi}(d\varphi - \omega dt)^2 - e^{2\mu_2}(dx^2)^2 - e^{2\mu_3}(dx^3)^2, \tag{7.118}$$

where φ denotes the azimuthal angle in the equatorial plane and $x^2(=r)$ and $x^3(=\theta)$ are the two remaining spatial coordinates. Also, in (7.118),

Chapter 7. The Kerr metric and its perturbations

ν, ψ, ω, μ_2, and μ_3 are functions of x^2 and x^3 only. Besides, we have the freedom of gauge to impose any coordinate condition we may choose on μ_2 and μ_3.

The two most important features of the Kerr spacetime, which makes it of special relevance to astronomy, are, first, its possession of an *event horizon* and, second, its *asymptotic flatness* – in other words, it represents a black hole. These are also the features that make for its uniqueness. In some ways, it would be most satisfying if one could derive the Kerr metric as the solution of Einstein's equations which satisfies the boundary conditions appropriate for a black hole. While such a derivation does not exist, it is not difficult to incorporate, from the outset, the possession of an event horizon in Carter's (1973) limited sense of a *Killing horizon: it is a convex smooth two-dimensional null-surface, in a stationary axisymmetric spacetime, which is spanned by the two Killing vectors $\partial/\partial t$ and $\partial/\partial\varphi$ of the assumed geometry.*

In accordance with the assumed stationarity and axisymmetry of the metric, let the equation of the event horizon be

$$N(x^2, x^3) = 0 \qquad (7.119)$$

The condition that it be null is

$$g^{ij} N_{,i} N_{,j} = 0. \qquad (7.120)$$

For the chosen form of the metric, (7.120) gives

$$e^{2(\mu_3 - \mu_2)} (N_{,r})^2 + (N_{,\theta})^2 = 0. \qquad (7.121)$$

Since we have the gauge freedom to impose any coordinate condition we may choose on μ_2 and μ_3, we shall suppose that

$$e^{2(\mu_3 - \mu_2)} = \Delta(r), \qquad (7.122)$$

where $\Delta(r)$ is some function of r which we shall leave unspecified for the present. From (7.121) it now follows that the equation of the null-surface is, in fact, given by

$$\Delta(r) = 0. \qquad (7.123)$$

The second condition, that the null-surface be spanned by the two Killing vectors $\partial/\partial t$ and $\partial/\partial\varphi$, requires that

$$e^{2(\psi+\nu)} = e^{2\beta} = 0 \quad \text{on } \Delta(r) = 0 \quad (\beta = \psi + \nu). \qquad (7.124)$$

Since we have left $\Delta(r)$ unspecified, we may now suppose that

$$e^\beta = \Delta^{1/2} f(r, \theta), \qquad (7.125)$$

where $f(r, \theta)$ is some function of r and θ which is regular on $\Delta(r)=0$ and on the axis $\theta=0$. We shall suppose, instead, that e^β has the somewhat more restricted form

$$e^\beta = \Delta^{1/2} f(\theta) \tag{7.126}$$

where $f(\theta)$ is a function of θ only; and we ask whether Einstein's equations allow such a separable solution.

Now, when one writes down the Einstein field equations, which are appropriate to the chosen form (7.118) of the metric, one finds that e^β must satisfy the equation (for details see Chandrasekhar and Friedman, 1972 and Chandrasekhar, 1978a)

$$[e^{\mu_3-\mu_2}(e^\beta)_{,2}]_{,2} + [e^{\mu_2-\mu_3}(e^\beta)_{,3}]_{,3} = 0. \tag{7.127}$$

For $e^{\mu_3-\mu_2}$ and e^β given by (7.122) and (7.125), the foregoing equation gives

$$[\Delta^{1/2}(\Delta^{1/2})_{,r}]_{,r} + f_{,\theta\theta}/f = 0. \tag{7.128}$$

A solution of this equation, compatible with the requirement of regularity at the poles, is determined by

$$\Delta_{,rr} = 2 \quad \text{and} \quad f = \sin\theta; \tag{7.129}$$

and the solution for Δ that is appropriate is

$$\Delta = r^2 + a^2 - 2Mr, \tag{7.130}$$

where a and M are certain constants.

Thus, with a choice of gauge that is consistent with the existence of a Killing horizon, we have the solution

$$e^{\mu_3-\mu_2} = \Delta^{1/2} \quad \text{and} \quad e^\beta = e^{\psi+\nu} = \Delta^{1/2} \sin\theta. \tag{7.131}$$

With the particular solutions for the two metric functions given in (7.131), we can rewrite the metric (7.118) in the 'canonical form'

$$ds^2 = (\Delta^{1/2} \sin\theta)\left[\chi(dt)^2 - \frac{1}{\chi}(d\varphi - \omega\, dt)^2\right] - \frac{e^{\mu_2+\mu_3}}{\Delta^{1/2}}[(dr)^2 + \Delta(d\theta)^2], \tag{7.132}$$

where

$$\chi = e^{-\psi+\nu}. \tag{7.133}$$

To complete the solution, we must specify the three remaining functions χ, ω, and $\mu_2+\mu_3$. And for these three functions, Einstein's equations provide the following equations (for details, see Chandrasekhar

Chapter 7. The Kerr metric and its perturbations

1978a)):

$$\tfrac{1}{2}(X+Y)[(\Delta X_{,2})_{,2}+(\delta X_{,3})_{,3}] = \Delta(X_{,2})^2+\delta(X_{,3})^2, \quad (7.134)$$

$$\tfrac{1}{2}(X+Y)[(\Delta Y_{,2})_{,2}+(\delta Y_{,3})_{,3}] = \Delta(Y_{,2})^2+\delta(Y_{,3})^2, \quad (7.135)$$

and

$$2(r-M)\frac{\partial}{\partial r}(\mu_2+\mu_3)+2\mu\frac{\partial}{\partial \mu}(\mu_2+\mu_3)-\frac{1}{\chi^2}(\Delta X_{,2}Y_{,2}-\delta X_{,3}Y_{,3})$$
$$+3\frac{M^2-a^2}{\Delta}-\frac{1}{\delta}=0, \quad (7.136)$$

where

$$X = \chi+\omega, \quad Y = \chi-\omega, \quad (7.137)$$

and

$$\delta = 1-\mu^2 = \sin^2\theta, \quad (7.138)$$

and the index 3 now refers to $\mu(=\cos\theta)$ instead of θ.

Equations (7.134) and (7.135) are the basic equations of the theory: on them depends the solution of Einstein's equation for stationary axisymmetric fields. For, once (7.134) and (7.135) have been solved for X and Y, the solution for $\mu_2+\mu_3$ can be obtained by solving a single ordinary differential equation (to which the solution of (7.136) can be reduced).

Now, the 'miracle' of the Kerr metric is that (7.134) and (7.135) are satisfied by

$$X = \frac{\Delta-a^2\delta}{\delta^{1/2}(\rho^2\Delta^{1/2}-2aMr\delta^{1/2})} \quad \text{and} \quad Y = \frac{\Delta-a^2\delta}{\delta^{1/2}(\rho^2\Delta^{1/2}+2aMr\delta^{1/2})}, \quad (7.139)$$

where

$$\rho^2 = r^2+a^2\cos^2\theta. \quad (7.140)$$

The verification that the foregoing expressions for X and Y do satisfy (7.134) and (7.135) *is* 'cumbersome' (but not unduly so!). The corresponding solutions for χ and ω are

$$\chi = e^{-\psi+\nu} = \frac{\rho^2\Delta^{1/2}}{\Sigma^2\delta^{1/2}} \quad \text{and} \quad \omega = \frac{2aMr}{\Sigma^2}, \quad (7.141)$$

where

$$\Sigma^2 = (r^2+a^2)^2-a^2\Delta\delta\left(=\frac{\rho^4\Delta-4a^2M^2r^2\delta}{\Delta-a^2\delta}\right). \quad (7.142)$$

The Kerr metric and the perturbation problem

With the solutions for X and Y given in (7.139), (7.136) reduces to

$$(r-M)\frac{\partial}{\partial r}(\mu_2+\mu_3)+\mu\frac{\partial}{\partial \mu}(\mu_2+\mu_3)=2-\frac{(r-M)^2}{\Delta}-2\frac{rM}{\rho^2}. \quad (7.143)$$

There is no difficulty in solving this elementary partial differential equation; and we find that

$$e^{\mu_2+\mu_3}=\rho^2\Delta^{-1/2}. \quad (7.144)$$

Collecting the various results ((7.131), (7.141), and (7.144)) we have the following solutions for the metric coefficients:

$$\left.\begin{array}{l} e^{2\nu}=\dfrac{\rho^2\Delta}{\Sigma^2}, \quad e^{2\psi}=\dfrac{\Sigma^2\sin^2\theta}{\rho^2}, \quad \omega=\dfrac{2aMr}{\Sigma^2}, \\[6pt] e^{2\mu_2}=\dfrac{\rho^2}{\Delta}, \quad \text{and} \quad e^{2\mu_3}=\rho^2; \end{array}\right\} \quad (7.145)$$

and this is Kerr's solution.

We first observe that when $a=0$ the Kerr metric, as we have written it, reduces to the Schwarzschild metric in its standard form. Also, from the asymptotic behaviours,

$$e^{2\nu}\to 1-\frac{2M}{r}, \quad e^{2\psi}\to r^2\sin^2\theta,$$

$$\omega\to\frac{2aM}{r^3}, \quad e^{-2\mu_2}\to 1-\frac{2M}{r}, \quad \text{and} \quad e^{2\mu_3}\to r^2, \quad (7.146)$$

of the various metric coefficients, it is clear that the Kerr metric approaches the Schwarzschild metric as $r\to\infty$. We conclude that the Kerr spacetime is asymptotically flat and, further, that the parameter M is to be identified with the mass of the black hole. Again, from the interpretation of ω as representing the 'dragging of the inertial frame' (i.e. the fact that, to an observer at infinity, one locally at rest will appear as rotating with an angular velocity differing from $d\varphi/dt$ by ω), we conclude from its asymptotic behaviour ($\omega\to 2aM/r^3$) that a is to be identified with the specific angular momentum of the black hole.

And finally, we shall write down the explicit forms of the matrices representing the covariant and the contravariant forms of the Kerr metric:

$$(g_{ij})=\begin{vmatrix} 1-2Mr/\rho^2 & 0 & 0 & 2aMr\sin^2\theta/\rho^2 \\ 0 & -\rho^2/\Delta & 0 & 0 \\ 0 & 0 & -\rho^2 & 0 \\ 2aMr\sin^2\theta/\rho^2 & 0 & 0 & -(r^2+a^2+2a^2Mr\sin^2\theta/\rho^2)\sin^2\theta \end{vmatrix},$$

$$(7.147)$$

Chapter 7. The Kerr metric and its perturbations

and

$$(g^{ij}) = \begin{vmatrix} \Sigma^2/\rho^2\Delta & 0 & 0 & 2aMr/\rho^2\Delta \\ 0 & -\Delta/\rho^2 & 0 & 0 \\ 0 & 0 & -1/\rho^2 & 0 \\ 2aMr/\rho^2\Delta & 0 & 0 & -(\Delta - a^2\sin^2\theta)/(\rho^2\Delta\sin^2\theta) \end{vmatrix}.$$

(7.148)

7.4.2 The separability of the Hamilton–Jacobi equation

The separability of the Hamilton–Jacobi equation in Kerr geometry, the first of its many remarkable features, was discovered by Carter (1968). We shall start with a demonstration of this fact, principally, for obtaining the equations governing the null-geodesics.

Now the Hamilton–Jacobi equation,

$$\frac{\partial S}{\partial \lambda} = \frac{1}{2} g^{ij} \frac{\partial S}{\partial x^i} \frac{\partial S}{\partial x^j},$$

(7.149)

for the principal function S, when g^{ij} has the form given in (7.148), becomes

$$2\frac{\partial S}{\partial \lambda} = \frac{\Sigma^2}{\rho^2\Delta}\left(\frac{\partial S}{\partial t}\right)^2 + \frac{4aMr}{\rho^2\Delta}\frac{\partial S}{\partial t}\frac{\partial S}{\partial \varphi} - \frac{\Delta - a^2\sin^2\theta}{\rho^2\Delta\sin^2\theta}\left(\frac{\partial S}{\partial \varphi}\right)^2$$

$$- \frac{\Delta}{\rho^2}\left(\frac{\partial S}{\partial r}\right)^2 - \frac{1}{\rho^2}\left(\frac{\partial S}{\partial \theta}\right)^2.$$

(7.150)

An alternative form of this equation is

$$2\frac{\partial S}{\partial \lambda} = \frac{1}{\Delta\rho^2}\left[(r^2+a^2)\frac{\partial S}{\partial t} + a\frac{\partial S}{\partial \varphi}\right]^2 - \frac{1}{\rho^2\sin^2\theta}\left[(a\sin^2\theta)\frac{\partial S}{\partial t} + \frac{\partial S}{\partial \varphi}\right]^2$$

(7.151)

$$- \frac{\Delta}{\rho^2}\left(\frac{\partial S}{\partial r}\right)^2 - \frac{1}{\rho^2}\left(\frac{\partial S}{\partial \theta}\right)^2.$$

We shall now verify that (7.151) allows a separable solution of the form

$$S = \tfrac{1}{2}m^2\lambda - Et + L_z\varphi + S_r(r) + S_\theta(\theta),$$

(7.152)

where m (the rest mass), E (the energy), and L_z (the angular momentum about the z-axis) are the three (obviously) conserved quantities, and S_r and S_θ (as the notation indicates) are functions of r and θ only. Inserting

The Kerr metric and the perturbation problem

this assumed form for S in (7.151), we find, after some rearrangements:

$$\left\{ m^2 r^2 - \frac{1}{\Delta} [(r^2 + a^2)E - aL_z]^2 + (L_z - aE)^2 + \Delta \left(\frac{dS_r}{dr}\right)^2 \right\}$$

$$+ \left\{ [a^2 m^2 + (L_z^2 \csc^2 \theta - a^2 E^2)] \cos^2 \theta + \left(\frac{dS_\theta}{d\theta}\right)^2 \right\} = 0. \quad (7.153)$$

We observe that the two expressions in curly brackets on the left-hand side of (7.153) are functions, separately, of r and θ only. Accordingly, the two expressions must separately be constants, of equal magnitude and of opposite signs. The occurrence of a new separation constant, $\mathcal{Q}$ (say), which we thus infer, corresponds to the existence of a new integral of the geodesic equations.

With the definitions

$$P = (r^2 + a^2)E - aL_z,$$
$$R = P^2 - \Delta[m^2 r^2 + (L_z - aE)^2 + \mathcal{Q}], \quad (7.154)$$
$$\Theta = \mathcal{Q} - [a^2(m^2 - E^2) + L_z^2 \csc^2 \theta] \cos^2 \theta,$$

we can now write

$$p_r = \frac{dS_r}{dr} = \frac{\sqrt{R}}{\Delta} \quad \text{and} \quad p_\theta = \frac{dS_\theta}{d\theta} = \sqrt{\Theta}, \quad (7.155)$$

where the signs of $\sqrt{R}$ and $\sqrt{\Theta}$ can be chosen independently of one another (but once made, must be retained consistently). And the solution for S is

$$S = \tfrac{1}{2} m^2 \lambda - Et + L_z \varphi + \int^r \frac{\sqrt{R}}{\Delta} dr + \int^\theta \sqrt{\Theta} \cdot d\theta. \quad (7.156)$$

By differentiating (7.156) with respect to m^2, $\mathcal{Q}$, E, and L_z, and setting each of the derivatives equal to zero, we obtain the equations

$$\int^\theta \frac{d\theta}{\sqrt{\Theta}} = \int^r \frac{dr}{\sqrt{R}},$$

$$\lambda = \int^r \frac{r^2}{\sqrt{R}} dr + a^2 \int^\theta \frac{\cos^2 \theta}{\sqrt{\Theta}} d\theta,$$

$$t = \int^r \frac{(r^2 + a^2)P}{\Delta \sqrt{R}} dr + a \int^\theta \frac{L_z - aE \sin^2 \theta}{\sqrt{\Theta}} d\theta, \quad (7.157)$$

$$\varphi = a \int^r \frac{P}{\Delta \sqrt{R}} dr + \int^\theta \frac{L_z \csc^2 \theta - aE}{\sqrt{\Theta}} d\theta.$$

Chapter 7. The Kerr metric and its perturbations

By writing the foregoing equations in differential form, we can obtain the following equations which represent the solution of the geodesic equations:

$$\rho^2 \frac{dr}{d\lambda} = \sqrt{R}, \qquad \rho^2 \frac{d\theta}{d\lambda} = \sqrt{\Theta}, \qquad \rho^2 \frac{d\varphi}{d\lambda} = (L_z \csc^2\theta - aE) + \frac{aP}{\Delta},$$

and

$$\rho^2 \frac{dt}{d\lambda} = \frac{(r^2 + a^2)P}{\Delta} - a(aE \sin^2\theta - L_z). \tag{7.158}$$

Equations (7.158) describe, equally, the null geodesics provided we set $m = 0$ in (7.154) defining the functions R and Θ.

7.4.3 The principal null congruences and the type-D character of the Kerr metric

Particularly simple classes of null-geodesics are selected by the equations derived in section 7.4.2, when

$$p^\theta = 0 \text{ (requiring } \Theta = 0\text{)} \quad \text{and} \quad L_z = aE \sin^2\theta. \tag{7.159}$$

(The second relation between the two constants of motion, L_z and E, is permissible since, by virtue of the first requirement, $d\theta/d\lambda = 0$.) From (7.154), we now find that under the assumptions (7.159) (and $m = 0$)

$$\mathscr{Q} = -a^2 E^2 \cos^4\theta, \quad P = \rho^2 E, \quad \text{and} \quad R = P^2 = \rho^4 E^2; \tag{7.160}$$

and the equations governing the null-geodesics become

$$\frac{dt}{d\lambda} = \frac{r^2 + a^2}{\Delta} E, \quad \frac{dr}{d\lambda} = \pm E, \quad \frac{d\theta}{d\lambda} = 0, \quad \text{and} \quad \frac{d\varphi}{d\lambda} = \frac{a}{\Delta} E \tag{7.161}$$

Equations (7.161) allow us to define the two null-vectors

$$(l^i) = \frac{1}{\Delta}(r^2 + a^2, +\Delta, 0, a)$$

and

$$(n^i) = \frac{1}{2\rho^2}(r^2 + a^2, -\Delta, 0, a), \tag{7.162}$$

which satisfy the orthogonality condition

$$l \cdot n = 1. \tag{7.163}$$

The null-congruence defined by l is clearly affinely parametrized; but the congruence defined by n is not, since we have re-scaled (by a factor

The Kerr metric and the perturbation problem

$\Delta/2\rho^2$) the vector, derived from (7.161) with the choice of the negative sign for $dr/d\lambda$, to satisfy the condition (7.163).

We can now supplement the vectors l and n by two additional, complex-conjugate null-vectors, m and $\bar{m}$ in order that we may have a complex null-frame as the basis for a description of Kerr geometry in a Newman–Penrose formalism. We readily verify that

$$(m^i) = \frac{1}{\bar{\rho}\sqrt{2}}(ia\sin\theta, 0, 1, i\csc\theta), \qquad (7.164)$$

where

$$\bar{\rho} = r + ia\cos\theta \quad \text{and} \quad \bar{\rho}^* = r - ia\cos\theta, \qquad (7.165)$$

satisfies the requisite conditions, namely,

$$m \cdot l = n \cdot l = 0 \quad \text{and} \quad m \cdot \bar{m} = -1. \qquad (7.166)$$

Since the congruences defined by l and n are geodesic, the spin coefficients κ and ν must vanish; and, besides, the spin-coefficient ε must also vanish since the congruence defined by l is, in addition, affinely parametrized. (These statements follow from the propositions established in section 7.3.2.)

We shall find in section 7.4.4 that in the frame $(l, n, m, \bar{m})$ the spin coefficients σ and λ also vanish. *The congruences defined by l and n are, therefore, shear-free as well*; and it follows from the Goldberg–Sachs theorem that *the Kerr metric is of Petrov-type D.*

Finally, we may note the covariant forms of the basis vectors

$$\left. \begin{aligned} (l_i) &= \frac{1}{\Delta}(\Delta, -\rho^2, 0, -a\Delta\sin^2\theta), \\ (n_i) &= \frac{1}{2\rho^2}(\Delta, +\rho^2, 0, -a\Delta\sin^2\theta), \\ (m_i) &= \frac{1}{\bar{\rho}\sqrt{2}}[ia\sin\theta, 0, -\rho^2, -i(r^2+a^2)\sin\theta]. \end{aligned} \right\} \qquad (7.167)$$

7.4.4 The description of Kerr geometry in the Newman–Penrose formalism

We shall choose the null-frame formed by the vectors,† $(l, n, m, \bar{m})$ defined in section 7.4.3 ((7.162) and (7.164)) as the basis for a Newman–Penrose formalism.

---

† This frame is commonly called the *Kinnersley frame* after Kinnersley (1969) who first derived it (from, however, different considerations).

Chapter 7. The Kerr metric and its perturbations

In the Newman–Penrose formalism, a geometry is described by the spin-coefficients and the Weyl scalars in the chosen frame. We now proceed to determine them for Kerr geometry.

As we have explained in section 7.1.2 the evaluation of the spin-coefficients, γ_{abc}, is best carried out in terms of the symbols, λ_{abc} (antisymmetric in a and c), defined in (7.23). The evaluation of the symbols λ_{abc} is a simple matter since they involve only the ordinary derivatives of the components of the basis vectors. We find that, in the chosen frame, the only non-vanishing λ_{abc} are

$$\lambda_{122} = -\frac{1}{\rho^4}[(r-M)\rho^2 - r\Delta]; \quad \lambda_{314} = -\frac{2ia\cos\theta}{\rho^2};$$

$$\lambda_{132} = +\sqrt{2}\frac{iar\sin\theta}{\rho^2\bar{\rho}}; \quad \lambda_{324} = -\frac{ia\Delta\cos\theta}{\rho^4}; \quad (7.168)$$

$$\lambda_{213} = -\sqrt{2}\frac{a^2\sin\theta\cos\theta}{\rho^2\bar{\rho}}; \quad \lambda_{334} = +\frac{(ia+r\cos\theta)\operatorname{cosec}\theta}{\bar{\rho}^2\sqrt{2}};$$

$$\lambda_{243} = -\frac{\Delta}{2\rho^2\bar{\rho}}; \quad \lambda_{341} = -\frac{1}{\bar{\rho}}.$$

The spin-coefficients are linear combinations of the λ_{abc} terms (cf. section 7.2.6) and we find

$$\kappa = \sigma = \lambda = \nu = \varepsilon = 0 \quad (7.169)$$

and

$$\tilde{\rho} = -\frac{1}{\bar{\rho}^*}, \quad \beta = \frac{\cot\theta}{\bar{\rho}2\sqrt{2}}, \quad \pi = \frac{ia\sin\theta}{(\bar{\rho}^*)^2\sqrt{2}}, \quad \tau = -\frac{ia\sin\theta}{\rho^2\sqrt{2}},$$

$$\mu = -\frac{\Delta}{2\rho^2\bar{\rho}^*}, \quad \gamma = \mu + \frac{r-M}{2\rho^2}, \quad \text{and} \quad \alpha = \pi - \beta^*. \quad (7.170)$$

We observe that, besides the vanishing of κ, ν, and ε (which, as we have explained in section 7.4.3, are consequences of the way the vectors l and n were selected), the spin-coefficients σ and λ, also, vanish. The vanishing of σ and λ implies (as we have already stated in section 7.4.3) that the Kerr metric is of Petrov-type D. We can, now further, conclude that in the chosen frame

$$\Psi_0 = \Psi_1 = \Psi_3 = \Psi_4 = 0. \quad (7.171)$$

The only non-vanishing Weyl-scalar is Ψ_2; and it is determined, apart

The Kerr metric and the perturbation problem

from a constant of proportionality, by the relations

$$D\Psi_2 = 3\tilde{\rho}\Psi_2, \quad \text{or} \quad \frac{\partial}{\partial r}\log\Psi_2 = -\frac{3}{\tilde{\rho}^*},$$

and

$$\delta\Psi_2 = 3\tau\Psi_2, \quad \text{or} \quad \frac{\partial}{\partial\theta}\log\Psi_2 = -\frac{3ia\sin\theta}{\tilde{\rho}^*}, \tag{7.172}$$

which follow from the Bianchi identities (7.64b) and (7.64f). Accordingly,

$$\Psi_2 = \text{constant}\,(\tilde{\rho}^*)^{-3}. \tag{7.173}$$

The constant of proportionality, in (7.173), can be determined from one of the other equations (e.g. (7.59n)). We find

$$\Psi_2 = -M(\tilde{\rho}^*)^{-3}. \tag{7.174}$$

Finally, we may note that the structure-constants (which are related to the spin-cofficients, as expressed in (7.56)) have the values

$$C^1_{21} = -\frac{r\Delta}{\rho^4} + \frac{r-M}{\rho^2}; \qquad C^1_{31} = 0;$$

$$C^1_{32} = 0; \qquad C^1_{43} = \frac{ia\Delta\cos\theta}{\rho^4};$$

$$C^2_{21} = 0; \qquad C^2_{31} = 0;$$

$$C^2_{32} = \sqrt{2}\frac{a^2\sin\theta\cos\theta}{\rho^2\bar{\rho}}; \qquad C^2_{43} = \frac{2ia\cos\theta}{\rho^2};$$

$$C^3_{21} = -\sqrt{2}\frac{iar\sin\theta}{\rho^2\bar{\rho}^*}; \qquad C^3_{31} = \frac{1}{\bar{\rho}}; \tag{7.175}$$

$$C^3_{32} = -\frac{\Delta}{2\rho^2\bar{\rho}}; \qquad C^3_{43} = \frac{-1}{(\bar{\rho}^*)^2\sqrt{2}}(\bar{\rho}^*\cot\theta - ia\sin\theta);$$

$$C^4_{21} = +\sqrt{2}\frac{iar\sin\theta}{\rho^2\bar{\rho}}; \qquad C^4_{31} = 0;$$

$$C^4_{32} = 0; \qquad C^4_{43} = \frac{+1}{(\rho)^2\sqrt{2}}(\bar{\rho}\cot\theta + ia\sin\theta).$$

Chapter 7. The Kerr metric and its perturbations

7.4.5 The perturbation problem and some preliminaries

There are diverse formulations of the general problem of the perturbations of a spacetime that represents a solution of Einstein's equations. For the special case of the Kerr black hole, one may formulate the perturbation problem as one which seeks to ascertain how the black hole affects (and is affected by) incident waves of different sorts: electromagnetic, gravitational, neutrino, or electron waves. One may expect that in all these cases what will happen is a reflection of a part, and an absorption (or transmission) of a part,† of the incident flux of energy, by the black hole. It would appear then that the prime information we seek from a theory of the perturbations of the Kerr black hole is the reflection and transmission coefficients for incident waves of specified nature and character.

The problem of incident gravitational waves differs in one important respect from the problem of the incidence of waves of other sorts: incident gravitational waves, unlike the others, will induce changes in the metric coefficients; and the determination of these changes is, from some points of view, the central problem in any mathematical theory of the perturbations of a spacetime. Nevertheless, the fact, that in Kerr geometry (by virtue of its type-D character) we can choose a complex null-frame in which $\Psi_0 = \Psi_1 = \Psi_3 = \Psi_4 = 0$, has the consequence that we can treat the problem of the reflection and transmission of gravitational waves in a manner that is quite identical with the treatments for waves of other sorts. And, as it happens, the determination of the induced changes in the metric coefficients is dependent on (and must await) the solution to the problem of reflection and transmission.

Consistent with the foregoing formulation of the perturbation problem, we shall suppose that the perturbations one considers have a t- and a φ-dependence given by

$$e^{i(\sigma t + m\varphi)}, \tag{7.176}$$

where σ denotes the frequency of the incident waves and m is an integer (positive, negative, or zero). The common factor (7.176) in all the quantities describing the perturbation will be suppressed; and the symbols representing them will be their amplitudes.

The basis vectors, $(l, n, m, \bar{m})$ (given in (7.162) and (7.164) when applied as tangent vectors to functions with a t- and a φ-dependence

---

† We shall find that, under certain circumstances, the black hole will reflect an amount of energy which is in excess of the incident energy. This is the phenomenon of *superradiance* which we shall consider in section 7.8.

given by (7.176), become the derivative operators,

$$l = D = \mathcal{D}_0, \quad n = \Delta = -\frac{\Delta}{2\rho^2}\mathcal{D}_0^\dagger,$$

$$m = \delta = \frac{1}{\bar{\rho}\sqrt{2}}\mathcal{L}_0^\dagger, \quad \text{and} \quad \bar{m} = \delta^* = \frac{1}{\bar{\rho}^*\sqrt{2}}\mathcal{L}_0, \tag{7.177}$$

where

$$\mathcal{D}_n = \partial_r + \frac{iK}{\Delta} + 2n\frac{r-M}{\Delta}, \quad \mathcal{D}_n^\dagger = \partial_r - \frac{iK}{\Delta} + 2n\frac{r-M}{\Delta},$$

$$\mathcal{L}_n = \partial_\theta + Q + n \cot \theta, \quad \mathcal{L}_n^\dagger = \partial_\theta - Q + n \cot \theta, \tag{7.178}$$

and

$$K = (r^2 + a^2)\sigma + am, \quad Q = a\sigma \sin \theta + m \operatorname{cosec} \theta. \tag{7.179}$$

It will be noted that while $\mathcal{D}_n$ and $\mathcal{D}_n^\dagger$ are purely radial operators, $\mathcal{L}_n$ and $\mathcal{L}_n^\dagger$ are purely angular operators.

The differential operators we have defined satisfy a number of elementary identities which we shall have occasions to use constantly in the subsequent analysis. For convenience, we collect them here as a series of lemmas.

Lemma 7.1

(i) $\quad \mathcal{L}_n(\theta) = -\mathcal{L}_n^\dagger(\pi - \theta); \quad \mathcal{D}_n^\dagger = (\mathcal{D}_n)^*;$

(ii) $\quad (\sin \theta)\mathcal{L}_{n+1} = \mathcal{L}_n \sin \theta; \quad (\sin \theta)\mathcal{L}_{n+1}^\dagger = \mathcal{L}_n^\dagger \sin \theta; \tag{7.180}$

(iii) $\quad \Delta \mathcal{D}_{n+1} = \mathcal{D}_n \Delta; \quad \Delta \mathcal{D}_{n+1}^\dagger = \mathcal{D}_n^\dagger \Delta.$

Lemma 7.2

$$\left(\mathcal{D} + \frac{m}{\bar{\rho}^*}\right)\left(\mathcal{L} + \frac{mia \sin \theta}{\bar{\rho}^*}\right) = \left(\mathcal{L} + \frac{mia \sin \theta}{\bar{\rho}^*}\right)\left(\mathcal{D} + \frac{m}{\bar{\rho}^*}\right), \tag{7.181}$$

where $\mathcal{D}$ can be any $\mathcal{D}_n$ or $\mathcal{D}_n^\dagger$ and $\mathcal{L}$ any $\mathcal{L}_{n'}$ or $\mathcal{L}_{n'}^\dagger$, and m is a constant (generally an integer, positive or negative).

Lemma 7.3

$$\mathcal{L}_{n+1}\mathcal{L}_{n+2}\cdots\mathcal{L}_{n+m}(f \cos \theta) = (\cos \theta)\mathcal{L}_{n+1}\mathcal{L}_{n+2}\cdots\mathcal{L}_{n+m}f$$
$$- (m \sin \theta)\mathcal{L}_{n+2}\cdots\mathcal{L}_{n+m}f, \tag{7.182}$$

Chapter 7. The Kerr metric and its perturbations

where f is any function of θ and the $\mathscr{L}$ terms can be replaced by $\mathscr{L}_n^\dagger$ terms.

Lemma 7.4

If $f(\theta)$ and $g(\theta)$ are any two (bounded) functions, regular at $\theta = 0$ and $\theta = \pi$, then

$$\int_0^\pi g(\mathscr{L}_n f)\sin\theta\,d\theta = -\int_0^\pi f(\mathscr{L}_{-n+1}^\dagger g)\sin\theta\,d\theta. \quad (7.183)$$

And finally, we may note the elementary relations

$$Q_{,\theta} + Q\cot\theta = 2a\sigma\cos\theta \quad \text{and} \quad K - aQ\sin\theta = \rho^2\sigma. \quad (7.184)$$

7.5 The solution of Maxwell's equations

7.5.1 The separation of the variables and the solutions for ϕ_0 and ϕ_2

The solution to any problem concerned with electromagnetic radiation in the vicinity of the Kerr black hole must ultimately depend on the solution of Maxwell's equations. And it will clearly suffice to consider solutions with a t- and a φ-dependence given in (7.176).

We have already written down in section 7.2.6 the form which Maxwell's equations take in the Newman–Penrose formalism. The equations appropriate to Kerr geometry, in the chosen frame, can be obtained by inserting in (7.70)–(7.73) the expressions for the spin coefficients and the directional derivatives given in (7.170) and (7.177). We find that the resulting equations take simple forms if they are written for

$$\Phi_0 = \phi_0, \quad \Phi_1 = \phi_1 \bar{\rho}^*\sqrt{2} \quad \text{and} \quad \Phi_2 = 2\phi_2(\bar{\rho}^*)^2. \quad (7.185)$$

We find

$$\left(\mathscr{L}_1 - \frac{ia\sin\theta}{\bar{\rho}^*}\right)\Phi_0 = \left(\mathscr{D}_0 + \frac{1}{\bar{\rho}^*}\right)\Phi_1, \quad (7.186)$$

$$\left(\mathscr{L}_0 + \frac{ia\sin\theta}{\bar{\rho}^*}\right)\Phi_1 = \left(\mathscr{D}_0 - \frac{1}{\bar{\rho}^*}\right)\Phi_2, \quad (7.187)$$

$$\left(\mathscr{L}_1^\dagger - \frac{ia\sin\theta}{\bar{\rho}^*}\right)\Phi_2 = -\Delta\left(\mathscr{D}_0^\dagger + \frac{1}{\bar{\rho}^*}\right)\Phi_1, \quad (7.188)$$

$$\left(\mathscr{L}_0^\dagger + \frac{ia\sin\theta}{\bar{\rho}^*}\right)\Phi_1 = -\Delta\left(\mathscr{D}_1^\dagger - \frac{1}{\bar{\rho}^*}\right)\Phi_0. \quad (7.189)$$

The solution of Maxwell's equations

It is evident that the commutativity of the operators $(\mathcal{D}_0 + 1/\bar{\rho}^*)$ and $(\mathcal{L}_0^\dagger + ia \sin\theta/\bar{\rho}^*)$ (cf. lemma 7.2) enables us to eliminate Φ_1 from (7.186) and (7.189) and to obtain a decoupled equation for Φ_0. We obtain

$$\left[\left(\mathcal{L}_0^\dagger + \frac{ia \sin\theta}{\bar{\rho}^*}\right)\left(\mathcal{L}_1 - \frac{ia \sin\theta}{\bar{\rho}^*}\right) + \Delta\left(\mathcal{D}_1 + \frac{1}{\bar{\rho}^*}\right)\left(\mathcal{D}_1^\dagger - \frac{1}{\bar{\rho}^*}\right)\right]\Phi_0 = 0. \tag{7.190}$$

A similar elimination of Φ_1 from (7.187) and (7.188) yields the equation

$$\left[\left(\mathcal{L}_0 + \frac{ia \sin\theta}{\bar{\rho}^*}\right)\left(\mathcal{L}_1^\dagger - \frac{ia \sin\theta}{\bar{\rho}^*}\right) + \Delta\left(\mathcal{D}_0^\dagger + \frac{1}{\bar{\rho}^*}\right)\left(\mathcal{D}_0 - \frac{1}{\bar{\rho}^*}\right)\right]\Phi_2 = 0. \tag{7.191}$$

On expanding (7.190) and (7.191), we are left with

$$[\Delta\mathcal{D}_1\mathcal{D}_1^\dagger + \mathcal{L}_0^\dagger\mathcal{L}_1 - 2i\sigma(r + ia \cos\theta)]\Phi_0 = 0, \tag{7.192}$$

and

$$[\Delta\mathcal{D}_0^\dagger\mathcal{D}_0 + \mathcal{L}_0\mathcal{L}_1^\dagger + 2i\sigma(r + ia \cos\theta)]\Phi_2 = 0. \tag{7.193}$$

These equations are clearly separable. Thus, with the substitutions,

$$\Phi_0 = R_{+1}(r)S_{+1}(\theta) \quad \text{and} \quad \Phi_2 = R_{-1}(r)S_{-1}(\theta), \tag{7.194}$$

(where $R_\pm(r)$ and $S_\pm(\theta)$ are, respectively, functions of r and θ only) (7.192) and (7.193) separate to give the two pairs of equations

$$(\Delta\mathcal{D}_1\mathcal{D}_1^\dagger - 2i\sigma r)R_{+1} = \lambda R_{+1}, \tag{7.195}$$

$$(\mathcal{L}_0^\dagger\mathcal{L}_1 + 2a\sigma \cos\theta)S_{+1} = -\lambda S_{+1}, \tag{7.196}$$

and

$$(\Delta\mathcal{D}_0^\dagger\mathcal{D}_0 + 2i\sigma r)R_{-1} = \lambda R_{-1}, \tag{7.197}$$

$$(\mathcal{L}_0\mathcal{L}_1^\dagger - 2a\sigma \cos\theta)S_{-1} = -\lambda S_{-1}. \tag{7.198}$$

where λ is a separation constant.

It will be noticed that we have not distinguished between the separation constants that derive from (7.192) and (7.193). The reason is the following. Considering (7.196), we first observe that λ is a characteristic-value parameter that is to be determined by the condition that $S_{+1}(\theta)$ be regular at $\theta = 0$ and $\theta = \pi$. On the other hand, since the operator acting on S_{-1} in (7.198) is the same as the operator acting on S_{+1} in (7.196), if we replace θ by $\pi - \theta$ it follows that a proper solution, $S_{+1}(\theta; \lambda)$, of

Chapter 7. The Kerr metric and its perturbations

(7.196), belonging to a characteristic value λ, provides a proper solution of (7.198), belonging to the same value λ, if we replace θ by $\pi - \theta$. In other words, (7.196) and (7.198) determine the same set of characteristic values for λ.

We readily verify that (7.195)–(7.198) are equivalent to those first derived by Teukolsky (1972); and his separation constant E is related to our characteristic-value parameter λ by

$$\lambda = E + a^2\sigma^2 + 2am\sigma. \tag{7.199}$$

By rewriting (7.195) in the form

$$(\Delta \mathcal{D}_0 \mathcal{D}_0^\dagger - 2i\sigma r)\Delta R_{+1} = \lambda \Delta R_{+1}, \tag{7.200}$$

and comparing it with (7.197), we conclude that if ΔR_{+1} is a solution of (7.200), its complex conjugate is a solution of (7.197).

While the solutions for Φ_0 and Φ_2 are given by $R_{+1}S_{+1}$ and $R_{-1}S_{-1}$, respectively, there still remains the question of their relative normalization. We consider this question in section 7.5.2.

7.5.2 The Teukolsky–Starobinsky identities

Some remarkable identities connecting the functions belonging to the spins +1 and −1 were discovered by Teukolsky (see Teukolsky and Press, 1972) and Starobinsky (1973; see also Starobinsky and Churilov, 1973). We shall enunciate these identities as they have been formulated by Chandrasekhar (1978b).

The proofs of the various identities, though essentially straightforward, require a fair amount of algebraic manipulations with the derivative operators $\mathcal{L}$ and $\mathcal{D}$ and involve frequent use of lemmas 7.1 and 7.3. The proofs will accordingly be omitted.

Theorem 7.1

$\Delta \mathcal{D}_0 \mathcal{D}_0 R_{-1}$ is a multiple of ΔR_{+1}, and $\Delta \mathcal{D}_0^\dagger \mathcal{D}_0^\dagger \Delta R_{+1}$ is a multiple of R_{-1}.

The theorem is equivalent to asserting the commutation relation

$$\Delta \mathcal{D}_0 \mathcal{D}_0 (\Delta \mathcal{D}_0^\dagger \mathcal{D}_0^\dagger + 2i\sigma r) = (\Delta \mathcal{D}_0 \mathcal{D}_0^\dagger - 2i\sigma r)\Delta \mathcal{D}_0 \mathcal{D}_0, \tag{7.201}$$

and its complex conjugate.

The solution of Maxwell's equations

Corollary 1

By a suitable choice of the relative normalization of the functions ΔR_{+1} and R_{-1} we can arrange that

$$\Delta \mathcal{D}_0 \mathcal{D}_0 R_{-1} = \mathscr{C} \Delta R_{+1} \qquad (7.202)$$

and

$$\Delta \mathcal{D}_0^\dagger \mathcal{D}_0^\dagger \Delta R_{+1} = \mathscr{C}^* R_{-1}, \qquad (7.203)\dagger$$

where $\mathscr{C}$ is a constant (which can be complex).

We can arrange the relative normalization in the way prescribed since ΔR_{+1} and R_{-1} satisfy complex-conjugate equations.

Corollary 2

The square of the absolute value of the constant $\mathscr{C}$ is given by

$$|\mathscr{C}|^2 = \lambda^2 - 4\alpha^2 \sigma^2 \quad \text{where} \quad \alpha^2 = a^2 + (am/\sigma). \qquad (7.204)$$

The relation (7.204) is a consequence of the identity

$$\Delta \mathcal{D}_0^\dagger \mathcal{D}_0^\dagger \Delta \mathcal{D}_0 \mathcal{D}_0 = |\mathscr{C}|^2 \text{ mod } \Delta \mathcal{D}_0^\dagger \mathcal{D}_0 + 2i\sigma r - \lambda = 0. \qquad (7.205)$$

Theorem 7.2

$\mathscr{L}_0 \mathscr{L}_1 S_{+1}$ *is a multiple of S_{-1} and $\mathscr{L}_0^\dagger \mathscr{L}_1^\dagger S_{-1}$ is a multiple of S_{+1}.*

This theorem is equivalent to asserting the commutation relation

$$\mathscr{L}_0 \mathscr{L}_1 (\mathscr{L}_0^\dagger \mathscr{L}_1^\dagger + 2a\sigma \cos\theta) = (\mathscr{L}_0 \mathscr{L}_1^\dagger - 2a\sigma \cos\theta)\mathscr{L}_0 \mathscr{L}_1, \qquad (7.206)$$

and its adjoint obtained by replacing θ by $\pi - \theta$.

Corollary 1

If S_{+1} and S_{-1} are simultaneously normalized, then

$$\mathscr{L}_0 \mathscr{L}_1 S_{+1} = D S_{-1} \qquad (7.207)$$

and

$$\mathscr{L}_0^\dagger \mathscr{L}_1^\dagger S_{-1} = D S_{+1}, \qquad (7.208)$$

where D is a real constant.

The fact that the same constant D occurs in both (7.207) and (7.208) follows from writing the normalization integral for S_{-1} (say), expressing

† We shall presently show that $\mathscr{C}$ is in fact real.

Chapter 7. The Kerr metric and its perturbations

the integrand as a product of S_{-1} and $\mathscr{L}_0\mathscr{L}_1 S_{+1}$ and, after two successive applications of lemma 7.4, making use of the relation (7.208).

Corollary 2

The constant D is given by

$$D^2 = \lambda^2 - 4\alpha^2\sigma^2 \quad \text{where} \quad \alpha^2 = a^2 + (am/\sigma). \tag{7.209}$$

The relation (7.209) is a consequence of the identity

$$\mathscr{L}_0^\dagger \mathscr{L}_1^\dagger \mathscr{L}_0 \mathscr{L}_1 = D^2 \mod \mathscr{L}_0^\dagger \mathscr{L}_1 + (\lambda + 2a\sigma \cos\theta) = 0. \tag{7.210}$$

It will be observed that

$$D^2 = |\mathscr{C}|^2. \tag{7.211}$$

We shall now show that $\mathscr{C}$ is in fact equal to D, and therefore, real.

In view of the commutativity of the operators $(\mathscr{D}_0 + 1/\bar{\rho}^*)$ and $(\mathscr{L}_0 + ia\sin\theta/\bar{\rho}^*)$, we can eliminate Φ_1 from (7.186) and (7.187) to obtain a relation directly between Φ_0 and Φ_2. We find

$$\left(\mathscr{L}_0 + \frac{ia\sin\theta}{\bar{\rho}^*}\right)\left(\mathscr{L}_1 - \frac{ia\sin\theta}{\bar{\rho}^*}\right)\Phi_0 = \left(\mathscr{D}_0 + \frac{1}{\bar{\rho}^*}\right)\left(\mathscr{D}_0 - \frac{1}{\bar{\rho}^*}\right)\Phi_2. \tag{7.212}$$

On expanding this relation, we are left with

$$\mathscr{L}_0 \mathscr{L}_1 \Phi_0 = \mathscr{D}_0 \mathscr{D}_0 \Phi_2. \tag{7.213}$$

On inserting the solutions for Φ_0 and Φ_2 given in (7.194) into (7.213) we obtain

$$(\mathscr{L}_0 \mathscr{L}_1 S_{+1})/S_{-1} = (\mathscr{D}_0 \mathscr{D}_0 R_{-1})/R_{+1}. \tag{7.214}$$

If we now suppose S_{+1} and S_{-1} are normalized, then it follows from (7.207) that

$$\mathscr{D}_0 \mathscr{D}_0 R_{-1} = D R_{+1}. \tag{7.215}$$

Comparison of (7.202) and (7.215) shows that *the relative normalization of ΔR_{+1} and R_{-1} implied in writing (7.202) and (7.203) must be associated with the simultaneous normalizations of the angular functions S_{+1} and S_{-1}*. It follows at the same time that

$$\mathscr{C} = \mathscr{C}^* = D. \tag{7.216}$$

The sign of D can be found by considering the spherical case ($a\sigma = 0$) when the angular functions reduce to the standard spin-weighted spherical harmonics (cf. Goldberg et al., 1967); and the comparison shows that we must take the positive square root in (7.209).

Finally, we may note that, with the stated normalizations of the radial and the angular functions, the solutions for ϕ_0 and ϕ_2 are given by (cf. (7.185))

$$\phi_0 = R_{+1}S_{+1} \quad \text{and} \quad \phi_2 = \frac{1}{2(\bar{\rho}^*)^2} R_{-1}S_{-1}. \tag{7.217}$$

7.5.3 The solution for ϕ_1

We shall now complete the solution of Maxwell's equations by showing how the remaining scalar ϕ_1 can be expressed in terms of Teukolsky's functions $R_{\pm 1}$ and $S_{\pm 1}$.

First, we define the functions (cf. Chandrasekhar, 1976b)

$$\left.\begin{aligned} g_{+1}(r) &= \frac{1}{D}(r\mathscr{D}_0 R_{-1} - R_{-1}), \\ f_{+1}(\theta) &= \frac{1}{D}[(\cos\theta)\mathscr{L}_1^\dagger S_{-1} + (\sin\theta)S_{-1}], \\ g_{-1}(r) &= \frac{1}{D}(r\mathscr{D}_0^\dagger \Delta R_{+1} - \Delta R_{+1}), \\ \text{and} \quad f_{-1}(\theta) &= \frac{1}{D}[(\cos\theta)\mathscr{L}_1 S_{+1} + (\sin\theta)S_{+1}], \end{aligned}\right\} \tag{7.218}$$

where D denotes the Teukolsky–Starobinsky constant defined in (7.209). It can be readily verified, with the aid of the Teukolsky–Starobinsky identities, that the functions we have defined satisfy the equations

$$\begin{gathered} \mathscr{D}_0 g_{+1} = rR_{+1}; \quad \Delta\mathscr{D}_0^\dagger g_{-1} = rR_{-1}; \\ \mathscr{L}_0^\dagger f_{+1} = S_{+1}\cos\theta; \quad \mathscr{L}_0 f_{-1} = S_{-1}\cos\theta. \end{gathered} \tag{7.219}$$

Now writing (7.186) in the form

$$\mathscr{D}_0(\bar{\rho}^*\Phi_1) = (\bar{\rho}^*\mathscr{L}_1 - ia\sin\theta)\Phi_0, \tag{7.220}$$

and substituting for Φ_0 its solution $R_{+1}S_{+1}$, we can rewrite it successively, with the aid of (7.218) and (7.219) and the Teukolsky–

Chapter 7. The Kerr metric and its perturbations

Starobinsky identities, in the manner

$$\mathcal{D}_0(\bar{\rho}^*\Phi_1) = (rR_{+1})\mathcal{L}_1 S_{+1} - iaR_{+1}[(\cos\theta)\mathcal{L}_1 S_{+1} + (\sin\theta)S_{+1}]$$
$$= (\mathcal{D}_0 g_{+1})\mathcal{L}_1 S_{+1} - ia(DR_{+1})f_{-1}$$
$$= (\mathcal{D}_0 g_{+1})\mathcal{L}_1 S_{+1} - iaf_{-1}\mathcal{D}_0\mathcal{D}_0 R_{-1}. \tag{7.221}$$

We thus obtain

$$\mathcal{D}_0(\bar{\rho}^*\Phi_1) = \mathcal{D}_0(g_{+1}\mathcal{L}_1 S_{+1} - iaf_{-1}\mathcal{D}_0 R_{-1}). \tag{7.222}$$

In a similar fashion, we find from (7.187):

$$\left.\begin{aligned}\mathcal{L}_0(\bar{\rho}^*\Phi_1) &= [(r - ia\cos\theta)\mathcal{D}_0 - 1]R_{-1}S_{-1} \\ &= (r\mathcal{D}_0 R_{-1} - R_{-1})S_{-1} - ia(\mathcal{D}_0 R_{-1})(\cos\theta)S_{-1} \\ &= g_{+1}\mathcal{L}_0\mathcal{L}_1 S_{+1} - ia(\mathcal{L}_0 f_{-1})\mathcal{D}_0 R_{-1},\end{aligned}\right\} \tag{7.223}$$

or

$$\mathcal{L}_0(\bar{\rho}^*\Phi_1) = \mathcal{L}_0(g_{+1}\mathcal{L}_1 S_{+1} - iaf_{-1}\mathcal{D}_0 R_{-1}). \tag{7.224}$$

From (7.222) and (7.224), we conclude that the required solution for Φ_1 is given by

$$\bar{\rho}^*\Phi_1 = g_{+1}(r)\mathcal{L}_1 S_{+1} - iaf_{-1}(\theta)\mathcal{D}_0 R_{-1}. \tag{7.225}\dagger$$

By treating (7.188) and (7.189) in an analogous fashion, we find that we have the following alternative form for the solution for Φ_1:

$$-\bar{\rho}^*\Phi_1 = g_{-1}(r)\mathcal{L}_1^\dagger S_{-1} - iaf_{+1}(\theta)\mathcal{D}_0^\dagger \Delta R_{+1}. \tag{7.226}$$

A comparison of the solutions (7.225) and (7.226) leads to the interesting identity

$$g_{+1}\mathcal{L}_1 S_{+1} + g_{-1}\mathcal{L}_1^\dagger S_{-1} - ia(f_{-1}\mathcal{D}_0 R_{-1} + f_{+1}\mathcal{D}_0^\dagger \Delta R_{+1}) = 0. \tag{7.227}$$

---

† Strictly, we should have added to the particular solution (7.225) a solution, P say, of the corresponding homogeneous equation, $\mathcal{D}_0(P) = \mathcal{L}_0(P) = 0$. But the solution for P, namely,

$$P = \text{constant} \exp[-i\sigma(r_* + ia\cos\theta)]\cot^m(\theta/2)$$

where r_* is defined by the equation

$$dr_*/dr = (r^2 + \alpha^2)/\Delta \quad (\alpha^2 = a^2 + am/\sigma)$$

is singular at $\theta = 0$ and $\theta = \pi$; and on this account we have not included it in the solution (7.224).

Gravitational perturbations

Finally, by combining (7.225) and (7.226), we can write the solution for ϕ_1 in the form

$$\phi_1 = \frac{\sqrt{2}}{4(\bar{\rho}^*)^2}[(g_{+1}\mathcal{L}_1 S_{+1} - g_{-1}\mathcal{L}_1^\dagger S_{-1}) - ia(f_{-1}\mathcal{D}_0 R_{-1} - f_{+1}\mathcal{D}_0^\dagger \Delta R_{+1})].$$

(7.228)

This completes the solution of Maxwell's equations in Kerr geometry. In section 7.8 we shall show how the equations governing $R_{\pm 1}$ are to be solved for determining the reflection and transmission coefficients for incident electromagnetic waves.

7.6 Gravitational perturbations

7.6.1 The Newman–Penrose equations that are already linearized

As we have seen, the type-D character of the Kerr metric has enabled us to choose a null-frame in which the spin coefficients, κ, σ, λ, and ν, as well as the Weyl scalars, Ψ_0, Ψ_1, Ψ_3, and Ψ_4, vanish. And it is a remarkable fact the Newman–Penrose equations provide a set of equations which are linear and homogeneous in these same quantities. Thus, four of the eight Bianchi identities (7.64*a, d, e,* and *h*) and two of the Ricci identities (7.59*b*) and (7.59*j*) give

$$\left.\begin{array}{l}(\delta^* - 4\alpha + \pi)\Psi_0 - (D - 2\varepsilon - 4\tilde{\rho})\Psi_1 = 3\kappa\Psi_2, \\ (\Delta - 4\gamma + \mu)\Psi_0 - (\delta - 4\tau - 2\beta)\Psi_1 = 3\sigma\Psi_2, \\ (D - \tilde{\rho} - \tilde{\rho}^* - 3\varepsilon + \varepsilon^*)\sigma - (\delta - \tau + \pi^* - \alpha^* - 3\beta)\kappa = \Psi_0,\end{array}\right\} \quad (7.229)$$

and

$$\left.\begin{array}{l}(D + 4\varepsilon - \tilde{\rho})\Psi_4 - (\delta^* + 4\pi + 2\alpha)\Psi_3 = -3\lambda\Psi_2, \\ (\delta + 4\beta - \tau)\Psi_4 - (\Delta + 2\gamma + 4\mu)\Psi_3 = -3\nu\Psi_2, \\ (\Delta + \mu + \mu^* + 3\gamma - \gamma^*)\lambda - (\delta^* + 3\alpha + \beta^* + \pi - \tau^*)\nu = -\Psi_4.\end{array}\right\} \quad (7.230)$$

The foregoing equations are already linearized in the sense that Ψ_0, Ψ_1, κ, and σ in (7.229) and Ψ_4, Ψ_3, λ, and ν in (7.230) are to be considered as quantities of the first order of smallness and that, accordingly, we may replace all the other quantities (including the derivative operators) which occur in these equations by their unperturbed values given in (7.170), (7.174), and (7.177). We consider (7.229) and (7.230) in section 7.6.2; but it is important to make here the following two observations. *First*, we have provided only six equations for the eight unknowns which

Chapter 7. The Kerr metric and its perturbations

occur in them; this implies that their solution must involve two arbitrary functions. *Second*, equations (7.229) governing Ψ_0, Ψ_1, κ, and σ are decoupled from equations (7.230) governing Ψ_4, Ψ_3, λ, and ν; this decoupling of the two sets of equations has, as we shall see, important consequences.

7.6.2 The reduction of the equations and their solution

As we have stated in section 7.6.1 we can replace the derivative operators, the various spin-coefficients (besides κ, σ, λ, and ν), and Ψ_2 in (7.229) and (7.230) by their values given in equations (7.170), (7.174), and (7.177). We find that the resulting equations take simple and symmetrical forms if we write them in terms of the variables

$$\Phi_0 = \Psi_0, \quad \Phi_1 = \Psi_1 \bar{\rho}^* \sqrt{2}, \quad k = \kappa/(\bar{\rho}^*)^2 \sqrt{2}, \quad \text{and} \quad s = \sigma \bar{\rho}/(\bar{\rho}^*)^2 \tag{7.231}$$

and

$$\Phi_4 = \Psi_4(\bar{\rho}^*)^4, \quad \Phi_3 = \Psi_3(\bar{\rho}^*)^3/\sqrt{2}, \quad l = \lambda \bar{\rho}^*/2, \quad \text{and} \quad n = \nu \rho^2/\sqrt{2}. \tag{7.232}$$

After the various replacements and substitutions, (7.229) and (7.230) reduce to the forms

$$\left(\mathscr{L}_2 - \frac{3ia \sin \theta}{\bar{\rho}^*}\right)\Phi_0 - \left(\mathscr{D}_0 + \frac{3}{\bar{\rho}^*}\right)\Phi_1 = -6Mk, \tag{7.233}$$

$$\Delta\left(\mathscr{D}_2^\dagger - \frac{3}{\bar{\rho}^*}\right)\Phi_0 + \left(\mathscr{L}_{-1}^\dagger + \frac{3ia \sin \theta}{\bar{\rho}^*}\right)\Phi_1 = +6Ms, \tag{7.234}$$

$$\left(\mathscr{D}_0 + \frac{3}{\bar{\rho}^*}\right)s - \left(\mathscr{L}_{-1}^\dagger + \frac{3ia \sin \theta}{\bar{\rho}^*}\right)k = +\frac{\bar{\rho}}{(\bar{\rho}^*)^2}\Phi_0, \tag{7.235}$$

and

$$\left(\mathscr{D}_0 - \frac{3}{\bar{\rho}^*}\right)\Phi_4 - \left(\mathscr{L}_{-1} + \frac{3ia \sin \theta}{\bar{\rho}^*}\right)\Phi_3 = +6Ml, \tag{7.236}$$

$$\left(\mathscr{L}_2^\dagger - \frac{3ia \sin \theta}{\bar{\rho}^*}\right)\Phi_4 + \Delta\left(\mathscr{D}_{-1}^\dagger + \frac{3}{\bar{\rho}^*}\right)\Phi_3 = +6Mn, \tag{7.237}$$

$$\Delta\left(\mathscr{D}_{-1}^\dagger + \frac{3}{\bar{\rho}^*}\right)l + \left(\mathscr{L}_{-1} + \frac{3ia \sin \theta}{\bar{\rho}^*}\right)n = +\frac{\bar{\rho}}{(\bar{\rho}^*)^2}\Phi_4. \tag{7.238}$$

Gravitational perturbations

It is evident that by making use of the commutation relation (7.181), we can eliminate Φ_1 from (7.233) and (7.234) by operating (7.233) by $(\mathscr{L}_{-1}^\dagger + 3ia \sin\theta/\bar\rho^*)$ and (7.234) by $(\mathscr{D}_0 + 3/\bar\rho^*)$ and adding. The right-hand side of the resulting equation is, apart from a factor $6M$, precisely the quantity which occurs on the left-hand side of (7.235). We, accordingly, obtain the decoupled equation,

$$\left[\left(\mathscr{L}_{-1}^\dagger + \frac{3ia \sin\theta}{\bar\rho^*}\right)\left(\mathscr{L}_2 - \frac{3ia \sin\theta}{\bar\rho^*}\right) + \left(\mathscr{D}_0 + \frac{3}{\bar\rho^*}\right)\Delta\left(\mathscr{D}_2^\dagger - \frac{3}{\bar\rho^*}\right)\right]\Phi_0 \quad (7.239)$$

$$= 6M\frac{\bar\rho}{(\bar\rho^*)^2}\Phi_0.$$

Similarly, we obtain from (7.236)–(7.238) the decoupled equation,

$$\left[\left(\mathscr{L}_{-1} + \frac{3ia \sin\theta}{\bar\rho^*}\right)\left(\mathscr{L}_2^\dagger - \frac{3ia \sin\theta}{\bar\rho^*}\right) + \Delta\left(\mathscr{D}_{-1}^\dagger + \frac{3}{\bar\rho^*}\right)\left(\mathscr{D}_0 - \frac{3}{\bar\rho^*}\right)\right]\Phi_4 \quad (7.240)$$

$$= 6M\frac{\bar\rho}{(\bar\rho^*)^2}\Phi_4.$$

On expanding (7.239) and (7.240), we find after some reductions (in which we make use of the elementary relations (7.184)) that we are left with

$$[\Delta\mathscr{D}_1\mathscr{D}_2^\dagger + \mathscr{L}_{-1}^\dagger\mathscr{L}_2 - 6i\sigma(r + ia \cos\theta)]\Phi_0 = 0 \quad (7.241)$$

and

$$[\Delta\mathscr{D}_{-1}^\dagger\mathscr{D}_0 + \mathscr{L}_{-1}\mathscr{L}_2^\dagger + 6i\sigma(r + ia \cos\theta)]\Phi_4 = 0. \quad (7.242)$$

Equations (7.241) and (7.242) clearly allow the separation of the variables. Thus, by the substitutions,

$$\Phi_0 = R_{+2}(r)S_{+2}(\theta) \quad \text{and} \quad \Phi_4 = R_{-2}(r)S_{-2}(\theta), \quad (7.243)$$

where $R_{\pm 2}$ and $S_{\pm 2}$ are functions, respectively, of r and θ only, we obtain the two pairs of separated equations:

$$(\Delta\mathscr{D}_1\mathscr{D}_2^\dagger - 6i\sigma r)R_{+2} = \lambda R_{+2}, \quad (7.244)$$

$$(\mathscr{L}_{-1}^\dagger\mathscr{L}_2 + 6a\sigma \cos\theta)S_{+2} = -\lambda S_{+2}, \quad (7.245)$$

and

$$(\Delta\mathscr{D}_{-1}^\dagger\mathscr{D}_0 + 6i\sigma r)R_{-2} = \lambda R_{-2}, \quad (7.246)$$

$$(\mathscr{L}_{-1}\mathscr{L}_2^\dagger - 6a\sigma \cos\theta)S_{-2} = -\lambda S_{-2}, \quad (7.247)$$

Chapter 7. The Kerr metric and its perturbations

where λ is a separation constant. (We have not distinguished between the separation constants derived from (7.239) and (7.240) since the characteristic values of λ determined by (7.245) and (7.247) are the same, for reasons already explained in the context of (7.196) and (7.198)).

Further, by rewriting (7.244) in the alternative form

$$(\Delta \mathcal{D}_{-1} \mathcal{D}_0^\dagger - 6i\sigma r)\Delta^2 R_{+2} = \lambda \Delta^2 R_{+2}, \qquad (7.248)$$

and comparing it with (7.246), we conclude that if $\Delta^2 R_{+2}$ is a solution of (7.248), then its complex conjugate is a solution of (7.246).

It can be verified that (7.244)–(7.247) are equivalent to those first derived by Teukolsky (1972); and our present characteristic value parameter, λ, is related to his separation constant, E, by (cf. (7.199))

$$\lambda = E + a^2\sigma^2 + 2a\sigma m - 2. \qquad (7.249)$$

The solutions for the spin-coefficients κ, σ, λ, and ν and the choice of gauge

Quite generally, the linearized equations governing the perturbations must be consistent with the freedoms we have in the choice of the tetrad and in the choice of the coordinates. As we have explained in section 7.3.1 we have six degrees of freedom to make infinitesimal rotations of the local tetrad frame; and in addition we have four degrees of freedom to make infinitesimal coordinate transformations. We can exercise these available degrees of freedom to restrict the solutions of the linearized equations as convenience or occasion may dictate.

Returning to (7.233)–(7.238), we recall that the solutions of these equations must involve two arbitrary functions. Since $\Phi_0 (=R_{+2}S_{+2})$ and $\Phi_4 (=R_{-2}S_{-2})$ have already been determined (apart from normalization constants), we may let Φ_1 and Φ_3 be the two arbitrary functions. At this point, we may exercise four of the six available degrees of freedom in the choice of the local (perturbed) tetrad frame to make

$$\Phi_1 = \Phi_3 = 0 \quad \text{or, equivalently} \quad \Psi_1 = \Psi_3 = 0. \qquad (7.250)$$

In this gauge, the solutions for κ, σ, λ, and ν can be directly read off from (7.233)–(7.238); we have

Gravitational perturbations

$$\kappa = -\frac{\sqrt{2}}{6M}(\bar{\rho}^*)^2 R_{+2}\left(\mathcal{L}_2 - \frac{3ia\sin\theta}{\bar{\rho}^*}\right)S_{+2},$$

$$\sigma = +\frac{1}{6M}\frac{(\bar{\rho}^*)^2}{\bar{\rho}}S_{+2}\Delta\left(\mathcal{D}_2^\dagger - \frac{3}{\bar{\rho}^*}\right)R_{+2},$$

$$\lambda = +\frac{1}{6M}\frac{2}{\bar{\rho}^*}S_{-2}\left(\mathcal{D}_0 - \frac{3}{\bar{\rho}^*}\right)R_{-2},$$

and
$$\nu = +\frac{\sqrt{2}}{6M}\frac{1}{\bar{\rho}^2}R_{-2}\left(\mathcal{L}_2^\dagger - \frac{3ia\sin\theta}{\bar{\rho}^*}\right)S_{-2},$$

(7.251)

where it should be noted that we have not so far specified the relative normalization of the functions $R_{+2}S_{+2}$ and $R_{-2}S_{-2}$.

Finally, it should be noted that the solutions for Φ_0 and Φ_4 are in no way dependent on the choice of the arbitrary functions that occur in the general solutions of (7.233)–(7.238). The reason is that Φ_0 and Φ_4 (i.e. Ψ_0 and Ψ_4) are unchanged to quantities of the second order of smallness by infinitesimal rotations of the tetrad frame if Ψ_1 and Ψ_3 are initially zero (cf. (7.79) and (7.83)).

7.6.3 Is there a preferred gauge to express the solution?

In section 7.6.2 we have seen how we may choose a gauge in which $\Psi_1 = \Psi_3 = 0$. But one may ask if there is a preferred gauge for the problem. That there may be one such is suggested by the following considerations.

The set of equations (7.233)–(7.235) governing Φ_0, Φ_1, k, and s, for example, appears strangely truncated: the symmetry of the equations in Φ_0, k, and s is only partially present with respect to Φ_1, k, and s. Thus, as we have seen in section 7.6.2 the equations permit the elimination of Φ_1 from (7.233) and (7.234), while (7.235) provides exactly the 'right' relation between k and s to obtain a decoupled equation for Φ_0. We may, similarly, eliminate Φ_0 from (7.233) and (7.234) by virtue of the commutativity of the operators $(\mathcal{L}_2 - 3ia\sin\theta/\bar{\rho}^*)$ and $\Delta(\mathcal{D}_2^\dagger - 3/\bar{\rho}^*)$; but we do not have the 'right' relation between k and s to obtain a decoupled equation for Φ_1. But, exercising the freedom we have to subject the local (perturbed) tetrad frame to an infinitesimal rotation, we can rectify the situation by supplying (ad hoc?) the needed equation.

Chapter 7. The Kerr metric and its perturbations

Thus, with the additional equation

$$\Delta\left(\mathcal{D}_2^\dagger - \frac{3}{\bar{\rho}^*}\right)k + \left(\mathcal{L}_2 - \frac{3ia\sin\theta}{\bar{\rho}^*}\right)s = 2\frac{\bar{\rho}}{(\bar{\rho}^*)^2}\Phi_1, \qquad (7.252)$$

we can then complete the elimination of Φ_0, k, and s to obtain the decoupled equation

$$\left[\Delta\left(\mathcal{D}_2^\dagger - \frac{3}{\bar{\rho}^*}\right)\left(\mathcal{D}_0 + \frac{3}{\bar{\rho}^*}\right) + \left(\mathcal{L}_2 - \frac{3ia\sin\theta}{\bar{\rho}^*}\right)\left(\mathcal{L}_{-1}^\dagger + \frac{3ia\sin\theta}{\bar{\rho}^*}\right)\right]\Phi_1 \qquad (7.253)$$
$$= 12M\frac{\bar{\rho}}{(\bar{\rho}^*)^2}\Phi_1.$$

On expanding this equation, we obtain

$$[\Delta\mathcal{D}_2^\dagger\mathcal{D}_0 + \mathcal{L}_2\mathcal{L}_{-1}^\dagger - 6i\sigma(r + ia\cos\theta)]\Phi_1 = 0. \qquad (7.254)$$

This equations is clearly separable: by the substitution,

$$\Phi_1 = R_{+1}(r)S_{+1}(\theta) \qquad (7.255)$$

(we justify the designation of these functions with the subscript $+1$ presently) we obtain the pair of equations:

$$(\Delta\mathcal{D}_2^\dagger\mathcal{D}_0 - 6i\sigma r)R_{+1} = +(\lambda^{(1)} - 2)R_{+1} \qquad (7.256)$$

and

$$(\mathcal{L}_2\mathcal{L}_{-1}^\dagger + 6a\sigma\cos\theta)S_{+1} = -(\lambda^{(1)} - 2)S_{+1} \qquad (7.257)$$

where $\lambda^{(1)}$ is a separation constant. By making use of the readily verifiable identities

$$\Delta\mathcal{D}_2^\dagger\mathcal{D}_0 = \Delta\mathcal{D}_1\mathcal{D}_1^\dagger + 4ir\sigma - 2 \qquad (7.258)$$

and

$$\mathcal{L}_2\mathcal{L}_{-1}^\dagger = \mathcal{L}_0^\dagger\mathcal{L}_1 - 4a\sigma\cos\theta + 2, \qquad (7.259)$$

we can rewrite (7.256) and (7.257) in the forms

$$(\Delta\mathcal{D}_1\mathcal{D}_1^\dagger - 2i\sigma r)R_{+1} = \lambda^{(1)}R_{+1} \qquad (7.260)$$

and

$$(\mathcal{L}_0^\dagger\mathcal{L}_1 + 2a\sigma\cos\theta)S_{+1} = -\lambda^{(1)}S_{+1}. \qquad (7.261)$$

With the equations written in these forms, we recognize that R_{+1} and S_{+1} are the same radial and angular functions which describe a Maxwell field (for spin $s = +1$) in Kerr geometry (cf. (7.195) and (7.196)).

Gravitational perturbations

Similarly, supplementing (7.236)–(7.238) by the equation

$$\left(\mathcal{D}_0 - \frac{3}{\bar{\rho}^*}\right)n - \left(\mathcal{L}_2^\dagger - \frac{3ia\sin\theta}{\bar{\rho}^*}\right)l = 2\frac{\bar{\rho}}{(\bar{\rho}^*)^2}\Phi_3, \qquad (7.262)$$

we can obtain a decoupled equation for Φ_3 which on separation enables us to express Φ_3 as $R_{-1}(r)S_{-1}(\theta)$, where $R_{-1}(r)$ and $S_{-1}(\theta)$ satisfy the equations appropriate for a Maxwell field with spin $s = -1$. (cf. (7.197) and (7.198)).

Thus, *in a gauge which restores to (7.233)–(7.238) complete symmetry in the relationships among the various quantities which occur in them, the components Ψ_1 and Ψ_3 of the Weyl tensor are governed by the same equations as the components ϕ_0 and ϕ_2 of the Maxwell tensor.*

Should one, on the foregoing grounds, consider that gauge as a 'preferred one'? And if it is a preferred one, what is its physical meaning? It is found (see Chandrasekhar, 1978b) that, when one considers the perturbations of the charged Kerr–Newman black hole and writes the relevant equations in a gauge in which the perturbations in the Maxwell scalars ϕ_0 and ϕ_2 vanish, one is led to the same equations (7.252) and (7.262); and these equations survive, in fact, in the limit the charge vanishes. It is as if the Kerr black hole is in some sense 'aware' that a charged version of itself exists.

7.6.4 The Teukolsky–Starobinsky identities

The radial and the angular functions, $R_{\pm 2}$ and $S_{\pm 2}$, satisfy identities similar to those satisfied by $R_{\pm 1}$ and $S_{\pm 1}$; and they can be enunciated similarly.

Theorem 7.3

$\Delta^2 \mathcal{D}_0 \mathcal{D}_0 \mathcal{D}_0 \mathcal{D}_0 R_{-2}$ *is a multiple of* $\Delta^2 R_{+2}$ *and* $\Delta^2 \mathcal{D}_0^\dagger \mathcal{D}_0^\dagger \mathcal{D}_0^\dagger \mathcal{D}_0^\dagger \Delta^2 R_{+2}$ *is a multiple of* R_{-2}.

In view of equations (7.244) and (7.246) governing R_{-2} and $\Delta^2 R_{+2}$, the theorem is equivalent to asserting the commutation relation

$$\Delta \mathcal{D}_0 \mathcal{D}_0 \mathcal{D}_0 \mathcal{D}_0 (\Delta \mathcal{D}_{-1}^\dagger \mathcal{D}_0 + 6i\sigma r) = (\Delta \mathcal{D}_{-1} \mathcal{D}_0^\dagger - 6i\sigma r)\Delta^2 \mathcal{D}_0 \mathcal{D}_0 \mathcal{D}_0 \mathcal{D}_0, \qquad (7.263)$$

and its complex conjugate.

Chapter 7. The Kerr metric and its perturbations

Corollary 1

By a suitable choice of the relative normalization of the functions $\Delta^2 R_{+2}$ and R_{-2} we can arrange that

$$\Delta^2 \mathcal{D}_0 \mathcal{D}_0 \mathcal{D}_0 \mathcal{D}_0 R_{-2} = \mathscr{C} \, \Delta^2 R_{+2}, \qquad (7.264)$$

and

$$\Delta^2 \mathcal{D}_0^\dagger \mathcal{D}_0^\dagger \mathcal{D}_0^\dagger \mathcal{D}_0^\dagger \, \Delta^2 R_{+2} = \mathscr{C}^* R_{-2}, \qquad (7.265)$$

where $\mathscr{C}$ is some complex constant.

We can arrange the relative normalizations in the way prescribed since $\Delta^2 R_{+2}$ and R_{-2} satisfy complex-conjugate equations.

Corollary 2

The square of the absolute value of the constant $\mathscr{C}$ in (7.264) and (7.265) is given by

$$|\mathscr{C}|^2 = \lambda^2(\lambda+2)^2 - 8\sigma^2 \lambda[\alpha^2(5\lambda+6) - 12a^2] + 144\sigma^4 \alpha^4 + 144\sigma^2 M^2, \qquad (7.266)$$

where

$$\alpha^2 = a^2 + (am/\sigma). \qquad (7.267)$$

The relation (7.266) is a consequence of the identity

$$\Delta^2 \mathcal{D}_0^\dagger \mathcal{D}_0^\dagger \mathcal{D}_0^\dagger \mathcal{D}_0^\dagger \, \Delta^2 \mathcal{D}_0 \mathcal{D}_0 \mathcal{D}_0 \mathcal{D}_0 = |\mathscr{C}|^2 \mod \Delta \mathcal{D}_{-1}^\dagger \mathcal{D}_0 + 6i\sigma r - \lambda = 0, \qquad (7.268)$$

which directly follows from (7.264) and (7.265).

Theorem 7.4

$\mathscr{L}_{-1}\mathscr{L}_0\mathscr{L}_1\mathscr{L}_2 S_{+2}$ *is a multiple of* S_{-2} *and* $\mathscr{L}_{-1}^\dagger \mathscr{L}_0^\dagger \mathscr{L}_1^\dagger \mathscr{L}_2^\dagger S_{-2}$ *is a multiple of* S_{+2}.

In view of equations (7.245) and (7.247) governing S_{+2} and S_{-2}, the theorem is equivalent to asserting the commutation relation

$$\mathscr{L}_{-1}\mathscr{L}_0\mathscr{L}_1\mathscr{L}_2(\mathscr{L}_{-1}^\dagger \mathscr{L}_2 + 6a\sigma \cos\theta) = (\mathscr{L}_{-1}\mathscr{L}_2^\dagger - 6a\sigma \cos\theta)\mathscr{L}_{-1}\mathscr{L}_0\mathscr{L}_1\mathscr{L}_2, \qquad (7.269)$$

and its 'adjoint' obtained by replacing θ by $\pi - \theta$.

Corollary 1

If S_{+2} and S_{-2} are simultaneously normalized, then

$$\mathscr{L}_{-1}\mathscr{L}_0\mathscr{L}_1\mathscr{L}_2 S_{+2} = DS_{-2} \tag{7.270}$$

and

$$\mathscr{L}^\dagger_{-1}\mathscr{L}^\dagger_0\mathscr{L}^\dagger_1\mathscr{L}^\dagger_2 S_{-2} = DS_{+2}, \tag{7.271}$$

where D is some real constant.

Corollary 2

The constant D in the relations (7.270) and (7.271) is given by

$$D^2 = \lambda^2(\lambda+2)^2 - 8\sigma^2\lambda[\alpha^2(5\lambda+6) - 12a^2] + 144\sigma^4\alpha^4, \tag{7.272}$$

where $\alpha^2 = a^2 + (am/\sigma)$.

The relation (7.272) is a consequence of the identity

$$\mathscr{L}^\dagger_{-1}\mathscr{L}^\dagger_0\mathscr{L}^\dagger_1\mathscr{L}^\dagger_2\mathscr{L}_{-1}\mathscr{L}_0\mathscr{L}_1\mathscr{L}_2 = D^2 \bmod \mathscr{L}^\dagger_{-1}\mathscr{L}_2 + (\lambda + 6a\sigma \cos\theta) = 0. \tag{7.273}$$

It will be noticed that (cf. (7.268))

$$|\mathscr{C}|^2 = D^2 + 144\sigma^2 M^2, \tag{7.274}$$

a relation which could not have been anticipated.

It is important to observe that, while (7.266) determines the absolute value of the constant $\mathscr{C}$, we do not, as yet, know its real and imaginary parts, separately; and without this knowledge, we cannot effect the relative normalization of the radial functions, $\Delta^2 R_{+2}$ and R_{-2}, as required by the compatibility of (7.264) and (7.265). We need, in fact, even more than the knowledge of the real and the imaginary part of $\mathscr{C}$ to complete the solution; for we must specify what multiple of $R_{-2}S_{-2}$ is appropriate as a solution for Φ_4, when the solution for Φ_0 is taken as $R_{+2}S_{+2}$, in some chosen normalization (as the compatibility of (7.264) and (7.265)). These lacunae in our information must be filled eventually. But we may ask, meantime, the reason for the present incompleteness in our knowledge, in contrast to the completeness with which the solution for the Maxwell scalars, ϕ_0 and ϕ_2, was effected in sections 7.5.1 and 7.5.2. In these sections the set of equations we considered, namely (7.186)–(7.189), enabled us to obtain decoupled equations for ϕ_0 and ϕ_2 and provided an equation (namely (7.213)) which related them

Chapter 7. The Kerr metric and its perturbations

directly; it was this equation which enabled the completion of the solution. In contrast, the equations we have so far considered, namely (7.229) and (7.230), provide no similar equation relating Φ_0 and Φ_4, and in the absence of such an equation the solution of the problem cannot be completed. To obtain, then, an equation relating Φ_0 and Φ_4, we turn to the other Newman–Penrose equations in section 7.6.5.

7.6.5 The problem of gravitational perturbations and the manner of its solution

We begin with a statement of the problem of gravitational perturbations as it pertains to the Kerr metric.

A general perturbation of the Kerr metric, in the Newman–Penrose formalism, will be described by the two sets of amplitudes

$$\Psi_0, \Psi_1, \Psi_3, \Psi_4, \kappa, \sigma, \lambda, \text{ and } \nu \qquad (7.275)$$

and

$$\Psi_2^{(1)}, \tilde{\rho}^{(1)}, \mu^{(1)}, \tau^{(1)}, \pi^{(1)}, \alpha^{(1)}, \beta^{(1)}, \gamma^{(1)}, \varepsilon^{(1)}, l^{(1)}, n^{(1)}, m^{(1)}, \text{ and } \bar{m}^{(1)}, \qquad (7.276)$$

where the superscripts (1), in the second set of amplitudes, distinguish them from their unperturbed values. (We have suppressed the corresponding distinguishing superscripts, in the first set of amplitudes, for those quantities which vanish in the stationary state.)

In section 7.6.2 we obtained explicit solutions for the quantities listed in (7.275). In particular, the solution for the spin-coefficients, κ, σ, λ, and ν (in the gauge $\Psi_1 = \Psi_3 = 0$) are given in (7.251). And the solutions for Ψ_0 and Ψ_4 (which do not depend on the choice of gauge) are expressed in terms of Teukolsky's functions, $R_{\pm 2}$ and $S_{\pm 2}$, by

$$\Psi_0 = R_{+2} S_{+2} \quad \text{and} \quad \Psi_4 = R_{-2} S_{-2}/(\bar{\rho}^*)^4 \qquad (7.277)$$

where it should be remembered that, if the radial and the angular functions are so chosen that $\Delta^2 R_{+2}$ and R_{-2} are compatible with (7.264) and (7.265) and S_{+2} and S_{-2} are normalized to unity, then an additional numerical factor must be included in the expression for Ψ_4 – a factor which remains to be specified.

In many ways, the determination of the perturbations in the basis vectors, $(l, n, m, \bar{m})$ is the central problem in the theory of gravitational perturbations for the perturbations in the metric coefficients are deter-

Gravitational perturbations

mined by the perturbations in the basis vectors. Thus,

$$g^{ij(1)} = l^i n^{j(1)} + l^{i(1)} n^j + l^j n^{i(1)} + l^{j(1)} n^i \\ - m^i \bar{m}^{j(1)} - \bar{m}^{(i)} m^{j(1)} - m^j \bar{m}^{i(1)} - \bar{m}^j m^{i(1)}. \quad (7.278)$$

To represent the perturbations in the basis vectors, it is convenient to introduce an index notation. Thus, letting

$$l_1 = l, \quad l_2 = n, \quad l_3 = m, \quad \text{and} \quad l_4 = \bar{m}, \quad (7.279)$$

we can express the perturbation $l_a^{(1)}$, in the vector l_a, as a linear combination of the basis vectors l_b, in the manner,

$$l_a^{(1)} = A_a{}^b l_b. \quad (7.280)$$

The perturbations in the basis vectors are then fully described by the matrix $\mathbf{A}$.

Since l_1 and l_2 are real and l_3 and l_4 are complex conjugates, it follows that $A_1{}^1$, $A_1{}^2$, $A_2{}^1$, and $A_2{}^2$ are real, while all the remaining elements of $\mathbf{A}$ are complex; also, that the elements, in which the indices 3 and 4 are replaced, one by the other, are complex conjugates. It is, however, clear that the specification of the matrix will require sixteen real functions.

The quantities which have to be determined (listed in (7.275) and (7.276)) require ten real functions to specify the five (complex) Weyl scalars, twenty-four real functions to specify the twelve (complex) spin coefficients, and sixteen real functions to specify the matrix $\mathbf{A}$: all together fifty real functions. These fifty functions are subject to ten degrees of gauge freedom. These ten degrees of freedom arise from the six degrees of freedom we have in setting up the local tetrad-frame and the four degrees of coordinate freedom we have from the general covariance of the theory.

To determine the various quantities which we have enumerated, we have recourse to the linearized versions of the equations provided by the Newman–Penrose formalism. And, as we have seen in detail in section 7.2 the Newman–Penrose formalism provides three sets of equations: the Bianchi identities (7.64), the commutation relations ((7.52)–(7.55)), and the Ricci identities (7.59). By counting each complex equation as equivalent to two (real) equations, we have sixteen equations representing the Bianchi identities, twenty-four equations representing the commutation relations (one for each structure constant), and thirty-six equations representing the Ricci identities: all together seventy-six equations. These seventy-six equations are (ostensibly) written down for determining the fifty real functions we have enumerated and which are

Chapter 7. The Kerr metric and its perturbations

subject to ten degrees of gauge freedom. (It would, therefore, appear that among the seventy-six equations there can be no more than forty independent equations.)

Returning to the problem on hand, we take note that the quantities which remain to be determined are those listed in (7.276) (leaving aside the matters pertaining to the relative normalization of the functions R_{+2} and R_{-2} and the real and the imaginary parts of the constant $\mathscr{C}$). The elements of $\boldsymbol{A}$, the perturbations in the eight (remaining) spin coefficients, and the perturbation in Ψ_2, all together require thirty-four real functions for their specification.

The solution to the complete perturbation problem, as we have formulated, has recently been effected (Chandrasekhar, 1978c; and a further paper in press) by solving the linearized versions of the relevant Newman–Penrose equations. But the analysis is too long and involved to allow even a partially deductive account. A brief outline of the method and a simple statement of the principal results must suffice.

(a) The linearized Bianchi identities

We have already used four of the Bianchi identities (equations (7.64a, d, e, and h)) in obtaining the solutions for the quantities listed in (7.275). The remaining identities (equations (7.64b, c, f, and g)), in the gauge $\Psi_1 = \Psi_3 = 0$ (which is consistently adopted in the developments to be described) are

$$D\Psi_2 = 3\tilde{\rho}\Psi_2, \quad \Delta\Psi_2 = -3\mu\Psi_2, \quad \delta\Psi_2 = 3\tau\Psi_2, \quad \text{and} \quad \delta^*\Psi_2 = -3\pi\Psi_2,$$
(7.281)

where quantities of the second order of smallness (such as $\lambda\Psi_0$, $\kappa\Psi_4$, $\nu\Psi_0$, and $\sigma\Psi_4$) are neglected. The remarkable feature of equations (7.281) is that *they are formally the same as in the stationary state: they are valid inclusive of terms of the first-order of smallness.*

It is clear that the linearization of equations (7.281) will enable us to express the perturbations in the spin-coefficients, $\tilde{\rho}$, μ, τ, and π, directly in terms of the elements of $\boldsymbol{A}$ and the perturbation in Ψ_2.

(b) The linearized commutation relations

Using (7.280), we can write the linearized version of the equation (cf. (7.50))

$$[l_a, l_b] = C^f{}_{ab}l_f,$$
(7.282)

Gravitational perturbations

in the form

$$[A_a{}^f l_f, l_b] + [l_a, A_b{}^f l_f] = C^f{}_{ab} A_f{}^c l_c + c^f{}_{ab} l_f, \qquad (7.283)$$

where $c^f{}_{ab}$ is the perturbation in $C^f{}_{ab}$. The quantities $c^f{}_{ab}$ are directly related to the perturbations in the particular combinations of the spin coefficients which occur in the tabulation (7.56). By expanding (7.283), we obtain

$$\begin{aligned} A_a{}^f C^m{}_{fb} l_m + A_b{}^f C^m{}_{af} l_m + (l_a A_b{}^m) l_m - (l_b A_a{}^m) l_m \\ = C^f{}_{ab} A_f{}^m l_m + c^m{}_{ab} l_m, \end{aligned} \qquad (7.284)$$

or, since the ls are linearly independent, we can write

$$l_a A_b{}^m - l_b A_a{}^m = A_a{}^f C^m{}_{bf} - A_b{}^f C^m{}_{af} + C^f{}_{ab} A_f{}^m + c^m{}_{ab}. \qquad (7.285)$$

Equation (7.285) is, in many ways, the basic equation of this theory: it provides a set of twenty-four simultaneous, linear, inhomogeneous, partial differential equations for the elements of $\mathbf{A}$; and the inhomogeneous terms are directly related to the perturbations in the spin coefficients.

The twenty-four equations which follow from equation (7.285) can be combined and grouped to give three systems of eight equations each: a group of eight equations in which the inhomogeneous terms are directly related to the perturbations in the spin-coefficients, $\tilde{\rho}$, μ, τ, and π (which, as we have seen, can be expressed in terms of the elements of $\mathbf{A}$ and the perturbation in Ψ_2); a second group of eight equations in which the inhomogeneous terms are precisely κ, σ, λ, and ν (or their complex conjugates) for which we have explicit solutions in terms of Teukolsky's functions (leaving aside the unspecified relative normalization of R_{+2} and R_{-2}); and, finally, a third group of eight equations in which the inhomogeneous terms are directly related to the remaining spin-coefficients, α, β, γ, and ε. The first two groups of equations, together, provide sixteen equations for the elements of $\mathbf{A}$ and $\Psi_2^{(1)}$; and the last group of eight equations will suffice to determine α, β, γ, and ε, once we have found the solutions for the elements of $\mathbf{A}$ and $\Psi_2^{(1)}$.

(c) The linearized Ricci identities

Besides the Ricci identities (7.59*b*) and (7.59*j*), which we have already used in section 7.6.1, it is found that to complete the solution it is necessary to consider, as well, the linearized version of the equations (7.59*a*, *g*, *n*, and *p*).

Chapter 7. The Kerr metric and its perturbations

In a remarkable way, all the equations we have enumerated allow explicit solutions by processes which are almost wholly algebraic; and the principal results of the analysis are the following.

(1) The perturbation in Ψ_2 is left unspecified in accordance with the fact that it is subject to changes by (infinitesimal) coordinate transformations. We can accordingly use two of the available four degrees of coordinate-freedom to arrange for the identical vanishing of $\Psi_2^{(1)}$. Since we have already used up four of the six available degrees of tetrad-freedom to preserve the vanishing of Ψ_1 and Ψ_3, we can take advantage of the four remaining degrees of gauge-freedom to make the diagonal elements of $\mathbf{A}$ (i.e. A_1^1, A_2^2, A_3^3, and A_4^4) vanish.

(2) The analysis leads to a crucial integrability condition in which the solutions for Ψ_0 and Ψ_4 as well as those for κ, σ, λ, and ν all appear; and the satisfaction of the integrability condition, remarkably, determines the real and the imaginary parts of the constant $\mathscr{C}$ as well as the numerical factor that must be included in the solution for Ψ_4 given in (7.277). It is found that

$$\mathscr{C} = D - 12i\sigma M \qquad (7.286)$$

where D denotes the positive square root of the expression on the right-hand side of (7.272); and, further, that, in a relative normalization of the radial functions which is compatible with (7.264) and (7.265), the appropriate solutions for Ψ_0 and Ψ_4 are

$$\Psi_0 = R_{+2}S_{+2} \qquad (7.287)$$

and

$$\Psi_4 = \frac{1}{4(\bar{\rho}^*)^4} R_{-2}S_{-2}. \qquad (7.288)$$

(3) The solution for the elements of $\mathbf{A}$ (and, therefore, also for the perturbations of the metric coefficients) can all be expressed in terms of the single function

$$\Psi = \frac{\mathscr{C}_1 + \Gamma_1}{96M\sqrt{2}} \mathscr{R}(r)\mathscr{S}(\theta), \qquad (7.289)$$

where

Equations (7.290)

$$\mathscr{C}_1 = D, \quad \Gamma_1 = \lambda(\lambda+2) - 12\sigma^2\alpha^2, \quad \alpha^2 = a^2 + (am/\sigma),$$

$$\mathscr{R}(r) = \mathscr{R}_- + \frac{4\lambda\sigma}{\mathscr{C}_1 + \Gamma_1}\mathscr{R}_+, \quad \mathscr{S}(\theta) = \mathscr{S}_- + \frac{4\sigma(\lambda\alpha^2 - 6a^2)}{a(\mathscr{C}_1 + \Gamma_1)}\mathscr{S}_+,$$

$$\mathcal{R}_+ = \Delta^{1/2} \int \frac{rX}{\Delta^{3/2}} \, dr, \quad \mathcal{R}_- = \Delta^{1/2} \int \frac{Y}{\Delta^{3/2}} \, dr,$$

$$\mathcal{S}_+ = \int (S_{+2} + S_{-2}) \, d\theta, \quad \mathcal{S}_- = \int (S_{+2} - S_{-2}) \cos\theta \, d\theta,$$

$$X = \Delta^2 R_{+2} + R_{-2}, \quad \text{and} \quad Y = -i(\Delta^2 R_{+2} - R_{-2}).$$

Moreover, the functions $\mathcal{R}$ and $\mathcal{S}$ can be expressed directly in terms of the functions X and Y. (Also, it should be noted that the radial functions, $\Delta^2 R_{+2}$ and R_{-2}, are defined so as to be compatible with (7.264) and (7.265).)

Alternative treatments of the metric perturbations of the Kerr metric are due to Cohen and Kegeles (1974) and Chrzanowski (1975).

7.7 The solution of Dirac's equation

7.7.1 Dirac's equation in the Newman–Penrose formalism

In Penrose's (1968) spinor notation, the four components of the wave function which satisfy Dirac's equation are represented by two spinors P^A and Q^A (say); and Dirac's equation (in units in which $c = 1$ and $\hbar = 1$) is written in the form

$$\nabla_{AA'} P^A + i\mu_e \bar{Q}_{A'} = 0 \qquad (7.291)$$

and

$$\nabla_{AA'} Q^A + i\mu_e \bar{P}_{A'} = 0, \qquad (7.292)$$

where $2^{1/2} \mu_e$ is the mass of the electron (in the chosen units) and $\nabla_{AA'}$ is the symbol for covariant differentiation.

Written out explicitly in terms of the spin-coefficients and the directional derivatives of the standard Newman–Penrose formalism, (7.290) is equivalent to the pair of equations

$$(D + \varepsilon - \tilde{\rho})P^0 + (\delta^* + \pi - \alpha)P^1 = +i\mu_e \bar{Q}^{1'} \qquad (7.293)$$

and

$$(\delta + \beta - \tau)P^0 + (\Delta + \mu - \gamma)P^1 = -i\mu_e \bar{Q}^{0'}. \qquad (7.294)$$

Equation (7.292) provides a similar pair of equations for Q^0 and Q^1. Letting,

$$F_1 = P^0, \quad F_2 = P^1, \quad G_1 = \bar{Q}^{1'}, \quad \text{and} \quad G_2 = -\bar{Q}^{0'}, \qquad (7.295)$$

Chapter 7. The Kerr metric and its perturbations

we obtain the following set of four coupled equations

$$\left.\begin{aligned}(D+\varepsilon-\tilde{\rho})F_1+(\delta^*+\pi-\alpha)F_2 &= i\mu_e G_1,\\(\Delta+\mu-\gamma)F_2+(\delta+\beta-\tau)F_1 &= i\mu_e G_2,\\(D+\varepsilon^*-\tilde{\rho}^*)G_2-(\delta+\pi^*-\alpha^*)G_1 &= i\mu_e F_2,\\(\Delta+\mu^*-\gamma^*)G_1-(\delta^*+\beta^*-\tau^*)G_2 &= i\mu_e F_1.\end{aligned}\right\} \quad (7.296)$$

7.7.2 The separation of the variables in Kerr geometry†

Assuming that the four components of the wave function have the customary dependence on t and φ (given by (7.176)) and inserting for the spin coefficients and the directional derivatives the expressions given in (7.170) and (7.177), we find that equations (7.296) reduce to

$$\left.\begin{aligned}\left(\mathcal{D}_0+\frac{1}{\tilde{\rho}^*}\right)F_1+\frac{1}{\tilde{\rho}^*\sqrt{2}}\mathcal{L}_{1/2}F_2 &= +i\mu_e G_1,\\\frac{\Delta}{2\rho^2}\mathcal{D}^\dagger_{1/2}F_2-\frac{1}{\tilde{\rho}\sqrt{2}}\left(\mathcal{L}^\dagger_{1/2}+\frac{ia\sin\theta}{\tilde{\rho}^*}\right)F_1 &= -i\mu_e G_2,\\\left(\mathcal{D}_0+\frac{1}{\tilde{\rho}}\right)G_2-\frac{1}{\tilde{\rho}\sqrt{2}}\mathcal{L}^\dagger_{1/2}G_1 &= +i\mu_e F_2,\\\frac{\Delta}{2\rho^2}\mathcal{D}^\dagger_{1/2}G_1+\frac{1}{\tilde{\rho}^*\sqrt{2}}\left(\mathcal{L}_{1/2}-\frac{ia\sin\theta}{\tilde{\rho}}\right)G_2 &= -i\mu_e F_1.\end{aligned}\right\} \quad (7.297)$$

The form of equations (7.297) suggests that, in place of F_1 and G_2, we define

$$f_1 = \tilde{\rho}^* F_1 = (r-ia\cos\theta)F_1$$

and $\quad (7.298)$

$$g_2 = \tilde{\rho} G_2 = (r+ia\cos\theta)G_2.$$

Also writing f_2 and g_1 in place of F_2 and G_1 (for symmetry in form of the resulting equations) we find that equations (7.297) become

$$\left.\begin{aligned}\mathcal{D}_0 f_1+2^{-1/2}\mathcal{L}_{1/2}f_2 &= +(i\mu_e r+a\mu_e\cos\theta)g_1,\\\Delta\mathcal{D}^\dagger_{1/2}f_2-2^{+1/2}\mathcal{L}^\dagger_{1/2}f_1 &= -2(i\mu_e r+a\mu_e\cos\theta)g_2,\\\mathcal{D}_0 g_2-2^{-1/2}\mathcal{L}^\dagger_{1/2}g_1 &= (i\mu_e r-a\mu_e\cos\theta)f_2,\\\Delta\mathcal{D}^\dagger_{1/2}g_1+2^{+1/2}\mathcal{L}_{1/2}g_2 &= -2(i\mu_e r-a\mu_e\cos\theta)f_1.\end{aligned}\right\} \quad (7.299)$$

† The analysis in this section is taken from Chandrasekhar (1976c).

The solution of Dirac's equation

It is now apparent that the variables can be separated by writing

$$f_1(r, \theta) = R'_{-1/2}(r)S_{-1/2}(\theta), \quad f_2(r, \theta) = R_{+1/2}(r)S_{+1/2}(\theta),$$
$$g_1(r, \theta) = R_{+1/2}(r)S_{-1/2}(\theta), \quad \text{and} \quad g_2(r, \theta) = R_{-1/2}(r)S_{+1/2}(\theta),$$
(7.300)

where $R_{\pm 1/2}(r)$ and $S_{\pm 1/2}(\theta)$ are functions only of r and θ respectively. For with these forms for the solutions, equations (7.299) become

Equations (7.301)

$$(\mathcal{D}_0 R_{-1/2} - i\mu_e r R_{+1/2})S_{-1/2}$$
$$+ [2^{-1/2}\mathcal{L}_{1/2}S_{+1/2} - (a\mu_e \cos\theta)S_{-1/2}]R_{+1/2} = 0,$$
$$(\Delta\mathcal{D}^\dagger_{1/2} R_{+1/2} + 2i\mu_e R_{-1/2})S_{+1/2}$$
$$- [2^{+1/2}\mathcal{L}^\dagger_{1/2}S_{-1/2} - 2(a\mu_e \cos\theta)S_{+1/2}]R_{-1/2} = 0,$$
$$(\mathcal{D}_0 R_{-1/2} - i\mu_e r R_{+1/2})S_{+1/2}$$
$$- [2^{-1/2}\mathcal{L}^\dagger_{1/2}S_{-1/2} - (a\mu_e \cos\theta)S_{+1/2}]R_{+1/2} = 0,$$
$$(\Delta\mathcal{D}^\dagger_{1/2} R_{+1/2} + 2i\mu_e R_{-1/2})S_{-1/2}$$
$$+ [2^{+1/2}\mathcal{L}_{1/2}S_{+1/2} - 2(a\mu_e \cos\theta)S_{-1/2}]R_{-1/2} = 0.$$

These equations imply

Equations (7.302)

$$\mathcal{D}_0 R_{-1/2} - i\mu_e r R_{+1/2} = \lambda_1 R_{+1/2};$$
$$2^{-1/2}\mathcal{L}_{1/2}S_{+1/2} - (a\mu_e \cos\theta)S_{-1/2} = -\lambda_1 S_{-1/2},$$
$$\Delta\mathcal{D}^\dagger_{1/2} R_{+1/2} + 2i\mu_e R_{-1/2} = \lambda_2 R_{-1/2};$$
$$2^{+1/2}\mathcal{L}^\dagger_{1/2}S_{-1/2} - 2(a\mu_e \cos\theta)S_{+1/2} = +\lambda_2 S_{+1/2},$$
$$\mathcal{D}_0 R_{-1/2} - i\mu_e r R_{+1/2} = \lambda_3 R_{+1/2};$$
$$2^{-1/2}\mathcal{L}^\dagger_{1/2}S_{-1/2} - (a\mu_e \cos\theta)S_{+1/2} = +\lambda_3 S_{+1/2},$$
$$\Delta\mathcal{D}^\dagger_{1/2} R_{+1/2} + 2i\mu_e R_{-1/2} = \lambda_4 R_{-1/2};$$
$$2^{+1/2}\mathcal{L}_{1/2}S_{+1/2} - 2(a\mu_e \cos\theta)S_{-1/2} = -\lambda_4 S_{-1/2},$$

where $\lambda_1, \ldots, \lambda_4$ are four constants of separation. However, it is manifest that the consistency of the foregoing equations requires that

$$\lambda_1 = \lambda_3 = \tfrac{1}{2}\lambda_2 = \tfrac{1}{2}\lambda_4 = \lambda \quad \text{(say)}. \tag{7.303}$$

Chapter 7. The Kerr metric and its perturbations

We are thus left with the two pairs of equations

$$\mathcal{D}_0 R_{-1/2} = (\tilde{\lambda} + i\mu_e r)R_{+1/2},$$
$$\Delta \mathcal{D}_{1/2}^\dagger R_{+1/2} = 2(\tilde{\lambda} - i\mu_e r)R_{-1/2},$$
(7.304)

and

$$\mathcal{L}_{1/2} S_{+1/2} = -2^{1/2}(\tilde{\lambda} - a\mu_e \cos\theta)S_{-1/2},$$
$$\mathcal{L}_{1/2}^\dagger S_{-1/2} = +2^{1/2}(\tilde{\lambda} + a\mu_e \cos\theta)S_{+1/2}.$$
(7.305)

It is convenient at this stage to replace

$$2^{1/2}\tilde{\lambda} \text{ by } \tilde{\lambda}, \quad 2^{1/2}\mu_e \text{ by } m_e, \quad \text{and} \quad 2^{1/2}R_{-1/2} \text{ by } R_{-1/2}. \quad (7.306)$$

With these replacements, (7.304) and (7.305) become

$$\Delta^{1/2}\mathcal{D}_0 R_{-1/2} = (\tilde{\lambda} + im_e r)\Delta^{1/2} R_{+1/2},$$
$$\Delta^{1/2}\mathcal{D}_0^\dagger \Delta^{1/2} R_{+1/2} = (\tilde{\lambda} - im_e r)R_{-1/2},$$
(7.307)

and

$$\mathcal{L}_{1/2} S_{+1/2} = -(\tilde{\lambda} - am_e \cos\theta)S_{-1/2};$$
$$\mathcal{L}_{1/2}^\dagger S_{-1/2} = +(\tilde{\lambda} + am_e \cos\theta)S_{+1/2}.$$
(7.308)

From equations (7.307), we conclude that if $(\Delta^{1/2} R_{+1/2}, R_{-1/2})$ represents a solution, then so does $(R^*_{-1/2}, \Delta^{1/2} R^*_{+1/2})$. Similarly, we conclude from equations (7.308) that $[S_{+1/2}(\theta), S_{-1/2}(\theta)]$ represents a solution, then so does $[S_{-1/2}(\pi - \theta), S_{+1/2}(\pi - \theta)]$.

We can eliminate $\Delta^{1/2} R_{+1/2}$ from the pair of equations (7.307) to obtain an equation for $R_{-1/2}$; thus

$$\left[\Delta \mathcal{D}_{1/2}^\dagger \mathcal{D}_0 - \frac{im_e \Delta}{\tilde{\lambda} + im_e r}\mathcal{D}_0 - (\tilde{\lambda}^2 + m_e^2 r^2)\right] R_{-1/2} = 0; \quad (7.309)$$

and $\Delta^{1/2} R_{+1/2}$ satisfies the complex-conjugate equation.

Similarly, we can eliminate $S_{+1/2}$ from the pair of equations (7.308) to obtain an equation for $S_{-1/2}$; thus

$$\left[\mathcal{L}_{1/2}\mathcal{L}_{1/2}^\dagger + \frac{am_e \sin\theta}{\tilde{\lambda} + am_e \cos\theta}\mathcal{L}_{1/2}^\dagger + (\tilde{\lambda}^2 - a^2 m_e^2 \cos^2\theta)\right] S_{-1/2} = 0; \quad (7.310)$$

and $S_{+1/2}$ satisfies the 'adjoint' equation (obtained by replacing θ by $\pi - \theta$). Also, $\tilde{\lambda}$ now appears as a characteristic-value parameter determined by the condition that $S_{-1/2}$ (and, therefore, also $S_{+1/2}$) is regular at $\theta = 0$ and $\theta = \pi$.

The potential barriers round the Kerr black hole

It may be noted that by setting $M = 0$ in the foregoing equations, we obtain Dirac's equation separated in oblate spheroidal coordinates in Minkowskian space.

7.7.3 The equations governing the two-component neutrino

The equations appropriate to the two-component neutrino can be obtained by setting $m_e = 0$ in the equations of section 7.7.2. Thus, in place of equations (7.307)–(7.310), we have

$$\mathscr{D}_0 R_{-1/2} = \bar{\lambda} R_{+1/2}, \quad \Delta^{1/2} \mathscr{D}_0^\dagger \Delta^{1/2} R_{+1/2} = \bar{\lambda} R_{-1/2}, \quad (7.311)$$

$$\mathscr{L}_{1/2} S_{+1/2} = -\bar{\lambda} S_{-1/2}, \quad \mathscr{L}_{1/2}^\dagger S_{-1/2} = +\bar{\lambda} S_{+1/2}, \quad (7.312)$$

and

$$(\Delta \mathscr{D}_{1/2}^\dagger \mathscr{D}_0 - \bar{\lambda}^2) R_{-1/2} = 0, \quad (\Delta \mathscr{D}_{1/2} \mathscr{D}_0^\dagger - \bar{\lambda}^2) \Delta^{1/2} R_{1/2} = 0, \quad (7.313)$$

$$(\mathscr{L}_{1/2} \mathscr{L}_{1/2}^\dagger + \bar{\lambda}^2) S_{-1/2} = 0, \quad (\mathscr{L}_{1/2}^\dagger \mathscr{L}_{1/2} + \bar{\lambda}^2) S_{+1/2} = 0. \quad (7.314)$$

The foregoing equations are equivalent to those first derived by Unruh (1973, 1974) and Teukolsky (1973).

7.8 The potential barriers surrounding the Kerr black hole and the problem of reflection and transmission

7.8.1 A transformation of Teukolsky's equations

The equations governing the radial functions $R_{\pm s}$, for $s = \frac{1}{2}$, 1, and 2 (namely (7.313), (7.195), (7.197), (7.244), and (7.246)), can be written as special cases of the equations

$$[\Delta \mathscr{D}_{1-s} \mathscr{D}_0^\dagger - 2(2s-1)i\sigma r]\Delta^s R_{+s} = \lambda \Delta^s R_{+s}$$

and

$$[\Delta \mathscr{D}_{1-s}^\dagger \mathscr{D}_0 + 2(2s-1)i\sigma r]R_{-s} = \lambda R_{-s}, \quad (7.315)$$

where, for the case $s = \frac{1}{2}$, we have written $\bar{\lambda}$ in place of $\bar{\lambda}^2$ in (7.313). (Also, note that we are assigning only positive values to s.) Expanding the equations (7.315), we find

$$\left\{ \Delta^{-s} \frac{d}{dr} \Delta^{s+1} \frac{d}{dr} + \frac{1}{\Delta} [K^2 + 2isK(r-M) - \Delta(\lambda - 2s + 4is\sigma r)] \right\} R_{+s} = 0,$$

Chapter 7. The Kerr metric and its perturbations

and

$$\left\{\Delta^{+s}\frac{d}{dr}\Delta^{-s+1}\frac{d}{dr}+\frac{1}{\Delta}[K^2-2isK(r-M)-\Delta(\lambda-4is\sigma r)]\right\}R_{-s}=0;$$
(7.316)

these are the forms in which Teukolsky (1972) first derived them, and the forms in which they are generally written.

It would appear from (7.316) that the different m-modes will have to be treated separately, since m occurs explicitly in the equations (via K). We shall now show how this explicit occurrence of m can be eliminated by a different choice of variables.

First, we introduce, in place of r, a new independent variable r_* defined by

$$\frac{d}{dr_*}=\frac{\Delta}{\rho^2}\frac{d}{dr},$$
(7.317)

where

$$\rho^2=r^2+\alpha^2 \quad \text{and} \quad \alpha^2=a^2+(am/\sigma).$$
(7.318)†

In view of the admissibility of negative values for α^2 (when m is negative and $\sigma \to 0$), it is clear that the resulting $r_*(r)$-relation can, under certain circumstances, become double-valued. We shall postpone to sections 7.8.2 and 7.8.3 a consideration of the consequences of this possible double-valuedness of the $r_*(r)$-relation and restrict ourselves, for the present, to deducing only the formal consequences of this change of variable. *Second*, in addition to this change in the independent variable, we shall also let

$$Y_{+s}=|\rho^2|^{-s+1/2}\Delta^s R_{+s} \quad \text{and} \quad Y_{-s}=|\rho^2|^{-s+1/2}R_{-s}.$$
(7.319)

In writing the resulting equations for $Y_{\pm s}$, we shall find it convenient to use the operators,

$$\Lambda_\pm = \frac{d}{dr_*} \pm i\sigma \quad \text{and} \quad \Lambda^2 = \Lambda_+\Lambda_- = \Lambda_-\Lambda_+ = \frac{d^2}{dr_*^2} + \sigma^2.$$
(7.320)

In terms of these operators, we find that $Y_{\pm s}$ satisfies the equation

$$\Lambda^2 Y + P\Lambda_- Y - QY = 0,$$
(7.321)

---

† This different usage of ρ^2 should not be confused with the earlier usage (with the meaning $\rho^2 = r^2 + a^2 \cos^2\theta$). But α^2 has the same meaning as in the definitions of the Teukolsky–Starobinsky constants (see (7.209) and (7.260)).

where

$$P = \frac{2s}{\rho^4}[2r\Delta - \rho^2(r-M)] = \frac{d}{dr_*}\left(\ln\frac{|\rho|^{4s}}{\Delta^s}\right) \quad (7.322)$$

and

$$Q = \frac{\Delta}{\rho^4}\left\{\lambda - (2s-1)\left[\frac{\Delta - 2(s-1)r(r-M)}{\rho^2} + (2s-3)\frac{r^2\Delta}{\rho^4}\right]\right\}. \quad (7.323)$$

And Y_{-s} satisfies the complex-conjugate equation obtained by replacing Λ_- by Λ_+.

It will be observed that (7.321) contains no explicit reference to m: *the non-axisymmetric case differs from the axisymmetric case only by a^2 being replaced by α^2.* (It should, however, be remembered that Δ involves a^2; and we should not tamper with *this* a^2.)

Finally, we may note that for the three important cases, $s = \frac{1}{2}$, 1, and 2, the expressions for Q are (see Chandrasekhar, 1976a, Chandrasekhar and Detweiler, 1975, 1976a, b)

$$Q_{s=1/2} = \lambda\frac{\Delta}{\rho^4}, \quad (7.324)$$

$$Q_{s=1} = \frac{\Delta}{\rho^4}\left(\lambda - \alpha^2\frac{\Delta}{\rho^4}\right), \quad (7.325)$$

and

$$Q_{s=2} = \frac{\Delta}{\rho^4}\left\{\lambda - 3\left[\frac{\Delta - 2r(r-M)}{\rho^2} + \frac{r^2\Delta}{\rho^4}\right]\right\}$$

$$= \frac{\Delta}{\rho^8}[\lambda\rho^4 + 3\rho^2(r^2 - a^2) - 3r^2\Delta]. \quad (7.326)$$

7.8.2 The $r_*(r)$-relation

Integrating (7.317), we obtain the relation

$$r_* = r + \frac{2Mr_+ + (am/\sigma)}{r_+ - r_-}\ln\left(\frac{r}{r_+} - 1\right)$$

$$- \frac{2Mr_- + (am/\sigma)}{r_+ - r_-}\ln\left(\frac{r}{r_-} - 1\right) \quad (r > r_+), \quad (7.327)$$

where

$$r_\pm = M \pm (M^2 - a^2)^{1/2} \quad (7.328)$$

Chapter 7. The Kerr metric and its perturbations

denote the radii of the outer and the inner horizons of the Kerr metric.

When considering the (external) perturbations of the Kerr black hole we are indifferent to 'what happens' inside $r < r_+$; and we need be concerned with the $r_*(r)$-relation only for $r > r_+$.

It is now apparent from (7.327), that the $r_*(r)$-relation is single-valued only so long as

$$r_+^2 + \alpha^2 = 2Mr_+ - a^2 + \alpha^2 = 2Mr_+ + (am/\sigma) > 0; \qquad (7.329)$$

and when this inequality obtains

$$r_* \to \infty \text{ when } r \to \infty \quad \text{and} \quad r_* \to -\infty \text{ when } r \to r_+ + 0. \qquad (7.330)$$

Letting

$$\sigma_s = -am/2Mr_+ \quad \text{(for } m \text{ negative)}, \qquad (7.331)$$

and remembering that our convention with respect to σ is that it is positive, we conclude that *so long as $\sigma > \sigma_s$, the $r_*(r)$-relation is single-valued and the range of r_* is the entire interval $(+\infty, -\infty)$*. But if $0 < \sigma < \sigma_s$ and $r_+^2 + \alpha^2 < 0$, the $r_*(r)$-relation is double-valued: r_* attains a minimum value when $r = |\alpha|$ and $r_* \to +\infty$ both when $r \to \infty$ and $r \to r_+ + 0$. In the latter case, it can be readily shown that in the neighbourhood of $r = |\alpha|$, the $r_*(r)$-relation has the behaviour

$$r_* = r_*(|\alpha|) + \frac{|\alpha|}{\Delta_{|\alpha|}} (r - |\alpha|)^2 + O[(r - |\alpha|)^3]. \qquad (7.332)$$

Also, when $0 < \sigma < \sigma_s$ and $r = |\alpha|$, the functions P and Q in (7.321) become singular; accordingly, the equation must be considered, separately, in the two branches of the $r_*(r)$-relation, namely, for $\infty > r > |\alpha|$ and for $|\alpha| > r > r_+$.

We shall show in section 7.8.4 that *the interval, $(0 < \sigma < \sigma_s)$, in σ defines the range of frequencies in which the reflection coefficient, for incident waves of integral spins, exceeds unity and the phenomenon of super-radiance occurs.*

Finally, it should be noted that α^2 becomes negative already before σ_s: thus,

$$\alpha^2 \leq 0 \quad \text{for} \quad \sigma \leq \sigma_c = -m/a; \qquad (7.333)$$

and $\alpha = 0$ when σ is 'co-rotational'. In the interval $\sigma_s < \sigma \leq \sigma_c$, while α^2 is negative, $r_+^2 + \alpha^2 > 0$ and the $r_*(r)$-relation continues to be single-valued.

7.8.3 The reduction to a one-dimensional wave equation

We shall now show (following Chandrasekhar and Detweiler, 1975, 1976a, b and Chandrasekhar, 1976a) how (7.321) can be reduced to a one-dimensional wave equation of the form

$$\Lambda^2 Z = VZ \qquad (7.334)$$

where V is some potential function of r_* (or r) to be determined. If such a reduction can be effected, then, the problem would be reduced to one in the elementary theory of the penetration of potential barriers.

It is clear that Z (on the assumption that one such exists) must be a linear combination of Y and its derivative since Y satisfies a second-order differential equation. There is, therefore, no loss of generality in assuming that Z is related to Y in the manner

$$Y = f\Lambda_+\Lambda_+ Z + W\Lambda_+ Z, \qquad (7.335)\dagger$$

where f and W are certain functions of r_* (or r) which are, for the present, unspecified. An alternative form of this equation, obtained by making use of (7.334), is

$$Y = fVZ + (W + 2i\sigma f)\Lambda_+ Z. \qquad (7.336)$$

Now applying the operator Λ_- to (7.336), and making use, once again, of the fact that Z has been assumed to satisfy (7.334), we find that

$$\Lambda_- Y = -\frac{\Delta^s}{\rho^{4s}} M\beta Z + R\Lambda_+ Z, \qquad (7.337)$$

where

$$-\frac{\Delta^s}{\rho^{4s}} M\beta(r_*) = \frac{d}{dr_*} fV + WV, \qquad (7.338)$$

and

$$R = fV + \frac{d}{dr_*}(W + 2i\sigma f). \qquad (7.339)$$

We must now require that (7.336) and (7.337) are compatible with (7.321) satisfied by Y. For this purpose, we must eliminate Y from

---

† This form may appear an 'odd' way of expressing an expected relation, but it is suggested by what is known from an analysis of the perturbations of the Schwarzschild metric (see Chandrasekhar, 1975, equation (49)). Indeed, in many ways, the perfect adaptability to the Kerr metric of methods found successful in the context of the Schwarzschild metric is one of the striking aspects of the entire theory.

Chapter 7. The Kerr metric and its perturbations

(7.336) and (7.337). The required elimination can be effected by applying the operator Λ_- to (7.337), making use of the equation (cf. (7.321))

$$\Lambda_-\Lambda_-Y = -(2i\sigma + P)\Lambda_-Y + QY, \tag{7.340}$$

substituting for Y and Λ_-Y from (7.336) and (7.337), and, eventually, obtaining a relation, linear in Z and Λ_+Z, which must vanish identically. From the vanishing of the coefficients of Z and Λ_+Z in the relation so obtained, we obtain the pair of equations

$$RV - M\frac{\Delta^s}{\rho^{4s}}\frac{d\beta}{dr_*} = QfV \tag{7.341}$$

and

$$\frac{d}{dr_*}\left(\frac{\rho^{4s}}{\Delta^s}R\right) = \frac{\rho^{4s}}{\Delta^s}[Q(W + 2i\sigma f) - 2i\sigma R] + \beta M. \tag{7.342}$$

It can now be verified that (7.338)–(7.342) admit the integral

$$\frac{\rho^{4s}}{\Delta^s}RfV + \beta M(W + 2i\sigma f) = K = \text{constant}. \tag{7.343}$$

This integral enables us to invert (7.336) and (7.337); and we obtain

$$\frac{\Delta^s}{\rho^{4s}}KZ = RY - T\Lambda_-Y, \tag{7.344}$$

and

$$K\Lambda_+Z = \beta MY + \frac{\rho^{4s}}{\Delta^s}fV\Lambda_-Y, \tag{7.345}$$

where, for brevity, we have written

$$T = W + 2i\sigma f. \tag{7.346}$$

It can now be verified that *equations (7.336), (7.337), (7.344), and (7.345), by virtue of the relations (7.338), (7.339), (7.341), and (7.342), are necessary and sufficient for equation (7.321) to imply equation (7.334), and conversely.*

Since (7.338), (7.339), (7.341), and (7.342) allow the integral (7.343), it will suffice to consider the following equations

$$R - fV = \frac{dT}{dr_*}, \tag{7.347}$$

$$\frac{d}{dr_*}\left(\frac{\rho^{4s}}{\Delta^s}R\right) = \frac{\rho^{4s}}{\Delta^s}(QT - 2i\sigma R) + \beta M, \tag{7.348}$$

$$R\left(R - \frac{dT}{dr_*}\right) + \frac{\Delta^s}{\rho^{4s}}\beta MT = \frac{\Delta^s}{\rho^{4s}}K, \tag{7.349}$$

and

$$RV - QfV = M\frac{\Delta^s}{\rho^{4s}}\frac{d\beta}{dr_*}, \tag{7.350}$$

where, it may be noted that (7.349) is an alternative form of (7.343) in which fV has been replaced in accordance with (7.347).

Equations (7.347)–(7.350) provide four equations for the five functions f, β, R, T, and V. There is, accordingly, a considerable latitude in seeking useful solutions of the equations. But for the special values of Q given in (7.324)–(7.326), it will appear that we can write down explicit solutions of the equations, essentially, by inspection. The transformation of (7.321) to the form of a one-dimensional equation can, therefore, be effected and explicit expressions for the potential V can be found.

Assuming, then, that (7.321) can be transformed into a one-dimensional wave equation of the form (7.334), let Z_1 and Z_2 denote two distinct solutions; their Wronskian will, clearly, be a constant:

$$W_{r_*}(Z_1, Z_2) = \text{constant}, \tag{7.351}$$

where the subscript r_* to W signifies that the derivatives, in evaluating the Wronskian, are to be taken with respect to r_*.

Some care should be exercised in interpreting (7.351) when σ is in the super-radiant interval. For in that case, the potential V will be found to be singular (even as the functions P and Q in (7.321) are) at $r = |\alpha| > r_+$. In these cases, the equation (as we have indicated earlier) must be considered, separately, in the two branches of the $r_*(r)$-relation. The Wronskians of the two solutions will take constant values in the two branches separately; but they need not be the same. On the other hand, if the two solutions Z_1 and Z_2 correspond to two solutions of Teukolsky's equations, then the Wronskians of Z_1 and Z_2, in the two branches, must be related; and the relation between them can be found as follows.

Now if Y_1 and Y_2 correspond to the solutions Z_1 and Z_2, then

$$W_{r_*}(Y_1, Y_2) = Y_1 \Lambda_- Y_2 - Y_2 \Lambda_- Y_1, \tag{7.352}$$

and substituting for Y and $\Lambda_- Y$ in accordance with (7.336) and (7.337),

Chapter 7. The Kerr metric and its perturbations

we have

$$W_{r_*}(Y_1, Y_2) = \left(RfV + \frac{\Delta^2}{\rho^{4s}}\beta MT\right)(Z_1\Lambda_+ Z_2 - Z_2\Lambda_+ Z_1); \quad (7.353)$$

or, making use of the integral (7.343), we can write

$$W_{r_*}(Y_1, Y_2) = K\frac{\Delta^s}{\rho^{4s}} W_{r_*}(Z_1, Z_2). \quad (7.354)$$

But by the defining equations (7.319)

$$W_{r_*}(Y_1, Y_2) = \frac{1}{|\rho^2|^{2s-1}} W_{r_*}(\mathcal{R}_1, \mathcal{R}_2) = \frac{\Delta}{\rho^2|\rho^2|^{2s-1}} W_r(\mathcal{R}_1, \mathcal{R}_2), \quad (7.355)$$

where we have written $\mathcal{R}$ in place of $\Delta^2 R_{+s}$. By combining (7.354) and (7.355), we have

$$KW_{r_*}(Z_1, Z_2) = \Delta^{-s+1}\frac{\rho^{4s-2}}{|\rho^2|^{2s-1}} W_r(\mathcal{R}_1, \mathcal{R}_2). \quad (7.356)$$

Since the Teukolsky functions cannot have any singularity for any finite r (outside of the horizon), we conclude from (7.356) that

$$W_{r_*}(Z_1, Z_2)_{r<|\alpha|} = (-1)^{2s-1} W_{r_*}(Z_1, Z_2)_{r>|\alpha|}, \quad (7.357)$$

in case $\alpha^2 < 0$ and $r_+ < |\alpha|$, a situation which will occur when $0 < \sigma < \sigma_s$. Therefore, *for $s = 1$ and 2, only the sign of the Wronskian, $W_{r_*}(Z_1, Z_2)$ changes as we cross the singularity at $r = |\alpha|(>r_+)$ while for $s = \frac{1}{2}$, the Wronskian $W_{r_*}(Z_1, Z_2)$ retains its value.*

Our considerations, so far, have been restricted to (7.321). We now inquire how those considerations will be affected when they are applied to the complex-conjugate equation. When the occasion to use both equations arises, we shall find it convenient to distinguish Y and its complex conjugate by $Y^{(+\sigma)}$ and $Y^{(-\sigma)}$ (rather than by Y_{+s} and Y_{-s} as in (7.319)); and we shall write

$$\Lambda^2 Y^{(\pm\sigma)} + P\Lambda_\mp Y^{(\pm\sigma)} - QY^{(\pm\sigma)} = 0 \quad (7.358)$$

for the equation governing them. And when we seek to transform the equation governing $Y^{(-\sigma)}$ to a one-dimensional wave equation (for a function $Z^{(-\sigma)}$) by a sequence of transformations similar to those adopted in the context of $Y^{(+\sigma)}$, we find that the same equations apply if we reverse the sign of σ wherever it explicitly occurs;† and this remark applies, in particular, to equations (7.347)–(7.350).

† It should, however, be noted that α^2 remains unchanged since 'complex-conjugation' requires that the signs of σ and m are simultaneously reversed (e.g. $\mathcal{D}_n$ and $\mathcal{D}_n^\dagger$, as defined, are complex conjugates for this simultaneous reversal in the signs of σ and m).

7.8.4 The potential barriers appropriate for electromagnetic perturbations and the problem of reflection and transmission

The considerations of the preceding sections have ignored or side-stepped several questions which arise from the singular nature of the underlying transformations: the double-valuedness of the $r_*(r)$-relation in the super-radiant interval, $0 < \sigma < \sigma_s$, and the resulting singularities in the derived potentials at $r = |\alpha|(>r_+)$. Besides, additional questions arise from the plurality of the potentials which describe the same problem and which are often complex. We shall clarify these various questions in the context of electromagnetic perturbations; for, it will appear that the solutions of the relevant equations, in this case, are explicit and simple; and the different aspects of the problem can be isolated and understood, singly.

Now we find by simple inspection that

$$T = 2i\sigma, \quad \mathcal{R} = fV,$$

$$R = \mp 2\sigma\alpha\frac{\Delta}{\rho^4}, \quad \beta M = -2i\sigma\left(\lambda - \alpha^2\frac{\Delta}{\rho^4} \pm 2\sigma\alpha\right), \quad (7.359)$$

and

$$K = 4\sigma^2(\lambda \pm 2\sigma\alpha),$$

satisfy (7.347)–(7.349) when $s = 1$ and Q has the value given in (7.325). Equation (7.350) then gives

$$V = \frac{\Delta}{\rho^4}\left[\lambda - \alpha^2\frac{\Delta}{\rho^4} \mp i\alpha\rho^2\frac{d}{dr}\left(\frac{\Delta}{\rho^4}\right)\right]. \quad (7.360)$$

With the solutions given by the foregoing equations, equations (7.335), (7.336), (7.344), and (7.345), relating Y and Z, become

$$Y = \mp 2\sigma\alpha\frac{\Delta}{\rho^4}Z + 2i\sigma\Lambda_+ Z, \quad (7.361)$$

$$\Lambda_- Y = 2i\sigma\left(\lambda - \alpha^2\frac{\Delta}{\rho^4} \pm 2\sigma\alpha\right)Z \mp 2\sigma\alpha\frac{\Delta}{\rho^4}\Lambda_+ Z, \quad (7.362)$$

and

$$KZ = \mp 2\sigma\alpha Y - 2i\sigma\frac{\rho^4}{\Delta}\Lambda_- Y, \quad (7.363)$$

$$K\Lambda_+ Z = -2i\sigma\left(\lambda - \alpha^2\frac{\Delta}{\rho^4} \pm 2\sigma\alpha\right)Y \mp 2\sigma\alpha\Lambda_- Y. \quad (7.364)$$

Chapter 7. The Kerr metric and its perturbations

The potential V given by (7.360) vanishes on the horizon (in fact, exponentially in r_* as $r \to r_+ + 0$) and falls off like r^{-2} as $r \to \infty$. Therefore, so long as σ is outside of the super-radiant interval, the potential is of short range: its integral over r_* is finite and always real. In the super-radiant interval, the potential has a singularity at $r = |\alpha| > r_+$ (see (7.394) below); but its behaviour at infinity and at the horizon are unaffected. Hence, in all cases, the solutions of the wave equation have the asymptotic behaviours

$$Z \to e^{\pm i\sigma r_*} \quad \text{for } r \to \infty \text{ and } r \to r_+ + 0. \tag{7.365}$$

By inserting these behaviours of Z in equation (7.361), we can infer the corresponding asymptotic behaviours of Y (and, therefore, also of the Teukolsky function).

The two solutions, which we shall distinguish by Z_+ and Z_- and which follow from (7.361)–(7.364) by choosing the upper or the lower sign, are simply related. Thus, by substituting in the relation,

$$K_+ Z_+ = -2\sigma\alpha Y - 2i\sigma \frac{\rho^4}{\Delta} \Lambda_- Y, \tag{7.366}$$

appropriate for Z_+, the expressions for Y and $\Lambda_- Y$ relating them to Z_-, we find that

$$K_+ Z_+ = 4\sigma^2 \left(\lambda - 2\sigma\alpha - 2\alpha^2 \frac{\Delta}{\rho^4} \right) Z_- - 8i\alpha\sigma^2 \Lambda_+ Z_-. \tag{7.367}$$

Therefore, given any solution Z_-, appropriate for the potential V_-, we can derive a solution Z_+, appropriate for the potential V_+. From (7.367) we conclude that

$$K_+ Z_+ \to K_- Z_- - 8i\alpha\sigma^2 \Lambda_+ Z_- \quad \text{for } r \to \infty \text{ and } r \to r_+ + 0. \tag{7.368}$$

From this last relation, it follows that solutions for Z_-, which have the asymptotic behaviours,

$$Z_- \to e^{-i\sigma r_*} \quad \text{and} \quad Z_- \to e^{+i\sigma r_*} \quad \text{for } r \to \infty \text{ and } r \to r_+ + 0. \tag{7.369}$$

lead to solutions for Z_+ which have, respectively, the asymptotic behaviours

$$Z_+ \to \frac{K_-}{K_+} e^{-i\sigma r_*} \quad \text{and} \quad Z_+ \to e^{+i\sigma r_*} \quad \text{for } r \to \infty \text{ and } r \to r_+ + 0. \tag{7.370}$$

(a) The distinction between $Z^{(+\sigma)}$ and $Z^{(-\sigma)}$

Distinguishing Y and its complex conjugate by $Y^{(+\sigma)}$ and $Y^{(-\sigma)}$ (as in (7.358)) and the functions satisfying the associated one-dimensional wave equations by $Z^{(+\sigma)}$ and $Z^{(-\sigma)}$, we find, by substitutions analogous to equations (7.359) in (7.347)–(7.350) with the signs of σ reversed, that $Z^{(+\sigma)}$ and $Z^{(-\sigma)}$ satisfy wave equations with the same potential (7.360):

$$\Lambda^2 Z^{(\pm\sigma)} = V Z^{(\pm\sigma)}. \tag{7.371}$$

Since V can be complex, it follows that $Z^{(+\sigma)}$ and $Z^{(-\sigma)}$, unlike $Y^{(+\sigma)}$ and $Y^{(-\sigma)}$, need not always be complex conjugate functions. Nevertheless, the asymptotic behaviours of $Z^{(-\sigma)}$, for $r \to \infty$ and $r \to r_+ + 0$, are the same as for $Z^{(+\sigma)}$.

If we select for $Z^{(+\sigma)}$ the solution which we had earlier distinguished by Z_+, we have the relations:

$$T^{(\pm\sigma)} = \pm 2i\sigma, \quad R^{(\pm\sigma)} = \mp 2\sigma\alpha \frac{\Delta}{\rho^4},$$

$$M\beta^{(\pm\sigma)} = \mp 2i\sigma\left(\lambda - \alpha^2 \frac{\Delta}{\rho^4} \pm 2\sigma\alpha\right), \quad \text{and} \quad K^{(\pm\sigma)} = 4\sigma^2(\lambda \pm 2\sigma\alpha); \tag{7.372}$$

and the equations relating $Z^{(\pm\sigma)}$ and $Y^{(\pm\sigma)}$ are

$$Y^{(\pm\sigma)} = \mp 2\sigma\alpha \frac{\Delta}{\rho^4} Z^{(\pm\sigma)} \pm 2i\sigma \Lambda_\pm Z^{(\pm\sigma)},$$

$$\Lambda_\mp Y^{(\pm\sigma)} = \pm 2i\sigma \frac{\Delta}{\rho^4}\left(\lambda - \alpha^2 \frac{\Delta}{\rho^4} \pm 2\sigma\alpha\right) Z^{(\pm\sigma)} \mp 2\sigma\alpha \frac{\Delta}{\rho^4} \Lambda_\pm Z^{(\pm\sigma)}, \tag{7.373}$$

and

$$K^{(\pm\sigma)} Z^{(\pm\sigma)} = \mp 2\sigma\alpha Y^{(\pm\sigma)} \mp 2i\sigma \frac{\rho^4}{\Delta} \Lambda_\mp Y^{(\pm\sigma)},$$

$$K^{(\pm\sigma)} \Lambda_\pm Z^{(\pm\sigma)} = \mp 2i\sigma\left(\lambda - \alpha^2 \frac{\Delta}{\rho^4} \pm 2\sigma\alpha\right) Y^{(\pm\sigma)} \mp 2\sigma\alpha \Lambda_\mp Y^{(\pm\sigma)}. \tag{7.374}$$

From these relations it can be shown, by a procedure analogous to that we used earlier to relate the solutions Z_+ and Z_-, that $Z^{(+\sigma)}$ is related to the complex conjugate of $Z^{(-\sigma)}$ by (cf. (7.367))

$$K^{(+\sigma)} Z^{(+\sigma)} = 4\sigma^{+2}\left(\lambda - 2\sigma\alpha - 2\alpha^2 \frac{\Delta}{\rho^4}\right)[Z^{(-\sigma)}]^* - 8i\alpha\sigma^2 \Lambda_+ [Z^{(-\sigma)}]^*. \tag{7.375}$$

Chapter 7. The Kerr metric and its perturbations

In particular,

$$K^{(+\sigma)}Z^{(+\sigma)} \to K^{(-\sigma)}[Z^{(-\sigma)}]^* - 8i\alpha\sigma^2\Lambda_+[Z^{(-\sigma)}]^*$$

$$\text{for } r \to \infty \text{ and } r \to r_+ + 0. \quad (7.376)$$

From this last relation, it follows that solutions for $Z^{(-\sigma)}$, which have the asymptotic behaviours,

$$Z^{(-\sigma)} \to e^{+i\sigma r_*} \quad \text{and} \quad Z^{(-\sigma)} \to e^{-i\sigma r_*} \quad \text{for } r \to \infty \text{ and } r \to r_+ + 0. \quad (7.377)$$

lead to solutions for $Z^{(+\sigma)}$ which have, respectively, the asymptotic behaviours,

$$Z^{(+\sigma)} \to \frac{K^{(-\sigma)}}{K^{(+\sigma)}} e^{-i\sigma r_*} \quad \text{and} \quad Z^{(+\sigma)} \to e^{+i\sigma r_*} \quad \text{for } r \to \infty \text{ and } r \to r_+ + 0.$$

$$(7.378)$$

Inserting the foregoing asymptotic behaviours $Z^{(\pm\sigma)}$ in the first of the two equations (7.373), we can infer the corresponding asymptotic behaviours of $Y^{(\pm\sigma)}$ and, therefore, also of the Teukolsky functions ΔR_{+1} and R_{-1}.

Finally, we may note the following relation

$$K^{(+\sigma)}K^{(-\sigma)} = 16\sigma^4(\lambda^2 - 4\alpha^2\sigma^2) = 16\sigma^4\mathscr{C}^2, \quad (7.379)$$

where $\mathscr{C}$ is the Teukolsky–Starobinsky constant (7.209).

With the completion of the formal solution to the problem of reducing Teukolsky's radial equations (for $s = 1$) to the form of one-dimensional wave equations, we turn to the problem of how to use these equations to determine the reflection and transmission coefficients for incident electromagnetic waves. In discussing this latter problem, it will be instructive to distinguish the different cases which arise: $\sigma > \sigma_c (= -m/a)$ when $\alpha^2 > 0$; $\sigma_s < \sigma \le \sigma_c$ when $\alpha^2 < 0$ but $r_+ > |\alpha|$; and $0 < \sigma < \sigma_s$, when $\alpha^2 < 0$ and $r_+ < |\alpha|$.

(b) The case $\sigma > \sigma_c$ and $\alpha^2 > 0$

Since $\alpha^2 > 0$, the $r_*(r)$-relation is single-valued and the potential,

$$V_\pm = \frac{\Delta}{\rho^4}\left[\lambda - \alpha^2\frac{\Delta}{\rho^4} \mp i\alpha\rho^2\frac{\mathrm{d}}{\mathrm{d}r}\left(\frac{\Delta}{\rho^4}\right)\right], \quad (7.380)$$

is bounded and of short range. The wave equation (7.371) will therefore

The potential barriers round the Kerr black hole

admit solutions satisfying the boundary conditions

$$Z_+^{(+\sigma)} \to e^{i\sigma r_*} + A_+^{(+\sigma)} e^{-i\sigma r_*} \quad (r_* \to +\infty)$$
$$\to B_+^{(+\sigma)} e^{+i\sigma r_*} \quad (r_* \to -\infty), \tag{7.381}$$

and

$$Z_+^{(-\sigma)} \to e^{-i\sigma r_*} + A_+^{(-\sigma)} e^{+i\sigma r_*} \quad (r_* \to +\infty)$$
$$\to B_+^{(-\sigma)} e^{-i\sigma r_*} \quad (r_* \to -\infty). \tag{7.382}$$

In writing the foregoing behaviours, we have further distinguished the solutions by subscripts $+$ to indicate that we are, in this instance, considering solutions appropriate to the potential V_+.

Since $Z^{(+\sigma)}$ and $Z^{(-\sigma)}$ both satisfy wave equations with the *same* potential, the reflection and transmission coefficients defined in the manner,

$$\mathbb{R} = A_+^{(+\sigma)} A_+^{(-\sigma)} \quad \text{and} \quad \mathbb{T} = B_+^{(+\sigma)} B_+^{(-\sigma)}, \tag{7.383}$$

will satisfy the *conservation law*

$$\mathbb{R} + \mathbb{T} = 1. \tag{7.384}$$

On the other hand, by (7.377) and (7.378) relating the asymptotic behaviours of $Z_+^{(+\sigma)}$ and $[Z_+^{(-\sigma)}]^*$,

$$A_+^{(-\sigma)} = \frac{K_+^{(+\sigma)}}{K_+^{(-\sigma)}} [A_+^{(+\sigma)}]^* = \frac{\lambda + 2\alpha\sigma}{\lambda - 2\alpha\sigma} [A_+^{(+\sigma)}]^*$$

and

$$B_+^{(-\sigma)} = [B_+^{(+\sigma)}]^*. \tag{7.385}$$

The expressions (7.383) for $\mathbb{R}$ and $\mathbb{T}$ can, therefore, be rewritten in the forms

$$\mathbb{R} = \frac{\lambda + 2\alpha\sigma}{\lambda - 2\alpha\sigma} |A_+^{(+\sigma)}|^2 \quad \text{and} \quad \mathbb{T} = |B_+^{(+\sigma)}|^2. \tag{7.386}$$

These alternative forms for $\mathbb{R}$ and $\mathbb{T}$ show that, as defined in (7.383), they are indeed real. However, the importance of the expressions (7.386) for $\mathbb{R}$ and $\mathbb{T}$ consists in showing that for the purposes of evaluating the reflection and transmission coefficients, it is not necessary to integrate the wave equation twice in order to obtain the solutions with the different asymptotic behaviours (7.381) and (7.382): it will suffice to integrate the appropriate equation only once, for example, to the boundary conditions (7.381).

Chapter 7. The Kerr metric and its perturbations

The expression for $\mathbb{R}$ and $\mathbb{T}$ given in (7.386) apply for solutions of the wave equation with the potential V_+. If instead, we had considered solutions of the wave equation with the potential V_-, we should have found

$$\mathbb{R} = \frac{\lambda - 2\alpha\sigma}{\lambda + 2\alpha\sigma}|A_-^{(+\sigma)}|^2 \quad \text{and} \quad \mathbb{T} = |B_-^{(+\sigma)}|^2. \tag{7.387}$$

On the other hand, according to (7.369) and (7.370)

$$A_+^{(+\sigma)} = \frac{\lambda - 2\alpha\sigma}{\lambda + 2\alpha\sigma} A_-^{(+\sigma)} \quad \text{and} \quad B_+^{(+\sigma)} = B_-^{(+\sigma)}. \tag{7.388}$$

Therefore, (7.386) and (7.387) define the same reflection and transmission coefficients.

The foregoing discussion clarifies how the standard methods of treating the penetration of real one-dimensional potential barriers have to be modified when the potentials are complex and the solutions satisfying complex-conjugate boundary conditions are, themselves, not complex-conjugate functions.

(c) The case $\sigma_s < \sigma \leq \sigma_c$

Now $\alpha^2 < 0$, but $r_+ > |\alpha|$. Therefore, the $r_*(r)$-relation continues to be single-valued. On the other hand, since α is now imaginary, the potential (7.360) now becomes real:

$$V_\pm = \frac{\Delta}{\rho^4}\left[\lambda + |\alpha|^2 \frac{\Delta}{\rho^4} \pm |\alpha|\rho^2 \frac{d}{dr}\left(\frac{\Delta}{\rho^4}\right)\right], \tag{7.389}\dagger$$

where

$$\rho^2 = r^2 - |\alpha|^2 \quad \text{and} \quad |\alpha|^2 = -\alpha^2 \quad (r_+ > |\alpha|). \tag{7.390}$$

The potentials given by (7.389), under these circumstances, are bounded, real, and of short range. Consequently, the theory of the penetration of one-dimensional potential barriers, familiar in elementary quantum theory, now becomes directly applicable. It should, however, be noted that the potentials $V_\pm$ both yield the same reflection and transmission coefficients. The reason is that the corresponding amplitudes of the reflected and transmitted waves are related in the

---

† Notice that for $\sigma = \sigma_c$ (when $\alpha^2 = 0$), the potential is particularly simple: $V = \lambda\Delta/r^4$.

manner (cf. (7.388))

$$A_+ = \frac{\lambda - 2i|\alpha|\sigma}{\lambda + 2i|\alpha|\sigma} A_- \quad \text{and} \quad B_+ = B_-$$

therefore,

$$|A_+|^2 = |A_-|^2 \quad \text{and} \quad |B_+|^2 = |B_-|^2. \tag{7.391}$$

While the potentials given by (7.389) are bounded so long as $\sigma > \sigma_s$, they become singular at the horizon as $\sigma \to \sigma_s + 0$: the barrier presented to the incoming waves, accordingly, becomes increasingly difficult to tunnel as σ approaches σ_s. And we might, on this account, expect that

$$\mathbb{R} \to 1 \quad \text{and} \quad \mathbb{T} \to 0 \quad \text{as } \sigma \to \sigma_s + 0. \tag{7.392}$$

This argument, while it is consistent with the expectation that superradiance, with an accompanying reflection coefficient $\mathbb{R} > 1$, begins at $\sigma = \sigma_s$, is not rigorous (as will become manifest in section 7.8.5); the numerical evaluation of the reflection coefficients with the aid of the potentials (7.389) does confirm the expectation (see Table 7.1, p. 445).

(d) The case $0 < \sigma < \sigma_s$

As we have seen, when $\sigma < \sigma_s$ and $r_+^2 - |\alpha|^2 < 0$, the $r_*(r)$-relation attains a minimum value at $r = |\alpha|$ and tends to $+\infty$ both when $r \to \infty$ and when $r \to r_+ + 0$. Therefore, we must consider, separately, the solutions along the two branches of the $r_*(r)$-relation which, starting at $r = |\alpha|$, progresses either towards $r \to \infty$ or towards the horizon at r_+. This requirement of separate considerations of the solution along the two branches is further compelled by the fact that the potentials given by (7.389) have singularities at $r = |\alpha|$. Thus, by rewriting the expression (7.389) for V in the manner

$$V_\pm = \frac{\Delta}{\rho^4}\left[\lambda + \frac{\Delta}{\rho^4}|\alpha|(|\alpha| \mp 4r) \pm 2|\alpha|\frac{r-M}{\rho^2}\right], \tag{7.393}$$

we find that in the neighbourhood of $r = |\alpha|$, $V_\pm$ has the behaviour

$$V_\pm = (-3 \text{ or } +5)\frac{\Delta_{|\alpha|}^2}{16|\alpha|^2}\frac{1}{(r-|\alpha|)^4}. \tag{7.394}$$

Postponing for the present the manner in which the singularity in the potential at $r = |\alpha|$ is to be taken into account in the solution of the wave equation, we first observe that the boundary conditions with respect to

Chapter 7. The Kerr metric and its perturbations

which the wave equation has to be solved are the same as hitherto, namely,

$$Z \to e^{+i\sigma r_*} + A_\pm e^{-i\sigma r_*} \quad (r_* \to \infty \text{ and } r \to \infty)$$
$$\to \quad B_\pm e^{+i\sigma r_*} \quad (r_* \to \infty \text{ and } r \to r_+ + 0), \quad (7.395)$$

where the subscript $\pm$ distinguishes the solutions derived with the potentials V_+ and V_-. On the other hand, according to (7.357), *the sign of the Wronskian, $W_{r_*}(Z, Z^*)$, must be reversed as we cross the singularity*. Therefore, with the usual definitions,

$$\mathbb{R} = |A_\pm|^2 \quad \text{and} \quad \mathbb{T} = |B_\pm|^2, \quad (7.396)$$

we shall now find the *conservation law*

$$\mathbb{R} - \mathbb{T} = 1. \quad (7.397)$$

In other words, $\mathbb{R} > 1$; this is the phenomenon of super-radiance.

Finally, it remains to clarify how a solution of the wave equation satisfying the boundary conditions (7.395) can be obtained, duly allowing for the singularity (7.394) in the potentials at $r = |\alpha|$. For this purpose, we must examine the behaviour of Z near the singularity. A straightforward calculation shows that in the neighbourhood of $r = |\alpha|$, Z allows two independent solutions with the behaviours

$$Z_+ \sim |(r - |\alpha|)|^{3/2} \quad \text{and} \quad |(r - |\alpha|)|^{1/2} \quad \text{for } V_+$$

and

$$Z_- \sim |(r - |\alpha|)|^{5/2} \quad \text{and} \quad |(r - |\alpha|)|^{-1/2} \quad \text{for } V_-; \quad (7.398)$$

and a general solution of Z in the neighbourhood of $r = |\alpha|$ will be a linear combination of these.

A method of procedure for solving the wave equation satisfying the boundary conditions (7.395) will be the following. Start with a solution for Z_+ (for the potential V_+, say) with the behaviour

$$Z_+ \to e^{+i\sigma r_*} \quad \text{for } r_* \to \infty \text{ along the branch } r \to r_+ + 0; \quad (7.399)$$

and continue the integration forward from the horizon (but backward in r_*). As we approach the singularity at $r = |\alpha|$, from the left (in r), the solution will tend to a determinate linear combination of the independent solutions that obtain here; let the linear combination be (cf.

(7.398))

$$Z_+ \to C_1(|\alpha|-r)^{3/2} + C_2(|\alpha|-r)^{1/2} \quad \text{as } r \to |\alpha|-0, \quad (7.400)$$

where C_1 and C_2 are two constants that will be determined by the integration. The requirement that the Wronskian, $W_{r_*}(Z_+, Z_+^*)$, reverses its sign at $r = |\alpha|$, implies that as $r \to |\alpha|+0$ the appropriate linear combination is

$$Z_+ \to iC_1(r-|\alpha|)^{3/2} - iC_2(r-|\alpha|)^{1/2} \quad \text{as } r \to |\alpha|+0. \quad (7.401)$$

With this form for Z_+ we can continue the integration forward (in r and r_*) beyond $r = |\alpha|$ along the branch $r \to \infty$. By such forward integration we shall eventually find that as $r \to \infty$, the solution tends to the limiting form

$$Z_+ \to C_{\text{inc}} e^{+i\sigma r_*} + C_{\text{ref}} e^{-i\sigma r_*} \quad (r \to \infty, r_* \to \infty), \quad (7.402)$$

where C_{inc} and C_{ref} are certain constants that will be determined by the integration. Since the solution we started with corresponds to a transmitted wave of unit amplitude approaching the horizon, it is apparent that the required reflection and transmission coefficients will be given by

$$\mathbb{R} = |C_{\text{ref}}|^2 |C_{\text{inc}}|^{-2} \quad \text{and} \quad \mathbb{T} = |C_{\text{inc}}|^{-2}. \quad (7.403)$$

The coefficients $\mathbb{R}$ and $\mathbb{T}$ derived in this fashion will satisfy the conservation law (7.397).

In Table 7.1 we list the reflection and transmission coefficients for electromagnetic waves of various frequencies incident on a Kerr black hole (with $a = 0.95$).

Table 7.1. *Reflection coefficients for electromagnetic waves incident on a Kerr black hole with $a = 0.95$*

		($l = 1, m = -1$)			
σ	σ/σ_s	$\mathbb{R}$	σ	σ/σ_s	$\mathbb{R}$
0.325000	0.8979	1.02428	0.405593	1.1205	0.70998
0.345000	0.9531	1.01919	0.415593	1.1481	0.56810
0.350000	0.9669	1.01565	0.425593	1.1758	0.41943
0.365593	1.0100	0.99241	0.435593	1.2034	0.28686
0.375593	1.0376	0.96100	0.445593	1.2310	0.18435
0.385593	1.0653	0.90807	0.455593	1.2586	0.11332
0.395593	1.0929	0.82563			

Chapter 7. The Kerr metric and its perturbations

7.8.5 The reflection and transmission of neutrino waves by the Kerr black hole

Now turning to the equations appropriate for $s=\frac{1}{2}$, we again find by simple inspection that (cf. (7.359) applicable to the case $s=1$)

$$T = 2i\sigma, \quad R = fV,$$

$$R = \frac{\Delta^{1/2}}{\rho^2} q_\pm, \quad \beta M = -2i\sigma \left(\lambda \frac{\Delta^{1/2}}{\rho^2} - q_\pm \right),$$

$$K = -4\sigma^2 q_\pm \quad \text{and} \quad q_\pm = \pm 2i\sigma\lambda^{1/2}, \tag{7.404}$$

satisfy (7.347)–(7.349) when $s=\frac{1}{2}$ and Q has the value given in (7.374). Equation (7.350) then gives

$$V_\pm = \lambda \frac{\Delta}{\rho^4} \mp \lambda^{1/2} \frac{d}{dr_*} \left(\frac{\Delta^{1/2}}{\rho^2} \right). \tag{7.405}$$

With the solutions given by the foregoing equations, the relations (7.335), (7.336), (7.344), and (7.345), relating Y and Z, become

$$Y = \frac{\Delta^{1/2}}{\rho^2} q_\pm Z + 2i\sigma \Lambda_+ Z,$$

$$\frac{\rho^2}{\Delta^{1/2}} \Lambda_- Y = 2i\sigma \left(\lambda \frac{\Delta^{1/2}}{\rho^2} - q_\pm \right) Z + q_\pm \Lambda_+ Z, \tag{7.406}$$

and

$$KZ = q_\pm Y - 2i\sigma \frac{\rho^2}{\Delta^{1/2}} \Lambda_- Y,$$

$$K \Lambda_+ Z = -2i\sigma \left(\lambda \frac{\Delta^{1/2}}{\rho^2} - q_\pm \right) Y + q_\pm \Lambda_- Y. \tag{7.407}$$

The two solutions, which we shall distinguish by Z_+ and Z_- and which follow from (7.406) and (7.407) by choosing the upper or the lower sign, are simply related. Thus, by substituting in the relation,

$$-8i\sigma^3 \lambda^{1/2} Z_+ = 2i\sigma \lambda^{1/2} Y - 2i\sigma \frac{\rho^2}{\Delta^{1/2}} \Lambda_- Y, \tag{7.408}$$

appropriate for Z_+, the expressions for Y and $\Lambda_- Y$ relating them to Z_-, we find

$$Z_+ = \frac{i}{\sigma} \left[\frac{(\lambda \Delta)^{1/2}}{\rho^2} Z_- - \frac{dZ_-}{dr_*} \right]. \tag{7.409}$$

Therefore, given a solution Z_-, belonging to V_-, we can obtain a solution Z_+, belonging to V_+, with the aid of the relation (7.409). In particular, for solutions which are bounded at infinity and at the horizon,

$$Z_+ \to -\frac{i}{\sigma}\frac{dZ_-}{dr_*} \quad \text{for } r \to \infty \text{ and } r \to r_+ + 0. \tag{7.410}$$

There is clearly no ambiguity in using the one-dimensional wave equation, with either of the two potentials,

$$V_\pm = \lambda\frac{\Delta}{\rho^4} \mp \frac{(\lambda\Delta)^{1/2}}{\rho^4}\left[(r-M) - \frac{2r\Delta}{\rho^2}\right], \tag{7.411}$$

given by (7.405), so long as $\sigma > \sigma_s$; for, then, the underlying $r_*(r)$-relation is single-valued and the potentials are moreover bounded and of short range. We can, accordingly, seek solutions of the wave equation with the potential V_-, for example, which have the asymptotic behaviours,

$$Z_- \to e^{+i\sigma r_*} + A\, e^{-i\sigma r_*} \quad (r_* \to +\infty)$$
$$\to \quad B\, e^{+i\sigma r_*} \quad (r_* \to -\infty). \tag{7.412}$$

With the aid of the relation (7.409) we can derive from these solutions Z_-, solutions Z_+, of the wave equation with the potential V_+, which have the asymptotic behaviours,

$$Z_+ \to e^{+i\sigma r_*} - A\, e^{-i\sigma r_*} \quad (r_* \to +\infty)$$
$$\to \quad B\, e^{+i\sigma r_*} \quad (r_* \to -\infty). \tag{7.413}$$

Hence, both potentials will lead to the same reflection and transmission coefficients given by

$$\mathbb{R} = |A|^2 \quad \text{and} \quad \mathbb{T} = |B|^2, \tag{7.414}$$

and satisfying the conservation law

$$\mathbb{R} + \mathbb{T} = 1. \tag{7.415}$$

It can be verified that the reflection and transmission coefficients, as we have defined them in terms of the solutions of the wave equations governing Z, are in accord with the physical definition (cf. Unruh, 1973) of the number-current of the neutrinos flowing down the black hole. (For an explicit demonstration, see Chandrasekhar and Detweiler, 1976b.)

Chapter 7. The Kerr metric and its perturbations

While the potentials given by (7.411) are bounded for $\sigma < \sigma_s$ they become singular exactly at the horizon, when $\sigma \to \sigma_s + 0$; and the tunnelling of the potential barrier becomes increasingly 'difficult'. But the singularity in the potentials, in this limit at the horizon, is weaker for neutrinos than it is for photons (section 7.8.4) or gravitons (section 7.8.6). Thus, considering the integral over V as a measure of the barrier, we find that for both potentials given by (7.4.11),

$$\int_{-\infty}^{+\infty} V \, dr_* = -\frac{\lambda}{2|\alpha|} \ln\left[\frac{2Mr_+(1-\sigma_s/\sigma)}{(r_+ + |\alpha|)^2}\right]$$

(for $\sigma_s < \sigma \leq \sigma_c$ and $0 \leq |\alpha| < r_+$). (7.416)

This integral diverges logarithmically as $\sigma \to \sigma_s + 0$, in contrast to a divergence like $(1-\sigma_s/\sigma)^{-2}$ in the electromagnetic and gravitational cases. It appears that this weaker divergence in the barrier for neutrinos results in a *finite* transmission coefficient in the limit $\sigma \to \sigma_s + 0$. The numerical results given in Table 7.2 strongly suggest this fact, though it remains to be established rigorously. In any event, the finiteness of $\mathbb{T}$, in this limit, is consistent with the fact (which we shall presently prove) that neutrinos cannot manifest super-radiance with a reflection coefficient exceeding unity.

Table 7.2. *Reflection coefficients for neutrinos incident on a Kerr black hole with $a = 0.95$*

($l = 0.5$, $m = -0.5$)

σ	σ/σ_s	$\mathbb{R}$	σ	σ/σ_s	$\mathbb{R}$
0.181987	1.0055	0.97627	0.220987	1.2210	0.89187
0.182987	1.0111	0.97556	0.230987	1.2763	0.83457
0.183987	1.0166	0.97479	0.240987	1.3315	0.75356
0.184987	1.0221	0.97398	0.250000	1.3813	0.66005
0.185987	1.0276	0.97313	0.250987	1.3868	0.64885
0.186987	1.0332	0.97223	0.260987	1.4420	0.52777
0.187987	1.0387	0.97129	0.300000	1.6576	0.14125
0.188987	1.0442	0.97012	0.350000	1.9338	0.01622
0.189987	1.0497	0.96925	0.400000	2.2101	0.00197
0.200987	1.1105	0.95353	0.450000	2.4864	0.00027
0.210987	1.1658	0.92971			

For $0 < \sigma < \sigma_s$, the $r_*(r)$-relation becomes double-valued and the potentials further become singular at $r = |\alpha|$ ($>r_+$ for $\sigma < \sigma_s$); they have, in fact, the behaviour (cf. (7.394))

$$V_{\pm} = \lambda^{1/2}\frac{\Delta_{|\alpha|}^{3/2}}{4|\alpha|^2}\frac{1}{(r-|\alpha|)^3}, \qquad (7.417)$$

in the neighbourhood of $r = |\alpha|$. However, since the Wronskian, $W_{r_*}(Z, Z^*)$, in this case, must retain its value along the two branches of the $r_*(r)$-relation (in contrast to a reversal in its sign demanded by (7.357) for integral spins), it follows that, in this instance, the conservation law (7.415) will continue to hold in the interval, $0 < \sigma < \sigma_s$. In other words, *neutrinos will not manifest super-radiance*.

The foregoing treatment for the two-component neutrinos has been extended by Güven (1977; see also Page, 1971) to the Dirac electrons; and he has shown, in particular, that the Dirac electrons, like neutrinos, will not manifest super-radiance.

7.8.6 The potential barriers appropriate for gravitational perturbations

The solution of (7.347)–(7.350), for the case $s = 2$, is not so easily accomplished as for the cases $s = 1$ and $\frac{1}{2}$. But a number of trials suggested that we seek solutions of the form (Chandrasekhar and Detweiler, 1975, 1976a)

$$T = T_1 + 2i\sigma, \quad \beta M = \beta_1 + 2i\sigma\beta_2, \quad \text{and} \quad K = (\kappa - 4\sigma^2\beta_2) + 2i\sigma\kappa_2, \qquad (7.418)$$

where κ, κ_2, and β_2 are some constants (as yet unspecified) and R, f, and V are explicitly independent of σ in that they do not contain any term linear in $i\sigma$ (as βM and K). The assumptions that the solutions are of the forms stated *and* that, when these forms for the solutions are substituted in (7.347)–(7.350), we may equate separately the terms which occur with the factor $i\sigma$ and the terms which do not, lead one to the equations

$$R = Q + \frac{\Delta^2}{\rho^8}\beta_2 = \frac{\Delta^2}{\rho^8}(F + \beta_2), \quad \beta_1 = \kappa_2 - \beta_2 T_1. \qquad (7.419)$$

$$V = Q - \frac{dT_1}{dr_*}, \quad \text{and} \quad T_1 = \frac{1}{F - \beta_2}\left(\frac{dF}{dr_*} - \kappa_2\right), \qquad (7.420)$$

where (cf. (7.326))

$$F = \frac{\rho^8}{\Delta^2}Q = \frac{1}{\Delta}[\lambda\rho^4 + 3\rho^2(r^2 - a^2) - 3r^2\Delta]. \qquad (7.421)$$

Chapter 7. The Kerr metric and its perturbations

And, further, the consistency of the foregoing forms for the solutions requires that for some suitably chosen constants κ, κ_2, and β_2, the following identity holds:

$$\left(\frac{dF}{dr_*}\right)^2 - F\frac{d^2F}{dr_*^2} + \frac{\Delta^2}{\rho^8}F^3$$
$$= (\kappa_2^2 - 2\beta_2\kappa) + \left(\kappa + 2\frac{\Delta^2}{\rho^8}\beta_2^2\right)F - \frac{\beta_2^2}{F}\left(\frac{d^2F}{dr_*^2} + \beta_2^2\frac{\Delta^2}{\rho^8} - \kappa\right), \quad (7.422)$$

where, it should be remembered that F is a *known function*. Remarkably, the identity, in fact, holds with the choice

$$\kappa = \lambda(\lambda+2), \quad \beta_2 = \pm 3\alpha^2, \quad (7.423)$$

and

$$\kappa_2 = \pm\{36M^2 - 2\lambda[\alpha^2(5\lambda+6) - 12a^2] + 2\beta_2\lambda(\lambda+2)\}^{1/2}. \quad (7.424)$$

The solution for V, which follows from (7.421), is

$$V = \frac{\Delta}{\rho^8}\left[q - \frac{\rho^2}{(q-\beta_2\Delta)^2}\{(q-\beta_2\Delta)[\rho^2\Delta q'' - 2\rho^2 q - 2r(q'\Delta - \Delta' q)]\right.$$
$$\left. + \rho^2(\kappa_2\rho^2 - q' + \beta_2\Delta')(q'\Delta - \Delta' q)\}\right], \quad (7.425)$$

where

$$q = \lambda\rho^4 + 3\rho^2(r^2 - a^2) - 3r^2\Delta \quad (7.426)$$

and a prime denotes differentiation with respect to r. Since the constants β_2 and κ_2, as defined in (7.423) and (7.424), can be assigned signs independently of each other, (7.425) will, in general, provide *four different potentials* for the reduced one-dimensional wave equation.

As in sections 7.8.3 and 7.8.4, we can relate the solutions of the wave equations belonging to the different potentials; and we find that their asymptotic behaviours, at infinity and at the horizon, are related exactly as in (7.369) and (7.370). Again, as in section 7.8.3 we must distinguish between $Z^{(+\sigma)}$ and $Z^{(-\sigma)}$ which are derived from $Y^{(+\sigma)}$ and $Y^{(-\sigma)}$, respectively. And we find, as before, that $Z^{(+\sigma)}$ and $Z^{(-\sigma)}$ satisfy wave equations with the same potential; that the solutions $Z^{(+\sigma)}$ and the complex conjugate of the solutions $Z^{(-\sigma)}$ can be related; and, finally, that their asymptotic behaviours are related exactly as in (7.377) and (7.378).

The discussion of the various cases which arise can be carried out exactly as in section 7.8.3; and we shall find, in particular, that the

Concluding remarks

present requirement (by (7.357)) of a reversal in the sign of the Wronskian, $W_{r*}(Z^{(+\sigma)}, Z^{(-\sigma)})$, as we cross the singularity in the potential at $r = |\alpha|$ when $0 < \sigma < \sigma_s$, leads to super-radiance.

In Table 7.3, we list the reflection and transmission coefficients for gravitational waves incident on a Kerr black hole (with $a = 0.95$).

Table 7.3. *Reflection coefficients for gravitational waves incident on a Kerr black hole with $a = 0.95$*
$(l = 2, m = -2)$

σ	σ/σ_s	$\mathbb{R}$	σ	σ/σ_s	$\mathbb{R}$
0.50	0.6907	1.04422	0.77	1.0636	0.27057
0.55	0.7597	1.07031	0.78	1.0774	0.16754
0.60	0.8288	1.10693	0.79	1.0912	0.10007
0.65	0.8979	1.15358	0.80	1.1051	0.05844
0.70	0.9669	1.15101	0.81	1.1189	0.03372
0.73	1.0084	0.92530	0.82	1.1327	0.01932
0.74	1.0222	0.76882	0.83	1.1465	0.01103
0.75	1.0360	0.58828	0.84	1.1603	0.00628
0.76	1.0498	0.41376	0.85	1.1741	0.00357

Finally, we may note the following relation which is analogous to the one we found earlier for $s = 1$ (equation (7.379)):

$$\begin{aligned} K^{(+\sigma)}K^{(-\sigma)} &= [\lambda(\lambda+2) - 4\sigma^2\beta_2]^2 + 4\sigma^2\kappa_2^2 \\ &= \lambda^2(\lambda+2)^2 - 8\sigma^2\lambda[\alpha^2(5\lambda+6) - 12a^2] + 144\sigma^2(M^2 + \alpha^4\sigma^2) \\ &= |\mathscr{C}|^2 \end{aligned} \quad (7.427)$$

where $\mathscr{C}$ is the Teukolsky–Starobinsky constant (7.266).

In figure 7.1, the dependence of the reflection coefficient $\mathbb{R}$ on σ/σ_s is contrasted for the three cases $s = \frac{1}{2}, 1,$ and 2.

An alternative treatment of the problems considered in sections 7.8.3 and 7.8.5 has been given by Detweiler (1976, 1977).

7.9 Concluding remarks

It was stated in the introduction that the Kerr metric has 'many properties which have the aura of the miraculous about them'. What, we may now inquire, are these properties?

In many ways, the most striking feature is the separability of all the standard equations of mathematical physics in Kerr geometry. One

Chapter 7. The Kerr metric and its perturbations

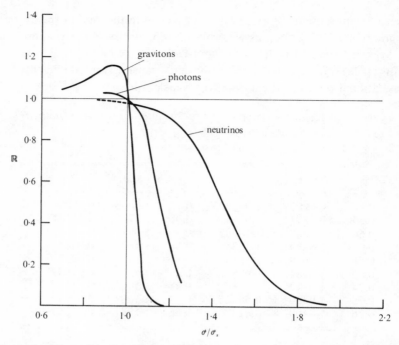

Figure 7.1. Illustrating the dependence of the reflection coefficient ($\mathbb{R}$) on σ/σ_s for neutrinos ($l = -m = 0.5$), photons ($l = -m = 1$), and gravitons ($l = -m = 2$) incident on a Kerr black hole ($a = 0.95$). The curves are distinguished by the massless particles to which they refer. Notice that while the photons and the gravitons manifest super-radiance ($\mathbb{R} > 1$) for $\sigma < \sigma_s$, the neutrinos do not (for reasons explained in the text). The dashed-line part of the curve referring to the neutrinos represents a 'visual' extrapolation of the computed part drawn as a full-line curve.

understands the separability of the Hamilton–Jacobi equation (in terms of the Kerr metric admitting a 'Killing tensor'); but one does not understand why the wave equations governing the fields of different spins ($s = \frac{1}{2}$, 1, and 2) separate. And there are other features besides. What, for example, is the origin of the 'ladder relations' connecting the functions – both radial and angular – belonging to opposite spins; and of the associated emergence of what we have called the Teukolsky–Starobinsky constants, $\mathscr{C}$ (for the radial functions) and D (for the angular functions)? Related questions are: Why are the constants $\mathscr{C}$ and D so curiously related: $\mathscr{C} = D$ for $s = 1$ and Re $\mathscr{C} = D$ for $s = 2$? Why must one have to go to the entire set of the Newman–Penrose equations to determine the real and the imaginary parts of $\mathscr{C}$ in the case $s = 2$? And why do these same constants again emerge when solving for the functions satisfying the one-dimensional wave equations?

Concluding remarks

Again, it is a matter for some astonishment that the entire set of the Newman–Penrose equations governing gravitational perturbations can be solved explicitly (without ever having to solve differential equations beyond those for the Teukolsky functions); and, further, that the solutions for the metric perturbations can be expressed in terms of two functions $\mathscr{R}(r)$ and $\mathscr{S}(\theta)$ whose relationships to the Teukolsky functions are so complicated. And, finally, there is the matter of the 'phantom gauge' in which the Weyl scalars, Ψ_1 and Ψ_3, are governed by the same equations as the Maxwell scalars, ϕ_0 and ϕ_2.

Finally, several crucial identities among the Teukolsky functions emerge when a complete solution of the entire set of the Newman–Penrose equations are undertaken; and the question arises whether the superfluity of the Newman–Penrose equations will enable us to discover new classes of identities among the special functions of mathematical physics when they occur as solutions of Einstein's equations.

The assessment of things as 'strange' or 'miraculous' are necessarily subjective; but it does seem that there are aspects of the Kerr metric which need understanding.

8. Black hole astrophysics

R. D. BLANDFORD AND K. S. THORNE†

8.1 Introduction
8.1.1 Historical remarks

Although pioneering theoretical work on black holes (Laplace, 1799; Chandrasekhar, 1934; Oppenheimer and Snyder, 1939) was motivated by astrophysical considerations, black holes were not taken seriously by practicing astronomers until fairly recently. One of us (KST) remembers being warned, when he was an undergraduate at Caltech in the late 1950s, that general relativity probably has no relevance for astronomy except in the big bang. 'All stars probably lose enough mass as they evolve,' he was told by an astronomy professor, 'to bring them below the Chandrasekhar limit' and thereby avoid the gravitational-collapse fate described in Oppenheimer's work.

This attitude prevailed within the astronomy community until two observational discoveries shook its foundations. The first was the discovery of quasars (Schmidt, 1963) and their huge energy outputs which suggested gravitational collapse as the driving mechanism. The second was the discovery of pulsars (Hewish *et al.*, 1968), and the demonstration a year later that they must be energized and regulated by rotating neutron stars that were probably formed in supernova explosions (Pacini, 1967; Gold, 1968; Cocke, Disney and Taylor, 1969). The reaction of the astronomy community to these discoveries is epitomized by the case of a well-known theoretical astrophysicist who, in the early 1960s, was occasionally heard to ask 'How many angels can dance on the head of a neutron star today?', but who in the late 1960s and early 1970s began doing astrophysical research not only on neutron stars, but also on black holes.

Some (but not all!) astronomers are now so willing to believe in black holes that one hears of holes being suggested as the *deus ex machina* for

† Supported in part by the National Science Foundation [AST76-20375 and AST76-80801 A01].

almost every bizarre observational phenomenon that is discovered – from Cygnus X-1 and other compact X-ray sources (Webster and Murdin, 1972; Bolton, 1972); to the case of the missing solar neutrinos (Stothers and Ezer, 1973; Clayton, Newman and Talbot, 1975); to the mystery of the missing mass required to stabilize spiral galaxies (Ostriker and Peebles, 1973; Ipser and Price, 1977), bind clusters of galaxies (e.g. Peebles, 1971), or even close the Universe (e.g. Thorstensen and Partridge, 1975; Carr, 1977a); to the nineteenth-century Siberian meteorite impact (Jackson and Ryan, 1973); to the disappearance of ships in the Bermuda Triangle (e.g. Berlitz, 1974). Things have become so extreme that it is a major task to winnow plausible hypotheses about roles of black holes from entanglement with implausible hypotheses.

There is now a fairly close interplay between theory and observation in black hole astrophysics. This was not always so. The pioneering works of Laplace (1799), and of Oppenheimer and Snyder (1939) were stimulated very little by observational astronomy. Subsequently, in the late 1950s and early 1960s, and without observational input, Wheeler (1959, 1963) forced his physicist colleagues to regard gravitational collapse to a singularity as 'one of the greatest crises of all time' for fundamental physics. It perhaps was Wheeler's influence that induced astrophysicists immediately, when quasars were discovered, to seek an explanation in gravitational collapse. And in the subsequent intellectual ferment, astrophysicists began asking themselves what other observational manifestations one might expect from collapsed objects. The modern era of black hole astrophysics, with its observational–theoretical intercourse, was now on its way, though the term 'black hole' was yet to be invented by Wheeler (1968).

8.1.2 On the nature of this chapter

In recent years a number of talented mathematical physicists have successfully moved from general relativity theory and quantum field theory into research on astrophysical problems, including black holes. This move is often extremely difficult. It entails learning much new physics, and, more importantly, learning an entirely new style of research.

This chapter is directed at those mathematical physicists who are tempted to follow their colleagues into astrophysics, at those who desire no change but are curious about the nature of life on the other side, at

Chapter 8. Black hole astrophysics

students who might wish to work in astrophysics but as yet have little training in it, and, finally, at those who are simply curious about what kinds of roles black holes might play in the astrophysical universe. This chapter is not directed at practicing astrophysicists.

Because of this direction (and out of laziness), we have refrained from performing a careful search of the black hole astrophysics literature, and from writing a thorough review of it. (Our indebtedness to earlier reviews, e.g. Eardley and Press (1975), will be apparent.) Instead we do the following. In section 8.2 we try to give the reader a rough feeling for the character of research in black hole astrophysics, i.e. the knowledge needed in pursuing it, and the style in which one thinks and computes. We then turn, in sections 8.3 to 8.7, to an overview of the most plausible current scenarios for the roles of black holes in the universe, and the observational data that impinge on those scenarios. For each scenario we list only a few references – those which we think might be most helpful to the novice in the field, or those from which the novice might find a fairly direct path into the other literature. Often this means that we list reviews rather than primary literature. Our scenarios are organized by section as follows:

8.3 Black holes of 'stellar mass' ($M \sim 1$ to $100\ M_\odot$) alone in the interstellar medium.
8.4 Black holes of stellar mass in binary systems.
8.5 Black holes that might form and reside in globular clusters.
8.6 Black holes in the nuclei of galaxies and quasars.
8.7 Black holes that might have formed before galaxies, either in the big bang itself, or in pregalactic condensations.

8.1.3 Some tacit assumptions of black hole astrophysics

In the literature and research of this field, several unproved assumptions are made from the outset, usually without explicit statement. These include:

(i) The 'hoop conjecture' – that a black hole forms when and only when a mass M gets compacted into a region whose circumference in *every* direction is $\mathscr{C} \lesssim 4\pi M$, so that a hoop of that circumference can be slipped over the region and rotated through 360° (see, e.g. Box 32.3 of Misner, Thorne and Wheeler (1973) – cited henceforth as MTW).

(ii) The 'hypothesis of cosmic censorship' – that when such compaction occurs to produce a black hole, and subsequently as holes evolve in the real universe, naked singularities never form (Penrose, 1969) except, perhaps, at the endpoint of black hole evaporation (chapters 13 and 15 of this book).

(iii) The 'rapid-loss-of-hair conjecture' – that in the absence of strong, dynamical, external gravitational perturbations, a black hole of mass M will settle down into a stationary state in a time

$$\Delta t \sim 100 \, M \sim 10^{-3} \, \text{s} \, (M/M_\odot)$$

(e.g. Box 32.2 of MTW, extended to generic situations).

(iv) The 'coalescence conjecture' – that when two black holes collide, they always coalesce to form a single black hole (e.g. Smarr (1977) extended to generic situations).

Of course, astrophysical research also relies on the powerful theorems and results about black holes that are reviewed in chapters 6 and 7 of this book.

8.2 On the character of research in black hole astrophysics

The fundamental theory of black holes, as laid out in chapters 6 and 7, is well posed, elegant, clean and self-contained. It follows inexorably and clearly from the fundamental laws of physics. The theory of black holes in an astrophysical environment is completely the opposite. Because it deals with the physics of matter in bulk – matter orbiting and accreting onto a hole – it is subject to all the dirty, complex uncertainties of the modern theory of the behavior of bulk matter. If thunderstorms and tornados on Earth have eluded accurate theoretical modeling, how can one expect to predict even qualitatively their analogues in the turbulent, magnetized plasmas that accrete onto a black hole in a close binary system? One cannot. The best that can be hoped for is to develop the crudest of models as to the gross behavior of matter in the vicinities of holes.

Fortunately, the resulting models have some modest hope of resembling reality. This is because the relative importance of physical processes near a hole can be characterized by dimensionless ratios that usually turn out to be very large, and consequently, the gross behavior of matter near a hole is dominated by a small number of processes. The task of the model builder is to identify the dominant processes in his given situation, and to construct approximate equations describing their macroscopic effects. Historically, to identify the dominant processes has

Chapter 8. Black hole astrophysics

not been easy. This is because a vast number of possible processes must be considered and the model builder often, out of ignorance, overlooks an important one.

Thus it is that research on black hole astrophysics involves large bodies of physical theory. Within each body of theory one must have at one's fingertips approximate formulae that characterize a long list of possibly relevant processes. The necessary bodies of theory include general relativity, the physics of equilibrium and non-equilibrium plasmas, the physics of radiative processes, and the physics of stellar dynamical systems. Those specific processes which had been identified by 1972 as important for black hole astrophysics are summarized in Novikov and Thorne (1973). Since 1972, theorists studying possible scenarios for black holes have encountered new important processes, including (i) the possibility that electrons and protons have different temperatures in plasmas around black holes (see Spitzer (1962) for fundamental theory, Shapiro, Lightman and Eardley (1976) for applications), (ii) electrodynamical processes near holes (Blandford and Znajek, 1977), and (iii) stellar dynamical processes near holes (reviewed in Lightman and Shapiro, 1978).

Research in black hole astrophysics also requires a good knowledge of the phenomenology of modern astronomy – the observed properties of stars, the main features of their evolution, the structure of the Galaxy, and the observed physical conditions in interstellar space. One way to learn such things is by reading the relevant portions of a good elementary textbook on astronomy (e.g. Abell, 1975; Jastrow and Thompson, 1972; Payne-Gaposchkin and Haramundanis, 1970; Menzel, Whipple and De Vaucouleurs, 1970) and by then browsing through the past 10 years or so of astronomy articles in *Scientific American*. Only after this is one really ready to absorb the review articles in *Annual Reviews of Astronomy and Astrophysics*.

The style of research in black hole astrophysics is very different from that in general relativity theory or other areas of fundamental theory. A crude approximation to the style is this:

(1) One poses a possible scenario for a specific role of black holes in the real universe. [*Example*: A black hole at rest in the interstellar medium will accrete interstellar gas; the infalling gas might heat up enough to radiate significantly before it reaches the horizon, and the radiation might be strong enough to observe at Earth. This scenario was first posed by Salpeter (1964) and Zel'dovich (1964), and has been studied in detail by Shvartsman (1971), Michel (1972), Shapiro

(1973a, b), Pringle, Rees and Pacholczyk (1973), Mészáros (1975b) and Bisnovatyi-Kogan (1978).]

(2) One guesses a macroscopic description of the physics of the scenario, and one solves the resulting macroscopic equations to get a first rough version of the model. [*Example*: One guesses that the accreting gas can be treated as a hydrodynamical perfect fluid with density $\rho(r)$, pressure $p(r)$, and radial inflow velocity $v(r)$, moving in the spherically symmetric Newtonian potential $\Phi = -GM/r$ of a hole with mass M. One guesses and hopes that the microscopic physics will ultimately yield a pressure which is 'polytropically' related to the density, $p = K\rho^\gamma$ for some constants, K and γ. This allows one to decouple the hydrodynamical equations from the equations for radiative cooling and heat transfer, and to delay considering cooling and heat transfer until later. One then solves the standard hydrodynamical equations with boundary conditions at infinity corresponding to 'typical' interstellar gas – an ionized, largely hydrogen plasma with temperature $T_\infty \sim 10^4$ K and with

$$v_\infty = 0, \quad \rho_\infty \sim 10^{-24} \text{ g cm}^{-3}, \quad (p_\infty/\rho_\infty)^{1/2} \sim 10 \text{ km s}^{-1}. \tag{8.1}$$

One discovers that the hole has little influence at radii $r \gg R_A \equiv$ ('accretion radius'), where R_A is given by

$$GM/R_A = p_\infty/\rho_\infty \approx (\text{speed of sound at } r = \infty)^2, \tag{8.2}$$
$$R_A \approx 10^{14} \text{ cm } (M/M_\odot),$$

but that near R_A the hole pulls the gas into infall and that at $r \ll R_A$ the gas is in free fall with

$$\tfrac{1}{2}v^2 \approx GM/r, \quad \rho \approx \rho_\infty (R_A/r)^{3/2}, \quad \text{in general}; \tag{8.3}$$
$$T \approx T_\infty (R_A/r)^{3(\gamma-1)/2} \quad \text{so long as } p \propto \rho T.$$

(Here and throughout M is the mass of an object as measured gravitationally, and $M_\odot = 2 \times 10^{33}$ g = 1.5 km is the mass of the Sun.) The total rate of mass accretion is

$$\dot M \approx 4\pi R_A^2 \rho_\infty (p_\infty/\rho_\infty)^{1/2} \sim (10^{11} \text{ g s}^{-1})(M/M_\odot)^2. \tag{8.4}$$

The details were first worked out by Hoyle and Lyttleton (1940) and by Bondi (1952); for a review see, e.g. § 4.2 of Novikov and Thorne (1973).]

(3) One next examines the microscopic physics of the scenario to see whether it is in accord with the macroscopic model. If it is not in accord,

Chapter 8. Black hole astrophysics

one guesses a modification of the macroscopic model and iterates. If it is in rough accord, one builds more detailed macroscopic equations that better approximate the microscopic physics, and can be solved to get an improved model. [*Example*: One asks, for the above solution, whether the hydrodynamical approximation is valid, at least at $r \sim R_A$ where the mass-accretion rate is being determined. To test the hydrodynamical approximation one compares the accretion radius R_A with the distance λ_p that a proton in the interstellar medium must travel before Coulomb scattering has deflected it substantially from straight-line motion:

$$\lambda_p \sim (7 \times 10^{12} \text{ cm})(\rho/10^{-24} \text{ g cm}^{-3})^{-1}(T/10^4 \text{ K})^2. \tag{8.5}$$

One discovers that

$$\lambda_p/R_A \sim 0.1 \, (M/M_\odot)^{-1}. \tag{8.6}$$

This is one of those rare circumstances when a dimensionless ratio comes out near unity. If λ_p/R_A had been $\ll 1$, the hydrodynamical approximation would have been valid and one would have proceeded happily forward with the model. If λ_p/R_A had been $\gg 1$ one would have started all over again with a new macroscopic model in which the protons are independent, non-interacting particles that carry electron clouds with them (see Begelman (1977) and references therein). But in this case neither extreme is true. In despair, one searches one's list of physical processes for something besides Coulomb scattering, which might save the day and make the hydrodynamical approximation valid. Two things come to mind: scattering of protons off plasmons (i.e. off collective excitations of the plasma), and the anchoring of protons to an interstellar magnetic field. Examination of dimensionless ratios of lengthscales shows proton–plasmon scattering to be marginally important at best, but magnetic-field anchoring to be extremely important:

$$\frac{\text{(Larmor radius of proton)}}{\text{(accretion radius)}} \approx \frac{(10^8 \text{ cm})(B/10^{-6} \text{ G})^{-1}}{(10^{14} \text{ cm})(M/M_\odot)} \ll 1. \tag{8.7}$$

If the magnetic field near R_A is tangled (i.e. inhomogeneous on scales $l_B \ll R_A$), then it will provide a physical coupling between protons; and the hydrodynamical approximation will be valid. One can then proceed with the hydrodynamical model, augmenting it with an appropriate macroscopic description of a magnetoturbulent plasma and appropriate equations for radiation emission by the synchrotron and bremsstrahlung processes; for details see Shvartsman (1971) and Shapiro (1973*a*, *b*).

On the other hand, if the magnetic field near R_A is highly homogeneous (i.e. $l_B/R_A \gg 1$), then one must begin all over again with a macroscopic model in which accreting, non-interacting protons drag the magnetic field smoothly in with them, producing ultimately an 'hour-glass-shaped' field configuration with plasma sliding down the field lines to the waist of the hour-glass. In the waist the plasma forms a disk, and gradually slips through the field lines toward the hole, getting heated strongly as it slips. For details see Bisnovatyi-Kogan and Ruzmaikin (1974, 1976) and Bisnovatyi-Kogan (1978).]

The above example illustrates the following features of research in black hole astrophysics:

(i) It involves an iteration back and forth between the equations of the macroscopic model and the microscopic physics which underlies those equations. One iterates until one obtains self-consistency.

(ii) One must search carefully, at each iterative stage, for overlooked processes that might be so important as to invalidate the model (anchoring to a homogeneous interstellar magnetic field in the above example).

(iii) One frequently encounters a 'branch point' where the model will take on two very different forms depending on what one assumes for the environment around the hole (homogeneous magnetic field versus tangled field in the above example), and where both branches might well occur in the real universe. This leads to a plethora of possible models, each corresponding to a different black hole environment and/or range of black hole masses.

The scenarios which we describe in the remainder of this chapter are all based on models which were built by iterative macroscopic–microscopic considerations. However, in describing the scenarios we shall give very few details of the models. The details would make this chapter much too long.

8.3 Isolated holes produced by collapse of normal stars

8.3.1 The deaths of stars and the births of holes

Stars are continually being born, and continually dying in our galaxy. It is known with confidence that they are born by gravitational contraction of dense interstellar gas clouds; that after the initial contraction stage, lasting $\sim 5 \times 10^7$ yr for $M \sim M_\odot$ and $< 10^5$ yr for $M > 10\, M_\odot$, they ignite

Chapter 8. Black hole astrophysics

hydrogen in their interiors; that they then live off nuclear burning for periods of $>10^{10}$ yr if $M < M_\odot$ and $< 2 \times 10^7$ yr if $M > 10 M_\odot$; and that they then die. (For detailed reviews see, e.g. §§ 6–5 to 6–7 of Clayton (1968) and Iben (1974).)

The details of the death throes are poorly understood. However, the possible remnants of the death are firmly established by theory. They include expanding gas clouds produced by disruption of part or all of the star (e.g. the Crab nebula), white dwarfs (hundreds of examples are known: see, e.g. Greenstein, 1976; Trimble and Greenstein, 1972), neutron stars (well established as the energizers and regulators of pulsars and as X-ray components of some binary X-ray sources: see, e.g. Manchester and Taylor, 1977; Lamb, 1977), and black holes. White dwarfs cannot be more massive than $M_{max}^{wd} \approx 1.4 M_\odot$ if slowly rotating (Chandrasekhar, 1931), or $\sim 3 M_\odot$ if rapidly and differentially rotating (Durisen, 1975). Neutron stars cannot exceed a limit $M_{max}^{n^*}$ that is variously estimated as between 1.3 and $2.5 M_\odot$ (e.g. Arnett and Bowers, 1977), and that is probably bounded by

$$M_{max}^{n^*} < (5 M_\odot)(\rho_0/\rho_{nuc})^{-1/2} \qquad (8.8)$$

even in the presence of rotation (e.g. Hartle, 1978). Here $\rho_{nuc} = 2 \times 10^{14}$ g cm^{-3} is the density of matter inside ordinary atomic nuclei, and $\rho_0 \sim (0.5 \text{ to } 5)\rho_{nuc}$ is the density at which one believes the current theory of bulk nuclear matter to be in serious error. The measured masses of white dwarfs and neutron stars are compatible with these theoretical limits (Weidemann, 1968; Greenstein, Oke and Shipman, 1971; Rappaport and Joss, 1977). Black holes formed by stellar death could have masses ranging from $M_{max}^{n^*}$ up to the mass of the heaviest normal stars $\sim 100 M_\odot$ (see, e.g. Clayton, 1968), but they probably cannot be less massive than $\sim M_{max}^{n^*}$ because nuclear forces resist the compression of such small amounts of matter to black hole densities.

The ultimate fate of a given star depends not only on its mass at birth, but also on how much mass it manages to shed between birth and final death. That significant mass loss must occur is clear from the presence of white dwarfs ($M < 1.4 M_\odot$) in the Hyades cluster of stars – a cluster measured to be so young that in it only stars of original mass $M > 2.1 M_\odot$ have died as yet (Auer and Woolf, 1965). Moreover, outflowing mass ('stellar wind') is observed spectroscopically in a wide variety of stars, and for some main-sequence (hydrogen-burning) stars of $M \gtrsim 20 M_\odot$ the measured winds are strong enough to reduce the mass substantially before the hydrogen fuel is exhausted (Lamers and Morton, 1976). Some current best guesses as to the amount of mass loss for various stars are reviewed by Weidemann (1977). However, these are only educated guesses. It is conceivable, though unlikely, that all stars *could* reduce their masses below $M_{max}^{n^*}$ before dying, thereby avoiding black hole fates; and it is also conceivable, though unlikely, that all stars exceeding $M_{max}^{n^*}$ at death might eject enough mass in their death throes to become neutron stars. Neither current observation nor current theory can rule out these possibilities. It is a sad commentary on the theory that if one ignores observational constraints, it even permits all stars to reduce themselves below M_{max}^{wd}, thereby avoiding neutron star formation. Indeed, theoretical work soon after pulsars were discovered (Arnett, 1969) suggested that neutron stars could never form in nature because a carbon-detonation nuclear explosion would disrupt all stars above M_{max}^{wd} before they could collapse. Only recently has the theory been adequately adjusted to account for neutron star formation (Mazurek, Meier and Wheeler (1977) and references therein).

At the other extreme it is perfectly plausible, on the basis of current theory and observation, that mass loss is negligible for a large fraction of all stars with $M > M_{max}^{n^*} \sim 2.5 M_\odot$. If so, then the observed stellar populations in the solar neighborhood imply a black hole formation rate of $\sim 1.5 \times 10^{-10}$ yr^{-1} (pc^2 of Galactic plane)$^{-1}$ (Ostriker, Richstone and Thuan, 1974). Extrapolating back in time over a Galaxy age of 10^9 yr, and assuming no substantial increase in the velocities of the holes during birth due to momentum ejection (Bekenstein, 1973, 1976), one infers that the nearest stellar-mass black hole to Earth today is at a distance of ~ 5 pc. Extrapolating this to the entire Galactic plane one infers a birth rate in our galaxy of $\sim$ one hole every five years and a total number of 2×10^9 holes. For comparison the supernova rate in our galaxy, as inferred from historical observations, is somewhere between one every six years and one every 25 years, and in external galaxies of our type and mass it is between one every 10 years and one every 40 years (Tammann, 1977); and the pulsar birth rate in our galaxy is estimated from observation to be between one every eight years and one every 40 years (Taylor and Manchester, 1977).

The above numbers for black holes must not be taken terribly seriously. Not only are there uncertainties due to mass loss, but also extrapolation backwards in time, or to other regions of the Galaxy, is very dangerous. The stellar populations long ago everywhere, and today in the central and outer regions of our galaxy, might be markedly

Chapter 8. Black hole astrophysics

different from the measured population today near Earth. See, e.g. van den Bergh (1975) and Tinsley and Larsen (1977).

Let us turn to the details of the births of black holes. Black holes cannot be formed by quasi-stationary contraction of a star, because the gravitational acceleration in a stationary star, at the point of horizon formation, would be infinite. Thus, it is certain that holes must form by dynamical stellar collapse. But almost nothing else is certain.

For the idealized case of spherical collapse there have been many computer simulations of black hole formation using realistic equations of state; see, e.g. May and White (1966). But unsolved issues of heat transfer by neutrinos produce enormous uncertainty in the true spherical story (e.g. van Riper, 1978), and when one permits magnetic fields, rotation, and other deviations from sphericity, almost nothing is known. Computer simulations of the non-spherical case are now underway in several research groups (Livermore, Harvard, Chicago, Texas), but they cannot hope to give reliable pictures in the near future. Among the important, wide-open issues are: Can black holes form at the cores of some supernova explosions? How much mass is ejected from the outer layers of the collapsing star? How much gravitational radiation is produced? How do the answers depend on the mass and angular momentum of the presupernova star? What are the roles of magnetic fields (Bisnovatyi-Kogan, Popov and Samochin, 1976; Meier et al., 1976)?

Perhaps one's best hope of getting a handle on these issues is through detection and study of gravitational radiation produced during the formation of holes. The waveforms of such radiation would give direct information about the dynamical behavior of the stellar core before and during horizon formation, and about the mass and angular momentum of the newborn hole. (The mass and angular momentum determine the complex eigenfrequencies of the hole's pulsational normal modes, and thereby might determine the behavior of the wave forms at late times; see chapter 7.) The strengths of the waves are very uncertain; but in the case of highly non-spherical collapse it seems likely that they will carry off a total energy equal to several per cent of the hole's mass (Detweiler, 1978; Detweiler and Szedenits, 1979) in a broad-band burst peaked near

$$\nu \sim (10 \text{ kHz})(M/M_\odot).$$

Such bursts from a source in our own galaxy (distance ~ 10 kpc) might barely have been detectable with the most sensitive gravitational-wave antennae operating in 1976. An ultimate goal of future generations of detectors is to monitor and study bursts from the distance of the M101 cluster of galaxies (~ 10 Mpc), within which there is roughly one supernova per year, and beyond which the number of supernovae increases as distance cubed. For further details see chapter 3, and also Thorne (1977).

8.3.2 Accretion of interstellar gas onto holes

Once a black hole has formed by the collapse of a normal star in our galaxy, there is negligible chance that it will ever (within $\sim 10^{10}$ yr) collide with any other hole or star. The only likely exceptions are black holes which form in binary systems (section 8.4), in globular clusters (section 8.5), or in the innermost few parsecs of a galactic nucleus (section 8.6). Also, a black hole alone in interstellar space has a negligible chance of acting as a gravitational lens for radiation from a more distant star (Refsdal, 1964). The only hope to detect such a hole, it seems, is through electromagnetic radiation produced by interstellar gas as it falls chaotically toward the hole's horizon.

Models for the accretion of interstellar gas onto an isolated hole have been developed in much detail during the past decade. The first few steps in such model building were described in section 8.2, (8.1) to (8.7). As we saw there, the interstellar magnetic field plays a crucial role. It turns out that, independently of whether the field is ordered or tangled on the scale of the accretion radius R_A, the field provides sufficient coupling of the protons to make the flow hydrodynamical at R_A. As a result, the rate of mass accretion is always given by (8.4) if the hole is at rest in the interstellar medium, and by a simple generalization of it in the case of a moving hole:

$$R_A \approx 10^{14} \text{ cm}(M/M_\odot)(T_\infty/10^4 \text{ K})^{-1}(1+\mu^2)^{-1},$$
$$\dot{M} \approx 4\pi R_A^2 \rho_\infty (p_\infty/\rho_\infty)^{1/2}(1+\mu^2)^{-1/2} \quad (8.9)$$
$$\approx 10^{11} \text{ g sec}^{-1}(M/M_\odot)^2 \xi,$$
$$\xi \equiv (\rho_\infty/10^{-24} \text{ g cm}^{-3})(T_\infty/10^4 \text{ K})^{-3/2}(1+\mu^2)^{-3/2}.$$

Here μ is the Mach number of the hole's motion through interstellar space

$$\mu = \frac{\text{(speed of hole)}}{(p_\infty/\rho_\infty)^{1/2}} = \left(\frac{\text{speed of hole}}{10 \text{ km s}^{-1}}\right)\left(\frac{10^4 \text{ K}}{T_\infty}\right)^{1/2}. \quad (8.10)$$

Note that at high Mach numbers, i.e. $\mu \gg 1$, (8.9) are obtained from the 'at rest' case by simply replacing the speed of sound $(p_\infty/\rho_\infty)^{1/2}$ by the

Chapter 8. Black hole astrophysics

speed of the hole. The $(1+\mu^2)$ factor is chosen to give a simple interpolation between the low-speed and high-speed cases (Bondi, 1952).

Below the accretion radius the gas flow approaches free fall with velocity $v \propto r^{-1/2}$, density $\rho \propto r^{-3/2}$, temperature $T \propto r^{-1}$, and (as a result of flux conservation) magnetic field $B \propto r^{-2}$; cf. (8.3). Outside and near R_A, magnetic pressure was negligible compared to thermal pressure; the magnetic field only served to couple protons to each other via Larmor gyrations. However, below R_A

$$B^2/8\pi \propto r^{-4}, \quad p \propto \rho T \propto r^{-5/2}, \quad (GM/r)\rho \propto r^{-5/2}, \qquad (8.11)$$

which means that magnetic pressure very quickly overwhelms thermal pressure, and soon thereafter magnetic energy density becomes comparable with gravitational energy density. The subsequent flow will be dominated by a competition between magnetic stresses and gravitational forces, and will deviate strongly from simple radial free fall.

A variety of different models has been constructed for the subsequent flow, corresponding to a variety of different conditions at infinity. When the interstellar field is homogeneous on scales $l_B \gg R_A$ (the most likely case), and if the magnetically impeded flow does not develop instabilities, one is led to the 'hour-glass model' described briefly in section 8.2 and developed in detail by Bisnovatyi-Kogan and Ruzmaikin (1974, 1976); see also Piddington (1970), Sturrock and Barnes (1972), Ozernoy and Usov (1973) and Bisnovatyi-Kogan (1978).

In the case of tangled field lines at infinity, or when instabilities produce tangling, the flow becomes magnetoturbulent. The models then assume, without good proof, that:

(i) The protesting magnetic fields slow the infall substantially, so that roughly half the gravitational energy released goes into infall energy and roughly half into magnetic and turbulent energies:

$$\tfrac{1}{2}\rho v^2 \sim \frac{B^2}{8\pi} + \tfrac{1}{2}\rho v_{\text{turb}}^2 \sim \tfrac{1}{2}\rho \frac{GM}{r}; \quad 4\pi r^2 \rho v = \dot{M} = \text{constant}. \qquad (8.12)$$

(ii) Field-line reconnection saves the magnetic field from mounting at the flux-conservation rate $B \propto r^{-2}$, and enables it to mount at the slower rate implied by (8.12), $B \propto r^{-5/4}$.

(iii) The magnetic energy lost in field-line reconnection – and turbulent energy lost due to turbulent viscosity – go into heat (i.e. into random kinetic energy of protons and electrons). The resulting heating rate per unit volume is

$$\Gamma \sim \frac{1}{2}\left(\frac{v}{r}\right)\left(\frac{B^2}{8\pi}+\frac{1}{2}\rho v_{\text{turb}}^2\right) \sim \left(\frac{GM}{r^3}\right)^{1/2} \rho\left(\frac{GM}{r}\right). \quad (8.13)$$

These assumptions seem reasonable, but the theory of strong magnetoturbulence is in such a primitive state that one cannot justify them rigorously.

Once these assumptions have been made, the magnetoturbulent models split up into several different cases depending on (i) the amount of angular momentum in the accreted gas (enough or too little to produce a centrifugal hangup in the radial infall and thereby create an accretion disk); (ii) the amount of cooling due to cyclotron and synchrotron emission (enough or too little to compete with the magnetoturbulent heating); and (iii) whether the infalling gas becomes optically thick to synchrotron and cyclotron emission. Which case applies depends on conditions in the accreting gas at infinity (ρ_∞, T_∞, B_∞, and vorticity). Models with centrifugal hangup (large vorticity at infinity) were discussed by Salpeter (1964) and Shvartsman (1971). The various cases without hangup are explored by Shvartsman (1971), Shapiro (1973a, b), and Mészáros (1975b).

The total luminosity from a model can be expressed in terms of the universal accretion rate, (8.9), and the model's efficiency ϵ for converting the gravitational energy released, $\dot{E}_{\text{grav}} = \dot{M}c^2$, into escaping radiation

$$L = \epsilon \dot{M} c^2 = (10^{-2} L_\odot)(\epsilon/0.5)(M/M_\odot)^2 \xi. \quad (8.14)$$

For the hour-glass model, and for the angular-momentum-hangup model, the efficiency is high: $\epsilon \sim (0.05$ to $0.5)$. For other models it is lower – e.g., Mészáros estimates $\epsilon \sim 10^{-4}$ and $L \sim 10^{-4} L_\odot$ for the case $M = 10 M_\odot$, $\rho_\infty = 10^{-24}$ g cm^{-3}, $T_\infty = 10^4$ K, $\xi = 1$, and magnetoturbulent flow without hangup.

In all models the outflowing radiation comes largely from cyclotron and synchrotron emission (electrons spiraling in the magnetic field). In most models the bulk of the radiation is emitted from near the Schwarzschild radius. If the emission from there is roughly black body, it must peak at a frequency ν_{max} given by

$$h\nu_{\text{max}} \approx kT_{\text{max}}, \quad 4\pi(3GM/c^2)^2 \sigma T_{\text{max}}^4 \approx L; \quad (8.15)$$

i.e.

$$h\nu_{\text{max}} \approx 60 \text{ eV } (\epsilon\xi)^{1/4}. \quad (8.16)$$

Chapter 8. Black hole astrophysics

This is in the ultraviolet part of the spectrum for $\epsilon\xi \sim 1$; in the visual for $\epsilon\xi \sim 10^{-4}$. If the spectrum is not black body, it will peak at or above 60 eV $(\epsilon\xi)^{1/2}$.

Most of the detailed models give spectral peaks near 60 eV $(\epsilon\xi)^{1/2}$, with exponential falloffs above there, and with gentle falloffs below there ($dL/d\nu \propto \nu^{0 \text{ to } 0.3}$ typically). Thus, a black hole of mass $M \sim$ a few $M_\odot$, accreting interstellar gas, should show up at optical frequencies as a 'star' of low but non-negligible luminosity, with a rather flat, featureless spectrum. Shvartsman (1971), the first model builder to reach this conclusion, noted the resemblance to DC white dwarfs (i.e. white dwarfs with no lines in their spectra) and suggested that some of the observed DC white dwarfs might, in fact, be black holes. He also called attention to a definitive 'signature' that the light from a hole might carry: it might fluctuate on timescales

$$\tau \sim (\text{a few}) \times (GM/c^3) \sim (\text{a few}) \times (10^{-5} \text{ s}) \times (M/M_\odot). \quad (8.17)$$

Unfortunately, to detect such fluctuations in an object of such low luminosity at reasonable distances (≥ 100 pc) requires a very large telescope, and the predictions are so uncertain that nobody has been willing to devote much large-telescope time to a search. Nevertheless, it is quite possible that future studies will identify such objects.

In addition to accretion from 'standard' interstellar gas, one can imagine other conditions of accretion onto an isolated black hole. At one extreme would be accretion in the galactic halo, where the interstellar density is far lower than 10^{-24} g cm^{-3} and where the resulting radiation is almost certainly unobservable unless the hole has $M \gg 100 M_\odot$. At the other extreme would be accretion from a very dense cloud of gas, e.g. gas ejected during the original formation of the hole, in which case a high-luminosity ($L \sim 10^5 L_\odot$) X-ray source might result; see Mészáros (1975c). The most extreme case would be a black hole that forms in the core of a massive star without ejecting the star's diffuse envelope. ('Conventional wisdom' says this never happens, but one can surely not rule it out.)

For the case of a hole inside a star, at least three research groups (Caltech, Cambridge, and Munich) have attempted to build steady-state models of the accretion, all without success. It seems plausible that the accretion might proceed in an exponential runaway manner, with the hole swallowing the entire star on a free-fall timescale. It also is plausible that the hole will eat the star only on a timescale of $\sim 10^7$ or 10^8 yr, that the accretion will proceed in an oscillatory, time-dependent way, and that a time-averaged luminosity $L_{\text{Edd}} \approx (4 \times 10^4 L_\odot) \times (M/M_\odot)$ will be produced by the accretion. Much work is needed on this problem. For details of previous failures and partial successes see Kafka and Mészáros (1976) and Begelman (1978).

8.4 Black holes in binary systems

8.4.1 Introduction

Roughly 50 per cent of all stars are born in binary systems; this is as true of big stars as of little. And in 50 per cent of all binaries, the stars are close enough together to interact substantially as they evolve. An interesting possibility is that the more massive star in a binary will exhaust its nuclear fuel and collapse to form a black hole, and that subsequently its less massive companion will dribble enough gas onto the hole to produce a very large luminosity.

A rough estimate of the maximum possible luminosity is the 'Eddington limit' – the value of L for which the pull of gravity on accreting gas is precisely counterbalanced by the outward force of photons that scatter off the gas's electrons:

$$\frac{GM\rho}{r^2} = \frac{L_{\text{Edd}}/c}{4\pi r^2} \sigma_\text{T} \frac{\rho}{m_\text{H}}, \tag{8.18}$$

i.e.

$$L_{\text{Edd}} = \frac{4\pi GcMm_\text{H}}{\sigma_\text{T}} = (1 \times 10^{38} \text{ erg s}^{-1})\left(\frac{M}{M_\odot}\right). \tag{8.19}$$

(Here σ_T is the Thomson cross-section for scattering of photons by electrons, m_H is the mass of a hydrogen atom, and ρ/m_H is the number density of electrons in the accreting gas, which we presume to be mainly ionized hydrogen.) If the bulk of the luminosity L comes from near the Schwarzschild radius, then the typical photons must have energies greater than, or of the order of, the black-body value:

$$h\nu \geqslant kT_{\text{BB}}, \quad 4\pi(3GM/c^2)^2 \sigma T_{\text{BB}}^4 \approx L; \tag{8.20}$$

i.e.

$$h\nu \geqslant 3 \text{ keV } (M/M_\odot)^{-1/2}(L/L_{\text{Edd}})^{1/4}. \tag{8.21}$$

Thus, a black hole accreting gas in a close binary system is a very promising source of high-luminosity X-rays. The same is true of a neutron star, and, to a lesser extent, a white dwarf.

Chapter 8. Black hole astrophysics

The idea to search for X-ray emission from black holes and neutron stars in binaries occurred to a large number of astrophysicists simultaneously in 1966 (see Burbidge (1972) for a partial description of the historical context; see Novikov and Zel'dovich (1966) and Shklovsky (1967) for early publications). The idea paid off in 1971 when Giacconi's research group flew the first Earth-orbiting X-ray telescope, Uhuru, and discovered pulsar-type X-ray emission with $L \sim L_{\text{Edd}}$ coming from several binary systems. (For a review see Giacconi and Gursky, 1974.) The pulsed X-rays were quickly and convincingly attributed to neutron stars (Pringle and Rees, 1972; Davidson and Ostriker 1973; Gnedin and Sunyaev, 1973; Lamb, Pethick and Pines, 1973); they could not come from black holes because theory requires holes to be axially symmetric and thereby prevents their rotation from acting as a clock to regulate precisely timed pulses.

On the other hand, if neutron stars can occur in binaries and can radiate X-rays by accretion, there is no obvious reason why black holes should fail to do the same. Thus, it became attractive to suppose that some of Uhuru's non-pulsed X-ray sources might be black holes. The most promising way to distinguish the holes from the neutron stars and white dwarfs was by weighing them; anything more massive than M_{max}^{n*} should be a hole. The weighing was done – albeit not terribly accurately – and it yielded one very promising black hole candidate: the X-ray source Cygnus X-1. In section 8.4.5 we shall describe in more detail the weighing of Cygnus X-1, and the other evidence that suggests (but does not yet prove!) that Cygnus X-1 and Circinus X-1 are both black holes.

But before discussing the observational situation, we shall outline a theoretical foundation for it by describing the evolution of massive binary systems in section 8.4.2, the theory of the dumping of gas onto a hole by its companion star in section 8.4.3, and models for the flow of the gas near the hole and its production of X-rays in section 8.4.4.

8.4.2 The evolution of close binary systems

The more massive a star is, the faster it will burn its nuclear fuel and the sooner it will die. This fact suggests that in binary systems the dead, accreting, X-ray-emitting object should be more massive than its live, mass-dribbling companion. However, just the opposite is true. In all measured cases the dead object is the less massive.

A simple explanation is provided by the theory of the evolution of close binaries. Consider two young stars in orbit around each other, with masses and radii M_1, R_1 and M_2, R_2, and with separation $D > R_1 + R_2$. The more massive star (star 1) is called the 'primary'; the less massive is the 'secondary'. When the primary exhausts the hydrogen at its center, it converts over to burning hydrogen in a shell; its center contracts somewhat, and its surface expands. As the hydrogen-burning shell gradually moves outward, the star's core contracts further, and its envelope continues to expand, gradually turning the star into a red giant. There is a critical radius in this expansion,

$$R_{\text{crit}} \approx [0.38 + 0.2 \log_{10}(M_1/M_2)]D \quad \text{if } 0.3 < M_1/M_2 < 20$$
$$\approx 0.462(1 + M_2/M_1)^{-1/3} D \quad \text{if } M_1/M_2 < 0.8, \tag{8.22}$$

at which the primary encroaches on the gravitational potential well of its companion. When R_1 reaches R_{crit}, the primary is said to 'fill its Roche lobe' and dumping begins. So much dumping occurs that, by the time dumping ceases, the primary is less massive than the secondary. If the primary has not by then reduced itself below the Chandrasekhar limit, $M_{\text{max}}^{\text{wd}}$, it will collapse to form a neutron star or black hole before the secondary has evolved much; and if mass ejection during collapse does not disrupt the binary system, the result will be a neutron star or black hole in orbit around a more massive, normal star.

Detailed models make this scenario seem plausible; see, e.g. van den Heuvel and Heise (1972). However, there are some difficulties with the models. In those models designed to produce the observed X-ray binaries, the initial separation D must not be much larger than $R_1 + R_2$. As a result, when the primary expands it dumps gas onto the secondary faster than the secondary can accept it; the secondary expands in response, coming into contact with the primary; mass continues to flow from primary to secondary in this contact system; but soon after contact is made, there is a breakdown of the computational techniques used by the model builders. The ultimate outcome cannot be predicted; it can only be inferred from the observed existence of X-ray binaries. For details of these model difficulties, and for speculations about the evolution of the contact system see, e.g. Flannery and Ulrich (1977), and Kippenhahn and Meyer-Hofmeister (1977). For reviews of the theory of binary evolution, see Thomas (1977) and references therein.

How massive must the primary be in order to wind up as a black hole? The answer will depend on the amount of mass transfer during the

Chapter 8. Black hole astrophysics

primary's giant stage, which in turn will depend on the initial mass ratio M_1/M_2 and separation D. At present this dependence is unknown because of uncertainties about the contact epoch. Also unknown is the likelihood that the binary will survive disruption when mass and gravitational waves are ejected by the hole's birth pangs (Gott, 1972). Thus, at present it is conceivable that a quarter of all the Galaxy's black holes were born in close binaries without disruption, and it is also conceivable that holes never occur in close binaries.

8.4.3 The dumping of gas onto a hole by a companion star

Consider a black hole that has managed to form in a close binary. So long as the hole's secondary companion remains a youthful 'main-sequence' star (i.e. so long as it is in the early stages of central hydrogen burning), it will dump very little mass onto the hole. However, as the secondary ages, two things happen. (i) It expands in size until ultimately it overflows its Roche lobe and starts dumping a heavy stream of gas onto the hole. (ii) It becomes more and more luminous, and its increasing luminosity drives an increasingly strong stellar wind of hot plasma off its surface. The hole can capture some of this wind. Once the hole is accreting somewhat from the wind or from Roche-lobe overflow, a third thing may happen. (iii) A strong flux of X-rays from the hole, impinging on the star's surface, may substantially modify the mass flow off the star, thereby modifying the X-ray flux and maybe even producing a self-regulated system. Of course, all three phenomena could occur for neutron star primaries as well as for black holes.

Ever since X-ray-emitting binaries were discovered, a debate has raged among theorists over the relative importance of Roche-lobe overflow, stellar wind, and self-regulation; see, e.g. Davidson and Ostriker (1973), van den Heuvel (1975), McCray and Hatchett (1975), Pratt and Strittmatter (1976), Savonije (1978), and references cited therein. Although the debate is not yet resolved, there seems to be tentative agreement on a few points:

(i) Self-regulation is probably not a dominant factor near the secondary star, though it may be important near the hole or neutron star.

(ii) A minimum accretion rate of

$$\dot{M}_{min} \sim 3(L/c^2) \sim 3 \times 10^{-10} M_\odot \, \text{yr}^{-1} \qquad (8.23)$$

is required to produce the observed X-ray luminosities of $L \gtrsim 3000 L_\odot$.

(iii) An accreting neutron star will not be able to accept a mass-accretion rate much in excess of

$$\dot M_{max} \sim 10 L_{Edd}/c^2 \sim (10^{-8} M_\odot \text{ yr}^{-1})(M/M_\odot). \tag{8.24}$$

A much larger mass-transfer rate will probably smother the X-ray source; see below.

(iv) Stellar winds large enough to produce $\dot M_{min}$ occur only in main-sequence stars bigger than $\sim 45 M_\odot$ (stars of spectral class Of), and in stars of $M \geqslant 20 M_\odot$ that have exhausted their central hydrogen and are expanding into the giant stage (stars that are classified spectroscopically as 'supergiants earlier than type B1'),

(v) The required winds last for only (a few) $\times 10^5$ yr in the rare main-sequence case, and only (a few) $\times 10^4$ yr in the more common 'giant' case.

(vi) Only in the earliest stages of Roche-lobe overflow can $\dot M$ be kept below $\dot M_{max}$; once the overflow gets going, an instability develops and drives mass off the secondary at a rate between $\sim 10^{-6} M_\odot \text{ yr}^{-1}$ (for $M_2 \approx 2 M_\odot$) and $\sim 10^{-3} M_\odot \text{ yr}^{-1}$ (for $M_2 \approx 20 M_\odot$).

(vii) The early, low-$\dot M$ stage of overflow will be prolonged if the secondary's rotation is slower than its orbital motion.

(viii) The low-$\dot M$ stage of overflow is longer in the closest of binaries, where overflow begins before central hydrogen is fully exhausted ('case A'), than in less close binaries, where it begins during hydrogen-shell burning ('case B').

(ix) In the more favorable case A, the total duration of overflow transfer at $\dot M_{min} \leqslant \dot M \leqslant \dot M_{max}$ is $\sim 10^6$ yr for $M_2 \approx 2 M_\odot$, and $\sim 10^4$ yr for $M_2 \approx 20 M_\odot$.

For point (i) see, e.g. McCray and Hatchett (1975). For (ii)–(vi) see, e.g. van den Heuvel (1975). For (vii)–(ix) see, e.g. Savonije (1978).

As discussed in section 8.4.2, the secondary will usually be much more massive than the hole itself, e.g. $M_2 \geqslant 10 M_\odot$. If so, then the above points suggest a lifetime in the X-ray binary stage of $\sim 10^4$ to 10^5 yr, compared to a secondary main-sequence life $\sim 10^7$ yr, and a current age for the galaxy of $\sim 10^{10}$ yr. This, plus the existence of two good black hole candidates among the observed X-ray binaries (section 8.4.5) suggests that our galaxy may contain $\geqslant 300$ black holes with young, close, main-sequence companions, and $\geqslant 3 \times 10^5$ dead black hole binaries. Where are they in the sky?

Holes with main-sequence companions are very difficult to identify observationally; see section 8.4.5. Dead X-ray binaries are probably

Chapter 8. Black hole astrophysics

even harder to identify, though one does not really know what to look for. The death of a black hole binary may begin when the hole gets buried by runaway Roche-lobe overflow from its companion. The hole may then find itself orbiting beneath the enlarged surface of its companion. It is not at all clear whether the hole will then spiral down into the center of its companion, producing a star with a black hole core which theorists have been unable to model (section 8.3.2), or whether the hole will catalyze an ejection of the companion's envelope but let the core evolve its merry way toward white dwarf, neutron star, or black hole extinction. In the first case the ultimate outcome would be a single, big black hole. In the second it would be a close binary made of two compact objects (a system like the 'binary pulsar'; Hulse and Taylor, 1975), which would ultimately spiral together and coalesce due to gravitational-radiation reaction. In either case, the death throes of the binary might prove observationally interesting – if one can figure out what to look for. Preliminary efforts to analyze these issues in the case of a neutron star binary have been made by Taam, Bodenheimer and Ostriker (1978), and by Thorne and Żytkow (1977). Little attention has been paid as yet to models for the black hole case.

8.4.4 Models for the flow of gas onto the hole, and the production of X-rays

Since 1971, when astronomers realized that Cygnus X-1 might be a black hole, astrophysicists have devoted much effort to modeling the flow of gas onto a hole and computing the properties of the X-rays it produces. Semi-quantitative agreement with the observations has been achieved. However, this does not guarantee that the models closely resemble reality, especially since very different models can be made to produce equally good agreement with the observations.

There are two main categories of models: radial-infall models and accretion-disk models. Which of these is correct depends on how much angular momentum the accreting gas possesses. When the angular momentum j per unit mass greatly exceeds

$$j_{\text{crit}} = (2GM_1/c^2)c, \qquad (8.25)$$

centrifugal forces will dominate the flow and produce an accretion disk at radii large compared to the horizon. (Here M_1 is the mass of the hole, denoted '1' because it was the original primary of the system.) When $j \lesssim j_{\text{crit}}$, centrifugal forces will have little influence, and the gas will fall quasi-radially onto the hole. The amount of angular momentum depends on the details of the mass ejection by the secondary star. In the case of Roche-lobe overflow, the gas has nearly as much specific angular momentum about the hole's center as does the secondary itself, so $j \gg j_{\text{crit}}$ and an accretion disk forms at very large radii. In the case of accretion from a wind of velocity v_w, the hole, moving supersonically through the wind, creates by its gravitational attraction a shock front of radius

$$R_s \approx 2GM_1/v_w^2. \qquad (8.26)$$

Material crossing this shock front falls onto the hole with net specific angular momentum

$$j \approx \left(\frac{R_A}{D}\right) R_A v_{\text{orb}} \qquad (8.27)$$

where D is the distance between the hole and the center of the secondary, and $v_{\text{orb}} \approx (GM_2/D)^{1/2}$ is the orbital velocity of the hole about the secondary. For reasonable values of the parameters (e.g. those corresponding to Cygnus X-1 where $M_1 \approx 10 M_\odot$, $M_2 \approx 25 M_\odot$, $D \approx 40 R_\odot$, $v_w \approx 2(GM_2/R_2)^{1/2} \approx 1000 \text{ km s}^{-1}$, $v_{\text{orb}} \approx 300 \text{ km s}^{-1}$, $r_A \approx 1 R_\odot$), j is of the order of j_{crit}, so that centrifugal forces become strong only near the hole's horizon, and the accretion disk, if any, forms not far from the horizon. However, this conclusion is very sensitive to the velocity of the wind near the hole ($j \propto R_A^2 \propto v_w^{-4}$), and that velocity is uncertain by at least a factor 4. If the wind is slower than expected, a big disk will form; if faster, no disk will form; if it fluctuates in time, a disk might form and be destroyed. For further detail see, e.g. Shapiro and Lightman (1976).

In the fast-wind case, gas that has passed through the shock falls quasi-radially onto the hole. It presumably will carry some magnetic field with itself, field that originated in the secondary star, and that is tangled up in the falling gas. This situation resembles mass accretion from the interstellar medium, except that the accretion rate $\dot{M}$ is far higher than in the interstellar case. Mészáros (1975b, c) has built models for the infall which are similar to the magnetoturbulent models of (8.12) et seq. The key difference is that, because of the larger $\dot{M}$, the density is higher, and cyclotron radiation is replaced as the dominant cooling process by bremsstrahlung and/or inelastic Compton scattering of low-energy cyclotron photons by high-energy electrons. The cooling is adequate to compete with magnetoturbulent heating near the horizon, and since a large fraction of the gravitational energy released goes into heating, (8.13), the efficiency is high for producing outpouring

Chapter 8. Black hole astrophysics

luminosity:

$$\epsilon = L/\dot{M}c^2 \gtrsim 0.2. \tag{8.28}$$

The temperature of the radiating electrons, as regulated by the balance of heating and cooling, comes out to be $T \sim 10^9$ K, and the radiation comes out largely as fairly hard X-rays.

By appropriate choices of the free parameters in this model, Mészáros achieves reasonable agreement with the observations of Cygnus X-1. However, (i) the agreement is no better than that achieved by disk models, and (ii) there are enough *ad hoc* (though reasonable) assumptions in the model, e.g. those discussed in the vicinity of (8.12) and (8.13), to make (even) an astrophysicist queasy.

When the secondary loses mass by Roche-lobe overflow or in a slow wind, j is large compared to j_{crit} and centrifugal forces will create an accretion disk at rather large radii.

An enormous amount of theoretical research has been done on accretion disks since the discovery of binary X-ray sources. (Our preprint and reprint files contain seven inches of publications on the subject.) A recent review of the literature is given by Pringle (1977). A pedagogical introduction is contained in the out-of-date review by Novikov and Thorne (1973).

The theory of the disk structure depends crucially on whether the disk is thick (h, the disk thickness at radius r, $\sim r$) or thin ($h \ll r$). For thin disks the theory is fairly reliable; for thick disks it is not. The best models for Cygnus X-1 involve disks that may be thick in their outer regions, but become thin after a few e-foldings of radius as one moves inwards, and then become thick again near the hole.

The thickness of the disk is governed by a balance between its internal pressure p, and vertical gravitational compression due to the tidal gravity (Riemann tensor) of the hole:

$$p/h \approx \rho(GM/r^3)h. \tag{8.29}$$

Since the internal energy density of the gas (thermal, plus magnetic, plus turbulent) is roughly equal to its pressure, this says

$$\frac{h}{r} \approx \left(\frac{p}{\rho GM/r}\right)^{1/2} \approx \left(\frac{\text{internal energy density}}{\text{gravitational energy density}}\right)^{1/2}, \tag{8.30}$$

which means that a disk is thin if, and only if, it is very efficient at radiating away its internal energy.

In a thin disk the gas orbits the hole in near-Keplerian orbits. ($p \ll \rho GM/r$ implies that radial pressure forces cannot much modify the Keplerian motion.) Consider two adjacent rings of gas in such a disk. The outer ring has a lower Keplerian angular velocity than the inner one. Consequently, viscous friction between the two rings speeds the outer one and slows the inner one, causing the outer ring to spiral outward while the inner spirals inward. In effect, viscosity produces a repulsion between the two rings. When one considers the interaction between all the rings in the disk, one finds a net inward spiraling everywhere, except in the outermost regions. The outer regions, left to their own devices, would move outward; but they are being bombarded by infalling gas from the companion star, producing a net inward flow, the details of which are very difficult to analyze. Fortunately, the enormous uncertainties of the outer regions are of only modest importance for binary X-ray models. This is because almost all of the gravitational energy release and X-ray production occurs near the hole.

In the language of angular-momentum conservation, viscous forces remove angular momentum from the gas in the inner regions of the disk, thereby permitting the gas to spiral inward. The angular momentum is transported mechanically (viscously) to the outer regions, where it is removed in an ill-understood way by interaction with the companion and with instreaming matter.

Suppose that in an annulus $r_1 < r < r_2$ the viscosity is too low to remove angular momentum at the required rate. Then gas will be fed into this annulus at r_2 by the stronger stresses outside it, but the annulus will not be able to pass the gas on in toward the hole at r_1. Gas will pile up in the annulus. Under most situations this added gas will increase the viscous stresses in the annulus until they become adequate to handle the mass transport. In this way a steady state will be achieved. (For mathematical details see Lightman (1974a, b), and Lynden-Bell and Pringle (1974).) However, in certain pathological situations the increase in gas density will trigger a decrease of viscosity, thereby causing the density in the annulus to run away while the density in adjacent regions falls to zero. It appears likely that this 'breakup-into-rings' instability occurs in thin accretion disks whenever the internal radiation pressure $p_{\text{rad}} = \frac{1}{3}aT^4$ exceeds the gas pressure $p_{\text{gas}} = \mathcal{R}\rho T$. See Lightman and Eardley (1974).

Another instability – one which develops faster than ring breakup and perhaps always accompanies it – is thermal runaway (Pringle et al., 1973; Pringle, 1976). This is triggered by viscous heating of the gas,

Chapter 8. Black hole astrophysics

which necessarily accompanies viscous transport of angular momentum. When radiative cooling fails locally to counterbalance viscous heating, the gas temperature rises. If higher temperature produces decreased heating or an increased ability to radiate, the temperature will fall back to equilibrium and the disk is thermally stable. But if higher temperature increases the disparity between heating and cooling, a thermal runaway occurs.

In actuality, there is not a completely clean distinction between these two instabilities (ring breakup and thermal runaway). They are analyzed in a unified fashion, along with others, in the definitive work of Shakura and Sunyaev (1976).

At present the most popular models for Cygnus X-1 (e.g. Shapiro et al., 1976) involve a thin accretion disk in which the viscosity is provided by turbulence and/or by magnetic stresses. In effect, these are disk analogues of Mészáros's magnetoturbulent, radial-inflow models; in both cases the heating is by field-line reconnection and turbulent viscosity. In these models the disk is optically thick, physically thin, and fairly cool ($T_\text{surface} < 10^7$ K) down to ~30 Schwarzschild radii. At that point burgeoning radiation pressure triggers both the ring-breakup and thermal-runaway instabilities, driving the innermost portion of the disk into a hot ($T \gtrsim 10^9$ K), thick ($h \sim r$), marginally optically thin state. Most of the observed X-rays are produced in this inner region by inelastic Compton scattering of lower-energy photons.

Thick-disk structures can result from huge ('super-Eddington') accretion rates, as well as from instabilities; and even in the thin-disk case, magnetohydrodynamic processes may generate thick, hot coronas above the disk. Thick-disk physics may in turn generate a strong wind that drives mass loss from the inner regions of the disk. For preliminary attempts to analyze super-Eddington accretion, coronas and winds, see Bisnovatyi-Kogan and Blinnikov (1977), Liang and Price (1977), Piran (1977), Icke (1977, 1978), Bardeen (1978). We shall return to these issues in section 8.6 in connection with black holes in galactic nuclei and quasars.

Research on accretion disks continues at an accelerating pace today, not because enormous progress is being made (it is not), but because astronomers have recently realized that accretion disks are a very common phenomenon in the universe: they surely occur around neutron stars and perhaps around black holes in X-ray binaries; they may occur around white dwarfs in dwarf novae and old novae (Smak, 1971, 1976; Osaki, 1974); they may occur around protostars in T Tauri systems

Black holes in binary systems

(Lynden-Bell and Pringle, 1974); and they probably occur elsewhere as well.

8.4.5 The observational search for black holes in binary systems

In 1965, long before X-ray binaries were discovered, Zel'dovich and Guseynov (1966) initiated a search in catalogs of binaries for systems that might contain a black hole. The likely candidates were single-line spectroscopic binaries with large secondary masses, i.e. systems from which only the light of one star is seen, and the spectral lines of that star exhibit a Doppler shift which oscillates periodically in time, and from the oscillations one infers a large mass for the object around which the star orbits. Unfortunately, all of the possible Zel'dovich–Guseynov candidates, and all of those found in a subsequent search by Trimble and Thorne (1969), could be explained fairly easily without invoking a black hole. In an attempt to rule out other explanations, Abt and Levy (1974) observed half (five) of the best Trimble–Thorne candidates and discovered that four of them had been misclassified: they were not binaries at all!

This illustrates the great difficulties that must plague any search for black hole binaries in the absence of a spectacular signature such as X-ray emission. Of the many candidates that have been suggested, e.g. ϵ Aurigae (Cameron, 1971), β Lyrae (Hack et al., 1974), and HD 108 (Bekenstein, 1976), none are convincing.

By contrast, a rather strong case can be made for Cygnus X-1, and a suggestive case for Circinus X-1.

When one observes Cygnus X-1 with an optical telescope, one sees a blue supergiant star of spectral type O9.7 Iab. The wavelengths of the absorption lines oscillate with a rather large amplitude, indicating that the star is in orbit around a massive companion with orbital period 5.6 d. One also sees emission lines with rather different oscillations. These are interpreted, convincingly, as light from gas that is flowing off the supergiant and onto its companion. One sees no evidence of light from the companion. These observations are summarized by Bolton (1975).

When one observes with an X-ray telescope, one sees an X-ray luminosity with roughly equal amounts of energy coming out in the bands $1 < E < 10$ keV and $10 < E < 100$ keV, and with substantial but smaller amounts above 100 keV. The X-rays of lowest energy (least penetration power) are partially eclipsed on occasion, and these eclipses generally occur when the supergiant is between us and its companion.

Chapter 8. Black hole astrophysics

One infers, convincingly, that the X-rays come from the vicinity of the companion, that the line of sight to Earth passes above the companion (no total eclipse), and that the X-ray source is sometimes partially eclipsed by the mass flowing off the supergiant and toward its companion. The X-rays fluctuate in intensity on all timescales $\geqslant 0.1$ s. (The data are not yet good enough to demonstrate convincingly or rule out faster fluctuations.) This means that at any one time the bulk of the X-ray energy comes from regions of size $\leqslant (0.1 \text{ light-seconds}) \approx 3 \times 10^4$ km. Bolton (1975) has summarized these observations and given relevant references.

The only reasonable explanation of all this is X-ray emission by gas falling onto a companion of size $\leqslant 3 \times 10^4$ km – a companion which, according to current theory, must be a white dwarf, a neutron star, or a black hole. (For unlikely but conceivable alternative explanations see §V of Bolton (1975) and references therein.) The obvious way to distinguish between white dwarf, neutron star, and black hole is to weigh the secondary.

There are many different ways to do the weighing. One involves combining Kepler's laws and elementary geometric considerations with the following data:
(1) the orbital velocity of the supergiant, as inferred from its oscillating spectral lines;
(2) an estimate of the radius of the supergiant, which comes from
   (a) its apparent visual brightness (erg cm^{-2} s^{-1} at Earth);
   (b) its distance, as inferred from
      (i) the amount of reddening of its light due to passage through interstellar dust (reddening determined by comparing the observed colors of its continuum with the intrinsic colors produced by atmospheres that have the temperature, density, and surface gravity inferred from the observed spectral lines),
      (ii) a curve of reddening versus distance for others stars near the line of sight to Cygnus X-1;
   (c) the absolute flux (erg cm^{-2} s^{-1}) of visual radiation off the supergiant's surface, as inferred from the atmosphere's inferred temperature, density, and surface gravity;
(3) the absence of X-ray eclipses, which places a limit on the inclination angle between our line of sight and the orbital plane.

Another method of weighing involves the amount by which the gravity of the companion deforms the supergiant, which one infers from a small

2.8 d variation of the supergiant's apparent brightness. Another (and somewhat dangerous) method makes use of a mass for the supergiant which one infers from its spectral type. Fortunately, all of the methods give agreement (see, e.g. Bolton, 1975): the mass of the dark companion lies between 8 and 18 $M_\odot$; it is above $M_{max}^{n^*}$ and M_{max}^{wd}; the companion can only be a black hole.

If this were a routine situation, astronomers would confidently accept the result. But since man's first discovery of a black hole hangs in the balance, and since firmer conclusions are sometimes destroyed by overlooked systematic errors, the astronomers are being cautious. Until additional, independent, confirming evidence is found – evidence of a positive rather than a negative 'what else can it be?' nature – they are not willing to conclude that Cygnus X-1 is definitely a black hole.

Current hopes for confirming evidence focus on rapid variability of the X-rays. Sunyaev (1973) has pointed out that if hot spots form in the accreting gas, then one might expect the X-rays to fluctuate quasi-periodically on timescales equal to the orbital periods of the hot spots (a few milliseconds near the hole). Hints of such fluctuations have been seen by large X-ray telescopes on rockets (Rothschild *et al.* 1974, 1977), but the statistics are poor, and the fluctuations might not be real (Weisskopf and Sutherland, 1978). If future observations with large telescopes in orbit reveal millisecond fluctuations, and if they show a sharp low-period cutoff, then the case for a black hole may become firm; and from the cutoff one may be able to infer the period of the last stable circular orbit around the hole. But for now these are only hopes.

Of the roughly one dozen other binary X-ray sources that are somewhat well studied, only one is regarded today as a likely black hole candidate: Circinus X-1. Unfortunately, the case for its candidacy is based entirely on a striking similarity between its X-ray spectrum and short-timescale variability, and those of Cygnus X-1 (Forman, Jones and Tananbaum, 1976; Buff *et al.*, 1977; Ostriker, 1977). Only in recent months have the optical and radio-stars associated with Circinus X-1 been identified (Whelan *et al.*, 1977). They are not yet well-enough studied to give useful information about its black hole candidacy.

8.5 Black holes in globular clusters

8.5.1 Theory of the evolution of star clusters

In section 8.3 we described the manner in which a massive, spherical star ages and dies. The aging is characterized by a gradual contraction of

Chapter 8. Black hole astrophysics

the core and expansion of the envelope. The death is triggered, presumably, by the collapse of the core to form a black hole.

Theory predicts a similar evolution for a spherical star cluster. In its core the cluster tries, by two-body hyperbolic-orbit encounters, to evolve toward a Maxwellian velocity distribution. But stars driven into the high-energy tail of the Maxwellian escape from the core into loosely bound, highly radial orbits, or escape from the cluster altogether ('evaporation'). As a result, the core gradually loses energy and contracts, and a halo of loosely bound, radially orbiting stars develops. Ultimately – if one ignores three-body encounters and physical star–star collisions – the core becomes unstable due to general relativistic effects, and collapses to form a supermassive black hole (Zel'dovich and Podurets, 1965; Ipser, 1969). Three-body encounters and star–star collisions make the ultimate outcome more complex and hard to analyze, but the formation of a supermassive hole is a reasonable possibility. See § VI of Lightman and Shapiro (1978) and references cited therein.

The timescale for noticeable core contraction is called the cluster's 'central relaxation time'. It is equal numerically to the time required for two-body encounters near the center of the cluster to change substantially a star's orbit. The impact parameter b for single encounters that change the orbit is given by

$$G\bar{m}/b = \tfrac{1}{2}\bar{v}^2 \tag{8.31}$$

where $\bar{v}$ is the root-mean-square velocity in the core and $\bar{m}$ is the average mass of a star. As in plasma physics, so also here, small-angle deflections have a net random-walk effect that is logarithmically larger than single encounters. Consequently, the central relaxation time is

$$t_{rc} \approx \frac{2(3/2)^{3/2}}{n_c(\pi b^2)\bar{v} \ln(0.5N)} = \frac{(3/2)^{3/2}\bar{v}^3}{2\pi G^2 m^2 n_c \ln(0.5N)}$$

$$= (1 \times 10^{13} \text{ yr}) \frac{(\bar{v}/10 \text{ km s}^{-1})^3}{(n_c/1 \text{ pc}^{-3})(m/M_\odot)^2 \ln(0.5N)}.$$

(8.32)

Here n_c is the central star density (number per cubic parsec) and N is the total number of stars in the cluster. Numerical and analytical studies show collapse of the cluster's core in roughly 100 t_{rc} (Spitzer and Thuan, 1972), and most stars more massive than $2\bar{m}$ sink to the center of the cluster after $\sim 30 t_{rc}$ (Spitzer and Shull, 1975).

Black holes in globular clusters

The timescales $100 t_{rc}$ and $30 t_{rc}$ are longer than the age of the universe for all types of clusters except two: globular clusters and (perhaps) the central cores of some galaxies. We consider globular clusters in section 8.5.2 and galaxy cores in section 8.6.

8.5.2 Observed features of globular clusters

Globular clusters of stars are spherical; they typically contain 10^4 to 10^6 stars within a cluster radius of 50 to 100 pc; their structure includes a central core with a radius of a few parsecs and star density $n_c \sim 10^3$ to 10^5 pc^{-3}, surrounded by a more diffuse halo. Astronomers can resolve individual stars in a cluster, and from the measured dispersion in radial velocities, the measured central star density, and the estimated mean star mass, they can compute central relaxation times. These range from $t_{rc} \sim 10^7$ yr to 10^{10} yr; see, e.g. Figure 1 of Bahcall and Ostriker (1975). For comparison, the actual age of all globular clusters in our galaxy (as inferred from the ages of the stars that are now just becoming giants) is 1×10^{10} yr, i.e. the age of the Galaxy itself. This strongly suggests that a sizeable fraction of the globular clusters with which our galaxy was born has died by core collapse before now, that roughly 50 more, out of the remaining ~200, will die in the next 10^9 yr, and that in those 50 doomed clusters, all stars heavier than $2 M_\odot$ have by now sunk to the center (Lightman, Press and Odenwald, 1978).

Where are the dead clusters? Nobody knows. And we don't know what they should look like, except for the fact that they might contain, in their collapsed cores, moderately massive black holes. How massive? Arguments about the effects of three-body encounters and star–star collisions in the dying core suggest that $M_{hole} \lesssim 1000 M_\odot$ and that much of the remaining mass of the cluster may have been evaporated away (Lightman and Fall, 1978).

And what of the doomed but live clusters, whose massive stars should already have sunk to the center? This question has motivated astronomers to scrutinize the cores of globular clusters more carefully in recent years. One sign of a central condensation – which might now be a massive black hole – would be its gravitational influence on stars in the cluster. The hole should preferentially pull stars toward itself, creating a cusp in the observed surface brightness of the cluster. Although several observed clusters have unresolved cores, the angular resolution is not yet good enough to prove or disprove the existence of a central black hole. The best one can do is place upper limits of $\sim 10^4 M_\odot$ on the

Chapter 8. Black hole astrophysics

masses of such holes (Bahcall and Hausman, 1977; Illingworth and King, 1977).

In 1974, when these issues of cluster death were in an early stage of study, X-ray sources were discovered unexpected in globular clusters. By now seven X-ray sources have position error boxes that contain globular clusters, and most of the seven presumably reside in those clusters. This means that, whereas globular clusters contain only $\sim 10^{-4}$ of the Galaxy's stellar mass, they contain several per cent of its compact X-ray sources. Moreover, all but one of the seven X-ray-emitting globular clusters are on the 'doomed list' with $100 t_{rc} < 10^{10}$ yr.

When these facts began to emerge, the conclusion seemed obvious (Bahcall and Ostriker, 1975; Silk and Arons, 1975): the X-ray sources are supermassive black holes ($M \sim 100$ to $10^4 M_\odot$) produced by the sinkage of heavy ($M \geq 2 M_\odot$) stars and holes to the cluster center, and the X-rays are produced by accretion of the cluster's interstellar gas.

Although this theory remains viable today, it has lost favor. The primary reason is that several of the globular-cluster X-ray sources have turned out to be 'bursters' (i.e. emitters of trains of imprecisely timed bursts), and most theorists today suspect that such bursting activity is associated with accretion of gas onto a magnetic neutron star. See, e.g. Lamb et al. (1977). On the other hand, the vagaries of astrophysical fashion may well bring the black hole model back into favor again before the final truth is found.

8.5.3 Tidal friction on globular clusters

As a globular cluster plunges through the disk of our galaxy, its strong gravitational field produces a wake in the disk's star distribution. This wake involves a density enhancement behind the cluster, which creates a gravitational drag ('tidal friction') on the cluster's motion. Tremaine (1976) has pointed out that the tidal friction is strong enough to drag the more massive of our galaxy's globular clusters into the Galaxy's center in a time less than 10^{10} yr. Whether this has already happened to many globulars is controversial (Thuan and Oke, 1976). However, if it has occurred, and if some of those clusters contained collapsed cores, one is invited to speculate about the observational consequences and fates of 10^2 to $10^4 M_\odot$ holes running around in the innermost few hundred parsecs of the Galaxy.

8.6 Black holes in quasars and active galactic nuclei

8.6.1 Introduction

As we remarked in section 8.1, many sceptics regard the black hole with considerable suspicion: as an easy, eye-catching panacea prescribed without too much thought for any astronomical ill. In the case of quasars (and henceforth in this article we shall also understand the term quasar to embrace any galactic nucleus showing evidence of non-stellar activity), it is well to remind ourselves that very soon after their discovery (Schmidt, 1963) they were interpreted as powerful examples of gravitational collapse (see especially the proceedings of the First Texas Symposium in 1963, Robinson, Schild and Schucking, 1965). From the start, black holes (e.g. Salpeter, 1964; Zel'dovich, 1964; Lynden-Bell, 1969) along with supermassive stars (Hoyle and Fowler, 1963; also called spinars by Morrison, 1969; or magnetoids by Ozernoy, 1966), dense star clusters (e.g. Colgate, 1967; Arons, Kulsrud and Ostriker, 1975) and a rich variety of ephemeral speculations have been actively discussed as energy sources for quasars. The reasons why black holes have only recently become the most popular candidate for the 'prime-mover' of a quasar are partly observational and partly theoretical. The theoretical problem was simply that, despite the depth of the potential well, the actual efficiency of extraction of *useful* energy from accreted gas by a massive ($\sim 10^8 M_\odot$) black hole was initially found to be extremely low. Black holes were then more readily associated with dead quasars (Lynden-Bell, 1969) than with quasars in the prime of life, especially since total gravitational collapse seemed the inevitable endpoint of most quasar models. We shall argue below that in fact high efficiency can be achieved fairly plausibly.

Detailed discussions of observations and theories of quasars are contained in the reports of the Eighth Texas Symposium on Relativistic Astrophysics (*Annals of the New York Academy of Sciences*, 1977) and in the proceedings of conferences held in Copenhagen and Cambridge in the summer of 1977 (*Physica Scripta*, **17**, 1977; Hazard and Mitton, 1979). It is not our plan to repeat these discussions; instead we shall confine our attention to giving a short review of the observed properties of quasars, emphasizing those that impinge directly on the nature of the primary energy source.

8.6.2 Radio-properties

It is nearly 25 yr since the radio-source Cygnus A was resolved by Jennison and Das Gupta (1953) into two comparably bright

Chapter 8. Black hole astrophysics

components, displaced by ~100 kpc, on opposite sides of what appears to be a giant-elliptical galaxy with an active nucleus. This observation effectively disposed of the earlier idea (Baade and Minkowski, 1954) that extragalactic radio-sources were galaxies in collision. It was also the first example of a source geometry (double structure or more generally linear structure) that has been repeated in radio-observations of strong sources ever since. In the physically largest known source, 3C 236 (Willis, Strom and Wilson, 1974), which is 6 Mpc in length, aligned double structure is observed on scales from 6 Mpc down to ≤ 200 pc (R. T. Schillizi, private communication), with the smallest structures located in the nucleus of the associated galaxy. There are now several examples of very small (≤ 1 pc across) compact sources displaying the same position angle on the sky as larger (≥ 100 kpc) associated extended sources (e.g. Readhead, Cohen and Blandford, 1978), although this is by no means always the case. Many compact sources are resolved into two components that appear to be separating with speeds greater than $2c$ (probably attributable to kinematical effects; e.g. Blandford, McKee and Rees, 1977). Here again successive outbursts usually occur in the same direction (e.g. Cohen et al., 1977).

Arguments independent of these observations (see the review by De Young, 1976) imply that power supplying the large sources must be generated continuously by the associated quasars and is probably supplied via two channels through the surrounding medium. Clearly, there must be some way for the central regions of a quasar to 'remember' a particular direction for the lifetime of a radio-source (typically 10^6–10^8 yr.). Arguably the best way to do this is through a spin axis, perhaps associated with a central dense star cluster or gas cloud, but more probably belonging to a single coherent object (black hole?) much smaller than a parsec in size.

A second argument for the occurrence of very dense structure in quasars is based on the observed timescale of *optical* luminosity variations of the compact 'central' sources, which in some quasars can be as short as ~1 d (e.g. Véron, 1975). In the absence of special relativistic kinematic effects, causality limits the size of the emission region to $\leq 3 \times 10^{15}$ cm ~ $100(M/10^8 M_\odot)$ Schwarzschild radii. On energetic grounds, quasars must typically supply $\geq 10^7 M_\odot c^2$ of energy over their lifetimes, which probably implies that the masses of their energy sources exceed $10^8 M_\odot$. If the rapidly fluctuating quasars are typical of the class as a whole, then the energy must be produced in a region fairly close to complete collapse. (It may be possible to circumvent this argument. If, for example, a quasar comprises a dense star cluster and each outburst corresponds to the formation of a new neutron star or black hole, then the cluster may extend over ≥ 1 pc. The constant rotation axis could be produced either by forming all the stars from a differentially rotating disk with parallel spins, or, more plausibly, by using an asymmetry in the surrounding gas cloud to shape the energy release. One of the problems with this type of idea, however, is that the energy associated with an individual quasar outburst does not seem to take on a particular value that can be associated with a $1 M_\odot$ object; in fact, it extends from $\leq 10^{-4} M_\odot c^2$ to $\geq 10^3 M_\odot c^2$.)

If quasars are fuelled by accretion of gas onto black holes, then there is a characteristic timescale (Salpeter, 1964) that is independent of the mass of the hole: when the Eddington luminosity, $L_{\text{Edd}} \sim 10^{46} (M/10^8 M_\odot)$ erg s^{-1}, is liberated by infalling gas with a (typical) efficiency ~ 0.1, i.e. when $L_{\text{Edd}} \sim 0.1 \dot{M} c^2$, then $M/\dot{M} \sim 5 \times 10^7$ yr. This timescale differs by less than an order of magnitude from other, independent estimates of quasar lifetimes.

The variable *radio*-emission from most quasars is plausibly attributed to synchrotron radiation emitted by relativistic electrons gyrating in magnetostatic fields. Synchrotron radiation theory predicts a minimum linear size, $l_{R,\text{min}}$, for the radio-components (e.g. Burbidge, Jones and O'Dell (1974) and references therein). The basis of this prediction is as follows. The energy of a relativistic electron radiating at frequency ν in a magnetic field of strength B is $\gamma m c^2 \sim [2\pi\nu/(eB/mc)]^{1/2} mc^2$. Now on thermodynamic grounds, we expect this electron energy to exceed the product of Boltzmann's constant and the 'brightness temperature' T_B of the observed radiation, and (by definition appropriate to the Rayleigh–Jeans part of a black-body spectrum) $T_B \propto l_R^{-2} L_R \nu^{-3}$ where L_R is the total radio-luminosity and l_R is the linear size of the source. Thus for a given radiation field we have a lower bound on the linear size of the source, $l_R \geq l_{R,\text{min}} \propto B^{1/4}$. Now it might be thought that l_R can be made arbitrarily small simply by decreasing B. However, this is not the case because, if the source becomes too compact, the energy density of radio-radiation $L_R/4\pi l_R^2 c$ exceeds the magnetic energy density $B^2/8\pi$; the relativistic electrons lose far more energy by 'inverse Compton' scattering the radio-photons than by synchrotron 'scattering' the virtual quanta in the magnetic field; and the total power requirements of the source become prohibitive. (Each inverse Compton scattering of a photon boosts its energy by a factor $\sim \gamma^2$.) The rule of thumb, which is fairly good in most instances, is that the radio-brightness temperature is

Chapter 8. Black hole astrophysics

unlikely to exceed 10^{12} K, and that correspondingly $l_{R, min}$ is ~ 1 pc (Kellerman and Pauliny-Toth, 1968).

The size of a radio-source is then typically $\geqslant 10^4$ times larger than that of a putative central black hole. The energy required to accelerate the emitting electrons must be transported from the hole to the components in a comparatively useful 'low-entropy' form such as bulk kinetic energy. It is no good dissipating all the energy close to the horizon in the form of freely escaping photons or relativistic particles if you want to explain the radio-observations. This is an important constraint on black hole models of quasars.

8.6.3 Optical and X-ray emission

A typical quasar radiates a total continuum power $\sim 10^{45}-10^{48}$ erg s^{-1}. The optical spectral index, defined as $\alpha = -\mathrm{d}\log S_\nu/\mathrm{d}\log \nu$ where S_ν is the energy flux per unit frequency, lies in the range $0 \leqslant \alpha \leqslant 4$, and most of the power can appear at either infra-red ($\alpha < 1$) or ultraviolet ($\alpha > 1$) wavelengths. (The ultraviolet flux is probably responsible for photo-ionization of the emission-line regions from which we obtain the redshifts, and hence the distances to the quasars.) If Seyfert galaxies are a reliable guide, in many quasars the bulk of the emergent power may even be in hard X-rays. (X-ray telescopes are not yet sensitive enough to determine whether or not this is generally so.)

At frequencies above the radio-range, it is by no means clear what the principal emission process is, and in view of the extreme diversity of the observed spectra, it is not unreasonable to expect that several competitive mechanisms may be involved. The most likely processes are synchrotron radiation, inverse Compton scattering (mentioned above), non-relativistic Compton scattering and bremsstrahlung. In non-relativistic Compton scattering (e.g. Katz, 1976) the frequency of a photon can be shifted upwards by an average fractional amount $\Delta\nu/\nu \sim (kT_e/mc^2)$ in each scattering off an electron of temperature T_e (assuming the downward Compton shift to be ignorable, i.e. $kT_e \gg h\nu$). If the source is optically thick to Compton scattering, i.e. if the electron density n_e exceeds (Thomson cross-section)$^{-1} \times$ (source size)$^{-1}$, then an individual photon can be scattered several times in escaping from the source region, and can, under some circumstances, have its energy more than doubled. Bremsstrahlung radiation liberated by hot, free electrons colliding with ions (see Novikov and Thorne (1973) for a thorough discussion) will dominate when the gas density near the hole is high.

X-ray lines radiated by highly ionized iron ions will probably be associated with thermal bremsstrahlung. If such lines exist, they may be detected in the near future by X-ray satellites like HEAO B.

The optical continuum emission is sometimes polarized, in particular in those sources showing rapid variability. The degree of polarization can occasionally rise as high as 30 per cent. The direction of the polarization vector may vary even more rapidly with time than the total intensity. High linear polarization is characteristic of the synchrotron process, though amounts $\leqslant 10$ per cent can be produced by Compton scattering in an aspherical geometry.

In the majority of sources, then, the emission could be produced by either non-relativistic or relativistic plasma mechanisms. Even the fact that the spectrum can extend over several decades of frequency doesn't rule out fundamentally thermal processes, for if there is a range of temperature in the source, then any spectrum with $\alpha > 0$ can be generated thermally.

Many calculations of theoretical spectra, making specific geometrical and gas dynamical assumptions, have been made. A characteristic feature of synchrotron models is that the lifetime of the optically emitting electrons against radiation reaction is much less than the time it takes light to cross the source. This implies that these electrons must be accelerated throughout the volume of the emitting region, and so again some means of transporting power from a hypothetical central black hole to the acceleration region would be required. Similar remarks apply, although less stringently, to Compton models.

8.6.4 Accretion onto black holes

Many models for the optical and X-ray emission of quasars have been based on quasi-spherical or disk accretion onto black holes (section 8.4). An important parameter in these models is the ratio of the infall time to the cooling time, $\tau_{\text{if}}/\tau_{\text{c}}$. If $\tau_{\text{if}}/\tau_{\text{c}}$ is much greater than or much less than unity, then the efficiency of energy release $\epsilon = L/\dot{M}c^2$ will be low, and the gravitational energy of the accreted material will be swallowed by the hole in the form of kinetic or thermal energy. If $\tau_{\text{if}}/\tau_{\text{c}} \sim 1$, then ϵ can be large. For quasi-spherical accretion most of the infalling gas might be in the form of cold clouds with low angular momentum. If (ideally) these clouds collide very close to the hole, where their relative velocities approach c, then shock waves will be driven into the clouds producing efficient dissipation (Fabian *et al.*, 1976; Rees, 1977). (We know from

Chapter 8. Black hole astrophysics

observations of galactic supernova remnants that shock waves with speeds $\sim 10^{-3}$–$10^{-2} c$ are fairly efficient at accelerating relativistic electrons, and so ultimate radiative efficiencies $\epsilon \geq 0.1$ are entirely plausible if this type of collision can in fact occur.) As we have described above, with disk accretion ϵ can also be ≥ 0.1.

The instabilities that plague models of binary X-ray sources are present in disk models of quasars with a vengeance. The innermost regions of a disk surrounding a $\sim 10^8 M_\odot$ hole accreting at the Eddington limit must have thermal temperatures $\geq 10^6$ K. This means that the ratio of radiation to gas pressures (see section 8.4) is large and that line cooling (e.g. Callahan, 1977) may be very important. Both of these factors are strongly conducive to thermal instability. (In any case, simple disk models without instabilities cannot reproduce the full range of the observed spectrum.) A similar instability can occur if magnetic field is accreted and synchrotron radiation by relativistic electrons is important in the inner part of the disk (Pringle, Rees and Pacholczyk, 1973).

Standard, instability-free assumptions about the disk accretion process, which may have some validity in binary X-ray sources, are then probably not relevant to quasars.

Two variants on the standard model of disk accretion have been discussed in the quasar context. Firstly, the disk may be sufficiently massive that its self-gravitation cannot be ignored. (The condition for this is that the matter density in the disk at radius r exceed the 'Roche limit' $\sim M/r^3$, where M is the mass of the hole (Paczyński, 1978).) In this case angular momentum may be transported outwards by gravitational interactions similar to those operating in spiral galaxies, rather than by a local viscous stress. Secondly, Abramowicz, Jaroszyński and Sikoru (1978) and Kozlowski, Jaroszyński and Abramowicz (1978) have investigated the nature of the equipotential surfaces near the event horizon of a Kerr hole. They find that if viscous and radiative stresses are not too important, then accreting matter can fill up the space bounded by the zero equipotential surface, and that the excluded volume may define the start of a pair of channels along which the energy needed to form a double radio-source can be focussed.

In the standard accretion disk, energy is dissipated locally in the form of radiation which escapes freely from the surface of the disk. A rather more promising idea in the context of quasars is that much of the energy is liberated near the hole in a non-radiative form, and that most of the continuum radiation is generated at some distance from the hole where the outflowing energy flux interacts with surrounding material.

One specific realization of this (e.g. Icke 1977, 1978; Liang and Price, 1977; Piran, 1977; Bardeen, 1978) is that the liberated gravitational energy is deposited in a 'corona' above the disk. The energy can be carried off in the form of a radiatively or thermally driven wind, a scaled-up version of the solar wind which carries off much of the energy deposited in the solar corona. Similarity solutions have been discovered in which a small fraction of the matter accreting in a disk is 'accepted' by the hole and is able to liberate luminosity $\sim L_{\rm Edd}$. The remaining matter is driven off by radiation pressure. It appears possible in this to produce flows collimated parallel and antiparallel to the spin axis.

In an alternative scheme (Blandford and Znajek, 1977; Blandford in Hazard and Mitton, 1979, and references therein) the energy and angular momentum of accreting gas are extracted by electromagnetic torques acting close to the hole. This, in fact, can be done with fairly high efficiency even in an axisymmetric geometry. Consider a magnetic field embedded in the disk. To a first approximation, the field will be frozen into the disk's orbiting matter. (This is because the electrical conductivity will be huge, which implies the 'perfect MHD condition' $\boldsymbol{E} + \boldsymbol{v} \times \boldsymbol{B}/c = \boldsymbol{0}$. The curl of this equation implies $\partial \boldsymbol{B}/\partial t = \boldsymbol{\nabla} \times (\boldsymbol{v} \times \boldsymbol{B})$, which is directly interpretable as the freezing of the magnetic field into the matter.) Magnetic field lines reaching out of the disk and frozen into the disk's orbiting matter will generate an electric field, as seen by locally non-rotating (stationary) observers. This electric field produces an electric potential difference across the innermost parts of the disk and, indeed, across the hole, just like that in a Faraday disk. This potential difference will cause currents to flow along the magnetic field lines out of the disk, establishing a magnetosphere around the hole. Finally these currents will generate a toroidal component of magnetic field, so that the field lines will be swept backwards by the motion of the matter. There will therefore be a resistive torque acting on any material near the hole, and this can lead to the transport of angular momentum (and energy) not outwards in the plane of the disk (as in conventional viscous models) but perpendicular to the disk in the form of an electromagnetic or hydromagnetic Poynting flux.

The same mechanism can lead to the extraction of spin energy from the hole itself. For a Kerr hole with specific angular momentum a, a fraction $1 - \{[1 + (1 - a^2/M^2)^{1/2}]/2\}^{1/2}$ (which varies from 0 to 29 per cent as a increases from 0 to M) of the hole's mass can, in principle, be

Chapter 8. Black hole astrophysics

extracted (Christodolou, 1970). However, for this to happen in practice, currents need to flow freely across the horizon. As particles must travel inwards at the horizon and presumably move outwards at large distances, there must be some source of current-carrying charges in the inner magnetosphere. This has to be supplied by a breakdown of the vacuum above the horizon just like in a lightning stroke. It turns out that under the anticipated conditions within the nucleus of a quasar, simple mechanisms exist that are capable of achieving this breakdown. This then provides an alternative method for liberating a significant fraction of the rest-mass energy of accreted material. In practice, any accreting magnetized gas is likely to be unstable so that much of the energy would be liberated in explosive flares (Shields and Wheeler, 1976) rather than by the idealized time-steady processes described here.

The end result of either the coronal or the electromagnetic mechanism is likely to be a collimated wind, quite probably moving with a relativistic speed. This is unlikely to be able to make a double radio-source, however, without additional focussing mechanisms far from the hole.

If it turns out that most of the radiant energy observed is of secondary origin and not generated very close to the hole, then it becomes much more difficult to distinguish a black hole model from one based on a supermassive star, for example. Probably the best hopes for seeing deepest down into the energy-generation region lie with X-ray observations of fairly distant quasars, and optical observations of 'Lacertids' (a class of quasar-like objects characterized by the absence of emission lines, and the presence of high polarization, steep spectrum and extremely rapid variability).

Unless it is surrounded by a particularly massive disk, a black hole in a quasar must be fed gaseous fuel at a rate $\sim 1-100 M_\odot \, \text{yr}^{-1}$. Two specific supplies have been postulated. First, a dense star cluster around a massive black hole should develop an even denser cusp at its center, by the same mechanisms as discussed in section 8.5 for black holes in globular clusters. Stars within this cusp can be destroyed by collisions or tidal forces, and a substantial fraction of the gaseous debris should fall onto the hole (e.g. Frank and Rees, 1976; Lightman and Shapiro, 1977). Provided that the hole is massive enough ($M \geqslant 10^6 M_\odot$) it should be capable of attracting a dense enough stellar cusp to account for the fuel requirements of even the most luminous quasars (Young, 1977). One might have thought that plasma supplied in this way would be unable to define a preferred axis. On the contrary, Bardeen and

Petterson (1975) have shown that, if the hole spins rapidly, then within ~ 100 Schwarzschild radii accreted plasma will settle into the hole's equatorial plane. It is interesting that a hole of mass $M_h \geqslant 10^8 M_\odot$ exerts a tidal acceleration, $\sim c^6 R_\odot (GM_h)^{-2}$, on solar-type stars that is *smaller* than their self gravity, $\sim GM_\odot R_\odot^{-2}$. This means that stars will not be tidally disrupted before crossing the horizon, and this may provide a means of switching off quasars when their masses grow larger than $\sim 10^8 M_\odot$ (Hills, 1975).

The second postulated source of material for black hole accretion is ambient gas in the neighborhood of the hole. Such gas might be generated by stellar processes (supernovae, planetary nebulae, etc.) in a hypothesized surrounding galaxy (e.g. Gisler, 1976), or might be torn out of a passing galaxy (e.g. Gunn in Hazard and Mitton, 1979).

The formation of the black hole in the first place provides ample opportunity for unconstrained theoretical speculation. As we discuss in section 8.7.1, the hole may be primordial and indeed may have something to do with seeding the condensation of a surrounding galaxy. However, it can also be a natural evolutionary product of a massive nuclear star cluster, a scaled-up version of the globular-cluster scenario described in section 8.5. As with globular clusters, it is only possible for the hole to swallow an appreciable fraction of the mass in the original cluster if there is an effective dissipation mechanism (e.g. stellar collision) that can increase the binding energy of the stars. This is discussed at greater length by Rees in Hazard and Mitton (1979).

8.6.5 Observing black holes in quasars

To conclude section 8.6, we remark that the black hole model of quasars does not seem to give an easily recognizable signature; competitive models frequently do. For instance, the detection of significant displacement of the radio-position from one outburst to the next would be most consistent with a star-cluster model; and the demonstration of a convincing periodicity in a quasar light curve (Ozernoy and Chertoprud, 1969) would be strongly suggestive of an uncollapsed supermassive star. As with binary X-ray sources, probably the best that can be hoped for in the near future is a chain of argument based on both theory and observation that leaves a black hole as the only plausible candidate for the central object.

Examples of such arguments have been given on radiative grounds by Lynden-Bell and Rees (1971) for the compact radio-source which

Chapter 8. Black hole astrophysics

resides at the precise center of our own galaxy. More recently Kellerman *et al.* (1977) find that a quarter of the compact radio-emission from the Galactic center comes from a region $\sim 10^{14}$ cm across which is only 100 Schwarzschild radii for a $3 \times 10^6 M_\odot$ hole. (From observations of a narrow infrared line in a region of linear size ≤ 0.1 pc, Wollman *et al.* (1977) were able to argue that the maximum mass of any central hole was $\leq 4 \times 10^6 M_\odot$.) As argued by Fabian *et al.* (1976), relatively unspectacular objects like Centaurus A might also contain a central black hole. In these cases the observed power would be limited by the gas supply, which would be much less than the Eddington limit.

Finally, Sargent *et al.* (1978) and Young *et al.* (1978) have examined the nucleus of the elliptical galaxy M87. From photometric and spectroscopic measurements, they find that $\sim 5 \times 10^9 M_\odot$ must be contained within the central ~ 100 pc, probably in a non-stellar form. If this is identified with a black hole, then it is interesting that the upper limit on the radius of the smallest component of the central radio-source is only ~ 50 Schwarzschild radii (Kellerman *et al.*, 1973). Perhaps we are almost resolving a quasar black hole after all.

8.7 Primordial black holes

8.7.1 Cosmological production of primordial holes

As well as being formed in the course of natural stellar or galactic evolution, black holes may also have been produced primordially, that is to say, at the very earliest epochs of cosmological time. In particular, if some portion of the early universe were sufficiently inhomogeneous, then masses much less than the maximum neutron star mass $M_{\max}^{n*}$ could collapse to form a black hole. An extreme form of inhomogeneous cosmology, studied by several authors (e.g. Misner, 1968; Rees, 1972; Barrow, 1977) has been termed 'chaotic'. There is a good reason why the universe might not have been isotropic and homogeneous back to the earliest times. This is that the mass within the particle horizon, essentially a measure of the amount of matter in causal contact, decreases towards zero as we approach the initial singularity. However, the present universe appears both isotropic and, on large enough scales, homogeneous, especially from observations of the isotropy of the microwave background. A tough problem for chaotic cosmologies, one which is still not satisfactorily resolved, is then to posit an efficient mechanism for smoothing things out at some intermediate epoch.

Primordial black holes

For a mass M to collapse and form a black hole in the early, high-temperature universe, its density ρ must be roughly twice that of its surroundings $\bar{\rho}$ when the universe has expanded enough that the particle horizon (length $\sim ct$ where $t=$ (age of universe)) is the size of the mass (length $\sim(M/\bar{\rho})^{1/3}$). Using the cosmological relation between density and age, $\bar{\rho}\sim(Gt^2)^{-1}$, we obtain the relation (Hawking, 1971) between the hole's mass M and the density $\bar{\rho}$ and time t at which it forms

$$M \sim (c^6/\bar{\rho}G^3)^{1/2} \sim 10(\bar{\rho}/10^{15}\text{ g cm}^{-3})^{-1/2}M_\odot \sim 10(t/10^{-4}\text{ s})M_\odot.$$
(8.33)

At times $t \leq 10^{-4}$ s when the temperature T exceeds $\sim 10^{12}$ K, cosmological ideas become highly speculative, mainly because of our uncertainty about the correct equation of state which can range from the extremely soft (e.g. Hagedorn 1965; Frautschi, 1971) to the maximally stiff (Zel'dovich 1962; Lin, Carr and Fall, 1976). Even if we were confident about the equation of state, our ignorance of the spectrum of fluctuations prevents us from making plausible *a priori* guesses as to the nature of the mass spectrum of primordial black holes (PBH henceforth). This is discussed further in Carr (1975). Some authors (Mészáros, 1974; Lin et al., 1976; Carr, 1977a) have argued that the minimum PBH mass is $\sim 1 M_\odot$. This would exclude the possibility of an observable density of holes radiating by the Hawking process (see below).

One initial worry about PBHs (Zel'dovich and Novikov, 1967) was that they might grow rapaciously during the radiation-dominated era of the universe, so as to produce holes of mass $\sim 10^{16} M_\odot$ by the time when electrons and protons recombined to form hydrogen at a temperature $T \sim 4000$ K. Carr and Hawking (1974) found, however, that a hole is only likely to double its mass after the time when it comes within its particle horizon, and so a wide spectrum of holes of mass $\ll 10^{16} M_\odot$ can survive the expansion of the universe without substantial growth. Curiously, with a maximally stiff equation of state (Zel'dovich, 1962), in which pressure equals mass–energy density and which may arise at times $t \ll 10^{-4}$ s, this rapid growth can occur and the sizes of the holes will keep pace with the horizon for as long as the stiff equation is valid (Lin et al., 1976).

Significant PBH formation in a hot universe is subject to one severe drawback that makes the idea seem somewhat unlikely (Novikov and Thorne, 1973). We can characterize the present PBH density by the parameter Ω_{PBH} which is the ratio of their mean mass density to the critical density $\rho_c = 3H_0^2/8\pi G$ (with H_0 the Hubble constant) that is

Chapter 8. Black hole astrophysics

necessary to close the universe. Now photons (both in the cosmic microwave background and in the spectral lines of distant quasars) are redshifted by the expansion of the universe so that their energies, at an epoch when the temperature of the background radiation was T, satisfy

$$h\nu = h\nu_0(T/T_0) = h\nu_0(1+z)^{-1}, \qquad (8.34)$$

where ν_0, T_0 are the present frequency and temperature of the background and z is the redshift. By contrast the energy in a black hole is not diminished by the expansion. This means that the density parameter at the formation redshift z_f must satisfy

$$\Omega_{\text{PBH}}(z_f) \leq 10^4 \Omega_{\text{PBH}}(0)(1+z_f)^{-1}, \qquad (8.35)$$

where we have used the fact that at present the microwave background accounts for $\sim 10^{-4}$ of the critical density. Since a typical PBH-formation redshift is $z_f \geq 10^{10}$, and since we know observationally that $\Omega_{\text{PBH}}(0) \leq 1$, we must conclude that $\Omega_{\text{PBH}}(z_f) \leq 10^{-6}$. There seems to be no particularly good reason why holes should form with a density $\Omega_{\text{PBH}}(z_f) \sim 10^{-8}$–$10^{-6}$ so as to be observable now. Indeed if primordial holes formed at all, it would seem more probable that they would form with $\Omega_{\text{PBH}}(z_f) \sim 1$ and thus violate present observational constraints by many orders of magnitude.

There is one way out of this and related difficulties, and that is to postulate a cold chaotic universe in which the background radiation is created at fairly recent times, i.e. at $z \ll 10^{10}$. Possible mechanisms for generating the necessary entropy (i.e., photons) 'recently' include shock waves (Rees, 1972), accretion by holes with $M \geq 1 M_\odot$ (Carr, 1977a; Barrow 1977) and evaporation of holes with $M \leq 10^{15}$ g (Hawking, 1974; Carr, 1976). Cosmologies of this type encounter problems when they attempt to account not only for the present entropy per baryon (Carr and Rees, 1977), but also for the thermalization of the microwave background (Zel'dovich and Starobinskii, 1976) and the production of helium and deuterium (Vainer and Nasel'skii, 1977; Zel'dovich et al., 1977); but such cosmologies cannot easily be ruled out.

Statistical fluctuations in the density of PBHs are not coupled to the surrounding radiation, and so may be able to grow on a cosmological timescale and thus 'seed' the formation of galaxies and clusters of galaxies. Mészáros (1974, 1975a) has argued that '$\sqrt{N}$' fluctuations of $1 M_\odot$ PBHs are adequate (but see Carr, 1977b), and Ryan (1972) and Barrow (1977) have considered the effects of larger masses. In view of the well-known difficulties associated with explanations of the formation of galaxies in the standard cosmology, the viability of PBH 'seeds' deserves further study.

8.7.2 Observability of primordial holes

As must be clear from the foregoing discussion, theoretical arguments in favor of a large number of primordial black holes are extremely speculative and far from compelling. What are the prospects for future detection, and what limits can be set from existing observations?

In view of the profound implications that their discovery would have for theoretical physics, it is not surprising that in the past few years attention has been focussed on the properties of holes of mass $M \sim 10^{15}$ g, which should evaporate by the Hawking (1974, 1975) process in a time comparable with the present age of the universe. Page (1976a, b, 1977) has computed the particle emission rates by the Hawking process, including the effects of the holes' rotation and charge fluctuations. He finds a value 5×10^{14} g $\leq M_H \leq 7 \times 10^{14}$ g for the 'Hawking mass' (i.e. the mass of a hole that evaporates in precisely the present age of the universe), with the precise value of M_H depending on the spin of the hole. He also finds that a significant fraction (≥ 0.2) of the evaporated luminosity should emerge in the form of γ-rays peaking at an energy ~ 100 MeV. This γ-radiation might be detectable in three ways:

(i) The integrated background from PBHs out to a redshift ~ 1 may be detectable. This can furnish an upper limit on the PBH density which Page and Hawking (1976) (cf. Chapline, 1975) computed to be $\bar{n}_H \leq 10^4$ pc^{-3} for holes of mass $\sim M_H$. This corresponds to $\Omega_{PBH}(M_H) \leq 10^{-8}$. A comparison between the predicted and the observed spectrum, in principle, could have provided a positive detection.

(ii) One might have hoped to detect γ-rays from individual, nearby PBHs. If holes are clustered around galaxies, then the local density n_H can exceed $\bar{n}_H$ by a clumping factor $C \leq$ (size of the universe)/(size of our galaxy) $\sim 10^6$. From this latter limit we can infer that one should not expect to see an *individual source* closer than

$$10^{15}(\bar{n}_H/10^4)^{-1/3}(C/10^6)^{-1/3} \text{ cm} \tag{8.36}$$

(i.e. outside the solar system). The γ-ray flux from a source at 10^{15} cm would be $\leq 10^{-9}$ photons cm^{-2} s^{-1}, which is totally unobservable.

We can put an upper limit on the permitted local density of heavier holes by requiring that $\Omega_{PBH} \leq 1$. This becomes

$$n(M) \leq 10^{11} C(M/10^{15} \text{ g})^{-1} \text{ pc}^{-3}. \tag{8.37}$$

Chapter 8. Black hole astrophysics

Thus the maximum allowed space density of PBHs is $n(M) \leq 10^9 C$ pc^{-3} for $M \sim 10^{17}$ g. (As the random velocity of holes is unlikely to exceed 100 km s^{-1}, the chances of one hitting the Earth (see Jackson and Ryan, 1973) this century are utterly negligible!)

(iii) It is with the third observational possibility, *black hole explosion*, that prospects of γ-ray detection seem most promising. As the hole's mass falls below M_H, the power radiated increases at least as fast as $(-t)^{-2/3}$, where t is time measured from the final disappearance of the hole. As a result, the last vestiges of the hole's rest mass are radiated explosively. A given type of particle starts to be radiated when the Schwarzschild radius of the hole shrinks to the particle's Compton wavelength. If, as originally proposed by Hagedorn (1965), the number of particle species with masses above the pion mass grows exponentially with mass, then $\sim 10^{34}$ erg of energy will be liberated in a time $\geq 10^{-7}$ s, mainly in the form of 250 MeV γ-rays (Carter *et al.*, 1976). On the other hand, for the harder equations of state consistent with the simplest quark models (to be more precise, for those models of high temperature matter which have a sound speed $\geq c/\sqrt{5}$ (see Carter *et al.*, 1976)) the energy release will be far from gradual:

$$E(>t) \sim 10^{29}(-t/1 \text{ s})^{1/3} \text{ erg}, \tag{8.38}$$

where $E(>t)$ is the energy released after time t. In this case, particle interactions can be ignored and explosion products should be observed with energies right up to the Planck mass $\sim (\hbar c/G)^{1/2} \sim 10^{28}$ eV. Independent of the high-temperature behavior of matter, the nearest explosion occurring in a month's observing time would be at a distance

$$\sim 2(\bar{n}_H/10^4 \text{ pc}^{-3})^{-1/3}(C/10^6)^{-1/3} \text{ pc} \tag{8.39}$$

from Earth. Even for the best case of a Hagedorn-like explosion and maximal PBH density, a flux of only 10^{-1}–10^{-2} γ-ray photons per square centimeter would be expected, requiring a detector with good time resolution and with an area ≥ 100 cm^2 (Carter *et al.*, 1976; Page and Hawking, 1976). Thus, although the observation of the explosion products from a high-temperature hole would, in principle, constitute a unique experiment in ultra-high-energy physics, the practical difficulties involved seem formidable unless PBHs of $\sim 10^{15}$ g have the maximum permitted density.

A more promising idea, due to Rees (1977a), relies on the fact that an exploding hole may also produce a significant yield of electron–positron pairs. These particles will soon be stopped by the interstellar magnetic field, but in so doing the field will be swept up and compressed to form a coherent low-frequency pulse of electromagnetic waves in a manner akin to that originally proposed by Colgate and Noerdlinger (1971) for supernovae.

If, in a best-case analysis, of order of half the hole's energy emerges as pairs with Lorentz factor γ, corresponding to an explosion temperature $\sim \gamma$ MeV, then $\sim 10^{43} \gamma^{-2}$ pairs should be produced. Provided that the duration of the explosion Δt is sufficiently short, then the pairs should form a thin, electrically conducting shell expanding radially into the surrounding field. As long as $\gamma \gg 1$, to an observer moving with the shell the magnetostatic field would look like electromagnetic radiation, and the virtual quanta in this field would be reflected by the shell. Transforming back into the frame of the explosion, we find that the energy flux of the 'scattered' radiation is $\sim \gamma^2$ times the incident magnetostatic energy flux (one factor of γ for each Lorentz transformation).

Thus the electrons and positrons will be decelerated when the energy of the explosion $E \sim 10^{37} \gamma^{-1}$ erg equals the radiated energy, i.e. when the shell has expanded out to a radius R satisfying

$$E \sim \gamma^2 \frac{B^2}{8\pi} \times \frac{4\pi}{3} R^3 \qquad (8.40)$$

where B is the ambient magnetic field. Now the incident virtual photons will have a wavelength $\sim R$ and so, Doppler shifting twice, the observed wavelength will be

$$\lambda \sim \gamma^{-2} R \sim (6E/\gamma^8 B^2)^{1/3}. \qquad (8.41)$$

Several physical criteria must be satisfied before an interaction like this can occur: the explosion must be rapid enough (i.e. $\Delta t \ll R/c\gamma^2$), the shell must have sufficient surface conductivity, the pairs must be able to avoid annihilation in the expanding fireball, etc. It turns out (Blandford, 1977; Rees, 1977a) that detectability is probably optimized if the hole explodes when its mass is $\sim 10^{11}$ g, producing $\sim 10^{32}$ erg of 100 GeV pairs. Such a hole, embedded in the interstellar magnetic field ($\sim 3 \mu$G) at the distance (10 kpc) of the Galactic center, should produce a linearly polarized radio-pulse of energy flux $\sim 10^{-23}$ erg cm^{-2} Hz^{-1} at a frequency $\lesssim 1$ GHz. The sensitivity required to detect this can be achieved with a fairly modest search effort. The existing upper limit on the explosion rate, assuming that most of the energy is produced in the 100 MHz – 1 GHz range, is $\sim 10^{-5}$ pc^{-3} yr^{-1} (Meikle, 1977). In view of the much greater energy sensitivity of a

Chapter 8. Black hole astrophysics

radio-telescope compared with a γ-ray detector, it is not surprising that this limit is $\sim 10^5$ better than the existing γ-ray limit. In principle, a dedicated search with a multiple-phased array using de-dispersing techniques could improve this limit to $\sim 10^{-12}\,\mathrm{pc}^{-3}\,\mathrm{yr}^{-1}$ (Meikle, 1977) corresponding to $\Omega_{\mathrm{PBH}}(M_{\mathrm{H}}) \sim 10^{-14}$. If the hole explodes at a temperature $\sim 10\,\mathrm{GeV}$, coherent optical emission can be produced. Slightly inferior upper limits occur in this case (Jelley, Baird and O'Morgain, 1977). However, in view of the uncertainties in the physics of the explosion and interaction with the surrounding medium, it must be re-emphasized that these limits correspond to detectability on the most optimistic of assumptions rather than firm upper limits on $\Omega_{\mathrm{PBH}}(M_{\mathrm{H}})$.

8.7.3 Limits on the density of primordial holes

It is even more difficult to set stringent limits on Ω_{PBH} for more massive holes than for holes near M_{H}. In fact it has been proposed by several authors that the 'missing mass' required to stabilize galaxies, bind clusters of galaxies, and indeed close the universe for those who deem this desirable, may take this form. (Conversely, if the arguments of, e.g. Gott *et al.* (1974) for a low-density universe are accepted, and the holes are clustered like galaxies, then obviously $\Omega_{\mathrm{PBH}} < \Omega_{\mathrm{total}} \sim 0.04$.)

One limit that does apply to holes (and, in fact, to any sufficiently compact object) in the mass range $10^4 M_\odot \leq M \leq 10^{15} M_\odot$ has been described by Press and Gunn (1973). This relies on the fact that a distant optical or radio-source can be focussed by the gravitational lens effect to form two similar images, if there is an intervening PBH (e.g. Refsdal, 1964; and Barnothy, 1965). If we consider rays from a source at a redshift $z \sim 1$, i.e. at a distance $\sim c/H_0$, then the distance of closest approach of rays passing on either side of the hole is $\sim (GM/c^2 \delta\theta) \sim (c\delta\theta/H_0) \sim (GM/cH_0)^{1/2}$, where $\delta\theta$ is the angle through which the ray is bent. In order to produce two images of comparable brightness, the two rays must be bent through similar angles $\delta\theta$, which means that the source must lie beyond the PBH, in a volume $\sim (c/H_0)(GM/cH_0)$. Hence the probability of observing a double image, in say a distant quasar, is roughly Ω_{PBH}. The expected angular separation is $\delta\theta \sim 10^{-6}(M/1M_\odot)^{1/2}$ seconds of arc, and provides a measure of the mass of the hole. The smallest measurable angular size using intercontinental radio-interferometry is $\sim 10^{-4}$ seconds of arc, which corresponds to $M \sim 10^4 M_\odot$. In fact double structure on the scale of 10^{-3} seconds of arc has been observed in several distant radio-sources, but this is believed to be unrelated to the gravitational lens effect (Blandford, McKee and Rees, 1977). Optical observations on seconds-of-arc angular scales indicate an upper limit $\Omega_{PBH}(10^{12}-10^{15}M_\odot) \leqslant 0.2$.

Holes with $M \geqslant 10^{15}M_\odot$ can be limited on dynamical grounds to $\Omega_{PBH} \leqslant (M/10^{16}M_\odot)^{-2}$ because of the absence of an observable gravitational effect on our galaxy (Press and Gunn, 1973). Similar limits can be placed on holes in galactic halos and in clusters of galaxies (van den Bergh, 1969; Rees, 1977b).

If we are prepared to make some assumptions about the nature of the intergalactic medium and the emissivity of plasma near the hole, as discussed in section 8.3, then some fairly interesting bounds can be placed on Ω_{PBH} for large masses (Dahlbacka, Chapline and Weaver, 1974; Rees, 1977b). For example, using (8.14) with $\xi = 10^{-11}$ and $\epsilon = 10^{-1}$, one obtains a predicted energy density in radiation from such holes

$$\sim \rho_c \frac{\Omega_{PBH}}{MH_0} L \sim 10^{-25} \Omega_{PBH} \frac{M}{M_\odot} \text{ erg cm}^{-3}. \qquad (8.42)$$

For a density $\Omega_{PBH} \sim 1$ in holes of mass $\geqslant 10^{12}M_\odot$ this background would probably be detectable in whichever waveband it appeared. A similar limit on the local density of PBHs with $M \sim 10^5 M_\odot$ has been suggested by Ipser and Price (1977); $10^5 M_\odot$ is an estimate of the mass of a cloud that can collapse at the epoch of recombination, and is therefore another possible characteristic PBH mass (Peebles and Dicke, 1968).

It is of course possible to determine more intricate limits on Ω_{PBH}. For instance, PBHs, like stars, may form gravitationally bound systems (binaries or clusters) which evolve by radiating gravitational waves. If a significant fraction of all PBHs are born in such systems, and if each system evolves by contraction and coalescence to a single hole of mass $\sim M^*$ at some redshift $\sim z^*$, there should result an isotropic gravitational-wave background of energy density $\sim 0.1\Omega_{PBH}\rho_c c^2/(1+z^*)$ centered around a frequency $\sim c^3/[6\pi GM^*(1+z^*)]$. If $\Omega_{PBH}(1+z^*) \gg 0.05\Omega_{total}^2$, then a continuous background rather than a series of outbursts will be seen. Near-future gravitational-wave experiments can put interesting bounds on such scenarios. Of course, if a comparable power is radiated into the electromagnetic background by these events (Eichler and Solinger, 1976), then this will be detected far more easily than a gravitational-wave background, provided that it has not been redshifted to frequencies $\leqslant 100$ MHz.

Chapter 8. Black hole astrophysics

In summary, although the notion of primordial black holes has generated much theoretical speculation and may well turn out to have stimulated the crucial unifying link between quantum mechanics and gravitation, and although PBHs have the potential for explaining many of the mysteries of the early universe, there is as yet no observational evidence for their existence. The probability of detecting them, either directly or indirectly through their radiative effects, does not seem high at present.

8.8 Concluding remarks

As must now be clear, in spite of many notable theoretical advances over the past 10 years and an explosion of new observational data throughout the whole electromagnetic spectrum, astrophysics has still not advanced to the stage at which we can be strongly confident that black holes exist at all. The best scientific case can be made for the X-ray source Cygnus X-1, and here the arguments, which are fairly compelling, have not changed substantially in five years. In the case of quasars, there is now more evidence for the occurrence of extremely compact structure, but it is still mainly theoretical considerations that lead one to the model described in section 8.6 with $\sim 1 M_\odot$ of gas accreting onto a $\sim 10^8 M_\odot$ black hole each year. The more optimistic hopes of five years ago for the future of black hole astrophysics have frankly not yet been fulfilled.

Nevertheless, the absence of direct observational confirmation has not dampened the enthusiasm of astronomers for speculating in what must surely be one of the most fascinating areas of modern theoretical physics. In addition, although the theoretical arguments for the existence of a significant density of primordial holes are somewhat weak, there is no serious alternative to the ultimate gravitational collapse of single stars of mass $M \geqslant 2.5 M_\odot$ and the collapse of sufficiently dense clusters of gas and stars; and the widespread theoretical prejudice that black holes do exist still seems justified.

What then are the prospects for improving the observational basis of general relativity using black holes in the next few years? The best hope probably lies with the X-ray detectors aboard HEAO B. These instruments have much greater sensitivity and time resolution than their predecessors, and might reveal the rapid, low-Q pulsations that are probably the best signature of black holes in quasars and binary X-ray sources. The chances of seeing a black hole either form (e.g. through a

Concluding remarks

burst of gravitational radiation) or explode (e.g. through a radio-pulse) are far from remote, but here again we must expect that any claimed detection will be fairly controversial.

This raises an interesting question. Are physical conditions near a black hole inevitably so messy that general relativistic calculations have no predictive power over and above Newtonian analyses terminated at, say, $r = 2M$? Fortunately (for relativists intent on making elaborate computations) there are already some examples of effects that are absent from Newtonian analyses, and whose influence may be crucial in interpreting future observations. Two examples are the Bardeen–Petterson (1975) process for driving an accretion disk into the equatorial plane of a rotating hole and the gravitational waves emitted in violent events involving black holes. Nevertheless, as we tried to emphasize in section 8.2, for most electromagnetic radiative effects, the inherent uncertainties in the microphysics are so extreme that we doubt that sophisticated relativistic calculations will soon play a crucial role in theoretical models of black holes and their environments. We hope that we are wrong.

9. The big bang cosmology – enigmas and nostrums†

R. H. DICKE AND P. J. E. PEEBLES

9.1 Introduction

'The time has come,' the Walrus said, 'To talk of many things: Of shoes – and ships – and sealing wax – Of cabbages – and kings – And why the sea is boiling hot – And whether pigs have wings.'   Lewis Carroll

The big bang cosmology that developed out of Einstein's ideas and Hubble's observations has stood the test of time and observation, but even the staunchest advocate would admit that it is at best only a reasonable first approximation that certainly does not tell the whole story. There are in particular some curious and enigmatic features of this cosmology that lead us to think that an important piece of the picture may be missing. It is useful to review and reconsider these curiosities from time to time because they certainly have something to teach us. But what is it?

Most of the conundrums and nostrums discussed here trace back, in one form or another, to the lively discussions of the early 1930s, before physical cosmology became encrusted with revealed truth. We have given references to original discussions from the 1930s where we know them. In many other cases we are reviewing and interpreting lore revealed to us when we came into the field.

9.2 Enigmas

The minister gave out his text and droned along monotonously through an argument that was so prosy that many a head by and by began to nod – and yet it was an argument that dealt in limitless fire and brimstone and thinned the predestined elect down to a company so small as to be hardly worth the saving.   Mark Twain, *The Adventures of Tom Sawyer*, Ch. 5

Einstein (1917) hit on the idea of a homogeneous closed world model as a way to satisfy Mach's principle, that the inertial properties of matter are determined by the presence of all the matter in the universe. He ruled out

† This research was supported in part by the National Science Foundation.

a bounded 'island universe' of matter in asymptotically flat space because a particle escaping from the island would move arbitrarily far from all other matter but yet retain all its standard inertial properties. He at first tried choosing boundary conditions so that the line element goes singular outside the assumed 'island' of matter, but then hit on a much more elegant solution: that a closed homogeneous world eliminates both boundaries and asymptotically flat space.

It is not clear how familiar Einstein was with the observational situation in astronomy, or how much attention he paid to it. There was at the time speculation that the spiral nebulae are island universes like the Milky Way galaxy, but there were also some good arguments that these objects must be only minor satellites. It was considered well established, from star counts, that the Milky Way star system is finite and bounded, shaped roughly like a flattened spheroid. The spiral nebulae seemed to be concentrated at the poles of this star system, which would suggest they are related to it. Also, by 1916 van Maanen had found the first tentative evidence of proper motions in some of the larger spirals (van Maanen, 1916). If valid, and if the internal velocities in these systems are less than the velocity of light, it would make them quite close and much smaller than the Milky Way.

Over the next two decades it became clear that these indications are misleading, the former because interstellar dust in the plane of the Milky Way obscures the galaxies and the latter because of observational problems. Einstein's vision was remarkably good. In 1924 Hubble showed, by the identification of variable stars of known intrinsic luminosity, that the spiral nebulae are well outside the Milky Way and at least comparable to it in size (Hubble, 1924). Hubble's surveys of the galaxy distribution, begun in 1926 (Hubble, 1926) and continuing through the 1930s, gave the first direct evidence of large-scale homogeneity and isotropy. This has been confirmed by recent observations of the precise isotropy of the radiation background – X-ray, microwave, and longer wavelength radio – which show that the matter distribution and motion integrated to the horizon is very close to isotropic. Of course the universe could be inhomogeneous yet isotropic about one point, but the observed universe has billions of galaxies, many apparently equally good homes, and the isotropy would be observed only from a special few, which seems unreasonable.

The concept of large-scale homogeneity has been with us so long that cosmologists tend to take it as a commonplace, but it is remarkable simply because it stands in such contrast to our experience that things

Chapter 9. The big bang cosmology – enigmas and nostrums

have structure – from the properties of subatomic particles on up to the organization of galaxies in great clouds. Milne (1935) was the first to recognize that homogeneity is a powerful principle in cosmology, and that without it, or some other simplifying principle, a theory of the universe would be intractable. But is it reasonable to believe that the universe is endowed with this simple structure only to make calculations easy?

The distant galaxies observed in well-separated parts of the sky are so far apart from each other that there is not time enough since the big bang for a signal to have traveled from one to the other. Observers on Earth can see and compare them, being about half-way in between, and in line with homogeneity it is found that the galaxies are quite similar. By comparing radiation background intensities across the sky it is also found that the temperature and expansion rate are precisely synchronized across the visible universe. Even though the separate parts of the visible universe are not visible to each other they are evolving in very precise unison.

Are structural relations between widely separated parts of the universe a problem? In the past these parts were much closer together. But close proximity in earlier times does not eliminate the problem. Assuming that causal relations require the transport of information from one place to another at a velocity not exceeding that of light, the zone of influence around any given object is not only smaller in the past (because the universe is younger) – it contains less matter.

The relationships of widely separated parts of the universe are not the only problem. There is a remarkable balance of mass density and expansion rate. In general relativity theory with $\Lambda = 0$ the two are related by the equation

$$H^2 = \left(\frac{1}{a}\frac{da}{dt}\right)^2 = \tfrac{8}{3}\pi G\rho(t) - \frac{c^2}{R^2 a^2}, \qquad (9.1)$$

where $a(t)$ is the expansion parameter, R is a constant, and $|R|a(t)$ is the magnitude of the space curvature (measured in a hypersurface of roughly constant galaxy proper number density, at fixed cosmic time t). The present relative value of the two terms on the right side of this equation is poorly known, because the mean mass density, ρ, is so uncertain, but it is unlikely that the first term is less than 3 per cent of the magnitude of the second. Since ρ varies as a^{-3} (or more rapidly if pressure is important) the mass term on the right-hand side dominates the curvature term when a is less than about 3 per cent of its present value. Tracing the expansion back in time, one finds that at $t \sim 1$ s, when much of the helium is thought to have been produced, the mass term is some 14 orders of magnitude larger than the curvature term. This means the expansion rate has been tuned to agree with the mass density to an accuracy better than 1 part in 10^{14}. In the limit $t \to 0$ this balance between the effective kinetic energy of expansion, measured by H^2, and the gravitational potential energy, measured by $\frac{8}{3}\pi G\rho$, is arbitrarily accurate.

As was first pointed out by Lemaître (1933a), (9.1) gives a reasonable approximation to the evolution of separate parts of an inhomogeneous universe, so this precise initial balance of density and expansion rate (with a well-synchronized start) must apply to each separate part. Otherwise the universe would run amuck. The inhomogeneities would produce large and irregular space curvature, leading to black holes of all sizes.† It has been cogently argued that small 'primeval' black holes could be beneficial in providing mass that is very difficult to see but could hold clusters of galaxies together. If so, one might want to relax the balance condition on small mass scales. But still it requires very careful regulation to ensure that we do not see such things as wholesale collapse of the part of the universe appearing in the Southern hemisphere, and general expansion in the other half.

The global H and ρ say that the present value of $c/a|R|$ is less than about 10^{-17} s^{-1}. It seems curious that such a small quantity should have been built into the universe at the big bang, and so it has often been suggested that the only 'reasonable' value is $R^{-2} = 0$. This was more or less the position of Einstein and de Sitter (1932). By then it was recognized that Λ and R^{-2} in (9.1) both could be negative as well as positive, and that by appropriate choices of these two parameters one could arrive at quite a variety of model universes. Einstein and de Sitter suggested that Λ be dropped, and that, until the observational measures of curvature improve, it is reasonable to concentrate on the simplest case, $R^{-2} = 0$.

We consider this argument from simplicity attractive but perhaps somewhat weakened by the fact that when (9.1) is applied to sections of the universe R^{-2} does not vanish. Matter does cluster (at least, that part

---

† More accurately, one must assume the pressure varies on scales large compared to the Jeans length $\leq c(G\rho)^{-1/2}$, but at high density ρ the Jeans length is very small so this is not a serious constraint. Also, if shear and vorticity are important (9.1) must be replaced with the Raychaudhuri equation, but that still requires a careful balance of gravity and motion if the universe is not to evolve into utter disorder.

Chapter 9. The big bang cosmology – enigmas and nostrums

in galaxies) on scales less than approximately two orders of magnitude smaller than the horizon. It has been hard to see how this clustering could have been produced from an initially strictly homogeneous distribution by non-gravitational forces. Gravitational instability could do it, but then one finds that the present clustering traces back to irregularities in the very early universe at least strong enough to produce roughly constant local fluctuations in space curvature (where the typical curvature R associated with irregularities on comoving scale x satisfies $|R| \ll x$, to avoid tying space in knots). So to account for the observed large-scale clumping of matter one must suppose that in the limit $t \to 0$ the local balance of expansion and gravity was extremely accurate but not exact!

A possible further complication is the origin of large-scale cosmic magnetic fields (in galaxies and radio-sources). It is often speculated that such fields were present at the big bang (though it is also possible, as argued by Parker (1975), that a dynamo can and must replenish the field in our galaxy). There is an interesting conceptual problem with this. A magnetic field, frozen in the ionized gas and presently disordered on a scale of a galaxy, would in the past have been more compact, its size scaling as the radius of the universe. But the information horizon, the limit of the observable universe, expands faster than the universe itself. At the time of helium formation the causally connected pieces of the universe are so small that the visible magnetic field would be close to uniform. Thus the field is then ordered on a scale that greatly exceeds the maximum distance for a causal relation.

Equally enigmatic is the role of antimatter. There must be very little of it in our galaxy, because there is no trace in the low-energy cosmic rays, but it is quite possible that the nearest large galaxy, the Andromeda Nebula, is pure antimatter. It has been proposed by Alfvén (1971) and Omnès (1969) that in the early universe matter and antimatter were well mixed, and later separated by non-gravitational processes. However to arrange for separation on the scale of a galaxy in the big bang model is a formidable and perhaps impossible problem. Apparently then, at the big bang either there were pockets of baryon number excess and deficit, or else the universe preferred matter.

Much of the problem in these discussions centers on the two assumptions that there was a singular big bang and that one can derive some detailed properties of it from observations of the present state of the universe. It is easy to find examples where the latter, a fairly detailed traceback, is not possible. Given a well-mixed cup of coffee there is no way one can determine the order in which the coffee, milk and sugar were added. On a more 'cosmic' level, the structure of a star is independent of many details of the initial conditions, and that is why it is possible to study stellar evolution even though there are great uncertainties in the theories of stellar formation. Could it be, as Misner (1967) has pointed out, that the details of conditions at the big bang have been similarly erased by thorough mixing? We suspect not, because (9.1) says to us that gravity plays a peculiarly dominant role here, that if the universe were chaotic it would tie itself up into knots.

The other point concerns the singularity at the big bang. The pioneering work of Penrose (1965) and Hawking (1966) has given compelling evidence that in Einstein's field equations there is at least one singularity at the big bang, and this precludes meaningful continuation of the solution to before the start of expansion. There is much discussion of mini versions of the singularity, in black holes. It is still not possible to point to a specific astronomical object where a black hole surely is forming, but again it is certain that in the standard theory there are many sites where relativistic collapse could be taking place. Many people take the relativistic singularities in a collapsing star and at the big bang to be a problem with the theory, not with the universe. Einstein (1945) clearly expressed this in the caution he added to the second edition of *The Meaning of Relativity*:

The doubts about the assumption of a 'beginning of the world' (start of the expansion) only about 10^9 years ago have roots of both an empirical and a theoretical nature. The astronomers tend to consider the stars of different spectral types as age classes of a uniform development, which process would need much longer than 10^9 years. Such a theory therefore actually contradicts the demonstrated consequences of the relativistic equations. It seems to me, however, that the 'theory of evolution' of the stars rests on weaker foundations than the field equations.

The theoretical doubts are based on the fact that for the time of the beginning of the expansion the metric becomes singular and the density, ρ, becomes infinite. In this connection the following should be noted: The present theory of relativity is based on a division of physical reality into a metric field (gravitation) on the one hand, and into an electromagnetic field and matter on the other hand. In reality space will probably be of a uniform character and the present theory be valid only as a limiting case. For large densities of field and of matter, the field equations and even the field variables which enter into them will have no real significance. One may not therefore assume the validity of the equations for very high density of field and of matter, and one may not conclude that the 'beginning of the expansion' must mean a singularity in the mathematical sense. All we have to realize is that the equations may not be continued over such regions.

Chapter 9. The big bang cosmology – enigmas and nostrums

The empirical problem has been relieved with the revision in the value of Hubble's constant. The second point according to current dogma is that something like a quantized gravity theory is needed to describe the big bang and what came before it. This problem has not been solved. We have only become very aware of the subsidiary puzzles that would be resolved if it were!

9.3 Nostrums and elixirs

The besetting evil of our age is the temptation to squander and dilute thought on a thousand different lines of inquiry.   Sir John Herschel

9.3.1 The universe as Phoenix

By 1933 Lemaître had become convinced that the universe must continue expanding forever (an opinion still widely held, and still perhaps premature), and he concluded 'from a purely esthetic point of view, one could perhaps regret it. The solutions where the universe successively expands and contracts, periodically reducing to an atomic system with the dimensions of the solar system, have an incontestable poetic charm and bring to mind the Phoenix of the legend' (Lemaître, 1933*b*).

In one version of the legend, 'when the Phoenix reaches the prescribed age of 500 years it ends its life on a fiery pyre. But from the dead ashes a new Phoenix arises, in the full freshness of youth and as brilliant in its red and gold plumage.' Does the universe have a similar fate?

An oscillating universe could provide an escape from some of the conceptual problems. As the universe ages, more and more of it becomes visible that earlier was beyond the horizon and presumably causally disconnected from us if first contact is reckoned from the big bang. How then are we to understand the remarkable familiarity of the objects just appearing on the horizon? Perhaps by tracing the evolution back through the big bang to an earlier collapsing phase. If the singularity associated with the collapse of the universe is an inadequacy of the (non-quantum) theory, some future and better theory might show that the collapse of the universe would lead to a 'bounce' instead of a singularity. Of course, since there is no fundamental theory of this we can only speculate on how such a 'bounce' might or ought to behave.

What might conditions be like as the universe collapses, assuming the mass density is high enough to make the universe collapse? When the mass density grows high enough it thermalizes the primeval fireball radiation with the starlight, neutrinos, cosmic rays and so on produced during the cycle. Since most of the production would have been during the expanding phase the photons and neutrinos have time to move through an appreciable part of the circumference of the universe, so that the radiation ought to be well mixed and nearly uniformly distributed. Much of the matter is in stars moving at peculiar velocities now in the range 300 km s^{-1} to 3000 km s^{-1}, in galaxies, clusters of galaxies, and superclusters. During the early part of the collapse these gravitationally bound systems remain stable. They are forced to contract as they start to overlap and as the radiation density becomes comparable to the mass density within the objects. During the contraction the peculiar velocities increase as $v \propto \langle \rho_m \rangle^\alpha$, $\frac{1}{4} \leq \alpha \leq \frac{1}{3}$. The stars moving at high speed through the increasing radiation density are torn apart by ablation (and less commonly by collisions), leaving the matter in streaks through the radiation. The high peculiar velocities of the stars and star clusters just before disruption thus somewhat smooth out the present strong irregularities in the matter distribution.

The mass concentration in a black hole (associated perhaps with a collapsing star or galaxy nucleus) also gains high peculiar velocity during the collapse. It captures radiation and matter, increasing the mass and decreasing the momentum per unit mass, but the effect is appreciable only when ct (time to the singularity) is comparable to the size of the black hole.† These objects thus remain relatively small and limited patches of 'advanced collapse'.

The bounce might be expected to preserve the smooth radiation field and the more clumpy matter distribution, which would agree with observation, and it must also somehow produce the precise balance of expansion rate and density. The black holes are a further problem. Are these regions of 'premature expansion', or are the black holes ironed out by the same mechanism that balances density and expansion rate?

The behavior of the total entropy through the bounce has attracted speculation ever since Tolman's pioneering work (Tolman, 1934). Experience would suggest the total entropy can only increase, though it certainly is conceivable that in the new physics of the bounce entropy is

---

† The radiation capture cross-section is $\sigma \sim (GM/c^2)^2$, where M is the black hole mass. If the radiation mass density ρ_r exceeds the matter density then $t \sim (G\rho_r)^{-1/2}$. Assuming the peculiar velocity is not highly relativistic the mass gained in a collapse time is $\delta M \sim \sigma \rho_r ct$, and $\delta M \sim M$ when $ct \sim GM/c^2$, that is, when the Schwarzschild radius is comparable to the collapse horizon.

Chapter 9. The big bang cosmology – enigmas and nostrums

eliminated, perhaps lost in black holes left over after the big bang. Under the former assumption, as Tolman pointed out, the universe could not be strictly cyclic, but one could imagine a series of ever larger hops as the amount of radiation increases.

The production of entropy in the present cycle of the universe can be roughly estimated as follows. The mean optical luminosity density is $j \sim 3 \times 10^8 \mathscr{L}_\odot \mathrm{Mpc}^{-3}$, and this can be expected to persist for a Hubble time, giving minimum radiation energy density on the order of $jH^{-1} \sim 6 \times 10^{-15}$ erg cm^{-3}. The emission outside the visible band and the enhanced activity at high redshift might bring the total to

$$10^{-14} \leq u_d \leq 10^{-13} \text{ erg cm}^{-3}. \tag{9.2}$$

(Energy production at high redshift may well be substantially stronger than it is now but it is relatively inefficient because the energy of each photon is diminished by the redshift factor.) The energy density in the primeval fireball radiation is $u_p = aT^4$ with $T = 2.7$ K and the ratio of the two densities is

$$0.03 \leq u_d/u_p \leq 0.3. \tag{9.3}$$

Future production during the expanding phase, even at diminished intensity, could appreciably increase the ratio (again because of the redshift). If the universe collapses, thermal evaporation of stars and nuclei tends to decrease $(u_d + u_p)/u_p$, but the effect must be negligible because the objects form when the universe has a much larger radius than when they decompose. Thus when the radiation is thermalized, at very high density, the entropy per baryon number (or better the entropy per comoving volume) is larger than at the start of expansion by the factor

$$f = (1 + u_d/u_p)^{3/4}, \quad 1.02 \leq f \leq 1.2. \tag{9.4}$$

If the magnitude of the increase were the same on each cycle the universe would have to be less than 100 cycles old. By this argument the Phoenix could reproduce only a very limited number of times.

An interesting alternative is to suppose that the total mass of matter in the universe increases on each cycle. This might happen through the creation of new and separate pockets of matter and antimatter, a sort of macroscopic extension of the creation of particle – antiparticle pairs, or perhaps through a net increase of the baryon number. The second course violates the observed precise conservation of baryons, but following Einstein we are not inclined to take this too seriously as the collapse certainly brings conditions well beyond the range of laboratory physics. One might also wonder whether this speculation violates energy conservation. Since general relativity theory does not admit a general analog of Newtonian energy conservation this is difficult to address even apart from the singularity at the bounce, but there is an interesting heuristic argument (whose history we have been unable to trace). In the homogeneous isotropic model R is conserved (9.1), that is, the proper radius of the universe is a fixed multiple of the mean distance between conserved particles, clearly independent of dissipative processes. Thus at the epoch of maximum expansion the proper radius and mass, r_m and M_m, satisfy

$$r_m \equiv R a_m, \quad M_m \sim \rho_m r_m^3, \quad GM_m/r_m c^2 \sim 1, \quad (9.5)$$

where the third equality follows from (9.1) with $\dot{a} = 0$. The 'total energy' of the universe, annihilation plus potential (at maximum expansion), should be

$$U \sim M_m c^2 - GM_m^2/r_m \sim M_m c^2 - M_m c^2 \approx 0. \quad (9.6)$$

This says the universe might be considered to have zero net energy.

Could the first cycle of our universe have developed out of 'nothing', as a zero-energy quantum fluctuation leading to the production of a few quanta? These few particles colliding in the subsequent collapse could generate more entropy and matter, at the expense of potential energy, leading through a series of ever larger universes to something fit to live in. It is also possible to imagine that the whole of our universe arose suddenly as a quantum fluctuation on a gigantic scale (Tryon, 1973).

One final aspect of the Phoenix deserves mention. In the early discussions Einstein and de Sitter took it as a matter of course that space in cosmology is static. (Perhaps that is a measure of the difficulty of breaking away from the rigid Cartesian notion of space.) One problem with the static Einstein universe was noted only after the expansion of the universe had been discovered. If stars have finite energy reservoirs they cannot shine forever, and if they do somehow shine forever space becomes filled with radiation in thermal equilibrium with the stars. Even the solar system is not completely stable; if left alone eventually some of the planets would be captured by the Sun (assuming it has not already exploded) and the rest would escape. Such relaxation processes were discussed by Jeans (1928). These remarks say that the universe could not be eternal; it must have a limited age. In the oscillating universe the scenario is altered because at each big bang the stars are disrupted, the elements decomposed, and the radiation thermalized. The universe is periodically refurbished.

Chapter 9. The big bang cosmology – enigmas and nostrums

9.3.2 The anthropomorphic universe

In medieval cosmology the Earth is a preferred spot, at the center of a mostly perfect spherical universe. Medieval philosophers speculated on the astrological influence of the universe on the behavior of people. Could it be just the opposite, that it is the presence of observers that determines the nature of the universe?

Imagine a game of Russian roulette played on a grand scale by many individuals using randomly distributed loaded and unloaded guns. At the conclusion of the deadly game a brilliant statistician after exhaustive statistical analyses concludes that there is a high probability of the randomly selected unloaded guns being drawn by the survivors of the game.

Imagine an ensemble of universes of all sorts. It should be no surprise that ours is not an 'average' one, for conditions on the average might well be hostile. We could only be present in a universe that happens to supply our needs.†

The exquisitely balanced adjustment of the initial expansion rate was noted above. Suppose, for example, that at $t = 1$ s, the epoch when helium production begins, the expansion rate has been reduced by 1 part in 10^6, making the gravitational potential energy exceed the kinetic energy of expansion by about this fractional amount. Then the universe would have stopped expanding at $t \sim 10^{12}$ s, when the radiation temperature is $T \sim 10^4$ K. Matter could not have combined to atoms and so it would have been tightly coupled to the radiation, preventing any stable clumping. Stars could not have formed, and there would have been no heavy elements out of which to make any reasonable facsimile of people. If at $t = 1$ s the expansion rate has been increased by 1 part in 10^6, making kinetic energy slightly exceed potential, then by the epoch of decoupling of matter and radiation, $T \sim 2000$ K, the kinetic energy of expansion would have so dominated gravity that minor density irregularities could not have collected into bound systems in which stars might form. Given the same initial spectrum of density irregularities as in our universe this one certainly would end up looking very different, and perhaps would not be the sort of universe that could have been observed.

In the oscillating universe picture a first bang with only a few quanta

---

† This 'anthropic principle' was first invoked as an explanation of the Dirac large numbers, and as an explanation for the value of physical constants like the gravitational and fine structure constant (Dicke, 1961a; Carter, 1968, 1974).

could not have been observed, and so, as J. A. Wheeler has emphasized, it is puzzling to know what is meant by the thought that it happens. If we suppose nonetheless that the 'quantum bang' happens, and that the particles multiply, perhaps we can imagine that eventually there is enough matter and radiation so that stars form and evolve and a favorable environment like the Earth appears.

Given that there apparently are observers in the present universe, what happens in the next oscillation? Are observers allowed, or needed? The entropy estimate suggests conditions might not be too different. Perhaps as long as the entropy per baryon does not grow out of hand there will be observers in each cycle, with an ever longer time between successive refurbishments.

9.3.3 Reproducing universes

We have been assuming for the most part that the universe is closed, fated to collapse back to the big bang. It is also possible that the universe is open, doomed to expand forever. (In our opinion the evidence either way is at best marginal.) In the latter case it is often said that the world suffers a heat death, all systems settling to the ground states, but that is not so. The Solar System is in a gravitationally bound galaxy, which probably is bound in the Local Group (scale $\sim 3 \times 10^6$ light years) in the Local Supercluster (scale $\sim 10^8$ light years). As the universe expands this complex remains more or less intact, and in the eternity available to it must lose energy, by evaporation of members and radiation of gravitational waves.

Eventually a good part of the Local Supercluster must collapse to a fiery death in a black hole (or in a hierarchy of them). What of a neutron star that evaporates from the Local Supercluster? Given all eternity, must it make the quantum jump across the barrier to the collapsing solution?

A collapsing black hole in an open universe presents us with much the same conundrum as a collapsing universe, but with the added complication that, viewed from the outside, the singularity never happens. However, a being riding in with the collapse would observe a singularity. Could this mini big bang bounce, perhaps adding entropy and matter? The result might look much like a Friedman–Lemaître model until the boundaries of a limited system like this became apparent. If the system had negative energy it might collapse again, be fattened at the bounce, and eventually develop into a suitable home for observers.

Chapter 9. The big bang cosmology – enigmas and nostrums

9.3.4 Variable strength of the gravitational interaction and oscillating universes

Mach's principle was an important element in Einstein's discovery of the homogeneous general relativity world model, but it can be argued that this model does not fully solve the problem because gravitation (inertia) remains constant as more and more of the universe comes into view (and, during the collapse, comes closer and closer together). The characteristic number

$$K = Gm_p^2/\hbar c,$$

where m_p is the proton mass, has figured in such discussions, and it has been proposed that K^{-1} is as large as it is because the universe is so massive (Dirac, 1938; Jordan, 1955; Sciama, 1953; Brans and Dicke 1961; Dicke, 1961*b*). Could the universe begin as a 'quantum fluctuation' with $K^{-1} \sim 1$, then through oscillations generate not only matter and entropy but also K^{-1}?

The scalar–tensor gravity theory is an interesting example. It provides no antidote to the singularity theorems. With the proper choice of units of measure (obtained by setting G, $\hbar$ and c equal to 1) Einstein's field equations are satisfied, and the energy density and pressure are positive, so the usual singularity theorems apply. The pressure contributed by the scalar field is equal to its energy density ρ_ϕ, so during a general collapse ρ_ϕ would rapidly increase, as $\rho_\phi \propto a^{-6}$, causing this originally minor energy to outstrip all others and dominate and hasten the final collapse. During the final stages ϕ may become strongly inhomogeneous as different parts race each other ever more rapidly to the singularities. As quantum effects become important scalar and tensor gravitons may become major ingredients of the cosmic soup. The chemistry of this soup could have some important consequences – perhaps through the scalar gravitons some of the abundant ϕ-energy is converted to other forms of matter; perhaps in these extreme conditions baryons are not conserved.

In the scalar–tensor cosmology K^{-1} increases with time, and goes to infinity at the singular collapse. In a sense this is what is wanted because matter is moving arbitrarily close together, but it raises a fresh puzzle. If the universe bounces what is K^{-1} during the new expansion? In the scalar–tensor theory if ϕ ($\phi > 0$) and $d\phi/dt$ come out of the bounce at not-too-large values the solution converges to a standard form which for a cosmologically flat model is

$$\phi \propto t^{2/(3\omega+4)}$$

where ω is the coupling constant in the theory. (The time t is expressed in atomic units.) We do not know how to compute the constant of proportionality. Perhaps it is larger on each bounce as the universe grows.

We mentioned the idea that our universe was 'selected' for features hospitable to observers. It could be objected that the universe was over-designed for such a modest purpose, that a star cluster or at most a large galaxy would do. Of course space must not curve up around the galaxy for then the world would be crushed much too soon as in a black hole. What of a single galaxy in asymptotically flat space? Einstein's original argument suggests an answer, that this is not allowed by Mach's principle. How then would a parsimonious creator develop our universe? Perhaps it began as a zero-energy quantum fluctuation, with $K \sim 1$, and involving only a few quanta. On each bounce matter is created in the cosmic soup, the energy being borrowed from gravity. On each oscillation K^{-1} increases. Finally gravity is weak enough to permit stellar evolution at a slow rate, that can provide a hospitable environment long enough to permit biological evolution. And here we are!

10. Cosmology and the early universe

YA. B. ZEL'DOVICH

10.1 Introduction

In a volume dedicated to the centenary of Einstein's birth, it is appropriate to begin this chapter with some remarks on his personal impact on cosmology. As early as 1917, immediately after the discovery of general relativity, Einstein wrote a paper entitled 'Kosmologische Betrachtungen zur allgemeinen Relativitätstheorie' (Einstein, 1917).

We are accustomed by now to the experience that every major advance in the scales investigated leads to new principles. So it was when on the atomic scale the quantum properties of matter became all-important. Perhaps now, in the seventies, one could add the quark-confinement theory operating on the subnuclear scale.

General relativity was conceived as a theory needed to cope with the large, perhaps infinite, scale of the universe as a whole. Therefore, its application to cosmology was quite in line with the internal logic of science. By the same token, no other specific theory of large scale was needed – it would have been superfluous because general relativity had no internal flaws and no disagreements with experiment. Einstein's initial paper on cosmology was not flawless, but as stated by Bohr, even the errors of a genius are of interest and their correction leads to important discoveries.

Einstein was assuming *a priori* the uniformity of the universe on a large scale: this belief is now confirmed on a scale greater than 1000 Mpc (compare with $c/H = 6000$ Mpc – the 'horizon', the limiting penetrating depth of any conceivable observation).

But he also assumed a static, non-evolving universe – and this proved to be wrong. The universe is in fact an expanding system with a definite evolution scenario. The disagreement with the actual situation also led Einstein into theoretical difficulties: he drew the conclusion that the Newtonian theory of gravitation is not applicable to cosmology and he changed the equations of general relativity introducing the 'cosmological term'. It is now well known that the difficulties of Newtonian theory (e.g.

Introduction

'the gravitational paradox') are prejudices. Exact logical analysis shows that the Newtonian approach to the cosmology of an expanding universe is impeccable; one needs only moral courage. In a simple form, such a theory was constructed by Milne (1934). For a review with refinement of the idea of uniformity see Zel'dovich (1977a).

Ironically, the simple Newtonian approach was formulated ten years after the much more complicated general-relativistic Friedmann (1922, 1924) solutions were found. People found a Newtonian proof when the answer was already known from the relativistic theory initiated by Einstein.

One question, unsolved even now, concerns the matter density in the universe. Even the question of whether the universe is open or closed is not solved with certainty. Even the rather wide constraints given by observational data lead to information about elementary particles which cannot be obtained by standard laboratory and accelerator-physics methods. This question is treated in section 10.2.

The establishment of the evolutionary theory including the dynamics of expansion gives the first approximation to the large-scale structure of the universe: the geometrical, kinematical and dynamical background. The next step consists in the elucidation of the physical content, the particles present, and the temperature of the medium.

Several attempts were made to explain the abundances of various chemical elements and isotopes by nucleosynthesis in uniformly expanding matter. They were based on overestimates of the Hubble constant, H, initially thought to be $550 \text{ km s}^{-1} \text{ Mpc}^{-1}$.

The resulting age $H^{-1} = 2 \times 10^9$ years, was shorter than or equal to the estimates of the age of Earth, Sun and globular clusters. Therefore the idea of cooking all the elements cosmologically seemed convincing.

Gamow (1946) tried to do it in a bath full of neutrons through absorption of neutrons by nuclei and subsequent beta decays. To prevent isotopes from reacting strongly with slow neutrons, Gamow (1952, 1956) introduced high initial temperatures and predicted for the present day a temperature of 6 K; amazingly close to the actually measured 2.7 K. All intermediate steps of Gamow's calculation failed: the age of the universe is 15 to 20×10^9 years, much greater than the Earth's and Sun's age ($\sim 5 \times 10^9$). Cosmological nucleosynthesis practically ends at He^4; the medium and heavy elements are cooked in stars.

Nevertheless, the hot big bang theory initiated by Gamow, was further developed by Alpher and Herman (1953), Hayashi (1950), Fermi and Turkevich (cited in Alpher and Herman, 1953), and others. In a paper

Chapter 10. Cosmology and the early universe

which remained unnoticed by radio-astronomers, Doroshkevich and Novikov (1964) pointed out that even at low temperature the cosmological radiation would overwhelm stars, radio-sources etc. in a definite wavelength band (e.g. with λ between 0.1 and 20 cm).

It is well known that Dicke *et al.* (1965) independently came to the same conclusion and moreover made an effort to find the cosmological radiation at 3 cm. Therefore, when Penzias and Wilson (1965) actually found a noise at 7.3 cm, Dicke and his team immediately understood its origin and announced that the hot big bang theory was proved. As an afterthought it was recognized that the inexplicable excitation of CN interstellar molecules observed by McKellar (1941) was a proof – and also a good measure of the cosmological radiation. The hot big bang theory leads to a natural division of the cosmological scenario. In order of increasing remoteness from our time and also of increasing complication let us consider:

(i) The lepton era; from $t = 0.01$ s, $z = 10^{11}$, $T = 3 \times 10^{11}$ K $= 30$ MeV, up to the present-day situation. The physical laws in this era are well known, cosmology helps to establish important details. The lepton era is considered in section 10.3.

(ii) The hadron era; from 10^{-42} s up to 0.01 s. The development of particle physics, including broken symmetries and quark confinement shows the possibility of many unexpected new qualitative features of the hot plasma in this region (section 10.4).

(iii) Last but not least, the Planckian unit of time 10^{-43} s is just characterized by the direct impact of quantum theory on cosmology, gravitation and metric. In fact other articles in this volume are dedicated to this topic. The evaporation of black holes is a quantum effect which is prolonged beyond the Planckian time. Some remarks on primordial black holes and quantum gravity are given in section 10.5.

10.2 The average matter density in the universe

The density of matter directly visible as stars and luminous gas clouds collected in galaxies averaged over all the volume, including dark intergalactic space, is of the order of 10^{-30} to 10^{-31} g cm^{-3}.

The energy density of electromagnetic radiation of cosmological origin (2.7 K black body relic radiation) is 5×10^{-13} erg cm^{-3}. Divided by c^2 it gives 5×10^{-34} g cm^{-3} which is now negligible compared with ordinary matter density.

The average matter density in the universe

Physics now knows about the existence of weakly interacting particles – neutrinos in the first place. The situation is exemplified by the search for solar neutrinos: their flux is thought to be 5 per cent of solar light fluxes, i.e. 10^5 erg cm^{-2} s^{-1} corresponding to energy density of 3×10^{-6} erg cm^{-3} and mass density of 3×10^{-27} g cm^{-3}, but they are still undetected. Electronic neutrinos of lower energy per particle and other types of neutrinos, as well as gravitons, are much more difficult to detect, as compared with the solar million-volt neutrinos.

This means that direct physical measurements in the laboratory are unable to detect neutrinos even in quantities which are already overwhelmingly important in cosmology. On the cosmological scale, gravitation is more sensitive than laboratory particle detectors to neutrinos (Zel'dovich and Smorodinsky, 1961).

The first and obvious result of an excess density would be a shortening of evolution time

$$t = \frac{1}{H} f(\Omega),$$

$$\Omega = \frac{\rho 8 \pi G}{3 H^2} = \frac{\rho}{\rho_c}, \quad f(0) = 1,$$

$f < 1$ for $\Omega > 0$; $f = (1 + \sqrt{\Omega})^{-1}$ for relativistic particles.

For $\Omega \gg 1$, t does not depend upon H asymptotically:

$$t \leq (3/8\pi G\rho)^{1/2}.$$

Taking the age of the solar system, 4.5×10^9 years, as a lower bound for t, one obtains $\rho < 10^{-28}$ g cm^{-3}. But this is a very generous estimate. With considerable confidence one can assume $\rho < 10^{-29}$ g cm^{-3}.

This allows us to give upper limits for the temperature of an equilibrium Fermi $\nu - \bar{\nu}$ gas or black body gravitational radiation: $T < 40$ K.

Another possibility is a degenerate neutrino sea (with negligible $\bar{\nu}$ admixture – or, vice versa, degenerate $\bar{\nu}$ without ν). This is compatible with the hot big bang theory scenario. The upper limit for the Fermi energy of the sea is of the order of 50 eV.

So much for the quantities of particles. Cosmology also gives information about the properties of neutrinos, i.e. about their rest-mass.

The assumption that neutrinos have a nonzero rest-mass is arbitrary, but it is hard to disprove it with the tools of nuclear physics. Experimentally, it is established that $m < 100$ eV for electronic neutrinos, $m < 5 \times 10^5$ eV for muonic neutrinos and $m < 5 \times 10^8$ eV for neutrinos associated with the new 'lepton' τ, a heavy ($m = 1.85 \times 10^9$ eV) analog of the

Chapter 10. Cosmology and the early universe

electron and muon. Gershtein and Zel'dovich (1966) were the first to use cosmological constraints. In the hot big bang model the number density of various neutrinos must be of the order of 200 cm^{-3} if their rest-mass is less than 10^6 eV so that they do not annihilate during expansion.

Using a generous estimate, $\rho < 10^{-28}$ g cm^{-3}, they obtained for the ν_e or ν_μ rest-masses an upper limit of approximately $m < 400$ eV. This was not important for the ν_e but very important for muonic neutrinos.

Cowsik and McLennand (1972) refined the theory, assuming $\rho \leqslant 5 \times 10^{-30}$, equal mass (if any) for all types of neutrinos and obtained the stronger result $m < 13$ eV. Szalay and Marx (1976) discuss the fate of heavy neutrinos in galaxies.

After the discovery of the third lepton, τ, many groups independently used cosmological evidence to exclude the range 10^6 eV $> m > 20$ eV of possible ν_τ rest-masses. But the $6 \times 10^8 > m > 10^6$ range poses new problems: the annihilation of ν_τ and $\bar{\nu}_\tau$ during expansion takes place when the temperature drops to $kT < m_\nu c^2$. The theory of annihilation during expansion, developed in connection with matter–antimatter (Chiu, 1966) and quark annihilation (Zel'dovich, Okun and Pikelner, 1965), was used.

Its important result is the algebraic (not exponential) dependence of final concentration upon the rest-mass.

The result was (Lee and Weinberg, 1977; Visotsky, Dolgov and Zel'dovich, 1977) the prohibition of the mass interval given above. Therefore the third neutrino rest-mass must also be less than 20 eV, and the strict zero value is very probable. The lifetime of massive neutrinos is discussed by Dicus, Kolb and Teplitz (1977). The case of quarks mentioned above was important for the development of elementary-particle theory. If free quarks can exist, their concentration after recombination would be equal to $(Gm^2/\hbar c)^{1/2} \sim 10^{-17}$ that of photons. The ratio of quarks to baryons would be of the order of 10^{-8}. The absence of such a concentration in ordinary matter was an argument in favor of quark confinement theory. This argument was stronger than the results of accelerator and cosmic-ray experiments. The possible properties of the Higgs meson are analyzed using cosmological arguments by Sato and Sato (1975).

10.3 The lepton era

The lepton phase, beginning from say 0.1 s or 1 s, gives important clues to exciting questions of physics and cosmology.

In the standard hot big bang scenario the equilibrium neutron–proton transformation is frozen at 15 per cent n, 85 per cent p, thereafter the standard primordial composition 30 per cent He^4, 70 per cent H is obtained by nucleosynthesis. On the other hand positrons annihilate with electrons, leaving only the e^- excess needed to compensate the charge of alpha particles and protons. The photons' equilibrium Planckian spectrum is maintained by absorption, radiation and scattering. Thereafter a long period ensues when the equilibrium is conserved because the expansion diminishes the temperature without perturbing the form of the electromagnetic spectrum.

At a minimum temperature, $T \sim 4000$ K, decoupling of matter (which is now neutral) and radiation occurs. Gravitational instability leads to the formation of galaxies beginning from small density perturbations. Every part of this process is sensitive to various perturbations of the standard scenario. Shvartsman (1969) pointed out that a population explosion among massless or light neutral particles (e.g. twenty types of neutrinos) would change the tempo of temperature change, therefore the $n:p$ ratio would change and the percentage of He^4 in primordial matter would increase. In a recent article Steigman et al. (1977) claim that the number of massless particle types is less than seven.

Particles decaying or annihilating later than $t = 10^4$ s (redshift, z, less than 10^7) would pump energy into radiation and spoil the Planckian spectrum. So would decaying acoustic oscillations if the initial perturbations on the appropriate scale are of greater than 10 per cent density contrast.

The sensitivity of the spectrum to energy input was studied by Zel'dovich, Illarionov and Sunyaev (1972). Due to photon spectrum rearrangement by Compton scattering, the input of 1 per cent of energy leads to an increase of order of one per cent of the average photon number in the most populated part of the spectrum, but there is a sensitive spectral region (in the Rayleigh–Jeans long wave part, wavelength ~ 20 cm) where the photon density and effective temperature are lowered by 10 to 20 per cent after energy input. These effects have not been found yet but due to the high sensitivity even their absence gives valuable upper limits on several types of perturbations superimposed on the idealized scenario.

10.4 The hadron era

At the high temperatures of the hadron era, theory predicts plenty of baryons and antibaryons – 'matter' and 'antimatter' in thermodynamic

Chapter 10. Cosmology and the early universe

equilibrium. Actually, in the lepton era there is a small admixture (10^{-8} to 10^{-9}) of baryons to the photons and neutrinos. Extrapolated to the hadron era this means that there is a slight asymmetry $1 + 10^{-8} : 1$ in the ratio of baryons:antibaryons in the past. This ratio seems very peculiar.

Evidence from galaxy formation and the background radiation (2.7 K) spectrum leads to conclusions about the spatial distribution of this ratio (extrapolated to the hadron era). On a large scale (megaparsecs of comoving coordinates†) the variations are less than 1 per cent of the excess, $1 + 1.01 \times 10^{-8}$ up to $1 + 0.99 \times 10^{-8}$.

On a small scale, but larger than one parsec, greater variations are possible, but still the ratio is never reversed ($r > 1$) so that no regions with antimatter excess exist.

One cannot disprove the existence of such regions on a scale even smaller than one parsec, provided that annihilation with baryon excess in the surroundings ends during the hadron era – but before the lepton era nucleosynthesis and spectral-sensitive periods.

The thermodynamical fluctuations with $\Delta N/N = 1/\sqrt{N}$ would give $r < 1$ in regions with $N \approx \bar{N} = 10^{16}$, $|N - \bar{N}| \sim 10^{8}$, corresponding to a comoving scale of the order of one kilometer. These mini-scale fluctuations must decay by diffusion long before the lepton era.

The information given above is a summary of the observations. No firm fundamental theory explaining the observed r and no theoretical statement about its spatial uniformity or variation exists to the author's knowledge.

A second topic concerning the hadron era is connected with theories of broken symmetry. It is assumed that one, or several wave fields Ψ have a peculiar dependence of potential energy on the amplitude, $V(\Psi)$, of the type $A - B\Psi^2 + C\Psi^4$ with A, B, $C > 0$. This function has minima at $\Psi = \pm(B/2C)^{1/2} = \pm\Psi_m$.

The cold vacuum corresponds to $\Psi = \pm\Psi_m \neq 0$. For example, T. D. Lee (1973) uses the existence of two solutions to construct a theory which is symmetric in its initial assumptions (including the symmetry $\Psi \to -\Psi$). But in a given region of space with, say, $\Psi = -\Psi_m$, the symmetry is broken, the properties of baryons and antibaryons are different (so called CP = charge and parity – symmetry violation). In other theories the masses of particles depend on Ψ_m being different from zero.

Kirzhnits (1972) pointed out (and his associate Linde elaborated further – Kirzhnits and Linde, 1972, 1974, 1976; Linde, 1974) that at a

---
† Measured by today's scale of distances, after expansion.

The hadron era

high enough temperature the symmetry of the vacuum is restored, $\bar{\Psi} \equiv 0$ everywhere. When the temperature decreases, a phase transition occurs from the symmetric phase $\Psi = 0$ into the asymmetric phase $|\bar{\Psi}| = \Psi_m$. But the signs of $\bar{\Psi}$ in distant regions (outside the horizon) are not correlated. Therefore boundary layers – 'walls' dividing $\bar{\Psi} = +\Psi_m$ and $\bar{\Psi} = -\Psi_m$ – must occur. The cosmological implications, including the mass of the walls and the perturbations connected with them, were considered by Zel'dovich, Kobzarev and Okun (1974). The result is rather negative for this type of theory. Other variants with an unstable vacuum in one state spontaneously decaying into another lower state are considered by Coleman (1977), Frampton (1977), Kobzarev, Okun and Voloshin (1974). The situation changes if Ψ is complex and $|\Psi|^2 = \Psi \bar{\Psi}$ is written instead of Ψ^2 in $V(\Psi)$.

Again the cold vacuum is non-symmetric, $|\bar{\Psi}| = \Psi_m$ and the hot vacuum is symmetric $\Psi \equiv 0$. But the cold vacuum degeneracy is now continuous: every

$$\Psi = \Psi_m e^{i\varphi}$$

with arbitrary phase φ is a candidate for the cold vacuum. The phase transition from hot $\bar{\Psi} = 0$ to the cold phase leaves vortex lines (the lines around which the phase is changed by 2π) instead of walls.

A detailed theory of vortex behaviour is not yet developed. This is not to blame the cosmologists for laziness. Clearly the particle physicists must first make their choice of the type of broken symmetry theory (if any), obtain a laboratory confirmation of this theory and find the parameters.

The moral of the above lies in the large number of qualitatively different variants. Some ten or twenty years ago many (including the author of this chapter) thought that the progress of particle physics could bring only quantitative changes in the equation of state (pressure as a function of density and entropy). Due to rapid attainment of equilibrium, this would mean that the lepton era does not depend on unknown details of the hadronic era (except for the ratio of leptons and baryons to values of the specific entropy).

The phase transitions are interesting because in principle they should be a source of perturbations leading later to galaxy formation in a cosmological model with ideal initial uniformity.

The first ideas about phase transitions were formulated by Omnès (1971a, b, c). He hypothesized that there is a temperature interval, $0.3 m_p c^2 < kT < n m_p c^2$ (with $n > 1$ unknown) where the homogeneous charge-symmetric mixture of baryons and antibaryons (as well as leptons

Chapter 10. Cosmology and the early universe

and mesons) in thermodynamic equilibrium is unstable. According to Omnès the baryons and antibaryons repel each other, two stable phases are formed – with excess matter and excess antimatter.

The latest considerations rather cast doubt on this particular type of phase transition (nuclear physics, Bogdanova and Shapiro, 1974; cosmological, 'case against antimatter', Steigman, 1971, 1973). The future of broken-symmetry theories is also unclear: whether they are only a mathematical tool for renormalization of the theory and will vanish later, or that the truth has been found and phase transitions of the Kirzhnits type really exist.

Also unexplored is the equation of state and qualitative features of a dense hot equilibrium mixture of quarks, antiquarks and gluons, with account of specific properties of gluons (see Chapline, 1975 – quarks at high densities).

It is known – or at least assumed – that free color-charge of quarks leads to long range gluonic forces, even stronger than electrostatic forces. Therefore, in equilibrium, the large-scale fluctuations of color are damped compared with baryon charge fluctuations. But we have no theory of the initial situation at the singularity. Therefore, one could ask what would occur if large-scale fluctuations of color are given *ab initio* in the singular state. The same question could be asked about electric charge and magnetic field (equivalent to currents) on a large scale. No answer is known and once more we see how rich is the choice of *a priori* cosmological models.

10.5 The quantum era and its effects

Quantum mechanics was never ignored in the development of cosmology. The properties of matter and radiation, spectral lines, light scattering, statistics of Bose or Fermi – all these topics were taken into account in the calculation of pressure, energy density, spectrum transport coefficients, etc. Therefore the right-hand side of Einstein's general relativity equations already included quantum effects.

Speaking about those effects now we emphasize the influence of spacetime curvature on particles and fields, as opposed to the usual physics of Minkowskian space.

The most interesting effect is the creation of particles by the gravitational field in vacuum. The reactions of the type $e^+ + e^- = g + g$, with g being gravitons, were considered and calculated in the thirties and forties. Taking many coherent gravitons, we obtain a classical gravitational wave.

The creation of e^+e^- pairs obviously occurs in colliding beams of (classical) gravitational waves, i.e. in a vacuum with time-dependent metric.

In a cosmological context the excitation of fields, i.e. the creation of field quanta (for example photons) in an expanding universe, was mentioned first by Schrödinger (1939, also 1970), and later by Utiyama and DeWitt (1962). A thorough investigation was made by Parker (1968, 1969). The general principle is that particle creation is due to the non-adiabatic behaviour of the corresponding field in a changing metric. The particles are created with a frequency of the order of the inverse characteristic time of the change of metric. This principle leads to the threshold of massive particle creation: no creation at $t > \hbar/mc^2$. By dimensional arguments the energy density of created particles is of the order of $\hbar/c^3 t^4$. The important qualitative features of particle creation are:

(i) The vacuum polarization effect – the appearance of energy–stress tensor components without real particles.† One example is the Casimir (1948) effect at zero temperature in the static situation. Here pure vacuum polarization occurs. In the case of a small localized perturbation, $g_{\mu\nu} = \eta_{\mu\nu} + h_{\mu\nu}$, where $h_{\mu\nu} = 0$ at $t < t_1$ and $t > t_2$, $h_{\mu\nu} \ll 1$ and smooth in the interval $t_1 < t < t_2$. The vacuum polarization is characteristically of first order in $h_{\mu\nu}$ and vanishes together with the metric perturbation for $t > t_2$. The creation of real particles is of second order $dm/dt \sim h^2$ (we omit the tensor indexes here); these particles – and their energy and stress – remain after the perturbation, at $t > t_2$.

(ii) The second important feature of the theory is its conformal invariance – at least in the limit of vanishing rest-masses (Chernikov and Tagirov, 1968; Penrose, 1964). The isotropic expansion is conformally equivalent to a static situation.

Therefore particle creation is abnormally small (proportional to $Gm^2/\hbar c = 10^{-38}$ in some positive power) in Friedmann isotropic and homogeneous cosmological models.‡

On the other hand (Grib and Mamaev, 1969), creation of particles in an anisotropic expansion gives divergent results if switched on at $t = 0$. If one (artificially) turns on the creation at $t = t_{pl} = 10^{-43}$ s, it destroys the

---

† Its importance for particle creation is shown by Zel'dovich and Pitaevsky (1971).
‡ An important exception are gravitons (Grishchuk, 1974). General relativity gives an unambiguous classical gravitational wave equation – with zero rest-mass but still not conformally invariant. To prevent creation of gravitons in the Friedmann model one needs an extra condition of zero curvature scalar $R = 0$, which in turn is the case if $\varepsilon = \frac{1}{3}p$ (where ε is the energy density and p the pressure) in the background matter. This condition is perhaps fulfilled in the actual universe near the singularity.

Chapter 10. Cosmology and the early universe

anisotropy during a time of the same order (Zel'dovich, 1970; Zel'dovich and Starobinsky, 1971, 1975, 1976; Lukash and Starobinsky, 1974). From these considerations it follows that cosmological solutions which are anisotropic at the singularity are prohibited. But we are not left solely with the Friedmann models. The quasi-isotropic solutions with isotropic expansion in the limit $t \to 0$ are also possible. They have an inhomogeneous comoving three-space metric. Therefore density perturbations as well as gravitational waves evolve from the quasi-isotropic solution. At the same time anisotropy also emerges – but sufficiently late that particle creation could be avoided if the three-space metric inhomogeneities are smooth enough.

The most important and beautiful result of the marriage between quantum theory and gravity is Hawking's (1974, 1975, 1977) theory of black hole evaporation. Its physical content is discussed in later chapters in this volume. Oversimplifying, one can say that collapse to a black hole is an event which does not end in a static situation (unlike the collapse of a white dwarf to a neutron star for example). The black hole formation leads to exponential reddening $\omega \sim e^{-ct/r_g}$ of all outgoing radiation. Therefore at all times the reddening is non-adiabatic for outgoing waves with frequency $\omega \sim c/r_g$ – so that the collapsing black hole emits particles with this frequency. This order of magnitude argument of Hawking was substantiated by arguments proving the detailed character of the radiation.

The evaporation is not important for stellar black holes, but only for primordial ones (PBH). The idea of PBH was first formulated by Zel'dovich and Novikov (1967).

If the initial metric and density distribution near the singularity are inhomogeneous enough it is quite possible that in some parts of the primordial plasma the expansion is followed by compression ending in early gravitational collapse. Limiting ourselves to quasi-isotropic solutions does not exclude the possibility of partial gravitational collapse, i.e., formation of PBH. Recently, numerical examples of PBH formation, from spherically symmetric smooth quasi-isotropic solutions were obtained by Nadjozhin, Novikov and Polnarev (1978).

Considering PBH in the sixties, Novikov and myself were not aware of quantum evaporation. We thought that the mass of every individual PBH after its formation could not decrease; due to accretion of surrounding matter it would increase monotonically. The ratio of final PBH mass to the initial mass was not clear; later analytical work of Carr and Hawking (1974), and still later calculations of Nadjozhin *et al.* (1978) show that the

The quantum era and its effects

ratio is near to unity. But despite there being no definite value of the ratio, it was quite clear from the very beginning (1966–7) that the initial density of PBHs must be very small compared with the surrounding expanding plasma – or perhaps no PBHs are formed at all.

The consequence of this reasoning was the statement that the initial metric of comoving three-dimensional space is smooth enough even at the smallest scale. One could not make this statement using the linear theory of perturbations, because in this theory the small-scale perturbations develop into oscillations and are totally damped very early, leaving no observable traces. The PBH formation is a phenomenon beyond the scope of linear theory. What are the changes in our views ten years later, after the discovery of PBH quantum evaporation?

If the mass of the PBH is greater than the Planckian mass $(\hbar c/G)^{1/2} = 10^{-5}$ g, i.e. if the PBH collapse occurs later than the Planckian time, 10^{-43} s, the formation and evaporation are separated in time, the evaporation does not affect formation. The connection between the initial metric perturbations and PBH formation remains. Detection methods and detection limits have been drastically changed. The PBHs with initial mass of the order of 3×10^{14} to 10^{16} g are still alive or died not long ago, they are (or were) strong emitters of hard γ-photons, electron–positron pairs and mesons. Such very active objects would give a major contribution to cosmic rays, X-ray and γ-ray background (Page and Hawking, 1976; Carr, 1976). In the mass range mentioned the detection sensitivity of active (evaporating) PBHs is 10^8 times better than the sensitivity of passive PBH detection by their gravitation, the observational limit for evaporating PBHs is much lower.

The PBH with evaporation times in the interval $100 \, \text{s} < t < 10^5$ years (before recombination) would spoil the Planckian spectrum of the background radiation. PBHs which evaporate earlier ($1 \, \text{s} < t < 100 \, \text{s}$) change the composition of nucleosynthesis (Veiner and Nasselsky, 1977; Zel'dovich et al. 1977). At $t < 1$ s the PBH will lead to an entropy increase; perhaps actually a part of entropy is due to this process – but one must not obtain more than the observed entropy per baryon, the contribution of PBH must be less than 10^9 dimensionless entropy units per baryon.

Our views on the cosmological importance of the lowest mass PBH, say 10^{-5} g to 1 g, have undergone real qualitative changes in light of developments in PBH evaporation theory. In the sixties they appeared to be prohibited even more strongly than the heavy PBH.

Chapter 10. Cosmology and the early universe

Now, in the context of evaporation theory, one can admit that their abundant initial formation followed the very early evaporation.† Strong departures from homogeneity are admitted on the smallest scale – from Planckian, 10^{-33} cm at $t = 10^{-43}$ s, up to 10^{-28} cm at $t = 10^{-38}$ s (corresponding to the scale $\sim 10^{-2}$ cm to 10^3 cm after expansion).

The formation and evaporation of PBH, leads to effective violation of the baryon conservation law. There is a trivial possibility that the energy of the vanishing baryons could be used to furnish the heat and entropy to the remaining baryons (Carr and Rees, 1977). But in this case the PBHs are not specific: it is possible, at least in principle, to construct an initial singularity with metric perturbations strong enough to give the needed entropy without PBH formation and/or baryon non-conservation. In the particular case of the stiff equation of state $p = \varepsilon$ this was shown by the author (Zel'dovich, 1972, 1973).

Another, non-trivial possibility is to obtain the observed small baryonic charge asymmetry $((B - \bar{B})/\gamma \sim 10^{-8})$ from a totally symmetric initial situation. Laboratory experiments have firmly established the existence of CP-violation, i.e. the absolute difference of some properties (decay branching ratios, but not masses) of particles and antiparticles. But in order to obtain a baryon excess from a symmetric state, baryon charge non-conservation is also needed. In earlier schemes (Kuzmin, 1970; Sakharov, 1967), where it was introduced on the elementary particle – quark – level, it was difficult to reconcile the hypothesis with the stability of ordinary matter (Pati and Salam, 1973, disagree!).

Hawking pointed out that PBH formation and evaporation could give the answer. Zel'dovich (1976) tried to give a particular variant of this idea. The evaporation itself is charge-symmetric, but if unstable particles and antiparticles resulting from the evaporation have different decay properties, the partial accretion of decay products could lead to an asymmetric universe. Indeed the hypothesis is very speculative. All processes are thought to occur in the earlier part of the hadron era.

The last topic possibly connected with quantization of the metric concerns the 'large numbers' $\hbar c/Gm_p^2 = 10^{37}$ or $\hbar c/Gm_e^2 = 10^{43}$, which are obtained by confrontation of particle physics with gravitation.

In a fascinating paper Dirac (1937) made the negative statement that it is impossible to obtain such numbers in any sound local theory.

---

† A comprehensive review of the admissible amount of PBHs is given in the report of Novikov, Starobinsky and Zel'dovich (1977), delivered at the General Relativity Conference (GR VIII) in Waterloo, Canada, August 1977.

The quantum era and its effects

Dirac's positive statement was that the large numbers are connected with the cosmological large numbers, for example the square root, $\sqrt{N}$, of particle number inside the horizon $N = (c/H)^3 n \sim 10^{80}$, or, the ratio of the age of universe $t \sim 10^{18}$ s to the characteristic hadronic time $\tau = \hbar/m_p c^2 = 10^{-24}$ s, $t/\tau = 10^{42}$. These rough coincidences, if taken seriously, suggest non-local, Machian physics. They also imply that the physical 'constants', or at least one of them, are time-dependent. Usually G is under suspicion.

The negative statement is of course valid for the usual perturbation-type theories. In those theories non-dimensional constants are powers of 2, 3, π, $\hbar c/e^2 = 137$, etc. – and Dirac's statement is well founded. It is crazy to build a theory with $(\hbar c/e^2)^{18}$.

But in the last few years examples of a new type of theory emerged due mostly to Polyakov (1975) and colleagues (Belavin et al., 1975). Changes of field topology are connected with intermediate states not obeying the classical equations. One is concerned with barrier penetration (tunneling).

In those cases exponentially small dimensionless numbers occur: $e^{-16\pi^2/g^2} \sim 10^{-430}$, for example.

In gravitational theory, no dimensionless charges exist. Still one could obtain numbers of the type $e^{8\pi^2} = 6 \times 10^{30}$. For example, one could speculate about wormholes† making space non-orientable: a right-hand helicity particle falls into the wormhole, a left-hand helicity particle emerges. The wormhole plays the role of rest-mass in a Dirac spin $\frac{1}{2}$ particle theory linking together two different helicities. In a space seeded with virtual wormholes of this type, every spin $\frac{1}{2}$ particle would acquire rest-mass. The first estimate is $m = (\hbar c/G)^{1/2} = 10^{-5}$ g (just by dimensional arguments), but if the wormholes are tunneling one could imagine an exponentially small dimensionless coefficient bringing the particles mass into an acceptable range. I feel that the particular example of a wormhole is very uncertain and full of difficulties (neutrino with zero rest-mass and parity violation; Zel'dovich and Novikov, 1967). But the general statement that new types of theories could give exponentially small (or large) numbers seems to be proved. The cosmological Machian explanation of large numbers is no longer necessary (Zel'dovich, 1977b).

The development of general relativity reveals an oscillatory pattern. Einstein began with spacetime curvature. Much effort went into

---

† The idea of wormholes as carriers of electric charge was advocated by J. Wheeler (Wheeler, 1962).

Chapter 10. Cosmology and the early universe

reformulating the theory as a more familiar non-linear field theory for a tensor of rank two in flat spacetime. Cosmology, with its characteristic interest in closed worlds was always the bastion of the geometrical approach. In the last decade with interest shifted to black holes and wormholes, one is more conscious of the beauty and efficiency of geometrical ideas, but mixed with quantum theory prophecies are dangerous. *Qui vivra, verra!*

11. Anisotropic and inhomogeneous relativistic cosmologies

M. A. H. MACCALLUM

11.1 Introduction

General-relativistic cosmology was for many years concerned almost exclusively with the simplest possible models, those obeying the assumptions that (i) the universe is the same at all points in space (spatial homogeneity) and (ii) all spatial directions at a point are equivalent (isotropy). Robertson and Walker proved that any Riemannian spacetime which is isotropic at every point is spatially-homogeneous and that its metric can be expressed as

$$ds^2 = -dt^2 + l^2(t)[dr^2 + f^2(r)(d\theta^2 + \sin^2\theta\, d\phi^2)] \tag{11.1}$$

where $f = \sin r$ if $k = 1$, $f = r$ if $k = 0$, and $f = \sinh r$ if $k = -1$, k being a parameter giving the curvature of the spacelike surfaces ($t = $ constant). The time-dependent function l remains to be determined; within Einstein's theory, one of the first to do this was Friedmann, and the models that arise are thus often called Friedmann–Robertson–Walker (FRW) models.

The Hubble relation between the apparent magnitudes and the redshifts of galaxies, published in 1929, was interpreted in terms of a cosmic expansion, but the FRW models then gave too short a timescale compared with the age of the Earth, unless there was a positive cosmological constant, $\Lambda > 0$. Einstein originally introduced Λ partly to obtain a model which was stationary (i.e. a model in which the universe looks the same at all times). This same desire motivated the 'steady-state' universe in which the expansion was balanced by continuous creation of new matter. There being rather few observationally known facts, the arguments concerning choice between cosmological models were largely of a philosophical character. Bondi (1960) gives an excellent review of this situation.

The present-day pre-eminence of general-relativistic models is based partly on the success of Einstein's theory as a local theory of gravity (which is described in chapter 2 by C. M. Will) and partly on the

Chapter 11. Anisotropic and inhomogeneous relativistic cosmologies

improvement over the past 25 years in our observational knowledge of the universe. It now appears that FRW models fit most features of the actual cosmos quite well. However, this does not mean that a suitable FRW model represents the complete and final answer to our search for an understanding of the universe (still less that general relativity is the final and complete theory of gravity). Einstein's theory gives no unique prediction for the universe: its equations must be supplemented by information about symmetries, matter content, boundary conditions 'at infinity' and other global conditions (Ellis and Sciama, 1971). Alternatives to the FRW models have been intensively studied over the past 15 years or so, and it is my purpose here to review this work.

The corpus of literature on anisotropic and inhomogeneous relativistic cosmologies is by now so great that justice cannot be done to all of it in a brief review. Moreover, although very comprehensive texts on relativity, such as that by Misner, Thorne and Wheeler (1973) (whose notation and conventions I shall generally follow with the exception that I take $8\pi G = 1$, rather than $G = 1$), are now available, they do not usually give any introduction to symmetry groups of spacetime. In section 11.2, I therefore give some space to an extremely abbreviated discussion of this topic. Sections 11.3 and 11.4 review the geometry of the models, and sections 11.5 and 11.6 examine their more physical aspects.

Since one of my selection principles has been to focus on results with experimental consequences, it is necessary to state briefly the presently known data on isotropy and homogeneity. (It is so difficult to reconcile a non-evolving universe with the Hubble law, the radio-source counts and the cosmic microwave background radiation that very few workers now espouse such a model; except for a special case discussed in section 11.6, I will therefore ignore such models.)

11.1.1 Evidence concerning isotropy

(1) The distribution of galaxies on the sky is isotropic to perhaps 30 per cent and has been investigated out to about $1100/h$ Mpc, where h is the Hubble constant in units of $100 \text{ km s}^{-1} \text{ Mpc}^{-1}$ (Peebles, 1971). There is clustering on scales up to $5/h$ Mpc (Peebles and Groth, 1975) and perhaps even up to $200/h$ Mpc. A 'supercluster' with a scale of 20–30 Mpc centred on the Virgo cluster has been positively suggested in a series of papers by de Vaucouleurs (e.g. de Vaucouleurs, 1976).

(2) The Hubble constant is isotropic to perhaps 25 per cent, but its interpretation has been challenged (for a review see Rowan-Robinson,

1976). There may be anisotropy associated with the Virgo supercluster. Rubin et al. (1976) found an anisotropy in measurements of ScI galaxies at distances around $50/h$ Mpc, interpreted as a velocity of our galaxy of 454 ± 125 km s^{-1} in the direction $l = 163°$, $b = -11°$. Schechter (1977) demonstrated the dependence of this result on the method of data analysis, and from an independent nearby sample found a velocity 346 ± 76 km s^{-1} in the direction $l = 72°$, $b = 28°$. Another group (Jaakkola et al., 1976) found similar anisotropy associated with the passage of light through clusters.

(3) The radio-source distribution is highly isotropic (Golden, 1974; Gillespie, 1975; Webster, 1976; Machalski, 1977), so much so as to suggest that, in general, a cluster of galaxies can contain only one strong radio-source. The limits on anisotropy are now below 5 per cent variation.

(4) The cosmic X-ray background is isotropic to less than 5 per cent (Silk, 1971).

(5) The cosmic microwave background is isotropic to fractions of a per cent on all scales (see, e.g. Peebles, 1971), but a dipole anisotropy, corresponding to a velocity for our galaxy relative to the microwave background of 603 ± 60 km s^{-1} in the direction $l = 261°$, $b = 33°$, has recently been found (Smoot, Gorenstein and Muller, 1977). Taken with the data of Rubin et al. this suggests a picture of large inhomogeneities with relative velocities up to several hundred km s^{-1} (Rowan-Robinson, 1977).

(6) A cosmic magnetic field would break isotropy. The effects of any such field are masked by fields in the sources and in our galaxy, and no firm conclusion seems possible yet (Kobolov, Reinhardt and Sazonov, 1976; Ruzmaikin and Sokoloff, 1977).

(7) Brown has presented evidence for systematic orientation of the axes of galaxies (Brown, 1968), which could be related to the cosmic magnetic field, if any (Reinhardt, 1971). Recent work has not confirmed this result (Hawley and Peebles, 1975), though there may still be some real effects (Borchkhadze and Kogoshvili, 1976; Fesenko, 1976; Thompson, 1976). Local alignments have also been found for radio-sources (Willson, 1972).

The above evidence contains tantalizing hints of anisotropy, and the recent microwave measurements look especially definite; however, all these could, of course, turn out like the reported anisotropies in the cosmic helium abundance, which played an important part in motivating early work on anisotropic models, but are no longer thought to exist.

Chapter 11. Anisotropic and inhomogeneous relativistic cosmologies

11.1.2 Evidence for spatial homogeneity

If isotropy is strictly true, only variations depending on radial distance from us are admissible. Such a geocentric picture of the universe conflicts with current faith in the 'Copernican principle' that the Earth is not in a specially favoured position. This faith itself is of comparatively recent origin: it was only in the 1920s that non-geocentricity of our galaxy was established, and only with the revision of the distance scale by Baade in 1952 that our galaxy ceased to be considered several times larger than all others.

Spatial-homogeneity is also supported by the counts of galaxies and the linearity of the Hubble law.

11.1.3 Reasons for considering non-FRW models

Apart from observational reasons suggested above, there are various theoretical considerations that have motivated study of anisotropic and inhomogeneous cosmologies. Among these are the following:

(1) Statistical fluctuations in FRW models cannot collapse fast enough to form the observed galaxies. This suggests that there must be real inhomogeneities at all stages in the universe. Moreover, some perturbations of FRW models are decaying modes which would have been more important in the past.

(2) Some kind of 'singularity' in our past is strongly indicated (Hawking and Ellis, 1973) if certain reasonable conditions hold. However, it could differ greatly from the type found in FRW models (Belinskii, Lifschitz and Khalatnikov, 1970).

(3) In FRW models we are now receiving microwave radiation from regions which have never had causal communication, but which are nevertheless exactly alike. This philosophically unsatisfactory situation motivated the programme of 'chaotic cosmology' which sought mechanisms to explain why the observed degree of isotropy and homogeneity should exist regardless of the initial conditions (Misner, 1969a).

11.2 Spacetime symmetries

Suppose we have a vector field Y, with components Y^μ, which has integral curves $\gamma(u)$, the parameter u being such that $Y = d/du$ (see Misner, Thorne and Wheeler (1973), p. 229). We can define *Lie transport* or 'dragging along' by the vector field Y by considering the map of the spacetime manifold onto itself produced by moving every point a fixed parameter distance u along an integral curve, so that if γ passes through a point $p = \gamma(u')$, p is mapped to $\gamma(u + u')$. This induces maps of vectors, tensors and other geometric objects, all of which we shall denote by Φ_u. The *Lie derivative* of a geometric object G with respect to the vector field Y is then defined as

$$\mathcal{L}_Y G = \lim_{u \to 0} \left(\frac{G - \Phi_u G}{u} \right).$$

Actual calculation of the components of $\mathcal{L}_Y G$ is usually straightforward. For small u, Φ_u maps a point $x'^\mu = x^\mu - u Y^\mu$ to x^μ (to lowest order in u). If G is a vector field Z, then, evaluating at x^μ,

$$(\Phi_u Z)^\mu = \frac{\partial x^\mu}{\partial x'^\nu} Z^\nu(x') = \left(\delta^\mu_\nu + u \frac{\partial Y^\mu}{\partial x^\nu} \right) Z^\nu(x')$$

by the usual formulae for maps of manifolds (see, e.g., Hawking and Ellis (1973), section 2.4). Hence

$$(\Phi_u Z)^\mu = \left(\delta^\mu_\nu + u \frac{\partial Y^\mu}{\partial x^\nu} \right) \left(Z^\nu - \frac{\partial Z^\nu}{\partial x^\kappa} u Y^\kappa \right)$$

and thence

$$(\mathcal{L}_Y Z)^\mu = \frac{\partial Z^\mu}{\partial x^\nu} Y^\nu - \frac{\partial Y^\mu}{\partial x^\nu} Z^\nu \qquad (11.2)$$

or, in coordinate-free notation,

$$\mathcal{L}_Y Z = [Y, Z],$$

where the right-hand side is just the commutator of the two vector fields, which is easily calculated by treating them as differential operators. Note that any three vector fields obey the Jacobi identity

$$[X, [Y, Z]] + [Y, [Z, X]] + [Z, [X, Y]] = 0. \qquad (11.3)$$

If a covariant derivative has been defined, (11.2) can be re-expressed using it. It is easy to generalize (11.2) to obtain Lie derivatives of 1-forms, tensors and so on. For a Riemannian metric g,

$$(\mathcal{L}_Y g)_{\mu\nu} = g_{\mu\nu,\kappa} Y^\kappa + Y^\kappa{}_{,\mu} g_{\nu\kappa} + Y^\kappa{}_{,\nu} g_{\mu\kappa}$$
$$= Y_{\mu;\nu} + Y_{\nu;\mu}$$

using the associated covariant derivative.

Chapter 11. Anisotropic and inhomogeneous relativistic cosmologies

Transformations Φ_u which drag along a particular geometric object G are of special interest. Symmetry of a relativistic spacetime is expressed by invariance of the metric. If $\mathscr{L}_Y g = 0$, then Y is called a *Killing vector* (field) and the Φ_u are called *isometries* or *motions*. The set of all isometries forms a *Lie group* and the set of all Killing vectors forms the associated *Lie algebra*, using the commutator as the necessary product. (One can readily check, using the symmetries of the Riemann curvature tensor, that if X and Y are Killing vectors, so is $[X, Y]$.)

The relation between Lie algebras and Lie groups is discussed in many texts (e.g. Cohn, 1957) and less rigorous introductions are also available (e.g. Gursey, 1964; MacCallum, 1973). The application to transformations of manifolds has been discussed by Eisenhart (1933). Some key results are:

(i) A Lie group is a differential manifold on which the group operations are differentiable maps.

(ii) Each Lie group is associated with a unique Lie algebra.

(iii) Each Lie algebra defines a unique largest Lie group, of which all other groups having the same Lie algebra are homomorphic images.

(iv) The dimensions of the group's manifold and Lie algebra are the same.

(v) Every element b of the Lie group is associated with transformations L_b (left translation) and R_b (right translation) of the group itself, defined by $R_b(a) = ab$, $L_b(a) = ba$ for all a in G, and the Lie algebra then represents the infinitesimal left translations.

The algebraic structure of a Lie group can be described in terms of the Lie algebra by taking a basis $\{X_A | A = 1, \ldots, r\}$ of the algebra and forming all commutators of the X_A. Since the Lie algebra is closed we must have

$$[X_A, X_B] = C^D{}_{AB} X_D$$

where $C^D{}_{AB}$ are constants known as the *structure constants* of the Lie algebra. The commutators of the X_A are a basis for the subalgebra generated by all commutators, which is called the (first) *derived algebra* and is clearly invariant.

As well as isometries we shall be interested in transformations with the property that $\mathscr{L}_Y g$ is a multiple of g; these are called *conformal motions*. In this case $\mathscr{L}_Y g = \alpha g$ where α is some scalar function. When α is constant, but nonzero, the transformation is called a *homothety* or *similarity*.

Theorem 11.1

A Riemannian manifold V_d of dimension d cannot admit an algebra of Killing vectors of dimension greater than $\frac{1}{2}d(d+1)$.

Proof. Suppose the dimension is $r > \frac{1}{2}d(d+1)$. Take a basis $\{X_A\}$. Then, for any constants C_A, $Y = C_A X_A$ is a Killing vector. Now we can impose on the C_A the $\frac{1}{2}d(d+1)$ conditions that at a chosen point p in V_d all components of Y and its covariant derivative vanish (the symmetric part of the latter is zero anyway); there will be nonzero C_A satisfying these conditions. Now the Φ_u associated with Y fix p and vectors at p, and they preserve lengths and geodesy (since the metric is preserved), and so a point at a certain distance from p along a geodesic with a certain tangent vector at p must remain at that point. Thus Φ_u moves no point and $Y = 0$, contrary to our supposition that not all C_A were zero. ∎

A transformation group G is said to be *transitive* on a manifold M if for any x and y in M there is a transformation T in G such that $Tx = y$. If we take a point p, its *orbit* is the set $\{Tp\}$ of all points Tp for T in G, and its stability group is the set of all T in G such that $Tp = p$. The stability group of p is a Lie group, whose Lie algebra consists of those Y which vanish at p. In the case of an isometry group, the stability group is called the *isotropy group*. If the group of transformations has dimension r, the orbit through p has dimension d, and the stability group of p has dimension s, then

$$r = d + s. \tag{11.4}$$

When $d = r$, the group is called *simply-transitive* on its orbits. Then for each orbit one can define d tangent vector fields $\{B_A\}$, $A = 1, \ldots, d$, by

$$[Y_A, B_D] = 0 \tag{11.5}$$

for all A and D. The integrability conditions for these equations are automatically satisfied by virtue of (11.3) for Y_A, Y_B, B_D. To fix B_D completely one must specify their values at one point in each orbit. One can write $[B_A, B_D] = D^E{}_{AD} B_E$, and (11.3) for Y_A, B_D, B_E then shows that the $D^E{}_{AD}$ are constants in the orbits. By writing $B_D = \psi_D{}^A Y_A$ one finds from (11.5) that

$$Y_A \psi_D{}^E + C^E{}_{AD} \psi_D{}^B = 0$$

and hence that

$$[B_A, B_D] = C^E{}_{BF} \psi_A{}^F \psi_D{}^B Y_E. \tag{11.6}$$

Chapter 11. Anisotropic and inhomogeneous relativistic cosmologies

Thus for the particular initial condition $\boldsymbol{B}_A = -\boldsymbol{Y}_A$, $D^A{}_{BF} = C^A{}_{BF}$; a more usual choice is $\boldsymbol{B}_A = \boldsymbol{Y}_A$, giving $D^A{}_{BF} = -C^A{}_{BF}$. The vector fields $\boldsymbol{B}_A$ generate a Lie group called the *reciprocal* group of the original transformation group. (In the simply-transitive case, each orbit is isomorphic to the group itself: consider an element T of the group being mapped to Tp for some fixed p. The reciprocal group is generated by infinitesimal right translations and is the algebraic dual of the original group.) The reciprocal group does not automatically share any special properties of the original group, e.g. the reciprocal group of a group of isometries is not, in general, an isometry group. If the orbits are submanifolds of an n-dimensional manifold ($n > d$) on which the group acts, one can extend the basis $\boldsymbol{B}_A$ to a basis $\{\boldsymbol{B}_\mu\}$, $\mu = 1, \ldots, n$ of the whole tangent space at each point, where the $\boldsymbol{B}_\mu$ obey (11.5).

In the applications considered here, we shall be interested mainly in groups of dimension $r \leq 4$. We denote a group of dimension r by G_r.

Theorem 11.2

Every G_4 contains a subgroup G_3. This theorem is credited to Egorov (Petrov, 1969) and Kantowski (1966). The proof that follows is adapted from Kantowski's (Collins, 1977).

Proof. Consider the derived algebra L'. If it has dimension $q \leq 3$, then a basis of L' (plus, if necessary, vectors in L' to make the dimension 3) generates a subgroup G_3. If $q = 4$, then

$$\text{rank}\,(C^D{}_{AB}) = \text{rank}\,(\varepsilon^{ABEF} C^D{}_{EF}) = 4$$

where $C^D{}_{AB}$ is considered as a 4×6 matrix with indices D and $[AB]$, and ε^{ABEF} is any nonzero completely skew tensor on the algebra. The kernel of $(\varepsilon^{ABEF} C^D{}_{EF})$ as a map on the six-dimensional space with indices $[AB]$ has dimension 2. But the four vectors $C^G{}_{BD}$ ($G = 1, \ldots, 4$) are linearly independent, and the Jacobi identity

$$C^D{}_{[EF} C^G{}_{B]D} = 0$$

yields

$$\varepsilon^{ABEF} C^D{}_{EF} C^G{}_{BA} = 0$$

on multiplying by ε_{DPQR}. Thus $\dim[\ker(\varepsilon^{ABEF} C^D{}_{EF})] \geq 4$. This is a contradiction. ∎

Spacetime symmetries

To enumerate the G_3 one can work either from the dimension of the derived algebra (Bianchi, 1897) or as follows (Estabrook, Wahlquist and Behr, 1968; Ellis and MacCallum, 1969). Take the completely skew object with components ε^{DFB} such that $\varepsilon^{123} = 1$. Let

$$\tfrac{1}{2}C^D{}_{BC}\varepsilon^{BCF} = N^{(DF)} + \varepsilon^{DFB}A_B.$$

The Jacobi identity yields

$$N^{AD}A_D = 0. \tag{11.7}$$

By a change of basis N^{AD} can be reduced to the form diag (N_1, N_2, N_3) where each N_A is ± 1 or 0, and A_D is $(A, 0, 0)$ so that $N_1 A = 0$. This produces the list of types shown in table 11.1. We refer to these as Bianchi types, although Bianchi's parametrization differed in types VI and VII (in type VI_h, Bianchi's q is such that $h = -[(1+q)/(1-q)]^2$ while in VII_h, Bianchi's q obeys $h = q^2/(4-q^2)$). Table 11.1 also shows, for each Bianchi type, the dimension of the derived algebra (dim L') and the dimension of the orbits in the six-dimensional space of constants satisfying (11.7) under the action of the general linear group for dimension 3 (which is a nine-dimensional group); this latter is denoted p, and its values were listed by Siklos (1976). In types VI and VII p is 5 for a fixed h, but 6 if varying h is considered. The further columns in table 11.1 are discussed in section 11.3.4.

Table 11.1. *The Bianchi types (for explanation see sections 11.2 and 11.3.4)*

Class	Type	N_1	N_2	N_3	A	dim L'	p	r	s
A	I	0	0	0	0	0	0	1	2
	II	1	0	0	0	1	3	2	5
	VI_0	0	1	-1	0	2	5	3	7
	VII_0	0	1	1	0	2	5	3	7
	VIII	1	1	-1	0	3	6	4	8
	IX	1	1	1	0	3	6	4	8
B	V	0	0	0	1	2	3	1	5
	IV	0	0	1	1	2	5	3	7
	III(VI_{-1})	0	1	-1	1	1	5	3	7
	$VI_h (h<0)$	0	1	-1	$\sqrt{-h}$	2	5(6)	3(4)	7(8)
	$VII_h (h>0)$	0	1	1	$\sqrt{h}$	2	5(6)	3(4)	7(8)
	$VI_h (h=-1/9)$	0	1	-1	1/3	2	5	4	7

Chapter 11. Anisotropic and inhomogeneous relativistic cosmologies

Finally, the G_2 have only one commutator to consider. This is either zero,
$$[Y_1, Y_2] = 0, \tag{11.8}$$
or nonzero, in which case we can choose the basis so that one vector lies in the derived algebra, and scale the other so that
$$[Y_1, Y_2] = Y_1. \tag{11.9}$$

11.3 Spatially-homogeneous anisotropic metrics

11.3.1 Introduction

The aim of this and the next section is to introduce the various metrics that have been used in cosmological discussions, and to make some remarks concerning their geometry, field equations, known exact solutions and general behaviour, while leaving aside temporarily the more astrophysical aspects.

Models invariant under a group of motions acting transitively on the whole spacetime are in effect 'steady-state' models. Exact solutions have been extensively investigated by Ozsvath in several papers (e.g. Ozsvath, 1965). We take the view that these are not satisfactory cosmological models. They have, however, the merit of simplicity in that, in a suitably chosen basis, the field equations are purely algebraic.

The next-simplest class, mathematically, are the spatially homogeneous models, in which the field equations can be reduced to ordinary differential equations, with time as the independent variable. In such cases the usual methods for qualitative treatment are available, e.g. those based on a Hamiltonian formulation or on reducing the equations to an autonomous system.

Spatially-homogeneous models admit a group G_r of isometries transitive on spacelike three-dimensional orbits. Since the (positive-definite) metric at a point is invariant under the rotation group, any isotropy group must be a subgroup of this. However, it has no subgroups of dimension 2, as is well known (two rotations about different axes produce a rotation about a third perpendicular axis). Thus since $d = 3$ and $s \neq 2$, $r \neq 5$. If $r = 6$, then $s = 3$ and we have the Robertson–Walker cases. Thus the anisotropic cases must have $s = 1$, $r = 4$ or $s = 0$, $r = 3$. These are discussed below.

Many forms of energy–momentum tensor have been used in discussing solutions of Einstein's equations. The most common is probably
$$T_{\mu\nu} = (p + \rho)u_\mu u_\nu + pg_{\mu\nu} \tag{11.10}$$

where u_μ is a unit timelike vector ($u^\mu u_\mu = -1$). This represents the energy–momentum of a perfect fluid flowing in the direction u^μ, with pressure p and energy density ρ. In the early epochs of a big bang cosmology one expects 'radiation' in thermal equilibrium

$$p = \rho/3, \tag{11.11}$$

and in the late stages, where galaxies move freely, one expects 'dust'

$$p = 0. \tag{11.12}$$

Many authors have used

$$p = (\gamma - 1)\rho \tag{11.13}$$

where $1 \leq \gamma \leq 2$, the lower limit being required for mechanical stability and the upper so that the sound speed is less than the speed of light. The principal addenda to (11.10) have been terms arising from an electromagnetic field, which contributes

$$T_{\mu\nu} = F_\mu{}^\kappa F_{\nu\kappa} - \tfrac{1}{4} g_{\mu\nu} (F^{\kappa\lambda} F_{\kappa\lambda}) \tag{11.14}$$

where $F_{\mu\nu}$ is the usual electromagnetic field tensor, and terms arising from viscous or other anisotropic stresses which add to (11.10) a term $\pi_{\mu\nu}$ where

$$\pi^\mu{}_\mu = 0, \quad \pi_{\mu\nu} u^\nu = 0, \quad \pi_{[\mu\nu]} = 0. \tag{11.15}$$

In all spatially-homogeneous models the unit normal to the orbits of the group is a geodesic vector field invariant under the group. It can thus be taken to be $n_\mu = -t_{,\mu}$ so that the orbits are surfaces ($t = $ constant) and the metric becomes

$$ds^2 = -dt^2 + g_{ij} \omega^i \omega^j \tag{11.16}$$

where ω^i are three 1-forms in the orbits. The proof of this is as follows. Take any point p in the orbit $\{Tp\}$. Erect the unit normal at p, and construct the geodesic with this as initial position and tangent vector. Let q lie on this geodesic and its orbit be $\{Tq\}$. Take any transformation T in the group of isometries. An orbit is invariant under T; hence any curve in an orbit is mapped to another curve in the orbit, and so tangent vectors in the orbit are mapped to tangent vectors in the orbit. The normal is thus mapped to a normal, since the metric is invariant. Thus the normal at p gives rise to a geodesic vector field n of unit magnitude and normal to $\{Tp\}$ at every point. Also, in $\{Tp\}$, $n_\mu Y_A{}^\mu = 0$ for all Killing vector fields Y_A of the group. Then

$$(n_\mu Y_A^\mu)_{,\nu} n^\nu = Y_A^\mu (n_{\mu;\nu} n^\nu) + Y_{A;\nu}^\mu n_\mu n^\nu = 0 \tag{11.17}$$

Chapter 11. Anisotropic and inhomogeneous relativistic cosmologies

as n is geodesic and Y_A is Killing. Thus n is also normal to $\{Tq\}$ and so is as stated above.

11.3.2 Kantowski–Sachs models

These are models with $r = 4$. We know that in all cases there is a subgroup G_3. If this acts simply-transitively, the cases with $r = 4$ are merely special cases of the $r = 3$ universes, which are called Bianchi models and are discussed below. It remains to search for cases where the G_3 acts multiply-transitively, and therefore on two-dimensional surfaces of maximal symmetry (by Theorem 11.1). These 2-spaces can have constant positive, zero or negative curvature K. The metrics are proportional to

$$d\theta^2 + f^2(\theta) d\phi^2 \tag{11.18}$$

where $f(\theta) = \sin \theta (K > 0)$, $\theta (K = 0)$ or $\sinh \theta (K < 0)$. They admit a G_3 with generators

$$\left. \begin{array}{l} X_1 = \sin \phi \partial_\theta + (f'/f) \cos \phi \partial_\phi \\ X_2 = \cos \phi \partial_\theta - (f'/f) \sin \phi \partial_\phi \\ X_3 = \partial_\phi \end{array} \right\} \tag{11.19}$$

where $\partial_x = \partial/\partial x$. The Bianchi type of this G_3 is IX$(K > 0)$, VII$_0(K = 0)$, or VIII$(K < 0)$.

The fourth Killing vector field X_4 must be nonzero everywhere (or the orbits of the G_4 are not three-dimensional). By an argument similar to that used in deducing (11.16) the unit normal in the hypersurface of homogeneity to the two-dimensional surfaces with metric (11.18) must be invariant under (11.19). Allowing for the time dependence, the metric of the spacetime must be

$$-dt^2 + A^2(t) dx^2 + B^2(t) g^2(x)(d\theta^2 + f^2(\theta) d\phi^2).$$

The Jacobi identities (11.3) give restrictions on the commutator of X_4 with the generators (11.19) which were investigated by Kantowski (see Collins, 1977). There are the following possibilities:

(i) $K > 0$, $[X_4, X_i] = 0$, $i = 1, 2, 3$;
(ii) $K < 0$, $[X_4, X_i] = 0$, $i = 1, 2, 3$;
(iii) $K = 0$, $[X_4, X_3] = CX_4$, $[X_4, X_1] = [X_4, X_2] = 0$;
(iv) $K = 0$, $[X_4, X_3] = 0$, $[X_4, X_1] = X_1, [X_4, X_2] = X_2$.

In cases (ii) to (iv) subalgebras of the following Bianchi types and generators exist, their generators are nowhere zero, and they are thus

simply-transitive on the hypersurfaces of homogeneity:
  (ii) (X_1+X_3, X_2, X_4),   Bianchi type III;
  (iii) (X_1, X_2, X_4),   Bianchi type I;
  (iv) (X_1, X_2, X_4),   Bianchi type V.

No such subalgebra is possible in case (i) since it would intersect the algebra generated by (11.19) in a two-dimensional subalgebra, and such a subalgebra does not exist (this is another way of saying that the rotation group SO(3) has no two-dimensional subgroup). Thus case (i) is the unique case of a spatially-homogeneous universe with no simply-transitive G_3 of isometries. It was given as case I in Kantowski and Sachs (1966) and is thus known as the Kantowski–Sachs metric (they erroneously included case (ii) above, but this was corrected by Kantowski (1966)).

In cases (i) and (ii) X_4 must be normal to the two-dimensional spaces with metric (11.18) since it is invariant under the isotropy at each point. Moreover, in all cases $A^{-1}g_{\mu\nu}X_4^\mu\delta_1^\nu$ is a constant, where $A^{-1}\delta_1^\nu$ are the components of the (geodesic) unit vector field normal to the 2-surfaces (11.18), by an argument similar to (11.17). Thus in cases (i) and (ii) X_4 is the normal to (11.18) and $g(x)$ is constant (and so can be taken to be 1). In case (iii) it is easy, by expressing X_4 as a linear combination of δ_1^ν and X_1, X_2, X_3 to verify, using the commutators given, that $C=0$ is the only possibility, and then again it follows that $g(x)=1$ may be chosen. These three cases have metrics

$$ds^2 = -dt^2 + A^2(t)\,dx^2 + B^2(t)(d\theta^2 + f^2(\theta)\,d\phi^2) \qquad (11.20)$$

and have been investigated by many authors, starting with Kompaneets and Chernov (1964). The field equations are readily calculated and give

$$T^1{}_1 = 2\ddot{B}/B + \dot{B}^2/B^2 + k/B^2 + \Lambda$$
$$T^2{}_2 = T^3{}_3 = \ddot{A}/A + \ddot{B}/B + \dot{A}\dot{B}/AB + \Lambda$$
$$T^4{}_4 = \dot{A}\dot{B}/AB + \dot{B}^2/B^2 + k/B^2 + \Lambda$$
$$T^\mu{}_\nu = 0 \quad (\mu \neq \nu)$$

where $k = +1, 0, -1$ if $K>0$, $K=0$, $K<0$, $(K=k/B^2)$, and the coordinates (x^1, x^2, x^3, x^4) are (x, θ, ϕ, t).

Exact solutions for the Kantowski–Sachs case, $k=1$, have been given for dust, radiation, cosmological constant, 'stiff' matter and combinations (Kantowski, 1966), for dust plus electromagnetic field (Doroshkevich, 1965; Shikin, 1966; Thorne, 1967), and for charged dust with magnetic

Chapter 11. Anisotropic and inhomogeneous relativistic cosmologies

field (Shikin, 1974). Vajk and Eltgroth (1970) give a compendium of solutions for all three k.

The $k=1$ cases are spherically symmetric, in that they contain orbits invariant under the rotation group. However, one must use the term with caution: at a given t, all the spheres have the same area and cannot be regarded as stacked inside one another. Even when they can be, as in the Schwarzschild solution, there may be no regular point of the Riemannian manifold at the centre. I shall reserve the term centro-symmetric for those cases where there truly is a centre.

11.3.3 The Bianchi metrics

These are the spatially-homogeneous models in which a three-dimensional isometry group acts simply-transitively on the hypersurfaces of homogeneity. Having found vector fields $\boldsymbol{B}_a$, $a = 1, 2, 3$, obeying (11.5) in each orbit, the metric (11.16) becomes

$$ds^2 = -dt^2 + g_{ab}\boldsymbol{B}^a\boldsymbol{B}^b \tag{11.21}$$

where $(dt, \boldsymbol{B}^a)$ are dual to $(\partial/\partial t, \boldsymbol{B}_a)$ and g_{ab} is a function of t alone.

A particular orbit is isomorphic to the group, and coordinates can thus be found directly from the canonical coordinates of the group (Cohn, 1957) with origin determined by the choice of isomorphism. Such coordinates were explicitly constructed by Siklos (1976). The choices of origin in neighbouring orbits are usually made by requiring the coordinates to be comoving with respect to some particular set of timelike curves. These can be the integral curves of $\boldsymbol{n}$, or, if there is matter present with 4-velocity $\boldsymbol{u} \neq \boldsymbol{n}$, of $\boldsymbol{u}$. In the latter case the t coordinate may be replaced by proper time τ along the $\boldsymbol{u}$ curves. If we replace the basis $(\boldsymbol{n}, \boldsymbol{B}_a)$ by the basis $(\boldsymbol{u}, \boldsymbol{B}_a)$, the dual 1-forms are $(d\tau, \hat{\boldsymbol{B}}^a)$ where $dt = u^4 d\tau$ and $\boldsymbol{B}^a = u^a d\tau + \hat{\boldsymbol{B}}^a$, the u^μ being such that $\boldsymbol{u} = u^4\boldsymbol{n} + u^a\boldsymbol{B}_a$. Then the metric (11.21) becomes

$$ds^2 = -d\tau^2 + 2g_{ab}u^a\,d\tau\,\hat{\boldsymbol{B}}^b + g_{ab}\hat{\boldsymbol{B}}^a\hat{\boldsymbol{B}}^b \tag{11.22}$$

where g_{ab} and u^a can be considered as functions of τ alone. The remaining freedom is in fixing the $\boldsymbol{B}_a$ by specifying them at one point in each orbit.

Let us denote by $\boldsymbol{E}_a$ a choice of $\boldsymbol{B}_a$ such that in each surface

$$[\boldsymbol{E}_a, \boldsymbol{E}_b] = -C^d{}_{ab}\boldsymbol{E}_d \tag{11.23}$$

where the structure constants take the canonical form of table 11.1. It should be noted that this basis is fixed only up to the stability group of the structure constants, which is a group of dimension $9-p$; since the same is true of the basis choice for the Lie algebra, and thus of the choice of the spatial coordinates mentioned above, we may always assume that at the origin of spatial coordinates $\boldsymbol{E}_a = \boldsymbol{Y}_a$ and the coordinates are canonical. Explicit expressions for $\boldsymbol{E}_a$, $\boldsymbol{Y}_a$, and $\boldsymbol{\omega}^a$, where the $\boldsymbol{\omega}^a$ are the 1-forms corresponding to the $\boldsymbol{E}_a$ in a dual basis, are given in table 11.1. We must have

$$d\omega^a = \tfrac{1}{2} C^a{}_{bd} \omega^b \wedge \omega^d, \tag{11.24}$$

and if t is used as the time coordinate and the space coordinates comove with $\boldsymbol{n}$, $[\boldsymbol{n}, \boldsymbol{E}_a] = 0$, while if τ is used as the time coordinate and the space coordinates comove with $\boldsymbol{n}$, $[\boldsymbol{u}, \boldsymbol{E}_a] = 0$.

The main alternative is to take an orthonormal basis. If this is chosen with $\boldsymbol{e}_4 = \boldsymbol{n}$, the spacelike vectors lie in the surfaces ($t = $ constant). Ryan and Shepley (1975) use $\boldsymbol{B}^i = b^i{}_j \omega^j$ where $b^i{}_j$ is the symmetric square-root matrix of g_{ab}. Ellis and MacCallum (1969) (and, in different terms, Estabrook, Wahlquist and Behr (1968)) used a basis $\boldsymbol{e}_a$ chosen so that the commutators

$$[\boldsymbol{e}_a, \boldsymbol{e}_b] = D^d{}_{ab} \boldsymbol{e}_d$$

have the form

$$D^d{}_{ab} = \varepsilon_{abf} n^{df} + \delta^d_b a_a - \delta^d_a a_b \tag{11.25}$$

where $n^{df} = \text{diag}(n_1, n_2, n_3)$ and $\boldsymbol{a} = (a, 0, 0)$. In this case $h = a^2/n_2 n_3$. The duals of the $\boldsymbol{e}_\mu$ can be denoted $\boldsymbol{e}^\mu$. They are determined up to some subgroup of the rotation group preserving the canonical form of the $\boldsymbol{n}$ and $\boldsymbol{a}$.

For models with matter content such that $\boldsymbol{u} \neq \boldsymbol{n}$ ('tilted' models) one can make a Lorentz transform of the $(\boldsymbol{n}, \boldsymbol{e}_a)$ basis, say to $(\boldsymbol{u}, \hat{\boldsymbol{e}}_a)$ where

$$\begin{aligned} \boldsymbol{u} &= \boldsymbol{n} \cosh \beta + \boldsymbol{c} \sinh \beta \\ \boldsymbol{n} &= \boldsymbol{u} \cosh \beta - \hat{\boldsymbol{c}} \sinh \beta, \end{aligned} \tag{11.26}$$

$\boldsymbol{c}$ being in the $\boldsymbol{e}_a$ plane and $\hat{\boldsymbol{c}}$ in the $\hat{\boldsymbol{e}}_a$ plane. This was used by King and Ellis (1973).

One can also use a group-invariant null-basis $(\boldsymbol{l}, \boldsymbol{k}, \boldsymbol{m}, \bar{\boldsymbol{m}})$ with $\boldsymbol{m}, \bar{\boldsymbol{m}}$ lying in the $\boldsymbol{e}_a$ plane, to define the Newman–Penrose variables and thus to study algebraically special cases (Siklos, 1976) and 'whimper' singularities (Siklos, 1978).

Chapter 11. Anisotropic and inhomogeneous relativistic cosmologies

Table 11.2. *Standard forms of the Y_a, E_a and ω^a (see text, section 11.3.3)*

Y_a	I	$\partial_1, \quad \partial_2, \quad \partial_3$
	II	$\partial_1, \quad \partial_2, \quad \partial_3 + x^2 \partial_1$
	IV	$\partial_1 - x^2 \partial_2 - (x^2 + x^3) \partial_3, \quad \partial_2, \quad \partial_3$
	V	$\partial_1 - x^2 \partial_2 - x^3 \partial_3, \quad \partial_2, \quad \partial_3$
	VI (III if $h=1$)	$\partial_1 + (x^3 - Ax^2) \partial_2 + (x^2 - Ax^3) \partial_3, \quad \partial_2, \quad \partial_3$
	VII	$\partial_1 + (x^3 - Ax^2) \partial_2 - (x^2 + Ax^3) \partial_3, \quad \partial_2, \quad \partial_3$
	VIII	$\partial_1, \quad -\sinh x^1 \tanh x^2 \partial_1 + \cosh x^1 \partial_2 - \sinh x^1 \operatorname{sech} x^2 \partial_3,$
		$\cosh x^1 \tanh x^2 \partial_1 - \sinh x^1 \partial_2 + \cosh x^1 \operatorname{sech} x^2 \partial_3$
	IX	$\partial_1, \quad \sin x^1 \tan x^2 \partial_1 + \cos x^1 \partial_2 + \sin x^1 \sec x^2 \partial_3,$
		$\cos x^1 \tan x^2 \partial_1 - \sin x^1 \partial_2 + \cos x^1 \sec x^2 \partial_3$
E_a	I	$\partial_1, \quad \partial_2, \quad \partial_3$
	II	$\partial_1, \quad \partial_2 + x^3 \partial_1, \quad \partial_3$
	IV	$\partial_1, \quad e^{-x^1}(\partial_2 - x^1 \partial_3), \quad e^{-x^1} \partial_3$
	V	$\partial_1, \quad e^{-x^1} \partial_2, \quad e^{-x^1} \partial_3$
	VI(III)	$\partial_1, \quad e^{-Ax^1}(\cosh x^1 \partial_2 + \sinh x^1 \partial_3),$
		$e^{-Ax^1}(\sinh x^1 \partial_2 + \cosh x^1 \partial_3)$
	VII	$\partial_1, \quad e^{-Ax^1}(\cos x^1 \partial_2 - \sin x^1 \partial_3), \quad e^{-Ax^1}(\sin x^1 \partial_2 + \cos x^1 \partial_3)$
	VIII	$\operatorname{sech} x^2 \cos x^3 \partial_1 - \sin x^3 \partial_2 - \tanh x^2 \cos x^3 \partial_3,$
		$\operatorname{sech} x^2 \sin x^3 \partial_1 + \cos x^3 \partial_2 - \tanh x^2 \sin x^3 \partial_3, \quad \partial_3$
	IX	$\sec x^2 \cos x^3 \partial_1 - \sin x^3 \partial_2 + \tan x^2 \cos x^3 \partial_3,$
		$\sec x^2 \sin x^3 \partial_1 + \cos x^3 \partial_2 + \tan x^2 \sin x^3 \partial_3, \quad \partial_3$
ω^a	I	$dx^1 \qquad dx^2 \qquad dx^3$
	II	$dx^1 - x^3 dx^2 \quad dx^2 \quad dx^3$
	IV	$dx^1 \qquad e^{x^1} dx^2 \qquad e^{x^1}(x^1 dx^2 + dx^3)$
	V	$dx^1 \qquad e^{x^1} dx^2 \qquad e^{x^1} dx^3$
	VI(III)	$dx^1, \quad e^{Ax^1}(\cosh x^1 dx^2 - \sinh x^1 dx^3),$
		$e^{Ax^1}(-\sinh x^1 dx^2 + \cosh x^1 dx^3)$
	VII	$dx^1, \quad e^{Ax^1}(\cos x^1 dx^2 - \sin x^1 dx^3),$
		$e^{Ax^1}(\sin x^1 dx^2 + \cos x^1 dx^3)$
	VIII	$\cosh x^2 \cos x^3 dx^1 - \sin x^3 dx^2, \cosh x^2 \sin x^3 dx^1 + \cos x^3 dx^2, \quad \sinh x^2 dx^1 + dx^3$
	IX	$\cos x^2 \cos x^3 dx^1 - \sin x^3 dx^1, \cos x^2 \sin x^3 dx^1 + \cos x^3 dx^2, \quad -\sin x^2 dx^1 + dx^3$

It should be noted that if for a particular a, $[n, e_a] = -\theta_a e_a$ (no sum on a) and $e^a([n, e_b]) = 0$ ($b \neq a$), then one may take $e_a = l_a E_a$ where $\dot{l}_a / l_a = \theta_a$ (no sum on a), and ˙ denotes d/dt.

To write out and follow through all the consequences of all these forms of the equations would take too much space. Unfortunately, no one form is really suitable for discussion of all aspects of the models. For the (u, E_a) basis, equations were given by Grishchuk (1970). For the (n, E_a) basis, the Ricci curvature is

$$R_{44} = -\dot{\theta} - \theta_{pq}\theta^{pq} \qquad (11.27)$$

$$R_{4q} = \theta^p{}_j C^j{}_{pq} + \theta^s{}_q C^r{}_{sr} \qquad (11.28)$$

$$R_{ab} = \dot{\theta}_{ab} + \theta\theta_{ab} - 2\theta_{ac}\theta^c{}_b + R^*{}_{ab} \qquad (11.29)$$

where $\theta_{ab} = \tfrac{1}{2}\dot{g}_{ab}$, and $R^*{}_{ab}$ is the curvature of the surfaces ($t = $ constant),

$$R^*{}_{ab} = \tfrac{1}{2}C^c{}_{ck}(C^r{}_{tb}g_{ar} + g_{br}C^r{}_{ta})g^{kt} - \tfrac{1}{2}C^c{}_{ka}(C^k{}_{cb} + g_{cm}g^{kt}C^m{}_{tb})$$
$$+ \tfrac{1}{4}C^m{}_{sk}C^r{}_{tc}g_{bm}g_{ar}g^{kt}g^{cs}.$$

The G_3 may be only part of the full group of motions. In this case there is an isotropy group acting in the spacelike hypersurfaces ($t = $ constant). Spacetimes with a spatial isotropy at every point are called locally rotationally symmetric (LRS) (Ellis, 1967). The LRS Bianchi spacetimes belong to types I, II, VII$_0$, VIII, IX, V, VII$_h$, and III. The LRS type I and VII$_0$, and type V and VII$_h$, are the same: they are in fact cases (iii) and (iv) of section 11.3.2. Bianchi type III LRS spaces with $n^a{}_a = 0$ are case (ii) of section 11.3.2. Bianchi type III LRS spaces with $n^a{}_a \neq 0$ are the same as LRS Bianchi VIII spaces. The Robertson–Walker spaces, with a three-dimensional isotropy group, are of types IX ($k = 1$), I (and VII$_0$) ($k = 0$), and V (and VII$_h$) ($k = -1$). For these results see Ellis and MacCallum (1969).

A considerable number of exact solutions are known; to avoid a long list of references I refer the reader to the table in MacCallum (1973). Some new solutions have become known since then. They are as follows.

Siklos (1976) has found algebraically special solutions of Petrov type N in Bianchi types VI$_h$, VII$_h$, and IV. The last of these has also been found by Harvey and Tsoubelis (1977). Barnes (1978) has found electromagnetic solutions of types I, II, VI$_0$ and VII$_0$ (some of which may duplicate known solutions). A class of Bianchi VI$_0$ fluid and electromagnetic solutions has been discussed by Dunn and Tupper (1976), see also Tupper (1977). An exact solution for Bianchi type VII$_h$ has been found by Lukash (1974). Solutions of Bianchi type VII$_h$ with dust and electromagnetic field have been considered by Melvin (1975). Collins, Glass and Wilkinson (unpublished work, 1977) have found some new fluid solutions for LRS type VIII models, although in these $p(\rho)$ is not based on physical considerations. Batakis and Cohen (1972) found a scalar field generalization of the electromagnetic LRS type IX solution.

Chapter 11. Anisotropic and inhomogeneous relativistic cosmologies

11.3.4 General properties of spatially-homogeneous models

The study of Bianchi models as cosmological models began with the work of Taub (1951), Raychaudhuri (1958), and Heckmann and Schucking (1962). The form of (11.29) for R_{ab} shows that the spatial curvature R^*_{ab} acts as a forcing term for θ_{ab}. Thus the simplest models are those where $R^*_{ab} = \frac{1}{3} R^* g_{ab}$. These are models of Bianchi types I ($R^* = 0$) and V ($R^* < 0$), and the FRW case of type IX ($R^* > 0$), together with special cases of types VII_0 and VII_h (the LRS cases and the Demianski–Grishchuk metric). These models were the earliest to be well studied (see e.g. Heckmann and Schucking, 1962). There has also been a considerable effort put into these types on the grounds that they contain FRW models (see below).

By now a considerable amount of information on all the Bianchi types and the Kantowski–Sachs model is available, and my aim here is to summarize it under various headings.

Generality

The values of p in table 11.1 show that, in terms of the arbitrariness of the stucture constants, types VI_h, VII_h, VIII and IX are the most general. The same conclusion applies when one considers assigning approximate symmetry to spacelike slices of inhomogeneous spacetimes (Spero and Baierlein, 1977).

Another aspect of this is the consideration of the degrees of freedom of the initial value (Cauchy) problem. Having chosen the canonical forms of table 11.1, the 12 variables g_{ab} and θ_{ab} are, for vacuum, related by four constraints given by (11.27) and (11.28). There are $9-p$ arbitrary degrees of freedom in choice of the E_a, and one in the origin of t. Thus in general there are $p-2$ real degrees of freedom. This has its maximum value (4) in types VIII, IX, VI_h and VII_h. In Bianchi types I, II and VI ($h = -1/9$) the four constraints following from (11.27) and (11.28) reduce, for vacuum, to 1, 3 and 3 respectively. Thus the number of essential parameters in a general vacuum solution for each type is that given as r in table 11.1. In a non-vacuum solution there will be additional data for the matter content. For example, for a perfect fluid with given equation of state $p = p(\rho)$, the additional data are initial values of ρ and three components of **u**. Thus the number of degrees of freedom in general becomes $s = r + 4$. However, Bianchi types I, II and VI ($h = -1/9$) are exceptional for the same reason as above. The values of s are given in table 11.1.

'Tilt' and rotation

A fluid model is said to have nonzero tilt when β in (11.26) is nonzero. From (11.20) and (11.28) it is found that the tilt must be zero for the Kantowski–Sachs metrics and Bianchi type I. In Bianchi types II and VI ($h = -1/9$) the tilt can be nonzero, but only two components of $\boldsymbol{u}$ can be chosen arbitrarily. King and Ellis (1973) have shown that the vorticity of the fluid is determined by

$$\omega_{ab} = -\tfrac{1}{2}\sinh\beta\varepsilon_{abc}(n^{cd}c_d + \varepsilon^{cde}c_d a_e)$$

in the notation of (11.25) and (11.26). Thus the Bianchi II models give zero vorticity, while in types VIII and IX tilted fluids must have vorticity. In class B, if $\boldsymbol{a}$ and $\boldsymbol{c}$ are parallel, $\omega_{ab} = 0$ (due to the Jacobi identities); otherwise, $\omega_{ab} \neq 0$ unless the group is Bianchi III. Moreover, the vorticity vector $\boldsymbol{\omega}$ cannot be orthogonal to $\boldsymbol{c}$ in types VIII and IX, but must be in type V. In type V and some type IV models $n^{ab}c_b = 0$ and $\boldsymbol{\omega} \neq 0$, and in these cases $\boldsymbol{\omega}$ is perpendicular to $\boldsymbol{n}$, $\boldsymbol{a}$ and $\boldsymbol{c}$. Finally, $\boldsymbol{\omega}$ is parallel to $\boldsymbol{c}$ only in class A models with $\boldsymbol{c}$ an eigenvector of n^{ab}. The only tilted LRS Bianchi models are of type V (or VII$_h$), i.e. case (iv) of section 11.3.2, and have no vorticity. For dust, the exact solutions were found by Farnsworth. The only tilted Bianchi model with $R^*{}_{ab} = \tfrac{1}{3}R^* g_{ab}$ other than this Bianchi V LRS case is the special solution of type VII$_0$ discussed by Demianski and Grishchuk, which has a non-physical equation of state.

Diagonalization

The simplest metrics of the form (11.21) with the choice (11.24) are those in which g_{ab} is a diagonal matrix. In itself, this condition merely leads to some odd restrictions on $T_{\mu\nu}$. For physical interest we must first restrict $T_{\mu\nu}$ in some way, and the most natural is to assume that $T_{\mu\nu}$ is simultaneously diagonal. Then any class A metric is possible and, except in types I and II, diagonality of $T_{\mu\nu}$ forces that of g_{ab} (MacCallum, Stewart and Schmidt, 1970). In class B, the diagonality conditions either give rise to an anisotropic stress violating the dominant-energy condition, or lead to the conditions that a_b is an eigenvector of the shear of the $\boldsymbol{n}$ congruence and $n^b{}_b \neq 0$ (MacCallum, 1972). The diagonality conditions for $T_{\mu\nu}$ alone force a_b to be a shear and stress eigenvector, except in type VI ($h = -1/9$), and so in this case

$$ds^2 = -dt^2 + g_{11}\omega^1\omega^1 + \sum_{2,3} g_{AB}\omega^A\omega^B.$$

Chapter 11. Anisotropic and inhomogeneous relativistic cosmologies

It should perhaps be noted that the reason why the diagonalization proof does not work in types I and II is because there may simultaneously be off-diagonal components of the shear of the **n** congruence and off-diagonal terms arising from the Fermi-rotation, i.e. the rotation of the axes relative to a dynamically non-rotating frame. In the other Bianchi types, these terms are coupled together in the T_{4a} equations and so the diagonality condition for $T_{\mu\nu}$ leads to their simultaneous elimination. The Fermi-rotation of axes should not be confused with the vorticity of a fluid matter content; it is non-physical in the sense that axes can always be chosen so that it vanishes, and it is retained in the Bianchi models in order to permit the specializations based on the canonical forms of the structure constants.

Lagrangians and Hamiltonians

It is extremely convenient to be able to write a Lagrangian (or Hamiltonian) form for the field equations of Bianchi models in which only the time-dependent functions are varied. Such a Lagrangian can formally be obtained by integrating the usual Lagrangian for Einstein's equations over the spatial variables. However, the resulting Lagrangian may not be correct. Rather than plough through the full computation, let me outline the source of the difficulty by a simple example.

Suppose we have to determine $y(x)$ from the Lagrangian integral

$$I = \int_{x_1}^{x_2} f(x, y, y') \, dx$$

where $y' = dy/dx$. On varying y to $y + \delta y$ we obtain

$$\delta I = \int_{x_1}^{x_2} \left(\frac{\partial f}{\partial y} \delta y + \frac{\partial f}{\partial y'} \delta y' \right) dx$$

$$= \left[\frac{\partial f}{\partial y'} \delta y \right]_{x_1}^{x_2} + \int_{x_1}^{x_2} \delta y \left(\frac{\partial f}{\partial y} - \frac{d}{dx} \left(\frac{\partial f}{\partial y'} \right) \right).$$

Then the usual Euler–Lagrange equation can be deduced,

$$\frac{d}{dx} \left(\frac{\partial f}{\partial y'} \right) - \frac{\partial f}{\partial y} = 0,$$

provided that the boundary terms

$$\left[\frac{\partial f}{\partial y'} \delta y \right]_{x_1}^{x_2} = 0.$$

Spatially-homogeneous anisotropic metrics

This is usually enforced either by taking $\delta y = 0$, or by insisting on the 'natural boundary conditions' $\partial f/\partial y' = 0$, at x_1 and x_2.

In the case of Einstein's equations the region of integration is four-dimensional and its boundary is three-dimensional. If the variations are spatially-homogeneous and vanish on the boundary, they must vanish everywhere. So to deduce correct equations from a spatially-homogeneous variation we must satisfy the natural boundary conditions. These arise from integrating a divergence, and the divergences involve the a_b of (11.25). Thus the class B spacetimes are the ones that give difficulty. Regarding the map of variations in $g_{\mu\nu}$ to variations of the action integral as a map from 10 dimensions to one dimension, it is clear that, at most, one equation is wrong.

The Lagrangian and Hamiltonian techniques were originally introduced by Misner for types I (1968) and IX (1969b, c). Hughston and Jacobs extended this to other Bianchi types (see Ryan, 1972a), but Estabrook and Wahlquist, and independently I, noted the failure in class B spaces, where only the special cases with $n^a{}_a = 0$ seemed to work (MacCallum, 1971). The explanation above was given by MacCallum and Taub (1972), following an idea of Hawking (1969). The Hamiltonian version was checked by Ryan (1974), who had been informed of the work by Ehlers. Unfortunately, MacCallum and Taub gave erroneous arguments about the number of incorrect equations and the possibility of amending the Lagrangian by adding constraints (which left unclear the way in which the $n^a{}_a = 0$ cases worked), while Ryan, as a result of a miscalculation, ascribed the result to the wrong cause. These points have recently been cleared up by Sneddon (1975), although there are still some open questions concerning amending the Lagrangian.

The Lagrangian for class A spaces with perfect-fluid matter content and $u = n$ can be written as follows. Take an E_a basis and let

$$g_{ab}(t) = e^{2\lambda}(e^{-2\beta})_{ab}$$

where λ is defined by $e^{6\lambda} = \det g_{ab}$ and β is a symmetric matrix with zero trace. In general β has five independent components, but in the case we are considering g_{ab} can be diagonalized and β then has only two independent components. It can be written

$$\beta_{ab} = \mathrm{diag}\,(\beta_1, -\tfrac{1}{2}\beta_1 + \tfrac{1}{2}\sqrt{3}\beta_2, -\tfrac{1}{2}\beta_1 - \tfrac{1}{2}\sqrt{3}\beta_2)$$

and then

$$\mathcal{L} = [R^* + 6\dot\lambda^2 - \tfrac{3}{2}(\dot\beta_1^2 + \dot\beta_2^2)]\,e^{3\lambda} + \mathcal{L}_M \qquad (11.30)$$

Chapter 11. Anisotropic and inhomogeneous relativistic cosmologies

where

$$R^* = -\tfrac{1}{2}e^{-2\lambda}[N_1^2 e^{4\beta_1} + e^{-2\beta_1}(N_2 e^{\sqrt{3}\beta_2} - N_3 e^{-\sqrt{3}\beta_2})^2$$
$$-2N_1 e^{\beta_1}(N_2 e^{\sqrt{3}\beta_2} + N_3 e^{-\sqrt{3}\beta_2})] \qquad (11.31)$$
$$+\tfrac{1}{2}N_1 N_2 N_3(1 + N_1 N_2 N_3)$$

and the term $\mathscr{L}_M$ is a Lagrangian for the matter. For perfect fluid with equation of state (11.13) the matter terms are of order $e^{-3(\gamma+1)\lambda}$ and are thus in general negligible as $\lambda \to \infty$. The potential (11.31) for types I and II is extremely simple. For types VI_0, VII_0, VIII and IX it is as shown in figure 11.1.

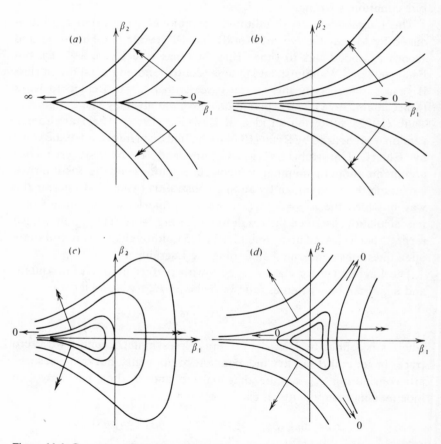

Figure 11.1. Contours of (11.31) for fixed λ: (a) type VI_0, (b) type VII_0, (c) type VIII, and (d) type IX. Double-headed arrows indicate exponential increase. Some asymptotic values are indicated.

For the class B models with $n^a{}_a = 0$ the R^* term contributes a potential

$$R^* = e^{-2\lambda}(-2(1-3h)/h)A^2 e^{2B/\sqrt{(1-3h)}} \qquad (11.32)$$

where

$$B = [-\beta_1 - \sqrt{(-3h)}\beta_2]/\sqrt{(1-3h)}.$$

Anisotropic stresses

There are three main forms of anisotropic stress that have been considered. The first of these is an electromagnetic field. For pure magnetic fields (as seen by an observer on the n congruence) Maxwell's equations, in the e_a basis, reduce to

$$n^{ba}H_a + \varepsilon^{bcd}a_c H_d = 0, \quad a_b H^b = 0,$$

i.e. H is orthogonal to the derived algebra of the e_a and to a. This allows the following number of free components of H: in type I, 3; in type II, 2; in types III, VI_0 and VII_0, 1; and in all other cases, 0 (Hughston and Jacobs, 1969; see also Jacobs, 1977). It is interesting that a type IX solution with magnetic field must also admit an electric field.

The second main form is a viscous term $\pi_{\mu\nu} = -2\eta\sigma_{\mu\nu}$ for a coefficient of viscosity η. This should be evaluated on the basis of a microscopic theory, such as relativistic kinetic theory (for which see Stewart, 1971) but simplifying assumptions, such as that $\eta \propto \rho^n$ for some power n, are often made.

Finally one may have a stress $\pi_{\mu\nu}$ which arises from the anisotropic expansion of a cloud of collisionless particles. In this case the particles are red- (or blue-) shifted by an amount dependent on their direction of motion, and so the distribution function becomes anisotropic. Such stresses have been considered particularly for massless particles – gravitons, neutrinos and photons. Even simple directed fluxes can be considered as an extreme case.

One can show formally (MacCallum, Stewart and Schmidt, 1970) that in the diagonal Bianchi metrics all these stresses can be represented as additional potentials. The actual form taken has been calculated for some cases. Hughston and Jacobs (unpublished work) showed that the magnetic fields contribute additional exponential terms in β to the potential. In type I this leads to a form like type II with no anisotropic stress; in type II it leads to a corner somewhat like the upper left part of

Chapter 11. Anisotropic and inhomogeneous relativistic cosmologies

the type VIII diagram (figure 11.1(c)); in types VI_0 and VII_0 the magnetic potential rises towards positive β_1, and in type III it reinforces the R^* term.

The free-massless-particle stresses due to an initial thermal-equilibrium isotropic distribution in Bianchi I models were calculated by Misner (1968); they lead to a potential similar to that in figure 11.1(d), but with the corners completely closed. The type VII_0 models can be regarded as Bianchi I models with standing gravitational waves giving rise to the potential in figure 11.1(b). This interpretation is based on the idea that the sinusoidal terms in the E_a (see table 11.2) represent a circularly polarized wave of constant amplitude with wavelength dependent on the value of β_1, whose initial value is of course arbitrary. Such an interpretation does involve any violation of exact homogeneity. A similar interpretation in terms of travelling waves can be given for type VII_h models (Lukash, 1974). Directed graviton fluxes are subject to similar restrictions to those on magnetic fields, and like them produce exponential potentials helping the R^* term to confine the model in the β plane (MacCallum and Taub, 1973). The treatment of more general massless-particle stresses is complicated. Matzner (1971) has discussed the case of a tilted type IX cosmology, and had previously (1969) made some progress with the case of type V.

In tilted models the energy–momentum of a perfect fluid flowing along the u congruence appears to an observer moving along the n congruence to contribute an anisotropic stress. The full five degrees of freedom of β may be required; the extra three can be parametrized by the Euler angles of the rotation required to make g_{ab} diagonal. The type IX tilted case has been very extensively considered by Ryan (1971). He finds that the potential in figure 11.1(d) is modified by the addition of potentials blocking off the corners (due to the additional β momenta) and the bisectors of the sides of the triangular well (due to the extra matter terms); these he calls the rotation and centrifugal terms respectively. The Euler angles only change significantly at collisions with the centrifugal potential wall.

Extensive discussions of the applications of the Hamiltonian methods with anisotropic stresses will be found in Ryan (1972a), and Ryan and Shepley (1975). The latter is a generally useful reference on spatially-homogeneous models (see also MacCallum, 1973).

The form of (11.20) allows only simple forms of anisotropic stress, and these can be treated in a very similar manner to the Bianchi cases; see, for example, Thorne (1967) and Ryan (1972a).

Spatially-homogeneous anisotropic metrics

Autonomous systems of equations

The differential equations governing spatially-homogeneous models take the form

$$\frac{dx}{dt} = f(x, t)$$

where x is an n-tuple of dependent variables and f is some n-tuple function of x and t. When f is independent of t, such a system is called autonomous. This can be formally (but not usefully) achieved by adjoining t as an extra component of x. The greatest progress is made when the system becomes a plane-autonomous system (i.e. x has only two components), because the Jordan curve theorem enables qualitative features to be readily discovered. Even in the general case the study of the critical points where $f = 0$ can be illuminating.

There are various spatially-homogeneous cosmologies whose governing equations can be reduced to a plane-autonomous system. The columns marked r and s in table 11.1 give some indication of these; one requires that the time dependence has exactly two degrees of freedom. The addition of pure constants, such as the cosmological constant, or constant coefficients in expressions for viscosity does not necessarily affect the possibility of reducing the system to plane-autonomous form. Imposition of the requirements of additional symmetry (making the model LRS), or of $u = n$ for a fluid model (which reduces s to $r+1$), is used to produce the necessary simplification.

The variables used in the reduced form are in most cases a new time variable λ (as used above), a parameter $\Omega = 3\rho/\theta^2$ describing the dynamical importance of matter (where $\theta = g^{ab}\theta_{ab} = 3\dot{\lambda}$), and a parameter $\beta' = d\beta/d\lambda = \sqrt{(\beta_1'^2 + \beta_2'^2)} = 2\sqrt{3}\sigma/\theta$ where $\sigma = \sqrt{(\frac{1}{2}\sigma^{ab}\sigma_{ab})}$ and $\sigma_{ab} = \theta_{ab} - \frac{1}{3}\theta g_{ab}$ is the shear. These parameters have been renamed here so that Ω is the same as the density parameter now used in studies of FRW models and λ has the time sense of evolution away from an initial big bang singularity or towards a fully expanded state. If the universe recollapses, as it does in the Kantowski–Sachs $k = 1$ cases, λ will be double valued.

The models which have been discussed as plane-autonomous systems are:

Type I with $\Lambda \neq 0$, Collins (1971), Kubo (1975)
      with magnetic field, Collins (1972)
      LRS with free neutrinos, Shikin (1972)
      with bulk viscosity, Belinskii and Khalatnikov (1975)

Chapter 11. Anisotropic and inhomogeneous relativistic cosmologies

Type II LRS, Collins (1971)
   vacuum, Collins (1971)
Type V Collins (1971)
   LRS tilted, Collins (1974), Shikin (1975)
Type VI$_0$ with $g_{22} = g_{33}$ in the diagonal basis, Collins (1971)
Type VI$_h$ $n^a{}_a = 0$ (including (11.20) with $k = -1$), Collins (1971)
Kantowski–Sachs models ((11.20) with $k = 1$), Goethals (1975), Collins (1977).

Except where otherwise stated the cases in this list were assumed to be filled with a perfect fluid with equation of state (11.13) and to have $u = n$. Bulk viscosity refers to a perfect fluid with $p = p(\rho) - K(\rho)\theta$ for some function K. It may be noted that the cases where the approach is successful are essentially those where the 'potentials' are just pure exponentials of β.

In general the solutions are structurally stable in that their qualitative features are unaltered by small changes in γ, except at $\gamma = 2$. This last case is exceptional because $\rho \propto e^{-6\lambda}$, and in regions where the potential is negligible $\sigma^2 \propto e^{-6\lambda}$ also. In the tilted LRS Bianchi V case the behaviour has structural instabilities at several values of γ.

The use of the qualitative techniques for more general cases has largely been confined to type IX where extensive studies have been made by Bogoyavlenskii and S. P. Novikov (1973); see also Bogoyavlenskii (1976) and S. P. Novikov (1972). Some of the results of these various studies are mentioned below.

Singularities

The spatially-homogeneous models in general have singularities. In fact, at least one timelike geodesic is incomplete in any spatially-homogeneous model in which $R_{\mu\nu}k^\mu k^\nu > 0$ for all timelike- and null-vectors k and the matter fields obey equations giving unique solutions to the Cauchy problem (theorem, p. 147 of Hawking and Ellis, 1973). Investigations have therefore concentrated on the nature of the singularities.

From (11.29), when neither the spatial curvature nor matter terms are important, the behaviour of g_{ab} approximates that of a vacuum Bianchi I model. The exact solutions for these models, due to Kasner, are

$$ds^2 = -dt^2 + t^{2p_1}(dx^1)^2 + t^{2p_2}(dx^2)^2 + t^{2p_3}(dx^3)^2 \quad (11.33)$$

where $\sum p_i = \sum p_i^2 = 1$. In agreement with table 11.1 there is just one free parameter. Except in the special case where one of the p_i is 1 and the others 0, the metric (11.33) has a real singularity at $t = 0$, and it is of 'cigar' type: i.e. a small spatial region which is spherical at some time $t_1 \neq 0$ becomes infinitely long and thin as $t \to 0$. In the special case the singularity is apparently of 'pancake' type where the spherical region becomes an infinitely thin disk. This turns out not to be a real singularity since the metric in this case is just a portion of flat space in unusual coordinates. However, when matter is present, so that (11.33) is only an approximation, Bianchi I universes have a singularity of infinite density. One naturally asks whether all Bianchi universes have singularities of this character. Ellis and King (1974) have proved that in all class A Bianchi models, and in all Bianchi models of class B with $u = n$, if the matter content is a perfect fluid its density becomes infinite at the singularity (or, in the case of type IX models, at the two singularities). However, in class B tilted models another possibility arises. This is that there is a Cauchy horizon, a null-surface bounding the region for which the equations give unique predictions, and that the matter entered the expanding universe across this horizon, which is a group orbit. The analytic continuation would be a universe admitting an isometry group of the same Bianchi type, but acting on timelike surfaces, i.e. a stationary solution of Einstein's equations. Associated with the Cauchy horizon is an 'intermediate singularity' into which the integral curves of n and the past null-geodesics on the Cauchy horizon run. (An 'intermediate singularity' is one such that Ricci components in an orthonormal frame parallelly propagated along a curve hitting the singularity are unbounded, but the components in some other orthonormal frame are bounded.) This situation is colloquially called a 'whimper'. Ellis and King, on the basis of study of the initial data on the Cauchy horizon, concluded that such singularities were possible in class B with the full freedom of table 11.1. Siklos (1978) has re-examined this argument. He finds that the methods were inappropriate because the usual quantities θ, n_{ab} and a_b are formally infinite, and argues that the Newman–Penrose basis is the appropriate one. He proves that class A whimpers are incompatible with $R_{\mu\nu}k^\mu k^\nu > 0$ on the Cauchy horizon, k being its null generator, and that perfect-fluid whimpers have two fewer parameters than the general case. Thus, even within the simplified class of spatially-homogeneous models, whimpers are not generic, whereas infinite-density singularities are. (Collins (1977) has proved that the Kantowski–Sachs $k = 1$ models have two singularities of infinite density; like type IX models, they recollapse.) A much fuller review of the singularities in spatially-homogeneous models has been given by Collins and Ellis (1977).

Chapter 11. Anisotropic and inhomogeneous relativistic cosmologies

The next question concerns the manner of approach of the model to its infinite-density singularity. First let us suppose that matter is not important in the Einstein equations near the singularity. In the models with a Lagrangian form we can see, for example from figures 11.1, that an approximate evolution can be constructed from three elements: the Kasner approximation for periods when the R^* terms are small, an approximation for a straight exponential wall potential, and an approximation for the regions such as that in the region $\beta_1 < 0$, β_2 small, in type VII_0 (figure 11.1(b)). The second of these is provided by the exact Bianchi type II vacuum solution, which was given by Taub (1951). This gives a law relating the values of the Kasner exponents for successive epochs. The third approximation required has been extensively studied in the context of both type VII_0 and type IX models; for this purpose types VIII and IX are equivalent. The result is that the model performs a substantial number of small oscillations, eventually reversing its motion into the corner.

These approximations can now be combined. For example, in a general type VII_0 model we see that the universe begins in a Kasner regime at large β_1, runs towards negative β_1, performs a number of oscillations and (in the vacuum case) eventually reverts to a Kasner regime. The type VII_h model, although it cannot be discussed in terms of a potential, behaves in a similar manner, essentially because the contributions of a_b to the vacuum field equations is an isotropic curvature R^*, and this is negligible near the singularity (Doroshkevich, Lukash and I. D. Novikov, 1973).

By such arguments we infer that in most spatially-homogeneous types of model, if the matter is dynamically negligible near the singularity, then the singularity will be of the cigar type, or in special cases of pancake type. The behaviour in types IV and VI_h does not seem to have been fully investigated. In types VIII and IX our qualitative arguments suggest an indefinitely oscillatory behaviour. This conclusion has been confirmed in the more rigorous analysis of Bogoyavlenskii and S. P. Novikov mentioned earlier, and by numerical experiments (see, e.g., Moser, Matzner and Ryan, 1973). There are also the special cases of the LRS type VII_0, VIII and IX models which are confined to the ordinate axis of figures 11.1(b)–(d). Here the vacuum solutions are exactly known: they are Kasner solutions with two of the p equal, for type VII_0, and for types VIII and IX are the Taub portions of the famous Taub–NUT metrics (Taub, 1951; Misner, 1967; Siklos, 1976). The Taub–NUT models have Cauchy horizons.

Spatially-homogeneous anisotropic metrics

We have also to consider whether matter is truly negligible near the singularity. If not, we may get a 'point' singularity like that in FRW models, where a spherical region shrinks in all directions, a 'barrel' singularity where it becomes infinitely thin but finitely long (Jacobs, 1968), or even a 'ribbon' singularity where it is infinitely long, finitely thick in one direction, and infinitesimally thin in the other (Evans, 1978). Except for perfect fluids with equation of state $p = \rho$, where it is clear that matter will have an effect near the singularity, it requires more than the simple analysis above to discover the influence of the matter terms.

Early examples where matter remains important near the singularity came from exact solutions (MacCallum, 1971). A more general illustration is provided by the LRS Bianchi II solutions, where the plane-autonomous system for a perfect fluid with equation of state (11.13), $1 \leq \gamma < 2$, gives rise to figure 11.2 (Collins, 1971). To interpret this diagram, note that models starting from $\Omega = 0$ are the cases where matter

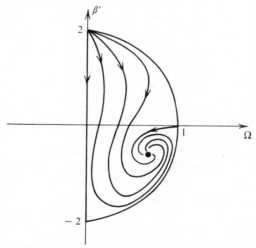

Figure 11.2. Evolution of LRS Bianchi type II models.

is dynamically negligible near the singularity, and the cases where matter is significant at early times are those which start from $\Omega \neq 0$. It may also be noted that crossing the β' axis corresponds to a 'bounce' off the exponential potential in which the shear is momentarily zero. Figure 11.2 shows that there are two special models, one starting from $(\Omega, \beta') = (1, 0)$, and the other at the focus of the curves, in which matter is important at early times. The diagrams for type VI_0 with $g_{22} = g_{33}$ and for type VI_h with $n^a{}_a = 0$ and $\gamma < 2(1-h)/(1-3h)$ have similar properties.

Chapter 11. Anisotropic and inhomogeneous relativistic cosmologies

The more complicated case of tilted type IX models has been considered by Belinskii, Khalatnikov and Ryan (see Ryan, 1972b). They have shown that the rotation and centrifugal potentials, which exist only if there is 'tilted' matter present, are important in the early universe in confining the model to one of the six congruent triangular regions that can be drawn in figure 11.1(d).

Isotropization

What one would like to mean by isotropization is the property of being as isotropic as the universe is now observed to be. This could be achieved for the instantaneous values of the anisotropy parameters merely by imposing limits on Cauchy data assigned on a 'present-day' surface ($t =$ constant). However, we observe some properties that depend on the run of anisotropy over long periods, e.g. the microwave-background isotropy. The various simplified conditions that have been applied, such as that $\beta \to$ constant as $\lambda \to \infty$ (MacCallum, 1971), are neither necessary nor sufficient to ensure agreement with observation (Barrow, 1976a). Doroshkevich *et al.* (1973) use 'isotropization' simply in the sense of an era when β' is very small (whether or not it again becomes significant later). For the Bianchi IX models S. P. Novikov (1972) uses the condition that $R^*{}_{ab}$ should be approximately isotropic at the momentum of maximum expansion.

Dustlike matter contributes terms of order $e^{-3\lambda}$ to the Einstein equations, while the anisotropic terms in (11.31) have an $e^{-2\lambda}$ dependence. In general, therefore, the anisotropic terms in $R^*{}_{ab}$ will dominate the dynamics in the late stages of expansion. Thus one infers that a spatially-homogeneous model only becomes and remains isotropic if it belongs to a Bianchi type admitting an FRW universe (Collins and Hawking, 1973). The only 'general' Bianchi types thus permitted are types IX and VII_h. Type IX models have been known for some time to recollapse due to the influence of the positive R^* (all other Bianchi types have negative R^*), and this has been proved rigorously for certain cases by S.P. Novikov (1972). A perturbation analysis of type VII_h models starting from the FRW case shows that they are unstable and do not remain isotropic (Collins and Hawking, 1973). Nevertheless, with a sufficiently high matter density to dominate the dynamics for a long time period, type VII_h models can be compatible with present-day observations (Doroshkevich, Lukash and I. D. Novikov, 1973, 1974).

The effect on isotropization of anisotropic stresses, and the implications for chaotic cosmology, are discussed in section 11.5.

11.4 Inhomogeneous metrics

11.4.1 Introduction

In principle, inhomogeneous models constitute some open set in a space of all solutions of Einstein's equations. However, a precise discussion of all such models is not possible: no general solution of the equations is known. Nevertheless, certain aspects of the general case have been discussed in a qualitative manner, especially the behaviour near the singularity (see section 11.5).

A number of more particular inhomogeneous models have been considered. The most highly symmetric of these admit a G_2 or G_3 of isometries acting on two-dimensional surfaces, and, in addition, a homothety; such models are known as (spatially) self-similar and are discussed in section 11.4.2.

Then come the inhomogeneous models with a G_3 of isometries. These include inhomogeneous models analogous to the Kantowski–Sachs class, spherically symmetric models and plane-symmetric models; they are discussed in section 11.4.3.

Next are the models with a G_2 of isometries; models in this class include the Gowdy models and cylindrically symmetric cosmologies, and are discussed in section 11.4.4.

The least symmetric exact solutions that have been considered as cosmological models are the Szekeres solutions, and these are discussed in section 11.4.5.

The most intensively discussed inhomogeneous cosmologies are perturbations either in the FRW models (i.e. with no perturbations of the geometry) or of the FRW models. These are used in the discussion of galaxy formation and possible turbulence mechanisms in the early universe (see, e.g. Jones, 1976), in considering variations in the primordial helium abundance (see, e.g. Schramm and Wagoner, 1977), and in evaluating the effect of inhomogeneity on measurable relationships such as the magnitude–redshift law (see, e.g. Weinberg, 1976; Roeder, 1975*a, b*). This last question has also been analysed by using the 'Swiss-cheese' model in which spherical regions of FRW metrics are replaced by portions of the well-known Schwarzschild solution (Kantowski, 1969; Dyer, 1976); these, as solutions of the equations, can be considered with the spherically symmetric solutions. The FRW perturbation analyses are beyond the scope of this article.

It is interesting to note that relativistic solutions with a shear-free perfect fluid form a very restricted class. There are no such solutions for dust unless the matter is either non-expanding or irrotational (Ellis, 1967). There are no shear-free, rotating, spatially-homogeneous solutions (King and Ellis, 1973), and there are strong indications that there are no rotating, expanding, shear-free solutions at all (King, 1974a).

11.4.2 Self-similar solutions

Eardley (1974) has considered solutions of Einstein's equations admitting a group of homotheties. Since every homothetic vector field gives

$$\mathcal{L}_Y g = \phi g$$

for some constant ϕ and $\mathcal{L}_Y$ is linear in Y, one can write the various ϕ in terms of a linear 1-form b on the vector fields such that $\phi = 2b(Y)$. The Y in the kernel of b are Killing vectors. Computation shows that the commutator of two homotheties is Killing. Moreover, except for a special case, a theorem of Defrise–Carter shows that the spacetime is conformally related to one in which all the Y are Killing. The conformal factor can be taken to be e^ψ where

$$\mathcal{L}_Y \psi = b(Y)$$

and $ds^2 = (ds_0^2) e^{2\psi}$, where in ds_0^2 all the homotheties become isometries. McIntosh (1976) has shown, in addition, that in a perfect-fluid solution any non-trivial homothetic vector (which is not an isometry) cannot be perpendicular to the fluid motion. The self-similar models thus all have 'tilt'.

Eardley considered the algebraic structure of the three-dimensional homothety groups. The derived algebra must have dimension 2 or less, and this must be orthogonal to b, so

$$b_a C^a{}_{bc} = 0. \tag{11.34}$$

If $b = |b| \neq 0$, but $A_c = 0$, then b is an eigenvector of N_{ab} with zero eigenvalue, where A_c, N_{ab} are as defined in section 11.2. The possibilities are like the class A types I, II, VI$_0$ and VII$_0$, with a new parameter f defined as $b^2/N_2 N_3$ when N_2 and N_3 are both nonzero; these cases Eardley calls class C. If both A and b are nonzero we have Eardley's class D. Either $A = b$ (class D$_0$) or not; if not, then the condition (11.34) easily shows $b = fA$ for some constant f.

In all the cases where $b \neq 0$, the orbits of the isometries are two-dimensional. Thus the self-similar spacetimes of classes C and D include some plane-symmetric and pseudo-spherically-symmetric solutions.

There is also the special case, analogous to the Kantowski–Sachs $k = 1$ metrics, of spherically symmetric self-similar solutions. These were examined by Cahill and Taub (1971).

Eardley has shown that a Lagrangian technique, with self-similar variations only, works for the class D_0 models. For the class D_0 Bianchi type VII_h models the potential is similar to that of class A type VII_0 (see figure 11.1(b)). The difference is that the Hamiltonian is singular at $\beta_2 = 0$, and this reflects the possibility of evolution to a Cauchy horizon, like the tilted Bianchi models of class B.

Luminet (1977) has recently discussed the spatial curvature scalar R^* for the self-similar models and shown that it can be positive, suggesting recollapse, in some class D cases.

11.4.3 Inhomogeneous metrics with a G_3 of motions

For these models the metric has the form

$$ds^2 = -A^2\,dt^2 + B\,dt\,dr + C^2\,dr^2 + D^2(d\theta^2 + f^2(\theta)\,d\phi^2)$$
(11.35)

where A, B, C, and D are functions of r and t alone, and $f(\theta)$ is as in (11.18). The argument to establish this is similar to those used in section 11.3.2 (see Schmidt, 1967). The metric in the (r, t) plane can be diagonalized, so that the B term can be eliminated; it may, however, be more convenient to use null-coordinates so that $A = C = 0$. There are four further simplifications that are often made by choice of coordinates (the integral curves of r and t remain fixed, so these choices amount to rescalings of r and t). They are to scale r so that $r = D$, to scale r so that $C = 1$, to scale r so that $D = rC$, and to scale t so that $A = 1$. One cannot, in a general case, perform more than one of these. When a fluid-filled universe is considered, it is common to take its world-lines as the t-curves and scale A to 1.

The universes are called spherically symmetric if $f(\theta) = \sin \theta$. They have a centre of symmetry only if D ranges down to zero and if the 'centrality' conditions are satisfied there (Ellis, Maartens and Nel, 1978). In this case D need not be a monotonic function of r for fixed t. There can be a second centre. For the choice of variable $\sin r = D$ one has the 'surface of revolution' cosmologies discussed by Ryan (1972c). If, with the choice $A = 1$, D is a function of t alone, the metric is of the 'inhomogenized' Kantowski–Sachs form (Ellis, 1967), becoming of the Kantowski–Sachs form (11.20) if C is also a function of t alone. The

Chapter 11. Anisotropic and inhomogeneous relativistic cosmologies

inhomogeneity present when C depends on r is of a very simple kind, since the spatial variation is in effect 'frozen in' (Lund, 1973). Other aspects of these models have been discussed by Ruban (1970).

The only models in this class that have been given extensive consideration are the centro-symmetric models with metric

$$ds^2 = -A^2\,dt^2 + C^2\,dr^2 + D^2(d\theta^2 + \sin^2\theta\,d\phi^2) \quad (11.36)$$

the centre being $r=0$. One can specify initial values as functions of r with constraints as usual. For the vacuum one has essentially only one parameter (the Schwarzschild mass), and the solution is singular at $r=0$. For a perfect fluid, with given equation of state, there are in effect two free functions of r representing the density and radial velocity. This enables a variety of solutions to be considered. It is even possible to have discontinuities, such as joining a Schwarzschild solution to an FRW solution, although rather careful treatment of the matching conditions is required (Korkina, 1975). A number of applications of these models have been made. Among these are the following:

(a) The study of the growth of local, but large, perturbations in an FRW universe, as an approach to the problems of galaxy formation. Self-similar spherically symmetric solutions have been used in a similar manner to study formation of black holes in the early universe (Carr and Hawking, 1974). The same idea has been applied to other problems involving inhomogeneity (see sections 11.4.1 and 11.6.1). Eisenstadt (1977) has pointed out that if the solution is to become exactly FRW for $r > r_0$, where r_0 is some fixed radius, the mass within r_0 must be exactly that in the relevant FRW model.

(b) Use of spherically symmetric models as global alternative models (e.g. Dodson, 1972; Ellis, Maartens and Nel, 1978). In particular this has been done to model a hierarchical universe. The idea is that, on a series of scales, any observer is within an effectively spherical inhomogeneity, although the complete spacetime is not of course spherically symmetric. A very extensive investigation on these lines has been made by Wesson (see, e.g. Wesson, 1975a). In the case of dust, one can choose the t-coordinate along the worldlines. The R_{14} equation gives

$$C = D'/f$$

where ' denotes $\partial/\partial r$ and f is an arbitrary function of r. The G_{11} equation then yields

$$2D\dot{D}^2 + 2D(1-f^2) = F$$

where ˙ denotes $\partial/\partial t$ and F is another arbitrary function of r. The G_{44} equation gives

$$\rho = F'/2D'D^2.$$

These are the models of Tolman (1934). Bonnor (1972) has considered a hierarchical universe of this type with $f = 1$ and ρ proportional to $(1 + kr^3)^{-1/2}$ where k is constant. Later (1974) he considered the approach of Tolman models to isotropy, and found that models with $f > 1$ can, and models with $f = 1$ must, approach FRW models, while $f < 1$ gives recollapse.

11.4.4 Spacetimes admitting a G_2 of motions

A G_2 must act on two-dimensional orbits (theorem 11.1). These can be flat, if the G_2 is (11.8), or of negative curvature, if (11.9). In the flat case one has the metrics which are sometimes described as cylindrically symmetric (though one has to check whether or not the spacetime really contains an axis for the cylinder). These have not really been given much consideration as cosmological models, apart from an examination of some types of singularity (King, 1974b).

The models in this class that have been the most studied are those due to Gowdy (1974). The 2-spaces are flat in these models, and it is also assumed that every orbit is compact and that there are invariant spacelike compact three-dimensional surfaces (which topologically must be the 3-torus T^3, $S^1 \times S^2$, or the 3-sphere S^3, or identifications thereon). Of these it is the T^3 case that has received the most attention. Its metric is

$$ds^2 = e^{-3B_+}(-e^{3\lambda}\, d\lambda^2 + e^{-\lambda}\, d\theta^2) + e^{2\lambda}(e^{2\sqrt{3}B_-}\, dx^2 + e^{-2\sqrt{3}B_-}\, dy^2)$$

where $B_\pm$ are functions of λ and θ alone. Spero and Baierlein (1978) have used this metric as an example for their approximate symmetry technique. With the obvious slicing, the approximate symmetry group is Bianchi type I, and these models can thus be regarded as generalizations of this spatially-homogeneous case. Near the singularity the behaviour is indeed Kasner-like. The metric can be interpreted as containing two types of gravitational wave, parametrized by the B, and from this point of view has been used in studies of graviton production (see section 11.5.4). It should perhaps be noted that if a different slicing than the obvious one is used, the approximate symmetry found by Spero and Baierlein is of Bianchi type VI_0.

Chapter 11. Anisotropic and inhomogeneous relativistic cosmologies

11.4.5 The Szekeres solutions

These are metrics of the form

$$ds^2 = -dt^2 + e^{2A}\,dx^2 + e^{2B}(dy^2 + dz^2)$$

where A and B are functions of (x, y, z, t). They were discussed by Szekeres (1975), and independently by Tomimura in unpublished work, for the case of dust. Szafron (1977) has given the most general discussion. The solutions have been invariantly characterized by the properties of principal null-directions of the Weyl tensor, which is of Petrov type D (Szafron, 1977; Wainwright, 1977), but a more interesting characterization from the cosmological viewpoint is the one due to Collins and Szafron (Spero and Szafron, 1978). This can be stated as follows.

Theorem 11.3

If a spacetime contains a perfect fluid and satisfies the conditions that (i) the fluid flow is geodesic and irrotational, (ii) the hypersurfaces orthogonal to the fluid flow are conformally flat, and (iii) both the hypersurface 3-space Ricci tensor and the expansion tensor of the fluid have two equal eigenvalues, then it is a solution to the Einstein equations if and only if it has the Szekeres form.

Proof. The proof is unpublished at the time of writing. ∎

Tedious calculations show that the Szekeres metric with perfect-fluid matter content always obeys

$$\frac{\partial^2 B}{\partial t\, \partial y} = \frac{\partial^2 B}{\partial t\, \partial z} = 0.$$

The remaining equations take the forms

$$e^B = R(t, x)/F(x, y, z)$$

$$\ddot{A} + 2\ddot{B} + \dot{A}^2 + 2\dot{B}^2 + \tfrac{1}{2}(\rho + 3p) = 0$$

$$2\ddot{R}/R + \dot{R}^2/R^2 + p + k(x)/R^2 = 0$$

and either (class I)

$$e^A = F(e^B)_{,x}$$

and

$$F = \alpha(x)(y^2 + z^2) + \beta(x)y + \gamma(x)z + \delta(x)$$

where α, β, γ and δ are arbitrary functions of x subject to

$$\alpha\delta - \beta^2 - \gamma^2 = \tfrac{1}{4}(1 + k(x)),$$

or (class II)

$$F = \tfrac{1}{2}[1 + k(y^2 + z^2)], \quad R = R(t)$$
$$e^A = E(x, y, z)R(t) + T(t, x)$$

where

$$EF = \varepsilon(x)(y^2 + z^2) + \zeta(x)y + \eta(x)z + \theta(x),$$
$$R\ddot{T} + \dot{T}\dot{R} + T\ddot{R} + TRp = \varepsilon + k\theta,$$

k is constant and ε, ζ, η and θ are arbitrary functions of x.

In general these metrics have no Killing vectors (Bonnor, Sulaiman and Tomimura, 1977), but various previously known solutions are included, most of which do have symmetry. Class I includes the Tolman–Bondi spherically symmetric models, some plane-symmetric models discussed by Eardley, Liang and Sachs (1971), and Schwarzschildian vacuum metrics, while class II includes the metric (11.20) and the LRS class II spaces of Ellis (1967).

The Szekeres models with $p \neq 0$ in general do not satisfy an equation of state in the usual sense: p is a function of t alone, but ρ is not, and there is no physical basis for the resulting relation between p and ρ. The algorithm for computations usually involves a choice of $p(t)$, computation of R and, only thence, of ρ. Spero and Szafron (1978) have shown that Szekeres solutions with $\rho = \rho(t)$ must be spatially-homogeneous; in fact, they are the FRW models, and the metrics (11.20). The cosmological relevance of the general Szekeres solutions is thus rather doubtful. However, they do give really inhomogeneous solutions with $p = 0$.

The cosmological investigation of these models has concentrated on their asymptotic behaviour in the far future and far past (Bonnor and Tomimura, 1976; Szafron, 1977; Szafron and Wainwright, 1977). Bonnor and Tomimura analysed the class II dust cases, while the two later papers discuss special cases, of class I and II respectively, with $k = 0$ and

$$p = 4q(1 - q)/3t^2,$$

where q is a constant.

A wide variety of possibilities arises in the class II dust case. The regime near the singularity is in all cases anisotropic, and in most cases inhomogeneous. $k = -1$ models tend to FRW metrics but the density remains

Chapter 11. Anisotropic and inhomogeneous relativistic cosmologies

inhomogeneous, while a subclass of $k = 0$ models approach the Einstein–de Sitter metric. Otherwise the future regime is anisotropic. The later papers consider not only the evolution along a worldline, but also the uniformity with which the limits are approached. In general the limits are isotropic, but not uniformly approached. Particular cases are discussed in which an initially anisotropic state with approximately homogeneous density distribution evolves through an inhomogeneous era to an approximately FRW form. The general conclusion is that the geometry will not necessarily become FRW in the far future, which echoes the results from the Bianchi models.

11.5 Physics of the models

11.5.1 Isotropization by anisotropic stresses

The behaviour of the anisotropic and inhomogeneous models in recent epochs involves, in general, very similar physical processes to those in the standard FRW models, and gives rise to various constraints on the parameters of the models (see section 11.6). Here we are more concerned with earlier stages of evolution.

In section 11.3.4 the anisotropic stresses in Bianchi models were mentioned. Misner (1968) claimed that viscous stresses due to neutrino–electron scattering were extremely effective in dissipating anisotropy of Bianchi I models. However, it was pointed out that the viscous approximation could only be valid if the collision timescale is short compared with the expansion timescale and this requires that the shear is already small (Doroshkevich, Zeldovich and I. D. Novikov, 1968; Stewart, 1968). A number of studies were made subsequently of the effectiveness of both viscous and collisionless particle stresses, and more precise approximations to the effect of particles with long, but finite, mean free paths (see, e.g. Matzner 1971, 1972, and section 11.6.3) were found.

Stewart initiated rather a general type of argument to show that dissipation must be limited. It relies on the fact that regular systems of ordinary differential equations have unique solutions for given initial data, provided the well-known Lipschitz conditions are met. This implies that arbitrarily large anisotropy can be set as an initial condition now, and a corresponding earlier state found for any past time. A more detailed study of the equations, using the 'positive pressure' condition that the total stress in any direction is positive, yields a limit on the viscous heating: in a radiation-dominated universe of any Bianchi type, except

IX, the heating cannot exceed one-half the adiabatic cooling. The weaker condition of 'dominant energy' leads to possibly greater, but still limited, dissipation (Collins and Stewart, 1971). The new condition enables the argument to encompass additional processes such as neutrino-antineutrino annihilation. The general conclusion is that in order to dissipate arbitrarily large anisotropy, some type of singular process must be invoked, such as quantum particle production, or a process that works all the way back to the initial big bang.

Nevertheless, some remarkably efficient processes are possible. Matzner's calculations (1972) show enhancements by a factor 10^6 over the adiabatic cooling, due to neutrinos with long mean free paths. He has also suggested that the collisionless particle stresses could be used to reduce the shear to zero and, the energy density of the particles being greatest at that instant, they could then have a much enhanced rate of annihilation to form an isotropic thermal distribution. This would leave no anisotropic stress and so the models could lose all their anisotropy. The approximations in this argument mean that the conclusion is not exactly true, but the process could still be important. It also, of course, ignores the restoring effect of the R^* terms on the shear. (See Doroshkevich *et al.* (1968) for another account of this argument.)

The calculations of viscosity depend on the cross-sections of the various interactions. Viscosity occurs just at the stage when decoupling is partial, in between a thermal-equilibrium phase and a collisionless phase. The arguments gave the interesting further possibility of universes which were not in equilibrium at very early times, or even, in extreme cases, at any time (Collins and Stewart, 1971).

The effects of magnetic stresses near the singularity have been considered and have the general tendency to force the model from cigar configurations towards pancake configurations, with the magnetic field direction as the distinguished axis (Thorne, 1967).

11.5.2 Perturbations and cosmogony

A number of authors have considered the stability of spatially-homogeneous models (e.g. Bonanos, 1972) especially those of Bianchi type I (Johri, 1972; Perko, Matzner and Shepley, 1972; Goorjian, 1975). In the latter case, the perturbations can be Fourier-analysed. There are, as in the FRW models, modes representing density perturbations, gravitational waves, and rotational perturbations. In the FRW case these are uncoupled at lowest order, but the curvature of the background space

Chapter 11. Anisotropic and inhomogeneous relativistic cosmologies

couples the density and gravity waves in anisotropic Bianchi I models. The general effect is to enhance the growth of perturbations, compared with the FRW models, but not by a large enough factor to account plausibly for galaxies; the growth does not become subject to an exponential rather than a power law.

The motion of matter in the type VII_0 models can, by interpretation of the sinusoidal terms in the $\boldsymbol{E}_a$ in a similar manner to that used in the gravitational-wave interpretation (see section 11.3.4), be regarded as vortex motion in a Bianchi I background (Lukash, 1976). This suggests a type VII_0 model as an example of the possible early stage of turbulent cosmologies.

The fact that many universes are approximated by the Kasner models at early stages, even if very near the singularity the R^* terms take over, has led to much use of test hydrodynamics in a Kasner background as a way of studying the combined effects of anisotropy and inhomogeneity (e.g. Tomita, 1973).

It may be noted that the presence of density inhomogeneities is naturally linked with that of anisotropies (Liang, 1975). Barrow and Carr (1978) recently considered black hole formation in anisotropic universes. They found that the shear inhibits black hole formation, sharing some of the characteristics of a fluid with $p = \rho$, as mentioned earlier. This, paradoxically, leads to the possibility of a much higher present-day black hole density than in the FRW cases, because the effect may suppress just those small black holes whose formation and subsequent explosion would lead to violations of the observed limits on the gamma-ray fluxes. The shear eventually ceases to dominate the dynamics and large black holes can then be formed in considerable numbers, and could play an important part in the later dynamics of the universe and in galaxy formation.

Perturbations of type IX universes are still more complicated than the type I cases, as there is an even greater degree of complicated mode mixing due to the couplings. Hu (1974) has considered the perturbations and their effect in particle production.

11.5.3 'Mixing'

The Kasner model with the pancake singularity and the Taub-like Bianchi IX models have the property that particle horizons are removed, thus enabling an observer to see an infinite comoving coordinate distance through the universe. This suggested that matter-filled models which approach these behaviours might have the same property. Such improved

Physics of the models

causal connection could be used to produce smoothing over large lengthscales, thereby resolving some of the difficulties of chaotic cosmology. This intriguing possibility was raised by Misner (1969b) for Bianchi IX universes.

The effect depends on the model running into the corner channels during the course of its evolution in the potential of figure 11.1(d). In this region it approximates a Taub model for long time periods. Study of the recurrence relation for the parameters of successive Kasner epochs and the necessary conditions for horizon removal reveals that in general the latter is not achieved (Lifschitz, Lifschitz and Khalatnikov, 1970; Doroshkevich, Lukash and I. D. Novikov, 1971) even when the tilted case is considered (Grishchuk, Doroshkevich and Lukash, 1971; Matzner and Chitre, 1971). There are a small proportion of models in which the horizon is removed for at least one axis, but even in these cases the light rays could have circumnavigated the universe only a few times since the end of the quantum era. The arguments are too lengthy to set out here, but Matzner has suggested the following nice intuitive picture: the potential well of figure 11.1(d) expands as we approach an initial big bang, while the achievement of horizon removal demands that one hit a region of fixed linear size at the entrances to the corner channels; the probability of success diminishes to zero fast enough that there is no hope of hitting each corner in turn, as required for full horizon removal, and only a small probability of hitting even one.

11.5.4 Quantum effects

Quantum effects are expected to be important near the singularity. In an FRW model, dimensional arguments suggest that quantum gravity is significant at $< 10^{-43}$ s from an initial singularity. Considerable effort has been put into 'quantum cosmology' where canonical quantization methods are applied to the gravitational fields of a limited class of models. The methods rely on Hamiltonian formulations of the equations, and the examples have largely been taken from among the spatially-homogeneous models (of class A, see section 11.3.4). Some of the hopes of this programme are that quantization might help avoid singularities, aid 'mixing', or allow the discussion of topology change. None of these has so far been realized, although there are some interesting features in the cases that have been studied (Ryan, 1972a; MacCallum, 1975).

The other possibility that has been investigated concerns the effects of quantum fields in curved spacetimes. Zel'dovich and his colleagues have

Chapter 11. Anisotropic and inhomogeneous relativistic cosmologies

investigated the production of particles near the big bang in anisotropic models. They proceed by assuming that no particles were produced during the quantum-gravity era, and start their calculation from the end of this era. The expectation value of the energy–momentum of the particles produced is used as a source term in the Einstein equations in order to obtain the back-reaction effect. The effect violates the classical energy–momentum conservation and turns the shear terms into energy. The effects are calculated to be quite dramatic; the universe is essentially reduced instantaneously to isotropy. These computations have been done for Bianchi I and VII_0 models, and have been reviewed by Lukash et al. (1976). Of course the conclusions depend critically on the validity of the generalization of quantum field theory which is used, and independent studies of this question have been made by Berger (1975b), Nariai (1977) and Pessa (1977). Similar effects in Taub models have been considered by Lapedes (1976).

The only inhomogeneous model where these questions have been considered at length is the Gowdy model of T^3 type. Berger (1974, 1975a) has studied the production of gravitons by methods similar to those used in quantum cosmology, while Winlove and Raine (1975) have used a different definition of particle number (based on vacuum–vacuum transition amplitudes rather than Hamiltonian diagonalization). Berger (1975b) has also considered a general universe with space sections of T^3 topology.

The details of the various calculations are complex. Some indication of the difficulties of combining quantum theory with relativity will be found elsewhere in this volume.

11.5.5 The classical singularities

Belinskii et al. (1970) have constructed an approximate solution of the Einstein equations for the neighbourhood of a spacelike singularity that has the full number of degrees of freedom expected of a general model. It has locally the nature of a Bianchi IX or VIII solution. The meaning of the series of approximations leading to this conclusion is rather unclear. One can regard the result as saying that if one takes a Bianchi VIII or IX universe and perturbs by making the free parameters (in the perfect fluid case, eight; see table 11.1) into free functions of spatial position, one obtains a general solution, in the sense of an open set in the space of all solutions.

There are a number of surprising features of the results. It is hard to see why the only type of singularity that can arise should be of the spatially-homogeneous type. This may arise from assumptions about the magnitudes of the spatial derivatives, or it may be necessitated by the spacelike nature of the supposed singular surface. In a general solution one would expect the space and time derivatives to have equal magnitudes, whereas the solution of Belinskii *et al.* in effect ignores all space derivatives except those giving the structure of the commutators of the spatial basis vectors. It could be that solutions in which other space derivatives matter naturally have singularities on null- or timelike surfaces.

The absence of types VI_h and VII_h among the paradigms for the singularity is also a little surprising, given the results in table 11.1. It could be due to the fact that variations in the eighth parameter (h) would have to have spatial derivatives of higher order than the leading terms, and so would not appear to give a freely disposable function at the lowest orders as required.

The behaviour near the singularity of the general case is thus oscillatory, while special cases, in which the $R^*{}_{ab}$ term in the field equations is negligible at early times, have Kasner-like singularities. These latter cases, the 'velocity-dominated' solutions for irrotational fluid cosmologies, have been investigated by Eardley *et al.* (1971) and by Liang (1972), and it has been found that the conclusions of Belinskii *et al.* are borne out. One of the examples discussed by Eardley *et al.* is the Tolman class of solutions (see section 11.4.3), where it is found that the time of the singularity $t_0(r)$ varies in space (cf. the papers cited in section 11.4.3).

The Kasner solutions have power-law asymptotic behaviour. Possible power-law behaviours, in which various terms can play the leading role, have been listed by Evans (1978) with the help of a Newtonian analogy. He has considered examples from among the Bianchi models. S. P. Novikov discovered a power-law behaviour that can occur in Bianchi IX models in addition to the FRW and Taub cases (see Bogoyavlenskii and S. P. Novikov, 1973). The oscillatory behaviour of type IX models can be modified to a power law on the introduction of a scalar field which mimics the energy–momentum tensor with $p = \rho$ (Belinskii and Khalatnikov, 1972; Ruban, 1976).

The oscillatory behaviour can be understood as an extension of the Lukash interpretation of the type VII metrics in terms of gravitational waves. To produce the type IX potential one would need on this argument two sets of waves moving on different axes. The generalization is to let these circularly polarised waves have variable amplitude. A rigorous study of the basis and conclusions of the argument of Belinskii *et al.* would be very welcome.

Chapter 11. Anisotropic and inhomogeneous relativistic cosmologies

Singularities other than the big bang type are thought to be implausible in cosmology (Clarke, 1974; King, 1975; Siklos, 1978).

11.6 Constraints and inferences

11.6.1 Observations of electromagnetic radiation

One can attempt to put limits on the anisotropy and inhomogeneity of the universe from purely local observations, i.e. of nearby galaxies. These do not give at all strong constraints (Kristian and Sachs, 1966). However, spherically symmetric cosmologies have been invoked to explain both the anomaly reported by Rubin *et al.* (see section 11.1) and the contrary view of de Vaucouleurs (Mavrides, 1976; Mavrides and Tarantola, 1977), by patching together three regions in a Tolman solution – the first FRW, then Schwarzschild, then an outer FRW. A similar idea was used by Fennelly (1977), who inferred the necessary parameters from the observations, and superposed the effect of peculiar velocities by using a self-similar model.

The main difficulty in discussing observations of more distant objects lies in integrating up the null geodesics. This is possible in metrics of the form (11.20) (Tomita, 1968) and Bianchi I models (Saunders, 1969), but in all other cases (except the central observer in a spherically symmetric model) has not, as far as I know, been carried through, except by approximations (MacCallum and Ellis, 1970). Nevertheless one can draw some interesting conclusions by taking small departures from FRW models and evaluating the effects. In the class A spatially-homogeneous models there is characteristically a quadrupole anisotropy, while in the class B models the variations are large in a small region around the *a* axis. Superposed on the purely geometric effects are the effects of the peculiar velocity of the cosmic matter, which produces dipole and other terms. Collins and Hawking (1972) used models of Bianchi types I, VII_0, IX, V, and VII_h in this way and applied the results to the microwave background to deduce limits on the shear, vorticity and peculiar velocity. These are given in table 11.3. The results for type VII depend on the scale taken for the initial value of $n^a{}_a$ (denoted x), with reasonable values in the range $(1/25, \infty)$, and on the density parameter Ω which determines the time at which the isotropic part of R^* becomes significant and hence the time at which the shear begins to be regenerated. Similar arguments from the X-ray background produce limits of the the order of 10^{-4} (Wolfe, 1970). A similar analysis of the geodesics has been carried out by Doroshkevich *et al.* (1974).

Table 11.3. *Limits on anisotropy deduced from the microwave and element-formation limits. The assumed redshift of the microwave background is denoted z. For other parameters, see section 11.6.1.*

Type			Microwave limits			Element limits
		Peculiar velocity	Shear σ/θ	Vorticity ω/θ		Shear σ/θ
I	$z = 1000$		6×10^{-8}	0		4.8×10^{-12}
	$z = 8$		10^{-4}	0		
VII_0 $x<1$	$z = 1000$	2×10^{-6}	$10^{-6}/x$	$2 \times 10^{-6}/x$		
	$z = 8$	3×10^{-4}	$2 \times 10^{-3}/x$	$3 \times 10^{-4}/x$		8×10^{-7}
$x>1$	$z = 1000$	10^{-6}	1.5×10^{-3}	$10^{-6}/x$		
	$z = 8$	1.5×10^{-4}	1.5×10^{-3}	$2.5 \times 10^{-4}/x$		
IX	$z = 1000$	10^{-3}	10^{-3}	10^{-11}		8×10^{-7}
	$z = 8$	1.5×10^{-4}	10^{-3}	1.5×10^{-8}		
V		10^{-3}	1.5×10^{-4}	4×10^{-4}		1.5×10^{-11}
VII_h $x<1$		10^{-3}	$10^{-5.5}\Omega^{-2/3}/x$	$10^{-3}/x$		$10^{-4.5}\Omega^{-2/3}$
$x>1$		10^{-3}	$1.5 \times 10^{-5.5}\Omega^{-2/3}$	4×10^{-4}		

It should be noted that these analyses derive their strong limits by assuming that the shear has in effect been decaying adiabatically since the last microwave scattering. Thorne (1967) remarked that one could accommodate the observed isotropy with a large present-day shear if the shear was oscillating and just happened to have small net cumulative effect at the present time. He regarded this as implausible, but it looks less so in the light of the various mechanisms for producing oscillating shear that have been discussed subsequently.

Ellis et al (1978) have proposed a quite different model – one with spherical symmetry and which is static. The singularity is now at a fixed r and gives off microwave radiation through a development in space similar to the time-evolution of FRW models. The basis of this idea is that we actually observe only our past light-cone and that the continuation off this could equally be on timelike surfaces as on spacelike hypersurfaces. In such a static model the redshift depends on the gravitational potential, and so, providing the emitting surface is at constant potential, the microwave background inevitably appears isotropic regardless of the actual geometric relationship between the observer and the singularity. To fit the numerical parameters, we would, in fact, still be near the centre of the model, and when the field equations are used it becomes very hard

Chapter 11. Anisotropic and inhomogeneous relativistic cosmologies

to fit the magnitude–redshift relation. Nevertheless, this model suggests some intriguing directions for future research.

An alternative way to investigate effects of anisotropy on the microwave background is to consider the effects on the polarization and the spectrum (Rees, 1968). These effects appear to be preserved from the scattering era (Anile and Breuer, 1977; Caderni *et al.*, 1978). They arise from Thomson scattering of a distribution function which is always slightly anisotropic due to the shear. The most detailed treatments, both for Bianchi I universes, seem to be those due to Anile (1974) for the polarization, and Rasband (1971) for the spectrum. The effects are well below present experimental limits, but suggest possible future experiments.

Small-scale effects in the microwave background are related to constraints on inhomogeneities. These are again quite strong, and are usually analysed by the use of perturbed FRW models, and so the details fall outside our present scope (see, e.g. Anile and Motta, 1976).

One more test that may be relevant is the argument of Gödel that in a tilted model the number counts of galaxies would appear anisotropic even if the universe is spatially-homogeneous; for recent use of this see Wesson (1975*b*).

11.6.2 Element formation

The effect of a different evolution rate on element production in big bang cosmologies is very marked, as was first noted by Hawking and Tayler (1966). Barrow (1976*a*) has analysed the limits imposed by the observed helium and deuterium abundances on anisotropy in the same Bianchi types as were considered by Collins and Hawking (1972). His results are quoted in table 11.3. The analysis is simplified by the fact that the element synthesis takes place in quite a short time, and so a simple approximation to the dynamics in this period is adequate. A refinement of this analysis has been performed by Olson (1978). In analogy with remarks in the previous section, Olson points out that the strong limits on present-day shear depend on the shear scaling like $e^{-3\lambda}$, and that this would not necessarily be so in those models with growing modes of shear perturbation. For a similar reason element abundances place no limit of interest on the vorticity. Nor do they limit inhomogeneities on scales bigger than the Galaxy, since the experimental data refer only to the solar neighbourhood.

A similar analysis for turbulent models has been done by Barrow (1977) by using inhomogeneous test hydrodynamics in a Kasner regime. A similar analysis by Tomita (1973) reached rather optimistic conclusions, whereas Barrow claims that such a model is ruled out by conflict between the abundance requirements and the microwave-background isotropy (which was not considered by Tomita) unless there are sources for the element abundances other than the big bang.

11.6.3 Entropy

The strongest constraint of all could come from the entropy generation in anisotropic and inhomogeneous models. This is usually expressed in terms of the extent to which the universe fails to be a heat bath at the temperature of the microwave background radiation, i.e. by taking the ratio of the black-body photon number density to the number density of baryons. This is (in units of Boltzmann's constant) $s \sim 10^8$ at the present day. While this may be thought a large number, it is in fact small compared with the numbers typical in numerological cosmology ($\sim 10^{39}$). If the initial 'anisotropy energy' is converted into photons, this imposes strong limits on the anisotropy dissipated (Barrow and Matzner, 1977). The later the dissipation, the less this problem is, and of course if the 'anisotropy energy' were to end up in some other form, the limit would be less stringent, but it is doubtful whether the general conclusion would be affected.

The actual contribution from various processes has been calculated. Rees (1972) gave order-of-magnitude estimates for dissipation of inhomogeneities and used $s \simeq 10^8$ to fix the maximum scale of inhomogeneity and thus the epoch at which it dissipated. Actually he assumed dissipation could only take place when the lengthscales were within the horizon size, but since most dissipative processes depend on local gradients this is not necessarily so. Fluctuations greater than the horizon size can be smoothed in every neighbourhood, and thus appear to be acausally smoothed over scales greater than the horizon. This merely reflects the fact that the data were set in a non-causal manner. Indeed, one could bring the 'chaotic-cosmology' philosophy full circle by saying that fluctuations of such non-causal character are never permitted. This leads to an FRW universe!

Klimek (1975) and, in more detail, Caderni and Fabbri (1978) have calculated entropy production by viscous processes in Bianchi I models (and in Klimek's case type V); cf. also the work mentioned in section

Chapter 11. Anisotropic and inhomogeneous relativistic cosmologies

11.5.1. Assuming that the viscous approximations are valid, a large amplification (10^6–10^{10}) of the entropy can be achieved, and a proportionately large decrease in the shear anisotropy. Barrow (1976*b*) has considered an inhomogeneous model, with test motions in a Kasner background, in which similarly the shear is limited to that level at which viscosity dominates the collisionless-particle type of stress. He also finds amplifications of order 10^6, and the model is compatible with the element-formation and microwave data. It is because the viscous processes require an initial input of 'radiation' that the results are expressed as amplifications.

11.6.4 Conclusion

It appears that the constraints on those time-evolving universes which are eventually like FRW models are very severe, and that unless quantum processes are invoked the 'chaotic-cosmology' hypothesis cannot be sustained in its most general form. There are indications that general models do not isotropize in the far future. Nevertheless, there are some interesting anisotropic and inhomogeneous models compatible with present observations, which may be of especial relevance near a big bang singularity. There may also be potential for development in the quite different cosmology suggested by Ellis *et al.* (1978). Certainly the non-FRW models offer much scope for further study, and may still hold some surprises in store.

12. Singularities and time-asymmetry

R. PENROSE

12.1 Introduction

It has been a source of worry to many people that the general theory of relativity – that supremely beautiful description of the geometry of the world – should lead to a picture of spacetime in which singularities are apparently inevitable. Einstein himself had fought against the inevitability of such seeming blemishes to his theory, suggesting different possible ways out, of considerable ingenuity (e.g. the Einstein–Rosen bridge,[1] the attempt at a black-hole-avoiding stable relativistic star cluster,[2] the idea that a non-singular 'bounce' of the universe might be achieved through irregularities,[3] even his attempts at modifying general relativity to obtain a singularity-free unified field theory[4,5]). Yet, the researches of theorists in more recent years have driven us more and more in the direction of having to accept, and face up to, the existence of such singularities as true features of the geometry of the universe.

This is not to say that some mathematically precise concept of 'singularity' should now form part of our description of physical geometry – though much elegant work has been done in this direction in recent years. Rather, it seems to be that the very notion of spacetime geometry, and consequently the physical laws as we presently understand them, are limited in their scope. Indeed, these laws are even *self*-limiting, as the singularity theorems[6,7,8] seem to show. But I do not feel that this is a cause for pessimism. There is a need for new laws in any case; while, in my opinion, the presence and the apparent structure of spacetime singularities contain the key to the solution to one of the long-standing mysteries of physics: the origin of the *arrow of time*.

The point of view I am going to present does not stem from any radical view of things. I shall adopt a basically conventional attitude on most issues – or so it would have seemed, were it not for the fact that my ultimate conclusions appear to differ in their basic essentials from those most commonly expressed on this subject! My arguments do not depend on detailed calculations, but on what seem to me to be certain 'obvious'

Chapter 12. Singularities and time-asymmetry

facts, whose very obviousness may contribute to their being frequently overlooked.

It was Einstein who was, after all, the supreme master at deriving profound physical insights from 'obvious' facts. I hope I may be forgiven for trying to emulate him on his hundredth birthday, for I feel sure that he would have cared not one fig for the supposed significance of so arbitrary an anniversary!

12.2 Statement of the problem

The basic issue is a familiar one.[9,10,11] The local physical laws we know and understand are all symmetrical in time. Yet on a macroscopic level, time-asymmetry is manifest. In fact, a number of apparently different such macroscopic arrows of time may be perceived. To these may be added the only observed time-asymmetry of particle physics – which features in the decay of the K^0-meson. And there is one further related issue, namely the interpretation of quantum mechanics, which I feel should not be banished prematurely from our minds in this connection. The conventional wisdom has it that, despite an initial appearance to the contrary, the framework of quantum mechanics contains no arrow.[9,12-16] I do not dispute this wisdom, but nevertheless believe, for reasons that I shall indicate, that the question must be kept alive.

Let me list, therefore, seven apparently independent arrows (or possible arrows) that have been discussed in the literature:[9] section 12.2.1, K^0-meson; section 12.2.2, quantum-mechanical observations; section 12.2.3, general entropy increase; section 12.2.4, retardation of radiation; section 12.2.5, psychological time; section 12.2.6, expansion of the universe; and section 12.2.7, black holes versus white holes. I shall discuss each of these in turn.

12.2.1 Decay of the K^0-meson

Can the asymmetry that is present in the decay rate of the K^0-meson have any remote connection with the other arrows? The effect, after all, is utterly minute. The T-violating component in the decay is perhaps only about one part in 10^9 of the T-conserving component[17-21] – and, in any case, the presence of this T-violation has to be inferred, rather than directly measured, from the presence of a minute CP-violation ($\sim 10^{-9}$) together with the observation that CPT-violation, if it exists, must be even smaller in this interaction ($\ll 10^{-9}$). A very weak interaction indeed (or very weak component of a weak interaction) seems to be involved, and it plays no significant role in any of the important processes that govern the behaviour of matter as we know it. Gravitation, of course, is even weaker, and does, in fact, dominate the motion of matter on a large scale. But the differential equations of general relativity are completely time-symmetric as are Maxwell's equations and, apparently, the laws of strong interactions and ordinary weak interactions.

Yet the tiny effect of an almost completely hidden time-asymmetry seems genuinely to be present in the K^0-decay. It is hard to believe that Nature is not, so to speak, 'trying to tell us something' through the results of this delicate and beautiful experiment, which has been confirmed several times.[20] One of the suggestions that was put forward early on was that the T-violating effect arose via some cosmological long-range interaction, whereby matter-to-antimatter imbalance provided the required asymmetry.[50] But subsequent analysis[19,20] has rendered this viewpoint implausible. It seems that the asymmetry is really present in the *local* dynamical laws. This is a matter that I shall return to later. I believe that it is a feature of key significance.

12.2.2 Quantum-mechanical observations

In standard quantum mechanics, the dynamical evolution of a state takes place according to Schrödinger's equation. Under time-reversal this equation is transformed to itself provided i is replaced by $-i$. But Schrödinger's equation must be supplemented by the procedure ('collapse of the wavefunction') whereby the current state vector is discarded whenever an 'observation' is made on a system, and is replaced by an eigenstate ψ_Q of the Hermitian operator Q which represents the observable being measured. As it stands, this procedure is time-asymmetric since the state of the system is an eigenstate of Q just after the observation is made, but not (normally) just before (figure 12.1(a)). However, this asymmetry of description is easily remedied:[14] in the time-reversed description, one simply regards this same eigenstate ψ_Q as referring, instead, to the state just before the observation and Schrödinger's equation is used to propagate backwards until the previous observation (with operator P) is reached. Thereupon this (backwards-evolved) state vector is discarded (according to a time-reversed version of the 'collapse of the wavefunction') and an eigenstate ψ_P of P (corresponding to the actual result of the observation P) is substituted (figure 12.1(b)).

Chapter 12. Singularities and time-asymmetry

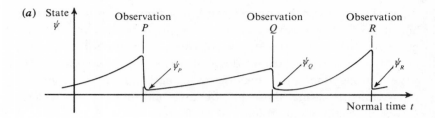

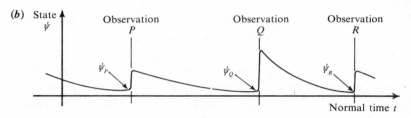

Figure 12.1. (a) Conventional (Schrödinger) picture of development of a wavefunction. (b) Essentially equivalent 'time-reverse' development of a wavefunction.

The relative probability of the observed Q-value, given the observed P-value, or of the observed P-value, given the observed Q-value, is the same in each mode of description, being

$$|\langle\psi_Q|U\psi_P\rangle|^2 = |\langle\psi_P|U^{-1}\psi_Q\rangle|^2,$$

where U is the unitary operator representing the evolution (according to Schrödinger's equation) of a state from the time of P to the time of Q. Since these probabilities are the only observational manifestation of the wavefunction ψ in any case, we see that the two modes of description are equivalent, and that the framework of quantum mechanics is time-symmetric.

It is often stated[16,22] that the actual value of the wavefunction at any time is not properly a description of physical reality. This is strikingly illustrated by figure 12.1, if time-symmetry is to be a feature of microscopic physics. But in any case there is a well-known difficulty, even with special relativity, concerning taking too strong a view that the wavefunction describes physical reality. For an 'observation' would seem to collapse the wavefunction into one of its eigenstates simultaneously with that observation – where 'simultaneous' presumably refers to the restframe of the one making the observation. This can lead to conceptual problems, when two spacelike-separated observations are carried out,

Statement of the problem

concerning the question of the *ordering* in which two collapses of the wavefunction take place. The difficulty is pinpointed particularly well in the famous thought experiment of Einstein, Podolsky and Rosen.[22]

There are, however, *other* situations in which it seems hard to maintain the view that the wavefunction (or state vector) does *not* give a proper description of physical reality in accordance with figure 12.1(*a*). Consider an isolated system on which observation P has just been made, giving a *conventional* description of a state ψ_P (eigenstate of P) which evolves forwards in time according to Schrödinger's equation to give a state $U\psi_P$ at some later time. Now (according to the conventional framework of quantum-mechanics, and assuming that no additional principles are incorporated) the operator UPU^{-1} is just as 'good' an observable as P. Furthermore, provided that the eigenvalue λ corresponding to ψ_P is simple, $U\psi_P$ has the property that it (or a nonzero multiple of it) is the *unique* state for which the probability is unity of giving the value λ for the observation UPU^{-1}. The isolated system cannot (in the ordinary way of looking at things) 'know' that the observation UPU^{-1} may be about to be performed upon it, but it must be prepared for that eventuality! So it seems that the information of $U\psi_P$ (up to phase) must be stored in the system, i.e. that the wavefunction *does* describe physical reality.

Of course UPU^{-1} may correspond to an utterly outlandish and totally impracticable experiment, as, for example, if in the case of the Schrödinger 'cat paradox'[24] we were to attempt to verify a resulting state $U\psi_P$: 'complex linear combination of dead cat and alive cat'. The very outlandishness of such an experiment suggests that $U\psi_P$ (and UPU^{-1}) may not, after all, 'really' refer to reality! But that is my whole point. There is something missing or something inappropriate about the laws of quantum mechanics, when applied to such situations. There is also something absurd about the whole idea of a collapsing wavefunction (or of any of the other essentially equivalent alternative ways of describing the same phenomenon, such as the conscious observer threading his way through an Everett-type[26–29] many-sheeted universe) as a description of physical reality. Yet what is a physical theory for if not to describe reality? In this I feel that I must align myself with many of the originators and main developers† of quantum mechanics – and, not least, with Einstein himself[23] – in believing that the resolution of the question of 'observations' is not to be found within the formalism of quantum mechanics itself. Some new (presumably nonlinear) theory seems to be required in

---

† E.g., in their different ways, Bohr,[30] Schrödinger,[24] Dirac,[31] Wigner.[25]

Chapter 12. Singularities and time-asymmetry

which quantum mechanics and classical mechanics each emerges as a separate limiting case.

The issue has importance here for two reasons. In the first place, the making of an observation seems to be associated with an *irreversible* process, depending upon an essential entropy increase. It is not obvious that what is missing (or wrong) about quantum mechanics is not some fundamentally *time-asymmetric* law. So the demonstration of time-symmetry in the formalism of quantum mechanics does not really settle the question of time-(a)symmetry in quantum-mechanical observations.

The second reason why the issue is important here has to do with the question of the role of quantum gravity. One must bear in mind the possibility that it might be the presence of a (measurable) *gravitational field* that takes the description of a physical system out of the realm of pure quantum physics.[32,33] And if a *new law* is needed, especially to cover situations in which both quantum and classical physics are being stretched in the extreme, then quantizers of gravity, beware!

I shall return to these questions in section 12.4. But, for the present, I propose to be wholly conventional in my attitude to quantum mechanics: it contains no manifest arrow, and the solution to the problem of macroscopic time-asymmetry must be sought elsewhere.

12.2.3 General entropy increase

The statistical notion of entropy is, of course, crucial for the discussion of time-asymmetry. And if the (important) local laws are all time-symmetric, then the place to look for the origin of statistical asymmetries is in the boundary conditions. This assumes that the local laws are of the form that, like Newtonian theory, standard Maxwell–Lorentz theory, Hamiltonian theory, Schrödinger's equation, etc., they determine the evolution of the system once boundary conditions are specified, it being sufficient to give such boundary conditions *either* in the past *or* in the future. (The boundary values are normally specified on a spacelike hypersurface.) Then the statistical arrow of time can arise via the fact that, for some reason, the *initial* boundary conditions have an overwhelmingly *lower* entropy than do the *final* boundary conditions.

There are several issues that must be raised here before we proceed further. In the first place, there is something rather unreasonable about determining the behaviour of a system by specifying boundary conditions at all, whether in the past *or* future. The 'unreasonableness' is particularly apparent in the case of future boundary conditions. Suppose I throw my watch on a stone floor so that it shatters irreparably, and then wait for 10 minutes. The future boundary condition is a mess of cogs and springs, but with minutely organized velocities of such incredible accuracy that when reversed in direction (i.e. with the clock run backwards) they suddenly reassemble my watch after a 10-minute period of apparent motionlessness. Though the models of physics that we are using (e.g. Newtonian theory) allow, in principle, such accuracy to be defined, we do not know (nor do we demand) that our models of physical laws correspond with such precision to reality. The problem is there also even for past boundary conditions, as stressed by Born.[34] And Feynman[35] has pointed out that, in Newtonian theory, if all the positions and velocities of a complex system are known to a certain accuracy, then all the accuracy is lost in less time than it takes to state that accuracy in words! Born (and Feynman) invoke this argument to demonstrate that classical mechanics is, in a sense, no more deterministic than quantum mechanics. Of course quantum mechanics has the additional problem of what happens when an 'observation' is made – and 'observations' seem to be necessary, in the normal view, to keep the wavefunctions from spreading throughout space.

I mention such things mainly to point out difficulties. But I shall ignore them henceforth and follow the conventional path that boundary conditions work! There is, however, a somewhat related question that needs further comment. Consider, again, my shattered watch as a future boundary condition. It will be seen that though this state has a higher entropy than that before my watch was shattered – and, therefore, in the normal way of looking at things, is a less 'unusual' state than the earlier one – the later state is nevertheless a very strange one indeed when one comes to examine it in minute detail, in view of the very precise correlations between the particle motions. But again I shall adopt a conventional 'macroscopic' view here. This strangeness is not of the kind that is described as 'low entropy'. And the 'reason' that I had a watch earlier is not that these precise correlations exist in the *future* boundary conditions, but that something in the *past* (say a watch factory) had a lower entropy than it might otherwise have had. Likewise, the 'reason' for the precise correlations in the particle motions of the shattered watch can be traced back to the factory, not the other way around.

I do not feel that this is begging the question of time-asymmetry. It could perfectly well have been the case, in a suitably designed universe, that some processes behave like my watch, while others (using the time-sense defined by my watch while it was still working!) indulge in apparently miraculous assembly procedures which suggest that special

Chapter 12. Singularities and time-asymmetry

(low-entropy) *future* boundary conditions should be invoked to provide the 'reason' for *their* behaviour. But our universe seems not to work that way.

An important related question that I have glossed over so far is that of coarse-graining.[36] What does entropy mean anyway? Is it a definite physical attribute of a system which, like energy–momentum, seems to be independent of the way that we look at it? In practice, entropy can normally be treated that way (for example, in physical chemistry). But for the most general definitions of entropy that might be expected to apply to a complicated system such as a watch, we need some apparently rather arbitrary (i.e. non-objective) way of collecting together physical states into larger classes (coarse-graining) where the members of each class are considered to be indistinguishable from one another. The entropy concept then refers to the classes of states and not to the individual states. Then, the (Boltzmann) entropy of a class containing N distinct (quantum) states is

$$S = k \log N$$

(where k is Boltzmann's constant). In fact, several conceptually different definitions of entropy are available.[36-38] But the whole question is clearly fraught with difficulties.† (The phenomenon of 'spin-echo' is one striking example that emphasizes these difficulties.[39] I am not even convinced that 'entropy increase' is at all an appropriate concept for describing the shattering of my watch. Probably taking a bath increases the entropy enormously more – while, in the case of my watch, the proportional increase in entropy must be quite insignificant.) The question of the objectivity of entropy will be returned to in section 12.4. But for the moment I hastily take refuge once more in conventionality: entropy is a concept that may be bandied about in a totally cavalier fashion!

12.2.4 Retardation of radiation

The question of boundary conditions is also intimately involved in the next of our arrows, that of retarded radiation. We may separate this phenemenon into two quite distinct aspects: the entropy question again, and the question of source-free or sink-free radiation. Retardation is not just a feature of electromagnetic radiation, of course, though it is usually

---

† I am leaving aside also such important questions as the 'H-theorem'[36-38] and 'branch systems'.[10,11] They do not *explain* the time-asymmetric origin of the total entropy imbalance.

Statement of the problem

discussed in that context. Imagine a stone thrown into a pond. We expect to see ripples expanding outwards from the point of entry to have their energy gradually dissipated, especially when they hit the bank. We do not expect to see, before the stone reaches the water, ripples being produced at the bank with such precise organization that they converge upon the point of entry at the exact moment that the stone enters the water. Still less do we expect to see such ripples converging inwards from the bank to some point in the middle of the pond, at which they entail the sudden ejection of a stone into the air! Such extraordinary behaviour is perfectly in agreement with the local physical laws. But its occurrence would require the sort of precise correlations in particle motions that could only be explained by some low-entropy future boundary condition.

I should emphasize, once more, a point made in the last section, since I feel that it is a key issue: correlations are present in the detailed particle motions in the future *because* the entropy was low in the past. Likewise, similar (but time-reversed) correlations are absent in the past *because* the entropy is high in the future. The latter statement is an unusual form of words, but I am trying to be unbiased with respect to time-ordering.† My point of view is that the correlations are not to be viewed as the 'reason' for anything; but low entropy (itself to be explained from some other cause) *can* provide the 'reason' for the correlations. (This way around we avoid the problem of over-extreme precision in physical laws.) And once more I stress that the 'specialness' of a state, due to its possessing intricate particle correlations of this kind, is *not* the type of 'specialness' that *at that time* corresponds to a low entropy. That is an essential point of coarse-graining.

So we see that the normal retarded behaviour of the ripples correspond to low entropy in the past and correlations in the future, while the two situations I have described, which seem to involve *advanced* behaviour of the ripples, involve some very precise correlations in the past of the kind leading to a reduction in entropy. Furthermore, an alternative hypothetical *retarded* situation, in which a stone is suddenly ejected from the pond accompanied by ripples propagating outwards towards the bank,‡ also involves such precise correlations (this time in the motions of the

---

† This leads to a logical reversal of the viewpoint expressed by O. Penrose and Percival[40] in their 'law of conditional independence', according to which the absence of initial correlations is *postulated*. The world-view of section 12.3.3 provides a certain justification for this law.

‡ The reader may notice the close relation between this situation and the 'zag' motion described by Gold.[9,41] Likewise, the converging waves meeting the falling stone are 'zaglike', while the other two are 'ziglike'.

Chapter 12. Singularities and time-asymmetry

particles near the stone at the bottom of the pond). Thus, in these situations we have no need to invoke any extra hypothesis to explain why the ripples are retarded. The entropy hypothesis is already sufficient to rule out the two advanced situations as overwhelmingly improbable – and it also rules out the above unreasonable retarded situation with the ejected stone (assuming the absence of any other agency responsible for ejecting the stone, such as an underwater swimmer, etc.).

The situation with electromagnetic radiation is, for the most part, similar to that of the ripples. There is a minor difference here, however, in the case of stars shining in a largely empty universe: it might well be that some of this radiation is never absorbed by any matter, but continues on indefinitely as the universe expands, or else terminates its existence on a spacetime singularity; likewise, there might be source-free radiation present in the universe, which had been produced directly in the big bang (or in a white hole) or possibly had come in from infinity from a previously collapsing phase of the universe.

I do not believe that these possibilities really make any essential difference to the discussion. I mention them mainly because a great deal has been written on the subject of 'the absorber theory of radiation'. In this view,[42,43] the contribution to the electromagnetic field due to each charged particle is taken to be half advanced and half retarded, while any additional source-free or sink-free radiation is regarded as 'undesirable'. By postulating the absence of such additional radiation, a link between the expansion of the universe and the retardation of radiation is obtained – though, in my opinion, not very convincingly. (And I have to confess to being rather out of sympathy with the whole programme, which strikes me as being unfairly biased against the poor photon, not allowing it the degrees of freedom admitted to all massive particles!)

In any case, the relevance of the entropy argument to the question of retardation seems to me to be quite independent of this.[44] The presence of free radiation coming in from infinity (or from the big bang singularity, say) which converges on a searchlight the moment it is switched on – or some other such absurdity – corresponds just as much to entropy-decreasing-type correlations in the initial state as would radiation coming in from sources. The only difference is that the correlations are just put directly into the photons themselves rather than into the particles producing the photons. So such correlations would be expected to be absent if the entropy in the future is to be high. Correspondingly, there is no objection to such correlations (in time-reversed form) being present in the *future* boundary conditions, because the entropy was low in the past – and this is, of course, necessary in order that the stars should shine.

The reader may be concerned about how one actually specifies boundary conditions at infinity, or on a spacetime singularity, in order to discuss such correlations in any detail. Of course serious technical problems can arise, particularly in the case of singularities. But the details of these problems should not substantially affect the foregoing discussion – at least if cosmic censorship holds true. I prefer to postpone these questions until section 12.3.2 except just to mention that under certain circumstances (e.g. in a big bang model in which the total charge within some observer's past light-cone is nonzero) there is necessarily a certain amount of source-free radiation present (and in other circumstances, a certain amount of sink-free radiation).[45] There is no reason to believe that this radiation should be correlated in any way which is incompatible with the entropy arguments. The stars still shine 'outwards', rather than 'inwards', whether or not there is some additional radiation permeating space – provided that this radiation has less than the intensity of a star and that it is not specially correlated.

12.2.5 Psychological time

The arrow most difficult to comprehend is, ironically, that which is most immediate to our experiences, namely the feeling of relentless forward temporal progression, according to which potentialities seem to be transformed into actualities. But since the advent of special relativity it has become clear that at least in *some* respects this feeling is illusory. One has the instinctive (or perhaps learnt) impression that one's own concept of temporal progression is universal, so that the transformation of potentialities into actualities that each one of us feels to be taking place ought to occur simultaneously for all of us. Special relativity tells us that this view of the world is false (and it is this lesson that had probably represented the major obstacle to the understanding and acceptance of the theory). Two people amble past one another in the street. What potential events are then becoming actual events on some planet in the Andromeda galaxy? According to the two people, there could be a discrepancy of several days!† (And adopting a view that events are becoming actual on, say, one's past or future light-cone – rather than

---

† To make the question seem more relevant, imagine that 'at that very moment' a committee is sitting on the planet, deciding the future of humanity!

Chapter 12. Singularities and time-asymmetry

using Einstein's definition of simultaneity – only makes the subjectivity of these occurrences even worse!)

So relativity *seems* to lead to a picture in which 'potentialities becoming actualities' is either highly subjective or meaningless. Nevertheless the feeling remains very strong within us that there is a very fundamental difference between the past and the future, namely that the past is 'actual' and unchangeable, whereas the future can yet be influenced and is somehow not really fixed. The usual view of the world according to relativity denies this, of course, presenting a rigid four-dimensional determinate picture and telling us that our instinctive feelings concerning the changeability of the future are illusory.

But I do not think that we should just dismiss such feelings out of hand. It is possible to envisage model universes in which the future is yet indeterminate, while the past is fixed. Imagine a continually branching universe, like that depicted in figure 12.2. One is to depict oneself located

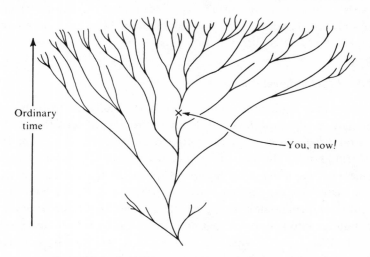

Figure 12.2. A model universe, branching into the future.

at some universe-point in the midst of it all, 'moving' up the picture as one's psychological time unfolds. In this model, the branching takes place only into the future. The path into the past from that point (i.e. the past history of the universe) is absolutely unique, whereas there are many alternative branches into the future (i.e. many alternative possible future histories for the universe, given the present state).

There are, in fact, (at least) two ways to make such a model relativistic. In the first (figure 12.3(*a*)), the branching takes place along the future

592

Statement of the problem

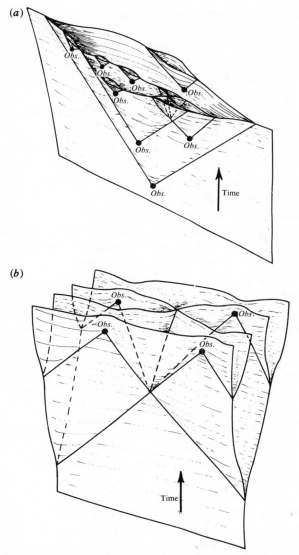

Figure 12.3. Two possible versions of a relativistic future-branching universe: *Obs.* means event at which an 'observation' is made.

light-cones of the points at which 'observations' (presumably quantum-mechanical) are made, and in the second (figure 12.3(*b*)), along the past light-cones of such points. This second case would seem to have considerably less plausibility than the first, since the universe has to 'know' that it has to branch in advance of the observation – which suggests

Chapter 12. Singularities and time-asymmetry

the presence of some sort of precise correlations in the past, like those leading to advanced-type radiation. The first model, however, is not altogether implausible. It is possible to envisage, for example, that the branching accompanies a kind of retarded collapse of the wavefunction, where on each branch the wavefunction starts out as a different eigenvector of the operator representing the observation.

Such a model may be referred to as an 'Everett-type' universe,[26-29] although it is by no means clear to me that this is really the kind of picture that the Everett formalism of quantum mechanics leads to. For, in the first place, if the Everett picture is to be essentially a reformulation of standard quantum mechanics, it ought to be time-reversible, i.e. there ought to be as much branching into the past as into the future. In the second place, in the Everett picture, one envisages a single wavefunction for the whole universe, which never itself 'collapses'. Instead, it becomes naturally represented as a linear combination of states, in each of which is a measuring apparatus and a physical system appearing to be in a separate eigenstate of the apparatus. Possibly, and with further assumptions as to, say, the nature of the Hamiltonian involved, etc., an Everett picture could, to a certain degree of approximation, be shown to lead to a picture somewhat resembling figures 12.2 and 12.3. But no-one appears to have done this. In particular, without some grossly time-asymmetrical additional assumption there would be nothing to rule out the time-reverses of figures 12.2 and 12.3, or innumerable possibilities in which branching occurs both to the past and to the future, with many branches temporarily separating and then coming together again.

Such considerations are clearly wildly speculative. But one is, in any case, groping at matters that are barely understood at all from the point of view of physics – particularly where the question of human (or non-human) 'consciousness' is implicitly involved, this being bound up with the whole question of psychological time, though not necessarily with the Everett picture. But spacetime models resembling figure 12.3(a), say, might well be worthy of study. They could be legitimate mathematical objects, e.g. four-dimensional Lorentzian manifolds subject to Einstein's field equations (say), but where the Hausdorff condition is dropped.[46] It could be argued that such a model is more in accordance with one's intuitive feelings of a determinate past and an indeterminate future than is our normal picture of a Hausdorff space-time. And one could claim that our feeling that 'time moves forwards into the future', rather than 'into the past', is natural in view of the fact that in the 'forward' direction potentialities become actualities, rather than the other way about! Such a model could be viewed as an 'objective' description of a world containing some strongly 'subjective' elements. One could envisage different conscious observers threading different routes through the myriads of branches (either by chance, say, or perhaps even by the exercise of some 'free will'). And each such observer would have a different 'subjective' view of the world.

This is not to say that I have any strong inclinations to believe in such a picture. I feel particularly uncomfortable about my friends having all (presumably) disappeared down different branches of the universe, leaving me with nothing but unconscious zombies to talk to! I have, in any case, strayed far too long from my avowed conventionality in this discussion, and no new insights as to the *origin* of time-asymmetry have, in any case, been obtained. I must therefore return firmly to sanity by repeating to myself three times: 'spacetime is a *Hausdorff* differentiable manifold; spacetime is a *Hausdorff*...'!

I have as yet made no real attempt to relate the *direction* of psychological time to the key question of entropy. It is clear that any attempted answer must necessarily be very incomplete, in view of our very rudimentary understanding of what constitutes psychological time. And I should emphasize that it is not just a question of the past being (apparently) more certainly *knowable* than the future. Indeed, it is not always the case that the past is easier to ascertain than the future, i.e. that 'retrodiction' is more certain than 'prediction'. At a meeting in Cornell in 1963, Gold[41] pointed out that it was far easier to predict the future behaviour of a recently launched Soviet satellite than to ascertain the time and place of launching! Furthermore, it is not very clear why the phenomenon of increasing entropy should, in any case, be compatible with ease in retrodiction and difficulty in prediction. If earlier states are more 'exceptional' and later ones more 'usual', then it would seem that we should be on safer ground predicting later ones than retrodicting earlier ones! Since, in practice, retrodiction is normally easier (or, at least, more accurate) than prediction (memory being more reliable than soothsaying), the precise relation between this and the entropy question is obscure.

But the issue is really rather different from this. It is not the ease in inferring the past that is relevant here, but the feeling that the past is unchangeable. Likewise, it is not the difficulty that we might have in guessing (or trying to calculate) the future that concerns us, but the feeling that we can affect the future. Thus, despite Gold's observation that the Soviet satellite's future was accurately predictable, one might have the

Chapter 12. Singularities and time-asymmetry

lingering unease that (had the technical expertise been available at that time) someone might have tried to intercept it. On the other hand, it seems totally inconceivable that any action could be applied *after* the satellite was in orbit to change the launching date! But if the future is to be, in principle, not essentially different from the past, one must entertain the awesome possibility[47] that even the 'fixed' past might conceivably, under suitable circumstances, be changeable! Is it just a question of 'money' that prevents this? (The science-fiction possibility suggests itself of a powerful government going one stage further than falsifying history: namely, actually *changing* the past!) I prefer to leave this question well alone.

There is a rather more helpful way of looking at the intuitive past–future distinction, however. One tends to view events in the past as providing 'causes' or 'reasons' for events in the future, not the other way around. This, at least, *is* compatible with the views I have been presenting in sections 12.2.3 and 12.2.4. My attitude has been that a low entropy at one time may be regarded as providing a 'reason' for precise correlations in particle motions at another, but that the presence or absence of such correlations should not be regarded as providing 'reasons' for anything else. We observe low entropy in the past and infer precise correlations in the future. Thus it is the presence of low entropy in the past that implies that the state of the past provides 'reasons' for the state of the future, not the other way around. This seems to me to be wholly sensible. If it had been the case in our world that collections of broken cogs and springs would sometimes spontaneously assemble themselves into working watches, then people would surely not be averse to attributing the 'causes' of such occurrences to events in the future. Such occurrences might co-exist with others of the more familiar kind, whose 'causes' could be attributed to events in the past. But our universe happens not to be quite like that! The 'causes' of things in *both* types of universe would be traced to situations of low entropy. And in *our* universe these low-entropy states turn out to be in the past.

So at least in this case our psychological feeling of a distinction between past and future can be directly linked to the entropy question. Perhaps the other aspects can too.

12.2.6 Expansion of the universe

I have implicitly indicated in section 12.2.4 that the expansion of the universe should not be regarded as directly responsible for the retardation of radiation, the latter phenomenon being simply one of the many consequences of an assumption that the initial state of the universe was of a far lower entropy than its final state will be (and, correspondingly, that entropy-decreasing correlations were absent in the initial state). I now wish to argue that the expansion of the universe cannot, in itself, be responsible for this entropy inbalance either.

For let us suppose that the contrary is the case and that, for some reason, increasing entropy is a necessary concomitant of an expanding universe. By time-reversal symmetry, this view would entail, correspondingly, that in a contracting universe the entropy should decrease.[48] There are two main situations to consider. First, it might be that the expansion of our actual universe will some day reverse and become a contraction, in which case, according to this view, the entropy would start decreasing again to attain a final low value. The second possibility is that the expansion will continue indefinitely and that the entropy will likewise continue increasing for ever – until a maximum entropy state is reached (ignoring the question of Poincaré cycles, etc.).

It seems that there are very serious objections to the idea that the trend of increasing entropy will reverse itself when the universe reaches maximum expansion. It is hard to see how such reversal could take place without some sort of thermal equilibrium state having been reached in the middle. Otherwise one would have to envisage, it seems to me, a middle state in which phenomena of the normal sort (e.g. retarded radiation and shattering watches) would co-exist with phenomena of the 'time-reversed' sort (e.g. advanced radiation and self-assembling watches). While it is possible to contemplate such situations 'for the purposes of argument', it is a different matter altogether for us to take them seriously for our *actual* universe. Furthermore, the moment of time-symmetry would be reached, it would seem, whilst normal retarded light from very distant galaxies is still coming in (those distant galaxies appearing still to be receding) and, at the same time, there would be advanced light behaving in the time-reversed way (i.e. specially correlated and converging on approaching galaxies). There would seem to be some serious self-consistency problems here[9] (though I am not claiming that they are totally insurmountable). I cannot find it in myself to take such a picture seriously – though some others have apparently not found their intuitions to be so constrained![48]

We might suppose, on the other hand, that the timescale for the reversal of the expansion is so enormously long that an effective thermal equilibrium can be achieved at maximum expansion. But such times that

Chapter 12. Singularities and time-asymmetry

one must contemplate for this are of a completely different order from the normal cosmological scales. *In effect*, then, such a universe does not recontract at all, and the situation can be considered alongside that of the indefinitely expanding universe-models.

One might think that these models would avoid the problems just considered, but this is not so. Let us envisage an astronaut in such a universe who falls into a black hole. For definiteness, suppose that it is a hole of $10^{10} M_\odot$ so that our astronaut will have something like a day inside, for most of which time he will encounter no appreciable tidal forces and during which he could conduct experiments in a leisurely way. In figure 12.4 the situation is depicted in a standard conformal diagram

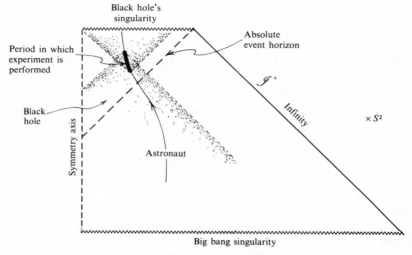

Figure 12.4. Conformal diagram of astronaut falling into black hole in Friedmann $k = -1$ universe.

(light-cones drawn at 45° and spherical symmetry assumed). Notice that the entire history of the astronaut beyond his point of crossing the absolute event horizon lies within the past domain of dependence of the black hole's singularity – and also within the future domain of dependence of the big bang singularity. Suppose that experiments are performed by the astronaut for a period while he is inside the hole. The behaviour of his apparatus (indeed, of the metabolic processes within his own body) is entirely determined by the conditions at the black hole's singularity (assuming that behaviour is governed by the usual hyperbolic-type differential equations) – as, equally, it is entirely determined by the

598

conditions at the big bang. The situation inside the black hole differs in no essential respect from that at the late stages of a recollapsing universe. If one's viewpoint is to link the local direction of time's arrow directly to the expansion of the universe, then one must surely be driven to expect that our astronaut's experiments will behave in an entropy-*decreasing* way (with respect to 'normal' time). Indeed, one should presumably be driven to expect that the astronaut would believe himself to be coming *out* of the hole rather than falling in (assuming his metabolic processes could operate consistently through such a drastic reversal of the normal progression of entropy).

I have to say that I cannot help regarding this possibility as even *more* of an absurdity than the entropy reversal at maximum expansion of a recollapsing universe! One would presumably not expect the entropy to reverse suddenly as the astronaut crosses the horizon (and thereafter, the collapsing aspect of his boundary conditions *enormously* outweighs the expanding one). So, strange behaviour in the entropy would have to occur well before the astronaut actually crosses the horizon – whereupon the astronaut could change his mind and accelerate outwards, so avoiding capture by the hole and being in a position to report his strange findings to the outside world! Indeed, the black hole argument can also be applied in the recollapsing universe. We do not need to wait for the whole universe to recollapse in order for the absurdities of this viewpoint to manifest themselves. (It may be that some holders of this viewpoint have a disinclination to accept the reality of black holes in any case. I have no desire to enter into the arguments – in my view very compelling – in favour of black holes here, but refer the reader to the literature.[7,51])

An argument could be put forward that the spacetime depicted in figure 12.4 is based on a too strong and unrealistic assumption of spherical symmetry. In fact this is not really the case; the dropping of spherical symmetry should make no essential qualitative difference to the picture, provided only that a (suitably strong) assumption of cosmic censorship is made. I prefer to postpone a more detailed discussion of this point until section 12.3.2. For the moment it is sufficient that I may fall back on conventionality again to conclude that the expansion of the universe is *not, in itself*, somehow responsible for the fact that the entropy of our universe is increasing.

This is not to say that I regard the correspondence between these two awesome facts as entirely fortuitous. Far from it. For I shall argue later on that both are consequences of the very special nature of the big bang – a special nature that is *not* to be expected in the singularities of recollapse.

Chapter 12. Singularities and time-asymmetry

12.2.7 Black holes versus white holes

General relativity is a time-symmetric theory. So, to any solution of its equations (with time-symmetric equations of state) that is asymmetric in time, there must correspond another solution for which the time-ordering is reversed.† One of the most familiar solutions is that representing (spherically symmetric) collapse of a star (described using, say, the T_{ab} of 'dust') to become a black hole.[54,55,7] In time-reversed form this represents what is referred to as a 'white hole' finally exploding into a cloud of matter. Spacetime diagrams for the two situations are given in figure 12.5.

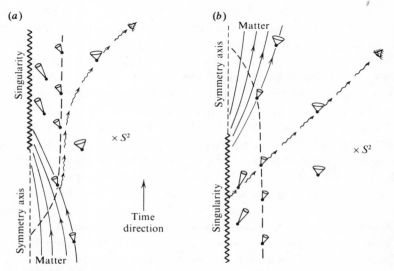

Figure 12.5. Black and white holes (Finkelstein-type picture): (*a*) collapse to a black hole, (*b*) explosion from a white hole. An observer's eye is at the top right.

Various authors[52,53] have attempted to invoke white holes as explanations for quasars or other violent astronomical phenomena (sometimes using the name of 'lagging cores'). Recall that in a (classical) collapse to a black hole, the situation starts out with a perfectly normal distribution of matter which follows deterministic evolutionary equations. At a certain stage a trapped surface may form, leading to the presence of an *absolute event horizon* into which particles can fall, but out of which none can escape. After all the available matter has been swallowed, the hole settles

---

† For this purpose, a 'solution of the Einstein equations' would be a Lorentzian 4-manifold with a *time-orientation* (and perhaps a space-orientation). 'Reversing the time-ordering' amounts to selecting the opposite time-orientation.

down and remains unchanging until the end of time (possibly at recollapse of the universe). (This ignores the quantum-mechanical effects of the Hawking process,[56] which I shall discuss in a moment.) A white hole, therefore, is created at the beginning of time (i.e. in the big bang) and remains in an essentially unchanging state for an indefinite period. Then it disappears by exploding into a cloud of ordinary matter. During the long quiescent period, the boundary of the white hole is a stationary horizon – the *absolute particle horizon* – into which no particle can fall, but out through which particles may, in principle, be ejected.

There is something that seems rather 'thermodynamically unsatisfactory' (or physically improbable) about this supposed behaviour of a white hole, though it is difficult to pin down what seems wrong in a definitive way. The normal picture of collapse to a black hole seems to be 'satisfactory' as regards one's conventional ideas of classical determinism. Assuming that (strong) cosmic censorship[57–59] holds true, the entire spacetime is determined to the *future* of some 'reasonable' Cauchy hypersurface, on which curvatures are small. But in the case of the white hole, there is no way of specifying such boundary conditions in the past because an initial Cauchy hypersurface has to encounter (or get very close to) the singularity. Put another way, the future behaviour of such a white hole does not, in any *sensible* way, seem to be determined by its past. In particular, the precise moment at which the white hole explodes into ordinary matter seems to be entirely of its own 'choosing', being unpredictable by the use of normal physical laws. Of course one could use future boundary conditions to retrodict the white hole's behaviour, but this (in our entropy-increasing universe) is the thermodynamically unnatural way around. (And, in any case, one can resort to memory as a more effective means of retrodiction!)

Related to this is the fact that while an external observer (using normal retarded light) cannot directly see the singularity in the case of a black hole, he can do so in the case of a white hole (figure 12.5). Since a spacetime singularity is supposed to be a place where the known physical laws break down it is perhaps not surprising, then, that this implies a strong element of indeterminism for the white hole. Causal effects of the singularity can, in this case, influence the outside world.

The presently accepted picture of the physical effects that are expected to accompany a spacetime singularity, is that *particle creation* should take place.[62,63,66] This is the general conclusion of various different investigations into curved-space quantum field theory. However, owing to the incomplete state of this theory, these investigations do not always agree

Chapter 12. Singularities and time-asymmetry

on the details of their conclusions. As applied to a white hole, two particular schools of thought have arisen. According to Zel'dovich[64] the white hole ought to be completely unstable to this process, evaporating away instantaneously, while Hawking[67] has put forward the ingenious viewpoint that the white hole evaporation ought to be much lower, and indistinguishable in nature from that produced, according to the Hawking process, by a black hole of the same mass – indeed, that a white hole ought itself to be indistinguishable from a black hole! This Hawking viewpoint is, in a number of respects, a very radical one which carries with it some serious difficulties. I shall consider these in a moment. The other viewpoint has the implication that white holes should not physically exist (though, owing to the tentative nature of the particle-creation calculations, this may not carry a great deal of weight; however, cf. also Eardley[65]).

There is another reason for thinking of white holes as antithermodynamic objects (though this reason, too, must be modified if one attempts to adopt the above-mentioned more radical of Hawking's viewpoints). According to the Bekenstein–Hawking formula,[56,69] the surface area A of a *black* hole's absolute event horizon is proportional to the intrinsic entropy S of the hole:

$$S = kAc^3/4\hbar G$$

(k being Boltzmann's constant and G being Newton's gravitational constant). The area principle of classical general relativity[7,70,71] tells us that A is non-decreasing with time in classical processes, and this is compatible with the thermodynamic time's arrow that entropy should be non-decreasing. Now, if a white hole is likewise to be attributed an intrinsic entropy, it is hard to see how the value of this entropy can be other than that given by the Bekenstein–Hawking formula again, but where A now refers to the absolute *particle* horizon. The time-reverse of the area principle then tells us that A is *non-increasing* in classical processes, which is the opposite of the normal thermodynamic time's arrow for entropy. In particular, the value of A will substantially *decrease* whenever the white hole ejects a substantial amount of matter, such as in the final explosion shown in figure 12.5. Thus, this is a strongly antithermodynamic behaviour.

It seems that there are two main possibilities to be considered concerning the physics of white holes. One of these is that there is a general principle that rules out their existence (or, at least, that renders them overwhelmingly improbable). The other possibility is contained in the

Statement of the problem

aforementioned line of argument due to Hawking, which suggests that because of quantum-mechanical effects, black and white holes are to be regarded as physically indistinguishable.[67] I shall discuss, first, Hawking's remarkable idea. Then I shall attempt to indicate why I nevertheless believe that this cannot be the true explanation, and that it is necessary that white holes do *not* physically exist.

Recall, first, the Hawking radiation that is calculated to accompany any black hole. The temperature of the radiation is inversely proportional to the mass of the hole, being of the general order of 10^{-7} K in the case of a black hole of 1 $M_\odot$. Of course this temperature is utterly insignificant for stellar-mass holes, but it could be relevant observationally for very tiny holes, if such exist. In an otherwise empty universe, the Hawking radiation would cause the black hole to lose mass, become hotter, radiate more, lose more mass, etc., the whole process accelerating until the hole disappears (presumably) in a final explosion. But for black hole of solar mass (or more) the process would take $>10^{53}$ Hubble times! And, so that the process could even begin, a wait of 10^7, or so, Hubble times would be needed to enable the expansion of the universe to reduce the present background radiation to below that of the hole – assuming an indefinitely expanding universe-model!

The absurdity of such figures notwithstanding, it is of some considerable theoretical interest to contemplate, as Hawking has done,[67,72] the state of thermal equilibrium that would be achieved by a black hole in a large container with perfectly reflecting walls. If the container is sufficiently large for a given total mass–energy content (case (a)), the black hole will radiate itself away completely (presumably) – after having swallowed whatever other stray matter there had been in the container – to leave, finally, nothing but thermal radiation (with perhaps a few thermalized particles). This final state will be the 'thermal equilibrium' state of maximum entropy (see figure 12.6(a)).

If the container is substantially smaller (case (c)) – or, alternatively, if the mass–energy content is substantially larger (though still not large enough to collapse the whole container) – the maximum-entropy state will be achieved by a single spherical black hole in thermal equilibrium with its surrounding radiation. Stability is here achieved because, if by a fluctuation the hole radiates a bit too much and consequently heats up, its surroundings heat up even more and cause it to absorb more than it emits and thus to return to its original size; if by a fluctuation it radiates less than it absorbs, its surroundings cool by more than it does and again it returns to equilibrium (figure 12.6(c)).

Chapter 12. Singularities and time-asymmetry

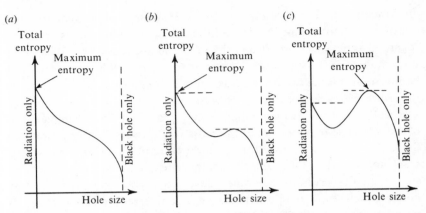

Figure 12.6. Hawking's black hole in a perfectly reflecting container: (*a*) large container, (*b*) intermediate container, (*c*) smallish container.

There is also a situation (case (*b*)) in which the container lies in an intermediate size range, for the given mass–energy content, and where the black hole is still stable, but only represents a *local* entropy maximum, the absolute maximum being a state in which there is only thermal radiation (and perhaps thermalized particles) but no black hole. In this case the black hole can remain in equilibrium with its surrounding radiation for a very long period of time. A large fluctuation would be needed in which a considerable amount of radiation is emitted by the hole, sufficient to get across the low entropy barrier between the two local maxima (figure 12.6(*b*)). With such a large mass loss from the hole, it is able to heat up by an amount greater than its surroundings are able to do; it then loses more mass, heats up more, and, as in case (*a*), radiates itself away completely, to give the required state of thermal radiation (plus occasional thermalized particles).

It should be pointed out that whereas we are dealing with processes that have absurdly long timescales,† these situations have a rather fundamental significance for physics. We are concerned, in fact, with the states of maximum entropy for *all* physical processes. In cases (*a*) and (*b*), the maximum-entropy state is the familiar 'heat death of the universe', but in case (*c*) we have something new: a black hole in thermal equilibrium with radiation. There are, of course, many detailed theoretical difficulties with this setup (e.g. the Brownian motion of the black hole would occasionally send it up against a wall of the container, whereupon

---

† If we allow (virtual) black holes of down to 10^{-33} cm (i.e. $\sim 10^{-20}$ of an elementary particle radius) then we obtain a picture[73] for which these timescales can be very short.

the container should be destroyed). Such problems will be ignored as irrelevant to the main issues! But also, since the relaxation times are so much greater than the present age of the universe, the interpretations of one's conclusions do need some care. Nevertheless, I feel that there are important insights to be gained here.

To proceed further with Hawking's argument, consider case (c). For most of the time the situation remains close to maximum entropy: a black hole with radiation. But occasionally, via an initial large fluctuation in which a considerable energy is emitted by the hole, a sequence like that just considered for case (b) will occur, where the black hole evaporates away to give thermal radiation. But then, after a further long wait, enough radiation (again by a fluctuation) collects together in a sufficiently small region for a black hole to form. Provided this hole is large enough, the system settles back into the maximum-entropy state again, where it remains for a very long while.

Cycles like this can also occur in case (b) (and even in case (a)), but with the difference that most of the time is spent in a state where there is no black hole. In case (c), a black hole is present most of the time. Hawking now argues that since the essential physical theories involved are time-symmetric (general relativity, Maxwell theory, neutrinos, possibly electrons, pions, etc., and the general framework of quantum mechanics), the equilibrium states ought to be time-symmetric also. But reversing the time-sense leads to white holes, not to black holes. Thus, Hawking proposes, white holes ought to be physically indistinguishable from black holes!

This identification is not so absurd as one might think at first. The Hawking radiation from the black hole becomes reinterpreted as particle creation near the singularity of the white hole (and hence Hawking proposes a rather slow rate of particle production at the white hole singularity). The swallowing of radiation by the black hole becomes time-reversed Hawking radiation from the white hole. One can, of course, envisage a black hole swallowing a complicated object such as a television set. How can this be thought of as time-reversed Hawking radiation? The argument is that Hawking radiation, being thermal,[67,74] produces all possible configurations with equal probability. It is *possible* to produce a television set as part of the Hawking radiation of a black hole, but such an occurrence is overwhelmingly improbable and would correspond to a large reduction in entropy. A black hole swallowing a television set only seems more 'natural' because we are used to situations in which the entropy is low in the initial state. We can equally well

Chapter 12. Singularities and time-asymmetry

envisage initial boundary conditions of low entropy for the time-reversed Hawking radiation – and this would be the case for a television set being annihilated as time-reversed Hawking radiation of a white hole.

So far, this all seems quite plausible, and there is even a certain unexpected elegance and economy in the whole scheme. But unfortunately is suffers from two (or perhaps three) very severe drawbacks which, in my opinion, rule it out as a serious possibility.

In the first place, whereas the geometry of the spacetime outside a stationary black hole's horizon is identical to that outside a stationary white hole's horizon, it is definitely *not* so that the exterior geometry of a black hole that forms by standard gravitational collapse and then finally disappears according to the Hawking process is time-symmetric. This time-asymmetry is made particularly apparent by use of conformal diagrams as shown in figure 12.7. A precise distinction between the

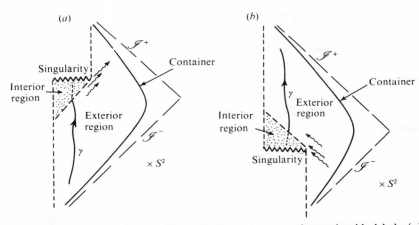

Figure 12.7. Conformal diagrams illustrating time-asymmetry of a transient black hole. (*a*) Classical collapse to a black hole followed by complete Hawking evaporation. (*b*) Hawking condensation to a white hole followed by its classical disappearance.

transient black hole and white hole external geometries can be made in terms of their TIP and TIF structures (cf. section 12.3.2). But intuitively, the distinction should be clear from the presence of timelike curves γ which, in the case of the black hole, 'leave' the exterior geometry to 'enter' the hole; and the other way around in the case of the white hole. The reason for this distinction is simply that the process of *classical* collapse is not the time-reverse of the *quantum* Hawking process. This should not really surprise us since each relies on quite different physical

Statement of the problem

theories (classical general relativity as opposed to quantum field theory on a fixed curved-space background).

A point of view adopted by Hawking which might avoid this difficulty is to regard the spacetime geometry as being somewhat observer-dependent. Thus, as soon as quantum mechanics and curved-space geometry have become essentially intertwined, so this viewpoint would maintain, one cannot consistently talk about a classically objective spacetime manifold. An observer who falls into a white hole to be evaporated away as its time-reversed Hawking radiation would, accordingly, believe the geometry to be, instead, that of a black hole whose horizon he crosses and inside which he awaits his 'classical' fate of final destruction by excessive tidal forces.

I have to say that I find this picture almost as hard to accept as those according to which the entropy starts decreasing when the observer's universe collapses about him. If one is considering black or white holes whose radius is of the order of the Planck length ($\sim 10^{-33}$ cm) – or even, possibly, of the order of an elementary particle size ($\sim 10^{-13}$ cm) – then such indeterminancy in the geometry might be acceptable. But for a black hole of solar mass (or more) this would entail a very radical change in our views about geometry, a change which would drastically affect almost any other application of general relativity to astrophysical phenomena. It is true that in section 12.2.5 I have briefly entertained the possibility of a world-view which allows for an element of 'observer-dependence' in the geometry. But I have as yet seen no way to relate such a view to the kind of indeterminacy in classical geometry that the physical identification of black holes with white holes seems to lead to.

But there are also other objections to attempting to regard classical gravitational collapse as being effectively the time-reverse of a quantum-mechanical particle creation process. One of these refers not so much to the attempted identification of the Hawking process with the time-reverse of the classical swallowing of matter by a black hole, but with the attempt to identify either of these processes with the phenomenon of particle production at regions of large spacetime curvature. Such a further identification *seems* to be an integral part of the time-symmetric view that I have been discussing, though it may well not really be what is intended. I am referring to the picture of Hawking radiation by a black hole as being alternatively regarded as a process of particle production near the singularity of a white hole. If, indeed, it can be so regarded, then this is not the 'normal' process of particle production at regions of large curvature that has many times been discussed in the literature.[62] For in

Chapter 12. Singularities and time-asymmetry

that process, particles are always produced in *pairs*: baryon with antibaryon, lepton with antilepton; positively charged particle with negatively charged particle. But the Hawking process is explicitly not of this form, as its thermal nature (for particles escaping to infinity) implies.[67,74,75]

The contrast is even more blatant if we try to relate this pairwise particle production near white hole singularities to the time-reverse of the destruction of matter at a black hole's singularity. For there are no constraints whatever on the type of matter that a black hole can classically absorb. And if strong cosmic censorship is accepted in classical processes, it seems that even individual charged particles have then to be separately destroyed at the singularity. (This point will be amplified in section 12.3.2.) There is no suggestion that the particles must somehow contrive to sort themselves into particle–antiparticle pairs before they encounter the singularity. The difficulty here is, perhaps, not so much directly to do with Hawking's identification of black holes with white holes, but with the whole idea of hoping to deal with the matter destruction–creation process in terms of known physics. Thus I maintain that, whereas it *may* be that matter creation at the big bang can be treated in terms of known (or at least partially understood) particle-creation processes, this seems *not* to be true of the destruction processes at black hole singularities – nor, if Hawking is right, of the creation processes at white hole singularities. Thus, the Hawking view would seem to lead to direct conflict with the often expressed hope that particle creation at the big bang can be understood in terms of processes of particle production by spacetime curvature. This relates to the question of whether time-symmetric physics can be maintained at spacetime singularities. This is a key issue that I shall discuss in more detail in section 12.3.

More directly related to Hawking's proposal is a difficulty which arises if we examine in detail the cycles whereby a solar-mass black hole (say), in stable equilibrium with its surroundings, in our perfectly reflecting container, may disappear and reform, owing to fluctuations, in cases (*b*) and (*c*) just discussed. What, in fact, is the most probable way for the black hole to evaporate completely? It might, of course, simply throw out its entire mass in one gigantic fluctuation. But the random Hawking process would achieve this only absurdly infrequently. Overwhelmingly *less* absurdly infrequent would be the emission, in one huge fluctuation, of that *fraction* of the hole's mass needed to raise its Bekenstein–Hawking temperature above that to which its surroundings would be consequently raised. From there on, the evaporation would proceed

Statement of the problem

'normally' needing no further improbable occurrence. But consider the final explosion according to which the hole 'normally' disappears. Electrons and positrons have been appearing, followed by pions; then, at the last moment, a whole host of unstable particles is produced which undergo complicated decays. Finally one expects many protons and antiprotons separately escaping the annihilation point. Only much later, by chance encounters as they move randomly about the container, would one expect the protons and antiprotons gradually to annihilate one another (or possibly occasionally to decay, themselves, by a Pati–Salam-type process[76] into, say, positrons and into electrons which would then mostly annihilate one another).

What, now, must we regard as the most probable way in which a black hole forms again in the container, to reach another point of stable equilibrium with its surroundings? Surely it is *not* the time-reverse of the above, according to which the protons and antiprotons must first (with long preparation) contrive to form themselves (quite unnecessarily) out of the background radiation before aiming themselves with immense accuracy at a tiny point, only to indulge in some (again unnecessary) highly contrived particle physics whereby they meet up with carefully aimed γ-rays, etc., to form various unstable particles, etc., etc.; and then (also with choreography of the utmost precision) other particles (pions, then electrons and positrons) must aim themselves inwards, having first formed themselves out of the background at the right moment and in the correct proportions. Only later does the background radiation itself fall inwards to form the bulk of the mass needed to form the hole.

My point is not that this curious beginning is necessarily the most improbable part of the process. I can imagine that it may well not be. But it is unnecessary. The essential part of the hole formation of a *black* hole would occur when the radiation itself collects together in a sufficiently small region to undergo what is, in effect, a standard gravitational collapse. In fact, it would seem that the tiny core that has been formed with such elaborate preparations ought somewhat to *inhibit* the subsequent inward collapse, owing to its excessive temperature!

So what has gone wrong? What time-asymmetric physics has been smuggled into the description of the 'most probable' mode of disappearance of the hole, that it should so disagree with the time-reverse of its 'most probable' mode of reappearance? Possibly none, if white holes, in principle, exist, and are simply *different* objects from black holes. While the above elaborate preparations are not necessary for the formation of a black hole, they could be for a white hole. After all, one has to contrive

Chapter 12. Singularities and time-asymmetry

some way of producing the white hole's initial singularity which, as is evident from figure 12.7, has a quite different structure from that of a black hole. It might be that the production of such a singularity is an extraordinarily delicate process, requiring particles of just the right kind and in an essentially right order to be aimed with high energy and with extraordinary accuracy. There is an additional seeming difficulty, however, in that one also has to conjure up a region of spacetime (namely that inside the horizon) which does not lie in the domain of dependence of some initial Cauchy hypersurface drawn before the white hole appears (figure 12.7(b)). Of course, the time-reverse of this problem also occurs for the (Hawking) disappearance of a black hole, but one is not in the habit of trying to retrodict from Cauchy hypersurfaces, so this seems less worrisome! An additional difficulty, if white holes are allowed, is that we now encounter the problem, mentioned earlier, of trying to predict what the white hole is going to emit, and when. As I have indicated, *if* time-symmetric physics holds at singularities, then 'normal' ideas of particle creation due to curvature will not do. Hawking's concept of 'randomicity'[67] might be nearer the mark, but it is unfortunately too vague to enable any calculations to be made – now that the essential guiding idea of an identification between black holes and white holes has been removed.

I find the picture of an equilibrium involving the occasional production of such white holes a very unpleasant one. And what other monstrous zebroid combinations of black and white might also have to be contemplated? I feel that such things have nothing really to do with physics (at least on the macroscopic scale). The only reason that we have had to consider white holes *at all* is in order to save time-symmetry! The consequent unpleasantness and unpredictability seems a high price to pay for something that is *not even true* of our universe on a large scale.

One of the consequences of the hypothesis that I shall set forth in the next section is that it rules out the white hole's singularity as an unacceptable boundary condition. The hypothesis is time-asymmetric, but this is necessary in order to explain the other arrows of time. When we add this hypothesis to the discussion of equilibrium within the perfectly reflecting container, we see explicitly what time-asymmetric physics has been 'smuggled' in. For the hypothesis is designed not to constrain the behaviour of black holes in any way, but it forbids white holes and therefore renders irrelevant the extraordinary scenario that we seem to need in order to produce one!

Singularities: the key?

I hope the reader will forgive me for having discussed white holes at such length only to end by claiming that they do not exist! But hypothetical situations can often lead to important understandings, especially when they border on the paradoxical, as seems to be the case here.

12.3 Singularities: the key?

What is the upshot of the discussion so far? According to sections 12.2.3, 12.2.4 and 12.2.5, the arrows of entropy and retarded radiation, and *possibly* of psychological time, can all be explained if a reason is found for the initial state of the universe (big bang singularity) to be of comparatively low entropy and for the final state to be of high entropy. According to section 12.2.6, some low-entropy assumption *does* need to be *imposed* on the big bang; that is, the mere fact that the universe expands away from a singularity is in no way sufficient. And according to section 12.2.7, we need some assumption on initial singularities that rules out those which would lie at the centres of white holes. On the other hand, the discussions in sections 12.2.1 and 12.2.2 were inconclusive, and I shall need to return to them briefly at the end.

But what is it in the nature of the big bang that is of 'low entropy'? At first sight, it would seem that the knowledge we have of the big bang points in the opposite direction. The matter (including radiation) in the early stages appears to have been completely thermalized (at least so far as this is possible, compatibly with the expansion). If this had not been so, one would not get correct answers for the helium abundance, etc.[77,78] And it is often remarked upon that the 'entropy per baryon' (i.e. the ratio of photons to baryons) in the universe has the 'high' value of $\sim 10^9$. Ignoring the contribution to the entropy due to black holes, this value has remained roughly constant since the very early stages, and then represents easily the major contribution to the entropy of the universe – despite all the 'interesting' processes going on in the world, so important to our life here on Earth, that depend upon 'small' further taking up of entropy by stars like our Sun. The answer to this apparent paradox – that the big bang thus *seems* to represent a state of *high* entropy – lies in the unusual nature of gravitational entropy. This I next discuss, and then show how this relates to the structure of singularities.

12.3.1 Gravitational entropy

It has been pointed out by many authors[79] that gravity behaves in a somewhat anomalous way with regard to entropy. This is true just as

Chapter 12. Singularities and time-asymmetry

much for Newtonian theory as for general relativity. (In fact, the situation is rather worse for Newtonian theory.) Thus, in many circumstances in which gravity is involved, a system may behave as though it has a negative specific heat. This is directly true in the case of a black hole emitting Hawking radiation, since the more it emits, the hotter it gets. But even in such familiar situations as a satellite in orbit about the Earth, we observe a phenomenon of this kind. For dissipation (in the form of frictional effects in the atmosphere) will cause the satellite to speed up, rather than slow down, i.e. cause the kinetic energy to increase.

This is essentially an effect of the universally attractive nature of the gravitational interaction. As a gravitating system 'relaxes' more and more, velocities increase and the sources clump together – instead of uniformly spreading throughout space in a more familiar high-entropy arrangement. With other types of force, their attractive aspects tend to saturate (such as with a system bound electromagnetically), but this is not the case with gravity. Only non-gravitational forces can prevent parts of a gravitationally bound system from collapsing further inwards as the system relaxes. Kinetic energy itself can halt collapse only temporarily. In the absence of significant non-gravitational forces, when dissipative effects come further into play, clumping becomes more and more marked as the entropy increases. Finally, maximum entropy is achieved with collapse to a black hole – and this leads us back into the discussion of section 12.2.7.

Consider a universe that expands from a 'big bang' singularity and then recollapses to an all-embracing final singularity. As was argued in section 12.2.6, the entropy in the late stages ought to be much higher than the entropy in the early stages. How does this increase in entropy manifest itself? In what way does the high entropy of the final singularity distinguish it from the big bang, with its comparatively low entropy? We may suppose that, as is apparently the case with the actual universe, the entropy in the initial *matter* is high. The kinetic energy of the big bang, also, is easily sufficient (at least on average) to overcome the attraction due to gravity, and the universe expands. But then, relentlessly, gravity begins to win out. The precise moment at which it does so, locally, depends upon the degree of irregularity already present, and probably on various other unknown factors. Then clumping occurs, resulting in clusters of galaxies, galaxies themselves, globular clusters, ordinary stars, planets, white dwarfs, neutron stars, black holes, etc. The elaborate and interesting structures that we are familiar with all owe their existence to this clumping, whereby the gravitational potential energy begins to be taken up and the entropy can consequently begin to rise above the *apparently* very high value that the system had initially. This clumping must be expected to increase; more black holes are formed; smallish black holes swallow material and congeal with each other to form bigger ones. This process accelerates in the final stages of recollapse when the average density becomes very large again, and one must expect a very irregular and clumpy final state.

There is a slight technical difficulty in that the concept of a black hole is normally only defined for asymptotically flat (or otherwise open) spacetimes. This difficulty could affect the discussion of the final stages of collapse when black holes begin to congeal with one another, and with the final all-embracing singularity of recollapse. But I am not really concerned with the location of the black holes' event horizons, and it is only in precisely defining these that the aforementioned difficulty arises. A black hole that is formed early in the universe's history has a singularity that is reached at early proper times for observers who encounter it;[57] for holes that are formed later, they can be reached at later proper times. On the basis of strong cosmic censorship (cf. section 12.3.2), one expects all these singularities eventually to link up with the final singularity of recollapse.[57] I do not require that the singularities of black holes be, in any clear-cut way, distinguishable from each other or from the final singularity of recollapse. The point is merely that the gravitational clumping which is characteristic of a state of high gravitational entropy should manifest itself in a very complicated structure for the final singularity (or singularities).

The picture is not altogether dissimilar for a universe that continues to expand indefinitely away from its big bang. We still expect local clumping, and (provided that the initial density is not altogether too low or too uniform for galaxies to form at all) a certain number of black holes should arise. For the regions inside these black holes, the situation is not essentially different from that inside a collapsing universe (as was remarked upon in section 12.2.6), so we expect to find, inside each hole, a very complicated singularity corresponding to a very high gravitational entropy. For those regions not inside black holes there will still be certain localized portions, such as rocks, planets, black dwarfs, or neutron stars, which represent a certain ultimate raising of the entropy level owing to gravitational clumping, but the gain in gravitational entropy will be relatively modest, though sufficient, apparently, for all that we need for life here on Earth.

Chapter 12. Singularities and time-asymmetry

I have been emphasizing a qualitative relation between gravitational clumping and an entropy increase due to the taking up of gravitational potential energy. In terms of spacetime curvature, the absence of clumping corresponds, very roughly, to the absence of Weyl conformal curvature (since absence of clumping implies spatial-isotropy, and hence no gravitational principal null-directions).[45] When clumping takes place, each clump is surrounded by a region of nonzero Weyl curvature. As the clumping gets more pronounced owing to gravitational contraction, new regions of empty space appear with Weyl curvature of greatly increased magnitude. Finally, when gravitational collapse takes place and a black hole forms, the Weyl curvature in the interior region is larger still and diverges to infinity at the singularity.

At least, that is the picture presented in spherically symmetric collapse, the magnitude of the Weyl curvature diverging as the inverse cube of the distance from the centre. But there are various reasons for believing that in generic collapse, also, the Weyl curvature should diverge to infinity at the singularity, and (at most places near the singularity) should dominate completely over the Ricci curvature.

This can be seen explicitly in the details of the Belinskii–Khalatnikov–Lifshitz analysis.[80] Moreover, one can also infer it on crude qualitative grounds. In the exact Friedmann models, it is true, the Ricci tensor dominates, the Weyl tensor being zero throughout. In these cases, as a matter world-line is followed into the singularity, it is approached isotropically by the neighbouring matter world-lines, so we have simultaneous convergence in three mutually perpendicular directions orthogonal to the world-line. In the case of spherically symmetrical collapse to a black hole, on the other hand, if we envisage some further matter falling symmetrically into the central singularity, it will normally converge in towards a given matter world-line only in *two* mutually perpendicular directions orthogonal to the world-line (and diverge in the third). This is the situation of the Kantowski–Sachs[81] cosmological model, giving a so-called 'cigar'-type singularity.[7] If r is the usual Schwarzschild coordinate, the volume gets reduced like $r^{3/2}$ near the singularity, so the densities are $\sim r^{-3/2}$. Thus, for a typical Ricci tensor component, $\Phi \sim r^{-3/2}$. However, in general, for a typical Weyl tensor component, $\Psi \sim r^{-3}$, showing that the Weyl tensor dominates near the singularity in these situations. Also, in the 'pancake' type of singularity, where there is convergence in only *one* direction orthogonal to a matter world-line, we again expect the Weyl tensor to dominate with $\Phi \sim r^{-1}$ and $\Psi \sim r^{-2}$ in this case.

Now the Friedmann type of situation, with simultaneous convergence of all matter from all directions at once, would seem to be a very special setup. If there is somewhat less convergence in one direction than in the other two, then a cigar-type configuration seems more probable very close to the singularity, while a pancake-type appears to result when the main convergence is only in one direction. Moreover, with a generic setup, a considerable amount of oscillation seems probable.[80] An oscillating Weyl curvature of frequency ν and complex amplitude Ψ, supplies an *effective* additional 'gravitational-energy' contribution to the Ricci tensor[61] of magnitude $\sim |\Psi|^2 \nu^{-2}$. If ν becomes very large so that many oscillations occur before the singularity is reached, then[49] $\nu^2 \gg \Phi^{-1}$, where Φ is a typical Ricci tensor component. Thus if, as seems reasonable in general, the 'energy content' of Ψ is to be comparable with Φ as the singularity is approached, we have $|\Psi|^2 \nu^{-2} \sim \Phi$, so $|\Psi| \gg \Phi$. These considerations are very rough, it is true, but they seem to concur with more detailed analysis[80] which indicates that in generic behaviour near singularities the contributions due to matter can be ignored to a first approximation and the solution treated as though it were a vacuum, i.e. that the Weyl part of the curvature dominates over the Ricci part.

The indications are, then, that a high-entropy singularity should involve a very large Weyl curvature, unlike the situation of the singularity in the Friedmann dust-filled universe or any other models of the Robertson–Walker class. At the time of writing, however, no clear-cut integral formula (say) which could be regarded as giving mathematical expression to this suggested relation between Weyl curvature and gravitational entropy has come to light. Some clues as to the nature of such a formula (if such exists at all) may be obtained, firstly, from the Bekenstein–Hawking formula for the entropy of a black hole and, secondly, from the expression for the particle number operator for a linear spin-2 massless quantized free field – since an estimate of the 'number of gravitons' in a gravitational field could be taken as a measure of its entropy.† Thus this entropy measures the number of *quantum* states that contribute to a given classical geometry.

There is one final point that should be mentioned in connection with the question of the entropy in the gravitational field. It was pointed out some time ago by Tolman[84] that a model universe containing matter that

---

† This point of view does not seem to agree with that of Gibbons and Hawking,[82] who apparently regard the gravitational entropy as being zero when black holes are absent. But an estimate of 'photon number' in a classical electromagnetic field gives a measure of its entropy[83] (without black holes). Gravity is presumably similar.

Chapter 12. Singularities and time-asymmetry

appeared to be in thermal equilibrium in its early stages can lead to a situation in which the matter gets out of equilibrium as the universe expands (a specific example of matter illustrating such behaviour being a diatomic gas which is capable of dissociating into its elements and recombining). Then, if such a model represents an expanding and recollapsing universe, the state of the matter during recollapse would differ from the corresponding state during expansion, where we make the correspondence at equal values of the universe radius R (or comoving radius R). In fact, the matter, during recollapse, would have acquired some energy out of the global geometry of the universe, the resulting difference in geometry showing up in the fact that $\dot{R}^2$ is greater, for given R, at recollapse than it is during the expansion. So the entropy of the system as a *whole* increases with time even though the matter *itself* is in thermal equilibrium during an initial stage of the expansion. There is, in fact, a contribution to the entropy from R (and $\dot{R}$), which must be regarded as a dynamical variable in the model. (This arises because of the phenomenon of *bulk viscosity*.[85])

One can view what is involved here as basically a transfer of potential energy from the global structure of the universe (gravitational potential energy) into the local energy of the matter, though there are well-known difficulties about defining energy in a precise way for models of this kind. But these difficulties should not concern us unduly here, since it is actually the entropy rather than the energy that is really relevant, and entropy has much more to do with probabilities and coarse-graining than it has to do with any particular definition of energy. In the example given by Tolman there is no state of maximum entropy, either achieved in any one specific model or throughout all models of this type. By choosing the value of R at maximum expansion to be sufficiently large (for fixed matter content), the total entropy can be made as large as we please. Tolman envisaged successive cycles of an 'oscillating' universe with gradually increasing maximum values of R. However, it is hard for us to maintain such a world-view now, because the singularity theorems[7,8] tell us that the universe cannot achieve an effective 'bounce' at minimum radius without violating the known† laws of physics.

From my own point of view, the situation envisaged by Tolman may be regarded as one aspect of the question of how the structure of the universe as a whole contributes to the entropy. It apparently concerns a somewhat different aspect of this question than does gravitational

† I am counting quantum gravity as 'unknown' whether or not it helps with the singularity problem!

Singularities: the key?

clumping, since the Weyl tensor is everywhere zero in Tolman's models. It is clear that this has also to be understood in detail if we are to perceive, fully, the role of gravitational entropy. Nevertheless, it appears that the entropy available in Tolman's type of situation is relatively insignificant[78] compared with that which can be obtained – and, indeed, *is* obtained – by gravitational clumping (cf. section 12.3.3).

The key question must ultimately concern the structure of the singularities. These singularities, in any case, provide the boundary conditions for the various cycles in Tolman's 'oscillating' universe. Moreover, as we shall see in a moment, if strong cosmic censorship holds true, the presence of irregularities should not alter the all-embracing nature of these singularities in the case of an expanding and recollapsing universe.

12.3.2 Cosmic censorship†

Though it is by no means essential, for the viewpoint that I am proposing, to suppose that naked singularities cannot occur, such an assumption of 'cosmic censorship' does nevertheless greatly simplify and clarify the discussion. It has been my stated intention to adopt basically conventional attitudes on most issues, so I should not be out of line here, also, were I simply to align myself with what appears to be a majority view and (at least for purposes of argument) adopt a suitable assumption of cosmic censorship. But in the following pages I shall also give some independent justification of this view.

A preliminary remark is required before considering the details of this, however. In the Hawking process of black hole evaporation, the (supposed) final disappearance of the hole produces, momentarily, a naked singularity. This is not normally considered to be a violation of cosmic censorship because the Hawking process is a quantum-mechanical process, whereas cosmic censorship is taken normally to refer only to classical general relativity. (In the words of Hawking,[68] cosmic censorship is 'transcended' rather than violated!) Nevertheless, the presence of actual naked singularities of this kind in the geometry of the world would make a certain difference to the discussion. But the difference seems unlikely to be of any great relevance to the problem at hand. The black holes that we have any clear reason to believe actually exist in the universe are all of the order of a solar mass or more – and we have seen

† Parts of this section are considerably more technical than the others, and it can be omitted without seriously affecting the thread of argument.

Chapter 12. Singularities and time-asymmetry

that their lifetimes are greater than 10^{53} Hubble times, so their final naked singularities (!) can be safely ignored. Moreover, an implication of the viewpoint I shall set out in section 12.3.3 seems to be that mini-holes are unlikely to exist, these being the only black holes (say of mass 10^{20} g or less) whose final naked singularity could occur early enough to be of any remote relevance to the discussion. But since, in any case, such holes would be of no greater than atomic dimension, they could be 'smoothed over' and need not be considered as constituting a significant part of the classical geometry.

It seems, then, that a discussion of cosmic censorship entirely within classical general relativity should be perfectly adequate for our purposes. So what form of statement should we adopt? The usual formulation involves some assertion such as:

A system which evolves, according to classical general relativity with reasonable equations of state, from generic non-singular initial data on a suitable Cauchy hypersurface, does not develop any spacetime singularity which is visible from infinity.

Something of this kind, forbidding naked singularities, appears to be required in order that the usual general discussion of black holes can be carried through (e.g. the area-increase principle, the merging of two black holes necessarily forming a third, the general macrostability of black holes – indeed the very physical existence of black holes at all, rather than something worse, in a generic collapse[51]). The statement is vague in several respects, and a considerable increase in precision would be required in order to obtain something capable of clear mathematical proof, or disproof.

But it is probably not too helpful simply to make the various conditions more precise in some way, without having a deeper idea of what is likely to be involved. For example, it seems to me to be quite unreasonable to suppose that the physics in a comparatively local region of spacetime should really 'care' whether a light ray setting out from a singularity should ultimately escape to 'infinity' or not. To put things another way, some observer (timelike world-line) might intercept the light ray and see the singularity as 'naked', though he be not actually situated at infinity (and no actual observer would be so situated in any case). The observer might be close by the singularity and possibly himself trapped, e.g. inside the usual black hole of figure 12.5(a). The unpredictability entailed by the presence of naked singularities which is so abhorrent to many people

Singularities: the key?

would be present just as much for this local observer – observing a 'locally naked' singularity – as for an observer at infinity.

It seems to me to be comparatively unimportant whether the observer himself can escape to infinity. Classical general relativity is a scale-independent theory, so if locally naked singularities occur on a very tiny scale, they should also, in principle, occur on a very large scale in which a 'trapped' observer could have days or even years to ponder upon the implications of the uncertainties introduced by his observation of such a singularity (compare the analogous discussion in section 12.2.6 of the astronaut inside a large black hole). Indeed, for inhabitants of recollapsing closed universes (as possibly we ourselves are) there is no 'infinity', so the question of being locally 'trapped' is one of degree rather than principle.

It would seem, therefore, that if cosmic censorship is a principle of Nature, it should be formulated in such a way as to preclude such *locally naked singularities*.[57,58,86] This viewpoint gains some support from, first of all, the standard picture of spherically symmetrical collapse inside a black hole as shown in figure 12.5(a). An observer who falls inside the hole cannot, in fact, see the singularity at all until he encounters it. This is perhaps clearer if we use a standard conformal diagram (with null-cones sloping at 45°) to depict the situation, as in figure 12.8, since then the spacelike nature of the singularity is brought out clearly and shows that it is *not* locally naked in the sense described above.

Secondly, there are some reasons for believing that generic perturbations away from spherical symmetry will not change the spacelike nature of the singularity (whose continued existence, in the perturbed case, is guaranteed by the singularity theorems[6–8]). The situation is slightly complicated, however, because the Schwarzschild–Kruskal singularity of figure 12.8 is, in fact, unstable. The introduction of a minute amount of angular momentum into the black hole to give a Kerr solution (with $a \ll m$) will actually change the singularity structure completely, and it ceases to be spacelike. Only when a further perturbation of a generic nature is made, may a spacelike structure of the singularity be expected to be restored.

It is somewhat easier to examine this behaviour if we consider adding charge rather than angular momentum, so that we get the Reissner–Nordstrom solution instead of the Kerr solution. The corresponding conformal diagram is shown in figure 12.9(a), and, indeed, it is evident that the singularity *is* now locally naked in the sense described above, since the observer whose world-line is γ can see the singularity. With a

Chapter 12. Singularities and time-asymmetry

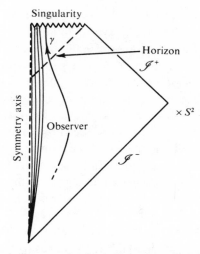

Figure 12.8. Conformal diagram of spherically symmetric collapse (with asymptotic flatness) illustrating the spacelike nature of the singularity.

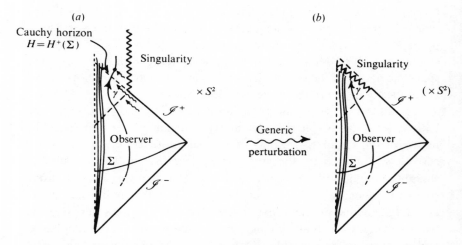

Figure 12.9. Conformal diagrams illustrating collapse to a black hole with a small charge: (a) spherically symmetric, (b) generically perturbed.

further perturbation of a generic nature one expects the situation more to resemble that of figure 12.9(b) in which the singularity is spacelike again (or perhaps null). The reason is that in the situation of figure 12.9(a) there is a null-hypersurface, H, in the spacetime, which is the *Cauchy horizon* of an (edgeless) spacelike hypersurface Σ extending all the way out to spatial infinity. (For terminology and notation, see reference 49.) An observer γ who crosses $H = H^+(\Sigma)$ will see, as he does so, the entire future history of the outside world flash by in an instant. If the data on Σ are perturbed in some mild way out near infinity, then this will lead to a drastic alteration of the geometry in the neighbourhood of H. This is because signals from the infinite regions of Σ will be blueshifted at H by an infinite amount. Indeed, the analysis of weak-field perturbations (or test fields)[87,88] explicitly indicates that these perturbations diverge along H. With the full nonlinear coupling, this would presumably give a curvature singularity in place of H. Furthermore, it *could* well be that nonlinear effects actually result in a spacelike rather than just a null-singularity, because the effects of large curvature might proliferate and reinforce one another further and further up along the singularity (cf. figure 12.9(*b*)).

In order to make precise the notion of a spacelike (or null) spacetime singularity it will be convenient to recall the concept of an *ideal point*[89] of a spacetime $\mathcal{M}$, defined in terms of the TIPs or TIFs (terminal indecomposable past-sets or future-sets) of $\mathcal{M}$. The ideal points may be thought of as some 'extra points' adjoined to the manifold $\mathcal{M}$, either 'singularities' or 'points at infinity', for which the timelike curves in $\mathcal{M}$ that are future-endless in $\mathcal{M}$ acquire future ideal endpoints (via TIPs), and those that are past-endless in $\mathcal{M}$ acquire past ideal endpoints (via TIFs).

For simplicity, assume that $\mathcal{M}$ is strongly causal. Let γ and γ' be two future-endless timelike curves in $\mathcal{M}$. Then γ and γ' have the same future ideal endpoint if and only if they have the same pasts, which, in standard notation, is written $I^-[\gamma] = I^-[\gamma']$. The TIPs of $\mathcal{M}$ are, in fact, the sets of the form $I^-[\gamma]$, with γ future-endless and timelike, and may be 'identified' with the future ideal points. Similarly, past-endless timelike curves η and η' have the same ideal past endpoints whenever their futures are the same: $I^+[\eta] = I^+[\eta']$, these sets being the TIFs of $\mathcal{M}$. (See figure 12.10). In each case, the TIP or TIF is said to be *generated* by the timelike curve in question. A simple criterion[57,58] that may be used to distinguish those TIPs representing points at infinity from those representing singular points is to define a TIP as an ∞-*TIP* if it is generated by some timelike curve of infinite proper length into the future, and as a *singular TIP* if it is generated by no such curve. The ∞-*TIFs* and *singular TIFs* are similarly defined. (One may also choose to call some of the ∞-TIPs and ∞-TIFs singular, in some appropriate sense, but I shall not bother with this in detail here.)

Next, a locally *naked* singularity can be defined as either a singular TIP contained in the past $I^-(q)$ of some point q of $\mathcal{M}$, or as a singular TIF

Chapter 12. Singularities and time-asymmetry

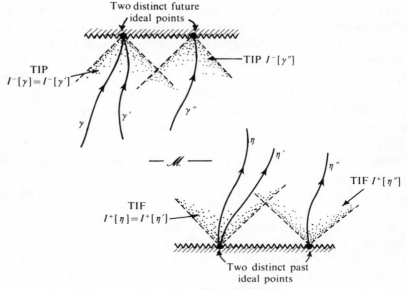

Figure 12.10. TIPs and TIFs defining ideal points of $\mathcal{M}$.

contained in the future $I^+(p)$ of some point p of $\mathcal{M}$ (figure 12.11). The upshot of such a definition is that in each case there is a timelike curve ζ (observer's world-line) from a point p to a point q, where the singularity lies to the future of p and to the past of q. (Take p in the TIP, in the first case, and q in the TIF, in the second.) The significance of this is that not

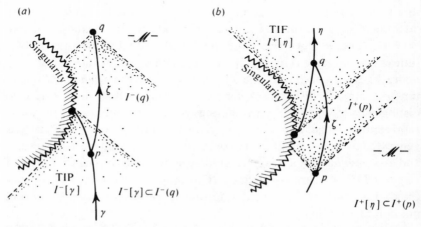

Figure 12.11. A locally naked singularity, lying to the future of p and to the past of q: (a) TIP definition, (b) TIF definition.

only can some observer (namely ζ) see the singularity (namely from q) but that at some earlier stage of his existence (namely at p) the singularity has yet to be produced. Thus, the usual all-embracing big bang does *not* qualify as locally naked, since no observer existed before it was produced.

One may say that $\mathcal{M}$ accords with a form of cosmic censorship if such locally naked singularities (as illustrated in figures 12.11(a) and (b)) do not occur.[57] However, it is convenient to go somewhat further than this and to exclude 'naked points at infinity' also, by dropping the condition that the TIP or TIF involved in figure 12.11 is necessarily a *singular* TIP or TIF. Indeed, it could be argued that such a naked point at infinity introduces as much indeterminacy into the future behaviour of a universe-model as does a naked singularity. But, in fact, naked points at infinity seem unlikely to arise unless they are also, in some appropriate sense, *singular* points (though still defined by ∞-TIPs or ∞-TIFs). The reason is that for a smooth conformal infinity $\mathcal{I}$, naked points at infinity can occur only if $\mathcal{I}$ is (in some places at least) timelike. And, with reasonable matter density, $\mathcal{I}$ can be timelike only if $\lambda < 0$, where λ is the cosmological constant.[90] However, the usual (Friedmann-type) $\lambda < 0$ models expand from, and recontract to, all-embracing singularities,[91] and hence possess no ∞-TIPs or ∞-TIFs, so (non-singular) naked points at infinity are not to be expected even in these cases.

Whether the TIPs or TIFs refer to singular points or to points at infinity, the situation of figure 12.11 is what characterizes ideal points as belonging to a boundary to $\mathcal{M}$ which is, in a sense, *timelike*. For if we extend ζ indefinitely into the future, in figure 12.11(a), to becomes a future-endless timelike curve ζ' we have

$$I^-[\gamma] \subset I^-(q), \quad q \in I^-[\zeta']$$

which is the condition[89] for the TIP $I^-[\gamma]$ to lie to the chronological (i.e. timelike) past of the TIP $I^-[\zeta']$. (The weaker condition $I^-[\gamma] \subset I^-[\zeta']$ obtains if $I^-[\gamma]$ lies to the *causal* past[89] of $I^-[\zeta']$.) Similarly, figure 12.11(b) gives us

$$I^+[\eta] \subset I^+(p), \quad p \in I^+[\zeta'']$$

with ζ'' past-endless, characterizing the TIF $I^+[\eta]$ as lying to the chronological future of the TIF $I^+[\zeta'']$. Thus, to rule out figure 12.11(a) configurations is to say that the future ideal points constitute an *achronal*[49] (i.e. roughly, spacelike or null) future boundary to $\mathcal{M}$, while to rule out those of figure 12.11(b) is to say that the past ideal points constitute an *achronal* past boundary to $\mathcal{M}$.

Chapter 12. Singularities and time-asymmetry

What is more, ruling out *either one* of these configurations throughout $\mathcal{M}$ is equivalent to ruling out the other, since each condition turns out to be equivalent[57] to the time-symmetric condition that $\mathcal{M}$ be *globally hyperbolic*.[7,49]

The proof of this fact is quite simple and is worth outlining here since it has not been spelled out in the standard literature. First, a globally hyperbolic spacetime is one with the property that for any two of its points p, q, the space of causal curves from p to q is compact. (A causal curve is a curve, not necessarily smooth, that everywhere proceeds locally within, or on, the future light-cone. Thus, causal curves are timelike curves or curves that are everywhere locally the C^0 *limits* of timelike curves.) Strong causality being assumed, global hyperbolicity is in fact equivalent to the statement that each $I^+(q) \cap I^-(p)$ has compact closure.[7,49] I also remark here that for a causal curve ζ, the set $I^-[\zeta]$ is a TIP if and only if ζ is future-endless,[89] with a similar result holding for TIFs.

Now suppose that $\mathcal{M}$ contains a point q and a future-endless timelike curve γ such that $I^-[\gamma] \subset I^-(q)$. It follows that $\mathcal{M}$ cannot be globally hyperbolic. For if p is a fixed point on γ, and $r_1, r_2, r_3, \ldots$ is a sequence of points proceeding indefinitely far up γ, we obtain a sequence $\zeta_1, \zeta_2, \zeta_3, \ldots$ of causal curves from p to q where ζ_i consists of that segment of γ from p to r_i together with a timelike curve from r_i to q (which exists because $r_i \in I^-[\gamma]$ and $I^-[\gamma] \subset I^-(q)$). If $\zeta_1, \zeta_2, \ldots$ had a limit causal curve ζ, from p to q, then $I^-[\zeta'] = I^-[\gamma]$, where $\zeta' = \zeta \cap I^-[\gamma]$ (as is easily seen), whereas ζ' cannot be future-endless, being continued as ζ to the future endpoint q. This would contradict the last statement of the preceding paragraph, showing that ζ does not exist and that $\mathcal{M}$ is consequently not globally hyperbolic.

The converse argument is slightly more technical. I use the notation and proposition numbers of reference 49. Suppose $\mathcal{M}$ is not globally hyperbolic. Then points p and q exist for which $I^+(p) \cap I^-(q)$ does not have compact closure. Hence (propositions 3.9, 5.20) $p \notin \text{int } D^-(\partial I^-(q))$, whereas $p \in I^-(q)$. Consequently (proposition 5.5h), $p' \notin D^-(\partial I^-(q))$, where $p' \in I^-(p)$, so there is a future-endless timelike curve γ from p' which does not meet $\partial I^-(q)$. But $p' \in I^-(q)$, whence $\gamma \subset I^-(q)$ and the TIP $I^-[\gamma]$ lies entirely within $I^-(q)$, as is required to prove.

My proposal[57,59] for a *strong cosmic censorship principle* is, therefore, that a physically reasonable classical spacetime $\mathcal{M}$ ought to have the property which can be stated in any one of the following equivalent forms: no TIP lies entirely to the past of any point in $\mathcal{M}$; no TIF lies entirely to the future of any point in $\mathcal{M}$; the TIPs form an achronal set; the

Singularities: the key?

TIFs form an achronal set; $\mathcal{M}$ is globally hyperbolic; there exists a Cauchy hypersurface for $\mathcal{M}$. (This last equivalence is a well-known result due to Geroch.[93]) The plausibility of this rests, of course, on what we expect of a 'physically reasonable' spacetime. Indeed, global hyperbolicity has tended to be regarded by many people as an over-strong restriction. Nevertheless, I believe that one *can* put forward plausibility arguments to support strong cosmic censorship. I next give an indication of this.

To begin with (except in the case of the big bang) it seems not unreasonable to restrict attention, in the first instance, to the case of vacuum only. The reason for this was indicated in section 12.3.1, namely that near 'generic' singularities we expect the Weyl curvature to dominate over the Ricci tensor. This is not totally satisfactory, however, because there are 'cumulative' effects due to the Ricci tensor (namely focussing) which the Weyl tensor can achieve only indirectly through nonlinearities. Nevertheless, the behaviour of vacuum solutions would seem to give a good first approximation near a generic singularity, which avoids the problems raised, for example, by the presence of apparently inessential 'shell-crossing' naked singularities[92] in idealized matter such as 'dust'. As a second approximation one could consider the Einstein–Maxwell theory, for example, which likewise avoids these problems. However, the big bang is a special situation (which relates to its low entropy) and the criterion of 'genericity' need not apply.† But strong cosmic censorship still seems to hold – for different reasons (cf. section 12.3.3). In the singularities of collapse, however, a high-entropy assumption of 'genericity' seems physically reasonable.

The sequence depicted in figures 12.8, 12.9(*a*) and 12.9(*b*) would seem to provide a plausible pattern for the general situation. In figure 12.8 (extended Schwarzschild solution), global hyperbolicity holds, but apparently fortuitously. Each singular TIP intersects a Cauchy hypersurface Σ in a compact region. The data in that region *alone* are all that are needed to imply the existence and nature of the singularity that the TIP represents. But perturb the solution slightly, so that it becomes that of Kerr and extend it maximally in the usual way (as in figure 12.9(*a*)); then the singularity disappears – in the sense that no singular TIP near to the original one now exists, but is replaced by the past $I^-(x)$ of a non-singular interior point x. Thus the original singularity apparently owes its existence to some very special aspect (e.g. the exact spherical symmetry) of the original initial data. Global hyperbolicity is violated in the slightly

† Indeed, it is perfectly in order for the big bang to be what, in the reverse direction in time, would be an *unstable* singularity (cf. section 12.3.3).

Chapter 12. Singularities and time-asymmetry

perturbed solution, but *because* of this violation there is now a Cauchy horizon $H = H^+(\Sigma)$. The past $I^-(y)$ of a point y of H intersects Σ in a region with *non-compact* closure (extending to infinity). We may take it that the structure of the spacetime (e.g. the curvature) at y is the result of some form of integral of the initial data over this region. If the data are perturbed generically, we are liable to get a *divergence* (owing to the non-compactness and resulting infinite blueshift) so that the non-singular point y changes, in effect, to a singular ideal point with (presumably) diverging curvature.

Suppose, now, that, instead of the asymptotically flat situation we have been examining, we consider a spacelike initial hypersurface Σ that is *compact*. It may still be that, as in figure 12.9(a), certain sets of data on Σ lead to a maximally extended vacuum spacetime which violates global hyperbolicity (e.g. Taub–NUT space $\mathcal{M}$), and a Cauchy horizon $H = H^+(\Sigma)$ is produced. Take $y \in H$, as before, and consider $I^-(y) \cap \Sigma$. This must now have compact closure (since Σ is compact) but it seems that, in a sense, it is liable to be *effectively non-compact* owing to an infinite wrapping round and round Σ by $I^-(y)$ – at least this is what appears to be indicated by an examination of the Bianchi IX models.[80,94] This effective non-compactness would seem to lead to a situation similar to the one just discussed, in which H gets converted to a curvature singularity upon generic perturbation, and strong cosmic censorship holds, because the integrals which define the perturbed curvature on H involve the same data on Σ over and over again, infinitely many times.

One is tempted to conjecture, therefore, that the singularities, like that of figure 12.8, which result from data on an effectively compact region are a 'measure zero' special case, and that while some perturbations give rise to Cauchy horizons, the points on these horizons are liable to be dependent upon effectively non-compact data regions, with infinite blueshift, so that the horizons should be unstable (like that of figure 12.9(a)). In this sense, strong cosmic censorship looks to be very plausibly true† – but the above argument is yet a long way from a proof.

We are thus presented with the picture of a globally hyperbolic universe, which starts from an achronal set $\partial \mathcal{M}$ of initial ideal points (the big bang), then remains topologically unchanging (an implication of global hyperbolicity[93]) – despite the presence of black holes – until the

---

† From many different viewpoints it is *mathematically* very desirable to be able to restrict attention to globally hyperbolic spacetimes. For example, most of the technical difficulties[89] concerning the topology and identifications for TIP and TIF structure now disappear.

Singularities: the key?

achronal set $\partial \hat{\mathcal{M}}$ of final ideal points is reached. (Strong cosmic censorship implies, in fact, that $\partial \check{\mathcal{M}}$ and $\partial \hat{\mathcal{M}}$ must be regarded as totally disjoint sets.) The initial ideal points are normally taken to be all singular points, but the final ideal points may be either points at infinity or singular points. Points at infinity would normally be thought to arise only in the case of an ever-expanding universe-model, in which case one would also expect singular final ideal points in black holes. But it is also conceivable (though in my view rather unlikely, for reasons similar to those just outlined above) that a universe which recollapses as a whole may nevertheless have certain limited portions that 'escape' to infinity in a (non-singular) ∞-TIP.

In figure 12.12, an indefinitely expanding universe-model is illustrated, showing how $\mathcal{M}$ can remain topologically unchanging – in the sense that $\mathcal{M} \cong \mathbb{R} \times \Sigma$, with each copy of $\mathbb{R}$ a timelike curve and each copy of Σ a spacelike Cauchy hypersurface[93] – even though there may be several

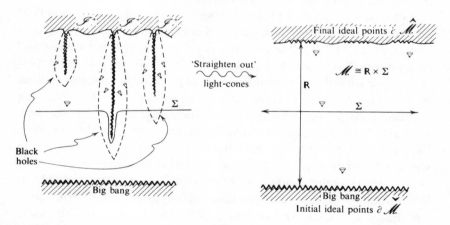

Figure 12.12. A universe-model subject to strong cosmic censorship. The indefinitely expanding case is illustrated. (The different $\mathscr{I}^+$ regions are actually all connected.)

black holes present.[57] The situation for a recollapsing universe-model is similar. The sets $\partial \check{\mathcal{M}}$ and $\partial \hat{\mathcal{M}}$ may or may not have the same topology as Σ, however. (For example, in the Einstein static universe, each of $\partial \check{\mathcal{M}}$ and $\partial \hat{\mathcal{M}}$ is a point, whereas Σ is an S^3.) According to the viewpoint in section 12.3.3, the big bang $\partial \check{\mathcal{M}}$ should, indeed, have the topology of Σ. But it is not at all clear that this should be so of $\partial \hat{\mathcal{M}}$. Another way of phrasing Misner's original hope that the generic Bianchi type IX 'mixmaster' empty models should be free of particle horizons[94] is that $\partial \check{\mathcal{M}}$ should be a

Chapter 12. Singularities and time-asymmetry

single point.† In time-reversed form, one might have correspondingly expected $\partial \hat{\mathcal{M}}$ to be a single point. But later analysis[95-98] has shown this kind of behaviour to be a very unlikely possibility. Nevertheless, one might anticipate that under certain circumstances $\partial \hat{\mathcal{M}}$ has, in an appropriate sense, fewer than three dimensions (as is the case for the two-dimensional $\partial \hat{\mathcal{M}}$ of a pancake singularity and for Taub space[7]).

If the big bang $\partial \check{\mathcal{M}}$ is, indeed, a smooth spacelike hypersurface with the topology of Σ (cf. section 12.3.3) then it constitutes a very satisfactory initial Cauchy hypersurface. But irrespective of the smoothness of three-dimensionality of $\partial \check{\mathcal{M}}$ and $\partial \hat{\mathcal{M}}$, each can be regarded as providing a home for the ultimate Cauchy data for $\mathcal{M}$ – the ultimate initial data in the case of $\partial \check{\mathcal{M}}$, and the ultimate final data in the case of $\partial \hat{\mathcal{M}}$. Each is, indeed, all-embracing in the sense of being intersected by every endless causal curve in $\mathcal{M}$, and is also achronal. Of course, the exact form that such Cauchy data would take is not yet clear in the general case, but the potentiality is certainly there in principle.

As a final consideration of this section, let us examine the possibility of an initial creation or the final destruction of a charged particle (cf. also section 12.2.7). If the initial creation, at $\partial \check{\mathcal{M}}$, can be regarded as the result of a more-or-less understood process[62,63,66] whereby the curvature itself creates particles, then we would expect not just a single charged particle, but a *pair* of oppositely charged particles. However, it might be that some totally unknown process is involved in particle production at the big bang, whereby charged particles could be produced singly.[67] In this situation, the particle's Coulomb field must also be conjured up at $\partial \check{\mathcal{M}}$, leading, as was remarked in section 12.2.4, to the existence of some effectively source-free incoming radiation.[45] Indeed, such a picture (at first sight, at least) would be presented as the time-reverse of particle destruction at $\partial \hat{\mathcal{M}}$. Imagine a single charge being swallowed by a large spherically symmetrical uncharged black hole. According to the geometry depicted in figure 12.8, this particle must be annihilated alone, on a singular part of $\partial \hat{\mathcal{M}}$. Is it conceivable that this one particle should so disturb the geometry of $\partial \hat{\mathcal{M}}$ that its destruction is held off until finally a particle of the opposite charge is swallowed by the hole and guided in to be annihilated by the first charge, before the energy of both can be absorbed by the singularity? It seems hardly credible. Yet the structure of generic singularities appears to be so delicate and elaborate that even this should not be dismissed out of hand. But the issue that I am attempting to raise here is whether or not

---

† This is also, and more explicitly, what is demanded for a Sachs–Budic 'deterministic' spacetime.[99]

the laws that dominate physical behaviour at a spacetime singularity can, after all, be time-symmetric. It seems to me that those who believe that they can be must face additional serious problems of principle!

12.3.3 A hypothesis and a world-view

Almost all the arguments that I have been presenting seem to focus themselves on one issue: what special geometric structure did the big bang possess that distinguishes it from the time-reverse of the generic singularities of collapse – and why? At this point I should mention the viewpoint of *chaotic cosmology*,[94] which has been in vogue for a number of years. According to this view, the big bang was not initially a very uniform singularity, but came to appear so because dissipative effects (e.g. neutrino viscosity, hadron collisions or particle creation[100-102]) served to iron out all its irregularities. One intended effect of this dissipation is to produce the observed 'high' entropy-per-baryon figure of $\sim 10^9$. Another is to produce the presently observed isotropy. And in order not to impose apparently arbitrary restrictions on the big bang, it was originally conjectured that the chaos of the initial singularity was, in some appropriate sense, *maximal*. In this way a supposedly canonical type of initial state was suggested, whose detailed implications could be worked out and compared with observation.

I have to say, however, that I regard the programme of chaotic cosmology – at least in its 'pure' form of maximal initial chaos – to be basically misconceived. For to assert that the initial chaos is maximal is presumably to assert that the initial *entropy* is maximal. And if this were the case, with time-symmetric physical laws, there would be no time's arrow and therefore no dissipation. (It is no good appealing to the expansion of the universe here – for reasons that were amply discussed in section 12.2.6!)

We might, however, entertain some milder version of chaotic cosmology in which the initial geometric chaos was not maximal but was constrained in some suitable way – the constraints being so adjusted that not only could the observed entropy per baryon be correctly obtained from early dissipation, but that density irregularities adequate for galaxy formation might also be obtainable as a result of a non-uniform initial geometry.[103,105] In my opinion, this version of chaotic cosmology is also quite unsatisfactory. The essential misconception seems to be to regard an entropy per baryon of $\sim 10^9$ as a *high* figure in a gravitational context. Consider a closed recollapsing universe containing, say, 10^{80} baryons.

Chapter 12. Singularities and time-asymmetry

When an $M_\odot$ black hole forms, it achieves, by the Bekenstein–Hawking formula, an entropy per baryon of 10^{18}, while a $10^{10} M_\odot$ galaxy with a $10^6 M_\odot$ black hole core has an entropy per baryon of 10^{20}. When finally the bulk of this galaxy collapses into the hole, the figure is 10^{28}. But as the collapse of the universe proceeds, these black holes unite into bigger and bigger ones, yielding an immense final entropy per baryon of $\sim 10^{40}$. The entropy resides in the irregularities of the geometry of the final singularity. Reversing the time-sense, we now see what a stupendous entropy would really have been available, had the Creator chosen to make use of it by an initial chaotic geometry. The supposedly 'high' figure of $\sim 10^9$ is small fry indeed! If we are to adopt 'mild' chaotic cosmology as an explanation for the figure 10^9, then we must ask why only the absurdly small fraction of at most $\sim 10^{-31}$ of the 'available chaos' was actually used! (Indeed, the figure would be $\sim 10^{-32}$ if the closed-universe value $\sim 10^{-8}$ for the entropy per baryon were used.)

It would seem that, with such an enormous discrepancy, we should not look for a gravitational explanation† for the figure 10^9. A more hopeful place to look might be in particle physics. I shall return to this question in section 12.4. Likewise it would seem that a *purely* gravitational explanation of the irregularities needed for galaxy formation will not be forthcoming. Again (though with considerably less confidence) I appeal to particle physics – in the very early stages of the expansion.

I propose,[57–60] then, that there should be a complete lack of chaos in the initial *geometry*. We need, in any case, some kind of low-entropy constraint on the initial state. But thermal equilibrium apparently held (at least very closely so) for the *matter* (including radiation) in the early stages. So the 'lowness' of the initial entropy was not a result of some special matter distribution, but, instead, of some very special initial spacetime geometry. The indications of sections 12.2.6 and 12.3.1, in particular, are that this restriction on the early geometry should be something like: *the Weyl curvature C_{abcd} vanishes at any initial‡ singularity.*[58–60]

This hypothesis is still a little vague, and is open to a number of different interpretations. We could require, for example, that the Weyl curvature tend to zero as the initial singularity is approached, or that it should do so at some preassigned faster rate, or, perhaps, that it should

---

† This is borne out by a recent detailed analysis by Barrow and Matzner[104] who come to the same essential conclusion.

‡ That is, with the notation of section 12.3.2, at $\partial \mathcal{M}$. Note that this includes the final singular TIF of a Hawking black hole explosion – at which the Weyl curvature indeed vanishes!

Singularities: the key?

only remain bounded (or merely dominated by the Ricci tensor near the singularity – so that the curvature tensor becomes, in the limit, *proportional* to one whose Weyl part vanishes). I have not examined the differences that might be entailed by adopting different versions of this hypothesis. My inclination is to try, first, the simplest of these, namely that $C_{abcd} \to 0$ (in, say, any parallelly propagated frame), as initial singularities are approached. I shall indicate, shortly, what the rough implications of this would be, though complete details have yet to be worked out.

Let me first explain my view of the role of this hypothesis, as it affects the 'selection' that the Creator might make of *one* particular universe out of the apparently infinite choice available, consistent with given physical laws. Imagine some vast manifold $\mathcal{U}_0$ (where I use the word 'manifold' in a rather loose sense) representing the different possible initial data for the universe, compatible with the physical laws. To select one universe, the Creator simply places a 'pin' somewhere in $\mathcal{U}_0$. But a viewpoint of this article is that we should not be biased in choosing initial rather than final data. So, equally well, the selection of the universe could be envisaged as the Creator's pin being placed in the manifold $\mathcal{U}_\infty$, representing all possible sets of final data compatible with the physical laws. Indeed, one could use any intermediate $\mathcal{U}_t$ representing the possible data 'at time t'.† All are equivalent, the equations of motion (which I am crudely assuming, for the purposes of the present discussion, to be of a classical determinate type) effecting canonical isomorphisms between $\mathcal{U}_0$, $\mathcal{U}_\infty$ and each $\mathcal{U}_t$. Thus we may envisage a *single* isomorphic abstract manifold $\mathcal{U}$ which represents any (or every) one of these, being the set of all possible universe-histories compatible with the physical laws.

We would like to be able to envisage that the Creator's pin is simply placed 'at random' in $\mathcal{U}$ (since if the pin is to be constrained in any further precise way, this would constitute another 'physical law'). But the notion of 'at random' requires that some appropriate *measure* be placed on $\mathcal{U}$. Is 'at random' the same concept when applied to initial conditions as when applied to final conditions? To put the question another way: is the phase-space measure that is naturally defined on $\mathcal{U}_0$ the same as that defined on $\mathcal{U}_\infty$ (or indeed on each $\mathcal{U}_t$) under the isomorphisms? Liouville's theorem tells us that it *is*, provided that we are adopting conventional Hamiltonian physics – and I am not proposing to be

---

† The concept 'time' is being used very loosely here. In the right-hand picture of figure 12.12 (where indefinite expansion need not be assumed) t could be some suitable parameter ranging from 0 to ∞ and measuring the 'height' up the picture.

Chapter 12. Singularities and time-asymmetry

'unconventional' in this respect here. I am, furthermore, going to ignore any difficulties† that might arise from a possible infinite-dimensionality, or infinite total measure, of the manifold $\mathcal{U}$ – not to mention all the very serious 'gauge' problems (etc.) which would arise in a proper general-relativistic treatment.

How are we to envisage the entropy of the universe according to this picture? A standard procedure is to coarse-grain each $\mathcal{U}_t$ by dividing it up into compartments, where the different members of any one compartment correspond to states macroscopically indistinguishable from one another at time t. If the Creator's pin pierces $\mathcal{U}_t$ at a point belonging to a compartment of phase-space volume V (measured in units where $\hbar = 1$) then the (Boltzmann) entropy at time t is $k \log V$ (cf. section 12.2.3). This entropy can fluctuate‡ or progressively change with time, because the coarse-grainings of the different $\mathcal{U}_t$ manifolds do not map to one another under the isomorphisms. Low entropy at one time (point in a small compartment) can correspond to high entropy at another (point in a large compartment), where the 'specialness' of the state has now become that of macroscopically indiscernible correlations.

In the present context, however, this description of entropy is not yet satisfactory. The points of $\mathcal{U}$ correspond only to those universe-histories that are compatible with the physical laws at *all* times. It may not be macroscopically discernible at a particular time t whether the laws are satisfied at all other times. I am hypothesizing, here, that there are in fact (local) physical laws which only become important near spacetime singularities, these being asymmetric in time and such as to force the Weyl curvature to vanish at any initial singular point (i.e. point of $\partial \mathcal{M}$). The effect of these laws is that each manifold $\mathcal{U}_t$ turns out to be much smaller than might otherwise have been expected. Only those motions and configurations at time t which are also compatible with the constraints ($C_{abcd} = 0$) at $t = 0$ are allowed. But since these implied constraints on each $\mathcal{U}_t$ are not macroscopically discernible, it is not reasonable, when calculating the entropy at time t, simply to use the phase-space volumes within $\mathcal{U}_t$. Instead, one must consider extended volumes within a certain

---

† I must apologize, particularly to the experts, for my crude and cavalier treatment of the delicate matters that I am embarking upon. My excuse is that for the questions with which I am now concerned, I do not believe that general-relativistic or thermodynamic sophistication is a key issue.

‡ If we prefer the *ensemble* picture of the world, we can to some extent avoid these fluctuations, envisaging that the Creator uses a *blunt* pin! So long as the diameter of the pin-point is large compared with the coarse-graining, fluctuations are smoothed over.

Singularities: the key?

larger manifold $\mathcal{W}_t$ – defined in the same way as $\mathcal{U}_t$ but for which these (initial) constraints are not required to hold.

The equations of motion again give isomorphisms between the different $\mathcal{W}_t$ manifolds at different times t (locally at least – and so long as the extra constraining laws remain physically insignificant) and there is a corresponding abstract manifold $\mathcal{W}$ representing the totality of unconstrained universe-histories. The imbedding of $\mathcal{U}$ in $\mathcal{W}$ has a very special relation to the coarse-graining at $t = 0$, because the vanishing of Weyl curvature *is* a macroscopically discernible property. Thus, no $t = 0$ compartment of $\mathcal{U}$ extends outside $\mathcal{U}$, into $\mathcal{W}$. But as t increases, the corresponding coarse-graining compartments of $\mathcal{U}$ extend more and more into $\mathcal{W}$, and accordingly acquire larger and larger extended volumes.† In this way, we regard the 'specialness' of the actual state of the universe (arising from its having started out with $C_{abcd} = 0$) as being more and more of the 'precise-correlations' kind, and less and less of the 'low-entropy' kind, as time progresses.

This is what gives us compatibility with the entropy-increase phenomenon of our universe. In this view, we do not impose any statistical low-entropy assumption at the big bang, but, instead, a precise *local condition* ($C_{abcd} = 0$). Aside from this constraint, there is to be complete randomness, that is, the Creator's pin is placed at random in the manifold $\mathcal{U}$. With this randomness assumption, we can attribute the 'reason' for the absence of initial correlations between particle motions in the initial state (i.e. for the *law of conditional independence*[40]) to the fact that at no time *other* than $t = 0$ is a new local constraint imposed (e.g. there is none imposed at the final singularity $t = \infty$). Correspondingly the 'reason' for the presence of increasing correlations as t increases, and for the increasing entropy, is the initial $C_{abcd} = 0$ constraint. In this way, the problem of time's arrow can be taken out of the realm of statistical physics and returned to that of determining what are the precise (local?) physical laws. I shall briefly discuss this question in section 12.4.

I should make mention, at this point, of the much discussed *anthropic principle*,[106] which is often invoked in connection with the matters I have been raising. This principle would, in effect, imply that the Creator's pin is placed in $\mathcal{U}$, not just at random, but with a weighting factor, weighted in

---

† Curiously, if it were not for extensions into $\mathcal{W}$, the volumes of the largest compartments would *decrease* with time owing to the profusion of differing macroscopic geometries that would be produced. This seems to correspond to the fact that the universe can get more 'interesting' even though the entropy increases!

Chapter 12. Singularities and time-asymmetry

favour of universes containing (many?) *conscious observers*. (Furthermore, the pin could also pierce other manifolds $\mathcal{V}, \mathcal{W}, \ldots$ corresponding to all the possible consistent alternative sets of pure-number physical constants, and to all the possible alternative laws of physics. I shall leave aside a discussion of this extended question as being 'beyond the scope of this article'!) Such an anthropic viewpoint has occasionally been invoked in an attempt to explain the entropy imbalance of the observed universe in terms of weighting in favour of a huge 'fluctuation' which might have been needed to produce the conditions necessary for life.[107] The trouble with this is that it is vastly 'cheaper' (in terms of negative entropy) simply to produce a few conscious beings out of some carefully organized particle collisions than it is to build, out of a fluctuation, such entropy imbalance as is familiar on Earth throughout the *entire*, apparently unending, universe – as revealed by the most powerful telescopes!

This is not to say that I believe 'randomness' for the Creator's pin will always remain the best explanation for the state of the world. But, with the addition of the initial $C_{abcd} = 0$ assumption, it seems to work remarkably well, and it saves our having to worry about what 'consciousness' means in physical terms – at least for the time being!

So what, indeed, are the implications of the world-view that I am proposing? With the Weyl curvature initially zero and thermal equilibrium for the matter (and radiation), we shall have something very close to spatial-isotropy and homogeneity for the initial state. Thus the discussions of Friedmann, Robertson and Walker hold good (initially), leading to a striking consistency with various remarkable observations: the uniformity of the 2.7 K black-body radiation (to one part in $\sim 10^3$),[108] the lack of measurable rotation of the universe relative to inertial frames ($<10^{-16}$ s^{-1}),[77] the large-scale gross uniformity of galactic clusters. The big bang singularity itself must have been closely of the Robertson–Walker type.† *Some* fluctuations in the matter distribution are allowed in this view – and, indeed, *must* occur – because the initial restrictions on R_{ab} are statistical, unlike those on C_{abcd}. However, the initial vanishing of C_{abcd} imposes considerable constraints on such initial density and velocity fluctuations. Yet, particle physics will have to be better understood before these fluctuations can be calculated in detail.[109]

White holes are excluded at all times, because their singularities are 'initial' singularities (i.e. points of $\partial \mathcal{M}$) which do not remotely satisfy the constraint $C_{abcd} = 0$. Black holes are allowed, of course, provided that

---
† This is independently supported by the accuracy of the helium-production calculations.[77,78]

they are formed in the normal way as a consequence of gravitational collapse of a massive body or bodies. But mini-holes are presumably *not* to be expected because they require an initial state with a chaotic geometry. The non-existence of such primordial black holes is consistent with present observations.[110]

Finally, the extraordinary observed behaviour of the entropy of our world – permeating so much of our everyday experience that we tend to take it quite for granted – is the major and most striking consequence of this world-view.

12.4 Asymmetric physics?

Some readers might feel let down by this. Rather than finding some subtle way that a universe based on time-symmetric laws might nevertheless exhibit gross time-asymmetry, I have merely asserted that certain of the laws are not in fact time-symmetric – and worse than this, that these asymmetric laws are yet unknown! But this is not so negative as it might seem. In particular, it tells us to look out for such asymmetries in other places in physics. Where do we look? Ultimately there must be some connection with gravity, since it is the Weyl tensor that describes gravitational degrees of freedom. Classical general relativity is a time-symmetric theory, but one may ask whether this time-symmetry will persist when finally the link with quantum mechanics is appropriately forged. Indeed, if one believes that virtual black holes at the Planck length (10^{-33} cm) are physically important,[73] then the arguments of section 12.2.7 suggest that the vacuum could be time-asymmetric in a significant quantum-gravitational way.

This is not greatly helpful, however, and there is another possible tentative connection between quantum mechanics and gravity that might be more relevant, namely the question of *quantum-mechanical observations*, that was left hanging in section 12.2.2. Certain associations with the Bekenstein–Hawking discussion of entropy should be pointed out. One regards a quantum-mechanical observation to have been 'made', after all, only when something 'irreversible' has taken place. But 'irreversible' here refers to the fact that an essential increase in the entropy has occurred. And entropy, as we have seen, seems to depend on the rather subjective concept of coarse-graining. For a quantum-mechanical observation 'actually' to take place, effecting a real change in the state of the world, one would seem to require *objectivity* for this entropy increase. Now recall that in the Bekenstein–Hawking formula, an entropy measure

Chapter 12. Singularities and time-asymmetry

is directly put equal to a precise feature of spacetime geometry, namely the surface area of a black hole. Is it that this geometry is now subjective, with the implication that *all* spacetime geometry (and therefore all physics) must in some measure be subjective? Or has the entropy, for a black hole, become objective? If the latter, then may not entropy also become objective in less extreme gravitational situations (cf. section 12.3.1)? Moreover, if it is only with (quantum) gravity that such a passage from subjectivity to objectivity of entropy can occur, then it is with (quantum) gravity that the linear superposition of von Neumann's chain[111] finally fails!

Perhaps, then, the new laws that seem to be needed to extend quantum mechanics (cf. section 12.2.2), so that observations can be incorporated within the theory, constitute some form of *quantum gravity* – by which I mean a theory having quantum mechanics and general relativity as two appropriate limits. I would contend, in any case, that the arguments I have been presenting (notably those of sections 12.2.7 and 12.3.3 which most directly relate to the Bekenstein–Hawking formula) point towards some new theory which is *time-asymmetric*. Accordingly, whatever nonlinear physics† eventually replaces suddenly collapsing wave functions may well turn out to involve an essential time-asymmetry.

At the present state of understanding, such considerations are highly speculative. Yet we know that there *is* a physical law which is time-asymmetric! Somewhere, hidden among the more familiar time-symmetric forces of Nature is one (or perhaps more than one) whose tiny effect has been almost completely masked by these others and has lain undetected in all physical processes bar one: that delicately poised decay of the K^0-meson. I am not suggesting that quantum gravity need be involved here. But, evidently, it is *not* one of Nature's inviolable rules that time-symmetry must always hold!

Moreover, the relative strength of the T-violating (or CP-violating) component in K^0-decay, perhaps 10^{-9}, is possibly suggestive. According to section 12.3.3, we need some explanation from particle physics for the observed entropy-per-baryon number of 10^9. It has, accordingly, been put forward[109,67,62] that perhaps in the early expansion there was a

---

† An interesting recent suggestion for a nonlinear modification of Schrödinger's equation is that of Bialynicki-Birula and Mycielski[112] according to which an additional term $(b \log |\psi|^2)\psi$ is incorporated. Though not time-asymmetric, there is the link with the present discussion that the constant b (a temperature) has the value $\sim 10^{-8}$ K, which is the Bekenstein–Hawking temperature for a 10 $M_\odot$ black hole. These are the smallest black holes that one has physical reason to believe in.

process of production of baryons and antibaryons, where baryons outnumbered antibaryons by about $1+10^{-9}:1$. Then, in the subsequent annihilation, not only would the (apparently) required nonzero net baryon number be produced, but also the observed entropy per baryon (i.e. photons per baryon). This initial production process, if it were to be explained as arising from something with the vacuum's quantum numbers, would need to violate baryon conservation and (for an imbalance to occur) both *CP*- and *C*-symmetries.† Gross violation of *C* has long been known to be a feature of weak interactions,[20] while *CP*-violation also occurs in the K^0-decay at the apparently required low level. Moreover, baryon non-conservation[76] is *somewhere* to be expected, on the basis of Hawking's black hole evaporation.[56,67]

But to explain the law that $C_{abcd} = 0$ initially, we would need something more, namely a violation of each of *T*, *PT*, *CT* and *CPT*. If, for example, *CPT* were not violated, we could take an allowed collapse to a singularity for which $C_{abcd} \neq 0$ and apply the *CPT*-symmetry to obtain a disallowed initial singularity. The symmetries *PT* and *CT* are maximally violated in weak interactions and *T* marginally violated in K^0-decay, but *CPT*-violations have not yet been detected. Of course there is the *CPT*-theorem[113] that lends some theoretical support to universal *CPT*-conservation. But it must be borne in mind that Poincaré covariance is an important assumption of that theorem, whereas I am envisaging situations (singularities, quantum gravity) for which this assumption is explicitly violated. I would contend that somehow, in experimental physics, a *CPT*-violating effect ought eventually to be discernable. But these are early days yet, and it is in no way surprising that such effects have not yet been seen.

I have presented, here, little in the way of quantitative detail. However, many of the phenomena I have been concerned with are of so gross and blatant a nature that much can yet be gleaned even without such detail. The most blatant phenomenon of all these is the statistical asymmetry of the universe. It is, to me, inconceivable that this asymmetry can be present without tangible cause. An explanation by way of the anthropic principle seems very wide of the mark (cf. section 12.3.3). So does an explanation of the 'symmetry-breaking'[114] type, according to which the most probable states of the universe might not share the symmetries of the laws that govern it. (It is difficult to see how our vast universe could

---

† Or, conceivably, *CPT*-violation in place of *CP*-violation, or *CT* in place of *C*, since there is a time-asymmetry in the universe expansion.

Chapter 12. Singularities and time-asymmetry

just 'flop' into one or the other of these states when it doesn't even know which temporal direction to start in!) In my own judgment, there remains the *one* ('obvious') explanation that the precise physical laws are actually *not* all time-symmetric!

The puzzle then becomes: why does Nature choose to hide this time-asymmetry so effectively? As we do not yet know the principles that govern Nature's choices of physical law, we cannot yet answer this question. But perhaps we should not be so surprised at a situation in which a fundamental asymmetry lies hidden deep beneath a facade of apparent symmetry. The fauna of this Earth, after all, exhibits with but few exceptions a superficial external bilateral symmetry. How could one have guessed that in the nucleus of every reproducing cell lies a helix whose structure governs the growth of these magnificent symmetrical creatures – yet every one of which is right-handed?

Acknowledgements

I am grateful to Dennis Sciama and, particularly, to Amelia Rechel-Cohn for critically reading the manuscript and drawing my attention to a number of references. My thanks go also to Stephen Hawking for several enlightening conversations.

13. Quantum field theory in curved spacetime

G. W. GIBBONS

13.1 Introduction

Einstein always regarded the fact that the right-hand side of his famous equations:

$$R_{\alpha\beta} - \tfrac{1}{2}g_{\alpha\beta}R + \Lambda g_{\alpha\beta} = 8\pi T_{\alpha\beta} \tag{13.1}$$

contains the stress tensor, an essentially non-geometrical entity, as a blemish of his theory. Since neither he, nor anyone else, has been able to do better, theorists have been forced to adopt non-geometrical models for the description of matter in general relativity. These have been predominantly classical models: hydrodynamical, classical Einstein–Maxwell theory and perhaps the most complete at the classical level – relativistic kinetic theory. However, in reality matter is governed by quantum mechanical laws and our most complete treatment of matter is that given by quantum field theory. The purpose of this article is to outline what the extension of such a treatment to curved space entails and to discuss what essentially new features arise when one takes into account the quantum mechanical nature of gravitating systems. I shall throughout assume a classical, unquantized gravitational field and confine the discussion to matter fields, although similar techniques and ideas may be applied to 'gravitons' – that is linearized perturbations of the metric propagating on some fixed, unperturbed, background.

The principal difference arises from one of Einstein's greatest discoveries, the equivalence between mass and energy. In classical general relativity this equivalence means gravitational potential energy contributes to the masses of bodies but nevertheless matter and gravitational energy are not completely interchangeable. In the classical theory matter cannot be completely converted into gravitational energy (Hawking, 1970). In the quantum theory however this process is possible – a sufficiently strong or rapidly varying gravitational field can create pairs of particles. *This means that the notion of a vacuum or no-particle state in a curved spacetime is inherently ambiguous.* This fact seems to have been

Chapter 13. Quantum field theory in curved spacetime

first realized by Schrödinger (1939) during his investigations of matter-wave propagation in Friedmann–Robertson–Walker (FRW) spacetimes, although an earlier paper of Rosenfeld (1930) seems to be the first to treat the quantum mechanical interaction of matter and gravity.

Over twenty years later Utiyama and DeWitt (1962) discussed renormalization in curved space and Lichnerowicz (1961) commutation relations and propagators. However it was not until the late sixties that Parker (1968) and Sexl and Urbantke (1967) following up Schrödinger's original lead began discussing the creation of particles during the initial rapidly expanding phase of the universe. In 1970 Zel'dovich suggested that not only could the kinetic energy of expansion of the universe be converted into particles but also the energy due to anisotropic expansion, thereby reducing any large-scale anisotropy. Since then he and his colleagues have worked extensively on these topics (Lukash, Novikov, Starobinsky and Zel'dovich, 1976). A further impulse to the subject came with Hawking's discovery (1974) of the spatial analogue of this process – that inhomogeneities of the gravitational field in the form of black holes can also be converted into particles leading to their 'evaporation'. That black holes should spontaneously radiate in super-radiant modes had been realized earlier by Zel'dovich (1971). Since then the literature has increased sharply and there already exists a number of reviews (B. DeWitt, 1975; Isham, 1977; Davies, 1976a; Parker, 1976a; Gibbons, 1977). In this account I shall not attempt a comprehensive survey of all that has been written but simply concentrate on what I consider to be the major features of the subject.

Conventions. I shall use the conventions of Misner, Thorne and Wheeler (1973), (MTW), and units such that $G = c = \hbar = k = 1$. Operators acting on the physical Hilbert spaces will be denoted by bold-face symbols. Standard definitions and notions from classical general relativity will be as in MTW or Hawking and Ellis (1973).

13.2 Basic notions

A quantum field theory on a fixed background consists of the following:
(1) Hilbert space of states $\mathfrak{H}$.
(2) A classical spacetime $\{\mathcal{M}, g_{\alpha\beta}\}$.
(3) A set of field operators $\{\boldsymbol{\phi}, \boldsymbol{\psi}\}$ acting on $\mathfrak{H}$ and defined on $\mathcal{M}$ as tensorial ($\boldsymbol{\phi}$, boson) or spinorial ($\boldsymbol{\psi}$, fermion) distributions (at least for linear theories).

Basic notions

(4) Wave equations satisfied by the fields $\{\phi, \psi\}$ which are linear if there are no mutual or self interactions between the particles and nonlinear otherwise.
(5) Commutation or anticommutation relations obeyed by the operators ('field algebra').
(6) Rules for constructing 'particle' observables and Fock bases for $\mathfrak{H}$.
(7) Rules for rendering formally divergent, nonlinear expressions in the fields (e.g. $\langle T_{\mu\nu}\rangle$) finite. That is 'renormalization' or 'regularization' schemes.

Let us treat these in turn.

Hilbert space $\mathfrak{H}$

We shall assume that to each physically measurable quantity there corresponds an 'observable' a self-adjoint linear map acting on $\mathfrak{H}$. Physical systems correspond to rays ('pure states') or to density matrices ρ ('mixed states'). I use the word 'state' for both. Density matrices describe systems associated with an additional degree of uncertainty beyond that due to quantum mechanics. For example they can describe an ensemble of systems in thermal equilibrium ('Gibbs state': $\rho = \exp(-\beta H)$, where β is inverse temperature and H the Hamiltonian). They can also describe situations in which one chooses, or is unable, to measure a 'complete set of commuting observables'. If $\{A\}$ and $\{B\}$ constitute such a complete set with normalized eigenvectors $|A_n\rangle$, $|B_n\rangle$ which span two Hilbert spaces $\mathfrak{H}_A$ and $\mathfrak{H}_B$ whose Cartesian product is the entire space, then the general normalized member of $\mathfrak{H}$ can be expressed as

$$|\psi\rangle = \sum_{m,n} \lambda_{mn} |A_m\rangle |B_n\rangle. \tag{13.2}$$

If an observable O acts only on $\mathfrak{H}_A$ its expectation value in $|\psi\rangle$ is

$$\langle\psi|O|\psi\rangle = \sum_{m,n} \rho_{nm} \langle A_n|O|A_m\rangle = \text{tr}\,\rho O \tag{13.3}$$

$$\rho = \sum_{n,m,s} |A_n\rangle \bar\lambda_{ns} \lambda_{ms} \langle A_m| \tag{13.4}$$

and the trace is taken over an orthonormal basis of $\mathfrak{H}$. The entropy of such a mixed state is S (from the point of view of observations made in the $\mathfrak{H}_A$ space) where

$$S = -\text{tr}\,\rho \ln \rho. \tag{13.5}$$

In the Heisenberg picture the states remain fixed for all time and the observables evolve in time. This picture is well suited to our purposes, since in general we have no invariant notion of time, and is almost always

Chapter 13. Quantum field theory in curved spacetime

used. The pure state $|\psi\rangle$ remains a pure state (with $S = 0$) for all time but loss of information about B gives rise to an apparent density matrix and with it nonzero entropy. In a similar way the total entropy of an initial mixed state remains constant but if only some observables are measured an increase of entropy occurs. Entropy looked at in this way thus depends on one's observations or measurements and hence on one's knowledge of the system.

The relevance of these remarks to our subject is that in curved space complete knowledge about all of spacetime is not always available to a given observer. If horizons are present then a degree of ignorance is forced upon such observers. This actually occurs when dealing with black holes for instance.

Finally we recall the alternative formulation of quantum mechanics due to Feynman in which one specifies the state of a system by a point in some configuration space $\{q'\}$ and writes the 'probability amplitude' for the system to move from a point q' at time t' to a point q at time t as

$$\langle q, t | q', t' \rangle = \int \mathscr{D}[q(t)] \exp[iI(t, t')] \qquad (13.6)$$

where $\int \mathscr{D}[q(t)]$ is a 'path integral' defined on the space of paths from q' to q and $I(t, t')$ is the classical action of such a path.

From the point of standard theory $|q, t\rangle$ is the eigenvector of $q(t)$ and (13.6) merely a compact and heuristic expression for the solution of the Heisenberg equations of motion $\dot{q} = i[H, q]$. However one may attempt to define these integrals directly (C. DeWitt, 1972; Alberverio and Hoegh-Krohn, 1977) or by means of analytic continuation in a suitable time variable (Gelfand and Yaglom, 1960) when the Feynman integral becomes a Wiener integral which is mathematically well understood. If this were possible in general one would have an alternative formulation of quantum theory. When applied to curved space this latter approach entails complexifying the spacetime. In this connection we recall that according to Cameron (1960) the functional integral can only be interpreted in measure theoretic terms if the resulting exponential is purely real, it is not sufficient merely to add an imaginary part to the action. This would seem to imply that only in Riemannian spaces can one have a true measure on the space of paths.

The spacetime

$g_{\alpha\beta}$ is a classical 'c number'. This approximation should be reasonable as long as:

(1) one ignores quantum gravitational fluctuations – i.e. at length scales greater than 10^{-33} cm,

(2) one does not take into account the back reaction to particle creation. It does not seem clear at present whether one can consistently consider a 'Hartree–Fock' form of Einstein's equations

$$R_{\alpha\beta} - \tfrac{1}{2} g_{\alpha\beta} R + \Lambda g_{\alpha\beta} = 8\pi \langle T_{\alpha\beta} \rangle \tag{13.7}$$

on length scales of the order of the Compton wavelength of elementary particles (10^{-10} cm to 10^{-13} cm) nevertheless it ought to be a good approximation at larger scales and should provide a reasonable first approximation for particle creation studies.

It is interesting to note that the curvatures necessary to create particles will necessarily build up during gravitational collapse or in the big bang. Since gravitational charge is always positive there is no possibility of these fields discharging before creation thresholds are reached, as presumably happens to electromagnetic fields. Because of this fact and also because of the intrinsically classical notion of a spacetime, external field theory seems better suited to gravitation than it is to the other three basic interactions.

In order to be able to solve wave equations, that is to be able to pose a Cauchy problem, $\{\mathcal{M}, g_{\alpha\beta}\}$ must be globally hyperbolic and time orientable (cf. Choquet-Bruhat, 1967). The fact that one may invariantly define 'left-handed' as the 'helicity of the massless fermion produced in the more abundant decay of the long lived neutral kaon' (Perkins, 1972) would seem to require that $\{\mathcal{M}, g_{\alpha\beta}\}$ be space-orientable although if parity violation is due to a spontaneous symmetry-breakdown mechanism (Lee, 1974) then it may be violated in different ways in different 'domains' (cf. Zel'dovich, Kobzarev and Okun, 1974) and no global definition of right-handed may exist. In any event it is simplest to assume space orientability.

These two assumptions will in turn imply the existence of a spinor structure (Geroch, 1968, 1970) and thus permit the existence of global spinor fields.

If $\{\mathcal{M}, g_{\alpha\beta}\}$ is not globally hyperbolic, then it may be that a physically interesting subset of it is (e.g. the exact Kerr solution). If on the other hand the region of physical interest contains singularities or Cauchy horizons it will not be possible to proceed until one augments the theory with extra boundary conditions. Without these the question of the creation of particles by 'naked singularities' for example, is meaningless.

Chapter 13. Quantum field theory in curved spacetime

Field operators

In principle at least these should be regarded as operator-valued distributions (linear maps from a space of test functions on $\mathcal{M}$ to the space of linear maps of $\mathfrak{H}$ into itself) and should only appear smeared with some test function $f: \phi[f]$, however this notation rapidly becomes cumbersome and is inappropriate to the level of rigour of this, and most other, accounts of field theory on curved spacetime. An exception to this rule are the papers of Wald (1975), Ashtekar and Magnon (1975) and Hajicek (1977). Nevertheless, at least for linear field theories, one can and should given in outline the construction of $\mathfrak{H}$ and the action of ϕ on it. This we will come to later.

Wave equations

Linear wave equations for arbitrary spins, exist in flat space but satisfactory generalizations to background gravitational or electromagnetic fields exist only for spin 0, $\frac{1}{2}$ and 1. The basic structure of the flat space equations is a set of propagation equations and subsidiary conditions which are evolved by the propagation equations. The subsidiary equations restrict the number of field components to $(2s+1)$. Minimal coupling or other generalizations to curved space can introduce extra-subsidiary conditions ('Buchdahl' conditions) or propagation equations whose characteristics lie outside the light-cone (cf. Aragone and Deser, 1971, for $s=2$; Gibbons, 1976, for $s=\frac{3}{2}$). The $s=\frac{3}{2}$ case can be dealt with in supersymmetric theories involving torsion (Deser and Zumino, 1976). No general theory exists so one must resort to a case by case analysis. The following equations pass the test.

Spin 0

$$(-\nabla_\alpha \nabla^\alpha + \xi R + m^2)\phi = 0, \qquad (13.8)$$

Lagrangian

$$-\tfrac{1}{2}[g^{\alpha\beta}\nabla_\alpha\phi\nabla_\beta\phi + (\xi R + m^2)\phi^2], \qquad (13.9)$$

Spin $\tfrac{1}{2}$

$$(\gamma^\alpha \nabla_\alpha + m)\psi = 0, \qquad (13.10)$$

Lagrangian

$$-\tfrac{1}{2}\bar\psi(\gamma^\alpha \nabla_\alpha + m)\psi, \qquad (13.11)$$

where

$$\gamma^\alpha\gamma^\beta + \gamma^\beta\gamma^\alpha = 2g^{\alpha\beta} \qquad (13.12)$$

if
$$\gamma_5 = \tfrac{1}{24}g^{1/2}\varepsilon_{\mu\nu\lambda\rho}\gamma^\mu\gamma^\nu\gamma^\lambda\gamma^\rho. \tag{13.13}$$

Then ψ decomposes into the right-handed, ψ_R, and left-handed, ψ_L, parts according to

$$\psi_L = \tfrac{1}{2}(1-\gamma_5)\psi, \tag{13.14}$$

$$\psi_R = \tfrac{1}{2}(1+\gamma_5)\psi. \tag{13.15}$$

γ_5 and $\gamma^\alpha \nabla_\alpha$ map right-handed into left-handed and vice versa. ψ_R and ψ_L may be regarded as 2-component spinors. If $m=0$ we have the Weyl equation together with the subsidiary condition

$$\psi_R = 0. \tag{13.16}$$

ψ obeys the following second-order equation

$$(-\nabla_\alpha \nabla^\alpha + \tfrac{1}{4}R + m^2)\psi = 0, \tag{13.17}$$

Spin 1

$$F_{\mu\nu} = \nabla_\mu A_\nu - \nabla_\nu A_\mu, \tag{13.18}$$

$$\nabla_\nu F^{\mu\nu} = -m^2 A^\mu, \tag{13.19}$$

$$\nabla_\mu A^\mu = 0, \tag{13.20}$$

which may be written as

$$(-\nabla_\alpha \nabla^\alpha + m^2)A^\sigma + R^\sigma{}_\beta A^\beta = 0, \tag{13.21}$$

$$\nabla_\mu A^\mu = 0. \tag{13.22}$$

Equation (13.21) is the propagation equation and (13.22) the subsidiary condition. The Lagrangian is (in unrationalized units)

$$-\frac{1}{16\pi}F_{\mu\nu}F^{\mu\nu} - \frac{m^2}{8\pi}A_\mu A^\mu. \tag{13.23}$$

For 2 complex solutions the following currents are conserved

$s=0$

$$J_\mu = i(\bar\phi_1 \nabla_\mu \phi_2 - \phi_2 \nabla_\mu \bar\phi_1), \tag{13.24}$$

$s=\tfrac{1}{2}$

$$J_\mu = \bar\psi_1 \gamma_\mu \psi_2, \tag{13.25}$$

$s=1$

$$J_\mu = i(\bar A_1^\sigma \nabla_\mu A_{2\sigma} - A_{2\sigma} \nabla_\mu \bar A_1^\sigma). \tag{13.26}$$

Chapter 13. Quantum field theory in curved spacetime

These currents generate hypersurface-independent inner products on the space of solutions by the formula

$$(\phi_1, \phi_2) = -\int J_\mu \, d\Sigma^\mu. \tag{13.27}$$

$(\, , \,)$ is antilinear in the first slot and linear in the second. For bosons it is indefinite, whereas for fermions it is positive-definite.

The massless equations, with $\xi = \frac{1}{6}$, possess the following symmetries

(1) Invariance under conformal rescaling. That is if $\{g_{\alpha\beta}, \gamma_\alpha, \phi, \psi, A_\alpha\}$ are solutions, then so are $\{\Omega^2 g_{\alpha\beta}, \Omega\gamma_\alpha, \Omega^{-1}\phi, \Omega^{-1}\psi, A_\alpha\}$.

(2) The massless Dirac equation possesses 'chiral invariance'. It is invariant under the replacement

$$\psi \to \exp(i\alpha\gamma_5)\psi$$

which mixes left- and right-handed components of ψ. This implies (by Noether's theorem) the conservation of the axial current.

$$J_\mu = \bar{\psi}\gamma_\mu\gamma_5\psi. \tag{13.28}$$

The conformal invariance implies that the symmetrized stress tensor is traceless. If spacetime admits a conformal Killing vector such that

$$k_{\alpha,\beta} + k_{\beta;\alpha} = 2\sigma g_{\alpha\beta} \tag{13.29}$$

then the current $J_{\alpha\beta}k^\beta$ will be conserved. If $k^\alpha = x^\alpha$ in flat space we have the 'dilation current'. In flat space one seeks theories which are 'dilation invariant' (e.g. Yang-Mills, $\lambda\phi^4$) since they have good renormalizability properties. Dilation invariance or the existence of a conformal Killing vector is not a generic property of spacetimes and the more appropriate symmetry is conformal rescaling. Both Yang-Mills and $\lambda\phi^4$ theory can be extended to curved space and both possess conformal rescaling (Weyl) invariance. Finally we note that the second-order operators governing the propagation can all be put in the form

$$D = -g^{\alpha\beta}\mathscr{D}_\alpha\mathscr{D}_\beta - \frac{1}{g^{1/2}}(g^{1/2}g^{\alpha\beta})_{,\beta}\mathscr{D}_\alpha + E \tag{13.30}$$

where E is a suitable matrix and

$$\mathscr{D}_\alpha = \partial_\alpha + W_\alpha \tag{13.31}$$

where W_α is a matrix of connection forms with curvature

$$W_{\alpha\beta} = \partial_\alpha W_\beta - \partial_\beta W_\alpha + W_\alpha W_\beta - W_\beta W_\alpha.$$

Commutation relations

The basic aim here is to pose canonical commutation or anticommutation relations on a given Cauchy Hypersurface Σ and show that the propagation equation propagates them off the surface. This is done by choosing a slicing of spacetime such that $t = $ constant surfaces are Cauchy surfaces, giving a $3+1$ decomposition of the field and exhibiting the equations in Hamiltonian form. One then replaces Poisson brackets by commutators or anticommutators. If $\delta_\Sigma^3(x, y)$ is a suitably defined delta function on Σ one has

$$\left[\boldsymbol{\phi}(x), \frac{\partial \boldsymbol{\phi}}{\partial t}(y)\right] = i\,\delta_\Sigma^3(x, y), \tag{13.32}$$

$$\{\bar{\boldsymbol{\psi}}(x), \boldsymbol{\psi}(y)\} = \delta_\Sigma^3(x, y), \tag{13.33}$$

$$\left[\boldsymbol{A}_\mu(x), \frac{\partial \boldsymbol{A}_\nu}{\partial t}(y)\right] = i\,\delta_\Sigma^3(x, y) g_{\mu\nu}. \tag{13.34}$$

By this means one may formally exhibit time evolution of observables $\boldsymbol{O}$ as the action of a one-parameter family of unitary maps $\boldsymbol{U}(t)$. I.e. $\boldsymbol{O}$ obeys

$$\mathscr{L}_{\frac{\partial}{\partial t}} \boldsymbol{O} = i[\boldsymbol{H}, \boldsymbol{O}] \tag{13.35}$$

with solutions

$$\boldsymbol{O}(t, x) = \boldsymbol{U}(t)\boldsymbol{O}(0, x)\boldsymbol{U}^{-1}(t) \tag{13.36}$$

where

$$\dot{\boldsymbol{U}} = i\boldsymbol{H}\boldsymbol{U}. \tag{13.37}$$

Here $\mathscr{L}$ denotes Lie derivative, $\boldsymbol{O}$ is a tensorial operator (it must be an even power of fermion operators if $\mathscr{L}$ is to be defined) and $\boldsymbol{H}(t)$ is the Hamiltonian given by

$$\boldsymbol{H}(t) = \int \boldsymbol{T}_{\alpha\beta} t^\alpha \, d\Sigma^\beta. \tag{13.38}$$

The integral being taken over a $t = $ constant slice, $t^\beta \, \partial/\partial x^\beta = \partial/\partial t$ and $T_{\alpha\beta}$ is the symmetrized stress tensor of the fields obtained by replacing the classical fields by operators and symmetrizing with respect to boson fields and the Hermitian conjugates and antisymmetrizing with respect to fermion fields and their Hermitian conjugates. All this is similar to the flat space except that $\boldsymbol{H}(t)$ depends upon time. Just as in flat space if $\{K_\alpha^A\}$ are Killing vectors we may formally construct the conserved momentum

Chapter 13. Quantum field theory in curved spacetime

operators by

$$K_\alpha^A = \int T_{\alpha\beta} K^{A\alpha} \, d\Sigma^\beta \qquad (13.39)$$

which will generate displacements along the orbits of K_α^A and whose commutator relations among one another are isomorphic to the corresponding Lie algebra of the Killing vector fields. If K_α^A is a globally timelike vector the corresponding operator is conserved Hamiltonian. Other examples would be the generators of rotations, which are angular-momentum operators for spacetimes with axi- or spherical symmetry. Many papers contain these formal developments, amongst them Urbantke (1969), Gibbons (1975a), Parker (1969, 1971), Kramers (1975, 1976). A general treatment of commutation relations in curved spacetime has been given by Lichnerowicz (1961).

Fock bases

In essence one wishes to construct $\mathfrak{H}$ and give it a Fock basis of 'particle states'. This is done by starting with the one-particle states which form a subspace of $\mathfrak{H}$, $\mathfrak{H}_1$ and building up the n particle states as totally symmetric (bosons) or totally antisymmetric (fermions) n-fold tensor products of $\mathfrak{H}_1$ with itself. $\mathfrak{H}_1$ is constructed from classical solutions of the wave equations. The scalar product in $\mathfrak{H}$ is just $(\ ,\ )$ suitably extended from $\mathfrak{H}_1$ to $\mathfrak{H}_n$. That is, $\mathfrak{H}_n$ consists of totally symmetric (boson) or antisymmetric (fermion) tensorial or spinorial functions of n spacetime variables and satisfying the relevant wave equation in each one separately.

The problem is that for bosons $(\ ,\ )$ is not positive-definite. For fermions $(\ ,\ )$ is positive-definite but one has no way of deciding what is a particle and what is an antiparticle. Both problems can be solved if one has a decomposition of arbitrary solutions of the wave equation into positive $f^+(x)$, and negative, $f^-(x)$, frequency solutions, where for bosons, the restriction of $(\ ,\ )$ to positive frequency functions must be positive-definite. One now builds up $\mathfrak{H}_n$ out of solutions which are positive frequency in each variable separately. For fermions one decomposes $\mathfrak{H}$, into $\mathfrak{H}^{(+)}$ and $\mathfrak{H}^{(-)}$, where $\mathfrak{H}^{(+)}$ contains the particle states and $\mathfrak{H}^{(-)}$ the antiparticle states. One can now define the positive-frequency part of $\phi[f]$ as an annihilation operator and the negative-frequency part as a creation operator in the sense that $\phi^-[f]$ maps $\mathfrak{H}_n \to \mathfrak{H}_{n+1}$ and $\phi^+[f]$ maps $\mathfrak{H}_n \to \mathfrak{H}_{n-1}$. The explicit action is in the creation case to multiply members of $\mathfrak{H}_n$ by $f^+(x)$ and totally symmetrize, and in the annihilation case to multiply by $f^-(x)$ and 'contract over x' using (,). The details may be found in Wald (1975) and parallel precisely the flat space case.

The major topic in curved space quantum field theory is what is positive frequency or what are 'particles'. If, as in flat space we had one agreed definition then there could of course be no particle creation. In globally static spacetimes for instance most people would accept as a reasonable definition that which is positive frequency with respect to the preferred time coordinate. Such 'particle states' would diagonalize the time independent Hamiltonian formed from the Killing vector and $T^{\mu\nu}$. Since the gravitational field is static one would in any case expect no particle creation. What is more interesting is to consider a spacetime which goes from a static stage in the past, $\mathcal{M}^-$ through a time varying stage $\mathcal{M}^0$ and then ends up static, in a region $\mathcal{M}^+$. Such a spacetime one might call a *'sandwich spacetime'*. This should on physical grounds yield some particle creation and this is reflected mathematically in the fact that negative frequency waves starting from $\mathcal{M}^-$ would propagate through $\mathcal{M}^0$ and arrive in $\mathcal{M}^+$ with both positive and negative frequencies. Since a negative-frequency wave corresponds to a state with no particles in it we see that 'particles have been created by the spacetime curvature in $\mathcal{M}^0$'. In order to make this statement however we are implicitly using a different definition of positive frequency in $\mathcal{M}^-$ from that being used in $\mathcal{M}^+$. Presumably this is reasonable because experiments carried out in $\mathcal{M}^-$ can conveniently be described in terms of the $\mathcal{M}^-$ definition and similarly for $\mathcal{M}^+$. The $\mathcal{M}^-$ particles would carry positive energy in $\mathcal{M}^-$ for instance. This is a simple example of a general point which is that *in curved spacetime one must use different definitions of particle appropriate to particular measurement processes at different times*. Put colloquially the notion of particle is 'observer dependent'.

A much deeper investigation of observer dependence has been carried out by Unruh (1976) and some of this analysis has been extended by Gibbons and Hawking (1977a). Unruh first investigated a simple model for a particle detector, thus stepping beyond the confines of the strictly non-interacting theory. The model consists of a quantum system carried by an observer along a timelike world-line. This is allowed to interact with the field with an interaction Hamiltonian of the form $ag\phi$ where a is a detector observable and g some coupling constant. The unperturbed detector Hamiltonian H_0 has energy levels E_n (the immediate vicinity of the world-line is assumed effectively static). The amplitude of a transition of the detector from a state $|n\rangle$ to a state $|m\rangle$ whilst the field changes from

Chapter 13. Quantum field theory in curved spacetime

a state $|i\rangle$ to a state $|f\rangle$ is, in perturbation theory, proportional to

$$\int dt \exp i(E_m - E_n)t \langle f|\phi|i\rangle\langle m|a|n\rangle, \quad (13.40)$$

where t is proper time along the observer's world-line. This expression shows that an 'Unruh detector' is sensitive to positive frequencies along its world-line. If one considers a uniformly accelerating world-line in Minkowski space with coordinates

$$T = \frac{1}{a} \sinh at, \quad (13.41)$$

$$Z = \frac{1}{a} \cosh at, \quad (13.42)$$

a being the acceleration one deduces that even in the standard Minkowski vacuum state the detector behaves as if a *thermal* distribution of particles is present with a temperature $T = a/2\pi$. The inequivalence of the standard vacuum definition to one based on time along these boost Killing orbits was first noticed by Fulling (1973) and was subsequently discussed by Davies (1975) after the discovery of black hole emission by Hawking (1974, 1975) which exhibits similar features. t and $l = 1/a$ may be used as coordinates in the region $|T| < Z$ and in these coordinates the metric of Minkowski space is

$$ds^2 = -l^2 dt^2 + dl^2 + dY^2 + dX^2.$$

This is a manifestly static and incomplete spacetime. The null surfaces $Z = |T|$ are boundaries of this 'Rindler space' and constitute past and future event horizons for the accelerating detector. Inertial observers, which pass through these horizons (e.g. Z = constant) will detect no particles in the Minkowski vacuum state.

Gibbons and Hawking (1977a) considered a similar example in de Sitter space. They constructed a de Sitter invariant state in which every *inertial* observer (each one of which has his own event horizon) sees particles coming from it with a thermal spectrum, even though these observers move with respect to one another with a relative velocity and one would thus expect, if the particles were observer independent, their observations to be redshifted.

Having given up the idea of a unique observer independent notion of particle we turn to considering the effect of changing one's definition – the subject of Bogoliubov transformations. Before doing so I should point out that the view advocated here has not found universal acceptance.

Other authors have sought to provide general and universally valid definitions. Amongst these are:

(1) Definitions based upon Feynman propagators – e.g. Mensky (1976). This is similar to Nachtmann's (1967) definition.
(2) Adiabatic particles (Parker and Fulling, 1974) and particles based on stationary-phase methods (Woodhouse, 1976).
(3) In and out particles defined using Feynman propagators (Rumpf, 1976; Rumpf and Urbantke, 1977).
(4) Particles defined by diagonalization of a time-dependent Hamiltonian (e.g. Castagnino, Verbeure and Weder, 1975). These seem to lead to infinite creation rates (Parker, 1969). They are used in cosmology but in a cosmological model with particle horizons it is not easy to see what measurement process these could correspond to.
(5) Particles defined on null surfaces (Frolov, Valovich and Zagrebnov, 1977; cf. Unruh, 1976).
(6) Particles, or rather creation rates defined by 'effective Lagrangians' (Raine and Winlove, 1975; Dowker and Critchley, 1976a). These formal constructions can yield probabilities exceeding unity for their creation.

A careful analysis of the notion of particle both in flat and curved space has been given by Hajicek (1976, 1977).

Bogoliubov transformations

Different Fock bases for $\mathfrak{H}$ should be related by unitary maps – 'S-matrices'. A Fock basis B^- is a set of states of the form

$$|n_1, n_2, \ldots, n_r\rangle = \frac{(a_1^+)^{n_1}}{\sqrt{n_1!}} \frac{(a_2^+)^{n_2}}{\sqrt{n_2!}} \cdots \frac{(a_r^+)^{n_r}}{\sqrt{n_r!}} |0_-\rangle \qquad (13.43)$$

and may be specified by giving a basis of positive frequency solutions $\{p_i\}$ of the relevant wave equation. These are often specified by giving the Cauchy data on some initial or final Cauchy surface. Here a_i^+ creates a particle corresponding to a wavefunction p_i. The a_i obey

$$a_i a_j^+ - a_j^+ a_i = \delta_{ij} \quad \text{for bosons,} \qquad (13.44)$$

$$a_i a_j^+ + a_j^+ a_i = \delta_{ij} \quad \text{for fermions.} \qquad (13.45)$$

If the new basis B^+ is $\{f_i\}$ then we have

$$f_i = \sum_j \alpha_{ij} p_j + \beta_{ij} \bar{p}_j \qquad (13.46)$$

Chapter 13. Quantum field theory in curved spacetime

where the bar denotes complex conjugation. α_{ij} and β_{ij} measure how much frequency mixing occurs. The field operator is in this notation

$$\phi = \sum_i a_i p_i + a_i^+ \bar{p}_i = \sum_i b_i f_i + b_i^+ \bar{f}_i.$$

To preserve the orthogonality relations (for bosons):

$$(p_i, p_j) = (f_i, f_j) = -(\bar{p}_i, \bar{p}_j) = -(\bar{f}_i, \bar{f}_j) = \delta_{ij} \tag{13.47}$$

$$(\bar{p}_i, p_j) = (\bar{f}_i, f_j) = 0. \tag{13.48}$$

α_{ij}, β_{ij} must obey (in the boson case)

$$\begin{pmatrix} \alpha, \beta \\ \bar{\beta}, \bar{\alpha} \end{pmatrix} \begin{pmatrix} \alpha^+, -\tilde{\beta} \\ -\beta^+, \tilde{\alpha} \end{pmatrix} = \begin{pmatrix} 1 & 0 \\ 0 & 1 \end{pmatrix}. \tag{13.49}$$

The tilde denotes transposition and $^+$ hermitean conjugate. The annihilation and creation operators transform as

$$b_i = \bar{\alpha}_{ij} a_j - \bar{\beta}_{ij} a_j^+. \tag{13.50}$$

The number of new particles in the old vacuum is

$$n_j^+ = \langle 0_-|b_j^+ b_j|0_-\rangle = \sum_i |\beta_{ij}|^2 \tag{13.51}$$

and the total number

$$N^+ = \sum n_i^+ = \operatorname{tr} \beta\beta^+ \tag{13.52}$$

we have

$$|0_-\rangle = S|0_+\rangle, \tag{13.53}$$

$$S = \exp(\gamma_{ij} b_i b_j - \bar{\gamma}_{ij} b_i^+ b_j^+), \tag{13.54}$$

$$\gamma = \alpha^{-1}\beta, \tag{13.55}$$

$$|\langle 0_+|0_-\rangle|^2 = |\det \alpha|^{-2}. \tag{13.56}$$

Equation (13.54) shows that particles are created in *pairs*. If $\langle 0_+|0_-\rangle$ vanishes then $|\langle 0_+|0_-\rangle|^2$ vanishes which can only occur if N diverges. In that case the two Fock bases cannot span the same Hilbert space – one has a 'unitarily inequivalent representation of the canonical commutation relations'. Physically this can occur because of an unsuitable choice of particle states or because the gravitational field is being allowed to create particles continuously without the back reaction being taken into account.

Basic notions

A simple and useful Bogoliubov transformation is (Parker, 1969)

$$p_{i\varepsilon} = \cosh \theta_i f_{i\varepsilon} + \sinh \theta_i \bar{f}_{i-\varepsilon}, \tag{13.57}$$

$$b_{i\varepsilon} = \cosh \theta_i a_{i\varepsilon} - \sinh \theta_i a^+_{i-\varepsilon}, \tag{13.58}$$

$$a_{i\varepsilon} = \exp(G) b_{i\varepsilon} \exp(-G), \tag{13.59}$$

$$G = \sum_i \tanh \theta_i (b^+_{i\varepsilon} b^+_{i-\varepsilon} - b_{i\varepsilon} b_{i-\varepsilon}), \tag{13.60}$$

$$|0_-\rangle = \prod_i (\cosh \theta_i)^{-1} \exp(\tanh \theta_i b^+_{i\varepsilon} b^+_{i-\varepsilon})|0_+\rangle \tag{13.61}$$

$$= \prod_i \sum_n (\cosh \theta_i)^{-1} (\tanh \theta_i)^n |n_{i\varepsilon}\rangle \otimes |n_{i-\varepsilon}\rangle. \tag{13.62}$$

Here $\varepsilon = \pm 1$ and labels two classes of particles (e.g. a momentum or angular momentum direction). Obviously particles are created in pairs – one from each class. If we were to ignore the particles in class $-\varepsilon$ we obtain a density matrix (in terms of ε variables)

$$\rho = \sum_n \prod_i p_n |n_{i\varepsilon}\rangle\langle n_{i\varepsilon}|, \tag{13.63}$$

$$p_n = \prod_i (\tanh \theta_i)^{2n_1} (\cosh \theta_i)^{-2}. \tag{13.64}$$

By suitable non-mixing (i.e. $\beta_{ij} = 0$) Bogoliubov transformations a general Bogoliubov transformation can be brought to this form. In the fermion case the minus signs become plus signs and the hyperbolic functions become trigonometric ones. This formalism has some simple applications. One to cosmology where – for instance in FRW with flat spatial sections ε represents the conserved momentum and so pairs are created with equal and opposite momentum (Parker, 1969). For spaces with horizons the two classes are particles which fall through the horizon and those that do not. If one cannot measure those that fall through one arrives at a density matrix. In fact for black holes one finds that

$$\cosh \theta = \exp(\pi\omega/\kappa) \tag{13.65}$$

where ω is the energy and κ the surface gravity of the hole. This yields a thermal density matrix (Wald, 1975; Parker, 1976b; Hawking, 1976b; Israel, 1976). Indeed Hawking (1976b) has argued that this foregoing of information on horizons and 'hidden surfaces' is not optional but obligatory. If this is so *the resulting loss of information, changing the initial pure state to an incoherent superposition is an intrinsically new feature of curved space quantum field theory.*

Chapter 13. Quantum field theory in curved spacetime

Regularization and propagator techniques

It is notorious that when one attempts to evaluate the expectation values of expressions which are nonlinear in the field operators, divergences result. The most elementary of these is the 'zero-point energy'. An unambiguous algorithm for rendering all such expressions finite is called a 'regularization' scheme. A 'renormalization' scheme is a device whereby solutions of an interacting system are expressed in the same form as the solutions of the non-interacting with certain constants given new, 'renormalized' values. It is well known that quantum electrodynamics may be 'regularized' using an infinite 'renormalization'. The situation with respect to quantum field theory on curved spacetimes seems to be that whilst regularization schemes exist none is a renormalization in our sense of the word. That is, one can in general express, for instance, the expectation value of the stress tensor as

$$\langle T_{\mu\nu} \rangle = \langle T_{\mu\nu}^{\text{ren}} \rangle + \langle T_{\mu\nu}^{\infty} \rangle \tag{13.66}$$

where $\langle T_{\mu\nu}^{\text{ren}} \rangle$ is finite and conserved and $\langle T_{\mu\nu}^{\infty} \rangle$ is infinite and in some sense

$$\langle T_{\mu\nu}^{\infty} \rangle = -2c_1(R_{\mu\nu} - \tfrac{1}{2}g_{\mu\nu}R) + c_0 g_{\mu\nu}, \tag{13.67}$$

where c_0, c_1 are infinite numbers. One could then write a 'Hartree–Fock' version of Einstein's equations

$$R_{\mu\nu} - \tfrac{1}{2}g_{\mu\nu} + \Lambda g_{\mu\nu} = 8\pi G \langle T_{\mu\nu}^{\text{ren}} \rangle$$

where Λ and G are the observed ('dressed') cosmological constant and Newton's constant related to the 'bare' values Λ_0, G_0 by

$$G = G_0(1 + 16\pi G_0 c_1)^{-1}, \tag{13.68}$$

$$\Lambda = (\Lambda_0 - 8\pi G_0 c_0)(1 + 16\pi G_0 c_1)^{-1}. \tag{13.69}$$

The divergences in $\langle T_{\mu\nu} \rangle$ have been examined in a number of ways and various regularization schemes proposed. These include
   (1) 'mode sums', e.g. Parker and Fulling (1974),
   (2) 'point splitting' (DeWitt, 1975; Christiensen, 1976; Davies, 1976*b*),
   (3) Pauli–Villars (Bernard and Duncan, 1977),
   (4) dimensional regularization (Brown, 1977; Brown and Cassidy, 1977; Duncan, 1977),
   (5) zeta-function regularization (Dowker and Critchley, 1976*a*),
   (6) the Fradkin–Vilkovisky (1977) approach.

Basic notions

Most of these schemes are based upon the construction of Feynman propagators and the DeWitt–Schwinger propertime formalism which we must first explain. For any state ρ we may define the generator $Z_\rho[J]$ as a functional of a classical source current by

$$Z_\rho[J] = \operatorname{tr}\left\{\rho \mathcal{T} \exp\left(i \int \phi(x)J(x)(-g)^{1/2}(x)\,d^4x\right)\right\} \quad (13.70)$$

(for fermions $J(x)$ is a classical anticommuting variable). Functionally differentiating leads to

$$\frac{\delta^n \ln Z_\rho}{\delta J(x)\ldots \delta J(x')\ldots,} = G_\rho(x,\ldots,x') \quad (13.71)$$

with

$$G_\rho(x,\ldots,x') = i^n \operatorname{tr}\{\rho \mathcal{T}\phi(x)\ldots\phi(x')\}. \quad (13.72)$$

$\mathcal{T}$ is the Wick time-ordering operator placing the field operators in chronological order from right to left with a minus sign for every fermion operator interchange. For linear fields these 'propagators' are Green's functions. For scalars, for example,

$$(-\nabla_\alpha \nabla^\alpha + \xi R + m^2)G_\rho(x,x') = \delta(x,x'). \quad (13.73)$$

The $G_\rho(x,\ldots,x')$ contain all the information about the theory. Formally at least one can construct stress tensor expectation values etc. by coincidence limits. The physical interpretation of $G_\rho(x,x')$ is the amplitude for a disturbance about the state to propagate from x' to x. In a globally static space ρ might be the obvious vacuum state and $G(x,x')$ describes the propagation of 'particles'. Alternatively G might be the associated finite temperature Gibbs state in which case the G_ρ are called thermal Green's functions (Gibbons and Perry, 1976, 1978). The propagators can either be constructed from Fock bases etc. or from other considerations, path integrals or analyticity (Hartle and Hawking, 1976; Mensky, 1976; Rumpf, 1976). In a path-integral formulation of curved space quantum field theory they should be directly constructable from the Feynman integral (cf. Hartle and Hawking, 1976). If we adopt a symbolic notation where I stands for the delta-function and A the differential operator defining the wave equation we have formally

$$AG_\rho = I \quad (13.74)$$

which has as a formal solution (DeWitt–Schwinger representation)

$$G_\rho = A^{-1} = i \int \exp(-iwA)\,dw \quad (13.75)$$

Chapter 13. Quantum field theory in curved spacetime

where $\exp(-iwA) = K(x, x', w)$ satisfies the Schrödinger-like equation

$$i\frac{\partial K}{\partial w} = AK, \quad K(x, x', 0) = I. \tag{13.76}$$

$K(x, x')$ will have a Feynman integral representation of the form

$$K(x, x', w) = \exp(-im^2 w) \int_0^w \mathcal{D}[x(w)] \exp(iI), \tag{13.77}$$

(see Dowker, 1974, for a review of path integrals for higher spin)

$$I = \tfrac{1}{4} \int_0^w g_{\alpha\beta} \frac{dx^\alpha}{dw} \frac{dx^\beta}{dw} dw. \tag{13.78}$$

One may also give $Z_\rho[J]$ a functional representation as

$$Z_\rho[J] = \int \mathcal{D}[\phi] \exp\left\{i\left[I[\phi] + \int \phi(x) J(x) g^{1/2}(x) d^4x\right]\right\} \tag{13.79}$$

where $I[\phi]$ is the classical action of ϕ, the field configuration. $\int \mathcal{D}[\phi]$ denotes a 'sum' over classical field configurations and formally

$$\operatorname{tr} \rho = (\det A)^{-1/2} = Z_\rho[0], \tag{13.80}$$

since

$$I[\phi] = \tfrac{1}{2} \int \phi A \phi g^{1/2} d^4x. \tag{13.81}$$

For fermions the sums are over anticommuting classical fields and (13.80) contains det A raised to the 1st power.

The general idea of these formal manipulations is that the state ρ and hence the boundary conditions obeyed by G_ρ are reflected in the class of paths over which these sums are taken. To obtain thermal Green's functions one sums over paths which are periodic in imaginary time with period β (cf. Gibbons and Hawking, 1977b).

B. DeWitt (1963, 1975) was the first person to introduce propagators in this way and he also introduced a power series expansion for $K(x, x', \Omega)$ which is closely related to the Hadamard expansion for retarded Green's functions. The expansion is of the form

$$K(x, x', \Omega) = \frac{-i}{(4\pi)^2} D^{1/2}(x, x') \exp\left[i\frac{\sigma(x, x')}{2w} - im^2 w\right] \sum_{n=0}^{\infty} a_n(x, x')(iw)^n. \tag{13.82}$$

$\sigma(x, x')$ is half the square of the geodesic distance between x' and x and D is constructed from $\sigma(x, x')$ by differentiation and the $a_n(x, x')$ obey a certain recurrence relation. This Hadamard–DeWitt expansion is closely related to an expansion for elliptic operators first written down by Minakshisundaram and which has subsequently been at the centre of a great deal of work on elliptic operators. Essentially if we write $iw = \Omega$ and suppose $g_{\alpha\beta}$ is a *Riemannian* metric (13.75) becomes the heat equation. If the Riemannian manifold were compact (with or without boundary) A would possess a discrete spectrum of eigenvalues λ_n, with eigenfunctions ϕ_n and we would have

$$K(x, x', \Omega) = \sum_{n=0}^{\infty} \exp[-\lambda_n \Omega] \phi_n(x) \phi_n(x'), \qquad (13.83)$$

$$Y(\Omega) = \int K(x, x, \Omega) g^{1/2}(x) \, d^4x = \sum_{n=0}^{\infty} \exp(-\lambda_n \Omega). \qquad (13.84)$$

Then it is known that $Y(\Omega)$ has an *asymptotic* expansion of the form

$$Y(\Omega) = \sum_{m=0}^{\infty} \frac{\Omega^{m/2}}{(4\pi\Omega)^2} \left\{ \int_M B_m(x) g^{1/2} \, d^4x + \int_{\partial M} C_m(x) h^{1/2} \, d^3x \right\} \qquad (13.85)$$

where $B_m(x)$ are local invariants of the curvature and $C_n(x)$ invariants of the intrinsic and extrinsic curvature of the boundary and $h^{1/2} d^3x$ the induced volume element on the boundary ∂M. The coefficients $B_m = a_{m/2}$ have been computed by many people in many different guises and they always crop up in discussing 'single closed loop processes' for gravitational or Yang–Mills theories. I propose calling them the Hadamard–Minakshisundaram–DeWitt = 'Hamidew' coefficients after the principal names associated with them.

$$E_2 = E - \frac{R}{6} + m^2, \qquad (13.86)$$

$$\begin{aligned} E_4 = {} & \tfrac{1}{30} R^{;\alpha}{}_{;\alpha} + \tfrac{1}{72} - \tfrac{1}{180} R_{\alpha\beta} R^{\alpha\beta} + \tfrac{1}{180} R_{\alpha\beta\gamma\delta} R^{\alpha\beta\gamma\delta} \\ & - \tfrac{1}{6} RE + \tfrac{1}{2} E^2 + \tfrac{1}{2} W_{\alpha\beta} W^{\alpha\beta} + \tfrac{1}{6} E^{;\alpha}{}_{;\alpha}. \end{aligned} \qquad (13.87)$$

E_n vanishes for n odd and E_3 is very long. If $\partial \mathcal{H} \neq 0$ there are C_1 and C_3 terms which for minimally coupled scalars are (McKean and Singer, 1967)

$$C_1 = \pm \tfrac{1}{4} (4\pi)^{1/2}, \qquad (13.88)$$

$$C_3 = \tfrac{1}{3} K \qquad (13.89)$$

where '+' denotes Dirichlet, '−' Neuman boundary conditions and K is the trace of the second fundamental form of the boundary. It should be

Chapter 13. Quantum field theory in curved spacetime

noted that Gilkey's results agree completely with earlier values given by DeWitt (1963). The relevance of the expansion for elliptic operators is that whereas the formal exponentiation etc. of elliptic operators can be rigorously justified (cf. Gilkey, 1974) the precise status of the integral representation for hyperbolic operators is unclear. At the very least one needs to give $\sigma(x, x')$ and $-m^2$ a small positive imaginary part to ensure the convergence of the integral. This has encouraged attempts (Hartle and Hawking, 1976; Hawking, 1977a) to define the path integrals using complexifications of spacetime. In simple cases, e.g. for de Sitter space (Gibbons and Hawking, 1977a), or for Schwarzschild (Hartle and Hawking, 1976), this has led to Riemannian spaces on which unique Feynman propagators could be defined. For these cases the asymptotic expansions are certainly relevant. We call such attempts to define path integrals by complexification, 'Riemmanization'. It requires that $\{\mathcal{M}, g^{\alpha\beta}\}$ admits a complexification as a 4-complex-dimensional manifold $\mathcal{M}_c$, together with a complex contravariant tensor field of type (2, 0) such that there exists 'real sections'. These are 2-complex-dimensional submanifolds on which, regarded as real manifolds, the restriction of $g_c^{\alpha\beta}$ to the cotangent space of the slice is a real form. They are called Lorentzian or Riemannian if the signature of $g^{\alpha\beta}$ is +++− or ++++ respectively. If a section is such that the restriction of $g_c^{\alpha\beta}$ has no real null covectors we call it 'quasi-Riemannian'. The general operator A will be elliptic on such a section. Riemannization is the analogue of Euclideanization in flat space quantum field theory.

We now need to define det A and hence Z. Once this is done one can define physical quantities by variation of $\ln Z$ with respect to a convenient parameter. For instance one has

$$\langle T_{\mu\nu}\rangle = \frac{\text{tr}\,\rho T_{\mu\nu}}{\text{tr}\,\rho} = \frac{2}{g^{1/2}}\frac{\delta \ln Z}{\delta g^{\mu\nu}} \qquad (13.90)$$

where $\delta/\delta g^{\mu\nu}$ denotes functional differentiation. This will guarantee a conserved expectation value. To define det A one may either start with the formal expression (DeWitt, 1975)

$$\delta \ln Z = -\tfrac{1}{2}\delta \ln \det A = -\tfrac{1}{2}\text{tr}\,\delta A A^{-1} = \tfrac{1}{2}\delta\,\text{tr}\int \frac{e^{-A\Omega}}{\Omega}\,d\Omega \qquad (13.91)$$

whence

$$\ln Z = \int_0^\infty \frac{Y(\Omega)}{\Omega}\,d\Omega = -L_{\text{eff}}. \qquad (13.92)$$

Now (13.91) as it stands does not converge so one can:

(1) change Ω to Ω^{1-s} and discard the poles at $s = 0$ or exhibit them as 'counter terms' in the effective Lagrangian (Dowker and Critchley, 1976a, b);

(2) write down a form of (13.91) valid in n dimensions and discard the poles at $n = 4$ (dimensional regularization – Brown, 1977; Brown and Cassidy, 1977) or regard them as 'counter terms' in the effective Lagrangian.

The B_0 and B_2 terms in L_{eff} that arise in this way can be thought of as renormalizing Λ and G but the B_4 term containing quadratic terms in the curvature does not 'renormalize' anything since terms of this form do not appear in the original Lagrangian. This was realized by DeWitt and Utiyama long ago.

An alternative approach suggested by Hawking, which is rather similar to Dowker and Critchley's method, is to consider the zeta-function of A. This is

$$\zeta(s) = \sum_{n=0}^{\infty} \frac{1}{\lambda_n^s} \qquad (13.93)$$

which will converge for Real $s > 2$ and its analytic continuation defines a meromorphic function of s analytic at $s = 0$, formally

$$\ln \det A = -\zeta'(0). \qquad (13.94)$$

Now the path integral expression leads to the formal expression

$$Z = \prod_{n=0}^{\infty} \left(\frac{2\pi\mu^2}{\lambda_n}\right)^{1/2} \qquad (13.95)$$

where μ has the dimensions of mass and is introduced to make Z dimensionless. This suggests adopting the definition

$$Z = (2\pi\mu^2)^{\zeta(0)/2} \exp \tfrac{1}{2}\zeta'(0) \qquad (13.96)$$

as the zeta-function definition of Z. In fact

$$\zeta(s) = \frac{1}{\Gamma(s)} \int_0^{\infty} \Omega^{s-1} Y(\Omega) \, d\Omega \qquad (13.97)$$

which shows the connection with Dowker and Critchley's approach. Since the n-dimensional 'Hamidew' expansion has the same form in all dimensions except that the factor in front of (13.85) changes from $(4\pi\Omega)^2$ to $(4\pi\Omega)^{n/2}$ we see that 'dimensional regularization' is also very closely

Chapter 13. Quantum field theory in curved spacetime

related. However there is one difference. The $\{B_n\}$ are obtained by taking dimension-dependent traces. On the other hand zeta-function regularization (which may be regarded as adding on $2s$ flat dimensions and then dimensionally regularizing) works with the 4-dimensional Hamidew expansion. This can, and does lead to different answers for the renormalized part of $T_{\mu\nu}$. This shows up in the trace anomaly for photons.

The third method of regularization which has been proposed which is based on the Schwinger–DeWitt expansion is 'point splitting'. Essentially one considers $T_{\mu\nu}$ as a coincidence limit of $T_{\mu\nu'}(x, x')$. Here $T_{\mu\nu'}(x, x')$ contains terms like $(\partial\phi(x)/\partial x^\mu)(\partial\phi(x')/\partial x'^{\nu'})$ and is constructed from G_ρ. Now the $\partial\phi(x)/\partial x^\mu$ terms are parallely propagated to x along the geodesic running from x' to x. If x' and x are close such that $x'^\mu - x^\mu = \varepsilon^\mu$ the expressions can be expanded in powers of ε^μ. One then discards the divergent or directionally dependent terms. Alternatively one subtracts off from the answer one has in a particular spacetime the infinities one expects to arise in general on the basis of the real-time Hamidew expansion. The final method suffers from the disadvantage that the notion of a direction-dependent term is ambiguous. For example

$$R_{\mu\nu}\varepsilon^\mu\varepsilon^\nu/\varepsilon_\sigma\varepsilon^\sigma \tag{13.98}$$

can be written as

$$G_{\mu\nu}\varepsilon^\mu\varepsilon^\nu - \tfrac{1}{2}G \tag{13.99}$$

with a different direction-dependent term. Proponents of point splitting claim that these ambiguities do not arise in their second method.

The regularized expressions for Z will depend (in zeta-function regularization) on μ, the 'renormalization mass', which is analogous to the subtraction point in conventional perturbation theory. This will also occur in dimensional regularization since one must write

$$\int (4\pi\mu^2\Omega)^{(4-n)/2} \frac{Y(\Omega)}{\Omega} d\Omega \tag{13.100}$$

in order to keep Z dimensionless. If $\zeta(0)$ which is given by the terms quadratic in the Riemann tensor by

$$(4\pi)\zeta(0) = \int_\mathcal{M} B_4 g^{1/2} d^4x + \int_{\partial\mathcal{M}} C_4 h^{1/2} d^3x \tag{13.101}$$

is not zero these will contribute to physical quantities. Another way of putting this is to say that since the B_2 terms are present the theory is not renormalizable and one must introduce extra terms in the Lagrangian to make the theory finite. From this point of view μ appears as a *new physical constant whose value is to be determined by experiment* (cf. DeWitt, 1975). Of course higher order quantum gravity effects may also introduce other corrections and other constants. Since the number of these is potentially infinite we face the basic problem that the non-renormalizability of gravitational theories leads to a breakdown in their predictability.

In conclusion we may say that if a state possesses a Feynman propagator which admits in some sense a Schwinger–DeWitt expansion together with some sort of (asymptotic?) Hamidew expansion then we may obtain regularized values for $\langle T_{\mu\nu} \rangle$. From the mathematical point of view it would be extremely useful to have a clearer idea of how general the Hamidew expansion really is and also what type of series it is. In the elliptic case this is well established but in the hyperbolic case much more needs to be done.

Anomalies. One speaks of an 'anomaly' in quantum field theory if the regulated quantities do not possess one of the symmetries of the classical theory. In curved space two such anomalies have been discovered.

(1) *Axial anomaly.* Following related work in quantum electrodynamics Kimura (1969), Delbourgo and Salam (1972), Eguchi and Freund (1976) independently discovered that chiral invariance was broken and the axial current acquires a divergence given by

$$\nabla_\mu(\bar{\psi}\gamma^\mu\gamma_5\psi) = \frac{1}{384\pi^2}\varepsilon^{\mu\nu\lambda\rho}R_{\alpha\beta\mu\nu}R^{\alpha\beta}{}_{\lambda\rho}. \qquad (13.102)$$

The integral of the right-hand side of (13.102) is in fact a topological invariant. That is, it is unchanged by continuous changes of the metric inside the region of integration. In fact for a compact Riemannian manifold one has the identity

$$\frac{P}{3} = \tau = \frac{1}{96\pi^2}\int R_{\alpha\beta\gamma\delta}R^{\alpha\beta}{}_{\lambda\rho}\varepsilon^{\gamma\delta\lambda\rho}g^{1/2}\,d^4x \qquad (13.103)$$

where τ is the 'signature' and P the 'Pontryagin number' of the manifold. Both are integers. Furthermore the Atiyah–Singer index theorem implies

$$N^+_{1/2} - N^-_{1/2} = -P/24 \qquad (13.104)$$

where $N^+_{1/2}$ and $N^-_{1/2}$ are respectively the number of right- and left-handed solutions of the equation

$$\left(-\nabla_\alpha\nabla^\alpha + \frac{R}{4}\right)\psi = 0. \qquad (13.105)$$

Chapter 13. Quantum field theory in curved spacetime

The relationship between these facts has been elucidated by Nielsen, Römer and Schroer (1977) and Hawking (1977b). It should be remarked that Kimura, Delbourgo and Salam, and Eguchi and Freund all obtained their results by a Feynman graph technique. Kimura also computed the anomaly using a form of point splitting in a background field, which was repeated by Nielsen, Römer and Schroer independently. Delbourgo and Salam obtained an answer which differed from (13.102) by a factor of 2 but this seems to have been an algebraic slip.

Riemannian manifolds with nonzero Pontryagin number are rare. Hawking (1977b) has given an example – a member of the Taub–NUT class of metrics, but it is not compact and boundary terms enter into the various theorems. The case given by Hawking (which has self dual curvature – popularly called the Instanton condition) has been discussed by Römer and Schroer (1977). The interest these matters have aroused is largely because of analogous effects in gauge theories – especially in quantum chromodynamics where the group is SU(2). The instanton configuration give rise to a variety of effects most noticeably baryon non-conservation and CP violation (t'Hooft, 1976).

(2) *Conformal or trace anomalies.* The breakdown of conformal invariance in gravitational fields was first noticed by Capper and Duff (1974) again using Feynman diagram techniques. Since then a considerable sub-literature has blossomed on the subject. In general terms the anomaly may be understood as a consequence of introducing a scale into the theory in order to regularize it. This occurs in different ways in different regularization methods but nevertheless there is a close agreement as to the resulting finite trace to the stress tensor which is the direct manifestation of this anomaly. Indeed one finds the following relation for the trace of $T_{\mu\nu}$

$$(4\pi)^2 180 T^{\sigma}{}_{\sigma} = \alpha R^2 + \beta R_{\alpha\beta}R^{\alpha\beta} + \gamma R_{\alpha\beta\gamma\delta}R^{\alpha\beta\gamma\delta} + \delta R^{;\alpha}{}_{;\alpha} \qquad (13.106)$$

with for conformally invariant scalars

$$\alpha = 0, \quad \beta = -1, \quad \gamma = 1, \quad \delta = 1, \qquad (13.107)$$

neutrinos

$$\alpha = \tfrac{5}{4}, \quad \beta = 2, \quad \gamma = \tfrac{7}{4}, \quad \delta = 3 \qquad (13.108)$$

photons

$$\alpha = -25, \quad \beta = 88, \quad \gamma = -13, \quad \delta = -18. \qquad (13.109)$$

The photon coefficient of $R^{;\alpha}{}_{;\alpha}$ is the only contentious term. The values quoted are those obtained using zeta-function regularization (Dowker and Critchley, 1976*b*). Brown and Cassidy (1977) obtained a value of +12 for this term using their version of dimensional regularization. References and a discussion of the various determinations may be found in Duff (1978).

In zeta-function regularization the anomaly may be obtained by considering the behaviour of Z under rescaling. If

$$g_{\alpha\beta} \to \Lambda g_{\alpha\beta}, \qquad (13.110)$$

$$A \to \Lambda^{-1} A, \qquad (13.111)$$

$$\mu \to \tilde{\mu}, \qquad (13.112)$$

then

$$Z \to Z(2\pi\tilde{\mu}^2/\mu^2)^{\frac{1}{2}\zeta(0)} \Lambda^{\frac{1}{2}\zeta(0)}. \qquad (13.113)$$

Thus if $\tilde{\mu} = \mu$ and

$$\int T^\sigma{}_\sigma g^{1/2} \, d^4x = 2 \frac{\partial}{\partial \Lambda} \ln Z = \zeta(0) \qquad (13.114)$$

$$T^\mu{}_\mu = \frac{B_4}{(4\pi)^2}. \qquad (13.115)$$

These are the values quoted before. The photon values include a contribution from Feynman–DeWitt–Faddeev–Popov ghosts.

The importance of conformal anomalies is that, like the axial anomaly, they are believed to be the same for each state. This means that knowing $T^\sigma{}_\sigma$ reduces the number of components one needs to compute directly. In some cases it determines the stress tensor completely. For instance in de Sitter space the de Sitter invariant vacuum discussed by Gibbons and Hawking has

$$\langle T_{\mu\nu} \rangle = \Lambda^2 g_{\mu\nu}(\alpha - 6\beta)/540(4\pi)^2. \qquad (13.116)$$

In the case of black holes it reduces the number of required components to unity (Christensen and Fulling, 1977). Anomalies also give the scaling behaviour of Z.

13.3 Applications

Having completed this survey of the basic structure of the formalism I wish to go on to discuss some of the applications. It should be fairly clear

Chapter 13. Quantum field theory in curved spacetime

that effects like particle production can only become important in the late stages of gravitational collapse or in the early universe. In no other circumstance can particle creation or the other effects like anomalies play a significant role. For this reason effort has been concentrated on quantum field theory in cosmological models and in black hole background metrics, although some work has been done on other spacetimes to illustrate some aspects of the theory.

Isotropic (FRW) cosmological models

These were the subjects of Schrödinger's original investigations and have been studied extensively ever since (see Parker, 1976a, for reviews). Roughly speaking one expects creation to take place when curvatures are comparable with the inverse square of the wavelengths involved (e.g. the Compton wavelength $1/m$ for massive particles). However one basic result dominates this field and that is that the FRW spaces are conformally static. That is, the metric

$$ds^2 = -dt^2 + R^2(t) \, d\Omega_3^2 \tag{13.117}$$

can be written as

$$ds^2 = R^2(\eta)(-d\eta^2 + d\Omega_3^2) \tag{13.118}$$

where $d\Omega_3^2$ is the metric on a 3-space of constant curvature and η obeys

$$\frac{d\eta}{dt} = \frac{1}{R}. \tag{13.119}$$

For conformally invariant theories one can write the field equations etc. in terms of the rescaled metric (even in the interacting case provided the interactions are conformally invariant) and define a vacuum (the 'conformal vacuum') using functions which are positive frequencies with respect to η. In this vacuum there should be no particle production, since the notion of positive frequency is independent of η or t. Put another way the rescaled Hamiltonian is independent of η (cf. Parker, 1969). One could also consider the Gibbs state with respect to this vacuum (cf. Gibbons and Perry, 1976, 1978). This is also stable, with a constant temperature T_0. When one rescales this to obtain physical quantities it becomes time-varying with the standard redshift factor

$$T = T_0/R(t). \tag{13.120}$$

This could describe the 3 K black body background.

Applications

If masses are present then in the rescaled space they will appear to increase with time as

$$m = m_0 R(\eta) \tag{13.121}$$

(either the universe gets bigger or the sizes of particles get smaller).

The resulting time-varying terms in the rescaled Hamiltonian will cause particle creation to take place when $(dm/d\eta)/m \sim m$ or $t \sim 10^{-22}$ s. For times earlier than this the rate of creation is small since $m \to 0$ as $R \to 0$, the instant of the big bang. Thus one may estimate the total number of particles created since the big bang if there were none (according to the conformal definition) to begin with. This has been done by Mamaev, Mostepanenko and Starobinsky (1976).

Many other types of calculation have been done, from different standpoints. The problem in general is deciding what initial state to choose and what subsequent particle definition to choose in order to compute a creation rate. The first problem obviously cannot be answered within the formalism and must be decided on physical or theological grounds. The second point also raises problems. For instance, if one instantaneously diagonalizes the physical Hamiltonian on $t =$ constant surfaces the resulting creation rates diverge (Parker, 1969; Castagnino, Verbeure and Weder, 1975). Parker and Fulling attempted to get round this difficulty by introducing the concept of 'adiabatic particles' (Parker and Fulling, 1974). They then went on to investigate the back reaction by performing an 'adiabatic regularization' of the stress tensor to see whether spacetime singularities would result. They found a class of states for which they did not. However because of their non-covariant form the status of these results is unclear. DeWitt (1975) has sought to relate the regularization process to the Schwinger–DeWitt representation and has argued that in this situation it is identical to another non-covariant scheme introduced by Zel'dovich and Starobinsky (1972) called n-wave regularization.

In view of these uncertainties attempts have been made to treat a sandwich spacetime version of the FRW model where $R(t)$ becomes constant before some initial time t_1 and after some final time t_2. Starting with the natural vacuum before t_1 one can compute the number of out particles after time t_2. Parker (1976c) has done so for a class of examples which he regards as representative and finds the resulting Bogoliubov transformations to have the asymptotic form

$$\beta_k \approx \exp(-a|k|). \tag{13.122}$$

The particles obey the minimally coupled scalar wave equation and k is

Chapter 13. Quantum field theory in curved spacetime

the momentum (the spatial cross-sections are taken to be flat). This indicates that massive particles are created with a thermal spectrum in momentum space. However a pure state results unless one has a mechanism (e.g. collisional thermalization) to remove the correlations. Chitre and Hartle (1977) have found similar results for massive fields using a propagator technique with a particular choice of scale factor.

For the conformally invariant equations, the high symmetry together with the trace anomaly enables one to determine the regularized vacuum stress tensor as (Bunch and Davies, 1977a, b)

$$\langle T_{\mu\nu}\rangle = \frac{1}{(4\pi)^2}\frac{1}{180}\left\{(\beta-2\gamma)\,^{(3)}H_{\mu\nu} - \frac{\delta}{6}\,^{(1)}H_{\mu\nu}\right\}. \quad (13.123)$$

where

$$^{(3)}H_{\mu\nu} = R^{\alpha\beta}R_{\alpha\mu\beta\nu} + \tfrac{1}{12}R^2 g_{\mu\nu}, \quad (13.124)$$

$$^{(1)}H_{\mu\nu} = 2g^{1/2}\frac{\delta}{\delta g^{\mu\nu}}\int R^2 g^{1/2}\,d^4x. \quad (13.125)$$

Einstein static space. This has been extensively investigated by Dowker and colleagues at zero and finite temperature (e.g. Dowker and Critchley, 1977) and by Ford (1976) from the point of view of zero point fluctuations.

de Sitter space. This space can be cast in the form of an FRW model with spaces of positive zero or negative curvature. Its high symmetry group 0(4, 1), has encouraged many investigations. These have mainly been concerned with defining states and particles in de Sitter invariant ways. These observer independent notions of particles lead either to stable states or to a constant rate of creation in each mode and hence to an infinite total creation rate. Similarly the regularized vacuum stress tensor is proportional to the metric as noted before.

This approach may be contrasted with that of Gibbons and Hawking (1977a), which was an adaptation of Hartle and Hawking's earlier treatment of black holes. They first constructed a propagator, following the Feynman integral approach. To do this they were led to work on a complexified version of de Sitter space, in fact on the Riemannian space S^4 with radius $(3/\Lambda)^{1/2}$. That is they defined a state $|0\rangle$ whose Feynman propagator was obtained by analytically continuing the unique Green's function of the elliptic equations on S^4 back to de Sitter space. Then applying Unruh's particle definition to an arbitrary geodesic world-line, they demonstrated that an Unruh detector would detect a thermal spectrum of particles at a temperature $T = (1/2\pi)(\Lambda/3)^{1/2}$. As mentioned earlier this result shows that these particles are very much 'observer dependent'. The derivation of these results is very analogous to those for the black hole case and will be dealt with later.

Anisotropic cosmological models

We have seen that it is unlikely that there was much particle creation near the singularity of FRW models, because they are conformally static. General homogeneous, anisotropic models do not share this property. The anisotropy should give rise to particle creation which, it has been suggested by Zel'dovich, might react back on its source to reduce the effect – i.e. to isotropize the models. This could account for the present high degree of isotropy of the universe. No matter what, within reason, the initial conditions were the universe would always end up isotropic. In order to give mathematical expression to these ideas Zel'dovich and Starobinsky (1972) investigated spacetimes of Kasner and Bianchi I type whose metrics may be cast in the form

$$ds^2 = -dt^2 + a^2(t)\,dx^2 + b^2(t)\,dy^2 + c^2(t)\,dz^2. \tag{13.126}$$

In vacuum

$$a = t^{p_1}; \quad b = t^{p_2}; \quad c = t^{p_3} \tag{13.127}$$

with

$$\begin{aligned} p_1 + p_2 + p_3 &= 1, \\ p_1^2 + p_2^2 + p_3^2 &= 1. \end{aligned} \tag{13.128}$$

The vacuum solution is in fact a good approximation for a matter filled model near the singularity at $t = 0$. Because of this singularity at $t = 0$ Zel'dovich and Starobinsky were unable to choose a natural initial particle state and were forced to consider a model which violated the first equations and became static before some time t_1 and after some time t_2. That is a, b, and c had constant values before and after those times. In this sandwich spacetime one may compute Bogoliubov coefficients and thus particle creation rates. They found that as t_1 tended to zero the created energy density rose as t_1^{-4}. In later papers (see, for example, Lukash, Novikov, Starobinsky and Zel'dovich, 1976) they estimated the back reaction of this created matter in a classical calculation and showed that there was indeed a tendency to isotropy. Plausible as these results are

Chapter 13. Quantum field theory in curved spacetime

physically it cannot be said that they are very satisfactory from a mathematical point of view. Other work in the same direction has been done by Berger (1975), Hu, Fulling and Parker (1974), Hu (1974), and Raine and Winlove (1975). Taub-NUT which is a special case of Bianchi type IX and also a related type VIII metric have been discussed by Lapedes (1976).

He uses propagator techniques and is able in part to circumvent in this way some of the difficulties of working in spacetimes which violate causality. Taub-NUT has closed timelike lines. The instantons discussed by Hawking (1977*b*) are complexifications of Taub-NUT. This suggests the possibility that instanton effects – such as baryon non-conservation and parity violation may be relevant in the early universe.

Plane waves

It is a well-known result in quantum electrodynamics that plane electromagnetic waves do not 'polarize the vacuum' nor create particles. This is also true for gravitational waves (Gibbons, 1975*a*; Deser, 1975). This means that for an asymptotically flat system radiating gravitational waves there will be no ambiguity in the notion of particles at large distances. The asymptotic symmetry group of general relativity, the 'BMS' group (Bondi, Van den Burg and Metzner, 1962), is the mathematical expression of the ambiguity in choosing time coordinates at future or past null infinity in the presence of gravitational radiation. It is easy to see that BMS transformations leave invariant the notion of positive frequency with respect to affine parameters on future or past null infinity (cf. Hawking, 1975). This fact is obviously related to the non-creation by gravitational radiation. It means that one has a well-defined asymptotic definition of particles, which for zero rest-mass particles is equivalent to defining positive frequency to be with respect to an affine parameter on future or past null infinity (which constitute, together with conditions at timelike infinity, Cauchy surfaces for asymptotically flat spacetimes).

Black hole spacetimes

The production of particles by black holes is by now the best understood of all the cases of particles creation. Since the early work of Starobinsky (1973), Unruh (1974) and most importantly Hawking (1974, 1975) a considerable literature on this subject has blossomed. The basic result is that black holes radiate as if they were bodies of a finite temperature

Applications

$T = \kappa/2\pi$, where $\kappa (=\frac{1}{4}M$ for non-rotating, neutral black holes) is the 'surface gravity' of the hole. This has been confirmed and derived in a number of ways and many aspects have been considerably clarified. In what follows I shall assume a basic familiarity with black hole physics (for a review of which see Carter's article in this book).

To fix notation let us consider the simplest Schwarzschild solution. This is a solution of the vacuum Einstein equations defined on $S^2 \times R^2$ with the metric

$$ds^2 = 32\frac{M^3}{r}\exp(r/2M)(-dT^2 + dX^2) + r^2(d\theta^2 + \sin^2\theta\, d\phi^2), \tag{13.129}$$

here M is the mass and r is defined by

$$X^2 - T^2 = \left(\frac{r}{2M} - 1\right)\exp\left(\frac{r}{2M}\right). \tag{13.130}$$

It is invariant under boosts in the X, T coordinates which are generated by Killing vector $K = \partial/\partial t$ whose orbits are spacelike if $|T|>|X|$ and timelike future- or past-directed if $|X|>|T|$ and $X>0$ or $X<0$ respectively. On the 2 null surfaces $X^2 = T^2$, K is null and at $X = T = 0$ it has a non-degenerate fixed point. In region I defined where K is future-directed timelike the metric may be given the form

$$ds^2 = -(1 - 2M/r)\,dt^2 + \frac{dr^2}{(1-2M/r)} + r^2(d\theta^2 + \sin^2\theta\, d\phi^2), \tag{13.131}$$

where $|K|$ has been normalized to unity at infinity. The surfaces $X = T$, $T>0$ and $X = T$, $T<0$ constitute future and past horizons $\mathcal{H}^+$ and $\mathcal{H}^-$ for region I. They lie on the boundary of validity of the r, t, chart. We may also define Schwarzschild coordinates

$$u = t - \int dr(1 - 2M/r)^{-1} \tag{13.132}$$

and

$$v = t + \int dr(1 - 2M/r)^{-1} \tag{13.133}$$

which together cover region I, and Kruskal null coordinates

$$U = T - X, \tag{13.134}$$

$$V = T + X, \tag{13.135}$$

which cover the entire manifold. u is an affine parameter along the

Chapter 13. Quantum field theory in curved spacetime

generators of $\mathscr{I}^+$ ($v = \infty$) and v an affine parameter along $\mathscr{I}^-$ ($u = -\infty$) and U and V are affine parameters along $\mathscr{H}^-$ and $\mathscr{H}^+$ respectively and the relation to u and v is

$$U = -\exp(-\kappa u), \qquad (13.136)$$

$$V = \exp(\kappa v). \qquad (13.137)$$

The surfaces $\mathscr{H}^-$ and $\mathscr{I}^-$ together with I^- (points such that $t \to -\infty$ but $v \to -\infty$) constitute (in Penrose's conformal sense) a Cauchy surface for region I. In this space we may introduce the following partial Fock bases. $B_S^\pm(\mathscr{I}^\pm)$, $B_K^\pm(\mathscr{I}^\pm)$, $B_S^\pm(\mathscr{H}^\pm)$, $B_K^\pm(\mathscr{H}^\pm)$. The notation is as follows. $^+$ or $^-$ denotes future or past. ($\mathscr{I}$) and ($\mathscr{H}$) denote the surface on which the Cauchy data specifying the bases has nonzero support (does not vanish) and S or K denote the fact that the relevant positive frequency definition is in terms of Schwarzschild or Kruskal coordinates. Thus $B_K^-(\mathscr{H}^-)$ is a basis of functions whose past Cauchy data are zero on $\mathscr{I}^-$ and are holomorphic in the lower half U plane on $\mathscr{H}^-$. These definitions only apply to zero rest-mass fields. For massive fields, the behaviour must be specified at $I^\pm$. These basis define the following 'vacuum states'

$$|0_B^\pm\rangle \equiv B_S^\pm(\mathscr{H}^\pm) \cup B_S^\pm(\mathscr{I}^\pm),$$

$$|0_{H^2}^\pm\rangle \equiv B_K^\pm(\mathscr{H}^\pm) \cup B_K^\pm(\mathscr{I}^\pm),$$

$$|0_U^\pm\rangle = B_K^\pm(\mathscr{H}^\pm) \cup B_S^\pm(\mathscr{I}^\pm)$$

where $\equiv$ means corresponds to in the sense described earlier. This does not exhaust all possibilities but the remaining state has no great physical interest. The subscripts B, H^2 and U stand for Boulware (1975a, b), Hartle and Hawking (1976), and Unruh (1976) and associated with them are Feynman propagators which bear the same names. The present notation was introduced by Gibbons and Perry (1976, 1978), cf. Fulling (1977). In fact the notation ± may be dropped from the Boulware and Hartle–Hawking vacua – the definitions are in fact equivalent (Gibbons and Perry, 1976).

We must turn from the exact solution to the physical situation at hand, the gravitational collapse of a body to form a black hole. We shall assume the collapse is spherically symmetric. Then by Birkhoff's theorem the spacetime is exactly described by the Schwarzschild solution outside the body. Inside, the geometry is time varying. The spacetime geometry can be pictured using a Penrose diagram of the complete Kruskal manifold. In this manifold we may define a basis analoguous to $B_S(\mathscr{I}^-)$ which will specify a state completely because in this case $\mathscr{I}^-$ is a Cauchy surface. We may also define future bases on $\mathcal{H}^+$ and $\mathcal{I}^+$ analogously. In view of our earlier discussion of asymptotically flat spaces it seems reasonable to regard $B_S^\pm(\mathcal{I}^\pm)$ as bases of ingoing or outgoing particle states – far from the hole a wavefunction behaving like $\exp(-i\omega t)f(r)Y_{lm}(\theta,\phi)$ will have most of the properties of a flat space particle wavefunction. Thus we ask ourselves the question if, in the real collapse situations we started in the initial state corresponding to $B_S^-(\mathcal{I}^-)$, call it $|0_c^-\rangle$, how many 'out' particles will it contain at $\mathcal{I}^+$ when analysed in terms of a basis of the form $B^+(\mathcal{H}^+) \cup B_S^+(\mathcal{I}^+)$ where $B^+(\mathcal{H}^+)$ is any basis on the future horizon. The answer to this question was ingeniously provided by Hawking (1975). Before giving his argument we remark that if we ignore the horizon states or rather 'trace over them' we obtain a density matrix as far as the $\mathcal{I}^+$ states are concerned, as mentioned earlier. This density matrix is independent of what basis is used on $\mathcal{H}^+$ and is a thermal density matrix describing uncorrelated particles flowing radially outwards (Hawking, 1975, 1976a, b; Wald, 1975; Parker, 1976b). The probability per unit time for the emission of n particles with energy ω to $\omega + d\omega$ is (for bosons)

$$p_n = \frac{1-z}{[1-(1-\Gamma)z]}\left[\frac{z}{1-(1-\Gamma)z}\right]^n \qquad (13.138)$$

$$= \sum_{p=n}^{\infty} e^{-p\omega\beta}\Gamma^n(1-\Gamma)^{p-n}\frac{p!}{n!(p-n)!} \qquad (13.139)$$

(Hawking, 1976; Bekenstein, 1975) with

$$z = \exp[-\omega\beta] = \exp[-8\pi M\omega] \qquad (13.140)$$

and Γ is the classical absorption coefficient for the mode in question.

The way in which Hawking obtained these results was essentially by noting that for radiation at late times it is only necessary to consider, when computing Bogoliubov coefficients, modes with very high frequency. This is because such modes, which pass through the star and escape to $\mathcal{I}^+$, must just graze the horizon. In order to arrive at $\mathcal{I}^+$ with a finite frequency they must have started off with a very high frequency because of the redshifting effect. This means that for the emission at late times one may use a geometric optics approximation. Now introduce a surface $\mathcal{H}^-$ which is obtained by continuing null rays backward from the end point of the future horizon – this is analogous to the past horizon in the Kruskal manifold. If σ is an affine parameter on generators of $\mathcal{H}^-$ the waves propagating from $\mathcal{I}^-$ will arrive on $\mathcal{H}^-$ with positive frequency with

Chapter 13. Quantum field theory in curved spacetime

respect to σ. Now for wave packets which are sufficiently close to the horizon σ will bear the same relation to u as does U to u in the exact Kruskal manifold. Indeed we may replace the messy gravitational collapse manifold with the much simpler Kruskal manifold but where the state in the exterior regions is specified by its behaviour on $\mathcal{H}^-$ and $\mathcal{I}^-$ together, and where the appropriate initial vacuum state is the Unruh vacuum.

Note that without reference to the realistic collapse we have no way of deciding what boundary conditions to impose on the past horizon. Without such means the question which is sometimes asked 'does an eternal black hole radiate' is unanswerable. Physically this is not unreasonable. An eternal black hole is a white hole or naked singularity in the past and without some additional physical or theological input into the theory we have no way of deciding what such an object will do.

To conclude we may say that Hawking showed that gravitational collapse brings about a state which can be described in the exact Kruskal manifold by the Unruh vacuum. It is now in principle a simple matter to determine the Bogoliubov transformation connecting $|0_U^-\rangle$ with, say $|0_U^+\rangle$ and to evaluate the density matrix ρ, obtained by averaging over the horizon states. Let $f_S(\mathcal{I}^+)$, $p_K(\mathcal{H}^-)$ etc. be positive-frequency members of $B_S^+(\mathcal{I}^+)$, $B_K^-(\mathcal{H}^-)$ etc. We will also assume that the Schwarzschild wavefunctions have frequencies ω then we may write

$$f_S(\mathcal{I}^+, \omega) = x(\omega)p_S(\mathcal{H}^-) + y(\omega)p_S(\mathcal{I}^-) \tag{13.141}$$

where $x(\omega)$ and $y(\omega)$ are complex numbers depending on ω and obeying

$$1 = |x|^2 + |y|^2 \tag{13.142}$$

and clearly $|x|^2 = \Gamma(\omega)$ the absorption coefficient for the mode in question. We may also write

$$p_S(\mathcal{H}^-) = \cosh\theta p_K(\mathcal{H}^-) + \sinh\theta \bar{p}_K(\mathcal{H}^-) \tag{13.143}$$

where $p_K(\mathcal{H}^-)$ are appropriately defined positive and negative frequency functions, in terms of the Kruskal coordinate U. $f_S(\mathcal{H}^-)$ is proportional to $\exp(-i\omega u)$ for $-\infty < u < \infty$, that is for $U < 0$, and vanishes for $U > 0$. We thus need to Fourier analyze the function $\theta(-U)(-U)^{i\omega/\kappa}$. One finds that

$$\theta(-U)(-U)^{i\omega/\kappa} \int_0^\infty \{\alpha_{\omega\omega'} \exp(-i\omega' U) + \beta_{\omega\omega'} \exp(+i\omega' U)\} d\omega' \tag{13.144}$$

with

$$|\alpha_{\omega\omega'}| = \exp\left(\frac{\omega\pi}{\kappa}\right)|\beta_{\omega\omega'}| \tag{13.145}$$

Applications

and so
$$\tanh\theta = \exp\left(-\frac{\omega\pi}{\kappa}\right). \tag{13.146}$$

Clearly if $x = 1$ we have the standard thermal Bogoliubov transformation discussed earlier representing thermal radiation. The non-trivial absorption coefficients slightly modify the analysis, however the overall transformation

$$f_S(\mathscr{I}^+) = \frac{z}{(1-z)^{1/2}} f_K(\mathscr{H}^-) + y f_S(\mathscr{I}^-) + \frac{1}{(1-z)^{1/2}} \bar{f}_K(\mathscr{H}^-) \tag{13.147}$$

shows that
$$|\beta_{ij}|^2 = \frac{\Gamma(\omega)}{1 - \exp(-\omega\beta)}. \tag{13.148}$$

And the number of particles created is
$$N = \frac{\Gamma(\omega)}{1 - \exp(-\omega\beta)} = \frac{\Gamma(\omega)}{1-z}. \tag{13.149}$$

This is the number in each mode. With more care as to the exact definition of these modes which should be wave packets one may construct wave packets equi-spaced in time so in effect $\Gamma(\omega)/[1 - \exp(-\omega\beta)]$ represents an emission rate per unit time per unit bandwidth. This is essentially an application of the Golden Rule in elementary quantum mechanics.

For rotating and charged holes similar results are valid. Again one can introduce Kruskal-like coordinates to cover the horizon of the Kerr–Newman solution. However one must also change the angular coordinate and the electromagnetic gauge. Let us consider the Kerr solution, the angular coordinate ϕ must be changed to $\tilde{\phi}$ defined by

$$\tilde{\phi} = \phi - \Omega_H t \tag{13.150}$$

where Ω_H is the angular velocity of the black hole. $\tilde{\phi}$ is constant along the null generators of the horizon. (More details of these concepts may be found in Carter's chapter in this volume or in the 1972 Les Houches lectures on Black Holes, C. DeWitt and B. DeWitt, 1973.) One may introduce Kruskal coordinates U and V which are related in the same way as in (13.136), (13.137) to the null coordinates u and v and are respectively affine parameters on the horizons and null infinity. Of course κ in (13.136), (13.137) is now the surface gravity of the rotating black hole. One then finds that the analogous quantity to z in (13.149) is

$$z = \exp[-(\omega - \Omega_H m)2\pi/\kappa] \tag{13.151}$$

Chapter 13. Quantum field theory in curved spacetime

where m is the angular momentum of the mode concerned. That is the hole behaves as if $m\Omega_H$ is the chemical potential for a mode with angular momentum m. In a similar way for charged holes the corresponding chemical potential is $e\Phi_H$ where e is the charge and Φ_H the electrostatic potential of the hole. That is, the rate is

$$\frac{\Gamma}{1-\exp(\omega-\mu)\beta} \tag{13.152}$$

with

$$\mu = m\Omega_H + e\Phi_H. \tag{13.153}$$

For fermions one of course obtains the correct Fermi–Dirac thermal factor. The modes for which $\omega < \mu$ correspond to so called super-radiant modes for which, for bosons, Γ is negative. It was these modes which first suggested that black holes might emit particles because they are associated with the existence of an ergo region. That is a region where the Killing vector $\partial/\partial t$ which is timelike at infinity becomes spacelike. Such a negative energy region is reminiscent of the deep potential wells that occur in the Klein Paradox. The super-radiance is easily seen by noting that a mode proportional to $\exp(-i\omega t + im\phi)$ carries a flux through the horizon which is proportional to $j_\mu l^\mu$ where j_μ is the current defined by (13.24) and l^μ is the null generator of the horizon

$$l^\mu \frac{\partial}{\partial x^\mu} = \frac{\partial}{\partial t} + \Omega_H \frac{\partial}{\partial \phi}. \tag{13.154}$$

This in turn is proportional to $(\omega - \mu)$. Put another way modes of this form may be put in the form $\exp[-i(\omega - n\Omega_H)t + im\tilde{\phi}]$ and so have a frequency along the null generators of the horizon of $\tilde{\omega} = \omega - n\Omega_H$ (for charged fields the relevant frequency is $\omega - e\Phi$ where Φ is the potential) and so the effective frequency at the horizon is again $\omega - \mu$. Clearly modes of positive $\tilde{\omega}$ can have negative ω and this frequency mixing leads to emission in the super-radiant modes even from the state analogous to $|0_B^-\rangle$. The analogous Bogoliubov coefficient is just the super-radiant coefficient $\Gamma(\omega)$. This super-radiant emission was first discussed from the point of view of second quantization by Unruh (1974) for rotating black holes and by Gibbons (1975a) for charged black holes. A different more intuitive approach to these super-radiant modes has been worked out by Ruffini, Damour and Deruelle based on the analogy by the Klein Paradox and two overlapping Dirac seas of particles and antiparticles. The two seas are displaced by an amount 2μ and if 2μ exceeds the energy gap between positive and negative frequencies (2 times the rest-mass), which is called 'level crossing', spontaneous emission results. A review of this approach has been given by Damour (1977), see also Gibbons (1977). For other aspects of these super-radiant modes and the application of the ideas of stimulated and spontaneous emission see Wald (1976), Bekenstein and Meisels (1977).

The precise rate of emission depends upon the detailed form of the emission coefficients $\Gamma(\omega)$. These cannot be obtained analytically. Page (1976a, b) has done some extensive numerical computations of the emission rates. However on general grounds it is clear that the rate is inversely proportional to the square of the mass. Thus if the back reaction to the creation is to cause the black hole to lose mass its emission rate will increase and the black hole would eventually explode – hence the title of Hawking's original paper. However, the fact that the emission is thermal, together with the fact that these particle creation calculations fit in so well with the classical laws of black hole thermodynamics described in Carter's article leads to the idea of a thermal equilibrium of a black hole and its emitted products and it is to this topic that we now turn.

The classical second law of black hole mechanics

$$dM = \frac{\kappa \, dA}{8\pi} + \Omega_H \, dJ + \Phi_H \, dQ \qquad (13.155)$$

where A is area of the horizon, J the angular momentum and Q the charge, implies that if T is given by

$$T = \frac{\kappa}{2\pi} \qquad (13.156)$$

then the entropy $S = \tfrac{1}{4} A$. Using the Smarr formula

$$\tfrac{1}{2} M = \kappa A / 8\pi + \Omega_H J + \tfrac{1}{2} \Phi_H Q \qquad (13.157)$$

one also deduces that the thermodynamic potential of this system, W, with chemical potential terms Ω_H and Φ_H, is given by

$$W = M - TS - \Omega_H J - \Phi_H Q = \tfrac{1}{2}(M - \Phi_H Q). \qquad (13.158)$$

Using these expressions one easily deduces on thermodynamic grounds that for a suitable range of parameters a stable thermodynamic equilibrium is possible between a black hole and a gas of radiation. (Hawking, 1976b; Gibbons, 1977; Gibbons and Perry, 1978). It should then be possible to find a state, the Gibbs state, which describes this situation, at the level of external field theory. In fact this can be done. The Gibbs state is the Hartle–Hawking vacuum and the Hartle–Hawking propagator is

Chapter 13. Quantum field theory in curved spacetime

the associated thermal Green's function (Gibbons and Perry, 1976). To see this in a simple way recall the well-known fact that a thermal Green's function for a system at temperature T and chemical potential μ obeys the following relation:

$$G_\beta(x, x', t-t') = e^{-\mu\beta} G_\beta(x, x', t-t'+i\beta)(-1)^{2s} \qquad (13.159)$$

where s is the spin of the particles. That is for $\mu = 0$, $G_\beta(x, x', t-t')$ is periodic in the imaginary time coordinate $\tau = i(t-t')$. Now the Hartle–Hawking (1976) propagator was originally constructed by the Riemannization method. That is it is the analytically continued version of a propagator which satisfies elliptic differential equations on the 'Riemannian Schwarzschild solution'. This is the positive definite, complete Ricci flat space with topology $R^2 \times S^2$ and metric

$$ds^2 = \frac{32}{r} M^3 e^{r/2M} (dX^2 + dY^2) + r^2(d\theta^2 + \sin^2\theta \, d\phi^2) \qquad (13.160)$$

with

$$X^2 + Y^2 = \left(\frac{r}{2M} - 1\right) e^{r/2M}. \qquad (13.161)$$

(13.160) is clearly invariant under rotations about $X = Y = 0$, which is a fixed point of this group of transformations which is generated by the Killing vector $\partial/\partial\tau$ where

$$X = \left(\frac{r}{2M} - 1\right)^{1/2} e^{r/4M} \cos \kappa\tau, \qquad (13.162)$$

$$Y = \left(\frac{r}{2M} - 1\right)^{1/2} e^{r/4M} \sin \kappa\tau. \qquad (13.163)$$

Comparison with (13.129), (13.130) and (13.131) shows that $\tau = it$ and $Y = iT$. Evidently τ is an angular coordinate. Points whose τ coordinate differs by $2\pi/\kappa$ are identified. The point $r = 2M$, $X = Y = 0$ is just the 'axis' – a fixed point set of this group of isometries. The periodicity of the coordinate τ means that the analytically continued propagator has all the properties of a thermal Green's function. Using the analytic properties established by Hartle and Hawking one may establish that this propagator corresponds to the Hartle–Hawking vacuum as defined above (Gibbons and Perry, 1978). As was pointed out by Gibbons and Perry (1976) this correspondence should hold even in the presence of particle interactions, thus extending the linear results obtained hitherto.

Applications

It is this basic periodicity which ties together the geometrical properties of the horizon and the thermodynamic character of the Hawking emission. Not surprisingly the analysis can be extended to spin s fields and to rotating and charged black holes. (Hartle and Hawking, 1976; Gibbons and Perry 1976, 1978). It turns out to be necessary to first perform the coordinate transformation to coordinates t, $\tilde{\phi}$ which corotate with the hole (13.150). One then performs the Wick rotation $t = i\tau$. This leads to a complex Riemannian metric which near the horizon is quasi-Riemannian in the sense defined earlier. The complex metric is *analytic* if τ is again interpreted as an angle with period $2\pi/\kappa$. Thus the Hartle–Hawking propagator will be periodic in this coordinate. Now consider the propagator restricted to modes of a fixed angular momentum n. Call this $G^n_\beta(x, x', \tau)$. G^n_β will be of the form

$$G^n_\beta(x, x', t-t') = \exp[in(\tilde{\phi} - \tilde{\phi}')]\tilde{G}^n_\beta(r, \theta, r', \theta', t-t') \tag{13.164}$$

where $\tilde{G}^n_\beta$ is periodic in $t - t'$ with period $2\pi i/\kappa$. Re-introducing the coordinates t, ϕ using (13.150) we see that G^n_β obeys the relation

$$G^n_\beta(t - t' + 2\pi i/\kappa) = \exp(2\pi n \Omega_H/\kappa) G^n_\beta(t - t') \tag{13.165}$$

which once again establishes μ as the chemical potential of the mode and $2\pi/\kappa$ as the temperature.

Gibbons and Hawking (1977b) have extended this analysis. They note that in the functional approach thermal averages are obtained by summing over configurations which are periodic in imaginary time. Further the generating functional Z has the property of being a partition function in that case. Thus one has a means of evaluating thermodynamic expressions for this system directly. To obtain the partition black hole thermodynamic quantities, like the entropy, in this way involves a quantum gravity calculation which is beyond the scope of this article. The reader is referred to the original paper and to Hawking's chapter in this volume for more details. However if we restrict ourselves to the external field theory case we see that to evaluate the stress tensor of the fields in this Hartle–Hawking state we are led to the evaluation of the functional determinant of an *elliptic* operator. This can be treated with the zeta-function technique described earlier (Hawking, 1977a) and yields quite finite answers which however depend on μ^2, the renormalization mass, instead of the divergent quantity one would get if one simply computed the density of a gas at the Tolman redshifted temperature $T = T_H/g_{00}^{1/2}$ which would diverge at the horizon.

Chapter 13. Quantum field theory in curved spacetime

The extension of these results to other spaces with horizons, including the thermodynamic aspects is straightforward. For instance for de Sitter space one finds that each inertial observer (timelike geodesic) has an event horizon which arises because the cosmic repulsion forces all bodies to move apart with ever increasing speed. Each observer is contained in a local static region in which the metric takes the form

$$ds^2 = -dt^2(1-\Lambda r^2/3) + \frac{dr^2}{1-\Lambda r^2/3} + r^2(d\theta^2 + \sin^2\theta\, d\phi^2) \qquad (13.166)$$

$r = (\Lambda/3)^{1/2}$ is this cosmological horizon. The surface gravity is readily shown to be $(\Lambda/3)^{1/2}$. The complexification is of course the 4-sphere. $\tau = it$ has period $2\pi(3/\Lambda)^{1/2}$ and temperature $(\Lambda/3)^{1/2}/2\pi$ (Gibbons and Hawking, 1977a).

One interpretation of the creation of particles by black holes has been in terms of the following simple model: virtual pairs are continually being spontaneously created and annihilated. To conserve energy one must carry positive energy and the other negative energy. If a deep enough potential well is present, so that there exist timelike curves with negative energy, one member of the pair can move along these curves, remaining trapped in the well while the other member can escape to infinity with positive energy. This gives an intuitively pleasing picture of creation by horizons and ergo-regions. In both cases there are regions where $\partial/\partial t$ is spacelike and hence the energy $-K\mu m\, dx^\mu/d\lambda = E$ can become negative. The loss of information about the absorbed pair gives rise to a randomness in the emitted particle spectrum. One can also view this process as the quantum tunnelling of particles out from behind the horizon. To do so such particles would classically have to move on spacelike or past directed timelike paths – i.e. 'backwards in time'. One might ask to what extent this intuitive picture is supported by theoretical investigations. To some extent the Bogoliubov coefficient computations bear this view out, the original state is expressible as a superposition of pair states, one particle moving out to infinity and the other into the horizon. More support for this view comes from the Hartle–Hawking and Gibbons–Hawking propagator computations where explicit amplitudes for such quantum tunnelling processes are computed. However these give no local information as to 'where the particles come from'. One should not expect too much in this direction since the emitted particles have wavelengths comparable to the emitting region – the horizon. The localization of particles to such small regions is impossible even in flat space. Nevertheless attempts have been made to give a more detailed picture of the Hawking radiation by evaluating regularized stress tensors representing the emitted products. Most noticeable of attempts along those lines has been the work of Davies, Fulling and Unruh (1976) and Christensen and Fulling (1977). The former are 2-dimensional calculations and the latter are fully 4-dimensional. The first is a 2-dimensional point splitting calculation which provides a 2-dimensional stress tensor for the massless scalar field which is conserved and has a 2-dimensional trace anomaly. The flux carried to infinity is the positive thermal value and that through the horizon is indeed negative. In the 4-dimensional case $\langle T_{\mu\nu} \rangle$ is shown to be determined up to one function of radius by the condition that it be conserved, spherically symmetric, have the correct trace and be well behaved on the future horizon. Again a negative flux of energy passes through the horizon, as it must on general grounds.

13.4 Conclusion

In this article I have tried to show how one attempts to extend flat space quantum field theory, in particular linear theories with no mutual and self interactions, to curved space. One meets certain problems – the ambiguity of the no-particle state and the divergences. Nevertheless these problems can to some extent be circumvented and the theory yields interesting physical predictions. Most success has been attained in the case of the black hole emission process and to a lesser extent in the cosmological case. Other problems however remain – the non-renormalizability of the theory and the back reaction problem. Neither of these can be properly tackled without quantizing the gravitational field, and the former cannot even be solved in that way if one adopts standard procedures as is made clear in other articles in this volume. Thus a cloud of uncertainty hangs above the subject. Nevertheless it is difficult to believe that the broad features which I have sketched out and particularly the beautiful tie up between thermodynamics and black hole physics will not form an integral part of whatever final theory gains acceptance.

14. Quantum gravity: the new synthesis

BRYCE S. DEWITT†

14.1 Introduction

In an address delivered in 1920 at the University of Leyden, Einstein referred to the 'wonderful simplification of theoretical principles' achieved by H. A. Lorentz in 'taking from the ether its mechanical... qualities'. He described the situation thus:

> According to Lorentz... the ether, and not matter, is the exclusive seat of electromagnetic fields... The elementary particles alone are capable of carrying out movements; their electromagnetic activity is entirely confined to the carrying of electric charges.... It may be said of [the ether], somewhat jokingly, that immobility is the only mechnical property of which it has not been deprived by Lorentz. It may be added that the change in the conception of the ether brought about by the special theory of relativity consists solely in removing from the ether its last mechanical quality, its immobility.‡

Einstein pointed out that it was possible to assume the following:

> The ether does not exist at all. The electromagnetic fields are not states of a medium... but are independent realities... not reducible to anything else... like atoms. This conception suggests itself the more readily as electromagnetic radiation, like ponderable matter, carries momentum and energy and, according to the special theory of relativity, both matter and radiation are but special forms of distributed energy.

This was the point of view subsequently adopted by many theorists. Einstein himself resisted it, citing the following homely example:

> Think of waves on the surface of water. Here we can describe two entirely different things. Either we may observe how the undulatory surface forming the boundary between water and air alters in the course of time; or else – with the help of small floats – we can observe how the positions of separate particles of water change. If the existence of such floats for tracking the motion of the particles of a fluid were a fundamental impossibility in physics – if, in fact, nothing else whatever were observable than the shape of the space occupied by the water..., we should have no ground for assuming that water consists of movable particles. But all the same we could characterize it as a medium.

---

† This work has been supported in part by a research grant from the National Science Foundation, by a travel grant from the North Atlantic Treaty Organization, and by organized research funds of the University of Texas.
‡ The original German will be found in Einstein (1933).

Introduction

Einstein's meaning is clear. According to the general theory of relativity spacetime itself is the medium. 'To deny the ether is to assume that empty space has no physical qualities whatever.' General relativity not only restores dynamical properties to empty space but also ascribes to it energy, momentum and angular momentum. In principle, gravitational radiation could be used as a propellant. Since gravitational waves are merely ripples on the curvature of spacetime, an anti-etherist would have to describe a spaceship using this propellant as getting something for nothing – achieving acceleration simply by ejecting one hard vacuum into another. This example is not as absurd as it sounds. It is not difficult to estimate that a star undergoing asymmetric (octupole) collapse may achieve a net velocity change of the order of 100 to 200 km s^{-1} by this means.

But Einstein's conception went beyond this. According to him curvature is not the only texture that the ether possesses. It must have other, more subtle textures which, like curvature, are best described in the language of differential geometry and differential topology. He laid down the following challenge to his Leyden audience:

> As to the part the new ether is to play in the physics of the future we are not yet clear. We know that it determines the metrical relations in the spacetime continuum, ... but we do not know whether it has an essential share in the structure of the elementary particles ... It would be a great advance if we could succeed in comprehending the gravitational field and the electromagnetic field together as one unified conformation.

Einstein's attack on the unified field problem, and his failure, are well known. The equal failure of others of high stature – Weyl, Klein, Pauli, to name but three – led to a strong reaction among theorists and to a turning away from such problems for many years. Yet the dream never wholly died. Two features of Einstein's conception remained permanently compelling: the potential richness of a geometrically based reality, and the predictive power of theories based on local invariance groups. These ideas have survived the explosive arrival in 1926 of the golden age of quantum mechanics, the great era of quantum electrodynamics, the subsequent disillusionment with quantum field theory, and the turning in despair to the applied arts of dispersion theory.

In the 1960s, fundamental physics – in Einstein's sense – began to revive. At first, interest focused on global invariance groups such as the Poincaré group and the isotopic spin and SU(3) groups relevant to experimental hadron physics. But the basic weak-interaction coupling had become known, and parity violation had been discovered. Therefore

Chapter 14. Quantum gravity: the new synthesis

hadron ideas also began to be applied to leptons. Current algebra – a frank return to field-theory concepts – began to make headway. In the meantime activity of a different sort was taking place.

In 1954 Yang and Mills had converted the isotopic spin group to a local group by introducing three new vector fields coupled to each other and to matter in a way that straightforwardly generalizes the coupling of electrodynamics. It was quickly recognized that the Yang–Mills idea could be extended to any compact Lie group, and an infinite class of 'non-Abelian gauge theories' was born. Utiyama (1956) showed that Einstein's theory of gravity too could be viewed as a non-Abelian gauge theory, the chief difference being that the gravitational gauge group – the group of diffeomorphisms of spacetime – has a richer and more complicated structure than the Yang–Mills groups. Conversely, non-Abelian gauge theories can be expressed in geometrical, even metrical, terms (see DeWitt, 1965, problem 77).

Although there was no experimental evidence for Yang–Mills fields in the 1950s and early 1960s, the gravitational field still existed as a stubborn fact. Unfortunately, it was viewed as a kind of anomaly of Nature in all domains but astrophysics. Only a few individuals, sustained by a belief in the essential unity of Nature, regarded such an attitude as foolish. They undertook to draw gravity theory back to the central position in physics that it occupied in the time of Newton, by quantizing it. In 1979, one hundred years after Einstein's birth, it is possible to glimpse where this effort, which now attracts the interest of many, may be leading.

For at least a decade quantum gravity occupied itself almost exclusively with formal questions.† It was not until 1962 that the first major advance was made permitting reliable computations of purely quantum effects to be carried out. Feynman (1963) showed, in one-loop order, that the naive perturbation rules used in quantum electrodynamics violate unitarity when applied to quantum gravity and other non-Abelian gauge theories. To correct for this one must add to every closed-loop graph a compensatory graph involving a closed, integral-spin, fermion loop. Another half a decade passed before it was discovered how to extend Feynman's prescription to all orders (DeWitt, 1967b; Faddeev and Popov, 1967).

In the meantime a major advance was taking place in the theory of weak interactions. Using a mechanism invented by Higgs (1964) for inducing symmetry breaking, Weinberg (1967) and Salam (1968) proposed what has subsequently developed into the first successful unified

---

† For a brief account of the work of this early period see DeWitt (1967a).

The quantum ether

field theory. It is not the unified field theory of which Einstein was dreaming in Leyden in 1920, for it unites electromagnetism with the weak interactions rather than with gravity. Nevertheless, because of the central role it gives to local gauge groups, it vindicates Einstein's basic geometrical viewpoint.

Because the broken symmetry of the Weinberg–Salam theory arises solely from vacuum degeneracy (analogous to the degenercy of the ground state of a ferromagnet) it leaves the local gauge group, which is a Yang–Mills group, intact. Using this fact 't Hooft (1971) was able to prove that the theory is renormalizable and therefore makes computable predictions, which can be tested by experiment. In his proof 't Hooft used the calculational rules that had been discovered in the effort to quantize gravity. This is worth noting because it marks the first occasion in modern times in which research in gravity theory has had a direct impact on another area of physics. Since then gravity theory has been moving rapidly back into the main stream.

Near the end of his Leyden lecture Einstein uttered some words of caution: 'In contemplating the ... future of theoretical physics we ought not to reject the possibility that the facts comprised in the quantum theory may set bounds to field theory beyond which it cannot pass'. These words were spoken eight years before the birth of quantum field theory, and therefore Einstein could not know that the quantum theory, far from setting bounds on field theory, would transmute and enrich it. Yet he was right in supposing that it would introduce severe new problems. Chief among these is the problem of ambiguities associated with the divergences that arise in the course of perturbation–theoretical calculations, a problem that is particularly difficult in the case of the gravitational field. The ambiguities of quantum gravity cannot presently be removed solely on the basis of information secured (in principle) from a finite number of experiments. Technically, quantum gravity is non-renormalizable.

Some aspects of this problem will be reviewed in a later section. One cannot, however, look at it in proper perspective, that is to say, with the proper optimism, without first noting the ways in which the general theory of relativity and the quantum theory of fields, taken together in a new synthesis, have thus far enriched each other.

14.2 The quantum ether

One of the most striking examples of mutual enrichment is to be found in the impact that the ideas of general relativity have had upon the concept

Chapter 14. Quantum gravity: the new synthesis

of 'the vacuum', and, conversely, in the reinforcement that quantum field theory has given to the idea that the vacuum may be viewed as a textured ether. It has been known from the earliest days of quantum electrodynamics that field strengths, in the vacuum state, undergo random fluctuations completely analogous to the zero-point oscillations of harmonic oscillators, and that when couplings to the electron field are taken into account these fluctuations are accompanied by pair-creation and annihilation events. The vacuum is thus in a state of constant turmoil.

From Einstein's point of view it would be natural to regard field fluctuations as having their seat in the ether and contributing to it qualities additional to its geometrical properties. A mathematical description of the vacuum that effectively embodies this idea was given years ago by Schwinger (1951). In the presence of an external source, a quantized field initially in the vacuum state need not stay in that state. Schwinger showed that all physical properties of the field can be derived from a knowledge of how the probability amplitude for the field to remain in the vacuum state varies as the source is changed. Functional derivatives of the vacuum-to-vacuum amplitude, with respect to the source, are response functions that describe how the ether reacts to external stimuli. The ether itself thus contains a complete blueprint for the field dynamics.

The ether may be probed by other means than sources. One may vary boundary conditions and external fields. For example the vacuum-to-vacuum matrix element of the stress tensor $T^{\mu\nu}$, of any combination of fields, including the gravitational field, is given by the functional derivative

$$\langle \text{out, vac}|T^{\mu\nu}|\text{in, vac}\rangle = -2\mathrm{i}\frac{\delta}{\delta g_{\mu\nu}}\langle \text{out, vac}|\text{in, vac}\rangle. \qquad (14.1)$$

Here $|\text{in, vac}\rangle$ and $|\text{out, vac}\rangle$ are the initial and final vacuum state vectors respectively and $g_{\mu\nu}$ is an external metric field, frequently referred to as *background field*, which serves as an arbitrary zero point for the quantum fluctuations of the gravitational field, and which can be used to fix the topology of the spacetime manifold. It is assumed in (14.1) that the 'vacuum' states are unambiguously (although not necessarily uniquely) defined relative to the background, and that topological transitions can only be described (if indeed they occur at all) by allowing $g_{\mu\nu}$ to move into the complex plane. It is also assumed that renormalizations have been carried out to eliminate any divergences that may arise.

The analogous equation in quantum electrodynamics is

$$\langle \text{out, vac}|j^{\mu}|\text{in, vac}\rangle = -\mathrm{i}\frac{\delta}{\delta A_{\mu}}\langle \text{out, vac}|\text{in, vac}\rangle, \qquad (14.2)$$

where j^μ is the current vector, and A_μ is the vector potential of an external or background electromagnetic field. Expression (14.2) is generally non-vanishing whenever the background field is non-vanishing — a phenomenon known as *vacuum polarization*. Similarly, expression (14.1) is generally different from zero whenever the background geometry is curved. But curvature is not the only source of gravitational 'vacuum polarization'. *Topology also contributes.* This means that the properties of the ether depend on the whole manifold!

This fact is so striking that it is worth looking at in some detail. The phenomenon was first discovered in the Casimir effect. In the course of computing Van der Waals forces between very close molecules, Casimir (1948) found that the interaction energy could be expressed as a sum of terms involving, in addition to the molecular separation distance and internal molecular parameters, powers of the curvature of the molecular surface. One term, however, depended on neither the curvature nor the molecular details. Its presence implied that an attractive force must exist between any two parallel flat conducting surfaces in a vacuum. The effect was soon verified in the Philips laboratories. Because the force is so tiny, great care had to be taken to insure that the surfaces were absolutely clean, neutral, and microflat so that they could be brought nearly into contact without other effects intervening.

The relevant field in the Casimir effect is the electromagnetic field, and the manifold involved is the plane-parallel *slab* between the conducting surfaces. In the mathematical terminology this is an *incomplete* manifold. Nothing prevents one from doing physics on an incomplete manifold provided one is supplied with appropriate boundary conditions — perfect-conductor boundary conditions in this case. The properties of the ether between the conductors are entirely determined by the field Green's functions (response functions) appropriate to the slab.

Consider first the infinite Minkowski vacuum — the standard vacuum of particle physicists. $T^{\mu\nu}$, the operator describing the energy, momentum and stresses in the electromagnetic field, is formally a bilinear product of operator-valued distributions (the field operators) and hence is meaningless. It is given meaning by a subtraction process that sets the expectation value $\langle T^{\mu\nu} \rangle$ equal to zero in the Minkowski vacuum. This subtraction corresponds to ignoring the zero-point energy of the field oscillators, but it is by no means arbitrary. $\langle T^{\mu\nu} \rangle$ *must* vanish in an empty Minkowski spacetime if quantum field theory is to be ultimately consistent with general relativity. The Minkowski vacuum serves as a standard to which all other vacua, whenever possible, are to be compared.

Chapter 14. Quantum gravity: the new synthesis

Now suppose a single, infinite plane conductor is introduced into the Minkowski vacuum. One may imagine it to be brought adiabatically from infinity so that the field suffers no excitation but remains in its ground state. The manifold of interest has become an infinite half-space. Introduce Minkowski coordinates x^μ, $\mu = 0, 1, 2, 3$, oriented so that x^3-axis is perpendicular to the plane of the conductor. By considerations of symmetry it is clear that $\langle T^{\mu\nu}\rangle$ must be diagonal and independent of x^0, x^1 and x^2. Moreover, because a perfect conductor remains a perfect conductor in any state of motion parallel to its surface, the vacuum stresses in its vicinity must look the same no matter how rapidly one is skimming over its surface. That is to say, the ether always keeps its relativistic properties, and hence $\langle T^{\mu\nu}\rangle$ must be invariant under Lorentz transformations that correspond to boosts parallel to the (x^1, x^2)-plane. This means that the first three rows and columns of $\langle T^{\mu\nu}\rangle$ must be proportional to the metric tensor of a $(2+1)$-dimensional Minkowski space, namely diag $(-1, 1, 1)$. If, to this inference, one adds the observation that $T_\mu{}^\mu = 0$ in the case of the electromagnetic field, one concludes that $\langle T^{\mu\nu}\rangle$ has the form

$$\langle T^{\mu\nu}\rangle = f(x^3) \times \text{diag}\,(-1, 1, 1, -3). \tag{14.3}$$

But that is not all. The form of the function $f(x^3)$ too may be deduced. For this one invokes the conservation law $\langle T^{\mu\nu}{}_{,\nu}\rangle = \langle T^{\mu\nu}\rangle_{,\nu} = 0$. In particular

$$0 = \langle T^{3\nu}\rangle_{,\nu} = -3f'(x^3), \tag{14.4}$$

which implies that f is a constant, independent of x^3. Now $\langle T^{\mu\nu}\rangle$ has the dimensions of energy density. The only fundamental constants that enter into the theory are $\hbar$ and c. To get a constant having the dimensions of energy density one needs also a unit of length, mass, or time. No natural units with these dimensions exist in the present problem. Therefore one can only conclude that $f = 0$ and hence $\langle T^{\mu\nu}\rangle = 0$ in an infinite half-space.

This conclusion is, in fact, confirmed by explicit calculation. The Green's function for an infinite half-space is readily constructed from the Minkowski Green's function by the method of images. $\langle T^{\mu\nu}\rangle$ is then obtained by appropriately differentiating this Green's function and bringing into coincidence the spacetime points on which it depends. The result, of course, diverges and must be 'renormalized' by subtraction of the corresponding Minkowski result. This is equivalent to subtracting the Minkowski Green's function from the half-space Green's function. Although the resulting 'renormalized' Green's function does not itself vanish it nevertheless yields $\langle T^{\mu\nu}\rangle = 0$.

All the above arguments concerning the form of $\langle T^{\mu\nu}\rangle$ hold equally well for the slab manifold, except that there is now a natural unit of length – the separation distance, a, between the parallel conductors. In the region between the conductors, therefore, we expect

$$\langle T^{\mu\nu}\rangle = f(a) \times \text{diag}\,(-1, 1, 1, -3). \tag{14.5}$$

The form of the function $f(a)$ may be determined by considering the work required to separate the conductors adiabatically. From the infinite half-space analysis one knows that the conductors experience no forces from the outside. There is an internal force, however, of amount $3f(a)$ per unit area, tending to pull them together. If the conductors are moved a distance da farther apart an amount of work $dW = 3f(a)\,da$, per unit area must be supplied. This must show up as an increase in the energy per unit area. $E = -af(a)$. Setting $dW = dE$ and integrating, one immediately obtains

$$f(a) = A/a^4, \tag{14.6}$$

where A is some universal constant.

The form (14.6) may also be inferred by dimensional analysis. The only combination in which $\hbar$, c and a can be united to yield an energy density is $\hbar c/a^4$. We shall henceforth set $\hbar = c = 1$. The constant A is then a pure number.

The evaluation of A requires explicit computation. Again the Green's function can be constructed by the method of images and again the 'renormalized' Green's function is obtained by subtracting off the Minkowski Green's function. The renormalized $\langle T^{\mu\nu}\rangle$ no longer vanishes. The anticipated form (14.5), (14.6) is confirmed and A is found to have the value $\pi^2/720$. Within expected errors this value is in agreement with experiment.

It will be observed that the energy density in the ether between the conductors is *negative*. It is a tiny energy, too small by many orders of magnitude to produce a gravitational field that anybody is going to measure. Yet one can easily construct gedanken experiments in which the law of conservation of energy is violated unless this energy is included in the source of the gravitational field. It turns out that the energy density in the quantum ether is often negative. The quantum theory therefore violates the hypotheses of the famous Hawking–Penrose theorems concerning the inevitability of singularities in spacetime, which imply the ultimate breakdown of classical general relativity.

Chapter 14. Quantum gravity: the new synthesis

14.2.1 Other topologies

In order for $\langle T^{\mu\nu} \rangle$ to be non-vanishing in a flat empty spacetime it is not necessary for the manifold to be incomplete, as in the Casimir effect. Complete manifolds also exhibit the phenomenon: for example, the manifolds $R \times \Sigma$, where the 'slices', Σ, are flat, spacelike Cauchy hypersurfaces having any one of the following topologies: $R^2 \times S^1$, $R \times T^2$, T^3, $R \times K^2$, etc., where T^n is the n-torus, K^2 is the 2-dimensional Klein bottle, etc. The case $\Sigma = R^2 \times S^1$ has the closest resemblance to the Casimir example. The only difference is that, instead of imposing conductor boundary conditions on the faces of a slab, one imposes periodic boundary conditions. $\langle T^{\mu\nu} \rangle$ again takes the form (14.5), (14.6), where a is now the period of the coordinate x^3. Again the renormalized Green's function is easily computed, and one finds in this case $A = \pi^2/45$. The cases $\Sigma = R \times T^2$ and $\Sigma = T^3$ are more complicated. Although $\langle T^{\mu\nu} \rangle$ is coordinate-independent its form is no longer given by (14.5) and (14.6) but depends on the ratios of the various coordinate periodicities. In the case $\Sigma = R \times K^2$, $\langle T^{\mu\nu} \rangle$ is not even coordinate-independent but is itself periodic (and smooth).†

One advantage in studying quantum field theory on complete manifolds is that any field may be selected. One does not have to worry about the question: what boundary conditions are analogous to perfect-conductor boundary conditions for the electromagnetic field? Scalar and spinor fields, and even the gravitational field, may be introduced. The spinor field is of particular interest because on some manifolds, for example on $\Sigma = R^2 \times S^1$, one may introduce spinor fields that are homotopically inequivalent. This means that more than one 'vacuum' state can be defined.

Situations of this type have been studied in recent years in connection with kinks, solitons and instantons (for a concise review see Jackiw, 1977). General relativity, with the richness of alternative topologies that it allows, increases the variety and complexity of these situations. Moreover, as a model for other (usually simpler) field theories, it has drawn attention to a fact that had often been overlooked earlier, namely that the configuration space of any set of interacting fields is itself a Riemannian manifold or, more generally (if fermion fields are involved), a graded Riemannian manifold‡ possessing a metric that is determined

---

† The author is indebted to Charles Hart for information about these cases, in all of which the energy density is negative.
‡ For an introduction to graded manifolds see Kostant (1977).

(at least in part) by the field Lagrangian. The topology of this manifold need by no means be trivial.†

Without sacrificing its Riemannian character one may also view the configuration space as a (graded) fiber bundle over spacetime. The topology of this fiber bundle need not be trivial. The homotopic inequivalence of classes of spinor fields on $\Sigma = R^2 \times S^1$ is an example. A still simpler example on $R^2 \times S^1$ has been suggested by Isham (1978). Consider a neutral scalar field. Classically it is a mapping from spacetime to the real line, R. Its configuration space is therefore a fiber bundle where the fibers are copies of R. Suppose the bundle is 'twisted' so that the field, instead of being periodic over $\Sigma = R^2 \times S^1$, is antiperiodic. The Green's function must reflect this antiperiodicity and the renormalized vacuum stress tensor must change accordingly. The computation is straightforward and, for a massless field, one finds that $\langle T^{\mu\nu} \rangle$ continues to have the form (14.5) and (14.6) but that now $A = -7\pi^2/720$. In this case the energy density is *positive*. In the untwisted case it is negative, as for the electromagnetic field.

In all the above examples the background manifold is flat; the energy and stresses in the ether are entirely due to topology. Not all topologies admit flat metrics – for example $\Sigma = S^3$ or $\Sigma = R \times S^2$. In these cases too, $\langle T^{\mu\nu} \rangle$ is non-vanishing. However, the 'vacuum polarization' is no longer exclusively produced by topology; curvature plays a role. Curvature also complicates the renormalization algorithm and gives rise to the phenomenon of *trace anomalies*: the formal identity $T_\mu{}^\mu = 0$, valid for conformally invariant classical field theories, fails for quantized fields when curvature is present.‡ These problems will be discussed briefly in a later section. We remark here only that trace anomalies are similar in a number of respects to the *axial-vector current anomalies* of weak-interaction theories in which, again, a formal identity (a divergence identity) fails in the quantum theory.

The axial-vector current anomalies play a role in the theory of π^0-decay and in the analysis of constraints that must be imposed on unified

---

† For example, in the case of the sine-Gordon field. Another interesting example is provided by the gravitational field itself. For a description of its configuration space as a 'stratified' infinite dimensional, Riemannian manifold see Fischer (1970). For a study of some of its topological properties see DeWitt (1967a and 1970).

‡ Capper and Duff (1974) were the first to discover a trace anomaly. General forms were not known, however, until the work of Davies, Fulling and Unruh (1976), Deser, Duff and Isham (1976), and Dowker and Critchley (1976). The exact coefficients were not fully known until the computations of Christensen and Fulling (1977), Bunch and Davies (1977), and Brown and Cassidy (1977a).

Chapter 14. Quantum gravity: the new synthesis

field theories of the Weinberg–Salam type. They have been known for many years and originally were not thought to have any connection with general relativity. It was dicovered by Kimura (1969), however, that they in fact have an intimate relation to the topology of spacetime. Kimura's discovery and its connection with the Pontryagin number and the Atiyah–Singer index (Atiyah and Singer, 1968; Atiyah, Bott and Patodi, 1973) have been subsequently amplified by Delbourgo and Salam (1977), Eguchi and Freund (1976), and Jackiw and Rebbi (1977). Its possible relevance to quantum-mechanical tunneling has been noted by Hawking (1977) based on work by 't Hooft (1976a, b) involving background Yang-Mills fields. It is too early to assess the significance of these developments, but it is clear that they, together with all the other examples that have been cited, guarantee a permanently important role for topology in the quantum field theory of the future.

14.2.2 Curved boundaries; acceleration

In the analysis of the Casimir effect, two incomplete manifolds were introduced – the half-space and the slab. The boundaries of these manifolds are flat. What happens when the boundaries are curved? Consider first the case in which the boundary consists of two non-parallel plane conductors joined along the line of intersection. The curvature may be regarded as concentrated on this line. The method of images, as always, is available for the construction of the Green's function. The renormalized $\langle T^{\mu\nu} \rangle$ is found to depend on the intersection angle and to vary as the inverse fourth power of the distance from the intersection line (Dowker and Kennedy, 1978). More generally, $\langle T^{\mu\nu} \rangle$ has this behavior near the intersection line of every dihedral angle of any polyhedral boundary surface.

The surface of a curved conductor may be regarded as the limit of a sequence of polyhedral surfaces, and $\langle T^{\mu\nu} \rangle$ may be approximated by taking such a limit. David Deutsch† has been able to show that, near the conductor, $\langle T^{\mu\nu} \rangle$ is proportional to the sum of the reciprocals of the two principal radii of curvature, and varies inversely as the cube of the distance from the conductor. As the conductor is approached, the energy density in the ether tends to $-\infty$ on the concave side and to $+\infty$ on the convex side.

† Private communication.

The inverse-cube law actually represents a breakdown of the perfect-conductor approximation. It implies not only that the total integrated energy in the ether, per unit conductor area, is infinite ($-\infty$ on the concave side and $+\infty$ on the convex side) but also that the conducting surface itself experiences infinite stresses. A real conductor does not remain perfect at arbitrarily high photon frequencies. An effective energy cut-off or skin depth, which depends on the atomic details of the conductor, ultimately comes into play. This is in agreement with the results that Casimir found in his study of Van der Waals forces. Those terms in the interaction energy that depend on the curvature of the molecular surface also depend on the molecular details, i.e. on such fundamental constants as e and m.

In the above examples the conducting surfaces are assumed to be at rest or in uniform motion. The curvature of the boundary of the relevant incomplete spacetime manifold is therefore purely spatial. If the conductor is made to accelerate then the manifold boundary possesses a *curvature in time* also. A particularly simple case is that of an infinite plane conductor undergoing acceleration perpendicular to its surface. If the acceleration varies with time the conductor will generally emit or absorb photons, i.e. exchange energy with the ether, but if the acceleration is constant then an equilibrium can be established on the concave side of the motion (acceleration directed *toward* the incomplete manifold). The reason for this is that a timelike Killing-vector field can be introduced on the concave side, relative to which the conductor is at rest. In all the previous examples at least one *geodesic* timelike Killing-vector field exists, and the 'vacuum' has always been tacitly assumed to be defined relative to that field, in terms of a standard mode decomposition into positive and negative frequencies. The geodesic character of the field, however, is not essential. A mode-function decomposition can be carried out, and a corresponding vacuum defined, relative to any timelike Killing-vector field whether geodesic or not.

In the case of the plane conductor undergoing constant acceleration this vacuum has some remarkable properties. Let a_0 be the acceleration of the conductor and p a spacetime point a distance s from it. Candelas and Deutsch (1977) have shown that for values of s small compared to a_0^{-1} the renormalized vacuum stress tensor $\langle T^{\mu\nu} \rangle$ at p, viewed in a local frame at rest with respect to the Killing-vector field, varies like $1/s^3$, just as it does near an unaccelerated curved conductor. In fact, with a_0^{-1} playing the role of a radius of curvature, the formulae in the two cases are

Chapter 14. Quantum gravity: the new synthesis

identical – numerical coefficients, signs and all. The local energy density in particular is negative.

Even more remarkable is the form that $\langle T^{\mu\nu}\rangle$ takes when $s \gg a_0^{-1}$. It becomes the *negative* of what it would be if the ether were an accelerating photon gas in thermal equilibrium. The local temperature varies in the standard relativistic manner: inversely as the length of the local Killing vector. In the present case this means that the temperature is proportional to the local acceleration. In units for which $k = 1$, the constant of proportionality turns out to be $1/2\pi$.

The mathematical analysis of this system is simplified if the acceleration, a_0, is made to tend to infinity. It is then convenient to introduce 'Rindler coordinates' τ and ζ, related to the Minkowski coordinates x^0 and x^3 by (Rindler, 1966)

$$x^0 = a^{-1} e^{a\zeta} \sinh a\tau, \qquad x^3 = a^{-1} e^{a\zeta} \cosh a\tau. \tag{14.7}$$

Here the acceleration (of the Killing vector $\partial/\partial\tau$) is in the x^3-direction, having a local value equal to a on the hypersurface $\zeta = 0$ and $a\,e^{-a\zeta}$ elsewhere. The conductor has been pushed to the edge of the manifold: $\zeta = -\infty$ (where its presence is actually irrelevant). The manifold is known as the Rindler 'wedge' and comprises the region $x^3 > |x^0|$.

The local temperature of the equivalent photon gas is now given by the formula $T = a\,e^{-a\zeta}/2\pi$ throughout the Rindler wedge. This means that in a local frame at rest with respect to $\partial/\partial\tau$ the stress tensor varies inversely as the fourth power of the proper distance from the edge of the wedge. This may be compared with the behavior of $\langle T^{\mu\nu}\rangle$ near the line of intersection of two conducting planes. Here, however, one is confronted with a temperature concept whose meaning needs to be elucidated. In addition, the following question must be answered: how is one to understand the *negativity* of the energy density?

It looks as if the ground state of the Rindler wedge (the so-called Rindler vacuum) is somehow *below* the absolute zero of temperature! One must *add* photons to it to bring its energy up to that of the Minkowski vacuum, and these photons must be added in a thermal distribution!

Bizarre as it may seem, such an interpretation is correct. This and many other examples discovered in the last few years have brought about major changes in our ways of thinking about 'particles' and in our ways of defining them and their associated vacuum states. As has happened before in relativity theory, *and* in the quantum theory, one has had to fall back on operational definitions. Here, for example, one must ask: How would a given particle detector respond in a given situation?†

To see how the answer to this question resolves the negative-energy puzzle consider first a particle detector in ordinary uncluttered Minkowski spacetime. For simplicity let the electromagnetic field be replaced by a massless scalar field ϕ and let the coupling between detector and field be of a simple monopole type, described by an interaction Lagrangian of the form

$$L_{int} = m(\tau)\phi(x(\tau)). \tag{14.8}$$

Here the functions $x^\mu(\tau)$ define the world-line of the detector (idealized as a pointlike object) and the operator $m(\tau)$ represents its monopole moment at proper time, τ. Suppose the detector has a discrete set of internal energy eigenstates described by vectors $|E\rangle$, where $E = 0$ corresponds to the ground state. The detector response will then involve the matrix elements

$$\langle E|m(\tau)|E'\rangle = \langle E|m(0)|E'\rangle\, e^{i(E-E')\tau}. \tag{14.9}$$

Suppose the detector is initially in its ground state and the field is in a state described by the symbol Ψ. Then the amplitude for the combined system to undergo a transition from the state $(0, \Psi)$ to a state (E, Ψ') is given by

$$A(E, \Psi'|0, \Psi) = \langle E, \Psi'|T\left(\exp\left(i\int_{-\infty}^{\infty} L_{int}\, d\tau\right)\right)|0, \Psi\rangle, \tag{14.10}$$

where T denotes the chronological ordering operator. If $E > 0$ and the (renormalized) monopole moment is small enough so that radiative corrections to the interaction Lagrangian can be neglected, this amplitude is adequately represented by first-order perturbation theory:

$$A(E, \Psi'|0, \Psi) \approx i\langle E, \Psi'|\int_{-\infty}^{\infty} m(\tau)\phi(x(\tau))\, d\tau|0, \Psi\rangle$$
$$= i\langle E|m(0)|0\rangle \int_{-\infty}^{\infty} e^{iE\tau}\langle\Psi'|\phi(x(\tau))|\Psi\rangle\, d\tau. \tag{14.11}$$

The total probability for the detector to wind up in an excited state of energy, E, is then

$$P(E) = \sum_{\Psi'} |A(E, \Psi'|0, \Psi)|^2$$
$$= |\langle E|m(0)|0\rangle|^2 \int_{-\infty}^{\infty} d\tau \int_{-\infty}^{\infty} d\tau'\, e^{-iE(\tau-\tau')}\langle\Psi|\phi(x(\tau))\phi(x(\tau'))|\Psi\rangle. \tag{14.12}$$

† The first person to undertake a systematic study of this question was W. G. Unruh (1976).

Chapter 14. Quantum gravity: the new synthesis

The detector response is seen to depend jointly on the monopole-moment matrix element and the Fourier transform (along the world-line) of the Wightman function of the field.

If the field is in the standard Minkowski vacuum state and the detector moves along a geodesic world-line, then the Wightman function has positive frequencies only, the Fourier transform vanishes, and the detector remains in its ground state $(P(E)=0)$. If the detector suffers an acceleration, $P(E)$ no longer generally vanishes; excitation transitions take place in a statistically predictable manner.

Conventionally these transitions are not regarded as signaling the absorption of photons by the detector, since no (Minkowski) photons are there to be absorbed! Rather the detector is viewed as *emitting* photons. If the detector were inert, possessing no internal degrees of freedom, the pattern of photon emission would be that of a given accelerating source. When internal degrees of freedom are present this pattern is altered: occasionally an emitted photon is softer than it would otherwise be. The detector has stolen some of its energy and passed to an excited state.

When the detector undergoes uniform acceleration it is possible to adopt an alternative viewpoint. If the magnitude of the acceleration is a then the detector may be regarded as moving along a line $x^1 = $ const, $x^2 = $ const, $\zeta = 0$, in a Rindler coordinates system (14.7). The boundary of the Rindler wedge is a horizon for the detector. As far as the detector is concerned the wedge is its universe. In this wedge it is possible for the field to be in the Rindler vacuum state. The Wightman function in (14.12) then has only positive frequencies *with respect to the Rindler time*, τ. No 'Rindler photons' are present, and the detector remains in its ground state.

It is evident that Rindler photons and Minkowski photons are not the same thing. A uniformly accelerating detector responds to Rindler photons; an unaccelerated one responds to Minkowski photons. Both kinds of photons contribute to $\langle T^{\mu\nu} \rangle$ but they refer to different ground states. The Minkowski vacuum is full of Rindler photons, although it is devoid of Minkowski photons. The energy carried by the Rindler photons brings $\langle T^{00} \rangle$ up from its negative value in the Rindler vacuum to its vanishing value in the Minkowski vacuum.

The (negative) thermality of $\langle T^{\mu\nu} \rangle$ in the Rindler vacuum can be understood from the fact that the distribution function of the Rindler photons in the Minkowski vacuum is thermal. The thermality may be demonstrated in several ways. One way is to compute the Wightman function of (14.12) for the case of uniformly accelerated motion in the

Minkowski vacuum. The result turns out to be

$$\langle\Psi|\phi(x(\tau))\phi(x(\tau'))|\Psi\rangle = -\frac{(a/2\pi)^2}{4\sinh^2\frac{1}{2}a(\tau-\tau'-i0)}. \quad (14.13)$$

This apparently pure-state Wightman function can be shown to be identical to the mixed-state Wightman function tr $[\rho\phi(x(\tau))\phi(x(\tau'))]$, where ρ is the density operator describing a thermal bath of Rindler photons at temperature $T = a/2\pi$ relative to the world-lines $\zeta = 0$:

$$\rho = \frac{e^{-2\pi H/a}}{\text{tr}(e^{-2\pi H/a})} \quad \begin{array}{l}(H = \text{Rindler Hamiltonian} \\ \text{relative to the world-lines } \zeta = 0).\end{array} \quad (14.14)$$

Another way to demonstrate thermality – a way that explains how a pure state can look like a mixed state – is to introduce a complete set of Rindler mode functions, not only in the original Rindler wedge but also in the other three quadrants ($x^0 > |x^3|$, $x^3 < -|x^0|$, and $x^0 < -|x^3|$) as well. This set may be used to define a generalized Rindler vacuum and corresponding Rindler photons in all four quadrants. One may then compute the Bogoliubov coefficients connecting the Minkowski and Rindler mode functions, which permits a direct determination of the distribution function of Rindler photons in the Minkowski vacuum in all four quadrants. It is characteristic of the Bogoliubov decomposition that each vacuum is representable as a distribution of statistically independent photon *pairs* relative to the other vacuum. In the present case it turns out that, in addition to being thermally distributed, the members of each Rindler pair correspond to mode functions having disjoint supports in spacetime. No two members are to be found in the same quadrant. All the Rindler photons in the original Rindler wedge, therefore, are statistically independent (incoherent), which means that the Minkowski vacuum is locally identical to a mixed thermal Rindler state.

14.2.3 Black holes

Thermal states play a particularly important role in the theory of black holes. This is seen most easily in the case of the Schwarzschild geometry. Let t be the standard Schwarzschild time coordinate. Then $\partial/\partial t$ is a timelike Killing vector everywhere outside the horizon and a vacuum state can be defined with respect to it. At large distances from the black hole this vacuum is indistinguishable from the ordinary Minkowski vacuum. In particular, $\langle T^{\mu\nu}\rangle$ (renormalized) vanishes there. Near the

Chapter 14. Quantum gravity: the new synthesis

horizon, on the other hand, this vacuum shares many of the features of the Rindler vacuum: a particle detector at rest with respect to $\partial/\partial t$ remains in its ground state. Furthermore $\langle T^{00} \rangle$ in a local Lorentz frame becomes negatively infinite on the horizon just as it does on the boundary of the Rindler wedge in the Rindler vacuum.

There is another 'vacuum' state that may be imposed on the Schwarzschild geometry – a state for which $\langle T^{\mu\nu} \rangle$ (in a local frame) remains finite on the horizon. This state is fixed by the requirement that a freely falling detector undergoes no stimulated transitions in the vicinity of the horizon. In this region the state is evidently similar to the Minkowski vacuum. At infinity, on the other hand, it has a thermal character that may be determined by the following reasoning.

Let M be the mass of the black hole. Choose units for which the gravity constant, G, is unity. Let a detector be placed at rest with respect to $\partial/\partial t$ at a position $r = 2M + \varepsilon$, $\varepsilon \ll 2M$, where r is the conventional Schwarzschild radial coordinate. In order to stay in this position the detector must experience an absolute acceleration equal to $(2M/\varepsilon)^{1/2}/4M$. Because the state has the local properties of the Minkowski vacuum near $r = 2M$ it follows that the detector must react, in the high-frequency range at least, as if it were immersed in a thermal photon bath at temperature $T_\varepsilon = (2M/\varepsilon)^{1/2}/8\pi M$. The photons in this bath correspond to mode functions based on $\partial/\partial t$. They are *real* because they carry energy that raises $\langle T^{00} \rangle$ from $-\infty$ to a finite value on the horizon. Furthermore they are able to escape to infinity where, because of a redshift factor $(\varepsilon/2M)^{1/2}$, they wind up with a temperature

$$T = \frac{1}{8\pi M}. \tag{14.15}$$

This is the temperature made famous by Hawking (1974). He first encountered it in the study of black holes formed by collapse. The collapse metric yields a well-posed Cauchy problem for the mode functions of the radiation field. Therefore $\langle T^{\mu\nu} \rangle$, if initially non-singular, must remain non-singular until the geometrical singularity at $r = 0$ is reached. In particular it must be smooth on the horizon. The black hole considered in the preceding paragraph was assumed to be an 'eternal' black hole, with a horizon that consists of both future and past parts. A black hole formed by collapse has only a future horizon. The condition that $\langle T^{\mu\nu} \rangle$ be smooth on this horizon leads (starting from a pre-collapse state with no particles at infinity) to a final state for which the photons at infinity are thermal, with $T = 1/8\pi M$, but are outgoing only. In the case of the eternal black hole the state at infinity is that of a thermal bath in equilibrium with the black hole, with as many photons being absorbed as emitted.

Hawking's association of a temperature with a black hole closed a major theoretical gap in the study of the statistical mechanics of black holes. He had earlier proved the important theorem (Hawking, 1971) that, in classical general relativity, the area of the future horizon of a black hole can never decrease. Led by the analogy of this result to the second law of thermodynamics, Bekenstein (1972) attempted to see whether the area could be regarded as a measure of entropy. By studying the information lost to the outside world when matter or radiation is trapped inside a black hole, he found that the area would do very nicely and that the proportionality constant between it and the conventional thermodynamic entropy must be very nearly equal to unity, in units for which $\hbar = c = G = k = 1$.

But Bekenstein's suggestion had one glaring defect. Classically a black hole can be in thermal equilibrium with its surroundings only if the surroundings are at the absolute zero of temperature, which means that the black hole itself must be at absolute zero. If the entropy, S, is to be regarded as proportional to the area then, according to the thermodynamical formula $dE = T\,dS$, an infinite change in the area would appear to be required to effect a finite change in the energy, E (or mass, M) of the black hole. This is in sharp disagreement with the elementary relation between mass and area: $A = 16\pi M^2$ (for a non-rotating black hole).

If, however, the black hole is assigned the temperature given by (14.15) then the thermodynamical relation takes the simple form $dM = \tfrac{1}{4}T\,dA$, permitting the immediate identification†

$$S = \tfrac{1}{4}A. \qquad (14.16)$$

Hawking's discovery not only establishes the proportionality constant between entropy and area but also shows in a most striking way that general relativity *must* be wedded to the quantum theory if consistency with statistical mechanics is to be assured.

---

† Equation (14.16) holds also for charged rotating black holes. For these the thermodynamical formula generalizes to $dM = \tfrac{1}{4}T\,dA + \Omega\,dJ + \Phi\,dQ$ where Ω, J, Φ, Q are respectively the angular velocity, angular momentum, electrostatic potential and charge. If such black holes are formed by collapse then the radiation emitted to infinity is not purely thermal but includes a spontaneous emission component arising from the ergosphere-super-radiance phenomenon in the neutral case (see Zel'dovich, 1971, 1972; Press and Teukolsky, 1972), and from both that and the Klein-paradox phenomenon in the charged case.

Chapter 14. Quantum gravity: the new synthesis

The generality of the above results deserves emphasis. Although the temperature (14.15) was originally derived from studies of *linear* quantized fields propagating in given black hole geometries, it holds equally well for interacting fields. This follows most clearly from the work of the Cambridge school (see chapters 13 and 15 by Gibbons and Hawking in this volume) who have shown that an important property of the chronological Green's functions for any field theory (interacting or not) in the black hole background may be derived simply by analytically continuing the metric to complex values of the time, t. One finds that if these functions are to be well behaved on the horizon, and have desirable properties at infinity, they must be periodic in t (antiperiodic in the fermion case) with imaginary period i/T. But this is a well-known property of thermal Green's functions (see Fetter and Walecka, 1971). It is displayed, for example, in (14.13), where the period is $2\pi i/a$.

14.3 The back reaction

All fundamental fields, including the gravitational field, contribute to the thermal radiation emitted by a black hole. Therefore one may correctly speak of the black hole itself as being quantized. A quantized black hole is evidently a dynamical object that may exchange energy and entropy with its surroundings. The exchange always respects the laws of statistical mechanics, provided these laws are understood as applying to the whole system: black hole plus surroundings. But the fact that a black hole can lose entropy means that, in the quantum theory, its area can sometimes decrease.

From the known expressions for area and temperature one easily infers that a black hole formed by collapse has a luminosity of the order of M^{-2}. If it is permissible to assume that the luminosity maintains its proportionality to M^{-2} under secular changes in M, then M steadily decreases until finally the black hole either disappears in a flash or else stabilizes itself by some as yet unknown quantum effect. The lifetime for this process is of the order of M^3.

The luminosity gives information about $\langle T^{\mu\nu} \rangle$, the stress tensor, only at infinity. In order to study the detailed temporal behavior of the geometry near (and even inside) the horizon one must determine $\langle T^{\mu\nu} \rangle$ in that region also. To do this one must surmount several practical difficulties: (1) The mode functions for the Schwarzschild geometry are not accurately known; they are even less accurately known for the Kerr geometry. (2) To renormalize $\langle T^{\mu\nu} \rangle$ it no longer suffices to subtract the Minkowski divergence; other divergences, involving the curvature, make their appearance. (3) Any renormalization scheme that is adopted must be covariant and applicable in a general context; it must not be tailored specifically to the black hole geometry.

In the following sections one of the standard methods for renormalizing $\langle T^{\mu\nu} \rangle$ will be outlined. First, however, a more precise statement is needed concerning the role that $\langle T^{\mu\nu} \rangle$ is envisaged as playing. Let ψ denote collectively any matter and/or radiation fields that are assumed to be present. In the quantum theory ψ is a quantum operator. The metric tensor too is an operator, $\mathbf{g}_{\mu\nu}$. Let it be separated into a (presently arbitrary) classical background, $g_{\mu\nu}$, and an operator remainder, $\boldsymbol{\phi}_{\mu\nu}$, both regarded as formal tensors. The conventional action, S, for the combined matter, radiation and gravitational fields involves $g_{\mu\nu}$ and $\boldsymbol{\phi}_{\mu\nu}$ only in the combination $g_{\mu\nu} + \boldsymbol{\phi}_{\mu\nu}$. Introduce the alternative functional

$$\bar{S}[g, \boldsymbol{\phi}, \boldsymbol{\psi}] \stackrel{\text{def}}{=} S[g+\boldsymbol{\phi}, \boldsymbol{\psi}] - S[g, 0]$$

$$-\int (\delta S[g, 0]/\delta g_{\mu\nu}) \boldsymbol{\phi}_{\mu\nu} \, \mathrm{d}^n x, \tag{14.17}$$

where (generalizing to n dimensions) $\mathrm{d}^n x$ denotes $\mathrm{d}x^0 \mathrm{d}x^1 \cdots \mathrm{d}x^{n-1}$. If no external sources are present, so that $\psi = 0$ is a solution of the classical matter/radiation equations, and if $g_{\mu\nu}$ is chosen to satisfy the classical empty-space equation $\delta S[g, 0]/\delta g_{\mu\nu} = 0$,† then the operator field equations may be written in the forms

$$\delta \bar{S}[g, \boldsymbol{\phi}, \boldsymbol{\psi}]/\delta \boldsymbol{\psi} = 0 \tag{14.18}$$

and

$$F^{\mu\nu\sigma\tau} \boldsymbol{\phi}_{\sigma\tau} = -\tfrac{1}{2} \boldsymbol{T}^{\mu\nu}, \tag{14.19}$$

where

$$F^{\mu\nu\sigma\tau} \boldsymbol{\phi}_{\sigma\tau} \stackrel{\text{def}}{=} \int \frac{\delta^2 S[g, 0]}{\delta g_{\mu\nu}(x) \delta g_{\sigma\tau}(x')} \boldsymbol{\phi}_{\sigma\tau}(x') \, \mathrm{d}x', \tag{14.20}$$

$$\boldsymbol{T}^{\mu\nu} \stackrel{\text{def}}{=} 2 \delta \bar{S}[g, \boldsymbol{\phi}, \boldsymbol{\psi}]/\delta g_{\mu\nu}. \tag{14.21}$$

† It is often convenient to change the zero points of the matter and radiation fields so that they too contribute to the background. If ψ_B denotes the matter–radiation background then the combined classical background must satisfy

$$\delta S[g, \psi_B]/\delta g_{\mu\nu} = 0, \quad \delta S[g, \psi_B]/\delta \psi_B = 0.$$

The ψ_Bs corresponding to fermion fields must be understood as anti-commuting Grassmann numbers.

Chapter 14. Quantum gravity: the new synthesis

$F^{\mu\nu\sigma\tau}$ is a covariant self-adjoint second-order differential operator that depends only on the background metric. (We shall not need its explicit form.) All the nonlinearity of the gravitational field equation is concentrated on the right-hand side of (14.19), which expresses a feedback principle: the field operator $\phi_{\mu\nu}$, through $T^{\mu\nu}$, serves partly as its own source. One may regard $\phi_{\mu\nu}$ and ψ as being formally on a common footing. Each interacts with the other and with itself while propagating in a background geometry described by $g_{\mu\nu}$. Under physical conditions for which it is permissible to regard $\phi_{\mu\nu}$ and ψ as 'small', both $\bar{S}[g, \phi, \psi]$ and $T^{\mu\nu}$ may be expressed as functional power series in $\phi_{\mu\nu}$ and ψ, starting with quadratic terms. These power series form the basis for a perturbation theory.

When (14.18) and (14.19) hold, $T^{\mu\nu}$ satisfies

$$T^{\mu\nu}{}_{;\nu} = 0, \tag{14.22}$$

the covariant derivative being defined in terms of the background metric. $T^{\mu\nu}$ has all the properties of a stress tensor and may be so regarded.† Its expectation value (renormalized) is the quantity of interest in the back-reaction problem. It appears in the relation obtained by taking the expectation value of (14.19):

$$F^{\mu\nu\sigma\tau}\langle\phi_{\sigma\tau}\rangle = -\tfrac{1}{2}\langle T^{\mu\nu}\rangle. \tag{14.23}$$

Several remarks must be made about this equation. Firstly, it will be convenient in all that follows to replace the conventional expectation value by the 'Schwinger average':

$$\langle A \rangle \stackrel{\text{def}}{=} \frac{\langle\text{out, vac}|T(A)|\text{in, vac}\rangle}{\langle\text{out, vac}|\text{in, vac}\rangle}, \tag{14.24}$$

where T denotes the chronological ordering operator. If A is a local operator, so that the T symbol may be omitted, and if there is no particle production by the background field, so that the 'in' and 'out' vacuum-state vectors are identical, then the Schwinger average coincides with the conventional vacuum-expectation value. More generally the two averages differ, but usually by a finite amount even when the operator A is unrenormalized. In studying renormalization questions therefore we may confine our attention to the Schwinger average. We postpone for the present the question how to define the Schwinger average when 'in' and 'out' states cannot be introduced in conventional ways because timelike

† When defined by (14.21), $T^{\mu\nu}$ is actually a tensor *density* of unit weight. It will be convenient throughout this review to take the stress tensor always in this density form.

The back reaction

Killing vectors are absent, or when they are ambiguous for other reasons.

Secondly, it must be noted that although the operators $\phi_{\mu\nu}$ and $T^{\mu\nu}$ may be regarded as transforming like ordinary tensors with respect to coordinate transformations of the background metric, they are *not* invariant under the gauge group of quantum gravity theory. Under gauge transformations the background metric is held fixed, and when the quantum metric tensor suffers a gauge transformation (see (14.95)) the full burden of change is carried by $\phi_{\mu\nu}$. The expectation values in (14.23) are therefore ambiguous. We shall introduce later a special way of removing the ambiguity (see (14.127), (14.132) and (14.133)) which consists of taking a Gaussian average over special gauges. When this averaging procedure is carried out, extra terms should actually appear in (14.23) (see (14.158)), but for the approximations of interest in the immediately following sections the equation is adequate as it stands.

It is worth remarking that this averaging procedure is analogous in several ways to the spacetime averaging method introduced by Isaacson (1968) in the classical theory to obtain a well-defined stress tensor. As was emphasized long ago by Bohr and Rosenfeld (1933), only spacetime averages of field quantities have physical meaning. An averaged stress tensor can be 'observed' only via measurements of spacetime averages of the 'local component' of the Riemann tensor in its vicinity, in effect, via measurements of the left side of (14.23). (A particle detector does *not* measure the stress tensor!) If one carries out a Bohr–Rosenfeld type analysis on the gravitational field (DeWitt, 1962) one finds that averages of the local Riemann tensor can be measured in principle to an accuracy well within the quantum domain, as long as the averaging region is larger than the Planck length. Equation (14.23) therefore has genuine operational significance.

A mathematically precise statement of its meaning can be obtained by introducing the definitions

$$\varphi_{\mu\nu} \stackrel{\text{def}}{=} g_{\mu\nu} + \langle \phi_{\mu\nu} \rangle = \langle g_{\mu\nu} \rangle, \quad \psi \stackrel{\text{def}}{=} \langle \psi \rangle. \tag{14.25}$$

It will be shown later that there exists a functional $\Gamma[\varphi, \psi]$ of $\varphi_{\mu\nu}$ and ψ such that (14.23) (corrected) and the Schwinger average of (14.18) may be recast in the respective forms

$$\delta\Gamma/\delta\varphi_{\mu\nu} = 0, \tag{14.26}$$

$$\delta\Gamma/\delta\psi = 0. \tag{14.27}$$

Chapter 14. Quantum gravity: the new synthesis

Γ is called the *effective action*. It describes the dynamical behavior of coherent large-amplitude gravitational, electromagnetic and matter fields. In this description $\varphi_{\mu\nu}$, rather than $g_{\mu\nu}$, plays the role of the 'classical' fiducial metric.

Because $\varphi_{\mu\nu}$ is governed by Γ rather than by S, its dynamical behavior includes the back reaction. Use of $\varphi_{\mu\nu}$ in place of $g_{\mu\nu}$ permits a *decaying* black hole to be described in quasi-classical terms (at least as long as the imaginary part of $\varphi_{\mu\nu}$ remains small). Since the value of $\varphi_{\mu\nu}$ is determined by (14.23) the role of $\langle T^{\mu\nu} \rangle$ is now evident.

When particle production becomes very large, as in the final stages of black hole decay, the imaginary part of $\varphi_{\mu\nu}$ is not small, and the quasi-classical picture breaks down. The correct interpretation of a *complex* $\varphi_{\mu\nu}$ remains to be discovered. Whether it can be comprehended under the general analytic continuation philosophy, or has a role to play in instanton theory and topological problems, is presently unknown. It is worth remarking that Γ has in any case a meaning in terms of transition amplitudes. The full S-matrix of the theory, for example, is given by the 'tree amplitudes' of Γ, and these may be defined relative to any asymptotically flat background. (See DeWitt, 1965.)

14.4 The one-loop approximation

In lowest approximation the right-hand side of (14.23) may be evaluated by inserting into the expression for $T^{\mu\nu}$ appropriate operator solutions of the equations $F^{\mu\nu\sigma\tau}\phi_{\sigma\tau} = 0$ and $(\delta \bar{S}[g, 0, \psi]/\delta\psi)_{\text{LIN}} = 0$, where 'LIN' means 'keep only the part linear in ψ'. In this approximation the fields $\phi_{\mu\nu}$ and ψ do not interact with each other (nor with themselves) but propagate linearly in the background $g_{\mu\nu}$. For consistency (at this level) one should keep only the terms quadratic in $\phi_{\mu\nu}$ and ψ in the series expansion of $T^{\mu\nu}$. The resulting value for $\langle T^{\mu\nu} \rangle$ is known as the *one-loop approximation*. This approximation is implicit in most current discussions of the stress tensor. It underlies the known results quoted in previous sections, out of which, as has been seen, quite a bit of reasonable sense can be made. Unfortunately, because quantum gravity is technically non-renormalizable, the best method for making sense out of the theory in higher orders is presently unknown.

In this review, renormalization of $\langle T^{\mu\nu} \rangle$ will be described in one-loop approximation only. Although our understanding of quantum gravity will be correspondingly incomplete it should be noted that the one-loop approximation in field theory is equivalent to the WKB approximation of

The one-loop approximation

ordinary quantum mechanics. It may therefore reasonably be expected, just as in ordinary quantum mechanics, to yield at least *some* significant insights into the exact theory.

In the study of the one-loop approximation it is convenient to introduce a functional $W[g]$ of the background metric, defined by

$$e^{iW} \stackrel{\text{def}}{=} \langle \text{out, vac} | \text{in, vac} \rangle. \tag{14.28}$$

Equations (14.1), (14.24) and (14.21) yield

$$\langle T^{\mu\nu} \rangle = 2\, \delta W / \delta g_{\mu\nu}. \tag{14.29}$$

Renormalization of $\langle T^{\mu\nu} \rangle$ may evidently be achieved by renormalizing W. In fact, this is the only foolproof procedure. Remembering that the background metric, $g_{\mu\nu}$, satisfies the empty-space field equation, one sees that (14.23) and (14.29) together imply

$$\frac{\delta S[g, 0]}{\delta g_{\mu\nu}} + F^{\mu\nu\sigma\tau} \langle \boldsymbol{\phi}_{\sigma\tau} \rangle + \frac{\delta W[g]}{\delta g_{\mu\nu}} = 0. \tag{14.30}$$

In one-loop order this is equivalent to

$$\frac{\delta S[\varphi, 0]}{\delta \varphi_{\mu\nu}} + \frac{\delta W[\varphi]}{\delta \varphi_{\mu\nu}} = 0, \tag{14.31}$$

which permits, in this order, identification of the sum $S[\varphi, 0] + W[\varphi]$ with that part of the effective action that is independent of ψ. Since it is the effective action that describes the *real* physics of the theory it follows that neither S nor W alone is physically relevant, but only the sum $S + W$. If W contains terms that are similar to those contained in S, then only the sums of the corresponding coefficients can be experimentally determined as observable 'coupling constants'. If any of the terms of W possess divergent coefficients these must be cancelled by 'counterterms' in S. In general there will also be finite terms that must be cancelled.

In one-loop order the divergences of the operator stress tensor, $T^{\mu\nu}$, are the same as those of its Schwinger average and hence are c-numbers. The renormalized stress tensor is defined by

$$T^{\mu\nu}_{\text{ren}} \stackrel{\text{def}}{=} T^{\mu\nu} - 2\, \delta \Delta W / \delta g_{\mu\nu}, \tag{14.32}$$

where ΔW is the maximal piece of W that can be cancelled by counterterms in S, subject to the following conditions: (1) only those types of counterterms are to be used that are needed to render $T^{\mu\nu}_{\text{ren}}$ free of

Chapter 14. Quantum gravity: the new synthesis

divergences; (2) $\langle T^{\mu\nu}_{\text{ren}} \rangle$ must vanish in flat, empty spacetime; (3) ΔW must be coordinate-invariant; (4) $\delta \Delta W/\delta g_{\mu\nu}$ must depend only locally on the background geometry. The coordinate invariance of ΔW guarantees that (14.22) will remain valid under the renormalization procedure. This is one of the most important reasons for renormalizing W first and then obtaining $T^{\mu\nu}_{\text{ren}}$ as a derived quantity. If $T^{\mu\nu}$ is tackled directly there is always a danger that (14.22) will be violated.

14.4.1 Formal relations; the role of the Feynman propagator

To subtract a coordinate-invariant divergent piece from W one must first unambiguously isolate it. This requires the introduction of a generally covariant regularization scheme. To set up such a scheme one must begin with some formal relations. Assume for the moment that the field of interest is a scalar field, ϕ. Then, $\langle T^{\mu\nu} \rangle$ is formally obtained by adding together appropriate derivatives of the Schwinger average $\langle \phi(x)\phi(x') \rangle$ and taking the limit $x' \to x$. If x' is required to approach x from a spacelike direction this average may be related to the 'Feynman propagator' $G(x, x')$:†

$$\langle \phi(x)\phi(x') \rangle = -\mathrm{i} G(x, x'). \tag{14.33}$$

The Feynman propagator is a Green's function satisfying

$$FG(x, x') = -\delta(x, x'), \tag{14.34}$$

where F is the differential operator appearing in the field equation $F\phi = 0$. Of crucial importance in the theory are the boundary conditions that define this Green's function. When the background geometry possesses a global timelike Killing vector, with respect to which the vacuum states are defined (in this case $|\text{in, vac}\rangle = |\text{out, vac}\rangle$), $G(x, x')$ is expressible as a sum of corresponding mode functions. Regarded as a function of x, with x' held fixed, it has the property of being decomposable into purely positive-frequency mode functions when x lies to the future of x' and purely negative-frequency mode functions when x lies to the past.‡ The same is true when $G(x, x')$ is regarded as a function of x', with x held fixed, for it is a symmetric function of its arguments.

---

† More generally, the chronological product should appear on the left side of (14.33).
‡ When x and x' are separated by a spacelike interval the decomposition may be performed either way. If zero-frequency modes are present (e.g. when the Universe is compact and the field is massless) a true Feynman propagator does not exist and the state-vector space cannot be constructed as a strict Fock space. The techniques needed for handling this case will not be considered in this review.

The one-loop approximation

These properties can be shown to follow from the following formal choice of solution to (14.34):

$$G = -\frac{1}{F + i0}. \qquad (14.35)$$

Here the labels x and x' have been suppressed and the symbol i0 means 'add a small positive imaginary multiple of the unit operator to F and take the limit as this added bit goes to zero'. An alternative way of obtaining the same Green's function is to rotate the time coordinate x^0 clockwise through 90° in the complex plane so that F, which in the scalar case has the form $g^{1/2}(\Box - \xi E - m^2)$,† becomes a negative definite operator having a single unique Green's function that vanishes when x and x' are infinitely separated. The Feynman propagator is obtained by analytically continuing this Green's function back to physical values of the time.

When there are no global timelike Killing vectors but only 'in' and 'out' regions, the Feynman propagator is defined by requiring it to be decomposable (with respect to either of its arguments) into negative-frequency mode functions in the 'in' region and positive-frequency mode functions in the 'out' region. A globally hyperbolic spacetime possessing 'in' and 'out' regions can be generated from one possessing a global timelike Killing vector by integrating an infinite sequence of infinitesimal variations, $\delta g_{\mu\nu}$, in the metric. The required boundary conditions are maintained if, under each of these variations, G suffers the change

$$\delta G = G \, \delta F \, G, \qquad (14.36)$$

where δF is the corresponding change in F. Because F, and hence δF, is self-adjoint, this leaves the symmetry of $G(x, x')$ in its arguments intact. More importantly, it leaves the representation (14.35) intact.

One naturally asks whether a Green's function exists, satisfying (14.35) and (14.36), and the symmetry property, even when there are no 'in' and 'out' regions. A general answer to this question has not yet been determined, but it is known that such a Green's function exists in cases in which global analytic continuation and/or conformal transformation techniques can be employed, *including* cases in which geometrical singularities are present. For example, the expectation value of the chronological product $iT(\phi(x)\phi(x'))$, in the natural thermal state associated with a black hole, has the representation (14.35). In this case the function is

---

† Here $g \stackrel{\text{def}}{=} -\det(g_{\mu\nu})$, $\Box$ is the Laplace–Beltrami operator, ξ is a constant, R is the curvature scalar, m is the field mass, and the metric signature is $- + + \cdots$.

Chapter 14. Quantum gravity: the new synthesis

unique. Whether existence implies uniqueness in general is a question which also has not yet been fully resolved.†

The representation (14.35) may be rewritten in the following form:

$$g^{1/4} G g^{1/4} = i \int_0^\infty \exp(ig^{-1/4} F g^{-1/4} s)\, ds, \qquad (14.37)$$

where the factors $g^{-1/4}$ have been inserted to render the integrand generally covariant. The question of the existence of G may be replaced by the question whether this integral exists. With the labels x and x' restored (14.37) is equivalent to

$$G(x, x') = i \int_0^\infty g^{-1/4}(x) K(x, x', s) g^{-1/4}(x')\, ds, \qquad (14.38)$$

where the function K satisfies the differential equation

$$\frac{\partial}{i\,\partial s} K(x, x', s) = g^{-1/4} F g^{-1/4} K(x, x', s) \qquad (14.39)$$

together with the 'initial' condition

$$K(x, x', 0) = \delta(x, x'). \qquad (14.40)$$

When spacetime is incomplete, K must also satisfy conditions on the boundary. In the Casimir problem these conditions are elementary, but if the incompleteness is due to geometrical singularities they may involve analytic continuation and/or conformal transformation of the metric.

Now the scalar contribution to the classical action may be written in the form

$$S_{sc}[\phi] = \tfrac{1}{2} \int \phi F \phi\, d^n x. \qquad (14.41)$$

Since $T^{\mu\nu} = 2\, \delta S_{sc}/\delta g_{\mu\nu}$, it follows from (14.29) that under a variation $\delta g_{\mu\nu}$ in the metric, with a corresponding change δF in F, the functional W suffers the change

$$\delta W = \tfrac{1}{2} \int \langle \phi\, \delta F\, \phi \rangle\, d^n x = -\frac{i}{2} \operatorname{tr}(G\, \delta F), \qquad (14.42)$$

in which use is made of (14.33), and integration over spacetime is replaced by the trace symbol, in passing to the final form. The operator δF is a local operator and hence the trace involves a coincidence limit (from a

---

† Existence does imply uniqueness when the spacetime manifold is complete and the associated mode functions are required to be bounded.

The one-loop approximation

spacelike direction) of the suppressed labels, x and x'. Because $F\phi = 0$ (14.42) may be equally well be written

$$\delta W = \tfrac{1}{2} \int \langle g^{1/4}\phi \delta(g^{-1/4}Fg^{-1/4})g^{1/4}\phi\rangle \, d^n x \qquad (14.42')$$
$$= -\frac{i}{2}\operatorname{tr}[g^{1/4}Gg^{1/4}\delta(g^{-1/4}Fg^{-1/4})],$$

which permits use of the representation (14.37):

$$\delta W = \delta\left[\tfrac{1}{2}\operatorname{tr}\int_0^\infty \frac{1}{is}\exp(ig^{-1/4}Fg^{-1/4}s)\,ds\right]. \qquad (14.43)$$

Integration of this variational equation is immediate. In terms of the function K one has formally

$$W = \tfrac{1}{2}\int_0^\infty \frac{1}{is} K(s)\,ds + \text{const}, \qquad (14.44)$$

$$K(s) \stackrel{\text{def}}{=} \int K(x, x, s)\,d^n x. \qquad (14.45)$$

14.4.2 Representations of $K(x, x', s)$

When x is close to x' it is convenient to introduce a representation for K that is suggested by its form in flat spacetime:

$$K(x, x', s) = \frac{iD^{1/2}(x, x')}{(4\pi i s)^{n/2}} e^{(i/2s)\sigma(x,x') - im^2 s} \Lambda(x, x', s). \qquad (14.46)$$

Here n is the dimensionality of spacetime, $\sigma(x, x')$ is half the square of the geodesic distance between x and x', $D(x, x')$ is the $n \times n$ determinant

$$D(x, x') \stackrel{\text{def}}{=} -\det\left[-\frac{\partial^2}{\partial x^\mu \partial x'^\nu}\sigma(x, x')\right], \qquad (14.47)$$

and $\Lambda(x, x', s)$ satisfies the boundary condition

$$\Lambda(x, x, 0) = 1. \qquad (14.48)$$

Inserting (14.46) into (14.39), remembering that $F = g^{1/2}(\Box - \xi R - m^2)$, and making use of the following equations satisfied by σ and D (see DeWitt, 1965),

$$\sigma_{;\mu}\sigma_{;}{}^\mu = \sigma_{;\mu'}\sigma_{;}{}^{\mu'} = 2\sigma, \quad (D^{1/2}\sigma_{;}{}^\mu)_{;\mu} = n, \qquad (14.49)$$

Chapter 14. Quantum gravity: the new synthesis

one finds that

$$\frac{\partial}{i\,\partial s}\Lambda + \frac{1}{is}\sigma_{;}{}^{\mu}\Lambda_{;\mu} = D^{-1/2}(\Box - \xi R)(D^{1/2}\Lambda). \quad (14.50)$$

When the spacetime manifold is incomplete (14.50) and the boundary condition (14.48) together do not suffice to determine Λ. However, they suffice to determine an asymptotic expansion of Λ valid when s is small and x is close to x':

$$\Lambda(x, x', s) \sim \sum_{r=0}^{\infty} a_r(x, x')(is)^r. \quad (14.51)$$

The coefficients $a_r(x, x')$ are generated by the differential recursion relations

$$\left.\begin{array}{l} a_0(x, x') = 1, \\ \sigma_{;}{}^{\mu} a_{r;\mu} + r a_r = D^{-1/2}(\Box - \xi R)(D^{1/2} a_{r-1}), \quad r = 1, 2, 3 \cdots \end{array}\right\} \quad (14.52)$$

which follow from (14.48) and (14.50). By repeatedly differentiating equations (14.49) and using $\sigma_{;\mu} \xrightarrow[x' \to x]{} 0$, $\sigma_{;\mu\nu'} \xrightarrow[x' \to x]{} -g_{\mu\nu}$, one finds that these recursion relations imply the coincidence limits†

$$a_1(x, x) = (\tfrac{1}{6} - \xi)R, \quad (14.53)$$

$$a_2(x, x) = \tfrac{1}{2}(\tfrac{1}{6} - \xi)^2 R^2 + \tfrac{1}{180}(-R_{\mu\nu}R^{\mu\nu} + R_{\mu\nu\sigma\tau}R^{\mu\nu\sigma\tau}) + \tfrac{1}{6}(\tfrac{1}{5} - \xi)\Box R. \quad (14.54)$$

Equations (14.45), (14.46) and (14.51) yield

$$K(s) \sim \frac{i}{(4\pi is)^{n/2}} e^{-im^2 s} \sum_{r=0}^{\infty} A_r (is)^r, \quad (14.55)$$

where the A_r are formal integrals:

$$A_r \stackrel{\text{def}}{=} \int g^{1/2} a_r(x, x)\,d^n x. \quad (14.56)$$

From this one sees that the integral in (14.44) diverges like

$$\tfrac{1}{2}(4\pi)^{-n/2}\left[\frac{2}{n} A_0 (is)^{-n/2} + \frac{2}{n-2}(A_1 - m^2 A_0)(is)^{-n/2+1} + \cdots\right]$$

at the lower limit.

If the spacetime manifold has a non-singular non-null boundary, (14.55) is not quite correct. $\Lambda(x, x, s)$ then contains terms of the form

† For $a_3(x, x)$ see Gilkey (1975).

exp $[if(x)/s]$ where $f(x)$, when x lies close to the boundary, is proportional to the square of the distance to the boundary. These terms have essential singularities at $s = 0$ and are not picked up by the asymptotic expansion (14.51).† However, when inserted into the integral (14.45) they yield two kinds of additional terms to the sum in (14.55): terms involving half-odd-integral powers of is, and terms involving integral powers like those already present. Terms of the first kind arise from a 'crowding' effect near the boundary, which reduces the effective dimensionality of spacetime by 1 there. Terms of the second kind involve the extrinsic curvature (second fundamental form) of the boundary and are most easily derived by 'doubling' the manifold, i.e. completing it by joining to it, along the boundary, a metrical copy of itself (see McKean and Singer, 1967). Except for the boundary-crowding effect the doubled manifold displays all the basic features of the original manifold, but because it is complete (14.55) may be applied to it as is. The connection components (in a natural coordinate system) suffer discontinuities at the 'join' along the original boundary, and the Riemann tensor behaves like a delta function there. This delta function is what produces the extrinsic-curvature corrections to (14.55).

We have seen that curved boundaries give rise to certain unrealistic and non-physical effects already in flat spacetime, e.g. infinite energy densities at the boundary. In the general case these become entwined with the standard infinities and complicate the renormalization problem in an unrealistic way. From now on we shall exclude manifolds with boundaries, unless the boundaries arise in a natural way by analytic continuation or conformal transformation from boundaryless manifolds. Equation (14.55) then needs no modification.

Another useful representation of $K(x, x', s)$ can be obtained by introducing a complete set of eigenfunctions $v(\alpha, x)$ of the operator $-\Box + \xi R$:

$$(-\Box + \xi R)v(\alpha, x) = \lambda(\alpha)v(\alpha, x). \tag{14.57}$$

Without loss of generality the $v(\alpha, x)$ may be taken real and orthonormalized:

$$\int g^{1/2}(x) v(\alpha, x) v(\alpha', x) \, d^n x = \delta_{\alpha\alpha'},$$
$$\sum_\alpha g^{1/4}(x) v(\alpha, x) v(\alpha, x') g^{1/4}(x') = \delta(x, x'). \tag{14.58}$$

† Other terms, containing factors of the form $\exp(i \times \text{const}/s)$, are also frequently present. Such terms, which can occur even when the manifold is boundaryless (e.g. whenever the Cauchy hypersurfaces are compact), are likewise missed by the asymptotic expansion, but they make no contribution to (14.55) nor to the divergences of the integral (14.44).

Chapter 14. Quantum gravity: the new synthesis

If spacetime is complete (and globally hyperbolic) the function v must be bounded; if incomplete they must satisfy appropriate conditions on the boundary. Here α stands for a complete set of labels, and the symbols $\delta_{\alpha\alpha'}$ and $\sum$ are to be understood as including respectively delta functions of, and integrations with respect to, any of the labels that are continuous. In terms of the function v one may write

$$K(x, x', s) = \sum_\alpha g^{1/4}(x) v(\alpha, x) v(\alpha, x') g^{1/4}(x') e^{-i[\lambda(\alpha)+m^2]s}, \quad (14.59)$$

$$K(s) = \sum_\alpha e^{-i[\lambda(\alpha)+m^2]s}. \quad (14.60)$$

The Feynman propagator may be expressed similarly:

$$G(x, x') = \sum_\alpha \frac{v(\alpha, x) v(\alpha, x')}{\lambda(\alpha) + m^2 - i0}. \quad (14.61)$$

14.4.3 The generalized zeta function

The properties of the ether (at least those properties for which the scalar field is responsible) are entirely determined by the operator F and the topology of spacetime. In analyzing the interplay of these two (for example in studying the distribution of the eigenvalues of F) mathematicians have found it convenient to introduce a generalization of Riemann's zeta function, defined as follows:

$$\zeta(x, z) \stackrel{\text{def}}{=} (g^{1/4} G g^{1/4})^z (x, x) = \sum_\alpha \frac{g^{1/2}(x) [v(\alpha, x)]^2}{[\lambda(\alpha) + m^2 - i0]^z}, \quad (14.62)$$

$$\zeta(z) \stackrel{\text{def}}{=} \operatorname{tr}(g^{1/4} G g^{1/4})^z = \int \zeta(x, z) \, d^n x = \sum_\alpha \frac{1}{[\lambda(\alpha) + m^2 - i0]^z}. \quad (14.63)$$

The series in (14.62) is to be evaluated in regions of the complex z-plane where it converges, and $\zeta(x, z)$ is then defined elsewhere by analytic continuation. $\zeta(z)$ itself is to be regarded as a formal integral over spacetime.†

The definitions (14.62) and (14.63) are strictly valid only if one is using 'unitless' units (e.g. units in which $\hbar = c = G = 1$), so that $g^{1/4} G g^{1/4}$, $\lambda(\alpha)$, m^2, etc. are dimensionless. Otherwise a scale parameter must be inserted into the definitions. We omit the scale parameter here and indicate briefly later what modifications (if any) are necessary when it is included.

† In compact manifolds (and some other cases as well) the integral actually exists.

The one-loop approximation

The zeta functions are readily seen to be related to the function K:

$$\zeta(x,z) = \frac{1}{\Gamma(z)} \int_0^\infty (is)^{z-1} K(x,x,s)\,ids$$

$$= \frac{i}{(4\pi)^{n/2}} \frac{g^{1/2}}{\Gamma(z)} \int_0^\infty (is)^{z-1-n/2} e^{-im^2 s} \Lambda(x,x,s)\,ids, \quad (14.64)$$

$$\zeta(z) = \frac{1}{\Gamma(z)} \int_0^\infty (is)^{z-1} K(s)\,ids$$

$$= \frac{i}{(4\pi)^{n/2}} \frac{1}{\Gamma(z)} \int_0^\infty (is)^{z-1-n/2} e^{-im^2 s} \Lambda(s)\,ids, \quad (14.65)$$

$$\Lambda(s) \stackrel{\text{def}}{=} \int g^{1/2}(x) \Lambda(x,x,s)\,d^n x \sim \sum_{r=0}^\infty A_r (is)^r. \quad (14.66)$$

These relations furnish information about the analytic structure of $\zeta(x,z)$. Integration by parts combined with analytic continuation from $\operatorname{Re} z > n/2$ yields

$$\zeta(x,z) = \begin{cases} \dfrac{i}{4\pi} \dfrac{g^{1/2}}{(z-1)\Gamma(z+1)} \displaystyle\int_0^\infty (is)^z \left(\dfrac{\partial}{i\partial s}\right)^2 [e^{-im^2 s} \Lambda(x,x,s)]\,ids \\ \hspace{6cm} (n=2), \quad (14.67a) \\[2pt] \dfrac{i}{(4\pi)^{3/2}} \dfrac{g^{1/2}}{(z-\tfrac{1}{2})(z-\tfrac{3}{2})\Gamma(z)} \displaystyle\int_0^\infty (is)^{z-1/2} \left(\dfrac{\partial}{i\partial s}\right)^2 [e^{-im^2 s} \Lambda(x,x,s)]\,ids \\ \hspace{6cm} (n=3), \quad (14.67b) \\[2pt] -\dfrac{i}{(4\pi)^2} \dfrac{g^{1/2}}{(z-1)(z-2)\Gamma(z+1)} \displaystyle\int_0^\infty (is)^z \left(\dfrac{\partial}{i\partial s}\right)^3 [e^{-im^2 s} \Lambda(x,x,s)]\,ids \\ \hspace{6cm} (n=4), \quad (14.67c) \\[2pt] \text{etc.} \end{cases}$$

from which one may conclude that when n is even, $\zeta(x,z)$ has simple poles at $z = 1, 2 \cdots n/2$, and, when n is odd, at $z = n/2 - r$, $r = 1, 2, 3 \cdots$. There are no other singularities. We note that

$$\zeta(x,0) = \begin{cases} 0 & (n \text{ odd}) \\ \dfrac{i}{(4\pi)^{n/2}} \dfrac{g^{1/2}}{(n/2)!} \left\{ \left(\dfrac{\partial}{i\partial s}\right)^{n/2} [e^{-im^2 s} \Lambda(x,x,s)] \right\}_{s=0} & (n \text{ even}). \end{cases} \quad (14.68)$$

Chapter 14. Quantum gravity: the new synthesis

It is of interest to examine this last relation in the special case $m = 0$ and $\xi = \frac{1}{4}(n-2)/(n-1)$. The scalar field equation is then conformally invariant and, under a change $\delta g_{\mu\nu} = g_{\mu\nu}\,\delta\lambda$ in the metric tensor, one can readily show that $\delta(g^{-1/4}Fg^{-1/4}) = -\frac{1}{2}\{g^{-1/4}Fg^{-1/4},\,\delta\lambda\}$, whence

$$\delta(g^{1/4}Gg^{1/4}) = \tfrac{1}{2}\{g^{1/4}Gg^{1/4},\,\delta\lambda\}, \quad \delta\zeta(z) = z\,\mathrm{tr}\,[(g^{1/4}Gg^{1/4})^z\,\delta\lambda]. \tag{14.69}$$

In particular, $\delta\zeta(0) = 0$. But (14.68) yields

$$\zeta(0) = \begin{cases} 0 & (n\text{ odd}), \\ \dfrac{\mathrm{i}}{(4\pi)^{n/2}} A_{n/2} & (n\text{ even}), \end{cases} \quad \left(m = 0,\ \xi = \frac{1}{4}\frac{n-2}{n-1}\right) \tag{14.70}$$

from which it follows that $A_{n/2}$ (n even) is a conformal invariant.†

14.4.4 Regularization and renormalization

In addition to expression (14.44) the variational equation (14.42′) has also the following formal integral:

$$W = -\frac{\mathrm{i}}{2}\mathrm{tr}\,\ln\,(g^{1/4}Gg^{1/4}) + \mathrm{const} \tag{14.71a}$$

$$= \frac{\mathrm{i}}{2}\sum_\alpha \ln\,[\lambda(\alpha) + m^2 - \mathrm{i}0] + \mathrm{const}. \tag{14.71b}$$

Comparison of (14.63) and (14.61b) yields yet another expression for W:

$$W = -\frac{\mathrm{i}}{2}\zeta'(0) + \mathrm{const}. \tag{14.72}$$

There is an important difference between expressions (14.44) and (14.72). Whereas the integral in (14.44) diverges at the lower limit, expression (14.72) is *finite*.‡ The finiteness may be checked by differentiating (14.65) and analytically continuing from $\mathrm{Re}\,z > n/2$ to $z = 0$.

---

† The author is indebted to J. S. Dowker for this proof.
‡ Or rather it is the formal spacetime integral of a finite geometrical quantity.

Dropping the constant of integration in (14.72) one finds

$$W = \begin{cases} \dfrac{1}{8\pi}\Big\{(\gamma+1)(A_1 - m^2 A_0) \\ \qquad - \displaystyle\int_0^\infty \ln(is)\left(\dfrac{\partial}{i\partial s}\right)^2 [e^{-im^2 s} \Lambda(s)]\, ids \Big\} & (n=2), \qquad (14.73a) \\[2ex] \dfrac{1}{12\pi^{3/2}} \displaystyle\int_0^\infty (is)^{-1/2} \left(\dfrac{\partial}{i\partial s}\right)^2 [e^{-im^2 s}\Lambda(s)]\, ids & (n=3), \qquad (14.73b) \\[2ex] \dfrac{1}{32\pi^2}\Big\{\left(\gamma+\dfrac{3}{2}\right)\left(A_2 - m^2 A_1 + \dfrac{1}{2} m^4 A_0\right) \\ \qquad - \dfrac{1}{2}\displaystyle\int_0^\infty \ln(is)\left(\dfrac{\partial}{i\partial s}\right)^3 [e^{im^2 s}\Lambda(s)]\, ids \Big\} & (n=4), \qquad (14.73c) \\ \text{etc.} \end{cases}$$

This method of obtaining a finite value for W has been exploited by Dowker and Critchley (1976) and by Hawking (1977b), and is known as *zeta-function regularization*. There is another popular method, known as *dimensional regularization*, that yields very similar results. In this method one inserts (14.46) into (14.45) and (14.44) and defines W by analytic continuation in the spacetime dimension n. When the physical dimension is odd the zeta-function and dimensional methods yield identical results. When the physical dimension is even the dimensional results differ from expressions (14.73) in the terms involving A_0, A_1, A_2, etc. To get the former one must replace the factors containing the Euler constant, γ, in (14.73) by certain functions of n, each having a simple pole at the corresponding physical dimension.

It is conventional to regard the explicitly appearing As as renormalization terms that are to be absorbed by counterterms in the classical action S. Both regularization schemes therefore yield the same *physical* results. Note that when $n=4$ a counterterm proportional to A_2 must be included in the classical action. Such a term demands the introduction of a new *non-classical* coupling constant. We shall comment presently on some ambiguities to which the presence of this term gives rise.

We do not obtain a fully renormalized W simply by dropping the renormalization terms in (14.73). Other terms still have to be eliminated. Consider first the case $m \neq 0$. Through integration by

Chapter 14. Quantum gravity: the new synthesis

parts it is straightforward to verify that

$$-\frac{1}{8\pi}\int_0^\infty \ln(is)\left(\frac{\partial}{i\partial s}\right)^2 [e^{-im^2 s}\Lambda(s)]\,ids$$

$$= W_{\rm ren} - \frac{1}{8\pi}(\gamma + \ln m^2)(A_1 - m^2 A_0) - \frac{1}{8\pi}A_1 \quad (n=2), \quad (14.74)$$

$$\frac{1}{12\pi^{3/2}}\int_0^\infty (is)^{-1/2}\left(\frac{\partial}{i\partial s}\right)^2 [e^{-im^2 s}\Lambda(s)]\,ids$$

$$= W_{\rm ren} - \frac{m}{12\pi}\left(\frac{3}{2}A_1 - m^2 A_0\right) \quad (n=3), \quad (14.75)$$

$$-\frac{1}{64\pi^2}\int_0^\infty \ln(is)\left(\frac{\partial}{i\partial s}\right)^3 [e^{-im^2 s}\Lambda(s)]\,ids$$

$$= W_{\rm ren} - \frac{1}{32\pi^2}(\gamma + \ln m^2)(A_2 - m^2 A_1 + \tfrac{1}{2}m^4 A_0)$$

$$-\frac{1}{32\pi^2}\left(\frac{3}{2}A_2 - \frac{1}{2}m^2 A_1\right) \quad (n=4), \quad (14.76)$$

where

$$W_{\rm ren} = \begin{cases} \dfrac{1}{8\pi}\int_0^\infty \dfrac{1}{(is)^2} e^{-im^2 s}[\Lambda(s) - A_0 - A_1(is)]\,ids & (n=2), \quad (14.77a) \\[2mm] \dfrac{1}{16\pi^{3/2}}\int_0^\infty \dfrac{1}{(is)^{5/2}} e^{-im^2 s}[\Lambda(s) - A_0 - A_1(is)]\,ids & (n=3), \\ & \quad (14.77b) \\[2mm] \dfrac{1}{32\pi^2}\int_0^\infty \dfrac{1}{(is)^3} e^{-im^2 s}[\Lambda(s) - A_0 - A_1(is) - A_2(is)^2]\,ids & (n=4). \\ & \quad (14.77c) \end{cases}$$

$W_{\rm ren}$ is the fully renormalized W. Note that it can be obtained directly from (14.44) in each case simply by subtracting from $K(s)$ as many of the initial terms of its asymptotic expansion (14.55) as are needed to render the integral convergent. These subtracted terms may be regarded as corresponding to the ΔW of (14.32).

This simple subtraction algorithm also works when $m=0$ and n is odd. It does not always work when $m=0$ and n is even, but must then be examined case by case. The difficulty is that the factor $e^{-im^2 s}$ is no longer present in the integrands of expressions (14.77) and hence there is nothing to control the convergence of the integrals at the upper limit. The problem does not come from $\Lambda(s)$, which is assumed to be well behaved at this limit, but from the subtracted term of highest power is, namely the term in $A_{n/2}$.

Consider first $n = 2$. In this case, by the Gauss–Bonnet formula, A_1 is proportional to the Euler-Poincaré characteristic of spacetime. It is therefore not merely conformally invariant but *metric*-invariant, and makes no contribution to the renormalized stress tensor, which is given by

$$\langle \boldsymbol{T}^{\mu\nu}_{\text{ren}} \rangle = 2\delta W_{\text{ren}}/\delta g_{\mu\nu}. \qquad (14.78)$$

This means that (14.77a) can be used formally as it stands, yielding

$$\langle \boldsymbol{T}^{\mu\nu}_{\text{ren}}(x) \rangle = \frac{1}{8\pi} \int_0^\infty \frac{1}{(is)^2} [T^{\mu\nu}(x,s) - g^{1/2}(x)g^{\mu\nu}(x)] i ds \quad (m = 0, n = 2),$$

$$(14.79)$$

$$T^{\mu\nu}(x,s) \stackrel{\text{def}}{=} 2 \delta \Lambda(s)/\delta g_{\mu\nu}(x). \qquad (14.80)$$

The quantity inside the square brackets in (14.79) is of order $(is)^2$ as $s \to 0$, and hence the integral converges.

The case $n = 4$ is not so simple. Suppose $\xi = \frac{1}{4}(n-2)/(n-1) = \frac{1}{6}$, so that the field equation is conformally invariant. Then $a_1(x, x) = 0$ (see (14.53)) and the term in A_1 disappears from (14.77c). The term in A_2, on the other hand, does not vanish. Moreover, although A_2 is conformally invariant it is not metric-invariant. Its functional derivative with respect to $g_{\mu\nu}$ does not vanish except in special cases. The most important special cases are those in which spacetime is either conformally flat (vanishing Weyl tensor) or Ricci-flat (vanishing Ricci tensor). In these cases (14.77c) *can* be used formally as it stands, yielding

$$\langle \boldsymbol{T}^{\mu\nu}_{\text{ren}}(x) \rangle = \frac{1}{32\pi^2} \int_0^\infty \frac{1}{(is)^3} [T^{\mu\nu}(x,s) - g^{1/2}(x)g^{\mu\nu}(x)] i ds \qquad (14.81)$$

($m = 0$, $n = 4$, $\xi = \frac{1}{6}$, spacetime conformally flat or Ricci-flat).

The quantity inside the square brackets is then of order $(is)^3$ as $s \to 0$.

If we do not have the conformally invariant case, or if spacetime is neither conformally flat nor Ricci-flat, (14.77c) cannot be used even formally. We must work directly with (14.73c), defining

$$W_{\text{ren}} = -\frac{1}{64\pi^2} \int_0^\infty \ln(\kappa^2 is)\left(\frac{\partial}{i\partial s}\right)^3 \Lambda(s) \, ids \quad (m = 0, n = 4). \qquad (14.82)$$

Here the arbitrary scale factor, κ, that should have been inserted into the definitions (14.62) and (14.63) at the outset, forces its appearance. It is

715

Chapter 14. Quantum gravity: the new synthesis

not difficult to verify that in all previous cases κ disappears from the theory upon renormalization. In the present case, however, it survives in the logarithm. We are faced here with an ambiguity. It is easy to see that a change in κ produces a change in W_{ren} proportional to A_2. Such a change can be absorbed by a counterterm, but that does not help us because we can never get rid of κ completely. Short of appealing to experiment (which is far beyond the realm of possibility for the tiny effects that are involved here) we appear to have no way of determining what κ should be.

This ambiguity is related to the non-renormalizability of quantum gravity. When $n = 4$ we have to introduce two non-classical quantities already in one-loop order: a coupling constant multiplying the A_2 that S must contain, and a scale factor. As we have just remarked, the two are not independent; a change in one can be compensated by a change in the other. This is a special case of a set of more general relations that in recent years have come to be known as *renormalization-group equations*. If quantum gravity were renormalizable one could write down a finite number of these equations, each with a finite number of terms. In actuality an infinity of terms is required, reflecting the ever-increasing new types of infinities (and hence ambiguities) that one encounters in two-loop and higher orders. (See chapter 16 by Weinberg, pp. 000–000.) From the point of view of conventional perturbation theory an infinity of experiments is required to fix the theory.

On the other hand, the results of a number of exploratory studies (DeWitt, 1964; Isham, Strathdee and Salam, 1971, 1972) give one strong reasons to believe that if reliable methods of approximately summing the whole quantum perturbation series could be found, quantum gravity would turn out to be *finite*, with all secondary quantities fully determined by Planck's constant and the classical constants G and c.† In particular, to

---

† These studies were deficient in that the summations performed, although involving infinite numbers of terms, were always partial and yielded results that, at best, were gauge invariant only up to a finite order. Moreover, ambiguities cropped up in the analytic continuation procedures that had to be invoked. Hawking (1977c) has suggested that such approaches may be inherently bad, arguing that black holes of Planckian mass will play a dominant role (as instantons) in the gravitational vacuum. However, black holes do not necessarily mean that summation techniques have to be abandoned, for it has long been known that the Schwarzschild metric can be built up from linearized gravity theory on a flat background by summing the classical perturbation series, provided one is allowed to introduce new coordinates and carry out an analytic extension after the summation has been performed (see Duff, 1973). Hawking's ideas are none the less valuable in that they focus attention on the analytic techniques that have proved useful in black hole theory. These techniques may possibly help to resolve the ambiguities in the earlier investigations, and the emphasis on black holes may point the way to gauge-invariant summation methods and better ways of evaluating the effective action.

The one-loop approximation

the extent that the one-loop approximation (14.82) may provide a rough approximation to the exact effective action, the scale factor, κ, would turn out to be of the order of the Planck mass $(\hbar c/G)^{1/2}$. From this point of view one may regard (14.82) as defective only in that we have not yet learned how to compute the proportionality constant between κ and $(\hbar c/G)^{1/2}$.

14.4.5 Conformal invariance and the trace anomaly

By functionally differentiating the differential equation (14.39) and making use of the representation (14.59) it is possible to show that

$$T^{\mu\nu}(x,s) = -2\mathrm{i}(4\pi)^{n/2}(\mathrm{i}s)^{n/2+1} \sum_{\alpha} \{T^{\mu\nu}(v(\alpha,x))$$
$$+ \tfrac{1}{2}g^{1/2}(x)g^{\mu\nu}(x)[\lambda(\alpha)+m^2][v(\alpha,x)]^2\} \, \mathrm{e}^{-\mathrm{i}[\lambda(\alpha)+m^2]s}, \tag{14.83}$$

where $T^{\mu\nu}(v(\alpha,x))$ is obtained from the conventional expression for the stress tensor by substituting $v(\alpha,x)$ for the scalar field, ϕ. Using (14.57) it is easy to verify that the covariant divergence of the quantity inside the curly brackets in (14.83) vanishes and hence that $T^{\mu\nu}(x,s)_{;\nu}=0$, which is, of course, consistent with (14.80). It is also straightforward to verify (see Brown and Cassidy, 1977*a*) that

$$T_\mu{}^\mu(x,s) = g^{1/2}(x)\left[n - 2\mathrm{i}s\left(m^2 + \frac{\partial}{\mathrm{i}\partial s}\right)\right][\mathrm{e}^{-\mathrm{i}m^2 s}\Lambda(x,x,s)]. \tag{14.84}$$

If expression (14.84) is inserted into (14.79), (14.81) or (14.82) one finds that when $m=0$ and $\xi=\tfrac{1}{4}(n-2)/(n-1)$ the trace of the renormalized stress tensor, unlike that of the classical stress tensor, does *not* vanish if n is even. This is known as the *trace anomaly*.

The trace anomaly at a given point depends on the local geometry there. It can be computed in each case with the aid of (14.84) or, more simply, as follows. Note that the second of equations (14.69) implies

$$g_{\mu\nu}(x)\frac{\delta\zeta(z)}{\delta g_{\mu\nu}(x)} = z\zeta(x,z), \tag{14.85}$$

which, upon differentiation with respect to z, yields

$$g_{\mu\nu}(x)\frac{\delta\zeta'(z)}{\delta g_{\mu\nu}(x)} = \zeta(x,z) + z\frac{\partial}{\partial z}\zeta(x,z). \tag{14.86}$$

Note furthermore that when W and W_{ren} are defined by the zeta-function

Chapter 14. Quantum gravity: the new synthesis

method they differ from one another by a quantity proportional to $A_{n/2}$ which, as we have seen, is a conformal invariant. Therefore, from (14.72) and (14.86) it follows that

$$\langle T_{\text{ren}\,\mu}{}^\mu(x)\rangle = 2g_{\mu\nu}(x)\,\delta W_{\text{ren}}/\delta g_{\mu\nu}(x) = 2g_{\mu\nu}(x)\,\delta W/\delta g_{\mu\nu}(x)$$

$$= -ig_{\mu\nu}(x)\,\delta\zeta'(0)/\delta g_{\mu\nu}(x) = -i\zeta(x,0)$$

$$= \begin{cases} 0 & (n \text{ odd}) \\ (4\pi)^{-n/2}g^{1/2}(x)a_{n/2}(x,x) & (n \text{ even}) \end{cases} \quad (14.87)$$

In the one-loop approximation the brackets $\langle\ \rangle$ can be removed from the left side of (14.87). This follows (as may be seen by setting up an arbitrary Fock space) from the fact that the matrix element of the trace between any two orthogonal states vanishes. The trace of the renormalized stress-tensor *operator* is therefore a c-number (multiple of the identity operator).

We shall see in the next section that in a number of important cases the trace anomaly determines, in a very simple way, the entire stress tensor. In other cases, unfortunately, the determination of $\langle T_{\text{ren}}^{\mu\nu}\rangle$ requires difficult computation and can be carried out successfully only when a great deal is known about the eigenfunctions of the operator F appearing in the field equation. Moreover, these eigenfunctions have to be handled in a sophisticated way. The naive mode sums that are often written for $\langle T^{\mu\nu}\rangle$ must be treated with extreme caution because they diverge and must be modified by subtractions that do not always obviously maintain the validity of (14.22). The difficulty is most acute when $m = 0$. It is useful in this case to have alternative sums that are already correctly renormalized. An example of such a sum is the following, which is obtained by substituting (14.83) into (14.81), expressing $\Lambda(x, x, s)$ as a sum over the $v(\alpha, x)$, and carrying out an integration by parts:

$$\langle T_{\text{ren}}^{\mu\nu}(x)\rangle = \int_0^\infty \sum_\alpha \{T^{\mu\nu}(v(\alpha, x)) + \tfrac{1}{4}\lambda(\alpha)g^{1/2}(x)g^{\mu\nu}(x)[v(\alpha, x)]^2\}\,e^{-i\lambda(\alpha)s}\,ds$$

$$+ \frac{1}{4(4\pi^2)}g^{1/2}(x)g^{\mu\nu}(x)a_2(x,x) \quad (14.88)$$

($m = 0$, $n = 4$, $\xi = \tfrac{1}{6}$, spacetime conformally flat or Ricci-flat).

This expression, in which the trace anomaly appears explicitly, has been checked by direct computation in simple cases and gives promise of being useful in a variety of more complicated settings. One must be careful not to interchange the order of summation and integration.

The one-loop approximation

14.4.6 Conformally flat spacetimes

It has been pointed out by Brown and Cassidy (1977b) that when $m=0$ and $\xi = \frac{1}{4}(n-2)/(n-1)$ the change in $\langle T_{\text{ren}}^{\mu\nu}\rangle$ resulting from a conformal variation in the metric is entirely determined by the trace anomaly. This follows from the readily verified identity

$$\left[g_{\mu\gamma}(x)\frac{\delta}{\delta g_{\nu\gamma}(x)}, \; g_{\alpha\beta}(x')\frac{\delta}{\delta g_{\alpha\beta}(x')}\right] = 0, \qquad (14.89)$$

which, when applied to W_{ren}, yields

$$\delta\langle T_{\text{ren}\,\mu}{}^{\nu}(x)\rangle = g_{\mu\gamma}(x)\frac{\delta}{\delta g_{\nu\gamma}(x)}\int \langle T_{\text{ren}\,\alpha}{}^{\alpha}(x')\rangle \,\delta\lambda(x')\,d^n x' \qquad (14.90)$$

where δ denotes the effect of varying the metric by the amount $\delta g_{\mu\nu} = g_{\mu\nu}\,\delta\lambda$.

It is straightforward to carry out the functional differentiation on the right-hand side of (14.90). If the metric is already conformally flat it is possible to integrate the resulting variational equation all the way from flat spacetime. In the case $n=2$, writing $g_{\mu\nu} = e^{\lambda}\eta_{\mu\nu}$ where $\eta_{\mu\nu}$ is the Minkowski metric, one finds

$$\langle T_{\text{ren}\,\mu}{}^{\nu}\rangle = \langle T_{\text{ren}\,\mu}{}^{\nu}\rangle_b + \frac{1}{24\pi}(\lambda_{,\mu}{}^{\nu} - \delta_{\mu}{}^{\nu}\lambda_{,\sigma}{}^{\sigma} - \tfrac{1}{2}\lambda_{,\mu}\lambda_{,}{}^{\nu} + \tfrac{1}{4}\delta_{\mu}{}^{\nu}\lambda_{,\sigma}\lambda_{,}{}^{\sigma}) \qquad (14.91)$$

$$(m=0, \; n=2, \; \xi=0, \; g_{\mu\nu}=e^{\lambda}\eta_{\mu\nu}),$$

where $\langle T_{\text{ren}\,\mu}{}^{\nu}\rangle_b$ is the renormalized stress tensor in flat spacetime and indices on the right are raised and lowered by means of the Minkowski metric.

In the case $n=4$, remembering that the Riemann tensor and the Bianchi identity are expressible entirely in terms of the Ricci tensor when spacetime is conformally flat, one finds by straightforward computation that (14.90) integrates to

$$\langle T_{\text{ren}\,\mu}{}^{\nu}\rangle = \langle T_{\text{ren}\,\mu}{}^{\nu}\rangle_b + \frac{g^{1/2}}{2880\pi^2}[-R_{\mu\alpha}R^{\nu\alpha} + RR_{\mu}{}^{\nu} - \tfrac{1}{3}R_{;\mu}{}^{\nu}$$
$$+ \delta_{\mu}{}^{\nu}(\tfrac{1}{2}R_{\alpha\beta}R^{\alpha\beta} - \tfrac{1}{3}R^2 + \tfrac{1}{3}R_{;\alpha}{}^{\alpha})] \qquad (14.92)$$

$$(m=0, \; n=4, \; \xi=\tfrac{1}{6}, \text{ spacetime conformally flat}).$$

When n is odd we have, of course,

$$\langle T_{\text{ren}\,\mu}{}^{\nu}\rangle = \langle T_{\text{ren}\,\mu}{}^{\nu}\rangle_b \qquad (14.93)$$

$(m=0, \; n \text{ odd}, \; \xi=\tfrac{1}{4}(n-2)/(n-1), \text{ spacetime conformally flat}).$

Chapter 14. Quantum gravity: the new synthesis

Note that $\langle T_{\text{ren}\,\mu}{}^\nu\rangle_b$ *need not be zero.* As we have seen earlier, it generally depends on topology and, if the flat reference-manifold is incomplete, on boundary conditions as well. In the present case the topology and boundary structure of the flat manifold are determined by its conformal relation to the curved manifold. Under the conformal transformation the representation (14.35) for the Feynman propagator is assumed to remain intact, and the integrated expressions (14.91), (14.92), and (14.93) thereby inherit a definition of the 'vacuum' from the flat manifold. Note that these expressions are purely real, which implies that W has no imaginary part. This means that particle production is absent and hence that $|\text{in, vac}\rangle$ and $|\text{out, vac}\rangle$ are identical.

Since every two-dimensional manifold is locally conformally flat, (14.91) is of universal validity. By mapping manifolds with curved boundaries conformally onto manifolds with straight boundaries one may express the stress tensor in complicated cases (e.g. the two-dimensional Rindler wedge) in terms of the stress tensor in simple cases (e.g. the half-space). The results are found to agree with results obtained by other methods *even when the Riemann tensor* vanishes. This means that the conformal anomaly is not just a consequence of curvature but is a manifestation of a deep consistency property of the theory.

The Robertson–Walker manifolds constitute an important class to which (14.92) may be applied. The simplest of these is de Sitter spacetime, which has a Riemann tensor given by $R_{\mu\nu\sigma\tau} = K(g_{\mu\sigma}g_{\nu\tau} - g_{\mu\tau}g_{\nu\sigma})$. Substitution into (14.92) and use of the boundary condition $\langle T^{\mu\nu}_{\text{ren}}\rangle \underset{K\to 0}{\to} 0$ immediately yield

$$\langle T^{\mu\nu}_{\text{ren}}\rangle = -\frac{K^2}{960\pi^2}g^{1/2}g^{\mu\nu} \quad (m=0,\ \xi=\tfrac{1}{6},\ \text{de Sitter spacetime}), \quad (14.94)$$

a result that has also been obtained, with much labor, by several other methods. The 'vacuum' state involved here is de Sitter invariant.

14.5 The full quantum theory

The methods of the preceding sections are as applicable to fields with spin as they are to the scalar field. A few new technical devices must be introduced, such as bitensors or bispinors to describe the parallel propagation of field polarizations along geodesics, and gauge-fixing terms and 'ghost' fields when gauge groups are present.† But basically the

---

† How these enter may be deduced from the full theory described in the following sections.

The full quantum theory

one-loop theory of $\langle T^{\mu\nu}_{\text{ren}} \rangle$ is the same for the electromagnetic, spin $-\frac{1}{2}$, Proca and gravitational fields as it is for the scalar field.†

As we have remarked earlier, the one-loop theory is the WKB approximation to the full theory to which we now turn. From this point on we shall be able to make fewer rigorous statements. Much of the time we shall be dealing with formal equations from which no one has yet learned how to extract reliable finite answers. The starting point for these equations, however, is the Feynman functional integral, which is widely believed to have a validity that transcends perturbation theory. The equations themselves, in one form or another, are therefore expected to survive future developments.

For simplicity we shall consider pure quantum gravity, but we shall introduce a compact notation that is equally applicable to all field theories. We begin by replacing the symbol $g_{\mu\nu}(x)$ for the quantum metric tensor by the symbol φ^i. The index i will be understood to stand for the combined set of labels μ, ν, x, where x identifies a spacetime point. The reason for including the continuous label x in this set is that much of the formalism of quantum field theory is purely combinatorial, with summations over dummy tensor indices being frequently accompanied by integrations over spacetime. In order to avoid having to write a lot of integral signs we lump the xs and the tensor indices together and adopt the convention that the repetition of a lower case Latin index implies a combined summation–integration. By extending the range of the index i one can include also the components of other fields in the symbol φ^i. Extra notation for keeping track of signs and factor orderings is needed if some of these fields are fermionic, but we ignore this complication for the present.

14.5.1 The gauge group

The only other kind of index that will be introduced is a lower case Greek index taken from the first part of the alphabet, called a *group* index. The way group indices arise is as follows. Consider the diffeomorphism that is generated by the displacement of spacetime points through an infinitesimal differentiable contravariant vector, $-\delta\xi$. In coordinate language one would write the corresponding transformation as $\bar{x}^\mu = x^\mu + \delta\xi^\mu$, and the effect of this transformation on $g_{\mu\nu}(x)$ would take the

† In supergravity theories the massless spin-3/2 field is also included.

Chapter 14. Quantum gravity: the new synthesis

form

$$\delta g_{\mu\nu}(x) = -\pounds_{\delta\xi} g_{\mu\nu}(x)$$
$$= \int [-g_{\mu\nu,\sigma}(x)\,\delta(x, x') - g_{\sigma\nu}(x)\,\delta_{,\mu}(x, x')$$
$$- g_{\mu\sigma}\,\delta_{,\nu}(x, x')]\,\delta\xi^\sigma(x')\,d^n x'. \tag{14.95}$$

The transformation (14.95) is called a *gauge transformation* and the diffeomorphism group is called the *gauge group* of quantum gravity.

It is not difficult to show that the quantity inside the brackets in the integral (14.95) is a bitensor density, transforming as a covariant tensor at the point x and as a covariant vector density of unit weight at the point x'. In the compact notation we replace this quantity by the symbol $Q^i_\alpha[\varphi]$ and rewrite (14.95) in the form

$$\delta\varphi^i = Q^i_\alpha[\varphi]\,\delta\xi^\alpha. \tag{14.96}$$

Here the labels μ, ν, x have been replaced by the index i, and the labels σ, x' have been replaced by the *group* index α.

Equation (14.96) expresses the action of the diffemorphism on the space of metric tensors φ^i. This action is a *realization* of the diffeomorphism. The Qs satisfy an important identity arising from the fact that diffeomorphisms form a group:

$$Q^i_{\alpha,j}Q^j_\beta - Q^i_{\beta,j}Q^j_\alpha = Q^i_\gamma c^\gamma_{\alpha\beta}. \tag{14.97}$$

Here (and from now on) commas are used to denote *functional* differentiation with respect to the φs. The cs are the structure constants of the gauge group and satisfy the cyclic identity†

$$c^\alpha_{\beta\varepsilon}c^\varepsilon_{\gamma\delta} + c^\alpha_{\gamma\varepsilon}c^\varepsilon_{\delta\beta} + c^\alpha_{\delta\varepsilon}c^\varepsilon_{\beta\gamma} = 0. \tag{14.98}$$

The range of the group indices can obviously be extended to include other gauge groups associated with other fields (e.g. Yang–Mills fields). The compact notation, in particular in (14.97) and (14.98), remains the same. By allowing for anticommutativity and generalized symmetries for the structure constants one can even include supergauge groups.

---

† In full notation the structure constants of the diffemorphism group are written $c^\mu_{\nu'\sigma''}$ and defined by

$$\int d^n x' \int d^n x''\, c^\mu_{\nu'\sigma''} X^{\nu'} Y^{\sigma''} = -[X, Y]^\mu$$

where X and Y are arbitrary contravariant vector fields and [,] denotes the Lie bracket. The $c^\mu_{\nu'\sigma''}$ are the components of a 3-point tensor density, transforming as a contravariant vector at x and as a covariant vector density of unit weight at both x' and x''. The cyclic identity that they satisfy is just the Jacobi identity for Lie brackets.

It is a remarkable fact, and one that can be frequently exploited, that the fields encountered in practice usually provide *linear* realizations of their gauge groups. This means that the functional derivatives $Q^i_{\alpha,j}$ are independent of the φ^i and, when regarded as continuous matrices (in i and j), yield a matrix *representation* of the Lie algebra associated with the group. Of course this simplicity is generally lost if the φs are replaced by nonlinear functions of themselves. But it is remarkable that there is usually a 'natural' set of field variables of which the Qs are linear functionals.†

14.5.2 The Feynman functional integral; the measure

Consider now a transition amplitude of the form $\langle \text{out} | \text{in} \rangle$ where the vectors $|\text{in}\rangle$ and $|\text{out}\rangle$ refer to states in which the field is maximally specified (in the quantum mechanical sense, for example, in terms of a complete set of commuting observables) in regions 'in' and 'out' respectively. These states need not be 'vacuum' states and the regions 'in' and 'out' need not refer to the infinite past and future respectively. If spacetime has geometrical singularities in the past and/or future, $|\text{in}\rangle$ and $|\text{out}\rangle$ may be defined not in terms of observables at all, but by some technical analytic continuation procedure. It will be assumed only that the 'in' and 'out' regions lie respectively to the past and future of the region of dynamical interest.

There are many ways of showing that the amplitude $\langle \text{out} | \text{in} \rangle$ can be expressed as a formal functional integral:

$$\langle \text{out} | \text{in} \rangle = N \int e^{iS[\varphi]} \mu[\varphi] \, d\varphi. \tag{14.99}$$

Here, N is a normalization constant, $S[\varphi]$ is the classical action functional, $\mu[\varphi]$ is a certain measure functional, and the integration is to be extended over all fields, φ, that satisfy the boundary conditions appropriate to the given 'in' and 'out' states. The integration also embraces as many topologies as can be reached by analytic continuation from the given background topology.

Expression (14.99) was first derived by Feynman (1948) in ordinary quantum mechanics, without gauge groups, and later (1950) applied by him to field theory. The full extension to field theories with gauge groups is the work of many people, and the reader is referred to the literature for

---

† In gravity theory there is a *family* of 'natural' fields, namely all tensor densities of the form $\mathbf{g}^r g^{\mu\nu}$ or $\mathbf{g}^{-r} g_{\mu\nu}$ where $r \neq 1/n$.

Chapter 14. Quantum gravity: the new synthesis

details.† In combination with the relation

$$\langle \text{out} | T(A[\varphi]) | \text{in} \rangle = N \int A[\varphi] \, e^{iS[\varphi]} \mu[\varphi] \, d\varphi \qquad (14.100)$$

(14.99) yields immediately such variational laws as (14.1), (14.2), (14.29), and (14.42).

The correct choice for the measure functional $\mu[\varphi]$ has been a subject of controversy for many years. There is no good reference that gives an overview of the problems involved, nor is there space in this review to do the job. Suffice it to say that the choice adopted by a given author depends on several factors: (1) how he chooses to think of the functional integral as being defined (e.g. by time-slicing or in terms of Green's functions and Feynman rules); (2) whether he works with a Lagrangian or a Hamiltonian scheme; (3) what interpretation he chooses to give to certain ambiguous expressions. Many of the ambiguities are related to factor-ordering problems and the problem of how to give meaning to local operators constructed from the basic fields φ^i. In renormalizable theories such ambiguities disappear upon adequate regularization, and there is no disagreement about how to carry out actual calculations. In the case of non-renormalizable theories, by contrast, conflicting interpretations are presently reminiscent of ancient arguments about angels and pins, because reliable finite answers do not yet exist.

We shall here adopt an uncompromisingly formal stance, as being the most likely to survive future developments, given the known tendency for pure formalism to acquire and maintain a consistency and logic of its own. We begin by noting that the amplitude (14.99) must be gauge-invariant. Since the action functional in the exponent of the integrand is already gauge-invariant we can secure gauge-invariance of the integral by requiring the 'volume element', $\mu \, d\varphi$, to be gauge-invariant. The easiest way to do this is to introduce a metric tensor, γ_{ij}, on the space of fields, φ^i, for which the actions of the gauge group are isometries. Explicitly we require

$$0 = \pounds_{Q\alpha} \gamma_{ij} = \gamma_{ij,k} Q^k{}_\alpha + \gamma_{kj} Q^k{}_{\alpha,i} + \gamma_{ik} Q^k{}_{\alpha,j}. \qquad (14.101)$$

We then choose

$$\mu = \text{const} \times [\det (\gamma_{ij})]^{1/2}, \qquad (14.102)$$

assuming the determinant can be defined in a suitable way. It is not difficult to verify that (14.97) is the integrability condition for (14.101)

---

† Useful modern references are Fadde'ev (1969), (1976), and Abers and Lee (1973).

and that (14.101) itself implies formally

$$(\mu Q^i{}_\alpha)_{,i} = 0. \tag{14.103}$$

This is an equation of 'divergenceless flow' that could in principle be used to select a suitable measure functional independently of a metric.

In pure quantum gravity, with $g_{\mu\nu}$ as the basic field variables, (14.101) translates into the statement that γ_{ij} must be a 2-point function, transforming under the diffeomorphism group, like a symmetric contravariant tensor density of unit weight at each point. Among all possible such bitensor densities there is a unique (up to a constant factor) 1-parameter family of them that may be characterized as *local*. These are given by

$$\gamma^{\mu\nu\sigma'\tau'} = \mathfrak{G}^{\mu\nu\sigma\tau}\delta(x, x'), \tag{14.104}$$

$$\mathfrak{G}^{\mu\nu\sigma\tau} \stackrel{\text{def}}{=} \tfrac{1}{2}g^{1/2}(g^{\mu\sigma}g^{\nu\tau} + g^{\mu\tau}g^{\nu\sigma} + \lambda g^{\mu\nu}g^{\sigma\tau}), \quad \lambda \neq -2/n. \tag{14.105}$$

The delta function appearing in (14.104) gives $\gamma^{\mu\nu\sigma'\tau'}$ a 'block' structure that permits its determinant to be expressed formally as

$$\det(\gamma^{\mu\nu\sigma'\tau'}) = \prod_x \mathfrak{G}(x), \tag{14.106}$$

where $\mathfrak{G}(x)$ is the determinant of the $\tfrac{1}{2}n(n+1) \times \tfrac{1}{2}n(n+1)$ matrix $\mathfrak{G}^{\mu\nu\sigma\tau}$. It is not a difficult computation to show that

$$\mathfrak{G} = (-1)^{n-1}\left(1 + \frac{n\lambda}{2}\right)g^{\frac{1}{4}(n-4)(n+1)} \tag{14.107}$$

where $g = -\det(g_{\mu\nu})$. In a four-dimensional spacetime $\mathfrak{G}$, and hence $\det(\gamma^{\mu\nu\sigma'\tau'})$, is seen to be a constant, independent of the $g_{\mu\nu}$. The measure, μ, may therefore be taken to be a constant; without loss of generality it may be chosen equal to 1. This will no longer be true in other dimensions or when other fields are present, if we stick to $g_{\mu\nu}$ as the basic variables. However, we can in principle replace the $g_{\mu\nu}$ by the variables $g^{-r}g_{\mu\nu}$ or $g^r g_{\mu\nu}$ ($r \neq 1/n$) and choose r so that μ remains constant. In practice, as we shall see later, this is superfluous. To set μ effectively equal to unity it turns out to be necessary only to choose basic fields that transform linearly under the gauge group. Equation (14.99) then becomes simply

$$\langle \text{out} | \text{in} \rangle = N \int e^{iS[\varphi]} d\varphi. \tag{14.108}$$

Chapter 14. Quantum gravity: the new synthesis

14.5.3 Globally valid gauge conditions; canonical coordinates

When a gauge group is present the integration in (14.108) is redundant. As φ ranges over a group *orbit* in the space of fields, the exponent in the integrand remains constant.† It is useful to eliminate this redundancy by adopting a gauge condition. In order to find a gauge condition that will be valid in contexts other than perturbation theory one must proceed with care.

Denote the space of all fields φ^i embraced in the integral (14.108) by Φ, and the gauge group by G. In pure gravity theory Φ is the space, Riem (M), of Riemannian metrics on the spacetime manifold M, and G is the group, Diff (M), of diffeomorphisms of M. The *physical* field-space is the quotient Φ/G, also called the *space of orbits*. In gravity theory the space of orbits, Riem $(M)/$Diff (M), is the space of *geometries* on M.

Consider a typical, i.e. *generic*, orbit. Modulo a possible discrete center it is a *copy* of the group manifold, G, because it provides a realization of G and has the same dimensionality. Not all orbits need have this dimensionality. There is often a class of degenerate orbits having fewer dimensions. Such orbits are *boundary points* of Φ/G and are generated by points in Φ that remain invariant under the action of non-trivial (continuous) subgroups of G. The greater the dimensionality of the invariance subgroup the smaller the dimensionality of the orbit. Fischer (1970) has shown that the orbits of given dimension may be assembled into boundary submanifolds so that the whole orbit space becomes a *stratified* manifold. In gravity theory the degenerate orbits are the symmetrical geometries, i.e. those that possess Killing vectors.‡ For ease of visualization one may think of Φ as being R^3, and G as being the group of rotations about a fixed axis. The orbits are then circles perpendicular to, and centered on, the axis, and the orbit space is a half-plane whose boundary points correspond to the points on the axis, which remain invariant under the group.

A *globally valid* gauge condition is a set of constraints that picks out a subspace in Φ, of codimension equal to the dimension of G, which intersects each orbit in precisely one point. In the case of the Yang–Mills field such a subspace does *not* exist if the gauge group corresponds to a

---

† The gauge invariance of $S[\varphi]$ may be expressed in the form $S_{,i}Q^i{}_\alpha \equiv 0$ which, in pure gravity theory, is the contracted Bianchi identity.

‡ Strictly speaking this is true only if spacetime is compact or has compact spatial sections. Minkowski spacetime, for example, is *not* a degenerate orbit, because the gauge parameters $\delta\xi^\sigma$ in (14.95) are required to vanish at infinity. The Poincaré isometries are therefore not contained in the gauge group.

twisted fiber bundle. Much less is known about the diffeomorphism group. Although its structure depends on the nature of the spacetime manifold, M, it cannot be viewed as a fiber bundle. Plausibility arguments can be adduced to suggest that globally valid gauge conditions *can* be introduced in the diffeomorphism case under a wide range of topologies for M. We shall here simply *assume* that a subspace with the desired properties exists. This subspace may be regarded as *representing* the orbit space Φ/G. Each orbit is represented by the point at which it intersects the subspace. To express this idea in equations one may think of the variables φ^i as being replaced by other variables $I^A[\varphi], P^\alpha[\varphi]$, where the I^A label individual orbits and are gauge-invariant, and the P^α label corresponding points *in* each orbit. The point on each orbit which is selected by the given gauge condition may be chosen as the origin of the 'coordinates' P^α in that orbit. The gauge condition is then simply $P^\alpha = 0$.

It will actually prove convenient to work with the continuum of gauge conditions

$$P^\alpha[\varphi] = \zeta^\alpha, \qquad (14.109)$$

where the ζ^α are constants (i.e. independent of the φ^i) whose values range from $-\infty$ to ∞. In order to embrace this range the Ps will have to constitute a very special set of coordinates on the orbits. Such coordinates may be obtained (in principle) as follows. Remembering that each (generic) orbit is a copy of G, let the origin, $P^\alpha = 0$, on each orbit be identified with the identity element of G. Then choose the Ps to be *canonical* group coordinates, i.e. normal coordinates generated by the 1-parameter Abelian subgroups of G. Noting that the action of the gauge group on each (generic) orbit mimics its action on itself, one is led to the functional differential equation

$$P^\alpha_{,i}[\varphi] Q^i_\beta[\varphi] = Q^\alpha_\beta[P[\varphi]], \qquad (14.110)$$

where the Q^α_β bear the same relation to the group manifold as the Q^i_α bear to Φ. In particular, they satisfy the analog of (14.97):

$$Q^\alpha_{\beta,\delta} Q^\delta_\gamma - Q^\alpha_{\gamma,\delta} Q^\delta_\beta = Q^\alpha_\delta c^\delta_{\alpha\beta}, \qquad (14.111)$$

where commas followed by Greek indices denote functional differentiation with respect to the Ps. Equation (14.111) is easily verified to be the integrability condition for (14.110).

Equations (14.110) and (14.111) hold for any coordinatization of the group. What renders canonical coordinates special is the form that Q^α_β

Chapter 14. Quantum gravity: the new synthesis

takes when they are used. One may show (e.g. DeWitt, 1965) that

$$Q^{-1}[P] = \frac{e^{c \cdot P} - I}{e^{c \cdot P}} \stackrel{\text{def}}{=} I + \frac{1}{2!} c \cdot P + \frac{1}{3!} (c \cdot P)^2 + \ldots \quad (14.112)$$

where I is the unit matrix (delta function) and

$$Q[P] \stackrel{\text{def}}{=} (Q^\alpha{}_\beta[P]), \quad c \cdot P \stackrel{\text{def}}{=} (c^\alpha{}_{\gamma\beta} P^\gamma). \quad (14.113)$$

The series (14.112) converges for *all* values of the P^α. For certain values the (continuous) matrix Q^{-1} may have vanishing roots which cause the right-hand side of (14.110) to blow up. At such values the canonical coordinate system becomes singular. However, singularities do not impair its usefulness. Canonical Ps are analogous to angular coordinates. As they vary over their allowed ranges (i.e. from $-\infty$ to ∞) each given orbit may be covered many (∞) times. But each set of Ps on a given orbit still defines a unique point on that orbit.

Equations (14.110) do not suffice completely to determine the functionals $P^\alpha[\varphi]$. Additional conditions are needed to 'line up' corresponding points on adjacent orbits. One possible way to do the lining up is as follows. Introduce into the field space Φ a non-singular metric γ_{ij} satisfying (14.101). Require that the matrix $\gamma_{ij} Q^i{}_\alpha Q^j{}_\beta$ be non-singular on all generic orbits so that a vector cannot be simultaneously tangent to and orthogonal to any of them. An associated metric can then be induced on the orbit space Φ/G by defining the distance between neighboring orbits in Φ/G to be the orthogonal distance between them in Φ. In pure gravity theory, if γ_{ij} is taken in the form (14.104), (14.105), with λ chosen to be -1, then $\gamma_{ij} Q^i_\alpha Q^j_\beta$ is a slight generalization of the Laplace–Beltrami operator for vector fields. When spacetime is asymptotically flat this operator is non-singular, so the required condition can be met in important practical cases. Another property of this special γ_{ij} is that when it is used, every pair of points in Φ can be connected by a unique geodesic. Methods for proving this will be found in De Witt (1967a), and the property will be assumed in what follows.

Choose a generic orbit and call it the *base orbit*. Call the identity on that orbit the *base point*. Let V be the subspace of Φ generated by the set of all geodesics emanating from the base point in directions orthogonal to the base orbit. It can be shown (De Witt, 1967a) that if a geodesic intersects one orbit orthogonally then it intersects every orbit in its path orthogonally and, moreover, traces out a geodesic curve in the orbit space Φ/G. Using the fact that every pair of points in Φ can be connected by a unique geodesic, and the fact that a geodesic cannot be simultaneously orthogonal to and tangent to an orbit, one can show that V ultimately intersects all orbits. To keep it from intersecting a given orbit more than once, one may terminate each of the generating geodesics as soon as it strikes a boundary point of Φ/G. V is then topologically (but not necessarily metrically) a copy of Φ/G.†

If another subspace, V', is constructed like V but starting from another point on the base orbit, it too will intersect all the orbits. Because the group operations are isometries of γ_{ij}, the P^α will be constants over V'. That is, once the identity points are 'lined up' all the other points are automatically lined up too.

The idea of using orthogonality to generate a representative subspace for Φ/G is an old one. It is usually applied in a rather crude manner, however, by choosing the functionals P^α to have the form

$$P^\alpha[\varphi] = Q^i{}_\alpha[g]\gamma_{ij}[g]\phi^j, \quad \phi^i \stackrel{\text{def}}{=} \varphi^i - g^i, \qquad (14.114)$$

where g^i is a background field. For example, if γ_{ij} has the form (14.104), (14.105) then, with the choice (14.114), the condition $P^\alpha = 0$ becomes

$$g^{1/2}(2\phi_\mu{}^\nu + \lambda \delta_\mu{}^\nu \phi_\sigma{}^\sigma)_{;\nu} = 0. \qquad (14.115)$$

With λ set equal to -1 this is a very popular gauge condition in gravity theory. (Here indices are raised and lowered by means of the background metric $g_{\mu\nu}$ and the covariant derivative is defined in terms of it.)

The condition $P^\alpha = 0$ with P^α taken in the form (14.114), or, more generally, the condition (14.109) with P^α taken in the form $P_{\alpha i}[g]\phi^i$ for some $P_{\alpha i}$, is known as a *linear* gauge condition. Linear gauge conditions are extremely convenient in perturbation theory, where the quantum field never gets very far from the background field. Covariant (with respect to the background field) gauge conditions like (14.115) are usually the best, but for some purposes non-covariant gauges (e.g. the Coulomb gauge in Yang–Mills theory) are more useful. In non-perturbative studies, however, linear gauge conditions have to be used with great care (see Gribov, 1977). This can be seen already with a condition

---

† To gain an appreciation of some of the metrical situations that can arise think of Φ as being R^3, and G as being the group of screw motions with fixed pitch about some axis. The orbits are then helices, and all are generic. If R^3 bears the Cartesian metric then the orbit space is topologically, but not metrically, a plane. Note that in this example there exist no surfaces that intersect all orbits orthogonally, although every plane not containing the axis is perpendicular to some orbit at its intersection point and is a surface like V, based on that orbit.

Chapter 14. Quantum gravity: the new synthesis

like (14.115) which defines a subspace that is orthogonal, approximately, only to those orbits that lie close to the background orbit through $\phi_{\mu\nu} = 0$. At least five things can go wrong with linear gauge conditions when applied globally: (1) The subspace defined by a linear condition has no boundary. Therefore it cannot represent Φ/G faithfully if there are degenerate orbits. (2) The subspace defined by a linear condition may intersect some orbits more than once. (3) There may be some orbits that it does not intersect at all. (4) Even if it intersects all orbits when the ζ^α in (14.109) have certain values, it may not intersect all the orbits when the ζ^α have other values. (5) When G is 'twisted' there are no globally valid gauge conditions at all, linear or otherwise.

It is possible in some cases to patch things up so that the advantages of linear gauge conditions can be maintained. This has been done in certain global studies in Yang–Mills theory. However, the diffeomorphism group of gravity theory is a much more complicated group than the Yang–Mills group, and both the difficulties to which it gives rise globally and the opportunities that it presents for technical innovation are almost unknown at the present time. In order to keep all options open we shall first develop the formal theory of the amplitude ⟨out | in⟩ using foolproof gauge conditions based on canonical coordinates and then make some remarks about how things might go when other gauge conditions are used.

14.5.4 Factoring out the gauge group

The way to remove the redundancy from the integral (14.108) by using a gauge condition was first shown by Faddeev and Popov (1967). We here adopt a refinement of their method. Let ξ be an arbitrary element of the gauge group (viewed as an abstract group) and let ξ^α be the canonical coordinates of ξ. Let φ^i be an arbitrary field in Φ. Denote by ${}^\xi\varphi^i$ the field to which φ^i is displaced under the action of ξ. Define

$$\Delta[\zeta, \varphi] \stackrel{\text{def}}{=} \int_G \delta[P[{}^\xi\varphi] - \zeta]\det Q^{-1}[\xi]\, d\xi, \quad d\xi \stackrel{\text{def}}{=} \prod_\alpha d\xi^\alpha, \quad (14.116)$$

where $\delta[\ ]$ is the delta *functional*, $P^\alpha[\varphi]$ are the functionals constructed by the canonical procedure described in the preceding section, Q^{-1} is the matrix defined by the series (14.112), and the integration extends over the entire gauge group! Because of the presence of the delta functional, the integrand 'switches on' only at one point in G, namely at that point for which ${}^\xi\varphi^i$ is equal to the unique field φ^i_ζ lying on the orbit containing φ^i

and picked out by the gauge condition (14.109):

$$P^\alpha[\varphi_\zeta] = \zeta^\alpha. \tag{14.117}$$

The determinant det $Q^{-1}[\xi]$ appearing in (14.116) is the well known right-invariant measure for G, satisfying

$$\det Q^{-1}[\xi\xi'] \, d(\xi\xi') = \det Q^{-1}[\xi] \, d\xi \quad \text{for all } \xi' \text{ in } G, \tag{14.118}$$

where $\xi\xi'$ is the group product of ξ and ξ'. Its presence renders the functional Δ gauge-invariant:

$$\Delta[\zeta, {}^{\xi'}\varphi] = \Delta[\zeta, \varphi] \quad \text{for all } \xi' \text{ in } G. \tag{14.119}$$

Because of this gauge invariance, evaluation of Δ is easy. One has only to shift φ^i to φ_ζ^i so that the integrand switches on at the identity element $\xi^\alpha = 0$. All quantities can then be expanded in power series in the ξ^α. For example, the argument of the delta functional becomes

$$\begin{aligned}P^\alpha[{}^\xi\varphi_\zeta] - \zeta^\alpha &= P^\alpha[\varphi_\zeta] - \zeta^\alpha + P^\alpha{}_{,i}[\varphi_\zeta] Q^i{}_\beta[\varphi_\zeta]\xi^\beta + \ldots \\ &= Q^\alpha{}_\beta[P[\varphi_\zeta]]\xi^\beta + \ldots, \end{aligned} \tag{14.120}$$

and hence

$$\begin{aligned}\Delta[\zeta, \varphi] &= \int_G \delta[Q[P[\varphi_\zeta]]\xi + \ldots](1 + \tfrac{1}{2} \text{tr } c \cdot \xi + \ldots) \, d\xi \\ &= (\det F[\varphi_\zeta])^{-1}, \end{aligned} \tag{14.121}$$

where F is the matrix with elements

$$F^\alpha{}_\beta[\varphi] \stackrel{\text{def}}{=} Q^\alpha{}_\beta[P[\varphi]]. \tag{14.122}$$

More generally, when the P^α are not chosen to satisfy the differential equation (14.110), F is defined by

$$F^\alpha{}_\beta[\varphi] \stackrel{\text{def}}{=} P^\alpha{}_{,i}[\varphi] Q^i{}_\beta[\varphi]. \tag{14.123}$$

Now insert unity into the integrand of (14.108), in the guise of $(\Delta[\zeta, \varphi])^{-1} \int_G \delta[P[{}^\xi\varphi] - \zeta] \det Q^{-1}[\xi] \, d\xi$, and interchange the order of integrations, obtaining

$$\langle \text{out} | \text{in} \rangle = N \int_G \det Q^{-1}[\xi] \, d\xi \int d\varphi \, e^{iS[\varphi]} (\Delta[\zeta, \varphi])^{-1} \delta[P[{}^\xi\varphi] - \zeta]. \tag{14.124}$$

Chapter 14. Quantum gravity: the new synthesis

We have seen earlier that the volume element $d\varphi$ is gauge-invariant. (Remember now μ = constant.) $S[\varphi]$ and $\Delta[\zeta, \varphi]$ are also gauge-invariant. Therefore a superscript ξ may be affixed to every φ in the integrand of (14.124) that does not already bear one. But every $^\xi\varphi$ is then a dummy, and hence all the ξs may be removed. Making use of (14.121) one immediately obtains

$$\langle \text{out} | \text{in} \rangle = N' \int e^{iS[\varphi]} \det F[\varphi] \delta[P[\varphi]-\zeta] \, d\varphi \qquad (14.125)$$

where $F[\varphi_\zeta]$ has been replaced by $F[\varphi]$ in the integrand because of the presence of the delta functional, and where

$$N' \stackrel{\text{def}}{=} N \int_G \det Q^{-1}[\xi] \, d\xi. \qquad (14.126)$$

The gauge group has now been factored out, and its 'volume' has been absorbed into the new normalization constant, N'. The integration in (14.125) is restricted to the subspace $P^\alpha[\varphi] = \zeta^\alpha$.

If the P^α are not chosen to satisfy (14.110) (e.g. if a linear gauge condition is adopted) expression (14.125) may have to be modified. For example, if the subspace $P^\alpha[\varphi] = \zeta^\alpha$ intersects certain orbits more than once then a factor $1/n[\zeta, \varphi]$ will have to be inserted into the integrand, $n[\zeta, \varphi]$ being the number of times the orbit containing φ^i is intersected. If some of the orbits are not intersected at all the problem is more difficult. $\Delta[\zeta, \varphi]$ vanishes when φ is on such an orbit and (14.124) breaks down. $\Delta[\zeta, \varphi]$ typically has a branch point behavior on the boundary between those orbits that are intersected and those that are not, and the matrix $F[\varphi_\zeta]$ blows up there. It is possible that an analytic continuation procedure could carry one around the branch point, particularly if the delta functional in (14.125) were to be represented as a Fourier integral. But this remains a program of research for the future.

The technique of confining the fields φ^i to a particular subspace can also be used to make operators well defined. Consider (14.100). It is strictly valid only if the functional $A[\varphi]$ is gauge-invariant. If $A[\varphi]$ is not gauge-invariant the integral is ambiguous, like $\int_{-\infty}^{\infty} x \, dx$. That is to say, matrix elements are definable only for gauge-invariant operators. However, given a non-gauge-invariant operator $A[\varphi]$, one can construct a gauge-invariant operator out of it by the following definition:

$$T(A[\varphi_\zeta]) \stackrel{\text{def}}{=} T((\Delta[\zeta, \varphi])^{-1} \int_G A[^\xi\varphi] \delta[P[^\xi\varphi]-\zeta] \det Q^{-1}[\xi] \, d\xi). \qquad (14.127)$$

The full quantum theory

The chronological ordering symbol is used here so that the non-commutativity of $A[^\xi\varphi]$ with both $(\Delta[\zeta,\varphi])^{-1}$ and the delta functional can be effectively ignored. Note that because the gauge group acts linearly on the φs there is no ambiguity about the symbol $^\xi\varphi$. Note, however, that diffeomorphisms in gravity theory can drag the field in very complicated ways. The chronological operation, which orders field operators solely by the value of the coordinate x^0, constantly rearranges the 'physical' fields in correspondingly complicated ways as the variable ξ in the integral (14.127) ranges over the group.

Inserting (14.127) into (14.100) and following the same reasoning as was used in passing from (14.124) to (14.125), one finds

$$\langle \text{out}| T(A[\varphi_\zeta])|\text{in}\rangle = N' \int A[\varphi] e^{iS[\varphi]} \det F[\varphi] - \zeta] d\varphi, \qquad (14.128)$$

valid for any functional $A[\varphi]$.

14.5.5 Gaussian averages

It is possible to develop a perturbation theory based on (14.125) and (14.128), but it is usually more convenient to work with a formalism from which the delta functionals have been eliminated. Note that although the parameters ζ^α appear on the right-hand side of (14.125), the amplitude $\langle \text{out}|\text{in}\rangle$ is actually independent of them. Therefore nothing changes if we integrate over these parameters, with a weight factor. A Gaussian weight factor turns out to be by far the most convenient.

Let $\gamma_{\alpha\beta}$ be an arbitrary symmetric non-singular continuous matrix. $\gamma_{\alpha\beta}$ is usually chosen to be local (i.e. proportional to the delta function) and to be covariantly dependent on a background field g^i. (g^i may, for example, be the base point of the geodesically generated subspace V that serves as the zero point for the functionals P^α when the latter are chosen to be canonical.) Suppressing indices one may write formally

$$\int e^{\frac{1}{2}i\zeta\gamma\zeta} d\zeta = C (\det \gamma)^{-1/2}, \quad d\zeta \stackrel{\text{def}}{=} \prod_\alpha \delta\zeta^\alpha, \qquad (14.129)$$

where C is a (divergent) constant. If $e^{\frac{1}{2}i\zeta\gamma\zeta}$, the Gaussian weight factor, is inserted into the integrand of (14.125) and an integration over the ζs is performed, one therefore gets

$$\langle \text{out}|\text{in}\rangle = N''(\det \gamma)^{1/2} \int e^{i(S[\varphi] + \frac{1}{2}P[\varphi]\gamma P[\varphi])} \det F[\varphi] d\varphi, \qquad (14.130)$$

Chapter 14. Quantum gravity: the new synthesis

$$N'' \stackrel{\text{def}}{=} N'/C, \qquad (14.131)$$

a result that was first obtained, with linear Ps, by the author (1967b). Note that if canonical Ps are used each orbit may be swept out many (∞) times in the ζ integration.

Equations (14.127) and (14.128) may also be replaced by their Gaussian averages. Defining

$$T(A[\varphi]) \stackrel{\text{def}}{=} C^{-1}(\det \gamma)^{1/2} \int T(A[\varphi_\zeta]) e^{\frac{1}{2}i\zeta\gamma\zeta} \, d\zeta, \qquad (14.132)$$

one may write

$$\langle \text{out} | T(A[\varphi]) | \text{in} \rangle = N''(\det \gamma)^{1/2} \int A[\varphi] \, e^{i(S[\varphi]+\frac{1}{2}P[\varphi]\gamma P[\varphi])} \det F[\varphi] \, d\varphi. \qquad (14.133)$$

Equation (14.133), and generalizations of it, will be used frequently in the following sections. Definitions (14.127) and (14.132) reveal precisely what kind of averaged quantum operator is associated with each classical functional $A[\varphi]$ in this formalism. Note that

$$T(P^\alpha[\varphi_\zeta]) = \zeta^\alpha, \quad \langle \text{out} | T(P^\alpha[\varphi]) | \text{in} \rangle = 0. \qquad (14.134)$$

Having so freely manipulated formal expressions we should now check that no inconsistencies have crept into our results, by verifying directly that the right-hand side of (14.130), for example, is truly independent of the choices we have made for the functionals $P^\alpha[\varphi]$ and the matrix $\gamma_{\alpha\beta}$. Obviously the right-hand side *will* be affected if we naively switch to Ps for which the subspaces $P^\alpha = \zeta^\alpha$ do not intersect all orbits, or else intersect them redundantly. However, it should be invariant under infinitesimal changes in the Ps and γs.

The way to verify this is as follows. Replace each φ^i in the integral by $\bar{\varphi}^i$, where†

$$\bar{\varphi}^i \stackrel{\text{def}}{=} \varphi^i + Q^i_{\ \alpha}[\varphi] \, \delta\xi^\alpha[\varphi], \qquad (14.135)$$

$$\delta\xi^\alpha[\varphi] \stackrel{\text{def}}{=} F^{-1\alpha}_{\ \ \beta}[\varphi](\delta P^\beta[\varphi] + \tfrac{1}{2}\gamma^{-1\beta\gamma} \delta\gamma_{\gamma\delta} P^\delta[\varphi]). \qquad (14.136)$$

Since the φs are just dummies in the integral this replacement has no effect. However, it is not difficult to show (see De Witt (1967b or 1972)

---

† For many choices of the Ps the matrix F is effectively a differential operator, and hence F^{-1} in (14.136) is a Green's function. This Green's function must satisfy the boundary conditions appropriate to the 'in' and 'out' states.

The full quantum theory

for details) (1) that the exponent suffers precisely the change it would experience if the Ps and γs were altered by the infinitesimal amounts δP^α and $\delta \gamma_{\alpha\beta}$ respectively; (2) that the same is true for the product $(\det \gamma)^{1/2} \det F[\varphi] \, d\varphi$ *provided* one is entitled to make the identifications

$$Q^i{}_{\alpha,i} = 0, \quad c^\beta{}_{\alpha\beta} = 0. \tag{14.137}$$

Subject to this proviso the invariance of the integral is therefore proved.

Equations (14.137) were not needed in the derivation of (14.130). Why are they needed now? The answer is that they are forced on us by the procedure of factoring out the gauge group. Our interchanging the orders of integration in arriving at (14.124), and our use of (14.126), amount to adopting the rule that the gauge group is to be treated formally as if it were *compact*. For consistency the associated Lie algebra must likewise be treated as compact. The generators of real representations of compact Lie algebras all have vanishing trace. Hence (14.137).

In Yang–Mills theories (14.137) hold automatically because the generating group is always compact. In gravity theory the situation is more subtle. Both $Q^i{}_{\alpha,i}$ and $c^\beta{}_{\alpha\beta}$, if one tries to compute them, are meaningless expressions involving derivatives of delta functions with coincident arguments. However, both are metric-independent covariant vector densities of unit weight. Any sensible regularization scheme *must* assign them the value zero, for otherwise spacetime would be endowed with a preferred direction even before a metric is imposed upon it. It will be observed that (14.137) and (14.103) are consistent when $\mu = 1$. However, the conclusion $Q^i{}_{\alpha,i} = 0$ is forced on us solely by the fact that the φ^i transform linearly under the diffeomorphism group and has nothing to do with the special choice (14.104), (14.105) for the metric γ_{ij}. This means that contributions to the functional intergral arising from the powers of g that survive in the determinant (14.106), (14.107) in other dimensions than 4, or for choices of field variables other than $g_{\mu\nu}$, must all be suppressed by any viable regularization scheme.†

14.5.6 Ghosts; the BRS transformation; the generating functional

Equations (14.130) and (14.133) lead to perturbation rules of standard type. The presence of the Ps in the exponent breaks the gauge symmetry and eliminates the redundancy that exists in the integration (14.108). An

---

† The surviving powers of g contribute a term $\text{const} \times \delta(0) \int \ln g \, d^n x$ to the exponent of the functional integral.

Chapter 14. Quantum gravity: the new synthesis

important symmetry nevertheless survives. It is most easily revealed by introducing two new *anticommuting* fields, χ_α and ψ^α, and making use of the formal rules for integrating with respect to such fields that were first introduced by Berezin (1966). These rules are analogous in many ways to the well known rules for ordinary definite integrals from $-\infty$ to ∞ with integrands that vanish asymptotically. For example, integrals of total derivatives vanish, and the position of the zero point may be shifted. On the other hand, with the Berezin rules, transformations of variables and evaluation of Gaussian integrals lead to determinants precisely inverse to those of standard theory. In particular, one gets

$$\int e^{i\chi_\alpha F^{\alpha\beta}[\varphi]\psi_\beta} \, d\chi \, d\psi = C' \det F[\varphi] \tag{14.138}$$

where C' is a (divergent) constant. This formula may be used to re-express the integral (14.130) in the form

$$\langle \text{out} | \text{in} \rangle = \bar{N}(\det \gamma)^{1/2} \int e^{i(S[\varphi] + \frac{1}{2}P[\varphi]\gamma P[\varphi] + \chi F[\varphi]\psi)} \, d\varphi \, d\chi \, d\psi, \tag{14.139}$$

$$\bar{N} \stackrel{\text{def}}{=} N''/C'. \tag{14.140}$$

It was discovered by Becchi, Rouet and Stora (BRS) (1975) that both the exponent and the volume element $d\varphi \, d\chi \, d\psi$ in (14.139) are invariant under a set of transformations whose infinitesimal forms are given by†

$$\left.\begin{aligned}
\delta\varphi^i &= Q^i{}_\alpha[\varphi]\psi^\alpha \, \delta\lambda, \\
\delta\chi_\alpha &= \gamma_{\alpha\beta} P^\beta[\varphi] \, \delta\lambda, \\
\delta\psi^\alpha &= -\tfrac{1}{2} c^\alpha{}_{\beta\gamma} \psi^\beta \psi^\gamma \, \delta\lambda,
\end{aligned}\right\} \tag{14.141}$$

where $\delta\lambda$ is an infinitesimal anticommuting constant. Using the anticommutativity of the χs, ψs and $\delta\lambda$, together with the identity (14.97) and the definition (14.123), one readily verifies the invariance of the exponent. By computing the Berezin Jacobian of the BRS transformation one finds that the volume element $d\varphi \, d\chi \, d\psi$ is likewise invariant, provided equations (14.137) are assumed to hold. It is also straightforward to verify that, if confined to the φs and ψs, the BRS transformations constitute an Abelian group. Inclusion of the χs destroys the group property unless $F^\alpha{}_\beta[\varphi]\psi^\beta = 0$.

---

† BRS transformations for quantum gravity in special gauges have been given by Dixon (1975) and by Delbourgo and Ramon-Medrano (1976). The BRS transformation has been applied to the quantization of higher-derivative gravity theories by Stelle (1977).

The full quantum theory

The fields χ_α and ψ^α are known as *ghost fields*. They do not give rise to physical quanta but play an important formal role in the theory, particularly when the so-called generating functional is introduced. What follows is an adaptation, applicable to the case in which the Ps are nonlinear of an account of the theory of the generating functional given by B. W. Lee (1976) and originally due to Zinn-Justin.

One begins by replacing the exponent in (14.139) by $\tilde{S}[\varphi, \chi, \psi, K, L, M] + J_i\varphi^i + \bar{J}^\alpha\chi_\alpha + \hat{J}_\alpha\psi^\alpha$, where

$$\tilde{S}[\varphi, \chi, \psi, K, L, M] \stackrel{\text{def}}{=} S[\varphi] + \tfrac{1}{2}P^\alpha[\varphi]\gamma_{\alpha\beta}P^\beta[\varphi] + \chi_\alpha F^\alpha{}_\beta[\varphi]\psi^\beta$$
$$+ K_i Q^i{}_\alpha[\varphi]\psi^\beta + \tfrac{1}{2}L_\alpha c^\alpha{}_{\beta\gamma}\psi^\beta\psi^\gamma + MP^\alpha[\varphi]\gamma_{\alpha\beta}F^\beta{}_\gamma[\varphi]\psi^\gamma \quad (14.142)$$

and by generalizing (14.133) to

$$\langle \text{out} | T(A[\varphi, \chi, \psi]) | \text{in} \rangle$$
$$\stackrel{\text{def}}{=} \bar{N}(\det \gamma)^{1/2} \int A[\varphi, \chi, \psi] \, e^{i(\tilde{S} + J\varphi + \bar{J}\chi + \hat{J}\psi)} \, d\varphi \, d\chi \, d\psi. \quad (14.143)$$

J_i, $\bar{J}^\alpha$, $\hat{J}_\alpha$, K_i, L_α and M are *external sources*, and the 'matrix element' (14.143) is a functional of them. J_i and L_α are bosonic sources; the rest are fermionic. If the functional A is replaced by unity one gets a generalization of the 'in–out' amplitude:

$$e^{iW[J,\bar{J},\hat{J},K,L,M]} \stackrel{\text{def}}{=} \langle \text{out} | \text{in} \rangle$$
$$\stackrel{\text{def}}{=} \bar{N}(\det \gamma)^{1/2} \int e^{i(\tilde{S} + J\varphi + \bar{J}\chi + \hat{J}\psi)} \, d\varphi \, d\chi \, d\psi. \quad (14.144)$$

This generalization, which is analogous in many ways to the partition function in statistical mechanics, is called the *generating functional*, because if it is expanded as a power series in the sources, the coefficients are the matrix elements of chronological products of field operators. The coefficient of zero order is the original amplitude (14.139).

The functional $\tilde{S}$ may be viewed as a generalized action functional. With the aid of (14.97) and (14.98) one may readily show that it is BRS invariant. Suppose the variables φ, χ, ψ in the integral (14.144) suffer a BRS transformation. Since these variables are dummies the integral remains unaffected. Therefore

$$0 = i\bar{N}(\det \gamma)^{1/2} \int (J_i Q^i{}_\alpha[\varphi]\psi^\alpha + \bar{J}^\alpha \gamma_{\alpha\beta} P^\beta[\varphi] - \tfrac{1}{2}\hat{J}_\alpha c^\alpha{}_{\beta\gamma}\psi^\beta\psi^\gamma)$$
$$\times e^{i(\tilde{S} + \hat{J}Q + \bar{J}\chi + \hat{J}\psi)} \, d\varphi \, d\chi \, d\psi \quad (14.145)$$

Chapter 14. Quantum gravity: the new synthesis

This result can be expressed in an alternative form through use of

$$\begin{aligned}0 &= \int \frac{\delta}{\delta\chi_\alpha}\{f[\varphi]\, e^{i(\tilde{S}+J\varphi+\bar{J}\chi+\hat{J}\psi)}\}\, d\varphi\, d\chi\, d\psi \\ &= i\int (F^\alpha{}_\beta[\varphi]\psi^\beta - \bar{J}^\alpha) f[\varphi]\, e^{i(\tilde{S}+J\varphi+\bar{J}\chi+\hat{J}\psi)}\, d\varphi\, d\chi\, d\psi, \quad (14.146)\end{aligned}$$

where f is any function of the φs. One obtains

$$0 = i\bar{N}(\det\gamma)^{1/2}\int\left[J_i\frac{\delta\tilde{S}}{\delta K_i} + \frac{\partial\tilde{S}}{\partial M} - \hat{J}_\alpha\frac{\delta\tilde{S}}{\delta L_\alpha}\right] e^{i(\tilde{S}+J\varphi+\bar{J}\chi+\hat{J}\psi)}\, d\varphi\, d\chi\, d\psi$$

$$= \left[J_i\frac{\delta}{\delta K_i} + \frac{\partial}{\partial M} - \hat{J}_\alpha\frac{\delta}{\delta L_\alpha}\right] e^{iW}, \quad (14.147)$$

or

$$J_i\frac{\delta W}{\delta K_i} + \frac{\partial W}{\partial M} - \hat{J}_\alpha\frac{\delta W}{\delta L_\alpha}. \quad (14.148)$$

Equation (14.148) is the expression of an important symmetry property of the generating functional and leads directly to the Ward–Takahashi identity to be discussed in section 14.5.8. Note that the derivative with respect to M is an ordinary one, not a functional one. M bears no indices and hence is independent of spacetime.

14.5.7 Many-particle Green's functions; the effective action

In this section important use will be made of the Schwinger average,

$$\langle A\rangle \stackrel{\text{def}}{=} \frac{\langle\text{out}|\, T(A)\,|\text{in}\rangle}{\langle\text{out}|\text{in}\rangle}. \quad (14.149)$$

Here A is an arbitrary functional of the operators φ^i, χ_α, ψ^α, and the numerator and denominator on the right are defined by (14.143) and (14.144) respectively. When 'in' and 'out' are vacuum states and the sources vanish this reduces to the Schwinger average (14.24). It will be convenient to define

$$\varphi^i \stackrel{\text{def}}{=} \langle\varphi^i\rangle, \quad \chi_\alpha \stackrel{\text{def}}{=} \langle\chi_\alpha\rangle, \quad \psi^a \stackrel{\text{def}}{=} \langle\psi^\alpha\rangle. \quad (14.150)$$

When the sources K_i, L_α, M, $\bar{J}^\alpha$ and $\hat{J}_\alpha$ vanish, the averages χ_α and ψ^a vanish, and when, in addition, J_i vanishes, φ^i reduces to the average introduced in (14.25). Note that although the symbols φ^i, χ_α and ψ^α have previously been used for integration variables, no confusion about their meaning will arise in practice.

It will also be convenient to denote the operators φ^i, χ_α, ψ^α collectively by $\hat\varphi^A$, their averages by φ^A, and the sources J_i, $\bar{J}^\alpha$, $\hat{J}_\alpha$ collectively by J_A. Let ΔJ_A be arbitrary finite increments in the sources. Then we may write

$$\sum_{n=0}^\infty \frac{i^n}{n!} \Delta J_{A_n} \ldots \Delta J_{A_1} \langle \text{out} | T(\hat\varphi^{A_1} \ldots \hat\varphi^{A_n}) | \text{in} \rangle$$

$$= \exp\left(\Delta J_A \frac{\delta}{\delta J_A}\right) \langle \text{out} | \text{in} \rangle = (e^{iW})_{J \to J + \Delta J}$$

$$= \exp\left(iW + i\Delta J_A \varphi^A + i \sum_{n=2}^\infty \frac{1}{n!} \Delta J_{A_n} \ldots J_{A_1} G^{A_1 \ldots A_n}\right), \quad (14.151)$$

where

$$\varphi^A = \langle \hat\varphi^A \rangle = e^{-iW} \frac{\delta}{i\delta J_A} e^{iW} = \frac{\delta W}{\delta J_A}, \quad (14.152)$$

$$G^{A_1 \ldots A_n} \stackrel{\text{def}}{=} \frac{\delta}{\delta J_{A_1}} \ldots \frac{\delta}{\delta J_{A_n}} W. \quad (14.153)$$

Dividing both sides of (14.151) by e^{iW} and comparing like powers of ΔJ_A, one obtains an infinite sequence of relations:

$$\langle \hat\varphi^A \hat\varphi^B \rangle = \varphi^A \varphi^B - iG^{AB}, \quad (14.154)$$

$$\langle \hat\varphi^A \hat\varphi^B \hat\varphi^C \rangle = \varphi^A \varphi^B \varphi^C - iP_3 \varphi^A G^{BC} + (-i)^2 G^{ABC}, \quad \text{etc.} \quad (14.155)$$

where P_N means 'sum over the N distinct permutations of indices, with a plus sign or a minus sign according to whether the permutation of the fermionic indices involved is even or odd'. G^{AB} is known as the *one-particle* propagator, and the $G^{A_1 \ldots A_n}$, $n \geq 3$, are known as *many-particle Green's functions*. In statistical mechanics they would be called *correlation functions*. They satisfy the boundary conditions specified by the 'in' and 'out' states.

Any functional of the sources J_A may be alternatively regarded as a functional of the averages φ^A. From (14.152) and (14.153) one sees that the one-particle propagator is the transformation matrix from one set of variables to the other:

$$G^{AB} = \frac{\delta \varphi^B}{\delta J_A}. \quad (14.156)$$

Chapter 14. Quantum gravity: the new synthesis

This fact may be used to establish an important relation between the functional W and the Schwinger average of the operator field equations. The latter is obtained from the formal functional identity

$$0 = -\mathrm{i}\, e^{-\mathrm{i}W}\bar{N}(\det \gamma)^{1/2} \int e^{\mathrm{i}(\tilde{S}+J\varphi+\bar{J}\chi+\bar{J}\psi)} \frac{\overleftarrow{\delta}}{\delta\varphi^A}\, \mathrm{d}\varphi\, \mathrm{d}\chi\, \mathrm{d}\psi$$

$$= \langle \tilde{S}_{,A} \rangle + J_A. \tag{14.157}$$

Here $\tilde{S}_{,A}$ is the operator corresponding to the functional $\tilde{S}_{,A}$ and one must now distinguish between left and right differentiation because of the presence of fermionic variables. When all sources vanish (14.157), with the index A set equal to i, reduces to

$$0 = \langle \tilde{S}_{,i} \rangle = \langle S_{,i} + P\gamma P_{,i} - \mathrm{i}(\ln \det F)_{,i} \rangle. \tag{14.158}$$

This is just (14.23) with the previously mentioned missing terms (involving P and det F) included.

If we differentiate (14.157) on the left with respect to J_B and make use of (14.156) we obtain

$$G^{BC}{}_{,C}\langle \tilde{S}_{,A} \rangle = -\delta^B{}_A. \tag{14.159}$$

Here the functional derivative inside the brackets $\langle\ \rangle$ is with respect to the field operator φ^A, and the functional derivative outside the brackets is with respect to the field average φ^C. $_C\langle \tilde{S}_{,A}\rangle$ is seen to be the operator of which the one-particle propagator G^{BC} is the Green's function. Because of its boundary conditions, G^{BC} may be shown to be both a left Green's function, as in (14.159), and a right Green's function as well. It has the symmetry

$$G^{AB} = (-1)^{AB} G^{BA} \tag{14.160}$$

which implies

$$_A\langle \tilde{S}_{,B} \rangle = (-1)^{A+B+AB} {}_B\langle \tilde{S}_{,A}\rangle, \tag{14.161}$$

where an index appearing in an exponent of (-1) is understood to have the value $+1$ or -1 according to whether it is bosonic or fermionic. But (14.161) is just the condition that there exist a functional $\tilde{\Gamma}[\varphi, \chi, \psi, K, L, M]$ such that

$$\tilde{\Gamma}_{,A} = \langle \tilde{S}_{,A}\rangle. \tag{14.162}$$

$\tilde{\Gamma}$ is known as the *effective action*. It satisfies the equations

$$\tilde{\Gamma}_{,A} = -J_A, \tag{14.163}$$

$$_A\tilde{\Gamma}_{,C}\, G^{CB} = -\delta_A{}^B, \tag{14.164}$$

The full quantum theory

and is related to the functional W by a Legendre transformation:

$$W = \tilde{\Gamma} + J_A \varphi^A. \tag{14.165}$$

This last relation may be verified through differentiation with respect to J_B and use of (14.163) in the form $_{,A}\tilde{\Gamma} = (-1)^A J_A$. Thus

$$\frac{\delta W}{\delta J_B} = \frac{\delta \varphi^A}{\delta J_B}[_{,A}\tilde{\Gamma} + (-1)^A J_A] + \varphi^B = \varphi^B, \tag{14.166}$$

which is just (14.152). Since $\tilde{\Gamma}$ is determined only up to an arbitrary constant of integration, (14.165) may be regarded as fixing it.

$\tilde{\Gamma}$ is also known as the *generating functional for proper vertices*. This stems from its relation to the many-particle Green's functions. By differentiating (14.164) one can relate functional derivatives of the one-particle propagator to derivatives of $\tilde{\Gamma}$. These relations yield, for example,

$$G^{ABC} = \frac{\delta}{\delta J_A} G^{BC} = G^{AD}{}_{,D}G^{BC}$$

$$= (-1)^{(B+C)D+(C+D)E+(D+E)F} G^{AD}G^{BE}G^{CF}{}_{,DEF}\tilde{\Gamma}. \tag{14.167}$$

If the propagators are represented by lines, and the third and higher derivatives of $\tilde{\Gamma}$ are represented by vertices, one easily sees that each new differentiation with respect to a source inserts a new line in all possible ways into the previous diagram. Each Green's function of a given order is thus representable as a sum of all the possible tree diagrams of that order. Because the S-matrix (when asymptotic regions exist) is expressible in terms of the chronological products appearing in (14.151), and because these products are expressible in terms of the Green's functions (14.154) and (14.155), it follows that when $\tilde{\Gamma}$ is used only tree diagrams are needed in the construction of the S-matrix. No closed loops appear. The vertices generated by $\tilde{\Gamma}$ are the *proper* vertices, already containing all quantum corrections. By noting that identical tree diagrams occur in classical perturbation theory, but with $\tilde{\Gamma}$ replaced by $\tilde{S}$, one can show that $\tilde{\Gamma}$ describes the quantum-corrected dynamics of coherent large-amplitude fields. The same must be true when no asymptotic region exists and there is no S-matrix.

14.5.8 The Ward–Takahashi identity

We now resume use of the symbols φ^i, χ_α, ψ^α, J_i, $\bar{J}^\alpha$, $\hat{J}_\alpha$ and rewrite (14.165) in the more explicit form

$$W[J, \bar{J}, \hat{J}, K, L, M] = \tilde{\Gamma}[\varphi, \chi, \psi, K, L, M] + J_i\varphi^i + \bar{J}^\alpha\chi_\alpha + \hat{J}_\alpha\psi^\alpha. \tag{14.168}$$

Chapter 14. Quantum gravity: the new synthesis

The averages $\varphi^i, \chi_\alpha, \psi^\alpha$ depend on all six sources, but because K_i, L_α, M do not participate in the Legendre transformation one may show that

$$\frac{\delta W}{\delta K_i} = \frac{\delta \tilde{\Gamma}}{\delta K_i}, \quad \frac{\delta W}{\delta L_\alpha} = \frac{\delta \tilde{\Gamma}}{\delta L_\alpha}, \quad \frac{\partial W}{\partial M} = \frac{\partial \tilde{\Gamma}}{\partial M}, \tag{14.169}$$

where the derivatives on the right refer only to the explicit dependence of $\tilde{\Gamma}$ on K_i, L_α, M. This result, combined with (14.163) in the form $_{,A}\tilde{\Gamma} = -(-1)^A J_A$, allows (14.148) to be rewritten as

$$-\frac{\delta \tilde{\Gamma}}{\delta \varphi^i} \frac{\delta \tilde{\Gamma}}{\delta K_i} + \frac{\partial \tilde{\Gamma}}{\partial M} - \frac{\delta \tilde{\Gamma}}{\delta \psi^\alpha} \frac{\delta \tilde{\Gamma}}{\delta L_\alpha} = 0. \tag{14.170}$$

This is the *Ward–Takahashi identity*.

The Ward–Takahashi identity has important implications for the structure of $\tilde{\Gamma}$. That it implies the existence of some sort of symmetry possessed by $\tilde{\Gamma}$ becomes evident when one notes that, because of its BRS-invariance, $\tilde{S}$ too satisfies the Ward–Takahashi identity. Unfortunately, to work from (14.170) *to* the symmetry possessed by $\tilde{\Gamma}$ is a much harder task. In principle one might do the following. Assume that $\tilde{\Gamma}$ can be expanded as a power series in $\chi_\alpha, \psi^\alpha, K_i, L_\alpha$ and M. Such an assumption has nothing *a priori* to do with perturbation theory, since the expansion is to be carried out *after* the functional integration (14.144) has been performed. It is based on the reasonable belief that (14.144) varies smoothly (at least after appropriate renormalizations) as $K_i, L_\alpha, M, \bar{J}^\alpha$ and $\hat{J}_\alpha$ (and hence χ_α and ψ^α) go to zero.

In determining the kinds of terms that can appear in the expansion it is useful to introduce the notion of 'ghost number'. If one assigns ghost numbers 1 to ψ^α and $\bar{J}^\alpha$; 0 to φ^i and J_i; -1 to $\chi_\alpha, \hat{J}_\alpha, K_i$ and M; and -2 to L_α, one easily sees that the integrand in (14.144) and the integral itself have total ghost number zero. Hence W and $\tilde{\Gamma}$ have total ghost number zero, and all the terms in an expansion have this property as well. Also the expansion can contain no terms in M of higher order than the first, since M is an anti-commuting constant.

If one inserts the expansion into (14.170) and groups together terms of like powers, one obtains an infinite sequence of subsidiary Ward–Takahashi identities relating the φ-dependent coefficients. Unfortunately there seems to be no easy way of drawing simple inferences from these identities *en gros*. So far (14.170) has been applied only to renormalizable models, order by order, in perturbation theory. There it has proved to be of great service in the practical details of the renormalization program as well as in the demonstration that the theory is indeed renormalizable to all orders and that unitarity is maintained.

The Ward–Takahashi identity must have an equally important role to play in quantum gravity. In particular, one expects it to yield the following result found to hold in renormalizable theories, namely, when all the sources vanish (14.163), with $A = i$, must reduce effectively to

$$\Gamma_{,i}[\varphi] = 0, \qquad (14.171)$$

where the functional Γ is gauge-invariant:

$$\Gamma_{,i} Q^i{}_\alpha \equiv 0. \qquad (14.172)$$

In quantum gravity (14.171) would be just (14.26), Γ being a *reduced effective action*. The demonstration that this is indeed so, however, remains a program for the future.

14.6 Conclusion

In any review of quantum gravity one always ends with a keen sense of how much remains unknown, of how much there is left to do. We still have only a very imperfect understanding of the structure of the diffeomorphism group and of the kinds of global difficulties that it can lead us into. It may turn out to be relatively harmless or it may lead to staggering complications. As an example of how wary one must be, consider the introduction of canonical coordinates as a device to achieve (in principle) globally valid gauge conditions. Canonical coordinates are based on the 1-parameter Abelian subgroups. Unlike the situation with most continuous groups, the 1-parameter Abelian subgroups of the diffeomorphism group do not fill a neighborhood of the identity (see Freifeld, 1968). There are some C^∞ diffeomorphisms, arbitrarily close to the identity, that cannot be reached from the identity by exponentiation. Very likely this fact does no harm. The structure of any covariant theory and, indeed, of the diffeomorphism group itself, is almost certainly determined already by those diffeomorphisms that *can* be exponentiated from the identity. And we should welcome anything that cuts the group down to size.

One may ask: why waste time on the group at all? Why not go directly for the invariants? They are where the physics lies. The reply to that is: Wonderful – if it can be done. But it won't be easy. We cannot deal with local fields (tensors, spinors), for example, without introducing the gauge

Chapter 14. Quantum gravity: the new synthesis

group, and so far no one has discovered how to set up an action principle without local fields.

These remarks are not meant to cast a shadow on the future of quantum gravity theory. In spite of the difficulties, everything that has been discovered so far has the right 'feel' about it. The results obtained are too beautiful not to be believed. Hawking's discovery of quantum black holes has already shown that this theory, like no other theory before it, unites relativity theory, the theory of quanta, and statistical mechanics into one harmonious whole. It is remarkable, and fitting, that it is in precisely these areas that Einstein's imperishable legacy to physics lies.

Ironically, Einstein would probably look on the whole effort with amusement, as a misguided attempt to unite the unjoinable. Quantum theory, for him, was not something to be imposed upon general relativity but rather was to be superseded someday by a nonlinear deterministic theory built, perhaps, on the lines of general relativity itself. And yet, with his intense perception of the deepest issues of physics, who knows?

That quantum gravity touches these deepest issues is undeniable, and it is a pity that Einstein could not live to see at least the first buds of its flowering. We have left, until the end, mention of one of the deepest issues of all. Most of the effects discussed in the first part of this review (Casimir effect, effects of topology and curvature on $\langle T^{\mu\nu} \rangle$, thermal radiation) are exceedingly small and become dominant only under conditions so extreme as to stagger the mind: arising from increases in the density of normal matter by over eighty powers of ten. These are endpoint conditions of gravitational collapse, by masses ranging in magnitude from those of stars up to that of the entire universe. This is quantum theory on a macroscopic scale with a vengeance.

If the whole universe was at one time, or will be again, a quantum object, what becomes of the objectivity that sustains us as scientists? If we are part of the wave function, how can we observe reality as we do? There is only one answer to these questions that has so far had any chance of being successful, without at the same time altering the structure of quantum mechanics as we know it, and that is the answer proposed by Everett (1957). Over the years since Everett's paper first quietly appeared, his conception has attracted an increasing number of distinguished adherents. It is a conception that maintains the experimentally verified absolute statistical laws of quantum mechanics but restores determinism globally by accepting the grand wave function for what it is: the description of a reality which, if taken literally, itself staggers the mind.

Conclusion

When one tries to determine the dynamical behavior of the universe in its early moments by computing the effective action (Hartle, 1977), or looks for isotropization mechanisms in the early universe by computing particle production (Zel'dovich, 1970; Hu and Parker, 1978), one is ultimately confronted with the interpretation issue and the Everett conception. It is general relativity, combined with the quantum theory in a new synthesis, that has forced the issue.

15. The path-integral approach to quantum gravity

S. W. HAWKING

15.1 Introduction

Classical general relativity is a very complete theory. It prescribes not only the equations which govern the gravitational field but also the motion of bodies under the influence of this field. However it fails in two respects to give a fully satisfactory description of the observed universe. Firstly, it treats the gravitational field in a purely classical manner whereas all other observed fields seem to be quantized. Second, a number of theorems (see Hawking and Ellis, 1973) have shown that it leads inevitably to singularities of spacetime. The singularities are predicted to occur at the beginning of the present expansion of the universe (the big bang) and in the collapse of stars to form black holes. At these singularities, classical general relativity would break down completely, or rather it would be incomplete because it would not prescribe what came out of a singularity (in other words, it would not provide boundary conditions for the field equations at the singular points). For both the above reasons one would like to develop a quantum theory of gravity. There is no well defined prescription for deriving such a theory from classical general relativity. One has to use intuition and general considerations to try to construct a theory which is complete, consistent and which agrees with classical general relativity for macroscopic bodies and low curvatures of spacetime. It has to be admitted that we do not yet have a theory which satisfies the above three criteria, especially the first and second. However, some partial results have been obtained which are so compelling that it is difficult to believe that they will not be part of the final complete picture. These results relate to the conection between black holes and thermodynamics which has already been described in chapters 6 and 13 by Carter and Gibbons. In the present article it will be shown how this relationship between gravitation and thermodynamics appears also when one quantizes the gravitational field itself.

There are three main approaches to quantizing gravity:

Introduction

1 The operator approach

In this one replaces the metric in the classical Einstein equations by a distribution-valued operator on some Hilbert space. However this would not seem to be a very suitable procedure to follow with a theory like gravity, for which the field equations are non-polynomial. It is difficult enough to make sense of the product of the field operators at the same spacetime point let alone a non-polynomial function such as the inverse metric or the square root of the determinant.

2 The canonical approach

In this one introduces a family of spacelike surfaces and uses them to construct a Hamiltonian and canonical equal-time commutation relations. This approach is favoured by a number of authors because it seems to be applicable to strong gravitational fields and it is supposed to ensure unitarity. However the split into three spatial dimensions and one time dimension seems to be contrary to the whole spirit of relativity. Moreover, it restricts the topology of spacetime to be the product of the real line with some three-dimensional manifold, whereas one would expect that quantum gravity would allow all possible topologies of spacetime including those which are not products. It is precisely these other topologies that seem to give the most interesting effects. There is also the problem of the meaning of equal-time commutation relations. These are well defined for matter fields on a fixed spacetime geometry but what sense does it make to say that two points are spacelike-separated if the geometry is quantized and obeying the Uncertainty Principle?

For these reasons I prefer:

3 The path-integral approach

This too has a number of difficulties and unsolved problems but it seems to offer the best hope. The starting point for this approach is Feynman's idea that one can represent the amplitude

$$\langle g_2, \phi_2, S_2 | g_1, \phi_1, S_1 \rangle,$$

to go from a state with a metric g_1 and matter fields ϕ_1 on a surface S_1 to a state with a metric g_2 and matter fields ϕ_2 on a surface S_2, as a sum over all field configurations g and ϕ which take the given values on the surfaces S_1

Chapter 15. The path-integral approach to quantum gravity

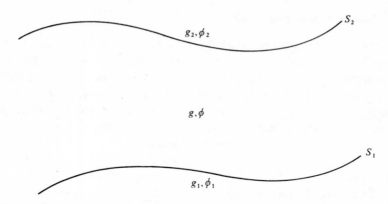

Figure 15.1. The amplitude $\langle g_2, \phi_2, S_2 | g_1, \phi_1, S_1 \rangle$ to go from a metric g_1 and matter fields ϕ_1, on a surface S_1 to a metric g_2 and matter fields ϕ_2 on a surface S_2 is given by a path integral over all fields g, ϕ which have the given values on S_1 and S_2.

and S_2 (figure 15.1). More precisely

$$\langle g_2, \phi_2, S_2 | g_1, \phi_1, S_1 \rangle = \int D[g, \phi] \exp (iI[g, \phi]),$$

where $D[g, \phi]$ is a measure on the space of all field configurations g and ϕ, $I[g, \phi]$ is the action of the fields, and the integral is taken over all fields which have the given values on S_1 and S_2.

In the above it has been implicitly assumed either that the surfaces S_1 and S_2 and the region between them are compact (a 'closed' universe) or that the gravitational and matter fields die off in some suitable way at spatial infinity (the asymptotically flat space). To make the latter more precise one should join the surfaces S_1 and S_2 by a timelike tube at large radius so that the boundary and the region contained within it are compact, as in the case of a closed universe. It will be seen in the next section that the surface at infinity plays an essential role because of the presence of a surface term in the gravitational action.

Not all the components of the metrics g_1 and g_2 on the boundary are physically significant, because one can give the components $g^{ab}n_b$ arbitrary values by diffeomorphisms or gauge transformations which move points in the interior, M, but which leave the boundary, ∂M, fixed. Thus one need specify only the three-dimensional induced metric h on ∂M and that only up to diffeomorphisms which map the boundary into itself.

In the following sections it will be shown how the path integral approach can be applied to the quantization of gravity and how it leads to the concepts of black hole temperature and intrinsic quantum mechanical entropy.

15.2 The action

The action in general relativity is usually taken to be

$$I = \frac{1}{16\pi G} \int (R - 2\Lambda)(-g)^{1/2} d^4x + \int L_m(-g)^{1/2} d^4x, \quad (15.1)$$

where R is the curvature scalar, Λ is the cosmological constant, g is the determinant of the metric and L_m is the Lagrangian of the matter fields. Units are such that $c = \hbar = k = 1$. G is Newton's constant and I shall sometimes use units in which this also has a value of one. Under variations of the metric which vanish and whose normal derivatives also vanish on ∂M, the boundary of a compact region M, this action is stationary if and only if the metric satisfies the Einstein equations:

$$R_{ab} - \tfrac{1}{2} g_{ab} R + \Lambda g_{ab} = 8\pi G T_{ab}, \quad (15.2)$$

where $T^{ab} = \tfrac{1}{2}(-g)^{-1/2}(\delta L_m/\delta g_{ab})$ is the energy-momentum tensor of the matter fields. However this action is not an extremum if one allows variations of the metric which vanish on the boundary but whose normal derivatives do not vanish there. The reason is that the curvature scalar R contains terms which are linear in the second derivatives of the metric. By integration by parts, the variation in these terms can be converted into an integral over the boundary which involves the normal derivatives of the variation on the boundary. In order to cancel out this surface integral, and so obtain an action which is stationary for solutions of the Einstein equations under all variations of the metric that vanish on the boundary, one has to add to the action a term of the form (Gibbons and Hawking, 1977a):

$$\frac{1}{8\pi G} \int K(\pm h)^{1/2} d^3x + C, \quad (15.3)$$

where K is the trace of the second fundamental form of the boundary, h is the induced metric on the boundary, the plus or minus signs are chosen according to whether the boundary is spacelike or timelike, and C is a term which depends only on the boundary metric h and not on the values of g at the interior points. The necessity for adding the surface term (15.3) to the action in the path-integral approach can be seen by considering the

Chapter 15. The path-integral approach to quantum gravity

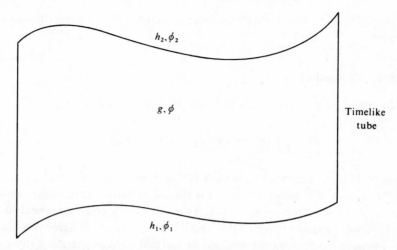

Figure 15.2. Only the induced metric h need be given on the boundary surface. In the asymptotically flat case the initial and final surfaces should be joined by a timelike tube at large radius to obtain a compact region over which to perform the path integral.

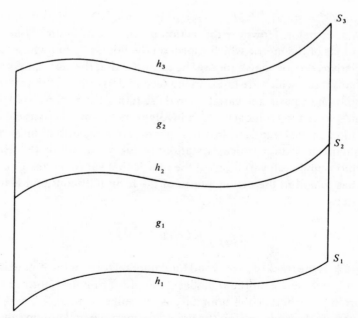

Figure 15.3. The amplitude to go from the metric h_1 on the surface S_1 to the metric h_3 on the surface S_3 should be the sum of the amplitude to go by all metrics h_2 on the intermediate surface S_2. This will be true only if the action contains a surface term.

situation depicted in figure 15.3, where one considers the transition from a metric h_1, on a surface S_1, to a metric h_2 on a surface S_2 and then to a metric h_3 on a later surface S_3. One would want the amplitude to go from the initial to the final state to be obtained by summing over all states on the intermediate surface S_2, i.e.

$$\langle h_3, S_3 | h_1, S_1 \rangle = \sum_{h_2} \langle h_2, S_2 | h_1, S_1 \rangle \langle h_3, S_3 | h_2, S_2 \rangle. \tag{15.4}$$

This will be true if and only if

$$I[g_1 + g_2] = I[g_1] + I[g_2], \tag{15.5}$$

where g_1 is the metric between S_1 and S_2, g_2 is the metric between S_2 and S_3, and $[g_1 + g_2]$ is the metric on the regions between S_1 and S_3 obtained by joining together the two regions. Because the normal derivative of g_1 at S_2 will not in general be equal to that of g_2 at S_2, the metric $[g_1 + g_2]$ will have a δ-function in the Ricci tensor of strength $2(K^1_{ab} - K^2_{ab})$, where K^1_{ab} and K^2_{ab} are the second fundamental forms of the surface S_2 in the metrics g_1 and g_2 respectively, defined with respect to the future-directed normal. This means that the relation (15.5) will hold if and only if the action is the sum of (15.1) and (15.3), i.e.

$$I = \frac{1}{16\pi G} \int (R - 2\Lambda)(-g)^{1/2} \, d^4x + \int L_m (-g)^{1/2} \, d^4x$$

$$+ \frac{1}{8\pi G} \int K(\pm h)^{1/2} \, d^3x + C. \tag{15.6}$$

The appearance of the term C in the action is somewhat awkward. One could simply absorb it into the renormalization of the measure $D[g, \phi]$. However, in the case of asymptotically flat metrics it is natural to treat it so that the contribution from the timelike tube at large radius is zero when g is the flat-space metric, η. Then

$$C = -\frac{1}{8\pi G} \int K^0 (\pm h)^{1/2} \, d^3x, \tag{15.7}$$

where K^0 is the second fundamental form of the boundary imbedded in flat space. This is not a completely satisfactory prescription because a general boundary metric h cannot be imbedded in flat space. However in an asymptotically flat situation one can suppose that the boundary will become asymptotically imbeddable as one goes to larger and larger radii. Ultimately I suspect that one should do away with all boundary surfaces and should deal only with closed spacetime manifolds. However, at the

Chapter 15. The path-integral approach to quantum gravity

present state of development it is very convenient to use non-compact, asymptotically flat metrics and to evaluate the action using a boundary at large radius.

A metric which is asymptotically flat in the three spatial directions but not in time can be written in the form

$$ds^2 = -(1 - 2M_t r^{-1}) dt^2 + (1 + 2M_s r^{-1}) dr^2$$
$$+ r^2 (d\theta^2 + \sin^2\theta \, d\phi^2) + O(r^{-2}). \quad (15.8)$$

If the metric satisfies the vacuum Einstein equations ($\Lambda = 0$) near infinity then $M_t = M_s$, but in the path integral one considers all asymptotically flat metrics, whether or not they satisfy the Einstein equation. In such a metric it is convenient to choose the boundary ∂M to be the t-axis times a sphere of radius r_0. The area of ∂M is

$$\int (-h)^{1/2} d^3x = 4\pi r_0^2 \int (1 - M_t r_0^{-1} + O(r_0^{-2})) dt. \quad (15.9)$$

The integral of the trace of the second fundamental form of ∂M is given by

$$\int K(-h)^{1/2} d^3x = \frac{\partial}{\partial n} \int (-h)^{1/2} d^3x, \quad (15.10)$$

where $\partial/\partial n$ indicates the derivative when each point of ∂M is moved out along the unit normal. Thus

$$\int K(-h)^{1/2} d^3x = \int (8\pi r_0 - 4\pi M_t - 8\pi M_s + O(r_0^{-2})) dt. \quad (15.11)$$

For the flat space metric, η, $K^0 = 2r_0^{-1}$. Thus

$$\frac{1}{8\pi G} \int (K - K^0)(-h)^{1/2} d^3x = \frac{1}{2G} \int (M_t - 2M_s) dt. \quad (15.12)$$

In particular for a solution of the Einstein equation with mass M as measured from infinity, $M_s = M_t = M$ and the surface term is

$$-\frac{M}{2G} \int dt + O(r_0^{-1}). \quad (15.13)$$

15.3 Complex spacetime

For real Lorentzian metrics g (i.e. metrics with signature $-+++$) and real matter fields ϕ, the action $I[g, \phi]$ will be real and so the path integral will oscillate and will not converge. A related difficulty is that to find a field

Complex spacetime

configuration which extremizes the action between given initial and final surfaces, one has to solve a hyperbolic equation with initial and final boundary values. This is not a well-posed problem: there may not be any solution or there may be an infinite number, and if there is a solution it will not depend smoothly on the boundary values.

In ordinary quantum field theory in flat spacetime one deals with this difficulty by rotating the time axis 90° clockwise in the complex plane, i.e. one replaces t by $-i\tau$. This introduces a factor of $-i$ into the volume integral for the action I. For example, a scalar field of mass m has a Lagrangian

$$L = -\tfrac{1}{2}\phi_{,a}\phi_{,b}g^{ab} - \tfrac{1}{2}m^2\phi^2. \tag{15.14}$$

Thus the path integral

$$Z = \int D[\phi] \exp(iI[\phi]) \tag{15.15}$$

becomes

$$Z = \int D[\phi] \exp(-\hat{I}[\phi]), \tag{15.16}$$

where $\hat{I} = -iI$ is called the 'Euclidean' action and is greater than or equal to zero for fields ϕ which are real on the Euclidean space defined by real τ, x, y, z. Thus the integral over all such configurations of the field ϕ will be exponentially damped and should therefore converge. Moreover the replacement of t by an imaginary coordinate τ has changed the metric η^{ab} from Lorentzian (signature $-+++$) to Euclidean (signature $++++$). Thus the problem of finding an extremum of the action becomes the well-posed problem of solving an elliptic equation with given boundary values.

The idea, then, is to perform all path integrals on the Euclidean section (τ, x, y, z real) and then analytically continue the results anticlockwise in the complex t-plane back to Lorentzian or Minkowski section (t, x, y, z real). As an example consider the quantity

$$Z[J] = \int D[\phi] \exp - (\tfrac{1}{2}\phi A\phi + J\phi)\, dx\, dy\, dz\, d\tau, \tag{15.17}$$

where A is the second-order differential operator $-\Box + m^2$, $\Box$ is the four-dimensional Laplacian and $J(x)$ is a prescribed source field which dies away at large Euclidean distances. The path integral is taken over all fields ϕ that die away at large Euclidean distances. One can write $Z[J]$

Chapter 15. The path-integral approach to quantum gravity

symbolically as

$$Z[J] = \exp\left(\tfrac{1}{2}JA^{-1}J\right) \int D[\phi] \exp\left(-\tfrac{1}{2}(\phi - A^{-1}J)A(\phi - A^{-1}J)\right), \quad (15.18)$$

where $A^{-1}(x_1, x_2)$ is the unique inverse or Green's function for A that dies away at large Euclidean distances,

$$A^{-1}J(x) = \int A^{-1}(x, x')J(x')\, d^4x' \quad (15.19)$$

$$JA^{-1}J = \iint J(x)A^{-1}(x, x')J(x')\, d^4x\, d^4x'. \quad (15.20)$$

The measure $D[\phi]$ is invariant under the translation $\phi \to \phi - A^{-1}J$. Thus

$$Z[J] = \exp\left(\tfrac{1}{2}JA^{-1}J\right)Z[0]. \quad (15.21)$$

Then one can define the Euclidean propagator or two-point correlation function

$$\langle 0|\phi(x_2)\phi(x_1)|0\rangle = \left.\frac{\delta^2 \log Z}{\delta J(x_1)\, \delta J(x_2)}\right|_{J=0}$$

$$= A^{-1}(x_2, x_1). \quad (15.22)$$

One obtains the Feynman propagator by analytically continuing $A^{-1}(x_2, x_1)$ anticlockwise in the complex $t_2 - t_1$-plane.

It should be pointed out that this use of the Euclidean section has enabled one to define the vacuum state by the property that the fields ϕ die off at large positive and negative imaginary times τ. The time-ordering operation usually used in the definition of the Feynman propagator has been automatically achieved by the direction of the analytic continuation from Euclidean space, because if $\text{Re}\,(t_2 - t_1) > 0$, $\langle 0|\phi(x_2), \phi(x_1)|0\rangle$ is holomorphic in the lower half $t_2 - t_1$-plane, i.e. it is positive-frequency (a positive-frequency function is one which is holomorphic in the lower half t-plane and which dies off at large negative imaginary t).

Another use of the Euclidean section that will be important in what follows is to construct the canonical ensemble for a field ϕ. The amplitude to propagate from a configuration ϕ_1 on a surface at time t_1 to a configuration ϕ_2 on a surface at time t_2 is given by the path integral

$$\langle \phi_2, t_2|\phi_1, t_1\rangle = \int D[\phi]\exp(iI[\phi]). \quad (15.23)$$

Using the Schrödinger picture, one can also write this amplitude as

$$\langle \phi_2|\exp(-iH(t_2-t_1))|\phi_1\rangle.$$

Complex spacetime

Put $t_2 - t_1 = -i\beta$, $\phi_2 = \phi_1$ and sum over a complete orthonormal basis of configurations ϕ_n. One obtains the partition function

$$Z = \sum \exp(-\beta E_n) \qquad (15.24)$$

of the field ϕ at a temperature $T = \beta^{-1}$, where E_n is the energy of the state ϕ_n. However from (15.23) one can also represent Z as a Euclidean path integral

$$Z = \int D[\phi] \exp(-i\hat{I}[\phi]), \qquad (15.25)$$

where the integral is taken over all fields ϕ that are real on the Euclidean section and are periodic in the imaginary time coordinate τ with period β. As before one can introduce a source J and obtain a Green's function by functionally differentiating $Z[J]$ with respect to J at two different points. This will represent the two-point correlation function or propagator for the field ϕ, not this time in the vacuum state but in the canonical ensemble at temperature $T = \beta^{-1}$. In the limit that the period β tends to infinity, this thermal propagator tends to the normal vacuum Feynman propagator.

It seems reasonable to apply similar complexification ideas to the gravitational field, i.e. the metric. For example, supposing one was considering the amplitude to go from a metric h_1 on a surface S_1 to a metric h_2 on a surface S_2, where the surfaces S_1 and S_2 are asymptotically flat, and are separated by a time interval t at infinity. As explained in section 15.1, one would join S_1 and S_2 by a timelike tube of length t at large radius. One could then rotate this time interval into the complex plane by introducing an imaginary time coordinate $\tau = it$. The induced metric on the timelike tube would now be positive-definite so that one would be dealing with a path integral over a region M on whose boundary the induced metric h was positive-definite everywhere. One could therefore take the path integral to be over all positive-definite metrics g which induced the given positive-definite metric h on ∂M. With the same choice of the direction of rotation into the complex plane as in flat-space Euclidean theory, the factor $(-g)^{1/2}$ which appears in the volume element becomes $-i(g)^{1/2}$, so that the Euclidean action, $\hat{I} = -iI$, becomes

$$\hat{I} = -\frac{1}{16\pi G} \int (R - 2\Lambda)(g)^{1/2} d^4x - \frac{1}{8\pi G} \int (K - K^0)(h)^{1/2} d^3x$$

$$- \int L_m (g)^{1/2} d^4x. \qquad (15.26)$$

Chapter 15. The path-integral approach to quantum gravity

The problem arising from the fact that the gravitational part of this Euclidean action is not positive-definite will be discussed in section 15.4.

The state of the system is determined by the choice of boundary conditions of the metrics that one integrates over. For example, it would seem reasonable to expect that the vacuum state would correspond to integrating over all metrics which were asymptotically Euclidean, i.e. outside some compact set as they approached the flat Euclidean metric on R^4. Inside the compact set the curvature might be large and the topology might be different from that of R^4.

As an example, one can consider the canonical ensemble for the gravitational fields contained in a spherical box of radius r_0 at a temperature T, by performing a path integral over all metrics which would fit inside a boundary consisting of a timelike tube of radius r_0 which was periodically identified in the imaginary time direction with period $\beta = T^{-1}$.

In complexifying the spacetime manifold one has to treat quantities which are complex on the real Lorentzian section as independent of their complex conjugates. For example, a charged scalar field in real Lorentzian spacetime may be represented by a complex field ϕ and its complex conjugate $\bar{\phi}$. When going to complex spacetime one has to analytically continue $\bar{\phi}$ as a new field $\tilde{\phi}$ which is independent of ϕ. The same applies to spinors. In real Lorentzian spacetime one has unprimed spinors λ_A which transform under SL(2, C) and primed spinors $\mu_{A'}$ which transform under the complex conjugate group $\overline{SL(2, C)}$. The complex conjugate of an unprimed spinor is a primed spinor and vice versa. When one goes to complex spacetime, the primed and unprimed spinors become independent of each other and transform under independent groups SL(2, C) and $\widetilde{SL}(2, C)$ respectively. If one analytically continues to a section on which the metric is positive-definite and restricts the spinors to lie in that section, the primed and unprimed spinors are still independent but these groups become SU(2) and $\widetilde{SU}(2)$ respectively. For example, in a Lorentzian metric the Weyl tensor can be represented as

$$C_{AA'BB'CC'DD'} = \psi_{ABCD}\varepsilon_{A'B'}\varepsilon_{C'D'} + \bar{\psi}_{A'B'C'D'}\varepsilon_{AB}\varepsilon_{CD}. \quad (15.27)$$

When one complexifies, $\bar{\psi}_{A'B'C'D'}$ is replaced by an independent field $\tilde{\psi}_{A'B'C'D'}$. In particular one can have a metric in which $\psi_{ABCD} \neq 0$, but $\tilde{\psi}_{A'B'C'D'} = 0$. Such a metric is said to be conformally self-dual and satisfies

$$C_{abcd} = {}^*C_{abcd} = \tfrac{1}{2}\varepsilon_{abef}C^{ef}{}_{cd}. \quad (15.28)$$

The metric is said to be self-dual if

$$R_{abcd} = {}^*R_{abcd}$$

which implies

$$R_{ab} = 0, \quad C_{abcd} = {}^*C_{abcd}. \tag{15.29}$$

A complexified spacetime manifold M with a complex self-dual or conformally self-dual metric g_{ab} may admit a section on which the metric is real and positive definite (a 'Euclidean' section) but it will not admit a Lorentzian section, i.e. a section on which the metric is real and has a signature $-+++$.

15.4 The indefiniteness of the gravitational action

The Euclidean action for scalar or Yang–Mills fields is positive-definite. This means that the path integral over all configurations of such fields that are real on the Euclidean section converges, and that only those configurations contribute that die away at large Euclidean distances, since otherwise the action would be infinite. The action for fermion fields is not positive-definite. However, one treats them as anticommuting quantities (Berezin, 1966) so that the path integral over them converges. On the other hand, the Euclidean gravitational action is not positive-definite even for real positive-definite metrics. The reason is that although gravitational waves carry positive energy, gravitational potential energy is negative because gravity is attractive. Despite this, in classical general relativity it seems that the total energy or mass, as measured from infinity, of any asymptotically flat gravitational field is always non-negative. This is known as the *positive energy conjecture* (Brill and Deser, 1968; Geroch, 1973). What seems to happen is that whenever the gravitational potential energy becomes too large, an event horizon is formed and the region of high gravitational binding undergoes gravitational collapse, leaving behind a black hole of positive mass. Thus one might expect that the black holes would play a role in controlling the indefiniteness of the gravitational action in quantum theory and there are indications that this is indeed the case.

To see that the action can be made arbitrarily negative, consider a conformal transformation $\tilde{g}_{ab} = \Omega^2 g_{ab}$, where Ω is a positive function which is equal to one on the boundary ∂M.

$$\tilde{R} = \Omega^{-2} R - 6\Omega^{-3} \Box \Omega \tag{15.30}$$

$$\tilde{K} = \Omega^{-1} K + 3\Omega^{-2} \Omega_{;a} n^a, \tag{15.31}$$

Chapter 15. The path-integral approach to quantum gravity

where n^a is the unit outward normal to the boundary ∂M. Thus

$$\hat{I}[\tilde{g}] = -\frac{1}{16\pi G}\int_M (\Omega^2 R + 6\Omega_{;a}\Omega_{;b}g^{ab} - 2\Lambda\Omega^4)(g)^{1/2}\,d^4x$$
$$-\frac{1}{8\pi G}\int_{\partial M}\Omega^2(K-K^0)(h)^{1/2}\,d^3x. \quad (15.32)$$

One sees that $\hat{I}$ may be made arbitrarily negative by choosing a rapidly varying conformal factor Ω.

To deal with this problem it seems desirable to split the integration over all metrics into an integration over conformal factors, followed by an integration over conformal equivalence classes of metrics. I shall deal separately with the case in which the cosmological constant Λ is zero but the spacetime region has a boundary ∂M, and the case in which Λ is nonzero but the region is compact without boundary.

In the former case, the path integral over the conformal factor Ω is governed by the conformally invariant scalar wave operator, $A = -\Box + \frac{1}{6}R$. Let $\{\lambda_n, \phi_n\}$ be the eigenvalues and eigenfunctions of A with Dirichlet boundary conditions, i.e.

$$A\phi_n = \lambda_n\phi_n, \ \phi_n = 0 \text{ on } \partial M.$$

If $\lambda_1 = 0$, then $\Omega^{-1}\phi_1$ is an eigenfunction with zero eigenvalue for the metric $\tilde{g}_{ab} = \Omega^2 g_{ab}$. The nonzero eigenvalues and corresponding eigenfunctions do not have any simple behaviour under conformal transformation. However they will change continuously under a smooth variation of the conformal factor which remains positive everywhere. Because the zero eigenvalues are conformally invariant, this shows that the number of negative eigenvalues (which will be finite) remains unchanged under a conformal transformation Ω which is positive everywhere.

Let $\Omega = 1 + y$, where $y = 0$ on ∂M. Then

$$\hat{I}[\tilde{g}] = -\frac{6}{16\pi g}\int (yAy + 2Ry)(g)^{1/2}\,d^4x + \hat{I}[g]$$

$$= -\frac{6}{16\pi G}\int \{(y - A^{-1}R)A(y - A^{-1}R)\}(g)^{1/2}\,d^4x$$

$$+ \frac{6}{16\pi G}RA^{-1}R + \hat{I}[g]$$

$$= \frac{6}{16\pi G}RA^{-1}R + \hat{I}[g] - \frac{6}{16\pi G}\int \gamma A\gamma(g)^{1/2}\,d^4x, \quad (15.33)$$

where $\gamma = (y - A^{-1}R)$.

Thus one can write

$$\hat{I}[\tilde{g}] = I^1 + I^2,$$

where I^1 is the first and second term on the right of (15.33) and I^2 is the third term.

I^1 depends only on the conformal equivalence class of the metric g, while I^2 depends on the conformal factor. One can thus define a quantity X to be the path integral of $\exp(-I^2)$ over all conformal factors in one conformal equivalence class of metrics.

If the operator A has no negative or zero eigenvalues, in particular if g is a solution of the Einstein equations, the inverse, A^{-1}, will be well defined and the metric $g'_{ab} = (1 + A^{-1}R)^2 g_{ab}$ will be a regular metric with $R' = 0$ everywhere. In this case I^1 will equal $\hat{I}[g']$, which in turn will be given by a surface integral of K' on the boundary. It seems plausible to make the *positive action conjecture*: any asymptotically Euclidean, positive-definite metric with $R = 0$ has positive or zero action (Gibbons, Hawking and Perry, 1978). There is a close connection between this and the positive energy conjecture in classical Lorentzian general relativity. This claims that the mass or energy as measured from infinity of any Lorentzian, asymptotically flat solution of the Einstein equations is positive or zero if the solution develops from a non-singular initial surface, the mass being zero if and only if the metric is identically flat. Although no complete proof exists, the positive energy conjecture has been proved in a number of restricted cases or under certain assumptions (Brill, 1959; Brill and Deser, 1968; Geroch, 1973; Jang and Wald, 1977) and is generally believed. If it held also for classical general relativity in five dimensions (signature $-++++$), it would imply the positive action conjecture, because a four-dimensional asymptotically Euclidean metric with $R = 0$ could be taken as time-symmetric initial data for a five-dimensional solution and the mass of such a solution would be equal to the action of the four-dimensional metric. Page (1978) has obtained some results which support the positive action conjecture. However he has also shown that it does not hold for metrics like the Schwarzschild solution which are asymptotically flat in the spatial directions, but are not in the Euclidean time direction. The significance of this will be seen later.

Let g_0 be a solution of the field equations. If I^1 increases under all perturbations away from g_0 that are not purely conformal transformations, the integral over conformal classes will tend to converge. If there is some non-conformal perturbation, δg, of g_0 which reduces I^1,

Chapter 15. The path-integral approach to quantum gravity

then in order to make the path integral converge one will have to integrate over the metrics of the form $g_0 + i\delta g$. This will introduce a factor i into Z for each mode of non-conformal perturbations which reduces I^1. This will be discussed in the next section. For metrics which are far from a solution of the field equation, the operator A may develop zero or negative eigenvalues. When an eigenvalue passes through zero, the inverse, A^{-1}, will become undefined and I^1 will become infinite. When there are negative eigenvalues but not zero eigenvalues, A^{-1} and I^1 will be well defined, but the conformal factor $\Omega = 1 + A^{-1}R$, which transforms g to the metric g' with $R' = 0$, will pass through zero and so g' will be singular. This is very similar to what happens with three-dimensional metrics on time-symmetric initial surfaces (Brill, 1959). If h is a three-dimensional positive-definite metric on the initial surface, one can make a conformal transformation $\tilde{h} = \Omega^4 h$ to obtain a metric with $\tilde{R} = 0$ which will satisfy the constraint equations. If the three-dimensional conformally invariant operator $B = -\Delta + R/8$ has no zero or negative eigenvalues (which will be the case for metrics h sufficiently near flat space) the conformal factor Ω needed will be finite and positive everywhere. If, however, one considers a sequence of metrics h for which one of the eigenvalues of B passes through zero and becomes negative, the corresponding Ω will first diverge and then will become finite again but will pass through zero so that the metric $\tilde{h}$ will be singular. The interpretation of this is that the metric h contained a region with so much negative gravitational binding energy that it cut itself off from the rest of the universe by forming an event horizon. To describe such a situation one has to use initial surfaces with different topologies.

It seems that something analogous may be happening in the four-dimensional case. In some sense one could think that metrics g for which the operator A had negative eigenvalues contained regions which cut themselves off from the rest of the spacetime because they contained too much curvature. One could then represent their effect by going to manifolds with different topologies. Anyway, metrics for which A has negative eigenvalues are in some sense far from solutions of the field equations, and we shall see in the next section that one can in fact evaluate path integrals only over metrics near solutions of the field equations.

The operator A appears in I^2 with a minus sign. This means that in order to make the path integral over the conformal factors converge at a solution of the field equations, and in particular at flat space, one has to take γ to be purely imaginary. The prescription, therefore, for making the path integral converge is to divide the space of all metrics into conformal

The indefiniteness of the gravitational action

equivalence classes. In each equivalence class pick the metric g' for which $R' = 0$. Integrate over all metrics $\tilde{g} = \Omega^2 g'$, where Ω is of the form $1 + i\xi$. Then integrate over conformal equivalence classes near solutions of the field equations, with the non-conformal perturbation being purely imaginary for modes which reduce I^1.

The situation is rather similar for compact manifolds with a Λ-term. In this case there is no surface term in the action and no requirement that $\Omega = 1$ on the boundary. If $\tilde{g} = \Omega^2 g$,

$$\hat{I}[\tilde{g}] = -\frac{6}{16\pi G} \int (\Omega^2 R + 6\Omega_{;a}\Omega_{;b}g^{ab} - 2\Lambda\Omega^4)(g)^{1/2} \, d^4x. \quad (15.34)$$

Thus quantum gravity with a Λ-term on a compact manifold is a sort of average of $\lambda\phi^4$ theory over all background metrics. However unlike ordinary $\lambda\phi^4$ theory, the kinetic term $(\nabla\Omega)^2$, appears in the action with a minus sign. This means that the integration over the conformal factors has to be taken in a complex direction just as in the previous case.

One can again divide the space of all the positive-definite metrics g on the manifold M into conformal equivalence classes. In each equivalence class the action will have one extremum at the vanishing metric for which $\Omega = 0$. In general there will be another extremum at a metric g' for which $R' = 4\Lambda$, though in some cases the conformal transformation $g' = \Omega^2 g$, where g is a positive-definite metric, may require a complex Ω. Putting $\tilde{g} = (1+y)^2 g'$, one obtains

$$\hat{I}[\tilde{g}] = -\frac{\Lambda V}{8\pi G} - \frac{6}{16\pi G} \int (6y_{;a}y_{;b}g^{ab} - 8y^2\Lambda - 8y^3\Lambda - 2y^4\Lambda)(g')^{1/2} \, d^4x, \quad (15.35)$$

where $V = \int (g')^{1/2} \, d^4x$.

If Λ is negative and one neglects the cubic and quartic terms in y, one obtains convergence in the path integral by integrating over purely imaginary y in a similar manner to what was done in the previous case. It therefore seems reasonable to adopt the prescription for evaluating path integrals with Λ-terms that one picks the metric g' in each conformal equivalence class for which $R' = 4\Lambda$, and one then integrates over conformal factors of the form $\Omega = 1 + i\xi$ about g'.

If Λ is positive, the operator $-6\Box - 8\Lambda$, which acts on the quadratic terms in ξ, has at least one negative eigenvalue, $\xi = $ constant. In fact it seems that this is the only negative eigenvalue. Its significance will be discussed in section 15.10.

Chapter 15. The path-integral approach to quantum gravity

15.5 The stationary-phase approximation

One expects that the dominant contribution to the path integral will come from metrics and fields which are near a metric g_0, and fields ϕ_0 which are an extremum of the action, i.e. a solution of the classical field equations. Indeed this must be the case if one is to recover classical general relativity in the limit of macroscopic systems. Neglecting for the moment, questions of convergence, one can expand the action in a Taylor series about the background fields g_0, ϕ_0,

$$\hat{I}[g, \phi] = \hat{I}[g_0, \phi_0] + I_2[\bar{g}, \bar{\phi}] + \text{higher-order terms}, \quad (15.36)$$

where

$$g_{ab} = g_{0ab} + \bar{g}_{ab}, \quad \phi = \phi_0 + \bar{\phi},$$

and $I_2[\bar{g}, \bar{\phi}]$ is quadratic in the perturbations $\bar{g}$ and $\bar{\phi}$. If one ignores the higher-order terms, the path integral becomes

$$\log Z = -\hat{I}[g_0, \phi_0] + \log \int D[\bar{g}, \bar{\phi}] \exp(-I_2[\bar{g}, \bar{\phi}]). \quad (15.37)$$

This is known variously as the stationary-phase, WKB or one-loop approximation. One can regard the first term on the right of (15.37) as the contribution of the background fields to $\log Z$. This will be discussed in sections 15.7 and 15.8. The second term on the right of (15.37) is called the one-loop term and represents the effect of quantum fluctuations around the background fields. The remainder of this section will be devoted to describing how one evaluates it. For simplicity I shall consider only the case in which the background matter fields, ϕ_0, are zero. The quadratic term $I_2[\bar{g}, \bar{\phi}]$ can then be expressed as $I_2[\bar{g}] + I_2[\bar{\phi}]$ and

$$\log Z = -\hat{I}[g_0] + \log \int D[\bar{\phi}] \exp(-I_2[\bar{\phi}]) + \log \int D[\bar{g}] \exp(-I_2[\bar{g}]). \quad (15.38)$$

I shall consider first the one-loop term for the matter fields, the second term on the right of (15.38). One can express $I_2[\bar{\phi}]$ as

$$I_2[\bar{\phi}] = \tfrac{1}{2} \int \bar{\phi} A \bar{\phi} (g_0)^{1/2} d^4x, \quad (15.39)$$

where A is a differential operator depending on the background metric g_0. In the case of boson fields, which I shall consider first, A is a second-order differential operator. Let $\{\lambda_n, \phi_n\}$ be the eigenvalues and the corresponding eigenfunctions of A, with $\phi_n = 0$ on ∂M in the case

The stationary-phase approximation

where there is a boundary surface. The eigenfunctions, ϕ_n, can be normalized so that

$$\int \phi_n \phi_m \cdot (g_0)^{1/2} \, d^4x = \delta_{mn}. \tag{15.40}$$

One can express an arbitrary field ϕ which vanishes on ∂M as a linear combination of these eigenfunctions:

$$\phi = \sum_n y_n \phi_n. \tag{15.41}$$

Similarly one can express the measure on the space of all fields ϕ as

$$D[\phi] = \prod_n \mu \, dy_n. \tag{15.42}$$

Where μ is a normalization factor with dimensions of mass or (length)$^{-1}$. One can then express the one-loop matter term as

$$Z_\phi = \int D[\phi] \exp(-I_2[\phi])$$

$$= \prod_n \int \mu \, dy_n \exp(-\tfrac{1}{2}\lambda_n y_n^2)$$

$$= \prod_n (2\pi \mu^2 \lambda_n^{-1})^{1/2}$$

$$= (\det(\tfrac{1}{2}\pi^{-1} \mu^{-2} A))^{-1/2}. \tag{15.43}$$

In the case of a complex field ϕ like a charged scalar field, one has to treat ϕ and the analytic continuation $\tilde{\phi}$ of its complex conjugate as independent fields. The quadratic term then has the form

$$I_2[\phi, \tilde{\phi}] = \tfrac{1}{2} \int \tilde{\phi} A \phi (g_0)^{1/2} \, d^4x. \tag{15.44}$$

The operator A will not be self-adjoint if there is a background electromagnetic field. One can write $\tilde{\phi}$ in terms of eigenfunctions of the adjoint operator $A^\dagger$:

$$\tilde{\phi} = \sum_n \tilde{y}_n \tilde{\phi}_n. \tag{15.45}$$

The measure will then have the form

$$D[\phi, \tilde{\phi}] = \prod_n \mu^2 \, dy_n \, d\tilde{y}_n. \tag{15.46}$$

Chapter 15. The path-integral approach to quantum gravity

Because one integrates over y_n and $\tilde{y}_n$ independently, one obtains

$$Z_\phi = (\det (\tfrac{1}{2}\pi^{-1}\mu^{-2}A))^{-1}. \tag{15.47}$$

To treat fermions in the path integrals one has to regard the spinor ψ and its independent adjoint $\tilde{\psi}$ as anticommuting Grassman variables (Berezin, 1966). For a Grassman variable x one has the following (formal) rules of integration

$$\int dx = 0, \quad \int x \, dx = 1. \tag{15.48}$$

These suffice to determine all integrals, since x^2 and higher powers of x are zero by the anticommuting property. Notice that (15.48) implies that if $y = ax$, where a is a real constant, then $dy = a^{-1} dx$.

One can use these rules to evaluate path integrals over the fermion fields ψ and $\tilde{\psi}$. The operator A in this case is just the ordinary first-order Dirac operator. If one expands $\exp(-I_2)$ in a power series, only the term linear in A will survive because of the anticommuting property. Integration of this respect to $d\psi$ and $d\tilde{\psi}$ gives

$$Z_\psi = \det (\tfrac{1}{2}\mu^{-2}A). \tag{15.49}$$

Thus the one-loop terms for fermion fields are proportional to the determinant of their operator while those for bosons are inversely proportional to determinants.

One can obtain an asymptotic expansion for the number of eigenvalues $N(\lambda)$ of an operator A with values less than λ:

$$N(\lambda) \sim \tfrac{1}{2}B_0\lambda^2 + B_1\lambda + B_2 + O(\lambda^{-1}), \tag{15.50}$$

where B_0, B_1 and B_2 are the 'Hamidew' coefficients referred to by Gibbons in chapter 13. They can be expressed as $B_n = \int b_n (g_0)^{1/2} d^4x$, where the b_n are scalar polynomials in the metric, the curvature and its covariant derivatives (Gilkey, 1975). In the case of the scalar wave operator, $A = -\Box + \xi R + m^2$, they are

$$b_0 = \frac{1}{16\pi^2} \tag{15.51}$$

$$b_1 = \frac{1}{16\pi^2}((1/6 - \xi)R - m^2) \tag{15.52}$$

$$b_2 = \frac{1}{2880\pi^2}(R^{abcd}R_{abcd} - R_{ab}R^{ab} + (6-30\xi)\Box R + \tfrac{5}{2}(6\xi - 1)^2 R^2$$
$$+ 30m^2(1 - 6\xi)R + 90m^4). \tag{15.53}$$

When there is a boundary surface ∂M, this introduces extra contributions into (15.50) including a $\lambda^{1/2}$-term. This would seem an additional reason for trying to do away with boundary surfaces and working simply with closed manifolds.

From (15.50) one can see that the determinant of A, the product of its eigenvalues, is going to diverge badly. In order to obtain a finite answer one has to regularize the determinant by dividing out by the product of the eigenvalues corresponding to the first two terms on the right of (15.50) (and those corresponding to a $\lambda^{1/2}$-term if it is present). There are various ways of doing this – dimensional regularization (t'Hooft and Veltman, 1972), point splitting (DeWitt, 1975), Pauli–Villars (Zeldovich and Starobinsky, 1972) and the zeta function technique (Dowker and Critchley, 1976; Hawking, 1977). The last method seems the most suitable for regularizing determinants of operators on a curved space background. It will be discussed further in the next section.

For both fermion and baryon operators the term B_0 is $(nV/16\pi^2)$, where V is the volume of the manifold in the background metric, g_0, and n is the number of spin states of the field. If, therefore, there are an equal number of fermion and boson spin states, the leading divergences in Z produced by the B_0-terms will cancel between the fermion and boson determinants without having to regularize. If in addition the B_1-terms either cancel or are zero (which will be the case for zero-rest-mass, conformally invariant fields), the other main divergence in Z will cancel between fermions and bosons. Such a situation occurs in theories with supersymmetry, such as supergravity (Deser and Zumino, 1976; Freedman, van Nieuwenhuizen and Ferrara, 1976) or extended supergravity (Ferrara and van Nieuwenhuizen, 1976). This may be a good reason for taking these theories seriously, in particular for the coupling of matter fields to gravity.

Whether or not the divergences arising from B_0 and B_1 cancel or are removed by regularization, the net B_2 will in general be nonzero, even in supergravity, if the topology of the spacetime manifold is non-trivial (Perry, 1978). This means that the expression for Z will contain a finite number (not necessarily an integer) of uncancelled eigenvalues. Because the eigenvalues have dimensions (length)$^{-2}$, in order to obtain a dimensionless result for Z each eigenvalue has to be divided by μ^2, where μ is the normalization constant or regulator mass. Thus Z will depend on μ. For renormalizable theories such as $\lambda\phi^4$, quantum electrodynamics or Yang–Mills in flat spacetime, B_2 is proportional to the action of the field. This means that one can absorb the μ-dependence into an effective

Chapter 15. The path-integral approach to quantum gravity

coupling constant $g(\mu)$ which depends on the scale at which it is measured. If $g(\mu) \to 0$ as $\mu \to \infty$, i.e. for very short length scales or high energies, the theory is said to be asymptotically free.

In curved spacetime however, B_2 involves terms which are quadratic in the curvature tensor of the background space. Thus unless one supposes that the gravitational action contains terms quadratic in the curvature (and this seems to lead to a lot of problems including negative energy, fourth-order equations and no Newtonian limit (Stelle, 1977, 1978)) one cannot remove the μ-dependence. For this reason gravity is said to be unrenormalizable because new parameters occur when one regularizes the theory.

If one tried to regularize the higher-order terms in the Taylor series about a background metric, one would have to introduce an infinite sequence of regularization parameters whose values could not be fixed by the theory. However it will be argued in section 15.9 that the higher-order terms have no physical meaning and that one ought to consider only the one-loop quadratic terms. Unlike $\lambda\phi^4$ or Yang–Mills theory, gravity has a natural length scale, the Planck mass. It might therefore seem reasonable to take some multiple of this for the one-loop normalization factor μ.

15.6 Zeta function regularization

In order to regularize the determinant of an operator A with eigenvalues and eigenfunctions $\{\lambda_n, \phi_n\}$, one forms a generalized zeta function from the eigenvalues

$$\zeta_A(s) = \sum \lambda_n^{-s}. \tag{15.54}$$

From (15.50) it can be seen that ζ will converge for Re $s > 2$. It can be analytically extended to a meromorphic function of s with poles only at $s = 2$ and $s = 1$. In particular it is regular at $s = 0$. Formally one has

$$\zeta_A'(0) = -\sum \log \lambda_n. \tag{15.55}$$

Thus one can *define* the regularized value of the determinant of A to be

$$\det A = \exp(-\zeta_A'(0)). \tag{15.56}$$

The zeta function can be related to the kernel $F(x, x', t)$ of the heat or diffusion equation

$$\frac{\partial F}{\partial t} + A_x F = 0, \tag{15.57}$$

where A_x indicates that the operator acts on the first argument of F. With the initial condition

$$F(x, x', 0) = \delta(x, x'), \tag{15.58}$$

F represents the diffusion over the manifold M, in a fifth dimension of parameter time t, of a point source of heat placed at x' at $t = 0$. The heat equation has been much studied by a number of authors including DeWitt (1963), McKean and Singer (1967) and Gilkey (1975). A good exposition can be found in Gilkey (1974).

It can be shown that if A is an elliptic operator, the heat kernel $F(x, x', t)$ is a smooth function of x, x', and t, for $t > 0$. As $t \to 0$, there is an asymptotic expression for $F(x, x, t)$:

$$F(x, x, t) \sim \sum_{n=0}^{\infty} b_n t^{n-2}, \tag{15.59}$$

where again the b_n are the 'Hamidew' coefficients and are scalar polynomials in the metric, the curvature and its covariant derivatives of order $2n$ in derivatives of the metrics.

One can represent F in terms of the eigenfunctions and eigenvalues of A

$$F(x, x', t) = \sum \phi_n(x) \phi_n(x') \exp(-\lambda_n t). \tag{15.60}$$

Integrating this over the manifold, one obtains

$$Y(t) = \int F(x, x, t)(g_0)^{1/2} \, d^4x = \sum \exp(-\lambda_n t). \tag{15.61}$$

The zeta function can be obtained from $Y(t)$ by an inverse Mellin transform

$$\zeta(s) = \frac{1}{\Gamma(s)} \int_0^\infty Y(t) t^{s-1} \, dt. \tag{15.62}$$

Using the asymptotic expansion for F, one sees that $\zeta(s)$ has a pole at $s = 2$ with residue B_0 and a pole at $s = 1$ with residue B_1. There would be a pole at $s = 0$ but it is cancelled by the pole in the gamma function. Thus $\zeta(0) = B_2$. In a sense the poles at $s = 2$ and $s = 1$ correspond to removing the divergences caused by the first two terms in (15.50).

If one knows the eigenvalue explicitly, one can calculate the zeta function and evaluate its derivative at $s = 0$. In other cases one can obtain some information from the asymptotic expansion for the heat kernel. For example, suppose the background metric is changed by a constant scale

Chapter 15. The path-integral approach to quantum gravity

factor $\tilde{g}_0 = k^2 g_0$, then the eigenvalues, λ_n, of a zero-rest-mass operator A will become $\bar{\lambda}_n = k^{-2}\lambda_n$. Thus

$$\zeta_{\tilde{A}}(s) = k^{2s}\zeta_A(s)$$

and

$$\zeta'_{\tilde{A}}(0) = 2 \log k \zeta_A(0) + \zeta'_A(0), \tag{15.63}$$

therefore

$$\log(\det \tilde{A}) = -2\zeta(0)\log k + \log(\det A). \tag{15.64}$$

Because B_2, and hence $\zeta(0)$, are not in general zero, one sees that the path integral is not invariant under conformal transformations of the background metric, even for conformally invariant operators A. This is known as a conformal anomaly and arises because in regularizing the determinant one has to introduce a normalization quantity, μ, with dimensions of mass or inverse length. Alternatively, one could say that the measure $D[\phi] = \prod \mu \, dy_n$ is not conformally invariant.

Further details of zeta function regularization of matter field determinants will be found in Hawking (1977), Gibbons (1977c), and Lapedes (1978).

The zeta function regularization of the one-loop gravitational term about a vacuum background has been considered by Gibbons, Hawking and Perry (1978). I shall briefly describe this work and generalize it to include a Λ-term.

The quadratic term in the fluctuations $\bar{g}$ about a background metric, g_0, is

$$I_2[\bar{g}] = \tfrac{1}{2} \int \bar{g}^{ab} A_{abcd} \bar{g}^{cd} (g_0)^{1/2} \, d^4x, \tag{15.65}$$

where

$$g^{ab} = g_0^{ab} + \bar{g}^{ab} \tag{15.66}$$

and

$$16\pi A_{abcd} = \tfrac{1}{4}g_{cd}\nabla_a\nabla_b - \tfrac{1}{4}g_{ac}\nabla_d\nabla_b + \tfrac{1}{8}(g_{ac}g_{bd} + g_{ab}g_{cd})\nabla_e\nabla^e + \tfrac{1}{2}R_{ad}g_{bc}$$
$$- \tfrac{1}{4}R_{ab}g_{cd} + \tfrac{1}{16}Rg_{ab}g_{cd} - \tfrac{1}{8}Rg_{ac}g_{bd} - \tfrac{1}{8}\Lambda g_{ab}g_{cd} + \tfrac{1}{4}\Lambda g_{ac}g_{bd}$$
$$+(a \leftrightarrow b) + (c \leftrightarrow d) + (a \leftrightarrow b, c \leftrightarrow d). \tag{15.67}$$

One cannot simply take the one-loop term to be $(\det(\tfrac{1}{2}\pi^{-1}\mu^{-1}A))^{1/2}$, because A has a large number of zero eigenvalues corresponding to the fact that the action is unchanged under an infinitesimal diffeomorphism

Zeta function regularization

(gauge transformation)

$$x^a \to x^a + \varepsilon \xi^a$$
$$g_{ab} \to g_{ab} + 2\varepsilon \xi_{(a;b)}.$$
(15.68)

One would like to factor out the gauge freedom by integrating only over gauge-inequivalent perturbations $\bar{g}$. One would then obtain an answer which depended on the determinant of A on the quotient of all fields $\bar{g}$ modulo infinitesimal gauge transformations. The way to do this has been indicated by Feynman (1972), DeWitt (1967) and Fade'ev and Popov (1967). One adds a gauge-fixing term to the action

$$I_f = \tfrac{1}{2} \int \bar{g}^{ab} B_{abcd} \bar{g}^{cd} (g_0)^{1/2} \, d^4 x. \tag{15.69}$$

The operator B is chosen so that for any sufficiently small perturbation $\bar{g}$ which satisfies the appropriate boundary condition there is a unique transformation, ξ^a, which vanishes on the boundary such that

$$B_{abcd}(\bar{g}^{cd} + 2\xi^{(c;d)}) = 0. \tag{15.70}$$

I shall use the harmonic gauge in the background metric

$$16\pi B_{abcd} = \tfrac{1}{4} g_{bd} \nabla_a \nabla_c - \tfrac{1}{8} g_{cd} \nabla_a \nabla_b - \tfrac{1}{8} g_{ab} \nabla_c \nabla_d$$
$$+ \tfrac{1}{16} g_{ab} g_{cd} \Box + (a \leftrightarrow b) + (c \leftrightarrow d) + (a \leftrightarrow b, c \leftrightarrow d).$$
(15.71)

The operator $(A+B)$ will in general have no zero eigenvalues. However, $\det(A+B)$ contains the eigenvalues of the arbitrarily chosen operator B. To cancel them out one has to divide by the determinant of B on the subspace of all $\bar{g}$ which are pure gauge transformations, i.e. of the form $\bar{g}^{ab} = 2\xi^{(a;b)}$ for some ξ which vanishes on the boundary. The determinant of B on this subspace is equal to the square of the determinant of the operator C on the space of all vector fields which vanish on the boundary, where

$$16\pi C_{ab} = -g_{ab}\Box - R_{ab}. \tag{15.72}$$

Thus one obtains

$$\log Z = -\hat{I}[g_0] - \tfrac{1}{2} \log \det (\tfrac{1}{2}\pi^{-1}\mu^{-2}(A+B)) + \log \det (\tfrac{1}{2}\pi^{-1}\mu^{-2}C). \tag{15.73}$$

The last term is the so-called ghost contribution.

In order to use the zeta function technique it is necessary to express $A+B$ as $K-L$ where K and L each have only a finite number of negative

Chapter 15. The path-integral approach to quantum gravity

eigenvalues. To do this, let

$$A + B = -F + G, \qquad (15.74)$$

where

$$F = -\tfrac{1}{16}(\nabla_a \nabla^a + 2\Lambda), \qquad (15.75)$$

which operates on the trace, ϕ, of $\bar{g}$, $\phi = \bar{g}^{ab} g_{0ab}$

$$G_{abcd} = -\tfrac{1}{8}(g_{ac}g_{bd} + g_{ad}g_{bc})\nabla^e \nabla_e - \tfrac{1}{4}(C_{dcab} + C_{dbac}) + \tfrac{1}{6}\Lambda g_{ab}g_{cd}, \qquad (15.76)$$

which operates on the trace-free part, $\tilde{\phi}$, of $\bar{g}$, $\tilde{\phi}^{ab} = \bar{g}^{ab} - \tfrac{1}{4}g_0{}^{ab}\phi$.

If $\Lambda \leq 0$, the operator F will have only positive eigenvalues. Therefore in order to make the one-loop term converge, one has to integrate over purely imaginary ϕ. This corresponds to integrating over conformal factors of the form $\Omega = 1 + i\xi$. if $\Lambda > 0$, F will have some finite number, p, of negative eigenvalues. Because a constant function will be an eigenfunction of F with negative eigenvalue (in the case where there is no boundary), p will be at least one. In order to make the one-loop term converge, one will have to rotate the contour of integration of the coefficient of each eigenfunction, with a negative eigenvalue to lie along the real axis. This will introduce a factor of i^p into Z.

If the background metric g_0 is flat, the operator G will be positive-definite. Thus one will integrate the trace-free perturbations $\tilde{\phi}$ along the real axis. This corresponds to integrating over real conformal equivalence classes. However for non-flat background metrics, G may have some finite number, q, of negative eigenvalues because of the Λ and Weyl tensor terms. Again one will have to rotate the contour of integration for these modes (this time from real to imaginary) and this will introduce a factor of i^{-q} into Z.

The ghost operator is

$$C_{ab} = -g_{ab}(\nabla^e \nabla_e + \Lambda). \qquad (15.77)$$

If $\Lambda > 0$, C will have some finite number, r, of negative eigenvalues. Because it is the determinant of C that appears in Z rather than its square root, the negative eigenvalues will contribute a factor $(-1)^r$.

One has

$$\log Z = -\hat{I}[g_0] + \tfrac{1}{2}\zeta'_F(0) + \tfrac{1}{2}\zeta'_G(0) - \zeta'_C(0)$$
$$+ \tfrac{1}{2}\log(2\pi\mu^2)(\zeta_F(0) + \zeta_G(0) - 2\zeta_C(0)). \qquad (15.78)$$

From the asymptotic expansion for the heat kernel one has to evaluate the zeta functions at $s = 0$. From the results of Gibbons and Perry (1979) one has

$$\zeta_F(0) + \zeta_G(0) - 2\zeta_C(0) = \int \left(\frac{53}{720\pi^2} C_{abcd} C^{abcd} + \frac{763}{540\pi^2} \Lambda^2 \right) (g_0)^{1/2} d^4x. \tag{15.79}$$

From this one can deduce the behaviour of the one-loop term under scale transformations of the background metric. Let $\tilde{g}_{0ab} = k^2 g_{0ab}$, then

$$\log \tilde{Z} = \log Z + (1 - k^2) \hat{I}[g_0] + \tfrac{1}{2} \gamma \log k, \tag{15.80}$$

where γ is the right-hand side of (15.79). Providing $\hat{I}[g_0]$ is positive, $\tilde{Z}$ will be very small for large scales, k. The fact that γ is positive will mean that it is also small for very small scales. Thus quantum gravity may have a cut-off at short length scales. This will be discussed further in section 15.10.

15.7 The background fields

In this section I shall describe some positive-definite metrics which are solutions of the Einstein equations in vacuum or with a Λ-term. In some cases these are analytic continuations of well-known Lorentzian solutions, though their global structure may be different. In particular the section through the complexified manifold on which the metric is positive-definite may not contain the singularities present on the Lorentzian section. In other cases the positive-definite metrics may occur on manifolds which do not have any section on which the metric is real and Lorentzian. They may nevertheless be of interest as stationary-phase points in certain path integrals.

The simplest non-trivial example of a vacuum metric is the Schwarzschild solution (Hartle and Hawking, 1976; Gibbons and Hawking, 1977a). This is normally given in the form

$$ds^2 = -\left(1 - \frac{2M}{r}\right) dt^2 + \left(1 - \frac{2M}{r}\right)^{-1} dr^2 + r^2 d\Omega^2. \tag{15.81}$$

Putting $t = -i\tau$ converts this into a positive-definite metric for $r > 2M$. There is an apparent singularity at $r = 2M$ but this is like the apparent singularity at the origin of polar coordinates, as can be seen by defining a new radial coordinate $x = 4M(1 - 2Mr^{-1})^{1/2}$. Then the metric becomes

$$ds^2 = \left(\frac{x}{4M}\right)^2 d\tau^2 + \left(\frac{r^2}{4M^2}\right)^2 dx^2 + r^2 d\Omega^2.$$

Chapter 15. The path-integral approach to quantum gravity

This will be regular at $x = 0$, $r = 2M$, if τ is regarded as an angular variable and is identified with period $8\pi M$ (I am using units in which the gravitational constant $G = 1$). The manifold defined by $x \geq 0$, $0 \leq \tau \leq 8\pi M$ is called the Euclidean section of the Schwarzschild solution. On it the metric is positive-definite, asymptotically flat and non-singular (the curvature singularity at $r = 0$ does not lie on the Euclidean section).

Because the Schwarzschild solution is periodic in imaginary time with period $\beta = 8\pi M$, the boundary surface ∂M at radius r_0 will have topology $S^1 \times S^2$ and the metric will be a stationary-phase point in the path integral for the partition function of a canonical ensemble at temperature $T = \beta^{-1} = (8\pi M)^{-1}$. As shown in section 15.2, the action will come entirely from the surface term, which gives

$$\hat{I} = \tfrac{1}{2}\beta M = 4\pi M^2. \tag{15.82}$$

One can find a similar Euclidean section for the Reissner–Nordström solution with $Q^2 + P^2 < M^2$, where Q is the electric charge and P is the magnetic monopole charge. In this case the radial coordinate has the range $r_+ \leq r < \infty$. Again the outer horizon, $r = r_+$, is an axis of symmetry in the $r - \tau$-plane and the imaginary time coordinate, τ, is identified with period $\beta = 2\pi\kappa^{-1}$, where κ is the surface gravity of the outer horizon. The electromagnetic field, F_{ab}, will be real on the Euclidean section if Q is imaginary and P is real. In particular if $Q = iP$, the field will be self-dual or anti-self-dual,

$$F_{ab} = \pm {}^*F_{ab} = \tfrac{1}{2}\varepsilon_{abcd}F^{cd}, \tag{15.83}$$

where ε_{abcd} is the alternating tensor. If F_{ab} is real on the Euclidean section, the operators governing the behaviour of charged fields will be elliptic and so one can evaluate the one-loop terms by the zeta function method. One can then analytically continue the result back to real Q just as one analytically continues back from positive-definite metrics to Lorentzian ones.

Because $R = 0$, the gravitational part of the action is unchanged. However there is also a contribution from the electromagnetic Lagrangian, $-(1/8\pi)F_{ab}F^{ab}$. Thus

$$\hat{I} = \tfrac{1}{2}\beta(M - \Phi Q + \psi P), \tag{15.84}$$

where $\Phi = Q/r_+$ is the electrostatic potential of the horizon and $\psi = P/r_+$ is the magnetostatic potential.

In a similar manner one can find a Euclidean section for the Kerr metric provided that the mass M is real and the angular momentum J is imaginary. In this case the metric will be periodic in the frame that co-rotates with the horizon, i.e. the point (τ, r, θ, ϕ) is identified with $(\tau+\beta, r, \theta, \phi+i\beta\Omega)$ where Ω is the angular velocity of the horizon (Ω will be imaginary if J is imaginary). As in the electromagnetic case, it seems best to evaluate the one-loop terms with J imaginary and then analytically continue to real J. The presence of angular momentum does not affect the asymptotic metric to leading order to that the action is

$$\hat{I} = \tfrac{1}{2}\beta M \quad \text{with } \beta = 2\pi\kappa^{-1},$$

where κ is the surface gravity of the horizon.

Another interesting class of vacuum solutions are the Taub–NUT metrics (Newman, Unti and Tamburino, 1963; Hawking and Ellis, 1973). These can be regarded as gravitational dyons with an ordinary 'electric' type mass M and a gravitational 'magnetic' type mass N. The metric can be written in the form

$$ds^2 = -V\left(dt + 4N \sin^2 \frac{\theta}{2} d\phi\right)^2 + V^{-1} dr^2 + (r^2 + N^2)(d\theta^2 + \sin^2\theta \, d\phi^2), \tag{15.85}$$

where $V = 1 - (2Mr + N^2)/(r^2 + N^2)$. This metric is regular on half-axis $\theta = 0$ but it has a singularity at $\theta = \pi$ because the $\sin^2(\theta/2)$ term in the metric means that a small loop around the axis does not shrink to zero length as $\theta = \pi$. This singularity can be regarded as the analogue of a Dirac string in electrodynamics, caused by the presence of a magnetic monopole charge. One can remove this singularity by introducing a new coordinate

$$t' = t + 4N\phi. \tag{15.86}$$

The metric then becomes

$$ds^2 = -V\left(dt' - 4N \cos^2 \frac{\theta}{2} d\phi\right)^2 + V^{-1} dr^2 + (r^2 + N^2)(d\theta^2 + \sin^2\theta \, d\phi^2). \tag{15.87}$$

This is regular at $\theta = \pi$ but not at $\theta = 0$. One can therefore use the (t, r, θ, ϕ) coordinates to cover the north pole ($\theta = 0$) and the (t', r, θ, ϕ) co-ordinates to cover the south pole ($\theta = \pi$). Because ϕ is identified with period 2π, (15.86) implies that t and t' have to be identified with period $8\pi N$. Thus if ψ is a regular field with t-dependence of the form $\exp(-i\omega t)$, then ω must satisfy

$$4N\omega = \text{an integer}. \tag{15.88}$$

Chapter 15. The path-integral approach to quantum gravity

This is the analogue of the Dirac quantization condition and relates the 'magnetic' charge, N, of the Taub–NUT solution to the 'electric' charge or energy, ω, of the field ψ. The process of removing the Dirac string singularity by introducing coordinates t and t' and periodically identifying, changes the topology of the surfaces of constant r from $S^2 \times R^1$ to S^3 on which $(t/2N)$, θ and ϕ are Euler angle coordinates.

The metric (15.85) also has singularities where $V = 0$ or ∞. As in the Schwarzschild case $V = \infty$ corresponds to an irremovable curvature singularity but $V = 0$ corresponds to a horizon and can be removed by periodically identifying the imaginary time coordinate. This identification is compatible with the one to remove the Dirac string if the two periods are equal, which occurs if $N = \pm iM$. If this is the case, and if M is real, the metric is real and is positive-definite in the region $r > M$ and the curvature is self-dual or anti-self-dual

$$R_{abcd} = \pm {}^*R_{abcd} = \pm \tfrac{1}{2}\varepsilon_{abef}R^{ef}{}_{cd}. \qquad (15.89)$$

The apparent singularity at $r = M$ becomes a single point, the origin of hyperspherical coordinates, as can be seen by introducing new radial and time variables

$$x = 2(2M(r-M))^{1/2},$$
$$\psi = -\frac{it}{2M}. \qquad (15.90)$$

The metric then becomes

$$ds^2 = \frac{Mx^2}{2(r+M)}(d\psi + \cos\theta\, d\phi)^2$$
$$+ \frac{r+M}{2M}dx^2 + \frac{x^2(r+M)}{8M}(d\theta^2 + \sin^2\theta\, d\phi^2). \qquad (15.91)$$

Thus the manifold defined by $x \geq 0$, $0 \leq \psi \leq 4\pi$, $0 \leq \theta \leq \pi$, $0 \leq \phi \leq 2\pi$, with ψ, θ, ϕ interpreted as hyperspherical Euler angles, is topologically R^4 and has a non-singular, positive-definite metric. The metric is asymptotically flat in the sense that the Riemann tensor decreases as r^{-3} as $r \to \infty$ but it is not asymptotically Euclidean, which would require curvature proportional to r^{-4}. The surfaces of the constant r are topologically S^3 but their metric is that of a deformed sphere. The orbits of the $\partial/\partial\psi$ Killing vector define a Hopf fibration π; $S^3 \to S^2$, where the S^2 is parametrized by the coordinates θ and ϕ. The induced metric on the S^2 is that of a

2-sphere of radius $(r^2 - M^2)^{1/2}$, while the fibres are circles of circumference $8\pi M V^{1/2}$. Thus, in a sense the boundary at large radius is $S^1 \times S^2$ but is a twisted product.

It is also possible to combine self-dual Taub–NUT solutions (Hawking, 1977). The reason is that the attraction between the electric type masses M is balanced by the repulsion between the imaginary magnetic type masses N. The metric is

$$ds^2 = U^{-1}(d\tau + \boldsymbol{\omega} \cdot d\boldsymbol{x})^2 + U\, d\boldsymbol{x} \cdot d\boldsymbol{x}, \qquad (15.92)$$

where

$$U = 1 + \sum \frac{2M_i}{r_i}$$

and

$$\operatorname{curl} \boldsymbol{\omega} = \operatorname{grad} U. \qquad (15.93)$$

Here r_i denotes the distance from the ith 'NUT' in the flat, three-dimensional metric $d\boldsymbol{x} \cdot d\boldsymbol{x}$. The curl and grad operations refer to this 3-metric, as does the vector v. Each NUT has $N_i = iM_i$.

The vector fields $\boldsymbol{\omega}$ will have Dirac string singularities running from each NUT. If the masses M_i are all equal, these string singularities and the horizon-type singularities at $r_i = 0$ can all be removed by identifying τ with period $8\pi M$. The boundary surface at large radius is then a lens space (Steenrod, 1951). This is topologically an S^3 with n points identified in the fibre S^1 of the Hopf fibration $S^3 \to S^2$, where n is the number of NUTs.

The boundary surface cannot be even locally imbedded in flat space so that one cannot work out the correction term K^0 in the action. If one tries to imbed it as nearly as one can, one obtains the value of $4\pi n M^2$ for the action, the same as Schwarzschild for $n = 1$ (Davies, 1978). In fact the presence of a gravitational magnetic mass alters the topology of the space and prevents it from being asymptotically flat in the usual way. One can, however, obtain an asymptotically flat space containing an equal number, n, of NUTs ($N = iM$) and anti-NUTs ($N = -iM$). Because the NUTs and the anti-NUTs attract each other, they have to be held apart by an electromagnetic field. This solution is in fact one of the Israel–Wilson metrics (Israel and Wilson, 1972; Hartle and Hawking, 1972). The gravitational part of the action is $8\pi n M^2$, so that each NUT and anti-NUT contributes $4\pi M^2$.

I now come on to positive-definite metrics which are solutions of the Einstein equations with a Λ-term on manifolds which are compact

Chapter 15. The path-integral approach to quantum gravity

without boundary. The simplest example is an S^4 with the metric induced by imbedding it as a sphere of radius $(3\Lambda^{-1})^{1/2}$ in five-dimensional Euclidean space. This is the analytic continuation of de Sitter space (Gibbons and Hawking, 1977b). The metric can be written in terms of a Killing vector $\partial/\partial\tau$:

$$ds^2 = (1 - \tfrac{1}{3}\Lambda r^2)\, d\tau^2 + (1 - \tfrac{1}{3}\Lambda r^2)^{-1}\, dr^2 + r^2\, d\Omega^2. \tag{15.94}$$

There is a horizon-type singularity at $r = (3\Lambda^{-1})^{1/2}$. This is in fact a 2-sphere of area $12\pi\Lambda^{-1}$ which is the locus of zeros of the Killing vector $\partial/\partial\tau$. The action is $-3\pi\Lambda^{-1}$.

One can also obtain black hole solutions which are asymptotically de Sitter instead of asymptotically flat. The simplest of these is the Schwarzschild–de Sitter (Gibbons and Hawking, 1977b). The metric is

$$ds^2 = V\, d\tau^2 + V^{-1}\, dr^2 + r^2\, d\Omega^2, \tag{15.95}$$

where

$$V = 1 - 2Mr^{-1} - \tfrac{1}{3}\Lambda r^2.$$

If $\Lambda < (9M^2)^{-1}$, there are two positive values of r for which $V = 0$. The smaller of these corresponds to the black hole horizon, while the larger is similar to the 'cosmological horizon' in de Sitter space. One can remove the apparent singularities at each horizon by identifying τ periodically. However, the periodicities required at the two horizons are different, except in the limiting case $\Lambda = (9M^2)^{-1}$. In this case, the manifold is $S^2 \times S^2$ with the product metric and the action is $-2\pi\Lambda^{-1}$.

One can also obtain a Kerr–de Sitter solution (Gibbons and Hawking, 1977b). This will be a positive-definite metric for values of r lying between the cosmological horizon and the outer black hole horizon, if the angular momentum is imaginary. Again, one can remove the horizon singularities by periodic identifications and the periodicities will be compatible for a particular choice of the parameters (Page, 1978). In this case one obtains a singularity-free metric on an S^2 bundle over S^2. The action is $-0.9553\,(2\pi\Lambda^{-1})$.

One can also obtain Taub–de Sitter solutions. These will have a cosmological horizon in addition to the ordinary Taub–NUT ones. One can remove all the horizon and Dirac string singularities simultaneously in a limiting case which is CP^2, complex, projective 2-space, with the standard Kaehler metric (Gibbons and Pope, 1978). The action is $-\tfrac{9}{4}\pi\Lambda^{-1}$.

One can also obtain solutions which are the product of two two-dimensional spaces of constant curvature (Gibbons, 1977b). The case of

$S^2 \times S^2$ has already been mentioned, and there is also the trivial flat torus $T^2 \times T^2$. In the other examples the two spaces have genera g_1 and $g_2 > 1$ and the Λ-term has to be negative. The action is $-(2\pi/\Lambda)(g_1 - 1)(g_2 - 1)$.

Finally, to complete this catalogue of known positive-definite solutions on the Einstein equations, one should mention $K3$. This is a compact four-dimensional manifold which can be realized as a quartic surface in CP^3, complex projective 3-space. It can be given a positive-definite metric whose curvature is self-dual and which is therefore a solution of the Einstein equation with $\Lambda = 0$ (Yau, 1977). Moreover $K3$ is, up to identifications, the only compact manifold to admit a self-dual metric. The action is 0.

There are two topological invariants of compact four-dimensional manifolds that can be expressed as integrals of the curvature:

$$\chi = \frac{1}{128\pi^2} \int R_{abcd} R_{efgh} \varepsilon^{abef} \varepsilon^{cdgh} (g)^{1/2} \, d^4x, \qquad (15.96)$$

$$\tau = \frac{1}{96\pi^2} \int R_{abcd} R^{ab}{}_{ef} \varepsilon^{cdef} (g)^{1/2} \, d^4x. \qquad (15.97)$$

χ is the Euler number of the manifold and is equal to the alternating sum of the Betti numbers:

$$\chi = B_0 - B_1 + B_2 - B_3 + B_4. \qquad (15.98)$$

The pth Betti number, B_p, is the number of independent closed p-surfaces that are not boundaries of some $p+1$-surface. They are also equal to the number of independent harmonic p-forms. For a closed manifold, $B_p = B_{4-p}$ and $B_1 = B_4 = 1$. If the manifold is simply connected, $B_1 = B_3 = 0$, so $\chi \geq 2$.

The Hirzebruch signature, τ, has the following interpretation. The B_2 harmonic 2-forms can be divided into B_2^+ self-dual and B_2^- anti-self-dual 2-forms. Then $\tau = B_2^+ - B_2^-$. It determines the gravitational contribution to the axial-current anomaly (Eguchi and Freund, 1976; Hawking, 1977; Hawking and Pope, 1978).

S^4 has $\chi = 2$ and $\tau = 0$; CP^2 has $\chi = 3$, $\tau = 1$; the S^2 bundle over S^2 has $\chi = 4$, $\tau = 0$; $K3$ has $\chi = 24$, $\tau = 16$ and the product of two-dimensional spaces with genera g_1, g_2 has $\chi = 4(g_1 - 1)(g_2 - 1)$, $\tau = 0$.

In the non-compact case there are extra surface terms in the formulae for χ and τ. Euclidean space and the self-dual Taub–NUT solution has $\chi = 1$, $\tau = 0$ and the Schwarzschild solution has $\chi = 2$, $\tau = 0$.

Chapter 15. The path-integral approach to quantum gravity

15.8 Gravitational thermodynamics

As explained in section 15.3, the partition function

$$Z = \sum \exp(-\beta E_n)$$

for a system at temperature $T = \beta^{-1}$, contained in a spherical box of radius r_0, is given by a path integral over all metrics which fit inside the boundary, ∂M, with topology $S^2 \times S^1$, where the S^2 is a sphere of radius r_0 and the S^1 has circumference β. By the stationary-phase approximation described in section 15.5, the dominant contributions will come from metrics near classical solutions g_0 with the given boundary conditions. One such solution is just flat space with the Euclidean time coordinate identified with period β. This has topology $R^3 \times S^1$. The action of the background metric is zero, so it makes no contribution to the logarithm of the partition function. If one neglects small corrections arising from the finite size of the box, the one-loop term also can be evaluated exactly as Z_g

$$\log Z_g = \frac{4\pi^5 r_0^3 T^3}{135}. \tag{15.99}$$

This can be interpreted as the partition function of thermal gravitons on a flat-space background.

The Schwarzschild metric with $M = (8\pi T)^{-1}$ is another solution which fits the boundary conditions. It has topology $R^2 \times S^2$ and action $\hat{I} = \beta^2/16\pi = 4\pi M^2$. The one-loop term has not been computed, but by the scaling arguments given in section 15.6 it must have the form

$$\frac{106}{45} \log\left(\frac{\beta}{\beta_0}\right) + f(r_0 \beta^{-1}) \tag{15.100}$$

where β_0 is related to the normalization constant μ. If $r_0 \beta^{-1}$ is much greater than 1, the box will be much larger than the black hole and one would expect $f(r_0 \beta^{-1})$ to approach the flat-space value (15.99). Thus f should have the form

$$f(r_0 \beta^{-1}) = \frac{4\pi^5 r_0^3}{135 \beta^3} + O(r_0^2 \beta^{-2}). \tag{15.101}$$

From the partition function one can calculate the expectation value of the energy

$$\langle E \rangle = \frac{\sum E_n \exp(-\beta E_n)}{\exp(-\beta E_n)}$$

$$= -\frac{\partial}{\partial \beta} \log Z. \tag{15.102}$$

Gravitational thermodynamics

Applying this to the contribution $(-\beta^2/16\pi)$ to $\log Z$ from the action of the action of the Schwarzschild solution, one obtains $\langle E \rangle = M$, as one might expect. One can also obtain the entropy, which can be defined to be

$$S = -\sum p_n \log p_n, \tag{15.103}$$

where $p_n = Z^{-1} \exp(-\beta E_n)$ is the probability that the system is in the nth state. Then

$$S = \beta \langle E \rangle + \log Z. \tag{15.104}$$

Applying this to the contribution from the action of the Schwarzschild metric, one obtains

$$S = 4\pi M^2 = \tfrac{1}{4} A, \tag{15.105}$$

where A is the area of the event horizon.

This is a remarkable result because it shows that, in addition to the entropy arising from the one-loop term (which can be regarded as the entropy of thermal gravitons on a Schwarzschild background), black holes have an intrinsic entropy arising from the action of the stationary-phase metric. This intrinsic entropy agrees exactly with that assigned to black holes on the basis of particle-creation calculations on a fixed background and the use of the first law of black hole mechanics (see chapters 6 and 13 by Carter and Gibbons). It shows that the idea that gravity introduces a new level of unpredictability or randomness into physics is supported not only by semi-classical approximation but by a treatment in which the gravitational field is quantized.

One reason why classical solutions in gravity have intrinsic entropy while those in Yang–Mills or $\lambda \phi^4$ do not is that the actions of these theories are scale-invariant, unlike the gravitational action. If g_0 is an asymptotically flat solution with period β and action $\hat{I}[g_0]$, then $k^2 g_0$ is a solution with a period $k\beta$ and action $k^2 \hat{I}$. This means that the action, $\hat{I}$, must be of the form $c\beta^2$, where c is a constant which will depend on the topology of the solution. Then $\langle E \rangle = 2c\beta$, $\beta \langle E \rangle = 2c\beta^2$, while $\log Z = -\hat{I} = -c\beta^2$. Thus $S = c\beta^2$. The reason that the action $\hat{I}$ is equal to $\tfrac{1}{2}\beta \langle E \rangle$ and not $\beta \langle E \rangle$, as one would expect for a single state with energy $\langle E \rangle$, is that the topology of the Schwarzschild solution is not the same as that of periodically identified flat space. The fact that the Euler number of the Schwarzschild solution is 2 implies that the time-translation Killing vector, $\partial/\partial \tau$, must be zero on some set (in fact a 2-sphere). Thus the surfaces of a constant τ have two boundaries: one at the spherical box of radius r_0 and the other at the horizon $r = 2M$. Consider now the region of

Chapter 15. The path-integral approach to quantum gravity

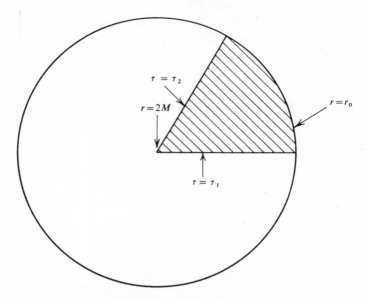

Figure 15.4. The τ–r plane of the Schwarzschild solution. The amplitude $\langle \tau_2 | \tau_1 \rangle$ to go from the surface τ_1 to the surface $\tau = \tau_2$ is dominated by the action of the shaded portion of the Schwarzschild solution.

the Schwarzschild solution bounded by the surfaces $\tau = \tau_1$, $\tau = \tau_2$ and $r = r_0$ (figure 15.4). The amplitude $\langle \tau_2 | \tau_1 \rangle$ to go from the surface τ_1 to the surface τ_2 will be given by a path integral over all metrics which fit inside this boundary, with the dominant contribution coming from the stationary-phase metric – which is just the portion of the Schwarzschild solution bounded by these surfaces. The action of this stationary-phase metric will be given by the surface terms because $R = 0$. The surface terms from the surfaces $\tau = \tau_1$ and $\tau = \tau_2$ will cancel out. There will be a contribution of $\frac{1}{2}M(\tau_2 - \tau_1)$ from the surface $r = r_0$. However there will also be a contribution from the 'corner' at $r = 2M$ where the two surfaces $\tau = \tau_1$ and $\tau = \tau_2$ meet, because the second fundamental form, K, of the boundary will have a δ-function behaviour there. By rounding off the corner slightly one can evaluate this contribution, and it turns out to be $\frac{1}{2}M(\tau_2 - \tau_1)$. Thus the total action is $\langle E \rangle (\tau_2 - \tau_1)$ and $\langle \tau_2 | \tau_1 \rangle = \exp(-\langle E \rangle (\tau_2 - \tau_1))$, as one would expect for a single state with energy $E = \langle E \rangle$. However, if one considers the partition function one simply has the boundary at $r = r_0$ and so the action equals $\frac{1}{2}\beta E$ rather than βE. This difference, which is equal to $\frac{1}{4}A$, gives the entropy of the black hole.

From this one sees that qualitatively new effects arise from the fact that the gravitational field can have different topologies. These effects would not have been found using the canonical approach, because such metrics as the Schwarzschild solution would not have been allowed.

The above derivation of the partition function and entropy of a black hole has been based on the use of the canonical ensemble, in which the system is in equilibrium with an infinite reservoir of energy at temperature T. However the canonical ensemble is unstable when black holes are present because if a hole were to absorb a bit more energy, it would cool down and would continue to absorb more energy than it emitted. This pathology is reflected in the fact that $\langle \Delta E^2 \rangle = \langle E^2 \rangle - \langle E \rangle^2 = (1/Z)(\partial^2 Z/\partial \beta^2) - (\partial \log Z/\partial B)^2 = -1/8\pi$, which is negative. To obtain sensible results with black holes one has to use the micro-canonical ensemble, in which a certain amount of energy E is placed in an insulated box and one considers all possible configurations within that box which have the given energy. Let $N(E)\,dE$ be the number of states of the gravitational field with energies between E and $E+dE$ in a spherical box of radius r_0. The partition function is given by the Laplace transform of $N(E)$,

$$Z(\beta) = \int_0^\infty N(E) \exp(-\beta E)\, dE. \qquad (15.106)$$

Thus, formally, the density of states is given by an inverse Laplace transform,

$$N(E) = \frac{1}{2\pi i} \int_{-i\infty}^{i\infty} Z(\beta) \exp(E\beta)\, d\beta. \qquad (15.107)$$

For large β, the dominant contribution to $Z(\beta)$ comes from the action of the Schwarzschild metric, and is of the form $\exp(-\beta^2/16\pi)$. Thus the right-hand side of (15.107) would diverge if the integral were taken up the imaginary β-axis as it is supposed to be. To obtain a finite value for (15.107) one has to adopt the prescription that the integral be taken along the real β-axis. This is very similar to the procedure used to evaluate the path integral in the stationary-phase approximation, where one rotated the contour of integration for each quadratic term, so that one would obtain a convergent Gaussian integral. With this prescription the factor $1/2\pi i$ in (15.107) would give an imaginary value for the density of states $N(E)$ if the partition function $Z(\beta)$ were real. However, as mentioned in section 5.6, the operator G which governs non-conformal or trace-free perturbations has one negative eigenvalue in the Schwarzschild metric. This contributes a factor i to the one-loop term for Z. Thus the partition function is purely imaginary but the density of space is real. This is what

Chapter 15. The path-integral approach to quantum gravity

one might expect: the partition function is pathological because the canonical ensemble is not well defined but the density of states is real and positive because the micro-canonical ensemble is well behaved.

It is not appropriate to go beyond the stationary-phase approximation in evaluating the integral in (15.107) because the partition function, Z, has been calculated in this approximation only. If one takes just the contribution $\exp(-\beta^2/16\pi)$ from the action of the background metric, one finds that a black hole of mass M has a density of states $N(M) = 2\pi^{-1/2} \exp(4\pi M^2)$. Thus the integral in (15.106) does not converge unless one rotates the contour integration to lie along the imaginary E-axis. If one includes the one-loop term Z_g, the stationary-phase point in the β integration in (15.107) occurs when

$$E = \frac{-\partial \log Z_g}{\partial \beta} \tag{15.108}$$

for the flat background metric, and

$$E = \frac{\beta}{8\pi} - \frac{\partial \log Z_g}{\partial \beta} \tag{15.109}$$

for the Schwarzschild background metric. One can interpret these equations as saying that E is equal to the energy of the thermal graviton and the black hole, if present. Using the approximate form of Z_g one finds that if the volume, V, of the box satisfies

$$E^5 < \frac{\pi^2}{15}(8354.5)V, \tag{15.110}$$

the dominant contribution to N comes from the flat-space background metric. Thus in this case the most probable state of the system is just thermal gravitons and no black hole. If V is less than the inequality (15.110), there are two stationary-phase points for the Schwarzschild background metric. The one with a lower value of β gives a contribution to N which is larger than that of the flat-space background metric. Thus the most probable state of the system is a black hole in equilibrium with thermal gravitons. These results confirm earlier derivations based on the semi-classical approximations (Hawking, 1976; Gibbons and Perry, 1978).

15.9 Beyond one loop

In section 15.5 the action was expanded in a Taylor series around a background field which was a solution of the classical field equations. The path integral over the quadratic terms was evaluated but the higher-order terms were neglected. In renormalizable theories such as quantum electrodynamics, Yang–Mills or $\lambda\phi^4$ one can evaluate these higher or 'interaction' terms with the help of the differential operator A appearing in the quadratic or 'free' part of the action. One can express their effect by Feynman diagrams with two or more closed loops, where the lines in the diagram represent the propagator or Green's function A^{-1} and the vertices correspond to the interaction terms, three lines meeting at a cubic term and so on. In these renormalizable theories the undetermined quantities which arise from regularizing the higher loops turn out to be related to the undetermined normalization quantity, μ, of the single loop. They can thus all be absorbed into a redefinition of the coupling constant and any masses which appear in the theory.

The situation in quantum gravity is very different. The single-loop term about a flat or topologically trivial vacuum metric does not contain the normalization quantity, μ. However, about a topologically non-trivial background one has $\log Z_g$ proportional to $(106/45)\chi \log \mu$, where Z_g is the one-loop term and χ is the Euler number. One can express this as an addition to the action of an effective topological term $-k(\mu)\chi$, where $k(\mu)$ is a scale-dependent topological coupling constant. One cannot in general provide such a topological interpretation of the μ-dependence of the one-loop term about a background metric which is a solution of the field equations with nonzero matter fields. However one can do it in the special case where the matter fields are related to the gravitational field by local supersymmetry or spinor-dependent gauge transformations. These are the various supergravity and extended supergravity theories (Freedman, Van Nieuwenhuizen and Ferrara, 1976; Deser and Zumino, 1976).

Two loops in supergravity, and maybe also in pure gravity, do not seem to introduce any further undetermined quantities. However it seems likely that, both in supergravity and in pure gravity, further undetermined quantities will arise at three or more loops, though the calculations needed to verify this are so enormous that no-one has attempted them. Even if by some miracle no further undetermined quantities arose from the regularization of the higher loop, one would still not have a good procedure for evaluating the path integral, because the perturbation expansion around a given background field has only a very limited range of validity in gravity, unlike the case in renormalizable theories such as Yang–Mills or $\lambda\phi^4$. In the latter theory the quadratic or 'free' term in the action $\int (\nabla\phi)^2 \, d^4x$ bounds the interaction term $\lambda \int \phi^4 \, d^4x$. This means

Chapter 15. The path-integral approach to quantum gravity

that one can evaluate the expectation value of the interaction term in the measure $D[\phi] \exp(-\int (\nabla \phi)^2 \, d^4 x)$ or, in other words, using Feynman diagrams where the lines correspond to the free propagator. Similarly in quantum electrodynamics or Yang–Mills theory, the interaction term is only cubic or quartic and is bounded by the free term. However, in the gravitational case the Taylor expansion about a background metric contains interaction terms of all orders in, and quadratic in derivatives of, the metric perturbations. These interaction terms are not bounded by the free, quadratic term so their expectation values in the measure given by the quadratic term are not defined. In other words, it does not make any sense to represent them by higher-order Feynman diagrams. This should come as no surprise to those who have worked in classical general relativity rather than in quantum field theory. We know that one cannot represent something like a black hole as a perturbation of flat space.

In classical general relativity one can deal with the problem of the limited range of validity of perturbation theory by using matched asymptotic expansions around different background metrics. It would therefore seem natural to try something similar in quantum gravity. In order to ensure gauge-invariance it would seem necessary that these background metrics should be solutions of the classical field equations. As far as we know, in a given topology and with given boundary conditions there is only one solution of the field equations or, at the most, a finite-dimensional family. Thus solutions of a given topology could not be dense in the space of metrics of that topology. However the Einstein action, unlike that of Yang–Mills theory, does not seem to provide any barrier to passing from fields of one topology to another.

One way of seeing this is to use Regge calculus (Regge, 1961). Using this method, one decomposes the spacetime manifold into a simplical complex. Each 4-simplex is taken to be flat and to be determined by its edge (i.e. 1-simplex) lengths. However the angles between the faces (i.e. 2-simplices) are in general such that the 4-simplices could not be joined together in flat four-dimensional space. There is thus a distortion which can be represented as a δ-function in the curvature concentrated on the faces. The total action is $(-1/8\pi) \sum A_i \delta_i$ taken over all 2-simplices, where A_i is the area of the ith 2-simplex and δ_i is the deficit angle at that 2-simplex, i.e. δ_i equals 2π minus the sum of the angles between those 3-simplices which are connected by the given 2-simplex.

A complex in which the action is stationary under small variations of the edge length can be regarded as a discrete approximation to a smooth solution of the Einstein equations. However, one can also regard the

Regge calculus as defining the action of a certain class of metrics without any approximations. This action will remain well defined and finite even if the edge lengths are chosen so that some of the simplices collapse to simplices of lower dimension. For example if a, b, c are the edge lengths of a triangle (a 2-simplex) then they must satisfy the inequalities $a < b + c$ etc. If $a = b + c$, the 2-simplex collapses to a 1-simplex. In general, the simplical complex will not remain a manifold if some of the simplices collapse to lower dimensions. However the action will still be well defined. One can then blow up some of the simplices to obtain a new manifold with a different topology. In this way one can pass continuously from one metric topology to another.

The idea is, therefore, that there can be quantum fluctuations of the metric not only within each topology but from one topology to another. This possibility was first pointed out by Wheeler (1963) who suggested that spacetime might have a 'foam-like' structure on the scale of the Planck length. In the next section I shall attempt to provide a mathematical framework to describe this foam-like structure. The hope is that by considering metrics of all possible topologies one will find that the classical solutions are dense in some sense in the space of all metrics. One could then hope to represent the path integral as a sum of background and one-loop terms from these solutions. One would hope to be able to pick out some finite number of solutions which gave the dominant contributions.

15.10 Spacetime foam

One would like to find which topologies of stationary-phase metrics give the dominant contribution to the path integral. In order to do this it is convenient to consider the path integral over all compact metrics which have a given spacetime volume V. This is not to say that spacetime actually is compact. One is merely using a convenient normalization device, like periodic boundary conditions in ordinary quantum theory: one works with a finite volume in order to have a finite number of states and then considers the values of various quantities per unit volume in the limit that the volume is taken to infinity.

In order to consider path integrals over metrics with a given 4-volume V one introduces into the action a term $\Lambda V/8\pi$, where Λ is to be regarded as a Lagrange multiplier (the factor $1/8\pi$ is chosen for convenience). This term has the same form as a cosmological term in the action but the motivation for it is very different as is its value: observational evidence

Chapter 15. The path-integral approach to quantum gravity

shows that any cosmological Λ would have to be so small as to be practically negligible whereas the value of the Lagrange multiplier will turn out to be very large, being of the order of one in Planck units.

Let

$$Z[\Lambda] = \int D[g] \exp\left(-\hat{I}[g] - \frac{\Lambda}{8\pi} V[g]\right), \quad (15.111)$$

where the integral is taken over all metrics on some compact manifold. One can interpret $Z[\Lambda]$ as the 'partition function' for what I shall call the *volume canonical ensemble*, i.e.

$$Z[\Lambda] = \sum_n \left\langle \phi_n \left| \exp-\left(\frac{\Lambda V}{8\pi}\right) \right| \phi_n \right\rangle, \quad (15.112)$$

where the sum is taken over all states $|\phi_n\rangle$ of the gravitational field. From $Z[\Lambda]$ one can calculate $N(V) dV$, the number of the gravitational fields with 4-volumes between V and $V+dV$:

$$N(V) = \frac{1}{16\pi^2 i} \int_{-i\infty}^{i\infty} Z[\Lambda] \exp(\Lambda V) d\Lambda \quad (15.113)$$

In (15.113), the contour of integration should be taken to the right of any singularities in $Z[\Lambda]$ on the imaginary axis.

One wants to compare the contributions to N from different topologies. A convenient measure of the complexity of the topology is the Euler number χ. For simply connected manifolds it seems that χ and the signature τ characterize the manifold up to homotopy and possibly up to homeomorphisms, though this is unproved. In the non-simply connected case there is no possible classification: there is no algorithm for deciding whether two non-simply connected 4-manifolds are homeomorphic or homotopic. This would seem a good reason to restrict attention to simply connected manifolds. Another would be that one could always unwrap a non-simply connected manifold. This might produce a non-compact manifold, but one would expect that one could then close it off at some large volume V with only a small change in the action per unit volume.

By the stationary-phase approximation one would expect the dominant contributions to the path integral Z to come from metrics near solutions of the Einstein equations with a Λ-term. From the scaling behaviour of the action it follows that for such a solution

$$\Lambda = -8\pi c V^{-1/2}, \quad \hat{I} = -\frac{8\pi c^2}{\Lambda} \quad (15.114)$$

where c is a constant (either positive or negative) which depends on the solution and the topology, and where the action $\hat{I}$ now includes the Λ-term. The constant c has a lower bound of $-(\frac{3}{8})^{1/2}$ which corresponds to its value for S^4. An upper bound can be obtained from (15.96) and (15.97) for χ and τ. For solutions of the Einstein equations with a Λ-term these take the form

$$\chi = \frac{1}{32\pi^2} \int (C_{abcd}C^{abcd} + 2\tfrac{2}{3}\Lambda^2)(g)^{1/2} d^4x, \qquad (15.115)$$

$$\tau = \frac{1}{48\pi^2} \int C_{abcd}{}^*C^{abcd}(g)^{1/2} d^4x. \qquad (15.116)$$

From (15.115) one sees that there can be a solution only if χ is positive. However this will be the case for simply connected manifolds because then $\chi = 2 + B_2$, where B_2 is the second Betti number. Combining (15.115) and (15.116) one obtains the inequality

$$2\chi - 3|\tau| \geq \frac{32c^2}{3}. \qquad (15.117)$$

From (15.115) one can see that, for large Euler number, at least one of the following must be true:

(a) c^2 is large

(b) $\int C_{abcd}C^{abcd}(g)^{1/2} d^4x$ is large.

In the former case c must be positive (i.e. Λ must be negative) because there is a lower bound of $-(\frac{3}{8})^{1/2}$ on c. In the latter case the Weyl tensor must be large. As in ordinary general relativity, this will have a converging effect on geodesics similar to that of a positive Ricci tensor. However, between any two points in space there must be a geodesic of minimum length which does not contain conjugate points. Therefore, in order to prevent the Weyl curvature from converging the geodesics too rapidly, one has to put in a negative Ricci tensor or Λ-term of the order of $-C_{abcd}C^{abcd}L^2$, where L is some typical length scale which will be of the order of $V^{1/4}\chi^{-1/4}$, the length per unit of topology. One would then expect the two terms in (15.115) to be of comparable magnitude and c to be of the order of $d\chi^{1/2}$, where $d \leq 3^{1/2}/4$.

This is borne out by a number of examples for which I am grateful to N. Hitchin. For products of two-dimensional manifolds of constant curvature one has $d = \frac{1}{4}$. For algebraic hypersurfaces one has $2^{1/2}/8$.

Chapter 15. The path-integral approach to quantum gravity

Hitchin has obtained a whole family of solutions lying between these limits. In addition, if the solution admits a Kähler structure one has the equality

$$3\tau + 2\chi = 32c^2. \tag{15.118}$$

One can interpret these results as saying that one has a collection of the order of χ 'gravitational instantons' each of which has action of the order of L^2, where L is the typical size and is of the order of $V^{1/4}\chi^{-1/4}$. One also has to estimate the dependence of the one-loop curve Z_g on Λ and χ. The dependence on Λ comes from the scaling behaviour and is of the form

$$Z_g \propto \Lambda^{-\gamma},$$

where

$$\gamma = \int \left(\frac{53}{720\pi^2} C_{abcd} C^{abcd} + \frac{763}{540\pi^2} \Lambda^2 \right) (g_0)^{1/2} \, d^4x. \tag{15.119}$$

One can regard γ as the number of extra modes from perturbations about the background metric, over and above those for flat space. From (15.119) one can see that it is of the same order as χ. One can therefore associate a certain number of extra modes with each 'instanton'.

From the above it seems reasonable to make the estimate

$$Z[\Lambda] = \left(\frac{\Lambda}{\Lambda_0}\right)^{-\gamma} \exp(b\chi\Lambda^{-1}), \tag{15.120}$$

where $b = 8\pi d^2$ and Λ_0 is related to the normalization constant μ. Using (15.120) in (15.113), one can do the contour integral exactly and obtain

$$N(V) = \Lambda_0^\gamma \left(\frac{8\pi b\chi}{V}\right)^{1-\gamma/2} I_{\gamma-1}\left(\frac{Vb\chi}{2\pi}\right) \tag{15.121}$$

for $V \geq 0$.

However the qualitative dependence on the parameters is seen more clearly by evaluating (15.113) approximately by the stationary-phase method. In fact it is inappropriate to do it more precisely because $Z[A]$ has been evaluated only in the stationary-phase approximation. The stationary-phase point occurs for

$$\Lambda_s = 4\pi \frac{\gamma \pm (\gamma^2 + Vb\chi/2\pi)^{1/2}}{V}. \tag{15.122}$$

Because the contour should pass to the right of the singularity at $\Lambda = 0$, one should take the positive sign of the square root.

Spacetime foam

The stationary-phase value of Λ is always positive even though $Z[\Lambda]$ was calculated using background metrics which have negative Λ for large Euler number. This means that one has to analytically continue Z from negative to positive Λ. This analytic continuation is equivalent to multiplying the metric by a purely imaginary conformal factor, which was necessary anyway to make the path integral over conformal factors converge.

From the stationary-phase approximation one has

$$N(V) = Q(\Lambda_s) \equiv \left(\frac{\Lambda_s}{\Lambda_0}\right)^{-\gamma} \exp\left(b\chi \Lambda_s^{-1} + \frac{V\Lambda_s}{8\pi}\right). \tag{15.123}$$

The dominant contribution to $N(V)$ will come from those topologies for which $dQ/d\chi = 0$. If one assumes $\gamma = a\chi$, where a is constant, one finds that this is satisfied if

$$-a \log\left(\frac{\Lambda_s}{\Lambda_0}\right) + b\Lambda_s^{-1} = 0. \tag{15.124}$$

If $\Lambda_0 \geqslant 1$, this will be satisfied by $\Lambda_s \approx \Lambda_0$. If $\Lambda_0 < 1$, $\Lambda_s \approx \Lambda_0^{a/b}$. Equation (15.122) then implies that $\chi = hV$, where the constant of proportionality, h, depends on Λ_0. In other words the dominant contribution to $N(V)$ comes from metrics with one gravitational instanton per volume h^{-1}.

What observable effects this foam-like structure of spacetime would give rise to has yet to be determined, but it might include the gravitational decay of baryons or muons, caused by their falling into gravitational instantons or virtual black holes and coming out again as other species of particles. One would also expect to get non-conservation of the axial-vector current caused by topologies with non-vanishing signature τ.

16. Ultraviolet divergences in quantum theories of gravitation

STEVEN WEINBERG[†]

16.1 Introduction

Ever since physicists began to think about a quantum theory of gravitation it has been clear that ultraviolet divergences would present problems. Simple dimensional analysis[1] tells us that if the coupling constant of a field theory has dimensions $[\text{mass}]^d$ (taking $\hbar = c = 1$), then the integral for a Feynman diagram of order N will behave at large momenta like $\int p^{A-Nd} \, dp$, where A depends on the process in question but not on N. Hence, the dangerous interactions are those with $d < 0$; in this case, the integrals for any process will diverge at sufficiently high order. But Newton's constant has dimensionality $d = -2$ (for $\hbar = c = 1$, $G = 6.7 \times 10^{-39} \, \text{GeV}^{-2}$) so general relativity is a prime example of a theory with dangerous interactions.

This intuitive argument became precise with the development of covariant rules for calculating Feynman diagrams in the quantum theory of gravitation.[2] Inspection of these rules showed immediately that general relativity fails the usual tests for renormalizability. This was confirmed by detailed calculations:[3] the one-loop integral for vacuum fluctuations in a classical background field $g_{\mu\nu}$ was found to have divergent terms[4] proportional to R^2 and $R^{\mu\nu}R_{\mu\nu}$. (These of course vanish if $g_{\mu\nu}$ satisfies the vacuum Einstein equation, $R_{\mu\nu} = 0$, but not in the presence of matter fields with $T_{\mu\nu} \neq 0$.) More generally, on dimensional grounds one would expect that diagrams with L loops should have divergent terms proportional to $L + 1$ powers of the curvature tensor. As in any non-renormalizable theory, the cancellation of these ultraviolet divergences would require introduction of an infinite number of terms in the Lagrangian, proportional to arbitrarily high powers of the curvature tensor and its covariant derivatives, and if these terms were introduced the divergences would be encountered even earlier.

[†] This research was supported in part by the National Science Foundation under Grant No. PHY77-22864.

Introduction

Of course, most of the business of physics can go on perfectly well without an understanding of quantum gravity and its ultraviolet divergences. However, it is deeply disturbing that the fusion of two such fundamental theories as quantum mechanics and general relativity should lead to an apparent contradiction. Further, the effort to construct a unified gauge theory of weak, electromagnetic, *and* strong interactions may force us to consider energies as high as 10^{19} GeV,[5] energies at which gravitation is a strong interaction. If so, then progress in developing such a 'superunified' theory will have to wait until we understand how to deal with the ultraviolet divergences in quantum gravity.

It is possible that this problem has arisen because the usual flat-space formalism of quantum field theory simply cannot be applied to gravitation. After all, gravitation is a very special phenomenon, involving as it does the very topology of space and time. It is also possible that a way may yet be found to describe gravitation, together with suitable matter fields, by an ordinary renormalizable quantum field theory. Several approaches along this line are reviewed in section 16.2. However, this paper will be chiefly concerned with another possibility, that a quantum field theory which incorporates gravitation may satisfy a generalized version of the condition of renormalizability known as *asymptotic safety*.

A theory is said to be asymptotically safe if the 'essential' coupling parameters approach a fixed point as the momentum scale of their renormalization point goes to infinity. This condition is introduced in section 16.3 as a means of avoiding unphysical singularities at very high energy. From the observed properties of second-order phase transitions, one can infer that for some, and probably all, fixed points the condition of asymptotic safety acts much like the condition of renormalizability, in fixing all but a finite number of the essential coupling parameters of a theory. For some fixed points, asymptotic safety even requires that the theory be renormalizable in the usual sense.

In section 16.4 we consider the behaviour of an asymptotically safe theory at ordinary energies, far below the Planck mass[6] $M_P = G^{-1/2} = 1.2 \times 10^{19}$ GeV. It appears that such a theory would be effectively renormalizable in the usual sense, with all non-renormalizable interactions suppressed by powers if E/M_P. The one exception is gravitation which, although incredibly weak, can nevertheless be detected because its coherence and long range allow us to observe its macroscopic effects; gravitation would be described at distances much longer than the Planck length, $M_P^{-1} = 1.6 \times 10^{-33}$ cm, by the Einstein Lagrangian,[7] $-\sqrt{g}R/16\pi G$. Thus from this point of view it is quite natural that the

Chapter 16. Ultraviolet divergences in quantum gravity

weak, electromagnetic, and strong interactions of microscopic physics should be described so well by renormalizable quantum field theories, and that the only force that seems to require a non-renormalizable description should be the one force that is observed purely through its macroscopic effects, the force of gravitation.

The question remains, whether there actually are any theories of gravitation which are asymptotically safe. In section 16.5 we show how to use dimensional continuation to give a tentative answer to this question. This method is applied to gravitation in section 16.6, and it is concluded that there is an asymptotically safe theory of pure gravity in $2+\varepsilon$ dimensions (here $0 < \varepsilon \ll 1$). Matter fields may or may not change this conclusion, depending on their type and number. Unfortunately, the problem of continuation from $2+\varepsilon$ to 4 dimensions remains unsolved.

As the reader will discover, the analysis presented here is extremely 'soft' – much is conjectured, but little is actually proved about quantum gravity. I am taking the opportunity afforded by this volume to present my own point of view on the most likely direction for progress in dealing with the ultraviolet divergences in the quantum theory of gravitation, rather than to exhibit a finished formalism.

One other purpose that I had here was to review some mathematical techniques that have been developed in the theories of critical phenomena and of elementary particles, which go under the general rubric of the 'renormalization group'.[8] Very little of this material is new, and almost none of it is due to me, but I hoped that my bringing it together in this article would provide some readers with a background to the renormalization group that they might find useful in future work on the quantum theory of gravitation.

16.2 Renormalizable theories of gravitation

As far as I know, there are just three approaches by which one might hope to construct a renormalizable theory of gravitation:

(a) Extended theories of gravitation

It could be that by combining gravitation with suitable matter fields, and imposing suitable symmetries on the theory, all ultraviolet divergences would be found to cancel, aside from a finite number which could be absorbed into a renormalization of the parameters of the theory. The extended theory of matter and gravitation would then be renormalizable in the usual sense.

As remarked in section 16.1 dimensional analysis suggests that a quantum field theory of gravitation will have ultraviolet divergencies everywhere. Experience has generally shown that if an ultraviolet divergence is expected on dimensional grounds, and is not ruled out by some symmetry of the theory, then the ultraviolet divergence will actually occur. The question of whether an ultraviolet divergence is ruled out by a symmetry is equivalent to the question of whether the term in the Lagrangian which would be needed to provide a counterterm for this divergence is forbidden by the symmetry. Thus it seems that the only hope for an extended theory of gravitation and matter that is renormalizable in the usual sense is that the symmetries of the theory should exclude all but a finite number of types of interaction from the Lagrangian.

This is certainly not possible for ordinary symmetries. If the symmetries allow the construction of a set of invariant interactions $\sqrt{g}\mathscr{L}_i$, then we can construct an infinite number of other allowed interactions by forming the products $\sqrt{g}\mathscr{L}_i\mathscr{L}_j$, $\sqrt{g}\mathscr{L}_i\mathscr{L}_j\mathscr{L}_k$, etc. and every one of the infinite number of corresponding ultraviolet divergences may be expected to occur.

However, it may be that the symmetries of the theory do not allow the construction of *any* invariant interactions, but only interactions $\sqrt{g}\mathscr{L}_i$ whose variations $\delta\sqrt{g}\mathscr{L}_i$ under the transformations of the symmetry group are total derivatives. Then the contributions $\int d^4x \sqrt{g}\mathscr{L}_i$ of these interactions to the action are invariant, so these are allowed interactions. However, the variations in the products $\sqrt{g}\mathscr{L}_i\mathscr{L}_j$, $\sqrt{g}\mathscr{L}_i\mathscr{L}_j\mathscr{L}_k$, etc. are in general not total derivatives, so these interactions could be forbidden by the symmetries of the theory, and the corresponding ultraviolet divergences would not occur.

The most interesting examples of a theory of this type are provided by supersymmetry theories,[9] in which particles of different spin are put together in irreducible representations. The Lagrangian is not actually invariant under supersymmetry transformations, but is a particular member of a supermultiplet which transforms into a total derivative. Supersymmetry imposes such stringent conditions on the Lagrangian that in some 'supergravity' theories[10] there are no allowed interactions of the sort that would be needed to cancel divergences in diagrams with one or two loops;[11] in consequence, no ultraviolet divergences are encountered in these theories in low orders of perturbation theory. However, there is as yet no reason to expect that there would not be an infinite number of allowed

Chapter 16. Ultraviolet divergences in quantum gravity

interactions in these theories, which will be needed as counterterms in higher orders of perturbation theory.

(b) Resummation

It is an old hope that a non-renormalizable theory could be made effectively renormalizable by a rearrangement of the perturbation series.[12] Certain infinite subclasses of graphs would have to be summed first, and if these sums vanished sufficiently rapidly at large momenta, then they could be used as building blocks in a new set of more convergent Feynman rules. However, it is difficult to show that such a resummation leads to a more reliable perturbation series, or indeed, to justify it on any grounds other than the elimination of ultraviolet divergences. In fact, in many cases we can see directly that terms of finite order in the resummed perturbation series contain what appear to be unphysical singularities.

The simplest example of this sort of resummation in the quantum theory of gravitation is provided by a theory based on the Lagrangian[13]

$$\mathscr{L} = -\frac{1}{16\pi G}\sqrt{g}R - f\sqrt{g}R^2 - f'\sqrt{g}R_{\mu\nu}R^{\mu\nu}, \qquad (16.1)$$

Normally one would treat the last two terms as a perturbation, and expand in powers of f and f' as well as G. In this way, of course, one would encounter horrible ultraviolet divergences, but there would be no unphysical singularities in any finite order of perturbation theory. However, we can sum up the contributions of the terms in R^2 and $R^{\mu\nu}R_{\mu\nu}$ which are quadratic in the gravitational field $h_{\mu\nu}$ (where as usual $g_{\mu\nu} = \eta_{\mu\nu} + \sqrt{32\pi G}h_{\mu\nu}$). This gives a modified graviton propagator, in which the usual $1/k^2$ is replaced with

$$\frac{1}{k^2} + \frac{1}{k^2}\alpha fGk^4\frac{1}{k^2} + \frac{1}{k^2}\alpha fGk^4\frac{1}{k^2}\alpha fGk^4\frac{1}{k^2} + \cdots = (k^2 - \alpha fGk^4)^{-1}, \qquad (16.2)$$

where α is a dimensionless function of f'/f. (Alternatively, we could simply regard the quadratic part of the R^2 and $R_{\mu\nu}R^{\mu\nu}$ terms as part of the zeroth-order Lagrangian, and obtain (16.2) directly as the 'bare' propagator for this theory.) This vanishes fast enough at infinity so that the Lagrangian (16.1) now passes the usual power-counting tests for renormalizability. The trouble is, of course, that (16.2) now has a pole at $k^2 = -1/\alpha fG$, with a residue, $-1/\alpha^2 f^2 G^2$, of the wrong sign to be consistent with unitarity.[14]

The problem encountered here is a special case of a more general impasse. Under the usual analyticity assumptions, any resummation would in general replace the $1/k^2$ in the graviton propagator with an integral[15]

$$\int \frac{\rho(\mu)}{\mu^2 + k^2} d\mu^2. \tag{16.3}$$

Unitarity requires that $\rho(\mu) \geq 0$ (except possibly for the contribution of loop graphs containing gravitons or Faddeev–Popov ghosts in Lorentz covariant gauges). As long as $\rho(\mu) \geq 0$, the integral (16.3) cannot vanish faster than $1/k^2$ as $k^2 \to \infty$.

We can see this again in a resummation based in the '$1/N$' approximation.[16] Suppose that gravity is coupled to N types of matter field, with gravitational constant, G, of the order of $1/N$. For large N, the dominant diagrams for the complete propagator consists of a chain of self-energy bubbles, each constructed from a single loop of matter lines, and connected with bare graviton propagators. (The two factors of $1/\sqrt{N}$ introduced by the graviton coupling to each matter loop are cancelled by a factor N from the sum over types of fields in the matter loop, so these diagrams are of zeroth order in $1/N$.) The matter-loop graphs are logarithmically divergent, so to cancel these divergences we must add terms proportional to $\sqrt{g}R^2$ and $\sqrt{g}R^{\mu\nu}R_{\mu\nu}$ in the Lagrangian, and the quadratic parts of these terms must be summed along with the matter loops to all orders. When this is done, the $1/k^2$ in the graviton propagator is found to be replaced with[17]

$$k^{-2} \to [k^2 + GNk^4 \alpha \ln(k^2/\mu^2)]^{-1}, \tag{16.4}$$

where α is a dimensionless constant, and μ is a renormalization mass which determines the values of f and f'. This vanishes fast enough as $k \to \infty$ so that no new ultraviolet divergences are encountered in graphs of higher order in $1/N$. However, the propagator (16.4) now has singularities at complex values of k^2. This certainly violates our usual ideas of analyticity, though it is not clear that complex singularities of this sort necessarily lead to a conflict with fundamental physical principles.[18]

(c) Composite gravitons

There is no renormalizable theory of the 3–3 resonance field, but no one worries about this – the 3–3 resonance is believed to be a three-quark bound state arising in an underlying renormalizable theory of strong

Chapter 16. Ultraviolet divergences in quantum gravity

interactions known as quantum chromodynamics. In the same way, the graviton may be a composite particle of mass zero and spin two which arises in a renormalizable quantum field theory of some sort.

If the graviton is a mere bound state, then why should it be described by a theory which is so elegantly geometrical as general relativity? A possible answer can be found within the framework of flat-space special relativity and quantum mechanics. It is very difficult to incorporate massless particles of spin two into any Lorentz-invariant quantum theory of long-range forces; it is necessary to couple such a particle to a conserved energy–momentum tensor which includes the gravitons themselves. It has in fact been shown[19] that massless particles of spin two must be described by an effective field theory satisfying the Principle of Equivalence. To the extent that we are only interested in effects of gravitation at long distances, the unique such theory is general relativity.[20]

An illuminating example of such effective field theories is provided by the theory of soft pions known as chiral dynamics.[21] In the limit of vanishing u and d quark masses, quantum chromodynamics tells us that the strong interactions have a global symmetry, chiral $SU(2) \otimes SU(2)$. This symmetry is spontaneously broken, giving rise to a massless Goldstone boson, the pion. Even though the pion is a quark–antiquark composite, its interactions are described by an effective field theory,[22] in which the pion is represented by a field $\vec{\pi}$ which transforms according to the nonlinear three-dimensional realization of $SU(2) \otimes SU(2)$. With one convenient definition of the pion field, the chiral dynamics Lagrangian takes the form

$$\mathscr{L} = -\frac{1}{2F_\pi^2} D_\mu \vec{\pi} \cdot D^\mu \vec{\pi} - f(D_\mu \vec{\pi} \cdot D^\mu \vec{\pi})^2 \cdots, \qquad (16.5)$$

where $D_\mu \vec{\pi}$ is a 'covariant derivative',

$$D_\mu \vec{\pi} = \partial_\mu \vec{\pi}/(1 + \vec{\pi}^2), \qquad (16.6)$$

and $F_\pi \simeq 190$ MeV and $f = O(1)$ are constants whose value must be taken from experiment. Suppose we use this theory to calculate any one of the invariant amplitudes $M(E)$ for pion–pion scattering at fixed angle and energy E. For E sufficiently small, the matrix element is given by a term of first order in the quartic part of the first term in (16.5), so $M(E) \propto E^2/F_\pi^2$. To next order in E we have to take into account the one-loop graphs constructed from the first term in (16.5), and the first-order effects of the second term in (16.5). The one-loop graphs are of course divergent, but the divergence can be absorbed in a renormalization of f, yielding a matrix element

$$M(E) = \frac{aE^2}{F_\pi^2} + \frac{bE^4}{F_\pi^4} \ln\left(\frac{E}{\mu}\right) + cf(\mu)\frac{E^4}{F_\pi^4} + O(E^6 \ln^2 E). \quad (16.7)$$

Here a, b, and c are known dimensionless quantities, which depend on the angle and isospin variables; $f(\mu)$ is a renormalized value of f; and μ is a unit of mass used in the definition of $f(\mu)$. Even though the effective theory is non-renormalizable, and in principle involves an infinite number of unknown parameters, we can calculate the E^2 and $E^4 \ln E$ terms in pion–pion scattering in terms of the single constant, F_π.

In much the same way, whether or not the graviton is a bound particle, the requirements of Lorentz invariance and quantum mechanics constrain the effective gravitational Lagrangian to take the form[19]

$$\mathcal{L}_G = -\frac{1}{16\pi G}\sqrt{g}R - f\sqrt{g}R^2 - f'\sqrt{g}R_{\mu\nu}R^{\mu\nu} - 16\pi G f''\sqrt{g}R^3 - \cdots \quad (16.8)$$

with a matter Lagrangian which depends on $g_{\mu\nu}$ in such a way as to be generally covariant. This is to be used to generate a perturbation series in powers of GE^2 or G/r^2 (where E and r are an energy and a length that are characteristic of the process under study), so the unperturbed Lagrangian must be taken as the quadratic part of only the first term, $-\sqrt{g}R/16\pi G$. Since the quadratic parts of the second and third terms must here be regarded as higher-order perturbations, we do not encounter the unphysical singularities discussed in section 16.2(b). We shall see in section 16.4 that the full Einstein term, $-\sqrt{g}R/16\pi G$, can be used in the tree approximation to calculate the leading terms in all multi-graviton Green's functions, and can even be used in the one-loop approximation to calculate the quantum corrections of relative order $GE^2 \ln E$ or $(G/r^2) \ln r$.

The main objection to viewing the graviton as a composite particle is simply that it is not clear why a bound state of spin two should have precisely zero mass. It is possible that this could be understood either dynamically,[23] or in terms of a supersymmetry[10] which puts the graviton in a multiplet with other composite particles which have to be massless because of chiral or other symmetries. However, this discussion has shown that if all we want is to study the low-energy or long-range properties of pions or gravitons, it is not necessary to know anything about the mechanism by which these particles are bound, or even whether pions and gravitons are composite particles at all. Section 16.4 will deal in

Chapter 16. Ultraviolet divergences in quantum gravity

a somewhat more systematic way with the low-energy limit of quantum theories of gravitation, without needing to raise the question whether gravitons are composite or elementary.

16.3 Asymptotic safety

Let us suppose for the sake of argument that none of the approaches outlined in the previous section can be successfully applied to gravitation. Suppose that no combination of gravity with matter fields yields a theory which is renormalizable in the usual sense, and that no resummation procedure yields a renormalizable perturbation theory consistent with unitarity, and that the graviton cannot be reinterpreted as a composite particle arising in an underlying renormalizable field theory. If gravitation is to be described by a flat-space quantum field theory at all, we would then have to face the prospect of dealing with a theory that is not renormalizable in the usual sense. The present section will describe a generalized version of the condition of renormalizability, which might still be applicable to gravitation in this case.

In any non-renormalizable theory, ultraviolet divergences occur in all Green's functions (and to all orders in their external momenta) so that in order to provide counterterms for these infinities, we must include in the Lagrangian all possible interactions allowed by the symmetries of the theory. Thus the gravitational Lagrangian would have to include not only the Einstein term $-\sqrt{g}R/16\pi G$, but also terms proportional to $\sqrt{g}R^2$, $\sqrt{g}R^\mu{}_\nu R^\nu{}_\mu$, $\sqrt{g}R^3$, etc., plus terms involving arbitrary powers of matter as well as gravitational fields.

It might be thought that the inclusion of terms proportional to $\sqrt{g}R^2$ and $\sqrt{g}R^\mu{}_\nu R^\nu{}_\mu$ would lead to unphysical singularities, of the sort discussed in section 16.2(*b*). However, these unphysical singularities are not encountered in perturbation theory unless the parts of these interactions quadratic in the fields are summed to all orders, as in (16.2). There is in general no justification for this partial summation – the apparent poles occur at momenta of the order of the Planck mass,[6] $G^{-1/2} = 1.2 \times 10^{19}$ GeV, and at such momenta gravitation is so strong that perturbation theory breaks down altogether. A Lagrangian with $\sqrt{g}R^2$ and $\sqrt{g}R^\mu{}_\nu R^\nu{}_\mu$ terms will *not* lead to unphysical singularities at energies $E \ll 10^{19}$ GeV, where perturbation theory can be trusted, and we cannot use perturbation theory to tell whether there will be unphysical singularities at energies of order 10^{19} GeV in such a theory, or indeed in general relativity itself. The question of possible unphysical singularities at very high energy is nevertheless an important one, and it is in fact the key issue which will lead us to the requirement of asymptotic safety.

In order to search for such singularities, we use the method of the renormalization group.[8] Let $g_i(\mu)$ denote the full set of all renormalized coupling parameters of a theory, defined at a renormalization point with momenta characterized by an energy scale μ. If $g_i(\mu)$ has the dimensions of $[\text{mass}]^{d_i}$ we replace it with a dimensionless coupling,

$$\bar{g}_i(\mu) = \mu^{-d_i} g_i(\mu). \tag{16.9}$$

Any sort of partial or total reaction rate R may be written in the form

$$R = \mu^D f\left(\frac{E}{\mu}, X, \bar{g}(\mu)\right) \tag{16.10}$$

where D is the ordinary dimensionality of R (e.g., for total cross sections, $D = -2$); E is some energy characterizing the process; and X stands for all other dimensionless physical variables, including all ratios of energies. The central idea of the renormalization group method is simply to recognize that the physical quantity R cannot depend on the arbitrary choice of renormalization point μ at which the couplings are defined, so that we can take μ in (16.10) to be anything like, and in particular to be $\mu = E$, in which case (16.10) becomes

$$R = E^D f(1, X, \bar{g}(E)). \tag{16.11}$$

Thus, apart from the trivial scaling factor E^D, the high-energy behavior of reaction rates depends on the behavior of the couplings $\bar{g}(\mu)$ as $\mu \to \infty$.

There is a technicality here which deserves a word of explanation at this point. Among the coupling parameters $g_i(\mu)$ are the particle masses $m(\mu)$, with dimensionality $d = +1$. The corresponding dimensionless parameters (16.9) are of the form $m(\mu)/\mu$, and may reasonably be expected to vanish[24] for $\mu \to \infty$. However, reaction rates will in general contain singularities for zero mass, corresponding to the infrared divergences present in a massless quantum field theory. Thus, it is not possible to evaluate the high-energy behavior of arbitrary reaction rates by simply setting the masses equal to zero. It is for this reason that the renormalization group method has historically been used to calculate the high-energy behavior of Green's functions far off the mass shell, where no infrared divergences occur even for zero mass. However, mass singularities can be eliminated from physical reaction rates themselves by performing suitable sums over certain sets of initial and final states.[25] In what follows, we will tacitly assume that this has been done.

Chapter 16. Ultraviolet divergences in quantum gravity

Our emphasis here on reaction rates rather than off-shell Green's functions has a very important advantage. Mass-shell matrix elements and reaction rates do not depend on how the fields are defined, so they are functions only of 'essential' coupling parameters, i.e. of those combinations of the coupling parameters in the Lagrangian that do not change when we subject the fields to a point transformation (as for instance $\phi \to \phi + \phi^2$ for a scalar field ϕ). In contrast, the off-shell Green's functions will of course reflect the definition of the fields involved, and will therefore be functions of all the coupling parameters in the Lagrangian, including those 'inessential' coupling parameters (like the field renormalization constants, Z) which are not invariant under redefinitions of the fields. It will always be understood here that the $g_i(\mu)$ comprise only the *essential* coupling parameters of the theory.

(There is a well-known test which can be used to identify the inessential coupling parameters in any theory. When we change any unrenormalized coupling parameter γ_0 by an infinitesimal amount ε the whole Lagrangian changes by

$$\mathscr{L} \to \mathscr{L} + \varepsilon \frac{\partial \mathscr{L}}{\partial \gamma_0}.$$

Suppose we try to produce this change by a mere redefinition of fields

$$\psi_n(x) \to \psi_n(x) + \varepsilon F_n(\psi(x), \partial_\mu \psi(x), \ldots).$$

The change in $\mathscr{L}$ induced thereby is

$$\delta \mathscr{L} = \varepsilon \sum_n \left[\frac{\partial \mathscr{L}}{\partial \psi_n} F_n + \frac{\partial \mathscr{L}}{\partial (\partial_n \psi_n)} \partial_n F_n + \cdots \right]$$

$$= \varepsilon \sum_n \left[\frac{\partial \mathscr{L}}{\partial \psi_n} - \partial_\mu \left(\frac{\partial \mathscr{L}}{\partial (\partial_\mu \psi_n)} \right) + \cdots \right] F_n + \text{total derivative terms}.$$

Thus a change $\delta \mathscr{L} = \varepsilon \, \partial \mathscr{L} / \partial \gamma_0$ in the Lagrangian can be brought about by a redefinition of the fields if and only if we can find functions, F_n, of the fields and their derivatives such that

$$\frac{\partial \mathscr{L}}{\partial \gamma_0} = \sum_n \left[\frac{\partial \mathscr{L}}{\partial \psi_n} - \partial_\mu \frac{\partial \mathscr{L}}{\partial (\partial_\mu \psi_n)} + \cdots \right] F_n(\psi, \partial_\mu \psi, \ldots) + \text{total derivative terms}.$$

In other words, *the coupling parameter γ_0 is inessential if and only if $\partial \mathscr{L}/\partial \gamma_0$ vanishes or is a total derivative when we use the Euler–Lagrange equations*

$$0 = \frac{\partial \mathscr{L}}{\partial \psi_n} - \partial_\mu \frac{\partial \mathscr{L}}{\partial (\partial_\mu \psi_n)} + \cdots.$$

For instance, in the renormalizable scalar field theory with Lagrangian

$$\mathscr{L} = -\tfrac{1}{2}Z(\partial_\mu\phi\,\partial^\mu\phi + m^2\phi^2) - \tfrac{1}{24}\lambda Z^2\phi^4$$

the field renormalization constant is an inessential coupling, because we can write

$$\frac{\partial\mathscr{L}}{\partial Z} = \tfrac{1}{2}\phi(\Box^2\phi - m^2\phi - \tfrac{1}{6}\lambda Z\phi^3) - \tfrac{1}{2}\partial_\mu(\phi\,\partial^\mu\phi)$$

and the first term vanishes when we use the field equation

$$\Box^2\phi = m^2\phi + \tfrac{1}{6}\lambda Z\phi^3.$$

On the other hand, neither the mass, m, nor the coupling, λ, nor any combination of m and λ are inessential. In this example, the field redefinition, $\phi \to \phi + \varepsilon F$, associated with the single inessential coupling Z is a simple rescaling, with $F \propto \phi$, but this is just because the theory is constrained to be renormalizable; more complicated transformations with F a nonlinear function of ϕ and its derivatives would have produced non-renormalizable terms in $\mathscr{L}$. In non-renormalizable theories, we have to consider all possible redefinitions of the fields consistent with their symmetry properties, and in consequence there is an infinite number of inessential as well as essential couplings.)

As we shall see, by working only with essential couplings we will be able to formulate the condition of asymptotic safety in a concise way. In addition, it fits in well with the 'background field method',[26] in which the renormalization of the coupling parameters is determined by calculating the matrix element between the 'in' and 'out' vacua in a classical background field which satisfies the Euler–Lagrange field equations.

Now let us return to the problem of determining the behavior of the essential couplings $\bar{g}_i(\mu)$. The change in $\bar{g}_i(\mu)$ under a given fractional change in μ is a dimensionless quantity, and can therefore depend on all the $\bar{g}_i(\mu)$, but not separately on μ itself, because there are no other dimensional parameters here with which μ can be compared. Thus the rate of change of $\bar{g}_i(\mu)$ with respect to rescaling of the renormalization point μ may be rewritten as a generalized Gell-Mann–Low equation[8]

$$\mu\frac{\mathrm{d}}{\mathrm{d}\mu}\bar{g}_i(\mu) = \beta_i(\bar{g}(\mu)). \tag{16.12}$$

Each specific theory is characterized by a *trajectory* in coupling constant space, generated by the solution of (16.12) with given initial conditions.

Chapter 16. Ultraviolet divergences in quantum gravity

The function $\beta(\bar{g})$ can be calculated as a power series in $\bar{g}_i$, but in general this will not help us determining the behavior of $\bar{g}(\mu)$ for $\mu \to \infty$. However, we can identify one general class of theories in which unphysical singularities are almost certainly absent. If the couplings $\bar{g}_i(\mu)$ approach a 'fixed point', g^*, as $\mu \to \infty$, then (16.11) gives a simple scaling behavior, $R \propto E^D$, for $E \to \infty$. In order for $\bar{g}(\mu)$ to approach g_i^* as $\mu \to \infty$, it is necessary that $\beta_i(\bar{g})$ vanish at this point

$$\beta_i(g^*) = 0 \tag{16.13}$$

and also that the couplings lie on a trajectory $\bar{g}_i(\mu)$ which actually hits the fixed point. The surface formed of such trajectories will be called the *ultraviolet critical surface*. The generalized version of renormalizability that we wish to propose for the quantum theory of gravitation is that the coupling constants must lie on the ultraviolet critical surface of some fixed point. Such theories will be called *asymptotically safe*.[27]

(Incidentally, the demand that the coupling parameters approach a fixed point for $\mu \to \infty$ cannot, in general, be met if we include inessential as well as essential coupling parameters. For instance, the renormalization group equations cannot change their form when we multiply each field by an independent constant; hence, if the field-renormalization constants $Z_r(\mu)$ satisfy these equations, then so do the $Z_r(\mu)$ times arbitrary constants, and therefore the equations for the $Z_r(\mu)$ must take the form

$$\mu \frac{d}{d\mu} Z_r(\mu) = Z_r(\mu) \gamma_r(\bar{g}(\mu)).$$

In general, there is no reason why $\gamma_r(g^*)$ should vanish or diverge, so the solution for $\mu \to \infty$ will be of the form

$$Z_r(\mu) \propto \mu^{\gamma_r(g^*)}.$$

This introduces corrections to scaling in off-shell Green's functions. However, reaction rates do not depend on the Zs, so they can exhibit 'naive' scaling $R \sim E^D$ even if $Z_r(\mu) \to \infty$ for $\mu \to \infty$.)

We do not really know that a theory which is not asymptotically safe will have unphysical singularities – the assumption of asymptotic safety is just one way of being reasonably sure that unphysical singularities do not occur. As an example of what can happen when a theory is *not* asymptotically safe, let us consider the sample differential equation[28]

$$\mu \frac{d}{d\mu} \bar{g}_i = a_i \sum_j (\bar{g}_j - g_j^*)^2 \tag{16.14}$$

where a_i and g_i^* are a set of arbitrary constants. The trajectories which hit the fixed point g^* are evidently those with initial values along the line

$$\bar{g}_i = g_i^* - a_i\xi, \quad \xi > 0, \tag{16.15}$$

so here the critical surface is one-dimensional. If $\bar{g}(\mu)$ is on this surface, with $\xi = \xi_0 > 0$ at $\mu = \mu_0$, then at greater values of μ, $\bar{g}$ is given by (16.15), with

$$\xi = \xi_0 \left[1 + \xi_0 \sum_i a_i^2 \ln(\mu/\mu_0) \right]^{-1}.$$

We see that in this case $\bar{g}_i$ smoothly approaches g_i^* as $\mu \to \infty$, and the theory is asymptotically safe. On the other hand, if $\bar{g}_i$ does not lie on the line (16.15), then it will diverge at a *finite* value of μ. For $\bar{g}_i \to \infty$, the solution to (16.14) becomes

$$\bar{g}_i \to a_i [-\sum a_j^2 \ln(\mu/\mu_\infty)]^{-1},$$

and we see that $\bar{g}_i \to \infty$ as μ approaches μ_∞. If we assume that an infinity in coupling parameters would produce an unphysical singularity in reaction rates, then we can conclude here that a theory which does not lie on the ultraviolet critical surface (16.15) will develop unphysical singularities at the energy $E = \mu_\infty$.

Of course, the question of whether or not an infinity in coupling constants betokens a singularity in reaction rates depends on how the coupling constants are defined. We could always adopt a perverse definition (e.g. $\bar{g}' \equiv (\bar{g} - g^*)^{-1}$) such that reaction rates are finite even at an infinity of the coupling parameters. This problem is avoided if we define the coupling constants as coefficients in a power series expansion of the reaction rates themselves around some physical renormalization point. In lowest-order perturbation theory, this procedure is indistinguishable from the usual renormalization procedure, in which the $\bar{g}_i(\mu)$ are defined in terms of a power series expansion of the Green's functions around some off-shell renormalization point. It seems worth exploring whether it would be possible to carry out a consistent definition of renormalized coupling parameters in terms of reaction rates rather than Green's functions, but this will not be attempted here.

The number of free parameters in an asymptotically safe theory is simply equal to the dimensionality of the ultraviolet critical surface. If the critical surface is infinite-dimensional, then the demand that the physical theory lie on this surface leaves us with an infinite number of undetermined parameters, and little has been learned. At the other extreme, if

Chapter 16. Ultraviolet divergences in quantum gravity

the critical surface is zero-dimensional, then the demand for asymptotic safety cannot be satisfied at all. The hope is that the dimensionality of the critical surface is some finite number, C; in this case, the theory will have C free parameters, of which $C-1$ are dimensionless parameters which identify a particular trajectory on the C-dimensional critical surface, and one is a dimensional parameter, which tells us the value of μ at which some given point on this trajectory is reached. Best of all of course would be the case $C=1$; in this case, physics would have no free parameters aside from one dimensional constant which merely defines our units of mass or length. As long as C is finite, the condition of asymptotic safety will play a role for us similar to that of the condition of renormalizability in quantum electrodynamics: it serves to fix all but a finite number of the parameters of the theory. In fact, as we shall see, the condition of asymptotic safety will in some cases require that a theory be renormalizable, in the usual sense.

The dimensionality of the critical surface can be determined from the behavior of the $\beta_i(\bar{g})$ functions near the fixed point. In the neighborhood of g^*, (16.12) may be written

$$\mu \frac{d}{d\mu} \bar{g}_i(\mu) \to \sum_j B_{ij}[\bar{g}_j(\mu) - g_j^*], \qquad (16.16)$$

where

$$B_{ij} \equiv (\partial \beta_i(\bar{g})/\partial \bar{g}_j)_{\bar{g}=g^*}. \qquad (16.17)$$

The general solution is

$$\bar{g}_i(\mu) \to \sum_K C_K V_i^K \mu^{\lambda_K} + g_i^* \qquad (16.18)$$

where V^K is the eigenvector of B_{ij} with eigenvalue λ_K,

$$\sum_j B_{ij} V_j^K = \lambda_K V_i^K, \qquad (16.19)$$

and the C_K are arbitrary coefficients. Clearly, the condition for $\bar{g}_i(\mu)$ to approach g_i^* as $\mu \to \infty$ is that C_K should vanish for all positive eigenvalues $\lambda_K > 0$. (The possibility of zero eigenvalues is a nuisance here, to which we shall return later.) The dimensionality of the ultraviolet critical surface then is the number of remaining C_K parameters, i.e. the number of negative eigenvalues of B_{ij}.

The crucial problem then is to determine how many eigenvalues of the B-matrix are negative. In all cases of which I know, this number is finite. This may be understood by a suggestive, though highly non-rigorous, argument. Recall that

$$\bar{g}_i(\mu) \equiv \mu^{-d_i} g_i(\mu),$$

where d_i is the dimensionality of the un-rescaled renormalized coupling $g_i(\mu)$ in powers of mass. The μ-dependence of $g_i(\mu)$ arises from the dependence of loop graphs on the momenta at the renormalization point, so that

$$\beta_i = -d_i \bar{g}_i + \text{loop contributions} \tag{16.20}$$

and

$$B_{ij} = -d_i \delta_{ij} + \text{loop contributions}. \tag{16.21}$$

Now, adding derivatives or powers of fields to an interaction will always *lower* the dimensionality d_i, so at most a finite number of d_i can be positive, and all but a finite number are below any given negative value. In the absence of the 'loop contributions', the eigenvalues of B_{ij} would be just the quantities $-d_i$, of which all but a finite number are positive. The 'loop contributions' can of course change the signs of some of the eigenvalues of B_{ij}, but if these contributions are bounded then they cannot change the sign of the infinite number of large positive eigenvalues, and only a finite number can be negative. We thus may guess that the ultraviolet critical surface will in general be of finite dimensionality.

This conclusion is in some cases empirically verified by the observed existence of second-order phase transitions. A second-order phase transition will in general occur at values of the parameters in a theory at which masses vanish or correlation lengths diverge, so that physical quantities can exhibit scaling at large distances or small momenta. By repeating the arguments of this section, we see that such scaling must be associated with a fixed point g^* where $\beta_i(g^*)$ vanishes; the phase transition occurs when the parameters of the theory take values on an *infrared* critical surface,[8] consisting of trajectories (16.12) which hit the fixed point for $\mu \to 0$. From (16.18), we see that the number of parameters which have to be adjusted in order to put the theory on the infrared critical surface is equal to the number of negative eigenvalues of B_{ij}, and hence is just equal to the dimensionality of the *ultraviolet* critical surface. But in all cases we know, this number is finite: only a finite number of parameters (temperature, pressure, magnetic field) have to be adjusted to induce a second-order phase transition. *Thus, at least for those fixed points associated with second-order phase transitions, we can be sure that the ultraviolet critical surface is of finite dimensionality.*

Chapter 16. Ultraviolet divergences in quantum gravity

We have already remarked that for a finite-dimensional critical surface, the condition of asymptotic safety works much like the condition of renormalizability in limiting the free parameters in physical theories. In fact, we can now see that the connection between asymptotic safety and renormalizability is even closer. Any theory will always have a fixed point at the origin, $g^* = 0$. (If the essential couplings vanish at one renormalization scale μ, they will vanish at all μ, so $\beta_i(\bar{g})$ always vanishes at $\bar{g} = 0$.) Suppose that for some theory this is the only suitable fixed point with an ultraviolet critical surface of nonzero dimensionality, so that asymptotic safety requires that the couplings lie on this surface. The 'loop contribution' in (16.21) vanishes at $\bar{g}_i = 0$, so the B-matrix for this fixed point is just

$$B_{ij} = -d_i \delta_{ij}. \tag{16.22}$$

In order for the trajectory $\bar{g}i(\mu)$ to hit $g^* = 0$ as $\mu \to \infty$, it is thus necessary that all g_i with $d_i < 0$ should vanish. But these are precisely the non-renormalizable interactions,[1] so this sort of theory must be renormalizable in the usual sense.

Strictly speaking, renormalizability may not be enough. The B-matrix for $g^* = 0$ will in general have some zero eigenvalues, corresponding to those 'strictly' renormalizable interactions with $d_i = 0$, and it is also necessary that these interactions have $\beta_i/g_i < 0$ near $\bar{g}_i = 0$, in order that $\bar{g}_i(\mu)$ should vanish for $\mu \to \infty$. Hence the ultraviolet critical surface of the fixed point $g^* = 0$ actually consists of all theories that are renormalizable *and* asymptotically free.[29] However, from a practical point of view a renormalizable theory like quantum electrodynamics can be regarded as asymptotically safe even though it is not asymptotically free, because the growth of a strictly renormalizable coupling like $e(\mu)$ is only logarithmic, and any unphysical singularities would only be encountered at exponentially high energies. This would not be the case if the theory involved non-renormalizable interactions, like Fermi interactions.

In some cases, asymptotic safety can lead to renormalizability even though the fixed point is not at $g^* = 0$. The seminal paper[8] of Gell–Mann and Low suggested the existence of a possible fixed point $e^* \neq 0$ in quantum electrodynamics, at which $\beta_e(e^*) = 0$. If there is such a fixed point in quantum electrodynamics, then it is a fixed point for the most general field theory of photons and electrons, but there is no reason to expect that trajectories with non-vanishing values for non-renormalizable couplings would hit this fixed point. Thus, barring other fixed points, the effect of the condition of asymptotic safety here would be just to require renormalizability of the usual sort.

The problem that we face in dealing with quantum gravity is that there may not be any theories that are renormalizable, let alone asymptotically free. Therefore we must search for other fixed points, away from $g^* = 0$. In general, there is no particular reason why a fixed point with $g^* \neq 0$ should have g^* small, so perturbation theory will not be of much use to us in searching for such fixed points, or in exploring their properties.

This is much the same as the problem encountered in the theory of critical phenomena: the failure of mean field theory showed that the phase transitions are not associated with a fixed point at zero coupling, and it became necessary to search for other fixed points. In this case, the problem could be solved[30] by a continuation in the spatial dimensionality of the system, the 'ε-expansion'. It therefore seems reasonable to try a continuation in spacetime dimensionality in our present problem.

Suppose we could find some spacetime dimensionality $D_r < 4$ at which there exists a renormalizable and asymptotically free theory of gravitation. As we have seen, this would imply that the fixed point at $g^* = 0$ has an ultraviolet critical surface of finite dimensionality. If we then increase the dimensionality, D, the theory will become non-renormalizable, but continuity would lead us to expect that the fixed point which is at $g^* = 0$ for $D = D_r$ will move smoothly away from the origin, and that for at least a finite range of D above D_r its ultraviolet critical surface will retain the same dimensionality. Hence for $D = D_r + \varepsilon$, we can expect to find a fixed point with g^* of order ε, and to learn the properties of this fixed point by an expansion in powers of ε.

This approach has already been applied[31] to the nonlinear σ-model (which is renormalizable and asymptotically free at $D_r = 2$), though from a somewhat different point of view. Its application to gravitation will be described in section 16.6, using techniques to be discussed in section 16.5.

In carrying out actual calculations, it is very useful to recognize that although the β_i-functions and the B_{ij}-matrix depend on the details of our renormalization procedure and on how the coupling parameters g_i are defined, the *eigenvalues* of the B_{ij}-matrix do not.[32] There is a wide variety of ways in which we might change the definition of the $g_i(\mu)$, as for example:

(a) We might simply choose a different set of renormalization points, in which case the rescaled coupling parameters (16.9) would become functions, $\tilde{g}_i(\tilde{\mu})$, of the scale, $\tilde{\mu}$, of the momenta at the new renormalization points.

Chapter 16. Ultraviolet divergences in quantum gravity

(*b*) We can use dimensional regularization,[33] in which case the 'renormalized' couplings can be taken as the constant terms in a Laurent expansion of the unrenormalized couplings, g_{i0}, around spacetime dimensionality $D = 4$:

$$g_{i0} \xrightarrow[D \to 4]{} \tilde{\mu}^{d_i(D)} \left[\tilde{g}_i(\tilde{\mu}) + \sum_{\nu=1}^{\infty} \frac{b_{\nu i}(\tilde{g}(\tilde{\mu}))}{(D-4)^\nu} \right], \quad (16.23)$$

with $\tilde{\mu}$ a 'unit of mass' introduced to make $\tilde{g}_i(\tilde{\mu})$ dimensionless.

(*c*) We can introduce an ordinary ultraviolet cut-off at momentum $\tilde{\mu}$, and take the rescaled unrenormalized couplings as functions $\tilde{g}_i(\tilde{\mu})$, with a cut-off dependence chosen so that the reaction rates are cut-off independent.[34]

In these or other cases, the new couplings $\tilde{g}_i$ must be expressible as functions of the old couplings $\bar{g}_j(\mu)$ and of the only other dimensionless quantity $\tilde{\mu}/\mu$:

$$\tilde{g}_i(\tilde{\mu}) = \tilde{g}_i\left(\frac{\tilde{\mu}}{\mu}, \bar{g}_i(\mu)\right).$$

But the new couplings do not depend on how the old couplings are defined, so they are independent of μ:

$$0 = \mu \frac{d\tilde{g}_i}{d\mu} = -\tilde{\mu} \frac{\partial \tilde{g}_i}{\partial \tilde{\mu}} + \sum_j \frac{\partial \tilde{g}_i}{\partial \bar{g}_j} \mu \frac{d\bar{g}_j}{d\mu}.$$

In other words, we may define a new β_i-function

$$\tilde{\beta}_i(\tilde{g}(\tilde{\mu})) \equiv \tilde{\mu} \frac{\partial \tilde{g}_i}{\partial \tilde{\mu}} \quad (16.24)$$

which is related to the old β-function by the transformation rule of a contravariant vector

$$\tilde{\beta}_i(\tilde{g}) = \sum_j \frac{\partial \tilde{g}_i}{\partial \bar{g}_j} \beta_j(\bar{g}). \quad (16.25)$$

We see that the existence of a fixed point is an invariant concept: if $\beta_i(g^*) = 0$ then $\tilde{\beta}_i(\tilde{g}(g^*)) = 0$. The β-functions themselves are not invariant, and neither are their derivatives

$$\frac{\partial \tilde{\beta}_i}{\partial \tilde{g}_j} = \sum_{kl} \frac{\partial^2 \tilde{g}_i}{\partial \bar{g}_k \partial \bar{g}_l} \frac{\partial \bar{g}_l}{\partial \tilde{g}_j} \beta_k(\bar{g}) + \sum_{kl} \frac{\partial \tilde{g}_i}{\partial \bar{g}_k} \frac{\partial \beta_k(\bar{g})}{\partial \bar{g}_l} \frac{\partial \bar{g}_l}{\partial \tilde{g}_j}.$$

But at a fixed point the first term vanishes, so the new B-matrix is related to the old one by

$$\tilde{B}_{ij} = \sum_{kl} A_{ik} B_{kl} A^{-1}{}_{lj} \qquad (16.26)$$

$$A_{ik} \equiv \left(\frac{\partial \tilde{g}_i(\bar{g})}{\partial \bar{g}_k}\right)_{\bar{g}=g^*}. \qquad (16.27)$$

This is a similarity transformation, so the eigenvalues of $\tilde{B}$ are the same as the eigenvalues of B. (These eigenvalues are known as *critical exponents*; they depend only on the nature of the degrees of freedom of a system, and on no other physical inputs.) In particular, the question of asymptotic safety is one that can be addressed in any of the formalisms (*a*), (*b*), (*c*) outlined above, with confidence that the answer will be the same.

16.4 Physics at ordinary energies

The condition of asymptotic safety entails the appearance of a fundamental energy scale M: it is the value of μ at which the trajectory $\bar{g}_i(\mu)$ approaches to within some definite distance of the fixed point $g_i{}^*$. We will see below that this characteristic energy should be of the order of the Planck mass,[6] $M_P = 1.2 \times 10^{19}$ GeV. The problem to which we now turn is the description of physical phenomena at ordinary energies, which are vastly less than M. In particular, we want to inquire why gravitational interactions are so well described at macroscopic distances by the Einstein Lagrangian, $-\sqrt{g}R/16\pi G$, and why the weak, strong, and electromagnetic interactions should be so well described at ordinary energies by renormalizable quantum field theories.[35]

Let us first consider the case of pure gravity, with a Lagrangian of form (16.8). We assume that the theory is asymptotically safe, so the infinite set of couplings are constrained to lie on a trajectory $\bar{g}_i(\mu)$ which hits some fixed point $g_i{}^*$ for $\mu \to \infty$. The renormalization-group arguments of the previous section show that physics at low energies is governed by the behavior of the couplings $\bar{g}_i(\mu)$ as $\mu \to 0$. The simplest possibility is for $\bar{g}_i(\mu)$ to approach another fixed point in this limit. Now, we saw in the previous section that in addition to $g_i{}^*$, there is always a fixed point at the origin. Also, every term in the Lagrangian (16.8) is non-renormalizable, so every eigenvalue of the matrix $\partial \beta_i / \partial \bar{g}_j$ at $\bar{g}_i = 0$ is positive. This means that the fixed point at the origin is entirely repulsive for $\mu \to \infty$, but by the same token, it is entirely attractive for $\mu \to 0$: whatever direction we go from the origin, we remain in the infrared critical surface. It follows that there is at least a finite region around the origin, within which *every*

Chapter 16. Ultraviolet divergences in quantum gravity

trajectory is attracted to the origin for $\mu \to 0$. We do not know whether the ultraviolet fixed point, g_i^*, lies within this region, but there is nothing unreasonable in supposing that it does. Under this assumption, we can conclude that all $\bar{g}_i(\mu)$ vanish for $\mu \to 0$.

(We have tacitly assumed here that the Lagrangian does not contain a cosmological constant term $\Lambda\sqrt{g}$. Such a term would be super-renormalizable, and would therefore correspond to an infrared-repulsive eigenvector of $\partial\beta_i/\partial\bar{g}_j$ at $\bar{g}_i = 0$. A cosmological constant is not needed as a counterterm in pure gravity, at least in the dimensional regularization scheme. However, theories with massive particles will in general require a cosmological constant counterterm, and it is somewhat of a mystery why Λ is not some forty orders of magnitude larger than the observed upper limit.)

For $\bar{g}_i(\mu)$ near zero, the 'loop contribution' in (16.20) is negligible, and the solution of (16.12) becomes simply

$$\bar{g}_i(\mu) \approx (M_i/\mu)^{d_i} \quad \text{for } \mu \ll M_i,$$

with M_i a set of unknown integration constants. The M_i are related to each other by the asymptotic safety condition, that $\bar{g}(\mu)$ must lie on the ultraviolet critical surface of a fixed point g^*. For a one-dimensional critical surface, the condition of asymptotic safety leaves us with only one free parameter, the fundamental energy scale M, so all M_i must be of order M. The same is true even if the dimensionality C of the critical surface is greater than unity, provided that the $C-1$ parameters which determine the orbit of $\bar{g}_i(\mu)$ do not take exceptionally large or exceptionally small values. We conclude then that the original couplings, before the rescaling (16.9), have the order of magnitude

$$g_i(\mu) \approx M^{d_i} \quad \text{for } \mu \ll M. \tag{16.28}$$

In particular, (16.28) gives a gravitational constant G of order M^{-2}, so M must be of the order of the Planck mass

$$M \approx M_P \equiv G^{-1/2} = 1.2 \times 10^{19} \text{ GeV}. \tag{16.29}$$

We can now see why gravitational phenomena are so well described by Einstein's theory at macroscopic distances. Consider a connected Green's function for a set of gravitational fields at points with typical spacetime separations r. Ultraviolet divergencies are removed by the renormalization of the infinite set of coupling parameters, but it is essential here that the renormalized coupling parameters $g_i(\mu)$ be defined at renormalization points with momenta of order $1/r$, not M_P. In this way, the integrals in the Feynman diagrams will begin to converge at momenta of order $1/r$, so r will be the only dimensional parameter in the theory other than the coupling parameters themselves. Equations (16.28) and (16.29) show that the coupling constants in a graph with $\mathcal{N}_i$ vertices of type i yield a factor proportional to N powers of $G^{1/2}$ or M_P^{-1}, where

$$N = -\sum_i \mathcal{N}_i d_i, \tag{16.30}$$

so ordinary dimensional analysis tells us that the contribution of such a graph will be suppressed by a factor

$$(G^{1/2}/r)^N = (rM_P)^{-N}. \tag{16.31}$$

The leading graphs for r much larger than the Planck length $M_P^{-1} = 1.6 \times 10^{-33}$ cm will thus be those with the smallest values of N. The dimensionality, d_i, is given by

$$d_i = 4 - p_i - g_i, \tag{16.32}$$

where p_i is the number of derivatives and g_i the number of graviton fields in the interaction of type i. To calculate N, we use the well known topological relations

$$\mathcal{L} = \ell - \sum_i \mathcal{N}_i + 1 \tag{16.33}$$

$$2\ell + \mathcal{E} = \sum_i \mathcal{N}_i g_i \tag{16.34}$$

where $\mathcal{L}$ is the number of loops; ℓ is the number of internal lines; and $\mathcal{E}$ is the number of external lines in the graph. Together with (16.30) and (16.32), these give

$$N = \sum_i \mathcal{N}_i (p_i - 2) + 2\mathcal{L} + \mathcal{E} - 2. \tag{16.35}$$

Leaving aside any super-renormalizable cosmological constant term, $\Lambda\sqrt{g}$, the gravitational interactions with the smallest number of derivatives are those derived from the Einstein term, $-\sqrt{g}R/16\pi G$, all of which have $p_i = 2$. Hence for any given Green's function with a given number $\mathcal{E}$ of external lines, the leading graphs for $r \gg M_P^{-1}$ will be the tree graphs ($\mathcal{L} = 0$) constructed purely from the Einstein Lagrangian. Summing these tree graphs is tantamount to solving the classical field equations;[36] in particular, the one-graviton Green's function in the presence of a classical background distribution of energy and momentum satisfies the classical Einstein field equations for the gravitational field produced by this energy–momentum tensor. The next corrections will arise both from the

Chapter 16. Ultraviolet divergences in quantum gravity

graphs in pure general relativity with one loop, and equally from the tree graphs containing *one* vertex arising from the higher interactions, $\sqrt{g}R^2$ or $\sqrt{g}R^{\mu\nu}R_{\mu\nu}$, for which $p_i = 4$. (The ultraviolet divergence in the loop graph is cancelled by the counterterm provided by the $\sqrt{g}R^2$ and $\sqrt{g}R^{\mu\nu}R_{\mu\nu}$ interactions.) Equation (16.35) shows that these quantum corrections to classical general relativity are suppressed by a factor of order $(rM_P)^{-2}$, and hence are utterly negligible at macroscopic distances.

Incidentally, (16.35) shows that even the classical tree-graph contributions to a given Green's function are suppressed by a factor proportional to $G^{1/2}$ for each extra external graviton line. When we calculate the metric produced by a mass, m, we also pick up another factor of $G^{1/2}m$ for each coupling of these external lines to the mass. The reason why the exchange of *trees* of graviton lines with $\mathscr{E} > 2$ external lines has a detectable effect on planetary motion is just that the solar mass, $m_\odot$, is so large that $Gm_\odot/r$ is not an utterly negligible quantity.

The above discussion can be immediately extended to theories of gravitation and matter, at least in the case that there are no masses and no super-renormalizable or asymptotically free renormalizable interactions among the matter fields. (For instance, this would be the case for the theory of gravitons, photons, and massless electrons.) In such a theory, the fixed point at zero coupling is still entirely attractive for $\mu \to 0$, so according to our previous arguments, the couplings at ordinary energy will have the order of magnitude

$$g_i(\mu) = O(M_P^{d_i}) \quad \text{for } \mu \ll M_P.$$

Physical phenomena at ordinary energies $E \ll M_P$ will be entirely governed by the renormalizable interactions, for which $d_i = 0$, because all non-renormalizable interactions with $d_i < 0$ are suppressed by powers of E/M_P and hence entirely undetectable. Only gravitation is an exception: as we have seen, the fact that gravitation couples coherently to every particle in a large body like the sun allows us to observe its macroscopic effects despite its intrinsic weakness.

Theories of gravitation and matter that involves masses or asymptotically free renormalizable interactions require a bit more attention. First, there may be particles with mass of order M_P – for instance, there are intermediate vector bosons almost this heavy in some superunified theories of weak, electromagnetic, and strong interactions.[5] Such particles cause no problems here; in Green's functions at ordinary energy, $E \ll M_P$, any internal line with mass M of order M_P may be replaced with a series of non-renormalizable interactions, obtained by expanding the heavy-particle propagators

$$\frac{1}{q^2+M^2} = \frac{1}{M^2} - \frac{q^2}{M^4} + \frac{(q^2)^2}{M^6} - \cdots.$$

(Again, it is essential here that the momentum scales, μ, of all renormalization points be taken of order E, not M, so that integrals converge at momenta $q \approx E \ll M$.) In this way, we obtain an effective field theory,[37] involving only the light particles with masses much less than M_P. These of course include all the particles with which we are familiar – gravitons, photons, leptons, quarks, intermediate vector bosons, Higgs bosons, and gluons.[38]

Now suppose that this effective field theory involves either some small masses $m \ll M_P$, or some asymptotically free, renormalizable interactions acting among the light particles, or both. In this case, the fixed point at the origin in coupling constant space is no longer entirely attractive for $\mu \to 0$. However, we can now identify an infrared-attractive *area*, if not an infrared-attractive *point*. Let us denote rescaled renormalizable couplings (including masses) as $\bar{f}_a(\mu)$, and non-renormalizable couplings as $\bar{F}_A(\mu)$. Since renormalizable couplings never generate infinities which require non-renormalizable counterterms, the renormalization group equations for $\bar{F}_A(\mu) \ll 1$ must take the form

$$\mu \frac{d}{d\mu} \bar{f}_a(\mu) \approx \beta_a(\bar{f}(\mu))$$

$$\mu \frac{d}{d\mu} \bar{F}_A(\mu) \approx \sum_B B_{AB}(\bar{f}(\mu)) \bar{F}_B(\mu).$$

Furthermore, for $\bar{f}(\mu) \ll 1$ the loop contributions to B_{AB} become negligible, so that

$$B_{AB}(\bar{f}) \approx -d_A \delta_{AB} \quad (\bar{f} \ll 1).$$

By definition, F_A is a non-renormalizable interaction, so $d_A > 0$. Thus there is at least a finite area $\mathcal{A}$ on the surface $\bar{F}_A = 0$ within which all eigenvalues of the matrix B_{AB} are positive. This is an infrared-attractive area – every trajectory in at least a finite slab $\mathcal{R}_\mathcal{A}$ around $\mathcal{A}$ is attracted to it for $\mu \to 0$. Once again, we do not know whether the ultraviolet fixed point g_i^* lies within the region $\mathcal{R}_\mathcal{A}$, but it is not unreasonable to suppose that it does. Indeed, this seems to be experimentally verified: there is an enormous range of renormalization scales, $1 \text{ GeV} \ll \mu \ll 10^{19} \text{ GeV}$, within which the rescaled non-renormalizable gravitational couplings,

Chapter 16. Ultraviolet divergences in quantum gravity

$\bar{F}_A(\mu)$, are all quite small, and the renormalizable weak, electromagnetic, and 'strong' interactions also have fairly small couplings, so that B_{AB} has positive eigenvalues $\approx -d_A$. On the assumption that the ultraviolet fixed point $g_i{}^*$ does lie within the region $\mathscr{R}_{\mathscr{A}}$, we can conclude again that the effect of non-renormalizable couplings becomes negligible at ordinary energies. More specifically, within the range of renormalization scales in which all $\bar{g}_i(\mu)$ are small, the unrescaled couplings $g_i(\mu)$ will again be of order M^{d_i}, as in (16.28), with M a common integration constant related to the renormalization scale at which $\bar{g}_i(\mu)$ approaches the fixed point, $g_i{}^*$. The conventional values of the gravitational couplings, G, and other non-renormalizable couplings are specified at a renormalization point with μ of the order of the mass of a typical light particle, where the rescaled renormalizable couplings (including m/μ) are just beginning to be of order unity, so these conventional couplings will also be of order M^{d_i}. We can thus identify M with the Planck mass, as in (16.29), and conclude again that effects of non-renormalizable couplings at ordinary energies, $E \ll M_P$, will be suppressed by powers of E/M_P.

16.5 Dimensional continuation

It was emphasized in section 16.3 that the existence of a fixed point and the dimensionality of its ultraviolet critical surface do not depend on whether we define the coupling parameters by ordinary renormalization, or by dimensional regularization, or by a 'floating' ultraviolet cut-off. However, experience has shown that the method of dimensional regularization[33] is by far the most convenient for actual calculation.[39] Somewhat suprisingly, dimensional regularization also turns out to provide a very convenient basis for the study of fixed points at arbitrary, non-integral dimensionality;[40] we remarked in section 16.3 that such dimensional continuation provides one method by which perturbation theory can be used in the study of fixed points.

Dimensional regularization allows us to calculate all Feynman integrals in finite form for non-rational spacetime dimensionality D, but the integrals will have poles as D approaches various rational values. Let us concentrate on some particular set of rational values, D_s, of the spacetime dimensionality, and suppose that the unrenormalized coupling constants $g_{i0}(D)$ have poles which will cancel the poles at $D = D_s$ in Feynman integrals, yielding reaction rates which are finite at $D = D_s$. (In the original formulation[39] of this approach, the set of D_s consisted of the single physical spacetime dimensionality $D = 4$; the generalization to several D_s is presented here because it requires essentially no extra work, and might turn out to be useful.) In order to express $g_{i0}(D)$ in terms of renormalized coupling parameters which are dimensionless for all D, we must introduce a unit of mass μ and write a Laurent expansion for the rescaled coupling, $g_{i0}(D)\mu^{-d_i(D)}$, instead of the ordinary unrenormalized coupling, $g_{i0}(D)$. (As usual, $d_i(D)$ is the dimensionality of $g_{i0}(D)$ in powers of mass, at a space-time dimensionality D). In addition to the poles at $D = D_s$, this expansion will have a remainder term which is analytic in D, and which we simply *define* as the dimensionless renormalized coupling, $g_i(\mu, D)$. The coefficients, $b_{\nu i}^{(s)}$, of the poles of order ν in $g_{i0}(D)\mu^{-d_i(D)}$ at $D = D_s$ will depend on μ and D only through their dependence on the renormalized coupling $g_i(\mu, D)$, because there is no dimensional parameter with which μ could be compared, and because any separate dependence of $b_{\nu i}^{(s)}$ on D would merely change the analytic terms and the lower-order poles in D. Thus the Laurent expansion may be written in the form

$$g_{i0}(D)\mu^{-d_i(D)} = g_i(\mu, D) + \sum_s \sum_{\nu=1}^{\infty} (D-D_s)^{-\nu} b_{\nu i}^{(s)}(g(\mu, D)). \quad (16.36)$$

(We now drop the tilde used in section 16.3 to distinguish $g_i(\mu, D)$ from the conventional rescaled renormalized coupling $\tilde{g}_i(\mu)$.)

To calculate the Gell–Mann–Low functions, $\beta_i(g, D)$, we first differentiate with respect to μ, and find

$$-d_i(D)\left[g_i + \sum_{s,\nu}(D-D_s)^{-\nu}b_{\nu i}^{(s)}(g)\right]$$

$$= \beta_i(g, D) + \sum_{s,\nu,j}(D-D_s)^{-\nu}b_{\nu ij}^{(s)}(g)\beta_j(g, D), \quad (16.37)$$

where

$$\beta_i(g(\mu, D), D) \equiv \mu\frac{\partial}{\partial\mu}g_i(\mu, D) \quad (16.38)$$

and

$$b_{\nu ij}^{(s)}(g) \equiv \partial b_{\nu i}^{(s)}(g)/\partial g_j. \quad (16.39)$$

Again, we write β_i as a function of all the $g_j(\mu, D)$ and also of D, but not of μ, because there is no dimensional quantity here with which μ could be compared.

Now, the dimensionality, $d_i(D)$, of g_{i0} is always a linear function of the spacetime dimensionality D, which we shall write as[39]

$$d_i(D) = \sigma_i + \rho_i D. \quad (16.40)$$

Chapter 16. Ultraviolet divergences in quantum gravity

The left-hand side of (16.37) can then be re-written as

$$-\rho_i g_i D - \left[\sigma_i g_i + \sum_s b_{1j}^{(s)}(g)\rho_i\right]$$

$$-\sum_{s,\nu}(D-D_s)^{-\nu}[\rho_i b_{\nu+1,i}^{(s)}(g) + (\sigma_i + \rho_i D_s)b_{\nu,i}^{(s)}(g)] \quad (16.41)$$

the sum over ν running as before from 1 to ∞. Since the highest power of D in the analytic part here is of first order, the same must be true on the right-hand side of (16.37). However, all poles in D are supposed to be eliminated when we express g_{i0} in terms of g_i, so $\beta_i(g, D)$ should be analytic in D. In order that the analytic part of the right-hand side of (16.37) contain no terms of higher than first order in D, it is necessary then that $\beta_i(g, D)$ be *linear* in D:

$$\beta_i(g, D) = \beta_i^{(1)}(g)D + \beta_i^{(0)}(g).$$

Equating terms of first and zeroth order in D in (16.41) and the right-hand side of (16.37) gives then

$$-\rho_i g_i = \beta_i^{(1)}(g)$$

$$-\sigma_i g_i - \sum_s b_{1i}^{(s)}(g)\rho_i = \beta_i^{(0)}(g) + \sum_{sj} b_{1ij}^{(s)}(g)\beta_j^{(1)}(g),$$

and therefore

$$\beta_i(g, D) = -\rho_i g_i D - \sigma_i g_i - \sum_s b_{1i}^{(s)}(g)\rho_i + \sum_{sj} b_{1ij}^{(s)}(g)\rho_j g_j. \quad (16.42)$$

It is both remarkable and convenient that the β_i-functions are linear in D; the D-dependence arises entirely from the term $-d_i(D)g_i$ in β_i, and the loop contributions in (16.20) are D-independent.

(It should be mentioned in passing that the comparison of pole terms in (16.37) leads to further relations,[39] which determine the $b_{\nu i}^{(s)}$ for $\nu > 1$ in terms of $b_{1i}^{(s)}$. Using (16.42), the right-hand side of (16.37) may be put in the form

$$-\rho_i g_i D - \sigma_i g_i - \sum_s b_{1i}^{(s)}(g)\rho_i$$

$$+ \sum_{s,\nu,j}(D-D_s)^{-\nu}[b_{\nu ij}^{(s)}(g)\beta_j(g, D_s) - b_{\nu+1,ij}^{(s)}(g)\rho_j g_j].$$

Equating the pole terms here with those in (16.41), we obtain the

816

recursion relation

$$\rho_i b_{\nu+1,i}^{(s)}(g) - \sum_j \rho_j g_j b_{\nu+1,ij}^{(s)}(g)$$
$$= -(\sigma_i + \rho_i D_s) b_{\nu i}^{(s)}(g) - \sum_j b_{\nu ij}^{(s)}(g) \beta_j(g, D_s) \quad (16.43)$$

for all $\nu \geq 1$.)

Further useful information about the residue functions can be obtained from dimensional considerations. One of the peculiarities of dimensional regularization is that the poles in Green's functions arise only from logarithmic ultraviolet divergences, not from quadratic or higher divergences. But a logarithmic divergence can occur in a given quantity only if the dimensionality of this quantity just equals the dimensionality of the coupling constants to which the quantity is proportional. It follows that $b_i^{(s)}(g)$ can contain a term of order $g_a g_b g_c \cdots$ only if the dimensionality of g_{i0} equals the total dimensionality of $g_{a0} g_{b0} g_{c0} \cdots$ at spacetime dimensionality D_s:

$$d_i(D_s) = d_a(D_s) + d_b(D_s) + \cdots . \quad (16.44)$$

From (16.42) we see that the same is then true of the D_s terms in β_i. For instance, in a Euclidean scalar field theory with symmetry under the transformation $\phi \to -\phi$ the general Lagrangian is

$$\mathcal{L} = -\frac{1}{2}\partial_\mu \phi \, \partial_\mu \phi - \frac{1}{2}g_{10}\phi^2 - \frac{1}{4!}g_{20}\phi^4 - \frac{1}{6!}g_{30}\phi^6 - \frac{1}{8!}g_{40}\phi^8 \cdots . \quad (16.45)$$

(We use our freedom to make point transformations with $\delta\phi$ a linear combination of $\phi, \phi^3, \Box\phi, \phi^5, \phi^2\Box\phi, \Box^2\phi$, etc. to eliminate terms such as $\phi^3\Box\phi, (\Box\phi)^2, \phi^5\Box\phi, \phi^2(\Box\phi)^2, (\Box^2\phi)^2$, etc., and to adjust the coefficient of $-\frac{1}{2}\partial_\mu\phi \, \partial_\mu\phi$ to unity.) In four dimensions, the g_{i0} have dimensionalities

$$d_1 = 2, \quad d_2 = 0, \quad d_3 = -2, \quad d_4 = -4, \cdots . \quad (16.46)$$

Hence if we define the renormalized couplings to eliminate only the poles at $D_s = 4$, the β-functions will have the structure

$$\beta_1 = g_1 F_1 \quad \beta_2 = F_2$$
$$\beta_3 = g_3 F_3 + g_1 g_4 F_{14} + \cdots \quad (16.47)$$
$$\beta_4 = g_4 F_4 + g_3^2 F_{33} + \cdots$$

where the Fs are functions only of the variables $g_2, g_1 g_3, g_1^2 g_4, \ldots$.

This formalism allows a very compact and convenient analysis of the properties of certain fixed points. Let us suppose from now on that the g_i

Chapter 16. Ultraviolet divergences in quantum gravity

are defined to eliminate only the poles at a single spacetime dimensionality D_s. The β_i for any non-renormalizable or super-renormalizable coupling must always be proportional to one or more powers of non-renormalizable or super-renormalizable couplings, respectively. Hence a set of couplings g_i^* with vanishing values for all interactions that are non-renormalizable or super-renormalizable at D_s will represent a fixed point, provided only that the $\beta_i(g^*)$ corresponding to the renormalizable couplings should vanish. Further, the only terms in $\partial \beta_i / \partial g_j$ which do not vanish at this fixed point are those with $d_i = d_j$, so the matrix B_{ij} is diagonal in the dimensionality of the couplings, and the eigenvalues of this matrix can be obtained by diagonalizing the submatrices connecting couplings of the same dimensionality. For instance, in the theory described by the Lagrangian (16.45), all the β-functions (16.46) will vanish at the point

$$g_1 = 0, \quad g_2 = g_2^*, \quad g_3 = 0, \quad g_4 = 0, \ldots \qquad (16.48)$$

provided only that g_2^* satisfies the condition

$$\beta_2(0, g_2^*, 0, \ldots) = 0. \qquad (16.49)$$

In addition, the B-matrix for this fixed point is diagonal, with non-vanishing elements

$$B_{11} = F_1(g^*) \quad B_{22} = (\partial F_2 / \partial g_2)^* \quad B_{33} = F_3(g^*) \cdots . \qquad (16.50)$$

These diagonal elements are then all eigenvalues of the B-matrix. In general, a fixed point of this sort may be found by working in a strictly renormalizable theory (even though the actual theory may not be renormalizable at all) and the eigenvalues of the B-matrix, the 'critical exponents', can be obtained by treating all super-renormalizable and non-renormalizable interactions as first-order perturbations.

It should perhaps be emphasized that even though the definition (16.36) of renormalized couplings gives the β-function a trival dependence on the spacetime dimensionality, D, the critical exponents have precisely the same complicated D-dependence that they would have with any other definition of the renormalized couplings. To illustrate this point, let us return to the Lagrangian (16.45), with $D_s = 4$. For reasons discussed above, we can find a fixed point by working with a truncated, strictly-renormalizable theory

$$\mathcal{L} = -\tfrac{1}{2}\partial_\mu \phi \, \partial_\mu \phi - \tfrac{1}{24}\lambda_0 \phi^4. \qquad (16.51)$$

Collins[41] has calculated the poles in λ_0 that are needed to cancel the poles in Green's functions as $D \to 4$ from below; to the two-loop order he finds

$$\lambda_0 \mu^{D-4} = \lambda + (D-4)^{-1}\left(\frac{-3\lambda^2}{16\pi^2} + \frac{17\lambda^3}{6(16\pi^2)^2} + \cdots\right)$$

$$+ (D-4)^{-2}\left(\frac{\lambda^3}{(16\pi^2)^2} + \cdots\right) + \cdots. \qquad (16.52)$$

The residue function for $\nu = 1$ is thus

$$b_{1\lambda} = \frac{-3\lambda^2}{16\pi^2} + \frac{17}{6}\frac{\lambda^3}{(16\pi^2)^2} + \cdots. \qquad (16.53)$$

The dimensionality, $\sigma_\lambda + \rho_\lambda D$, of λ_0 is $4-D$, so $\sigma_\lambda = 4$, $\rho_\lambda = -1$, and (16.42) gives the β-function here as

$$\beta_\lambda(\lambda, D) = (D-4)\lambda + b_{1\lambda} - \lambda\, \partial b_{1\lambda}/\partial\lambda \qquad (16.54)$$

or, using (16.53),

$$\beta_\lambda(\lambda, D) = (D-4)\lambda + \frac{3\lambda^2}{16\pi^2} - \frac{17}{3}\frac{\lambda^3}{(16\pi^2)^2} + \cdots. \qquad (16.55)$$

The fixed point where $\beta_\lambda(\lambda^*, D)$ vanishes can therefore be obtained as a power series in $D-4$:

$$\lambda^* = 16\pi^2[\tfrac{1}{3}(4-D) + \tfrac{17}{9}(4-D)^2 + \cdots]. \qquad (16.56)$$

(This is known as the Wilson–Fisher fixed point.[30] It has a physical sign $\lambda^* > 0$ only for $D < 4$.) Also, we easily calculate the critical exponent

$$\left(\frac{\partial \beta_\lambda}{\partial \lambda}\right)_{\lambda = \lambda^*} = -(4-D) + \frac{6\lambda^*}{16\pi^2} - \frac{17\lambda^{*2}}{(16\pi^2)^2} + \cdots$$

$$= (4-D) + \tfrac{85}{9}(4-D)^2 + \cdots. \qquad (16.57)$$

We note that this is positive for at least a finite range of D below $D = 4$. All other critical exponents are also positive in a neighborhood of $D = 4$, except for the one corresponding to the super-renormalizable coupling $-\tfrac{1}{2}m_0^2\phi^2$. Collins[41] has also calculated the poles in m_0^2 needed to cancel the singularities at $D = 4$ introduced by this coupling; to two-loop order, his result is

$$m_0^2 \mu^{-2} = m^2\left[1 + (D-4)^{-1}\left(\frac{-\lambda}{16\pi^2} + \frac{5}{12}\frac{\lambda^2}{(16\pi^2)^2} + \cdots\right)\right.$$

$$\left. + (D-4)^{-2}\frac{2\lambda^2}{(16\pi^2)^2} + \cdots\right]. \qquad (16.58)$$

Chapter 16. Ultraviolet divergences in quantum gravity

Hence the residue function for $\nu = 1$ and $D_s = 4$ is

$$b_{1,m^2} = m^2 \left[\frac{-\lambda}{16\pi^2} + \frac{5}{12} \frac{\lambda^2}{(16\pi^2)^2} + \cdots \right]. \tag{16.59}$$

The dimensionality, $\sigma + \rho D$, of m_0^2 is $+2$, so $\sigma_{m^2} = +2$, $\rho_{m^2} = 0$, and (16.42) gives the β-function for m^2 as

$$\beta_{m^2} = -2m^2 - \lambda \frac{\partial b_{1,m^2}}{\partial \lambda} \tag{16.60}$$

or using (16.59),

$$\beta_{m^2} = m^2 \left[-2 + \frac{\lambda}{16\pi^2} - \frac{5}{6} \frac{\lambda^2}{(16\pi^2)^2} + \cdots \right]. \tag{16.61}$$

The corresponding critical exponent is usually denoted $-\nu^{-1}$:

$$-\nu^{-1} \equiv \left(\frac{\partial \beta_{m^2}}{\partial m^2} \right)_{\lambda = \lambda^*, m^2 = 0} = -2 + \frac{\lambda^*}{16\pi^2} - \frac{5}{6} \frac{\lambda^{*2}}{(16\pi^2)^2} + \cdots \tag{16.62}$$

or, using (16.56),

$$\nu = \tfrac{1}{2} + \tfrac{1}{12}(4-D) + \tfrac{7}{162}(4-D)^2 + \cdots \tag{16.63}$$

in agreement with known results.[42] The fact that only one eigenvalue, $-\nu^{-1}$, of B_{ij} is negative for $D = 4 - \varepsilon$ implies that the Wislon–Fisher fixed point has a one-dimensional ultraviolet critical surface, and that there is just one parameter that need be adjusted to produce a second-order phase transition.

We saw in section 16.3 that the existence of a theory which is renormalizable and asymptotically safe at a spacetime dimensionality D_r indicates the existence of a fixed point near $g^* = 0$ with a finite-dimensional critical surface for at least a finite range of spacetime dimensionalities D above D_r. This can be seen very conveniently by using the method of dimensional continuation described in this section. Suppose for simplicity that the theory which is (strictly) renormalizable at $D = D_r$ has just a single coupling parameter, λ_0, with dimensionality $-(D - D_r)\rho$, where $\rho > 0$. Define the dimensionless renormalized coupling parameter $\lambda(\mu)$ so as to eliminate all poles in reaction rates at $D = D_r$. For the reasons discussed above in this section, we can find a fixed point of the whole theory for any D by setting all couplings equal to zero which would be non-renormalizable or super-renormalizable for $D = D_r$, and looking for a fixed point of the truncated theory. Having done this, the renormalization group equation satisfied by the coupling $\lambda(\mu)$ will be of the form

$$\mu \frac{d}{d\mu} \lambda(\mu) = \beta(\lambda(\mu), D) = [D - D_r]\rho\lambda(\mu) + \beta(\lambda(\mu), D_r). \quad (16.64)$$

The second term on the right arises from loop diagrams, so its power series will generally begin with terms of second order

$$\beta(\lambda(\mu), D_r) = -b\lambda^2(\mu) + O(\lambda^3(\mu)). \quad (16.65)$$

In order for the theory to be asymptotically free when $D = D_r$ (and $\lambda(\mu) > 0$) it is necessary that b be *positive*. For $D - D_r$ positive and sufficiently small, there will then be a fixed point

$$\lambda^* = (D - D_r)\rho/b + O((D - D_r)^2). \quad (16.66)$$

All critical exponents are positive, except for the one associated with λ

$$\left(\frac{\partial \beta_\lambda}{\partial \lambda}\right)_{\lambda = \lambda^*} = (D - D_r)\rho - 2b\lambda^* + O(\lambda^{*2})$$

$$= -(D - D_r)\rho + O((D - D_r)^2) < 0, \quad (16.67)$$

and those associated with any masses or couplings that would be super-renormalizable at $D = D_r$. Thus the ultraviolet critical surface is finite-dimensional, consisting of just those theories which would be renormalizable at $D = D_r$.

This should not be interpreted as a statement that these asymptotically safe theories are renormalizable as usual for $D > D_r$. The method of dimensional regularization is a bit misleading here — it eliminates ultraviolet divergences in *all* theories at non-rational values of the spacetime dimensionality D, at the price of introducing poles at rational values of D. With any other regularization scheme, there is a plethora of ultraviolet divergences at $D > D_r$, and these must be eliminated by including in the Lagrangian all possible interactions allowed by the symmetries of the theory. The occurrence of a fixed point (16.66) with a finite-dimensional critical surface in the dimensional regularization formalism ensures that there *is* a fixed point, with an ultraviolet critical surface of the same dimensionality, in the more conventional renormalization schemes, but the fixed point there will in general have non-vanishing values for *all* couplings, renormalizable *and* non-renormalizable, and the asymptotically safe theories will not even appear to be renormalizable in the usual sense.

Why then should we leave the elegant formalism of dimensional regularization, in which asymptotically safe theories appear so simple?

Chapter 16. Ultraviolet divergences in quantum gravity

The reason is just that we must in the end concern ourselves with a physical spacetime dimensionality $D = 4$ which is greater than D_r, and in continuing from $D = D_r$ up to $D = 4$ we must avoid the poles at intervening rational values of D and at $D = 4$ itself, which would be present with dimensional regularization. The conventional renormalization scheme offers a possible way of carrying out this continuation, at the price of giving up the appearance of renormalizability. However, the first step in assuring ourselves that there is a fixed point with a finite dimensional critical surface for D just above D_r is one that can be most easily accomplished by the method of dimensional regularization.

16.6 Gravity in $2 + \varepsilon$ dimensions

At last we come back to gravitation. We want to know whether it is possible to demand that the quantum theory of gravitation is asymptotically safe, and how many free parameters there would be in such a theory. This depends on whether there is a fixed point g^*, and on the dimensionality of its critical surface.

To address this question, we use the technique of dimensional continuation discussed in the previous section. In two dimensions there is a unique, strictly renormalizable theory of pure gravity, based on the Einstein Lagrangian $-\sqrt{g}R/16\pi G$. (The integral $\int d^2x \sqrt{g}R$ is dimensionless in two dimensions, so that G must be dimensionless in order for the action $\int d^2x \mathscr{L}$ to have no dimensions.) The theory remains renormalizable if we add matter fields with a minimal coupling to gravitation, though it may then be necessary to add couplings of matter fields with each other.

Of course, general relativity is not much of a theory in two dimensions. The Lagrangian $\sqrt{g}R$ is a total derivative for $D = 2$, and in consequence the left-hand side of the Einstein field equations, $R_{\mu\nu} - \frac{1}{2}g_{\mu\nu}R$, vanishes identically.[43] This raises a problem in using the method of dimensional regularization. Suppose that when we calculate invariant amplitudes in $2 + \varepsilon$ dimensions, we find that some invariant amplitude has a pole at $\varepsilon = 0$, and that in order to cancel this pole we have to add a term to the Lagrangian proportional to $\sqrt{g}R/\varepsilon$. Can we ignore such counterterms, on the grounds that $\int d^2x \sqrt{g}R$ vanishes for $\varepsilon = 0$? The answer appears to be no:[44] if we did not include a counterterm proportional to $\sqrt{g}R/\varepsilon$ where needed to cancel poles in invariant amplitudes at $\varepsilon = 0$, the Green's functions we calculate might be finite at $\varepsilon = 0$, but they would not be analytic functions of ε at $\varepsilon = 0$ in $2 + \varepsilon$ dimensions, as assumed in the dimensional-continuation formulation of the renormalization group.

Gravity in $2+\varepsilon$ dimensions

We conclude then that the gravitational coupling constant which appears in the Einstein Lagrangian, $-\sqrt{g}R/16\pi G$, is subject to renormalization in $2+\varepsilon$ dimensions, even for $\varepsilon \to 0$. We will return later to the question of whether G is an *essential* coupling, which cannot be altered by a suitable redefinition of fields.

The unrenormalized gravitational constant, $G_0(\varepsilon)$, in $2+\varepsilon$ dimensions has dimensionality [mass]$^{-\varepsilon}$, so in the notation of section 16.5, $\rho_G = -1$, $\sigma_G = 0$. The singularity structure of $G_0(\varepsilon)$ for $\varepsilon \to 0$ is given here by (16.36) as

$$G_0(\varepsilon)\mu^\varepsilon \to G(\mu) + \sum_{\nu=1}^{\infty} \varepsilon^{-\nu} b_\nu(G(\mu)). \qquad (16.68)$$

Also (16.38) and (16.42) yield a renormalization group equation for finite ε

$$\mu \frac{\mathrm{d}}{\mathrm{d}\mu} G(\mu) = \beta(G(\mu), \varepsilon) \qquad (16.69)$$

with

$$\beta(G, \varepsilon) = \varepsilon G + b_1(G) - G b_1'(G). \qquad (16.70)$$

For small G, we expect

$$b_1(G) = bG^2 + O(G^3) \qquad (16.71)$$

so (16.70) gives

$$\beta(G, \varepsilon) = \varepsilon G - bG^2 + O(G^3). \qquad (16.72)$$

The crucial question is whether b is positive; if so, then there is a fixed point

$$G^* = \varepsilon/b + O(\varepsilon^2) \qquad (16.73)$$

and as shown in section 16.5, it has an ultraviolet critical surface of finite dimensionality.

The calculation of b was carried out for a number of special cases by several different groups.[45] Their results up to mid-1977 can be summarized in the statement that the singularities in all purely gravitational Green's functions in $2+\varepsilon$ dimensions at $\varepsilon = 0$ are canceled in one-loop order if we suppose that the bare gravitational constant has the pole

$$G_0 \mu^\varepsilon \to G + bG^2/\varepsilon, \qquad (16.74)$$

with

$$b = \tfrac{38}{3} + 4N_V - \tfrac{1}{3}N_F - \tfrac{2}{3}N_S. \qquad (16.75)$$

Chapter 16. Ultraviolet divergences in quantum gravity

The four terms here arise from one-loop graphs, whose internal lines are respectively either graviton lines or matter lines of spin $1, \frac{1}{2}$, or 0; N_V and N_S are the number of real vector and scalar fields, and N_F is the number of Majorana fermion fields. It therefore appears at first sight that $b > 0$ and hence general relativity is asymptotically safe in $2 + \varepsilon$ dimensions, provided there are enough gauge fields to balance any scalar or fermion fields.

However, before reaching any such conclusion, we have to pay careful attention to the physical interpretation of (16.75). In pure general relativity, the trace of the vacuum Einstein field equations gives $R = 0$ for any spacetime dimensionality, so as explained in section 16.3, the coefficient $1/16\pi G$ of this Lagrangian is not an essential coupling, and there is no reason why it should be required to approach a fixed point for $\mu \to \infty$. The same is true if we add to the theory any number of 'photon' fields with purely gravitational interactions; in this case the Einstein field equations give $\sqrt{g} R$ proportional to $\sqrt{g} \sum F_{\mu\nu} F^{\mu\nu}$, but Maxwell's equations allow us to re-write this as the total derivative $\partial_\mu [\sqrt{g} \sum A_\nu F^{\mu\nu}]$, so here $\partial \mathscr{L}/\partial G$ is a total derivative, and G is again not an essential coupling. (The simplest theory which is renormalizable in two dimensions and which *does* have essential couplings is the Einstein–Yang–Mills[46] theory. In this case, reaction rates depend on the single essential coupling $e^{2\varepsilon} G^{2-\varepsilon}$, where e is the gauge coupling constant. However, this quantity is dimensionless for all ε, and therefore satisfies the trivial renormalization-group equation $d(e^{2\varepsilon} G^{2-\varepsilon})/d\mu = 0$.)

Recently a new interpretation of these calculations has been proposed by Gastmans, Kallosh and Truffin[47] (GKT). Their starting point is a reconsideration of the structure of the Lagrangian of general relativity. Gibbons and Hawking[48] have emphasized that in applying the functional formalism to general relativity, we do not actually use the Einstein Lagrangian $-\sqrt{g}R/16\pi G_0$, but rather

$$\mathscr{L}_G = -\frac{1}{16\pi G_0}[\sqrt{g}R - \Phi], \qquad (16.76)$$

with Φ a total derivative designed so that $\mathscr{L}_G$ is a function only of $g_{\mu\nu}$ and its *first* derivatives

$$\Phi = \frac{\partial}{\partial X^\mu}\left\{\frac{\sqrt{g}}{2}\left[g^{\lambda\nu}g^{\mu\kappa}\frac{\partial g_{\lambda\nu}}{\partial X^\kappa} - g^{\lambda\nu}g^{\mu\kappa}\frac{\partial g_{\kappa\nu}}{\partial X^\lambda} - g^{\lambda\nu}g^{\mu\kappa}\frac{\partial g_{\lambda\kappa}}{\partial X^\nu} + g^{\lambda\mu}g^{\nu\kappa}\frac{\partial g_{\nu\kappa}}{\partial X^\lambda}\right]\right\}. \qquad (16.77)$$

Of course, adding a total derivative has no effect if we restrict our attention to metrics which vanish sufficiently fast for $|X| \to \infty$, but in the functional formulation of quantum gravity we must sum over *all* metrics in Euclidean spacetime, and Φ makes an important contribution to the action for some of these metrics, such as the Euclidean Schwarzschild metric.

Now, when we calculate one-loop graphs in $2+\varepsilon$ dimensions, we must expect to find $1/\varepsilon$ poles as $\varepsilon \to 0$ which require independent counterterms proportional both to $\sqrt{g}R$ and to Φ. GKT argue that the Lagrangian should therefore then be written as the sum of two independent terms

$$\mathscr{L}_G = -\frac{1}{16\pi G_0}[\sqrt{g}R - \Phi] - \frac{1}{16\pi F_0}\sqrt{g}R. \tag{16.78}$$

By using the trace of the Einstein field equations, we can express $\sqrt{g}R$ in terms of matter fields, so its coefficient is not an independent essential coupling. However, the gravitational coupling, G_0, appears in the coefficient of a different term $\sqrt{g}R - \Phi$, which is *not* given by the field equations in terms of matter fields, so G_0 is an independent essential coupling. In other words, the earlier calculations[45] gave the counterterms in $1/G_0 + 1/F_0$ correctly, but the physically interesting essential coupling is $1/G_0$, and its counterterms must be calculated anew.

GKT have calculated the poles in G_0 required to cancel the $1/\varepsilon$ poles in gravitational Green's functions in a theory with N_S and N_V real scalar and vector fields and N_F and N_Δ Majorana fields of spin $\tfrac{1}{2}$ and $\tfrac{3}{2}$. Their results can be summarized in the statement that G_0 must have a pole (16.74), with b now given by

$$b = \tfrac{2}{3}[1 + \tfrac{15}{2}N_\Delta - N_F - N_S], \tag{16.79}$$

the first term arising from graviton loops. Very recently, Christensen and Duff[49] (CD) have carried out a calculation along the same lines, and find a formula for b with different fermion contributions:

$$b = \tfrac{2}{3}[1 - N_\Delta + N_F - N_S]. \tag{16.80}$$

In either case, there is an asymptotically safe theory of pure gravity in $2+\varepsilon$ dimensions, with a one-dimensional critical surface. Asymptotic safety is also preserved when we add matter fields, provided we add fields of spin $\tfrac{3}{2}$ (GKT) or $\tfrac{1}{2}$ (CD) to balance the contributions of fields of spin zero and $\tfrac{1}{2}$ (GKT) or $\tfrac{3}{2}$ (CD), and providing also that the couplings of the matter fields with themselves do not raise problems.

It may be noted that (16.79) or (16.80) and (16.75) give the same result for the contribution of scalar particles to b, so in this case the earlier

Chapter 16. Ultraviolet divergences in quantum gravity

calculations[45] did actually give the counterterms proportional to $\sqrt{g}R - \Phi$, not $\sqrt{g}R$. According to both GKT and CD, the new feature introduced by the distinction between $\sqrt{g}R - \Phi$ and $\sqrt{g}R$ is that the contribution of particles of arbitrary spin to b is simply proportional to the number of degrees of freedom of their fields, but (according to CD) with an extra minus sign for fermions. (The factor $\frac{15}{2}$ in (16.79) is puzzling here.) For instance, a symmetric traceless tensor field $h_{\mu\nu}$ has $\frac{1}{2}D(D+1)-1$ independent components, of which D are eliminated by the gauge condition which specifies $\partial_\mu h^\mu{}_\nu$, and $D-1$ are eliminated by our freedom to make further gauge transformations $\delta h_{\mu\nu} = \partial_\mu \phi_\nu + \partial_\nu \phi_\mu$ with $\partial_\mu \phi^\mu = 0$ and $\Box \phi^\mu = 0$; hence the number of degrees of freedom of the gravitational field in D dimensions is

$$\tfrac{1}{2}D(D+1) - 1 - D - (D-1) = \tfrac{1}{2}D(D-3).$$

This is -1 for $D = 2$, so the contribution of the graviton to b should be equal and opposite to the contribution of a single spinless particle. On the other hand, a vector field A_μ has D components, of which one is eliminated by the gauge condition which specifies $\partial_\mu A^\mu$, and one is eliminated by our freedom to make further gauge transformations $\delta A_\mu = \partial_\mu \phi$ with $\Box \phi = 0$; hence the number of degrees of freedom of the photon field in D dimensions is $D - 2$. This vanishes for $D = 2$, so photons make no contributions to b.

Since the evaluation of the contribution of scalar fields to b does not apparently raise problems of distinguishing between $\sqrt{g}R$ and $\sqrt{g}R - \phi$, and since this contribution sets the scale for the contributions of particles of nonzero spin, it may be of interest to see one more calculation of this quantity. A calculation based on the methods of reference 15 is presented in the Appendix.

It is amusing to apply these results to extended supergravity theories,[10] in which the graviton appears in a multiplet with fields of lower spin. In a four-dimensional theory with $n \leq 7$ supersymmetry generators, the helicity $+2$ graviton will appear in a multiplet with $\begin{bmatrix} n \\ r \end{bmatrix}$ massless particles of helicity $2 - r/2$, with $r = 1, 2, \ldots n$, and there is a separate multiplet containing the helicity -2 graviton and $\begin{bmatrix} n \\ r \end{bmatrix}$ massless particles of helicity $-2 + r/2$. For $n = 8$, there is a single multiplet containing gravitons of helicity ± 2, and $\begin{bmatrix} 8 \\ r \end{bmatrix}$ massless particles of helicity $2 - r/2$. We suppose that for arbitrary spacetime dimensionality, the number of fields of a given

Gravity in $2+\varepsilon$ dimensions

spin S is equal to the number of fields of that spin in four dimensions, and hence equal to the number of states in four dimensions with helicity $+S$ (or $-S$, but not both), even though the supersymmetry is actually present only for four dimensions. The numbers N_Δ, N_V, N_F, N_S of fields of spin $\frac{3}{2}$, $1, \frac{1}{2}, 0$ in $O(n)$-extended supergravity are given in Table 16.1, along with values of b obtained from (16.79) or (16.80): According to GKT's results, $b>0$ for pure supergravity ($n=1$), and also for $0(n)$-extended supergravity with $n \leq 5$. However, if CD are correct, then a two-loop calculation is needed to settle the question of asymptotic safety in pure supergravity, although we will always have $b>0$ for $n=1$ if we add enough 'vector' supermultiplets with spins 1 and $\frac{1}{2}$. According to CD's results, it appears to be impossible to have $b>0$ in $0(n)$-extended supergravity with $n \geq 2$, whether or not we add additional matter supermultiplets.

Table 16.1. *Numbers of field types and values of b in extended supersymmetry theories with n supersymmetry generators*

n	N_Δ	N_V	N_F	N_S	$3b$ (GKT)	$3b$ (CD)
0	0	0	0	0	2	2
1	1	0	0	0	17	0
2	2	1	0	0	32	−2
3	3	3	1	0	45	−2
4	4	6	4	2	50	−2
5	5	10	11	10	35	−6
6	6	16	26	30	−20	−18
7	8	28	56	70	−130	−42
8	8	28	56	70	−130	−42

The really important question here is that of continuation to four dimensions. In this respect, the dimensional regularization formalism in $2+\varepsilon$ dimensions may be somewhat misleading. It is true that the asymptotically safe theory of gravitation in this formalism is based on Lagrangians that must be renormalizable in two dimensions, and these Lagrangians do not contain any counterterms that would cancel the poles in Feynman diagrams at $D=4$ spacetime dimensions. However, the presence of these poles indicates that the perturbation expansion in powers of ε will have broken down long before we reach $\varepsilon = 2$.

Chapter 16. Ultraviolet divergences in quantum gravity

A better view of the possibilities of continuation from $2+\varepsilon$ to 4 dimensions may be provided by the conventional renormalization scheme with which we started in section 16.3. In this formalism, integrals are regulated with some sort of large ultraviolet cut-off Λ, and the cut-off dependence is removed for $\Lambda \to \infty$ by cancellation with the counterterms provided by the unrenormalized coupling constants. There are an infinite number of counterterms required in $D > 2$ dimensions, and such a theory of gravitation cannot be said to be renormalizable in the usual sense, but there are also no new singularities encountered when D approaches 4. The one result of the dimensional regularization method that can be taken over directly into the conventional renormalization scheme is that in $2+\varepsilon$ dimensions for sufficiently small ε there exists a fixed point with an ultraviolet critical surface of finite dimensionality.

Acknowledgements

It is a pleasure to thank S. Coleman, S. Deser, and M. J. Duff for their frequent and valuable help in the preparation of this report. I am also grateful to L. Brown, S. Christensen, J. C. Collins, B. DeWitt, E. S. Fradkin, G. 't Hooft, B. Lee, P. Martin, D. Nelson, A. Salam, H. Schnitzer, L. Smollin, H.-S. Tsao, P. van Nieuwenhuizen, K. Wilson, E. Witten and B. Zumino for informative conversations on various special topics.

16.7 Appendix. Calculation of b

This appendix will present a calculation of the residue of the pole in the bare gravitational constant at $D = 2$, using the method of reference 15.

We introduce a gravitational field $h_{\mu\nu}$ with

$$g_{\mu\nu} = \eta_{\mu\nu} + (32\pi G)^{1/2} h_{\mu\nu} \tag{A.1}$$

and work in a gauge with

$$\partial_\mu h^{\mu\nu} = 0. \tag{A.2}$$

(Indices here are raised and lowered with $\eta_{\mu\nu}$ not $g_{\mu\nu}$.) The bare graviton propagator in D spacetime dimensions is then

$$\langle T\{h_{\mu\nu}(x), h_{\lambda\rho}(0)\}\rangle_0 = \int \frac{d^D q}{(2\pi)^D} e^{iq\cdot x} \Delta_{\mu\nu,\lambda\rho}(q) \tag{A.3}$$

Appendix

$$\Delta_{\mu\nu,\lambda\rho}(q) = \frac{1}{2q^2}\left[L_{\mu\rho}(q)L_{\nu\lambda}(q) + L_{\mu\lambda}(q)L_{\nu\rho}(q) - \frac{2}{D-2}L_{\mu\nu}(q)L_{\lambda\rho}(q)\right] \quad \text{(A.4)}$$

$$L_{\mu\nu}(q) \equiv \eta_{\mu\nu} - q_\mu q_\nu/q^2. \quad \text{(A.5)}$$

In one-loop order, the propagator becomes

$$\Delta'_{\mu\nu,\lambda\rho}(q) = \Delta_{\mu\nu,\lambda\rho}(q) + \Delta_{\mu\nu,\mu'\nu'}(q)\Pi^{\mu'\nu',\lambda'\rho'}(q)\Delta_{\lambda'\rho',\lambda\rho}(q), \quad \text{(A.6)}$$

where Π is the graviton vacuum polarization tensor, defined by

$$\langle T\{T^{\mu\nu}(x), T^{\lambda\rho}(0)\}\rangle_0 = \frac{-i}{8\pi G}\int \frac{d^D q}{(2\pi)^D} e^{iq\cdot x}\Pi^{\mu\nu,\lambda\rho}(q). \quad \text{(A.7)}$$

Since the energy–momentum tensor is conserved, Π can be written

$$\Pi^{\mu\nu,\lambda\rho}(q) = (q^2)^2 A(q^2)L^{\mu\nu}(q)L^{\lambda\rho}(q)$$
$$- q^2 B(q^2)[L^{\mu\lambda}(q)L^{\nu\rho}(q) + L^{\mu\rho}(q)L^{\nu\lambda}(q) - 2L^{\mu\nu}(q)L^{\lambda\rho}(q)] \quad \text{(A.8)}$$

with $A(q^2)$ and $B(q^2)$ free of poles at $q^2 = 0$. It is straightforward then to calculate the corrected propagator as

$$\Delta'^{\mu\nu,\lambda\rho}(q) = \frac{1}{2q^2}(1 - 2B(q^2))\left[L^{\mu\rho}(q)L^{\nu\lambda}(q)\right.$$
$$\left. + L^{\mu\lambda}(q)L^{\nu\rho}(q) - \frac{2}{D-2}L^{\mu\nu}(q)L^{\lambda\rho}(q)\right]$$
$$+ \frac{A(q^2)}{(D-2)^2}L^{\mu\nu}(q)L^{\lambda\rho}(q). \quad \text{(A.9)}$$

We see that the renormalized gravitational constant which measures the strength of long-range graviton exchange is

$$G = G_0(1 - 2B(0)), \quad \text{(A.10)}$$

where G_0 is the bare gravitational constant. Hence if $B(0)$ has a pole at $D = 2$ of the form

$$B(0, D) \to \frac{bG}{2}\left(\frac{1}{D-2}\right) \quad \text{(A.11)}$$

we must introduce a pole in G_0 of the form

$$\mu^{D-2}G_0 \xrightarrow[D\to 2]{} G + bG^2\left(\frac{1}{D-2}\right). \quad \text{(A.12)}$$

Chapter 16. Ultraviolet divergences in quantum gravity

Thus the quantity b in (A.11) is the same as the b in (16.71)–(16.73). Our task is to calculate the residue of the pole in $B(q^2, D)$ at $D = 2$ and determine b by comparison with (A.11).

For this purpose, we introduce the spectral functions ρ_A, ρ_B by the formula

$$\sum_n \delta^D(p_n - p)\langle 0|T^{\mu\nu}(0)|n\rangle\langle 0|T^{\rho\sigma}(0)|n\rangle^*$$
$$= (2\pi)^{-D+1}\theta(p^0)[\rho_A(-p^2)(p^2)^2 L^{\mu\nu}(p)L^{\rho\sigma}(p)$$
$$-p^2\rho_B(-p^2)\{L^{\mu\rho}(p)L^{\nu\sigma}(p)+L^{\mu\sigma}(p)L^{\nu\rho}(p)-2L^{\mu\nu}(p)L^{\mu\rho}(p)\}]. \quad \text{(A.13)}$$

Apart from possible subtractions, (A.7), (A.8), and (A.13) give

$$A(q^2) = 8\pi G \int_0^\infty [q^2 + \mu^2 - i\varepsilon]^{-1}\rho_A(\mu^2)\,d\mu^2 \quad \text{(A.14)}$$

$$B(q^2) = 8\pi G \int_0^\infty [q^2 + \mu^2 - i\varepsilon]^{-1}\rho_B(\mu^2)\,d\mu^2. \quad \text{(A.15)}$$

Now let us consider the contribution of a state consisting of a pair of identical neutral spinless particles of mass m and momenta $\vec{k}$, $\vec{k}'$ to ρ_A and ρ_B. In lowest-order perturbation theory, for any D,[50]

$$\langle 0|T^{\mu\nu}(0)|\vec{k}, \vec{k}'\rangle$$
$$= -(2\pi)^{-(D-1)}(2\omega)^{-1/2}(2\omega')^{-1/2}[k^\mu k'^\nu + k'^\mu k^\nu + \eta^{\mu\nu}(-k^\lambda k'_\lambda + m^2)$$
$$- \frac{(D-2)}{2(D-1)}\{(k+k')^\mu(k+k')^\nu - (k+k')^2\eta^{\mu\nu}\}], \quad \text{(A.16)}$$

where

$$k^0 = \omega = (\vec{k}^2 + m^2)^{1/2} \quad k^{0\prime} = \omega' = (\vec{k}'^2 + m^2)^{1/2}.$$

Equation (A.13) then gives

$$\rho_A(\mu^2) = \tfrac{1}{8}(2\pi)^{-D+1}\Omega_D k^{D-3}\mu^{-5}\left[\frac{12k^4}{D^2-1} - \frac{2k^2\mu^2}{(D-1)^2} + \frac{\mu^4}{4(D-1)^2}\right] \quad \text{(A.17)}$$

$$\rho_B(\mu^2) = \tfrac{1}{2}(2\pi)^{-D+1}\Omega_D k^{D+1}\mu^{-3}/(D^2-1), \quad \text{(A.18)}$$

where Ω_D is the surface area of a unit sphere in $D-1$ spatial dimensions,

$$\Omega_D = 2\pi^{(D-1)/2}/\Gamma\!\left(\frac{D-1}{2}\right),$$

and
$$k \equiv \left(\frac{\mu^2}{4} - m^2\right)^{1/2}.$$

Now we can calculate b. The function $B(q^2, D)$ is given by (A.15) and (A.18) as

$$B(q^2, D) = \frac{4\pi G \Omega_D}{(D^2-1)(2\pi)^{D-1}} \int_{4m^2}^{\infty} \frac{k^{D+1} \, d\mu^2}{\mu^3(\mu^2+q^2)}. \tag{A.19}$$

The integral is well defined for $D < 2$, and can be analytically continued to $D > 2$, with a pole at $D = 2$

$$B(q^2, D) \xrightarrow[D \to 2]{} \frac{G}{3}\left(\frac{1}{2-D}\right). \tag{A.20}$$

Comparing with (A.11), we see that

$$b = -2/3 \tag{A.21}$$

in agreement with (6.79) and (6.80) for $N_s = 1$.

This method of calculation has the advantage of allowing us to draw general conclusions about the sign and other properties of spectral functions and pole residues at various dimensionalities for intermediate states of arbitrary spin. We take p in (A.13) to lie in the 'time' direction $p = (0, \ldots, 0, \mu)$, and contract with $a_\mu b_\nu a_\rho b_\sigma$, where a and b are purely spatial vectors separated by an angle ϕ; this gives

$$(1 + \tan \phi)^2 \rho_A(\mu^2)\mu^4 - 4 \tan \phi \rho_B(\mu^2)\mu^2 \geq 0 \tag{A.22}$$

for all ϕ and μ. For $D = 2$ we must of course take $\phi = 0$, so this gives only the condition

$$\rho_A(\mu^2) \geq 0. \tag{A.23}$$

For integer dimensionalities $D \geq 3$, we can pick ϕ freely; by choosing it to minimize the left-hand side of (A.22), we find

$$0 \leq \rho_B(\mu^2) \leq \rho_A(\mu^2)\mu^2. \tag{A.24}$$

Finally, for a traceless energy–momentum tensor, (A.13) gives

$$0 = \rho_A(\mu^2)\mu^2(D-1) + \rho_B(\mu^2)(4-2D). \tag{A.25}$$

In particular, $\rho_A(\mu^2) = 0$ for $D = 2$. Even if finite masses give the energy–momentum tensor a non-vanishing trace, (A.25) will be asymptotically valid for $\mu \to \infty$, so the integral (A.14) for $A(q^2)$ will not have a pole at $D = 2$.

References

Chapter 1 An introductory survey

Bardeen, J., Carter, B. and Hawking, S. W. (1973). *Commun. Math. Phys.*, **31**, 161.
Christodoulou, D. (1970). *Phys. Rev. Lett.*, **25**, 1596.
Clarke, C. J. S. (1975). *Commun. Math. Phys.*, **41**, 65.
Einstein, A. (1915). *Sitzungsber. Preuss. Akad. Wiss.*, 778, 779, 844.
Einstein, A. (1939). *Ann. Math. (Princeton)*, **40**, 922.
Hawking, S. W. (1971). *Phys. Rev. Lett.*, **26**, 1344.
Hawking, S. W. (1974). *Nature*, **248**, 30.
Hawking, S. W. and Penrose, R. (1970). *Proc. R. Soc. Lond.*, **A314**, 529.
Hilbert, D. (1915), *Nachr. Ges. Wiss. Göttingen*, 395.
Holton, G. (1972). *Am. J. Phys.*, **37**, 968.
Lifshitz, E. M. and Khalatnikov, I. M. (1963). *Sov. Phys.: Uspekhi*, **6**, 495.
Lifshitz, E. M. and Khalatnikov, I. M. (1970). *Phys. Rev. Lett.*, **24**, 76.
Penrose, R. (1965). *Phys. Rev. Lett.*, **14**, 57.
Smarr, L. (1977). *Ann. N.Y. Acad. Sci.*, **302**, 569.

Chapter 2 The confrontation between theory and experiment

Alley, C. O., Cutler, L. S., Reisse, R., Williams, R., Steggerda, C., Rayner, J., Mullendore, J. and Davis, S. (1977). Atomic-clock measurements of the General Relativistic time differences produced by aircraft flights using both direct and laser-pulse remote time comparison. (In preparation.)
Anderson, J. D. (1974). Lectures on physical and technical problems posed by precision radio tracking. In *Experimental Gravitation: Proceedings of Course 56 of the International School of Physics 'Enrico Fermi'*, ed. B. Bertotti, pp. 163–99. Academic Press: London, New York.
Anderson, J. D., Esposito, P. B., Martin, W., Thornton, C. L. and Muhleman, D. O. (1975). Experimental test of General Relativity using time-delay data from Mariner 6 and Mariner 7. *Astrophys. J.*, **200**, 221–33.
Anderson, J. D., Colombo, G., Friedman, L. D. and Lau, E. L. (1977). An arrow to the Sun. In *Proceedings of the International Meeting on Experimental Gravitation*, ed. B. Bertotti, pp. 393–422. Accademia Nazionale dei Lincei: Rome.
Anderson, J. D., Keesey, M. S. W., Lau, E. L., Standish, E. M., Jr and Newhall, XX (1977). Tests of General Relativity using astrometric and

Chapter 2. References

radiometric observations of the planets. In *Proceedings of the Third International Space Relativity Symposium* (27th Congress, International Astronautical Federation), in press.

Anderson, J. L. (1967). *Principles of Relativity Physics.* Academic Press: London, New York.

Baierlein, R. (1967). Testing General Relativity with laser ranging to the Moon. *Phys. Rev.*, **162**, 1275–88.

Balbus, S. A. and Brecher, K. (1976). Tidal friction in the binary pulsar system PSR 1913+16. *Astrophys. J.*, **203**, 202–5.

Barker, B. M. (1978). General Scalar-Tensor theory of gravity with constant G. *Astrophys. J.*, **219**, 5–11.

Barker, B. M. and O'Connell, R. F. (1975). Relativistic effects in the binary pulsar PSR 1913+16. *Astrophys. J.*, **199**, L25–6.

Baum, W. A. and Florentin-Nielsen, R. (1976). Cosmological evidence against time variation of the fundamental atomic constants. *Astrophys. J.*, **209**, 319–29.

Bekenstein, J. D. (1977). Are particle rest masses variable? Theory and constraints from solar system experiments. *Phys. Rev.*, **D15**, 1458–68.

Bender, P. L., Currie, D. G., Dicke, R. H., Eckhardt, D. H., Faller, J. E., Kaula, W. M., Mulholland, J. D., Plotkin, H. H., Poultney, S. K., Silverberg, E. C., Wilkinson, D. T., Williams, J. G. and Alley, C. O. (1973). The lunar laser ranging experiment. *Science*, **182**, 229–38.

Bergmann, P. G. (1968). Comments on the Scalar–Tensor theory. *Int. J. Theor. Phys.*, **1**, 25–36.

Bernacca, P. L., Ciatti, F., Guzzi, L., Sedmak, G., Campisi, I. E. and Treves, A. (1965). Search for an optical counterpart of the binary pulsar PSR 1913+16. *Astron. Astrophys.* **40**, 327–9.

Bertotti, B., Brill, D. R. and Krotkov, R. (1962). Experiments on gravitation. In *Gravitation: An Introduction to Current Research*, ed. L. Witten, pp. 1–48. Wiley: New York.

Bessel, F. W. (1832). *Poggendorff's Ann.*, **25**, 401.

Bisnovatyi-Kogan, G. S. and Komberg, B. V. (1976). Possible evolution of a binary-system radio pulsar as an old object with a weak magnetic field. *Sov. Astron. Lett.*, **2**, 130–2.

Blandford, R. and Teukolsky, S. A. (1975). On the measurement of the mass of PSR 1913+16. *Astrophys. J.*, **198**, L27–9.

Blandford, R. and Teukolsky, S. A. (1976). Arrival-time analysis for a pulsar in a binary system. *Astrophys. J.*, **205**, 580–91.

Braginsky, V. B. and Ginzburg, V. L. (1974). Possibility of measuring the time dependence of the gravitational constant. *Sov. Phys.: Dokl.*, **19**, 290–1.

Braginsky, V. B. and Panov, V. I. (1972). Verification of the equivalence of inertial and gravitational mass. *Sov. Phys.: JETP*, **34**, 463–6.

Braginsky, V. B., Caves, C. M. and Thorne, K. S. (1977). Laboratory experiments to test relativistic gravity. *Phys. Rev.*, **D15**, 2047–68.

Brans, C. and Dicke, R. H. (1961). Mach's principle and a relativistic theory of gravitation. *Phys. Rev.*, **124**, 925–35.

Brault, J. W. (1962). The gravitational redshift in the solar spectrum. Ph.D. Thesis, Princeton University.

Chapter 2. References

Brecher, K. (1975). Some implications of period changes in the first binary radio pulsar. *Astrophys. J.*, **195**, L113–5.

Brill, D. R. (1973). Observational contacts of General Relativity. In *Relativity, Astrophysics, and Cosmology*, ed. W. Israel, pp. 127–52. Reidel: Dordrecht.

Brumberg, V. A., Zel'dovich, Ya. B., Novikov, I. D. and Shakura, N. I. (1975). Component masses and inclination of binary systems containing a pulsar, determined from relativistic effects. *Sov. Astron. Lett.*, **1**, 2–4.

Caporaso, G. and Brecher, K. (1977). Neutron-star mass limit in the bimetric theory of gravitation. *Phys. Rev.*, **D15**, 3536–42.

Caves, C. M. (1977). Cosmological observations as tests of relativistic gravity: Rosen's bimetric theory. In *Proceedings of the 8th International Conference on General Relativity and Gravitation* (unpublished), 104. University of Waterloo, Canada.

Chanan, G., Middleditch, J. and Nelson, J. (1975). An upper limit on optical pulsations from PSR 1913 + 16. *Astrophys. J.*, **199**, L167–8.

Chapman, P. K. and Hanson, A. J. (1971). An Eötvös experiment in Earth orbit. In *Proceedings of the Conference on Experimental Tests of Gravitation Theories*, ed. R. W. Davies, pp. 228–35. NASA–JPL Technical Memorandum 33-499.

Chin, C.-W. and Stothers, R. (1976). Limit on the secular change of the gravitational constant based on studies of solar evolution. *Phys. Rev. Lett.*, **36**, 833–5.

Chiu, H.-Y. and Hoffman, W. F. (1964). Introduction. In *Gravitation and Relativity*, eds. H.-Y. Chiu and W. F. Hoffman, pp. xiii–xxxv. Benjamin: New York.

Counselman, C. C., III, Kent, S. M., Knight, C. A., Shapiro, I. I., Clark, T. A., Hinteregger, H. F., Rogers, A. E. E. and Whitney, A. R. (1974). Solar gravitational deflection of radio waves measured by very-long-baseline interferometry. *Phys. Rev. Lett.*, **33**, 1621–3.

Damour, T. and Ruffini, R. (1974). Sur certaines vérifications nouvelles de la Relativité Générale rendues possibles par la découverte d'un pulsar membre d'un système binaire. *C.R. Acad. Sci. (Paris)*, **279**, A971–3.

Davidsen, A., Margon, B., Liebert, J., Spinrad, H., Middleditch, J., Chanan, G., Mason, K. O. and Sanford, P. W. (1975). Optical and X-ray observations of the PSR 1913 + 16 field. *Astrophys. J.*, **200**, L19–21.

Davies, P. C. W. (1972). Time variation of the coupling constants. *J. Phys.*, **A5**, 1296–1304.

Dearborn, D. S. and Schramm, D. N. (1974). Limits on variation of G from clusters of galaxies. *Nature*, **247**, 441–3.

Demianski, M. and Shakura, N. I. (1976). A secular relativistic change in the period of a binary pulsar. *Nature*, **263**, 665–6.

Deser, S. and Laurent, B. (1973). Linearity and parametrization of gravitational effects. *Astron. Astrophys.*, **25**, 327–8.

de Sitter, W. (1916). On Einstein's theory of gravitation and its astronomical consequences. *Mon. Not. R. Astron. Soc.*, **77**, 155–84.

Dicke, R. H. (1964a). Experimental relativity. In *Relativity, Groups and Topology*, eds. C. DeWitt and B. DeWitt, pp. 165–313. Gordon and Breach: New York.

Chapter 2. References

Dicke, R. H. (1964b). Remarks on the observational basis of General Relativity. In *Gravitation and Relativity*, eds. H.-Y. Chiu and W. F. Hoffman, pp. 1–16. Benjamin: New York.

Dicke, R. H. (1969). *Gravitation and the Universe*. American Philosophical Society: Philadelphia.

Dicke, R. H. (1974). The oblateness of the Sun and relativity. *Science*, **184**, 419–29.

Dicke, R. H. and Goldenberg, H. M. (1974). The oblateness of the Sun. *Astrophys. J. Suppl.*, **27**, 131–82.

Drever, R. W. P. (1961). A search for anisotropy of inertial mass using a free precession technique. *Philos. Mag.*, **6**, 683–7.

Duff, M. J. (1974). On the significance of perihelion shift calculations. *Gen. Relativ. Grav.*, **5**, 441–52.

Dyson, F. J. (1972). The fundamental constants and their time variation. In *Aspects of Quantum Theory*, eds. A. Salam and E. P. Wigner, pp. 213–36. Cambridge University Press.

Eardley, D. M. (1975). Observable effects of a scalar gravitational field in a binary pulsar. *Astrophys. J.*, **196**, L59–62.

Eardley, D. M., Lee, D. L. and Lightman, A. P. (1973). Gravitational-wave observations as a tool for testing relativistic gravity. *Phys. Rev.*, **D10**, 3308–21.

Eardley, D. M., Lee, D. L., Lightman, A. P., Wagoner, R. V. and Will, C. M. (1973). Gravitational-wave observations as a tool for testing relativistic gravity. *Phys. Rev. Lett.*, **30**, 884–6.

Eddington, A. S. (1922). *The Mathematical Theory of Relativity*. Cambridge University Press.

Eddington, A. S. and Clark, G. L. (1938). The problem of n bodies in General Relativity theory. *Proc. R. Soc. Lond.*, **166**, 465–75.

Ehlers, J., Rosenblum, A., Goldberg, J. N. and Havas, P. (1976). *Astrophys. J.*, **208**, L77–81.

Eötvös, R. V., Pekár, V. and Fekete, E. (1922). Beitrage zum Gesetze der Proportionalität von Trägheit und Gravität. *Ann. Phys. (Leipzig)*, **68**, 11–66.

Epstein, R. (1977). The binary pulsar: post-Newtonian timing effects. *Astrophys. J.*, **216**, 92–100.

Epstein, R. and Wagoner, R. V. (1975). Post-Newtonian generation of gravitational waves. *Astrophys. J.*, **197**, 717–23.

Esposito, L. W. and Harrison, E. R. (1975). Properties of the Hulse–Taylor binary pulsar system. *Astrophys. J.*, **196**, L1–2.

Everitt, C. W. F. (1974). The gyroscope experiment. I. General description and analysis of gyroscope performance. In *Experimental Gravitation: Proceedings of Course 56 of the International School of Physics 'Enrico Fermi'*, ed. B. Bertotti, pp. 331–60. Academic Press: London, New York.

Fairbank, W. M., Witteborn, F. C., Madey, J. M. J. and Lockhart, J. M. (1974). Experiments to determine the force of gravity on single electrons and positrons. In *Experimental Gravitation: Proceedings of Course 56 of the International School of Physics 'Enrico Fermi'*, ed. B. Bertotti, pp. 310–30. Academic Press: London, New York.

Chapter 2. References

Faulkner, J. (1971). Ultrashort-period binaries, gravitational radiation, and mass transfer. I. The standard model, with applications to WZ Sagittae and Z Camelopardalis. *Astrophys. J.*, **170**, L99–104.

Flannery, B. P. and van den Heuvel, E. P. J. (1975). On the origin of the binary pulsar PSR 1913 + 16. *Astron. Astrophys.*, **39**, 61–7.

Fomalont, E. B. and Sramek, R. A. (1975). A confirmation of Einstein's General Theory of Relativity by measuring the bending of microwave radiation in the gravitational field of the Sun. *Astrophys. J.*, **199**, 749–55.

Fomalont, E. B. and Sramek, R. A. (1976). Measurements of the solar gravitational deflection of radio waves in agreement with General Relativity. *Phys. Rev. Lett.*, **36**, 1475–8.

Fomalont, E. B. and Sramek, R. A. (1977). The deflection of radio waves by the Sun. *Comment. Astrophys.*, **7**, 19–33.

Fujii, Y. (1971). Dilaton and possible non-Newtonian gravity. *Nature (Phys. Sci.)*, **234**, 5-7.

Fujii, Y. (1972). Scale invariance and gravity of hadrons. *Ann. Phys. (N.Y.)*, **69**, 494–521.

Gilvarry, J. J. and Muller, P. M. (1972). Possible variation of the gravitational constant over the elements. *Phys. Rev. Lett.*, **28**, 1665–8.

Hafele, J. C. and Keating, R. E. (1972a). Around-the-world atomic clocks: Predicted relativistic time gains. *Science*, **177**, 166–8.

Hafele, J. C. and Keating, R. E. (1972b). Around-the-world atomic clocks: Observed relativistic time gains. *Science*, **177**, 168–70.

Hari Dass, N. D. and Radhakrishnan, V. (1975). The new binary pulsar and the observation of gravitational spin precession. *Astrophys. Lett.*, **16**, 135–9.

Haugan, M. P. (1978). Energy conservation and the principle of equivalence. *Ann. Phys. (N.Y.)*, in press.

Haugan, M. P. and Will, C. M. (1976). Weak interactions and Eötvös experiments. *Phys. Rev. Lett.*, **37**, 1–4.

Haugan, M. P. and Will, C. M. (1977). Principles of equivalence, Eötvös experiments, and gravitational redshift experiments: The free fall of electromagnetic systems to post-post-Coulombian order. *Phys. Rev.*, **D15**, 2711–20.

Hellings, R. W. and Nordtvedt, K., Jr (1973). Vector-metric theory of gravity. *Phys. Rev.*, **D7**, 3593–602.

Hill, H. A. (1971). Light deflection. In *Proceedings of the Conference on Experimental Tests of Gravitation Theories*, ed. R. W. Davies, pp. 89–91. NASA–JPL Technical Memorandum 33–499.

Hill, H. A. and Stebbins, R. T. (1975). The intrinsic visual oblateness of the Sun. *Astrophys. J.*, **200**, 471–83.

Hill, H. A., Clayton, P. D., Patz, D. L., Healy, A. W., Stebbins, R. T., Oleson, J. R. and Zanoni, C. A. (1974). Solar oblateness, excess brightness and relativity. *Phys. Rev. Lett.*, **33**, 1497–500.

Hill, J. M. (1971). A measurement of the gravitational deflection of radio waves by the Sun. *Mon. Not. R. Astron. Soc.*, **153**, 7P–11P.

Hjellming, R. M. and Gibson, D. M. (1975). An interferometric search for the Hulse–Taylor binary pulsar. *Astrophys. J.*, **199**, L165–6.

Chapter 2. References

Hughes, V. W., Robinson, H. G. and Beltran-Lopez, V. (1960). Upper limit for the anistropy of inertial mass from nuclear resonance experiments. *Phys. Rev. Lett.*, **4**, 342–4.

Hulse, R. A. and Taylor, J. H. (1975). Discovery of a pulsar in a binary system. *Astrophys. J.*, **195**, L51–3.

Jaffe, J. and Vessot, R. F. C. (1976). Feasibility of a second-order gravitational redshift experiment. *Phys. Rev.*, **D14**, 3294–300.

Jenkins, R. E. (1969). A satellite observation of the relativistic Doppler shift. *Astron. J.*, **74**, 960–3.

Jones, B. F. (1976). Gravitational deflection of light: solar eclipse of 30 June 1973. II. Plate reductions. *Astron. J.*, **81**, 455–63.

Koester, L. (1976). Verification of the equivalence of gravitational and inertial mass for the neutron. *Phys. Rev.*, **D14**, 907–9.

Kreuzer, L. B. (1968). Experimental measurement of the equivalence of active and passive gravitational mass. *Phys. Rev.*, **169**, 1007–12.

Kristian, J., Clardy, K. D. and Westphal, J. A. (1976). Upper limits for the visible counterpart of the Hulse–Taylor binary pulsar. *Astrophys. J.*, **206**, L143–4.

Lee, D. L., Lightman, A. P. and Ni, W.-T. (1974). Conservation Laws and variational principles in metric theories of gravity. *Phys. Rev.*, **D10**, 1685–1700.

Lee, D. L., Caves, C. M., Ni, W.-T. and Will, C. M. (1976). Theoretical frameworks for testing relativistic gravity. V. Post-Newtonian limit of Rosen's theory. *Astrophys. J.*, **206**, 555–8.

Levi-Civita, T. (1937). Astronomical consequences of the relativistic two-body problem. *Am. J. Math.*, **59**, 225–34.

Levi-Civita, T. (1964). *The n-Body Problem in General Relativity*. Reidel: Dordrecht.

Lightman, A. P. and Lee, D. L. (1973a). Restricted proof that the Weak Equivalence Principle implies the Einstein Equivalence Principle. *Phys. Rev.*, **D8**, 364–76.

Lightman, A. P. and Lee, D. L. (1973b). New two-metric theory of gravity with prior geometry. *Phys. Rev.*, **D8**, 3293–302.

Lipa, J. A., Fairbank, W. M. and Everitt, C. W. F. (1974). The gyroscope experiment. II. Development of the London-moment gyroscope and of cryogenic technology for space. In *Experimental Gravitation: Proceedings of Course 56 of the International School of Physics 'Enrico Fermi'*, ed. B. Bertotti, pp. 361–80. Academic Press: London, New York.

Long, D. R. (1976). Experimental examination of the gravitational inverse square law. *Nature*, **260**, 417–18.

McGuigan, D. F. and Douglass, D. H. (1977). Clocks based upon high mechanical Q single crystals. In *Proceedings of the 31st Annual Frequency Control Symposium* (in press).

Mansouri, R. and Sexl, R. U. (1977a). A test theory of Special Relativity. I. Simultaneity and clock synchronization. *Gen. Relativ. Grav.*, **8**, 497–513.

Mansouri, R., and Sexl, R. U. (1977b). A test theory of Special Relativity. II. First order tests. *Gen. Relativ. Grav.*, **8**, 515–24.

Chapter 2. References

Marchant, A. and Mansfield, V. (1977). Evolution of dynamical systems with time-varying gravity. *Nature*, **270**, 699–700.

Masters, A. R. and Roberts, D. H. (1975). On the nature of the binary system containing the pulsar PSR 1913+16. *Astrophys. J.*, **195**, L107–11.

Merat, P., Pecker, J. C., Vigier, J. P. and Yourgrau, W. (1974). Observed deflection of light by the Sun as a function of solar distance. *Astron. Astrophys.*, **32**, 471–5.

Mikkelson, D. R. (1977). Very massive neutron stars in Ni's theory of gravity. *Astrophys. J.*, **217**, 248–51.

Mikkelson, D. R. and Newman, M. J. (1977). Constraints on the gravitational constant at large distances. *Phys. Rev.*, **D16**, 919–26.

Misner, C. W., Thorne, K. S. and Wheeler, J. A. (1973). *Gravitation*. Freeman: San Francisco. Referred to in text as MTW.

Morrison, D. and Hill, H. A. (1973). Current uncertainty in the ratio of active-to-passive gravitational mass. *Phys. Rev.*, **D8**, 2731–3.

Morrison, L. V. (1973). Rotation of the Earth from AD 1663–1972 and the constancy of G. *Nature*, **241**, 519–20.

Morrison, L. V. and Ward, C. G. (1975). An analysis of the transits of Mercury: 1667–1973. *Mon. Not. R. Astron. Soc.*, **173**, 183–206.

Muhleman, D. O. and Reichley, P. (1964). Effects of General Relativity on planetary radar distance measurements. *JPL Space Programs Summary 4 No. 37–39*, 239–41.

Muhleman, D. O., Ekers, R. D. and Fomalont, E. B. (1970). Radio interferometric test of the general relativistic light bending near the Sun. *Phys. Rev. Lett.*, **24**, 1377–80.

Nather, R. E., Robinson, E. L., Van Citters, G. W. and Hemenway, P. D. (1977). An upper limit to optical pulses from the binary pulsar PSR 1913+16. *Astrophys. J.*, **211**, L125–7.

Newman, R., Pellam, J., Schultz, J. and Spero, R. (1977). Experimental test of the gravitational inverse square law at laboratory distances. In *Proceedings of the 8th International Conference on General Relativity and Gravitation* (unpublished), 268. University of Waterloo, Canada.

Newton, I. (1687). *Philosophiae Naturalis Principia Mathematica*. London.

Ni, W.-T. (1972). Theoretical frameworks for testing relativistic gravity. IV. A compendium of metric theories of gravity and their post-Newtonian limits. *Astrophys. J.*, **176**, 769–96.

Ni. W.-T. (1963). A new theory of gravity. *Phys. Rev.*, **D7**, 2880–3.

Ni, W.-T. (1977). Equivalence principles and electromagnetism. *Phys. Rev. Lett.*, **38**, 301–4.

Nordtvedt, K., Jr (1968a). Equivalence principle for massive bodies. I. Phenomenology. *Phys. Rev.*, **169**, 1014–16.

Nordtvedt, K., Jr (1968b). Equivalence principle for massive bodies. II. Theory. *Phys. Rev.*, **169**, 1017–25.

Nordtvedt, K., Jr (1968c). Testing relativity with laser ranging to the Moon. *Phys. Rev.*, **170**, 1186–7.

Nordtvedt, K., Jr (1969). Equivalence principle for massive bodies including rotational energy and radiation pressure. *Phys. Rev.*, **180**, 1293–8.

Chapter 2. References

Nordtvedt, K., Jr (1970a). Solar-system Eötvös experiments. *Icarus*, **12**, 91–100.

Nordtvedt, K., Jr (1970b). Post-Newtonian metric for a general class of Scalar–Tensor gravitational theories and observational consequences. *Astrophys. J.*, **161**, 1059–67.

Nordtvedt, K., Jr (1970c). Gravitational and inertial mass of bodies of interacting electrical charges. *Int. J. Theor. Phys.*, **3**, 133–9.

Nordtvedt, K., Jr (1971a). Equivalence principle for massive bodies. IV. Planetary orbits and modified Eötvös-type experiments. *Phys. Rev.*, **D3**, 1683–9.

Nordtvedt, K., Jr (1971b). Tests of the equivalence principle and gravitation theory using solar system bodies. In *Proceedings of the Conference on Experimental Tests of Gravitation Theories*, ed. R. W. Davies, pp. 32–7. NASA-JPL Technical Memorandum 33–499.

Nordtvedt, K., Jr (1972). Gravitation theory: Empirical status from solar-system experiments. *Science*, **178**, 1157–64.

Nordtvedt, K., Jr (1973). Post-Newtonian gravitational effects in lunar laser ranging. *Phys. Rev.*, **D7**, 2347–56.

Nordtvedt, K., Jr (1975a). Quantitative relationship between clock gravitational redshift violations and non-universality of free-fall rates in non-metric theories of gravity. *Phys. Rev.*, **D11**, 245–7.

Nordtvedt, K., Jr (1975b). Anisotropic gravity and the binary pulsar PSR 1913+16. *Astrophys. J.*, **202**, 248–9.

Nordtvedt, K., Jr (1976). Anisotropic parametrized post-Newtonian gravitational metric field. *Phys. Rev.*, **D14**, 1511–17.

Nordtvedt, K., Jr (1977). A study of one- and two-way Doppler tracking of a clock on an arrow toward the Sun. In *Proceedings of the International Meeting on Experimental Gravitation*, ed. B. Bertotti, pp. 247–56. Accademia Nazionale dei Lincei: Rome.

Nordtvedt, K., Jr and Haugan, M. P. (1978). Paper in preparation.

Nordtvedt, K., Jr and Will, C. M. (1972). Conservation laws and preferred frames in relativistic gravity. II. Experimental evidence to rule out preferred-frame theories of gravity. *Astrophys. J.*, **177**, 775–92.

O'Hanlon, J. (1972). Intermediate-range gravity: A generally covariant model. *Phys. Rev. Lett.*, **29**, 137–8.

Ozernoi, L. M. and Reinhardt, M. (1975). Companion of the binary-system radio pulsar 1913+16. *Sov. Astron. Lett.*, **1**, 75–6.

Ozernoi, L. M. and Shishov, V. I. (1975). Upper limit on the electron density near the binary-system pulsar 1913+16. *Sov. Astron. Lett.*, **1**, 55–6.

Pagel, B. E. J. (1977). On the limits to past variability of the proton-electron mass ratio set by quasar absorption lines. *Mon. Not. R. Astron. Soc.*, **179**, 81P–85P.

Paik, H. J. (1977). Response of a disk antenna to scalar and tensor gravitational waves. *Phys. Rev.*, **D15**, 409–15.

Paik, H. J., Mapoles, E. and Fairbank, W. M. (1977). Private communication.

Peebles, P. J. E. (1962). The Eötvös experiment, spatial isotropy, and generally covariant field theories of gravity. *Ann. Phys. (N.Y.)*, **20**, 240–60.

Chapter 2. References

Peebles, P. J. E. and Dicke, R. H. (1962). Significance of spatial isotropy. *Phys. Rev.*, **127**, 629–31.

Peters, P. C. and Mathews, J. (1963). Gravitational radiation from point masses in Keplerian orbit. *Phys. Rev.*, **131**, 435–40.

Potter, H. H. (1923). Some experiments on the proportionality of mass and weight. *Proc. R. Soc. Lond.*, **104**, 588–610.

Pound, R. V. and Rebka, G. A., Jr (1960). Apparent weight of photons. *Phys. Rev. Lett.*, **4**, 337–41.

Pound, R. V. and Snider, J. L. (1965). Effect of gravity on gamma radiation, *Phys. Rev.*, **140**, B788–803.

Press, W. H., Wiita, P. J. and Smarr, L. L. (1975). Mechanism for inducing synchronous rotation and small eccentricity in close binary systems. *Astrophys. J.*, **202**, L135–7.

Rastall, P. (1976). A theory of gravity. *Can. J. Phys.*, **54**, 66–75.

Rastall, P. (1977a). A note on a theory of gravity. *Can. J. Phys.*, **55**, 38–42.

Rastall, P. (1977b). Conservation laws and gravitational radiation. *Can. J. Phys.*, **55**, 1342–8.

Rastall, P. (1977c). The maximum mass of a neutron star. *Astrophys. J.*, **213**, 234–8.

Reasenberg, R. D. and Shapiro, I. I. (1976). Bound on the secular variation of the gravitational interaction. In *Atomic Masses and Fundamental Constants*, vol. 5, eds. J. H. Sanders and A. H. Wapstra, pp. 643–9. Plenum: New York.

Reasenberg, R. D. and Shapiro, I. I. (1977). Solar-system tests of General Relativity. In *Proceedings of the International Meeting on Experimental Gravitation*, ed. B. Bertotti, pp. 143–60. Accademia Nazionale dei Lincei: Rome.

Renner, J. (1935). *Mat. Termeszettud. Ert.*, **53**, 542.

Richard, J.-P. (1975). Tests of theories of gravity in the solar system. In *General Relativity and Gravitation*, eds. G. Shaviv and J. Rosen, pp. 169–88. Wiley, New York.

Riley, J. M. (1973). A measurement of the gravitational deflection of radio waves by the Sun during October 1972. *Mon. Not. R. Astron. Soc.*, **161**, 11P–14P.

Ritter, R. C., Beams, J. W. and Lowry, R. A. (1976). A laboratory experiment to measure the time variation of Newton's gravitational constant. In *Atomic Masses and Fundamental Constants*, vol. 5, eds. J. H. Sanders and A. H. Wapstra, pp. 629–35. Plenum: New York.

Roberts, D. H., Masters, A. R. and Arnett, W. D. (1976). Determining the stellar masses in the binary system containing the pulsar PSR 1913+16: Is the companion a helium main-sequence star? *Astrophys. J.*, **203**, 196–201.

Robertson, H. P. (1938). The two-body problem in General Relativity. *Ann. Math.*, **39**, 101–4.

Robertson, H. P. (1962). Relativity and cosmology. In *Space Age Astronomy*, eds. A. J. Deutsch and W. B. Klemperer, pp. 228–35. Academic Press: London, New York.

Rochester, M. G. and Smylie, D. E. (1974). On changes in the trace of the Earth's inertia tensor. *J. Geophys. Res.*, **79**, 4948–51.

Chapter 2. References

Roll, P. G., Krotkov, R. and Dicke, R. H. (1964). The equivalence of inertial and passive gravitational mass. *Ann. Phys. (N.Y.)*, **26**, 442–517.

Rose, R. D., Parker, H. M., Lowry, R. A., Kuhlthau, A. R. and Beams, J. W. (1969). Determination of the gravitational constant G. *Phys. Rev. Lett.*, **23**, 655–8.

Rosen, N. (1973). A bi-metric theory of gravitation. *Gen. Relativ. Grav.*, **4**, 435–47.

Rosen, N. (1974). A theory of gravitation. *Ann. Phys. (N.Y.)*, **84**, 455–73.

Rosen, N. (1977a). Bimetric gravitation and cosmology. *Astrophys. J.*, **211**, 357–60.

Rosen, N. (1977b). Is there gravitational radiation? *Nuovo Cimento Lett.*, **19**, 249–50.

Rosen, N. (1978). Bimetric theory on a cosmological basis. *Gen. Relativ. Grav.*, **9**, 339–45.

Rosen, N. and Rosen, J. (1975). The maximum mass of a cold neutron star. *Astrophys. J.*, **202**, 782–7.

Rudolph, E. (1977). Relativistic observable effects in the binary pulsar PSR 1913+16. In *Proceedings of the International School of Physics 'Enrico Fermi' 1976*. Academic Press: London, New York. (In press.)

Schiff, L. I. (1960a). On experimental tests of the General Theory of Relativity. *Am. J. Phys.*, **28**, 340–3.

Schiff, L. I. (1960b). Motion of a gyroscope according to Einstein's theory of gravitation. *Proc. Nat. Acad. Sci. USA*, **46**, 871–82.

Schiff, L. I. (1960c). Possible new test of General Relativity Theory. *Phys. Rev. Lett.*, **4**, 215–17.

Schiff, L. I. (1967). Comparison of theory and observation in General Relativity. In *Relativity Theory and Astrophysics. I. Relativity and Cosmology*, ed. J. Ehlers, pp. 105–16. American Mathematical Society: Providence.

Seielstad, G. A., Sramek, R. A. and Weiler, K. W. (1970). Measurement of the deflection of 9.602-GHz radiation from 3C279 in the solar gravitational field. *Phys. Rev. Lett.*, **24**, 1373–6.

Shapiro, I. I. (1964). Fourth test of General Relativity. *Phys. Rev. Lett.*, **13**, 789–91.

Shapiro, I. I. (1967). New method for the detection of light deflection by solar gravity. *Science*, **157**, 806–8.

Shapiro, I. I. (1968). Fourth test of General Relativity: preliminary results. *Phys. Rev. Lett.*, **20**, 1265–9.

Shapiro, I. I. (1972). Testing General Relativity: progress, problems and prospects. *Gen. Relativ. Grav.*, **3**, 135–48.

Shapiro, I. I., Ash, M. E., Ingalls, R. P., Smith, W. B., Campbell, D. B., Dyce, R. B., Jurgens, R. F. and Pettengill, G. H. (1971). Fourth test of General Relativity: new radar result. *Phys. Rev. Lett.*, **26**, 1132–5.

Shapiro, I. I., Smith, W. B., Ash, M. E., Ingalls, R. P. and Pettengill, G. H. (1971). Gravitational constant: experimental bound on its time variation. *Phys. Rev. Lett.*, **26**, 27–30.

Shapiro, I. I., Pettengill, G. H., Ash, M. E., Ingalls, R. P., Campbell, D. B. and Dyce, R. B. (1972). Mercury's perihelion advance: determination by radar. *Phys. Rev. Lett.*, **28**, 1594–7.

Chapter 2. References

Shapiro, I. I., Counselman, C. C., III and King, R. W. (1976). Verification of the principle of equivalence for massive bodies. *Phys. Rev. Lett.*, **36**, 555–8.

Shapiro, I. I., Reasenberg, R. D., MacNeil, P. E., Goldstein, R. B., Brenkle, J., Cain, D., Komarek, T., Zygielbaum, A., Cuddihy, W. F. and Michael, W. H., Jr (1977). The Viking relativity experiment. *J. Geophys. Res.*, **82**, 4329–34.

Shapiro, S. L. and Terzian, Y. (1976). Galactic rotation and the binary pulsar. *Astron. Astrophys.*, **52**, 115–18.

Shapiro, S. L. and Teukolsky, S. A. (1976). On the maximum gravitational redshift of white dwarfs. *Astropohys. J.*, **203**, 697–700.

Shlyakhter, A. I. (1976). Direct test of the constancy of fundamental nuclear constants. *Nature*, **264**, 340.

Smarr, L. L. and Blandford, R. (1976). The binary pulsar: physical processes, possible companions, and evolutionary histories. *Astrophys. J.*, **207**, 574–88.

Smoot, G. F., Gorenstein, M. V. and Muller, R. A. (1977). Detection of anisotropy in the cosmic blackbody radiation. *Phys. Rev. Lett.*, **39**, 898–901.

Snider, J. L. (1972). New measurement of the solar gravitational redshift. *Phys. Rev. Lett.*, **28**, 853–6.

Solheim, J.-E., Barnes, T. G., III and Smith, H. J. (1976). Observational evidence against a time variation in Planck's constant. *Astrophys. J.*, **209**, 330–4.

Sramek, R. A. (1971). A measurement of the gravitational deflection of microwave radiation near the Sun, 1970 October. *Astrophys. J.*, **167**, L55–60.

Sramek, R. A. (1974). The gravitational deflection of radio waves. In *Experimental Gravitation: Proceedings of Course 56 of the International School of Physics, 'Enrico Fermi'*, ed. B. Bertotti, pp. 529–42. Academic Press: London, New York.

Stein, S. R. (1974). The superconducting-cavity stabilized oscillator and an experiment to detect time variation of the fundamental constants. Ph.D. Thesis, Stanford University.

Stein, S. R. and Turneaure, J. P. (1975). Superconducting–cavity stabilized oscillators with improved frequency stability. *IEEE Proc.*, **63**, 1249–50.

Taylor, J. H. (1975). Discovery of a pulsar in a binary system. *Ann. N.Y. Acad. Sci.*, **262**, 490–2.

Taylor, J. H., Hulse, R. A., Fowler, L. A., Gullahorn, G. E. and Rankin, J. M. (1976). Further observations of the binary pulsar PSR 1913+16. *Astrophys J.*, **206**, L53–8.

Texas Mauritanian Eclipse Team (1976). Gravitational deflection of light: solar eclipse of 30 June 1973. I. Description of procedures and final results. *Astron. J.*, **81**, 452–4.

Thorne, K. S. (1977). The generation of gravitational waves. V. Multipole-moment formalisms. Preprint, Cornell University, CRSR 663.

Thorne, K. S. and Will, C. M. (1971). Theoretical frameworks for testing relativistic gravity. I. Foundations. *Astrophys. J.*, **163**, 595–610.

Chapter 2. References

Thorne, K. S., Will, C. M. and Ni, W.-T. (1971). Theoretical frameworks for testing relativistic gravity – A review. In *Proceedings of the Conference on Experimental Tests of Gravitation Theories*, ed. R. W. Davies, pp. 10–31. NASA–JPL Technical Memorandum 33-499.

Thorne, K. S., Lee, D. L. and Lightman, A. P. (1973). Foundations for a theory of gravitation theories. *Phys. Rev.* **D7**, 3563–78.

Turneaure, J. P. and Stein, S. R. (1976). An experimental limit on the time variation of the fine structure constant. In *Atomic Masses and Fundamental Constants*, vol. 5, eds. J. H. Sanders and A. H. Wapstra, pp. 636–42. Plenum: New York.

Turneaure, J. P. and Will, C. M. (1975). A null gravitational redshift experiment. *Bull. Am. Phys. Soc.*, **20**, 1488.

Van Citters, G. W. and Rybski, P. M. (1977). Area photometry in the region of the pulsar 1913+16. *Astrophys. J.*, **214**, 233–4.

van den Berg, S. (1975). The binary pulsar 1913+16. *Astrophys. Lett.*, **16**, 75.

Van Flandern, T. C. (1975a). A determination of the rate of change of G. *Mon. Not. R. Astron. Soc.*, **170**, 333–42.

Van Flandern, T. C. (1975b). Recent evidence for variations in the value of G. *Ann. N.Y. Acad. Sci.*, **262**, 494–5.

Van Flandern, T. C. (1976). Is gravity getting weaker? *Scientific American*, **234**, no. 2, 44–52.

Van Horn, H. M., Sofia, S., Savedoff, M. P., Duthie, J. G. and Berg, R. A. (1975). Binary pulsar PSR 1913+16: model for its origin. *Science*, **188**, 930–3.

Van Patten, R. A. and Everitt, C. W. F. (1976). Possible experiment with two counter-orbiting drag-free satellites to obtain a new test of Einstein's General Theory of Relativity and improved measurements in geodesy. *Phys. Rev. Lett.*, **36**, 629–32.

Vessot, R. F. C. (1974). Lectures on frequency stability and clocks and on the gravitational redshift experiment. In *Experimental Gravitation: Proceedings of Course 56 of the International School of Physics 'Enrico Fermi'*, ed. B. Bertotti, pp. 111–62. Academic Press: London, New York.

Vessot, R. F. C. and Levine, M. W. (1976). A preliminary report on the gravitational redshift rocket-probe experiment. In *Proceedings of the 2nd Frequency Standards and Metrology Symposium*, ed. H. Hellwig, pp. 659–88. National Bureau of Standards: Boulder, Colorado.

Wagoner, R. V. (1970). Scalar–tensor theory and gravitational waves. *Phys. Rev.*, **D1**, 3209–16.

Wagoner, R. V. (1975). Test for the existence of gravitational radiation. *Astrophys. J.*, **196**, L63–5.

Wagoner, R. V. (1976). A new test of General Relativity. *Gen. Relativ. Grav.*, **7**, 333–7.

Wagoner, R. V. and Malone, R. C. (1974). Post-Newtonian neutron stars. *Astrophys. J.*, **189**, L75–8.

Wagoner, R. V. and Paik, H. J. (1977). Multi-mode detection of gravitational waves by a sphere. In *Proceedings of the International Meeting on Experimental Gravitation*, ed. B. Bertotti, pp. 257–66. Accademia Nazionale dei Lincei: Rome.

Wahr, J. M. and Bender, P. L. (1976). Determination of PPN parameters from Earth–Mercury distance measurements. Preprint.
Warburton, R. J. and Goodkind, J. M. (1976). Search for evidence of a preferred reference frame. *Astrophys. J.*, **208**, 881–6.
Ward, W. R. (1970). General Relativistic light deflection for the complete celestial sphere. *Astrophys. J.*, **162**, 345–8.
Webbink, R. F. (1975). PSR 1913 + 16: Endpoints of speculation. A critical discussion of possible companions and progenitors. *Astron. Astrophys.*, **41**, 1–8.
Weiler, K. W., Ekers, R. D., Raimond, E. and Wellington, K. J. (1974). A measurement of solar gravitational microwave deflection with the Westerbork synthesis telescope. *Astron. Astrophys.*, **30**, 241–8.
Weiler, K. W., Ekers, R. D., Raimond, E. and Wellington, K. J. (1975). Dual-frequency measurement of the solar gravitational microwave deflection. *Phys. Rev. Lett.*, **35**, 134–7.
Weinberg, S. (1972). *Gravitation and Cosmology*. Wiley: New York.
Wheeler, J. C. (1975). Timing effects in pulsed binary systems. *Astrophys. J.*, **196**, L67–70.
Wheeler, J. C. (1976). The binary pulsar: preexplosion evolution. *Astrophys. J.*, **205**, 578–9.
Whitehead, A. N. (1922). *The Principle of Relativity*. Cambridge University Press.
Wilkins, D. C. (1970). General equation for the precession of a gyroscope. *Ann. Phys. (N.Y.)*, **61**, 277–93.
Will, C. M. (1971a). Theoretical frameworks for testing relativistic gravity. II. Parametrized post-Newtonian hydrodynamics and the Nordtvedt effect. *Astrophys. J.*, **163**, 611–28.
Will, C. M. (1971b). Relativistic gravity in the solar system. I. Effect of an anisotropic gravitational mass on the Earth–Moon distance. *Astrophys. J.*, **165**, 409–12.
Will, C. M. (1971c). Theoretical frameworks for testing relativistic gravity. III. Conservation laws, Lorentz invariance, and values of the PPN parameters. *Astrophys. J.*, **169**, 125–40.
Will, C. M. (1971d). Relativistic gravity in the solar system. II. Anisotropy in the Newtonian gravitational constant. *Astrophys. J.*, **169**, 141–55.
Will, C. M. (1972). Einstein on the firing line. *Physics Today*, **25**, No. 10, 23–9.
Will, C. M. (1973). Relativistic gravity in the solar system. III. Experimental disproof of a class of linear theories of gravitation. *Astrophys. J.*, **185**, 31–42.
Will, C. M. (1974a). The theoretical tools of experimental gravitation. In *Experimental Gravitation: Proceedings of Course 56 of the International School of Physics 'Enrico Fermi'*, ed. B. Bertotti, pp. 1–110. Academic Press: London, New York. Referred to in text as TTEG.
Will, C. M. (1974b). Gravitation theory. *Scientific American*, **231**, No. 5, 25–33.
Will, C. M. (1974c). Gravitational redshift measurements as tests of non-metric theories of gravity. *Phys. Rev.*, **D10**, 2330–7.
Will, C. M. (1975). Periastron shifts in the binary system PSR 1913 + 16: Theoretical interpretation. *Astrophys. J.*, **196**, L3–5.

Chapter 3. References

Will, C. M. (1976a). Active mass in relativistic gravity: Theoretical interpretation of the Kreuzer experiment. *Astrophys. J.*, **204**, 224–34.

Will, C. M. (1976b). A test of post-Newtonian conservation laws in the binary system PSR 1913+16. *Astrophys. J.*, **205**, 861–7.

Will, C. M. (1977a). Clocks and experimental gravitation: A null gravitational redshift experiment, laboratory tests of post-Newtonian gravity, and gravity-wave detection by spacecraft tracking. *Metrologia*, **13**, 95–8.

Will, C. M. (1977b). Gravitational radiation from binary systems in alternative metric theories of gravitation: Dipole radiation and the binary pulsar. *Astrophys. J.*, **214**, 826–39.

Will, C. M. and Eardley, D. M. (1977). Dipole gravitational radiation in Rosen's theory of gravity: Observable effects in the binary system PSR 1913+16. *Astrophys. J.*, **212**, L91–4.

Will, C. M. and Nordtvedt, K., Jr (1972). Conservation laws and preferred frames in relativistic gravity. I. Preferred-frame theories and an extended PPN formalism. *Astrophys. J.*, **177**, 757–74.

Williams, J. G., Dicke, R. H., Bender, P. L., Alley, C. O., Carter, W. E., Currie, D. G., Eckhardt, D. H., Faller, J. E., Kaula, W. M., Mulholland, J. D., Plotkin, H. H., Poultney, S. K., Shelus, P. J., Silverberg, E. C., Sinclair, W. S., Slade, M. A. and Wilkinson, D. T. (1976). New test of the equivalence principle from lunar laser ranging. *Phys. Rev. Lett.*, **36**, 551–4.

Wolfe, A. M., Brown, R. L. and Roberts, M. S. (1976). Limits on the variation of fundamental atomic quantities over cosmic time scales. *Phys. Rev. Lett.*, **37**, 179–81.

Worden, P. W., Jr (1976). A cryogenic test of the Equivalence Principle. Ph.D. Thesis, Stanford University.

Worden, P. W., Jr and Everitt, C. W. F. (1974). The gyroscope experiment. III. Tests of the equivalence of gravitational and inertial mass based on cryogenic techniques. In *Experimental Gravitation: Proceedings of Course 56 of the International School of Physics 'Enrico Fermi'*, ed. B. Bertotti, pp. 381–402. Academic Press: London, New York.

Zel'dovich, Ya. B. and Shakura, N. I. (1975). Relativistic irregularity in the rotation of a pulsar moving in an elliptic orbit. *Sov. Astron. Lett.*, **1**, 222–3.

Chapter 3 Gravitational-radiation experiments

Allen, W. D. and Christodoulides, C. (1975). *J. Phys.*, **A8**, 1726.

Anderson, J. A. (1971). *Nature*, **229**, 547.

Anderson, J. A. (1977). *Proceedings of the International Meeting on Experimental Gravitation*, ed. B. Bertotti. Accademia Nazionale dei Lincei: Rome.

Anderson, J. D. (1974). Lectures on physical and technical problems posed by precision radio tracking. In *Experimental Gravitation: Proceedings of Course 56 of the International School of Physics 'Enrico Fermi'*, ed. B. Bertotti, pp. 163–99. Academic Press: London, New York.

Barnes, J. A. *et al.* (1971). *IEEE Trans. Instrum. Meas.*, **IM-20**, 105.

Chapter 3. References

Billing, H., Kafka, P., Maischberger, K., Meyer, F. and Winkler, W. (1975). *Lett. Nuovo Cimento*, **12**, 111.
Boriakoff, V. (1976). *Astrophys. J.*, **208**, L43.
Braginsky, V. B. (1966). *Sov. Phys.: Uspekhi*, **8**, 513.
Braginsky, V. B. (1977a). *Proceedings of the International School of Cosmology and Gravitation*, ed. V. De Sabbata and J. Weber. Plenum: New York.
Braginsky, V. B. (1977b). *Proceedings of the International Meeting on Experimental Gravitation*, ed. B. Bertotti. Accademia Nazionale dei Lincei: Rome.
Braginsky, V. B. (1977c). *8th International Conference on General Relativity and Gravitation*, August 7–12, 1977. Waterloo, Ontario, Canada.
Braginsky, V. B. and Manukin, A. B. (1977). *Measurement of Weak Forces in Physics Experiments*, ed. D. H. Douglass, p. 116. University of Chicago Press.
Braginsky, V. B., Manukin, A. B., Papov, E. I., Rudenko, V. N. and Khorev, A. A. (1972). *Zh. Eksp. Teor. Fiz. Pisma Red.*, **16**, 157. (English translation in *Sov. Phys.: JETP Lett.*, **16**, 108.)
Brickhill, A. (1975). *Mon. Not. R. Astron. Soc.*, **170**, 405.
Clark, J. P. A. and Eardley, D. M. (1977). *Astrophys. J.*, **215**, 311.
Clark, J. P. A., van den Heuvel, E. P. J. and Sutantyo, W. (1978). *Astron. Astrophys.*, in press.
Davis, M., Ruffini, R., Press, W. H. and Price, R. H. (1971). *Phys. Rev. Lett.*, **27**, 1466.
DeBra, D. B. (1971). *Proceedings of the Cal Tech Conference on Experimental Tests of Gravitation Theories*, ed. R. W. Davies. NASA–JPL Technical Memorandum 33-499.
Detweiler, S. L. (1975a). *Astrophys. J.*, **197**, 203.
Detweiler, S. L. (1975b). *Astrophys. J.*, **201**, 440.
Detweiler, S. L. (1977). *8th International Conference on General Relativity and Gravitation*, August 7–12, 1977. Waterloo, Ontario, Canada.
Doroshkevich, A. G., Zel'dovich, Ya. B. and Novikov, I. D. (1967). *Astron. Zh.*, **44**, 295 (1967). (English translation in *Sov. Astron.*, **11**, 233.)
Douglass, D. H. (1971). *Proceedings of the Conference on Experimental Tests of Gravitational Theories*, ed. R. W. Davies. NASA–JPL Technical Memorandum 33-499.
Douglass, D. H. (1976). *Proceedings of the International Meeting on Experimental Gravitation*, ed. B. Bertotti. Accademia Nazionale dei Lincei: Rome.
Douglass, D. H. (1978). To be published.
Douglass, D. H., Gram, R. Q., Tyson, J. A. and Lee, R. W. (1975). *Phys. Rev. Lett.*, **35**, 480.
Drever, R. W. P. (1977a). *Q. J. R. Astron. Soc.*, **18**, 9–27.
Drever, R. W. P. (1977b). *8th International Conference on General Relativity and Gravitation*, August 7–12, 1977. Waterloo, Ontario, Canada.
Drever, R. W. P., Hough, J., Bland, R. and Lessonoff, G. W. (1973). *Nature*, **246**, 340.
Eddington, A. S. (1922). *The Mathematical Theory of Relativity*. Cambridge University Press.
Einstein, A. (1916). *Sitzungsber Preuss. Akad. Wiss.*, 688.

Chapter 3. References

Einstein, A. (1918). *Sitzungsber Preuss. Akad. Wiss.*, 154.
Estabrook, R. B. and Wahlquist, H. D. (1975). *Gen. Relativ. Grav.*, **6**, 439.
Faulkner, J., Flannery, B. P. and Warner, B. (1972). *Astrophys. J.*, **175**, L79.
Forward, R. L., Moss, G. E. and Miller, L. R. (1971). *Appl. Opt.*, **10**, 2495.
Gaposchkin, S. (1958). *Encyclopedia of Physics*, ed. S. Flügge, vol. L, *Astrophysics I – Stellar Surfaces–Binaries*. Springer-Verlag: Berlin.
Giffard, R. P. (1976). *Phys. Rev.*, **D14**, 2478.
Gursky, H. and Schreier, E. (1975). IAU Symposium No. 67, *Variable Stars and Stellar Evolution*. Reidel: Dordrecht.
Gusev, A. V. and Rudenko, V. N. (1977). *Zh. Eksp. Teor, Fiz.*, **72**, 1217. (English translation in *Sov. Phys.: JETP*, **45**, No. 4.)
Hack, M. (1963). *Proceedings of the International School of Physics 'Enrico Fermi' Stellar Evolution* (Course 28), ed. L. Gratton. Academic Press: London, New York.
Hansen, C. J., Cox, J. P. and Van Horn, H. M. (1977). *Astrophys. J.*, **217**, 151.
Hawking, S. W. (1971). *Phys. Rev. Lett.*, **26**, 1344.
Hawking, S. W. (1972). *Contemp. Phys.*, **13**, 273.
Hirakawa, H. and Narihara, K. (1975). *Phys. Rev. Lett.*, **35**, 330.
Hirakawa, H., Tsubono, K. and Fujimoto, M. (1977). *Phys. Rev.*, **D17**, 1919.
Hough, J., Pugh, J. R., Edelstein, W. A. and Martin, W. (1977). *J. Phys.*, **E10**, 997.
Isaacson, R. A. (1968). *Phys. Rev.*, **166**, 1272.
Kraft, R. (1962). *Astrophys. J.*, **136**, 312.
Lattimer, J. M. and Schramm, D. N. (1976). *Astrophys. J.*, **216**, 549.
Levine, J. L. and Garwin, R. L. (1973). *Phys. Rev. Lett.*, **31**, 173.
McGuigan, D. F., Lam, C. C., Gram, R. Q., Hoffman, A. W., Douglass, D. H. and Gutche, G. W. (1978). *J. Low Temp. Phys.*, **30**, 621.
Misner, C. W. (1974). *IAU Symposium No. 64, Gravitational Radiation and Gravitational Collapse*, ed. C. Dewitte-Morette. Reidel: Dordrecht.
Misner, C. W., Thorne, K. S. and Wheeler, J. A. (1973). *Gravitation*. W. H. Freeman: San Francisco.
Mironovskii, V. N. (1966). *Astron. Zh.*, **42**, 977. (English translation in *Sov. Astron.*, **9**, 762.)
Novikov, I. D. (1975). *Astron. Zh.*, **55**, 657. (English translation in *Sov. Astron.*, **19**, 398.)
Osaki, Y. and Hansen, C. H. (1973). *Astrophys. J.*, **185**, 277.
Papini, G. (1974). *Can. J. Phys.*, **52**, 880.
Payne-Gaposchkin, C. (1957). *The Galactic Novae*. North-Holland: Amsterdam.
Peebles, P. J. E. and Dicke, R. H. (1968). *Astrophys. J.*, **154**, 891.
Peters, P. C. (1964). *Phys. Rev.*, **136**, B1224.
Peters, P. C. and Mathews, J. (1963). *Phys. Rev.*, **131**, 435.
Pines, D., Shaham, J. and Ruderman, M. (1974). *IAU Symposium No. 53, Physics of Dense Matter*, ed. C. Hansen, p. 189. Reidel: Dordrecht.
Pizzella, G. (1975). *Riv. Nuovo Cimento*, **5**, 369.
Prentice, A. F. R. and ter Haar, D. (1969). *Mon. Not. R. Astron. Soc.*, **146**, 423.

Chapter 3. References

Press, W. H. and Thorne, K. S. (1972), *Annu. Rev. Astron. Astrophys.*, **10**, 338.
Price, W. H. and Thorne, K. S. (1969). *Astrophys. J.*, **144**, 201.
Rees, M. J. (1974). *Colloques Internationaux CNRS No. 220 Ondes et Radiations Gravitationelles*, ed. Y. Choquet-Bruhat. CNRS: Paris.
Robinson, E. L. (1977). *Proceedings of the Los Alamos Conference on Solar and Stellar Pulsations*, eds. A. Cox and R. Deupree, p. 98. Los Alamos Report, LA-6544-C.
Ruffini, R. and Wheeler, J. A. (1971). *Proceedings of the ESRO Colloquium, The Significance of Space Research for Fundamental Physics*. ESRO SP-52: Paris.
Saeny, R. A. and Shapiro, S. L. (1978). *Astrophys. J.*, **221**, 286.
Savedoff, M. P., Van Horn, H. H. and Vila, S. C. (1969). *Astrophys. J.*, **155**, 221.
Sciama, D. W. (1972). *Gen. Relativ. Grav.*, **3**, 149.
Shapiro, S. L. (1977). *Astrophys. J.*, **214**, 566.
Smarr, L. (1976). *8th Texas Symposium on Relativistic Astrophysics.*, *Ann. N.Y. Acad. Sci.*, **302**, 569.
Stein, S. R. (1975). *Proceedings of 29th Annual Symposium on Frequency Control*. Electronic Industries Association: Washington, D.C.
Stein, S. R. and Turneaure, J. P. (1973). *Proceedings of 27th Annual Symposium on Frequency Control*, p. 414. Electrical Industry Association: Washington, D.C.
Stockman, H. S., Angel, J. R. P., Noukck, R. and Woodgate, R. E. (1973). *Astrophys. J.*, **183**, L63.
Strand, K. A. (1948). *Astrophys. J.*, **107**, 106.
Talbot, R. J. (1976). *Astrophys. J.*, **205**, 535.
Tamman, G. A. (1974). *Supernovae and Supernovae Remnants*, ed. C. B. Cosmovici. Reidel: Dordrecht.
Taylor, J. H. and Manchester, R. N. (1977). *Annu. Rev. Astron. Astrophys.*, **15**, 19–44.
Thorne, K. S. (1969a). *Astrophys. J.*, **158**, 1.
Thorne, K. S. (1969b). *Astrophys. J.*, **158**, 997.
Thorne, K. S. (1978). In *Theoretical Principles in Astrophysics and Relativity*, eds. N. Lebovitz, W. Reid and P. Vandervoot. University of Chicago Press.
Thorne, K. S. and Braginsky, V. B. (1976). *Astrophys. J.*, **204**, L1.
Thorne, K. S. and Campolattaro, A. (1967). *Astrtrophys. J.*, **149**, 591.
Thuan, T. X. and Ostriker, J. P. (1974). *Astrophys. J.*, **191**, L105.
Tyson, J. A. (1973). *Phys. Rev. Lett.*, **31**, 326.
Unruh, W. G. (1977a). *Phys. Rev.*, in press.
Unruh, W. G. (1977b). *8th International Conference on General Relativity and Gravitation*, August 7–12, 1977. Waterloo, Ontario, Canada.
van den Heuvel, E. P. J. (1976). *IAU Symposium No. 73, Structure and Evolution of Close Binary Systems*, eds. D. Eggleton, S. Mitton and J. A. J. Whelan. Reidel: Dordrecht.
Van Horn, H. M. and Savedoff, M. P. (1977). *Proceedings of Los Alamos Conference on Solar and Stellar Pulsations*, eds. A. Cox and R. Deupree, p. 109. Los Alamos Report, LA-6544-C, 109.

Chapter 4. References

Vasilevskis et al. (1975). *Publ. Lick. Obs.*, **22**, part 5.
Vessot, R. F. C. (1976). *Methods of Experimental Physics*, 12, *Astrophysics*, part C, p. 198. Academic Press: London, New York.
Vessot, R. F. C., Levine, M. W. and Nordtvedt, K. L. (1977). *28th Astronautical Congress*. Academic Press: London, New York.
Wahlquist, H. D., Anderson, J. D., Eastabrook, F. B. and Thorne, K. S. (1976). *Proceedings of the International Meeting on Experimental Gravitation*, ed. B. Bertotti. Accademia Nazionale dei Lincei: Rome.
Warner, B. (1976). *IAU Symposium No. 73, Structure and Evolution of Close Binary Systems*, eds. D. Eggleton, S. Mitton and J. A. J. Whelan. Reidel: Dordrecht.
Weber, J. (1961). *Phys. Rev.*, **117**, 306.
Weber, J. (1964). *General Relativity and Gravitational Waves*, Interscience: London.
Weber, J. (1970a). *Phys. Rev. Lett.*, **22**, 1320.
Weber, J. (1970b). *Phys. Rev. Lett.*, **24**, 276.
Weber, J. (1970c). *Phys. Rev. Lett.*, **25**, 180.
Weber, J. (1971). *Lett. Nuovo Cimento*, **4**, 653.
Weber, J. (1972). *Nuovo Cimento*, **4B**, 197.
Weber, J., Gretz, D. J., Lee, M., Rydbek, G., Trimble, V. L. and Steppel, S. (1973). *Phys. Rev. Lett.*, **31**, 779.
Weiss, R. (1972). *MIT Quarterly Progress Report, Research Laboratory of Electronics*, **105**, 54.
Winkler, W. (1976). *Proceedings of the International Meeting on Experimental Gravitation*, ed. B. Bertotti. Accademia Nazionale dei Lincei: Rome.
Woo, R. (1975). *Astrophys. J.*, **201**, 238.
Zel'dovich, Ya. B. and Novikov, I. D. (1971). *Relativistic Astrophysics*, vol. 1, eds. K. S. Thorne and W. D. Arnett. University of Chicago Press.
Zel'dovich, Ya. B. and Polnarev, A. G. (1974). *Astron. Zh.*, **51**, 30. (English translation in *Sov. Astron.*, **18**, 17.)
Zimmerman, M. 1977. *8th International Conference on General Relativity and Gravitation*, August 7–12, 1977. Waterloo, Ontario, Canada.
Zimmerman, M. (1978). *Nature*, **271**, 524.

Chapter 4 The initial value problem and the dynamical formulation of general relativity

We have not referred in the text to all the references given here. This list is intended to be a reasonable core for the topics discussed in this paper.

Abraham, R. and Marsden, J. (1978). *Foundations of Mechanics*, Second Edition. W. A. Benjamin: New York.
Arms, J. (1977a). Linearization stability of the Einstein–Maxwell system. *J. Math. Phys.*, **18**, 830–3.
Arms, J. (1977b). Linearization stability of coupled gravitational and gauge fields. Thesis, Berkeley.
Arms, J., Fischer, A. and Marsden, J. (1975). Une approche symplectique pour des théorèmes de décomposition en géometrie ou relativité générale, *C. R. Acad. Sci. (Paris)*, **281**, 571–20.

Chapter 4. References

Arnowitt, R., Deser, S. and Misner, C. W. (1959). Dynamical structure and definition of energy in general relativity. *Phys. Rev.*, **116**, 1322–30.

Arnowitt, R., Deser, S. and Misner, C. W. (1960). Note on positive-definiteness of the energy of the gravitational field. *Ann. Phys.*, **11**, 116–21.

Arnowitt, R., Deser, S. and Misner, C. W. (1962). The dynamics of general relativity. In *Gravitation: an Introduction to Current Research*, ed. L. Witten, pp. 227–65. Wiley: New York.

Avez, A. (1963). Essais de géométrie riemannienne hyperbolique global – applications a la relativité générale. *Ann. Inst. Fourier (Grenoble)*, **13**, 105–90.

Avez, A. (1967). Le probleme des conditions initiales. In *Fluides et champs gravitationnels en relativité générale*, pp. 163–7. Colloq. Intern. CRNS **170**: Paris.

Bancel, D. and Choquet-Bruhat, Y. (1973). Existence, uniqueness and local stability for the Einstein–Maxwell–Boltzmann system. *Commun. Math. Phys.*, **33**, 83–96.

Bancel, D. and Lacaze, J. (1976). Espaces de fonctions avec conditions asymptotiques et sections d'espace maximal non compact. *C. R. Acad. Sci. (Paris)*, **283**, 405–7.

Barbance, C. (1964). Decomposition d'un tenseur symmetrique sur en espace d'Einstein. *C. R. Acad. Sci. (Paris)*, **258**, 5336; **264**, 515–16.

Berezdivin, J. (1974). The analytic noncharacteristic Cauchy problem for nonlightlike isometrics in vacuum spacetimes. *J. Math. Phys.*, **15**, 1963–6.

Berger, M. (1965). Sur les varietes d'Einstein compactes. In *Comptes Rendu, IIIe Reunion Math. Expression Latine Namur*. Centre Belge de Recherches Mathématiques.

Berger, M. (1970). Quelques formules de variation pour une structure riemannienne, *Ann. Scient. Ec. Norm. Sup.*, **3**, 285–94.

Berger, M. and Ebin, S. (1969). Some decompositions of the space of symmetric tensors on a Riemannian manifold. *J. Differ. Geom.*, **3**, 379–92.

Bergmann, P. G. (1958). Conservation Laws in general relativity as the generators of coordinate transformations. *Phys. Rev.*, **112**, 287–9.

Bourguignon, J. P. (1975). Une stratification de l'espace des structures riemanniennes, *Comp. Math.*, **30**, 1–41.

Bourguignon, J. P., Ebin, D. and Marsden, J. (1976). Sur le noyau des operateurs pseudo-differentiels a symbole surjective et non-injective. *C. R. Acad. Sci. (Paris)*, **282**, 867–70.

Brezis, H. (1973). *Operateurs maximaux monotones*. North-Holland: Amsterdam.

Brill, D. (1977). Maximal surfaces in closed and open spacetimes. *Proceedings of the Marcel Grossman Meeting, July 1975*, ed. R. Ruffini. North-Holland: Amsterdam.

Brill, D. and Deser, S. (1968). Variational methods and positive energy in energy in general relativity. *Ann. Phys. (N.Y.)*, **50**, 548–70. (See also Brill, Deser and Fadeev (1968). *Phys. Lett.*, **26A**, 538.)

Brill, D. and Deser, S. (1973). Instability of closed spaces in general relativity. *Commun. Math. Phys.*, **32**, 291–304.

Chapter 4. References

Budic, R., Isenberg, J., Lindblom, L. and Yasskin, P. (1978). On the determination of Cauchy surfaces from intrinsic properties. *Commun. Math. Phys.*, **61**, 87–101.

Cantor, M. (1975). Spaces of functions with asymptotic conditions on R^n. *Indiana University Math. J.*, **24**, 897–902.

Cantor, M. (1977). The existence of non-trivial asymptotically flat initial data for vacuum spacetimes. *Commun. Math. Phys.*, **57**, 83–96.

Cantor, M. (1978). Some problems of global analysis on asymptotically simple manifolds. *Comp. Math.*, in press.

Cantor, M., Fischer, A., Marsden, J., O'Murchadha, N. and York, J. (1976). On the existence of maximal slices. *Commun. Math. Phys.*, **49**, 897–902.

Caricato, C. (1969). Sur le problème de Cauchy intrinsèque pour les equations de Maxwell–Einstein dans le vide. *Ann. Inst. H. Poincare*, Sect. A, **11**, 373–92.

Chernoff, P. and Marsden, J. (1974). *Properties of Infinite-Dimensional Hamiltonian Systems.* Springer Lecture Notes No. 425.

Choquet (Foures)-Bruhat, Y. (1948). Sur l'intégration des équations d'Einstein. *C. R. Acad. Sci. (Paris)*, **226**, 1071. (See also *J. Ration. Mech. Anal.*, **5** (1956) 951–66.)

Choquet-Bruhat, Y. (1952). Théorème d'existence pour certains systèmes d'équations aux derivées partielles non linéaires. *Acta Math*, **88**, 141–225.

Choquet-Bruhat, Y. (1958). Théorèmes d'existence an mécanique des fluides rélativistes. *Bull. Soc. Math. France*, **86**, 155–75.

Choquet-Bruhat, Y. (1960). Fluides rélativistes de conductivité infinie. *Astron. Acta*, **6**, 354–65.

Choquet-Bruhat, Y. (1962). Cauchy problem. In *Gravitation: an Introduction to Current Research*, ed. L. Witten. Wiley: New York.

Choquet-Bruhat, Y. (1968a). Espaces-temps einsteiniens généraux chocs gravitationels. *Ann. Inst. H. Poincaré*, **8**, 327–38.

Choquet-Bruhat, Y. (1968b). Etude des equations des fluides chargés rélativistes inductifs et conducteurs. *Commun. Math. Phys.*, **3**, 334–57.

Choquet-Bruhat, Y. (1970). Mathematical problems in general relativity. *Proc. Int. Congress, Nice.* French Mathematics Society.

Choquet-Bruhat, Y. (1971a). New elliptic system and global solutions for the constraints equations in general relativity. *Commun. Math. Phys.*, **21**, 211–8.

Choquet-Bruhat, Y. (1971b). Solutions C^∞ d'equations hyperbolique non linéaires. *C. R. Acad. Sci. (Paris)*, **272**, 386–88.

Choquet-Bruhat, Y. (1971c). Probleme de Cauchy pour le système integro-differentiel d'Einstein–Liouville. *Ann. Inst. Fourier XXI*, **3**, 181–201.

Chouquet-Bruhat, Y. (1972a). Stabilité de solutions d'équations hyperboliques non linéaires. Application a l'espace–temps de Minkowski en rélativité générale. *C. R. Acad. Sci. (Paris)*, **274**, Ser A, 843. (See also Uspeskii *Math. Nauk.* XXIX(2) **176**, 314–22.)

Choquet-Bruhat, Y. (1972b). C^∞ solutions of hyperbolic non linear equations. *Gen. Relativ. Gravit.*, **2**, 359–62.

Choquet-Bruhat, Y. (1973). Global solutions of the equations of constraints in general relativity on closed manifolds. *Symposia Math.*, **XII**, 317–25.

Chapter 4. References

Choquet-Bruhat, Y. (1975a). Sous-variétés maximales, ou a courbure constante, de variétés lorentziennes. *C. R. Acad. Sci. (Paris)*, **280**, 169–71.

Choquet-Bruhat, Y. (1975b). Quelques propriétés des sous-variétés maximales d'un variété lorentzienne. *C. R. Acad. Sci. (Paris)*, **281**, 577–80.

Choquet-Bruhat, Y. (1976a). The problem of constraints in general relativity, solution of the Lichnerowicz equation. In *Differential Geometry and Relativity*, eds. M. Cahen and M. Flato. Reidel: Dordrecht.

Choquet-Bruhat, Y. (1976b). Maximal submanifolds and manifolds with constant mean, extrinsic curvature of a Lorentzian manifold. *Ann. Scuola Norm. Pisa, Serie IV*, vol. III (in honour of J. Leray), p. 361.

Choquet-Bruhat, Y., and Deser, S. (1972). Stabilité initiale de l'espace temps de Minkowski. *C. R. Acad. Sci. (Paris)*, **275**, 1019–27.

Choquet-Bruhat, Y. and Geroch, R. (1969). Global aspects of the Cauchy problem in general relativity. *Commun. Math. Phys.*, **14**, 329–35.

Choquet-Bruhat, Y., and Lamoureau-Brousse, L. (1973). Sur les equations de l'elasticite relativiste. *C. R. Acad. Sci. (Paris)*, **276**, 1217–1320.

Choquet-Bruhat, Y. and Marsden, J. (1976). Solution of the local mass problem in general relativity. *C. R. Acad. Sci. (Paris)*, **282**, 609–12; *Commun. Math. Phys.*, **51**, 283–96.

Choquet-Bruhat, Y. and York, J. W. (1979). In *Einstein Centenary Volume*, eds. P. Bergmann, J. Goldberg and A. Held, in press.

Choquet-Bruhat, Y., Fischer A. and Marsden, J. (1978). Maximal hypersurfaces and positivity of mass. *Proc. Enrico Fermi Summer School of the Italian Physical Society Varenna*, ed. J. Ehlers, in press.

Christodoulou, D. and Francaviglia, F. (1978). The thin sandwich conjecture. *Proc. Enrico Fermi Summer School Varenna*, ed. J. Ehlers, in press.

Coll, B. (1975). Sur la détermination, par des donnés de Cauchy, des champs de Killing admis par un espace-temps, d'Einstein-Maxwell. *C. R. Acad. Sci. (Paris)*, **281**, 1109–12.

Coll, B. (1976). Sur la stabilité linéaire des équations d'einstein du vide. *C. R. Acad. Sci. (Paris)*, **282**, 247–50.

Coll, B. (1978). On the evolution equations for Killing fields. *J. Math. Phys.*, **18**, 1918–22.

Cordero, P. and Teitelboim, C. (1976). Hamiltonian treatment of the spherically symmetric Einstein–Yang–Mills system. *Ann. Phys. (N.Y.)*, **100**, 607–31.

Courant, R. and Hilbert, D. (1962). *Methods of Mathematical Physics*, vol. II. Interscience: New York.

D'Eath, P. D. (1976). On the existence of perturbed Robertson–Walker universes. *Ann. Phys. (N.Y.)*, **98**, 237–63.

Deser, S. (1967). Covariant decomposition of symmetric tensors and the gravitational Cauchy problem. *Ann. Inst. H. Poincare*, **7**, 149–88.

Deser, S. (1975). Gravitational energy–momentum on nonmaximal surfaces. *Phys. Rev.*, **D12**, 943–5.

Deser, S. and Teitelboim, C. (1976). Supergravity has positive energy. *Phys. Rev. Lett.*, **39**, 249–52.

DeWitt, B. (1967). Quantum theory of gravity, I. Canonical Theory. *Phys. Rev.*, **160**, 1113–48.

Chapter 4. References

DeWitt, B. (1972). Covariant quantum geometrodynamics. In *Magic Without Magic*, ed. J. R. Klauder. W. H. Freeman: San Francisco.

Dionne, P. (1962). Sur les problemes de Cauchy bien posés. *J. Anal. Math., Jerusalem*, **10**, 1–90.

Dirac, P. A. M. (1950a). Generalized Hamiltonian dynamics. *Can. J. Math.*, **2**, 129–48.

Dirac, P. A. M. (1950b). The Hamiltonian form of field dynamics, *Can. J. Math.*, **3**, 1–23.

Dirac, P. A. M. (1958a). Generalized Hamiltonian dynamics. *Proc. R. Soc. Lond.*, **A246**, 326–32.

Dirac, P. A. M. (1958b). The theory of gravitation in Hamiltonian form. *Proc. R. Soc. Lond.*, **A246**, 333–43.

Dirac, P. A. M. (1959). Fixation of coordinates in the Hamiltonian theory of gravitation. *Phys. Rev.*, **114**, 924–30.

Dirac, P. A. M. (1964). *Lectures on Quantum Mechanics*. Belfer Graduate School of Science, Monograph Series No. 2. Yeshiva University: New York.

Dorroh, J. R. and Marsden, J. E. (1975). Differentiability of nonlinear semigroups. (Unpublished notes.)

Ebin, D. (1970). On the space of Riemannian metrics. *Symp. Pure Math., Am. Math. Soc.*, **15**, 11–40.

Ebin, D. and Marsden, J. (1970). Groups of diffeomorphisms and the motion of an incompressible fluid. *Ann. Math.*, **92**, 102–63.

Eells, J. and Sampson, J. (1964). Harmonic maps of Riemannian manifolds. *Am. J. Math.*, **86** (1964) 109–60.

Einstein, A. (1961a). Hamiltonsches Princip und allgemeine Relativitatstheorie. *Sitzungsber. Preuss. Akad. Wiss.*, 1111–16.

Einstein, A. (1961b). Naherungsweise Integration der Feldgleichungen der Gravitation. *Sitzungsber. Preuss. Akad. Wiss.*, 688–96.

Einstein, A. (1918). Der Energiesatz in der allgemeinen Relativitatstheorie *Sitzungsber Preuss. Akad. Wiss.*, 448.

Fadeev, L. D. (1970). Symplectic structure and quantization of the Einstein gravitation theory. *Actes du Congres Intern. Math.*, **3**, 35–40.

Fischer, A. (1970). The theory of superspace. In *Relativity*, eds. M. Carmelli, S. Fickler and L. Witten. Plenum Press: New York.

Fischer, A. and Marsden, J. (1972a). The Einstein equations of evolution – a geometric approach. *J. Math. Phys.*, **13**, 546–68.

Fischer, A. and Marsden, J. (1972b). The Einstein evolution equations as a first-order symmetric hyperbolic quasilinear system. *Commun. Math. Phys.*, **28**, 1–38.

Fischer, A. and Marsden, J. (1973a). Linearization stability of the Einstein equations. *Bull. Am. Math. Soc.*, **79**, 995–1001.

Fischer, A. and Marsden, J. (1973b). New theoretical techniques in the study of gravity. *Gen. Relativ. Grav.*, **4**, 309–317.

Fischer, A. and Marsden, J. (1974). Global analysis and general relativity. *Gen. Relativ. Grav.*, **5**, 89–93.

Fischer, A. and Marsden, J. (1975a). Linearization stability of nonlinear partial differential equations. *Proc. Symp. Pure Math., Am. Math. Soc.*, **27**, Part 2, 219–63.

Fischer, A. and Marsden, J. (1975b). Deformations of the scalar curvature. *Duke Math. J.*, **42**, 519–47.
Fischer, A. and Marsden, J. (1976). A new Hamiltonian structure for the dynamics of general relativity. *Gen. Relativ. Grav.*, **12**, 915–20.
Fischer, A. and Marsden, J. (1977). The manifold of conformally equivalent metrics. *Can. J. Math.*, **29**, 193–209.
Fischer A. and Marsden, J. (1978a). Topics in the dynamics of general relativity. In *Proceedings Enrico Fermi Summer School of the Italian Physical Society Varenna*, ed. J. Ehlers, in press.
Fischer, A. and Marsden, J. (1978b). Hamiltonian field theories on spacetime. (In preparation.)
Fischer, A. and Marsden, J. (1978c). The space of gravitational degrees of freedom. (In preparation.)
Fischer, A., Marsden, J. and Moncrief, V. (1978). The structure of the space solutions of Einstein's equations. (In preparation.)
Frankl, F. (1937). Uber das Angangswertproblem fur lineare und nichtlineare partielle Differentialgleichungen zweiter Ordnung. *Mat. Sb.*, **2**(44), 814–68.
Friedman, A. (1969). *Partial differential equations.* Holt: New York.
Friedrichs, K. O. (1954). Symmetric hyperbolic linear differential equations. *Commun. Pure Appl. Math.*, **8**, 345–92.
Friedrichs, K. (1974). On the laws of relativistic magneto-fluid dynamics. *Commun. Pure Appl. Math.*, **27**, 749–808.
Garding, L. (1957). Cauchy's problem for hyperbolic equations. *Lecture Notes*. University of Chicago.
Garding, L. (1964). Energy inequalities for hyperbolic systems. In *Differential Analysis*, Bombay Colloquium, 209–25. Oxford University Press.
Geroch, R. (1975). General relativity. *Proc. Symp. Pure Math.*, vol. 27 (part 2), pp. 401–14; *J. Math. Phys.*, **13** (1972) 956; *Ann. N. Y. Acad. Sci.*, **224** (1973) 108.
Hanson, A., Regge, T. and Teitelboim, C. (1976). Constrained Hamiltonian systems. *Accademia Nazionale dei Lincei, Rome*. No. 22, 1–135.
Hawking, S. W. and Ellis, G. F. R. (1973). *The Large Scale Structure of Space–time.* Cambridge University Press.
Hermann, R. (1975). *Gauge Fields and Cartan–Ehresman Connections.* Math-Sci Press: Brookline, Massachusetts.
Hille, E. and Phillips, R. (1967). *Functional Analysis and Semigroups.* American Mathematical Society: Providence, Rhode Island.
Hormander, L. (1966). Pseudo-differential operators and non-elliptic boundary value problems. *Ann. Math.*, **83**, 129–209.
Hughes, T., Kato, T. and Marsden, J. (1977). Well-posed quasi-linear second-order hyperbolic systems with applications to nonlinear elastodynamics and general relativity. *Arch. Ration. Mech. Anal.*, **63**, 273–94.
Kato, T. (1966). *Perturbation Theory for Linear Operators.* Springer: Berlin, Heidelberg, New York.
Kato, T. (1970). Linear evolution equations of 'hyperbolic' type. *J. Fac. Sci. Univ. Tokyo*, **17**, 241–58.

Chapter 4. References

Kato, T. (1973). Linear evolution equations of 'hyperbolic' type, II. *Math. Soc. Japan*, **25**, 648–66.

Kato, T. (1975a). The Cauchy problem for quasi-linear symmetric hyperbolic systems. *Arch. Ration. Mech. Anal.*, **58**, 181–205.

Kato, T. (1975b). Quasi-linear equations of evolution with applications to partial differential equations. *Springer Lecture Notes*, **448**, 25–70.

Kato, T. (1977). *Linear and Quasilinear Equations of Evolution of Hyperbolic Type*. Bressanone Lectures. Centro Internazionale Matematico Estivo: Rome.

Kazdan, J., and Warner, F. (1975). A direct approach to the determination of Gaussian and scalar curvature functions. *Invent. Math.*, **28**, 227–30.

Kijowski, J., and Szczyrba, W. (1976). A canonical structure for classical field theories. *Commun. Math. Phys.*, **46**, 183–205.

Kobayashi, S. (1974). *Transformation Groups in Differential Geometry*. Springer: Berlin, Heidelberg, New York.

Kryzanski, M., and Schauder, J. (1934). Quasilineare Differentialgleichungen zweiter Ordnung von hyperbolischen Typus. *Gemischte Randwertaufgaben, Studia, Math.*, 162–89.

Kuchař, K. (1972). A bubble-time canonical formalism for geometrodynamics. *J. Math. Phys.*, **13**, 768–81.

Kuchař, K. (1974). Geometrodynamics regained: a Lagrangian approach. *J. Math. Phys.*, **15**, 708–15.

Kuchař, K. (1976a). Geometry of hyperspace. I. *J. Math. Phys.*, **17**, 777–91.

Kuchař, K. (1976b). Kinematics of tensor fields in hyperspace. II. *J. Math. Phys.*, **17**, 792–800.

Kuchař, K. (1976c). Dynamics of tensor fields in hyperspace. III. *J. Math. Phys.*, **17**, 801–20.

Kuchař, K. (1977). Geometrodynamics with tensor sources. IV. *J. Math. Phys.*, **18**, 1589–97.

Kuchař, K. (1978). On equivalence of parabolic and hyperbolic super-Hamiltonians. *J. Math. Phys.*, **19**, 390–400.

Lanczos, C. (1922). Ein verinfachendes Koordinatensystem für die Einsteinschen Gravitationsgleichungen, *Phys. Z.*, **23**, 537–9.

Lax, P. (1955). Cauchy's problem for hyperbolic equations and the differentiability of solutions of elliptic equations. *Commun. Pure Appl. Math.*, **8**, 615–33.

Leray, J. (1953). *Hyperbolic Differential Equations*. Institute for Advanced Study. (Notes.)

Lichnerowicz, A. (1939). *Problèmes globaux en mécanique relativiste*. Hermann.

Lichnerowicz, A. (1944). L'intégration des équations de la gravitation problème des *n* corps. *J. Math. Pures Appl.*, **23**, 37–63.

Lichnerowicz, A. (1955). *Théories relativistes de la gravitation et de l'électromagnetism*. Masson: Paris.

Lichnerowicz, A. (1961). Propagateurs et commutateurs en relativité générale. Publications Mathématiques, No. 10, pp. 293–344. Institute des Hautes Etudes Scientifiques: Paris.

Lichnerowicz, A. (1967), *Relativistic Hydrodynamics and Magnetohydramics: Lectures on the Existence of Solutions*. W. A. Benjamin: New York.

Lichnerowicz, A. (1969). Ondes de choc et hypothèses de compressibilité en MHD rélativiste. *Commun. Math. Phys.*, **12**, 145–74.
Lichnerowicz, A. (1972). Ondes de choc gravitationnelles et électromagnetiques. *Symposia Math. Inst. Naz. di Alta Matematica*, Rome, **12**, 93–110.
Lichnerowicz, A. (1976). Shock waves in relativistic magnetohydrodynamics under general assumptions. *J. Math. Phys.*, **17**, 2135–42.
Lions, J. L. (1969). *Quelques methodes de résolution des problèmes nonlineares*, Dunod: Paris.
Marsden, J. (1974). *Applications of Global Analysis in Mathematical Physics*. Publish or Perish: Boston.
Marsden, J. and Fischer, A. (1972). On the existence of nontrivial, complete, asymptotically flat spacetimes. *Publ. Univ. Lyon*, **4** (fasc. 2), 182–93.
Marsden, J. and Hughes, T. (1978). Topics in Mathematical Foundations of Elasticity. In *Non-linear Analysis and Mechanics*, vol. II, ed. R. J. Knops. Pitman.
Marsden, J. E. and Tipler, F. J. Maximal hypersurfaces and foliations of constant mean curvature in general relativity. (Preprint).
Marsden, J. and Weinstein, A. (1974). Reduction of symplectic manifolds with symmetry. *Rep. Math. Phys.*, **5**, 121–310.
Marsden, J., Ebin, D. and Fischer, A. (1972). Diffeomorphism groups, hydrodynamics, and general relativity. *Proc. 13th Bienniel Seminar of the Canadian Math. Cong.*, Montreal, 135–279.
Misner, C. W., Thorne, K. and Wheeler, J. A. (1974). *Gravitation*. W. H. Freeman: San Francisco.
Moncrief, V. (1975a). Spacetime symmetries and linearization stability of the Einstein equations, I. *J. Math. Phys.*, **16**, 493–8.
Moncrief, V. (1975b). Decompositions of gravitational perturbations. *J. Math. Phys.*, **16**, 1556–60.
Moncrief, V. (1975c). Gauge-invariant perturbations of Reissner-Nordstrom black holes. *Phys. Rev.*, **D12**, 1526–37.
Moncrief, V. (1976). Spacetime symmetries and linearization stability of the Einstein equations, II. *J. Math. Phys.* **17**, 1893–1902.
Moncrief, V. (1977). Gauge symmetries of Yang-Mills fields, *Ann. Phys. (N.Y.)*, **108**, 387–400.
Müller zum Hagen, H. and Seifert, H. J. (1976). On characteristic initial-value and mixed problems. *Gen. Relativ. Grav.*, in press.
Nijenhuis, A. (1951). X_{n-1} forming sets of eigenvectors. *Proc. Kon. Ned. Akad., Amsterdam*, **54**, 200–212.
Nirenberg, L. (1959). On elliptic partial differential equations. *Ann. Scuola. Norm. Sup. Pisa*, **13**, 115–62.
Nirenberg, L. and Walker, H. (1973). The null space of elliptic partial differential operators in R^n. *J. Math. Anal. Appl.*, **47**, 271–301.
O'Murchadha, N. and York, J. W. (1974a). The initial-value problem of general relativity. *Phys. Rev.*, **D10**, 428–46.
O'Murchada, N. (1974b). Gravitational energy. *Phys. Rev.*, **D10**, 2345–57.
Palais, R. (1959). Natural operations on differential forms. *Trans. Am. Math. Soc.*, **92**, 125–41.

Chapter 4. References

Palais, R. (1965). *Seminar on the Atiyah-Singer Index Theorem.* Princeton University Press.
Palais, R. (1968). *Foundations of Global Nonlinear Analysis.* Benjamin: New York.
Petrovskii, I. (1937). Über das Cauchysche Problem für lineare und nichlineare hyperbolische partielle Differentialgleichungen. *Mat. Sb.,* **2,** 814-68.
Pham Mau Quan (1965). Magneto-hydrodynamique relativiste. *Ann. Inst. H. Poincaré, nouv. Serie* **t.2,** 151-65.
Regge, T. and Teitelboim, C. (1974). Role of surface integrals in the Hamiltonian formulation of general relativity. *Ann. Phys. (N.Y.),* **88,** 286-318.
Rund, H. and Lovelock, D. (1972). Variational principles in the general theory of relativity. *Uber Deutsch. Math-Verein.,* **74,** 1-65.
Sachs, R. and Wu, H. (1977). *General Relativity and Cosmology for Mathematicians.* Springer-Verlag Graduate Texts in Mathematics. Springer: Berlin, Heidelberg, New York.
Schauder, J. (1935). Das Anfangswertproblem einer quasi-linearen hyperbolischen Differentialgleichung zweiter Ordnung in beliebiger Anzahl von unabhangigen Veranderlichen. *Fund. Math.,* **24,** 213-246.
Schwartz, J. (1969). *Nonlinear Functional Analysis.* Gordon and Breach: New York.
Szczryba, W. (1977). On geometric structure of the set of solutions of Einstein equations. *Diss. Math.,* **150,** 1-87.
Segal, I. (1974). Symplectic structures and the quantization problem for wave equations. *Symp. Math.,* **14,** 99-117.
Sobolev, S. S. (1939). On the theory of hyperbolic partial differential equations. *Mat. Sb.,* **5,** 71-99.
Sobolev, S. S. (1963). *Applications of Functional Analysis in Mathematical Physics.* Translations of Math. Monographs **7.** American Mathematical Society: Providence, Rhode Island.
Souriau, J. M. (1970). *Structure des systemes dynamique.* Dunod: Paris.
Taub, A. (1970). *Variational Principles in General Relativity,* Bressanone Lectures, pp. 206-300. Centro Internazionale Matematico Estivo: Rome.
Terng, C. L. (1976). Natural Vector Bundles and Natural Differential Operators. Thesis, Brandeis University, Waltham, Massachusetts.
Weinstein, A. (1977). *Lectures on Symplectic Manifolds,* CBMS No. 29. American Mathematical Society: Providence, Rhode Island.
Wheeler, J. A. (1964). Geometrodynamics and the issue of the final state. In *Relativity, Groups, and Topology,* eds. C. M. Dewitt and B. S. Dewitt. Gordon and Breach: New York.
Wheeler, J. A. (1962). *Geometrodynamics.* Academic Press: New York, London.
York, J. W. (1971). Gravitational degrees of freedom and the initial-value problem. *Phys. Rev. Lett.,* **26,** 1656-8.
York, J. W. (1972a). Mapping onto solutions of the gravitational initial-value problem. *J. Math. Phys.,* **13,** 125-30.
York, J. W. (1972b). Role of conformal three-geometry in the dynamics of gravitation. *Phys. Rev. Lett.,* **28,** 1082-5.

York, J. W. (1973). Conformally invariant orthogonal decomposition of symmetric tensors on Riemannian manifolds and the initial-value problem of general relativity. *J. Math. Phys.*, **14**, 456–64.

York, J. W. (1974). Covariant decompositions of symmetric tensors in the gravitation. *Ann. Inst. H. Poincaré*, **21**, 319–32.

Yosida, K. (1974). *Functional Analysis*, Fourth edition. Springer: Berlin, Heidelberg, New York.

Chapter 5 Global structure of spacetime

(1) Hawking, S. W. and Ellis, G. F. R. (1973). *The Large Scale Structure of Space–Time.* Cambridge University Press. Penrose, R. (1972). *Techniques of Differential Topology in Relativity.* SIAM: Philadelphia. Geroch, R. (1971). In *General Relativity and Cosmology*, p. 71. Academic Press: New York, London. Penrose, R. (1968). In *Battelle Rencontres*, eds. C. M. deWitt and J. A. Wheeler, p. 121, Benjamin: New York.

(2) Kobayashi, S. and Nomizu, K. (1963, 1969). *Foundations of Differential Geometry*, vols. I and II. Interscience: New York.

(3) Steenrod, N. E. (1951). *The Topology of Fibre Bundles.* Princeton University Press.

(4) Markus, L. (1955). *Ann. Math.*, **62**, 411.

(5) Avez, A. (1963). *Ann. Inst. M. Fourier*, **13**, 105.

(6) Abraham, R. and Robbin, J. (1967). *Transversal Mappings and Flows.* Benjamin: New York.

(7) Geroch, R. (1970). *J. Math. Phys.*, **11**, 343.

(8) Wallace, A. H. (1963). *An Introduction to Algebraic Topology.* Pergamon: Oxford.

(9) Christenson, J. H. et al. (1964). *Phys. Rev. Lett.*, **13**, 138. Wu, C. S. (1957). *Phys. Rev.*, **105**, 1413. Garwin, R. L., Lederman, L. M. and Weinrich, M. (1957). *Phys. Rev.*, **105**, 1415.

(10) Geroch, R. (1967). Ph.D. Thesis, Princeton University.

(11) Kronheimer, E. H. and Penrose, R. (1967). *Proc. Cam. Phil. Soc.*, **63**, 481.

(12) Havas, P. (1968). *Synthese*, **18**, 75. Reichenbach, H. (1958). *The Philosophy of Space and Time*, Dover Publications: New York. Tipler, F. J. (1976). *Phys. Rev. Lett.*, **37**, 879. Tipler, F. J. (1977). *Ann. Phys. (N.Y.)*, **108**, 1.

(13) Hawking, S. W. (1969). *Proc. R. Soc. Lond.*, **A308**, 433.

(14) Hawking, S. W. (1966). Ph.D. Thesis, Cambridge University.

(15) Geroch, R. (1967). *J. Math. Phys.*, **8**, 782.

(16) Courant, R. and Hilbert, D. (1965). *Methods of Mathematical Physics*, vol. 2. Interscience: New York.

(17) Geroch, R. (1970). *J. Math. Phys.*, **11**, 437.

(18) Choquet-Bruhat, Y. (1962). In *Gravitation: An Introduction to Current Research*, ed. L. Witten, chapter 4. Wiley: New York. Choquet-Bruhat, Y. and Geroch, R. (1969). *Commun. Math. Phys.*, **14**, 329. O'Murchadha, N. and York, J. W. (1974). *Phys. Rev.*, **D10**, 428.

(19) Ashtekar, A. (1979). In *Einstein Centenary Volume*, ed. A. Held and P. Bergmann. Plenum Press: New York.

Chapter 5. References

(20) Schmidt, B. G. (1971). *Gen. Relativ. Grav.*, **1**, 269.
(21) Penrose, R. (1965). *Phys. Rev. Lett.*, **14**, 57. Hawking, S. W. (1967). *Proc. R. Soc. Lond.*, **A300**, 187. Hawking, S. W. and Penrose, R. (1970). *Proc. R. Soc. Lond.*, **A314**, 529.
(22) Geroch, R. (1968). *J. Math. Phys. Phys.*, **9**, 450. Hawking, S. W., unpublished essay submitted for Adam's Prize, Cambridge University, 1966.
(23) Johnson, R. A. (1977). *J. Math. Phys.*, **18**, 898.
(24) Penrose, R. (1969). *Rev. Nuovo Cimento*, **1**, 252.
(25) Muller zum Hagen, H. *et al.* (1974). *Commun. Math. Phys.*, **37**, 29.
(26) Thorne, K. S. (1972). In *Magic Without Magic*, ed. J. R. Klauder, p. 231. Freeman: San Francisco. Liang, E. (1974). *Phys. Rev.*, **D10**, 447.
(27) Demianski, M. and Lasota, J. P. (1968). *Astrophys. Lett.*, **1**, 205.
(28) Penrose, R. (1965). *Proc. R. Soc. Lon.*, **A284**, 159. Geroch, R. and Horowitz, G. T. (1978). *Phys. Rev. Lett.*, **40**, 203; erratum: *Phys. Rev. Lett.*, **40**, 483.
(29) Hawking, S. W. (1973). In *Black Holes*, eds. C. DeWitt and B. S. DeWitt, p. 5. Gordon and Breach: New York.
(30) Penrose, R. (1973). *Ann. N.Y. Acad. Sci.*, **224**, 125. Gibbons, G. W. (1972). *Commun. Math. Phys.*, **27**, 87.
(31) Jang, P. S. and Wald, R. (1977). *J. Math. Phys.*, **18**, 41.
(32) Del la Cruz, V. *et al.* (1970). *Phys. Rev. Lett.*, **24**, 423. Price, R. (1972). *Phys. Rev.*, **D5**, 2419.
(33) Ryan, M. P. and Shepley, L. C. (1975). *Homogeneous Relativistic Cosmologies*. Princeton University Press.
(34) Simpson, M. and Penrose, R. (1973). *Int. J. Theor. Phys.*, **7**, 183. McNamara, J. M. (1978). *Proc. R. Soc. Lond.*, **A358**, 499.
(35) Budic, R. *et al.* (1978). *Commun. Math. Phys.*, **61**, 87.

Chapter 6   The general theory of black hole properties

Bardeen, J. M. (1973). Rapidly rotating stars, discs, and black holes. In *Black Holes*, eds. B. DeWitt and C. DeWitt. Gordon and Breach: New York.
Bardeen, J. H., Carter, B. and Hawking, S. W. (1973). *Commun. Math. Phys.*, **31**, 181.
Beckenstein, J. (1973a). *Phys. Rev.*, **D7**, 949.
Beckenstein, J. (1973b). *Phys. Rev.*, **D7**, 2333.
Blandford, R. D. and Znajek, R. L. (1977). *Mon. Not. R. Astron. Soc.*, **179**, 433.
Birkhoff, G. (1923). *Relativity and Modern Physics*. Harvard University Press.
Bose, S. K. and Wang, H. Y. (1974). *Phys. Rev.*, **D10**, 1675.
Carter, B. (1968a). *Phys. Rev.*, **174**, 1559.
Carter, B. (1968b). *Commun. Math. Phys.*, **10**, 280.
Carter, B. (1969). *J. Math. Phys.*, **10**, 70.
Carter, B. (1970). *Commun. Math. Phys.*, **17**, 233.
Carter, B. (1971a). *Gen. Relativ. Grav.*, **1**, 349.
Carter, B. (1971b). *Phys. Rev. Lett.*, **26**, 331.
Carter, B. (1973a). *Nature, Phys. Sci.*, **238**, 71.

Chapter 6. References

Carter, B. (1973b). *Commun. Math. Phys.*, **30**, 261.
Carter, B. (1973c). General theory of stationary black hole states. In *Black Holes*, eds. B. DeWitt and C. DeWitt. Gordon and Breach: New York.
Carter, B. (1975). *Ann. Phys. (N.Y.)*, **95**, 53.
Carter, B. (1976). In *Journées Relativistes*, eds. M. Cahen, R. Debever and J. Geheniau. Univ. Libre de Bruxelles.
Carter, B. (1978). *Gen. Relativ. Grav.*, **9**, 437.
Carter, B. (1979). In *Active Galactic Nuclei*, eds. S. C. Hazard and S. Mitton. Cambridge University Press.
Catenacci, R. and Diaz Alonso, J. (1976). *J. Math. Phys.*, **17**, 2232.
Chandrasekhar, S. (1976). *Proc. R. Soc. Lond.*, **A349**, 571.
Choqet-Bruhat, Y. (1968). *Geometrie differentielle et systemes exterieurs.* Dunod: Paris.
Damour, T. (1975). *Ann. N.Y. Acad. Sci.*, **262**, 113.
Damour, T. (1978). *Phys. Rev.* (in press). Preprint 1977. Observatoire de Paris: Meudon.
Ernst, F. J. (1968a). *Phys. Rev.*, **167**, 175.
Ernst, F. J. (1968b). *Phys. Rev.*, **167**, 1175.
Gibbons, G. W. (1972). *Commun. Math. Phys.*, **27**, 87.
Gibbons, G. W. (1974). *Commun. Math. Phys.*, **35**, 13.
Hajicek, P. (1975). *J. Math. Phys.*, **16**, 518.
Hajicek, P. (1976). In *Proceedings of the Marcel Grossmann Meeting on General Relativity*, ed. R. Ruffini. North-Holland: Amsterdam.
Hartle, J. B. (1973a). *Phys. Rev.*, **D8**, 1010.
Hartle, J. B. (1973b). In *Relativity, Astrophysics and Cosmology*, ed. W. Israel. Reidel: Dordrecht.
Hartle, J. B. and Hawking, S. W. (1972a). *Commun. Math. Phys.*, **26**, 87.
Hartle, J. B. and Hawking, S. W. (1972b). *Commun. Math. Phys.*, **27**, 283.
Hawking, S. W. (1957). *Proc. R. Soc. Lond.* **A300**, 187.
Hawking, S. W. (1972). *Commun. Math. Phys.*, **25**, 152.
Hawking, S. W. (1973a). The event horizon. In *Black Holes*, eds. B. DeWitt and C. DeWitt. Gordon and Breach: New York.
Hawking, S. W. (1973b). *Commun. Math. Phys.*, **33**, 325.
Hawking, S. W. (1975). *Commun. Math. Phys.*, **43**, 199.
Hawking, S. W. and Ellis, G. F. R. (1973). *The Large Scale Structure of Space-Time.* Cambridge University Press.
Hawking, S. W. and Penrose, R. (1970). *Proc. R. Soc. Lond.*, **A314**, 529.
Hicks, N. J. (1971). *Notes on Differential Geometry.* Van Nostrand: New York.
Hoffman, B. (1933). *Q. J. Math.*, **4**, 179.
Ipser, J. R. (1971). *Phys. Rev. Lett.*, **27**, 529.
Israel, W. (1967). *Phys. Rev.*, **164**, 1776.
Israel, W. (1968). *Commun. Math. Phys.*, **8**, 245.
Kerr, R. (1963). *Phys. Rev. Lett.*, **11**, 237.
King, A. R. and Lasota, J. P. (1977). *Astron. Astrophys.*, **58**, 175.
Komar, A. (1959). *Phys. Rev.*, **113**, 934.
Lichnerowicz, A. (1955). *Theories relativistes de la gravitation et de l'électromagnetism.* Masson: Paris.
Linet, B. (1977). *Phys. Lett.*, **A60**, 395.

Chapter 6. References

Misner, C. W. (1972). *Phys. Rev. Lett.*, **28**, 994.
Misner, C. W., Thorne, K. S. and Wheeler, J. A. (1973). *Gravitation*. Freeman: San Francisco.
Morse, M. and Heinze, H. (1945). *Ann. Math.*, **96**, 625.
Müller zum Hagen, H. Robinson, D. C. and Seifert, H. J. (1973). *Gen. Relativ. Gravit.*, **4**, 53.
Müller zum Hagen, H., Robinson, D. C. and Seifert, H. J. (1974). *Gen. Relativ. Gravit.*, **5**, 59.
Müller zum Hagen, H. (1970). *Proc. Camb. Phil. Soc.*, **68**, 199.
Newman, E. T. and Penrose, R. (1962). *J. Math. Phys.*, **3**, 566.
Newman, E. T. *et al.* (1965). *J. Math. Phys.*, **6**, 918.
Page, R. (1976). *Phys. Rev.*, **D14**, 1509.
Papapetrou, A. (1966). *Ann. Inst. H. Poincare*, **A4**, 83.
Penrose, R. (1965). *Phys. Rev. Lett.*, **14**, 57.
Penrose, R. (1967). In *Battelle Rencontres*, eds. C. DeWitt and J. A. Wheeler. Benjamin: New York.
Penrose, R. (1969). *Riv. Nuovo Cimento I*, **I**, 252.
Penrose, R. (1974). *Ann. N.Y. Acad. Sci.*, **224**, 125.
Pirani, F. (1965). In *Lectures on General Relativity (Brandeis Summer Institute, 1964)*, eds. S. Deser and K. W. Ford. Prentice-Hall: New Jersey.
Pollock, M. D. (1976). *Proc. R. Soc. Lond.*, **A350**, 239.
Pollock, M. D. and Brinkmann, W. P. (1977). *Proc. R. Soc. Lond.*, **A356**, 351.
Press, W. H. (1972). *Astrophys. J.*, **175**, 243.
Press, W. H. and Teukolsky, S. A. (1972). *Nature*, **238**, 211.
Press, W. H. and Teukolsky, S. A. (1973). *Astrophys. J.*, **185**, 649.
Prior, C. R. (1977). *Proc. R. Soc. Lond.*, **A355**, 1.
Robinson, D. C. (1974). *Phys. Rev.*, **D10**, 458.
Robinson, D. C. (1975). *Phys. Rev. Lett.*, **34**, 905.
Robinson, D. C. (1977). *Gen. Relativ. Grav.*, **8**, 695.
Sachs, R. K. (1962). *Proc. R. Soc. Lond.*, **A270**, 103.
Smarr, L. (1973). *Phys. Rev. Lett.*, **30**, 71.
Stewart, J. (1975). *Proc. R. Soc. Lond.*, **A344**, 65.
Teukolsky, S. A. (1972). *Phys. Rev. Lett.*, **29**, 114.
Teukolsky, S. A. (1973). *Astrophys. J.*, **185**, 635.
Tipler, F. J. (1976). *Phys. Rev. Lett.*, **37**, 879.
Trautman, A. (1965). In *Lectures on General Relativity (Brandeis Summer Institute, 1964)*, eds. S. Deser and K. W. Ford. Prentice-Hall: New Jersey.
Wald, R. (1972). Static axially symmetric electromagnetic fields in Kerr spacetime. Ph.D. thesis, Princeton University.
Vishveshwara, C. V. (1968). *J. Math. Phys.*, **9**, 1319.
Vishveshwara, C. V. (1970). *Phys. Rev.*, **D1**, 2870.
Wilkins, D. and Hartle, J. B. (1974). *Commun. Math. Phys.*, **38**, 47.
Yodzis, P., Seifert, H. J. and Muller zum Hagen, H. (1974). *Commun. Math. Phys.*, **37**, 29.
Zel'dovich, Ya. B. (1972). *J. Exp. Theor. Phys.*, **62**, 2076.
Zel'dovich, Ya. B. and Novikov, I. D. (1971). Translated by K. Thorne. *Relativistic Astrophysics*. Chicago University Press.

Znajek, R. L. (1977). *Mon. Not. R. Astron. Soc.*, **179**, 457.
Znajek, R. L. (1978). *Mon. Not. R. Astron. Soc.*, **182**, 639.

Chapter 7 The Kerr metric and its perturbations

Boyer, R. H. and Lindquist, R. W. (1967). *J. Math Phys.*, **8**, 265.
Carter, B. (1968). *Phys. Rev.*, **174**, 1559.
Carter, B. (1971). *Phys. Rev. Lett.*, **26**, 331.
Chandrasekhar, S. (1975). *Proc. R. Soc. Lond.*, **A343**, 289.
Chandrasekhar, S. (1976a). *Proc. R. Soc. Lond.*, **A348**, 39.
Chandrasekhar, S. (1976b). *Proc. R. Soc. Lond.*, **A349**, 1.
Chandrasekhar, S. (1976c). *Proc. R. Soc. Lond.*, **A349**, 571.
Chandrasekhar, S. (1977a). *Proc. R. Soc. Lond.*, **A358**, 405.
Chandrasekhar, S. (1977b). *Proc. R. Soc. Lond.*, **A358**, 421.
Chandrasekhar, S. (1977c). *Proc. R. Soc. Lond.*, **A358**, 441.
Chandrasekhar, S. and Detweiler, S. (1975). *Proc. R. Soc. Lond.*, **A345**, 145.
Chandrasekhar, S. and Detweiler, S. (1976a). *Proc. R. Soc. Lond.*, **A350**, 165.
Chandrasekhar, S. and Detweiler, S. (1976b). *Proc. R. Soc. Lond.*, **A352**, 325.
Chandrasekhar, S. and Friedman, J. L. (1972). *Astrophys. J.*, **175**, 379.
Chrzanowski, P. L. (1975). *Phys. Rev.*, **D11**, 2042.
Cohen, J. M. and Kegeles, L. S. (1974). *Phys. Rev.*, **D10**, 1070.
Detweiler, S. (1976). *Proc. R. Soc. Lond.*, **A349**, 217.
Detweiler, S. (1977). *Proc. R. Soc. Lond.*, **A352**, 381.
Goldberg, J. N., Macfarlane, A. J., Newman, E. T., Rohrlich, F. and Sudarshan, E. C. G. (1967). *J. Math. Phys.*, **8**, 2155.
Güven, R. (1977). *Phys. Rev.*, **D16**, 1706.
Kerr, R. P. (1963). *Phys. Rev. Lett.*, **11**, 237.
Kinnersley, W. (1975). In *General Relativity and Gravitation*, eds. G. Shaviv and J. Rosen, p. 109. John Wiley: New York.
Landau, L. D. and Lifschitz, E. M. (1975). *Classical Fields*. Pergamon: Oxford.
Newman, E. T. and Penrose, R. (1962). *J. Math. Phys.*, **3**, 566.
Page, D. N. (1977). *Phys. Rev.*, **D16**, 2402.
Penrose, R. (1968). *Battelle Rencontres*, eds. C. M. DeWitt and J. A. Wheeler, p. 121. W. A. Benjamin: New York.
Robinson, D. C. (1975). *Phys. Rev. Lett.*, **34**, 905.
Schwarzschild, K. (1916). *Berliner Sitzungsber.*, 189.
Starobinsky, A. A. (1973). *Zh. Eksp. Teor. Fiz.*, **64**, 48; (also *Sov. Phys.: JETP* (1974), **37**, 28).
Starobinsky, A. A. and Churilov, S. M. (1973). *Zh. Eksp. Teor. Fiz.*, **65**, 3; (also *Sov. Phys.: JETP* (1974), **38**, 1).
Teukolsky, S. A. (1972). *Phys. Rev. Lett.*, **29**, 114.
Teukolsky, S. A. (1973). *Astrophys. J.*, **185**, 635.
Teukolsky, S. A. and Press, W. H. (1974). *Astrophys. J.*, **193**, 443.
Unruh, W. (1973). *Phys. Rev. Lett.*, **31**, 1265.
Unruh, W. (1974). *Phys. Rev.*, **D2**, 10, 3194.

Chapter 8. References

Chapter 8 Black hole astrophysics

Abell, G. O. (1975). *Exploration of the Universe*, 3rd edition. Holt, Reinhart and Winston: New York.
Abramowicz, M., Jarosyński, M. and Sikrou, M. (1978). *Astron. Astrophys.*, **63**, 221.
Abt, H. A. and Levy, S. G. (1974). *Astrophys. J.*, **188**, 291.
Arnett, W. D. (1969). *Astrophys. Space Sci.*, **5**, 180.
Arnett, W. D. and Bowers, R. L. (1977). *Astrophys. J. Suppl. Ser.*, **33**, 415.
Arons, J., Kulsrud, R. M. and Ostriker, J. P. (1975). *Astrophys. J.*, **198**, 687.
Auer, L. H. and Woolf, N. J. (1965). *Astrophys. J.*, **142**, 182.
Baade, W. and Minkowski, R. (1954). *Astrophys. J.*, **119**, 206.
Bahcall, N. A. and Hausman, M. A. (1977). *Astrophys. J.*, **213**, 93.
Bahcall, J. N. and Ostriker, J. P. (1975). *Nature*, **256**, 23.
Bardeen, J. M. (1978). Paper in preparation.
Bardeen, J. M. and Petterson, J. A. (1975). *Astrophys. J. (Lett.)*, **195**, L65.
Barnothy, J. M. (1965). *Astron. J.*, **70**, 666.
Barrow, J. D. (1977). *Nature*, **267**, 117.
Begelman, M. C. (1977). *Mon. Not. R. Astron. Soc.*, **181**, 347.
Begelman, M. C. (1978). *Mon. Not. R. Astron. Soc.*, **184**, 53.
Bekenstein, J. D. (1973). *Astrophys. J.*, **183**, 657.
Bekenstein, J. D. (1976). *Astrophys. J.*, **210**, 544.
Berlitz, C. F. (1974). *The Bermuda Triangle*. Doubleday: Garden City, New York.
Bisnovatyi-Kogan, G. S. (1978). *Rev. Nuovo Cimento*, in press.
Bisnovatyi-Kogan, G. S. and Blinnikov, S. I. (1977). *Astron. Astrophys.*, **59**, 111.
Bisnovatyi-Kogan, G. S. and Ruzmaikin, A. A. (1974). *Astrophys. Space Sci.*, **28**, 45.
Bisnovatyi-Kogan, G. S. and Ruzmaikin, A. A. (1976). *Astrophys. Space Sci.*, **42**, 401.
Bisnovatyi-Kogan, G. S., Popov, Yu. P. and Samochin, A. A. (1976). *Astrophys. Space Sci.*, **41**, 287.
Blandford, R. D. (1977). *Mon. Not. R. Astron. Soc.*, **181**, 489.
Blandford, R. D. and Znajek, R. L. (1977). *Mon. Not. R. Astron. Soc.*, **179**, 433.
Blandford, R. D., McKee, C. F. and Rees, M. J. (1977). *Nature*, **267**, 211.
Bolton, C. T. (1972). *Nat. Phys. Sci.*, **240**, 124.
Bolton, C. T. (1975). *Astrophys. J.*, **200**, 269.
Bondi, H. (1952). *Mon. Not. R. Astron. Soc.*, **112**, 195.
Buff, J., Jernigan, G., Laufer, B., Bradt, H., Clark, G. W., Lewin, W. H. G., Matilsky, T., Mayer, W. and Primini, F. (1977). *Astrophys. J.*, **212**, 768.
Burbidge, G. R. (1972). *Comments Astrophys. Space. Sci.*, **4**, 105.
Burbidge, G. R., Jones, T. W. and O'Dell, S. L. (1974). *Astrophys. J.*, **193**, 43.
Callahan, P. S. (1977). *Astron. Astrophys.*, **59**, 127.
Cameron, A. G. W. (1971). *Nature*, **229**, 178.
Carr, B. J. (1975). *Astrophys. J.*, **201**, 1.
Carr, B. J. (1976). *Astrophys. J.*, **206**, 8.

Chapter 8. References

Carr, B. J. (1977a). *Mon. Not. R. Astron. Soc.*, **181**, 293.
Carr, B. J. (1977b). *Astron. Astrophys.*, **56**, 377.
Carr, B. J. and Hawking, S. W. (1974). *Mon. Not. R. Astron. Soc.*, **168**, 399.
Carr, B. J. and Rees, M. J. (1977). *Astron. Astrophys.*, **61**, 705.
Carter, B., Gibbons, G. W., Lin, D. N. C. and Perry, M. J. (1976). *Astron. Astrophys.*, **52**, 427.
Chandrasekhar, S. (1931). *Astrophys. J.*, **74**, 81.
Chandrasekhar, S. (1934). *Observatory*, **57**, 373, last paragraph.
Chapline, G. F. (1975). *Nature*, **253**, 251.
Christodolou, D. (1970). *Phys. Rev. Lett.*, **25**, 1596.
Clayton, D. D. (1968). *Principles of Stellar Evolution and Nucleosynthesis*. McGraw-Hill: New York.
Clayton, D. D., Newman, M. J. and Talbot, R. J. Jr (1975). *Astrophys. J.*, **201**, 489.
Cocke, W. J., Disney, M. J. and Taylor, D. J. (1969). *Nature*, **221**, 525.
Cohen, M. H., Kellermann, K. I., Shaffer, D. B., Linfield, R. P., Moffet, A. T., Romney, J. D., Seielstad, G. A., Pauliny-Toth, I. I. K., Preuss, E., Witzel, A., Schilizzi, R. T. and Geldzahler, B. J. (1977). *Nature*, **268**, 405.
Colgate, S. A. (1967). *Astrophys. J.*, **150**, 163.
Colgate, S. A. and Noerdlinger, P. D. (1971). *Astrophys. J.*, **165**, 509.
Dahlbacka, G. H., Chapline, G. F. and Weaver, T. A. (1974). *Nature*, **250**, 36.
Davidson, K. and Ostriker, J. P. (1973). *Astrophys. J.*, **179**, 585.
De Young, D. S. (1976). *Annu. Rev. Astron. Astrophys.*, **14**, 447.
Detweiler, S. L. (1978). *Astrophys. J.*, in press.
Detweiler, S. L. and Szedenits, E., Jr (1979). *Astrophys. J.*, in press.
Durisen, R. H. (1975). *Astrophys, J.*, **199**, 179.
Eardley, D. M. and Press, W. H. (1975). *Annu. Rev. Astron. Astrophys.*, **13**, 381.
Eichler, D. and Solinger, A. (1976). *Astrophys. J.*, **203**, 1.
Fabian, A. C., Maccagni, D., Rees, M. J. and Stoeger, W. R. (1976). *Nature*, **260**, 683.
Flannery, B. P. and Ulrich, R. K. (1977). *Astrophys. J.*, **212**, 533.
Forman, W., Jones, C. and Tananbaum, H. (1976). *Astrophys. J.*, **208**, 849.
Frank, J. H. and Rees, M. J. (1976). *Mon. Not. R. Astron. Soc.*, **176**, 633.
Frautschi, S. (1971). *Phys. Rev.* **D3**, 2821.
Giacconi, R. and Gursky, H. (1974). *X-Ray Astronomy*. Reidel: Dordrecht.
Gisler, G. R. (1976). *Astron. Astrophys.*, **51**, 137.
Gnedin, Yu. N. and Sunyaev, R. A. (1973). *Astron. Astrophys.*, **25**, 233.
Gold, T. (1968). *Nature*, **218**, 731.
Gott, J. R., III (1972). *Astrophys. J.*, **173**, 227.
Gott, J. R., III, Gunn, J. E., Schramm, D. N. and Tinsley, B. M. (1974). *Astrophys. J.*, **194**, 543.
Greenstein, J. L. (1976). *Astron. J.*, **81**, 323.
Greenstein, J. L., Oke, J. B. and Shipman, H. L. (1971). *Astrophys. J.*, **169**, 563.
Hack, M., Hutchings, J. B., Kondo, Y., McCluskey, G. E., Plavec, M. and Polidan, R. S. (1974). *Nature*, **249**, 534.

Chapter 8. References

Hagedorn, R. (1965). *Nuovo Cimento Suppl.*, **3**, 147.
Hartle, J. B. (1978). *Phys. Rep.*, in press.
Hawking, S. W. (1971). *Mon. Not. R. Astron. Soc.*, **152**, 75.
Hawking, S. W. (1974). *Nature*, **248**, 30.
Hawking, S. W. (1975). *Comm. Math. Phys.*, **43**, 189.
Hazard, C. and Mitton, S. A. (eds.) (1979). *Active Galactic Nuclei*. Cambridge University Press.
Hewish, A., Bell, S. J., Pilkington, J. D. H., Scott, P. F. and Collins, R. A. (1968). *Nature*, **217**, 709.
Hills, J. G. (1975). *Nature*, **254**, 295.
Hoyle, F. and Fowler, W. A. (1963). *Nature*, **197**, 533.
Hoyle, F. and Lyttleton, R. A. (1940). *Proc. Camb. Philos. Soc.*, **36**, 325 and 424.
Hulse, R. A. and Taylor, J. H. (1975). *Astrophys. J. (Lett.)*, **195**, L51.
Iben, I., Jr (1974). *Annu. Rev. Astron. Astrophys.*, **12**, 215.
Icke, V. (1977). *Nature*, **266**, 699.
Icke, V. (1978). *Astron. Astrophys.*, in press.
Illingworth, G. and King, I. R. (1977). *Astrophys. J. (Lett.)*, **218**, L109.
Ipser, J. R. (1969). *Astrophys. J.*, **158**, 17.
Ipser, J. R. and Price, R. H. (1977). *Astrophys. J.*, **216**, 578.
Jackson, A. A. and Ryan, M. P., Jr (1973). *Nature*, **245**, 88.
Jastrow, R. and Thompson, M. H. (1972). *Astronomy: Fundamentals and Frontiers*. Wiley: New York.
Jelley, J. V., Baird, G. A. and O'Mongain, E. (1977). *Nature*, **267**, 499.
Jennison, R. C. and Das Gupta, M. K. (1953). *Nature*, **172**, 996.
Kafka, P. and Mészáros, P. (1976). *Gen. Relativ. Grav.*, **7**, 841.
Katz, J. I. (1976). *Astrophys. J.*, **206**, 910.
Kellermann, K. I. and Pauliny-Toth, I. I. K. (1968). *Annu. Rev. Astron. Astrophys.*, **6**, 417.
Kellermann, K. I., Clark, B. G., Cohen, M. H., Shaffer, D. B., Broderick, J. J. and Jauncey, D. L. (1973). *Astrophys. J. (Lett.)*, **179**, L141.
Kellermann, K. I., Shaffer, D. B., Clark, B. G. and Geldzahler, B. J. (1977). *Astrophys. J. (Lett.)*, **214**, L61.
Kippenhahn, R. and Meyer-Hofmeister, E. (1977). *Astron. Astrophys.*, **54**, 539.
Kozlowski, M., Jarosyński, M. and Abramowicz, M. A. (1978). *Astron. Astrophys.*, **63**, 209.
Lamb, F. K. (1977). In Proceedings of the Eighth Texas Symposium on Relativistic Astrophysics, *Ann. N.Y. Acad. Sci.*, **302**, 482.
Lamb, F. K., Fabian, A. C., Pringle, J. E. and Lamb, D. Q. (1977). *Astrophys. J.*, **217**, 197.
Lamb, F. K., Pethick, C. J. and Pines, D. (1973). *Astrophys. J.*, **184**, 271.
Lamers, H. J. G. L. M. and Morton, D. C. (1976). *Astrophys. J. Suppl. Ser.*, **32**, 715.
Laplace, P. S. (1799). *Allgemeine geographische Ephemeriden*, vol. 1, ed. F. X. von Zach. English translation in S. W. Hawking and G. F. R. Ellis (1973). *The Large Scale Structure of Space–Time*, appendix A. Cambridge University Press.
Liang, E. P. T. and Price, R. H. (1977). *Astrophys. J.*, **218**, 247.
Lightman, A. P. (1974a). *Astrophys. J.*, **194**, 419.

Lightman, A. P. (1974b). *Astrophys. J.*, **194**, 429.
Lightman, A. P. and Eardley, D. M. (1974). *Astrophys. J. (Lett.)*, **187**, L1.
Lightman, A. P. and Fall, M. (1978). *Astrophys. J.*, **221**, 567.
Lightman, A. P. and Shapiro, S. L. (1977). *Astrophys. J.*, **211**, 244.
Lightman, A. P. and Shapiro, S. L. (1978). *Rev. Mod. Phys.*, **50**, 437.
Lightman, A. P., Press, W. H. and Odenwald, S. F. (1978). *Astrophys. J.*, **219**, 629.
Lin, D. N. C., Carr, B. J. and Fall, S. M. (1976). *Mon. Not. R. Astron. Soc.*, **177**, 51.
Lynden-Bell, D. (1969). *Nature*, **223**, 690.
Lynden-Bell, D. and Pringle, J. E. (1974). *Mon. Not. R. Astron. Soc.*, **168**, 603.
Lynden-Bell, D. and Rees, M. J. (1971). *Mon. Not. R. Astron. Soc.*, **152**, 461.
McCray, R. and Hatchett, S. P. (1975). *Astrophys. J.*, **199**, 196.
Manchester, R. N. and Taylor, J. H. (1977). *Pulsars*. W. H. Freeman: San Francisco.
May, M. M. and White, R. H. (1966). *Phys. Rev.*, **141**, 1232.
Mazurek, T. J., Meier, D. L. and Wheeler, J. C. (1977). *Astrophys. J.*, **213**, 518.
Meier, D. L., Epstein, R. I., Arnett, W. D. and Schramm, D. N. (1976). *Astrophys. J.*, **204**, 869.
Meikle, W. P. S. (1977). *Nature*, **269**, 41.
Menzel, D. H., Whipple, F. L. and De Vaucouleurs, G. (1970). *Survey of the Universe*. Prentice-Hall: Englewood Cliffs, New Jersey.
Mészáros, P. (1974). *Astron. Astrophys.*, **37**, 225.
Mészáros, P. (1975a). *Astron. Astrophys.*, **38**, 5.
Mészáros, P. (1975b). *Astron. Astrophys.*, **44**, 59.
Mészáros, P. (1975c). *Nature*, **258**, 583.
Michel, F. C. (1972). *Astrophys. Space Sci.*, **15**, 153.
Misner, C. W. (1968). *Battelle Rencontres in Mathematics and Physics*, eds. C. DeWitt and B. DeWitt. Benjamin Press: New York.
Misner, C. W., Thorne, K. S. and Wheeler, J. A. (1973). *Gravitation*. W. H. Freeman: San Francisco. Cited in text as MTW.
Morrison, P. (1969). *Astrophys. J. (Lett.)*, **157**, L73.
Novikov, I. D. and Thorne, K. S. (1973). In *Black Holes*, ed. C. DeWitt and B. DeWitt, p. 343. Gordon and Breach: New York.
Novikov, I. D. and Zel'dovich, Ya. B. (1966). *Nuovo Cimento Suppl.*, **4**, 810, addendum 2.
Oppenheimer, J. R. and Snyder, H. (1939). *Phys. Rev.*, **56**, 455.
Oppenheimer, J. R. and Volkoff, G. (1939). *Phys. Rev.*, **55**, 374.
Osaki, Y. (1974). *Publ. Astron. Soc. Jap.*, **26**, 429.
Ostriker, J. P. (1977). In Proceedings of the Eighth Texas Symposium on Relativistic Astrophysics, *Ann. N.Y. Acad. Sci.*, **302**, 229.
Ostriker, J. P. and Peebles, P. J. E. (1973). *Astrophys. J.*, **186**, 467.
Ostriker, J. P., Richstone, D. O. and Thuan, T. X. (1974). *Astrophys, J. (Lett.)*, **188**, L87.
Ozernoy, L. M. (1966). *Sov. Astron.*, **10**, 241.
Ozernoy, L. M. and Chertoprud, V. E. (1969). *Sov. Astron.*, **13**, 738.
Ozernoy, L. M. and Usov, V. V. (1973). *Astrophys. Space Sci.*, **25**, 149.
Pacini, F. (1967). *Nature*, **216**, 567.

Chapter 8. References

Paczynski, B. (1978). *Acta Astron.*, in press.
Page, D. N. (1976a). *Phys. Rev.*, **D13**, 198.
Page, D. N. (1976b). *Phys. Rev.*, **D14**, 3260.
Page, D. N. (1977). *Phys. Rev.*, **D16**, 2402.
Page, D. N. and Hawking, S. W. (1976). *Astrophys. J.*, **206**, 1.
Payne-Gaposchkin, C. and Haramundanis, K. (1970). *Introduction to Astronomy.* Prentice-Hall: Englewood Cliffs, New Jersey.
Peebles, P. J. E. (1971). *Physical Cosmology.* Princeton University Press: New Jersey.
Peebles, P. J. E. and Dicke, R. H. (1968). *Astrophys. J.*, **154**, 891.
Penrose, R. (1969). *Nuovo Cimento 1*, special number, 252.
Piddington, J. N. (1970). *Mon. Not. R. Astron. Soc.*, **148**, 131.
Piran, T. (1977). *Mon. Not. R. Astron. Soc.*, **180**, 45.
Pratt, J. P. and Strittmatter, P. A. (1976). *Astrophys. J. (Lett.)*, **204**, L29.
Press, W. H. and Gunn, J. E. (1973). *Astrophys. J.*, **185**, 397.
Pringle, J. E. (1976). *Mon. Not. R. Astron. Soc.*, **177**, 65.
Pringle, J. E. (1977). In Proceedings of the Eighth Texas Symposium on Relativistic Astrophysics, *Ann. N.Y. Acad. Sci.*, **302**, 6.
Pringle, J. E. and Rees, M. J. (1972). *Astron. Astrophys*, **21**, 1.
Pringle, J. E., Rees, M. J. and Pacholczyk, A. G. (1973). *Astron. Astrophys.*, **29**, 179.
Rappaport, S. A. and Joss, P. C. (1977). In Proceedings of the Eighth Texas Symposium on Relativistic Astrophysics, *Ann. N.Y. Acad. Sci.*, **302**, 460.
Readhead, A. C. S., Cohen, M. H. and Blandford, R. D. (1978). *Nature*, **272**, 131.
Rees, M. J. (1972). *Phys. Rev. Lett.*, **28**, 1660.
Rees, M. J. (1977a). *Nature*, **266**, 333.
Rees, M. J. (1977b). In Proceedings of the Eighth Texas Symposium on Relativistic Astrophysics, *Ann. N.Y. Acad. Sci.*, **302**, 613.
Refsdal, S. (1964). *Mon. Not. R. Astron. Soc.*, **128**, 295.
Robinson, I., Schild, A. and Schucking, E. L. (1965). *Quasi-Stellar Sources and Gravitational Collapse.* University of Chicago Press: Chicago.
Rothschild, R. E., Boldt, E. A., Holt, S. S. and Serlemitsos, P. J. (1974). *Astrophys. J. (Lett.)*, **189**, L13.
Rothschild, R. E., Boldt, E. A., Holt, S. S. and Serlemitsos, P. J. (1977). *Astrophys. J. (Lett.)*, **213**, 818.
Ryan, M. P., Jr (1972). *Astrophys. J. (Lett.)*, **177**, L79.
Salpeter, E. E. (1964). *Astrophys. J.*, **140**, 796.
Sargent, W. L. W., Young, P. J., Boksenberg, A., Shortridge, K., Lynds, C. R. and Hartwick, F. D. A. (1978). *Astrophys. J.*, **221**, 731.
Savonije, G. J. (1978). *Astron. Astrophys.*, **62**, 317.
Schmidt, M. (1963). *Nature*, **197**, 1040.
Shakura, N. I. and Sunyaev, R. A. (1976). *Mon. Not. R. Astron. Soc.*, **175**, 613.
Shapiro, S. L. (1973a). *Astrophys. J.*, **180**, 531.
Shapiro, S. L. (1973b). *Astrophys. J.*, **185**, 69.
Shapiro, S. L. and Lightman, A. P. (1976). *Astrophys. J.*, **204**, 555.
Shapiro, S. L., Lightman, A. P. and Eardley, D. M. (1976). *Astrophys. J.*, **204**, 187.

Chapter 8. References

Shields, G. A. and Wheeler, J. C. (1976). *Astrophys. Lett.*, **17**, 69.
Shklovsky, I. S. (1967). *Astrophys. J. (Lett.)*, **148**, L1.
Shvartsman, V. F. (1971). *Sov. Astron.*, **15**, 377.
Silk, J. and Arons, J. (1975). *Astrophys. J. (Lett.)*, **200**, L131.
Smak, J. (1971). New Directions and New Frontiers in Variable Star Research, IAU Colloquium No. 15, *Veröffentlichung der Remeis-Sternwarte, Bamberg*, **9**, 248.
Smak, J. (1976). *Acta Astron.*, **26**, 277.
Smarr, L. (1977). In Proceedings of the Eighth Texas Symposium on Relativistic Astrophysics, *Ann. N.Y. Acad. Sci.*, **302**, 569.
Spitzer, L. Jr (1962). *Physics of Fully Ionized Gases*, 2nd edition. Interscience: New York.
Spitzer, L. Jr and Shull, J. M. (1975). *Astrophys. J.*, **201**, 773.
Spitzer, L. Jr and Thuan, T. X. (1972). *Astrophys. J.*, **175**, 31.
Stothers, R. and Ezer, D. (1973). *Astrophys. Lett.*, **13**, 45.
Sturrock, P. A. and Barnes, C. (1972). *Astrophys. J.*, **176**, 31.
Sunyaev, R. A. (1973). *Sov. Astron.*, **16**, 941.
Taam, R. E., Bodenheimer, P. and Ostriker, J. P. (1978). *Astrophys. J.*, **222**, 269.
Tammann, G. A. (1977). In Proceedings of the Eighth Texas Symposium on Relativistic Astrophysics, *Ann. N.Y. Acad. Sci.*, **302**, 61.
Taylor, J. H. and Manchester, R. N. (1977). *Annu. Rev. Astron. Astrophys.*, **15**, 19.
Thomas, H.-C. (1977). *Annu. Rev. Astron. Astrophys.*, **15**, 127.
Thorne, K. S. (1978). In *Theoretical Principles in Astrophysics and Relativity*, ed. N. Lebovitz, W. Reid and P. Vandervoort. University of Chicago Press: Chicago.
Thorne, K. S. and Żytkow, A. (1977). *Astrophys. J.*, **212**, 832.
Thorstensen, J. R. and Partridge, R. B. (1975). *Astrophys. J.*, **200**, 527.
Thuan, T. X. and Oke, J. B. (1976). *Astrophys. J.*, **205**, 360.
Tinsley, B. M. and Larsen, R. B. (eds.) (1977). *The Evolution of Galaxies and Stellar Populations*. Yale University Observatory: New Haven.
Tremaine, S. D. (1976). *Astrophys. J.*, **203**, 345.
Trimble, V. and Greenstein, J. L. (1972). *Astrophys. J.*, **177**, 441.
Trimble, V. and Thorne, K. S. (1969). *Astrophys. J.*, **156**, 1013.
Vainer, B. V. and Nasel'skii, P. D. (1977). *Pisma Astron. Zh.*, **3**, 147.
van den Bergh, S. (1969). *Nature*, **224**, 891.
van den Bergh, S. (1975). *Annu. Rev. Astron. Astrophys.*, **13**, 217.
van den Heuvel, E. P. J. (1975). *Astrophys. J. (Lett.)*, **198**, L109.
van den Heuvel, E. P. J. and Heise, J. (1972). *Nature, Phys. Sci.*, **239**, 67.
van Riper, K. (1978). *Astrophys. J.*, **221**, 304.
Véron, M. P. (1975). *Astron. Astrophys.*, **41**, 423.
Webster, B. L. and Murdin, P. (1972). *Nature*, **235**, 37.
Weidemann, V. (1968). *Annu. Rev. Astron. Astrophys.*, **6**, 351.
Weidemann, V. (1977). *Astron. Astrophys.*, **59**, 411.
Weisskopf, M. C. and Sutherland, P. G. (1978). *Astrophys. J.*, **221**, 228.

Chapter 9. References

Wheeler, J. A. (1958). In *La structure et l'evolution de l'univers*, chapter by B. K. Harrison, M. Wakano and J. A. Wheeler, p. 124. Editions Stoops: Brussels.
Wheeler, J. A. (1964). In *Relativity, Groups, and Topology*, ed. C. M. DeWitt and B. S. DeWitt, p. 317. Gordon and Breach: New York.
Wheeler, J. A. (1968). *Am. Sci.*, **56**, 1.
Whelan, J. A. J., Mayo, S. K., Wickramasinghe, D. T., Murdin, P. G., Peterson, B. A., Hawarden, T. G., Longmore, A. J., Haynes, R. F., Goss, W. M., Simons, L. W., Caswell, J. L., Little, A. G. and McAdam, W. B. (1977). *Mon. Not. R. Astron. Soc.*, **181**, 259.
Willis, A. G., Strom, R. G. and Wilson, A. S. (1974). *Nature*, **250**, 625.
Wollman, E. R., Geballe, T. R., Lacy, J. H., Townes, C. H. and Rank, D. M. (1977). *Astrophys. J. (Lett.)*, **218**, L103.
Young, P. J. (1977). *Astrophys. J.*, **215**, 36.
Young, P. J., Westphal, J. A., Kristian, J., Wilson, C. P. and Landauer, F. P. (1978). *Astrophys. J.*, **221**, 721.
Zel'dovich, Ya. B. (1962). *Sov. Phys.: JETP*, **16**, 1163.
Zel'dovich, Ya. B. (1964). *Sov. Phys.: Dokl.*, **9**, 195.
Zel'dovich, Ya. B. and Guseynov, O. H. (1966). *Astrophys. J.*, **144**, 840.
Zel'dovich, Ya. B. and Novikov, I. D. (1967). *Relativistic Astrophysics*. Nauka: Moscow. English translation (1971). *Relativistic Astrophysics*, vol. 1, *Stars and Relativity*. University of Chicago Press: Chicago.
Zel'dovich, Ya. B. and Podurets, M. A. (1965). *Astron. Zh.*, **42**, 963; (1966). *Sov. Astron.*, **9**, 742.
Zel'dovich, Ya. B. and Starobinskii, A. A. (1976). *Sov. Phys.: JETP Lett.*, **24**, 571.
Zel'dovich, Ya. B., Starobinskii, A. A., Khlopov, M. Yu. and Checketkin, P. M. (1977). *Sov. Astron. Lett.*, **3**, 208.

Chapter 9   The big bang cosmology – enigmas and nostrums

Alfvén, H. (1971). *Physics Today*, Feb., p. 28.
Brans, C. and Dicke, R. H. (1961). *Phys. Rev.*, **124**, 925.
Carter, B. (1968). Unpublished preprint, Institute of Astronomy, Cambridge.
Carter, B. (1974). *Confrontation of Cosmological Theories with Observational Data*, ed. M. S. Longair, p. 291. Reidel: Dordrecht.
Dicke, R. H. (1961*a*). *Phys. Rev.*, **126**, 1875.
Dicke, R. H. (1961*b*). *Nature*, **192**, 440.
Dirac, P. A. M. (1938). *Proc. R. Soc. Lond.*, **A165**, 199.
Einstein, A. (1917). *Sitzungsber. Preuss. Akad. Wiss.*, 142.
Einstein, A. (1945). *The Meaning of Relativity*. Princeton University Press.
Einstein, A. and de Sitter, W. (1932). *Proc. Nat. Acad. Sci. U.S.A.*, **18**, 213.
Hawking, S. W. (1966). *Proc. R. Soc. Lond.*, **A294**, 511.
Hubble, E. (1924). *Annual Report of the Mount Wilson Observatory*, p. 94.
Hubble, E. (1926). *Astrophys. J.*, **64**, 321.
Jeans, J. (1928). *Astronomy and Cosmogony*. Cambridge University Press.
Jordan, P. (1955). *Schwerkraft und Weltall*. Vieweg und Sohn: Braunschweig.
Lemaître, G. (1933*a*). *C.R. Acad. Sci. (Paris)*, **196**, 903, 1085.
Lemaître, G. (1933*b*). *Ann. Soc. Sci. Brussels*, **A53**, 85.

Milne, E. A. (1935). *Relativity, Gravitation and World Structure*. Clarendon Press: Oxford.
Misner, C. W. (1967). *Nature*, **214**, 40.
Omnès, R. (1969). *Phys. Rev. Lett.*, **23**, 38.
Parker, E. N. (1975). *Astrophys. J.*, **202**, 523.
Penrose, R. (1965). *Phys. Rev. Lett.*, **14**, 57.
Sciama, D. W. (1953). *Mon. Not. R. Astron. Soc.*, **113**, 34.
Tolman, R. C. (1934). *Relativity, Thermodynamics and Cosmology*, p. 443. Clarendon Press: Oxford.
Tryon, E. P. (1973). *Nature*, **246**, 396.
van Maanen, A. (1916). *Astrophys. J.*, **44**, 210.

Chapter 10 Cosmology and the early universe

Alpher, R. and Herman, R. (1953). *Annu. Rev. Nucl. Sci.*, **2**, 1.
Belavin, A. A., Polyakov, A. A., Schwartz, A. S. and Tyupkin, Ya. S. (1975). *Phys. Lett.*, **B59**, 85.
Bogdanova, L. M. and Shapiro, I. S. (1974). *Zh. Eksp. Teor. Fiz. Pisma Red.*, **20**, 217; (English translation (1974) *Sov. Phys.: JETP Lett.*, **20**, 94).
Carr, B. J. (1976). *Astrophys. J.*, **206**, 8.
Carr, B. J. and Hawking, S. W. (1974). *Mon. Not. R. Astron. Soc.*, **168**, 399.
Carr, B. B. and Rees, M. (1977). Preprint.
Casimir, H. B. C. (1948). *Proc. Nederl. Acad. Wetensch.*, **60**, 793.
Chapline, G. F. (1975). *Nature*, **253**, 1.
Chernikov, N. A. and Tagirov, E. A. (1968). *Ann. Inst. H. Poincaré*, **9**, 109.
Chiu, H. Y. (1966). *Phys. Rev. Lett.*, **17**, 712.
Coleman, S. (1977). *Phys. Rev.*, **D15**, 2929.
Cowsik, R. and McClelland, J. (1972). *Phys. Rev. Lett.*, **29**, 669.
Dicke, R. H., Peebles, P. J. E., Roll, P. G. and Wilkinson, D. T. (1965). *Astrophys. J.*, **142**, 414.
Dicus, D. A., Kolb, E. M. and Teplitz, V. L. (1977). *Phys. Rev. Lett.*, **39**, 168.
Dirac, P. A. M. (1937). *Nature*, **139**, 323.
Doroshkevich, A. G. and Novikov, I. D. (1964). *Dokl. Acad. Nauk USSR*, **154**, 809; (English translation (1964) *Sov. Phys.: Dokl.*, **9**, 111).
Einstein, A. (1917). *Sitzungsber. Preuss. Akad. Wiss.*, **1**, 142; see also *Sobranie Nauchnykh Trudov*, vol. 1, p. 601. Nauka: Moscow (1965). (In Russian.)
Frampton, P. H. (1977). *Phys. Rev.*, **D15**, 2922.
Friedman, A. (1922). *Z. Physik*, **10**, 377.
Friedman, A. (1924). *Z. Physik*, **21**, 326.
Gamow, G. (1946). *Phys. Rev.*, **70**, 572.
Gamow, G. (1952). *Phys. Rev.*, **86**, 251.
Gamow, G. (1956). In *Vistas in Astronomy*, ed. X. Beer, vol. 2, p. 1726. Pergamon Press: Oxford, New York.
Gershtein, S. S. and Zeldovich, Ya. B. (1966). *Zh. Eksp. Teor. Fiz. Pisma Red.*, **4**, 174; (English translation (1966) *Sov. Phys.: JETP Lett.*, **4**, 120).
Grib, A. A. and Mamaev, S. G. (1969). *Yad. Phys.*, **10**, 1276. (In Russian.)

Chapter 10. References

Grishchuk, L. P. (1974). *Zh. Eksp. Theor. Fiz.*, **67**, 825; (English translation (1975) *Sov. Phys.: JETP*, **40**, 409).
Hawking, S. W. (1974). *Nature*, **248**, 30.
Hawking, S. W. (1975). *Commun. Math. Phys.*, **43**, 199.
Hawking, S. W. (1976). Quantum gravity and path integrals. (Talk given at the *8th International Conference on General Relativity and Gravitation*, Waterloo, Ontario, August 9th).
Hayashi C. (1950). *Progr. Theor. Phys.*, **5**, 224.
Kirzhnitz, D. A. (1972). *Zh. Eksp. Teor. Fiz. Pisma Red.*, **15**, 745; (English translation (1972) *Sov. Phys.: JETP Lett.*, **15**, 529).
Kirzhnitz, D. A. and Linde, A. D. (1972). *Phys. Lett.* **B42**, 471.
Kirzhnitz, D. A. and Linde, A. D. (1974). *Zh. Eksp. Teor. Fiz.*, **67**, 1263; (English translation (1975). *Sov. Phys.: JETP*, **40**, 628).
Kirzhnitz, D. A. and Linde, A. D. (1976). *Ann. Phys. (N.Y.)*, **101**, 195.
Kobzarev, I. Yu., Okun, L. B. and Voloshin, M. B. (1974). *Yad Phys.*, **20**, 1229.
Kuzmin, V. A. (1970). *Zh. Eksp. Teor. Fiz. Pisma. Red.*, **12**, 335; (English translation (1970). *Sov. Phys.: JETP Lett.*, **13**).
Lee, B. W. and Weinberg, S. (1977a) In Fermilab–Pub–77/41–THY.
Lee, B. W. and Weinberg, S. (1977b). *Phys. Rev. Lett.*, **39**, 165.
Lee, T. D. (1973). *Phys. Rev.*, **D8**, 1226.
Linde, A. D. (1974). *Zh. Eksp. Teor. Fiz. Pisma Red.*, **19**, 320; (English translation (1974). *Sov. Phys.: JETP Lett.*, **19**, 183).
Lukash, V. N. and Starobinsky, A. A. (1974). *Zh. Eksp. Teor. Fiz.*, **66**, 1515; (English translation (1974) *Sov. Phys.: JETP*, **39**, 742).
McKellar, A. (1941). *Publ. Dominion Astrophys. Observ.*, **7**, No. 15.
Milne, E. (1934). *Q. J. Math.*, **5**, 64.
Nadjozhin, D. K., Novikov, I. D. and Polnarev, A. G. (1978). *Astron. J. USSR*, **55**, No. 2. (In Russian).
Novikov, I. D., Starobinsky, A. A. and Zel'dovich, Ya. B. (1977). Primordial Black Holes. *8th International Conference on General Relativity and Gravitation*, Waterloo, Ontario).
Omnes, R. (1971a). *Astron. Astrophys.*, **10**, 228.
Omnes, R. (1971b). *Astron. Astrophys.*, **11**, 450.
Omnes, R. (1971c). *Astron. Astrophys.* **15**, 273.
Page, D. K. and Hawking, S. W. (1976). *Astrophys. J.*, **206**, 1.
Parker, L. (1968). *Phys. Rev. Lett.*, **21**, 562.
Parker, L. (1969). *Phys. Rev.*, **183**, 1057.
Pati, J. S. and Salam, A. (1973). *Phys. Rev. Lett.*, **31**, 661.
Penrose, R. (1964). In *Relativity, Groups and Topology*, eds. C. M. DeWitt and B. S. DeWitt, p. 565. Gordon and Breach: New York.
Penzias, A. A. and Wilson, R. W. (1965). *Astrophys. J.*, **142**, 419.
Polyakov, A. M. (1975). *Phys. Lett.*, **B59**, 78.
Sakharov, A. D. (1967). *Zh. Eksp. Teor. Fiz. Pisma Red.*, **5**, 32; (English translation (1967) *Sov. Phys.: JETP Lett.*, **5**, 24).
Sato, K. and Sato, H. (1975). *Prog. Theor. Phys.*, **54**, 912.
Schrödinger, E. (1939). *Physica*, **6**, 899.
Schrödinger, E. (1970). *Proc. R. Irish Acad.*, **46**, 25.

Shvartsman, V. F. (1969). *Zh. Eksp. Teor. Fiz. Pisma. Red.*, **9**, 315; (English translation (1969) *Sov. Phys.: JETP Lett.*, **9**, 184).
Steigman, G. (1971). In *General Relativity and Cosmology*, ed. R. K. Sachs. Academic Press: London, New York.
Steigman, G. (1973). In *Proc. IAU Symp.*, No. 63, Poland.
Steigman, G., Schramm, D. N. and Gunn, J. E. (1977). *Phys. Lett.*, **B66**, 202.
Szalay, A. S. and Marx, G. (1976). *Astron. Astrophys.*, **49**, 437.
Utiyama, R. and DeWitt, B. S. (1962). *J. Math. Phys.*, **3**, 608.
Veiner, B. V. and Nasselsky, P. D. (1977). *Astron. J. USSR Lett.*, **3**, 147; (English translation (1977) *Sov. Astron. Lett.*, **3**, 76).
Visotsky, M. I., Dolgov, A. D. and Zel'dovich, Ya. B. (1977). *Zh. Eksp. Teor. Fiz. Pisma Red.*, **26**, 200; (English translation (1977) *Sov. Phys.: JETP Lett.*, **26**).
Wheeler, J. A. (1962). *Geometrodynamics*. Academic Press: London, New York.
Zel'dovich, Ya. B. (1970). *Zh. Eksp. Teor. Fiz. Pisma Red.*, **12**, 443; (English translation (1970) *Sov. Phys.: JETP Lett.*, **12**, 307).
Zel'dovich, Ya. B. (1972). *Mon. Not. R. Astron. Soc.*, **160**, 1L.
Zel'dovich, Ya. B. (1973). *Zh. Eksp. Teor. Fiz.*, **64**, 58; (English translation (1973) *Sov. Phys.: JETP*, **37**, 33).
Zel'dovich, Ya. B. (1976). *Zh. Eksp. Teor. Fiz. Pisma Red.*, **24**, 25; (English translation (1976) *Sov. Phys.: JETP Lett.*, **24**).
Zel'dovich, Ya. B. (1977a). *Annu. Rev. Fluid Mech.*, **9**, 215.
Zel'dovich, Ya. B. (1977b). *Usp. Phys. Nauk USSR*, **123**, 487; (English translation (1977) *Sov. Phys.: Uspekhi*).
Zel'dovich, Ya. B. and Novikov, I. D. (1967). *Astron. J. USSR*, **44**, 663; (English translation (1967) *Sov. Astron.: AJ*, **11**, 526).
Zel'dovich, Ya. B. and Pitaevsky, L. P. (1971). *Commun. Math. Phys.*, **23**, 185.
Zel'dovich, Ya. B. and Smorodinsky, Ya. A. (1961). *Zh. Eksp. Teor. Fiz.*, **41**, 907; (English translation (1962) Sov. Phys.: *JETP*, **14**, 647).
Zel'dovich, Ya. B. and Starobinsky, A. A. (1971). *Zh. Eksp. Teor. Fiz.*, **61**, 2161; (English translation (1972). *Sov. Phys.: JETP*, **34**, 1159).
Zel'dovich, Ya. B. and Starobinsky, A. A. (1975). In *Problems of Nuclear Physics and Elementary Particles Physics*; p. 141. Nauka: Moscow. (In Russian.)
Zel'dovich, Ya. B. and Starobinsky, A. A. (1976). *Zh. Eksp. Teor. Fiz. Pisma Red.*, **24**, 616 (English translation (1976) *Sov. Phys.: JETP Lett.*, **24**).
Zel'dovich, Ya. B., Okun, C. B. and Pikelner, S. B. (1965). *Usp. Phys. Nauk USSR*, **87**, 113; (English translation (1966) *Sov. Phys.: Uspekhi*, **8**, 702).
Zel'dovich, Ya. B., Illarionov, A. F. and Sunyaev, R. A. (1972). *Zh. Eksp. Teor Fiz.*, **62**, 1217; (English translation (1972) *Sov. Phys.: JETP*, **35**, 643).
Zel'dovich, Ya. B., Kobzarev, I. Yu. and Okun, L. B. (1974). Preprint IPM No. 15; (1974) *Zh. Eksp. Teor. Fiz.*, **67**, 3; (English translation (1974) *Sov. Phys.: JETP*, **40**, 1).
Zel'dovich, Ya. B., Starobinsky, A. A., Khlopov, M. Yu. and Chechetkin, V. M. (1977). *Astron. J. USSR Lett.*, **3**, 208. (In Russian.)

Chapter 11 Anisotropic and inhomogeneous relativistic cosmologies

Anile, A. M. (1974). Anisotropic expansion of the universe and the anisotropy and linear polarisation of the cosmic microwave background. *Astrophys. Space Sci.*, **29**, 415.

Anile, A. M. and Breuer, R. (1977). Polarisation transport in anisotropic universes. *Astrophys. J.*, **217**, 353.

Anile, A. M. and Motta, S. (1976). Perturbation of the general Robertson–Walker metric and angular variations of the cosmic microwave background. *Astrophys. J.*, **207**, 685.

Barnes, A. (1978). A class of homogeneous Einstein–Maxwell fields. *J. Phys.*, **A11**, 1303.

Barrow, J. (1976a). Light elements and the isotropy of the universe. *Mon. Not. R. Astron. Soc.*, **175**, 359.

Barrow, J. (1976b). A chaotic cosmology. *Nature*, **267**, 117.

Barrow, J. (1977). The synthesis of light elements in turbulent cosmologies. *Mon. Not. R. Astron. Soc.*, **178**, 625.

Barrow, J. D. and Carr, B. J. (1978). Primordial black hole formation in an anisotropic universe. *Mon. Not. R. Astron. Soc.*, **182**, 537.

Barrow, J. D. and Matzner, R. A. (1977). The homogeneity and isotropy of the universe. *Mon. Not. R. Astron. Soc.*, **181**, 719.

Batakis, N. and Cohen, J. M. (1972). Closed anisotropic cosmological models. *Ann. Phys. (N.Y.)*, **73**, 578.

Belinskii, V. A. and Khalatnikov, I. M. (1972). On the influence of scalar and vector fields on the cosmological singularity character. *Zh. Eksp. Teor. Fiz.*, **63**, 1121. (English translation in *Sov. Phys.: JETP*, **36**, 591.)

Belinskii, V. A. and Khalatnikov, I. M. (1975). Effect of viscosity on the nature of the cosmological solution. *Zh. Eksp. Teor. Fiz.*, **69**, 401 (English translation in *Sov. Phys.: JETP*, **42**, 205.)

Belinskii, V. A., Lifschitz, E. M. and Khalatnikov, I. M. (1970). The oscillatory regime near the singularity in relativistic cosmology. *Usp. Fiz. Nauk.*, **102**, 463. (English translation in *Sov. Phys.: Usp.*, **13**, 475; and *Adv. Phys.*, **19**, 525.)

Berger, B. K. (1974). Quantum gravitation creation in a model universe. *Ann. Phys. (N.Y.)*, **83**, 203.

Berger, B. K. (1975a). Quantum cosmology: exact solution for the Gowdy T^3 model. *Phys. Rev.*, **D11**, 2770.

Berger, B. K. (1975b). Scalar particle creation in an anisotropic universe. *Phys. Rev.*, **D12**, 368.

Bianchi, L. (1897). Sugli spazii a tre dimensioni che ammettono un gruppo continuo di movimenti. *Mem. di Mat. Soc. Ital. Sci.* **11**, 267. Reprinted (1952) in *Opere*, vol. IX, ed. A. Maxia. Editioni Cremonese, Rome.

Bogoyavlenskii, O. I. (1976). Some properties of the type IX cosmological model with moving matter. *Zh. Eksp. Teor. Fiz.*, **70**, 361. (English translation in *Sov. Phys.: JETP.*, **43**, 187.)

Bogoyavlenskii, O. I. and Novikov, S. P. (1973). Singularities of the cosmological model of the Bianchi IX type according to the qualitative theory of differential equations. *Zh. Eksp. Teor. Fiz.*, **64**, 1475. (English translation in *Sov. Phys.: JETP*, **37**, 747.)

Chapter 11. References

Bonanos, S. (1972). Stability of homogeneous universes. *Commun. Math. Phys.*, **26**, 259.

Bondi, H. (1960). *Cosmology*. Cambridge University Press.

Bonnor, W. B. (1972). A non-uniform relativistic cosmological model. *Mon. Not. R. Astron. Soc.*, **159**, 261.

Bonnor, W. B. (1974). The evolution of inhomogeneous cosmological models. *Mon. Not. R. Astron. Soc.*, **167**, 55.

Bonnor, W. B. and Tomimura, N. (1976). Evolution of Szekeres's cosmological models. *Mon. Not. R. Astron. Soc.*, **175**, 85.

Bonnor, W. B., Sulaiman, A. H. and Tomimura, N. (1977). Szekeres's space-times have no Killing vectors. *Gen. Relativ. Grav.*, **8**, 549.

Borchkhadze, T. M. and Kogoshvili, N. G. (1976). The anisotropy of spiral galaxy orientation. *Astron. Astrophys.* **53**, 431.

Brown, F. G. (1968). The forms and position angles of 4287 galaxies in Hydra, Ursa Major, Virgo and Eridanus. *Mon. Not. R. Astron. Soc.*, **138**, 327.

Caderni, N. and Fabbri, R. (1978). Production of entropy and viscous damping of anisotropy in homogeneous cosmological models: Bianchi type I spaces. *Nuovo Cimento.* **44B**, 228.

Caderni, N., Fabbri, R., Melchiori, B., Melchiori, F. and Natale, V. (1978). Polarisation of the microwave background radiation: I. Anisotropic cosmological expansion and evolution of the polarisation states. *Phys. Rev.*, **D17**, 1901.

Cahill, M. E. and Taub, A. H. (1971). Spherically symmetric similarity solutions of the Einstein field equations for a perfect fluid. *Commun. Math. Phys.*, **21**, 1.

Carr, B. J. and Hawking, S. W. (1974). Black holes in the early universe. *Mon. Not. R. Astron. Soc.*, **168**, 399.

Clarke, C. J. S. (1974). Singularities in globally hyperbolic space–times. *Commun. Math. Phys.*, **41**, 65.

Cohn, P. M. (1957). *Lie Groups*. Cambridge University Press.

Collins, C. B. (1971). More qualitative cosmology. *Commun. Math. Phys.*, **23**, 137.

Collins, C. B. (1972). Qualitative magnetic cosmology. *Commun. Math. Phys.*, **27**, 37.

Collins, C. B. (1974). Tilting at cosmological singularities. *Commun. Math. Phys.*, **39**, 131.

Collins, C. B. (1977). Global structure of the Kantowski–Sachs cosmological models. *J. Math. Phys.*, **18**, 2116.

Collins, C. B. and Ellis, G. F. R. (1979). Singularities in Bianchi cosmologies. In *Bianchi Universes and Relativistic Cosmology*, ed. R. Ruffini, in press.

Collins, C. B. and Hawking, S. W. (1972). The rotation and distortion of the universe. *Mon. Not. R. Astron. Soc.*, **162**, 307.

Collins, C. B. and Hawking, S. W. (1973). Why is the universe isotropic? *Astrophys. J.*, **180**, 317.

Collins, C. B. and Stewart, J. M. (1971). Qualitative cosmology. *Mon. Not. R. Astron. Soc.*, **153**, 419.

de Vaucouleurs, G. (1976). Supergalactic studies V. *Astrophys. J.*, **205**, 13.

Chapter 11. References

Dodson, C. T. J. (1972). A spherically symmetric inhomogeneous cosmological model. *Astrophys. J.*, **172**, 1.

Doroshkevich, A. G. (1965). Model of a universe with a uniform magnetic field. *Astrofizika*, **1**, 255. (English translation in *Astrophysics*, **1**, 138.)

Doroshkevich, A. G., Lukash, V. N. and Novikov, I. D. (1971). Impossibility of mixing in a cosmological model of the Bianchi IX type. *Zh. Eksp. Teor. Fiz.*, **60**, 1201. (English translation in *Sov. Phys.: JETP*, **33**, 649.)

Doroshkevich, A. G., Lukash, V. N. and Novikov, I. D. (1973). Isotropisation of homogeneous cosmological models. *Zh. Eksp. Teor. Fiz.*, **64**, 1457. (English translation in *Sov. Phys.: JETP*, **37**, 739.)

Doroshkevich, A. G., Lukash, V. N. and Novikov, I. D. (1974). Primordial radiation in a homogeneous but anisotropic universe. *Astron. Zh.*, **51**, 940. (English translation in *Sov. Astron.*, **18**, 554.)

Doroshkevich, A. G., Zel'dovich, Ya. B. and Novikov, I. D. (1968). Weakly interacting particles in the anisotropic cosmological model. *Zh. Eksp. Teor. Fiz.*, **53**, 644. (English translation in *Sov. Phys.: JETP*, **26**, 408.)

Dunn, K. A. and Tupper, B. O. J. (1976). A class of Bianchi type VI cosmological models with electromagnetic field. *Astrophys. J.*, **204**, 322.

Dyer, C. C. (1976). The gravitational perturbation of the cosmic background radiation by density concentrations. *Mon. Not. R. Astron. Soc.*, **175**, 429.

Eardley, D. M. (1974). Self-similar spacetimes; geometry and dynamics. *Commun. Math. Phys.*, **37**, 287.

Eardley, D., Liang, E. and Sachs, R. K. (1971). Velocity-dominated singularities in irrotational dust cosmologies. *J. Math. Phys.*, **13**, 99.

Eisenhart, L. P. (1933). *Continuous Groups of Transformations.* Princeton University Press: Princeton, New Jersey. (Reprinted (1961) by Dover: New York.)

Eisenstadt, J. (1977). Density constraint on local inhomogeneities of a Robertson–Walker universe. *Phys. Rev.*, **D16**, 927.

Ellis, G. F. R. (1967). Dynamics of pressure-free matter in general relativity. *J. Math. Phys.*, **8**, 1171.

Ellis, G. F. R. and King, A. R. (1974). Was the big bang a whimper? *Commun. Math. Phys.*, **38**, 119.

Ellis, G. F. R. and MacCallum, M. A. H. (1969). A class of homogeneous cosmological models. *Commun. Math. Phys.*, **12**, 108.

Ellis, G. F. R. and Sciama, D. W. (1971). Global and non-global problems in cosmology. In *Studies in Relativity (Synge Festschrift)*, ed. L. O'Raiffeartaigh. Oxford University Press.

Ellis, G. F. R., Maartens, R. and Nel, S. D. (1978). The expansion of the universe. *Mon. Not. R. Astron. Soc.*, **184**, 439.

Estabrook, F. B., Wahlquist, H. D. and Behr, C. G. (1968). Dyadic analysis of spatially homogeneous world models. *J. Math. Phys.*, **9**, 497.

Evans, A. B. (1978). Correlation of relativistic and Newtonian cosmology. *Mon. Not. R. Astron. Soc.*, **183**, 727.

Fennelly, A. J. (1977). Anisotropy in the Hubble parameter and large-scale cosmological imhomogeneity. *Mon. Not. R. Astron. Soc.*, **181**, 121.

Fesenko, B. I. (1976). Organisation of the position angles of galaxies belonging to systems. *Astron. Zh.*, **53**, 1153. (English translation in *Sov. Astron.*, **20**, 650.)

Chapter 11. References

Gillespie, A. R. (1975). Investigations into reported anisotropies in radio source counts and spectra at 1421 MHz. *Mon. Not. R. Astron. Soc.*, **170**, 541.

Goethals, M. (1975). Modeles cosmologiques du type de Kantowski and Sachs I. *Ann. Soc. Sci. de Bruxelles*, **89**, 50.

Golden, L. M. (1974). Observational selection in the identification of Quasars and claims for anisotropy. *Observatory*, **94**, 122.

Goorjian, P. (1975). Electromagnetic plane wave perturbations in Kasner cosmologies. *Phys. Rev.*, **D12**, 2978.

Gowdy, R. H. (1974). Vacuum spacetimes with two-parameter spacelike isometry groups and compact invariant hypersurfaces: Topologies and boundary conditions. *Ann. Phys. (N.Y.)*, **83**, 203.

Grishchuk, L. P. (1970). On spatially homogeneous gravitational fields. *Dokl. AN SSR*, **190**, 1066. (English translation in *Sov. Phys.: Dokl.*, **15**, 130.)

Grishchuk, L. P., Doroshkevich, A. G. and Lukash, V. N. (1971). Model of mixmaster universe with arbitrarily moving matter. *Zh. Eksp. Teor. Fiz.*, **61**, 3. (English translation in *Sov. Phys.: JETP*, **34**, 1.)

Gursey, F. (1964). Introduction to group theory. In *Relativity, Groups and Topology*, ed. C. DeWitt and B. DeWitt. Blackie and Son: London and Glasgow.

Harvey, A. and Tsoubelis, D. (1977). Exact Bianchi IV cosmological model. *Phys. Rev.*, **D15**, 2734.

Hawking, S. W. (1969). On the rotation of the universe. *Mon. Not. R. Astron. Soc.*, **142**, 129.

Hawking, S. W. and Ellis, G. F. R. (1973). *The Large Scale Structure of Space-Time*. Cambridge University Press.

Hawking, S. W. and Tayler, R. J. (1966). Helium production in an anisotropic big-bang cosmology. *Nature*, **209**, 1278.

Hawley, D. L. and Peebles, P. J. E. (1975). Distribution of observed orientations of galaxies. *Astron. J.*, **80**, 477.

Heckmann, O. and Schucking, E. L. (1962). In *Gravitation: An Introduction to Current research*, ed. L. Witten. Wiley: New York.

Hu, B. L. (1974). Scalar waves in the mixmaster universe: II. Particle creation. *Phys. Rev.*, **D9**, 3263.

Hughston, L. P. and Jacobs, K. C. (1969). Homogeneous electromagnetic and massive-vector-meson fields in Bianchi cosmologies. *Astrophys. J.*, **160**, 147.

Jaakkola, T., Karoji, H., Le Denmat, G., Moles, M., Nottale, L., Vigier, J.-P. and Pecker, J.-C. (1976). Additional evidence and possible interpretation of angular redshift anisotropy. *Mon. Not. R. Astron. Soc.*, **177**, 191.

Jacobs, K. C. (1968). Spatially homogeneous and Euclidean cosmological models with shear. *Astrophys. J.*, **153**, 661.

Jacobs, K. C. (1977). Source-free Bianchi electromagnetic fields. Preprint *MPI-PAE-Astro 121*. Max Planck Institut: Munich.

Johri, V. B. (1972). Gravitational stability in an anisotropic universe. *Tensor*, **25**, 241.

Jones, B. J. T. (1976). The origin of galaxies: a review of theoretical developments and their confrontation with observation. *Rev. Mod. Phys.*, **48**, 107.

Chapter 11. References

Kantowski, R. (1966). Some relativistic cosmological models. Ph.D. thesis. University of Texas.

Kantowski, R. (1969). Corrections in the luminosity redshift relations of the homogeneous Friedmann models. *Astrophys. J.*, **155**, 89.

Kantowski, R. and Sachs, R. K. (1966). Some spatially homogeneous anisotropic relativistic cosmological models. *J. Math. Phys.*, **7**, 443.

King, A. R. (1974a). Generalised shear-free singularities. *Gen. Relativ. Grav.*, **5**, 371.

King, A. R. (1974b). New types of singularity in general relativity: the general cylindrically symmetric stationary solution. *Commun. Math. Phys.*, **38**, 157.

King, A. R. (1975). Instability of intermediate singularities in general relativity. *Phys. Rev.*, **D11**, 763.

King, A. R. and Ellis, G. F. R. (1973). Tilted homogeneous cosmological models. *Commun. Math. Phys.*, **31**, 209.

Klimek, Z. (1975). Dissipation in early universe: I. Bianchi type I and V models. *Acta Astron.*, **25**, 79.

Kobolov, V. M., Reinhardt, M. and Sazonov, V. N. (1976). A test of the isotropy of the universe. *Astrophys. Lett.*, **17**, 183.

Kompaneets, A. S. and Chernov, A. S. (1964). Solution of the gravitation equations for a homogeneous anisotropic model. *Zh. Eksp. Teor. Fiz.*, **47**, 1939. (English translation in *Sov. Phys.: JETP*, **20**, 1303.)

Korkina, M. P. (1975). Conditions for fitting a Friedman universe to empty space. *Astron. Zh.*, **52**, 299. (English translation in *Sov. Astron.*, **19**, 185.)

Kristian, J. and Sachs, R. K. (1966). Observations in cosmology. *Astrophys. J.*, **143**, 679.

Kubo, M. (1975). Evolution and age of anisotropic cosmological models with cosmological constant. *Publ. Astron. Soc. Jap.*, **26**, 355.

Lapedes, A. S. (1976). Thermal particle production in two Taub-NUT spacetimes. *Commun. Math. Phys.*, **51**, 121.

Liang, E. P. T. (1972). Velocity-dominated singularities in irrotational hydrodynamic cosmological models. *J. Math. Phys.*, **13**, 386.

Liang, E. P. T. (1975). Anisotropy and large scale density inhomogeneities in non-rotating cosmologies. *Phys. Lett.*, **A51**, 141.

Lifschitz, E. M., Lifschitz, I. M. and Khalatnikov, I. M. (1970). Asymptotic analysis of oscillatory mode of approach to a singularity in homogeneous cosmological models. *Zh. Eksp. Teor. Fiz.*, **59**, 322. (English translation in *Sov. Phys.: JETP*, **32**, 173.)

Lukash, V. N. (1974). Gravitational waves that conserve the homogeneity of space. *Zh. Eksp. Teor. Fiz.*, **67**, 1594. (English translation in *Sov. Phys.: JETP*, **40**, 792.)

Lukash, V. N. (1976). Physical interpretation of homogeneous cosmological models. *Nuovo Cimento*, **B35**, 268.

Lukash, V. N., Novikov, I. D., Starobinskii, A. A. and Zel'dovich, Ya. B. (1976). Quantum effects and evolution of cosmological models. *Nuovo Cimento*, **B35**, 293.

Luminet, J. P. (1977). Courbure scalaire intrinseque dans les espaces–temps spatialement homothetiques. *C. R. Acad. Sci. (Paris)*, **A285**, 821.

Lund, F. (1973). Effective homogeneity of some inhomogeneous cosmologies. *Phys. Rev.*, **D8**, 4229.
MacCallum, M. A. H. (1971). A class of homogeneous cosmological models: III. Asymptotic behaviour. *Commun. Math. Phys.*, **20**, 57.
MacCallum, M. A. H. (1972). On 'diagonal' Bianchi cosmologies. *Phys. Lett.*, **A40**, 385.
MacCallum, M. A. H. (1973). Cosmological models from a geometric point of view. In *Cargese Lectures*, vol. 6, ed. E. Schatzman. Gordon and Breach: New York.
MacCallum, M. A. H. (1975). Quantum cosmological models. In *Quantum Gravity: an Oxford symposium*, ed. C. J. Isham, R. Penrose and D. W. Sciama. Oxford University Press.
MacCallum, M. A. H. and Ellis, G. F. R. (1970). A class of homogeneous cosmological models: II. Observations. *Commun. Math. Phys.*, **19**, 31.
MacCallum, M. A. H. and Taub, A. H. (1972). Variational principles and spatially-homogeneous universes, including rotation. *Commun. Math. Phys.*, **25**, 173.
MacCallum, M. A. H. and Taub, A. H. (1973). The averaged Lagrangian and high-frequency gravitational waves. *Commun. Math. Phys.*, **30**, 153.
MacCallum, M. A. H., Stewart, J. M. and Schmidt, B. G. (1970). Anisotropic stresses in homogeneous cosmologies. *Commun. Math. Phys.*, **17**, 343.
Machalski, J. (1977). A new statistical investigation of the problem of isotropy in radio source populations at 1400 MHz: I. Spatial density, source counts and spectral index distributions in a new GB sky survey. *Astron. Astrophys.*, **56**, 53.
McIntosh, C. B. G. (1976). Homothetic motions in general relativity. *Gen. Relativ. Grav.*, **7**, 199.
Matzner, R. A. (1969). The evolution of anisotropy in non-rotating Bianchi type V cosmologies. *Astrophys. J.*, **157**, 1085.
Matzner, R. A. (1971). Closed rotating cosmologies containing matter described by the kinetic theory. A. Formalism. *Ann. Phys. (N.Y.)*, **65**, 438; B. Small anisotropy calculations, application to observations. *Ann. Phys. (N.Y.)*, **65**, 482.
Matzner, R. A. (1972). Dissipative effects in the expansion of the universe: II. A multicomponent model for neutrino dissipation of anisotropy in the early universe. *Astrophys. J.*, **171**, 433.
Matzner, R. A. and Chitre, D. M. (1971). Rotation does not enhance mixing in the mixmaster universe. *Commun. Math. Phys.*, **22**, 173.
Mavrides, S. (1976). Anomalous Hubble expansion and inhomogeneous cosmological models. *Mon. Not. R. Astron. Soc.*, **177**, 709.
Mavrides, S. and Tarantola, A. (1977). Local supercluster and anomalous Hubble expansion. *Gen. Relativ. Grav.*, **8**, 665.
Melvin, M. A. (1975). Homogeneous axial cosmologies with electromagnetic fields and dust. *Ann. N.Y. Acad. Sci.*, **262**, 253.
Misner, C. W. (1967). Taub–NUT space as a counter-example to almost anything. In *Relativity Theory and Astrophysics*, vol. 1, ed. J. Ehlers. American Mathematical Society: Providence, Rhode Island.
Misner, C. W. (1968). The isotropy of the universe. *Astrophys. J.*, **151**, 431.

Chapter 11. References

Misner, C. W. (1969a). Relativistic fluids in cosmology. In *Colloques Internationaux de C.N.R.S. No. 220.*

Misner, C. W. (1969b). The mixmaster universe. *Phys. Rev. Lett.*, **22**, 1071.

Misner, C. W. (1969c). Quantum cosmology I. *Phys. Rev.*, **186**, 1319.

Misner, C. W., Thorne, K. S. and Wheeler, J. A. (1973). *Gravitation.* W. H. Freeman and Sons: San Francisco.

Moser, A. R., Matzner, R. A. and Ryan, M. P. Jr (1973). Numerical solutions for symmetric Bianchi type IX universes. *Ann. Phys. (N.Y.)*, **79**, 558.

Nariai, H. (1977). On a quantised scalar field in some Bianchi type I universes. *Prog. Theor. Phys.*, **58**, 560.

Novikov, S. P. (1972). On some properties of cosmological models. *Zh. Eksp. Teor. Fiz.*, **62**, 1977. (English translation in *Sov. Phys.: JETP*, **35**, 1031.)

Olson, D. W. (1978). Helium production and limits on the anisotropy of the universe. *Astrophys. J.*, **219**, 777.

Ozsvath, I. (1965). New homogeneous solutions of Einstein's field equations with incoherent matter obtained by a spinor technique. *J. Math. Phys.*, **6**, 590.

Peebles, P. J. E. (1971). *Physical Cosmology.* Princeton University Press: Princeton, New Jersey.

Peebles, P. J. E. and Groth, E. J. (1975). Statistical analysis of catalogs of extragalactic objects: V. Three point correlation function for the galaxy distribution in the Zwicky catalog. *Astrophys. J.*, **196**, 1.

Perko, T. E., Matzner, R. A. and Shepley, L. C. (1972). Galaxy formation in anisotropic cosmologies. *Phys. Rev.*, **D6**, 969.

Pessa, E. (1977). Scalar particle production near the singularity in an anisotropic universe: I. Scalar field theory. *Nuovo Cimento.*, **B37**, 155; II. Mass creation by gravitational fields. *Nuovo Cimento.*, **B41**, 99.

Petrov, A. Z. (1969). *Einstein Spaces.* Pergamon: London. (Also in German, (1964) *Einstein-Raume.* Akademie-Verlag, Berlin; and in Russian, (1966) Nauka.)

Rasband, S. N. (1971). Expansion anisotropy and the spectrum of the cosmic background radiation. *Astrophys. J.*, **170**, 1.

Raychaudhuri, A. (1958). An anisotropic cosmological solution in general relativity. *Proc. Phys. Soc.*, **72**, 263.

Rees, M. J. (1968). Polarisation and spectrum of the primeval radiation in an anisotropic universe. *Astrophys. J.*, **153**, 1.

Rees, M. J. (1972). Origin of the cosmic microwave background in a chaotic universe. *Phys. Rev. Lett.*, **28**, 1669.

Reinhardt, M. (1971). Orientation of galaxies and a magnetic 'urfield'. *Astrophys. Space Sci.*, **10**, 363.

Roeder, R. C. (1975a). Significance of the angular diameter redshift relation. *Nature*, **255**, 124.

Roeder, R. C. (1975b). Apparent magnitudes, redshifts and inhomogeneities in the universe. *Astrophys. J.*, **196**, 671.

Rowan-Robinson, M. (1976). Quasars and the cosmological distance scale. *Nature*, **262**, 97.

Rowan-Robinson, M. (1977). Aether drift detected at last. *Nature*, **270**, 9.

Chapter 11. References

Ruban, V. A. (1970). Spherically symmetric T-models in the general theory of relativity. *Zh. Eksp. Teor. Fiz.*, **56**; 1914. (English translation in *Sov. Phys.: JETP*, **29**, 1027.)

Ruban, V. A. (1976). On the influence of massless scalar and vector electromagnetic fields on the singularity character in anisotropic cosmology. Preprint 291. Leningrad Nuclear Physics Institute.

Rubin, V. A., Thonnard, N., Ford, W. K. and Roberts, M. S. (1976). Motion of the galaxy and the local group determined from the velocity anisotropy of distant ScI galaxies: II. The analysis for the motion. *Astron. J.*, **81**, 719.

Ruzmaikin, A. A. and Sokoloff, D. D. (1977). The interpretation of rotation measures of extragalactic radio sources. *Astron. Astrophys.*, **58**, 247.

Ryan, M. P. Jr (1971). Qualitative cosmology: Diagrammatic solutions for Bianchi type IX universes: I. The symmetric case. *Ann. Phys. (N.Y.)*, **65**, 506; II. The general case. *Ann. Phys. (N.Y.)*, **68**, 541.

Ryan, M. P. Jr (1972a). Hamiltonian cosmology. *Lecture Notes in Physics*, vol. 13. Springer-Verlag: Berlin.

Ryan, M. P. Jr (1972b). The oscillatory regime near the singularity in Bianchi type IX universes. *Ann. Phys. (N.Y.)*, **70**, 301.

Ryan, M. P. Jr (1972c). Surface-of-revolution cosmology. *Ann. Phys. (N.Y.)*, **72**, 584.

Ryan, M. P. Jr (1974). Hamiltonian cosmology: death and transfiguration. *J. Math. Phys.*, **15**, 812.

Ryan, M. P. Jr and Shepley, L. C. (1975). *Homogeneous Relativistic Cosmologies*. Princeton University Press: Princeton, New Jersey.

Saunders, P. T. (1969). Observations in some simple cosmological models with shear. *Mon. Not. R. Astron. Soc.*, **142**, 213.

Schechter, P. L. (1977). On the solar motion with respect to external galaxies. *Astron. J.*, **82**, 569.

Schmidt, B. (1967). Isometry groups with surface-orthogonal trajectories. *Zs. Naturfor.*, **22a**, 1351.

Schramm, D. N. and Wagoner, R. V. (1977). Element production in the early universe. *Annu. Rev. Nucl. Sci.*, **27**.

Shikin, I. S. (1966). A uniform anisotropic cosmological model with a magnetic field. *Dokl. AN SSR*, **171**, 73. (English translation in *Sov. Phys.: Dokl.*, **11**, 944.)

Shikin, I. S. (1972). Study of gravitational fields in an anisotropic model with matter and neutrinos. *Zh. Eksp. Teor. Fiz.*, **63**, 1529. (English translation in *Sov. Phys.: JETP*, **36**, 811.)

Shikin, I. S. (1974). Investigation of a class of gravitational fields of a charged dustlike medium. *Zh. Eksp. Teor. Fiz.*, **67**, 433. (English translation in *Sov. Phys.: JETP*, **40**, 215.)

Shikin, I. S. (1975). Anisotropic cosmological model of Bianchi type V in general (axially symmetric) case with moving matter. *Zh. Eksp. Teor. Fiz.*, **68**, 1563. (English translation in *Sov. Phys.: JETP*, **41**, 794.)

Siklos, S. T. C. (1976). Singularities, invariants and cosmology. Ph.D. thesis. University of Cambridge.

Siklos, S. T. C. (1978). Occurrence of whimper singularities. *Commun. Math. Phys.*, **58**, 255.

Chapter 11. References

Silk, J. (1971). Diffuse cosmic X and gamma radiation: the isotropic component. *Space Sci. Rev.*, **11**, 671.

Smoot, G. F., Gorenstein, M. V. and Muller, R. A. (1977). Detection of anisotropy in the cosmic blackbody radiation. *Phys. Rev. Lett.*, **39**, 898.

Sneddon, G. E. (1975). Hamiltonian cosmology: a further investigation. *J. Phys.*, **A9**, 229.

Spero, A. and Baierlein, R. (1977). Approximate symmetry groups of inhomogeneous metrics. *J. Math. Phys.*, **18**, 1330.

Spero, A. and Baierlein, R. (1978). Approximate symmetry groups of inhomogeneous metrics: examples. *J. Math. Phys.*, **19**, 1324.

Spero, A. and Szafron, D. (1978). Spatial conformal flatness in homogeneous and inhomogeneous cosmologies. *J. Math. Phys.*, **19**, 1536.

Stewart, J. M. (1968). Neutrino viscosity in cosmological models. *Astrophys. Lett.*, **2**, 133.

Stewart, J. M. (1971). Non-equilibrium relativistic kinetic theory. *Lecture Notes in Physics*, vol. 10. Springer-Verlag, Berlin.

Szafron, D. A. (1977). Inhomogeneous cosmologies: new exact solutions and their evolution. *J. Math. Phys.*, **18**, 1673.

Szafron, D. A. and Wainwright, J. (1977). A class of inhomogeneous perfect fluid cosmologies. *J. Math. Phys.*, **18**, 1608.

Szekeres, P. (1975). A class of inhomogeneous cosmological models. *Commun. Math. Phys.*, **41**, 55.

Taub, A. H. (1951). Empty space–times admitting a three-parameter group of motions. *Ann. Math.*, **53**, 472.

Thompson, L. A. (1976). The angular momentum properties of galaxies in rich clusters. *Astrophys. J.*, **209**, 22.

Thorne, K. S. (1967). Primordial element formation, primordial magnetic fields and the isotropy of the universe. *Astrophys. J.*, **148**, 51.

Tolman, R. (1934). Effect of inhomogeneity on cosmological models. *Proc. Nat. Acad. Sci. USA*, **20**, 169.

Tomita, K. (1968). Theoretical relations between observable quantities in an anisotropic and homogeneous universe. *Prog. Theor. Phys.*, **40**, 264.

Tomita, K. (1973). Element formation in a chaotic early universe. *Prog. Theor. Phys.*, **50**, 1285.

Tupper, B. O. J. (1977). Conductivity in type VI_0 cosmologies with electromagnetic field. *Astrophys. J.*, **216**, 192.

Vajk, J. P. and Eltgroth, P. G. (1970). Spatially homogeneous anisotropic cosmological models containing relativistic fluid and magnetic field. *J. Math. Phys.*, **11**, 2212.

Wainwright, J. (1977). Characterisation of the Szekeres inhomogeneous cosmologies as algebraically special solutions. *J. Math. Phys.*, **18**, 672.

Webster, A. (1976). The clustering of radio sources: I. The theory of power-spectrum analysis. *Mon. Not. R. Astron. Soc.*, **175**, 61; II. The 4C, GB and MC1 surveys. *Mon. Not. R. Astron. Soc.*, **175**, 71.

Weinberg, S. (1976). Apparent luminosities in a locally inhomogeneous universe. *Astrophys. J. Lett.*, **208**, L.1.

Wesson, P. S. (1975a). Relativistic hierarchical cosmology: I. Derivation of a metric and dynamical equations. *Astrophys. Space Sci.*, **32**, 273; II. Some classes of model universes. *Astrophys. Space Sci.*, **32**, 305; III. Comparison with observational data. *Astrophys. Space Sci.*, **32**, 315.

Wesson, P. S. (1975b). Discrete source counts and the rotation of the local universe in hierarchical cosmology. *Astrophys. Space Sci.*, **37**, 235.

Willson, M. A. G. (1972). The relative orientation of radio sources. *Mon. Not. R. Astron. Soc.*, **155**, 275.

Winlove, C. P. and Raine, D. J. (1975). Pair creation and the Gowdy model. *Ann. Phys. (N.Y.)*, **93**, 116.

Wolfe, A. M. (1970). New limits on the shear and rotation of the universe from the X-ray background. *Astrophys. J. Lett.*, **159**, L.61.

Chapter 12 Singularities and time-asymmetry

(1) Einstein, A. and Rosen, N. (1935). *Phys. Rev.* ser. 2, **48**, 73.
(2) Einstein, A. (1939). *Ann. Math.*, **40**, 922.
(3) Einstein, A. (1931). *Berl. Ber.*, 235.
(4) Einstein, A. (1945). *Ann. Math.*, ser. 2, **46**, 578.
(5) Einstein, A. and Straus, E. G. (1946). *Ann. Math.*, ser. 2, **47**, 731.
(6) Penrose, R. (1965). *Phys. Rev. Lett.*, **14**, 57.
(7) Hawking, S. W. and Ellis, G. F. R. (1973). *The Large Scale Structure of Space–time*. Cambridge University Press.
(8) Hawking, S. W. and Penrose, R. (1970). *Proc. R. Soc. Lond.*, **A314**, 529.
(9) Davies, P. C. W. (1974). *The Physics of Time Asymmetry*. Surrey University Press.
(10) Grünbaum, A. (1963). *Philosophical Problems of Space and Time*. Knopf: New York.
(11) Reichenbach, H. (1956). *The Direction of Time*. University of California Press: Berkeley.
(12) Einstein, A., Tolman, R. C. and Podolsky, B. (1934). *Phys. Rev.*, ser. 2, **37**, 780.
(13) Watanabe, S. (1955). *Rev. Mod. Phys.*, **27**, 179; (1965). *Prog. Theor. Phys.*, Suppl. (Extra No.), 135.
(14) Aharanov, Y., Bergmann, P. G. and Lebowitz, J. L. (1964). *Phys. Rev.*, **B134**, 1410.
(15) Penfield, R. H. (1966). *Am. J. Phys.*, **34**, 422.
(16) D'Espagnat, B. (1971). *The Conceptual Foundations of Quantum Mechanics*. Benjamin: Menlo Park, California.
(17) Christenson, J. H., Cronin, J. W., Fitch, V. L. and Turlay, R. (1964). *Phys. Rev. Lett.*, **13**, 138.
(18) Casella, R. S. (1968). *Phys. Rev. Lett.*, **21**, 1128; (1969), **22**, 554.
(19) Kabir, P. K. (1968). *The CP Puzzle*. Academic Press: London, New York.
(20) Commins, E. D. (1973). *Weak Interactions*. McGraw-Hill: New York.
(21) Wolfenstein, L. (1964). *Phys. Rev. Lett.*, **13**, 562.
(22) Einstein, A., Podolsky B. and Rosen, N. (1935). *Phys. Rev.*, ser. 2, **47**, 777.
(23) Einstein, A. (1974). *Ideas and Opinions*. Souvenir Press: London.
(24) Schrödinger, E. (1935). *Naturwiss.*, **23**, 807, 823, 844.
(25) Wigner, E. P. In *Symmetries and Reflections*, (1970). MIT Press; and in *The Scientist Speculates*, (1961) ed. I. J. Good. Heineman: London.
(26) Everett, H. III. (1957). *Rev. Mod. Phys.*, **29**, 454.

Chapter 12. References

(27) Wheeler, J. A. (1957). *Rev. Mod. Phys.*, **29**, 463.
(28) DeWitt, B. S. (1971). In *Foundations of Quantum Mechanics*, ed. B. d'Espagnat. Academic Press: London, New York.
(29) Clarke, C. J. S. (1974). *Phil. Sci.*, **41**, 317.
(30) Bohr. N. (1934, 1961). *Atomic Theory and the Description of Nature.* Cambridge University Press.
(31) Dirac, P. A. M. (1973). In *The Physicist's Conception of Nature*, ed. J. Mehra, p. 14. Reidel: Dordrecht.
(32) Komar, A. (1969). *Int. J. Theor. Phys.*, **2**, 157.
(33) Karolyhazy, F. (1966). *Nuovo Cimento*, **A42**, 390.
(34) Born, M. (1927). *Nature*, **119**, 354.
(35) Feynman, R. P., Leighton, R. B. and Sands, M. (1965). *The Feynman Lectures on Physics*, vol. III. Addison-Wesley: Reading, Massachusetts.
(36) Ehrenfest, P. and Ehrenfest, T. (1959). *The Conceptual Foundations of the Statistical Approach in Mechanics.* Cornell University Press.
(37) Tolman, R. C. (1938). *The Principles of Statistical Mechanics.* Oxford University Press.
(38) Penrose, O. (1970). *Foundations of Statistical Mechanics.* Pergamon: Oxford.
(39) Hahn, E. L. (1950). *Phys. Rev.*, **80**, 580.
(40) Penrose, O. and Percival, I. C. (1962). *Proc. Phys. Soc.*, **79**, 509.
(41) Gold, T. (1967). *The Nature of Time.* Cornell University Press.
(42) Hogarth, J. (1962). *Proc. R. Soc. Lond.*, **A267**, 365.
(43) Sciama, D. W. (1963). *Proc. R. Soc. Lond.*, **A273**, 484.
(44) Einstein, A. and Ritz, W. *Physikalische Zeitschrift*, **9** Jahrgang, no. 25, 903; **10** Jahrgang, no. 6, 185; **10** Jahrgang, no. 9, 322.
(45) Penrose, R. (1964). In *Relativity, Groups and Topology*, eds. C. M. DeWitt and B. S. DeWitt. Gordon & Breach: New York; and in Gold, ref. 41.
(46) Geroch, R. P. (1968). In *Battelle Rencontres*, eds. C. M. DeWitt and J. A. Wheeler. Benjamin: New York.
(47) Cf. Lord Dunsany. *The King That Was Not*, in *Time and the Gods*.
(48) Gold, T. (1962). *Am. J. Phys.*, **30**, 403.
(49) Penrose, R. (1972). *Techniques of Differential Topology in Relativity.* S.I.A.M.: Philadelphia.
(50) Bell, J. S. and Perring, J. K. (1964). *Phys. Rev. Lett.*, **13**, 348.
(51) Penrose, R. (1969). *Rivista Nuovo Cimento*, Num. Spec. I, **1**, 252.
(52) Novikov, I. D. (1964). *Astron. Zh.*, **41**. (English translation in *Sov. Astron.: Astron. J.*, **8**, 857, 1075.)
(53) Ne'eman, Y. (1965). *Astrophys. J.*, **141**, 1303.
(54) Oppenheimer, J. R. and Synder, H. (1939). *Phys. Rev.*, **56**, 455.
(55) Misner, C. W., Thorne, K. S. and Wheeler, J. A. (1973). *Gravitation.* Freeman: San Francisco.
(56) Hawking, S. W. (1975). *Commun. Math. Phys.*, **43**, 199.
(57) Penrose, R. (1974). In *Confrontation of Cosmological Theories with Observational Data.* (IAU Symp. 63), ed. M. S. Longair. Reidel: Boston.
(58) Penrose, R. (1978). In *Theoretical Principles in Astrophysics and Relativity*, eds. N. R. Lebovitz, W. H. Reid and P. O. Vandervoort. University of Chicago Press.

Chapter 12. References

(59) Penrose, R. (1977). In *Proceedings of the First Marcel Grossmann Meeting on General Relativity, ICTP Trieste*, ed. R. Ruffini, North-Holland: Amsterdam.
(60) Penrose, R. (1977). In *Physics and Contemporary Needs*, ed. Riazuddin. Plenum: New York.
(61) Penrose, R. (1966). In *Perspectives in Geometry and Relativity*, ed. B. Hoffmann. Indiana University Press.
(62) Parker, L. (1977). In *Asymptotic Structure of Space-Time*, eds. F. P. Esposito and L. Witten. Plenum: New York.
(63) Sexl, R. U. and Urbantke, H. K. (1969). *Phys. Rev.*, **179**, 1247.
(64) Zel'dovich, Ya. B. (1974). In *Gravitational Radiation and Gravitational Collapse* (IAU Symp. 64), ed. C. M. DeWitt. Reidel: Boston.
(65) Eardley, D. M. (1974). *Phys. Rev. Lett.*, **33**, 442.
(66) Zel'dovich, Ya. B. and A. A. Starobinsky (1971). *Zh. Eksp. Teor. Fiz.*, **61**, 2161. (English translation in *Sov. Phys.*: *JETP*, **34**, 1159.)
(67) Hawking, S. W. (1976). *Phys. Rev.*, **D13**, 191; **14**, 2460.
(68) Hawking, S. W. (1977). Report to GR8 meeting, Waterloo, Ontario.
(69) Bekenstein, J. D. (1973). *Phys. Rev.*, **D7**, 2333; (1974) **9**, 3292.
(70) Penrose, R. and Floyd, R. M. (1971). *Nature Phys. Sci.*, **229**, 177.
(71) Hawking, S. W. (1972). *Commun. Math. Phys.*, **25**, 152.
(72) Gibbons, G. W. and Perry, M. J. (1978). *Proc. R. Soc. Lond.*, **A358**, 467.
(73) Wheeler, J.A. (1962). *Geometrodynamics*. Academic Press: London, New York.
(74) Wald, R. M. (1975). *Commun. Math. Phys.*, **45**, 9.
(75) Parker, L. (1975). *Phys. Rev.*, **D12**, 1519.
(76) Pati, J. C. and Salam, A. (1973). *Phys. Rev. Lett.*, **31**, 661; *Phys. Rev.*, **D8**, 1240; (1974) **10**, 275.
(77) Sciama, D. W. (1971). *Modern Cosmology*. Cambridge University Press.
(78) Weinberg, S. (1972). *Gravitation and Cosmology*. Wiley: New York.
(79) Lynden-Bell, D. and Lynden-Bell, R. M. (1977). *Mon. Not. R. Astron. Soc.*, **181**, 405.
(80) Belinskii, V. A., Khalatnikov, I. M. and Lifshitz, E. M. (1970). *Adv. Phys.*, **19**, 525.
(81) Kantowski, R. and Sachs, R. K. (1967). *J. Math. Phys.*, **7**, 443.
(82) Gibbons, G. W. and Hawking, S. W. (1977). *Phys. Rev.*, **D15**, 2752.
(83) Zel'dovich, Ya. B. Personal Communication.
(84) Tolman, R. C. (1934). *Relativity, Thermodynamics and Cosmology*. Clarendon Press: Oxford.
(85) Israel, W. (1963). *J. Math. Phys.*, **4**, 1163.
(86) Liang, E. P. T. (1973). *Lett. Nuovo Cimento*, **7**, ser. 2, 599.
(87) Simpson, M. and Penrose, R. (1973). *Int. J. Theor. Phys.*, **7**, 183.
(88) McNamara, J. M. (1978). *Proc. R. Soc. Lond.*, **A358**, 499.
(89) Geroch, R., Kronheimer, E. H. and Penrose, R. (1972). *Proc. R. Soc. Lond.*, **A327**, 545.
(90) Penrose, R. (1965). *Proc. R. Soc. Lond.*, **A284**, 159.
(91) Bondi, H. (1952). *Cosmology*. Cambridge University Press.
(92) Yodzis P., Seifert, H.-J. and Muller zum Hagen, H. (1973). *Commun. Math. Phys.*, **34**, 135; (1974) **37**, 29.

Chapter 13. References

(93) Geroch, R. P. (1970). *J. Math. Phys.*, **11**, 437.
(94) Misner, C. W. (1968). *Astrophys. J.*, **151**, 431; (1969), *Phys. Rev. Lett.*, **22**, 1071.
(95) Chitre, D. M. (1972). *Phys. Rev.*, **D6**, 3390.
(96) Doroshkevich, A. G. and Novikov, I. D. (1970). *Astron. Zh.*, **47**, 948. (English translation in *Sov. Astron.*, **14**, 763.)
(97) Lifschitz, E. M., Lifschitz, I. M. and Khalatnikov, I. M. (1970). *Zh. Eksp. Teor. Fiz.*, **59**, 322. (English translation in *Sov. Phys.: JETP*, **32**, 173.)
(98) MacCallum, M. A. H. (1971). *Nature, Phys. Sci.*, **230**, 112.
(99) Budic, R. and Sachs, R. K. (1976). *Gen. Relativ. Gravit.*, **7**, 21.
(100) Matzner, R. A. and Misner, C. W. (1972). *Astrophys J.*, **171**, 415.
(101) Parker, L. (1976). *Nature*, **261**, 20.
(102) Zel'dovich, Ya. B. (1972). In *Magic Without Magic: John Archibald Wheeler*. Freeman: San Francisco.
(103) Zel'dovich, Ya. B. and Novikov, I. D. (1971). *Relativistic Astrophysics II*. University of Chicago Press.
(104) Barrow, J. D. and Matzner, R. A. (1977). *Mon. Not. R. Astron. Soc.*, **181**, 719.
(105) Barrow, J. D. (1977). *Nature*, **267**, 117.
(106) Carter, B. (1974). In *Confrontation of Cosmological Theories with Observational Data* (IAU Symp. 63), ed. M. S. Longair. Reidel: Boston.
(107) Boltzmann, L. (1895). *Nature*, **51**, 413.
(108) Partridge, R. B. and Wilkinson, D. T. (1967). *Phys. Rev. Lett.*, **18**, 557.
(109) Harrison, E. R. (1973). *Annu. Rev. Astron. Astrophys.*, **11**, 155.
(110) Carr, B. J. (1975). *Astrophys. J.*, **201**, 1.
(111) von Neumann, J. (1955). *Mathematical Foundations of Quantum Mechanics*. Princeton University Press.
(112) Bialynicki-Birula, I. and Mycielski, J. (1976). *Ann. Phys. (N.Y.)*, **100**, 62.
(113) Lüders, G. (1957). *Ann. Phys.*, **2**, 1.
(114) Taylor, J. C. (1976). *Gauge Theories of Weak Interactions*. Cambridge University Press.

Chapter 13 Quantum field theory in curved spacetime

Albeverio, S. A. and Hoech-Krohn, R. J. (1977). *Mathematical Theory of Feynman Path Integrals*. Lecture Notes in Maths 523. Springer-Verlag: New York.
Aragone, C. and Deser, S. (1971). *Nuovo Cimento*, **3A**, 709.
Ashtekar, A. and Magnon, A. (1975). Quantum fields in curved space time. *Proc. R. Soc. Lond.*, **A346**, 375–94.
Bekenstein, J. D. (1975). Statistical black hole thermodynamics. *Phys. Rev.*, **D12**, 3077.
Bekenstein, J. D. and Meisels, A. (1977). Einstein A and B coefficients for a black hole. *Phys. Rev.*, **D15**, 2775–81.
Berger, B. (1975). Scalar particle creation in an anisotropic universe. *Phys. Rev.*, **D12**, 368–75.

Chapter 13. References

Bernard C. and Duncan, A. (1977). Regularization and renormalization of quantum field theory in curved space–time. *Annu. Phys.* (*N.Y.*), **107**, 201–21.

Bondi, H., Van den Burg, M. G. J. and Metzner, A. W. K. (1962). *Proc. R. Soc. Lond.*, **A269**, 21.

Boulware, D. G. (1975a). Quantum field theory in Schwarzschild and Rindler spaces. *Phys. Rev.*, **D11**, 1404–23.

Boulware, D. G. (1975b). Spin $\frac{1}{2}$ quantum field theory in Schwarzschild space. *Phys. Rev.*, **D12**, 350–67.

Boulware, D. G. (1976). Hawking radiation and thin shells. *Phys. Rev.*, **D13**, 2169–86.

Brown, L. S. (1977). Stress-tensor trace anomalies in a gravitational metric: Scalar fields. *Phys. Rev.*, **D15**, 1469.

Brown, L. S. and Cassidy, J. P. (1977). Stress tensors and their trace anomalies in conformally flat space–times. *Phys. Rev.*, **D15**, 2810–29.

Bunch, T. S. and Davies, P. C. W. (1977a). Covariant point-splitting regularization for a scalar field in a Robertson–Walker universe with spatial curvature. *Proc. R. Soc. Lond.*, **A357**, 381–94.

Bunch, T. S. and Davies, P. C. W. (1977b). Stress tensor and conformal anomalies for massless fields in a Robertson–Walker universe. *Proc. R. Soc. Lond.*, **A356**, 596–74.

Cameron, R. H. (1960). *J. Math. Phys.*, **39**, 126.

Capper, D. M. and Duff, M. J. (1974). Trace anomalies in dimensional regularization. *Nuovo Cimento*, **23A**, 173–83.

Castagnino, M., Verbeure, A. and Weder, R. A. (1975). Catastrophies in the canonical quantization in an expanding universe. *Nuovo Cimento*, **26B**, 396–408.

Choquet-Bruhat, Y. (1967). Hyperbolic partial differential equations on a manifold. *Battelle Rencontres*, eds. C. DeWitt and J. A. Wheeler. W. A. Benjamin: New York.

Chitre, D. M. and Hartle, J. B. (1977). Path integral quantization and cosmological particle production: an example. *Phys. Rev.*, **D16**, 251–60.

Christensen, S. M. (1976). Vacuum expectation values of stress tensor in arbitrarily curved background – the covariant point separation method. *Phys. Rev.*, **D14**, 2490.

Christiensen, S. M. and Fulling, S. A. (1977). Trace anomalies and the Hawking effect. *Phys. Rev.*, **D15**, 2088–104.

Damour, T. (1977). Klein paradox and vacuum polarization. In *Recent Developments in the Fundamentals of General Relativity*, ed. R. Ruffini. North-Holland: Amsterdam.

Davies, P. C. W. (1975). Scalar particle production in Scwarzschild and Rindler metrics. *J. Phys.*, **A8**, 609–16.

Davies, P. C. W. (1976). Quantum field theory in curved space–time. *Nature*, **263**, 377.

Davies, P. C. W. (1977). Stress tensor calculations and conformal anomalies. Paper presented to the Eighth Texas Symposium on Relativistic Astrophysics. *Ann. N.Y. Acad. Sci.*, **302**, 166–85.

Davies, P. C. W., Fulling, S. A. and Unruh, W. G. (1976). Energy-momentum tensor near an evaporating black hole. *Phys. Rev.*, **D13**, 2720–3.

Chapter 13. References

Delbourgo, R. and Salam, A. (1972). The gravitational correction to PCAC. *Phys. Lett.*, **40B**, 381–2.

Deser, S. (1975). Plane waves do not polarize the vacuum. *J. Phys.*, **A8**, 1972–4.

Deser, S. and Zumino, B. (1976). Consistent supergravity. *Phys. Lett.*, **62B** 335–6.

DeWitt, B. S. (1963). In *Les Houches 1963 Lectures*, eds. B. S. DeWitt and C. DeWitt. Gordon and Breach: London.

DeWitt, B. S. (1975). Quantum field theory in curved space–time. *Phys. Rep.*, **19C**, 297–357.

DeWitt, C. (1972). Feynman's path integral. Definition without limiting procedure. *Commun. Math. Phys.*, **28**, 47–67.

DeWitt, C. and DeWitt, B. S. (1973). *Black Holes*. Gordon and Breach: London.

Dowker, J. S. (1974). *Functional Integration*, ed. A. M. Arthur. Oxford University Press.

Dowker, J. S. and Critchley, R. (1976a). Effective Langrangian and energy–momentum tensor in de Sitter space. *Phys. Rev.*, **D13**, 3224–32.

Dowker, J. S. and Critchley, R. (1976b). The stress tensor conformal anomaly for scalar and spinor fields. Preprint. University of Manchester.

Dowker, J. S. and Critchley, R. (1977). Vacuum stress tensor in Einstein Universe: Finite temperature effects. *Phys. Rev.*, **D15**, 1484.

Duff, M. J. (1978). Observations on conformal anomalies. *Nucl. Phys.*, **B125**, 334–47.

Duncan, A. (1977). Conformal anomalies in curved space–time. *Phys. Lett.*, **66B**, 170–2.

Eguchi, T. and Freund, P. G. O. (1976). Quantum gravity and world topology. *Phys. Rev. Lett.*, **37**, 1251–4.

Ford, L. H. (1976). Quantum vacuum energy in a closed universe. *Phys. Rev.*, **D14**, 3304–14.

Fradkin, E. S. and Vilkovisky, G. A. (1977). On the renormalization of quantum field theory in curved space–time. *Lett. Nuovo Cimento*, **19**, 47–54.

Frolov, V. P., Volovich, I. V. and Zagrebnov, V. A. (1977). On a new definition of vacuum-state in a gravitational field $\mathcal{H}$-vacuum. Preprint no. 13, High Energy Physics and Cosmic Rays. Academy of Sciences of the USSR, Lebedev Physical Institute, Moscow.

Fulling, S. A. (1973). Nonuniqueness of canonical field quantization in Riemannian space–time. *Phys. Rev.*, **D7**, 2850–62.

Fulling, S. A. (1977). Alternative vacuum states in static space–times with horizons. *J. Phys.*, **A10**, 917–52.

Gelfand, I. M. and Yaglom, A. M. (1960). Integration in functional spaces. *J. Math. Phys.*, **1**, 48–9.

Geroch, R. (1968, 1970). Spinor structure of space–times in general relativity: I and II. *J. Math. Phys.*, **9**, 1739, and **11**, 343.

Gibbons, G. W. (1975a). Vacuum polarization and the spontaneous loss of charge by black holes. *Commun. Math. Phys.*, **44**, 245–64.

Gibbons, G. W. (1975b). Quantized fields propagating on plane wave space–times. *Commun. Math. Phys.*, **45**, 191–202.

Chapter 13. References

Gibbons, G. W. (1976). A note on the Rarita–Schwinger equation in a gravitational background. *J. Phys.*, **A9**, 145–8.

Gibbons, G. W. (1977). Quantum processes near black holes. In *Proceedings of the Marcel Grossman meeting on Recent Advances in the Fundamentals of General Relativity*, ed. R. Ruffini. North Holland: Amsterdam.

Gibbons, G. W. and Hawking, S. W. (1977a). Cosmological event horizons, thermodynamics and particle creation. *Phys. Rev.*, **D15**, 2738–51.

Gibbons, G. W. and Hawking, S. W. (1977b). Action integrals and partition functions in quantum gravity. *Phys. Rev.*, **D15**, 2752–6.

Gibbons, G. W. and Perry, M. J. (1976). Black holes in thermal equilibrium. *Phys. Rev. Lett.*, **36**, 985.

Gibbons, G. W. and Perry, M. J. (1978). Black holes and thermal Green's functions. *Proc. R. Soc. Lond.*, **A358**, 467–94.

Gilkey, P. B. (1974). *The Index Theorem and the Heat Equation*. Mathematics Lecture Series No. 4. Publish and Perish: Boston, Massachusetts.

Gilkey, P. B. (1975). The spectral geometry of a Riemannian manifold. *J. Differ. Geom.*, **10**, 601–18.

Hajicek, P. (1976). On quantum field theory in curved space–time. *Nuovo Cimento*, **33B**, 597–612.

Hajicek, P. (1977). I. Theory of particle detection in curved space–time. *Nuovo Cimento Lett.*, **18**, 251–4; II. On particle detection in de Sitter space–time. *Phys. Rev.*, **D15**, 2757–74.

Hartle, J. B. and Hawking, S. W. (1976). Path-integral derivation of black-hole radiance. *Phys. Rev.*, **D13**, 2188–203.

Hawking, S. W. (1970). The conservation of matter in general relativity. *Commun. Math. Phys.*, **18**, 301.

Hawking, S. W. (1974). Black hole explosions. *Nature*, **248**, 30–1.

Hawking, S. W. (1975). Particle creation by black holes. *Commun. Math. Phys.*, **43**, 199–220.

Hawking, S. W. (1976a). Breakdown of predictability in gravitational collapse. *Phys. Rev.*, **D14**, 2460.

Hawking, S. W. (1976b). Black holes and thermodynamics. *Phys. Rev.*, **D13**, 191–7.

Hawking, S. W. (1977a). Zeta function regularization of path integrals in curved space–time. *Commun. Math. Phys.*, **56**, 133–48.

Hawking, S. W. (1977b). Gravitational instantons. *Phys. Lett.*, **A60**, 81.

Hawking, S. W. and Ellis, G. F. R. (1973). *The Large Scale Structure of Space–Time*. Cambridge University Press.

Hu, B. L. (1974). Scalar waves in the mixmaster universe: II. Particle creation. *Phys. Rev.*, **D9**, 3263.

Hu, B. L., Fulling, S. A. and Parker, L. (1974). Quantized scalar fields in a closed anisotropic universe. *Phys. Rev.*, **D8**, 2377–85.

Isham, C. J. (1977). Quantum field theory in curved space-times – an overview. Paper presented to the Eighth Texas Symposium on Relativistic Astrophysics. *Ann. N.Y. Acad. Sci.*, **302**, 114–57.

Israel, W. (1976). Thermo-field dynamics of black holes. *Phys. Lett.*, **57A**, 107–10.

Chapter 13. References

Kimura, T. (1969). Divergence of axial-vector current in the gravitational field. *Prog. Theor. Phys.*, **42**, 1191.

Kramers, D. (1975). Quantization of the electromagnetic field in Riemannian spaces. *Acta Phys. Pol.*, **B6**, 467–78.

Kramers, D. (1976). Quantized scalar field in curved space–time. *Acta. Phys. Pol.*, **B7**, 237–45; Quantization of the Dirac field in Riemannian spaces. *Acta. Phys. Pol.*, **B7**, 4, 227–36; Canonical quantization on general spacelike hypersurfaces. *Acta. Phys. Pol.*, **B7**, 117–25.

Lapedes, A. S. (1976). Thermal particle production in two Taub–NUT type spacetimes. *Commun. Math. Phys.*, **51**, 121–33.

Lee, T. D. (1974). CP non conservation and spontaneous symmetry breaking. *Phys. Rep.*, **9C**, 145.

Lichnerowicz, A. (1961). Spineurs harmonique. *C. R. Acad. Sci. (Paris)*, **257**, 7–9.

Lukash, V. N., Novikov, I. D., Starobinsky, A. A. and Zel'dovich, Ya. B. (1976). Quantum effects and evolution of cosmological models. *Nuovo Cimento*, **35B**, 293–307.

McKean, H. P. and Singer, I. M. (1967). Curvature and the eigenvalues of the Laplacian. *J. Diff. Geom.*, **1**, 43–69.

Mamaev, S. G., Mostepanenko and Starobinsky, A. A. (1976). Particle creation from vacuum near a homogenous isotropic singularity. *Sov. Phys.: JETP*, **43**, 823–9.

Mensky, M. B. (1974). Feynman quantization and the S-matrix for spinning particles in Riemannian spacetime. *Teor. Mat. Fiz.*, **18**, 190–202. (English translation: *Sov. Phys. Theor. Math. Phys.*, **18**, 136–44.)

Mensky, M. B. (1976). Relativistic quantum theory without quantized fields. *Commun. Math. Phys.*, **47**, 97–108.

Misner, C. W., Thorne, K. S. and Wheeler, J. A. (1973). *Gravitation*. W. H. Freeman: San Francisco.

Nachtmann, O. (1967). Quantum theory in de Sitter space. *Commun. Math. Phys.*, **6**, 1–16.

Nielsen, N. K., Römer, H. and Schroer, B. (1977). Classical anomalies and local version of the Atiyah–Singer theory. *Phys. Lett.*, **70B**, 445–8.

Page, D. N. (1976a). Particle emission rates from a black hole: Massless particles from an uncharged non-rotating hole. *Phys. Rev.*, **D13**, 198–206.

Page, D. N. (1976b). Particle emission rates from a black hole: III. Charged leptons from a non-rotating hole. *Phys. Rev.* (to appear).

Parker, L. (1968). Particle creation in expanding universes. *Phys. Rev. Lett.*, **21**, 562–4.

Parker, L. (1969). Quantized fields and particle creation in expanding universes I. *Phys. Rev.*, **183**, 1057–68.

Parker, L. (1971). Quantized fields and particle creation in expanding universes II. *Phys. Rev.*, **D3**, 346–56.

Parker, L. (1975). Quantized fields and particle creation in curved space–time. Lectures presented at the Second Latin American Symposium on Relativity and Gravitation, Caracas, Venezuela, December 8–13.

Parker, L. (1976a). The production of elementary particles by strong gravitational fields. In *Proceedings of the Symposium on Asymptotic*

Properties of Space–Time, ed. F. P. Exposito and L. Witten. Plenum: New York. (In press).

Parker, L. (1976b). Probability distribution of particles created by a black hole. *Phys. Rev.*, **D12**, 1519–25.

Parker, L. (1976c). Thermal radiation produced by the expansion of the Universe. *Nature*, **261**, 20–3.

Parker, L. and Fulling, S. A. (1974). Adiabatic regularization of the energy–momentum tensor of a quantized field in homogeneous spaces. *Phys. Rev.*, **D9**, 341–54.

Perkins, D. H. (1972). *Introduction to High Energy Physics*. Addison Wesley: Reading, Massachusetts.

Raine, D. J. and Winlove, C. P. (1975). Pair creation in expanding universes. *Phys. Rev.*, **D12**, 946–51.

Römer, H. and Schroer, B. (1977). Fractional topological configurations and surface effects. *Phys. Lett.*, **71B**, 182–4.

Rosenfeld, L. (1930). Zur quantelung der Wellenfelder. *Ann.–Phys. (Leipz.)*, 5th Ser., **5**, 113.

Rumpf, H. (1976). Covariant description of particle creation in curved spacetimes. *Nuovo Cimento*, **35B**, 321; Covariant treatment of particle creation in curved space–time. *Phys. Lett.*, **61B**, 272.

Rumf, H. and Urbanthe, H. K. (1977). Covariant 'in–out' formalism for creation by external fields. Preprint. University of Wien.

Schrödinger, E. (1939). *Physica*, **6**, 899.

Sexl, R. U. and Urbantke, H. K. (1967). Cosmic particle creation processes. *Acta. Phys. Austriaca*, **26**, 339–56.

Starobinsky, A. A. (1973). Amplification of waves during reflection from a rotating 'black hole'. *Zh. Eksp. Teor. Fiz.*, **64**, 48–57.

t'Hooft, G. (1976). Computation of the quantum effects due to a 4-dimensional pseudo particle. *Phys. Rev.*, **D14**, 3432.

Unruh, W. G. (1974). Second quantization in the Kerr metric. *Phys. Rev.*, **D10**, 3194–204.

Unruh, W. G. (1976). Notes on black hole evaporation. *Phys. Rev.*, **D14**, 870.

Urbantke, H. K. (1969). Remark on noninvariance groups and field quantization in curved space. *Nuovo Cimento Ser. X*, **63B**, 203–14.

Utiyama, R. and DeWitt, B. S. (1962). Renormalization of a classical gravitational field interacting with quantized matter fields. *J. Math. Phys.*, **3**, 608–18.

Wald, R. M. (1975). On particle creation by black holes. *Commun. Math. Phys.*, **45**, 9–34.

Wald, R. M. (1976). Stimulated emission effects in particle creation near black holes. *Phys. Rev.*, **D13**, 3176–82.

Woodhouse, N. M. J. (1976). Particle creation by gravitational fields. *Phys. Rev. Lett.*, **36**, 999–1001.

Zel'dovich, Ya. B. (1971). Generation of waves by a rotating body. *Zh. Eksp. Teor. Fiz.*, **14**, 270–2. (English translation in *Sov. Phys.: JETP Lett.*, **14**, 180–1.)

Zel'dovich, Ya. B., Kobzarev, I. Yu., and Okun, L. B., (1974). Cosmological consequences of a spontaneous breakdown of a discrete symmetry, *Zh. Eksp. Teor. Fiz.*, **67**, 3. (English translation in *Sov. Phys.: JETP*, **40**, 1.)

Chapter 14. References

Zel'dovich, Ya. B. and Starobinsky, A. A. (1972). Particle production and vacuum polarization in an anisotropic gravitational field. *Zh. Eksp. Teor. Fiz.*, **6**, 2161–640. (English translation in *Sov. Phys.: JETP*, **34**, 1159–412.)

Chapter 14 Quantum gravity: the new synthesis

Abers, E. S. and Lee, B. W. (1973). *Phys. Rep.*, **9C**, 1.
Atiyah, M. F. and Singer, I. M. (1968). *Ann. Math.* **87**, 546.
Atiyah, M. F., Bott, R. and Patodi, V. K. (1973). *Invent. Math.*, **19**, 279.
Becchi, C., Rouet, A. and Stora, R. (1975). *Commun. Math. Phys.*, **42**, 127.
Bekenstein, J. D. (1973). *Phys. Rev.*, **D7**, 2333.
Berezin, F. A. (1966). *The Method of Second Quantization*, Academic Press: London, New York.
Bohr, N. and Rosenfeld, L. (1933). *Kgl. Danske Videnskab. Selskab. Mat.-fys. Med.*, **12**, No. 8.
Brown, L. S. and Cassidy, J. P. (1977a). *Phys. Rev.*, **D15**, 2810.
Brown, L. S. and Cassidy, J. P. (1977b). *Phys. Rev.*, **D16**, 1712.
Bunch, T. S. and Davies, P. C. W. (1977). *Proc. R. Soc. Lond.*, **A356**, 569.
Candelas, P. and Deutsch, D. (1977). *Proc. R. Soc. Lond.*, **A354**, 79.
Capper, D. M. and Duff, M. J. (1974). *Nuovo Cimento*, **23A**, 173.
Casimir, H. B. G. (1940). *Proc. Kon. Ned. Akad. Wetenschap.*, **51**, 793.
Christiensen, S. M. and Fulling, S. A. (1977). *Phys. Rev.*, **D15**, 2088.
Davies, P. C. W., Fulling, S. A. and Unruh, W. G. (1976). *Phys. Rev.*, **D13**, 2720.
Delbourgo, R. and Ramon Medrano, M. (1976). *Nucl. Phys.*, **B110**, 467.
Delbourgo, R. and Salam, A. (1972). *Phys. Lett.*, **40B**, 381.
Deser, S., Duff, M. J. and Isham, C. J. (1976). *Nucl. Phys.*, **B111**, 45.
DeWitt, B.S. (1962). In *Gravitation: An Introduction to Current Research*, ed. L. Witten. Wiley: New York.
DeWitt, B. S. (1964). *Phys. Rev. Lett.*, **13**, 114.
DeWitt, B. S. (1965). *Dynamical Theory of Groups and Fields*. Gordon and Breach: New York.
DeWitt, B. S. (1967a). *Phys. Rev.*, **160**, 1113.
DeWitt, B. S. (1967b). *Phys. Rev.*, **162**, 1195, 1239.
DeWitt, B. S. (1970). In *Relativity: Proceedings of the Relativity Conference in the Midwest*, eds. M. Carmeli, S. I. Fickler and L. Witten. Plenum Press: New York.
DeWitt, B. S. (1972). In *Magic Without Magic*, ed. J. Klauder. W. H. Freeman: San Francisco.
Dixon, J. (1975). D. Phil. thesis. University of Oxford.
Dowker, J. S. and Critchley, R. (1976). *Phys. Rev.*, **D13**, 3224.
Dowker, J. S. and Kennedy, G. (1978). *J. Phys.*, **11A**, 895.
Duff, M. J. (1973). *Phys. Rev.*, **D7**, 2317.
Eguchi, T. and Freund, P. G. O. (1976). *Phys. Rev. Lett.*, **37**, 1251.
Einstein, A. (1933). *Mein Weltbild*. Querido Verlag: Amsterdam.
Everett, H. (1957). *Rev. Mod. Phys.*, **29**, 454. Reprinted, together with several other papers on the same subject, in *The Many-Worlds*

Chapter 14. References

Interpretation of Quantum Mechanics, eds. B. S. DeWitt and R. N. Graham. Princeton University Press (1973).
Fadde'ev, L. D. (1969). *Theor. Math. Phys.*, **1**, 3.
Fadde'ev, L. D. (1976). In *Methods in Field Theory*, eds. R. Balian and J. Zinn-Justin (1975 Les Houches Lectures). North-Holland: Amsterdam.
Fadde'ev, L. D. and Popov, V. N. (1967). *Phys. Lett.*, **25B**, 29.
Fetter, A. L. and Walecka, J. D. (1971). *Quantum Theory of Many-Particle Systems*. McGraw-Hill: New York.
Feynman, R. P. (1948). *Rev. Mod. Phys.*, **20**, 267.
Feynman, R. P. (1950). *Phys. Rev.*, **80**, 440.
Feynman, R. P. (1963). *Acta Phys. Polon.*, **24**, 697. See also in *Proceedings of the 1962 Warsaw Conference on the Theory of Gravitation*, PWN-Editions Scientifiques de Pologne, Warszawa (1964).
Fischer, A. E. (1970). In *Relativity: Proceedings of the Relativity Conference in the Midwest*, eds. M. Carmeli, S. I. Fickler and L. Witten, Plenum Press: New York.
Freifeld, C. (1968). In *Battelle Rencontres* (1967 Lectures in Mathematics and Physics) eds. C. M. DeWitt and J. A. Wheeler. Benjamin: New York.
Gilkey, P. B. (1975). *J. Differ. Geom.*, **10**, 601.
Gribov, V. N. (1977). Lecture at the Twelfth Winter School of the Leningrad Nuclear Physics Institute. (SLAC translation No. 176.)
Hartle, J. B. (1977). *Phys. Rev. Lett.*, **39**, 1373.
Hawking, S. W. (1971), *Phys. Rev. Lett.*, **26**, 1344.
Hawking, S. W. (1974). *Nature*, **248**, 30.
Hawking, S. W. (1977a). *Phys. Lett.*, **60A**, 81.
Hawking, S. W. (1977b). *Commun. Math. Phys.*, **55**, 133.
Hawking, S. W. (1978). *Phys. Rev.*, **D18**, 1447.
Higgs, P. W. (1964). *Phys. Rev. Lett.*, **12**, 132; **13**, 508. (See also *Phys. Rev.*, **145**, 1156 (1966).)
Hu, B. L. and Parker, L. (1978). *Phys. Rev.*, **D17**, 933.
Isaacson, R. A. (1968). *Phys. Rev.*, **166**, 1263, 1272.
Isham, C. J. (1978). *Proc. R. Soc. Lond.*, **A362**, 383.
Isham, C. J., Strathdee, J. and Salam, A. (1971). *Phys. Rev.*, **D3**, 1805.
Isham, C. J., Strathdee, J. and Salam, A. (1972). *Phys. Rev.*, **D5**, 2584.
Jackiw, R. (1977). *Rev. Mod. Phys.*, **49**, 681.
Jackiw, R. and Rebbi, C. (1977). *Phys. Rev.*, **D16**, 1052.
Kimura, T. (1969). *Prog. Theor. Phys.*, **42**, 1191.
Kostant, B. (1977). In *Differential Geometrical Methods in Mathematical Physics* (Proceedings of the July 1–4, 1975 Symposium in Bonn), eds. K. Bleuler and A. Reetz. Lecture Notes in Mathematics No. 570. Springer: Berlin.
Lee, B. W. (1976). In *Methods in Field Theory* (1975 Les Houches Lectures), eds. R. Balian and J. Zinn-Justin. North-Holland: Amsterdam.
McKean, H. P. and Singer, I. M. (1967). *J. Differ. Geom.*, **1**, 43.
Press, W. and Teukolsky, S. (1972). *Nature*, **238**, 211.
Rindler, W. (1966). *Am. J. Phys.*, **34**, 1174.
Schwinger, J. (1951). *Proc. Nat. Acad. Sci. USA*, **37**, 452, 455.
Stelle, K. S. (1977). *Phys. Rev.*, **D16**, 953.
't Hooft, G. (1971). *Nucl. Phys.*, **B35**, 167.

Chapter 15. References

't Hooft, G. (1976a). *Phys. Rev. Lett.*, **37**, 8.
't Hooft, G. (1976b). *Phys. Rev.*, **D14**, 3422.
Unruh, W. G. (1976). *Phys. Rev.*, **D14**, 870.
Utiyama, R. (1956). *Phys. Rev.*, **101**, 1597.
Yang, C. N. and Mills, R. (1954). *Phys. Rev.*, **96**, 191.
Zel'dovich, Ya. B. (1970). *Pisma v. Zh. Eksp. Teor. Fiz.*, **12**, 443 (English translation, *Sov. Phys.*: *JETP Lett.*, **12**, 307.)
Zel'dovich, Ya. B. (1971). *Pisma v. Zh. Eksp. Teor. Fiz.*, **14**, 270 (English translation, *Sov. Phys.*: *JETP Lett.*, **14**, 180.)
Zel'dovich, Ya. B. (1972). *Zh. Eksp. Teor. Fiz.*, **62**, 2076. (English translation, *Sov. Phys.*: *JETP*, **35**, 1085.)

Chapter 15 The path-integral approach to quantum gravity

Berezin, F. A. (1966). *The Method of Second Quantization.* Academic Press: London, New York.
Brill, D. R. (1959). *Ann. Phys. (N.Y.)*, **7**, 46.
Brill, D. R. and Deser, S. (1968). *Ann. Phys. (N.Y.)*, **50**, 548.
Davis, L. R. (1978). Unpublished report.
Deser, S. and Zumino, B. (1976). *Phys. Lett.*, **63B**, 335–6.
DeWitt, B. S. (1963). In *Relativity Groups and Topology*, eds. B. S. and C. DeWitt. Gordon and Breach: New York.
DeWitt, B. S. (1967). *Phys. Rev.*, **162**, 1195–1239.
DeWitt, B. S. (1975). *Phys. Rep.*, **196**, 297–357.
Dowker, J. S. and Critchley, R. (1976). *Phys. Rev.*, **D13**, 3224–32.
Eguchi, T. and Freund, P. G. O. (1976). *Phys. Rev. Lett.*, **37**, 1251–4.
Fade'ev, L. D. and Popov, V. N. (1967). *Phys. Lett.*, **25B**, 697.
Ferrara, S. and van Nieuwenhuizen, P. (1976). *Phys. Rev. Lett.*, **37**, 669.
Feynman, R. P. (1972). In *Magic Without Magic*, ed. J. Klauder. W. H. Freeman: San Francisco.
Freedman, D. Z., van Nieuwenhuizen, P. and Ferrara, S. (1976). *Phys. Rev.*, **D13**, 3214.
Geroch, R. P. (1973). *Ann. N.Y. Acad. Sci.*, **224**, 108.
Gibbons, G. W. (1977a). *Phys. Lett.*, **60A**, 385–6.
Gibbons, G. W. (1977b). *Phys. Lett.*, **61A**, 3–5.
Gibbons, G. W. and Hawking, S. W. (1977a). *Phys. Rev.*, **D15**, 2752–6.
Gibbons, G. W. and Hawking, S. W. (1977b). *Phys. Rev.*, **D15**, 2738–51.
Gibbons, G. W. and Perry, M. J. (1978). *Proc. R. Soc. Lond.*, **A358**, 467–94.
Gibbons, G. W. and Perry, M. J. (1979). To be published.
Gibbons, G. W. and Pope, C. N. (1978). *Commun. Math. Phys.*, **61**, 239–48.
Gibbons, G. W., Hawking, S. W. and Perry, M. J. (1978). *Nucl. Phys.*, **B138**, 141–50.
Gilkey, P. B. (1974). *The Index Theorem and the Heat Equation.* Publish or Perish: Boston.
Gilkey, P. B. (1975). *J. Differ. Geom.* **10**, 601–18.
Hartle, J. B. and Hawking, S. W. (1972). *Commun. Math. Phys.*, **26**, 87–101.
Hartle, J. B. and Hawking, S. W. (1976). *Phys. Rev.*, **D13**, 2188–2203.
Hawking, S. W. (1976). *Phys. Rev.*, **D13**, 191–7.

Hawking, S. W. (1977a). *Phys. Lett.*, **60A**, 81.
Hawking, S. W. (1977b). *Commun. Math. Phys.*, **55**, 133–48.
Hawking, S. W. and Ellis, G. F. R. (1973). *The Large Scale Structure of Space–Time.* Cambridge University Press.
Hawking, S. W. and Pope, C. N. (1978). *Phys. Lett.*, **73B**, 42–4.
Israel, W. and Wilson, G. A. (1972). *J. Math. Phys.*, **13**, 865–7.
Jang, Pong Soo and Wald, R. M. (1977). *J. Math. Phys.*, **18**, 41–4.
Lapedes, A., (1978). To be published.
McKean, M. P. and Singer, I. M. (1967). *J. Differ. Geom.*, **1**, 43–69.
Newman, E. T., Unti, T. and Tamburino, L. (1963). *J. Math. Phys.*, **4**, 915.
Page, D. N. (1978). *Phys. Rev.*, **D**, in press.
Perry, M. J. (1978). *Nucl. Phys.*, **B**, in press.
Regge, T. (1961). *Nuovo Cimento*, **19**, 558–71.
Steenrod, N. (1951). *The Topology of Fibre Bundles.* Princeton University Press.
Steelle, K. (1977). *Phys. Rev.*, **D16**, 953–69.
Stelle, K. (1978). To be published.
t'Hooft, G. and Veltman, M. (1972). *Nucl. Phys.*, **B44**, 189.
Wheeler, J. (1963). In *Relativity Groups and Topology*, eds. B. S. and C. M. DeWitt. Gordon and Breach, New York.
Yau, S. T. (1977). *Proc. Nat. Acad. Sci. USA*, **74**, 1798–9.
Zel'dovich, Ya. B. and Starobinsky, A. (1972). *Zh. Eksp. Teor. Fiz.*, **6**, 2161–2640. (English translation in *Sov. Phys.: JETP*: **34**, 1159–1412.)

Chapter 16 Ultraviolet divergences in quantum gravity

(1) Long before the modern development of renormalization theory, Heisenberg proposed a classification of elementary-particle interactions into those with dimensionless couplings and those whose couplings have the dimensions of a negative power of mass, and he suggested that the mass scale which enters in the latter couplings would set a bound to the applicability of existing theories (see W. Heisenberg (1938) *Z. Physik*, **110**, 251). He also noted that in Fermi's theory of beta decay, the coupling constant had the dimensions of [mass]$^{-2}$, and he suggested that dynamical effects might be associated with energies of order $G_F^{-1/2}$, as for instance in cosmic ray showers (see W. Heisenberg (1938) *Z. Physik*, **101**, 251; (1939) *Z. Physik*, **113**, 61). After the development of renormalization theory, it was noted that non-renormalizable theories are in general just those whose couplings have the dimensionality of negative powers of mass, and that reaction rates would grow rapidly with energy in such theories (see S. Sakata, H. Umezawa and S. Kamefuchi (1952) *Prog. Theor. Phys.*, **7**, 327).

(2) R. P. Feynman (1963) *Acta Phys. Polon.*, **24**, 697; B. S. DeWitt (1967) *Phys. Rev.*, **162**, 1195, 1239 (erratum (1968) *Phys. Rev.*, **171**, 1834); L. D. Faddeev and V. N. Popov (1967) *Phys. Lett.*, **B25** (1967) 29; S. Mandelstam (1968) *Phys. Rev.*, **175**, 1604; E. S. Fradkin and I. V. Tyutin (1970), *Phys. Rev.*, **2**, 2841. For a derivation of covariant rules from the canonical formalism see E. S. Fradkin and G. A. Vilkovsky (1975) *Phys. Lett.*, **55B**, 224; *Nuovo Cimento*, **13**, 187. For a review see M. J. Duff (1975) in

Chapter 16. References

Quantum Gravity, eds. C. J. Isham, R. Penrose and D. W. Sciama (Oxford University Press).
(3) G. 't Hooft (1973) *Nucl. Phys.*, **B62**, 444; G. 't Hooft and M. Veltman (1974) *Ann. Inst. Poincaré*, **20**, 69; S. Deser and P. van Nieuwenhuizen (1974) *Phys. Rev. Lett.*, **32**, 245; S. Deser and P. van Nieuwenhuizen (1974) *Phys. Rev.*, **D10**, 401, 411; S. Deser, H-S Tsao and P. van Nieuwenhuizen (1974) *Phys. Lett.*, **50B**, 491. Infrared divergences pose no problem here; see S. Weinberg (1965) *Phys. Rev.*, **140**, B516.
(4) The need for counterterms proportional to $R_{\mu\nu}R^{\mu\nu}$ and R^2 was pointed out very early by R. Utiyama and B. S. DeWitt (1962) *J. Math. Phys.*, **3**, 608.
(5) H. Georgi, H. Quinn and S. Weinberg (1974). *Phys. Rev. Lett.*, **33**, 451.
(6) The Planck mass is given (for $\hbar = c = 1$) in terms of the Newton Constant G, as $G^{-1/2} = 1.2 \times 10^{19}$ GeV. See M. Planck (1899) *Sitz. Deut. Akad. Wiss.* (Berlin), 440.
(7) In the notation used in this article, $\hbar = c = 1$; the flat-space metric $\eta_{\mu\nu}$ has diagonal elements $+, +, +, -$; and sign conventions for $R, R_{\mu\nu}$, etc. are the same as in S. Weinberg (1972) *Gravitation and Cosmology – Principles and Applications of the General Theory of Relativity* (Wiley: New York).
(8) The renormalization group idea was introduced in its modern form into elementary particle physics by M. Gell-Mann and F. E. Low (1954) *Phys. Rev.*, **95**, 1300. Similar concepts were also discussed by E. C. G. Stueckelberg and A. Petermann (1953) *Helv. Phys. Acta*, **26**, 499. For surveys of the applications to the theory of critical phenomena see the following reviews: K. G. Wilson and J. Kogut (1974) *Phys. Rep.*, **12C**, No. 2; M. E. Fisher (1974) *Rev. Mod. Phys.*, **46**, 597; E. Brézin, J. C. LeGuillou and J. Zinn-Justin (1975) in *Phase Transitions and Critical Phenomena*, eds. C. Domb and M. S. Green (Academic Press: London, New York); F. J. Wegner (1975) in *Trends in Elementary Particle Theory*, p. 171 (Springer-Verlag, Berlin); K. Wilson (1975) *Rev. Mod. Phys.*, **47**, 773; Shang-Keng Ma (1976) *Modern Theory of Critical Phenomena* (W. A. Benjamin, New York).
(9) Supersymmetry was introduced by J. Wess and B. Zumino (1974) *Nucl. Phys.*, **B70**, 34; *Nucl. Phys.*, **B78**, 1; *Phys. Lett.*, **49B**, 52. Similar ideas had also been explored by Yu. A. Gol'fand and E. P. Likhtman (1971) *Sov. Phys.: JETP Lett.*, **13**, 323; D. V. Volkov and V. P. Akulov (1973) *Phys. Lett.*, **46B**, 109. A 'superspace' formulation was given by A. Salam and J. Strathdee (1974) *Nucl. Phys.*, **B76**, 477. For a review, see P. Fayet and S. Ferrara (1977) *Phys. Rep.*, **32C**, 249.
(10) 'Supergravity' is the supersymmetric theory of the multiplet consisting of the graviton plus a massless Majorana particle of spin 3/2. It was introduced by D. Z. Freedman, P. van Nieuwenhuizen and S. Ferrara (1976) *Phys. Rev.*, **D13**, 3214; S. Deser and B. Zumino (1976) *Phys. Lett.*, **62B**, 335. For reviews with discussions of 'extended supergravity' theories see B. Zumino (1977) CERN preprint; S. Deser (1978) Brandeis preprint; D. Z. Freedman and P. van Nieuwenhuizen (1978) *Rev. Mod. Phys.*, to be published. For supergravity theories based on a generalization of the Lagrangian $\sqrt{g}R^2$, see M. Kaku, P. K. Townsend and P. van Nieuwenhuizen (1977) *Phys. Rev. Lett.*, **39**, 1109; S. Ferrara and P. van Nieuwenhuizen (1978) Ecole Normale Superieur preprint 78/14.

(11) The cancellation of some divergences in globally supersymmetric field theories of matter was noted by B. W. Lee (1974) (unpublished; cited in Wess and Zumino, *infra*); J. Wess and B. Zumino (1974) *Phys. Lett.*, **49B**, 52; J. Iliopoulos and B. Zumino (1974) *Nucl. Phys.*, **B76**, 310 (1974); S. Ferrara, J. Iliopoulos and B. Zumino (1974) *Nucl. Phys.*, **B77**, 413. The cancellation of divergences in gravitational Green's functions in a theory of supersymmetric matter fields was pointed out by B. Zumino (1975) *Nucl. Phys.*, **B89**, 535. This led to the conjecture that a supersymmetric theory of gravitation could be renormalizable; see B. Zumino (1974) in *Proceedings of the XVIII International Conference on High Energy Physics*, ed. J. R. Smith (Rutherford Laboratory, Chilton, Didcot, Oxfordshire). Cancellation of one-loop divergences in pure supergravity theory were noted by D. Z. Freedman, P. van Nieuwenhuizen and S. Ferrara (1976) *Phys. Rev.* **D13**, 3214; *Phys. Rev.*, **D14**, 912; S. Deser and B. Zumino (1976) *Phys. Lett.*, **62B**, 335; M. T. Grisaru, P. van Nieuwenhuizen and J. A. M. Vermaseren (1976) *Phys. Rev. Lett.*, **37**, 1662; S. Deser, J. H. Kay and K. S. Stelle (1977) *Phys. Rev. Lett.*, **38**, 527. One-loop divergences in theories with supergravity coupled to matter were found in explicit calculations by P. van Nieuwenhuizen and J. A. M. Vermaseren (1976) *Phys. Lett.*, **65B**, 263, and were explained in general terms by Deser, Kay and Stelle (1977) *Phys. Rev. Lett.*, **38**, 527. However, the finiteness of one-loop graphs for $O(n)$ extended supergravity has been shown by Grisaru, van Nieuwenhuizen and Vermaseren (1976) *Phys. Rev. Lett.*, **37**, 1662, for $n=2$; by P. van Nieuwenhuizen and J. A. M. Vermaseren (1977) *Phys. Rev.*, **D16**, 298, for $n=3$ and $n=4$; and by M. Fischler and P. van Nieuwenhuizen (1978) to be published, for $n=8$. In two-loop order, pure supergravity was shown to be free of ultraviolet divergences by M. T. Grisaru (1977) *Phys. Lett.*, **66B**, 75; E. Tomboulis (1977) *Phys. Lett.*, **67B**, 417; Deser, Kay, and Stelle (1977) *Phys. Rev. Lett.* (The same was found to be true in one- and two-loop order for $O(n)$-extended supergravity with n supersymmetry generators.) However, it was pointed out by Deser, Kay, and Stelle (1977), *op. cit.*, and Ferrara and Zumino (1978), CERN preprint, that there are possible ultraviolet divergences in pure supergravity in three-loop order which cannot yet be ruled out by symmetry arguments. Similar problems have been found in three-loop order in $O(2)$-extended supergravity; S. Deser and J. H. Kay (1978), private communication. The possible ultraviolet divergences in supergravity have been analyzed for all orders of perturbation theory, by S. Ferrara and P. van Nieuwenhuizen (1978), Ecole Normale Superieur preprint 78/14. Possible divergences are found in every order beyond two loops, but it is not yet known whether any of their coefficients are non-zero.

(12) Various methods of resummation in general relativity or analogous theories have been studied by B. S. DeWitt (1964) *Phys. Rev. Lett.*, **13**, 114; I. B. Khriplovich (1966) *Yadernaya Fizika*, **3**, 575 (English translation in *Sov. J. Nucl. Phys.*, **3**, 415); A. Salam and J. Strathdee (1970) *Nuovo Cimento Lett.*, **4**, 101; C. J. Isham, A. Salam and J. Strathdee (1971) *Phys. Rev.*, **D3**, 867; (1972) *Phys. Rev.*, **D5**, 2548; A. Salam (1963), *Phys. Rev.*, **130B**, 1287; J. Strathdee (1964) *Phys. Rev.*, **135B**, 1428; A. Salam and R. Delbourgo (1964) *Phys. Rev.*, **135B**, 1398; R. Delbourgo (1977), University of Tasmania preprint; R. Delbourgo and P. West (1977) *J. Phys.*, **A10**, 1049.

Chapter 16. References

(13) Renormalizable theories of gravitation based on Lagrangians which include $\sqrt{g}R^2$ and $\sqrt{g}R^\mu{}_\nu R^\nu{}_\mu$ terms were proposed by S. Deser (1975) in *Proceedings of the Conference on Gauge Theories and Modern Field Theory*, eds. R. Arnowitt and P. Nath (MIT Press, Cambridge, Massachusetts); S. Weinberg (1974) in *Proceedings of the XVII International Conference on High Energy Physics*, ed. J. R. Smith (Rutherford Laboratory, Chilton, Didcot, Oxfordshire), III-59. A general study of such theories has been carried out by K. S. Stelle (1977) *Phys. Rev.* **D16**, 953.

(14) A resummation of the theory which might eliminate this pole has been suggested by A. Salam and J. Strathdee (1978) Trieste preprint. Similar ideas have been pursued by S. Deser (1978), private communication.

(15) G. Källén (1952) *Helv. Phys. Acta*, **25**, 417; H. Lehmann (1954) *Nuovo Cimento*, **11**, 342. (Subtractions might be needed in the dispersion relation for the propagator, but this would make its asymptotic behavior even worse than (16.3).)

(16) The 'large N' approximation was developed in statistical physics by H. E. Stanley (1968) *Phys. Rev.*, **176**, 718; E. Brézin and D. J. Wallace (1973), *Phys. Rev.*, **B7**, 1967; K. G. Wilson (1973) *Phys. Rev.*, **D7**, 2911; L. Dolan and R. Jackiw (1974) *Phys. Rev.*, **D9**, 3320; and in relativistic quantum field theory by R. Abe (1972) *Prog. Theor. Phys.*, **48**, 1414; G. Parisi and L. Peliti (1972) *Phys. Lett.*, **41A**, 331; M. Suzuki (1972) *Phys. Lett.*, **54A**, 5; R. A. Ferrel and D. J. Scalapino (1972) *Phys. Rev. Lett.*, **29**, 413; S. Ma (1972) *Phys. Rev. Lett.*, **29**, 1361. Unphysical 'tachyon' poles were found in some theories by S. Coleman, R. Jackiw, and H. D. Politzer (1974) *Phys. Rev.*, **D10**, 2491; D. J. Gross and A. Neveu (1974) *Phys. Rev.*, **D10**, 3235. However, it was pointed out that these poles are absent in the stable solutions of the theory, by L. F. Abbott, J. S. Kang and H. J. Schnitze (1976) *Phys. Rev.*, **D13**, 2212. Fixed points in non-renormalizable field theories were studied in the large N approximation by G. Parisi (1975) *Nucl Phys.*, **B100**, 368.

(17) E. Tomboulis (1977) Princeton University preprint.

(18) T. D. Lee and G. C. Wick (1969) *Nucl. Phys.*, **B9**, 209; *Nucl. Phys.* **B10**, 1; (1970) *Phys. Rev.*, **D2**, 1033.

(19) S. Weinberg (1964) *Phys. Lett.*, **9**, 357; (1964) *Phys. Rev.*, **B135**, 1049; (1965) *Phys. Rev.*, **B138**, 988; and (1965) in *Lectures on Particles and Field Theory*, eds. S. Deser and K. Ford, p. 988 (Prentice-Hall: New Jersey). The program of deriving classical general relativity from quantum mechanics and special relativity was completed by D. Boulware and S. Deser (1975) *Ann. Phys.*, **89**, 173. I understand that similar ideas were developed by R. Feynman in unpublished lectures at Cal Tech.

(20) Einstein derived his field equations as the unique generally covariant equations in which each term (on the left) would involve just two derivatives of the metric. General covariance is just a convenient way of implementing the Principle of Equivalence, and the limitation to two derivatives picks out those terms in the most general generally covariant field equations which are relevant at long distances. For a discussion of the Einstein field equations along these lines, see S. Weinberg (1972) *Gravitation and Cosmology – Principles and Applications of the General Theory of Relativity*, Section 7.1 (Wiley: New York).

(21) For reviews, see S. Weinberg (1968) *Proceedings of the XIV International Conference on High Energy Physics* (CERN, Geneva), 253; S. Weinberg (1970) in *Lectures on Elementary Particles and Quantum Field Theory – 1970 Brandeis Summer Institute in Theoretical Physics*, eds. S. Deser, M. Grisaru and H. Pendleton (MIT Press: Cambridge, Mass.); B. W. Lee (1972) *Chiral Dynamics* (Gordon and Breach: New York).
(22) S. Weinberg (1967) *Phys. Rev. Lett.*, **18**, 507; J. Schwinger (1967) *Phys. Lett.*, **24B**, 473; S. Weinberg (1968) *Phys. Rev.*, **166**, 1568; S. Coleman, J. Wess, and B. Zumino (1968), *Phys. Rev.* **177**, 2239; C. Callan, S. Coleman, J. Wess, and B. Zumino (1968), *Phys. Rev.*, **177**, 2247.
(23) Photon pairing instabilities have been proposed as a possible origin for gravitation by S. L. Adler, J. Lieberman, Y. J. Ng and H.-S. Tsao (1976) *Phys. Rev.*, **D14**, 359; S. L. Adler (1976) *Phys. Rev.*, **D14**, 379. For other theories with composite gravitons, see P. R. Phillips (1966) *Phys. Rev.*, **146**, 966; A. D. Sakharov (1967) *Dokl. Akad. Nauk. SSSR*, **177**, 70; H. C. Ohanian (1969) *Phys. Rev.*, **184**, 1305; H. P. Dürr (1973) *Gen. Relativ. Grav.*, **4**, 29; D. Atkatz (1977) Stony Brook preprint ITP-SB-77-59; H. Teragawa, Y. Chikashige, K. Akama and T. Matsuki (1977) *Phys. Rev.*, **D15**, 1181; T. Matsuki (1978) *Prog. Theor. Phys.*, **59**, 235; K. Akama, Y. Chikashige and T. Matsuki (1978) *Prog. Theor. Phys.*, **59**, 653; K. Akama, Y. Chikashige, T. Matsuki and H. Terazawa (1977), INS-Report-304; K. Akama (1978), Saitama preprint.
(24) The condition that $m(\mu)/\mu$ should vanish for $\mu \to \infty$ is discussed by S. Weinberg (1973) *Phys. Rev.*, **D8**, 3497.
(25) T. Kinoshita (1962) *J. Math. Phys.*, **3**, 650; T. D. Lee and M. Nauenberg (1964) *Phys. Rev.*, **133**, B1549. For a recent application see G. Sterman and S. Weinberg (1977) *Phys. Rev. Lett.*, **39**, 1436.
(26) B. S. DeWitt (1967) *Phys. Rev.*, **162**, 1195, 1239; (1975) *Phys. Rep.*, **19**, 295; G. 't Hooft and M. Veltman (1974) ref. 3; J. Honerkamp (1972) *Nucl. Phys.*, **B48**, 269; R. Kallosh (1974) *Nucl. Phys.*, **B78**, 293; M. T. Grisaru, P. van Nieuwenhuizen and C. C. Wu (1975) *Phys. Rev.*, **D12**, 3203.
(27) For earlier discussions, see S. Weinberg (1977), invited talk at the Eighth International Conference on General Relativity, Waterloo, Ontario, Canada (unpublished). Related ideas are discussed by E. S. Fradkin and G. A. Vilkovsky (1976) *Proceedings of the XVIII International Conference on High Energy Physics*, vol. 2, Sec. T28 (JINR, Dubna, 1977); (1976), Berne preprint; (1978), I.A.S. preprint 778-IPP.
(28) This is a generalization of an example suggested to me by E. Witten. Of course, this example is specifically chosen to have trajectories with the desired behavior, that they either hit a fixed point for $\mu \to \infty$ or go to infinity for finite values of μ. However, this example does show that this behavior can arise for β-functions that are not at all pathological.
(29) The asymptotic freedom of non-Abelian gauge theories was discovered by H. D. Politzer (1973) *Phys. Rev. Lett.*, **30**, 1346; D. J. Gross and F. Wilczek (1973) *Phys. Rev. Lett.*, **30**, 1343.
(30) K. G. Wilson and M. E. Fisher (1972) *Phys. Rev. Lett.*, **28**, 240; K. G. Wilson (1972) *Phys. Rev. Lett.*, **28**, 548.
(31) W. A. Bardeen, B. W. Lee and R. E. Shrock (1976) *Phys. Rev.*, **D14**, 985; E. Brézin, J. Zinn-Justin and J. C. Le Guillou (1976) *Phys. Rev.*, **D14**,

Chapter 16. References

2615; also see A. M. Polyakov (1975) *Phys. Rev. Lett.*, **59B**, 79; A. Midgal (1975) *Zh. Eksp. Teor. Fiz.*, **69**, 1457.

(32) This is shown for the theory of critical phenomena by F. J. Wegner (1974) *J. Phys., C. Solid State Physics*, **7**, 2098. This work emphasizes that the invariance of the critical exponents applies only to those eigenvectors associated with essential couplings. In statistical physics, couplings that are not essential are called 'redundant'. I use the term 'inessential' here instead, because in referring to its opposite, it is easier to say 'essential' than 'irredundant'.

(33) G. 't Hooft and M. Veltman (1972) *Nucl. Phys.*, **B48**, 189; C. G. Bollini and J. J. Giambiagi (1972) *Phys. Lett.*, **40B**, 566; J. F. Ashmore (1972) *Nuovo Cimento Lett.*, **4**, 289.

(34) This is the method used in most applications of the renormalization group to critical phenomena. See, for example, Kogut and Wilson (1974) ref. 8.

(35) For a review of modern renormalizable gauge theory of weak and electromagnetic interactions, see E. S. Abers and B. W. Lee (1973) *Phys. Rep.*, **9**, 1; J. C. Taylor (1976) *Gauge Theories of Weak Interactions* (Cambridge University Press).

(36) This is shown for general field theories by Y. Nambu (1968) *Phys. Lett.*, **26B**, 626; D. G. Boulware and L. S. Brown (1968) *Phys. Rev.*, **172**, 1628; L. V. Prokhorov (1969) *Phys. Rev.*, **183**, 1515, and specifically for general relativity in low orders by M. J. Duff (1973) *Phys. Rev.*, **D7**, 2317. Quantum loop connections are considered by M. J. Duff (1974) *Phys. Rev.*, **D9**, 1837. The applications of these results to the motion of large masses is considered by D. Boulware and S. Deser (1975) *Ann. Phys.*, **89**, 173.

(37) This is a generalization of a result of T. Applequist and J. Carazzone (1975) *Phys. Rev.*, **D11**, 2856. They showed that, in a renormalizable theory of 'light' and (much heavier) 'heavy' particles, the interactions among the 'light' particles is given by an effective renormalizable field theory in which 'heavy' particles do not appear.

(38) For a discussion, see pp. 49–50 of S. Weinberg (1977) *Physics Today*, **30**, No. 4; also E. Gildener and S. Weinberg (1976) *Phys. Rev.*, **D13**, 3333.

(39) The use of dimensional regularization to construct a new set of renormalization group equations is due to G. t'Hooft (1973), *Nucl. Phys.*, **B61**, 455; *Nucl. Phys.*, **B82**, 444. The derivation given here is a somewhat simplified version of the one given by 't Hooft.

(40) In ref. 39, the renormalized coupling constant was defined strictly at $D = 4$, as the constant term in a Laurent expansion around $D = 4$. The extension of this definition and the corresponding renormalization group equation to arbitrary irrational spacetime dimensions has been implicity used for some time by the Saclay group (see Brézin *et al.* (1976), ref. 42) and was given explicitly bt D. J. Gross (1976), in *Methods in Field Theory*, eds. R. Balian and J. Zinn-Justin, section 4 (North-Holland: Amsterdam).

(41) J. C. Collins (1974) *Phys. Rev.*, **D10**, 1213; and University of Cambridge thesis (unpublished.)

(42) E. Brézin, J. C. LeGuillou, J. Zinn-Justin and B. G. Nickel (1973) *Phys. Lett.*, **44A**, 227; E. Brézin, J. C. LeGuillou and J. Zinn-Justin (1973) *Phys. Rev.*, **D8**, 2418; Kogut and Wilson (1974), ref. 8, Table 8.1.

(43) See, for example, S. Weinberg (1972), ref. 6, Section 6.7.
(44) A. Duncan (1977) *Phys. Lett.*, **66B**, 170.
(45) For spin 0 contributions see L. Brown (1977) *Phys. Rev.*, **D15**, 1469. For spin $\frac{1}{2}$ contributions see D. M. Capper and M. J. Duff (1974) *Nucl. Phys.*, **B82**, 147. For spin 1 contributions see D. M. Capper, M. J. Duff and L. Halpern (1974) *Phys. Rev.*, **D10**, 461. For graviton contributions, see H.-S. Tsao (1977) *Phys. Lett.*, **66B**, 79.
(46) C. N. Yang and R. L. Mills (1954) *Phys. Rev.*, **96**, 141.
(47) R. Gastmans, R. Kallosh and C. Truffin (1977), Lebedev Institute preprint, to be published. Also see E. S. Fradkin and G. A. Vilkovsky, ref. 27.
(48) G. W. Gibbons and S. W. Hawking (1977) *Phys. Rev.*, **D15**, 2752. Also see S. W. Hawking (1977) *Phys. Lett.*, **60A**, 81 (1977). In ref. 49, Christensen and Duff relate the extra term in (16.76) back to early work of R. Arnowitt, S. Deser, and C. W. Misner (1962), in *Gravitation: An Introduction to Current Research*, ed. L. Witten (Wiley: New York); J. W. York (1972) *Phys. Rev. Lett.*, **28**, 1082; C. W. Misner, K. S. Thorne and J. A. Wheeler (1973) *Gravitation*, Chapter 21 (Freeman: San Francisco).
(49) S. M. Christensen and M. J. Duff (1978), Brandeis preprint.
(50) In this formula, it is assumed that the form of the energy–momentum tensor is 'improved' for arbitrary D in the manner suggested by C. Callan, S. Coleman and R. Jackiw (1970), *Ann. Phys.*, **59**, 42, for $D = 4$. However, this only affects $A(q^2)$, not $B(q^2)$.

Index

absolute space and absolute time, 1
absorber theory of radiation, 590
accelerated conductor, renormalized electromagnetic stress tensor near, 691
accelerated observer, vacuum for, 692–5, 650
accretion (black hole), 458–61, 465–9, 472–81, 484, 487, 489–93, 496
achronal past (future) boundary, 623–8
achronal slices, 245–6, 291–2
action
    for Bianchi models, 552–5
    effective, 650, 658, 702, 717, 740, 745
    Euclidean, 753–61, 773–7
    field, 749
    functional for quantized fields, 699, 713
    generalized, 737
    for gravity in general Hamiltonian formalism, 152–3
    for gravity in Lagrangian formulation, 142
    for gravity in Regge formalism, 784
    for quantized scalar field, 706
    for tensor field, 144
    surface term for Einstein action, 158, 749
ADM formalism, *see* Einstein field equations, in Dirac–ADM form; Hamiltonian formalism, for the Einstein field equations in Dirac–ADM form
aether, 1–2
    Einstein's views on, 680
    quantum, 683–98, 710
    see also vacuum
algebraically special fields, 387–9
angular momentum, of Kerr black hole, 395
angular momentum flux, ratio to energy flux, 331–4
anisotropic cosmological models, particle creation in, 527–8
anisotropic homogeneous cosmological models, 550–62

group of motions, 542
particular metrics, 563–70
reasons for considering them, 536
anisotropic stresses, *see* energy–momentum tensor; isotropization of cosmologies
anisotropy energy, 579–80
anomalous perihelion shifts, 60
anomaly
    axial-vector current, 661, 689
    trace, 662, 679, 689, 717–20, 768
anthropic principle, 633
antimatter, in cosmology, 508, 512
apparent horizon, 300
area
    decrease for white holes, 602
    and entropy of a black hole, 602, 635
area theorem (black hole), 17–18, 300–1, 602, 619
arrow of time, 581
    see also time-asymmetry
asymptotic freedom (quantum gravity), 22, 764, 806, 812
asymptotic predictability, 217
asymptotic safety, 22, 791, 798–828
    critical exponents, 809, 818, 821
    dimensional continuation, 792, 807, 814–28
    infrared critical surface, 805, 809
    ultraviolet critical surface, 802–7, 810, 814, 820–1, 823, 825
asymptotically de Sitter solution, *see* Schwarzschild–de Sitter solution; Kerr–de Sitter solution; Taub–de Sitter solutions
asymptotically Euclidean metrics, in positive action conjecture and fall-off of Riemann tensor, 774
    see also positive action conjecture
asymptotically flat spacetime, 10, 277–9
Atiyah–Singer index theorem, 661, 690
axial anomaly, 661, 689
axial-vector current anomaly, 689

back reaction due to particle creation, 643, 652, 675, 679, 698–702

Index

back reaction (*cont.*)
  near the big bang in anisotropic models, 574
background field, method of, 698–704, 801
background (metric) fields, as stationary-phase points in path-integrals, 771–7
'barrel'-type singularity, 561
baryon excess in universe, 524, 636–7
baryon non-conservation in gravitational collapse, 20, 530
  in big-bang particle production, 637
  and instantons, 662
Bekenstein–Hawking formula (for entropy of black hole), 17, 602, 615, 630, 635, 636
Berger–Ebin splitting, *see* reduction of phase space
Bergmann–Wagoner–Nordtvedt theory, 48, 69
Berkeley criticism of Newtonian theory, 12
Betti number, *see* topological invariants
Bialynicki–Birula–Mycielski nonlinear modification of Schrödinger's equation, 636 (footnote)
Bianchi cosmological models, 550–62
Bianchi identities
  and constraints, 159
  in Newman–Penrose formalism, 381
  in tetrad formalism, 374–5
Bianchi metrics, 546–9
Bianchi types, classification, 541
big-bang theory, 13–15, 504–17, 520, 543
  see also singularities
bimetric theories of gravity, *see* Rosen theory
binary pulsar, 70–84
  gravitational waves from, 78–81
binary systems
  binary cataclysmic variables, 103, 104 (table 3.5)
  black holes as members of, 100, 469–81
  classical binaries, list of, 101, 102 (table 3.4)
  death of, 100, 473–4
  doubly compact, 106–7, 115
  evolution of, 471–2
  generation of gravitational waves by, 97–108, 98 (table 3.2), 101 (figure 3.1)
  maximum rotational frequency, 100, 100 (table 3.3)
  and novae, 110–11, 471–4

black hole(s), 4, 15–21
  accretion of gas onto, 458–61, 465–9, 472, 484, 487, 489–93, 496
  angular velocity, 318–21
  area theorem, 17–18, 300–1, 602, 619
  astrophysical aspects, 454–503, 511
  birth of, 461–4; *see also* gravitational collapse
  chemical potential, 674, 675, 677
  CP-violation, 530
  collisions, 17, 457, 465
  cosmic censorship, *see* cosmic censorship
  detection of, 464, 467, 470
  effective surface current, 339–42
  effective surface viscosity, 313
  electric potential, 322–3
  electrodynamics, 313–14, 322–3, 336–52
  energy extraction, 6, 351
  entropy, 7, 17–20, 602–5, 636 (footnote), 675, 697
  ergo region, 674, 678
  eternal, 672, 696
  evaporation, 18–21, 528–30, 603, 698, 702
  event horizon, 280–2
  evolution of horizon, 302–14
  formation in early universe, 566, 572, 608–9, 612
  generalized Smarr formula, 353
  global characterization, 296–300
  in globular clusters, 17, 481–4
  gravitational waves from collapse to, 98 (table 3.2), 114 (figure 3.5), 114–18
  gravitational waves from fall of matter into, 118–19
  gravitational waves from two colliding black holes, 118
  Hartle–Hawking formula, 310–12
  horizon, *see* horizon
  Kerr metric, *see* Kerr black hole geometry
  luminosity, 698
  mass and angular momentum integrals, 352–6
  mechanics: first law, 356–9; second law, 301, 675, 697; third and zeroth laws, 323–7
  as member of binary, 100, 469–81, 501
  mini, 618, 635
  no-hair theorem (uniqueness of Kerr), 4, 17, 359–69, 370
  particle creation by, 7, 18–21, 528, 601, 640, 668–79, 695–8
  Penrose process, 6
  primordial, *see* primordial black holes

904

black hole(s) (*cont.*)
  in quasars and galactic nuclei, 17, 485–93
  in radiation bath, 603–5, 608–10
  rigidity, 320
  Schwarzschild, *see* Schwarzschild spacetime
  singularity, 16, 598, 613, 617–28, 272; *see also* singularities
  Smarr formula, 675
  stationary, 314–27
  stress tensor, 662
  super-radiance, 6, 335, 351, 640, 674
  surface gravity, 18, 307, 313, 323–7
  tacit assumptions, 456–7
  temperature, 7, 18, 20, 313, 668, 675, 677, 696, 772
  thermal equilibrium, 603–5, 608–10, 675–7
  thermal radiation from, 18–21
  thermodynamics, 7, 16–19, 301, 323–7, 356–9, 675, 696–7
  thermodynamic potential, 675
  and time-asymmetry, 600–11
  3-velocity, 321–2
  versus white holes, 600–11
  violation of baryon conservation in, 20, 530, 662
  and X-ray sources, 17, 488–9, 474–8
BMS (Bondi–Metzner–Sachs) group, 668
Bogoliubov transformation, 651–3, 665, 667, 671, 678, 695
Boltzmann entropy, 588, 632
Bondi–Metzner–Sachs (BMS) group, 668
branch systems, 589 (footnote)
Brans–Dicke theory, 48, 54, 55, 86, 88, 516
broken symmetries, in hadron era, 524–5
bulk viscosity, 616

C-violation, in weak interactions and in cosmological particle production, 637
canonical energy–momentum tensor, 328–331
canonical momentum, in ADM formalism, 145
canonical quantization of gravity and 'quantum cosmology', 573–4
  see also quantization of gravity
Casimir effect, 18, 685–7, 690, 744
cat paradox, 585
Cauchy horizon, 254–5, 291–2, 559, 565, 620, 626
Cauchy problem, 138–211
  for Einstein equations, 8–10, 185–93, 284–8
  for Einstein vacuum equations: global formulation, 192–3; local existence theorem, 185–9; local uniqueness theorem, 190–1
  for linear evolution equations, 168–75
  for Maxwell equations, 249
  for nonlinear evolution equations, 175–83
  in presence of a white hole, 601
Cauchy stability of vacuum solution, 190
Cauchy surface, 252–5, 291–2
causal curve, 624, 628
causal past (future), 623
causal structure, 232–55
causal violations, 238–43
  see also time-asymmetry
causality, stable, 210–11, 241–3
Cavendish experiment, 58
Chandrasekhar limit, 454, 462
chaotic cosmology, 13, 494, 496, 536, 572–3, 579, 580, 581, 629–35
chiral dynamics, 796–7
chiral invariance, 646
chromodynamics, 796
chronological past (future), 623
'cigar'-type singularity, 559, 560
CN interstellar molecules, excitation by cosmic microwave radiation, 520
collapse, *see* black hole; gravitational collapse; singularities
collapse of the wavefunction, 584–6, 636
commutation relations
  canonical, 647
  in quantum gravity, 747
commutator of intrinsic derivatives, 374
compact manifold, 219
complex projective space, 777
complexification of spacetime, 658, 666, 668, 676–8, 753–6
conductor, renormalized stress tensor near, 691
conformal anomaly, 662, 679, 689, 717–20, 768
conformal diagram, 619
  for Reissner–Nordström solution, 619
conformal equivalence classes of metrics, 760
conformal infinity, 277–82, 623
conformal transformation of Einstein action, 758
conformally invariant field theory, 646, 664, 689, 712, 715
conformally self-dual metric, 757
congruences of timelike and null curves, kinematics, 302–14
conservation laws
  energy–momentum, 9–10

Index

conservation laws (*cont.*)
  gravitational energy, 10
constraint equations
  in Dirac–ADM formalism, 147
  in Hamiltonian formalism, 138, 152
  linearization stability, 574–9
constraint manifold
  as divergence constraint manifold, 166
  Hamiltonian constraint manifold, 163
  and symplectic reduced space, 206–7
  and symplectic symmetries (*J*-invariances), 208–9
  for vacuum Einstein system, 158, 166
'continuous creation' (in cosmology), 14, 533
cosmic censorship, 4, 16, 270–88, 297–8, 456–7, 601, 608, 613, 617–29
  strong, 624–7
cosmic gamma-ray background, 21
cosmic microwave background, 3, 14, 520
cosmological constant, as counterterm in effective Lagrangian, 654, 659, 810
cosmological density parameter, 86
cosmological models, 533–80
  admitting a G_2 of motions, 567–70
  anisotropic, 13, 527–8
  de Sitter, 12–13, *see* de Sitter spacetime
  Einstein, 12, 628, 666
  Ellis, 577, 580
  entropy generation, 579–80
  Friedmann, 2, 4, 13, 527, 533, 614
  Gowdy, 563, 567, 574
  helium production, 4, 578, 634
  homogeneous, 504–8
  homogeneous anisotropic, 13, 550–62
  inhomogeneous, 494, 507, 563–70
  island universe, 504, 516
  isotropization by viscous effects, 13, 562, 570–1
  isotropization by particle creation, 6, 527, 573–4
  Kantowski–Sachs, 544–6, 563, 565–6, 615
  Kasner, 558–60, 567, 572–3, 575, 580
  locally rotationally symmetric (LRS), 549
  nucleosynthesis, *see* nucleosynthesis in early universe
  particle creation in, 6, 526–8, 573–4, 608, 628–9, 640, 643, 664–8, 745
  quantum effects, 526–32, 573–4
  quasi-isotropic, 528
  self-similar, 564–5
  singularities in, 4, 15–16, 255–89, 509, 611, 614–15; *see* singularities
  steady-state, 14, 533, 542, 577, 580
  Swiss-cheese, 563
  Tolman, 566–7, 575, 615–17
  turbulent, 579
  see also cosmology; universe
cosmological term, in Einstein's field equations, 8, 11, 12–13
cosmological tests of gravitational theories, 86, 88
cosmology,
  Brans–Dicke theory, 516–17
  chaotic, 13, 494, 496, 536, 572–3, 579, 580, 626–30
  continuous creation, 14, 533
  Dirac theory of 'large numbers', 530–1
  Einstein's first paper on, 518
  Newtonian, 11–12, 518–19
  relativistic, 12–13
  see also cosmological models; universe
covering space, universal, 229–32
CP-violation and black hole evaporation, 530
CPT-theorem, 637
CPT-violation in an expanding universe with *CP*-violation, 637 (footnote)
Crab Nebula, 70
cut-off for quantum gravity, 771
  see also quantum gravity
Cygnus X-1, 5, 17, 470, 475, 476, 478, 479–81

deformation of Lorentz metric on manifold, 222–4
deformation (infinitesimal) of vacuum Einstein solution, 196
density matrix, 641–2
  for Bogoliubov-transformed vacuum, 653
  for Hawking radiation, 671
  for Rindler vacuum, 695
dependence, domain of, 247–55, 291–2
Deser's splitting, *see* reduction of phase space
de Sitter spacetime, 12, 776
  complexification, 658
  particle creation, 666, 678
  particle definition, 650
  renormalized stress tensor, 663, 720
DeWitt metric, 148
DeWitt–Schwinger proper-time formalism, 655–8, 704–10
Dicke framework for experimental gravity, 27, 45
dilation invariance, 646
dimensional regularization, 654, 659, 663, 713, 765, 810, 814–21
Dirac canonical commutation relations, 162

Index

Dirac cosmological theory, 40, 58–9, 530–1
Dirac equation
  in curved spacetime, 644
  in Kerr geometry, 426
  in Minkowski space, 429
  in Newman–Penrose formalism, 425
Dirac string, analogue in general relativity, 773–4, 775
Dirac–ADM evolution equations, *see* Einstein field equations, in Dirac–ADM form
direction field on manifold, 219–20, 223–4
directional derivative, 372
divergences of quantum field theory, 654–61, 703–17, 790–831
divergences in quantum gravity, 21–2, 660, 716, 790–8
domain of dependence, 247–55, 291–2
domain of influence, 232–43, 290–1
domain of outer communications, 296–7
dominant-energy condition, 551, 571
Doppler ranging and gravitational waves, 133–5, 137
Doppler-shift, binary (pulsar) system, 76–7
dragging of inertial frames in Kerr geometry, 395
dynamical representation, 207
dyons, gravitational, 773–5
  see also Taub–NUT spacetime

early universe, 518–32
  chaotic, 13, 496, 536, 572–3, 577 (table 11.3), 579, 629; primordial black holes, 19, 494–502; quarks in, 522, 526
  decoupling of matter and radiation, 523
  galaxy formation, 496, 523, 525
  hadron era, 520, 523–6
  helium production, 4, 578, 611, 634–5
  isotropization of, 6, 13, 562–3, 570–1; by creation of particles, 6, 527–8, 573–4, 667
  lepton era, 520, 522–3
  neutrinos in, 521–2
  nucleosynthesis, 519, 523, 578, 611, 634
  particle creation in, 6, 526–30, 608, 628–9, 640, 643, 663–8, 744–5
  quantum effects in, 526–32, 744–5
  quarks in, 522, 526
  uniformity of, 518
eclipse expedition (1919), 2
effective action, 702, 717, 740, 745

Einstein, 90, 91, 504–5, 509
  on aether, 680
  on his field equations, 639
  first paper on cosmology, 12, 518
  geodesic postulate, 9
  on gravitational radiation, 91
  on quantum theory, 6, 744
  on unified field theory, 681
  principle of equivalence, 8, 26–7, 28, 35–9, 54–5, 84–8
Einstein field equations, 8–11
  with cosmological term, 749
  in Dirac–ADM form, 146–7
  in Euler–Lagrange form, 142–4
  in Hamiltonian form, 138, 151
  linearization stability, 195–203
  quantized form, 699
Einstein–Podolsky–Rosen experiment, 685
Einstein static universe, 12, 627
  particle creation in, 666
Einstein's theory, *see* general relativity; special relativity
electromagnetic waves
  reflection and transmission by Kerr black hole, 370, 402, 429, 437–45
element formation (big-bang cosmologies), 4, 519, 523, 578–9, 580, 611, 634–5
Ellis cosmological model, 577, 580
energy, gravitational, 10–11, 639
energy condition, 259–60, 315–17
energy integral (stationary system) 352–6
energy loss from pulsars, 80
energy–momentum tensor
  canonical and metric 328–31
  local conservation of, 9, 69
  pseudo-tensor for gravitation, 10
  in quantum field theory, *see* stress tensor
energy of universe, 513
entropy
  behaviour in oscillating universe, 616
  of black hole, 7, 17–20, 602–5, 636 (footnote), 675, 697, 779
  Bekenstein–Hawking formula, 602
  generation by viscous effects, 616
  of gravitational field, 611–17, 778–82
  of mixed state, 642
  observer-dependence of, 642
  of universe, 511–12, 579–80, 611–17, 629–30, 636–7
  of Yang–Mills field, 779
Eötvös–Dicke–Braginsky experiment, 8, 30, 54
equivalence principle, 8
  strong, 84–8

907

Index

equivalence principle (*cont.*)
  weak, 8, 26–7, 28, 35–9, 54, 55, 84
ergo region, particle creation in, 674–5, 678–9
  see also Kerr black hole geometry
ether, *see* aether; vacuum
Euclidean action
  definiteness of the Euclidean Yang–Mills action, 757
  for de Sitter space, 776
  for gravity, 755
  indefiniteness of Euclidean gravitational action, 757–61
  for $K3$-space, 777
  for Kerr solution, 772
  for Kerr–de Sitter solution, 776
  for NUT and anti-NUT solution, 775
  for Reissner–Nordström solution, 772
  for scalar field, 753
  for Schwarzschild solution, 771
  for Schwarzschild–de Sitter solution, 776
  for Taub–de Sitter solution or CP^2, 776
Euclidean propagator, for scalar field, 754
Euclidean section,
  of curved spacetime, 757
  of flat spacetime, 753
  of Kerr solution, 772
  of Reissner–Nordström solution, 772
  of Schwarzschild solution, 771
  of Taub–NUT spacetime, 775
Euler number, *see* topological invariants
event horizon, 7, 16–17
  see also black hole, event horizon; horizon
Everett-type many-sheeted universe, 585, 594, 744
expansion of congruence, 386
experimental tests of gravitational theories
  binary pulsar, 70–84
  cosmological, 84–8
  gravitational radiation, 62–70, 78–81, 90–137
  solar system, 5, 9, 24–62
extended quantum mechanics (Penrose), 636

Faddeev–Popov ghost field, 735–8
Fermi rotation, 552
Feynman path-integral, *see* path-integral
Feynman propagator, 655–6, 661, 704–10, 739
  in black hole spacetime, 675–7
  in de Sitter spacetime, 666
  graviton, 794, 797, 810–13, 825, 828
  for scalar field, 754
  in Taub–NUT spacetime, 668
Feynman rules, for quantum gravity, 21–2
  see also quantum gravity
fibre bundle
  associated with gravitational and Yang–Mills field, 140–1
  Hopf bundle, 774
field equations, *see* Einstein field equations
foam-like structure of spacetime, 785, 789
Fock bases, 648–52, 718
  in Schwarzschild spacetime, 670
  see also Bogoliubov transformation
Friedmann cosmological models, 2, 4, 13, particle creation in, 526–7
  perturbations of, 13
Friedmann–Robertson–Walker (FRW) models, 533, 614–15
fully conservative theory, 44
future and past of a point, 232–42

galaxy formation, 496–7, 523, 525
gauge-fixing term in Euclidean–Einstein action, 769
gauge invariance, identities generated by vacuum gravitational fields, 159
gauge theories
  non-Abelian, 682
  quantization, 720–43
Gell-Mann–Low equation, 801
Gell-Mann–Low functions, 815
general relativity,
  discovery, 2
  foundations, 8–11
generic spacetime, 264
geodesic focussing and crossing, 258–60, 299–301
geodesic hypothesis, 9
geodesic precession, 60
geodesically incomplete spacetime, 257–8
ghost contribution to the Euclidean–Einstein action, 769
ghost field, 735–8, 769
global structure of spacetime, 212–93
globally hyperbolic manifold, 145, 624–6
globular cluster,
  black holes in, 116, 132, 481–4
  central relaxation time, 482
  evolution of, 481–4
  gravitational waves from, 114 (figure 3.5), 116
  tidal friction, 484
Gödel spacetime 238, 243
Goldberg–Sachs theorem, 389–91
  applied to Kerr geometry, 399

Index

Goldstone boson, 796
Gowdy model (of universe), 563, 567, 574
Grassmann variables, in path-integral, 764
gravitational collapse, 3, 4, 15–17
  cosmic censorship, 4, 16, 270–88, 297–8, 601, 608, 613, 617–29
  event horizon, 280, 298
  particle creation in, 643, 670–5, 696–8
  singularity 'crisis', 455, 509–10
  singularity theorems, 258–66
  source of gravitational waves, 95, 114–18, 137, 464–5
gravitational constant
  constancy of, 40, 86, 516, 531–2
  variations in locally measured value, 58–9
gravitational dyons, 773–5
  see also Taub–NUT spacetime
gravitational energy, 10–11, 54, 83
gravitational energy–momentum tensor, 10
gravitational entropy, 611–17, 778–82
gravitational field, universal coupling, 8
gravitational instantons, 788
gravitational magnetic-type mass in Taub–NUT spacetime, 773
gravitational radiation, 5–6, 10–11, 17, 62–70, 78–81, 90–137
  black-body cosmological, 98 (table 3.2)
  bursts, 94–6, 131–2, 134; sources of, 114–19
  continuous, 94, 127–8, 134–5; sources of, 97–112
  detection of, 92, 119–37; from binary pulsar, 78–81
  detectors: Brownian-motion noise, 123–4, 125, 136; and cryogenics, 136, 136 (figure 3.9); cylindrical-bar, 93 (figure 3.2), 120, 121–2, 123, 128; Doppler ranging, 133, 137; free-mass and almost free-mass, 129–35; idealized resonant antenna (two masses on a spring), 93 (figure 3.2), 120; laser ranging, 129–32; mass quadrupole, 92; mechanically resonant, 120–2, 128, 135; prospects for the future, 78–81, 135–6; quantum non-demolition, 127, 136; research groups, 130, 135, 136; sensor noise, 124–5, 136; space experiments, 129, 132, 133, 137
  dimensionless amplitude (h), 94, 96, 101 (figure 3.3), 114, 114 (figure 3.5)
  dipole, 68–9, 80–1, 91
  efficiency factor, 94–5, 117

  Einstein on, 91
  frequency bands, 98 (table 3.2)
  generation of, 91–2
  linearized theory of, 90
  luminosity formulae, 91–2, 94, 95, 99
  man-made, 98 (table 3.2)
  polarization, 63–7
  as a propellant, 681
  quadrupole nature, 91, 92
  reflection, transmission by Kerr black hole, 370, 402, 449–51
  Rosen time-symmetric theory, 78
  sources of: binary pulsar, 78–81; binary systems, 67–70, 501; fall of matter into black hole, 117–18; gravitational collapse, 114–18, 131–2, 134, 464–5; gravitational scattering of two bodies, 119; neutron starquakes, 98 (table 3.2), 118; novae, 110–11; pulsation of neutron star, 111–12, 115; rotation of a star, 112; supernovae, 92, 110, 115–16, 131–2, 135, 464–5; two colliding black holes, 118; vibrations of a neutron star, 108, 111–12, 115, 118; vibrations of a white dwarf, 98 (table 3.2), 101 (figure 3.3), 109–10; white dwarf quakes, 19
  speed of gravitational waves, 67
  weak-field theory, 64–7, 91–100
gravitational redshift experiment, 32
gravitational theories, 5, 26–9, 34–50
  Dicke framework, 27–9
  metric and non-metric theories, 35–40
  PPN formalism, 42–5
  Rastall, 49
  Rosen, 48–9, 75, 81, 87, 88
  scalar–tensor (Brans–Dicke), 48, 54–5, 64, 68–9, 86, 88, 516
  Schiff's conjecture, 34–5
  stratified, 49, 88
  vector–tensor, 48
gravitational thermodynamics, 7, 17–20, 611–17, 778–82
  see also black hole(s), entropy; entropy
gravitational waves, *see* gravitational radiation
gravitons, composite, 795–8
gravity, experimental, *see* experimental tests of gravitational theories; gravitational radiation
Green's function
  for Casimir effect, 685–7, 689
  for field in curved spacetime, 655–8, 704–10
  graviton, 794, 797, 811, 825, 828
  in black hole spacetime, 675–7

Index

Green's function (*cont.*)
  many-particle, 655, 739
  renormalized, 686
  thermal, 655, 676–7, 698 *see also*
    Feynman propagator
gyroscope experiments, 42, 60–1

H-theorem, 588 (footnote)
hadron era, 520, 523–6
  phase transitions in, 525–6
  symmetry breaking in, 524
Hamidew (Gilkey) expansion, 656, 659, 661, 764–5, 766
Hamilton–Jacobi equation for orbits in Kerr geometry, 396
Hamiltonian for field theory in curved spacetime, 647
Hamiltonian formalism
  for the Einstein equations in Dirac–ADM, 138, 144, 146–7, 151–2
  for the field equations of Bianchi models, 552–5
  for scalar field coupled with gravity, 157–8
Hamiltonian vector field, 154
harmonic coordinates, 185
Hartle–Hawking formula (black hole), 310–13
Hausdorff condition for spacetime, 594, 595
Hawking area theorem 16–17, 301
Hawking–Penrose singularity theorems, 4, 258–66, 299–300, 616
Hawking radiation (black holes), 7, 18–21, 528–30, 603–6, 637, 640, 668–79, 695–8, 743
heat death of the universe, 604
heat equation and zeta function, 766–7
helicity states of the graviton, 63
helium production, 4, 578–9, 634 (footnote)
Hilbert space for quantum field theory, 641–2
Hirzebruch signature, *see* topological invariants
homogeneous cosmological models, *see* spatially-homogeneous cosmological models
homotheties of solutions of Einstein equations, 564–5
Hopf bundle, 774–5
horizon
  apparent, 300
  Cauchy, 254–5, 291–2
  connection with black hole entropy and particle creation, 642, 650, 653, 675, 677–9, 695

  in cosmology, 494, 505–6, 508, 510
  in de Sitter spacetime, 678
  event: definition, 280–2; stationary, 314–18; *see also* black hole(s)
  inner, perturbations of, 283
  in Kerr spacetime, 391, 431, 673
  particle, 572–3, 601, 602, 606, 627
  in Rindler space, 650, 694, 696
  in Schwarzschild spacetime, 669
Hubble expansion parameter, 86–7, 534
Hubble redshift, 2
  isotropy, 534–5
Hubble relation, 533
Hughes–Drever experiment, 27, 29, 38

ideal points, 606–7, 621–6
incompleteness, geodesic, 257–8
index theorem, Atiyah–Singer, 661, 690
infinity, conformal, 277–82
influence, domain of, 232–43, 290–1
inhomogeneous cosmological models, 563–70
initial value problem, 138–211
  for Einstein's field equations, 8–10
  for gravitational field, 249–50, 284–8
  for Maxwell field, 249
  see also Cauchy problem
instanton, 668, 688, 690, 702
  condition, 662
intermediate singularity, 547, 559
interstellar gas, 458–9
  accretion onto black hole, 458–61, 465–9
intrinsic derivative, 373
isolation theorem (in geometry), 195
isotropization of cosmological models
  by particle creation, 6, 526–7, 573–4, 667
  by viscous effects, 13, 562–4, 570–1
isotropy (cosmology)
  of cosmic magnetic field, 535
  of cosmic microwave and X-ray backgrounds, 535
  of distribution of galaxies, 534
  and local condition on the Weyl tensor at big bang, 634
  as principle, 533
  of radio-source distribution, 535
  of Virgo supercluster, 534

Kaehler metrics, 776, 778
Kantowski–Sachs cosmological models, 544–6, 563, 565, 614
Kasner cosmological model, 558–9, 560, 567, 572, 573, 575, 580

Index

Kerr black hole geometry, 4, 17, 370–453, 619–21, 625–6, 773
  derivation, 391–6
  event horizon, 392, 431–2
  extraction of spin energy from, 6, 351, 491
  Maxwell's equations in, 404–11
  in Newman–Penrose formalism, 375–82, 399
  perturbations of, 370, 402
  potential barriers, 429–51
  principal null-directions for, 398
  reflection and transmission of waves, 370, 402, 429, 437–45
  reflection–transmission coefficients, 440–51
  separability of (field) equations in, 452
  spin-coefficients for, 375–7, 400
  structure constants for, 378, 401
  super-radiance, 432, 444, 449, 451
  type-D character in Petrov classification, 398–9, 400, 402
  uniqueness as external field of black hole, 4, 17, 359–70
Kerr–de Sitter geometry, 776
Kerr–Newman geometry, 4, 17, 366
  perturbations of, 417
Killing horizon, 314–18, 392
Killing vector congruences, 305
Killing vectors,
  and commutator of homotheties, 564–5
  in de Sitter space, 776
  and isometries (motions), 538
  and Szekeres models, 569
  in Taub–NUT solution, 774
  on generic symplectic manifold (Killing form), 202
Komar's integrals (stationary system), 352
Kreuzer experiment, 61
Kruskal coordinates, 670

Lagrangian
  of chiral dynamics, 796
  counterterms, 659, 660, 822, 825–6
  'effective', 656, 659, 701–2, 716, 740
  of Euclidean scalar field, 817
  of general relativity, 791, 809
  for particle detector, 693
  of quantum gravity, *see* quantum gravity
  of spin 0, $\frac{1}{2}$, 1 fields, 644–5
  of ϕ^4 theory, 818
Lagrangian formalism
  for field equations of Bianchi models, 552–5

  in classical field theory for fields coupled to gravity, 140–4
  see also action
Lamb shift, 18
Laplace–Beltrami operator, behaviour of the eigenvalues under conformal transformations, 758–60
lapse functions (ADM), 146, 160, 206
large numbers (cosmology), 530–1
  see also Dirac cosmological theory
laser ranging and gravitational waves, 129–32
law of conditional independence, 589, 633
lens effect, 465, 500–1
lens space, 775
Lense–Thirring effect, 44, 60–1
lepton era, 520, 522–3
Lie derivative
  Cartan formula, 304
  definition, 537
  and isometries, 538
Lifshitz–Khalatnikov–Belinsky analysis (of the big bang singularity), 4, 14, 574–5, 614, 615
light-deflection test, 42, 50–2
linearization stability, 194–9
local conservation of energy–momentum tensor, 9, 69
local inertial coordinates, 8
locally rotationally symmetric (LRS) cosmological models, 549
Lorentz metric on manifolds, 219–24, 289–90
lunar-laser-ranging tests, 42, 54–5, 60, 82

Mach's principle, 12, 84, 504, 516–17
manifold
  admitting direction field, 219–20
  admitting flat metric, 220–1
  admitting positive-definite metric, 218–19
  causal structure, 232–55
  compact, 219; in gravitational path-integral, 785; with self-dual Weyl tensor ($K3$), 777
  definition, 218
  examples, 289–90
  geodesically incomplete, 257–8
  globally hyperbolic, 145, 624–6
  with Lorentz metric, 219
  orientable, 224–232
  simply connected (and Poincaré conjecture), 786
  spacetime, 140
  strongly causal, 621, 624
  see also singularities; spacetime

Index

Maxwell's equations
  in curved spacetime, 645
  in Kerr geometry, 404–11
  in Newman–Penrose formalism, 382
Maxwell tensor in Newman–Penrose formalism, 382, 384
meson (K^0), time-asymmetry in decay-rate, 582–3
metric
  as classical background field, 643, 790
  as quantum operator, 699–702
metric and non-metric theories of gravity, 37, 40
metric meshing law, 37
Michelson–Morley experiment, 1–2
microwave background, 3, 14, 85, 88, 520, 576–8, 579, 634
mild chaotic cosmology, 630
mini-hole, 19–20, 494–502, 618, 635
minimal coupling, 144
mixed state, *see* density matrix
'mixing' models, 572–3
momentum operators, canonical, 647
Moncrief's splitting, *see* reduction of phase space
Mössbauer effect for redshift experiments, 32
motion, equations of
  approximation schemes, 9
  test bodies, 9
motions
  as isometries, 538
  conformal, 538
  group motions, G_2, 540–1
  see also Bianchi types
multipolarity of gravitational waves, 67–70, 85

naked points at infinity (ideal points), 623
naked singularity, 20, 270–88
  see also singularities
Neumann's chain, 636
neutrino(s)
  cosmological, 521–2
  massive, 522
  solar, 521
neutrino equation
  in curved spacetime, 645
  in Kerr geometry, 429
neutrino reflection and transmission by Kerr black hole, 446–9
neutrino super-radiance, absence of, 449
neutrino waves, 370
neutron stars, 4, 5, 77
  accretion of gas onto, 473, 484
  binary pulsar, 70–84
  as member of binary systems, 100 (table 3.3), 470, 480
  birth of, 461–2
  birthrate, 115–16
  collapse to, as source of gravitational waves, 115 (figure 3.5), 114
  mass limit, 462
  model for pulsars, 454
  pulsation of, as source of gravitational waves, 111–12, 115
  quake of, as source of gravitational waves, 98 (table 3.2), 118
  rotation of, as source of gravitational waves, 112
Newman–Penrose formalism, 64, 375–91
  Kerr metric in, 399–401
Newtonian concept of space and time, 1
Newtonian cosmology 11–12, 518–19
Newtonian gravitational potential, 42
Newtonian gravitational theory, 1
Newtonian limit, 30
Noether identities, 328–31
Noether's theorem, 203
no-hair theorem, 4–5, 17, 359–69
non-localizability of gravitational energy, 10
non-polynomial field equation, 747
Nordtvedt effect, 54
  lunar-laser-ranging tests, 42, 54–5, 60, 82
Nordtvedt gedanken experiment, 38
novae
  and binary stars, 110, 111
  as source of gravitational waves, 109
  dwarf novae, 110–11
nucleosynthesis in early universe, 519, 523, 578–9, 580, 611, 634–5
null-basis, 375–7
  for Kerr geometry, 398–9
  perturbations of, 420–1
  rotations of, 383–5
null-congruence, kinematics, 302–14
null-infinity, 277–9
null-surfaces, 11

observer-dependence
  of entropy, 642
  of particle concept, 649–51
  of spacetime geometry, 607
Olbers' paradox, 12, 13
one-loop approximation, 797, 812, 829
  general formulation, 762
  ghost fields, 682
  Hamidew coefficients, 657, 659–60, 661, 707–8, 764–5, 767
  in $2+\varepsilon$ dimensions, 823
  quantum gravity, 702–20

Index

one-loop approximation (*cont.*)
  ultraviolet divergence, 790, 793
  see also quantum field theory, in curved spacetime
one-loop term
  contribution to path-integral, 762
  dependence on topology, 788
  for gravitational field on a vacuum background metric, 768–71
  for quantized field on background metric, 762
operator quantization of gravity, *see* quantization of gravity
Oppenheimer limit, 3
optical scalars, 386
orientable spacetime, 224–32, 643
outer communications, domain of, 296–7

PT-violation, maximally in weak interaction, 637
'pancake'-type singularity, 559, 560, 614, 628
parametrized post-Newtonian (PPN) formalism, 42–5
parity violations and orientability of the universe, 229
particle, definitions of, 649–51, 668, 670, 692
particle annihilation and creation operators, 648, 651
particle creation, 639, 652, 664, 700, 720
  back reaction, 643, 652, 675, 679, 698–702; singularity avoidance by, 665; isotropization by, 667
  by black holes, 7, 18–21, 528, 601, 640, 668–79, 698–702, 744; by primordial black holes, 528–30
  in conformally static spacetime, 664
  in cosmological models, 6, 526–7, 573–4, 608, 628, 640, 643, 664–7, 720, 745
  in de Sitter space, 666, 678
  in early universe, 526–30
  in gravitational collapse, 643, 670–2, 696
  by gravitational wave, 668
  intuitive picture, 678–9
  by naked singularities, 643
  in sandwich spacetime, 649
particle detector, 649, 669, 693–4, 696
particle states, *see* Fock bases; Boguliubov transformation
partition function, 737
  definition, 755, 778
  as Euclidean path-integral, 75
  and generating functional, 677

  of thermal gravitons on a flat spacetime background, 778
  when black holes are present, 781–2
  see also density matrix
past and future of a point, 232–42
path-integral
  BRS transformation, 736, 742
  for fermions (rules of integration), 735–6, 764
  formulation: of quantized gauge theories, 723–43; of quantum field theory, 642, 655; of quantum gravity, 22, 723–43
  generating functional, 737
  for gravity and matter fields, 747, 762
  and propagators, 655, 658
  for pure gravity, 768–9
  for (charged) scalar field, 753–4
  and topologies of stationary-phase metrics, 785–9
  see also action; quantum gravity
Pati–Salam process, 609
pendulum experiment, 30
Penrose conformal diagram, 277–9
Penrose energy extraction process, 6, 351
Penrose–Hawking singularity theorems, 4, 258–66, 299–300, 616
periastron shift in binary pulsar, 74–6
perihelion shift
  measurements, 42, 55–7
  anomalous, 60
perturbations
  of Bianchi type I, 571–2
  of Bianchi type VII$_h$, 562
  of Bianchi type IX, 572, 574
  of FRW models, 536, 563, 566
  of Kerr metric, 370–453
  of Schwarzschild solution, 625
  of spherical symmetry, 619
Petrov classification, 387–9
  of Kerr geometry, 398–9, 400, 402
Petrov type
  type D, 568
  type N, 549
phase transition, second-order, 805, 820
physical constants, variation of, 40, 58, 86, 530–2
Planck length, 607, 635, 791
Planck mass, 20, 717, 766, 791, 798, 809, 810–14
planetary tests of preferred-frame effects, 42, 57–60
plane-wave spacetime, no particle creation in, 668
point-splitting regularization, 654, 660, 662, 679
point-type singularity, 561

Index

points at infinity (ideal points), 606, 621–4
Poisson bracket, 155–6, 159
polarization of gravitational waves, 63–7
positive action conjecture, 759
positive energy conjecture, 10, 757
positive frequency
   in conformally static spacetime, 664
   in curved spacetime, 648–53, 704
   invariance under BMS transformations, 668
   on Kerr horizon, 674
   in Schwarzschild spacetime, 670
'positive pressure' condition (in Bianchi models), 570
post-Newtonian conservation laws, experimental test for, 81–4
Pound–Rebka–Snider experiment, 32
precession
   of periastron, 74–6
   of perihelion, 42, 55–7
   of spin axis, 60; binary pulsar, 77
predictability, asymptotic, 297
primordial black holes, 19–20, 494–502, 528–30, 618, 635
   cosmological production of, 494–7, 507
   detection of, 497–500
   evaporation of, 496, 497
   particle creation by, 528–30
   seeds for galaxies, 496
principal null-directions, 387–9
   for Kerr geometry, 398–9
Principle of Equivalence, 8, 26–7, 28, 35–9, 54–5, 84
propagator, Feynman, *see* Feynman propagator
proper-time formalism, 655–8, 704–10
pseudo-tensor for gravitational energy, 10
pulsar(s), 5
   as source of gravitational waves, 98 (table 3.2), 101 (figure 3.3), 112, 113 (table 3.7), 128 (Crab pulsar), 132 (Crab and Vela pulsars), 78–81 (binary pulsars)
   binary pulsar PSR 1913+16, 70–804
   birthrate, 115–16
   rotating neutron star model, 454

quadrupole moment and generation of gravitational waves, 91, 92
quantization of gauge theories, 721–43
quantization of gravity
   comparison of different approaches, 747–9
   see also action; particle creation; path-integrals; quantum gravity; regularization; renormalization

quantum chromodynamics, 796
   see also strong interaction
quantum effects, in early universe, 6, 509, 526–32, 573–4, 608, 628–9, 640, 643, 663–8, 744
quantum electrodynamics, 661, 668, 682, 804, 806
   Casimir effect, 18, 685–7, 690, 744
quantum field theory
   in curved spacetime, 6–7, 639–79
   divergences of, 654–61, 703–17, 790–831
   on incomplete manifolds and non-Euclidean topologies, 685–90
   of massless spin 2 particles, 796
   path-integral formulation, 642, 655, 723–43
quantum fluctuations, 513, 516
quantum gravity, 6, 7, 21–3, 680–831
   approaches to renormalization, 792–7
   background-field method, 699–702, 768–71
   divergences of, 21–2, 660, 716, 790–8
   finite? 716
   gauge group, 701, 722
   non-renormalizability, 661, 716, 790–1
   one-loop approximation, 744, 762, 797, 811–12, 823–4, 829
quantum theory
   Einstein's views on, 6, 744
   Everett interpretation of, 585, 592–4
quarks, in early universe, 522, 526
quasars, 600
   and black hole events in galactic nuclei, 116, 485–94
   and gravitational collapse, 454
   discovered, 3
   gravitational waves from, 114 (figure 3.5), 134
   optical and X-ray emission, 488–9
   radio-properties, 485–8

radiation, gravitational, *see* gravitational radiation
radio-source, counts, 534
Rastall theory, 49
Raychaudhuri equation, 309
reaction rates, 799, 808, 814, 820, 824
   scaling, 802
redshift
   cosmological, 2, 533
   gravitational: experiments, 32; universality of, *see* universality of gravitational redshift (UGR)
reduction of phase space
   definition of the reduced spaces, 203–4

reduction of phase space (*cont.*)
  splitting theorem for a generic symplectic manifold, 204–5
  splitting theorems for relativity, 206–8
Regge calculus, 784
regularization, 641, 654–61, 704, 712–17
  adiabatic, 665
  in conformally flat spacetime, 715, 718, 720
  dimensional, 654, 659, 663, 713
  n-wave, 665
  Pauli–Villars, 765
  point-splitting, 654, 660, 662, 679, 765
  zeta-function, 654, 659–61, 663, 677, 710–12, 713, 717, 766–71
  see also renormalization
Reissner–Nordström solution, 619, 722
renormalization, 641, 654, 689, 698, 709, 712–16, 796
  background-field method, 801
  group, 716, 792, 799, 813, 820, 822
  of quantum gravity, 766, 783, 790–8
  of stress tensor, *see* stress tensor
  of Weinberg–Salam theory, 683
  of ϕ^4 and Yang–Mills theory, 766, 783–4
  see also regularization
'ribbon'-type singularity, 561
Ricci identities
  in Newman–Penrose formalism, 379–80
  in tetrad formalism, 374–5
Ricci rotation-coefficients, 372–4
  and spin-coefficients, 375–7
Ricci tensor, representation in Newman–Penrose formalism, 377–8
Riemannization, 658, 666, 668, 676–7
Rindler space, 692, 720
  particle detector in, 694–5
Robertson–Walker universe, 2, 4, 13, 527, 533, 613–14
  particle creation in, 664–7
Robinson's theorem (black holes), 5, 365–6
Roche lobe, 471–3
rocket-redshift experiment, 34, 38
Rosen theory, 48–9, 75, 81, 85, 87, 88
Rubin–Ford effect, 535

S-matrix, 651, 702, 741
Sachs–Budic 'deterministic' spacetime, 628 (footnote)
sandwich formulation, 148
scalar-tensor theories of gravity, 48, 64, 68, 69, 85, 86, 88, 516–17
scale parameter
  and Planck mass, 766

in quantum gravity and Yang–Mills theories, 763–6, 770, 783
Schiff's conjecture, 27, 35, 36–7, 38
Schwarzschild spacetime, 563, 566, 619, 625, 771
  action for, 825
  complexification, 658, 676–7
  Kruskal coordinates, 670
  limiting case of Kerrmetric, 370, 395
  particle creation in, 669–73, 676–7, 695–8
  perturbations of, 433 (footnote)
Schwarzschild–de Sitter spacetime, 776
second fundamental form, 751
secular acceleration in gravitational theories, 84
Segal (or Kostant–Souriau) quantization formalism, 210
self-acceleration of centre of mass of binary system, 81
self-dual Maxwell tensor, 772
self-dual metrics
  definition, 757
  Taub–NUT, 775
  $K3$, 777
self-similar solutions, 564–5
semi-conservative theory, 44, 81
shear
  (cosmology), 551, 557, 576, 578
  of congruence, 308–9, 386–7
shell-crossing singularity, 272–3, 297–8, 625
shift vector field (ADM), 146, 160, 206
simplicial complex, *see* Regge calculus
singularities, 255–88, 293, 611
  Belinskii–Khalatnikov–Lifshitz analysis, 574–6, 615
  in cosmology and gravitational collapse, 2, 4, 13–16, 509, 574–6, 610–11, 613–15
  definitions of, 255–8, 267–9
  dynamical, 139
  of 'high entropy' (black hole singularities), 613
  as ideal points, 606, 621–5
  locally naked, 619, 621–3
  of 'low entropy' (big bang singularities), 611–17, 625, 628, 634
  naked, 270–88, 456–7, 617–19
  particle creation at, *see* particle creation
  quantum field theory in presence of, 643, 705, 723
  quantum gravity, avoidance by? 510–11, 616, 665, 687
  shell-crossing, 272–3, 297–8, 625
  and time-asymmetry, 581–638

Index

singularities (*cont.*)
  Weyl tensor, behaviour at, 614
  white hole (and pair creation), 607, 608, 634
  'measure zero'? 626
singularity theorems, 4, 15, 16, 139, 258–66, 299–300, 536, 581, 616, 619, 687
slices of spacetime, 243–7, 291–2
  achronal, 245–6
solar quadrupole moment, 56–7
soliton, 688
space experiments, 34, 50, 52–7, 60–61
  for detection of gravitational waves, 129, 132, 133, 137
spacetime
  asymptotically flat, 277–82
  causal structure, 232–55
  complexification, 658, 666, 668, 676–7, 752–7
  Euclidean section of, *see* Euclidean section
  generic, 264
  geodesically incomplete, 257–8
  geometry observer-dependent? 607
  global structure, 212–93
  orientability, 224–32, 643
  sandwich, 649, 665, 667
  singular, *see* singularities
  slices, 243–7, 291–2
  topology and quantum field theory, 661, 684, 688–90, 710, 720, 723, 744
  see also cosmological models; manifold
spatial homogeneity of universe, 4
  from counts of galaxies and linearity of the Hubble law, 536
  as principle, 533
spatial isotropy of universe, 4
  see also isotropy
spatially-homogeneous cosmological models, 13, 504
  anisotropic stresses, 555–6
  diagonalization, 551–2
  differential equations governing them, 557–8
  generality, 550
  isotropization, 562
  Lagrangians and Hamiltonians, 552–5
  metric, 543, 546
  singularities, 558–62
  tilt and rotation, 551
special relativity, discovery, 2
speed of gravitational waves, 66 (table 2.8), 67
spin-coefficients, 375–7, 400
  and structure constants, 378, 401
  effect of tetrad transformation on, 384–5

  for Kerr geometry, 400
spin-echo, 588
spinors, in complex spacetime, 756
stability of spatially-homogeneous models, 571–2
stability, linearization, 195–203
stable causality, 241–3, 290–1
stars
  birth and death of, 461–5
  rotation of, as source of gravitational waves, 112
  starquake, 118
  stellar wind, 462, 467, 468, 475, 491
  supermassive, 485
  vibrations of, as source of gravitational waves, 108–12
  see also binary systems; neutron stars; pulsars; white dwarf stars
stationary axisymmetric system, mass and angular momentum, 352–6
steady-state and static models of universe, 14, 533, 542, 577–8
stellar-system tests for gravitational theories, 70–84
stratified theories of gravity, 49, 88
stress tensor,
  from action, 749
  as quantum operator, 698–702, 706
  regularization, 654–61, 685–720
  renormalized, *see* stress tensor, renormalized
  trace anomaly, 662, 679, 689, 717–20
stress tensor, renormalized
  in black hole spacetime, 695–9
  Casimir effect, 18, 685–7, 690
  near conductor, 691
  in conformally flat spacetimes, 719–20
  in curved spacetime, 689
  in de Sitter spacetime, 663, 720
  in non-Euclidean topology spacetimes, 688–90
  in one-loop approximation, 702–20
  in Ricci-flat spacetimes, 715, 719
  in Robertson–Walker universe, 666, 720
  trace, 662–3, 717–18
strong causality principle, 621–4
strong equivalence principle, 84–88
strong interaction, 681, 791, 795, 809, 812
structure constants (commutation coefficients), 374
  for Kerr geometry, 401
  and spin-coefficients, 378
supergravity, 7, 22, 765, 783–4, 793
supernovae
  birthrate, 116

916

supernovae (cont.)
　and formation of pulsars, 454
　as source of gravitational waves, 92, 110, 115–16, 131–2, 135, 464–5
super-radiance, 6, 335, 432, 444–5, 674–5, 678
　see also Kerr black hole geometry, super-radiance
superspace, 210
supersymmetry, 793, 797, 826
surface at infinity, 623
　in Einstein action, 748, 751
surface gravity (black hole), 18, 307, 313, 323–7
　and periodicity of Euclidean section, 772
'surface of revolution' cosmologies, 565
surface term in Einstein action, 158, 749–51, 824
Swiss-cheese cosmological models, 563
symmetry-breaking
　in hadron era, 524
　spontaneous, 643, 682, 796
symplectic manifold
　and Dirac–ADM evolution equations, 148–9
　as space of degrees of freedom for fields and gravity, 211
　as space of gravitational degrees of freedom, 140
　see also constraint manifold
symplectic structure,
　for a generic manifold, 202
　for relativity, 149–50
　on $S_2 \times S_d^2$, 209
symplectic symmetry (J-invariance), 208, 209
Szekeres solutions, 563, 568–70

T-violation, 582–3, 637–8
Taub–de Sitter solutions, 776
Taub–NUT spacetime, 238, 248–9, 282–3, 566, 626–8
　as gravitational dyon, 773–5
　as instanton, 662
　particle creation in, 668
temperature of black hole, 7, 18–20, 313, 668–9, 675–6, 677, 695–8
temperature of quantum vacuum state, 650–3, 691–5
tetrad formalism, 371–2ff
　and Newman–Penrose formalism, null-basis, 376–7ff
tetrad transformations, 383–5
Teukolsky–Starobinsky identities, 406–9, 410, 417–19

$TH\varepsilon\mu$ formalism, 35–9
thermal Green's function and black holes, 676–7, 698, 755
thermodynamics of black holes, 7, 16–19, 301, 323–7, 356–9, 675, 696–7
tidal dissipation effects (binary pulsar), 79
tilt of cosmological model, 551, 564
tilted cosmological models, 547, 551, 562, 571–3, 578
time-asymmetry, 581–638
　and asymmetric laws of physics, 635–8
　black holes versus white holes, 606–8, 610, 611
　CP-violation (or T-violation) in the decay of the K^0-meson, 582–3, 636, 637
　entropy, increase of, 586
　expansion of the universe, 596–9
　increase of entropy, 586
　psychological time, 592–4
　quantum-mechanical observations, 583–6, 635, 636
　retardation of radiation, 590, 591
　see also causal structure
time-delay experiments, 42, 52–3
time-function, 242
timelike and null congruences, kinematics, 302–14
time-orientable spacetime, 224–32
time-orientation, 600 (footnote)
TIP and TIF (terminal indecomposable past sets or future sets), structure of spacetime, 606, 621–5
Tolman cosmological model, 566–7, 575, 615–17
topological coupling constant, 783
topological invariants
　Betti number, 777, 787
　Euler number, 777, 786–8
　Hirzebruch signature (Pontryagin number), 777, 786–8
topology
　of Einstein universe, 627
　of Euclidean sections of complex manifolds, 772–7, 778
　of globally hyperbolic manifold, 627
　of Gowdy cosmological models, 567
　of our universe, 217–54
　of spacetime and Casimir effect, 688–9
torsion-balance experiment (Eötvös–Dicke–Braginsky), 8, 30, 54
trace anomaly, 662, 679, 689, 717–20, 768
trapped surface, 16, 265–6, 299–300
tree graphs
　Green's functions, 797, 810–11
　S-matrix, 702, 741

Index

turbulent cosmological models, 579
twist, 386–7

ultraviolet divergence of quantum gravity, 660, 716, 790–831
unitarity, 682, 743, 794, 798
universal coupling of the gravitational field, 8
universal covering space, 229–32
universality of gravitational redshift (UGR), 29, 32–4, 35, 37, 38, 39, 84
universe
 age of, 519
 antimatter, 508, 512
 baryon excess, 524, 636–7
 big bang model, 504–17
 causal structure, 232–55
 causal violations, 238–43
 collapse of, 510–11
 cosmological horizons, 494, 505–6, 508, 510
 entropy of, 511–12, 579–80, 610–17, 629–30, 636–7; reversal at maximum expansion, 596–9
 early, see early universe
 expansion, 2, 13; rate of, 506f, 514
 galaxy formation, 13, 496
 homogeneity of, 494, 504–10, 535
 island universe, 505, 517
 isotropy of, 494–504, 534–5, 576–8 (table 11.3)
 'large numbers' (Dirac theory), 40, 86, 530–1
 mass density, 506–7, 520–2
 microwave background, 3, 14, 519, 535, 576–8 (table 11.3)
 Olbers' paradox, 12, 13, 14
 oscillating, 510–15, 612–13, 616
 radiation density, 520
 singular origin, 13–15, 255–88, 509, 574–6, 610–11, 614–16
 space- and time-orientability, 228–31
 steady state, 14
 time-asymmetry, 581–638
 topology of, 217–54
 total energy, 513
 see also cosmological models; cosmology

vacuum (quantum field theory), 680, 683–98, 710
 for accelerated observer, 649–50, 692–5
 ambiguity in curved spacetime, 639, 688
 conformal, 664

 degeneracy, 683
 in de Sitter spacetime, 650, 720
 fluctuations, 7, 684
 Minkowski, 650, 685, 694
 polarization, 526–7, 685–98
 for quantum gravity, 755–6
 in Schwarzschild spacetime, 670, 672, 695
 and time-asymmetry, 635–6
vacuum polarization,
 in gravitational field, 527
 negative energy in, 678, 687, 692–5
variation of Einstein action, 142, 153, 749
 for Bianchi models, 552–5
variation of physical constants, 40, 58, 86, 530–2
vector–tensor theories of gravitation, 48
Virgo cluster, gravitational waves from supernova in, 67, 131–2, 135, 465
Virgo supercluster, 534–5
virtual black holes, 635
viscosity (cosmology), 555, 570–1
volume canonical ensemble, 786
vorticity (of fluid), 551, 576

Ward–Takahashi identity, 738, 742–3
wave equation in curved spacetime, 644–6
 Cauchy problem, 643
 formal solution of, 655
weak equivalence principle, 8, 26–7, 28, 30–1, 34–9, 54, 55, 84
weak-field theory (gravitational waves), 64–7, 91–100
weak interactions, 637, 681, 791, 809, 812
Weinberg–Salam theory, 683, 690
Weyl neutrino equation in curved spacetime, 645
Weyl (conformal) tensor
 behaviour in a periodic universe, 614
 behaviour near singularities, 614–15, 630, 634
 in a Lorentzian metric, 756
 representation in the Newman–Penrose formalism, 377–8
whimper singularity, 547, 559
white dwarf stars
 as member of binary systems, 100 (table 3.3), 480
 birth of, 462
 mass limit, 462
 quakes, as source of gravitational waves, 119

918

white dwarf stars (*cont.*)
  vibration of, as source of gravitational waves, 98, 101 (figure 3.3), 109–10
white holes, 590, 600, 601–10
Whitehead theory, 50
Will–Nordtvedt version of PPN formalism, 45
wormholes, 531

X-ray binaries, 103, 105 (table 3.6), 469, 470, 471–81

X-ray sources, 5, 17
  and black holes, 455, 469, 474, 479–84, 488–9
  see also Cygnus X-1

Yang–Mills fields, 682, 690, 729, 735, 824
  and minimal coupling, 144

zag-motion, 589 (footnote)
zeta-function, generalized, 710–12
zeta-function regularization, 654, 659–61, 663, 677, 710–12, 713, 717, 766–71